Methods in Enzymology

Volume 191
BIOMEMBRANES
Part V
Cellular and Subcellular Transport:
Epithelial Cells

METHODS IN ENZYMOLOGY

EDITORS-IN-CHIEF

John N. Abelson   Melvin I. Simon

DIVISION OF BIOLOGY
CALIFORNIA INSTITUTE OF TECHNOLOGY
PASADENA, CALIFORNIA

FOUNDING EDITORS

Sidney P. Colowick and Nathan O. Kaplan

Methods in Enzymology

Volume 191

Biomembranes

Part V

Cellular and Subcellular Transport:
Epithelial Cells

EDITED BY

Sidney Fleischer
Becca Fleischer

DEPARTMENT OF MOLECULAR BIOLOGY
VANDERBILT UNIVERSITY
NASHVILLE, TENNESSEE

Editorial Advisory Board

Ronald Kaback
Yasuo Kagawa
Martin Klingenberg
Robert L. Post

George Sachs
Antonio Scarpa
Widmar Tanner
Karl Ullrich

ACADEMIC PRESS, INC.
Harcourt Brace Jovanovich, Publishers
San Diego   New York   Boston
London   Sydney   Tokyo   Toronto

This book is printed on acid-free paper. ∞

Copyright © 1990 by Academic Press, Inc.
All Rights Reserved.
No part of this publication may be reproduced or transmitted in any form or by any means, electronic or mechanical, including photocopy, recording, or any information storage and retrieval system, without permission in writing from the publisher.

Academic Press, Inc.
San Diego, California 92101

United Kingdom Edition published by
Academic Press Limited
24-28 Oval Road, London NW1 7DX

Library of Congress Catalog Card Number: 54-9110

ISBN 0-12-182092-0 (alk. paper)

Printed in the United States of America
90 91 92 93   9 8 7 6 5 4 3 2 1

Table of Contents

CONTRIBUTORS TO VOLUME 191 ix
PREFACE xiii
VOLUMES IN SERIES. xv

1. Epithelial Transport: An Introduction — KARL JULIUS ULLRICH — 1
2. Determination of Paracellular Shunt Conductance in Epithelia — G. KOTTRA AND E. FRÖMTER — 4

Section I. Kidney

A. Intact Organ

3. Isolated Perfused and Nonfiltering Kidney — PATRICIO SILVA — 31
4. Multiple Indicator Dilution and the Kidney: Kinetics, Permeation, and Transport *in Vivo* — CHARLES J. LUMSDEN AND MELVIN SILVERMAN — 34
5. Micropuncture Techniques in Renal Research — IHAB M. SHAFIK AND GARY A. QUAMME — 72
6. Kidney: Microperfusion–Double-Perfused Tubule *in Situ* — KARL JULIUS ULLRICH AND GERHARD RUMRICH — 98

B. Glomerular Filtration

7. Transcapillary Fluid Transport in the Glomerulus — AIDA YARED AND IEKUNI ICHIKAWA — 107
8. Preparation and Study of Isolated Glomeruli — DETLEF SCHLONDORFF — 130
9. Isolation and Study of Glomerular Cells — D. A. TROYER AND J. I. KREISBERG — 141
10. Isolation and Culture of Juxtaglomerular and Renomedullary Interstitial Cells — E. ERIC MUIRHEAD, WILTON A. RIGHTSEL, JAMES A. PITCOCK, AND TADASHI INAGAMI — 152

C. Isolated Tubules

11. Comparative Kidney Tubule Sources, Isolation, Perfusion, and Function — KLAUS W. BEYENBACH AND WILLIAM H. DANTZLER — 167

12. Microdissection of Kidney Tubule Segments	P. A. WRIGHT, M. B. BURG, AND M. A. KNEPPER	226
13. Measurement of Transmural Water Flow in Isolated Perfused Tubule Segments	JAMES C. WILLIAMS, JR., AND JAMES A. SCHAFER	232
14. Identification and Study of Specific Cell Types in Isolated Nephron Segments Using Fluorescent Dyes	GEORGE J. SCHWARTZ AND QAIS AL-AWQATI	253
15. Functional Morphology of Kidney Tubules and Cells *in Situ*	BRIGITTE KAISSLING AND MICHEL LE HIR	265
16. An Electrophysiological Approach to the Study of Isolated Perfused Tubules	RAINER GREGER	289
17. Hormonal Receptors in the Isolated Tubule	FRANÇOIS MOREL AND DANIEL BUTLEN	303
18. Metabolism of Isolated Kidney Tubule Segments	GABRIELE WIRTHENSOHN AND WALTER G. GUDER	325
19. Endocytosis and Lysosomal Hydrolysis of Proteins in Proximal Tubules	THOMAS MAACK AND C. HYUNG PARK	340
20. Flux Measurements in Isolated Perfused Tubules	JAMES A. SCHAFER AND JAMES C. WILLIAMS, JR.	354
21. Measurements of Volume and Shape Changes in Isolated Tubules	WILLIAM B. GUGGINO, DIANE MARKAKIS, AND L. MARIO AMZEL	371

D. Isolated Renal Cells

22. Transport in Isolated Cells from Defined Nephron Segments	ROLF K. H. KINNE	380
23. Primary Culture of Isolated Tubule Cells of Defined Segmental Origin	MICHAEL F. HORSTER AND MASAYOSHI SONE	409
24. Tissue Culture of Established Renal Cell Lines	N. L. SIMMONS	426
25. Giant MDCK Cells: A Novel Expression System	H. OBERLEITHNER, A. SCHWAB, H.-J. WESTPHALE, B. SCHURICHT, B. PÜSCHEL, AND H. KOEPSELL	437

E. Isolation of Membrane Vesicles

26. Isolation of Lumenal and Contralumenal Plasma Membrane Vesicles from Kidney	EVAMARIA KINNE-SAFFRAN AND ROLF K. H. KINNE	450

27. Transport Studies by Optical Methods	G. SACHS, E. RABON, AND S. J. D. KARLISH	469
28. Stoichiometry of Coupled Transport Systems in Vesicles	R. JAMES TURNER	479
29. Phosphate Transport in Established Renal Epithelial Cell Lines	J. BIBER, K. MALMSTRÖM, S. RESHKIN, AND H. MURER	494
30. ATP-Driven Proton Transport in Vesicles from the Kidney Cortex	IVAN SABOLIĆ AND GERHARD BURCKHARDT	505

F. Hormonal Modulation

31. Aldosterone	DIANA MARVER	520
32. The Cellular Action of Antidiuretic Hormone	DENNIS BROWN, ALAN VERKMAN, KARL SKORECKI, AND DENNIS AUSIELLO	551

G. Reconstitution of Transport Proteins

33. Purification and Reconstitution of Epithelial Chloride Channels	DONALD W. LANDRY, MYLES A. AKABAS, CHRISTOPHER REDHEAD, AND QAIS AL-AWQATI	572
34. Reconstitution and Fractionation of Renal Brush Border Transport Proteins	HERMANN KOEPSELL AND STEFAN SEIBICKE	583

Section II. Stimulus Secretion Coupling in Epithelia

35. Receptor Identification	RAKESH VINAYEK AND JERRY D. GARDNER	609
36. cAMP Technologies, Functional Correlates in Gastric Parietal Cells	CATHERINE S. CHEW	640
37. Stimulus–Secretion Coupling: General Models and Specific Aspects in Epithelial Cells	HOWARD RASMUSSEN	661
38. Metabolism and Function of Phosphatidylinositol-Derived Arachidonic Acid	LOWELL E. HOKIN	676
39. Measurement of Intracellular Free Calcium to Investigate Receptor-Mediated Calcium Signaling	CARL A. HANSEN, JONATHAN R. MONCK, AND JOHN R. WILLIAMSON	691
40. Two-Stage Analysis of Radiolabeled Inositol Phosphate Isomers	K. A. WREGGETT, D. J. LANDER, AND ROBIN F. IRVINE	707

Section III. Pharmacological Agents in Epithelial Tansport

41. Pharmacological Agents of Gastric Acid Secretion: Receptor Antagonists and Pump Inhibitors — BJÖRN WALLMARK AND JAN FRYKLUND — 721

42. Cation Transport Probes: The Amiloride Series — THOMAS R. KLEYMAN AND EDWARD J. CRAGOE, JR. — 739

43. Photoaffinity-Labeling Analogs of Phlorizin and Phloretin: Synthesis and Effects on Cell Membranes — DONALD F. DIEDRICH — 755

44. Diuretic Compounds Structurally Related to Furosemide — SCOTT M. O'GRADY, MARK W. MUSCH, AND MICHAEL FIELD — 781

45. Chloride Channel Blockers — RAINER GREGER — 793

Section IV. Targeting and Intracellular Trafficking in Epithelial Cells

46. *In Vitro* Recovery of Exocytic Transport Vesicles from Polarized MDCK Cells — MARK K. BENNETT, ANGELA WANDINGER-NESS, ANDRÉ W. BRÄNDLI, AND KAI SIMONS — 813

47. Biogenesis of the Rat Hepatocyte Plasma Membrane — JAMES R. BARTLES AND ANN L. HUBBARD — 825

48. Oligomerization and Protein Transport — ROBERT W. DOMS — 841

AUTHOR INDEX 855

SUBJECT INDEX 895

Contributors to Volume 191

Article numbers are in parentheses following the names of contributors.
Affiliations listed are current.

MYLES A. AKABAS (33), *Department of Medicine, College of Physicians & Surgeons, Columbia University, New York, New York 10032*

QAIS AL-AWQATI (14, 33), *Department of Medicine, College of Physicians & Surgeons, Columbia University, New York, New York 10032*

L. MARIO AMZEL (21), *Department of Biophysics, Johns Hopkins University School of Medicine, Baltimore, Maryland 21205*

DENNIS AUSIELLO (32), *Renal Unit, Massachusetts General Hospital and Department of Medical Services, Harvard Medical School, Boston, Massachusetts 02114*

JAMES R. BARTLES (47), *Department of Cell, Molecular, and Structural Biology, Northwestern University Medical School, Chicago, Illinois 60611*

MARK K. BENNETT (46), *Department of Biological Sciences, Stanford University, Stanford, California 94305*

KLAUS W. BEYENBACH (11), *Section of Physiology, Cornell University, Ithaca, New York 14853*

J. BIBER (29), *Institute of Physiology, University of Zürich-Irchel, CH-8057 Zürich, Switzerland*

ANDRÉ W. BRÄNDLI (46), *Department of Biochemistry and Biophysics, University of California-San Francisco, San Francisco, California 94143*

DENNIS BROWN (32), *Renal Unit, Massachusetts General Hospital, and Department of Pathology, Harvard Medical School, Boston, Massachusetts 02114*

GERHARD BURCKHARDT (30), *Max-Plank-Institut für Biophysik, D-6000 Frankfurt/Main 70, Federal Republic of Germany*

M. B. BURG (12), *Laboratory of Kidney and Electrolyte Metabolism, National Institutes of Health, Bethesda, Maryland 20814*

DANIEL BUTLEN (17), *Laboratoire de Physiologie Cellulaire, Collège de France, 75231, Paris Cedex 05, France*

CATHERINE S. CHEW (36), *Department of Physiology, Morehouse School of Medicine, Atlanta, Georgia 30310*

EDWARD J. CRAGOE, JR. (42), *Department of Medicine, University of Pennsylvania, Philadelphia, Pennsylvania 19104, and Merck, Sharp and Dohme Research Laboratories, West Point, Pennsylvania 19486*

WILLIAM H. DANTZLER (11), *Department of Physiology, University of Arizona College of Medicine, Tucson, Arizona 85724*

DONALD F. DIEDRICH (43), *Department of Pharmacology and Center for Membrane Sciences, University of Kentucky College of Medicine, Lexington, Kentucky 40506*

ROBERT W. DOMS (48), *Laboratory of Viral Diseases, National Institute of Allergy and Infectious Diseases, National Institutes of Health, Bethesda, Maryland 20892*

MICHAEL FIELD (44), *Department of Medicine and Physiology, College of Physicians & Surgeons, Columbia University, New York, New York, 10032*

E. FRÖMTER (2), *Zentrum der Physiologie, Johann Wolfgang Goethe-Universität, D-6000 Frankfurt/Main 70, Federal Republic of Germany*

JAN FRYKLUND (41), *AB Hässle, Department of Biology, S-431 83 Mölndal, Sweden*

JERRY D. GARDNER (35), *Digestive Diseases Branch, National Institutes of Health, Bethesda, Maryland 20892*

R. GREGER (16, 45), *Physiologisches Institut, Albert-Ludwigs-Universität, D-7800 Freiburg, Federal Republic of Germany*

WALTER G. GUDER (18), *Institute of Clinical Chemistry, Bogenhausen City Hospital, D-8000 München 81, Federal Republic of Germany*

WILLIAM B. GUGGINO (21), *Department of Physiology, Johns Hopkins University School of Medicine, Baltimore, Maryland 21205*

CARL A. HANSEN (39), *Department of Biochemistry and Biophysics, University of Pennsylvania School of Medicine, Philadelphia, Pennsylvania 19104*

MICHEL LE HIR (15), *Zentrum für Lehre und Forschung, Kantonsspital, 4031 Basel, Switzerland*

LOWELL E. HOKIN (38), *Department of Pharmacology, University of Wisconsin Medical School, Madison, Wisconsin 53706*

MICHAEL F. HORSTER (23), *Institute of Physiology, University of Munich, D-8000 Munich, Federal Republic of Germany*

ANN L. HUBBARD (47), *Department of Cell Biology and Anatomy, Johns Hopkins University School of Medicine, Baltimore, Maryland 21205*

IEKUNI ICHIKAWA (7), *Pediatric Nephrology, Vanderbilt University, Nashville, Tennessee 37232*

TADASHI INAGAMI (10), *Department of Biochemistry, Vanderbilt University School of Medicine, Nashville, Tennessee 37235*

ROBIN F. IRVINE (40), *Institute of Animal Physiology and Genetics Research, Cambridge Research Station, Cambridge CB2 4AT, England*

BRIGITTE KAISSLING (15), *Anatomisches Institut der Universität, University of Zürich-Irchel, CH-8057 Zürich, Switzerland*

S. J. D. KARLISH (27), *Department of Biochemistry, Weizmann Institute, Rehovoth, Israel*

ROLF K. H. KINNE (22, 26), *Max-Plank-Institut für Systemphysiologie, 4600 Dortmund, Federal Republic of Germany*

EVAMARIA KINNE-SAFFRAN (26), *Max-Plank-Institut für Systemphysiologie, Rheinlanddamm 201, 4600 Dortmund, Federal Republic of Germany*

THOMAS R. KLEYMAN (42), *Department of Medicine, University of Pennsylvania, Philadelphia, Pennsylvania 19104*

M. A. KNEPPER (12), *Laboratory of Kidney and Electrolyte Metabolism, National Institutes of Health, Bethesda, Maryland 20814*

H. KOEPSELL (25, 34), *Max-Planck-Institut für Biophysik, D-6000 Frankfurt/Main 70, Federal Republic of Germany*

G. KOTTRA (2), *Zentrum der Physiologie, Klinikum der Johann Wolfgang Goethe-Universität, D-6000 Frankfurt/Main 70, Federal Republic of Germany*

J. I. KREISBERG (9), *Departments of Pathology and Medicine, The University of Texas Health Science Center at San Antonio, San Antonio, Texas 78284*

D. J. LANDER (40), *Institute of Animal Physiology and Genetics Research, Cambridge Research Station, Cambridge CB2 4AT, England*

DONALD W. LANDRY (33), *Department of Medicine, College of Physicians & Surgeons, Columbia University, New York, New York 10032*

CHARLES J. LUMSDEN (4), *Membrane Biology Group and Department of Medicine, University of Toronto, Toronto, Ontario M5S 1A8, Canada*

THOMAS MAACK (19), *Department of Physiology, Cornell University Medical College, New York, New York 10021*

K. MALMSTRÖM (29), *Institute of Physiology, University of Zürich-Irchel, CH-8057 Zürich, Switzerland*

DIANE MARKAKIS (21), *Department of Physiology, Johns Hopkins University School of Medicine, Baltimore, Maryland 21205*

DIANA MARVER (31), *Department of Internal Medicine, University of Texas Southwestern Medical Center, Dallas, Texas 75235*

JONATHAN R. MONCK (39), *Department of Physiology and Biophysics, Mayo Clinic, Rochester, Minnesota 55902*

FRANÇOIS MOREL (17), *Laboratoire de Physiologie Cellulaire, Collège de France, 75231, Paris Cedex 05, France*

E. ERIC MUIRHEAD (10), *Department of Pathology, Baptist Memorial Hospital, Memphis, Tennessee 38146*

H. MURER (29), *Institute of Physiology, University of Zürich-Irchel, CH-8057 Zürich, Switzerland*

MARK W. MUSCH (44), *Department of Medicine, The University of Chicago, Chicago, Illinois 60637*

SCOTT M. O'GRADY (44), *Department of Veterinary Biology, University of Minnesota, St. Paul, Minnesota 55108*

H. OBERLEITHNER (25), *Department of Physiology, University of Würzburg, D-8700 Würzburg, Federal Republic of Germany*

C. HYUNG PARK (19), *Department of Physiology, Cornell University Medical College, New York, New York 10021*

JAMES A. PITCOCK (10), *Department of Pathology, Baptist Memorial Hospital, Memphis, Tennessee 38146*

B. PÜSCHEL (25), *Max-Plank-Institut für Biophysik, D-6000 Frankfurt/Main 70, Federal Republic of Germany*

GARY A. QUAMME (5), *Division of Nephrology, Department of Medicine, University of British Columbia, Vancouver, British Columbia V6T 1W5, Canada*

E. RABON (27), *Department of Physiology and Medicine, University of California-Los Angeles, Wadsworth VA Medical Center, Los Angeles, California 90073*

HOWARD RASMUSSEN (37), *Departments of Internal Medicine, Cell Biology, and Physiology, Yale University School of Medicine, New Haven, Connecticut 06510*

CHRISTOPHER REDHEAD (33), *Department of Medicine, College of Physicians & Surgeons, Columbia University, New York, New York 10032*

S. RESHKIN (29), *Institute of Physiology, University of Zürich-Irchel, CH-8057 Zürich, Switzerland*

WILTON A. RIGHTSEL (10), *Department of Pathology, Baptist Memorial Hospital, Memphis, Tennessee 38146*

GERHARD RUMRICH (6), *Max-Planck-Institut für Biophysik, D-6000 Frankfurt/Main 70, Federal Republic of Germany*

IVAN SABOLIĆ (30), *Department of Physiology, University of Zagreb, 41000 Zagreb, Yugoslavia*

G. SACHS (27), *Department of Physiology and Medicine, University of California-Los Angeles, Wadsworth VA Medical Center, Los Angeles, California 90073*

JAMES A. SCHAFER (13, 20), *Nephrology Research and Training Center, Departments of Physiology and Biophysics, and Medicine, University of Alabama at Birmingham, Birmingham, Alabama 35294*

DETLEF SCHLONDORFF (8), *Renal Division, Department of Medicine, Albert Einstein College of Medicine, Bronx, New York 10461*

B. SCHURICHT (25), *Department of Physiology, University of Würzburg, D-8700 Würzburg, Federal Republic of Germany*

A. SCHWAB (25), *Department of Physiology, University of Würzburg, D-8700 Würzburg, Federal Republic of Germany*

GEORGE J. SCHWARTZ (14), *Departments of Pediatrics and Physiology/Biophysics, Albert Einstein College of Medicine, Bronx, New York 10462*

STEFAN SEIBICKE (34), *Max-Planck-Institut für Biophysik, D-6000 Frankfurt/Main 70, Federal Republic of Germany*

IHAB M. SHAFIK (5), *Division of Nephrology, Department of Medicine, University of British Columbia, Vancouver, British Columbia V6T 1W5, Canada*

PATRICIO SILVA (3), *Department of Medicine, Joslin Diabetes Center, New England Deaconess Hospital and Harvard Medical School, Boston, Massachusetts 02215*

MELVIN SILVERMAN (4), *Membrane Biology Group and Department of Medicine, University of Toronto, Toronto, Ontario M5S 1A8, Canada*

N. L. SIMMONS (24), *Department of Physiological Sciences, The Medical School, University of Newcastle-upon-Tyne, Newcastle-upon-Tyne NE2 4HH, England*

KAI SIMONS (46), *European Molecular Biology Laboratory, D-6900 Heidelberg, Federal Republic of Germany*

KARL SKORECKI (32), *Toronto General Hospital, Division of Nephrology, Toronto, Toronto, Ontario M5G 1L7, Canada*

MASAYOSHI SONE (23), *The Fourth Department of Medicine, Tokyo Women's Medical College, Tokyo, Japan*

D. A. TROYER (9), *Department of Pathology, The University of Texas Health Science Center at San Antonio, San Antonio, Texas 78284*

R. JAMES TURNER (28), *Clinical Investigations and Patient Care Branch, National Institute of Dental Research, National Institutes of Health, Bethesda, Maryland 20892*

KARL JULIUS ULLRICH (1, 6), *Max-Planck-Institut für Biophysik, D-6000 Frankfurt/ Main 70, Federal Republic of Germany*

ALAN VERKMAN (32), *Department of Medicine and Division of Nephrology, Cardiovascular Research Unit, University of California, San Francisco, San Francisco, California 94143*

RAKESH VINAYEK (35), *Digestive Diseases Branch, National Institutes of Health, Bethesda, Maryland 20892*

BJÖRN WALLMARK (41), *AB Hässle, Department of Biology, S-431 83 Mölndal, Sweden*

ANGELA WANDINGER NESS (46), *European Molecular Biology Laboratory, D-6900 Heidelberg, Federal Republic of Germany*

H.-J. WESTPHALE (25), *Department of Physiology, University of Würzburg, D-8700 Würzburg, Federal Republic of Germany*

JAMES C. WILLIAMS, JR. (13, 20), *Department of Anatomy and Cell Biology, Medical University of South Carolina, Charleston, South Carolina 29425*

JOHN R. WILLIAMSON (39), *Department of Biochemistry and Biophysics, University of Pennsylvania School of Medicine, Philadelphia, Pennsylvania 19104*

GABRIELE WIRTHENSOHN (18), *Institute of Clinical Chemistry, Bogenhausen City Hospital, D-8000 München 81, Federal Republic of Germany*

K. A. WREGGETT (40), *Institute of Animal Physiology and Genetics Research, Cambridge Research Station, Cambridge CB2 4AT, England*

P. A. WRIGHT (12), *Laboratory of Kidney and Electrolyte Metabolism, National Institutes of Health, Bethesda, Maryland 20814*

AIDA YARED (7), *Division of Pediatric Nephrology, Vanderbilt University School of Medicine, Nashville, Tennessee 37232*

Preface

Biological transport is part of the Biomembranes series of *Methods in Enzymology*. It is a continuation of methodology concerned with membrane function. This is a particularly good time to cover the topic of biological membrane transport because there is now a strong conceptual basis for its understanding. The field of transport has been subdivided into five topics.

1. Transport in Bacteria, Mitochondria, and Chloroplasts
2. ATP-Driven Pumps and Related Transport
3. General Methodology of Cellular and Subcellular Transport
4. Cellular and Subcellular Transport: Eukaryotic (Nonepithelial) Cells
5. Cellular and Subcellular Transport: Epithelial Cells

Topic 1 covered in Volumes 125 and 126 initiated the series. Topic 2 is covered in Volumes 156 and 157, Topic 3 in Volumes 171 and 172, and Topic 4 in Volumes 173 and 174. The remaining topic 5 is now covered in Volumes 191 and 192.

Topic 5 is divided into two parts: this volume (Part V) which covers transport in kidney, hormonal modulation, stimulus secretion coupling, pharmacological agents and targeting, and intracellular trafficking in epithelial cells, and Volume 192 (Part W) which deals with gastrointestinal and a diversity of other epithelial cells.

We are fortunate to have the good counsel of our Advisory Board. Their input insures the quality of these volumes. The same Advisory Board has served for the complete transport series. Valuable input on the outlines of the five topics was also provided by Qais Al-Awqati, Ernesto Carafoli, Halvor Christensen, Isadore Edelman, Joseph Hoffman, Phil Knauf, and Hermann Passow. Additional valuable input for Volumes 191 and 192 was obtained from Michael Berridge, Eberhard Frömter, Ari Helenius, and Heine Murer.

The names of our advisory board members were inadvertently omitted in Volumes 125 and 126. When we noted the omission, it was too late to rectify the problem. For volumes 125 and 126, we are also pleased to acknowledge the advice of Angelo Azzi, Youssef Hatefi, Dieter Oesterhelt, and Peter Pedersen.

The enthusiasm and cooperation of the participants have enriched and made these volumes possible. The friendly cooperation of the staff of Academic Press is gratefully acknowledged. We are pleased to acknowledge Ms. Laura Taylor for her tireless efforts and secretarial skills.

These volumes are dedicated to Professor Sidney Colowick, a dear friend and colleague, who died in 1985. We shall miss his wise counsel, encouragement, and friendship.

SIDNEY FLEISCHER
BECCA FLEISCHER

METHODS IN ENZYMOLOGY

VOLUME I. Preparation and Assay of Enzymes
Edited by SIDNEY P. COLOWICK AND NATHAN O. KAPLAN

VOLUME II. Preparation and Assay of Enzymes
Edited by SIDNEY P. COLOWICK AND NATHAN O. KAPLAN

VOLUME III. Preparation and Assay of Substrates
Edited by SIDNEY P. COLOWICK AND NATHAN O. KAPLAN

VOLUME IV. Special Techniques for the Enzymologist
Edited by SIDNEY P. COLOWICK AND NATHAN O. KAPLAN

VOLUME V. Preparation and Assay of Enzymes
Edited by SIDNEY P. COLOWICK AND NATHAN O. KAPLAN

VOLUME VI. Preparation and Assay of Enzymes (*Continued*)
Preparation and Assay of Substrates
Special Techniques
Edited by SIDNEY P. COLOWICK AND NATHAN O. KAPLAN

VOLUME VII. Cumulative Subject Index
Edited by SIDNEY P. COLOWICK AND NATHAN O. KAPLAN

VOLUME VIII. Complex Carbohydrates
Edited by ELIZABETH F. NEUFELD AND VICTOR GINSBURG

VOLUME IX. Carbohydrate Metabolism
Edited by WILLIS A. WOOD

VOLUME X. Oxidation and Phosphorylation
Edited by RONALD W. ESTABROOK AND MAYNARD E. PULLMAN

VOLUME XI. Enzyme Structure
Edited by C. H. W. HIRS

VOLUME XII. Nucleic Acids (Parts A and B)
Edited by LAWRENCE GROSSMAN AND KIVIE MOLDAVE

VOLUME XIII. Citric Acid Cycle
Edited by J. M. LOWENSTEIN

VOLUME XIV. Lipids
Edited by J. M. LOWENSTEIN

VOLUME XV. Steroids and Terpenoids
Edited by RAYMOND B. CLAYTON

VOLUME XVI. Fast Reactions
Edited by KENNETH KUSTIN

VOLUME XVII. Metabolism of Amino Acids and Amines (Parts A and B)
Edited by HERBERT TABOR AND CELIA WHITE TABOR

VOLUME XVIII. Vitamins and Coenzymes (Parts A, B, and C)
Edited by DONALD B. MCCORMICK AND LEMUEL D. WRIGHT

VOLUME XIX. Proteolytic Enzymes
Edited by GERTRUDE E. PERLMANN AND LASZLO LORAND

VOLUME XX. Nucleic Acids and Protein Synthesis (Part C)
Edited by KIVIE MOLDAVE AND LAWRENCE GROSSMAN

VOLUME XXI. Nucleic Acids (Part D)
Edited by LAWRENCE GROSSMAN AND KIVIE MOLDAVE

VOLUME XXII. Enzyme Purification and Related Techniques
Edited by WILLIAM B. JAKOBY

VOLUME XXIII. Photosynthesis (Part A)
Edited by ANTHONY SAN PIETRO

VOLUME XXIV. Photosynthesis and Nitrogen Fixation (Part B)
Edited by ANTHONY SAN PIETRO

VOLUME XXV. Enzyme Structure (Part B)
Edited by C. H. W. HIRS AND SERGE N. TIMASHEFF

VOLUME XXVI. Enzyme Structure (Part C)
Edited by C. H. W. HIRS AND SERGE N. TIMASHEFF

VOLUME XXVII. Enzyme Structure (Part D)
Edited by C. H. W. HIRS AND SERGE N. TIMASHEFF

VOLUME XXVIII. Complex Carbohydrates (Part B)
Edited by VICTOR GINSBURG

VOLUME XXIX. Nucleic Acids and Protein Synthesis (Part E)
Edited by LAWRENCE GROSSMAN AND KIVIE MOLDAVE

VOLUME XXX. Nucleic Acids and Protein Synthesis (Part F)
Edited by KIVIE MOLDAVE AND LAWRENCE GROSSMAN

VOLUME XXXI. Biomembranes (Part A)
Edited by SIDNEY FLEISCHER AND LESTER PACKER

VOLUME XXXII. Biomembranes (Part B)
Edited by SIDNEY FLEISCHER AND LESTER PACKER

VOLUME XXXIII. Cumulative Subject Index Volumes I–XXX
Edited by MARTHA G. DENNIS AND EDWARD A. DENNIS

VOLUME XXXIV. Affinity Techniques (Enzyme Purification: Part B)
Edited by WILLIAM B. JAKOBY AND MEIR WILCHEK

VOLUME XXXV. Lipids (Part B)
Edited by JOHN M. LOWENSTEIN

VOLUME XXXVI. Hormone Action (Part A: Steroid Hormones)
Edited by BERT W. O'MALLEY AND JOEL G. HARDMAN

VOLUME XXXVII. Hormone Action (Part B: Peptide Hormones)
Edited by BERT W. O'MALLEY AND JOEL G. HARDMAN

VOLUME XXXVIII. Hormone Action (Part C: Cyclic Nucleotides)
Edited by JOEL G. HARDMAN AND BERT W. O'MALLEY

VOLUME XXXIX. Hormone Action (Part D: Isolated Cells, Tissues, and Organ Systems)
Edited by JOEL G. HARDMAN AND BERT W. O'MALLEY

VOLUME XL. Hormone Action (Part E: Nuclear Structure and Function)
Edited by BERT W. O'MALLEY AND JOEL G. HARDMAN

VOLUME XLI. Carbohydrate Metabolism (Part B)
Edited by W. A. WOOD

VOLUME XLII. Carbohydrate Metabolism (Part C)
Edited by W. A. WOOD

VOLUME XLIII. Antibiotics
Edited by JOHN H. HASH

VOLUME XLIV. Immobilized Enzymes
Edited by KLAUS MOSBACH

VOLUME XLV. Proteolytic Enzymes (Part B)
Edited by LASZLO LORAND

VOLUME XLVI. Affinity Labeling
Edited by WILLIAM B. JAKOBY AND MEIR WILCHEK

VOLUME XLVII. Enzyme Structure (Part E)
Edited by C. H. W. HIRS AND SERGE N. TIMASHEFF

VOLUME XLVIII. Enzyme Structure (Part F)
Edited by C. H. W. HIRS AND SERGE N. TIMASHEFF

VOLUME XLIX. Enzyme Structure (Part G)
Edited by C. H. W. HIRS AND SERGE N. TIMASHEFF

VOLUME L. Complex Carbohydrates (Part C)
Edited by VICTOR GINSBURG

VOLUME LI. Purine and Pyrimidine Nucleotide Metabolism
Edited by PATRICIA A. HOFFEE AND MARY ELLEN JONES

VOLUME LII. Biomembranes (Part C: Biological Oxidations)
Edited by SIDNEY FLEISCHER AND LESTER PACKER

VOLUME LIII. Biomembranes (Part D: Biological Oxidations)
Edited by SIDNEY FLEISCHER AND LESTER PACKER

VOLUME LIV. Biomembranes (Part E: Biological Oxidations)
Edited by SIDNEY FLEISCHER AND LESTER PACKER

VOLUME LV. Biomembranes (Part F: Bioenergetics)
Edited by SIDNEY FLEISCHER AND LESTER PACKER

VOLUME LVI. Biomembranes (Part G: Bioenergetics)
Edited by SIDNEY FLEISCHER AND LESTER PACKER

VOLUME LVII. Bioluminescence and Chemiluminescence
Edited by MARLENE A. DELUCA

VOLUME LVIII. Cell Culture
Edited by WILLIAM B. JAKOBY AND IRA PASTAN

VOLUME LIX. Nucleic Acids and Protein Synthesis (Part G)
Edited by KIVIE MOLDAVE AND LAWRENCE GROSSMAN

VOLUME LX. Nucleic Acids and Protein Synthesis (Part H)
Edited by KIVIE MOLDAVE AND LAWRENCE GROSSMAN

VOLUME 61. Enzyme Structure (Part H)
Edited by C. H. W. HIRS AND SERGE N. TIMASHEFF

VOLUME 62. Vitamins and Coenzymes (Part D)
Edited by DONALD B. MCCORMICK AND LEMUEL D. WRIGHT

VOLUME 63. Enzyme Kinetics and Mechanism (Part A: Initial Rate and Inhibitor Methods)
Edited by DANIEL L. PURICH

VOLUME 64. Enzyme Kinetics and Mechanism (Part B: Isotopic Probes and Complex Enzyme Systems)
Edited by DANIEL L. PURICH

VOLUME 65. Nucleic Acids (Part I)
Edited by LAWRENCE GROSSMAN AND KIVIE MOLDAVE

VOLUME 66. Vitamins and Coenzymes (Part E)
Edited by DONALD B. MCCORMICK AND LEMUEL D. WRIGHT

VOLUME 67. Vitamins and Coenzymes (Part F)
Edited by DONALD B. MCCORMICK AND LEMUEL D. WRIGHT

VOLUME 68. Recombinant DNA
Edited by RAY WU

VOLUME 69. Photosynthesis and Nitrogen Fixation (Part C)
Edited by ANTHONY SAN PIETRO

VOLUME 70. Immunochemical Techniques (Part A)
Edited by HELEN VAN VUNAKIS AND JOHN J. LANGONE

VOLUME 71. Lipids (Part C)
Edited by JOHN M. LOWENSTEIN

VOLUME 72. Lipids (Part D)
Edited by JOHN M. LOWENSTEIN

VOLUME 73. Immunochemical Techniques (Part B)
Edited by JOHN J. LANGONE AND HELEN VAN VUNAKIS

VOLUME 74. Immunochemical Techniques (Part C)
Edited by JOHN J. LANGONE AND HELEN VAN VUNAKIS

VOLUME 75. Cumulative Subject Index Volumes XXXI, XXXII, XXXIV–LX
Edited by EDWARD A. DENNIS AND MARTHA G. DENNIS

VOLUME 76. Hemoglobins
Edited by ERALDO ANTONINI, LUIGI ROSSI-BERNARDI, AND EMILIA CHIANCONE

VOLUME 77. Detoxication and Drug Metabolism
Edited by WILLIAM B. JAKOBY

VOLUME 78. Interferons (Part A)
Edited by SIDNEY PESTKA

VOLUME 79. Interferons (Part B)
Edited by SIDNEY PESTKA

VOLUME 80. Proteolytic Enzymes (Part C)
Edited by LASZLO LORAND

VOLUME 81. Biomembranes (Part H: Visual Pigments and Purple Membranes, I)
Edited by LESTER PACKER

VOLUME 82. Structural and Contractile Proteins (Part A: Extracellular Matrix)
Edited by LEON W. CUNNINGHAM AND DIXIE W. FREDERIKSEN

VOLUME 83. Complex Carbohydrates (Part D)
Edited by VICTOR GINSBURG

VOLUME 84. Immunochemical Techniques (Part D: Selected Immunoassays)
Edited by JOHN J. LANGONE AND HELEN VAN VUNAKIS

VOLUME 85. Structural and Contractile Proteins (Part B: The Contractile Apparatus and the Cytoskeleton)
Edited by DIXIE W. FREDERIKSEN AND LEON W. CUNNINGHAM

VOLUME 86. Prostaglandins and Arachidonate Metabolites
Edited by WILLIAM E. M. LANDS AND WILLIAM L. SMITH

VOLUME 87. Enzyme Kinetics and Mechanism (Part C: Intermediates, Stereochemistry, and Rate Studies)
Edited by DANIEL L. PURICH

VOLUME 88. Biomembranes (Part I: Visual Pigments and Purple Membranes, II)
Edited by LESTER PACKER

VOLUME 89. Carbohydrate Metabolism (Part D)
Edited by WILLIS A. WOOD

VOLUME 90. Carbohydrate Metabolism (Part E)
Edited by WILLIS A. WOOD

VOLUME 91. Enzyme Structure (Part I)
Edited by C. H. W. HIRS AND SERGE N. TIMASHEFF

VOLUME 92. Immunochemical Techniques (Part E: Monoclonal Antibodies and General Immunoassay Methods)
Edited by JOHN J. LANGONE AND HELEN VAN VUNAKIS

VOLUME 93. Immunochemical Techniques (Part F: Conventional Antibodies, Fc Receptors, and Cytotoxicity)
Edited by JOHN J. LANGONE AND HELEN VAN VUNAKIS

VOLUME 94. Polyamines
Edited by HERBERT TABOR AND CELIA WHITE TABOR

VOLUME 95. Cumulative Subject Index Volumes 61–74, 76–80
Edited by EDWARD A. DENNIS AND MARTHA G. DENNIS

VOLUME 96. Biomembranes [Part J: Membrane Biogenesis: Assembly and Targeting (General Methods; Eukaryotes)]
Edited by SIDNEY FLEISCHER AND BECCA FLEISCHER

VOLUME 97. Biomembranes [Part K: Membrane Biogenesis: Assembly and Targeting (Prokaryotes, Mitochondria, and Chloroplasts)]
Edited by SIDNEY FLEISCHER AND BECCA FLEISCHER

VOLUME 98. Biomembranes (Part L: Membrane Biogenesis: Processing and Recycling)
Edited by SIDNEY FLEISCHER AND BECCA FLEISCHER

VOLUME 99. Hormone Action (Part F: Protein Kinases)
Edited by JACKIE D. CORBIN AND JOEL G. HARDMAN

VOLUME 100. Recombinant DNA (Part B)
Edited by RAY WU, LAWRENCE GROSSMAN, AND KIVIE MOLDAVE

VOLUME 101. Recombinant DNA (Part C)
Edited by RAY WU, LAWRENCE GROSSMAN, AND KIVIE MOLDAVE

VOLUME 102. Hormone Action (Part G: Calmodulin and Calcium-Binding Proteins)
Edited by ANTHONY R. MEANS AND BERT W. O'MALLEY

VOLUME 103. Hormone Action (Part H: Neuroendocrine Peptides)
Edited by P. MICHAEL CONN

VOLUME 104. Enzyme Purification and Related Techniques (Part C)
Edited by WILLIAM B. JAKOBY

VOLUME 105. Oxygen Radicals in Biological Systems
Edited by LESTER PACKER

VOLUME 106. Posttranslational Modifications (Part A)
Edited by FINN WOLD AND KIVIE MOLDAVE

VOLUME 107. Posttranslational Modifications (Part B)
Edited by FINN WOLD AND KIVIE MOLDAVE

VOLUME 108. Immunochemical Techniques (Part G: Separation and Characterization of Lymphoid Cells)
Edited by GIOVANNI DI SABATO, JOHN J. LANGONE, AND HELEN VAN VUNAKIS

VOLUME 109. Hormone Action (Part I: Peptide Hormones)
Edited by LUTZ BIRNBAUMER AND BERT W. O'MALLEY

VOLUME 110. Steroids and Isoprenoids (Part A)
Edited by JOHN H. LAW AND HANS C. RILLING

VOLUME 111. Steroids and Isoprenoids (Part B)
Edited by JOHN H. LAW AND HANS C. RILLING

VOLUME 112. Drug and Enzyme Targeting (Part A)
Edited by KENNETH J. WIDDER AND RALPH GREEN

VOLUME 113. Glutamate, Glutamine, Glutathione, and Related Compounds
Edited by ALTON MEISTER

VOLUME 114. Diffraction Methods for Biological Macromolecules (Part A)
Edited by HAROLD W. WYCKOFF, C. H. W. HIRS, AND SERGE N. TIMASHEFF

VOLUME 115. Diffraction Methods for Biological Macromolecules (Part B)
Edited by HAROLD W. WYCKOFF, C. H. W. HIRS, AND SERGE N. TIMASHEFF

VOLUME 116. Immunochemical Techniques (Part H: Effectors and Mediators of Lymphoid Cell Functions)
Edited by GIOVANNI DI SABATO, JOHN J. LANGONE, AND HELEN VAN VUNAKIS

VOLUME 117. Enzyme Structure (Part J)
Edited by C. H. W. HIRS AND SERGE N. TIMASHEFF

VOLUME 118. Plant Molecular Biology
Edited by ARTHUR WEISSBACH AND HERBERT WEISSBACH

VOLUME 119. Interferons (Part C)
Edited by SIDNEY PESTKA

VOLUME 120. Cumulative Subject Index Volumes 81–94, 96–101

VOLUME 121. Immunochemical Techniques (Part I: Hybridoma Technology and Monoclonal Antibodies)
Edited by JOHN J. LANGONE AND HELEN VAN VUNAKIS

VOLUME 122. Vitamins and Coenzymes (Part G)
Edited by FRANK CHYTIL AND DONALD B. MCCORMICK

VOLUME 123. Vitamins and Coenzymes (Part H)
Edited by FRANK CHYTIL AND DONALD B. MCCORMICK

VOLUME 124. Hormone Action (Part J: Neuroendocrine Peptides)
Edited by P. MICHAEL CONN

VOLUME 125. Biomembranes (Part M: Transport in Bacteria, Mitochondria, and Chloroplasts: General Approaches and Transport Systems)
Edited by SIDNEY FLEISCHER AND BECCA FLEISCHER

VOLUME 126. Biomembranes (Part N: Transport in Bacteria, Mitochondria, and Chloroplasts: Protonmotive Force)
Edited by SIDNEY FLEISCHER AND BECCA FLEISCHER

VOLUME 127. Biomembranes (Part O: Protons and Water: Structure and Translocation)
Edited by LESTER PACKER

VOLUME 128. Plasma Lipoproteins (Part A: Preparation, Structure, and Molecular Biology)
Edited by JERE P. SEGREST AND JOHN J. ALBERS

VOLUME 129. Plasma Lipoproteins (Part B: Characterization, Cell Biology, and Metabolism)
Edited by JOHN J. ALBERS AND JERE P. SEGREST

VOLUME 130. Enzyme Structure (Part K)
Edited by C. H. W. HIRS AND SERGE N. TIMASHEFF

VOLUME 131. Enzyme Structure (Part L)
Edited by C. H. W. HIRS AND SERGE N. TIMASHEFF

VOLUME 132. Immunochemical Techniques (Part J: Phagocytosis and Cell-Mediated Cytotoxicity)
Edited by GIOVANNI DI SABATO AND JOHANNES EVERSE

VOLUME 133. Bioluminescence and Chemiluminescence (Part B)
Edited by MARLENE DELUCA AND WILLIAM D. MCELROY

VOLUME 134. Structural and Contractile Proteins (Part C: The Contractile Apparatus and the Cytoskeleton)
Edited by RICHARD B. VALLEE

VOLUME 135. Immobilized Enzymes and Cells (Part B)
Edited by KLAUS MOSBACH

VOLUME 136. Immobilized Enzymes and Cells (Part C)
Edited by KLAUS MOSBACH

VOLUME 137. Immobilized Enzymes and Cells (Part D)
Edited by KLAUS MOSBACH

VOLUME 138. Complex Carbohydrates (Part E)
Edited by VICTOR GINSBURG

VOLUME 139. Cellular Regulators (Part A: Calcium- and Calmodulin-Binding Proteins
Edited by ANTHONY R. MEANS AND P. MICHAEL CONN

VOLUME 140. Cumulative Subject Index Volumes 102–119, 121–134

VOLUME 141. Cellular Regulators (Part B: Calcium and Lipids)
Edited by P. MICHAEL CONN AND ANTHONY R. MEANS

VOLUME 142. Metabolism of Aromatic Amino Acids and Amines
Edited by SEYMOUR KAUFMAN

VOLUME 143. Sulfur and Sulfur Amino Acids
Edited by WILLIAM B. JAKOBY AND OWEN W. GRIFFITH

VOLUME 144. Structural and Contractile Proteins (Part D: Extracellular Matrix)
Edited by LEON W. CUNNINGHAM

VOLUME 145. Structural and Contractile Proteins (Part E: Extracellular Matrix)
Edited by LEON W. CUNNINGHAM

VOLUME 146. Peptide Growth Factors (Part A)
Edited by DAVID BARNES AND DAVID A. SIRBASKU

VOLUME 147. Peptide Growth Factors (Part B)
Edited by DAVID BARNES AND DAVID A. SIRBASKU

VOLUME 148. Plant Cell Membranes
Edited by LESTER PACKER AND ROLAND DOUCE

VOLUME 149. Drug and Enzyme Targeting (Part B)
Edited by RALPH GREEN AND KENNETH J. WIDDER

VOLUME 150. Immunochemical Techniques (Part K: *In Vitro* Models of B and T Cell Functions and Lymphoid Cell Receptors)
Edited by GIOVANNI DI SABATO

VOLUME 151. Molecular Genetics of Mammalian Cells
Edited by MICHAEL M. GOTTESMAN

VOLUME 152. Guide to Molecular Cloning Techniques
Edited by SHELBY L. BERGER AND ALAN R. KIMMEL

VOLUME 153. Recombinant DNA (Part D)
Edited by RAY WU AND LAWRENCE GROSSMAN

VOLUME 154. Recombinant DNA (Part E)
Edited by RAY WU AND LAWRENCE GROSSMAN

VOLUME 155. Recombinant DNA (Part F)
Edited by RAY WU

VOLUME 156. Biomembranes (Part P: ATP-Driven Pumps and Related Transport: The Na,K-Pump)
Edited by SIDNEY FLEISCHER AND BECCA FLEISCHER

VOLUME 157. Biomembranes (Part Q: ATP-Driven Pumps and Related Transport: Calcium, Proton, and Potassium Pumps)
Edited by SIDNEY FLEISCHER AND BECCA FLEISCHER

VOLUME 158. Metalloproteins (Part A)
Edited by JAMES F. RIORDAN AND BERT L. VALLEE

VOLUME 159. Initiation and Termination of Cyclic Nucleotide Action
Edited by JACKIE D. CORBIN AND ROGER A. JOHNSON

VOLUME 160. Biomass (Part A: Cellulose and Hemicellulose)
Edited by WILLIS A. WOOD AND SCOTT T. KELLOGG

VOLUME 161. Biomass (Part B: Lignin, Pectin, and Chitin)
Edited by WILLIS A. WOOD AND SCOTT T. KELLOGG

VOLUME 162. Immunochemical Techniques (Part L: Chemotaxis and Inflammation)
Edited by GIOVANNI DI SABATO

VOLUME 163. Immunochemical Techniques (Part M: Chemotaxis and Inflammation)
Edited by GIOVANNI DI SABATO

VOLUME 164. Ribosomes
Edited by HARRY F. NOLLER, JR. AND KIVIE MOLDAVE

VOLUME 165. Microbial Toxins: Tools for Enzymology
Edited by SIDNEY HARSHMAN

VOLUME 166. Branched-Chain Amino Acids
Edited by ROBERT HARRIS AND JOHN R. SOKATCH

VOLUME 167. Cyanobacteria
Edited by LESTER PACKER AND ALEXANDER N. GLAZER

VOLUME 168. Hormone Action (Part K: Neuroendocrine Peptides)
Edited by P. MICHAEL CONN

VOLUME 169. Platelets: Receptors, Adhesion, Secretion (Part A)
Edited by JACEK HAWIGER

VOLUME 170. Nucleosomes
Edited by PAUL M. WASSARMAN AND ROGER D. KORNBERG

VOLUME 171. Biomembranes (Part R: Transport Theory: Cells and Model Membranes)
Edited by SIDNEY FLEISCHER AND BECCA FLEISCHER

VOLUME 172. Biomembranes (Part S: Transport Membrane Isolation and Characterization)
Edited by SIDNEY FLEISCHER AND BECCA FLEISCHER

VOLUME 173. Biomembranes [Part T: Cellular and Subcellular Transport: Eukaryotic (Nonepithelial) Cells]
Edited by SIDNEY FLEISCHER AND BECCA FLEISCHER

VOLUME 174. Biomembranes [Part U: Cellular and Subcellular Transport: Eukaryotic (Nonepithelial) Cells]
Edited by SIDNEY FLEISCHER AND BECCA FLEISCHER

VOLUME 175. Cumulative Subject Index Volumes 135–139, 141–167

VOLUME 176. Nuclear Magnetic Resonance (Part A: Spectral Techniques and Dynamics)
Edited by NORMAN J. OPPENHEIMER AND THOMAS L. JAMES

VOLUME 177. Nuclear Magnetic Resonance (Part B: Structure and Mechanism)
Edited by NORMAN J. OPPENHEIMER AND THOMAS L. JAMES

VOLUME 178. Antibodies, Antigens, and Molecular Mimicry
Edited by JOHN J. LANGONE

VOLUME 179. Complex Carbohydrates (Part F)
Edited by VICTOR GINSBURG

VOLUME 180. RNA Processing (Part A: General Methods)
Edited by JAMES E. DAHLBERG AND JOHN N. ABELSON

VOLUME 181. RNA Processing (Part B: Specific Methods)
Edited by JAMES E. DAHLBERG AND JOHN N. ABELSON

VOLUME 182. Guide to Protein Purification
Edited by MURRAY P. DEUTSCHER

VOLUME 183. Molecular Evolution: Computer Analysis of Protein and Nucleic Acid Sequences
Edited by RUSSELL F. DOOLITTLE

VOLUME 184. Avidin-Biotin Technology
Edited by MEIR WILCHEK AND EDWARD A. BAYER

VOLUME 185. Gene Expression Technology
Edited by DAVID V. GOEDDEL

VOLUME 186. Oxygen Radicals in Biological Systems (Part B: Oxygen Radicals and Antioxidants)
Edited by LESTER PACKER AND ALEXANDER N. GLAZER

VOLUME 187. Arachidonate Related Lipid Mediators
Edited by ROBERT C. MURPHY AND FRANK A. FITZPATRICK

VOLUME 188. Hydrocarbons and Methylotrophy
Edited by MARY E. LIDSTROM

VOLUME 189. Retinoids (Part A: Molecular and Metabolic Aspects)
Edited by LESTER PACKER

VOLUME 190. Retinoids (Part B: Cell Differentiation and Clinical Applications)
Edited by LESTER PACKER

VOLUME 191. Biomembranes (Part V: Cellular and Subcellular Transport: Epithelial Cells)
Edited by SIDNEY FLEISCHER AND BECCA FLEISCHER

VOLUME 192. Biomembranes (Part W: Cellular and Subcellular Transport: Epithelial Cells)
Edited by SIDNEY FLEISCHER AND BECCA FLEISCHER

VOLUME 193. Mass Spectrometry
Edited by JAMES A. MCCLOSKEY

VOLUME 194. Guide to Yeast Genetics and Molecular Biology
Edited by CHRISTINE GUTHRIE AND GERALD R. FINK

VOLUME 195. Adenylyl Cyclase, G Proteins, and Guanylyl Cyclase (in preparation)
Edited by ROGER A. JOHNSON AND JACKIE D. CORBIN

VOLUME 196. Molecular Motors and the Cytoskeleton (in preparation)
Edited by RICHARD B. VALLEE

VOLUME 197. Phospholipases (in preparation)
Edited by EDWARD A. DENNIS

VOLUME 198. Peptide Growth Factors (Part C) (in preparation)
Edited by DAVID BARNES, J.P. MATHER, AND GORDON H. SATO

[1] Epithelial Transport: An Introduction

By KARL JULIUS ULLRICH

Multicellular organisms exchange material, mostly solutes and water, with their environment. This is achieved by transport through their outer and inner body surfaces, skin, and intestinal and urinary tract. All are covered by epithelia, i.e., sheets of cells, which are held together by terminal bars (Fig. 1). Depending on the number of strands which form the terminal bars, one can discriminate between leaky and tight epithelia. In tight epithelia the transepithelial electrical resistance is high and most of the transepithelial transport proceeds transcellularly. Tight epithelia are able to create considerable concentration differences since passive backflux, which predominantly occurs paracellularly, is largely prevented. Such a situation is found (1) in the terminal parts of the intestine (colon and rectum), (2) in the distal parts of the urinary tract (collecting duct and urinary bladder), (3) in the excretory ducts of the various glands (sweat glands and salivary glands), and (4) in integuments (frog skin). By changing the rates of ion (Na^+, K^+, Ca^{2+}, Mg^{2+}, H^+/HCO_3^-, Cl^-, etc.) and water transport under the influence of hormones these epithelia are able to regulate the body salt and water content. In leaky epithelia, on the other hand, the transepithelial electrical resistance, and in particular the paracellular resistance, is low. Even with small driving forces large quantities of solutes and water can be transported paracellularly. This is advantageous where a large mass flow is required, as in (1) the small intestine to reabsorb NaCl, (2) the proximal tubule to reabsorb most of the filtered primary urine, and (3) the acini of glands to form a plasma-like primary secretion. Here, transcellular active transport processes create small osmotic gradients, which are sufficient to promote passive water flux followed by bulk flux of small solutes.

The key questions of research in the field of epithelial transport are as follows: (1) What is transported along which pathways and across which barriers? (2) How does the transport occur?

Logically, all analyses start with intact epithelial cell layers. These preparations allow the black box approach, i.e., variation of the outer compartments, mucosal–serosal or lumenal–contralumenal, and measurement of the respective transport rates, but do not yet provide answers to the above questions. By the black box approach it is possible to discriminate between (1) passive transport, which is driven by external driving forces, and (2) active transport, i.e., net transport, which is still present

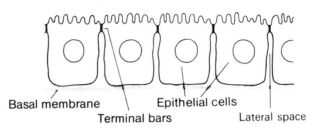

Fig. 1. Scheme of an epithelial cell layer (renal proximal tubule or small intestine).

when the external driving forces vanish. Active transport can be classified as primary active transport, driven by ATPases (Na^+, K^+-ATPase, H^+-ATPase, Ca^{2+}-ATPase), and as secondary active transport, driven by passive flux coupling of one substance with a second substance that undergoes primary active transport (for instance, transepithelial transport of D-glucose which is coupled to primary active Na^+ transport) (see Fig. 2).

To identify transport routes or transport mechanisms requires more sophisticated techniques. For example, by using voltage scanning or vibrating probe techniques it is possible to identify transport across individual cells or to separate cellular and paracellular pathways. The paracellular pathway proceeds through the strands of the terminal bars, the biochemical composition and regulation of which are still unknown, and through the lateral intercellular space. The transcellular pathway proceeds through the lumenal and the contralumenal cell membrane, usually in a polarized manner such that net reabsorption or net secretion results. It is obvious that separate analyses of lumenal and of contralumenal transport processes are required to understand their concerted interaction in achieving net transepithelial transport. In other words: The physiological function of transepithelial transport can be understood only if the molecular events at each cell side are elucidated by biochemical and biophysical methods. Thus, each method has its strength and its limitations. A few examples are as follow: An experimenter, who impales a cell with an intracellular microelectrode and applies electric circuit analysis, usually knows in what cell type the electrode is located. He or she has, however, to consider the whole electrical circuit, which means both cell membranes and paracellular resistance. On the other hand, an experimenter who intends to work with membrane vesicles from a certain cell side will be able to impose quite easily different driving forces for solute transport, but must bear the uncertainty created by working with an admixture of vesicles from undefined locations. Nevertheless, he or she is able to study almost all transport

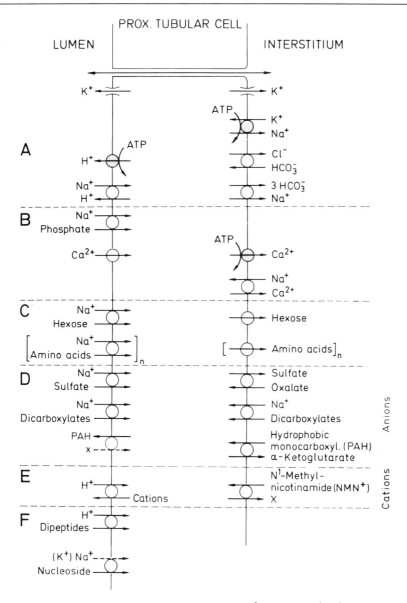

FIG. 2. Transporters in the renal proximal tubular cells: ATPase; channel; uniporter; exchanger; cotransporter. n, Several amino acid transport systems; PAH, p-aminohippurate.

processes by using tracer or fluorescence indicator techniques, as well as any which are not electrogenic. In a third instance an experimenter who works with the patch-clamp method can get signals from a cell in its natural surrounding by a cell-attached patch pipet. The worker can, however, also impose different driving forces and channel modifiers on an excised patch, but is restrained to transport processes which proceed through channels and are electrogenic. To analyze the molecular events of each transport process it is necessary to identify and separate the different transport proteins. This is achieved by affinity or photoaffinity labeling and the general procedure of protein separation and purification. Furthermore, reconstitution of a functional transport unit into liposomes is obligatory to ensure the retention of a protein subunit or lipid constituent during the separation procedure just mentioned. Since for some solutes, for instance Na^+, HCO_3^-, and Ca^{2+}, several transporters participate in transepithelial transport, the quantitative contribution of each of them is important. Thus, once the transporters and their location within the cell are known a balance sheet would be helpful. It certainly helps to prevent confusion, if (as in the proximal renal tubule) more than 15 different transporters are located at each cell side, half of them being Na^+ cotransporters (Fig. 2). In future studies those methods which identify the transporters *in situ* might be of great value, for instance, electron microscopic localization with monoclonal antibodies. Studies of transporters at the molecular level might utilize recombinant DNA techniques, introduction or attachment of molecules emitting molecular signals, production of mixed detergent–transporter crystals, and X-ray crystallography.

[2] Determination of Paracellular Shunt Conductance in Epithelia

By G. KOTTRA and E. FRÖMTER

Introduction

Ever since it became clear that epithelia do not simply consist of two cell membranes in series but may have significant shunt pathways between the cells[1-4] researchers have been concerned with identifying the structural

[1] H. H. Ussing and E. E. Windhager, *Acta Physiol. Scand.* **61**, 484 (1964).
[2] T. W. Clarkson, *J. Gen. Physiol.* **50**, 695 (1967).
[3] E. L. Boulpaep, in "Electrophysiology of Epithelial Cells" (G. Giebisch, ed.), p. 91. Schattauer, Stuttgart, Federal Republic of Germany, 1971.
[4] E. Frömter, *J. Membr. Biol.* **8**, 259 (1972).

equivalent of the shunt and quantifying the shunt permeability in relation to the permeability of the transcellular transport pathway. In epithelia (such as that in *Necturus* gall bladder) which exhibit a relatively simple microscopic structure and may be easily investigated with microelectrodes, the problem has been largely solved — at least so far as ion permeation is concerned — but in others, such as frog skin, the identification and quantification of the shunt path and the determination of its transport properties remains a considerable challenge.

The classical approach to quantify the shunt conductance is to measure and compare transepithelial and cell membrane resistances. Since the latter resistances are often difficult to determine a number of more or less indirect approaches have been devised. These are based on specific assumptions, the validity of which is usually difficult to ascertain.

In this chapter we attempt to critically review the electrophysiological methods which have been used in the past to quantify epithelial shunt conductances. Emphasis is on flat sheet epithelia. A number of specific problems involved with shunt analysis in tubular epithelia has been discussed in a methodological review[5] and will be mentioned only briefly here.

The methods to be discussed differ in their ability to represent the electrical properties of given epithelia appropriately. All are based on equivalent circuit representations of epithelia which, for various reasons, are gross oversimplifications. This is true, for example, for the circuits of Fig. 1a–c (so-called "lumped" models) in which the epithelium is essentially represented by three elements: two resistors (R_a and R_{bl} for the apical and basolateral cell membrane, respectively) arranged in series and an additional resistor (R_{sh} for the paracellular shunt) arranged in parallel. These models yield erroneous results whenever the resistance of the lateral space cannot be neglected. To account for lateral space effects, the "distributed" model (Fig. 1d) of the lateral space and lateral cell membrane has been developed.[4] In this model the shunt pathway consists of the tight junction resistance (R_j) and of a chain of resistors for the lateral intercellular space (r_{lis}, with $R_{lis} = \Sigma r_{lis}$). The distributed model is particularly important for leaky epithelia with low-resistance tight junctions.

There are other more general shortcomings, however. Virtually all equivalent circuits treat all resistors as constant, although, in reality, the current voltage relation of cell membranes is usually nonlinear even in the narrow range of physiological membrane voltages (-90 to $+40$ mV). This is not yet problematic so long as the analysis is restricted to resistance measurements (as in models 1a and 1c–d of Fig. 1) and so long as the

[5] E. Frömter, *Kidney Int.* **30**, 216 (1986).

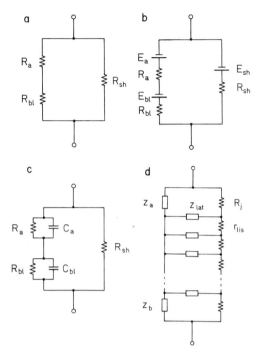

FIG. 1. (a) Simple lumped model. (b) Lumped model with electromotive forces representing zero-current potentials E_i for each membrane element. (c) Lumped model including capacitances C_a and C_{bl} to represent the apical and basal cell membrane impedance. (d) Distributed model; the Z elements denote parallel RC units for cell membrane impedances.

measurements are restricted to the linear range (determination of so-called "small signal" conductance or "slope" conductances at the normal operating point of the system). The analysis becomes problematic, however, if (slope) resistances and zero current potentials E_0 are used together in one circuit as in Fig. 1b and in methods 2.2, 2.3, and 2.4 below. If the resistance is nonlinear, extrapolation of the slope conductance to the zero current membrane voltage does not yield the proper E_a or E_{bl} that may be equated with ion concentration ratios on the basis of the Nernst equation.[6,7] It is not surprising, therefore, that these methods have led to a number of inconsistent results.

Another uncertainty comes up in impedance measurements (see methods 3.1 and 3.2 below). It relates to the fact that in reality biological membranes cannot be represented by ideal capacitors (e.g., C_a and C_{bl})

[6] S. I. Helman and S. M. Thompson, *Am. J. Physiol.* **243**, F519 (1982).
[7] S. M. Thompson, *Am. J. Physiol.* **250**, C333 (1986).

because they may exhibit dielectric loss. In addition it may be sometimes difficult to separate membrane capacitances from polarization effects of unstirred layers, but the latter problem arises only at low-frequency currents (≈ 1 to 10 Hz).

1. Methods Based on Equivalent Circuit 1a

1.1 Quantitative Comparison of Direct Current (DC) Cell Membrane Resistances and DC Transepithelial Resistance

This method requires that the transepithelial resistance, R_T, and the cell membrane resistances, R_a and R_{bl}, be measured and compared. In contrast to R_T, the determination of R_a and R_{bl} is usually quite difficult. Since most epithelial cells are electrically coupled with their neighbors,[8] a current that is injected into a single cell does not only pass across R_a and R_{bl} into the apical and basal fluid compartments, respectively, but spreads also radially within the epithelial cell layer. A theoretical analysis of this problem, the so-called two-dimensional cable analysis,[4,9] predicts that radial voltage (V) attenuation within the cell layer follows a first order Bessel function K_0 which can be defined by two parameters: a proportionality coefficient A and an apparent length constant λ:

$$V(x) = AK_0(x/\lambda) \qquad (1)$$

where x is the radial distance. From these parameters two resistances can be calculated: R_z, which describes the resistance of the parallel arrangement of R_a and R_{bl}, and R_x, which represents the coupling resistance for horizontal current flow in the cell layer:

$$R_z = [(1/R_a) + (1/R_{bl})]^{-1} = (2\pi A \lambda^2)/i_0 \qquad (2)$$
$$R_x = (2\pi A)/i_0 \qquad (3)$$

where i_0 is the injected current.

On first view, to define the voltage attenuation within the cell layer may appear quite demanding and time consuming. However, once it has been verified that the voltage as function of distance x from the point of current injection obeys the proper Bessel function, it is sufficient to determine the current-induced voltage change at only two distances x, since two data points suffice to define the Bessel function and hence to determine A and λ. In practice it may be preferable to collect more data points in order to average out individual measuring errors.

[8] W. R. Loewenstein, *Physiol. Rev.* **61**, 829 (1981).
[9] L. Reuss and A. L. Finn, *J. Gen. Physiol.* **64**, 1 (1974).

Alternatively, one can also try to circumvent cable analysis, and determine R_z by measuring the input resistance of an individual cell after electrical uncoupling. In practice, however, this approach is of little value, since uncoupling by depolarizing current[10] will certainly alter R_a and R_{bl}, thus invalidating the entire concept and the same might happen with chemical uncouplers such as high CO_2 pressure.

Once R_z has been determined, R_a and R_{bl} can be calculated with the help of α, the direct current (DC) voltage divider ratio (VDR), which in the simple models of Fig. 1a–c equals R_a/R_{bl}. Finally the shunt resistance, R_{sh}, is obtained by comparing the sum of R_a and R_{bl} with R_T:

$$R_{sh} = [(R_a + R_{bl})R_T]/(R_a + R_{bl} - R_T) \qquad (4)$$

The method can also be applied to tubular structures where the voltage attenuation along the tubular lumen and along the cell column has to be measured separately. Both follow double-exponential functions from which the three resistances R_a, R_{bl}, R_{sh}, as well as R_T can be derived by data-fitting techniques.[5,11–13]

Comments. The method is straightforward but technically rather involved since it requires the insertion of at least two (or better three) microelectrodes simultaneously into nearby cells, which must be recognized under the microscope and the distance between which must be accurately measured. In tubular epithelia an even greater number of measurements is necessary, again including intracellular recordings that must be combined with accurate distance measurements.

The advantage of the method is that it does not rely on unproved assumptions such as constancy of zero-current potentials, which is a precondition for most of the other methods discussed below. On the other hand, the method has limited power, because it is based on the "lumped" model equivalent circuit. The problem manifests itself in the use of the voltage divider ratio, α, which is taken to equal R_a/R_{bl}, while in fact it may be lower if R_{bl} was overestimated due to a significant contribution of the lateral space resistance, R_{lis}. As a consequence the estimate of R_{sh} will be in error, as has already been discussed in the literature.[14] Problems with the appropriateness of the measured α values have indeed been noticed in various epithelial preparations, including frog skin,[15] which is commonly held to obey the relation $R_j \gg R_{lis}$, with R_{lis} being negligible.

[10] S. A. Lewis, D. C. Eaton, and J. M. Diamond, *J. Membr. Biol.* **28**, 41 (1976).
[11] T. Anagnostopoulos, J. Teulon, and A. Edelman, *J. Gen. Physiol.* **75**, 553 (1980).
[12] T. Hoshi, K. Kawahara, R. Yokoyama, and K. Suenaga, *Adv. Physiol. Sci.* **2**, 403 (1981).
[13] W. B. Guggino, E. E. Windhager, E. L. Boulpaep, and G. Giebisch, *J. Membr. Biol.* **67**, 143 (1982).
[14] E. L. Boulpaep and H. Sackin, *Curr. Top. Membr. Transp.* **13**, 169 (1980).
[15] W. Nagel, J. F. Garcia Diaz, and A. Essig, *Pfluegers Arch.* **399**, 336 (1983).

1.2. Paired Measurements of R_T and VDR in Response to Alteration of One Membrane Resistance

This method consists of measuring R_T and α while one surface of the epithelium is suddenly perfused with a solution that alters one or more ion transport mechanisms in the respective cell membrane by partial or total ion substitution, or by application of inhibitors or transport facilitators. The method is easier to apply than the above discussed method 1.1, since it requires only one intracellular microelectrode (for determination of α). On the other hand, it is less reliable, since it depends on the precondition that the solution change affects only one single resistor of the circuit, leaving the others unchanged. If this precondition is fulfilled, and if R_a was changed from R_a to R_a', four independent data are available from which R_a, R_{bl}, and R_{sh} as well as R_a' can be determined. The respective equations which are presented here in the form published previously[16] were first derived by Reuss and Finn[9] (see also Refs. 17 and 18):

$$R_a = [\alpha(\alpha - \alpha')R_T R_T']/(1 + \alpha)(1 + \alpha')(R_T V - R_T') \quad (5)$$

$$R_{bl} = [(\alpha - \alpha')R_T R_T']/(1 + \alpha)(1 + \alpha')(R_T - R) \quad (6)$$

$$R_{sh} = [(\alpha - \alpha')R_T R_T']/[(1 + \alpha)R_T' - (1 + \alpha')R_T] \quad (7)$$

where $\alpha = R_a/R_{bl}$, $\alpha' = R_a'/R_{bl}$, and R_T' is the transepithelial resistance measured after the pertubation.

The same principles have been applied to tubular epithelia where R_T must be determined by cable analysis. In that case, however, the determination of VDR presents a problem, because usually $\alpha \neq R_a/R_{bl}$, but changes from an overestimate of R_a/R_{bl} near the point of current injection to an underestimate at large distances from the current source. As analyzed previously[19] the error has a minimum close to $x = \lambda$, i.e., at around one length constant from the point of current injection.

Comments. While technically speaking the method is comparatively simple and the calculations are straightforward the quantitative validity of the analysis depends on two conditions: (1) the accuracy of α as a measure of R_a/R_{bl} (cf. lumped versus distributed model!) and (2) whether indeed only one resistor of the equivalent circuit of Fig. 1a was altered and the others remained constant. Hence, when performing such experiments the following precautions must be taken.

1. The solution change must be very fast (depending on cell size, in the order of seconds) and it must be ascertained that only instantaneous

---

[16] E. Frömter and B. Gebler, *Pfluegers Arch.* **371**, 99 (1977).
[17] R. A. Frizzell and S. G. Schultz, *J. Gen. Physiol.* **59**, 318 (1972).
[18] T. Anagnostopoulos, *J. Physiol. (London)* **233**, 375 (1973).
[19] D. I. Cook and E. Frömter, *Pfluegers Arch.* **403**, 388 (1985).

changes of R_T and α are recorded. If solution change and recording are too slow, intracellular ion concentrations may change and this may secondarily alter the cell membrane resistances, or even the shunt resistance.

2. For each individual epithelium the proper experimental maneuver must be selected.

In tight epithelia, such as frog skin, urinary bladder, or cornea, for example, amiloride can be given to suddenly increase R_a, or amphotericin B or gramicidin D to decrease it.[10,16,20,21] Although such measurements yield valuable estimates of the lumped circuit parameters, some uncertainty remains with regard to the quantitative reliability of the analysis because of possible side effects on R_{bl} and/or R_{sh}. For example, if the basolateral cell membrane contains voltage-dependent conductances, hyper- or depolarization of this membrane might change R_{bl}. On the other hand, ionophores might not be restricted to the apical cell membrane but might also be taken up into the basolateral cell membrane. Moreover, as amiloride blocks tight junctions in leaky epithelia,[22,23] it has been postulated that it exerts similar effects in tight epithelia.[15,24,25]

In leaky epithelia, such as renal proximal tubule, small intestine, or gallbladder, the situation is similar. Here sugars and/or amino acids have been given to activate rheogenic Na^+–substrate cotransport and thereby increase the apical cell membrane conductance as well as Ba^{2+} or high K^+ concentrations to increase or, respectively, decrease R_a or R_{bl}.[26-28] In the former case, however, besides R_a, also R_{bl} could have changed because of voltage-dependent conductances and/or R_{lis} could have changed in response to changes in the rate of fluid transport. In the latter case, besides possible effects on R_{bl} from voltage-dependent conductances, R_j could have decreased by the presence of high K^+ or increased[29] in response to Ba^{2+}.

We have tried to quantify the possible errors that can be made by neglecting such side effects in Appendix A. Generally speaking one can conclude that the greater the change of α during an experimental maneuver the more accurate the determination of R_{sh} becomes. Also, the greater

[20] N. K. Wills, S. A. Lewis, and D. C. Eaton, *J. Membr. Biol.* **45**, 81 (1979).
[21] S. A. Lewis and N. K. Wills, *J. Membr. Biol.* **67**, 45 (1982).
[22] R. S. Balaban, L. J. Mandel, and D. J. Benos, *J. Membr. Biol.* **49**, 363 (1979).
[23] G. Kottra and E. Frömter, *J. Exp. Biol.* **106**, 217 (1983).
[24] R. G. O'Neil and S. I. Helman, *Am. J. Physiol.* **231**, 164 (1976).
[25] L. G. M. Gordon, *J. Membr. Biol.* **52**, 61 (1980).
[26] E. Frömter, *Pfluegers Arch.* **393**, 179 (1982).
[27] R. Greger and E. Schlatter, *Pfluegers Arch.* **396**, 315 (1983).
[28] E. Bello Reuss, *J. Physiol. (London)* **370**, 25 (1986).
[29] G. Kottra and E. Frömter, *Pfluegers Arch.* **415**, 718 (1990).

the value of R_{sh} in comparison to R_a plus R_{bl}, the more uncertain its accuracy becomes.

1.3. Plot of G_T versus VDR in Response to Alteration of One Cell Membrane Resistance

A variant of technique 1.2 was described by Lewis et al.[10] It is based on the following equation which is equivalent to Eqs. (5–7) derived from the circuit in Fig. 1a:

$$G_T = G_{sh} + G_{bl}(1 + \alpha)^{-1} \qquad (8)$$

with $G_T = 1/R_T$, $G_{bl} = 1/R_{bl}$, and $G_{sh} = 1/R_{sh}$. This equation can be applied to experiments in which R_T and α are recorded while the apical cell membrane conductance is progressively changed by apical application of an ionophore, such as nystatin. If the data are plotted as G_T versus $(1 + \alpha)^{-1}$, Eq. (8) predicts a straight line which intercepts with the conductance axis at G_{sh} and has a slope of G_{bl}.

Comments. The method is not superior to the approach defined by Eqs. (5–7) above, since the same problems may arise that have been commented on before. In particular it seems questionable whether R_{bl} does indeed remain constant during the progressive action of the ionophore as the theory requires, since the experimental design does not fulfill the criterion of instantaneous action and measurement. In fact, data which have been obtained with this method and were published in conformation with Eq. (8) might as well be interpreted as exhibiting a systematic curvilinear relationship, rather than the postulated straight line.[20]

2. Methods Based on Equivalent Circuit 1b

2.1. Measurements of R_T, VDR, and of Transepithelial and Cell Membrane Potentials in Response to Alteration of One Membrane Resistance

This method does not differ much from method 1.2. Again a solution change is performed which should affect the ionic conductance of only *one* cell membrane, and the effect on α is recorded as well as the change in transepithelial potential difference, ΔV_T, and the change in membrane potential across the contralateral cell membrane, ΔV_a or ΔV_{bl}. In addition R_T must be known. If, for example, the apical membrane is altered but E_{bl}, E_{sh}, R_{bl} and R_{sh} remain constant, we obtain

$$R_{sh} = [(1 + \alpha + \beta)/(1 + \alpha)]R_T \qquad (9)$$

where $\alpha = R_a/R_{bl}$ and $\beta = \Delta V_{bl}/\Delta V_T$.

This approach differs from method 1.2, in that β instead of α' is used for the calculations. However, since α' is also known, the results of both approaches can be directly compared.

Comments. Essentially the same preferences and restrictions apply and the same precautions should be observed which have been discussed with regard to method 1.2 above. With method 2.1, however, it is particularly important to make fast solution changes and to record the potential changes instantaneously, since the additional restriction applies that some of the zero-current potentials be constant. The method has been used in *Necturus* urinary bladder in which the Na^+ channel blocker amiloride was applied to the apical surface.[16] A comparison of methods 1.2 and 2.1 in those experiments gave virtually identical results.

Alternatively, combining methods 1.2 and 2.1 one obtains five data points from which a total of five circuit parameters can be determined. Hence, if a solution change affects R_a and R_{sh} simultaneously, both R'_a and R'_{sh} can be determined.[30] However, the validity of such calculations rests entirely on whether or not R_{bl} and E_{bl} remain constant. This assumption, however, cannot be tested independently unless more sophisticated techniques are applied (see below).

2.2. Measurements of R_T, VDR, and of the Relationship between Cell Membrane Potential and Short Circuit Current (I_{sc})

Disregarding the fundamental problems with the circuit of Fig. 1b, various methods have been devised for determination of shunt conductances in tight epithelia such as amphibian skin and urinary bladder which generate significant short circuit currents (I_{sc}). If an epithelium is short circuited, current flow across R_{sh} vanishes, and $V_a = V_{bl}$. If I_{sc} is now altered by applying, for example, amiloride apically the simplified treatment of circuit 1b yields the following equation:

$$V_a = V_{bl} = E_{bl} - R_{bl}I_{sc} \tag{10}$$

which predicts a straight-line relationship between V_{bl} and I_{sc} provided E_{bl} and R_{bl} remain constant.[31,32] From the slope of this relation R_{bl} can be directly determined. If in addition R_T and α are measured in the same experiments, R_a and R_{sh} can be readily calculated. In analogy to that, I_{sc} can be altered by applying ouabain basolaterally to determine R_a.

Comments. The method is technically rather simple, but the validity of the data depends critically on whether or not the precondition (constancy

---

[30] S. Henin, D. Cremaschi, T. Schettino, G. Meyer, C. L. L. Donin, and F. Cotelli, *J. Membr. Biol.* **34,** 73 (1977).
[31] S. G. Schultz, R. A. Frizzell, and H. N. Nellans, *J. Theor. Biol.* **65,** 215 (1977).
[32] S. I. Helman and R. S. Fisher, *J. Gen. Physiol.* **69,** 571 (1977).

of E_{bl}) is fulfilled or, in other words, whether the conductances are constant in the entire range between open circuit and short circuit. By applying increasing concentrations of amiloride Nagel[33] attempted to derive the current–voltage *(I–V)* relation of the basolateral cell membrane with this approach; however, the results showed marked nonlinearity and in some cases even slopes with negative resistances. In addition, comparison of the shunt resistances calculated with methods 1.2 and 2.2 from the same experiments showed discrepancies of between 0.185- and 1.13-fold.[33] This indicates that E_{bl} was not constant or that the circuit of Fig. 1b did not adequately describe the data.

2.3 Plots of Transepithelial Conductance versus I_{sc} in Response to Alteration of One Cell Membrane Resistance

While the previously discussed methods require intracellular measurements with one or two microelectrodes it was desirable that methods be found which allow the shunt conductance to be determined exclusively from macroscopic measurements. From the viewpoint of principle this is, of course, impossible since one cannot expect to obtain insight into the structure of an unknown membrane by simply studying its input/output relations. On the other hand, it might have been possible that in some cases enough information was already available to justify a purely macroscopic approach.

Based on present-day knowledge of epithelial transport a method has been devised to determine the shunt resistance of tight epithelia by appropriately varying the apical cell membrane resistance and simply measuring $G_T = 1/R_T$ and the short circuit current, I_{sc}. Provided R_{bl}, R_{sh}, E_a, and E_{bl} remain constant the equivalent circuit of Fig. 1b yields

$$G_T = G_{sh} + (E_a + E_{bl})^{-1} I_{sc} \qquad (11)$$

Accordingly, if the preconditions are fulfilled, plotting the data as G_T versus I_{sc} should show a linear relationship with an intercept on the conductance axis equal to $G_{sh} = 1/R_{sh}$ (see Fig. 2a). Alternatively the same information can be obtained from a plot of V_T versus R_T according to the following equation where, in the ideal case, the linear relation between V_T and R_T intercepts with the resistance axis at R_{sh}:

$$1 = [V_T/(E_a + E_{bl})] + (R_T/R_{sh}) \qquad (12)$$

The introduction of this concept goes back to a paper by Yonath and Civan,[34] who apparently used the word shunt in the sense of comprising all

---

[33] W. Nagel, *J. Membr. Biol.* **42**, 99 (1978).
[34] J. Yonath and M. M. Civan, *J. Membr. Biol.* **5**, 366 (1971).

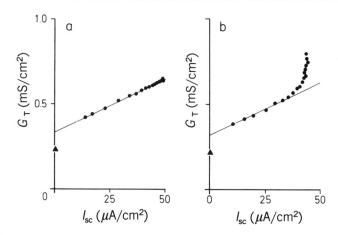

FIG. 2. G_T vs I_{sc} plots obtained with 1 mM novobiocin on two frog skins. The triangles denote the G_{sh} values determined with amiloride. (From Rick et al.[39])

ion flux mechanisms in parallel with the Na$^+$ pump; however, the concept was later reinterpreted by Lewis and collaborators,[10] who also derived Eq. (12) as a method for determination of the paracellular shunt conductance.[20] This method has already been widely used in tight epithelia such as frog skin, urinary bladder, colon, and others.[20,21,35-39] R_a was either increased by increasing concentrations of amiloride or decreased by ionophores such as amphotericin B, gramicidin D, nystatin, novobiocin, silver ions, or by stimulation with vasopressin.

Comment. The method is subject to the general criticism of the underlying model circuit (Fig. 1b; see Introduction). In addition, its application is restricted to epithelia whose properties are already largely known. It is *not* capable of identifying paracellular shunts in epithelia which have not yet been further investigated. Furthermore, the quantitative validity of the derived data must be considered with precaution, because usually the measurements do not fulfill the criteria of fast perturbation and instantaneous recordings. Indeed, deviations from a straight-line relationship between G_T and I_{sc} have often been observed (see Fig. 2b) and R_{sh} values derived either by decreasing R_a with ionophores or increasing R_a with amiloride usually differ from each other[20] (see also Fig. 2).

[35] L. C. Isaacson, *J. Membr. Biol.* **30**, 301 (1977).
[36] P. U. Feig, G. D. Wetzel, and H. S. Frazier, *Am. J. Physiol.* **232**, F448 (1977).
[37] W. S. Marshall and S. D. Klyce, *J. Membr. Biol.* **73**, 275 (1983).
[38] N. K. Wills and C. Clausen, *J. Membr. Biol.* **95**, 21 (1987).
[39] R. Rick, A. Dörge, and E. Sesselmann, *Pfluegers Arch.* **411**, 243 (1988).

In Appendix B we report some model calculations which define the validity limits of the method quantitatively. From this we deduce that the method is restricted to the simple model of tight Na^+-absorbing epithelia in which R_a is *exclusively* selective for Na^+ and to the use of ionophores that are also purely selective for Na^+. Alternatively, if the cell membrane is permeable to more than one ion species present, an ionophore must be used whose relative selectivity properties are identical with the relative ionic permeabilities of the respective cell membrane in the control state. *Note:* Even a straight-line relationship does not prove that the estimate of G_{sh} from the intercept is correct; also, the method is not able to distinguish a paracellular shunt from a transcellular Cl^- shunt or K^+ shunt.

A special case of method 2.3 is the determination of R_{sh} by blocking the apical cell membrane conductance with amiloride. If the apical cell membrane was exclusively permeable to Na^+, R_T in the presence of amiloride should equal R_{sh}. This approach has been widely used in frog skin and other tight epithelia, particularly in Cl^--free solutions.[40] On the other hand, it is also known that K^+ channels are present in the apical cell membrane of the skin of at least some frog species[41,42] and the question about a possible effect of amiloride on R_j is still open.

2.4. Steady State Measurements of V_T, R_T, VDR, and of the Intracellular Activity of an Ion Species for Which One of the Cell Membranes Is Permselective

This method was published recently.[43] It is based on the analysis of current flow in an equivalent circuit such as that depicted in Fig. 1b, with the restriction that all current across one cell membrane (say membrane *a*) is exclusively carried by one ion species (*i*). If the intracellular activity of *i* can be measured the following equation allows calculation of R_{sh}:

$$R_{sh} = R_T \left[1 - \frac{V_T}{[V_a + (RT/z_i F)] \ln(a_{i,c}/a_{i,o})} \frac{\alpha}{1+\alpha} \right] \quad (13)$$

where z_i, F, R, and T are valence, Faraday constant, gas constant, and absolute temperature and $a_{i,c}$ and $a_{i,o}$ are the intra- and extracellular activities of the permeating ion species. A similar equation has been derived for the short-circuited state in which the I_{sc} appears rather than V_T.

Comment. The main advantage of the method is the fact that only steady state measurements are required. This eliminates a number of

---

[40] J. Warncke and B. Lindemann, *J. Membr. Biol.* **86**, 255 (1985).
[41] W. Zeiske and W. Van Driessche, *J. Membr. Biol.* **47**, 77 (1979).
[42] W. Nagel and W. Hirschmann, *J. Membr. Biol.* **52**, 107 (1980).
[43] N. J. Willumsen and R. C. Boucher, *Am. J. Physiol.* **256**, C1054 (1989).

problems discussed above. On the other hand, as with the preceding two methods (2.2 and 2.3), this method relies on the assumption that membrane resistances are linear in the range between open and closed circuit, which normally is not true. Other, more technical, problems arise if other ion conductances must be blocked to establish the proper measuring conditions or if ion-selective microelectrode readings must be corrected for interferences.

2.5 Determination of R_{sh} from Break Point in the Transepithelial Current Voltage Relation

This method is based on a special form of the equivalent circuit 1b. When studying transepithelial current–voltage *(I–V)* relations in frog skin, Helman and Fischer[32] observed that the *I–V* curve consisted of linear segments with distinct break points in between, where the slope suddenly changed to a new constant value. These observations suggested that the apical cell membrane rectified and hence could be modeled by a parallel arrangement of two resistors of different magnitude, each having an ideal diode in series to allow either only inflow or outflow of current. If this model is correct the transition point between two segments in the *I–V* curve should then be the point at which current flow across the apical cell membrane and hence across the entire cell vanishes. One can conclude then that under those conditions all transepithelial current flows across the shunt path so that R_{sh} can be directly calculated by dividing the break point voltage through the applied current.

Comments. Although this method is extremely simple and requires only transepithelial measurements it has only rarely been used in the past, probably because the break points in the *I–V* relations are not always well-enough defined. This may be related to the fact that in principle the applicability of the method may be restricted to epithelia having an apical cell membrane that is exclusively selective for Na^+. Any other parallel ion conductances may be expected to reduce the sharpness of the break point.

3. Methods Based on Equivalent Circuits 1c and d (Impedance Techniques)

Unfortunately, measurements of the frequency-dependent transepithelial impedance alone do not allow one to recognize and quantify the presence of a shunt path. This derives from the impossibility of identifying parallel elements inside a two-terminal network by simply studying its input–output relations. (In contrast, it is well known that serial *RC* elements inside a two-pole network can be recognized provided they cause

3.1. Measurements of Transepithelial Impedance and of VDR in the Time Domain

If a constant current pulse is passed across an epithelium, the equivalent circuit in Fig. 1c predicts that the voltage response will rise with a double-exponential time course toward a new steady value[44,45]:

$$\Delta V_T(t) = [K_1(1 - e^{-t/\tau_1}) + K_2(1 - e^{-t/\tau_2})]I \tag{14}$$

where I is the transepithelial current and the parameters K_1, K_2, τ_1, and τ_2 are functions of all circuit elements. Note that the time constants τ_1 and τ_2 differ from the intrinsic time constants of the individual cell membranes ($\tau_a = R_a C_a$ and $\tau_{bl} = R_{bl} C_{bl}$) except in the case $G_{sh} = 0$ (for details see Ref. 44). The analysis of this time course yields four independent parameters, which do not suffice to determine the five elements of circuit 1c. Consequently a further independent measurement is required, which — in addition — must contribute information from inside the cell layer. Such a measurement would be the DC voltage divider ratio. Hence, with these rather simple measurements the circuit is already fully defined. In reality, however, the approach is less promising because fitting such data with two exponential functions is usually difficult unless the time constants differ by nearly one order of magnitude or more.

The technique can be improved if not only the DC voltage divider ratio is recorded, but the full rising flank of the intracellular voltage in response to the transepithelial constant current step:

$$\Delta V_{bl}(t) = [K_3(1 - e^{-t/\tau_1}) + K_4(1 - e^{-t/\tau_2})]I \tag{15}$$

This time course of the intracellular potential response, which also obeys a double-exponential function and has the same time constants τ_1 and τ_2, allows the data to be fitted more easily. This is particularly true for the voltage response across the cell membrane with the smaller RC product, which exhibits a transient overshoot resulting from superposition of a fast positive and a slow negative exponential ($K_3 < 0$, $K_4 > 0$).

Comments. In spite of the simple and straightforward principle, impedance measurements in the time domain cannot be recommended for determination of the shunt resistance. The main problem is associated with fitting two exponentials in the data analysis (see above). From a technical

---

[44] K. Suzuki, G. Kottra, L. Kampmann, and E. Frömter, *Pfluegers Arch.* **394**, 302 (1982).
[45] S. A. Lewis and J. L. C. DeMoura, *J. Membr. Biol.* **82**, 123 (1984).

point of view the method is quite demanding, since analysis of the intracellular voltage rise time requires shielded intracellular microelectrodes to be used. In addition, compared to the experimental investment the results are of limited use since the underlying equivalent circuit (lumped model) is generally too simple. Once the experimental set-up is available it would always be preferable to perform measurements in the frequency domain, which yields more explicit information.

3.2. Measurements of the Transepithelial Impedance and of the VDR in the Frequency Domain

In these experiments, instead of a square wave pulse, a series of sine wave currents of constant amplitude is passed across the epithelium, either as preprogrammed single-frequency bursts,[40] or as pseudo-random binary sequence (PRBS) noise,[46] or as multifrequency superimposed sine wave bursts (MSSB), obtained by inverse Fourier transform from a selected sequence of logarithmically spaced frequencies.[47] The relative advantages of the different methods have been discussed.[47] In the latter two cases the spectral content of the voltage response is obtained by fast Fourier transform techniques: By dividing the cross-power spectrum through the autopower spectrum of the current signal the frequency-dependent impedance is obtained, which can be represented either as a Bode[48] diagram (logarithm of amplitude and phase versus logarithm of frequency) or as a Nyquist plot[47] (imaginary versus real component of impedance). The data are then fitted with the appropriate equations which describe the impedance of circuit 1c and 1d from which quantitative estimates of the circuit parameters can be derived.[48,49]

As in the case of the time domain analysis mentioned above, transepithelial data of this type alone are not sufficient to determine the five elements of the equivalent circuit of Fig. 1c. However, under specific conditions ($R_j \rightarrow \infty$) and after inclusion of additional information (e.g., the DC voltage divider ratio) it is possible to determine not only the elements of Fig. 1c but to analyze even the distributed equivalent circuit of Fig. 1d. This means that the lateral intercellular space resistance can be resolved.[38,50,51]

Further progress and confirmation of the appropriateness of the analysis and the validity of the results were obtained by combining the transepithelial impedance analysis with simultaneous recordings of the fre-

[46] C. Clausen and J. M. Fernandez, *Pfluegers Arch.* **390**, 290 (1981).
[47] G. Kottra and E. Frömter, *Pfluegers Arch.* **402**, 409 (1984).
[48] C. Clausen, S. A. Lewis, and J. M. Diamond, *Biophys. J.* **26**, 291 (1979).
[49] G. Kottra and E. Frömter, *Pfluegers Arch.* **402**, 421 (1984).
[50] C. Clausen, P. S. Reinach, and D. C. Marcus, *J. Membr. Biol.* **91**, 213 (1986).
[51] J. R. Pappenheimer, *J. Membr. Biol.* **100**, 137 (1987).

quency-dependent voltage divider ratio.[47] In experiments on *Necturus* gallbladder such measurements have clearly indicated that the distributed model (Fig. 1d) is more appropriate to represent the epithelium than the lumped model, or in other words that R_{lis} cannot be neglected in comparison with R_j in this tissue.[49]

A problem with the impedance analysis of leaky epithelia, which remained unsolved thus far, was the proper determination of R_a. According to our experience R_a values showed a large scatter with skewed distribution and did not respond consistently to perturbations that might have been expected to increase R_a. This problem seems to reflect an unfavorable distribution of resistance values in the *Necturus* gallbladder equivalent circuit under control conditions.[52] Indeed, in gallbladder the major current flows across the shunt path and the small amount of current that passes across the cells produces only minor voltage changes across the cell membranes which may be lost in the scatter. However, further progress was achieved by combining the transepithelial impedance and frequency-dependent voltage divider measurements with a three-point two-dimensional cable analysis.[29] Together with the voltage divider ratio the latter data define R_a and R_{bl} more accurately and thereby restrict the variability of the fit to the impedance data.

Comments. The combined transepithelial and intracellular impedance analysis has the following advantages.

1. Compared to all other methods discussed above it not only yields a lumped value for the shunt resistance but allows one to differentiate between the resistance of the tight junction R_j and the resistance of the lateral space R_{lis}. It thereby enables one to study variations of R_j, for example in response to neurovegetative stimuli, which was not possible previously due to overlapping changes in R_{lis}.

2. It allows resistance changes of *all* circuit parameters to be followed with time resolution of a few seconds, for example during apical or basolateral solution changes. This enables the experimenter to test the assumptions on which some of the above-discussed methods to calculate R_{sh} are based.

3. It also allows changes in cell membrane capacitances to be followed that might occur in response to neurovegetative stimuli (insertion/removal of proteins and lipids into the cell membranes).

The method has the following drawbacks: (1) From a technical point of view it is extremely involved; (2) due to the specific distribution of cell membrane and shunt resistances in leaky epithelia, the estimated cell membrane resistances exhibit large scatter so that the measurements have

[52] P. Weskamp, *Pfluegers Arch.* **402**, R10 (1984).

to be supplemented with simultaneously collected two-dimensional cable measurements; and (3) a particular shortcoming is that the method cannot simply be transferred to tubular epithelia, since DC current injection into the tubular lumen via a microelectrode leads to a double-exponential voltage decay along the tubular axis, which complicates the analysis of time or frequency-dependent signals enormously.

Conclusions

In this chapter we have reviewed the methods used in the past to determine the paracellular shunt conductance in epithelia. These methods differ in their technical complexity and in their appropriateness to represent the electrical properties of given epithelia correctly. Some even rely on assumptions which are difficult to prove so that the results can only be used with great reservation.

Among the methods based on the lumped model circuits of Fig. 1a–c, the first method, direct comparison of cell membrane and transepithelial resistances, is the most reliable one, since it is not based on unproved assumptions. On the other hand, this method involves two-dimensional cable analysis with intracellular voltage measurements and microscopic distance measurements which must be performed with low enough current density and extreme care to yield reliable results.

All methods based on the equivalent circuit of Fig. 1b are technically less demanding but suffer from the uncertainty of whether or not the preconditions (perturbation of only one circuit element while all others remain constant) are indeed fulfilled. In addition, some of these methods are subject to the restriction of being applicable only to tight epithelia which perform rheogenic active transport. On the other hand, they offer the advantage of allowing the shunt conductance to be determined from simple transepithelial measurements without using even a single microelectrode. The applicability of those "macroscopic" methods, however, is restricted to epithelia which fulfill certain criteria (such as permselectivity of the apical cell membrane for Na^+), which must be ascertained before use.

The impedance techniques, finally, have the advantage that they are able to define a distributed equivalent circuit model of epithelia (Fig. 1d) which allows the shunt resistance to be separated into its contributions from tight junctions and lateral intercellular spaces, and thereby also provides more reliable values for the apical and basal cell membrane resistances; but they have their own limitations and are technically rather involved.

In summary, we conclude that an ideal, simple, and reliable method applicable to all epithelia does not exist. In order to obtain meaningful data the more complicated techniques have to be applied and the results from two or three different approaches must be compared before the data can be accepted as valid.

Appendix A

Method 1.2 for the calculation of cell membrane and shunt resistances depends on the assumption that only one resistance (generally R_a) changes after the experimental perturbation of the system. If this condition is not fulfilled, and hence R_{sh} and/or R_{bl} change simultaneously, the calculation leads to erroneous values of all circuit parameters. The possible errors in the determination of R_{sh} were in part analyzed in the past.[9]

In this Appendix we analyze the errors made when either R_{sh} or R_{bl} do not remain constant. The general case that both R_{bl} and R_{sh} change simultaneously with R_a leads to unpredictable results and hence will not be considered further here.

Case I: R_{sh} Changes, R_{bl} Remains Constant

The following notations will be used. The true control values are denoted as R_a, R_{bl}, R_{sh}, and $\alpha = R_a/R_{bl}$; the experimental values are denoted as R'_a, R'_{bl}, R'_{sh} and $\alpha' = R'_a/R'_{bl}$; and the (erroneously) calculated control values are denoted as R^*_a, R^*_{bl}, R^*_{sh} with $\alpha = R^*_a/R^*_{bl}$. The ratio p of the change in G_{sh} and G_T will be denoted as

$$p = \Delta G_{sh}/\Delta G_T \tag{A1}$$

R^*_a can be calculated from Eq. (5) as

$$R^*_a = \{[\alpha(\alpha' - \alpha)]/[(1 + \alpha)(1 + \alpha')]\} (1/\Delta G_T) \tag{A2}$$

To calculate the true value of R_a, however, the change of G_{sh} must be subtracted from the change in G_T:

$$R_a = \{[\alpha(\alpha' - \alpha)]/(1 + \alpha)(1 + \alpha')]\}[1/(\Delta G_T - \Delta G_{sh})] \tag{A3}$$

Dividing Eq. (A2) by Eq. (A3) leads to

$$R^*_a/R_a = (\Delta G_T - \Delta G_{sh})/\Delta G_T = 1 - p \tag{A4}$$

Since $\alpha = R_a/R_{bl} = R^*_a/R^*_{bl}$ it follows that

$$R^*_{bl}/R_{bl} = 1 - p \tag{A5}$$

Equations (A4) and (A5) show that a change of G_{sh} into the same direction as the change of G_T (i.e., $0 < p < 1$) will lead to an underestimation of the membrane resistances, while a change of G_{sh} into the opposite direction (i.e., $p < 0$) will overestimate R_a and R_{bl}.

The error in the determinations of R_{sh} can be calculated from

$$1/R_T = (1/R_{sh}) + [1/(R_a + R_{bl})] = (1/R^*_{sh}) + [1/(R^*_a + R^*_{bl})] \quad (A6)$$

Inserting $R^*_a = R_a(1 - p)$ and $R^*_{bl} = R_{bl}(1 - p)$ into Eq. (A6) and rearranging leads to

$$\frac{R^*_{sh}}{R_{sh}} = \frac{1}{[R_{sh}/(R_a + R_{bl})]\{1 - [1/(1 - p)]\} + 1} \quad (A7)$$

An analysis of Eq. (A7) shows that the error in the determination of R_{sh} does not only depend on the factor p, but also on the ratio $R_{sh}/(R_a + R_{bl})$. When this ratio is small, as in the case of leaky epithelia, the error becomes small.

Case II: R_{bl} Changes, R_{sh} Remains Constant

Using the same notation as above in the control state we have R_a, R_{bl}, R_{sh}, and $\alpha = R_a/R_{bl}$; in the experimental state we have R'_a, $R'_{bl} = pR_{bl}$, R_{sh}, and $\alpha' = R'_a/R'_{bl}$; and the (erroneously) calculated control values are R^*_a, R^*_{bl}, R^*_{sh}, and $\alpha = R^*_a/R^*_{bl}$ where p denotes the fractional change of R_{bl}. The transepithelial resistances in the control and experimental state are

$$1/R_T = (1/R_{sh}) + [1/(R_a + R_{bl})] = (1/R_{sh}) + \{1/[R_{bl}(1 + \alpha)]\} \quad (A8)$$

and

$$1/R'_T = (1/R_{sh}) + [1/(R'_a + R'_{bl})] = (1/R_{sh}) + \{1/[pR_{bl}(1 + \alpha')]\} \quad (A9)$$

From Eqs. (A8) and (A9) R_{bl} can be expressed as

$$R_{bl} = \{[p(1 + \alpha') - (1 + \alpha)]/[p(1 + \alpha)(1 + \alpha')]\}/\Delta G_T \quad (A10)$$

On the other hand, R^*_{bl} can be calculated using the assumption that $p = 1$:

$$R^*_{bl} = \{(\alpha' - \alpha)/[(1 + \alpha)(1 + \alpha')]\}/\Delta G_T \quad (A11)$$

Dividing Eq. (A11) by Eq. (A10) and rearranging the results yields

$$R^*_{bl}/R_{bl} = [p(\alpha' - \alpha)]/\{[p(1 + \alpha')] - (1 + \alpha)\} \quad (A12)$$

Since $\alpha = R_a/R_{bl} = R^*_a/R^*_{bl}$ the same expression is also true for the apical membrane:

$$R^*_a/R_a = [p(\alpha' - \alpha)]/\{[p(1 + \alpha')] - (1 + \alpha)\} \quad (A13)$$

The error R^*_{sh}/R_{sh} can be calculated from Eq. (A7), case I; however, the expression $(1 - p)$ must be replaced by the right-hand side of Eq. (A12):

$$R^*_{sh}/R_{sh} = \frac{1}{[R_{sh}/(R_a + R_{bl})]\{1 - [p(1 + \alpha') - (1 + \alpha)]/[p(\alpha' - \alpha)]\} + 1} \quad (A14)$$

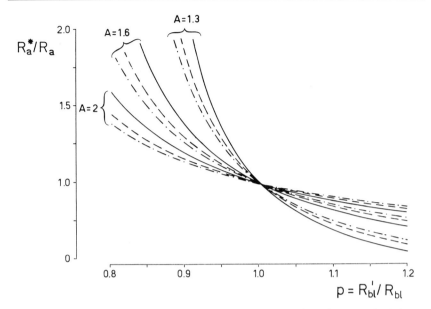

FIG. A1. Relative error of the calculated value of R_a as a function of an unnoticed change in R_{bl}. Values of α are as follow: 2 for linetype 1, 4 for linetype 2, 8 for linetype 3. The change of VDR is denoted as $A = \alpha'/\alpha$. $1(—), 2(---), 3(-\cdot-)$.

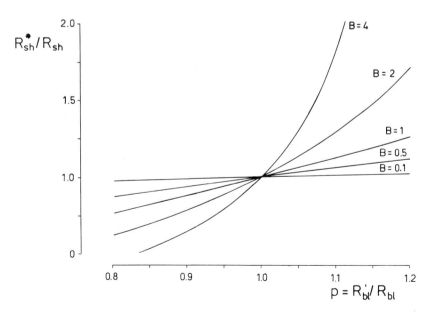

FIG. A2. Relative error of the calculated value of R_{sh} as a function of an unnoticed change in R_{bl}. The "tightness" of the epithelium is described with $B = R_{sh}/(R_a + R_{bl})$. For all lines: $\alpha = 4, \alpha' = 8$.

Numerical calculations of the error in R_a as a function of an unnoticed change in R_{bl} are depicted in Fig. A1 and similar calculations of the error in R_{sh} as a function of unnoticed changes in R_{bl} are depicted in Fig. A2.

Appendix B

This Appendix analyzes the theoretical basis of method 2.2 for the determination of the shunt conductance under the constraint that $I-V$ relations of all cell membranes are linear and zero-current potentials constant. In this method the resistance of the apical cell membrane R_a is varied and the shunt conductance $G_{sh} = 1/R_{sh}$ is derived from a plot of the transepithelial conductance (G_T) versus the short circuit current (I_{sc}). In particular we address the following questions.

1. Under what conditions is the relationship between G_T and I_{sc} linear? What does this imply for the applicability of the method?
2. What is the meaning of the extrapolated intercept with the conductance axis? Under what conditions does it or does it not equal G_{sh}?

For the sake of clarity we limit our discussion to a system which contains two ion species, 1 and 2, which can penetrate all membrane elements passively, but ion 1 is thought to be pumped across the basolateral cell membrane by active transport to generate a short circuit current. This leads to the model of Fig. B1, where R_{a1}, R_{a2}, E_{a1}, and E_{a2} are resistances and zero-current potentials for ions 1 and 2 at the apical membrane, R_{bl} and E_{bl} are the total resistance and total zero-current potential of the basolateral cell membrane, including active pumps, and R_{sh} is the paracellular shunt conductance. In essence the method consists of changing $R_a = [(1/R_{a1}) + (1/R_{a2})]^{-1}$ by some ionophore that produces an extra conductance pathway in the apical cell membrane. The latter can be either (1) specific for one ion, say ion 1, in that it lowers R_{a1} but leaves R_{a2} constant, (2) semispecific for ion 1, in that both R_{a1} and R_{a2} are lowered but the fractional reduction of R_{a2} is less, or (3) nonspecific, in that the fractional reduction of R_{a1} and R_{a2} is identical (the relative selectivity of the ionophore for ions 1 and 2 must in that case equal the relative conductances of ions 1 and 2 in the apical cell membrane).

In the following we shall discuss these three cases using the following formalism: application of the ionophore changes R_{a1} to $x R_{a1}$ with $0 < x < 1$, and simultaneously R_{a2} may change to $[1 + k(x - 1)]R_{a2}$, where k indicates the selectivity of the ionophore for ion 2 in relation to its selectivity for ion 1 normalized for the apical membrane conductances of ions 1 and 2.

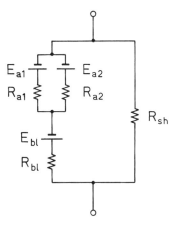

FIG. B1. Model of the epithelium with two distinct apical permeation pathways. For details see text.

Case A corresponds to $k = 0$ (the ionophore transfers only ion 1, but R_{a2} does not change).

Case B corresponds to $0 < k < 1$ (the ionophore also has a small selectivity for ion 2; however, R_{a1} and R_{a2} change differently).

Case C corresponds to $k = 1$ (both R_{a1} and R_{a2} are lowered by the same factor x).

We now consider the general case and derive G_T and I_{sc} for the equivalent circuit of Fig. B1.

The transepithelial conductance is

$$G_T = G_{cl} + G_{sh} = \frac{xR_{a1} + [1 + k(x-1)]R_{a2}}{x[1 + k(x-1)]R_{a1}R_{a2} + R_{bl}\{xR_{a1} + [1 + k(x-1)]R_{a2}\}} + G_{sh} \quad (B1)$$

where G_{cl} is the conductance of the *transcellular* permeation pathway. For the short circuit current we obtain

$$I_{sc} = G_{cl}(E_a + E_{bl}) \quad (B2)$$

where the zero-current potential across the apical cell membrane takes the form

$$E_a = \frac{E_{a1}[1 + k(x-1)]R_{a2} + E_{a2}xR_{a1}}{xR_{a1} + [1 + k(x-1)]R_{a2}} \quad (B3)$$

Now we ask at what point the extrapolated relationship of G_T versus I_{sc} intersects with the conductance axis. This point at $I_{sc} = 0$ will be called G_x. If we do not want to exclude a possible curvilinear relation of G_T and I_{sc} from the discussion a priori, we may define G_x by linear extrapolation of the slope at the point that lies closest to the G_T (and I_{sc}) axis. This is, of course, the control state with $x = 1$. The coordinates of the control state $(G_T)_{x=1}$ and $(I_{sc})_{x=1}$ are

$$(G_T)_{x=1} = (R_{a1} + R_{a2}/[R_{a1}R_{a2} + R_{bl}(R_{a1} + R_{a2})] + G_{sh} \tag{B4}$$

$$(I_{sc})_{x=1} = E_{a1}R_{a2} + E_{a2}R_{a1} + E_{bl}(R_{a1} + R_{a2})]/[R_{a1}R_{a2} + R_{bl}(R_{a1} + R_{a2})] \tag{B5}$$

and for G_x we obtain

$$G_x = (G_T)_{x=1} - (dG_T/dI_{sc})_{x=1} (I_{sc})_{x=1} \tag{B6}$$

As both G_T and I_{sc} change with x the slope dG_T/dI_{sc} can be obtained by parametric differentiation:

$$dG_T/dI_{sc} = [dG_T(x)/dx]/[dI_{sc}(x)/dx] \tag{B7}$$

Inserting Eqs. (B1) and (B2) into Eq. (B7) and rearranging leads to

$$\frac{dG_T}{dI_{sc}} = \frac{kx^2 R_{a1} + [1 + k(x-1)]^2 R_{a2}}{(1-k)(E_{a1} - E_{a2})R_{bl} + kx^2(E_{a2} + E_{bl})R_{a1} + [1 + k(x-1)]^2(E_{a1} + E_{bl})R_{a2}} \tag{B8}$$

This result shows that in the general case the slope of the relationship between G_T and I_{sc} depends on x and hence on I_{sc}, or in other words that the relationship is nonlinear

Inserting now Eqs. (B4), (B5), and (B6) and rearranging we obtain for the intercept G_x

$$G_x = G_{sh} + \frac{(1-k)(E_{a1} - E_{a2})}{E_{a1}(R_{a2} + (1-k)R_{bl}) + E_{a2}[kR_{a1} - (1-k)R_{bl}] + E_{bl}(kR_{a1} + R_{a2})} \tag{B9}$$

As can be seen, in the general case $G_x \neq G_{sh}$ in contrast to the initial expectation of the above-discussed method 2.3 to determine the shunt conductance. Note also that the deviation which will be denoted as ΔG_x may be either positive or negative.

Let us now consider the special cases A and C introduced above, in which k is either zero or unity. For $k = 0$ (ionophore highly selective for the actively transported ion) Eq. (B8) reduces to

$$dG_T/dI_{sc} = R_{a2}/[E_{a1}(R_{a2} + R_{bl}) - E_{a2}R_{bl} + E_{bl}R_{a2}] \quad (B10)$$

This expression is independent of x and hence indicates that the relationship between G_T and I_{sc} is linear. The same holds also for $k = 1$ (nonselective ionophore), in which case Eq. (B8) reduces to

$$dG_T/dI_{sc} = (R_{a1} + R_{a2})/[E_{a1}R_{a2} + E_{a2}R_{a1} + E_{bl}(R_{a1} + R_{a2})] \quad (B11)$$

The situation is different, however, if we consider the error ΔG_x in determining the shunt conductance G_{sh} by linear extrapolation. In the case of $k = 0$ (highly selective ionophore), ΔG_x can be calculated from Eq. (B9) to yield

$$(\Delta G_x)_{k=0} = (E_{a1} - E_{a2})/[R_{a2}(E_{a1} + E_{bl}) + R_{bl}(E_{a1} - E_{a2})] \quad (B12)$$

which reduces to

$$(\Delta G_x)_{k=0} = 1/R_{bl} \quad (B13)$$

if $E_{a1} = -E_{bl}$, while in case of $k = 1$ (non-selective ionophore)

$$(\Delta G_x)_{k=1} = 0 \quad (B14)$$

This result indicates that $(G_x)_{k=0} \neq G_{sh}$, while $(G_x)_{k=1} = G_{sh}$, even though the relationship between G_T and I_{sc} was linear in both cases.

In summary, we conclude the following about method 2.2:

1. It yields accurate data only when the ionophore and the cell membrane have the same relative selectivities or respective permeabilities for all permeable ions (this includes, of course, membranes that are permeable for only one ion species in case they are investigated with an ionophore that is highly selective for this ion). In addition it is mandatory that the basic preconditions be fulfilled: constancy of R_{sh}, R_{bl}, E_a, and E_{bl}, which implies, of course, constancy of intracellular ion concentrations.

2. A straight-line relationship between G_T and I_{sc} does not prove that the derived intercept with the conductance axis is a true measure of G_{sh}.

Acknowledgment

The secretarial assistance of Mrs. U. Merseburg is gratefully acknowledged.

Section I

Kidney

A. Intact Organ
Articles 3 through 6

B. Glomerular Filtration
Articles 7 through 10

C. Isolated Tubules
Articles 11 through 21

D. Isolated Renal Cells
Articles 22 through 25

E. Isolation of Membrane Vesicles
Articles 26 through 30

F. Hormonal Modulation
Articles 31 and 32

G. Reconstitution of Transport Proteins
Articles 33 and 34

[3] Isolated Perfused and Nonfiltering Kidney

By PATRICIO SILVA

Among the many advantages of the isolated perfused kidney is that because of the nature of the preparation glomerular filtration can be prevented, thus allowing the investigator to study renal functions independently of glomerular filtration and the associated cellular work related to tubular reabsorption of salt and other organic and inorganic compounds. The nonfiltering isolated kidney preparation can be used to determine whether the site of uptake into renal tubular cells is basolateral, independent of glomerular filtration, or lumenal, dependent on glomerular filtration. To my knowledge, the first reported use of the preparation was that of Johnson and Maack[1] where they used a nonfiltering isolated perfused rat kidney to study the metabolism by the kidney of growth hormone. Since then, it has been used by many other investigators to examine peptide or protein catabolism, uptake of toxic substances, renin secretion, or to determine physiological parameters such as basal oxygen consumption.[1-5]

There is an additional advantage in the use of the nonfiltering mode of the isolated kidney preparation perfused with an artificial solution over that of the filtering mode, and that is that the nonfiltering preparation never develops the morphological lesions in the thick ascending limb of the loop of Henle[6] associated with perfusion regularly found in the filtering one.[7,8]

Several different methods have been used to prevent glomerular filtration: ureteral ligation, reduction in the pressure of perfusion, increased oncotic pressure of the perfusate, or a combination of these. Ureteral ligation prevents the excretion of urine and, because of the increased back pressure along the urinary tract, it reduces glomerular filtration; however, it does not prevent it completely. While electrolyte and water reabsorption continue, predominantly in the proximal tubule, glomerular filtration continues as well, albeit at a much reduced rate. Another way of reducing

---

[1] V. Johnson and T. Maack, *Am. J. Physiol.* **233**, F185 (1977).
[2] S. T. Kau and T. Maack, *Am. J. Physiol.* **233**, F445 (1977).
[3] K. Miura, R. S. Goldstein, D. A. Pasino, and J. B. Hook, *Toxicology* **44**, 147 (1987).
[4] A. J. Cohen, K. Spokes, R. S. Brown, J. S. Stoff, and P. Silva, *Circ. Res.* **50**, 400 (1982).
[5] R. D. Swartz, P. Silva, R. Hallac, and F. H. Epstein, *Curr. Probl. Clin. Biochem.* **8**, 123 (1978).
[6] M. Brezis, S. Rosen, P. Silva, and F. H. Epstein, *Kidney Int.* **25**, 65 (1984).
[7] D. Alcorn, K. R. Emslie, B. D. Ross, G. B. Ryan, and J. D. Tange, *Kidney Int.* **19**, 638 (1981).
[8] H. J. Schurek and W. Kriz, *Lab. Invest.* **53**, 145 (1985).

glomerular filtration is to reduce the pressure of perfusion to a level where the hydrostatic pressure in the glomerular capillary is lower than the algebraic sum of the oncotic pressure of the perfusate and the proximal tubular pressure. The problem associated with reducing the pressure of perfusion is that the flow of perfusate, and hence the delivery of nutrients and oxygen to the renal tubular cells, may become rate limiting. A much better approach is to increase the oncotic pressure of the perfusate. This can be readily accomplished by perfusing the kidney with a concentration of bovine serum albumin calculated to provide an oncotic pressure that is high enough to counterbalance the filtration pressure at the glomerulus. Because the flow of perfusate through the isolated kidney perfused with an artificial solution is generally very high, of the order of 20-22 ml/min/g, there is little or no decline in the pressure of filtration along the glomerular capillary (in the filtering kidney this translates into a very low filtration fraction, of the order of 2 to 3%, and an absence of filtration pressure equilibrium; in other words, there is filtration all along the glomerular capillary). There is also no change in the oncotic pressure. Therefore, the balance between filtration pressure and oncotic pressure remains constant along the length of the glomerular capillary.

The glomerular capillary pressure in the isolated perfused rat kidney is comparable to that measured *in vivo*.[9] The proximal tubular pressure has been either measured directly or estimated in different isolated perfused rat kidney preparations and found to be either comparable to that measured *in vivo* or slightly higher.[9-11] Thus, the filtration pressure in the glomerular capillary of the isolated perfused kidney depends on the oncotic pressure of the perfusate.

The technique used is the same as that of the filtering perfused kidney preparation. A number of techniques have been developed over the years, but for the purposes of a nonfiltering kidney, perfusion of rat kidneys with bovine serum albumin in Krebs-Henseleit solution is the most convenient.

The preparation that we use is that developed by Nishiitsutsuji-Uwo *et al.*[12] and first reported in 1967. The rat is anesthetized with pentobarbital, 60 mg/kg intraperitoneally. Heparin (1000 U) is injected into the femoral vein. The peritoneal cavity is then opened and the right ureter dissected free and catheterized with a polyethylene PE-10 catheter. The ureter is then cut below the site of catheterization. The superior mesenteric artery, the right renal artery, and the right adrenal artery (which usually comes off the right renal artery about 3 mm from the origin of the renal artery) are

[9] H. M. Brink, W. M. Moons, and J. F. Slegers, *Pfluegers Arch.* **397**, 48 (1983).
[10] M. Bullivant, K. O. Hicks, and B. H. Smaill, *Pfluegers Arch.* **389**, 251 (1981).
[11] G. De Mello, and T. Maack, *Am. J. Physiol.* **231**, 1699 (1976).
[12] J. M. Nishiitsutsuji-Uwo, B. D. Ross, and H. A. Krebs, *Biochem. J.* **103**, 852 (1967).

dissected free. Care is taken not to exert any traction on the right renal artery to avoid trauma-induced vasoconstriction. The right adrenal artery is ligated. The superior mesenteric artery is ligated about 1 cm from its aortic origin. Loose ligatures are placed around the origin of the superior mesenteric artery and the renal artery. An incision is made in the mesenteric artery about 5 mm from its origin and the perfusion cannula inserted into the artery. Perfusion is started at this time. The cannula is then threaded across the aorta into the renal artery and both the mesenteric and renal arterial ligatures are tied. This procedure allows perfusion of the kidney without interruption of flow. The renal vein is cut and the kidney removed from the animal and placed in the perfusion cabinet. In this preparation the venous effluent drains over the kidney into the reservoir. If sampling or derivation of the venous effluent is required, the vein can be separately catheterized.

The perfusion medium we use consists of a Krebs–Henseleit solution of the following composition (mM): sodium, 145; potassium, 4; calcium, 2.5; magnesium, 1.2; chloride, 103; sulfate, 0.8; phosphate, 1.2; bicarbonate, 25. The pH of the perfusate is maintained at 7.4 by gassing with a mixture of 95% O_2 and 5% CO_2. With this gas mixture the O_2 content of the perfusate is maintained at 400–500 mmHg. The albumin is prepared as a stock solution of 13 g% of bovine serum albumin in Krebs–Henseleit solution. The albumin is dialyzed in a counterflow dialysis system against Krebs–Henseleit solution for a period of 24 hr. One pump runs the perfusate through and another the Krebs–Henseleit solution.

The concentration of albumin necessary to produce an oncotic pressure high enough to counterbalance the filtration pressure at the level of the glomeruli can be calculated from the Landis–Pappenheimer equation, where c is the concentration of albumin in grams per deciliter:

$$2.8c + 0.18(c^2) + 0.012(c^3) = \text{oncotic pressure}$$

The temperature of the system is maintained at 37°. The easiest way to attain this is to keep the whole system in a warm, thermostatted cabinet. Another way is to warm the perfusate. This can be done by using a water-jacketed oxygenator in which the water is heated by a circulator. This approach uses much less space. Evaporation of water, particularly during long perfusions, must be compensated by the continuous addition of distilled water at a rate sufficient to maintain the osmolality of the perfusate constant.

The glassware used consists of a reservoir that holds the kidney on top and collects the venous effluent on the bottom. It must have a side opening allowing the ureter catheter through so that the urine can be collected. It can also be made out of plastic. We use a bubble oxygenator, consisting of several glass bubbles attached one on top of the other with five openings.

At the top is the perfusate inlet; at the bottom is the perfusate outlet; a side arm at the bottom allows a small amount, 4 to 5 ml, to remain at the bottom of the oxygenator at all times (this is the overflow that goes back directly to the reservoir); a side arm slightly higher than the previous one is the oxygen/CO_2 inlet; a side arm in the top bubble is the gas outlet.

Two pumps are used. It does not matter whether they are of the peristaltic or continuous type. One pump takes the perfusate from the bottom reservoir, through two Millipore (Bedford, MA) filters arranged in parallel, to the top of the oxygenator. The filters used are Millipore type LC 10 μm with a prefilter type AP Cat. #AP20 02200. Both filters are contained in stainless steel holders. The second pump takes the perfusate from the bottom of the oxygenator to the kidney. The arterial line has an in-line Brooks flowmeter with a tantalum float, the only float that is capable of reading the large flows found in this preparation. The arterial line also has a pressure gauge attached to it by a T-type connector. The arterial line has in addition a derivation to the reservoir. A needle valve in it permits adjustment of the pressure/flow through the kidney.

We use glass arterial cannulas, but stainless steel needles are also adequate. The glass cannulas are preferable because they allow one to see whether there is air inside as the cannulation is performed. The problem with glass cannulas is threefold: First, the glass available in the United States is thick walled, therefore the cannulas that are thin enough to insert into the artery easily have high resistance. Second, they break. Third, they must be calibrated to determine their intrinsic resistance, that is, the pressure drop across the cannula. This is necessary to determine the perfusion pressure of the kidney and can be done by establishing a pressure-flow relation for each cannula.

[4] Multiple Indicator Dilution and the Kidney: Kinetics, Permeation, and Transport *in Vivo*

By CHARLES J. LUMSDEN and MELVIN SILVERMAN

Introduction

In human kidneys almost 200 liters of water and solutes are filtered daily from the blood at the level of the glomerulus, but only 1 liter is finally voided as concentrated urine. The glomerular ultrafiltrate must therefore be subject to extensive modification prior to excretion from the body. This is achieved by transcellular movement of ligands between the renal microcirculation and the ultrafiltrate pool. Through the expenditure of metabolic energy, active transport ultimately retrieves metabolically important substances such as glucose and amino acids from the ultrafiltrate, along with 99% of the filtered water. Additional waste products are added, making the final urine product (reviewed in Ref. 1).

The microcirculation and the extravascular regions of the kidney are coupled via solute and water fluxes. These fluxes move ligands between the renal capillaries and the extracellular space, between the extracellular space and renal parenchymal cells, and between the cells and the tissue compartments holding the ultrafiltered plasma. At each stage of ultrafiltration, secretion, and reabsorption, the couplings and exchanges involve the interactions of mobile molecules with macromolecules fixed in the extracellular space or on the plasma membranes of the renal parenchymal cells (transporters). The specificity of interaction may reflect generalized properties of molecular size, shape, and charge, as in the ultrafiltration step, or the unique recognition of specific ligands by membrane transport receptors, as in secretion and reabsorption. Renal physiological and pathophysiological states are therefore defined in fundamental terms by the specific physical mechanisms of molecular recognition, transport, and mobility within kidney tissue.

Both physiological and biochemical work on these mechanisms employ subcellular fractionation to isolate and characterize the structural and functional behavior of biomolecules. This reductionist approach has yielded deep insights into the specificity and kinetics of membrane transport at the molecular level in the kidney.[2-4] Beginning with these microscopic descriptions, the challenge that faces modern renal studies is to reconstitute the complex matrix of transport reactions representing the cooperative and regulatory behavior of the intact tissue. It is worth remembering that the turnover rate of enzymes and transport proteins is in the micro- to millisecond range. Thus the rate-limiting step for *in vitro* reactions is determined solely by the diffusional delivery rate of ligand to the active site of the biologic macromolecules (transporters or carriers). But *in vivo*, catalysis and transport take place within highly structured mem-

---

[1] E. Koushanpour and W. Kriz, "Renal Physiology: Principles, Structure, and Function," 2nd ed. Springer-Verlag, New York, 1986.
[2] M. Silverman and R. J. Turner, *Biomembranes* **10**, 1 (1979).
[3] M. Silverman and R. J. Turner, *Handb. Physiol. Sect. 8: Renal Physiol.* (in press) (1990).
[4] M. Silverman, *Hosp. Pract.* **24**, 180 (1989).

brane arrays. Such an arrangement favors the emergence of organized reaction chains in which ligands are channeled from one intracellular compartment (or organelle) to another without the necessity of having to traverse long diffusion pathways. In other words, the organizational geometry (anatomy) of the renal cell and indeed the whole kidney is dedicated to efficient transport kinetics.

Since the normal operation of a transport mechanism under these circumstances depends upon a complex pattern of interacting macromolecular constituents, we cannot understand renal physiology on the basis of *in vitro* studies alone. Analyses of purified receptors or transport proteins must lead to the quantitative study of the constituents' function *in vivo*. The quantitative analysis of the *in vivo* transport kinetics underlying organ physiology, in the kidney and elsewhere, is the province of *multiple indicator dilution* (MID), a dynamic tracer method uniquely suited to bridging from microscopic mechanism to macroscopic function.

In MID a pulse of chemical signals is introduced in tracer concentration into the microcirculation via the arterial inflow of an organ and the timed fractional recovery for each chemical signal measured in the venous outflow. Provided appropriate reference tracers are included it is possible to extract from the temporal pattern of fractional recoveries considerable information about transport kinetics at the capillary, interstitial, and cell surface within an organ, and utilize these data to obtain a comprehensive quantitative description of transport and metabolic events under *in vivo* conditions.

Introduced originally to study transcapillary exchange, MID has developed into a powerful approach to study cellular uptake, transcellular transport, and ligand–receptor kinetics. The MID literature on transcapillary exchange and capillary permeability in the kidney and other organs is enormous and excellent summaries are available elsewhere.[5-12] In this chapter we are interested in describing how MID is now used in the kidney to analyze ultrafiltration, secretion, and reabsorption in terms of quantita-

[5] L. H. Peterson (ed.) *Circ. Res.* **10**, 377 (1962).
[6] K. L. Zierler, *Handb. Physiol. Sect. 2: Cardiovasc. Syst.* **1**, 585 (1962).
[7] F. P. Chinard, R. Effros, W. Perl, and M. Silverman, *in* "Compartments, Pools, and Spaces in Medical Physiology" (P.-E. E. Bergner and C. C. Lushbaugh, eds.), At. Energy Comm. Symp. **11**, p. 381. Clearinghouse Fed. Sci. Tech. Inf., Springfield, Virginia, 1967.
[8] D. A. Bloomfield, "Dye Curves: The Theory and Practice of Indicator Dilution." Univ. Park Press, Baltimore, Maryland, 1974.
[9] J. B. Bassingthwaighte and C. A. Goresky, *Handb. Physiol. Sect. 2: Cardiovasc. Syst.* **4**, 549 (1984).
[10] C. P. Rose and C. A. Goresky, *Handb. Physiol. Sect. 2: Cardiovasc. Syst.* **4**, 781 (1984).

tive hypotheses (models) about kinetic mechanism.[13-16] Two advances have produced renal MID in this modern form. First, basic modifications to experimental MID protocols have eliminated virtually all effects on tracer outflow except those arising from a single passage of the tracer impulse through the target tissue.[15] And second, mathematical models have been formulated that express specific kinetic schemes for ultrafiltration, secretion, etc., in terms of binding and transport within the unique geometric organization of renal tissue. The models predict patterns of whole-organ response in terms of cellular and subcellular events. After outlining our approach to experimental protocol and mathematical hypothesis in renal MID we will discuss the method's potential by highlighting recent work, where we have evaluated the kinetic mechanisms associated with major steps in renal ultrafiltration, ligand–receptor interaction, and transcellular transport.

It is remarkable that MID, a "whole-organ method," provides detailed specific information about intraorgan cellular transport and metabolism. Three properties of the method lead to this capacity. First, MID is dynamic. The impulse response of organs as measured by the MID technique takes place over a time period of seconds at sampling rates that are of the same order as the kinetics of cellular uptake and transcapillary exchange *in vivo*. Thus the fractional recovery curves embody the true dynamics of intraorgan exchange processes and therefore contain extensive information about the kinetics of solute exchange between the microcirculation and the parenchymal cells. Second (and see below, Mathematical Approaches to Renal MID), the simultaneous presentation of multiple tracers to the tissue allows the otherwise confounding effects of mechanisms antecedent and subsequent to the target kinetics to be peeled away, exposing the *in vivo* response of the target kinetic system to detailed analysis.

Third, MID is not to be confused with the "black box" compartmental approach to whole-organ physiology. In the latter approach the physiological input and output are related in very coarse terms, usually via an average or lumped kinetic scheme that mixes together the individual contributions of the organ's many specific transport and exchange steps (Fig. 1, top). The

---

[11] M. Silverman, C. I. Whiteside, and C. Trainor, *Fed. Proc., Fed. Am. Soc. Exp. Biol.* **43**, 171 (1984).
[12] C. I. Whiteside and C. J. Lumsden, *Am. J. Physiol.* **256**, F882 (1989).
[13] W. Perl and F. P. Chinard, *Circ. Res.* **22**, 273 (1968).
[14] M. Silverman and C. Trainor, *Fed. Proc., Fed. Am. Soc. Exp. Biol.* **41**, 3054 (1982).
[15] M. Silverman, in "Carrier Mediated Transport of Solutes from Blood to Tissue" (D. L. Yudilevich and G. E. Mann, eds.), p. 51. Longman, London, 1985.
[16] C. J. Lumsden and M. Silverman, *Am. J. Physiol.* **251**, F1073 (1986).

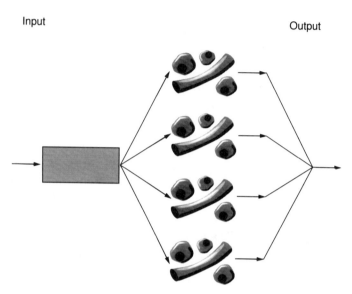

FIG. 1. *Top:* Whole-organ modeling with a black box. The box contains *ad hoc* mathematical representations of transport and metabolism in the organ and relates input to output in a purely phenomenological manner. *Bottom:* Whole-organ modeling in multiple indicator dilution (MID) theory. The organ is modeled as a hierarchical system of basic functional units, each of which is in turn composed of specified arrangements of microcirculation, interstitial elements, and parenchymal cells linked by physical fluxes of matter, energy, and information. A schematic representation of several functional units, each built around a capillary and parenchymal cells, is shown here. Multiple indicator dilution theory predicts whole-organ responses to experimental (tracer physiological) manipulations given this detailed structural blueprint of the organ.

hypothesis underlying MID is very different: organs are not black boxes but intrinsically hierarchical arrangements of functional subunits, each supplied and drained by a portion of microcirculation and each independent, to first order, from the rest (Fig. 1, bottom). These functional subunits in turn consist of capillaries, cells, and interstitial regions whose geometric arrangement is to be specified precisely. Finally, the individual kinetic steps of the transport mechanisms are organized within the spatial

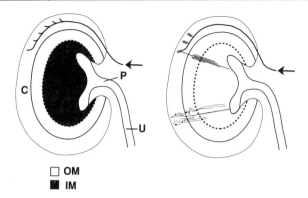

FIG. 2. *Left:* Rat kidney showing the two major anatomical regions: cortex (C) and medulla, divided into outer medulla (OM) and inner medulla (IM), including the pelvis (P) and ureter (U). *Right:* Arrangement of arterial vessels (black) and tubules (dotted). An arcuate artery (arrow), running along the corticomedullary border, gives rise to three interlobular arteries, which ascend into the renal cortex. They branch into afferent arterioles, which run to the capillary beds of individual glomeruli. Each efferent arteriole, leaving a glomerular capillary bed, runs to the peritubular capillary bed of the nephron tubule within the cortex. The efferent arteriole of a juxtamedullary glomerulus splits into a descending vasa recta supplying the renal medulla. In addition, three nephrons are illustrated: a cortical, a short looped, and a long looped nephron, along with their collecting duct. (After Koushanpour and Kriz,[1] p. 42.)

complex of each functional subunit. With appropriate assumptions about geometry and the thermodynamic forces driving transport and exchange in each functional subunit, and about the interrelation among the subunits defined by their anatomical positions on the organ's vascular tree, the kinetics of a binding or transport mechanism can be related explicitly to the whole-organ pattern of tracer fractional recovery. The result is an systematic, self-consistent approach to quantitative *in vivo* kinetics.

The Renal MID Method

The hierarchical structure of the kidney begins with its principal anatomical regions (Fig. 2, left) — the overlying cortex and the underlying zones of the medulla. In the cortex and the medulla, function is divided between the microcirculation and the renal tubules. The tubules, through which urine flows, course back and forth from cortex to medulla. The renal microcirculation is composed of three capillary beds lying in series (Fig. 2, right). Two occupy the renal cortex.[17] The first cortical bed is the glomerulus, a tuft of capillaries in which a hydrostatic pressure gradient ultrafilters

[17] R. Beeuwkes III, *Annu. Rev. Physiol.* **42**, 531 (1980).

blood plasma across the glomerular capillary wall. From there it enters the postglomerular tubule (there are approximately one million glomeruli per human kidney cortex). Under normal conditions the ultrafiltrate contains only water and small molecules. Plasma proteins are for the most part held back.

Blood moves from the glomerular capillary tuft into the peritubular capillary, which invests the proximal part of the tubule during its descent into the medulla. In this second capillary bed water, biomolecules, and ions are reabsorbed from the filtrate by the transport activity of epithelial cells forming the tubule walls. The proximal tubule is also the site of secretory activities that move ligands from the peritubular microcirculation into the urine. Under normal conditions 95% of the renal blood flows through the glomerular and peritubular capillaries. Five percent goes to the vasa recta capillaries forming the medullary microcirculation. Because of its extraordinary circulatory perfusion and its key location in the renal pathway of filtration, secretion, and reabsorption, the renal cortex has become the focus of our MID studies.

Pioneering work by Chinard in the early 1950s laid the basis for MID and its subsequent development in the kidney and other organs. In Chinard's original description of the MID protocol for kidney studies, a pulse injection of approximately 0.3 ml of a solution containing several tracers was made into the left renal artery of an anesthetized mongrel dog and timed serial samples obtained separately from the corresponding venous and urine outflows.[18-20] Only a fraction of total left renal vein outflow was sampled and the renal vein curves were corrected for recirculation by extrapolation of the first linear portion of the downslope on a semilogarithmic plot. As will become evident below, the dual output design of renal MID allows separation of glomerular from postglomerular events.

A major strength of the method as originally conceived by Chinard resides in the use of *multiple* simultaneously injected indicators. By employing this strategy of multiple tracers it becomes possible to include reference compounds that serve as markers of functionally meaningful intraorgan compartments. In the kidney two such probes are regularly employed: albumin as a plasma reference, and creatinine or L-glucose as extracellular markers. The physiologically relevant information about the intrarenal handling of test solutes is obtained by comparing the renal vein and urine outflow curves for a given substrate relative to those of simultaneously injected albumin and creatinine.

[18] F. P. Chinard, *Am. J. Physiol.* **180,** 617 (1955).
[19] F. P. Chinard, W. R. Taylor, M. F. Nolan, and T. Enns, *Am. J. Physiol.* **196,** 535 (1959).
[20] F. P. Chinard, T. Enns, C. A. Goresky, and M. F. Nolan, *Am. J. Physiol.* **209,** 243 (1965).

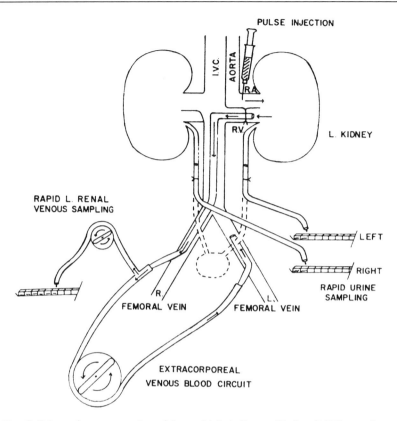

FIG. 3. Schematic representation of the multiple indicator dilution (MID) experimental method. See text for details. IVC, inferior vena cava; RA, renal artery; RV, renal vein. (Reprinted by permission from Silverman.[15])

Our experimental design (see Fig. 3) is in many ways identical to Chinard's original description with several notable refinements.[15,21] Total venous outflow from the left kidney is diverted via the right femoral vein through an extracorporeal circuit controlled by a perfusion pump, and returned into the left femoral vein. A portion of the outflow is extracted from a side arm by means of a Sigma motor pump and led to the blood sampling rack. Total blood flow is measured continuously by an electromagnetic flow probe located immediately distal to the sampling site. The volume of the extracorporeal circuit is just large enough so that there is no

[21] C. I. Whiteside and M. Silverman, *Am. J. Physiol.* **245**, F485 (1983).

recirculation of injected tracer during the renal vein sampling period. This modification eliminates one of the major sources of experimental error in MID by removing the effects of tracer molecules that have passed through the kidney, left it, passed through the systemic circulation and other organs, and reentered the kidney. The urine sampling rate is usually 10 to 20 sec/sample and the right side is used as a control for the left. Blood outflow from the renal vein is samples at much higher rates (0.3–0.6 sec/sample). The concentration of each of the indicators is determined in all blood and urine samples. By means of internal standards, the renal vein and urine outflow data can be expressed as plots of fractional recovery versus time. The availability of an independent measurement of blood flow enables us to correct for partial injection of the indicator bolus into the renal artery, so that the exact amount of the injection bolus entering the kidney is known in absolute terms.

There is sufficient volume per sample to permit a single analytical measurement for each of the indicators. By optimizing counting statistics and pipetting techniques we achieve reproducibility of better than 5% for each experimental point. Corrections for any distorting effects of the sampling catheters on the pattern of fractional recovery versus time are routinely built into the data reduction scheme. Chromatographic techniques are used to monitor the chemical identity of labeled substrates relative to the injectate. Some situations necessitate anaerobic sampling in order to ensure metabolic conservation of indicator.

Glomerular and Postglomerular Events

Consider the situation where albumin and creatinine are injected in pulse fashion into the kidney and timed serial samples are obtained simultaneously from renal vein and ureters as we have just described (Fig. 3). Because albumin is restricted to plasma it emerges in the renal venous outflow with a short mean transit time $\bar{t} = \Sigma_i t_i f_{obs}(t_i)/\Sigma_i t_i$, $t_{i=1, \ldots, K}$ the sampling times. There is no loss across the glomerulus. Creatinine, on the other hand, is freely filtered at the glomerulus (roughly 25% of that injected moves into the tubular lumen so that its total renal vein recovery is only 75% of that of albumin). Moreover, in contrast to albumin it readily exchanges across the peritubular capillary wall. Thus it emerges with a larger renal vein mean transit time than albumin.

Consider now a homologous set of macromolecular weight markers such as neutral dextrans, prepared in homogeneous fractions and radiolabeled to serve as tracer probes of graded size.[21,22] The neutral dextrans do

[22] C. I. Whiteside and M. Silverman, *Am. J. Physiol.* **245**, F496 (1983).

not interact with the lumenal or contralumenal surfaces of the tubular epithelium and therefore are neither secreted nor reabsorbed. The urine fractional recovery curves for neutral dextran tracers will depend only on glomerular extraction, whereas the renal vein curves reflect both glomerular and postglomerular events. Thus if any of these dextran probes has a renal vein mean transit time greater than that of the plasma reference (albumin), then some solute molecules must have left and later returned to the vascular space at the level of the postglomerular circulation (Fig. 4, right). Moreover, because no secretion or reabsorption takes place in the case of neutral dextrans, the extent of glomerular extraction can be directly estimated either from quantitative urine recovery or alternatively as a fraction of total renal vein recovery relative to albumin.

As the dextran size decreases, the available postglomerular distribution volume increases and eventually the renal vein transit curve for the dextran probe superimposes on creatinine. At the other extreme, as the size of dextran increases, the postglomerular volume of distribution becomes vanishingly small and ultimately the renal vein mean transit time for a test solute equals that of the plasma reference. At this point the curve for test solute and albumin will be perfectly symmetrical but the area under the dextran curve will be less, reflecting loss from the glomerular circulation (Fig. 4, left). We have examined the relative renal vein transit times for a range of homologous neutral dextrans and compared their renal vein mean transit times to that of albumin.[22] For dextrans below 15,500 Da the albumin transit time is less than that for dextrans, indicating a finite postglomerular distribution. Above 15,500 Da the two become equal and the postglomerular extraction is negligible.

Prior to these experiments the capability of MID to distinguish glomerular from postglomerular events had been the subject of controversy.[23] Our observations establish that the analysis of renal vein curves for fractional recovery, when done *in parallel* with an analysis of the urine fractional recovery curves, provides extensive information on the respective roles of glomerular and postglomerular events in shaping the patterns of tracer recovery relative to those of simultaneously injected reference tracers. This capacity is further enhanced when mathematical models of the MID process are applied to the renal vein and urine data.

Mathematical Approaches to Renal MID

Consider a point $\mathbf{x}$ in the renal cortex at time t during an MID experiment. If $c_j(\mathbf{x},t)$ denotes the concentration of substance C_j at that position

---

[23] See, for example, the discussion following F. P. Chinard, *Alfred Benzon Symp.* **2**, 32 (1970).

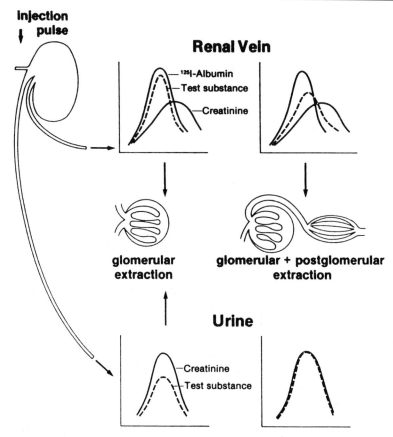

FIG. 4. Schematic representation of the separation of glomerular and postglomerular events by means of the MID experimental method as applied to dog kidney *in vivo*. [^{125}I]Albumin is the plasma reference; creatinine is the extracellular (interstitial) reference. Dashed line represents the test marker (e.g., a neutral dextran), injected in tracer quantities simultaneously along with the tracer quantities of albumin and creatinine. *Left panels:* The renal vein and urine fractional recovery curves expected from a test marker with negligible postglomerular extraction and limited filtration at the glomerulus. *Right panels:* The corresponding patterns for a test marker that is completely extracted at the glomerulus (i.e., is ultrafiltered in a manner identical to water) and, in addition, has a finite postglomerular distribution. (Reprinted by permission from Whiteside and Silverman.[21])

and time, then the local rate of change in $c_j(\mathbf{x},t)$ may be due to diffusion, convection, or chemical reaction:

$$\frac{\partial}{\partial t} c_j(\mathbf{x},t) = \nabla \cdot [\mathbb{D} \cdot \nabla c_j] - \mathbf{K} \cdot \nabla c_j + R_j(c_1, \ldots, c_N | \alpha_1, \ldots, \alpha_M) \quad (1)$$

where[24] $\mathbb{D}$ is the diffusive mobility tensor, $\mathbf{K}$ is the convective mobility, and $R_j(\cdot | \cdot)$ is the (generally nonlinear) chemical reaction rate of change in the local volume element immediately surrounding the point $\mathbf{x}$. A total of N species are germane to the kinetics of C_j and $\alpha_1, \ldots, \alpha_M$ denote parameters (such as temperature, pressure, and local abundances of inhibitor ligands) controlled by the experimenter.

In order to relate the transport kinetics expressed by Eq. (1) directly to the patterns of renal vein and urine tracer fractional recovery it is necessary to solve (1) for the individual functional units of the cortex (glomeruli, tubules) and then integrate over their behavior, modeling the whole-organ response. To the observed fractional recoveries $f_{\text{obs}}(t)$ there then correspond theoretically predicted recoveries $f_{\text{Model}}(t|\mathbb{D}, \mathbf{K}, \mathbf{R})$ where $\mathbf{R}$ is the kinetic vector $(R_1, \ldots, R_N)$. Given this correspondence one can attempt to fit a tracer pattern of fractional recovery to, for example, a secretive or reabsorptive mechanism of tubular function. This process of classification is pushed much further when optimization techniques are used to quantify the individual reaction rates in $\mathbf{R}$ and the magnitude of the mobility coefficients comprising $\mathbb{D}$ and $\mathbf{K}$. The question of the norms, or measures of best fit, most suited to relating MID models to data is a subject in its own right (for further discussion see Ref. 9).

Analytical solutions to Eq. (1) can of course be obtained only for the simplest models, which we will describe in somewhat more detail before turning to our results. But even these simple models capture significant properties of filtration, reabsorption, and secretion in the renal cortex and therefore have been immensely informative about the basic organization and quantitative kinetics of specific transport pathways. In principle the lack of analytical solutions for more complex models poses no problem given sufficient computational resources, and efficient algorithms for numerically implementing Eq. (1) for tracer MID are being devised.[9,16] Since MID uses ligands at the physiological tracer concentrations $c_j^* \ll c_j$, the assumed steady state of the kidney is not disturbed by the passage of the multiple indicator impulse through the organ and Eq. (1) immediately

---

[24] J. Bear, "Dynamics of Fluids in Porous Media." Elsevier, New York, 1972.

reduces to the linear problem

$$\frac{\partial}{\partial t} c_j^*(\mathbf{x},t) = \nabla \cdot [\mathbb{D} \cdot \nabla c_j^*] - \mathbf{K} \cdot \nabla c_j^* + \sum_{k=1}^{N} X_{jk} c_k^* \qquad (2)$$

where $X_{jk} \equiv \partial R_j(\,\cdot\,|\,\cdot\,)/\partial c_k$. Measuring the dependence of the tracer transport coefficients on the steady state concentrations of ligands competing for (or otherwise modifying) specific steps in the transport mechanism leads to *in vivo* estimates of the binding, dissociation, and inhibition constants controlling the movement of ligands through the functional unit.

How in practice is the complex general problem posed by Eqs. (1) and (2) translated into schemes for interpreting renal MID data? Consider the three diagrams in Fig. 5, which illustrates the steps taken to build renal MID models from an understanding of cortical anatomy and physiology. The functional unit depicted in this illustration is the proximal tubule and we are interested in the kinetics by which ligands are transported across the two opposing surfaces of the tubular epithelial cell [the contralumenal or basolateral membrane (BM) and the lumenal or bush border membrane (LM; BBM)]. Epithelial cells, comprising a physiological space C contact the urine space Tu on their BBM side, and an interstitium Int on their BM side. The Int separates C from the microcirculation or blood space Bl, which exchanges with Int via the capillary endothelium. Metabolic events within C may consume or sequester tracer molecule indicators (small downward pointing arrows). Blood moves through the peritubular capillary, carrying tracers in direction x with a specified convective velocity W_{Bl}. The convective velocity of urine flow is W_{Tu}. The tracer relationships between Bl and Tu are governed by the splitting of the tracer input bolus at the glomerulus, upstream from the tubular epithelium. Glomerular filtration produces the equivalent of two input boluses to each exchange unit. One enters the arterial end of the postglomerular capillary and one forms at $x = 0$ in Tu. Convection in Bl and Tu moves these boluses toward $x = L$.

The assumptions used in reducing a mathematical description of this system [Eqs. (1) and (2)] to manageable proportions can be organized into four sets: (1) assumptions pertinent to the modeling of MID at the level of single functional units generally, (2) assumptions relevant to the construction of models applying MID in the renal cortex, (3) assumptions specific to epithelial transport in the proximal tubule, and (4) assumptions about how the individual functional units (intact proximal tubules) are arranged in the whole organ. A large number of assumptions are involved in laying out such a model completely. Extensive reviews of their details and justification can be found elsewhere.[9,16,25] Here we will mention only the most

[25] C. A. Goresky, W. H. Ziegler, and G. G. Bach, *Circ. Res.* **27**, 739 (1970).

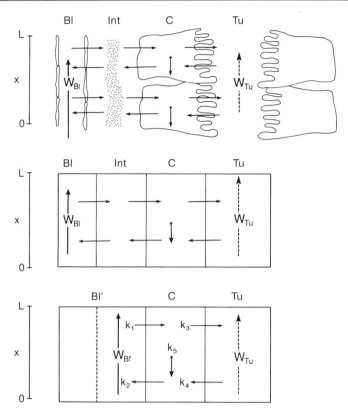

FIG. 5. Steps in modeling indicator exchange between the renal tubule (Tu) and the microcirculation (Bl) of the peritubular capillary. *Top:* Schematic representation of the structural relationship between Bl, Tu, and the indicator fluxes joining them. *Bottom:* Final model for the passage of flow-limited indicators through the system. See text for details.

important assumptions from each of the four sets:

1. Aside from the tracers, the system is steady state. Tracers move between the spaces Bl, Int, C, and Tu (Fig. 5, middle, horizontal arrows). The rate of transport or flux across a barrier is proportional to the concentration difference across the barrier and the barrier's permeability to the tracer. The lateral diffusion of the tracers is infinitely rapid between the barriers. Convective flow within the capillary is plug flow and tracer diffusion parallel to the convective flow is negligible. Tracer concentration in each space depends only on the time t and on x, the distance along the capillary. The tracer bolus introduced at $x = 0$ at $t = 0$ approximates a

Dirac delta function. The amount of tracer entering a capillary is proportional to the flow through it. Metabolic consumption or irreversible sequestration can occur intracellularly.

2. In the intact kidney the medulla receives some 5 to 10% of the total blood flow. The contribution of medullary structures to the pattern of single-unit and whole-organ indicator dilution is therefore taken to be negligible. Each proximal tubule interacts with one capillary, although the nephron as a whole might consist of a series of tubular segments, each fed by a different vessel.

3. Tracer movement across the peritubular capillary wall is flow limited. Under this condition the transcapillary permeabilities are very large and the combined space Bl' = Bl + Int functions as one virtual convective space[25] for pulses of tracer that enter Bl at $x = 0$. Tracers appear at $x = L$, the capillary outflow, as though they convected through Bl with an effective velocity related to actual blood velocity by the inverse of the ratio of transcapillary fluxes. The natural reference indicator for subsequent processes of epithelial exchange into the spaces C and Tu is then the extracellular marker of the flow-limited process.

4. Diffusive or convective coupling among the functional units is zero (but see ref. 17) and the whole-organ response to tracer input is a summation of single-unit responses, taking into account the heterogeneities introduced by the distribution of times a tracer molecule spends moving *to* a functional unit (arterial input), passing *through* it, and moving *from* the unit to the collection point (venous output). The tracer molecules can move out of the vascular pool only within the functional units; the arterial and vein walls have negligible permeability to the tracer over the time course of the MID experiment. For the kidney, the transit times along the nonexchanging vessels can to a good approximation be taken as equal for all the routes tracer substrates can follow through the organ.[16] Different total vascular transit times are therefore caused by differences in length among the peritubular capillaries. Similarly, differences in the urine transit times of glomerular markers, which are neither secreted nor reabsorbed, arise from differences in length of the tubules. Each peritubular capillary of length L is paired with a tubule of equal length.

With those assumptions Eqs. (1) and (2) reduce to

$$\partial u(x,t)/\partial t = -W_{Bl'}\partial u/\partial x - k_1 \gamma u + k_2 \gamma v$$
$$\partial v(x,t)/\partial t = k_1 u - (k_1 + k_3 + k_5)v + k_4 z \quad (3)$$
$$\partial z(x,t)/\partial t = -W_{Tu}\partial z/\partial x + k_3 \gamma v/\theta - k_4 \gamma z/\theta$$

for each tracer in the final model shown in Fig. 5 (bottom) where $u(x,t)$, $v(x,t)$, and $z(x,t)$ are the concentrations of the tracer at distance x down the capillary and time t in the compartments Bl', C, and Tu, respectively, $k_1, \ldots, k_4$ are the unidirectional flux coefficients (permeability surface area products per unit length) for movement across the BM and LM (Fig. 5, bottom), k_5 is the intracellular metabolic/sequestration rate constant, and γ and θ are the volumes of C and Tu per unit length of peritubular capillary relative to the vascular volume of tracer distribution. The solution of equations such as Eq. (3) is discussed elsewhere[9,16] and illustrated in Fig. 6 (top). Integration of the outflow $u(L,t)$ across all functional units gives the renal vein recovery curve (Fig. 6, bottom); since the vascular tracer does not leave the capillaries its renal vein fractional recovery curve $B_{Bl}(t)$ is the statistical distribution of vascular transit times through the organ. Similarly, the extracellular reference recovery curve $B_{Int}(t)$ is the transit time distribution for flow-limited tracers. It then follows that for any tracer crossing the tubular epithelial cells between Bl' and Tu the renal vein outflow curve is

$$B(t) = B(t|k_1, \ldots, k_5) = \int_0^\infty u(W_{Bl}\tau, t - t_0|k_1, \ldots, k_5) B_{Int}(\tau) d\tau \quad (4)$$

where t_0 is the delay time of the nonexchanging vessels. An equivalent relationship holds for the urine outflow curve.

In Eq. (4) we have made explicit the parametric dependence of $B(t)$ [and $u(x,t)$] on the kinetic coefficients for transcellular transport. Given fractional recovery data, the values of k_j are adjusted to obtain the best fit between model and data.

Glomerular Permselectivity

Ultrafiltration at the glomerular capillary wall differentiates solutes on the basis of size, charge, and shape.[1,12] Although the investigation of all three determinants may be carried out using MID in the kidney, our quantitative, model-based studies have thus far focused on the kinetics of size and charge selectivity. This work[12,21] is of particular general interest because it represents the first time MID has been used to measure the permselectivity characteristics of an ultrafiltrative microcirculation. Historically MID has been applied only under conditions when net capillary ultrafiltration is zero. In other words, although transcapillary flux of solute in the direction perpendicular to blood flow is the sum of convective and diffusive fluxes, the net convective efflux from capillary to interstitium is assumed to be much smaller than the diffusional flux. We have demonstrated how MID can be used to study transcapillary exchange in a strongly ultrafiltering capillary bed—the renal glomerulus.

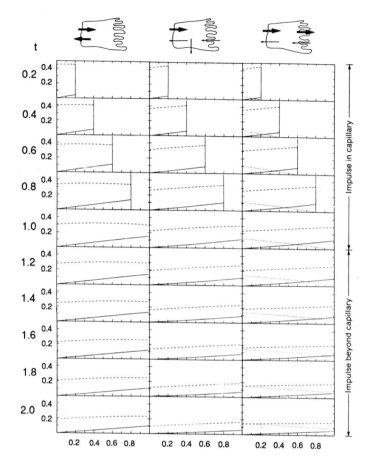
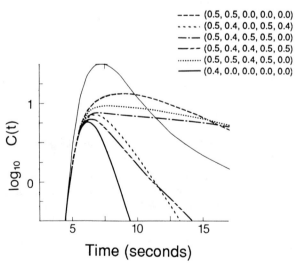

Size. Above a certain size, the kinetics of steric interaction (collision) between a permeating solute and the fixed macromolecular structures of the glomerular capillary wall [endothelial fenestrae, glomerular basement membrane, glomerular epithelial cells, and "slit pores" (structural overview in Ref. 1)] lead to the exclusion of a certain fraction of the molecules attempting to pass through the wall and enter the urine space.[21] These steric kinetics are characterized by the hydraulic reflection coefficient σ_s. When convective motion dominates the ultrafiltration process, $1 - \sigma_s$ of the permeating species enters and passes through the capillary wall.

In general, solute flux across any membrane can be considered as the result of both diffusive and convective forces. We examined the role of diffusive forces by testing for flow dependence in the unidirectional extraction of neutral dextrans from the glomerular microcirculation in anesthetized dogs maintained in autoregulatory range and under mannitol diuresis.[21] If diffusion represented a significant component of total flux, then with decreasing flow and consequently decreasing glomerular filtration rate, a disproportionate increase in glomerular extraction of dextrans should occur relative to simultaneously filtered extracellular reference. In contrast to findings in the rat[26,27] no such change was observed when the renal plasma flow was lowered two- to threefold.

For steady state ultrafiltration through channels permitting transmembrane movement by diffusion and convection but no chemical reaction,

[26] R. L. S. Chang, C. R. Robertson, W. M. Deen, and B. M. Brenner, *Biophys. J.* **15,** 861 (1975).
[27] R. L. S. Chang, I. F. Ueki, J. L. Troy, W. M. Deen, C. R. Robertson, and B. M. Brenner, *Biophys. J.* **15,** 887 (1975).

FIG. 6. *Top:* Spatiotemporal patterns of indicator concentration within the exchange units depicted in Fig. 5. Solid line, indicator concentration $u(x,t)$ in the effective vascular space Bl'; dashed line, intracellular concentration $v(x,t)$; dotted line, indicator concentration in the tubular lumen. Capillaries have length 1 unit and $W_{Bl'}$ is 1 unit/sec. Thus for $t > 1$ sec the indicator bolus from the injection (spike at the front of the concentration waves depicted in the panel) has left the capillary and the indicator outflow is determined by the residual indicator in the unit. Convective flow in the tubule is assumed to be much slower than $W_{Bl'}$ in the examples shown here, so $W_{Tu} \approx 0$. The three columns correspond to three patterns of relative values of the kinetic constants for tubule epithelial cell uptake, release, and sequestration of indicator as defined by the set of values $(k_1, k_2, k_3, k_4, k_5)$ (see Fig. 5): (0.5, 0.5, 0, 0, 0) ("antilumenal equilibration"), (0.5, 0.4, 0, 0.5, 0.4) ("antilumenal drive"), and (0.5, 0.4, 0.5, 0.5, 0) ("net secretion"). *Bottom:* Whole-organ outflow curves for this model, plotted in semilogarithmic coordinates and shown in terms of their dependence on the rate constants $(k_1, k_2, k_3, k_4, k_5)$. The thin solid line that peaks above the other curves is the outflow pattern for creatinine (reference indicator). (Reproduced by permission from Lumsden and Silverman.[16])

solution of Eq. (1) gives

$$C_2/C_1 = (1 - \sigma_s)/(1 - \sigma_s e^{-k}) \quad (5)$$

where C_2/C_1 is the relative urine-to-plasma concentration of permeating solute and $k = (1 - \sigma_s)J_V/P_S$ with J_V the volume flux and P_S the permeability constant.[28,29] The left-hand side of Eq. (5) is equivalent to the fractional clearance $[(U/P)_D/(U/P)_I]$ or fractional extraction E_D/E_I, measured by MID,[11,21] where the subscript D refers to the tracer dextran and I refers to the glomerular reference (inulin). When diffusion is negligible, $C_2/C_1 \rightarrow 1 - \sigma_s$.

Under these conditions the sieving characteristics of the glomerular filtration barrier can be defined experimentally and the data can be used to calculate *in vivo* reflection coefficients. Our results are shown in Fig. 7. Agreement with σ_s values obtained by classical steady state tracer clearance methods[26,27,30] is excellent and corresponds to a capillary wall containing pores of radius approximately 55 Å.[31]

Charge. The charge-selective properties of the glomerular capillary wall are maintained by anionic proteins, the glomerular polyanions (the GPA), that restrict the passage of negatively charged molecules more than neutral ones of the same molecular size and facilitate the transport of positively charged molecules.[12] Clearance studies[30] have suggested that the GPA provide a free energy barrier preventing excessive ultrafiltration of anionic plasma proteins. But because of their steady state nature, clearance methods are unsuited to kinetic studies of the binding events ligands experience while passing through the glomerular wall. This kinetic information is, however, expressed directly in the dynamics of the MID fractional recovery curves for the urine outflow. We therefore turned to MID using anionic and cationic probes (dextrans), together with *in vitro* studies of isolated glomerular binding, in order to elucidate the action of the GPA. When compared to neutral dextrans of the same equivalent Stokes–Einstein radius the anionic dextrans consistently showed lower urine recovery. We attributed this lowered recovery to the effects of the GPA free energy barrier on the negatively charged anionic solutes.

Cationic dextran tracers introduced into the glomerular microcirculation via MID showed incomplete urine recovery (Fig. 8); however, in this case *excess* urine recovery of cationic tracer is obtained when excess nontracer unlabeled cationic dextran is added to the injection solution along

---

[28] J. R. Pappenheimer, *Physiol Rev.* **33**, 387 (1953).
[29] E. H. Bresler and L. S. Groome, *Am. J. Physiol.* **241**, F469 (1982).
[30] W. M. Deen, C. R. Bridges, and B. M. Brenner, *J. Membr. Biol.* **71**, 1 (1983).
[31] C. J. Lumsden, unpublished observations (1989).

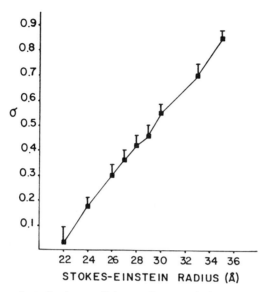

FIG. 7. Glomerular reflection coefficient σ for neutral dextrans calculated from the corresponding mean fractional extraction relative to inulin during single-pass MID *in vivo* in the dog. (Reprinted by permission from Whiteside and Silverman.[21])

with the radioactively labeled tracer cationic dextran.[12] These observations suggest a kinetic mechanism in which tracer cationic dextrans bind reversibly to anionic sites in the glomerular wall during ultrafiltration. We applied Eq. (1) to this mechanism by means of the following reasoning. The probability of cationic dextran tracer entering the glomerular barrier is determined partly by molecular size. Once inside the barrier, however, the cationic dextran may take a noninteractive, convective route through the wall and exit simultaneously with tracer inulin (glomerular filtration marker) or it may bind reversibly to polyanionic sites distributed within the glomerular capillary wall and more slowly emerge in the urine output. The capillary wall of each glomerulus was modeled as a convectively well-mixed compartment (strong ultrafiltration) in which the binding sites are uniformly distributed and react with cationic solute independently of each other. Because the mass bound is tracer quantity, the sites are not saturated and the amount of tracer remaining in the wall at any time t can be described as an exponential function with a decay rate λ for tracer disappearance into the urine. λ depends on the effective volume of distribution available to the tracer, the binding constant for cationic tracer–anionic site interaction, the saturation limit for the binding process, and

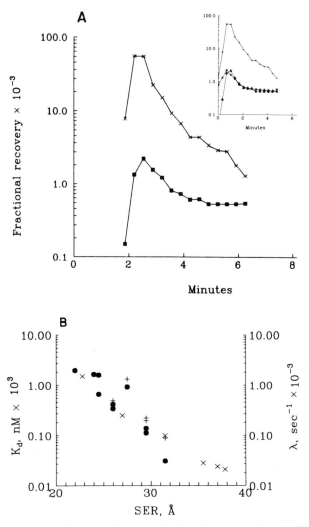

FIG. 8. (A) Multiple indicator dilution urine fractional recovery curves for [^{14}C]inulin (×) and cationic [^{3}H]dextran (■) of 29.5-Å Stokes–Einstein radius (SER). Incomplete recovery of the dextran is observed during tracer conditions. *Inset:* Comparison of observed dextran curve to the best fit obtained with the convection/binding model of glomerular capillary wall transit [see Eq. (6)]. (B) *In vitro* dissociation constant K_d and *in vivo* binding/washout rate constant 1 plotted as logarithmic functions of cationic [^{3}H]dextran SER. ×, K_d; ●, λ when $FF_{CDEX} = FF_{IN}$; +, λ when $FF_{CDEX} = FF_{NDEX}$. (Reprinted by permission from Whiteside and Lumsden.[12])

the rate of convective flow through the glomerular capillary wall. The glomeruli are considered to be identical to one another and to function independently. Under these conditions the fractional recovery $f(t)$ of tracer cationic dextran is related to that for insulin by

$$f(t) = x \frac{FF_{DEX}}{FF_{INU}} f_{INU}(t) + (1-x) \frac{FF_{DEX}}{FF_{INU}} \int_0^t \lambda \exp[-\lambda(t-\tau)] f_{INU}(\tau) \, d\tau \tag{6}$$

where λ is the rate constant, x is the tracer fraction taking the purely convective route through the glomerular wall, and the FF are filtration fractions.[12] This is a two-parameter (x,λ) representation of the urine recovery data. The parameters x and λ were obtained for each MID run by computer fitting Eq. (6) to the observed net fractional urine recoveries. We applied the least-squares norm to the minimization process and compared the resulting estimates of λ to the dissociation constants K_d for tracer cationic dextran binding to isolated glomeruli.

The model provided a good simulation of the urine fractional recovery curves for tracer cationic dextran (Fig. 8). Moreover, as suggested by the modeling assumptions, λ is highly correlated with K_d and its behavior across a graded series of cationic dextran sizes appears primarily to reflect the intensification of charge–charge interactions between dextrans and GPA as molecular size (and thus net effective charge of the uniformly substituted tracers) increases. In contrast to previous work on the glomerular wall, these studies lead directly to the conclusion that the mobility of cationic solutes within the ultrafiltration barrier is a sensitive function of solute charge as well as size. Hence, transglomerular cationic solute flux is mediated by a convection-binding mechanism that creates what is, in effect, a polyvalent charge barrier within the glomerular capillary wall.

Peritubular Capillary Permselectivity

In the past, the postglomerular (peritubular) capillary network has been relatively inaccessible to direct studies of transcapillary exchange. Quantification of postglomerular capillary permselectivity has required the use of renal lymphatic clearance measurements. Multiple indicator dilution provides a new and exciting empirical approach to peritubular capillary permeability. Just as the urine outflow curves for neutral, anionic, and cationic dextrans reflect the size, shape, and charge selectivity of the glomerular capillary wall, the renal vein outflow patterns for these tracers contain information about their kinetics of interaction with the peritubular capillary wall and the interstitium beyond.

Thus far in our work we have characterized these peritubular interactions by calculating the initial upslope ratios of neutral, anionic, and cationic dextran tracers relative to that of simultaneously injected plasma (albumin) and extracellular (inulin, creatinine) references.[22,32,33] The upslope method, which sets the stage for more detailed kinetic modeling, was pioneered by Crone[34] and gives the single passage extraction of a test molecule relative to the plasma reference. This extraction E is

$$E = 1 - f_{0T}/f_{0P} = 1 - e^{-PS/F} \qquad (7)$$

where f_0 denotes fractional recovery at the initial point, T and P refer to test and plasma tracer, respectively, PS is the permeability–surface area product for solute transit through the capillary wall, and F is the flow.

Extraction values obtained in this way are not corrected for the effects of water reabsorption (convective backflux) on macromolecules leaving and reentering the peritubular capillary. In addition, the permeability calculated by applying the right-hand side of Eq. (7) reflects only the behavior of the capillaries with the shortest transit time. Nevertheless, by a careful comparison of the single-pass dextran extractions across a broad range of size, charge, and molecular shapes we were able to obtain new and unique information about the peritubular capillary as a permeability barrier:

Size. From the upslope ratios of test neutral dextran markers relative to plasma references we determined that the peritubular capillary sterically impedes the efflux of macromolecular solutes.[22] The capillary wall acts as though it is pierced by small, rigid pores with a radius of approximately 55 Å.[31]

Charge. Our anionic dextrans had a *greater* peritubular capillary permeability than neutral dextran markers (Fig. 9), even though cationic probes confirmed the presence of anionic binding sites along the peritubular capillary.[32,33] Thus in contrast to microvascular barriers like the glomerular capillary wall, the peritubular wall *in vivo* may contain localized regions of positive charge as well as negatively charged sites.

Shape. Not all of the differences in the peritubular behavior of charged versus neutral dextrans can, however, be assigned to charge–charge interactions. In a recent series of studies we considered the role of charge in setting the effective shape and conformational plasticity of permeating macromolecules.[33] We were able for the first time to demonstrate that the elevated permeability of charged dextran tracers (anionic or cationic) rela-

---

[32] C. I. Whiteside and M. Silverman, *Am. J. Physiol.* **247**, F965 (1984).
[33] C. I. Whiteside and M. Silverman, *Am. J. Physiol.* **253**, F500 (1987).
[34] C. Crone, *Acta Physiol. Scand.* **58**, 292 (1963).

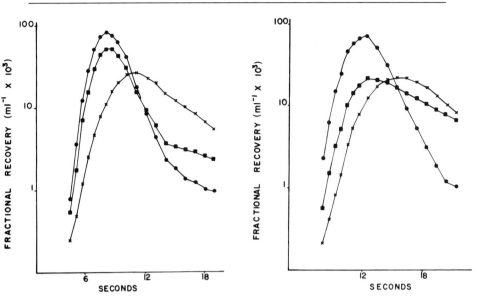

FIG. 9. Renal vein recovery curves for neutral (■, left) and sulfated (■, right) [^{3}H]dextran indicators of 28.0-Å SER. For both panels the reference indicators are [^{125}I]albumin (●) and [^{14}C]inulin (×). Inspection of the curves demonstrates that $t_{Neutral} \ll t_{Anionic}$ for indicators of the same equivalent SER, implying greater extraction of the anionic as compared to the neutral dextran probes. (Reprinted by permission from Whiteside and Silverman.[32])

tive to neutral dextrans in the peritubular capillary is due, at least in part, to shape differences among the test solutes (charged dextrans = relatively rigid rods; neutral dextrans = flexible chains).

Transport and Receptor Kinetics in the Proximal Tubule

We noted above that the renal tubule is a polarized epithelium in which lumenal and contralumenal surface exhibit unique (and different) morphological, biochemical, and functional properties. Distinguishing between the behavior of the two opposing surfaces is therefore crucial for an appropriate description of tubular binding and transport. A major advantage of our approach to renal MID is that this experimental separation can be achieved entirely *in vivo*. A sensitivity analysis of the components of Eq. (3) shows that the fractional recovery curves obtained from the renal vein reflect primarily events in the peritubular microcirculation.[35] For small

[35] C. J. Lumsden and N. V. Rau, *Am. J. Physiol.* (submitted).

molecules such as sugars or amino acids this means that the renal vein transit patterns contain information dominated by interactions that occur between test substrates and the antilumenal surface of the epithelium. In contrast, the urine outflow curves reflect primarily interactions that take place at the lumenal surface.

Peritubular extraction of a test substrate as monitored by comparing its time course of fractional recovery to that of the extracellular reference could in principle reflect an uptake process at the endothelial cell, basement membrane, renal interstitium, or contralumenal cell surface. In most instances it is straightforward to identify the pertinent barrier, leading to an appropriate model in mathematical form. Thus for example test substances that exhibit contralumenal interactions emerge in the renal vein (in terms of their fractional recoveries) lower than creatinine at early times, peak at lower values, and then generally cross over creatinine on the downslope. On the other hand, superposition of the outflow curve of a test substrate on that of creatinine in the renal vein outflow implies that no contralumenal interaction takes place.

To illustrate the features that have just been described, with the special purpose of defining the kinetics of solute interaction at the lumenal and contralumenal surfaces of the epithelium, let us consider some actual examples.

Enumeration of Transport Pathways. Figure 10 shows typical renal vein and urine fractional recovery curves for simultaneously injected T1824-albumin (plasma reference), creatinine (extracellular reference), and two test sugars, D-[^{14}C]fucose and D-[^{3}H]glucose, each in tracer quantities.[14] The runs were carried out under mannitol diuresis with a plasma glucose concentration of approximately 5 mM. In the urine D-fucose superimposes on the creatinine curve and therefore exhibits no evidence of lumenal interaction. In contrast the urine recovery of D-glucose is less than 2% of that filtered; 98% of filtered tracer D-glucose was extracted at the lumenal surface (Fig. 10, top). In the renal vein, D-glucose and D-fucose emerge lying below creatinine at early times, peak at lower values, and then cross over creatinine on the downslope. The total renal vein recovery is almost 100% of that injected. Inasmuch as less than 0.2% of the injected D-[^{14}C]glucose is metabolized to ^{14}CO$_2$ during the single pass and, in addition, there has been no rearrangement of the carbon skeleton,[19] we conclude that all reabsorbed glucose has been transported *transepithelially* and appears in the renal vein outflow. The renal vein curve for D-[^{14}C]fucose has a longer mean transit time than creatinine, which indicates a larger apparent volume of distribution in the postglomerular microcircula-

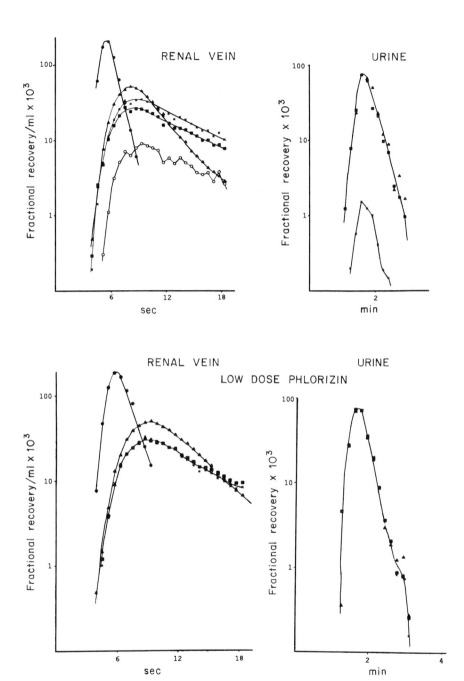

FIG. 10. Renal vein and urine fractional recovery curves for the monosaccharides D-fucose (■) and D-glucose (×). *Top:* Under control conditions D-fucose superimposes on creatinine (▲) in the urine, showing no evidence of lumenal interaction. D-Glucose is avidly reabsorbed by the tubular epithelial cells. *Bottom:* The presence of phlorizin blocks the lumenal interactions of D-glucose and the renal behavior of D-glucose is now identical to that of D-fucose. (●), T1824. (Reprinted by permission from Silverman and Trainor.[14])

tion. On the basis of the foregoing discussion we interpret this to mean bidirectional flux of D-fucose across the antilumenal nephron surface.

The same indicators can be compared after infusing a low dose of phlorizin—a known specific competitive inhibitor of D-glucose transport in the kidney (Fig. 10, bottom). D-Glucose now superimposes on the creatinine curve. Lumenal uptake has been completely blocked. But antilumenal uptake of both sugars is unaffected by phlorizin at these micromolar concentrations. In fact, after blockage of lumenal uptake, D-[^{14}C]fucose and D-[^{3}H]glucose behave identically in the renal vein effluent and the curves reflect cellular uptake of both sugars only at the contralumenal surface.

Our results imply that, under the existing steady state conditions employed in the experiment, the contralumenal uptake pathways for tracer D-glucose and D-fucose are indistinguishable. Therefore subtraction of the renal vein outflow curve of D-fucose from that of D-glucose in the absence of phlorizin should yield the renal vein appearance curve of transepithelially reabsorbed D-glucose.

With this experimental approach we have successfully defined the specificity of interaction of a homologous series of pyranosides and pyranoside derivatives with the two opposing surfaces of the tubular epithelium. By carrying out competitive inhibition experiments *in vivo* it has been possible to test which sugars share common transport systems. We have detected at least three different sugar carriers at the lumenal membrane (G, M, and Myo) and two at the contralumenal surface (G' and Myo'). Their specificity characteristics are reviewed comprehensively in Ref. 3. The G carrier may comprise two different glucose transport systems, but it is sufficient here to point out that the G carrier transports D-glucose, D-galactose, and several other sugars and is sensitive to phlorizin whereas the G' carrier transports D-glucose and 10 other sugars and is relatively insensitive to phlorizin, requiring 1000 times greater concentration to achieve inhibition.

In view of the very high circulatory perfusion received by the cortex, it is reasonable to conclude that the sugar carriers are located at the level of the proximal tubule (convoluted and/or straight segment). To test this hypothesis we turned to highly purified plasma membrane vesicle preparations,[36] one derived from the contralumenal membrane and the other from the brush border surface. Our studies of sugar uptake in the vesicle systems show that the phlorizin-sensitive G carrier is Na^+ dependent and exists exclusively in proximal tubule brush border membrane vesicles. The

---

[36] R. Kinne, H. Murer, E. Kinne-Saffran, M. Thees, and G. Sachs, *J. Membr. Biol.* **21**, 375 (1975).

phlorizin-insensitive carrier is Na$^+$ independent and localized to vesicles from the contralumenal membrane.[37,38] Preparation of vesicles from different regions of the cortex have refined the MID picture according to the number and types of different sugar carriers.[39] But the functional specification and localization of these carriers to lumenal versus contralumenal surface as deduced from MID holds true.

Quantification of Transport Kinetics. Two distinctive patterns of substrate handling are fundamental to the normal operation of the tubular epithelium. In net secretion, transport events at the opposing surfaces (BM, LM) of the epithelial cell create a vectorial displacement of substrate out of the microcirculation and into the urine. During net reabsorption the vector is reversed and substrate passes across the cell into the immediate vicinity of the tubular microcirculation. D-Glucose, discussed above, is a prime example of a strongly reabsorbed substrate. The available evidence indicates that, in contrast, organic anions undergo net secretion from the postglomerular microcirculation. Micropuncture studies on isolated perfused tubule preparations have shown that the first several millimeters of the proximal tubule are the principal secretion sites for those substrates.[40,41] The exact mechanisms and, in particular, the driving forces and transport kinetics have not been identified *in vivo*.

p-Aminohippuric acid (PAH) is the model substrate most often used to study organic anion transport in the renal proximal tubule. In a series of MID experiments we presented the kidneys of anesthetized mongrel dogs with bolus injections combining [^{125}I]albumin (vascular reference), [^{14}C]creatinine (extracellular reference), and [^{3}H]PAH. The experimental animals were maintained in a state of brisk, mannitol-induced diuresis.[42] Some animals were brought to specific steady state plasma concentrations of infused, unlabeled PAH prior to receiving the bolus of tracers. The renal vein and urine fractional recoveries were calculated (Fig. 11) and the renal vein PAH recoveries computer fitted with the model of transepithelial tracer exchange given by Eq. (3) in the limit $W_{Tu} \lll W_{Bl}$ (brisk diuresis). The renal vein recovery ratios of [^{3}H]PAH relative to [^{14}C]creatinine without PAH loading were well below unity, indicating net uptake of [^{3}H]PAH across the BM into the tubular cells. This avid postglomerular extraction is illustrated for a typical control run in Fig. 11 (left). The renal

---

[37] R. J. Turner and M. Silverman, *Biochim. Biophys. Acta* **507**, 305 (1978).
[38] R. J. Turner and M. Silverman, *Biochim. Biophys. Acta* **511**, 470 (1978).
[39] R. J. Turner and A. Moran, *Am. J. Physiol.* **242**, F406 (1982).
[40] M. Tune, M. B. Burg, and C. S. Patlak, *Am. J. Physiol.* **217**, 1057 (1969).
[41] A. Shimomura, A. M. Chonko, and J. J. Grantham, *Am. J. Physiol.* **240**, F430 (1981).

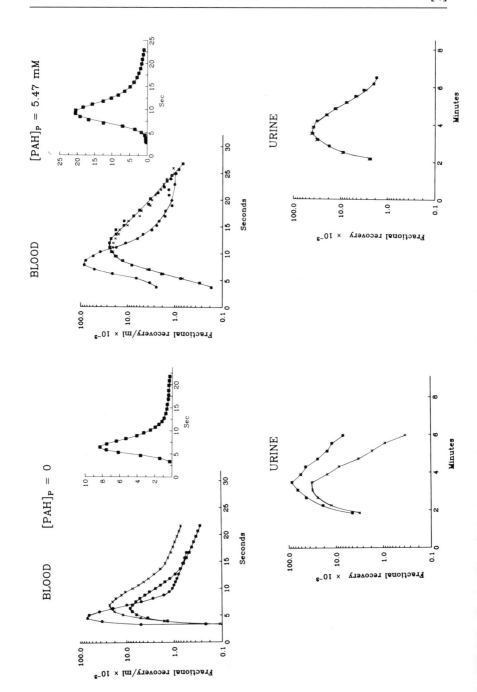

vein curve for [^{3}H]PAH falls below [^{14}C]creatinine. The urine outflow curves from the same control runs are sensitive indicators of LM events. For a typical control the large positive area between [^{3}H]PAH and [^{14}C]creatinine indicates [^{3}H]PAH flux from the tubular cells into the urine space.

As ambient PAH concentration was increased, BM uptake of [^{3}H]PAH from the interstitium decreased. The upslopes and peaks of the renal vein [^{3}H]PAH curves began to approach those of [^{14}C]creatinine. This behavior implies saturation of the [^{3}H]PAH uptake pathway across the BM. At the highest values of infused, unlabeled PAH (Fig. 11, right), saturation is complete and PAH is restricted to the space available to the extracellular reference. As this occurs, the urine outflow curve of [^{3}H]PAH approaches that of [^{14}C]creatinine, the glomerular reference.

Best estimates of unidirectional flux coefficients $k_1, \ldots, k_4$ were obtained across the full concentration range of infused, unlabeled PAH. Detailed discussion of their inhibition kinetics as a function of [PAH] infused can be found elsewhere. Of particular interest here is the picture we obtain for the BM and LM fluxes in the control (normal) state. In units of second^{-1}, the values are $k_1 = 0.32 \pm 0.03$, $k_2 = 0.21 \pm 0.04$, $k_3 = 0.34 \pm 05$, and $k_4 = 0.12 \pm 0.03$ (mean $\pm$ standard error, $N = 18$ runs). This defines a pattern of overall net secretion across the tubular epithelial cell, quantified to *in vivo* conditions. Our calculations indicate that at the BM ratio k_1/k_2 exceeds unity, consistent with active PAH transport at this surface. The LM unidirectional flux ratio k_3/k_4 is approximately 3–6, in keeping with passive transport of accumulated tracer from cytoplasm to lumen.

Many methods have been used to develop and test models of organic acid transport across the proximal tubular cell. The majority of these techniques have used *in vitro* preparations such as the isolated perfused tubule and isolated plasma membrane vesicles (reviewed in Ref. 42). The

[42] M. Silverman, C. I. Whiteside, C. J. Lumsden, and H. Steinhart, *Am. J. Physiol.* **256**, F255 (1989).

FIG. 11. Typical renal vein and urine fractional recovery curves for simultaneously injected [^{125}I]albumin (●), [^{14}C]creatinine (×), and [^{3}H]PAH (^{3}H-labeled *p*-aminohippuric acid) (■). The best fit of the renal vein fractional recoveries to the model of PAH handling by the tubular epithelial cell (see Fig. 5) is shown in the inset. *Left:* Recovery pattern in the absence of infused, unlabeled PAH, showing secretion of tracer-level [^{3}H]PAH into the urine. *Right:* Recovery pattern at high concentrations of infused, unlabeled PAH, for which the behavior of PAH now coincides with the extracellular reference, creatinine. (Reprinted by permission from Silverman *et al.*[42])

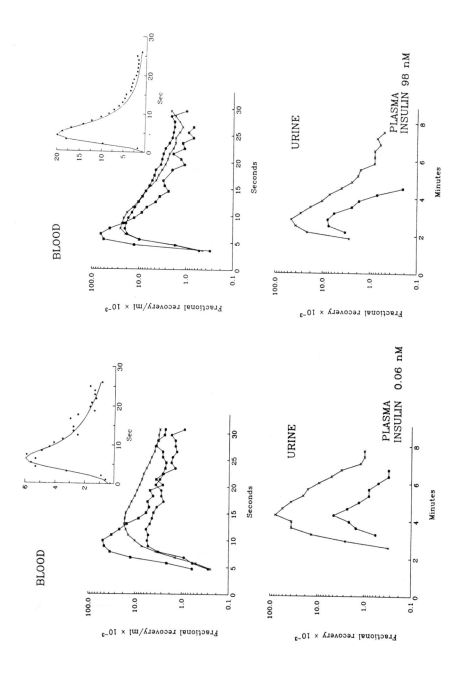

in vitro methodologies give rather strict control of the experimental environment. Although it is more difficult to manipulate the *in vivo* situation with similar efficacy, renal MID provides an equivalent separation of lumenal and antilumenal events, which together define the kinetic mechanism of PAH secretion at these surfaces in the normal intact state.

Ligand–Receptor Kinetics. The cells of the kidney are targets for a vast array of blood-borne effectors that induce hemoregulatory, growth, inflammatory, immune, and biomechanical responses. In nonepithelial tissues, insulin acts through specific membrane receptors. *In vitro* analyses of insulin interaction with isolated kidney membranes, isolated cortical tubules, and antilumenal vesicle preparations suggest that high-affinity insulin-specific receptors are present in the kidney, too.[43-45] The kinetics of *in vitro* insulin–receptor interaction are similar to those observed in adipocytes[46] and monocytes.[47] Insulin has an antinatriuretic and antiphosphaturic effect in the rat kidney,[48] suggesting that this peptide hormone may regulate sodium and water reabsorption by the tubular epithelium. However, evidence that insulin increases $Na^+–H^+$ exchange activity in epithelia is derived largely from rabbit proximal tubular cells in culture. We have recently applied MID to examine *in vivo* the interaction of ^{125}I-labeled insulin at the lumenal and contralumenal tubular epithelial surfaces of the dog kidney.[49] Since insulin does not undergo significant reabsorption into the postglomerular microcirculation during the time course of our single-pass MID experiment (approximately 15–20 sec), we were particularly interested in testing kinetic models of insulin interaction with the peritu-

[43] R. F. Blanchard, P. J. Davis, and S. D. Blas, *Diabetes* **27**, 88 (1978).
[44] K. Kurokawa and R. L. Lerner, *Endocrinology (Baltimore)* **106**, 655 (1980).
[45] Z. Talor, S. Emmanouel, and A. I. Katz, *J. Clin. Invest.* **69**, 1136 (1982).
[46] J. Olefsky, P. Jen, and G. Reaven, *Diabetes* **23**, 565 (1974).
[47] J. Olefsky and G. Reaven, *J. Clin Endocrinol. Metab.* **38**, 554 (1974).
[48] R. A. DeFronzo, A. M. Goldberg, and Z. S. Agus, *J. Clin. Invest.* **58**, 83 (1976).
[49] C. I. Whiteside, C. J. Lumsden, and M. Silverman, *Am. J. Physiol.* **255**, E357 (1988).

FIG. 12. Typical renal vein and urine fractional recovery curves for simultaneously injected [^{3}H]dextran (plasma reference) (■), [^{14}C]inulin (extracellular reference) (×), and [^{125}I]insulin (test substance) (●). *Left:* Recovery pattern at normal concentrations of nontracer unlabeled plasma insulin. *Right:* Recovery pattern at high steady state concentrations of infused nontracer unlabeled plasma insulin, showing approach of the tracer [^{125}I]insulin renal vein curve to that of the extracellular reference. *Insets:* Best fit of renal vein recovery data for [^{125}I]insulin to our computer-based mathematical model of tracer insulin binding to, uptake into, and release from an effective physiological space in series with the interstitium surrounding the peritubular capillary. (Reprinted by permission from Whiteside *et al.*[49])

bular interstitium. We compared the renal vein fractional recoveries expected from the alternative hypotheses of (1) irreversible binding to interstitial receptors or to the BM, (2) reversible binding to such sites, and (3) uptake and release of interstitial insulin from a third compartment defined either by morphological (e.g., intracellular) or kinetic ("enzymatic space") conditions.

Our renal vein data (Fig. 12, left) show that at normal plasma values of unlabeled (nontracer) insulin, the fractional recovery curve for tracer [^{125}I]insulin lies well below the [^{14}C]inulin (interstitial reference) curve, illustrating postglomerular capillary extraction of [^{125}I]insulin in the single-pass experiment. At increased values of infused unlabeled plasma insulin the urine outflow curves show an unchanging uptake of [^{125}I]insulin at the lumenal surface of the tubular epithelium. However, the renal vein curve of [^{125}I]insulin now approximates the [^{14}C]inulin marker, in keeping with saturable postglomerular uptake (Fig. 12, right).

Only the kinetic model of postglomerular tracer [^{125}I]insulin exchange based on uptake and release from a third compartment could explain the pattern of renal vein recoveries we observed (insets, Fig. 12). Uptake and release from this space corresponds to rates of 0.60 ± 0.08 and 0.16 ± 0.03 sec^{-1}, respectively, in the control state ($N = 9$, means $\pm$ standard error). The inhibition of tracer [^{125}I]insulin by infused unlabeled insulin in the steady state was compatible with simple competitive inhibition governed by a Michaelis constant equal to 15.44 nM (Fig. 13). Kinetic constants previously obtained from studies of [^{125}I]insulin binding to renal antilumenal membranes were carried out *in vitro* and under equilibrium conditions: dog BM $K_d = 1.2$ nM[50]; rabbit BM $K_d = 3$ nM.[45] Our result is also similar to the dissociation constant derived from insulin–receptor binding studies with rat hepatocyte $K_d = 3.5$ nM.[51]

Our insulin experiments give the first quantitative *in vivo* characterization of the kinetics of insulin interaction with the antilumenal renal tubular cell membrane. Further modeling indicates that the third compartment is indeed a virtual "enzyme" space created by the binding of insulin to an interstitial or BM receptor through a rate-limiting step that is slow relative to the rate of hemodynamic perfusion through the peritubular capillary.[52] Taken together these findings have exciting implications for the extension of our insulin experiments to other ligands involved in the kidney's complex system of receptor-mediated autoregulation.

[50] M. R. Hammerman and J. R. Gavin III, *Am. J. Physiol.* **247**, F408 (1984).
[51] S. Terris and D. F. Steiner, *J. Biol. Chem.* **250**, 8389 (1975).
[52] J. Booth, C. J. Lumsden, C. Whiteside, and M. Silverman, *Am. J. Physiol.* (submitted).

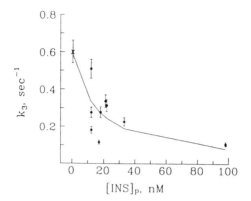

FIG. 13. Inhibition of the rate constant k_3 for tracer [^{125}I]insulin binding and uptake into an effective physiological space in series with the interstitium surrounding the peritubular capillary, as observed *in vivo* with MID in the dog. The independent variable is [INS]$_P$, the steady state plasma concentration of infused, unlabeled nontracer insulin. Each point is the best computer fit value of k_3 for the MID experiment carried out at the indicated value of [INS]$_P$. For the control state (×), $N = 9$; otherwise $N = 1$ in this series. The error bars denote ±10% uncertainty in the best fit estimates. The solid line is the least-squares fit of a simple competitive inhibition kinetics with inhibition constant K_d ($K_d = 15.44$ nM) to the points. (Reprinted by permission from Whiteside et al.[49])

Discussion

The potential of MID as an instrument for *in vivo* kinetics is evident. What, in terms of both method and model, so strongly recommends MID renal studies also applies to other tissues, and MID is being actively developed in virtually every major organ system in the body. While many of those applications deal with classical issues of capillary wall permeability an increasing number emphasize the *in vivo* study of cellular transport and receptor kinetics.[9,10,14,15,53-55] Despite these advances important limitations in method and theory remain, and these demand attention if the current rate of progress is to be maintained. We would like to conclude our report with a brief summary of these intriguing issues.

The first is MID applications to transient and time-dependent phenomena *in vivo*. Originally a tracer method used on steady state systems, MID

---

[53] A. J. Schwab, *Math. Biosci.* **71**, 57 (1984).
[54] N. Itoh, Y. Sawada, Y. Sugiyama, T. Iga, and M. Hanano, *Am. J. Physiol.* **251**, F103 (1986).
[55] J. B. Bassingthwaighte, C. Y. Wang, and I. S. Chan, *Circ. Res.* (in press) (1989).

has only recently been applied to time-dependent processes.[9] With a system in steady state it is difficult, if not impossible, to separate all the causal factors that contribute to the numerical magnitude of the steady state unidirectional flux coefficient k. For a ligand–receptor system these are factors like receptor size, number density, and ligand interaction energetics, which combine to determine the flux coefficient in individual cases. If the steady state of the system is disturbed transiently (e.g., via osmotic or pressure transients) the MID outflow dynamics may reflect the individual contributions of these individual causal factors more explicitly than in the normal steady state, and are amenable to direct modeling.[52] Moreover, kinetic parameters such as reflection coefficients, not normally accessible via steady state MID, become measurable.[9,56]

An important related issue involves physiological processes that are time dependent normally, rather than time independent or static, and vary on a time scale more or less equal to that of MID. In renal MID this time scale ranges from seconds to minutes—in other words, the time scales over which effector mechanisms regulate hemodynamics and ultrafiltration in the cortex and medulla of the kidney. Biomechanical events causing changes in, e.g., blood vessel diameter, mesangial cell size and shape, the glomerular epithelium, and hydrostatic driving forces provide the physical basis for these regulatory alterations. The biomechanical changes themselves are triggered by the interaction of a wide variety of primary messenger ligands with receptors embedded in renal tissue, as well as the activity of hemodynamic sensors in response to modifications in flow.[57-61] Mathematical models of MID in which tracers are caught up in dynamic alterations in flow and filtration remain to be developed and applied to the broad subject of cortical control systems.

Methodological restriction to (impulse) input–output measurements is a second principal limitation of MID. Even with detailed hierarchical models of organ structure, there are inevitable ambiguities in the average picture that data from one or two whole-organ outflow channels (in the kidney, renal vein, and urine) provide. Are our data on tracer PAH fractional recovery to be explained, for example, by one homogeneous popula-

[56] R. M. Effros, *J. Clin. Invest.* **54,** 935 (1974).
[57] P. M. Vanhoutte, *Mayo Clin. Proc.* **57,** 20 (1982).
[58] T. Sakai and W. Kriz, *Anat. Embryol.* **176,** 373 (1987).
[59] P. Andrews, *J. Electron Microsc. Tech.* **9,** 115 (1988).
[60] E. N. Ellis, S. M. Mauer, D. E. R. Sutherland, and M. W. Steffes, *Lab. Invest.* **60,** 231 (1989).
[61] J. W. U. Fries, D. J. Sandstrom, T. W. Meyer, and H. G. Rennke, *Lab. Invest.* **60,** 205 (1989).

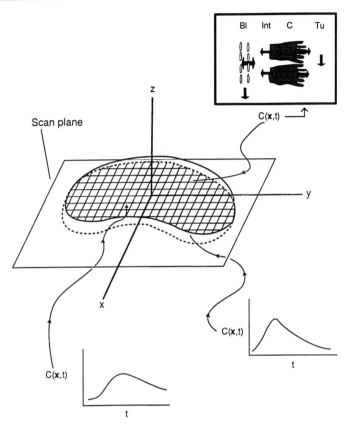

FIG. 14. Multiple indicator dilution modeling approach to the analysis of local tissue physiology and dynamic kidney imaging. The kidney is imaged at high spatiotemporal resolution in multiple planes of transection (e.g., dynamic computed tomography). In each scan plane, the organ's structure is resolved into contiguous volume elements, or voxels, $v(\mathbf{x})$, where x denotes three-dimensional spatial position. Each voxel contains a relatively large number of microcirculatory exchange units *(inset)*. The handling of indicator by these units, and its transport among units in adjacent voxels, is described mathematically in terms of volumes of distribution, barrier permeabilities, and thermodynamic driving forces acting between, e.g., the microcirculation (Bl), the interstitium (Int), the renal parenchymal cells (C), and the tubules (Tu) (compare Fig. 5). The time course $C(\mathbf{x},t)$ of indicator mass in each voxel $v(\mathbf{x})$ is determined from the digitized scans as a function of time t following bolus injection of radioopaque indicator into the renal artery. The rate constants of the microcirculatory exchange model are adjusted to give the best fit to the indicator curve $C(\mathbf{x},t)$ in each voxel. The resulting image is a map of the kidney in terms of the rate constants, permeabilities, and forces underlying tissue function.

tion of transporters spread the length of the proximal tubule, or several populations with varying dissociation constants, receptor densities, and spatial distribution along the tubule? A parsimonious model based on a homogeneous population gives a good account of itself (Fig. 11) but it is conceivable that our renal vein and urine data present a deceptively simple-looking picture with a more complex basis in truth (e.g., two or more types of receptors). In view of the residual experimental uncertainty (currently 5% per data point) ambiguities are inevitable.

Methods that localize measurements to the site of interest (e.g, segments of the proximal tubule) provide data not accessible to whole-organ input-output methods. Vesicle transport studies and micropuncture analyses are good examples of *in vitro* methods with the capacity to localize function. Recently we have begun to develop renal MID in a form that realizes the intrinsic information gains of localized tissue characterization while retaining the analytical power of whole-organ MID. Our strategy is to mathematically model dynamic CT (computed tomography[62,63]) images of the renal cortex responding to impulse injections of contrast agents (Fig. 14). As in whole-organ MID, we visualize (Fig. 15) the passage of several probe molecules (vascular reference, extracellular reference, test substance) through the system, but (in view of the imaging requirements of CT) the probes are introduced serially, one after another, rather than simultaneously. With rapid imaging (3-10 sections/sec) the indicator residue curve (the fractional retained mass, rather than fractional recovery as in whole-organ MID) is evaluated for the cortex on a point-by-point basis. Model-based analysis of these curves then leads to the assignment of kinetic parameters to that localized region and, by extension, across the entire structure of the renal cortex. Unlike the imaging of flow parameters[63] or, as in positron emission tomography (PET), metabolic/sequestra-

---

[62] M. D. Bentley, E. A. Hoffman, M. J. Eiksen-Olsen, F. G. Knox, E. I. Ritman, and C. Romero, *Am. J. Anat.* **181**, 77 (1988).

[63] G. K. von Schulthess, "Morphology and Function in MRI: Cardiovascular and Renal Systems." Springer-Verlag, New York, 1989.

FIG. 15. Dynamic spatial imaging of renal indicator dilution *in vivo* in the dog kidney. The pulse-injected indicator (renal artery) is a nonionic radiocontrast agent (interstitial marker). (A) Color-enhanced sections from early, middle, and late in the sequence of cortical filling by the indicator. Local mass of indicator increases along the color code blue, green, yellow, red, etc. Interlobar distribution of indicator arrival is clearly visible in the early frames. (B) Full sequence of images comprising a single run. The time interval between each scan is 0.1 sec. Our data were gathered using the Imatron CT scanner at the Buffalo Children's Hospital, Buffalo, New York.

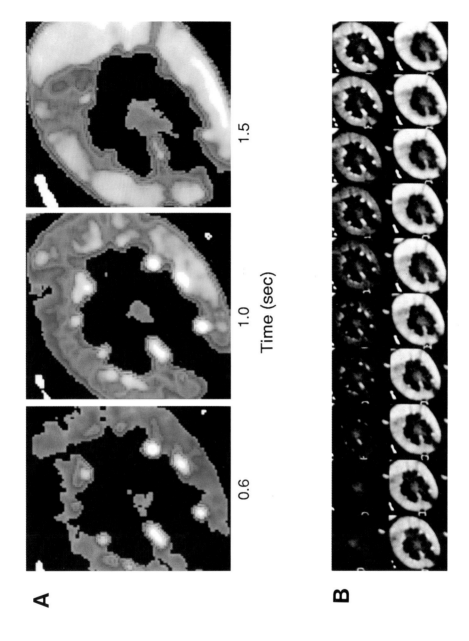

tion rate via the irreversible uptake of tracers into cellular spaces ("scalar physiology"), MID-based medical imaging visualizes the kinetics of biomolecular transport regionally within a tissue or organ ("vector physiology").

Realistic modeling for such MID initiatives necessarily begins with accurate representations of tissue structure. For the most part, however, previous work has approximated the morphology of blood vessels and the regions of tissue they perfuse by the simplest geometric forms—especially concentric cylinders, along which a central hemodynamic flow feeds surrounding regions of interstitium and parenchymal cells.[9,16,25,64] These models do surprisingly well in accounting for the patterns of fractional recovery observed among simultaneously injected tracers, though ultimately one desires to interpret transport kinetics (via the unidirectional flux coefficients) in terms of parameters fitted to morphometrically realistic models of tissue organization, not "equivalent cylinders." This is particularly so in the kidney, where MID output depends on two microcirculations in series, the second involving the exact spatial relation between two elongated convective compartments (peritubular capillary and tubule) pursuing complex paths through the cortex.[17,65] The move to regional studies in MID, augmenting input–output measurements at the whole-organ level, will further increase the need for MID models based on accurate and complete data (see, e.g., Ref. 66) about the size, shape, and spatial disposition of the microcirculation, interstitium, and parenchymal elements disposed within the target tissue.

The control for background given by the stepwise comparison of test substance ↔ extracellular reference ↔ vascular reference lies at the heart of MID. But as the number of kinetic steps separating the events traced by the test substance and the baseline events of extracellular distribution increase, the sensitivity of MID models decreases rapidly.[9] In our applications to the kidney, kinetic parameters can rarely be obtained with confidence if many more than three or four kinetic steps separate the test indicator from the reference indicator. As MID pushes past the frontier represented by the capillary wall and penetrates deeper into the receptor-based events at the cell surface, progress in modeling will have to be matched with the application of more complex injection solutions, where $N > 3$ reference indicators trace much further down the chains of kinetic steps by which solutes are recognized, processed, and taken across the plasma membrane. We consider this an exciting opportunity for the appli-

---

[64] W. C. Sangren and C. W. Sheppard, *Bull. Math. Biophys.* **15**, 387 (1953).
[65] M. J. Horacek, A. M. Earle, and J. P. Gilmore, *J. Anat.* **148**, 205 (1986).
[66] W. Pfaller, *Adv. Anat. Embryol. Cell Biol.* **70**, 1982.

cation of enzymologic and molecular engineering principles to design reference substances that optimally "peel away" irrelevant kinetic events and maximize the signal, in the form of net fractional recovery, from the test tracer. This is the analog in tracer physiology, still awaiting serious pursuit, of the histological problem solved by monoclonal antibodies and their fluorescent/electron-dense labels: select a specific structure (function) and label it in a way that minimizes the interfering effects of ancillary structures (functions).

Finally, we are struck by the degree to which normal function, and normal structure–function relationships, have occupied renal MID investigation over the past three decades. By comparison, no more than sparing attention has been given so far to MID studies of renal pathophysiology in well-documented animal models of kidney disease. This is a missed opportunity, explained partly by the relatively high cost of MID work in its traditional setting (the dog) and by the fact that many disease models are set in species (rat, rabbit) much smaller than the one for which the MID protocols were designed. Since dysfunctions in the kinetics of flow, filtration, and tubular physiology underlie virtually all the significant effects of renal disease, a new decade of renal MID research opens with the signal prospect of extending to pathophysiology the analytical potential of an *in vivo* kinetic method now firmly established in both theory and experiment.

Acknowledgments

The Medical Research Council of Canada, through its grants to our experimental (MRC Group Grant for the MRC Group in Membrane Biology, University of Toronto) and theoretical (MRC Grant MA-8635) investigations, has made possible the work reported here. Medical Research Council career awards (MRC Scholar, MRC Scientist) to C.J.L. are also gratefully acknowledged.

[5] Micropuncture Techniques in Renal Research

By IHAB M. SHAFIK and GARY A. QUAMME

Micropuncture studies of single nephrons have offered a large amount of information on many aspects of fundamental glomerular and tubular function. The methodology has been especially useful in determining the location and quantification of transport processes within the nephron.[1-6]

[1] E. E. Windhager, in "Molecular Biology and Medicine Series" (E. E. Bittar, ed.), p. 1. Butterworth, London, 1968.

With appropriate advances in chemical and electrical quantification our understanding of essential transport mechanisms has also been clarified. Micropuncture methodologies have also been used in other tissues such as tracheal, hepatic, and intestinal epithelia to evaluate physiological processes. Although this chapter will center on the kidney, the principles involved in micropuncture are generally applicable to other tissues. The aim of this chapter is to describe micropuncture methodologies and their applications to epithelial studies. We will also suggest avenues for future use of these techniques in the kidney and in other organs.

Micropuncture is the technique of inserting small pipets into tissue in order to gather chemical and physical information. As the name implies, these are performed at a microlevel, usually requiring special stereotaxis micromanipulators and stereoscopic microscopes. The technique was used some 60 years ago by Richards and Wearn and their colleagues to describe solute- and water-free ultrafiltration of the glomerulus and tubular reabsorption in the amphibian kidney.[7] Subsequently, workers have applied these micropuncture techniques to the mammalian kidney.[8,9]

Micropuncture may be performed *in vitro* or *in vivo*. *In vitro* micropuncture of isolated nonfiltering kidney or individual tubule segments will be considered in detail elsewhere in this volume. *In vivo* micropuncture experiments are performed on anesthetized animals maintained at constant body temperature of 37–38°. Sodium pentobarbital [sodium 5-ethyl-5-(1-methylbutyl)-barbiturate] and inactin [sodium 5-ethyl-5-(1-methylpropyl)-2-thiobarbiturate] are the two commonly used anesthetics, both being administered intraperitoneally. The jugular or femoral vein is usually cannulated for intravenous infusions and blood pressure is monitored by catheters located in either the carotid or femoral artery. Blood samples are obtained from this catheter throughout the duration of the study. The kidney is surgically exposed and immobilized in a Lucite cup in order to minimize transmission of respiratory movements to the kidney surface. Urine samples are collected via catheters inserted into the bladder or ureter.

---

[2] G. Giebisch (ed.), *Yale J. Biol. Med.* **45**, 187 (1972).
[3] C. W. Gottschalk and W. E. Lassiter, *Handb. Physiol., Sect. 8: Renal Physiol. 1973* p. 129 (1973).
[4] F. Lang, R. Greger, C. Lechene, and F. G. Knox, *Methods Pharmacol.* **4B**, 75 (1978).
[5] V. E. Andreucci (ed.), "Manual of Renal Micropuncture," Idelson, Naples, Italy, 1980.
[6] G. A. Quamme and J. H. Dirks, *Kidney Int.* **30**, 152 (1986).
[7] J. T. Wearn, *Physiologist* **23**, 1 (1980).
[8] A. M. Walker and J. Oliver, *Am. J. Physiol.* **133**, 562 (1941).
[9] A. M. Walker and P. A. Bott, *Am. J. Physiol.* **134**, 580 (1941).

Viewed through the microscope, the surface of the kidney appears as a mass of tubular segments, most of which are proximal convolutions. Distal convoluted segments appear on the kidney surface less frequently than proximal convolutions. Identification of different segments of superficial tubules is facilitated by the intravenous injection of a small amount of Lissamine Green or other water-soluble dye, whole renal excretion is rapid and is accomplished by glomerular filtration. Thus the dye highlights the path through the different nephron segments. At first, a diffuse green coloration of the kidney surface indicates passage of dye through surface blood vessels and into Bowman's space. This is taken as "zero time." As the dye in the tubules reaches the terminal segments of proximal convolutions, the tubules may be seen converging in groups on the surface (see Fig. 1). This marks the "proximal transit time" which averages 8–12 sec. Afterward, the kidney regains its normal red color for a short period of time (15–30 sec) corresponding to passage of dye through loops of Henle. The dye reappears on the kidney surface again as it passes through into the distal tubules before flowing out in the urine. Besides differentiating proximal and distal tubules, measurement of the transit time to each segment

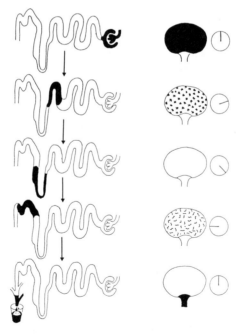

FIG. 1. Diagrammatic representation of time sequence of Lissamine Green transit time.

also helps assess tubular urine flow and the functional integrity of the kidney.

Glomerular Micropuncture

Studies using micropuncture methods have provided a direct description of water and solute ultrafiltration of accessible glomeruli. Glomerular function has often been inferred from findings on the whole kidney filtration rate because the glomeruli of the mammalian kidney normally lie below the surface of the kidney, unlike those of the frog, and are not available for micropuncture. However, an occasional glomerulus may be observed on the surface. Micropuncture of these occasionally available glomeruli provided evidence that an ultrafiltrate of plasma is elaborated by the glomerular wall.[9] More recently, the discovery in K. Thurau's laboratory of a strain of Wistar rats (called the Munich–Wistar strain) that frequently have glomeruli on the surface, has enabled the direct evaluation of glomerular function.[10]

Micropuncture of surface glomeruli in Munich–Wistar rats has been used extensively and has helped greatly in defining the nature of the glomerular ultrafiltration process, its biophysical determinants, and the various factors regulating and influencing this process. It also provided the most direct and accurate approach toward assessing the composition of the glomerular filtrate and estimating the percentage ultrafilterabilty of the different solutes reaching the glomerular capillary bed.

Net filtration, the initial step in the formation of urine, occurs at the glomerulus and results in the elaboration of a nearly ideal ultrafiltrate of plasma from the blood that flows through the glomerular capillary network. Understanding the principles of this ultrafiltration process is of utmost importance to the management of many disorders of renal function.

Two opposing forces operate across the glomerular capillary wall; the transcapillary hydraulic pressure (P), which favors filtration, and the transcapillary oncotic pressure (π), which opposes it. P, in turn, is the composite of the outward driving hydraulic pressure within the glomerular capillary (P_{GC}) and the inward driving hydraulic pressure within Bowman's space, which is almost equal to the pressure in the proximal tubules (P_T). Similarly, π is the composite of the oncotic pressure of plasma proteins within the glomerular capillaries (π_{GC}) and that of the filtrate within the Bowman's space (π_T). The net ultrafiltration pressure (P_{UF}) therefore equals the

[10] B. M. Brenner, J. L. Troy, and T. M. Daugharty, *J. Clin. Invest.* **50**, 1776 (1971).

difference between P and π, i.e.,

$$P_{UF} = P - \pi = (P_{GC}-P_T) - (\pi_{GC}-\pi_T)$$

The local rate of ultrafiltration at any one point of the glomerular capillary therefore should equal the product of P_{UF} and the local effect capillary wall permeability k. If *k is multiplied by the surface area available for filtration of the whole capillary network belonging to one glomerulus S*, the rate of ultrafiltration in that single nephron (SNGFR) can be calculated from the following equation:

$$\text{SNGFR} = kS[(P_{GC}-P_T)-(\pi_{GC}-\pi_T)] = K_f[P-\pi] = K_f P_{UF}$$

where K_f is the ultrafiltration coefficient and P_{UF} is the mean net ultrafiltration pressure for the whole capillary length.

Measurement of the glomerular capillary hydraulic pressure (P_{GC}) is a classic example of the ingenious use of micropuncture methodology in renal research. Brenner and colleagues have developed a number of micropuncture techniques for direct measurement of the mean glomerular capillary hydraulic pressure.[11] A value of 45–50 mmHg has been obtained in Munich–Wistar rats and primates. Similarly, the hydraulic pressure in Bowman's space was found to be 10–12 mmHg, giving a value of transcapillary hydraulic pressure P of about 33–40 mmHg. From this information much has been learned about the physiology and biophysics of the filtration process in health and disease.

The other area where glomerular micropuncture has provided valuable information is in determination of the composition of the glomerular filtrate (Table I). Various investigators have relied in the past on the use of artificial membranes for estimation of the composition of the glomerular filtrate. Inevitably, the reports showed wide variations of ultrafiltrability of different solutes. However, direct sampling of the glomerular filtrate from Bowman's capsule of Munich–Wistar rats has yielded much more consistent results.[12,13]

The glomerular filtrate is a protein-free fluid whose macromolecular concentration is determined by the respective protein binding and the Donnan effect across the glomerular membrane. Several factors that determine the glomerular filtration of macromolecules have been identified, including molecular size, charge, and configuration as well as glomerular hemodynamics. In terms of size selectivity, the normal glomerular capillary wall acts as an ultrafiltration membrane with pores of about 50-Å radii

[11] B. M. Brenner, J. L. Troy, T. M. Daugharty, W. M. Deen, and C. R. Robertson, *Am. J. Physiol.* **223**, 1184 (1972).
[12] C. Le Grimellec, P. Poujeol, C. De Rouffignac, *Pfluegers Arch.* **354**, 117 (1975).
[13] C. A. Harris, P. G. Baer, E. Chirito, and J. H. Dirks, *Am. J. Physiol.* **227**, 972 (1974).

TABLE I
COMPOSITION OF GLOMERULAR FILTRATE

	Plasma concentration (mM)	Glomerular fluid (mM)	Percentage ultrafilterability[a]
Na$^+$	140–145	139–145	100
Cl$^-$	111–115	120–124	104–108
K$^+$	4.1–4.5	3.60–4.01	80–90
Ca^{2+} (total)	2.2–2.6	1.28–1.51	60–66
Mg^{2+} (total)	0.6–0.7	0.48–0.52	70–80
P$_i$	3.0–3.2	3.2–3.6	109–120

[a] Ultrafilterability was determined by micropuncture of superficial glomeruli of Munich–Wistar rats and comparisons of the filtrate with plasma concentration.

preventing any larger macromolecule from moving across the membrane.[10,11] Filtration of macromolecules with lesser radii will inevitably be influenced by some other hemodynamic factors such as the mean transcapillary hydraulic pressure P, ultrafiltration coefficient, as well as the initial glomerular plasma flow rate and protein concentration. Molecular configuration, however, exerts little, if any, influence on the sieving of macromolecules. Finally, a good deal of information indicates that the glomerular membrane is a charge-selective barrier as the membrane is rich in negatively charged sialoproteins. Molecular charge is therefore another important determinant of macromolecular filtration. Significant electrostatic interactions between the charged solutes and the negatively charged components in the glomerular capillary wall will facilitate transport of polycations while retarding the filtration of polyanions.[12]

It is clear from this brief outline of the biophysical factors affecting the process of ultrafiltration that the glomerular capillary wall does not discriminate solely on the basis of molecular size. Hence, attempts at estimating the composition of the glomerular filtrate by the use of artificial membranes are bound to be less accurate than directly sampling the fluid in Bowman's capsule by glomerular micropuncture. Indeed, this has been directly tested by Le Grimellec *et al.*, who punctured Bowman's capsule in Munich–Wistar rats and compared the composition of the fluid obtained to that of an ultrafiltrate prepared *in vitro* using artificial membranes.[12]

As is shown in Table I, sodium is freely filterable by the glomerular membrane.[26] Chloride is 104–108% ultrafilterable as determined in glomeruli of Munich–Wistar rats.[12,13] Potassium concentration in the glomerular filtrate was found to be 80–90% of the plasma concentration.[13] Calcium and magnesium are significantly bound to plasma proteins and therefore only 60–66 and 70–80%, respectively, are found in the glomerular fluid.[12] About 20% of plasma phosphate is also protein bound and not freely filtered through the glomerular membrane. Glomerular micropuncture has also helped in assessing the suitability of the use of certain substances, such as inulin, as markers of the glomerular filtration rate. The propensity of being freely ultrafilterable through the glomerulus is a prerequisite for a suitable marker; accordingly, the demonstration by Le Grimellec et al.[12] and by Harris et al.[13] of a glomerular fluid/plasma concentration ratio of inulin (GF/P_{In}) of 1.00 indicates completeness of the ultrafiltration of this substance and the suitability of its use.

Micropuncture of Bowman's capsule is best performed with micropipets with tip diameters between 6 and 7 μm. It is also preferable to collect the fluid sample without introducing an oil block to avoid any alteration of the transcapillary hydrostatic pressure P. Collection therefore should proceed at a rate slower than the single-nephron GFR in order to avoid contamination of the sample by retrograde collection of fluid from the proximal tubule.

Tubular Micropuncture

Free-Flow Micropuncture

Free-flow micropuncture, as the names implies, involves the insertion of glass micropipets into the tubule lumen and collection of tubule fluid samples under "free-flow" conditions. A successful tubule fluid collection depends to a large extent on using a suitable, carefully prepared micropipet. Micropipets used for puncturing proximal convoluted tubules should have smooth, well-bevelled tips with outer diameters between 80 and 10 μm. Figure 2 shows a schematic illustration of a free-flow micropuncture experiment in a rat kidney. The collection micropipet is filled with mineral oil colored with Sudan Black, mounted on a micromanipulator, and lowered onto the surface of the kidney at an angle of about 20–30° until it is seen touching the wall of the selected tubule. With gentle forward pressure, the tubule wall may be easily penetrated without tearing. The ease with which tubules may be penetrated depends largely on the sharpness of the pipet tip and the angle of penetration. The kidney should be absolutely motionless, since slight pulsatory movement will lead to difficulty in penetration and, frequently, damage to the neighboring capillaries. It is good practice to attempt puncturing tubules only with a long enough length available on the surface, i.e., approximately four to five tubule diameters. On the one hand, this will facilitate visualizing part of the oil column during the collection (*vide infra*). On the other hand, if tubules are punctured perpendicular to their longitudinal axis there is a high chance of the

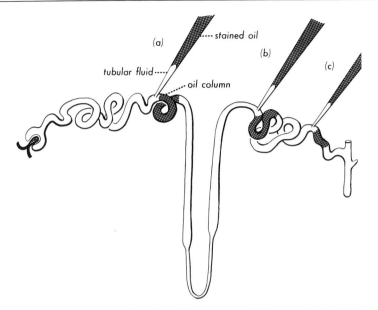

FIG. 2. Schematic aspects of free-flow micropuncture: (a) late proximal tubular site; (b) early distal tubular site; (c) late distal tubular site.

micropipet tip penetrating the opposite tubular wall and creating a fistulous tract with a neighboring tubule with subsequent leakage and contamination of samples.

Measurement of single-nephron glomerular filtration rate (SNGFR) in free-flow micropuncture necessitates complete collection of tubule fluid which reaches the pipet tip at a given time.[13] For this reason an oil block is introduced into the lumen prior to collection. The oil column prevents retrograde flow and subsequent contamination of the collected sample by fluid distal to the puncture site. An oil column of approximately three to five tubular diameters in length is commonly employed by most micropuncturists. As soon as the oil column moves downstream to the pipet tip a slight transient suction is applied to the pipet which starts collection of the tubule fluid. Excessive or continuous suction, however, should be avoided since this will alter the intratubular pressure and the SNGFR. As described earlier, segments or distal convoluted tubules may be easily identified with the aid of Lissamine Green. In addition to visualizing distal segments, the dye will also identify the direction of tubular fluid flow. The technical procedure for puncturing distal tubule segments is essentially the same as described for proximal tubules. Micropipets with small tip diameters (5–7 μm) are necessary to puncture the narrower diameter of distal tubules.

TABLE II
COMPOSITION OF EARLY AND LATE PROXIMAL AND EARLY AND LATE DISTAL TUBULE FLUID (MMOL/LITER)

	Early proximal site	Late proximal site	Early distal site	Late distal site
$TF/P_{In}{}^a$	1.01–1.1	2.5–3.0	5.0–7.5	10–20
Na^+	142–145	142–150	42–70	28–35
K^+	3.8–4.0	3.8–4.6	1.5–2.1	3.6–19.0
Cl^-	122–126	134–152	31–50	24–37
Ca^{2+}	1.4–1.5	1.5–1.8	0.4–0.7	0.3–0.4
Mg^{2+}	0.50–0.55	0.7–1.5	0.3–0.4	0.4–0.5
P_i	2.5–3.0	1.7–2.1	2.5–3.0	2.38–2.70
HCO_3^-	20–25	4.0–5.0	5.0–8.0	5.0–8.0
Urea	2–5	2–5	5–10	10–50
Glucose	3–5	0.7–1.2	0.1–0.5	0.5–1.0

[a] TF/P_{In} is the tubule fluid-to-plasma concentration ratio for inulin.

A unique modification of free-flow micropuncture has been the application of the recollection technique.[14] In this method, samples from the same site of the same tubule are collected under control and experimental conditions. This approach allowed investigators to perform an experimental maneuver and compare fluid composition from the same site on a "paired" basis during control and following an imposed experimental maneuver. The technique has contributed a great deal to our understanding of factors controlling water and solute transport.

Studies using free-flow micropuncture have provided a direct description of water and solute reabsorption in the nephron segments accessible to the micropuncture pipette. With the use of inulin and the appropriate chemical analysis, the concentration profile along the nephron may be determined (Table II). The accessible tubule segments which may be directly examined include the proximal convoluted tubule and distal convoluted tubule. The segments between these accessible sites, the loop of Henle, including portions of S_2 and S_3 of the proximal tubule, thin descending and thick ascending limbs, and the distal tubule may be inferred from the differences in solute deliveries to the respective collection sites. Also, some information may be inferred by comparing the solute deliveries into the final urine with those at the distal tubule. This ignores the possibility of functional internephron heterogeneity between superficial and deep nephrons.[15] To circumvent these inferences, a number of workers have directly sampled the descending and ascending limbs of the papillary loops

[14] H. G. Rennke, R. S. Cotran, and M. A. Venkatachalam, *J. Cell Biol.* **67**, 638 (1975).
[15] F. G. Knox and G. R. Marchand, *Methods Pharmacol.* **4A**, 73 (1976).

of Henle as well as the papillary collecting ducts.[16,17] Wirz first used micropuncture to sample fluid from the papilla of young rats (body wt < 100 (mmol/liter), which protrudes at this age, into the renal pelvis.[18] Another innovation has been developed to study the collecting duct system.[19-21] Jarausch and Ullrich[21] first used catheterization to collect tubule fluid from the collecting duct. In this approach, the papilla is exposed by an incision into the ureter. Then a fine (15–35 μm) polyethylene catheter is advanced several millimeters through the opening of a papillary collecting duct so that the proximal and distal portions of the collecting tree can be sampled.

Microinjection and Microinfusion

A number of variants of the micropuncture technique have been devised to investigate the mechanics of solute and water transport within the nephron. The microinjection technique entails the injection of a small volume of a solution containing radioactive [^{3}H]inulin and a labeled tracer of the substance under study into some part of a superficial nephron.[22,23] The urine is then monitored sequentially for both substances. The urinary recovery of the tracer is compared to the recovery of inulin, which allows for an estimation of the unidirectional fractional reabsorption of the injected substance beyond the microinjection site. The feasibility of the microinjection technique depends to a large extent on the ability to reproduce precisely equal volumes of injected solution for tubular injection and for standard determination. This necessitates the construction of a special type of micropipet in which a constriction is added to provide a constant volume. The resulting volume may be repeatedly filled with the injectable solution in a quantitatively reproducible manner. Constant-volume micropipets used for microinjection into proximal tubule segments usually have volumes ranging between 8 and 10 nl. For distal tubule microinjections, small volumes (5–7 nl) are needed. The total urinary recovery of [^{3}H]inulin and the tracer substance is expressed as fractional urinary recovery:

Fractional urinary recovery = (total urinary recovery/injected load) × 100

or as fractional unidirectional:

[16] J. H. Dirks, W. J. Cirksena, and R. W. Berliner, *J. Clin. Invest.* **44,** 1160 (1965).
[17] A. D. Baines, C. J. Baines, and C. De Rouffignac, *Pfluegers Arch.* **308,** 244 (1969).
[18] C. W. Gottschalk and M. Mylle, *Am. J. Physiol.* **196,** 927 (1959).
[19] C. De Rouffignac and F. Morel, *J. Clin. Invest.* **48,** 474 (1969).
[20] H. Wirz, *Helv. Pharmacol. Acta* **11,** 20 (1953).
[21] K. H. Jarausch and K. J. Ullrich, *Pfluegers Arch.* **264,** 88 (1956).
[22] H. H. Bengele, E. R. McNamara, and E. A. Alexander, *Am. J. Physiol.* **232,** F566 (1977).
[23] C. W. Gottschalk, F. Morel, and M. Mylle, *Am. J. Physiol.* **209,** 173 (1965).

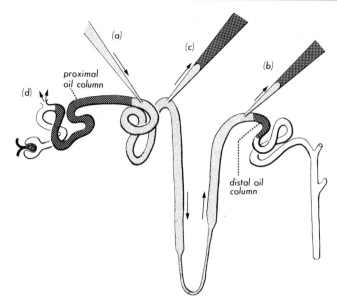

FIG. 3. Schematic aspects of *in vivo* microperfusion: (a) perfusion pipet in a late proximal site; (b) collection pipet in early distal site; (c) collection pipet in last visible proximal site; (d) a wide hole is made in a superficial early proximal segment to enable escape of glomerular filtrate.

$$\text{Fractional unidirectional reabsorption} = 100 - \text{fractional urinary recovery}$$

The microinfusion technique is a recent modification which involves continuous infusion of the solution using a microperfusion pump instead of the injection of a bolus.[24] The unidirectional outflux of the tracer substance from one single nephron may then be monitored continuously during control and experimental conditions.

Microperfusion

Another method of investigating solute flux is by continuous microperfusion of specific tubule segments and subsequent collection downstream to the perfusion site.[25] Figure 3 shows a schematic drawing of a continuous microperfusion experiment. An ordinary micropipet with tip diameter between 8 and 10 μm and filled with colored oil is inserted into an early

[24] C. De Rouffignac and F. Morel, *Nephron* **4**, 92 (1967).
[25] R. Greger, F. Lang, G. Marchand, and F. G. Knox, *Pfluegers Arch.* **369**, 111 (1977).

proximal segment. The more distal segments of the same nephron and the direction of tubular fluid flow within these segments are identified by the injection of small oil droplets and following their path. The perfusion pipet (a) containing the artificial solution to be studied is inserted downstream to pipet (d) in a segment where perfusion would be in direction with tubular fluid flow. An oil block is positioned between the two pipets. This oil block is necessary to isolate the perfusion solution from the glomerular filtrate, which is either collected in pipet (d) or allowed to escape from the hole left after removal of pipet (d). The collection pipet (also filled with colored oil) is inserted further downstream to the perfusion pipet in a proximal (c) or in a distal (b) tubule segment as appropriate, and a distal oil block is injected. This distal oil block will ensure complete collection of the perfusion solution and prevent retrograde collection of tubule fluid from more distal segments. The microperfusion pump can be set to deliver the artificial solution at constant flow rates (between 1 and 50 nl min^{-1}), and the collection initiated downstream to the perfusion site.

The advantage of this continuous microperfusion approach is that one may artificially alter the intralumenal fluid composition and control the flow rate into any given segment.[26] Also, by including labeled solute in the perfusate one can readily discern unidirectional solute movements by comparing the disappearance rate with the cold solute concentration.[26] The continuous microperfusion method has been and still is successfully applied to study the unidirectional and net fluxes of a variety of substances in the proximal tubule, loop of Henle, and distal tubules.

By adding a nonreabsorbable substance, such as inulin, to the perfusion solution and measuring its concentration in the collected fluid, the net fluid reabsorption (J_v) in the perfused segment may be calculated from

$$J_v = V_0[1 - (C/P)_{In}^{-1}] \tag{1}$$

where V_0 is the perfusion rate, C_{In} is the concentration of inulin in the collected sample, and P_{In} is the concentration of inulin in the perfusion solution.

Likewise, the net electrolyte flux ($J_n S$) may be calculated by the equation for any given segment (n).

$$J_n S = V_0[S]_0 - V_x[S]_x \tag{2}$$

where $[S]_0$ is the concentration of electrolyte in the perfusion solution, $[S]_x$ is the concentration of electrolyte in the collected fluid at some distance x from the perfusion site, and V_x is the collection rate.

The collection rate V_x may be calculated either by timing the collection

[26] H. Sonnenberg and P. Deetjen, *Pfluegers Arch.* **278**, 669 (1964).

of the sample, or from the equation

$$V_x = V_0[1/(C/P)_{\text{In}}] \tag{3}$$

Another approach for investigating water and solute flux within the nephron has been the use of lumenal fluid droplets injected into the tubule. Richards and Walker studied transport in the *Necturus* nephron by inserting a fluid droplet into the tubule and determining the alteration in size relative to the fluid composition. Two modifications of this droplet technique have been developed. First, Shipp and colleagues described a method whereby a droplet was inserted between two oil columns, left for a few seconds, and then collected for analysis of solute composition.[1] An unreabsorbable solute, such as raffinose, was added to the droplet fluid to prevent water absorption. This has been termed stationary microperfusion.[27,28] Second, Gertz and Ullrich developed the shrinking droplet method for determination of isotonic fluid reabsorption in the mammalian kidney.[29,30] A droplet of isotonic saline is injected into an oil column within the tubule, and the rate of disappearance of the droplet is measured by serial microphotometry. These approaches were used to measure, respectively, the development of limiting concentration gradients across the tubule wall and the absolute determination of fluid reabsorption. Further, Ullrich and colleagues have developed an intriguing approach. They have used a combination of the standing droplet technique and simultaneous perfusion of the peritubule capillaries to determine the disappearance rates of various solutes across the epithelial membranes. The concentration gradient generated by the half-time disappearance rate is a measure of transport rate. This novel approach has been extensively used to describe the kinetic properties of proximal transport. The above-mentioned techniques of microinjection, microperfusion, and standing droplet may also be applied to the peritubule capillary to alter the basolateral membrane environment and/or evaluate basolateral membrane transport.[30] Micropuncture of capillaries is more difficult than tubular micropuncture. Peritubular capillaries are smaller than tubules and blood in the capillaries tend to clot in and around the pipet tip. Special precautions are therefore necessary to prevent frequent blocking of the tip, such as siliconization or heparinization (5000 U/ml, aqueous heparin) of the micropipet before use.

Micropuncture techniques require careful animal preparation, meticulous collection of fluid samples and precise ultramicroanalyses; each step brings its specific difficulties and is susceptible to technical errors.[6,30] Al-

[27] F. Morel and Y. Murayama, *Pfluegers Arch.* **320**, 1 (1970).
[28] A. Z. Gjöry, *Pfluegers Arch.* **324**, 328 (1971).
[29] K. H. Gertz, J. A. Mangos, G. Braun, and H. D. Pagel, *Pfluegers Arch.* **285**, 360 (1965).
[30] K. J. Ullrich, H. Fasold, G. Rumrich, and S. Klöss, *Pfluegers Arch.* **400**, 241 (1984).

though it is not the purpose of this chapter to criticize methodologies, a comment should be made concerning the technical problems involved in micropuncture since, occasionally, not enough attention is paid to them. Exacting care to published details and knowledge of individual experiences have facilitated micropuncture and eliminated many of the interlaboratory differences that were common in the 1960s. Nevertheless, discrepancies will arise from time to time which may not be explained solely by species differences, the most readily used excuse for rationalization of technical error.

In the next several pages, we would like to review the contributions of the various approaches of micropuncture, indicating where appropriate some of the advantages and limitations of its application. The major contribution of micropuncture has been in our current understanding of the function of the accessible proximal tubule. Figure 4 illustrates the concentration profile of electrolytes along the superficial proximal tubule. Mammalian proximal tubules consist of two readily recognizable portions, a convoluted segment, pars convoluta, which begins abruptly at the urinary

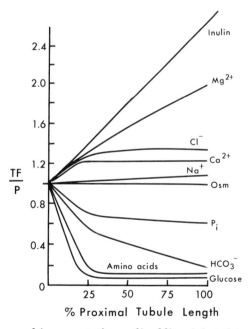

FIG. 4. Summary of the concentration profile of filtered electrolytes and solutes along the superficial proximal tubule. TF/P indicates the tubule fluid-to-plasma concentration ratio for each of the substances except Ca^{2+} and Mg^{2+}, where the ultrafilterable plasma concentration was used.

pole of the glomerulus, and a more distal, shorter straight segment, the pars recta, which is located underneath the kidney surface and is therefore not accessible to *in vivo* micropuncture. Only the early portions, delineated S_1 and parts of S_2, are available for micropuncture. While the recognized structural differences of the proximal tubule segments cannot always be related to specific functional variations, an examination of transport patterns along the proximal tubule suggests that such a relationship does indeed exist. A number of solutes, present in the glomerular filtrate, are preferentially reabsorbed along the very first portion of the proximal tubule. Whereas fluid reabsorption occurs isosmotically and without alteration in the lumenal concentration of sodium, bicarbonate ions are reabsorbed more avidly than chloride ions in the very first part of the proximal convoluted tubule.[31,32] The concentration of chloride in the tubule fluid therefore rises, and reaches its maximum in the first 25% of proximal tubular length. Other solutes are also more avidly reabsorbed in the early segments. For instance, at a point corresponding to 15-20% of proximal tubule length, almost total reabsorption of amino acids has taken place. In addition, the concentrations of glucose and lactic acid[6] have been drastically reduced by the time the fluid has traversed the first 30% of proximal tubular length. Furthermore, inorganic phosphate is more rapidly reabsorbed along the first part of the proximal tubule than in the later portions of this nephron segment.[33] Evidence has also been provided that the early proximal tubule is more effective in lowering the pH of the tubule fluid as measured by the steeper glycodiazine gradients that can be observed in this tubule segment. Although free-flow micropuncture is the most commonly used micropuncture technique, localization is important to correctly determine the concentration profile along the nephron.

The total amount of salt and water reabsorbed by the proximal tubules of the mammalian kidney is about 60-70% of the glomerular filtrate. Normally, the reabsorptive rate changes in parallel to alterations in glomerular load, i.e., the absolute quantity of salt and water than enters the proximal tubule per unit time. The proportional adaptation of proximal tubular reabsorption to changes in glomerular load is commonly referred to as "glomerular-tubular balance".[28,34] The basis for glomerular-tubular balance is thought to involve first an increase in protein concentration within the peritubule capillary commensurate with the change in filtration fraction, and second, an increase in intralumenal glucose, amino acids, carboxylic acids, etc., which have been shown to stimulate sodium uptake.

[31] R. Greger, F. Lang, F. G. Knox, and C. Lechene, *Methods Pharmacol.* **4B,** 75 (1978).
[32] F. C. Rector, Jr., N. W. Carter, and D. W. Seldin, *J. Clin. Invest.* **44,** 278 (1965).
[33] C. W. Gottschalk, W. Lassiter, and M. Mylle, *Am. J. Physiol.* **198,** 581 (1960).
[34] K. J. Ullrich, G. Rumrich, and S. Klöss, *Pfluegers Arch.* **372,** 269 (1977).

The use of micropuncture technology has provided an understanding of these influences. The convoluted, as well as straight, proximal tubules have a high permeability to sodium chloride and water which is consistent with the presence of significant paracellular ionic shunt pathways.[35] Ionic permeability have been studied by tracer techniques and by electrophysiological approaches.[35,36,37] These observations support the view that paracellular ion movement acts as a shunt path for electrolytes in parallel to transcellular ionic routes. The paracellular ionic movement is presumably passive whereas the transcellular movement is dependent, in part, on active processes. Windhager and associates provided direct evidence for active transport of sodium by the method of stop-flow microperfusion of proximal tubules of *Necturus* kidneys.[38] The most important advantage of stationary or stop-flow microperfusion is that relatively long contact times can be used, which allows the determination of equilibrium concentrations. Similar perfusion experiments have been performed on mammalian proximal tubules, which can establish limiting gradients of 30 to 50 mEq liter^{-1} between tubule fluid and blood plasma. The disadvantage of this technique is the volume of the test droplet is normally less than 1 nl. This poses problems with regard to chemical analysis. Additionally, the injected volume of 1 nl may alter local intracellular pools which are considerable smaller than 1 nl, thus the disappearance rates may not reflect epithelial transport in the normal state. However, in support of the stop-flow micropuncture studies, several investigators, using conventional micropuncture techniques under free-flow conditions, have also shown that the intact kidney of the rat can reabsorb sodium chloride against a concentration gradient under conditions of osmotic diuresis induced by the infusion of mannitol.[38] Reabsorptive net transport of sodium across proximal tubule epithelium is only a small fraction of unidirectional transepithelial sodium movement.[36] The rate of net absorption into peritubule capillaries amounts to only some 20% of the unidirectional sodium efflux from the tubular lumen. Unidirectional passive backflux of sodium is remarkably high. It is likely that most of this backflux occurs via paracellular channels and plays a major role in the regulation of net reabsorption of salt and water. Unidirectional flux studies require the use of radioactive tracers of high specific activity. Accordingly, care must be taken to ensure that these tracers are in the free form and thus available for transport.[24]

[35] K. J. Ullrich, G. Rumrich, and K. Baumann, *Pfluegers Arch.* **357**, 149 (1975).
[36] E. E. Windhager, in "Membrane Transport in Biology" (G. Giebisch, D. C. Tosteson, and H. H. Ussing, eds.), p. 143. Springer-Verlag, Berlin, 1973.
[37] T. E. Andreoli and J. A. Schafer, *Am. J. Physiol.* **237**, F89 (1979).
[38] E. E. Windhager, G. Whittembury, D. E. Oken, H. J. Schatzmann, and A. K. Solomon, *Am. J. Physiol.* **197**, 313 (1959).

Rector and colleagues first postulated that a sizable fraction of proximal tubular reabsorption of sodium is passive in nature.[39] These studies involved the use of continuous microperfusion of intact proximal tubules. Subsequent studies[39,40,41,42] have provided additional support for the notion of passive transport. Key elements in the development of this thesis were the observations that increases in peritubule fluid bicarbonate concentration augment proximal tubular reabsorption and that a change occurs in the polarity of the electrical profile along the proximal convoluted tubule.[42] The essential aspects of the concept of passive reabsorption are as follows. In the earliest segment of the convoluted proximal tubule the cotransport of sodium with glucose and amino acids generates a lumen-negative potential difference. In the same segment, bicarbonate is reabsorbed preferentially to chloride by a nonelectrogenic mechanism. This results in increased tubule fluid chloride concentrations at all points of the proximal tubule beyond the initial segment (approximately 1 mm from the glomerulus). Glucose, amino acids, and other organic solutes that might be cotransported with sodium are reabsorbed within the initial portion of the proximal tubules so that little substrate remains in the part in which the tubule fluid chloride concentration exceeds that in plasma water. Essential for passive sodium chloride flux is the generation of the chloride concentration gradient $[Cl^-]_{lumen} > [Cl^-]_{plasma}$. This gradient renders the lumen electrically positive by generating a chloride diffusion potential across the proximal tubule epithelium. Hence, a driving force for sodium reabsorption is created along that part of the proximal tubule where the chloride concentration has risen significantly above peritubule plasma levels.

A second driving force for passive sodium chloride reabsorption has been postulated to be the effective osmotic gradient of organic solutes such as glucose, amino acids, and bicarbonate.[40] The early removal of nearly all filtered organic compounds yields tubule fluid in the later segments that is isosmotic but effectively hypotonic to peritubule fluid. Accordingly, the establishment of different concentrations for organic solutes and chloride in the early segments allows for passive salt and water absorption. Rector, Seldin, Ullrich, and their respective collaborators have concluded on theoretical grounds that a major fraction of proximal tubular sodium reabsorption in rats is driven by the small lumen-positive potential difference and the effective osmotic driving force of bicarbonate across the main portion of the proximal tubule beyond its initial segments.

[39] E. E. Windhager and G. Giebisch, *Am. J. Physiol.* **200**, 581 (1961).
[40] F. C. Rector, Jr., F. P. Brunner, and D. W. Seldin, *J. Clin. Invest.* **45**, 590 (1966).
[41] E. Frömter and K. Luer, *Pfluegers Arch.* **243**, 47 (1973).
[42] D. L. Maude, *Kidney Int.* **5**, 253 (1974).

The use of continuous microperfusion has provided a clear understanding of the role of intralumenal bicarbonate/chloride and solute concentrations in the control of proximal fluid absorption. Although the technique is technically arduous, it has the advantages that one may artificially change the intralumenal composition, or simply add tracer to determine unidirectional flux rates, or uniformly alter flow rates by adjusting the setting of the microperfusion pump. Accordingly, transport coefficients and solute permeability may be readily assessed. The limitations are that the altered permeability at the perfusion site may overestimate unidirectional transport rates.[25] Micropuncture involves the insertion of a glass pipet across the epithelial cell wall, thus some damage may be expected in the vicinity of the micropuncture site and may be perceived as an increase in electrolyte permeability.[26] To avoid this error it is preferable to calculate transport fluxes from the disappearance along the perfused length rather than using the perfusion site in the flux equations.

In addition to elucidating some of the basic mechanisms, the use of micropuncture has defined some of the controls of salt and water transport within the proximal tubule. The question of how proximal tubular sodium reabsorption is regulated is best brought into focus by considering, first, the inhibition of reabsorption observed during expansion of the extracellular fluid volume and, second, the proportionate adjustment of proximal reabsorption to changes in glomerular filtration rate. Clearly, any hypothesis for the control of proximal tubular sodium reabsorption must explain both phenomena. Two micropuncture techniques have played an instrumental role in explaining these controls. First, recollection micropuncture has been used by Dirks and colleagues[16] to show that extracellular fluid expansion following saline infusion diminishes fluid absorption. Alteration in salt and water reabsorption is presumably a function of the balance between peritubule capillary plasma hydraulic and oncotic pressures. Recollection micropuncture, as the name implies, allows for the comparison of an experimental maneuver with the control state. The disadvantage of this approach is the limitation of time between control and experimental influence; most animal preparations are stable for only 2–3 hr, thus any influence acting beyond this time is not detected. Second, direct evidence for the existence of a peritubular capillary control mechanism was subsequently obtained in *in vivo* microperfusion studies of capillaries in which the peritubule oncotic pressure was controlled by varying the concentration of dextran in peritubular capillary perfusion fluid. An almost linear direct relationship was found between the oncotic pressure of the capillary perfusate and the rate of proximal tubule fluid reabsorption under both free-flow conditions and in split-droplet experiments. The results suggested that oncotic driving forces can modulate fluid reabsorption when acting from the capillary side. In contrast, no significant effect has been observed when oncotic forces are applied to the lumenal side of the proximal tubule epithelium. Similarly, hydraulic pressure has been shown to alter net salt and water absorption.[43] Accordingly, alterations in both parameters, peritubular capillary oncotic pressure or hydraulic pressure, have profound effects on net salt and water absorption. Both are sensitively controlled by pre- and postglomerular resistance. Besides offering the ability to independently control lumenal and peritubule fluid composition, the combined perfusion of tubules and peritubule capillaries allows for the careful control of the respective flow rates.

The mechanism of potassium handling in the proximal tubule remains controversial.[44] Free-flow micropuncture studies in rats and dogs have shown that the TF/P_{K^+} ratio remains near unity or slightly below 1.0 along the accessible convoluted tubule.[44] However, Le Grimellec and colleagues[45] have reported that the ultrafiltration of potassium across the glomerular membrane may be in the order of 80–90%, accordingly the tubule fluid-to-ultrafilterable potassium (TF/UF_{K^+}) value may be greater than unity. No clear association of TF/UF_{K^+} with TF/P_{In} has been demonstrated in the rat or dog. However, in a number of animal specials, including the chinchilla, the desert rodents *Psammomys obesus* and *Perognathus penicillatus,* and the rabbit the TF/UF_{K^+} value is clearly above unity.[6,44] This is particularly evident in *Perognathus* and the rabbit, where the TF/UF_{K^+} rises along the proximal tubule to reach 1.4 at the end of the accessible portion. From these data, Jamison and colleagues[44] have taken the view that potassium absorption is passive and driven by a diffusional flux down a concentration gradient that develops as a consequence of water absorption.

In all animal species studied to date, the proximal tubule fluid sodium concentration normally remains the same as plasma whereas calcium increases by 1.1- to 1.2-fold[6] with respect to the ultrafilterable value (Table II). Accordingly, 54 and 47%, respectively, of the filtered sodium and calcium are reabsorbed prior to the late proximal sampling site. The 10–20% elevation of tubule fluid calcium has not been adequately explained but is a consistent observations in all species reported to date. The formation of nonabsorbable intralumenal complexes of calcium has been suggested as one method whereby the concentration may increase above the ultrafilterable value. This remains to be explored further.

The lumenal magnesium concentration rises sharply along the proximal tubule with respect to the ultrafilterable magnesium concentration

[43] B. M. Brenner and R. W. Berliner, *Am. J. Physiol.* **217,** 6 (1969).
[44] R. L. Jamison, J. Work, and J. A. Schafer, *Am. J. Physiol.* **242,** F297 (1982).
[45] C. Le Grimellec, *Pfluegers Arch.* **354,** 113 (1975).

(Fig. 4). Morel and colleagues were the first investigators to demonstrate this phenomenon in the rat.[46] They found a mean TF/UF magnesium ratio of 1.7 at a point in the nephron where 50–55% of the sodium and water had been reabsorbed. This appears to be a common phenomenon in all species studied to date; it has been demonstrated to occur in rats, desert rodents (*Psammomys obesus, Perognathus penicillatus*), hamsters, dogs, monkeys, and rabbits.[47] Thus, the overall fractional magnesium reabsorption in the proximal tubule is some 20–30% of the filtered load, which is about 50–70% the fractional reabsorption rate of sodium. The evidence in the rat suggests that this is due to a relatively low permeability of the proximal tubule for magnesium so that the reabsorption rate is fractionally lower than sodium and the lumenal magnesium concentration rises as water is reabsorbed.[47] The profile for the TF/UF magnesium ratio relative to the inulin concentration remains largely unchanged in a variety of experimental circumstances such as volume expansion and volume contraction.[47] Accordingly, magnesium reabsorption in the proximal tubule closely follows changes in NaCl and water reabsorption, but at a fractionally lower rate.

There is agreement from free-flow micropuncture studies in the dog, rat, and other species that a larger fraction of filtered phosphate than of sodium and water is normally reabsorbed along the proximal tubule so that the TF/P_{Pi} ratio falls below unity, to about 0.7.[48] This TF/P ratio of 0.7 is already attained in the earliest accessible proximal tubule and remains virtually unchanged along the remainder of the available proximal tubule in the face of ongoing fluid absorption. The fluid TF/P ratio attained is highly dependent on the controlling factors for proximal phosphate absorption; greater than 0.7 with elevated parathyroid hormone levels, excess dietary phosphate, and the presence of metabolic acidosis.[49] Mean values below 0.7, to as low as 0.2, have been reported following thyroparathyroidectomy and dietary phosphate deprivation.[49] Evidence has been provided that the rate of phosphate reabsorption is greater in the early segments in contrast to the late proximal tubule. Further, phosphate flux is essentially transcellular movement dependent on active sodium transport. No firm evidence has been provided for tubular secretion of phosphate.

In addition to restoring a large fraction of the filtered sodium and water to the body, proximal tubular sodium transport also has an important role in several other tubular transport processes. First, since a fraction of proxi-

---

[46] F. Morel, N. Roinel, and C. Le Grimellec, *Nephron* **6**, 350 (1969).
[47] G. A. Quamme and J. H. Dirks, *Am. J. Physiol.* **239**, F393 (1980).
[48] C. Amiel, H. E. Kuntziger, and G. Richet, *Pfluegers Arch.* **317**, 92 (1970).
[49] A. Haramati, J. A. Haas, and F. G. Knox, *Am. J. Physiol.* **244**, F265 (1983).

mal sodium reabsorption is closely associated with that of bicarbonate, a substantial fraction of proximal tubular sodium transfer is involved in the maintenance of a normal acid-base balance.[32] Second, recent studies have clearly shown that sodium transport by proximal tubular epithelium plays an important role in the reabsorption of solutes such as glucose and amino acids by cotransport mechanisms. In such a process, downhill movement of sodium from tubule lumen into the cellular compartment, across the lumenal brush-border membranes of proximal epithelium, provides the driving force for uphill sugar and amino acid transport into the cell. Microperfusion studies have revealed that these coupled processes are highly substrate specific. Third, the primary active transport of sodium by the proximal tubule epithelium, which its concomitant transfer of water, also provides the driving force for the transfer of such nonelectrolytes as urea and other similar uncharged substances. Since the tubule wall has a lower permeability to urea than to water, the removal of fluid from the lumen, secondary to sodium reabsorption, results in a significant increase of urea concentration in tubule fluid, above the levels in peritubule fluid and plasma water. The favorable diffusion gradient thus produced results in a passive reabsorptive movement of urea from the tubule lumen into blood. Obviously, the energy-consuming step in proximal tubular urea reabsorption is the active transport of sodium, which is responsible for water transfer across the epithelium.

Some evidence has been obtained for transport processes located on the basolateral membrane of the proximal epithelial cell. These studies have been either of an indirect nature, through the elevation of plasma and presumably peritubule capillary concentration while holding the lumenal concentration constant, or of a direct nature by perfusion of the portal system of amphibian kidneys and peritubule capillaries.[9] Ullrich and colleagues[30] have recently used a stopped flow capillary perfusion to describe organic solute fluxes across the basolateral membrane *in situ*.

Reabsorption of salt and water by the loop of Henle are the means by which the kidney adjusts the urinary osmolality. The loop of Henle is composed of the descending limb, which is highly permeable to water, and the ascending limb, which is impermeable to water and actively transports sodium chloride. Micropuncture studies of the tip of juxtamedullary nephrons have considerably expanded our view of the descending limb.[50] It has also been an area of considerable controversy. These debates have revolved around the issues of permeability of the thin descending limb and the presence of active transport in the thin ascending limb of the loop.[50]

Wirz first performed micropuncture on the exposed papilla of hamsters.[20] These observations and those of Gottschalk and colleagues provided evidence which gave credence to the counter-current theory.[18,23]

Tubule fluid collected at the tip of the loop was found to be hypertonic, similar to the urine in the adjacent collecting duct tip. Salt, principally NaCl, but also containing significant amounts of HCO_3^-, potassium, calcium, and magnesium, makes up the major portion of the total fluid osmolality.[16] This has been confirmed in the rat using a method developed by Berliner and colleagues which allows micropuncture of inaccessible regions of rat medulla following partial nephrectomy. It was demonstrated that at comparable levels in the medulla, the contents of the thin ascending limb are significantly less concentrated than those of an adjacent descending limb. Data from hamster and *Psammomys* have provided evidence to support the notion that tubule fluid is concentrated in part by water abstraction and solute addition in the descending limb[17] and diluted by active sodium transport in the ascending limb.[17,50]

Micropuncture of the superficial distal tubule has provided some inferential data on the functional role of the ascending limb of the cortical nephron. Micropuncture may consist of random sampling of late proximal and early distal fluid or the collection may be of a paired nature from the same nephron as described by Amiel *et al.*[48] Superficial loops are short and devoid of thin ascending limbs, at least in the rat. Free-flow micropuncture studies have shown that 25–40% of the filtered sodium is normally reabsorbed in the loop at less than 15% of the filtered water; accordingly, the fluid issuing from the loop is dilute with respect to plasma. Active reabsorption of NaCl devoid of water provides the concentration–dilution potential of the kidney. As for the proximal tubule, a balance exists between the salt delivery to the loop and the reabsorption rate; however, the basis is different. Sophisticated and technically difficult microperfusion studies have provided the basis for our understanding of functional controls of the loop. First, the rate of reabsorption varies directly with the salt concentration in the lumen and the salt concentration gradient from lumen to intestinum. Second, Thurau, Schnermann, and their colleagues[51] have provided evidence from microperfusion studies to establish the notion of a tubuloglomerular feedback system whereby the individual glomeruli sense the flow rate and composition of the tubule fluid at the macula densa located near the end of the loop of Henle and respond appropriately. This has given the impetus for further studies on the mechanisms and disturbances of this sensitive medium.[52] More recently, antidiuretic hormone and other hormones mediated through cAMP have been

---

[50] R. L. Jamison, *in* "The Kidney" (B. Brenner and F. C. Rector, Jr., eds.), 2nd ed., p. 495. Saunders, Philadelphia, Pennsylvania, 1981.
[51] K. Thurau and J. Schnermann, *Klin. Wochenschr.* **43**, 410 (1965).
[52] L. G. Navar, *Am. J. Physiol.* **234**, F357 (1978).

implicated in control of NaCl transport by the loop.[53] Firm evidence has also been provided that alteration of NaCl transport is associated with concurrent changes of similar magnitudes in potassium, calcium, and magnesium handling.[53] These micropuncture studies suggest that these cations are dependent, in part, on NaCl movement in the thick ascending limb possibly due to a common dependence on the transepithelial electrical potential. Moreover, micropuncture studies have provided data to enhance our understanding of the action of diuretics within the loop.[54]

The distal convoluted tubule is applied to those segments between the macula densa and the first confluence with another distal tubule; accordingly, it is composed of a number of distinct cell types, including portions of the thick ascending limb. As the majority of the distal tubule is accessible to the micropuncture pipet, the tubular function of this segment has been extensively studied.

Net reabsorption of sodium ions occurs against an electrochemical gradient along the distal convoluted tubule.[55] Tubule fluid-to-plasma sodium ratios average 0.5 at the beginning and 0.2 at the end of the distal tubule.[56] Wright first demonstrated that the transtubular electrical potential difference increased along the length of the distal tubule from about −8 to −45 mV. Accordingly, sodium transport is active. Net sodium transport is only one-third to one-fifth that reabsorbed by the proximal convoluted tubule.[9] Absolute net sodium reabsorption is dependent on delivery and enhanced by mineralocorticoids.[57] The distal tubule is highly permeable to potassium ions, which determines, in the main, the amount of potassium appearing in the urine.[56] Micropuncture and microperfusion studies have also supplied ample evidence that calcium and magnesium are absorbed in the distal tubule and have enlightened us as to some of the factors controlling divalent ion balance.[58,59]

One of the means by which the terminal tubule segments have been investigated is by the microinjection technique.[22-24] A small amount of fluid with labeled substance of interest is injected into the superficial nephron and subsequentially recovered in the urine. The urinary recovery of the tracer is compared to that of inulin to calculate the fractional

[53] C. De Rouffignac, J. M. Elalouf, and N. Roinel, *Kidney Int.* **31,** 611 (1987).
[54] G. A. Quamme and J. H. Dirks, *in* "Reviews of Physiology, Biochemistry and Pharmacology" (K. Ullrich, ed.), p. 69. Springer-Verlag, Berlin, 1983.
[55] R. N. Khuri, M. Wiederholt, N. Strieder, and G. Giebisch, *Am. J. Physiol.* **228,** 1262 (1975).
[56] F. S. Wright, *Am. J. Physiol.* **220,** 624 (1971).
[57] K. Hierholzer, M. Wiederholt, and H. Stolte, *Pfluegers Arch.* **291,** 43 (1966).
[58] L. S. Costanzo and E. E. Windhager, *Am. J. Physiol.* **235,** F492 (1978).
[59] C. Bailly, N. Roinel, and C. Amiel, *Pfluegers Arch.* **403,** 28 (1985).

reabsorption of the substance. Only disappearance rates can be ascertained, i.e., unidirectional efflux rates. Moreover, the disappearance rates are influenced by alterations in fluid absorption which are not assessed by this technique. It does have the advantage in that it is relatively simple and inferential data may be obtained in segments not accessible to direct micropuncture.

Although the nephron segments beyond the later superficial distal tubule are not normally available to the micropuncture pipet, some data has been directly obtained by surgical exposure of the papilla with micropuncture and microcatheterization studies. The microcatheterization technique is unique in that it involves the collection of tubule fluid by a catheter advanced up the collecting duct from the tip of the papilla.[19] This avoids the issue of cellular damage entailed with micropuncture. However, it is limited to the collecting duct system. Moreover, it is difficult to ascertain if the catheter has been wedged into the side of the collecting duct thus impeding the flow. Sonnenberg has minimized this by using a pliable catheter.[60] The major benefit of this approach is the proximal and late portions of the collecting system are available for sampling. Direct micropuncture and microcatheterization of the medullary collecting duct have shown net reabsorption of sodium, chloride, urea, and water, and net secretion of hydrogen ions and ammonia, whereas potassium is neither secreted nor reabsorbed, at least in the antidiuretic state.[56] Additionally, calcium and phosphate undergo net reabsorption, whereas magnesium is poorly absorbed in this segment.[61] These methods are technically very difficult; however, they have provided direct information on normal functions within the collecting duct system.

Although the flow rate in nephrons is relatively rapid, 30–50 nl/min, the total volume at any given time is minute. For example, the proximal tubule, with radius of 13 μm and a length of 10 mm, contains 5 nl of tubule fluid. Smaller volumes are contained in the superficial distal segments, about 0.2–0.3 nl. Accordingly, collection rates may be long and the final collected volumes are relatively tiny. This has given impetus to efforts in developing ultramicroanalytical approaches to chemically define the composition of the tubule fluid. Ramsey and colleagues were the first to develop cryoscopic and electrometric methods for analyzing aqueous fluids for osmolality and chloride, respectively.[62] These sensitive techniques are still widely used today. Spectrophotometry,[63] fluorometry,[64] helium glow

---

[60] H. Sonnenberg, *Am. J. Physiol.* **228**, 565 (1975).
[61] M. G. Brunette, N. Vigneault, and S. Carrière, *Pfluegers Arch.* **373**, 229 (1978).
[62] J. A. Ramsay, R. H. J. Brown, and P. C. Croghan, *J. Exp. Biol.* **32**, 822 (1955).
[63] K. J. Ullrich and W. Hampel, *Pfluegers Arch.* **268**, 177 (1958).
[64] G. G. Vurek and R. L. Bowman, *Anal. Biochem.* **29**, 238 (1969).

photometry[65] and more recently electron microprobe spectrophotometry have been in tubule fluid analysis. Micropuncturists today are able to analyze most elements and solutes of interest, including CO_2, in ultramicrovolumes that are normally collected in the glass micropuncture pipet. New techniques are constantly being developed to extend these capabilities and facilitate the ease with which the aqueous fluids are analyzed.

In summary, the use of micropuncture methodologies have provided our basic understanding of tubular function. The above examples are intended to highlight the advantages and perhaps the disadvantages of direct micropuncture studies. Notably, with regard to deficiencies, the thick ascending limb and collecting system are unavailable for direct investigation. Nevertheless, innovative approaches have been developed to describe almost all portions of the nephron.

We have spent some time reviewing the methodology of micropuncture and the contributions of its use; now we would like to speculate on some of the new approaches currently being developed which may have future application. First, a solid tissue punch has been developed which, when used in conjunction with the electron microprobe, may provide information on total lumenal, intracellular, and interstitial elemental concentrations. Second, novel ion-selective resins are being developed which, with the use of micropuncture, will ultimately allow the description of transmembrane ionic activities, in addition to the electrical and pH distributions currently being extensively used. Third, the patch-clamp technique is now being widely used to assess specific membrane conductivity in cultured cells and intact tubules. It is only a matter of time before this approach will be routinely used in *in vivo* micropuncture.

Electron microprobe analyses of solid tissue samples obtained by rapid freezing have been used to determine total elemental composition.[66,67] Potentially, this approach offers a very sensitive method of determining elemental concentrations within the tubule lumen, cell, and interstitium of any region within the kidney. Unfortunately, to date it has been fraught with difficulties; principal among them is fixation of electrolytes during the tissue sampling and preparation. Ionic constituents diffuse very rapidly during these procedures. Nevertheless, with proper care this approach may prove beneficial to the micropuncture field.

Ionic activities are, of course, essential in describing fundamental fluxes across membranes. Ion-selective electrodes have been extensively used in

---

[65] G. G. Vurek and R. L. Bowman, *Science* **149**, 448 (1965).
[66] J. Mason, F. Beck, A. Dörge, R. Rick, and K. Thurau, *Kidney Int.* **20**, 61 (1981).
[67] N. Roinel and C. De Rouffignac, *Scanning Electron Microsc.* **3**, 1155 (1982).

micropuncture studies to determine hydrogen ion activities.[68] More recently, the development of novel ion selective resins has allowed the determination of Na^+, K^+, Cl^-, and Ca^{2+} concentrations within specific intra- and extracellular compartments. New ion-selective resins will allow the measurement of additional ions in the future. Additionally, many electrodes are currently being developed to measure a variety of solutes such as glucose and reducing substances. Rather than collection of fluid and external chemical analysis this approach allows the determination of ion concentrations within the intact functioning tubule. Thus, rapid, on the spot measurements can be made which provide immediate assessment of tubule function as opposed to retrospective evaluation. Accordingly, this approach may be considered micropuncture probing rather than micropuncture sampling.

Ionic channels are integral membrane proteins that facilitate the movement of ions, in a selective manner, across cell membranes. Changes in the protein conformation of these channels can turn the flow of ions on and off, thereby producing pulses of current. The patch-clamp technique can easily resolve these current pulses; accordingly, it is one of the most sensitive assays known for the study of protein conformational changes. The essential feature of the patch clamp is the isolation of a small patch of cell membrane within the tip of glass micropipet. The currents flowing into and out of the pipet across the membrane patch are measured, with the aim of detecting active single of multiple channels. The patch-clamp methodology has been used to study ionic current in both intact kidney and isolated cultured renal cells. Studies have also been directed at distinguishing the various channels, for instance the separation of sodium and calcium current. Akin to this approach is the technique of voltage clamping small cells or whole cell patch clamping. This application involves voltage clamping small cells with a relatively large low-resistance single electrode. It promises to be a major new tool in the armament of electrophysiologists because it allows the investigation of the electrical properties of a wide variety of cells, both in culture as well as in the intact tissue.

In conclusion, the general application of micropuncture techniques has provided a vast amount of information concerning the functions of the kidney. As with all technical approaches, the interest in micropuncture has waxed and waned over the four-odd decades in which it has been used. As a broad methodology, there are still many fundamental problems which can be addressed within the kidney; these will necessarily require the development of novel analytical techniques.

[68] N. W. Carter, in "Manual of Renal Puncture" (V. E. Andreucci, ed.), p. 227. Idelson, Naples, Italy, 1980.

[6] Kidney: Microperfusion–Double-Perfused Tubule *in Situ*

By KARL JULIUS ULLRICH and GERHARD RUMRICH

Introduction

To study epithelial function it is necessary to be able to rinse each cell side independently with artificial solutions. This goal can be achieved in whole kidneys of amphibia and birds by double perfusion of the renal artery, which supplies the glomeruli, and of the portal vein, which supplies the peritubular capillaries (see Chapter 3, this volume). However, in the intact mammalian kidney double perfusion is possible only by applying micropuncture techniques. It is obvious that by double perfusion a three-compartment (lumen–cell–interstitium) approach is applicable to the tubules *in situ:* Either in steady state or after rapid solution changes the concentrations of most substances in the compartments (i.e., the compartment volume) can be measured. The same holds for electrical parameters, potential difference, DC and AC resistances.

In the following we give a detailed description of a double-microperfusion technique of the rat kidney *in situ* as it is applied in our laboratory.

Animals and Experimental Set-Up

For standard experiments male Wistar rats of 200–250 g body weight fed on standard diet and tap water are used. If glomeruli or early proximal tubules are punctured the Munich–Wistar strain with surface glomeruli is chosen.[1] For techniques to load with or deprive the animals of minerals and hormones before the experiment a recent review should be consulted.[2] The rats are anesthetized by injecting Inactin [sodium 5-ethyl-5-(1'-methylpropyl)-2-thiobarbiturate; Byk Gulden, Konstanz, FRG] intraperitoneally at a dose of 120–150 mg/kg body weight. To prevent allergy of the experimenter to rat hairs it is recommended the skin be moistened with paraffin oil. For the same reason the hairs at the operation field are removed with scissors, not with an electrical cutter. The animals are placed on a heated operating table with a thermostat control set at 37° (Fig. 1). In addition the rectal temperature of the animal should be controlled during

---

[1] H. Hackbarth, D. Büttner, D. Jarck, M. Pothmann, C. Messow, and K. Gärtner, *Renal Physiol.* **6,** 63 (1983).

[2] K. J. Ullrich and R. Greger, in "The Kidney Physiology and Pathophysiology" (D. W. Seldin and G. Giebisch, eds.), p. 427. Raven, New York, 1985.

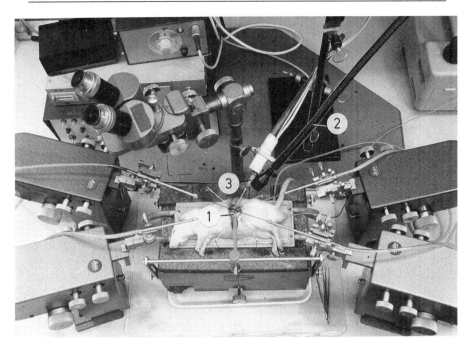

FIG. 1. Experimental set-up for micropuncture microperfusion. Shown is an anesthetized rat on a heated animal table. The Leitz stereomicroscope is set aside and the four Leitz micromanipulators are seen. 1, Lucite cap; 2, light conductor with focusing device; 3, tube to rinse the kidney surface with 38° paraffin oil.

the whole experimental period, which lasts for 1.5 to 3 hr. The jugular vein of the animals is cannulated with polyethylene tubes, when replacement of fluids and/or injection of drugs and colors is desired. The trachea is not opened because cutting the tissue and introducing a polyvinyl chloride tube causes secretion, which disturbs the experiment. By a subcostal incision in the left flank the kidney is exposed and freed from its perirenal attachments. The capsule is stripped off with microforceps and the kidney immobilized in a Lucite cup with cotton wool. The organ is covered and continuously superfused with paraffin oil heated to 37° using a feedback heating system. The organ is illuminated by a light conductor (Coldlight lamp Schott 1500 with focusing device) and inspected with a Leitz or Wild binocular microscope at ×100–×216. Micropuncture is performed by means of one to four Leitz micromanipulators, a separate one for each micropuncture pipet. The glass micropipets are drawn from Pyrex or an equivalent type of glass (Borosilikat glass of 0.9-mm diameter and 0.09-mm wall thickness; Glasmühle Hilgenberg, Malsfeld, FRG) using a horizontal puller to get shank angles of ≈17°. To prevent blocked tips the glass tubes must be cleaned carefully by sucking distilled H_2O-air-H_2O-air about 10 times, followed by absolutely clean acetone-air-acetone-air, again 10 times. The tips are bevelled for ca. 15 sec at an angle of 20° on a rotating lead-tin plate (Fig. 2A) which is covered by diamond paste (0.25 µm; E. Winter, Hamburg, FRG). In other laboratories the bevelling is performed with a rotating carborundum stone.[3] After grinding, the tips have to be cleaned by dipping them slightly into chromic-sulfuric acid and several times into three different distilled H_2O-filled beakers, keeping continuous pressure with a syringe-ethylene tube inside the glass micropipets to prevent acid and/or distilled water flowing into the glass pipets. The bevelled tips have a diameter of 6-8 µm. The glass micropipets are protected against dust by covering them with an upside-down 2-liter glass beaker. Rarely, double-barrelled pipets are used, made either by twisting pipets[4] or from Theta glass (see below).[5] The glass micropipets are filled with castor oil stained with Sudan Black (20 g/liter) or with perfusion solutions by means of syringes connected to the micropipets by heat-pulled polyethylene tubes. Before filling the pipets, a short piece of the end of the pipet is cut off and briefly softened by melting (fire polishing) the sharp glass rims in a microflame. Otherwise particles from the plexiglas and/or from the ethylene tubes will be scraped off and carried with the solutions to the tip during the filling procedure. To remove particles which may clog the tip it is necessary that the solutions be filtered or centrifuged before use. The filled puncture micropipets are mounted in micromanipulators by means of metal holders which are connected to 20-ml syringes by silicon tubes (i.d. 4 mm). Two syringes are mounted in a stage so that they can be manipulated with one hand (Fig. 2B). The system is filled with air. For quick release of negative or positive pressure, which is created by moving a lever connected to the syringe piston, a valve can be opened by tilting the lever. Alternatively, the puncture capillary is connected to a thermally insulated microperfusion pump (type III; W. Hampel, Neu Isenburg, FRG) which delivers 1-50 nl/min with a reproducibility < 1%.[6] In other laboratories Sage pumps[7] (Sage Instruments, Inc., White Plains, New York) are used.

[3] E. E. Windhager, "Micropuncture Techniques and Nephron Function," p. 24. Butterworth, London, 1968.
[4] K. J. Ullrich, E. Frömter, and K. Baumann, in "Laboratory Techniques in Membrane Biophysics" (H. Passow and R. Stämpfli, eds.), p. 106. Springer-Verlag, Berlin, 1969.
[5] E. Frömter, M. Simon, and B. Gebler, in "Progress in Enzyme and Ion Selective Electrodes" (D. W. Lübbers, eds.), p. 35. Springer-Verlag, Berlin, 1981.
[6] H. Sonnenberg, P. Deetjen, and W. Hampel, *Pfluegers Arch.* **278**, 669 (1964).
[7] R. Green, E. E. Windhager, and G. Giebisch, *Am. J. Physiol.* **226**, 265 (1974).

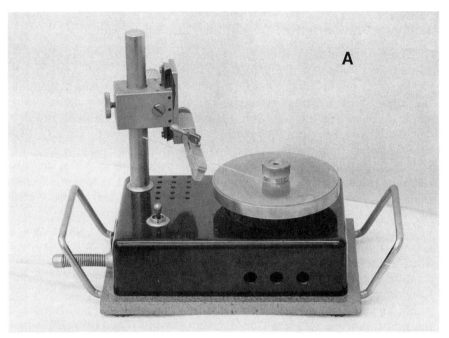

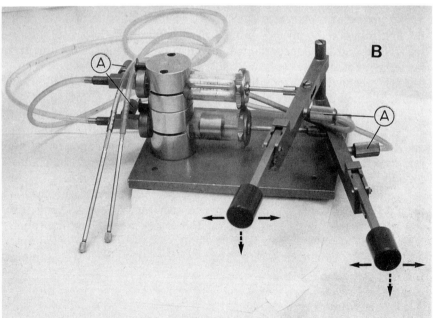

FIG. 2. (A) Device to sharpen the tips of glass pipets. The lead-tin plate which is slightly covered by diamond paste rotates at 1300 rpm. (B) Device to apply positive or negative pressure to the puncture pipets. This is achieved by pushing or pulling the levers (↔). If the levers are tilted downward (↓) a valve is opened so that the pressure in each system can be released immediately via bypassing tubes Ⓐ.

Lumenal Stationary and Continuous Microperfusion

The procedure starts with a puncture of a surface proximal tubule with an oil-filled capillary (A in Fig. 3). By injecting small oil droplets which are carried away by the tubular fluid, the location of the pipet with respect to the length of the proximal convolution can be judged. Alternatively this can be done by injecting a 0.02-ml bolus of 10% Lissamine Green into the cannulated jugular vein. The passage of Lissamine Green through the superficial proximal and subsequently distal loops can easily be judged. However, the problem arises that injected dyes or dyes added to the perfusate might change tubular function. [The apparent K_i value of Lissamine Green against contralumenal p-aminohippurate (PAH) transport was found to be 1.8 mmol/liter and that of Kiton Blue 1.4 mmol/liter (unpublished results).] If the tip of the oil-filled pipet is located in the segment needed, the whole segment is filled with colored oil. Thereafter, the oil-filled segment is punctured with a second pipet which contains the test solution (B in Fig. 3). This solution might be plasma ultrafiltrate or Ringer solution and is reabsorbed like the glomerular filtrate. Alternatively, 16 mmol/liter NaCl of the Ringer perfusate might be replaced by 32 mmol of the nonreabsorbed raffinose ("equilibrium solution," see below). In this case volume reabsorption ceases. Now two types of lumen perfusion can be applied, either stationary microperfusion or continuous microperfusion. In the first instance the injected droplet of ultrafiltrate or Ringer solution is reabsorbed and the concomitant lumenal volume change is documented by serial photography with a Robot photorecorder ("Gertz's shrinking droplet" method).[8,9] Alternatively the injected droplet of "equilibrium solution" remains in place and is removed, usually after 45 sec, by a third pipette (C in Fig. 3) for analysis ("lumenal stopped flow microperfusion"). For continuous microperfusion the tubule is perfused via the micropump and the solution is sampled in the same tubule at a distance of several millimeters downstream. At the perfusion rate of 16 nl/min the contact time to pass 1 mm is 1.2 sec. To avoid artifacts at the infusion site, several subsequent samplings at the same tubule are recommended proceeding from distal to proximal.[10] The distance between the infusion and puncture site is estimated simply by ocular micrometer or after filling the tubule with microfil (Canton Biomedical Products, Boulder, CO USA) and maceration of the tubule in 6 N HCl at 37° for 60 min.[11,12] The samples

---

[8] K. H. Gertz, *Pfluegers Arch.* **276**, 336 (1963).
[9] A. Z. Györy, *Pfluegers Arch.* **324**, 328 (1971).
[10] K. Loeschke and K. Baumann, *Pfluegers Arch.* **305**, 139 (1969).
[11] P. A. Bott, *Am. J. Physiol.* **168**, 107 (1952).
[12] G. Malnic, R. M. Klose, and G. Giebisch, *Am. J. Physiol.* **206**, 674 (1964).

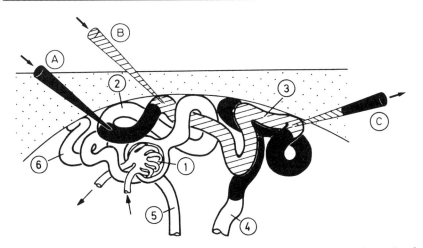

FIG. 3. Schematic drawing of a continuous microperfusion experiment: 1, glomerulus; 2, early proximal loop; 3, late proximal loop; 4, thick descending limb of loop of Henle (pars recta); 5, thick ascending limb; 6, distal convoluted tubule. An oil-filled pipet A is inserted into an early proximal loop. The perfusion pipet B is inserted downstream to pipet A. The collection pipet C is inserted farther downstream. A distal oil block is injected, the pump is set at the desired rate of delivery through pipet B, and the collection is started through pipet C.

obtained are placed under water-saturated mineral oil, sucked into constant-bore pipets for volume measurements, and analyzed as soon as possible. Since the scatter of the results gained from several tubules is very large, it is highly recommended that test and control samples be taken from the same tubule.[13,14] Recently a lumenal stopped flow stationary microperfusion modification was devised, which also avoids the measurement of sample volume prior to analysis. With this method an "equilibrium solution" with added radioactive test substance at different concentrations and radioactive inulin as volume marker is injected in the oil-blocked tubule. After 1-, 2-, 4-, and 10-s contact times the solution is removed by a sampling pipet and placed directly in a scintillation cocktail for radioactive counting. The disappearance curves relative to inulin can be fitted to kinetic models.[15] They give information about the transcellular and paracellular transport pathways. With this method one obtains from each animal more than a dozen samples so that the scatter from tubule to tubule can be balanced by a large number of observations.

[13] H. W. Radtke, G. Rumrich, S. Klöss, and K. J. Ullrich, *Pfluegers Arch.* **324**, 288 (1971).
[14] K. Sato, *Curr. Probl. Clin. Biochem.* **4**, 175 (1975).
[15] E. Sheridan, G. Rumrich, and K. J. Ullrich, *Pfluegers Arch.* **399**, 18 (1983).

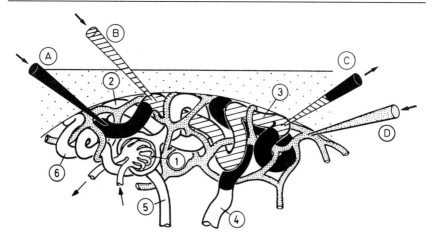

FIG. 4. Set-up of continuous microperfusion with stimultaneous capillary perfusion. The arrangement is the same as described in Fig. 3, but in addition a star vessel is punctured by pipet D. Through this pipet the peritubular capillaries are perfused.

Lumenal and Simultaneous Capillary Perfusion

After lumenal microperfusion was practised for some time it became clear that for rigorous changes of the lumenal perfusate it was also necessary to perfuse the peritubular capillaries with artificial solutions. Thus, it was not possible to test the sodium dependence of transtubular solute transport on proximal tubules *in situ* before the capillary perfusate was also sodium free.[16] Therefore, the lumenal stationary or continuous microperfusion was combined with capillary microperfusion (Fig. 4).

For perfusion of the peritubular blood capillaries a large star vessel at the surface is punctured with sharpened glass micropipets of 7- to 8-μm o.d. By applying 70–140 kPa pressure fluid is injected into the capillaries at a rate of approximately 1.0–3.0 μl/min. The blood stream is thereby interrupted and an area comprising four to eight tubular loops with a diameter of ~300 μm is artificially perfused. In the experiments with continuous microperfusion the peritubular capillaries are continuously perfused for up to 8 min. In all other experiments the individual perfusion lasts less than 1 min. For some experiments rapid changes of peritubular solutions are necessary. This is achieved by puncturing with sharpened double- or triple-channel micropipets of 7- to 12-μm tip diameter.[17] The latter pipets are pulled from glass capillary tubing containing one or two

---

[16] K. J. Ullrich, G. Rumrich, and S. Klöss, *Pfluegers Arch.* **351**, 35 (1974).
[17] B.-C. Burckhardt, K. Sato, and E. Frömter, *Pfluegers Arch.* **401**, 34 (1984).

separation walls (RD Scientific glass, Spencerville, MD). To prevent the formation of air bubbles during filling the pipets are vapor silanized on their inside with dimethyldichlorosilane (Serva, Heidelberg, FRG). Each channel is connected to fine polyethylene tubing, which contains a perfusion solution and is attached at the other end to an oil-filled gravity perfusion syringe, which allows us to apply pressure steps between 25 and 600 kPa (average 150 kPa) by placing calibrated weights on the stem. Rapid switching between two perfusates is effected by simultaneously turning three-way stopcocks (Hamilton, Darmstadt, FRG) which are inserted in both perfusion pathways after the appropriate pressure has been selected in a test perfusion. Since the perfusion system retains some volume elasticity between micropipet tip and stopcock, the pressure is almost completely released in nonperfusing channels by connecting these channels to a fluid column which is adjusted in height so as to counterbalance the pressure inside the blood vessel. The proper adjustment is tested by observing the position of a dye front inside the tip of the respective micropipet channel, since one of the perfusion solutions is stained with Kiton Blue (0.1%). To reduce fluid mixing in the micropipet tips the last 0.3 mm of the tips is pulled into long narrow channels of 15- to 30-μm diameter. As judged from the upstroke of the cell potential response this perfusion system allows one to effectively change fluid composition at the peritubular cell membrane within 100 to 200 msec.

Capillary Microperfusion Only

In order to study contralumenal transport, i.e., from the interstitium into cortical, mainly proximal tubular cells, a method was devised which causes the lumena to collapse so that only two compartments, extracellular and intracellular space, are present (Fig. 5).[18] For this purpose the renal artery and vein are isolated from the ureter so that they can be clamped underneath the plastic cup with a U-shaped hook which bends the blood vessels against the edge of the cup when a weight is attached to the hook. Clamping causes cessation of glomerular filtration and collapse of the proximal tubules within 30 sec, because their content is being reabsorbed. Now an oil-filled sharpened glass pipet (tip diameter about 5-6 μm) for sampling is impaled in a thick blood capillary. Another blood vessel is punctured with a filling pipet (tip diameter about 7 μm) at a distance of 100-130 μm from this first pipet. Through this second pipet, a test solution is quickly injected. The solution contains radiolabeled test substance

[18] G. Fritzsch, W. Haase, G. Rumrich, H. Fasold, and K. J. Ullrich, *Pfluegers Arch.* **400**, 250 (1984).

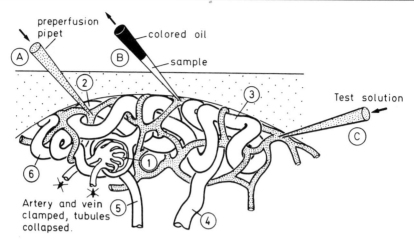

FIG. 5. Set-up for microperfusion of the peritubular capillaries in the renal cortex. Artery and vein are clamped and the tubules collapsed. Test solution is injected and withdrawn by different pipets which are impaled in adjacent capillary loops. If desired the capillaries are preperfused by a third micropipet. The numbering of tubules has the same meaning as given in the legend to Fig. 3 and 4.

—for instance sulfate, chloride, bicarbonate, or *p*-aminohippurate—as well as radiolabeled inulin, both with high specific activity (minimum ~15 mCi/mmol). The test solution is withdrawn by the sampling pipet after 1–10 sec as checked by an acoustic timing device. Filling and collecting takes not more than 0.2 sec altogether. The total clamping time of the renal vein and artery is at most 2 min. Another 2-min recovery period for the kidney is allowed before a new sample is taken at a different site. The area in which the blood capillaries plus interstitial space are filled has a diameter of about 500 μm. The sample which has a volume of 1–2 nl is diluted in 10 μl Ringer solution and is then blown directly into the scintillation fluid. The contralumenal concentration change of the test substance is evaluated relative to inulin; i.e., it is corrected for dilution by extracellular fluid, which is not pushed away by the injected test solution and/or sucked into the sampling pipet from neighboring blood capillaries. The dilution amounts to 10–20%. The disappearance of the test substance from the extracellular space, which is equivalent to the flow into the intracellular space, is plotted against time and the curve fitted by a computer program.[18] Saturation type or diffusion type or mixed curves are obtained. Since the morphological data are also available the data could be expressed as moles second^{-1} centimeter^{-1} of tubular length.

Composition of Perfusates

As standard solution for lumenal and peritubular perfusion Ringer bicarbonate solution is recommended, gassed with 95% O_2, 5% CO_2. The composition of this solution is (in mmol/liter): Na^+, 146; Cl^-, 130; HCO_3^-, 25; K^+, 4; Ca^{2+}, 1.5; Mg^{2+}, 1.0. The addition of substrates is not necessary for most experiments and may even complicate the analysis of the results. The so-called "equilibrium solution" for lumenal perfusion, where net electrolyte and water reabsorption has ceased, has in our laboratory the following composition (in mmol/liter): Na^+, 130.5; K^+, 4; Ca^{2+}, 1.5; Mg^{2+}, 1.0; Cl^-, 135.5; HCO_3^-, 4; raffinose, 31. If Na^+-free solutions are desired Na^+ can be replaced by Li^+, choline, or N-methyl-D-glucamine (Merck, Darmstadt, FRG). Chloride can be replaced by gluconate (Merck), whereby the free Ca^{2+} concentration must be carefully adjusted. As reabsorbable buffer bicarbonate can be replaced by glycodiazine (Redul, Schering, West Berlin). Other buffers, HEPES (N-2-hydroxyethylpiperazine-N'-2-ethanesulfonic acid), Tris [tris(hydroxymethyl)aminomethane], or TES [tris(hydroxymethyl)methyl-2-aminoethanesulfonic acid], should be used only if it is certain that they do not disturb the function which is to be studied. Bicarbonate-free solutions are gassed with pure O_2. For some experiments the solutions must be as simple as possible. Thus, for short-term contralumenal flux measurements, when the tubules are collapsed and the tight junction without function, Ca^{2+}- and Mg^{2+}-free solutions can be used (in mmol/liter): Na^+, 155; K^+, 4; gluconate, 159: However, when the lumena are filled and paracellular transport takes place, the paracellular permeability is drastically augmented when the perfusates contain no Ca^{2+}.

[7] Transcapillary Fluid Transport in the Glomerulus

By AIDA YARED and IEKUNI ICHIKAWA

Glomerular filtration rate (GFR) is determined by two factors: the number of filtering nephrons and the filtration rate in each nephron, or single-nephron glomerular filtration rate (SNGFR). Nephronogenesis occurs mainly during intrauterine life. It proceeds at different rates in different species. Thus, in humans, it is completed by 36 weeks of gestation,[1] while in rats it continues for 1 week,[2] in dogs for 3 weeks,[3] and in

---

[1] E. L. Potter and J. M. Craig, in "Pathology of the Fetus and the Infant," p. 436. Yearbook, Chicago, 1975.
[2] L. Larsson, A. Aperia, and P. Wilton, *Kidney Int.* **18**, 2 (1980).
[3] M. Horster and H. Valtin, *J. Clin. Invest.* **50**, 77 (1971).

guinea pigs through the sixth week of extrauterine life.[4] At maturity, the number of filtering nephrons per kidney is approximately one million in the human, and 30,000–36,000 in the albino rat.[5,6]

Mathematical Model of Glomerular Filtration

The Determinants of Single-Nephron Glomerular Filtration Rate

The determinants of SNGFR have been defined in a mathematical model.[7] Blood entering through the afferent arteriole courses through the glomerular capillary tuft and exits through the efferent arteriole. Along the glomerular capillary tuft, a portion of the plasma will be lost (as the filtrate) into Bowman's space, to be processed by the renal tubules and collecting ducts and leave the kidney as the final urine.

Figure 1 (upper part) depicts the glomerular capillary network in the idealized form of a cylindrical tube extending from afferent to efferent end. The filtration rate across its wall is determined by the same two factors that affect fluid flux across other capillaries, namely, properties of the capillary wall, and the prevailing Starling forces.

The capillary wall properties are represented by the ultrafiltration coefficient K_f, a product of the hydraulic conductivity per unit area of the capillary wall and the surface area available for filtration. Thus

$$K_f = k \times S \tag{1}$$

At any point along the glomerular capillary, the sum of the Starling forces, denoted P_{UF} or ultrafiltration pressure, is equal to the difference between the transcapillary hydraulic and oncotic pressures:

$$\begin{aligned} P_{UF} &= \Delta P - \Delta \pi \\ &= (P_{GC} - P_{BS}) - (\pi_{GC} - \pi_{BS}) \end{aligned} \tag{2}$$

where P_{GC} is the hydraulic pressure within the glomerular capillary, P_{BS} is the hydraulic pressure within Bowman's space (usually low), π_{GC} is the oncotic pressure within the glomerular capillary, and π_{BS} is the oncotic pressure within Bowman's space. Since glomerular filtrate is virtually protein free, $\pi_{BS} \approx 0$. Even in markedly proteinuric states, π_{BS} is negligible

---

[4] C. Merlet-Benichou and C. de Rouffignac, *Am. J. Physiol.* **232**, F178 (1977).
[5] M. Arataki, *Am. J. Anat.* **36**, 399 (1926).
[6] J. Strassberg, J. Paule, H. C. Gonick, M. H. Maxwell, and C. R. Kleeman, *Nephron* **4**, 384 (1967).
[7] W. M. Deen, J. L. Troy, C. R. Robertson, and B. M. Brenner, *J. Clin. Invest.* **52**, 1500 (1973).

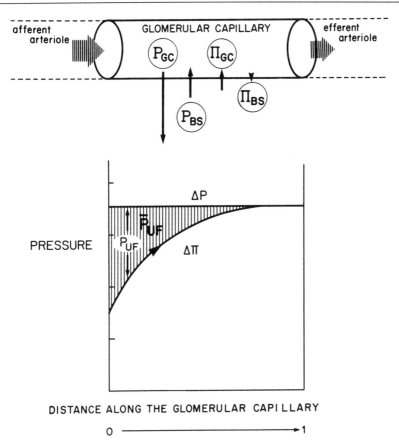

FIG. 1. Schematic portrayal of the process of glomerular filtration. $P_{UF} = \Delta P - \Delta \pi$, where P_{UF} is net ultrafiltration pressure, and ΔP and $\Delta \pi$ are the glomerular transcapillary hydraulic and colloid osmotic pressure difference, respectively. $\Delta P = P_{GC} - P_{BS}$ and $\Delta \pi = \pi_{GC} - \pi_{BS}$, where P_{GC} and P_{BS} are glomerular capillary and Bowman's space hydraulic pressure, and π_{GC} and π_{BS} are glomerular capillary and Bowman's space colloid osmotic pressure, respectively. π_{BS} is regarded as equal to zero. Resulting from loss of colloid-free solution from plasma as a filtrate, π_{GC}, hence $\Delta \pi$ increases progressively along the glomerular capillary, so that P_{UF} progressively decreases. Mean net ultrafiltration pressure ($\bar{P}_{UF}$) is represented by the shaded area. [Reproduced by permission from M. A. Holliday, T. M. Barratt, and R. L. Vernier (eds.), "Pediatric Nephrology," p. 47. Williams & Wilkins, Baltimore, Maryland, 1987.]

relative to π_{GC}, and therefore $\Delta \pi \approx \pi_{GC}$. Both P_{GC} and π_{BS} favor filtration, while P_{BS} and π_{GC} oppose it.

As seen in Fig. 1, $\Delta \pi$ (or π_{GC}) increases along the glomerular capillary. The profile of $\Delta \pi$ is determined by (1) its value at the afferent end of the glomerular capillary where filtration starts, π_A, determined by systemic plasma protein concentration, and (2) the degree of the increase in protein concentration along the capillary as a result of losing colloid-free solution as ultrafiltrate.

When the Starling forces are expressed as values averaged over the entire length of the glomerular capillary (denoted as overbars), SNGFR is given by the equations

$$\begin{align} \text{SNGFR} &= K_f \overline{P}_{UF} \\ &= K_f (\overline{\Delta P} - \overline{\Delta \pi}) \\ &= K_f [(\overline{P}_{GC} - P_{BS}) - (\overline{\pi}_{GC} - \pi_{BS})] \end{align} \tag{3}$$

where $\overline{\Delta P}$ and $\overline{\Delta \pi}$ are the *mean* glomerular transcapillary hydraulic and oncotic pressure differences; $\overline{P}_{UF}$, the *mean* ultrafiltration pressure, is the difference between $\overline{\Delta P}$ and $\overline{\Delta \pi}$, seen as the shaded area in Fig. 1.

Although not appearing in Eq. (3), SNGFR is also determined by the rate of glomerular plasma flow in the afferent arteriole, Q_A, as it modulates the profile of change in $\Delta \pi$ along the glomerular capillary (see "Glomerular plasma flow rate").

The four panels in Fig. 2 illustrate the various parameters influencing SNGFR as they affect ultrafiltration pressure (shaded area). Since $\overline{\Delta P} = (\overline{P}_{GC} - P_{BS})$, a change in either mean glomerular capillary pressure ($\overline{P}_{GC}$) or Bowman's space hydraulic pressure (P_{BS}) can affect $\overline{P}_{UF}$ and hence SNGFR (Fig. 2A). The level of oncotic pressure in systemic plasma, π_A, is also important in determining SNGFR, since it is the initial value of $\Delta \pi$, at the afferent end of the glomerular capillary (Fig. 2B). $\overline{\Delta \pi}$, the mean value of $\Delta \pi$ for the entire glomerular capillary, is a function not only of π_A, but also of the subsequent profile of $\Delta \pi$ along the capillary, a profile modulated by the glomerular plasma flow rate (Q_A) (Fig. 2C). When Q_A is decreased, a given amount of protein-free fluid filtered into Bowman's space will leave a higher intravascular protein concentration at any one point intralumenally; hence, the rise of $\Delta \pi$ along the capillary will be accelerated (from curve 2 to curve 3 in Fig. 2C). Thus, by modulating $\overline{\Delta \pi}$, and therefore $\overline{P}_{UF}$, Q_A is potentially capable of regulating SNGFR. The SNGFR is also a function of the glomerular capillary ultrafiltration coefficient, K_f.

Selective Effect of the Determinants of SNGFR

Mean Glomerular Transcapillary Hydraulic Pressure Difference ($\overline{\Delta P}$). Generally, $\overline{\Delta P}$ is considered not to play a major role in alteration in GFR, except under some extreme circumstances. This is in part because changes in $\overline{\Delta P}$ are predicted to cause directionally similar changes in $\overline{\Delta \pi}$ (Fig. 2A), so that the resulting net change in $\overline{P}_{UF}$ is minimal.

Since $\overline{\Delta P} = \overline{P}_{GC} - P_{BS}$, changes in $\overline{\Delta P}$ can be due to changes in the glomerular capillary pressure, $\overline{P}_{GC}$. $\overline{P}_{GC}$ is partly determined by the values

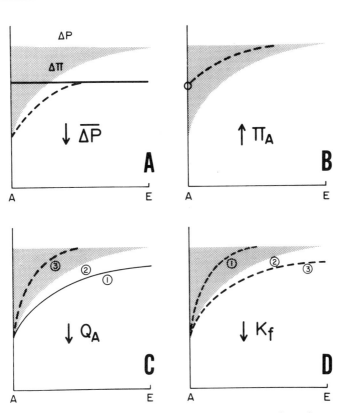

FIG. 2. Schematic Portrayal of the process of glomerular filtration for a glomerulus with (A) reduced mean transcapillary hydraulic pressure difference ($\overline{\Delta P}$), (B) increased systemic oncotic pressure (π_A), (C) reduced glomerular plasma flow rate (Q_A), and (D) reduced ultrafiltration coefficient (K_f). The shaded areas represent normal mean net ultrafiltration pressure ($\overline{P}_{UF}$), determined by the normal profiles of hydraulic (ΔP) and oncotic ($\Delta \pi$) pressure differences. The altered $\Delta\pi$ profile as a consequence of each of the above changes is given by an interrupted line in each panel. Curve 1 in (C) and curve 3 in (D) represent the condition of filtration pressure disequilibrium, while curve 3 in (C) and curve 1 in (D) represent equilibrium. The Starling equation (top) describes the determinants for single-nephron glomerular filtration rate (SNGFR). [Reproduced by permission from M. A. Holliday, T. M. Barratt, and R. L. Vernier (eds.), "Pediatric Nephrology," p. 47. Williams & Wilkins, Baltimore, Maryland, 1987.]

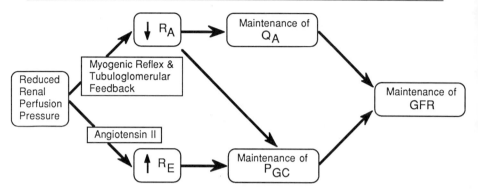

FIG. 3. Mechanisms contributing to the autoregulatory maintenance of renal blood flow and glomerular filtration rates in the face of a reduction in renal perfusion pressure. R_A, afferent (and R_E, efferent) arteriolar resistance; Q_A, glomerular plasma flow rate; P_{GC}, glomerular capillary hydraulic pressure; GFR, glomerular filtration rate. (Reproduced by permission from Badr and Ichikawa.[8])

of afferent (R_A) and efferent arteriolar resistances (R_E). The autoregulatory capacity of these arterioles helps maintain $\bar{P}_{GC}$ relatively constant during changes in systemic blood pressure, further minimizing the importance of ΔP in determining GFR. The capacity of renal autoregulation is largely determined by the ability of the arterioles (both afferent and efferent) to constrict or dilate, and thus blunt the change in $\bar{P}_{GC}$ that would occur in response to a change in renal perfusion pressure[8] (Fig. 3). The level of GFR in patients with mild to moderate hypertension is usually within the normal range.[9] Rats with spontaneous hypertension also have near normal levels of SNGFR.[10] This is because $\bar{P}_{GC}$ in moderately hypertensive states is maintained remarkably constant. The mechanism whereby the elevated systemic arterial pressure is not transmitted to the glomerular capillary is a markedly high afferent arteriolar resistance.[10]

Changes in ΔP can also be due to changes in Bowman's space hydraulic pressure, P_{BS}. These occur only under some exceptional circumstances, including acute obstruction of the urinary tract, where P_{BS} is elevated.[11] With prolonged severe obstruction, P_{BS} returns to near normal level, yet total kidney GFR remains low, due to atrophy of filtering nephrons,

[8] K. F. Badr and I. Ichikawa, *N. Engl. J. Med.* **319,** 623 (1988).
[9] G. M. London, M. E. Safar, J. A. Levenson, A. C. Simon, and M. A. Temmar, *Kidney Int.* **20,** 99 (1981).
[10] W. J. Arendhorst, *Circ. Res.* **44,** 344 (1979).
[11] A. Dal Canton, R. Stanziale, A. Corradi, V. E. Andreucci, and L. Migone, *Kidney Int.* **12,** 403 (1977).

mainly in the deeper cortex,[12] as well as renal vasoconstriction.[13] Thus, the level of GFR during urinary tract obstruction is influenced by variety of factors in addition to P_{BS}. An increase in P_{BS} also occurs in some forms of acute renal failure, secondary to intratubular obstruction by sloughed epithelial cells.[14,15]

Systemic Plasma Colloid Osmotic Pressure (π_A). The oncotic pressure of circulating plasma is primarily derived from the serum protein content, mainly albumin.

Since intraglomerular oncotic pressure opposes filtration, changes in π_A are theoretically expected to lead to opposite changes in SNGFR. Experiments in animals, however, have failed to demonstrate such an inverse relation between C_A (plasma protein concentration) and SNGFR. This appears to be due to changes in K_f occurring in parallel to the changes in C_A through an unknown mechanism.[16,17] A low level of K_f induced by hypoproteinemia may contribute to the normal or low GFR often seen in the human nephrotic syndrome.[18]

Hypoproteinemia associated with severe malnutrition is often accompanied by a decrease in GFR.[19] In rats with chronic protein malnutrition,[20] GFR and SNGFR are reduced, even in the absence of a major reduction in C_A. This is due to severe reductions in Q_A and K_f. The low K_f might reflect a reduced filtering surface area, since the size of the glomeruli is uniformly smaller in protein-malnourished rats.[20]

Glomerular Plasma Flow Rate (Q_A). Glomerular plasma flow rate, Q_A, influences SNGFR by modulating the rate of rise in $\Delta\pi$ along the glomerular capillary. When Q_A is low, or K_f is very high, filtration does not occur along the entire length of the glomerular capillary, but ceases at some point before its end. This condition, termed *filtration pressure equilibrium*, occurs when $\Delta\pi$ becomes equal to ΔP, i.e., $P_{UF} = 0$ by the end of the glomerular capillary.

Since glomerular filtrate is virtually protein free, the amounts of protein

---

[12] D. R. Wilson, *Kidney Int.* **7**, 19 (1975).
[13] I. Ichikawa and B. M. Brenner, *Am. J. Physiol.* **236**, F131 (1979).
[14] G. A. Tanner and M. Steinhausen, *Kidney Int.* **10**, S65 (1979).
[15] T. H. Hostetter and B. M. Brenner, in "Acute Renal Failure" (B. M. Brenner and J. M. Lazarus, eds.), 2nd ed., pp. 67–89. Churchill Livingstone, New York, 1988.
[16] C. Baylis, I. Ichikawa, W. T. Willis, C. B. Wilson, and B. M. Brenner, *Am. J. Physiol.* **232**, F58 (1977).
[17] R. C. Blantz, F. C. Rector, Jr., and D. W. Seldin, *Kidney Int.* **6**, 209 (1974).
[18] R. H. R. White, E. F. Glasgow, and R. J. Mills, *Lancet* **1**, 1353 (1970).
[19] S. Klahr and G. A. O. Alleyne, *Kidney Int.* **3**, 129 (1973).
[20] I. Ichikawa, M. L. Purkerson, S. Klahr, J. L. Troy, M. Martinez-Maldonado, and B. M. Brenner, *J. Clin. Invest.* **65**, 982 (1980).

entering and exiting the glomerular capillary network per unit time are essentially equal. Thus,

$$C_A Q_A = C_E (Q_A - \text{SNGFR}) \qquad (4)$$

where C_A and C_E are the protein concentrations at the afferent and efferent ends. Rearranging Eq. (4) yields

$$\text{SNGFR} = Q_A (1 - C_A/C_E) \qquad (5)$$

Single-nephron filtration fraction is equal to

$$\text{SNFF} = (1 - C_A/C_E) \qquad (6)$$

As long as both C_A and C_E remain constant, i.e., filtration pressure equilibrium is sustained (as between curves 2 and 3 in Fig. 2C), SNGFR will vary in direct proportion to the rate of glomerular plasma flow, Q_A.

In contrast, when Q_A increases beyond a certain level, a resulting deceleration in the rate of rise of $\Delta\pi$ along the glomerular capillary will lead to attainment of *filtration pressure disequilibrium;* i.e., $\Delta\pi$ remains lower than ΔP at the efferent end (from curve 2 to curve 1 in Fig. 2C). Under this condition, the value of C_E is less than its value under filtration pressure equilibrium, and the single-nephron filtration fraction (SNFF) is reduced. That is, SNGFR increases with an increase in Q_A, but to a proportionally lesser degree. Overall, the magnitude of the effect of changes in Q_A on SNGFR will depend upon whether filtration pressure equilibrium or disequilibrium prevails. While high K_f and C_A favor equilibrium, high $\overline{\Delta P}$ and Q_A favor disequilibrium. When an increase in Q_A is accompanied by an increase in K_f, such as occurs during postnatal development,[21] the impact of the increase in Q_A on SNGFR becomes much more pronounced, by maintaining SNFF.

When changes occur in renal blood and plasma flow rates, concomitant changes in other determinants of SNGFR are noted to variable degrees. The impact of a change in Q_A on SNGFR will therefore depend on the cause of the change in Q_A, i.e., whether the other determinants of SNGFR have concomitantly changed. For example, an increase in Q_A induced by an acute infusion of plasma is not accompanied by major changes in other SNGFR determinants, and SNGFR increases. A decrease in Q_A secondary to partial constriction of the aorta or the renal artery is accompanied by a substantial parallel decrease in $\overline{\Delta P}$, leading to a marked fall in SNGFR.

When potent vasodilators such as prostaglandin E_1, acetylcholine, bradykinin, or histamine are administered, SNGFR remains constant, despite

---

[21] I. Ichikawa, D. A. Maddox, and B. M. Brenner, *Am. J. Physiol.* **236**, F465 (1979).

an increase in Q_A, because of the opposing occurrence of a decrease in K_f.[22] The decrease in K_f represents an effect of these substances on the glomerular vasculature distinct from their known dilatory effects on the renal arterioles. Conversely, vasoconstrictors such as angiotensin II and norepinephrine are capable of producing substantial reductions in renal plasma flow. Again, little change results in GFR, due in this case to a marked compensatory increase in $\overline{P}_{GC}$, consequent to an increase in R_E.[23]

Congestive heart failure is clinically characterized by several hemodynamic changes, including markedly low cardiac output and renal plasma flow rate,[24] and a high filtration fraction (FF).[25] In rats with congestive heart failure following surgical ligation of a coronary artery, Q_A is markedly decreased.[26] Single-nephron glomerular filtration rate is also low, but to a much lesser extent than Q_A, so that SNFF is increased. K_f is also low. The preservation of GFR despite the marked reductions in Q_A and K_f is due to an increase in P_{GC}, owing to an increase in R_E.[26]

Clinically, renal hypoperfusion typically occurs in intravascular volume depletion. In experimental animals deprived of water for 24–48 hr, a marked decrease in Q_A is noted, along with a decrease in SNGFR.[27] This decrease in Q_A, together with a decrease in K_f and an increase in $\overline{\pi}_A$, accounts for the decrease in GFR. It is expected that a decrease in $\overline{\Delta P}$ must also contribute to decreasing GFR when volume depletion is so severe as to cause profound hypotension.

Glomerular Capillary Ultrafiltration Coefficient (K_f). The ultrafiltration coefficient, K_f, is a product of the surface area available for filtration (S), and the glomerular capillary permeability to water (k). Like $\overline{\Delta P}$, changes in K_f inevitably lead to directionally similar changes in $\overline{\Delta \pi}$ (Fig. 2D). Hence changes in K_f, unless extreme, do not cause major changes in SNGFR. In fact, when K_f is high enough to permit achievement of filtration pressure equilibrium ($\Delta \pi = \Delta P$ at the efferent end), a further increase in K_f will merely allow the pressure equilibrium to be reached more proximally along the glomerular capillary (from curve 2 to curve 1 in Fig. 2D). Since no filtration is taking place beyond this point, SNGFR will remain unchanged. Thus, a change in K_f will affect SNGFR only under filtration pressure desequilibrium (curve 3 in Fig. 2D). Whether filtration equilib-

---

[22] C. Baylis, W. M. Deen, B. D. Myers, and B. M. Brenner, *Am. J. Physiol.* **230**, 1148 (1976).
[23] B. D. Myers, W. M. Deen, and B. M. Brenner, *Circ. Res.* **37**, 101 (1975).
[24] A. P. Briggs, D. M. Fowell, W. F. Hamilton, J. W. Remington, N. C. Wheeler, and J. A. Winslow, *J. Clin. Invest.* **27**, 810 (1948).
[25] B. I. Heller and W. E. Jacobson, *Am. Heart J.* **39**, 188 (1950).
[26] I. Ichikawa, J. M. Pfeffer, M. A. Pfeffer, T. H. Hostetter, and B. M. Brenner, *Circ. Res.* **55**, 669 (1984).
[27] A. Yared, V. Kon, and I. Ichikawa, *J. Clin. Invest.* **75**, 1447 (1985).

rium or disequilibrium prevails is also determined by other factors, including Q_A (Fig. 2C). Nevertheless, a profound fall in K_f can lead to filtration pressure disequilibrium, and has been demonstrated to prevail in rats with various experimental diseases, including nephrotic syndrome, glomerulonephritis, some forms of acute renal failure, acute and chronic extracellular fluid depletion, and congestive heart failure;[28] under these circumstances, a reduction in K_f is one of the major contributory factors for decreased SNGFR.

A variety of hormones and vasoactive substances are capable of modulating SNGFR. Many, including vasopressin, angiotensin II, and epinephrine, do so by reducing K_f. The mechanism of this effect is thought to involve modulation of the contractility of mesangial cells. These cells, interposing the capillary network, could regulate S, the glomerular capillary filtering surface area, hence SNGFR.[29]

Assessment of GFR

Whole Kidney GFR: The Concept of Clearance

The renal clearance of a substance x, denoted as Cl_x, is defined as

$$Cl_x = (U_x/Pl_x) V \tag{7}$$

where U_x and Pl_x are the concentrations of x in urine and plasma, respectively, and V is urine flow rate. Clearance has units of volume per unit time, usually milliliters per minute for the whole kidney, and nanoliters per minute for the single nephron. If a substance x is freely permeable across the glomerular capillary, uncharged, biologically inert (neither synthesized nor degraded by the kidney), and neither secreted nor reabsorbed by the renal tubule, then the amount of x excreted per unit time $(U_x V)$ is equal to the amount filtered across the glomerulus:

$$Pl_x GFR = U_x V \tag{8}$$

Thus,

$$GFR = (U_x/Pl_x)V \tag{9}$$

i.e., the clearance rate of x.

Numerous substances have been studied for their suitability as markers of GFR. Inulin, a fructose polymer with a molecular weight of about 5200,

[28] V. Kon and I. Ichikawa, *Annu. Rev. Med.* **36**, 515 (1985).
[29] D. Schlondorff, *FASEB J.* **1**, 272 (1987).

under most circumstances fulfills the criteria enumerated above, in humans and other mammalian species, and comes close to being an ideal marker for glomerular filtration. Polyfructosan S, another fructose polymer of higher water solubility and smaller molecular weight than inulin, is also often used.

Whole kidney GFR is routinely measured in an experimental animal by infusing it with a solution of inulin (usually 5-10% in normal saline) following administration of a bolus. After an equilibration time of at least 20 min has elapsed, two to three timed urine collections (10-20 min each) are obtained via a catheter inserted into the ureter, and blood samples obtained midway through the collection. Plasma and urine concentrations of inulin are determined, after appropriate dilution and precipitation of the plasma proteins, by the macroanthrone method.[30] Alternatively, a radioactive marker may be administered, and its clearance rate similarly determined.

Single-Nephron GFR

Micropuncture Technique

Tubule Fluid Collection. Proximal tubule fluid collection: To evaluate GFR at the level of the single nephron, a marker of filtration that can be identified visually, chemically, or after radioactive tagging (e.g., inulin, polyfructosan S, ferrocyanide) is administered. Quantitation of SNGFR commonly uses the concept of clearance and Eq. (7), measuring the clearance of the marker across a single nephron in a fashion similar to whole kidney GFR. The standard procedure in rats is as follows.

After induction of anesthesia, rats are placed on a temperature-regulated micropuncture table. A femoral arterial catheter is placed for subsequent periodic blood sampling and monitoring of mean systemic arterial pressure ($\overline{AP}$). Venous catheters are also inserted for infusion of inulin and test agent solutions, and for replacement of plasma losses consequent to anesthesia and surgery.[31] An intravenous infusion of 5-10% inulin in 0.9% NaCl is then started at a rate of 1.2 ml/hr. The left kidney is exposed by a subcostal incision and separated from the adrenal gland and the surrounding perirenal fat. It is suspended on a Lucite holder, and its surface illuminated with a fiber optic light source and bathed with isotonic NaCl heated to 35-37°.

Micropuncture measurements are started ~60 min after onset of plasma infusion. Using micropipets (o.d. ~10 μm) exactly timed (1-2

[30] J. Führ, J. Kaczmarczyk, and C. D. Krüttgen, *Klin. Wochenschr.* **33,** 729 (1955).
[31] I. Ichikawa, D. A. Maddox, M. C. Cogan, and B. M. Brenner, *Renal Physiol.* **1,** 121 (1978).

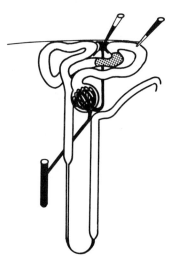

FIG. 4. Schematic representation of the nephron *in situ*. Pipets are shown inserted into the proximal tubule (which contains an oil drop downstream) and the distalmost segment of the efferent arteriole (star vessel). See text for details and applications.

min) samples of fluid are collected from surface proximal convolutions for determination of flow rate and inulin concentration (Fig. 4). Coincident with these tubule fluid collections, samples of femoral arterial blood are obtained in each period for determination of the plasma concentration of inulin. Measurements of SNGFR are obtained in several nephrons during one study period, and their values averaged as representative of the SNGFR prevailing during that period. The same technique can be adapted for guinea pigs,[32] dogs,[3] and other mammals.

Single-nephron glomerular filtration rate is calculated from Eq. (9), as applied to the single nephron:

$$\text{SNGFR} = (\text{TF}_{\text{In}}/\text{Pl}_{\text{In}})V_{\text{TF}} \qquad (10)$$

where TF_{In} and Pl_{In} refer to tubule fluid and plasma inulin concentrations, respectively, and V_{TF} to tubule fluid flow rate. The concentration of inulin in tubule fluid is measured chemically by the microfluorescence method.[33] The volume of fluid collected from individual proximal tubules (hence V_{TF}) is calculated from the length of the fluid column in a constant-bore capillary tube of known internal diameter.

Alternatively, radiolabeled inulin ([³H]inulin) may be administered.

[32] A. Spitzer and C. M. Edelmann, Jr., *Am. J. Physiol.* **221**, 1431 (1971).
[33] G. G. Vurek and S. E. Pegram, *Anal. Biochem.* **16**, 409 (1966).

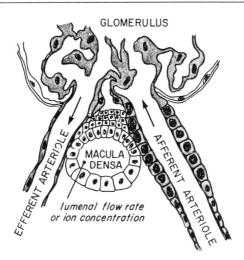

FIG. 5. Presumed site where tubuloglomerular feedback takes place. A stimulus related to lumenal flow rate or ion concentration in the loop of Henle is perceived at the macula densa and transmitted to the vascular structures of the nephron (glomerulus, afferent or efferent arteriole), thus influencing SNGFR. [Reproduced by permission of the American Heart Association from J. O. Davis, *Circ. Res.* **3**, 301 (1971).]

The radioactivity measured in the sample of tubule fluid collected will reflect the amount of inulin filtered over the same period of time, thus obviating the need for measuring tubule fluid sample volume and inulin concentration.

Distal tubule fluid collection—the tubuloglomerular feedback mechanism: The tubuloglomerular feedback mechanism, coupling distal nephron flow and glomerular filtration rates, is thought to contribute to the autoregulatory maintenance of GFR. In this system, a stimulus, related to lumenal flow rate or ion concentration in the loop of Henle, is perceived at the macula densa and transmitted to the vascular structures of the nephron (glomerulus, afferent or efferent arteriole) that partly control filtration rate (Fig. 5). This feedback system is geared to maintaining a constant rate of water and salt delivery to the distal segments of the nephron, specifically the macula densa, where reabsorption is regulated to maintain body fluid balance. Experimentally, changes in loop of Henle flow rate produced by microperfusion of a single nephron result in inverse changes in the filtration rate of that nephron.[34]

Because of the existence of this system, there has been concern that blockade of flow to the distal tubule, which is necessary during quantitative collections from the proximal tubule, would result in an undesirable in-

[34] J. Briggs, *Kidney Int.* **22**, S-143 (1982).

crease of SNGFR due to abolition of a flow-dependent signal at a distal sensor site.[35] Single-nephron glomerular filtration rate values have therefore been compared in proximal and distal tubules of the same nephron, again using timed collections and inulin clearance rates. Indeed, one group of investigators obtained SNGFR values by distal puncture on average 10 nl/min higher than those obtained by proximal puncture.[35] However, in the rat and under normotensive conditions, the sensitivity of this system is very weak[36]: its activation is not instantaneous, and varies with the experimental conditions, being more likely to occur if the animal is volume-contracted sodium deficient.[37] Thus, measurements obtained by proximal puncture, as well as period-to-period comparisons, remain valid under most experimental conditions.

Early proximal tubule fluid collection: Single-nephron glomerular filtration rate can also be measured as the rate of fluid flow in the earliest portion of the proximal tubule (EtPTFR), before tubule reabsorption has taken place (Fig. 6). This method obviates the need for injection or quantitation of a marker.

Pairs of surface glomeruli and their earliest proximal tubules are identified by injecting (using a pipet with o.d. 2–4 μm), and using modest pressure, a small quantity of 0.1% Lissamine Green into proximal convolutions in the vicinity of surface glomeruli.[38] Precisely timed (1–2 min) samples of fluid are collected from the earliest surface segment of proximal tubules thus identified for determination of SNGFR. For the purpose of this timed fluid collection, mineral oil stained with Sudan Black is injected into the earliest proximal tubule with a collection micropipet (o.d. 6–8 μm) to fill approximately three to five diameter lengths of the tubule lumen. Fluid is then collected with the minimum degree of suction required to prevent the oil block from flowing downstream.

A study comparing SNGFR values estimated from EtPTFR with those obtained by the quantitation of inulin in the tubule fluid discussed earlier revealed that EtPTFR measurement provides an accurate estimated value for SNGFR.[38]

Measurement of SNGFR by EtPTFR is useful in some forms of acute renal failure that involve injury or necrosis of the renal tubule, hence disruption of its functional integrity. Under such conditions, clearance of markers may yield an inadequate estimate of GFR as a result of backleak of GFR markers across the tubule epithelia. In that respect, it has been

---

[35] J. Schnermann, J. M. Davis, P. Wunderlich, D. Z. Levine, and M. Horster, *Pfluegers Arch.* **329**, 307 (1971).
[36] F. S. Wright and J. P. Briggs, *Physiol. Rev.* **59**, 958 (1979).
[37] B. Dev, C. Drescher, and J. Schnermann, *Pfluegers Arch.* **346**, 263 (1974).
[38] Y. Yoshida, A. Fogo, H. Shiraga, A. D. Glick, and I. Ichikawa, *Kidney Int.* **33**, 855 (1988).

FIG. 6. Measurement of SNGFR by collection of fluid from the earliest proximal tubule segment. See text for details.

shown that certain glomerular markers, including inulin, can be reabsorbed by the renal tubule of humans and experimental animals in the setting of acute renal injury.[39,40] Hence, one of the conditions for using the clearance of a substance as a marker for GFR would not be fulfilled.

Measurement of SNGFR by EtPTFR, sometimes after stripping off the overlying renal capsule, has been performed in rats as young as 17 days of age, seemingly the earliest age in this species at which all proximal tubules are patent,[41] and in newborn guinea pigs.[32]

Use of EtPTFR has been found to reliably allow repeated measurement of SNGFR in the same nephrons over days and weeks, making it a useful tool in following the progression of renal disease under conditions where functional and morphological analyses need to be made on the same glomeruli.[38,42] Note that inulin collection from a tubule does not allow the indentification of the specific glomerulus which belongs to the same nephron.

Efferent Arteriolar Blood Collection. Single-nephron glomerular filtration rate has also been calculated by measuring efferent arteriolar plasma flow rate (Q_E), derived from efferent arteriolar blood collection, as well as afferent and efferent glomerular plasma protein concentrations.[11] Single-nephron filtration fraction is calculated from Eq. (6); SNGFR is then calculated as

$$\text{SNGFR} = [Q_E/(1 - \text{SNFF})] - Q_E \tag{11}$$

In this way, no tubule fluid collection is required.

---

[39] B. M. Myers, F. Chui, M. Hilberman, and A. S. Michaels, *Am. J. Physiol.* **237**, F319 (1979).
[40] G. M. Eisenbach and M. Steinhausen, *Pfluegers Arch.* **343**, 11 (1973).
[41] A. Aperia and P. Herin, *Am. J. Physiol.* **221**, 1319 (1975).
[42] A. Fogo, Y. Yoshida, A. D. Glick, T. Homma, and I. Ichikawa, *J. Clin. Invest.* **82**, 322 (1988).

Timed collections of blood are performed from efferent arterioles of superficial nephrons, using oil-filled pipets (o.d. 12–15 μm). After insertion of the pipet tip, a small quantity of mineral oil is injected, displacing the blood within the capillary. As soon as blood flowing from the glomerulus reaches the pipet tip and the oil is displaced into peritubular capillaries, the blood collection is started using gentle suction (Fig. 4). The presence of the oil in the peritubular capillaries is evidence for a complete collection, and prevents retrograde collection of peritubular capillary blood. The volume of the specimen collected allows calculation of efferent blood flow rate, from which Q_E is derived. Plasma protein concentration is measured on efferent blood and on a specimen of systemic arterial blood, taken as representative of afferent blood.

SNGFR values measured by this technique were found to be indistinguishable from those obtained by the conventional method of tubule fluid collection, indicating that it alters neither the rate of filtration nor the intrinsic hemodynamic pattern of the glomerular microcirculation.[11] This technique is useful when an extremely high Bowman's space pressure prevails (e.g., acute urinary tract obstruction), which would be dissipated by suction of fluid from tubules for determination of SNGFR.

Ferrocyanide Injection Technique. Measurement of SNGFR by the standard micropuncture technique has the limitation of assessing only superficial nephrons. To detect variations in the SNGFR value across the renal cortex, some investigators have compared SNGFR values measured in superficial nephrons with SNGFR values estimated from the ratio of whole kidney GFR to number of nephrons per kidney.

Conversely, a visual marker which is freely filtered across the glomerulus, such as ferrocyanide, may be used as a qualitative index of filtration.[43] Briefly, in the rat, 0.8 ml of 10% ferrocyanide solution is rapidly injected into the aorta above the renal arteries. Fifteen seconds after the marker reaches the kidneys, the renal pedicle is clamped, the kidney excised and frozen in liquid nitrogen. The ferrocyanide remaining in the kidney is precipitated to Prussian Blue. This is followed by maceration of the kidney in HCl. Complete superficial and juxtamedullary proximal tubules are then dissected under a stereomicroscope, placed in a drop of distilled water, and inspected for the presence of Prussian Blue particles, i.e., the ferrocyanide precipitate. In the case of filtering glomeruli, the ferrocyanide should be entirely recovered in the tubules. Glomeruli are identified as superficial or juxtamedullary according to defined criteria, superficial occupying the outer 10% and juxtamedullary glomeruli the inner 15% of cortical volume.[43] This technique does not allow quantitation

---

[43] A. Aperia, O. Broberger, P. Herin, and I. Joelsson, *Acta Physiol. Scand.* **99**, 261 (1977).

of filtrate, but rather determination of whether a glomerulus is filtering or not; it has found usefulness in developmental studies of filtration, allowing demonstration that nephron maturation occurs in a centrifugal manner, from deep to superficial cortex.[43]

The same marker, radiolabeled with ^{14}C, has also been used to quantitate SNGFR, using the principle of clearance. Thus, during a continuous infusion of [^{14}C]ferrocyanide, a pulse of unlabeled marker is injected into the aorta. The renal pedicle is tied after precise timing (e.g., 8–10 sec) and an arterial sample collected. The kidney is removed and the ferrocyanide precipitated as above. After maceration, complete (superficial or juxtamedullary) nephrons are dissected and cut at the front of the visible blue unlabeled precipitate. These contain the [^{14}C]ferrocyanide filtered during the 8- to 10-sec interval. Single-nephron glomerular filtration rate is calculated by determining radioactivity of the plasma and the precipitate contained in the proximal tubule segment upstream of the front. This method has been shown to be in good agreement with SNGFR measured by inulin clearance in both adult and young rats.[44] It allows separate measurement of SNGFR in nephrons at different cortical levels.

Assessment of the Determinants of SNGFR

Micropuncture

Pressures and flows governing SNGFR can be evaluated in the Munich–Wistar strain of rats, uniquely endowed with some glomeruli at the surface of the kidney directly accessible to micropuncture techniques. By assessing preglomerular, glomerular, and postglomerular pressures and flows, Starling forces, glomerular capillary ultrafiltration coefficient, and afferent and efferent arteriolar resistances can be determined in control and experimental conditions.[7,31] Glomerular micropuncture, allowing the measurement of SNGFR and its determinants, is possible in the Munich–Wistar rat over 40 days of age.[21,45]

It should be recognized that these calculations assume homogeneity of hemodynamics among glomeruli. Several parameters, namely C_E, P_{GC}, P_{BS}, SNGFR, and efferent arteriolar pressure are measured in different nephrons. Thus, when internephron heterogeneity of glomerular hemodynamics prevails, as in chronic renal failure, this approach becomes invalid.

Hydraulic Pressures. Time-averaged pressures can be directly measured in surface glomerular capillaries ($\overline{P}_{GC}$), proximal tubules (P_T, taken as

---

[44] C. de Rouffignac and L. Monnens, *J. Physiol. (London)* **262**, 119 (1976).
[45] V. Kon, M. L. Hughes, and I. Ichikawa, *Kidney Int.* **25**, 391 (1984).

representative of P_{BS}), and efferent arterioles (P_E) with a continuous recording, servo-null micropipet transducer system. Micropipets with o.d. 2–3 μm and containing 2.0 M NaCL are used.

$\overline{P}_{GC}$ can also be indirectly estimated from "stop flow pressure," P_{SF}, i.e., the pressure required in the proximal tubule (hence in Bowman's space) to stop glomerular filtration. Under this condition, $\overline{P}_{UF}$ is presumably zero; rearranging Eq. (2), and substituting P_{SF} for P_{BS}:

$$\overline{P}_{GC} = P_{SF} + \pi_A \qquad (12)$$

For this purpose, the earliest accessible portion of the proximal tubule is identified by injecting a small quantity of 0.1% Lissamine Green in retrograde fashion, while applying modest pressure. A pipet ($\sim$7-μm o.d.) containing mineral oil stained with Sudan Black is then inserted into the tubule at a site at least three tubule diameters in length downstream from the earliest surface segment. The tubule is filled with oil, and proximal tubule fluid flow is stopped by advancing the injected oil toward Bowman's space and covering any previously punctured site(s). The P_{SF} is then measured using a 1-μm pressure pipet in the previously identified earliest surface segment of the nephron.[46] A specimen of systemic blood is simultaneously obtained for measurement of C_A, hence π_A.

Alternatively, proximal tubule flow can be blocked by injecting, downstream from the pressure pipet, a small amount of paraffin wax, using a pipet ($\sim$ 10-μm o.d.) attached to a perfusion pump.[47]

While the measurement of P_{SF} is useful in conditions when surface glomeruli are not accessible for direct P_{GC} measurement and gives an accurate value of the P_{GC} prevailing during the stop-flow condition, the experimental procedure of stop-flow by itself has been found to lead to a variable increase in P_{GC}, on average by $\sim$7 mmHg.[46] The possibility also exists that SNGFR may not be zero under "stop-flow" conditions, due to proximal fluid reabsorption which can occur proximally to the wax block.

Mean glomerular transcapillary hydraulic pressure difference is calculated as

$$\overline{\Delta P} = \overline{P}_{GC} - P_T \qquad (13)$$

Colloid Osmotic Pressures. Systemic blood, obtained through the femoral catheter, is considered representative of afferent arteriolar blood.

Efferent arteriolar blood is collected using microscopic visualization. The most distal segment of the postglomerular efferent arteriole (stellate vessel) is identified on the surface of the kidney (Fig. 4), and a sharpened

---

[46] I. Ichikawa, *Am. J. Physiol.* **242**, F580 (1982).
[47] J. P. Briggs and F. S. Wright, *Am. J. Physiol.* **236**, F40 (1979).

pipet (~15-μm o.d.) inserted. A drop of mineral oil sufficient to obliterate all the radiating capillary channels is injected, followed immediately by gentle suction to collect a blood sample of 50- to 150-nl volume. Suction pressure is adjusted to maintain the column of oil just distal to the site of puncture. The reliability of this technique is highly dependent on the investigator's technical skill.[48]

To estimate the colloid osmotic pressure of plasma entering and leaving glomerular capillaries, plasma protein concentrations are measured in femoral arterial (C_A) and the surface efferent arteriolar specimens (C_E). Assays for plasma protein concentration are carried out, usually in duplicate, by the method of Viets.[49] Colloid osmotic pressure (π) is calculated according to the equation

$$\pi = k_1 + k_2 C^2 \qquad (14)$$

where k_1 and k_2 equal 1.63 and 0.294, respectively, assuming an albumin/globulin ratio of one, the value found in rats under a variety of physiologic and pathophysiologic conditions.[7]

Glomerular Plasma Flow Rate (Q_A). Single-nephron filtration fraction is calculated using Eq. (6). Glomerular plasma flow rate, Q_A, can then be calculated as

$$Q_A = \text{SNGFR}/\text{SNFF} \qquad (15)$$

Glomerular Capillary Ultrafiltration Coefficient (K_f). The ultrafiltration coefficient, K_f, is calculated using a differential equation which gives the rate of change of protein concentration with distance along an idealized glomerular capillary. This equation, together with its derivation and the method for its solution, is given in detail elsewhere.[7]

Arteriolar Resistances. Other hemodynamic parameters are calculated from the equations below. Blood flow rate per single afferent arteriole or glomerulus:

$$\text{GBF} = Q_A(1 - \text{Hct}_A) \qquad (16)$$

where Hct_A, the hematocrit of afferent arteriolar blood, is taken as equal to femoral arterial hematocrit.

Efferent arteriolar blood flow rate (EABF) is calculated as

$$\text{EABF} = \text{GBF} - \text{SNGFR} \qquad (17)$$

Resistance per single afferent arteriole is calculated as

$$R_A = (\overline{\text{AP}} - \overline{P}_{GC})[(7.962 \times 10^{10})/\text{GBF}] \qquad (18)$$

[48] M. L. Hughes and I. Ichikawa, *Kidney Int.* **29**, 814 (1986).
[49] J. W. Viets, W. M. Deen, J. L. Troy, and B. M. Brenner, *Anal. Biochem.* **88**, 513 (1978).

where the factor 7.962×10^{10} is used to give resistance in units of (dynes·second)/centimeter^5 when AP and P_{GC} are expressed in millimeters Hg, and GBF in nanoliters per minute. Resistance per single efferent arteriole is calculated as

$$R_E = (\bar{P}_{GC} - P_E)[(7.962 \times 10^{10})/\text{EABF}] \tag{19}$$

Total arteriolar resistance (R_{TA}) for a single pre- to postglomerular vascular unit is calculated as

$$R_{TA} = R_A + R_E \tag{20}$$

Microperfusion

Juxtamedullary Glomeruli. A population of juxtamedullary nephrons is uniquely located on the inside cortical surface, in apposition to the renal pelvis. These nephrons are, therefore, accessible for *in situ* microperfusion study after stripping of the pelvic mucosa. Briefly, after the animal is anesthetized, the kidney is removed and longitudinally cut to expose the pelvic cavity, leaving the papilla intact. Tissues are removed to expose the glomeruli of interest, together with their related vascular structures and tubules. The kidney is perfused with oxygenated homologous blood.[50]

This preparation allows study of the hydraulic pressure profile along the glomerular vasculature. In that, in addition to the measurement of $\bar{P}_{GC}$, P_T, and P_E, hydraulic pressure can be obtained in afferent and arcuate arteries, superficial veins, peritubular capillaries, and ascending and descending vasa recta.[51] Single-nephron glomerular filtration rate can be measured by adding a marker of filtration to the perfusate and collecting proximal tubule fluid in a fashion similar to standard micropuncture techniques.

This technique has allowed segmental studies of the renal microvascular response to hormonal (e.g., infusion of angiotensin II or norepinephrine) and mechanical stimuli (e.g., change in renal perfusion pressure).[51] Supplemented by videometric techniques in conjunction with infusion of a fluorescent marker, it has allowed visual assessment and measurement of microvessel diameters in response to vasoactive drugs.[52]

Isolated Glomeruli. Direct assessment of SNGFR and its determinants is possible *in vivo* only in the superficial nephron of the Munich–Wistar rat. Investigators have, therefore, attempted to develop methods to study it *in vitro* using isolated glomeruli.

Isolated dog glomeruli have been perfused by one group of investiga-

---

[50] D. Casellas and L. G. Navar, *Am. J. Physiol.* **246**, F349 (1984).
[51] D. Casellas, P. K. Carmines, and L. G. Navar, *Kidney Int.* **28**, 752 (1985).
[52] P. K. Carmines, T. K. Morrison, and L. G. Navar, *Am. J. Physiol.* **251**, F610 (1986).

tors.[53] Briefly, after harvesting the kidney, glomeruli with intact afferent and efferent arterioles are dissected free under microscopy. Pipets are introduced into the arterioles to allow perfusion, pressure measurement, and sample collection. The afferent arteriole is perfused at a pressure (P_{GC}) of 30–35 mmHg. [^{125}I]Albumin is added to the perfusate.

Calculations for glomerular hemodynamics are derived from Eq. (3). In the isolated perfused glomerulus, $P_{BS} = 0$, hence

$$\text{SNGFR} = K_f(\bar{P}_{GC} - \bar{\pi}_{GC}) \tag{21}$$

The SNGFR is calculated as the difference between afferent (Q_A) and efferent (Q_E) arteriolar flow rates. Q_E is measured directly in the collection pipet, and Q_A is calculated from the formula

$$Q_A = Q_E/(1 - \text{FF}) \tag{22}$$

where FF is determined as

$$\text{FF} = 1 - ([^{125}\text{I}]\text{albumin}_p/[^{125}\text{I}]\text{albumin}_c) \tag{23}$$

where [^{125}I]albumin represents the counts per minute in perfusate (p) or collected fluid (c).

$\bar{\pi}_{GC}$ is calculated as the arithmetic mean of the afferent (π_A) and efferent (π_E) colloid osmotid pressures, an assumption which is valid under the condition of filtration disequilibrium. π_A is measured directly on an oncometer, and π_E from the efferent protein concentration (calculated from the FF). Thus, $\bar{P}_{GC}$ can be directly measured, SNGFR and $\bar{\pi}_{GC}$ determined, hence K_f can be derived.

Using this method, reproducible values for SNGFR and FF have been obtained, and remain stable over 120 min of perfusion. However, SNGFR values (hence K_f values) are uniformly lower than those obtained in *in vivo* studies. It is unclear whether these differences relate to a change in glomerulus permeability consequent to the procedure and manipulation, or whether it could be related to a difference in the viscosity of the perfusate fluid (absence of erythrocytes).

The theoretical advantages of this techniques would be the ability to study glomeruli from various cortical levels, as well as the ability to examine the effect of vasoactive substance on the glomerulus independent of systemic effects (hemodynamic, hormonal, or neural). Major problems, however, have been the technical difficulty in obtaining an adequate preparation (success rate 8/60 in the original studies), as well as scatter in Q_A values, possibly related to differences in glomerular size or arteriolar tone.[53]

---

[53] R. W. Osgood, M. Patton, M. J. Hanley, M. Venkatachalam, H. J. Reineck, and J. H. Stein, *Am. J. Physiol.* **244**, F349 (1983).

It is undoubted, however, that perfecting this technique would find widespread usefulness.

Incubation of Isolated Glomeruli

Incubation of isolated glomeruli has been used to estimate SNGFR and K_f in vitro.[54] Glomeruli are isolated by the differential sieving technique, and allowed to equilibrate for at least 10 min in a medium containing a high concentration (4 g/dl) of bovine serum albumin (BSA). Individual isolated glomeruli which appear intact and free of Bowman's capsule are selected. Each is held by gentle suction on a micropipet tip while being observed using a video monitor. The video image is recorded while the 4 g/dl BSA medium is rapidly changed to a medium containing 1 g/dl BSA. During the change in medium, the difference in protein concentration between the plasma within the glomerular capillaries and the new medium induces an oncotic gradient across the capillary wall and causes filtration of fluid into the capillaries. Filtration is evidenced by expansion of individual capillaries and of the glomerular image; glomerular size reaches a maximum within 0.1 to 0.2 sec after the medium change. Further filtration is evidenced by ejection of plasma and erythrocytes contained in the glomerulus. Glomerular volume is calculated as $\frac{4}{3}\pi(D/2)^3$, where D is the geometric mean of four diameters measured on the video image. Maximum filtration rate is estimated as the maximum rate of change in glomerular volume during a single $\frac{1}{60}$-sec interval in the first 0.1 sec of filtration. K_f is calculated as the quotient of the maximum filtration rate and the difference in oncotic pressures of the incubation and test media, as estimated by Eq. (14).

The advantages of K_f determination by this method are that it gives a more direct estimate, and that glomeruli at varying cortical levels can be studied and compared. On the other hand, it assumes that the permeability characteristics of isolated glomeruli in suspension are similar to those prevailing in vivo in viable glomeruli and that the forces determining "reverse filtration" in vitro are the same are those determining filtration in vivo.

Derivation from Macromolecular Sieving Data

Another approach to the study of glomerular hemodynamics has involved the use of differential solute clearances, a technique generally used to assess the permselectivity characteristics of the glomerular capillary wall to macromolecules.[55] This approach estimates various hemodynamic pa-

---
[54] V. J. Savin, C. Beason-Griffin, and W. P. Richardson, *Kidney Int.* **28,** 926 (1985).
[55] W. M. Deen, C. R. Bridges, B. M. Brenner, and B. D. Myers, *Am. J. Physiol.* **249,** F374 (1985).

rameters, including P_{UF}, P_{GC}, and K_f, by applying the notion that these parameters influence transport of macromolecules with different sizes in a mathematically predictable fashion. The models and equations used are detailed elsewhere.[56]

The experimental animal is infused with a marker of renal plasma flow rate (e.g., *p*-aminohippurate) and of GFR (e.g., inulin) that will allow measurement of these two parameters, hence of FF. It is then injected with a mixture of macromolecules of various sizes and molecular weights (e.g., polyvinylprolidine) radiolabeled with ^{125}I. Plasma and timed urine collections are then started. The urine and plasma samples are later fractionated according to the molecular size of the polymers, and the radioactivity of each fraction is measured. Clearances of macromolecules with different molecular radii, their fractional clearances (i.e., relative to the clearance of inulin, which is freely filtered), and hemodynamic and membrane parameters are calculated.[56]

The theoretical advantage of this method is that functional indices can be obtained noninvasively. This would make it a valuable tool for inferring glomerular hemodynamics when micropuncture cannot be performed, such as to study the effect of a drug or disease in humans. However, the models used and the assumptions on which they are based lead to such a wide degree of error in estimated values that the applicability of the technique remains limited.

Morphological Examination

Methods for the determination of the two components of K_f, namely surface area available for filtration and hydraulic conductivity of the capillary wall, remain rudimentary. Surface area has been measured in perfusion-fixed glomeruli as the area of capillary wall exposed to plasma flow, i.e., adjacent to fenestrated endothelium of patent capillaries (denoted as "peripheral" basement membrane, as opposed to glomerular basement membrane adjoining mesangial areas), measured by electron microscopy, and extrapolated to a whole glomerulus[57]; hydraulic conductivity has been estimated in some studies as the quotient of K_f and peripheral basement membrane area for each glomerulus.[54] Morphological measurements are at best approximate, since the actual area of the glomerular capillary wall defined in the Starling relationship and involved in fluid transfer is unknown.

[56] P. P. Lambert, R. Du Bois, P. Decoodt, J. P. Gassée, and A. Verniory, *Pfluegers Arch.* **359**, 1 (1975).
[57] M. A. Venkatachalam and H. G. Rennke, *Circ. Res.* **43**, 337 (1978).

[8] Preparation and Study of Isolated Glomeruli

By DETLEF SCHLONDORFF

Over the past 10 to 15 years, it has become increasingly apparent that the glomerulus is not just a passive filtration apparatus, but is also actively involved in regulating its characteristics for blood flow and filtration. Thus the glomerulus has been found to represent a target organ for a variety of hormones and vasoactive agents that control glomerular function.[1] In addition, the glomerulus itself is capable of generating a number of mediators, e.g., renin, prostaglandins, lipoxygenase products, and histamine, which in turn can influence its function.[2,3] This new body of information concerning the glomerulus has been gained by a combination of studies involving *in vivo* micropuncture of glomeruli,[1] *in vitro* microperfusion of isolated glomeruli,[4] *in vitro* biochemical and morphological experiments using isolated glomeruli,[5] and primary cell culture of glomerular endothelial, epithelial, and mesangial cells.[3,6] In the present chapter I will review some of the techniques used in our laboratory for isolation and incubation of glomeruli.

Techniques for Isolation of Glomeruli

Depending on the envisioned use of the glomeruli, a large number of isolation techniques have been employed. These range from microdissection to mechanical and enzymatic disruption of the renal cortex followed by enrichment of glomeruli based on physical characteristics.[5] Microdissection is tedious and has a low yield.[7] It is almost exclusively used for microperfusion requiring isolated glomeruli with attached afferent and efferent arterioles.[4] In order to obtain a good yield of glomeruli in a rapid and simple manner in most other methods the renal cortex is dissected, minced to a paste, forced through a sieve, and glomeruli and tubules are

---

[1] L. D. Dworkin, I. Ichikawa, and B. M. Brenner, *Am. J. Physiol.* **244,** F95 (1983).
[2] D. Schlondorff, *Am. J. Med.* **81,** 1 (1986).
[3] D. Schlondorff, *FASEB J.* **1,** 272 (1987).
[4] R. N. McCoy, T. A. Fried, R. W. Osgood, and J. H. Stein, *Kidney Int.* **27,** 297 (1985).
[5] D. Schlondorff, *Kidney Int.* **30,** 201 (1986).
[6] G. E. Striker and L. J. Striker, *Lab. Invest.* **53,** 122 (1985).
[7] M. Imbert, D. Chabardes, and F. Morel, *Mol. Cell. Endocrinol.* **1,** 295 (1974).
[8] P. Mandel, J. J. Helwig, and C. Bollack, *Exp. Cell Res.* **83,** 414 (1973).
[9] D. G. Taylor, R. G. Price, and D. Robinson, *Biochem. J.* **121,** 27P (1972).
[10] V. E. Torres, T. E. Northrup, R. M. Edward, S. V. Shah, and T. P. Dousa, *J. Clin. Invest.* **62,** 133 (1978).

then further separated either by density centrifugation,[8-10] sequential sieving,[11] or glomeruli are pulled out in a magnetic field after having been preloaded *in situ* with iron particles.[12] The most commonly used and simplest method is that of sequential sieving, as originally described by Krakower and Greenspon.[11] I shall give a detailed description of this method as presently employed in our laboratory for preparation of isolated glomeruli from rat kidneys. Modifications employed for different species will then be mentioned.

Isolation of Glomeruli from Rat Kidney

Animals

Adult rats (100–200 g body wt) are used. We mostly use Sprague–Dawley rats, but other strains can be used as well. Either male or female rats should be employed exclusively for any given study as there may be some sex differences, depending on the parameters to be examined in the glomeruli.

Blood-Free Perfusion of Kidneys

Animals are anesthetized by either ether or pentobarbital. The aorta, vena cava, and both kidneys are exposed through an abdominal midline incision. A 23-gauge butterfly needle is inserted into the abdominal aorta just below the level of the renal arteries. The abdominal aorta is ligated above the renal arteries and below the site of needle insertion. A cut is made in the vena cava and the kidneys are perfused by hand with 20–50 ml of ice-cold Krebs–Ringer's (KR) buffer until they blanch. They are then rapidly removed and placed into a beaker with KR buffer on ice. This procedure results in an almost blood-free kidney and can be performed within a few minutes.

Preparation without Prior Perfusion of Kidney

For this procedure, the rats can be either sacrificed by decapitation or exsanguination after prior anesthesia. Both kidneys are rapidly removed by midabdominal incision and placed on iced KR buffer. The actual isolation of glomeruli then proceeds in an identical fashion for both methods. We believe that the isolation of glomeruli without prior perfusion is simpler and eventually yields glomeruli that are just as free of blood-borne ele-

---

[11] C. A. Krakower and S. A. Greenspon, *Arch. Pathol.* **51**, 629 (1951).
[12] E. Meezan, K. Brendel, J. Ulreich, and E. C. Carlson, *J. Pharmacol Exp. Ther.* **187**, 332 (1973).

ments as those obtained from perfused kidneys. This is due to ejection of blood elements from the glomeruli during the subsequent isolation procedure. The isolation procedure is performed in solutions without oncotically active substances, i.e., albumin or protein. Initially, the glomeruli contain particulate blood elements (e.g., red cells, polys, platelets) as well as the oncotically active plasma. When exposed to the protein-free KR buffer (or normal saline) used during isolation, they act as reverse oncometers, drawing in fluid and thereby ejecting the blood elements. This reversed filtration has been used by Savin and Terreros[13] to estimate the glomerular ultrafiltration coefficient in isolated glomeruli. In any case, this process frees glomeruli of blood elements during the extensive isolation procedure, making prior perfusion of kidney *in situ* superfluous. We have, therefore, abandoned the blood-free perfusion of kidneys prior to glomerular isolation.

Cortical Dissection. The kidneys are gently stripped of their capsule by hand. The cortex is then cut off with a fine pair of surgical scissors and placed in a plastic Petri dish containing 1 ml of KR buffer. Using a razor blade, the cortical tissue is minced to a fine paste. These steps are performed at 4°.

Sieving of Glomeruli. The isolation of the glomeruli by sequential sieving is performed at room temperature, as the glomeruli are more deformable at room temperature than at 4° and hence suffer less damage during the mechanical sieving procedure. Three stainless steel sieves are placed on top of each other and over a sink. The three sieves are all 8 in. in diameter and are, from top to bottom: (1) Tyler equivalent 100 mesh, pore size 150 μm, (2) Tyler equivalent 150 mesh, pore size 106 μm, and (3) Tyler equivalent 200 mesh, pore size 75 μm. The sieves are prewetted with normal saline and the cortical tissue is placed on top of the first sieve. The cortical paste is forced through the sieve with rotatory movements using the bottom of a glass beaker and washing with 2–3 liters of normal saline (room temperature). This top sieve is then removed and the tissue on the underside is scraped off and added onto the next sieve (pore size 106 μm). The tissue remaining on the 150-μm pore size sieve (mostly vascular and tubular elements) is discarded. The tissue collected on the middle sieve is extensively washed with normal saline (3–5 liters) while being rubbed very gently (essentially no pressure applied) in rotatory movements with the bottom of a beaker. Rubbing too hard at this step will force unwanted tubular elements through the sieve, while extensive washing will allow glomeruli to pass through. The middle sieve is removed and the tissue remaining on it (mostly tubules) is discarded. The glomeruli have now

---

[13] V. J. Savin and D. A. Terreros, *Kidney Int.* **20,** 188 (1981).

collected on the bottom sieve, while very small tubular fragments will have passed through this sieve. The bottom sieve is again washed with 1-2 liters of normal saline and tilted to 30-45° so that the glomeruli accumulate at one edge. From there they are flushed into 50-ml conical polypropylene tubes using 50 ml of normal saline in a wash bottle.

Centrifugation and Washing of Glomeruli. The glomeruli are then spun down at 1000 rpm for 5 min in a clinical centrifuge, resuspended in 5 ml of cold KR buffer, and checked for purity under the microscope (see the next section). If there is more than a 5% admixture of tubular fragments, or if more than 20% of glomeruli still have Bowman's capsule, the tubular fragments and Bowman's capsule are disrupted by sucking the glomerular preparation up and down a long Pasteur pipet 5 to 10 times. To further purify glomeruli this is combined with a short period of collagenase digestion [10-20 min at 37° in Dulbecco's modified Eagle's medium (DMEM) with 1.5 mg/ml of collagenase type IV; Worthington, Inc., Freehold, NJ]. This treatment breaks up tubular fragments, removes Bowman's capsule, and results in better access of glomerular cells to, e.g., hormones during subsequent incubations.[14] After the collagenase digestion the glomeruli are collected on a 38-μm pore size, 400 mesh opening sieve, washed, and removed from the sieve as described above. The glomerular suspension is then centrifuged at 500 rpm for 2 min, the supernatant discarded, 5 ml of fresh buffer added to resuspend the glomeruli, and the centrifugation step repeated. The glomerular pellet is then resuspended in 5 ml of KR buffer and purity is reevaluated.

Evaluation of Purity and Glomerular Counting. One or two drops of the final glomerular preparation are applied to a microscope slide and observed under a phase-contrast microscope. Glomeruli free of Bowman's capsule appear as a "bunch of grapes," while those still encapsulated can be identified by the smooth contour of the capsule. Tubular fragments are easily identified. If glomerular counts are performed, it is important to assure even suspension of the glomeruli to removing an aliquot for counting in a hemocytometer chamber. Glomeruli will rapidly settle by gravity. In order to maintain them in suspension, the air from the collecting pipet is blown out at the bottom of the tube containing the glomeruli prior to removing aliquots.

This procedure will allow isolation of 20,000-30,000 glomeruli/rat within 20 min (corresponding to 2-3 mg/protein). In order to obtain good purity (greater than 95% glomeruli), it is advisable to use no more than two to four kidneys per sieving procedure and to employ large volumes of saline for washing, rather than vigorously rubbing the tissue through the

---

[14] J. B. Lefkowith and G. Schreiner, *J. Clin. Invest.* **80,** 947 (1987).

sieves. This will also avoid excessive mechanical trauma to the glomeruli. Following this procedure, we have observed that the overall glomerular architecture is well preserved, as evaluated either by scanning or transmission electron microscopy.

It is important that the sieves be meticulously cleaned before reuse as they otherwise become clogged by proteinaceous material. This can be achieved by immediately soaking them in hot water and scrubbing them with a bottle cleaner. After two to three uses, the sieves are also soaked overnight in 1 N NaOH to remove remaining cell material.

Modification of Procedure for Isolation of Glomeruli from Other Than Adult Rats

As the sieving method is based on size exclusion, the method has to be modified according to glomerular size. When young rats (50 g) are used, as is common practice for culture of glomeruli, glomeruli could be collected on a 38-μm pore size, 400 mesh opening sieve, rather than the one with 75-μm pore size used in adult rats. Isolation of glomeruli from mice is performed by using consecutive sieves of 150-, 106-, and 63-μm pore size. For preparation of glomeruli from rabbit, dog, or monkey, the following combination of sieves can be used: (1) 246-μm pore size, Tyler equivalent 60 mesh, (2) 180-μm pore size, Tyler equivalent 80 mesh, and (3) 125-μm pore size, Tyler equivalent 115 mesh. It should be noted that it has been a problem to obtain a pure glomerular preparation from rabbit kidneys. For human kidneys — surgical samples or donor kidneys found unsuitable for transplantation — the sieving method can be successfully employed with a good yield and purity by employing sieves with 246-μm pore size followed by 212-μm pore size and collecting glomeruli on a 106-μm pore size sieve. This can also be combined with a short collagenase digestion.

Alternative Methods of Glomerular Isolation

If a 90–95% purity of glomeruli is judged insufficient, an initial sieving step (either with the first sieve only, or with all three sieves) can be combined with a gradient centrifugation step using either sucrose or Ficoll.[8,10,15] This will result in further elimination of cells and tubular fragments, but is time consuming and reduces the yield. Closely following the procedures outlined, we no longer find any need for additional density centrifugation steps.

The isolated glomeruli can now be used for studies of, e.g., arachidonic acid metabolism,[16-20] phospholipid turnover,[21] hormone binding,[22-26] gen-

---

[15] D. Schlondorff, P. Yeo, and B. E. Alpert, *Am. J. Physiol.* **235**, F458 (1978).

eration and metabolism of cAMP[10,15,27,28] and cGMP,[28,29] histamine metabolism,[30,31] renin release,[32] contraction experiments,[33-35] or cell culture.[6] I shall briefly describe methods for incubating glomeruli for prostaglandin production and for measurements of glomerular contractility.

Incubation of Isolated Glomeruli in Test Tubes for Determination of Prostaglandin Synthesis

The final preparation of isolated glomeruli is kept on ice in KR buffer until the glomeruli have settled by gravity. The supernatant buffer is then removed and replaced by 10 ml of KR buffer at 37° and kept at 37° under an atmosphere of 95% O_2, 5% CO_2. After the glomeruli have again settled by gravity (usually 10 min) the supernatant buffer is removed and replaced by 2–4 ml of fresh KR buffer at 37°. The glomeruli are gently suspended by blowing through some air with the pipet to be used for aliquoting of

[16] A. Hassid, M. Konieczkowski, and M. J. Dunn, *Proc. Natl. Acad. Sci. U.S.A.* **76**, 1155 (1979).
[17] J. Sraer, J. D. Sraer, D. Chansel, F. Russo-Marie, B. Kouznetzova, and R. Ardaillou, *Mol. Cell. Endocrinol.* **16**, 29 (1979).
[18] D. Schlondorff, S. Roczniak, J. A. Satriano, and V. W. Folkert, *Am. J. Physiol.* **239**, F486 (1980).
[19] J. Sraer, M. Rigaud, M. Bens, H. Rabinovitch, and R. Ardaillou, *J. Biol. Chem.* **258**, 4325 (1983).
[20] K. Jim, A. Hassid, F. Sun, and M. J. Dunn, *J. Biol. Chem.* **257**, 10294 (1982).
[21] V. W. Folkert, M. Yunis, and D. Schlondorff, *Biochim. Biophys. Acta* **794**, 206 (1984).
[22] J. D. Sraer, J. Sraer, R. Ardaillou, and D. Mimoune, *Kidney Int.* **6**, 240 (1974).
[23] B. J. Ballermann, R. L. Hoover, M. J. Karnovsky, and B. M. Brenner, *J. Clin. Invest.* **76**, 2049 (1985).
[24] K. Kurokawa, F. J. Silverblatt, K. L. Klein, M. S. Wang, and R. L. Lerner, *J. Clin. Invest.* **64**, 1357 (1979).
[25] D. Chansel, T. P. Oudinet, M. P. Nivez, and R. Ardaillou, *Biochem. Pharmacol.* **31**, 367 (1982).
[26] T. R. Sedor and H. E. Abboud, *Kidney Int.* **26**, 144 (1984).
[27] J. Sraer, R. Ardaillou, N. Loreau, and J. D. Sraer, *Mol. Cell. Endocrinol.* **1**, 285 (1974).
[28] V. E. Torres, Y. S. F. Hui, S. V. Shah, T. E. Northrupt, and T. P. Dousa, *Kidney Int.* **14**, 444 (1978).
[29] J. J. Helwig, C. Bollack, P. Mandel, and C. Goridis, *Biochim. Biophys. Acta* **377**, 463 (1975).
[30] H. Abboud, *Kidney Int.* **24**, 534 (1983).
[31] J. I. Heald and T. M. Hollis, *Am. J. Physiol.* **230**, 1349 (1976).
[32] W. Beierwaltes, S. Schryver, P. S. Olson, and J. C. Romero, *Am. J. Physiol.* **239**, F602 (1980).
[33] L. A. Scharschmidt, J. G. Douglas, and M. J. Dunn, *Am. J. Physiol.* **250**, F348 (1986).
[34] L. A. Scharschmidt, E. Lianos, and M. J. Dunn, *Fed. Proc., Fed. Am. Soc. Exp. Biol.* **42**, 3058 (1983).
[35] R. Barnett, P. Goldwasser, L. A. Scharschmidt, and D. Schlondorff, *Am. J. Physiol.* **250**, F838 (1986).

glomeruli. It is important to avoid excessive mechanical agitation, such as vortexing. Aliquots of 0.5 ml of glomerular suspension are then immediately added to 1.5 ml Eppendorf conical test tubes containing 0.5 ml of either buffer only or buffer plus experimental agents maintained at 37° in a water bath. It is important to assure that the glomeruli have been resuspended in fresh buffer immediately (i.e., seconds) before the initiation of the experimental incubations. Incubations are carried out in duplicate for 10 min under an atmosphere of 95% O_2, 5% CO_2 in a shaking water bath (60 cpm). No more than six tubes should be incubated simultaneously in order to avoid excessive variations (greater than 15 sec) in incubation times. Incubations are terminated by a 10-sec spin in an Eppendorf microfuge and immediate removal of 0.5 to 0.8 ml of supernatant for subsequent determination of prostaglandins (PGs). These samples are stored frozen at $-20°$. The different prostaglandins are determined within 2 weeks by either radioimmunoassay (RIA)[36] or enzyme immunoassay.[37] To the glomeruli remaining in the test tubes, 1 ml of 1 N NaOH is added for subsequent determination of protein content. Glomerular PG synthesis is proportional to glomerular protein up to about 200 µg used per incubation and for incubation times up to 10 min. As two kidneys from a rat will yield from 2 to 4 mg of glomerular protein, this will allow for a total of 10 to 30 incubations to be carried out per rat.

Superfusion System of Isolated Glomeruli

In order to allow for more dynamic studies of, e.g., prostaglandin synthesis or renin release from isolated glomeruli, a superfusion system can be employed.[18]

Loading of Glomeruli onto Superfusion Chamber. A prewetted HA-Millipore filter (pore size 0.45 µm) is put on the bottom part of a disk filter holder (Swinnex 13- or 25-mm diameter; Millipore Co., Bedford, MA) horizontally held on a plastic support (Fig. 1). The rubber sealing ring is placed on the filter and a 5-ml syringe is connected to the outflow of the bottom part of the filter holder. A 1-ml aliquot of an isolated glomerular preparation (approximately 0.5–1 mg of glomerular protein) is slowly dripped onto the filter while the buffer of the suspension is sucked through by the syringe attached to the outflow of the filter holder. When the entire 1-ml sample of glomerular suspension has settled down onto the filter, the top part of the filter holder is screwed on and the filter chamber filled with buffer. It is important to assure that the top of the filter chamber is totally filled and free of air. The syringe and tubing connected to the outflow of

[36] F. B. Dray, B. Charbonnel, and J. Maclouf, *Eur. J. Clin. Invest.* **5,** 311 (1975).
[37] P. Pradelles, J. Grassi, and J. Maclouf, *Anal. Chem.* **57,** 1170 (1985).

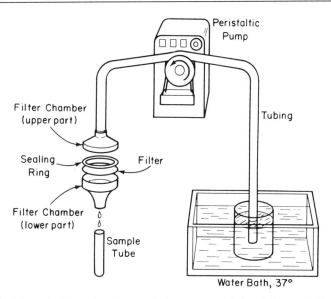

FIG. 1. Schematic illustration for superfusion system of isolated glomeruli. The isolated glomeruli are loaded onto the filter and sealed inside the filter chamber for continuous superfusion.

the filter chamber is removed and replaced by a stainless steel needle (18 gauge). The inflow of the filter holder is then connected to Tygon plastic tubing (i.d. $\frac{1}{16}$ in.; o.d. $\frac{1}{8}$ in.). The plastic tubing is inserted into a rotating peristaltic pump and placed into a beaker containing KR buffer equilibrated with 95% O_2, 5% CO_2 and maintained at 37°. Using the peristaltic pump, the glomeruli are then continuously superfused (usually 40 ml/hr), while the effluent is collected at 5- or 10-min intervals by placing tubes under the filter chamber outlet. At this rate of perfusion the temperature of the buffer in the filter chamber is maintained at 37°, i.e., the temperature present in the reservoir, so that the filter chamber does not have to be heated.

The effect of experimental agents in the superfusion buffer can be examined by simply switching the tube feeding the filter chamber back and forth from beakers containing either KR buffer only or KR plus the test agent. The time required for the buffer to travel from the reservoir to the superfusion chamber can be determined by timing, e.g., the transit time of a small air bubble through the tubing.

We routinely set up four superfusion chambers per glomerular preparation, superfusing them in parallel with the same peristaltic pump. Glomeruli in one chamber serve as a time control, while the others can be used for experimental agents. A time control is very important in this system as, e.g., prostaglandin synthesis or renin release may decline during the continuous superfusion. At the end of the experiment the filter with the glomeruli is removed from the filter holder and glomerular protein on each filter is determined. The filter holders and the tubing can be reused after extensive rinsing with normal saline.

Glomerular Contraction as Determined by Planar Surface Area

Glomeruli alter their filtration characteristics in response to a variety of vasoactive agents, such as angiotensin II, vasopressin, norepinephrine, leukotrienes, and platelet-activating factor.[1,2,34] This is thought to relate to contraction of mesangial cells, which have characteristics of modified smooth muscle cells.[34] *In vitro* contractility of isolated glomeruli can be determined by measuring their planar surface area, a direct reflection of their cross-sectional area.[33,34] Essentially all methods employed in such studies rely on the determination of planar surface area of isolated glomeruli incubated under different experimental conditions, using either photography followed by morphometric determinations[33] or direct video image analysis.[35]

Determination of Planar Glomerular Surface Area after Fixation. Glomeruli are isolated as described above. For these experiments, purity of the glomerular preparation is less crucial, but structural integrity is. Therefore, overly vigorous mechanical sieving or tricheration through Pasteur pipets should be avoided. If glomeruli are severely damaged, no or minimal contraction in response to, e.g., angiotensin will result. The glomerular preparation is resuspended in phosphate-buffered Robinson's solution containing 1 mM $CaCl_2$ at either room temperature or at 37°. Aliquots (0.5 ml) of the glomerular suspension are then added to plastic tubes containing either 0.5 ml of the above Robinson's buffer only (controls) or buffer plus experimental agents to be examined. The reaction is terminated by the addition of 0.1 ml of 10% glutaraldehyde and vortexing, resulting in fixation at 1% final concentration of glutaraldehyde. If experiments are carried out at room temperature, the fixation is performed after a 10- to 20-min incubation period, while at 37° a 2- to 5-min experimental incubation is sufficient. The planar surface area of the glomeruli is then analyzed by pipetting an aliquot (0.1 ml) of the glutaraldehyde-fixed glomerular suspension onto a microscope slide, which is observed under a phase-contrast microscope equipped with a video camera and connected to an automated image analyzer (e.g., Quantimet, Cambridge Instruments, Moncie, NY). The glomeruli on the slide are projected onto a video screen, usually about 10–20 per visual field. The image analyzer now allows exclusion of nonglomerular structures from analysis. The image analyzer will then measure the planar surface area of each glomerulus on the screen. The planar surface area of 40–60 glomeruli can be analyzed and recorded by the computerized morphometry within minutes by examining three to four different visual fields per slide. Aliquots of glomeruli from the different incubation conditions can thus be automatically analyzed without observer bias. Multiple aliquots from control incubations can be interspersed among experimental samples to assure reproducibility. The computerized morphometer is calibrated using beads of standard size under the same conditions. Changes in mean planar surface area of as little as 3–4% can reliably be determined by this method if enough glomeruli are analyzed (about 50 under each condition). If an automated image analyzer is not available, the glomeruli can be photographed under phase-contrast microscopy and their individual surface area determined from the photographic prints by either planimetry or measurements of diameters. Obviously, the latter methods are much more tedious, time consuming, and open to observer bias.

Determination of Glomerular Planar Surface Area during Continuous Superfusion. In this method, isolated glomeruli are superfused in a microperfusion chamber mounted on a microscope equipped with a video camera (Fig. 2) and their planar surface area is continuously recorded by an automated image analyzer (as above). A microscope cover slip is coated by immersing it in polylysine (1 mg/ml) for 10 min followed by rinsing with distilled water. An aliquot (0.05 ml) of the glomerular preparation is then applied to the coated cover slip, and placed upside down on a microperfusion chamber. The polylysine coating allows point attachment of glomeruli to the coverslip, thus preventing them from floating away during superfusion without interfering with their contraction. The microperfusion chamber consists of a microscope slide with a 5-mm-thick, 2×2-cm plastic block glued on. The plastic block has a central 1-cm-diameter bore hole connected to the outside by a 20-gauge stainless steel needle on each side. The inverted cover slip with the glomeruli is placed over the hole, sealed with vacuum grease, and secured by spring clamps. The chamber is continuously perfused with Robinson's buffer at 37° via a peristaltic pump and plastic tubing connected to one of the needles in the superfusion chamber. The other needle is also connected to a plastic tube and serves as outflow. The perfusion chamber is mounted on the microscope stage and the glomeruli on the inside surface of the cover slip bathed by the perfusion fluid can be observed under phase contrast on the video screen. About 5 to 10 glomeruli can be observed on a visual field and their planar surface areas continuously recorded by the computerized image analyzer. By using a superfusion system as described above for prostaglandin synthesis, the bathing fluid can be switched to contain the experimental test agents. This

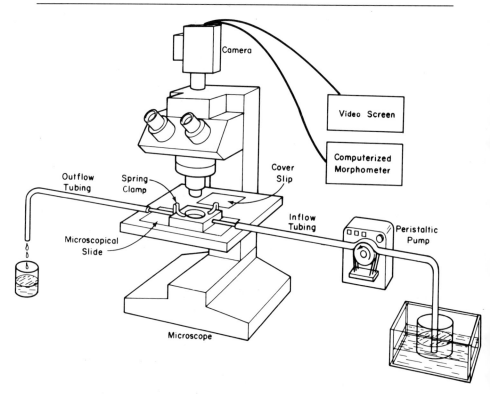

FIG. 2. Schematic illustration for microscopic and video recording system for superfusion of isolated glomeruli. The isolated glomeruli are adherent to the underside of the coverslip, which is fixed to the superfusion chamber on the microscope slide by two spring clamps.

system thus allows sequential determinations of planar surface areas of 5–10 isolated glomeruli by computerized image analysis. This has the advantage of allowing paired analysis of the same glomeruli and recording a detailed time course of glomerular surface changes. The disadvantage is that it is much more time consuming, thereby restricting the number of experimental conditions that can be examined on a given glomerular preparation.

[9] Isolation and Study of Glomerular Cells

By D. A. Troyer and J. I. Kreisberg

Techniques for Isolation and Culture of Glomeruli

This section describes methods to establish homogeneous cultures of four types of glomerular cells. The goal of these cell culture protocols is to isolate the cells of interest from the four or five cell types present in the mammalian glomerulus. Publications detailing established techniques for cell culture are available.[1,2] These texts also discuss the equipment needed for tissue culture.

Glomeruli from animals are obtained from kidneys that have been perfused with tissue culture medium or Hanks' balanced salt solution (HBSS) with Ca^{2+} and Mg^{2+} to remove the blood. Rats are anesthetized with intraperitoneal injection of Nembutal (50 mg/kg body wt). The abdomen is entered via a midline incision. The abdominal viscera are reflected to expose the retroperitoneal structures. Using cotton-tipped applicators to dissect away fat and serosa, the superior mesenteric artery is located just above the right renal artery. A ligature is placed just superior to the right renal artery, but not tied. The distal aorta is similarly dissected and two ligatures placed but not tied. A curved hemostat is then placed more distally to place tension on the aorta. A vascular clamp is then placed on the abdominal aorta at a sufficient distance from the distal ligature to allow for manipulation of a 20-gauge catheter into the aorta. A 20- or 21-gauge needle can be used if the point is smoothed with a file to avoid penetrating the posterior aorta. After nicking the distal aorta with iris scissors, the catheter is threaded into the aorta and the ligatures tied. After removal of the vascular clamp, the perfusion line is opened. The superior ligature is then tied, and the renal vein (either side) is then nicked to allow flow of perfusate (HBSS containing Ca^{2+} and Mg^{2+}) through the kidneys into the abdominal cavity. The kidneys should blanch rapidly and be completely pale within 1–2 min of perfusion with 25–50 ml. The kidneys are then removed to an iced bath of HBSS for brief storage until they can be dissected. A diagram of this procedure is presented in Fig. 1. The cortex is dissected and glomeruli are isolated from renal cortical tissue using a modification of a technique originally described by Krakower and Green-

---

[1] W. B. Jakoby and I. Pastan (eds.), "Methods in Enzymology," Vol. 58. Academic Press, New York, 1979.

[2] R. I. Freshney, "Culture of Animal Cells." Liss, New York, 1987.

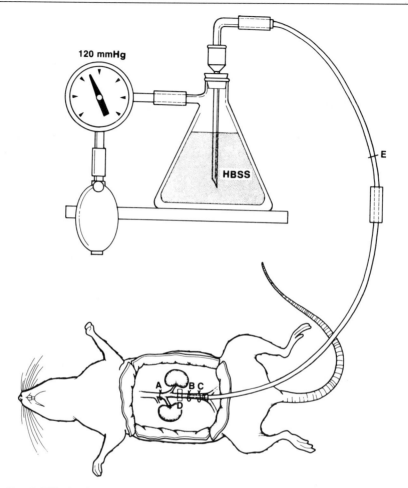

FIG. 1. This drawing shows the apparatus for perfusion of rat kidneys prior to nephrectomy. Ligatures are placed at A, B, and C. The vascular clamp is placed at D. A hemostat can be placed at E to regulate the flow of perfusate. HBSS, Hanks' balanced salt solution. [Adapted from L. D. Griffith, R. E. Bulger, and B. F. Trump, *Lab. Invest.* **16,** 220 (1967).]

spon.[3] Glomeruli are isolated by sieving through stainless steel screens (W. Tyler Co., Mentor, OH) of different pore size as shown for the rat in Fig. 2. The type of screen chosen to retain glomeruli depends on the species selected for study (Table I).[4] A spatula is used to press the renal cortical tissue through the largest screen only. By subsequent rinsing through the

[3] C. A. Krakower and S. A. Greenspon, *Arch. Pathol.* **58,** 401 (1954).
[4] S. R. Holdsworth, E. F. Glasgow, R. C. Atkins, and N. H. Thomson, *Nephron* **22,** 454 (1978).

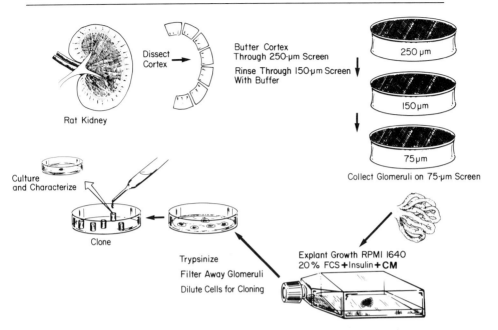

FIG. 2. This diagram illustrates the techniques used for isolation of rat glomeruli and cloning of glomerular cells: RPMI 1640 with 20% fetal calf serum (FCS), insulin, and conditioned medium (CM) are used to clone glomerular epithelial cells. [From J. I. Kreisberg, in "Cell Separation Methods and Selected Applications" (T. G. Pretlow II and T. P. Pretlow, eds.), Vol. 1, p. 247. Academic Press, New York, 1982.]

TABLE I
APPROPRIATE SIZE SCREEN TO RETAIN
GLOMERULI FROM DIFFERENT ANIMAL SPECIES[a]

Gauge (mesh)	Opening size (μm)	Species
80	184	Human
100	151	Human
120	128	Monkey, dog
150	103	Sheep, rabbit
200	75	Rat
300	53	Rat

[a] Adapted from Holdsworth et al.[4]

series of screens (without further use of the spatula), glomeruli can be obtained virtually free of tubule contamination and stripped of Bowman's capsule (Fig. 2). Contamination with fibroblasts has not been a problem. Human glomeruli are best obtained from surgically resected kidneys. In this regard, it is important that the tissue be as fresh as possible.

Two techniques are most often employed for growing glomerular cells: (1) Direct plating of the glomeruli into culture for explant growth of glomerular cells (glomeruli are plated into culture at approximately 10,000 glomeruli/28 cm^2 tissue culture dish): Once plated, the glomeruli should not be disturbed for 4 days to allow attachment to the dish. (2) Enzyme dissociation of the glomeruli and plating of the dissociated cells into tissue culture: Digestion with purified bacterial collagenase yields a population of cells that is considered to be epithelial,[5] while the partially digested glomerular cores remaining behind contain mesangial, monocytic, and endothelial cells. In either case, there are at least four cell types present in the culture and purification procedures are necessary to obtain homogeneity.

Several methods have been used to obtain homogeneous cultures of glomerular cell types. One is based upon the differential growth of glomerular epithelial and mesangial cells.[6] With this technique the first cell type to grow out from the glomerulus is "epithelial-like" (i.e., has a polygonal shape and cobblestone-like appearance after confluence). Some have isolated these cells within the first week to 10 days of growth and identified them as glomerular epithelial cells. Clearly, endothelial cells, mesangial cells, and monocytes could also be present. Additionally, if Bowman's capsule is not stripped from the glomeruli, parietal epithelial cells may also be present.[7] Mesangial cells, however, can be obtained in homogeneous culture with this technique by allowing the glomerular explants to grow for 1 month; at this time, mesangial cells represent almost 100% of the culture (Fig. 3a). The only reliable way to obtain homogeneous cultures of endothelial and epithelial cells is by cloning.[8] Cells are diluted with a sufficient volume of tissue culture medium to allow the addition of single cells to the culture. Colonies of cells, derived from a single cell, are individually subcultured and transferred to other dishes (Fig. 2). It should be noted, however, that when cells are plated at low densities, the survival rate falls. This can present a substantial problem with glomerular epithelial and endothelial cells where the plating efficiency is low. However, as will be discussed later in this chapter, attempts have been made to improve the

[5] G. E. Striker and L. J. Striker, *Lab. Invest.* **53,** 122 (1985).
[6] J. B. Foidart, C. A. Dechenne, P. Mahieu, C. A. Creutz, and J. De Mey, *Invest Cell Pathol.* **2,** 15 (1979).
[7] J. O. R. Norgaard, *Lab Invest.* **57,** 277 (1987).
[8] J. I. Kreisberg and M. J. Karnovsky, *Kidney Int.* **23,** 439 (1983).

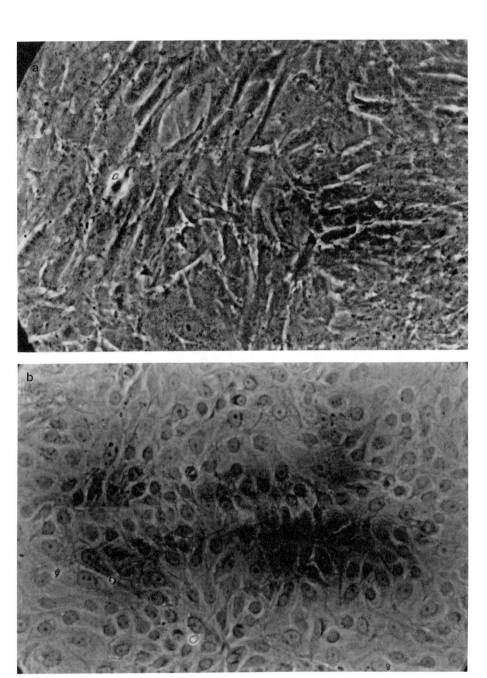

FIG. 3. (a) Photomicrograph showing a confluent culture of rat mesangial cells under phase-contrast microscopy (×76). (b) Photomicrograph showing cultured bovine glomerular endothelial cells. (Courtesy of Dr. Barbara Ballermann, Renal Division and Dept. of Medicine, Brigham and Women's Hospital, and the Harvard Center for the Study of Kidney Diseases, Harvard Medical School, Boston, MA) (×76). (c) Photomicrograph showing cultured glomerular epithelial cells under phase-contrast microscopy (×76).

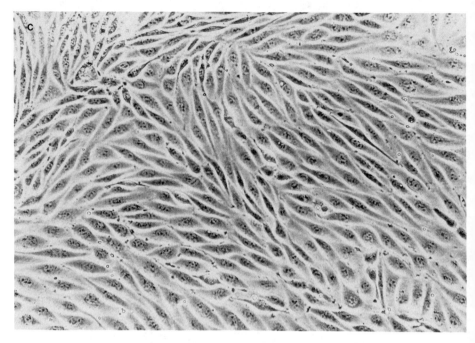

FIG. 3. *(Continued)*

plating efficiencies of these cells by coating plastic culture dishes with matrix elements that more closely resemble *in vivo* conditions. Cloning has been used to obtain homogeneous cultures of epithelial, endothelial, and mesangial cells from animal as well as human glomeruli.

Isolation and Culture of Mesangial Cells

Two cell types have been shown to reside in the glomerular mesangium: (1) the contractile mesangial cell and (2) a phagocytic bone marrow-derived monocyte which expresses the leukocyte common (LC) antigen.

Contractile Mesangial Cells

The contractile mesangial cell is the principal cell type residing in the glomerular mesangium. Mesangial cells are relatively easy to isolate and propagate in homogeneous culture and have been studied extensively following isolation from several species, including humans.[5,8] The growth medium most commonly used is Roswell Park Memorial Institute medium (RPMI) 1640 with 20% fetal calf serum (FCS) supplemented with antibiotics/antimycotics (penicillin, 100 U/ml, streptomycin, 100 μg/ml, and Fungizone, 0.25 μg/ml) and 0.66 IU of insulin/ml medium.[9] Mesangial cells are most often isolated from glomerular explants (see above for isolation of glomeruli). After 30 days of growth, the glomerular explants are nearly 100% mesangial (Fig. 3a). Upon passaging with trypsin–versene [0.5% trypsin and 0.53 mM ethylenediaminetetraacetic acid (EDTA) in HBSS without Ca^{2+} or Mg^{2+}], the cells are resuspended in complete culture medium, transferred to a sterile culture tube, the glomeruli allowed to sediment out at 1 g, and the suspended cells replated. Absolute purity can be achieved by subsequent cloning. Early passages of mesangial cells (i.e., within 10 passages) isotonically contract in response to vasoactive agonists such as angiotensin II and arginine vasopressin. After multiple passages when the cells develop tight adhesive characteristics,[10] isotonic contraction can be demonstrated after plating on slippery substrata.[11] The contractile qualities of these cells at higher passage can also be studied by evaluation of myosin light chain phosphorylation.[12] Additionally, only mesangial cells form hills and valleys (so-called hillocks[13]) after confluence, similar to smooth muscle cells in culture.[14]

Bone Marrow-Derived Cells

These phagocytic cells which contain the LC antigen, a membrane glycoprotein characteristic of hematopoietic cells, represent about 3–7% of the total cells present in the mesangium.[15] They are isolated as follows: Washed isolated glomeruli are digested in a cocktail composed of 1 mg/ml of trypsin (type III, Sigma Chemical Co., St. Louis, MO), 1 mg/ml collagenase (type CLS IV, Millipore Corp., Bedford, MA), and 0.01 mg/ml DNase (type III, Sigma Chemical Co., St. Louis, MO), in HBSS with Ca^{2+} and Mg^{2+} for 45 min at 37° on a rocking platform. The partially digested glomeruli are washed in HBSS without Ca^{2+} or Mg^{2+} and treated with 2 mM EDTA. After 10 min at 37°, the glomeruli are pelleted at 50 g for 10 min, the supernatant drawn off and stored at 4°, and the sediment consisting of partially digested glomerular fragments incubated for an additional 20 min at 37° in 2 mg/ml collagenase. This preparation is allowed to settle

---

[9] D. A. Ausiello, J. I. Kreisberg, C. Roy, and M. J. Karnovsky, *J. Clin. Invest.* **65**, 754 (1980).
[10] J. I. Kreisberg and M. A. Venkatachalam, *Am. J. Physiol.* **251**, C505 (1986).
[11] M. A. Venkatachalam and J. I. Kreisberg, *Am. J. Physiol.* **249**, C48 (1985).
[12] J. I. Kreisberg, M. A. Venkatachalam, R. A. Radnik, and P. Y. Patel, *Am. J. Physiol.* **249**, F227 (1985).
[13] R. B. Sterzel, D. H. Lovett, H. G. Foellmer, M. Perfetto, D. Biemesderfer, and M. C. Kashgarian, *Am. J. Pathol.* **125**, 130 (1986).
[14] M. A. Gimbrone and R. S. Cotran, *Lab Invest.* **33**, 16 (1975).
[15] G. F. Schreiner, R. S. Cotran, and E. R. Unanue, *Lab. Invest.* **51**, 524 (1984).

for 90 sec and the supernatant centrifuged at 400 g, washed in HBSS without Ca^{2+} or Mg^{2+}, and added to the previously stored supernatant. The combined suspended cells are centrifuged at 400 g and resuspended in HBSS with Ca^{2+} and Mg^{2+} containing 1 mg/ml albumin and 1 mM HEPES buffer. This procedure has been reported to yield $2-5 \times 10^6$ single cells per rat.[16] After plating into culture in RPMI 1640 with 10% FCS and antibiotics, the cells are allowed to adhere for 2 hr. The cultured cells are characterized as mononuclear phagocytes since they have receptors for C3b, Fc, contain the LC antigen, and are actively phagocytic.[15,16] Approximately 20% of the adherent cells are phagocytic and stain for the LC antigen.

Isolation and Culture of Glomerular Endothelial Cells

Three different methods have been described for isolation of glomerular endothelial cells; human, rat, and bovine glomerular endothelial cells have been isolated and propagated *in vitro*.[17-19]

Striker *et al.*[17] have described the isolation of endothelial cells by cloning explants of human glomeruli cultured in Waymouth's medium supplemented with 20% FCS and 2 ng/ml platelet-derived growth factor (PDGF). By cloning a dilution of the initial mixed outgrowth, a pure line of endothelial cells was obtained. The cells were characterized by assays for factor VIII antigen and angiotensin-converting enzyme activity.

Another method for isolation of glomerular endothelial cells involves filtration of dissociated glomeruli to yield a single cell suspension.[18] Glomeruli are isolated as described previously, treated with 0.2% trypsin for 20 min, followed by 0.1% collagenase (189 U/ml, Worthington Diagnostic Systems, Freehold, NJ) for 40 min at 37° in HBSS with Ca^{2+} and Mg^{2+} buffered with HEPES to pH 7.4. The dissociated glomeruli are filtered through a 15-μm Nitex filter to yield a single cell suspension. The resulting cells are plated at 1000 cells/100-mm dish in a 1:1 mixture of Taub and Sato K-1 defined medium containing 5% Nuserum (Collaborative Research, Waltham, MA) and conditioned medium. The conditioned medium is obtained from exponentially growing Swiss 3T3 cells grown in Dulbecco's modified Eagle's medium (DMEM) with 10% FCS. The K-1

---

[16] G. F. Schreiner, J. M. Kiely, R. S. Cotran, and E. R. Unanue, *J. Clin. Invest.* **68**, 920 (1981).

[17] G. E. Striker, C. Soderland, D. F. Bowen-Pope, A. M. Gown, G. Schmer, A. Johnson, D. Luchtel, R. Ross, and L. J. Striker, *J. Exp. Med.* **160**, 323 (1984).

[18] J. J. Castellot, R. L. Hoover, and M. J. Karnovsky, *Am. J. Pathol.* **125**, 493 (1986).

[19] B. J. Ballermann, *Am. J. Physiol.* **256**, C182 (1989).

medium of Taub and Sato contains a 50:50 mixture of DMEM and Hams F12 medium supplemented with 10 mM HEPES buffer, sodium bicarbonate (1.1 mg/ml), 10 nM sodium selenite ($Na_2SeO_3 \cdot 5H_2O$), insulin (5 μg/ml), PGE_1 (25 ng/ml), triiodothyronine (T_3; 5×10^{-12} M), hydrocortisone (5×10^{-8} M), and transferrin (5 μg/ml).[20] Selected colonies are removed and transferred to a well of a 24-well multiwell plate and cultured in the same medium. High levels of angiotensin-converting enzyme activity are used to identify the cells as endothelial cells. The method has a relatively low yield and it is difficult to generate sufficient cells for biochemical studies.[18]

Another possibly more successful method for culturing bovine endothelial cells combines cell-sorting techniques and selective media.[19] Glomeruli are collected by pressing 5–10 g of minced bovine kidney cortex through a 180-μm screen. The material collected from the undersurface is then sieved (rinsed) through stacked screens and collected on a 104-μm screen; the retained glomeruli are washed in HBSS with Ca^{2+} and Mg^{2+} and exposed to 1 mg/ml collagenase (type V, Sigma Chemical Co. St. Louis, MO) in HBSS with Ca^{2+} and Mg^{2+} buffered with 20 mM HEPES for 20 min at 37° with periodic mixing by pipetting using a 5-ml pipet. The volume is adjusted to 50 ml with RPMI 1640 containing 10% calf serum and glomerular remnants sedimented for 5 min at 500 g. The supernatant is then pelleted at 1000 g for 10 min. The resulting pellet is resuspended in RPMI 1640 medium containing 20% iron-supplemented bovine calf serum. The cells are plated on gelatin-coated dishes or plates overlaid with fibronectin. Coating of the plates is accomplished by dissolving gelatin (0.1 g/dl), (Sigma Cell Culture Reagents, 300 Bloom) in cell culture-grade distilled water, filtering the solution through a 0.2-μm syringe filter and spreading a thin film on the plates. After drying, the plates are overlaid with bovine fibronectin (5 μg/ml in HBSS with Ca^{2+} and Mg^{2+}) using 100 μl/cm² of surface area. After 45 min the plates are washed once with HBSS with Ca^{2+} and Mg^{2+} and used immediately to plate cells. Four hours after plating at a density of 50–100 cells/ml, nonadherent cells are washed off and endothelial cell medium added. The latter consists of RPMI 1640 medium containing 20% bovine plasma-derived serum, 50 μg/ml heparin sodium, and 300 μg/ml endothelial cell growth supplement (Biomedical Technologies, Stoughton, MA), penicillin (5 U/ml), and streptomycin (5 μg/ml). After 96 hr the growth medium is changed to RPMI 1640 without heparin and with iron-supplemented bovine calf serum (15%) in place of plasma-derived serum. For cloning, this medium is diluted 1:1 with 3T3

---

[20] M. Taub and G. Sato, *J. Cell. Phys.* **105**, 369 (1980).

cell-conditioned medium, consisting of DMEM with 10% iron-supplemented calf serum from 3T3 cells grown for 24 hr in log phase growth.

At this point either cloning or fluorescence-activated cell sorting is used to obtain pure endothelial cell cultures (Fig. 3b). The cloning technique utilizes transfer of cells following growth on fibronectin-coated microspheres laid down in wells containing viable colonies of endothelial cells. Microspheres (Lux Scientific Instruments, New York, NY) are coated with fibronectin by suspension in 10 vol of HBSS with Ca^{2+} and Mg^{2+} containing 5 µg/ml fibronectin for a minimum of 2 hr. Just before use, the microspheres are washed in HBSS with Ca^{2+} and Mg^{2+} and centrifuged at 500 g for 3 min. Sufficient coated microspheres are added to cover the bottom of a well of a 96-well culture dish. These wells should contain only one colony of endothelial cells and are utilized 10 days after plating of the initial cell isolation. After 3–4 days, when cells can be seen growing on the microspheres, they are transferred to 1 well of a gelatin-coated 24-well plate. After growth to near confluency, cells are trypsinized and transferred to gelatin-coated 60-mm dishes using RPMI 1640 supplemented with 15% iron-supplemented bovine calf serum and 50 µg/ml endothelial cell growth factor (Biomedical Technologies, Stoughton, MA). Cells are propagated thereafter at a 5:1 dilution every 5 days. Cells are identified by the presence of factor VIII antigen, and angiotensin I-converting enzyme activity. Alternatively, fluorescence-activated cell sorting using indocarbocyanine-conjugated acetylated low-density lipoprotein (LDL) is used to obtain cell lines. Endothelial cells are used for study up to passage 15. Loss of expression of angiotensin-converting enzyme activity is seen after more prolonged culture. Mesangial cell contamination can be recognized by light microscopy and the fact that mesangial cells (if present) overgrow and replace the endothelial cells within two or three passages. In summary, this detailed method allows for isolation and culture of endothelial cells by selective growth and cloning (Fig. 3b). The cloning efficiency averaged 48 ± 18%.

This method described by Ballermann provides the most detailed method available for obtaining glomerular endothelial cells. It is more complex than culturing mesangial cells, and not as widely applied. However, future studies may increase the use of this approach.

Isolation and Culture of Glomerular Epithelial Cells

A number of procedures have been employed to obtain homogeneous cultures of glomerular epithelial cells. All of the procedures require subsequent cloning to obtain homogeneity. In one technique, isolated glomeruli are collagenase digested, the undigested glomeruliar cores are removed by centrifugation, and the cells recovered in the supernatant are plated into Waymouth's medium containing 20% FCS.[21] In another technique, isolated glomeruli are plated for explant growth in RPMI 1640 containing 20% FCS diluted in half with sterile-filtered conditioned medium obtained from Swiss 3T3 cells in log phase growth. Insulin (0.66 IU/ml) and antibiotics/antimycotics are added.[22] The Swiss 3T3 cells are grown in DMEM with 10% FCS. After 10 days of growth, polygonal cells are seen growing out from the glomeruli which can be trypsinized and subcultured at clonal density.

An improved method has been described where 75% of the colonies remaining after low-density plating could be successfully established in long-term cultures.[23] Isolated glomeruli are treated with 0.2% trypsin for 20 min followed by a 40-min incubation with 0.1% collagenase (189 U/ml; Worthington Diagnostic Systems, Inc., Freehold, NJ) in HBSS with Ca^{2+} and Mg^{2+} buffered with HEPES, pH 7.4. Enzyme-treated glomeruli are resuspended in K1-3T3 medium consisting of one part Taub and Sato K-1 defined medium (see above, Isolation and Culture of Glomerular Endothelial Cells) containing 5% Nuserum and one part of sterile-filtered conditioned medium from Swiss 3T3 cells grown for 24 hr in DMEM plus 10% fetal calf serum. After 1 week, cells are trypsinized and plated at a concentration of 10^2 cells/ml onto collagen-coated plates in the K1-3T3 medium. Collagen coating is done by mixing collagen (Flow Laboratories, Inc., McLean, VA) 8:1:1 with 10× RPMI 1640 medium and 0.1 N NaOH. Three to four milliliters of this solution is added to a 100-mm culture dish and allowed to gel at 37° for 60 min. Colonies appear at about 7 days, and these can be retrieved by cutting around each colony with a hypodermic needle. The colony is placed in 1 ml of 0.2% collagenase until all collagen is digested. The cell pellet is resuspended in K1-3T3 medium, the cells pelleted for 5 min at 100 g, resuspended in 1 ml of K1-3T3 medium, and plated into 1 well of a 24-well cluster dish coated with collagen gel. Once confluent, cells can be transferred to larger dishes, always using collagen-coated plastic and K1-3T3 medium. Glomerular epithelial cells can be identified by their polygonal shape and cobblestone appearance of the monolayer when confluence is reached (Fig. 3c), by the ability to form domelike structures, which are morphological indicators for vectorial transport of salt and water, and also by intermediate filaments containing cytokeratin. A cytotoxic response to the aminonucleoside puromycin can

[21] G. E. Striker, P. Killen, and F. M. Farin, *Transplant Proc.* **12**, 88 (1980).
[22] J. I. Kreisberg, R. L. Hoover, and M. J. Karnovsky, *Kidney Int.* **14**, 21 (1978).
[23] P. A. Harper, J. M. Robinson, R. L. Hoover, T. C. Wright, and M. J. Karnovsky, *Kidney Int.* **26**, 875 (1984).

be observed in accordance with preferential injury of glomerular epithelial cells and not other glomerular cells *in vivo*.[24] Also, one can rule out contamination of cultures with other glomerular cell types by staining with specific markers (e.g., staining for factor VIII antigen to rule out the presence of contaminating endothelial cells).

[24] P. M. Andrews, *Lab Invest.* **36**, 183 (1977).

[10] Isolation and Culture of Juxtaglomerular and Renomedullary Interstitial Cells

By E. ERIC MUIRHEAD, WILTON A. RIGHTSEL, JAMES A. PITCOCK, and TADASHI INAGAMI

Two monolayer cell cultures having unique biologic attributes can be derived from the kidney by special techniques. The juxtaglomerular cells (JGCs) are derived from the renal cortex. The renomedullary interstitial cells (RICs) are derived from the renal medulla, especially its inner portion or renal papilla. Both cell lines exert local as well as systemic effects. The derivation of these cells and certain of their attributes are the subject of this chapter.

Juxtaglomerular Cell Line

Isolation and Culture

Juxtaglomerular cells were isolated from the kidneys of neonatal rats belonging to an inbred rat strain (Wistar/GM). The rats, about 24 hr old, were anesthetized with ether and the kidneys were excised. Care was taken to have the rats belong to one sex. The capsule, surrounding connective tissue, and fat were removed, and the cortices were dissected from each kidney, pooled, and minced with iris scissors. The cells were dissociated by treating the minced tissue with 100 ml of 0.25% Difco trypsin (Detroit, NJ) for 2 hr at 37°. Aliquots of the dispersed cells were removed at 30-min intervals, and were replaced with an equal volume of fresh 0.25% trypsin. Each aliquot was centrifuged for 3 min at 275 g, and the packed cells were held at 4° after the addition of 5 ml of the following tissue culture medium: Basal medium Eagle's (BME) containing 0.35% lactalbumin hydrolysate, 0.05% yeast extract, and 16% unfiltered fetal bovine serum at a final pH of 7.3. The cell suspension aliquots were pooled and recentrifuged. The packed cells were resuspended in tissue culture medium. Approximately

2×10^6 cells were inoculated into 75-cm^2 Wheaton tissue culture flasks (Scientific Products, Stone Mountain, GA) containing culture medium. The flasks were incubated at 37° in a stationary position until complete monolayers developed. Subcultures of the cortical cells were made by treating the culture with 0.25% trypsin at 37° for 30 min in order to remove the cell monolayer. The trypsinized cell suspension was centrifuged at 275 g for 3 min. The packed cells were washed twice with tissue culture medium. The final cell suspension was transferred to 136-cm^2 tissue culture flasks. The cells were subcultured at a ratio of 3:1 for three serial passages.

A cell suspension prepared from the fourth transfer was cloned by the limiting dilution method. Dilutions were made in order to obtain suspensions that theoretically contained 10 cells/ml and 1 cell/ml. By this limiting dilution technique, 1-ml aliquots of dilutions assumed to contain 10 cells/ml and 1 cell/ml were inoculated into replicate tissue culture tubes which were incubated at 37° in a stationary position. It has been assumed that each colony is a clone if the growth of cells could be observed in less than 20% of the tubes inoculated with a suspension calculated to contain 1 cell/ml, according to the Poisson distribution. Moreover, those tubes containing initial growth from one cell in less than 20% of the tubes were confirmed by daily microscopic observations. Thus, when colonies of cells developed in only the single area from a single cell in less than 20% of the tubes, the culture was transferred as a clone and assumed to be derived from a single cell. The clones were grown and transferred in tissue culture medium for at least 60 passages. Juxtaglomerular cells in early transfers were stored in the frozen state. A cell suspension containing 6×10^6 cells/ml was prepared in tissue culture medium to which was added 10% glycerol. Aliquots of 1 ml were placed in glass ampoules, sealed, and quick frozen in liquid nitrogen, using a Linde Bf-5 biological freezer. The ampoules were stored in a liquid nitrogen refrigerator. To initiate a new culture from the frozen cells, the ampoules were rapidly thawed at 37° and the entire contents inoculated into a 75-cm^2 Wheaton flask containing tissue culture medium.

Appearance

The cultured JGCs were examined by light and electron microscopy (LM and EM), using for the former a plastic cover slip preparation and for the latter a button of centrifuged cells.[1]

The JGCs were also examined by fluorescent microscopy. The indirect

---

[1] E. E. Muirhead, G. Germain, B. E. Leach, J. A. Pitcock, P. Stephenson, W. L. Brooks, E. G. Daniels, and J. W. Hinman, *Circ. Res.* **31** (Suppl. 2), 161 (1972).

fluorescent antibody technique was used for the purpose of demonstrating renin in the cytoplasmic granules. Two preparations were used, slides on which the cultured JGCs grew and frozen sections of the subcutaneous (sc) transplants of JGCs. The JGCs were grown in an eight-chamber tissue culture slide tray with lids obtained from Lab-Tek Products (Naperville, IL). Each chamber was inoculated with 5×10^5 cells. These were incubated at 37° in standard medium in an atmosphere of room air and 5% CO_2 for 2 days. The removed chamber cover slips were rinsed with phosphate-buffered saline (PBS).

The tissue culture slides and frozen section slides were processed in a similar manner by fixation in 95% methanol at −20° for 30 min, then air dried. These slides were stored frozen at −76°. To each slide of the fixed tissue culture cells or to the frozen section of the transplant was added a 1:5 or a 1:10 dilution of anti-rat renin antiserum made in the rabbit. This antiserum was shown to inhibit completely renin but not cathepsin D activity. It satisfied multiple criteria for purity.[2] The antibody gave a single precipitin band by double immunodiffusion. It did not cross-react with human renin, and reacted exclusively with JGCs when used for immunohistochemical localization of renin-containing cells in rat kidney section.[3]

The light microscopy of thick sections (0.6 μm) stained with toluidine blue emphasized the presence of many granules in the cytoplasm. The EM section at lower power indicated the presence of two types of granules, a dense granule and a pale granule. This mixture of granules has been described in JGCs within the kidney.[4]

Higher magnification of the cultured JGCs indicated the presence of features identical to those of JGCs *in situ* in the kidney (Fig. 1). These include secretory-type granules, lysosomal-type granules, rough endoplasmic reticulum containing a fluffy precipitate indicative of protein synthesis,[5] a prominent Golgi apparatus, peripheral dense bodies, and myofibrils indicative of a smooth muscle origin. Moreover, some granules were acid phosphatase positive, supporting their lysosomal nature. By this stain most granules were negative but displayed a finely stippled make-up and membrane binding. Some JGCs displayed crystalloid structures within the granules as described for JGCs *in situ* and in JGCs of JGC tumors. These findings are similar to those described by multiple authors for JGCs in the kidney.[6-11]

---

[2] T. Motaba, K. Murakami, and T. Inagami, *Biochim. Biophys. Acta* **526**, 560 (1978).
[3] K. Naruse, Y. Takii, and T. Inagami, *Proc. Natl. Acad. Sci. U.S.A.* **78**, (1981).
[4] L. Barajas and H. Latta, *Lab. Invest.* **12**, 1046 (1963).
[5] D. W. Fawcett, "An Atlas of Fine Structure: The Cell, Its Organelles and Inclusions." Saunders, Philadelphia, Pennsylvania, 1966.
[6] J. C. Lee, S. Hurley, and J. Hopper, Jr., *Lab. Invest.* **15**, 1459 (1966).
[7] E. R. Fisher, *Science* **152**, 1752 (1966).

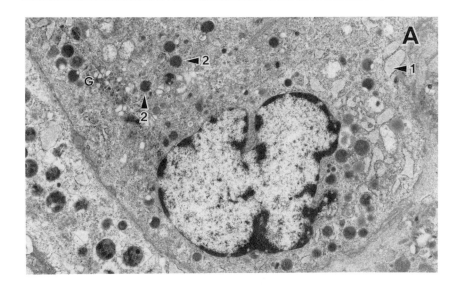

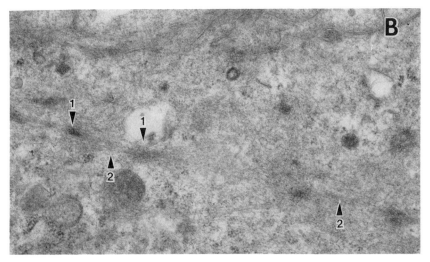

FIG. 1. Cultured juxtaglomerular cell. (A) Electron microscopy (×3570) emphasizes three features: Rough endoplasmic reticulum containing fine precipitate (arrowhead 1), granules (arrowhead 2), and Golgi apparatus (G). (B) Peripheral dense bodies are indicated by arrowheads 1 and myofibrils by arrowheads 2 (×3570). (Reproduced by permission from Muirhead et al.[19])

[8] L. Barajas and H. Latta, *Lab. Invest.* **12,** 257 (1963).
[9] L. Barajas and H. Latta, *Lab. Invest.* **12,** 1046 (1963).
[10] C. G. Biava and M. West, *Am. J. Pathol.* **49,** 679 (1966).
[11] P. M. Hartroft, *Anat. Rec.* **124,** 458 (1956) (abstr.).

Chromosome analysis yields diploid cells with a modal of about 42 chromosomes, as expected.

Functions

Renin–Angiotensin System. In order to determine the presence of various components of the renin–angiotensin system (RAS) within the cultured cells, the cells were washed and a cell extract was derived. First, JGCs were cultured in T75 Falcon flasks (Scientific Products, Stone Mountain, GA) in basal medium Eagle's containing 0.25% lactalgumin hydrolysate, 0.5% yeast extract, and 10% fetal calf serum with the pH maintained at 7.2–7.4 by means of 4 mmol/liter sodium bicarbonate and 25 mmol/liter N-2-hydroxyethylpiperazine-N'-2-ethanesulfonic acid (HEPES) buffer. At confluence (7–10 days) the medium was changed to either serum-free basal medium Eagle's or fresh serum-containing medium 24–28 hr before harvesting to determine the effect of the serum. Cells were detached from the flask with 0.25% trypsin and 1 mmol/liter ethylenediaminetetraacetic acid (EDTA), washed once in serum-containing medium to inhibit the trypsin, then twice with serum-free medium, and then suspended in water containing a mixture of 1 mmol/liter diisopropylfluorophosphate (DFP), 1 mmol/liter captopril (Squibb, Princeton, NJ), 5 μg/ml leupeptin (Protein Research Foundation, Osaka, Japan), and 5 mmol/liter EDTA. The cells were then lysed by five cycles of freeze–thawing and then treated with 0.1% Triton X-100. The cell extract used for angiotensin determination was prepared as above in the presence of 1 mol/liter pepstatin (Protein Research Foundation). For the assay of angiotensin I-converting enzyme, cells were disrupted in a similar manner but without the addition of EDTA and with captopril instead of the detergent Triton X-100. Nonidet P-40 was used for lysing the cells and releasing the enzyme.

The assay for renin activity was performed by taking aliquots of the cell extract and incubating these for 6 hr at 37° in plasma from 24-hr bilaterally nephrectomized rats as renin substrate in 0.2 mmol/liter sodium maleate buffer, pH 6.0, containing 5 mmol phenylmethanesulfonyl fluoride, 10 mmol/liter EDTA, and 0.1% neomycin. The reaction was stopped by heating in boiling water for 10 min. The angiotensin generated was determined by radioimmunoassay.[12] As control for nonspecific renin-like protease activity,[13,14] parallel experiments were run with cell extract preincubated with the specific antiserum to renin at 1:104 dilution for 16 hr at

---

[12] E. Haber, T. Koerner, L. Page, B. Klinman, and A. Purnode, *J. Clin. Endocrinol. Metab.* **29,** 1349 (1969).

[13] R. P. Day and I. A. Reid, *Endocrinology (Baltimore)* **99,** 93 (1976).

[14] E. Hackenthal, R. Hackenthal, and V. Hilgenfeld, *Biochim. Biophys. Acta* **522,** 561 (1978).

4°. At this dilution this antiserum inhibited completely 10 μg of renin but not cathepsin D of rat renal extract. The pH dependence of renin activity was determined in 0.2 mmol/liter buffers: sodium acetate (pH 3.5–5.0), sodium maleate (pH 5.0–6.5), and sodium phosphate (pH 6.5–7.5).

A modification of the De Pierre and Roth[15] method was used to determine the angiotensin I-converting enzyme activity. Solutions were incubated with the synthetic substrate hippuryl-His-Leu for 30 min at 37° in 0.4 ml of 50 mmol/liter phosphate/50 mmol/liter borate buffer, pH 8.0, containing 0.2 mmol/liter NaCl. The degree of hydrolysis of the hippuryl–histidyl peptide bond was measured by fluorometric assay using o-phthalaldehyde. Inhibition of the reaction by captopril (12 mmol/liter), EDTA (2.3 mmol/liter), and by omission of NaCl was demonstrated.

In determining the amount of angiotensin present, aliquots of cell extract were treated with 3 vol of acetone and centrifuged at 6000 rpm for 10 min. The supernatant was removed and dried under a nitrogen stream at room temperature, redissolved in 0.1 mol/liter Tris-acetate buffer, pH 7.4, heated in boiling water, and subjected to the radioimmunoassay for angiotensin I[12] and angiotensin II.[16] The angiotensin II antiserum had less than 1% cross-reactivity with angiotensin I but 100% cross-reactivity with the heptapeptide, hexapeptide, and pentapeptide (from Protein Research Foundation). More than 90% recovery of 1 ng of angiotensin I and angiotensin II/III was derived when these peptides were added to the extract and treated in the same way. Identity of angiotensins in the extract was confirmed by chromatography on cellulose plates (Analtech, Newark, NJ) by modification of the method of Semple et al.,[17] using sec-butanol–ammonia, 5:1 (v/v). After developing the plate, areas containing the angiotensins were identified by standards on a marker plate, eluted into methanol–ammonia, 2:1 (v/v), dried under a nitrogen stream, redissolved in Tris-acetate buffer, pH 7.4, heated in boiling water, and subjected to radioimmunoassay.

The fluorescent technique demonstrated the presence of granules yielding apple-green fluorescence when treated with the fluorescein isothiocyanate (FITC)-conjugated antirenin antibody from the rabbit. At least 50% of the cells fluoresced. Controls with normal rabbit serum gave no fluorescence. These results support the presence of renin in the cells.

Renin activity, neutralizable by specific renin antibody, angiotensin I and II/III, and converting enzyme was demonstrated in the JGC extract. Cells grown in serum-containing medium revealed a considerable reduc-

[15] D. De Pierre and M. Roth, *Enzyme* **19,** 65 (1975).
[16] D. J. Gocke, L. M. Sherwood, I. Oppenhoff, J. Gerten, and J. H. Laragh, *J. Clin. Endocrinol. Metab.* **28,** 1675 (1968).
[17] P. F. Semple, W. A. Macrae, and J. J. Morton, *Clin Sci.* **59,** 61S (1980).

tion in renin and angiotensin levels when washed with serum-free medium. Cell harvested 24–48 hr after the withdrawal of the serum had negligible amounts of renin that could be readily removed by repeated washing with serum-free medium. Activity not removed by repeated washing was considered to be produced by the cells. Withdrawal of serum from the culture medium for 48 hr caused a three- to fourfold and two- to threefold increase in renin and angiotensin II/III levels in the cells, respectively.

Much of the total renin-like activity, as indicated by generation of angiotensin I from rat angiotensin substrate, was inhibited by specific renin antibody. The antibody-sensitive portion was considered to be renin. The remainder was considered renin-like activity due to proteases.[13,14] The pH profile of the sum of the antibody-sensitive and antibody-insensitive activities revealed an optimum pH in the acidic area (pH < 4.0), but the activity inhibitable by the antibody had a pH optimum between 6.0 and 6.5, in agreement with rat renin derived from the kidney.[2]

Renin activity in the medium was 98.7 ± 1.3% inhibited by the antibody. This activity also shows a neutral pH optimum. These observations indicated that the renin-like activity found in the cells was specific renin.

The hippuryl-His-Leu hydrolysis was almost completely inhibited by the addition of EDTA or captopril, and by the removal of chloride. The findings support the presence of converting enzyme. The converting enzyme activity was not altered by the presence of fetal calf serum.

The substances detected by radioimmunoassay as angiotensin I and II/III had the same chromatographic mobility as synthetic angiotensin I and II/III.

The findings indicate the presence of renin, angiotensin I, converting enzyme, and angiotensin II in the cultured JGCs. They imply the presence of renin substrate. In other words, the cultured JGCs appear to contain the entire renin–angiotensin system.

Hypertension Induced by JGCs. Transplantation of cultured JGCs subcutaneously (45–100 million viable cells) into syngeneic recipients followed by reduction of the renal mass causes hypertension (Fig. 2).[18] The reduction of the renal mass can be accomplished by either uninephrectomy (50% reduction) or by removal of one kidney and the poles of the opposite kidney (~70% reduction).[19] The ultimate elevation of the blood pressure (BP) is comparable by these two means of reducing the renal mass but the

[18] E. E. Muirhead, W. A. Rightsel, J. A. Pitcock, T. Inagami, T. Okamura, Y. Takii, T. L. Goodfriend, J. E. Sealey, B. Brooks, and P. Brown, *Trans. Assoc. Am. Physicians* **95**, 110 (1982).

[19] E. E. Muirhead, W. A. Rightsel, J. A. Pitcock, and T. Inagami, *in* "Kidney Hormones" (J. W. Fisher, ed.), p. 247. Academic Press, Orlando, Florida, 1986.

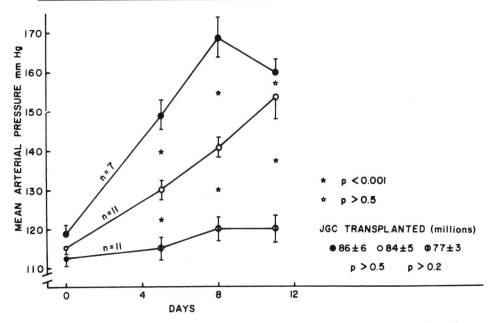

FIG. 2. Hypertension induced by transplant of cultured juxtaglomerular cells. Relation of renal mass to hypertension is depicted. Intact renal mass (◐) was attended by virtually no change of BP. Hypertension occurred when renal mass was reduced by 50% (○) or 70% (●), more rapidly by the latter. The number of transplanted juxtaglomerular cells is indicated in millions. (Reproduced by permission from Muirhead et al.[19])

rate of elevation is more rapid following the greater decrement of renal mass.

The necessity for renal mass reduction in order for the BP to elevate remains an enigma. It could be related to rate of metabolism of angiotensin II (see below), as renal tissue is very avid toward destruction of this molecule.[20]

This hypertensive state has two phases, namely, (1) a developmental phase lasting 2 to 3 weeks, during which time the plasma angiotensin II level is elevated (as high as 200–400 pg/ml), and (2) a maintenance phase beyond 3 weeks, during which time the plasma angiotensin II level is reduced to baseline (<30 pg/ml) (Fig. 3). Plasma renin concentration was within the normal range throughout, indicating secretion of angiotensin II by the transplanted JGCs. The dependence on the plasma angiotensin II level of the developmental phase is supported by the depression of the BP during this time by the angiotensin II competitive inhibitor saralasin when

[20] S. Oparil and M. D. Bailie, *Circ. Res.* **33**, 500 (1973).

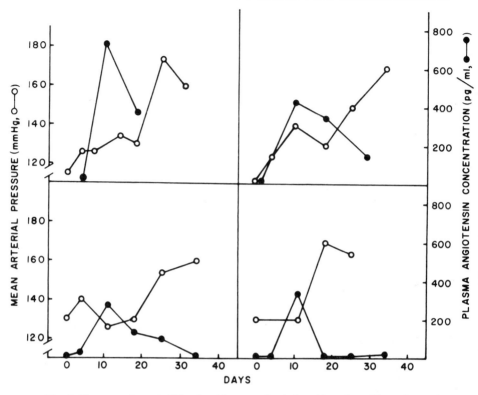

FIG. 3. Plasma angiotensin II level and hypertension induced by cultured juxtaglomerular cells. Plasma angiotensin II concentration elevated sharply over about 2 weeks (developmental phase), then receded to baseline (maintenance phase) as the hypertension evolved and remained. (Reproduced by permission from Muirhead et al.[19])

infused iv and by the prevention of the hypertension by the converting enzyme inhibitor captopril when started at the time of the JGC transplantation. During the maintenance phase saralasin has no effect on the BP, as expected (Fig. 4). The hypertension is also dependent on the transplant. The BP begins its rise at about 3 days posttransplant or at the time of vascularization of this structure. Removal of the transplant up to 12–14 days lowers the BP to baseline. Removal of the transplant after 12–14 days has no effect on the evolution of the hypertension (Fig. 5). These results indicate that the developmental phase is transplant and angiotensin II dependent while the maintenance phase is dependent on some other mechanism(s). In this respect, this hypertensive state resembles the one-kidney, one-clip Goldblatt-type hypertension of the rat.

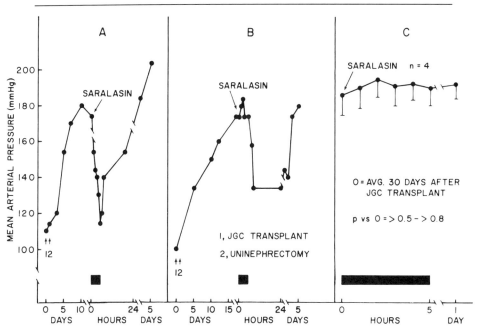

FIG. 4. Effect of saralasin, competitive inhibitor of angiotensin II, on the hypertension induced by juxtaglomerular cells. Saralasin lowered the BP during the developmental phase of hypertension (first 10–15 days) but had no effect during the maintenance phase (after 30 days). (Reproduced by permission from Muirhead et al.[19])

The JGC-induced hypertension is associated with degeneration of the RICs of the remaining kidney or renal nubbin. These cells become smaller, rounded, and have few lipid granules (Fig. 6). In view of the fact that the RICs exert an antihypertensive function (see the section, Function, below), one wonders if a deficiency of this function contributes to the pathogenesis of the hypertension (Fig. 7). Under circumstances wherein the plasma angiotensin II level is very high, malignant hypertension develops as indicated by the severe vascular disease (Fig. 8). Fibrinoid necrosis and musculomucoid intimal hyperplasia (onion skin lesion) of the visceral small arteries and arterioles are encountered. These are the morphologic expressions of malignant hypertension.

The hypertension induced by cultured JGCs constitutes a unique model of experimental hypertension. It offers possibilities in attempts to dissect out the role of the renin–angiotensin system in initiating and maintaining the hypertensive state.

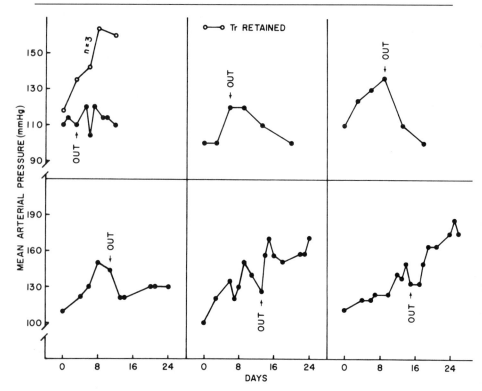

FIG. 5. Removal of transplant (Tr) of cultured juxtaglomerular cells versus the hypertension induced by these cells. Transplant removal at 3 days prevented hypertension. Transplant removal up to 10 days lowered the BP. Transplant removal by 12–14 days had no effect on the BP. (Reproduced by permission from Muirhead et al.[19])

Renomedullary Interstitial Cells

Isolation and Culture

Transplants of fragmented renal medulla were used originally to prevent the hypertensive state.[21] Subsequently, it was shown that these transplants not only prevent hypertension but also lower the BP of animals having an established hypertensive state.[22] These transplants, in time, consist mostly of RICs and a vasculature.[23,24] In addition, there is a mini-

[21] E. E. Muirhead, J. A. Stirman, and F. Jones, *J. Clin. Invest.* **39**, 266 (1960).
[22] E. E. Muirhead, G. B. Brown, G. S. Germain, and B. E. Leach, *J. Lab. Clin. Med.* **76**, 641 (1970).
[23] E. E. Muirhead, B. Brooks, J. A. Pitcock, and P. Stephenson, *J. Clin. Invest.* **51**, 181 (1972).
[24] E. E. Muirhead, B. Brooks, J. A. Pitcock, P. Stephenson, and W. L. Brosius, *Lab. Invest.* **27**, 192 (1972).

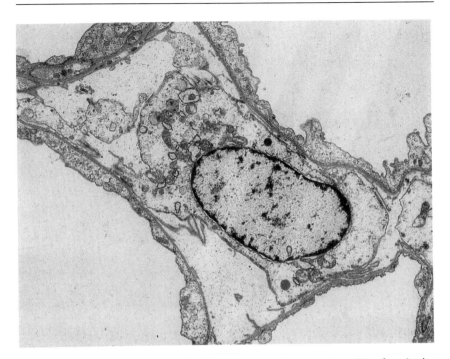

FIG. 6. Renomedullary interstitial cell (RIC) in remaining renal papilla of rat having hypertension induced by JGCs. The RIC has no cytoplasmic processes, few lipid granules, and clear cytoplasm (signs of degeneration). (Reproduced by permission from Muirhead et al.[19])

mum of connective tissue, collapsed basement membrane of resorbed tubules, a few macrophages, lymphocytes, and granulocytes. The RICs form clusters, suggesting hyperplasia of these elements. The concentration of RICs within these transplants makes them the source of the RIC monolayer cell culture.[25]

Either autotransplants or syngeneic transplants may be used as the source of RICs. Whereas originally transplants shown to exert an antihypertensive action were used, subsequently it became apparent that autotransplants in normal, nonhypertensive recipients could also serve as a source of these cells.[26–28]

[25] E. E. Muirhead, G. Germain, B. E. Leach, J. A. Pitcock, P. Stephenson, B. Brooks, W. L. Brosius, E. G. Daniels, and J. W. Hinman, *Circ. Res.* **31**, (Suppl. 2), 161 (1972).
[26] M. J. Dunn, R. S. Staley, and M. Harrison, *Prostaglandins* **12**, 37 (1976).
[27] R. M. Zusman and H. R. Keiser, *J. Biol. Chem.* **252**, 2069 (1977).
[28] F. Russo-Marie, M. Paing, and D. Duval, *J. Biol. Chem.* **254**, 8498 (1979).

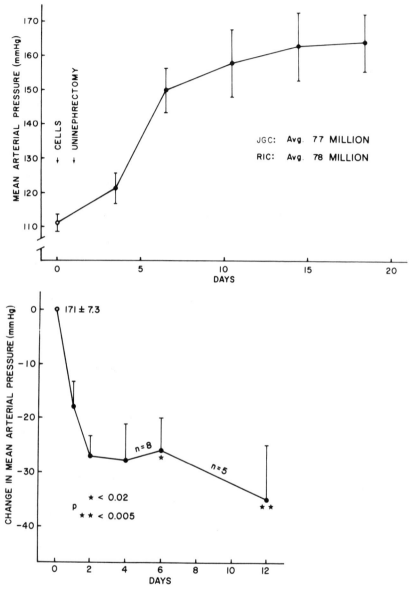

FIG. 7. Transplant of cultured RICs during the developmental and maintenance phases of hypertensive induced by JGCs. During the developmental phase of hypertension transplants of RICs had no effect on the JGC hypertension (upper panel). During the maintenance phase transplants of RICs lowered the BP significantly (lower panel). (Reproduced by permission from Muirhead et al.[19])

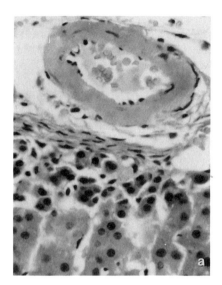 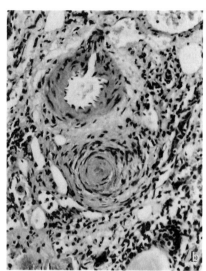

FIG. 8. Vascular disease in JGC-induced hypertension. *Left:* Atypical fibrinoid necrosis of artery in capsule of adrenal gland is shown. *Right:* Musculomucoid hyperplasia of arteriole (onion skin lesion) in the kidney is depicted. (Reproduced by permission from Muirhead *et al.*[19])

Transplants of fragmented total renal medulla (outer and inner portions) and transplant of the renal papilla (inner medullary portion) may be used. This results from the fact that the RICs are concentrated in the renal papilla, forming a ladder-like arrangement between the vasa recta, Henle's loop, and the collecting ducts.

Two features of the renomedullary transplantation are essential. (1) The tissue must be fragmented into fine particles 1 mm or less across, and (2) the transplantation must disperse the particles so as not to form a bolus of tissue. The first feature can be accomplished by mincing the tissue with iris scissors or a straight razor blade. A small Waring blender has also been used, activating it for about 5 sec to prevent homogenization. The second feature is accomplished by either scattering the particles intraperitoneally or by injecting them subcutaneously or intramuscularly as fine tracts. A long needle is inserted to the hilt and withdrawn slowly as a fine tract of tissue is described. Bolus formation causes necrosis in the center of the masses, which in turn induces inflammation that may destroy the transplant. Naturally, sterile precautions are essential throughout all of these procedures.

At the sites of transplantation elongated nodular masses form. These can be palpated subcutaneously. The nodules can be shelled out at operation.

Nodules of renomedullary transplants are minced with iris scissors and the fragments placed in a 75-cm^2 Corning disposable plastic tissue culture flask (Scientific Products, Stone Mountain, GA) containing 30 ml of modified Eagle's medium. Formulation of the medium is as follows: 500 ml of 10× basal medium Eagle's (BME) in Hanks' balanced salt solution; 250 ml of 20× stock lactalbumin hydrolysate (LAH)/yeast extract (YE) containing 50 g LAH and 10 g YE/liter; 150 ml of 5× stock sodium bicarbonate/sodium hydroxide containing 7.00 g sodium bicarbonate and 4.40 g sodium hydroxide/liter; and 4100 ml of sterile deionized water. All stock solutions are prepared under sterile conditions and stored at 4° until used. Growth medium consists of modified Eagle's medium to which filtered fetal bovine serum is added. Final serum concentration is 16%. Antibiotics are not routinely used. However, 100 U/ml penicillin and 100 μg/ml streptomycin may be used during initial isolation of renomedullary interstitial cells. Final pH of growth medium is 7.2–7.4. Culture medium is carefully removed and replaced with fresh medium every 5 days until cells evolve into a confluent sheet.

Once cells have formed a confluent monolayer, transfer to new bottles is accomplished by trypsinizing the cell sheet for 20 min at 37° using 0.25% Enzar-T in Hanks' balanced salt solution. The Enzar-T solution is brought to pH 7.6 with sterile stock sodium bicarbonate/sodium hydroxide solution. The trypsin-treated cell suspension is pelleted by centrifugation. The pellet is then used to inoculate two new bottles. After three to five transfers, 150-cm^2 Corning disposable plastic tissue culture flasks are used. Cell pellets are then used to inoculate three 150-cm^2 flasks. Cultures are transferred every 5 to 7 days.

Appearance

The original organ culture derived from the renomedullary transplant contains cells having all of the characteristics of the RICs as observed *in situ* within the renal papilla by light and electron microscopy. These are the cells that grow into sheets. At first the cells have cytoplasmic processes. Later they tend to round up.

The characteristic appearance includes lipid granules. These are intensely osmiophilic and stain with Oil Red O and Sudan Black B. Other organelles include perinuclear and cytoplasmic cisterns of the rough endoplasmic reticulum (ER) type, smooth ER, the Golgi apparatus, scattered mitochondria, and glycogen bodies. Normal karyotype for the rabbit and the rat, main species used, can be demonstrated.

Function

Cultured RICs, when transplanted into syngeneic hypertensive recipients, can be shown to exert a powerful antihypertensive action.[29] The BP drops to normal levels within hours to a few days and remains depressed for a period of days. When these transplants are removed the BP returns to its prior hypertensive level. Eventually the transplants disappear. This may require about 4 weeks.

These transplants consist mostly of RICs and a capillary vasculature. The cells exert a positive tropism toward the capillaries. Reculture of these transplants is possible.

[29] E. E. Muirhead, G. S. Germain, F. B. Armstrong, B. Brooks, B. E. Leach, L. W. Byers, J. A. Pitcock, and P. Brown, *Kidney Int.* **8,** s271 (1975).

[11] Comparative Kidney Tubule Sources, Isolation, Perfusion, and Function

By KLAUS W. BEYENBACH and WILLIAM H. DANTZLER

Introduction

"... the solution to a physiological or pathological problem often depends solely on the appropriate choice of the animal for the experiment so as to make the results clear and searching."

Claude Bernard[1] came to this conclusion 150 years ago, a conclusion which is still appropriate today. The history of physiology is full of examples where unique or seemingly bizarre features in the anatomy and physiology of some animals have been studied in order to gain insights of universal importance in biology. The history of renal physiology is no exception, drawing on nature for providing the most convincing tests of a hypothesis. Thus, it was only after work on the glomeruli of frogs, rats, and rabbits that Cushny, Starling, and Richards confirmed Ludwig's idea about ultrafiltration in Malpighian corpuscles. Likewise, it was only with the help

[1] C. Bernard, "An Introduction to the Study of Experimental Medicine" (H. C. Greene, transl.). Dover, New York, 1957.

of aglomerular fish that Marshall proved tubular secretion to be a fact. In modern times, the kidneys of the common rat, the desert rat, the dog, and the beaver have played important roles in our understanding of the mechanisms of urinary concentration in the renal medulla. Frog skins have been pivotal in providing a conceptual basis of renal salt transport; toad urinary bladders have been instrumental in our understanding of the regulation of water permeabilities; and gall bladders have been central in our appreciation of leaky epithelia.

After Burg developed the techniques for *in vitro* microperfusion of renal tubules, the need for model tissues representing the renal tubules of the kidney has subsided. However, the value of the comparative approach to physiological problems, as stated above by Bernard, has persisted. To this date, the renal tubules of different animals contribute importantly to the understanding of renal physiology in general and the details of tubular transport and its regulation in particular. Hence, a treatment of comparative kidney tubule sources, isolation, perfusion, and function must (1) emphasize the unique physiological systems of animals and (2) review the experimental techniques for studying them.

Accordingly we will summarize the unique renal tubular functions so far as they are known in an invertebrate and in nonmammalian vertebrates: in an insect, and in fish, amphibians, reptiles, and birds, in that order. As we move up the evolutionary tree we will point out at each stop the special renal functions from the perspective of comparative physiology and the experimental procedures to learn from these.

We have approached this task by covering the experimental work and methods with which we are most familiar. Accordingly, Dr. Dantzler reviews the work done on reptilian and avian tubules while focusing on the mechanisms of organic solute transport that have been studied with isotopic and biochemical methods. Dr. Beyenbach discusses the work done on the excretory tubules of insects, elasmobranchs, fish, and amphibians while focusing on the mechanisms of electrolyte transport that have been studied with electrophysiological methods. With this division we hope to present a survey of nonmammalian renal tubules and a balanced view of the experimental methods used.

Insect Malpighian Tubules

Unique Tubular Functions

The "kidney" of insects consists of a small number of Malpighian tubules. The blind end of these tubules are suspended by tracheal tubes in the hemolymph, the respiratory organ in insects. The open end of each

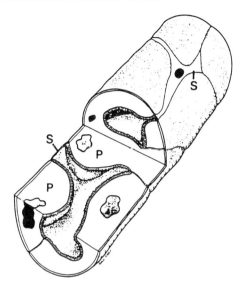

Fig. 1. Diagram of the upper portion of the Malpighian tubule in the mosquito *Aedes taeniorhynchus*. P, Primary cells; S, stellate cells. [From T. J. Bradley, A. M. Stuart, and P. Satir, *Tissue Cell* 14, 759 (1982).]

tubule empties its contents into the gut at the junction of the midgut and the hindgut.

The blind end of the tubule secretes fluid. The rate of fluid secretion and the composition of secreted fluid depend upon the nature of the challenges to the extracellular fluid compartment (hemolymph). After secretion at the blind end of the tubule, the secreted fluid may be modified on its passage through the remaining Malpighian tubule, the hindgut, and the rectum. Hence, renal excretion in insects entails both secretory and reabsorptive epithelial transport. Malpighian tubules have two types of epithelial cells, the prominent principal cells and the stellate cells (Fig. 1). It is believed that most epithelial transport functions are carried out by the principal cells.

Whereas glomerular filtration in the vertebrate kidney indiscriminately delivers an ultrafiltrate of the plasma to the tubule lumen, urine formation begins in insects with a highly specific and regulated epithelial transport function. The tubules secrete a K^+-rich fluid when the insect has ingested a K^+ load; the tubules secrete a Na^+-rich fluid when the insect has ingested a Na^+ load. The epithelial transport systems involved have been studied in detail, primarily in England but also in other countries.[2]

[2] V. B. Wigglesworth, "The Principles of Insect Physiology." Methuen, London, 1965.

Insect Sources

In the United States insects can be obtained from a number of sources. Carolina Biological Supply Company (Burlington, NC) and Wards Biology (Rochester, NY) are commercial suppliers of a variety of insects. Most any university department of entomology is a ready supplier. In the case of our studies on the yellow fever mosquito, *Aedes aegypti,* we originally acquired the yellow fever mosquito from the National Institutes of Health, and since then we have reared them ourselves as described by Shapiro and Hagedorn.[3]

Isolation of Malpighian Tubules

The mosquito is cold anesthetized at 5°. The rectum together with the gastrointestinal tract is removed under Ringer's solution by blunt dissection using fine forceps and a dissection microscope ($\times 10 - \times 40$). Malpighian tubules are then removed by freeing them from their attachments to the pylorus of the midgut and from the trachea.

Methods for Studying Malpighian Tubules in Vitro

The Ramsay Method. The Ramsay method[4] is the most widely practiced experimental technique for studying the functions of Malpighian tubules *in vitro*. The isolated Malpighian tubule is placed into a known volume (50–100 μl) of Ringer under paraffin oil (Fig. 2). The open end of the tubule is then pulled from the Ringer droplet into the paraffin oil using fine glass hooks fabricated with a microforge. In our laboratory we leave approximately 2 mm of the blind-ended tubule in the Ringer droplet. As the blind segment of the tubule secretes fluid, the hydrostatic pressure in the tubule lumen drives the fluid out of the open end where secreted fluid accumulates in the oil as a droplet separate from the incubating Ringer (Fig. 2).

The volume of this droplet (V) can be measured directly with nanoliter volumetric pipets or indirectly by measuring (in micrometers) the long (l) and short (s) axes of the droplet and treating the droplet as a prolate spheroid [Eq. (1)].

$$V = (\Pi s^2 l)/[6(10^6 \; \mu m^3/nl)] \tag{1}$$

Timed volume measurements yield secretion rates. Figure 3 illustrates the results of a typical Ramsay experiment. Cumulative volume of secreted fluid is plotted as a function of time. During the first hour of the experi-

[3] J. P. Shapiro and H. H. Hagedorn, *Gen. Comp. Endocrinol.* **46,** 176 (1982).
[4] J. A. Ramsay, *J. Exp. Biol.* **31,** 104 (1954).

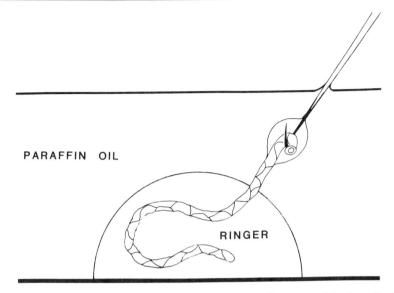

FIG. 2. The Ramsay method for the measurement of secretion rates and for the collection of secreted fluid. The blind end of the Malpighian tubule is incubated in insect Ringer's solution (20 μl), and the open end of the tubule pulled into the oil bath where fluid secreted by the tubule can accumulate as a droplet separate from the Ringer.

ment (control) the tubule secretes fluid at a constant rate of 0.9 nl/min (slope of regression line). Thereafter, cAMP is added to the incubating Ringer to yield a 1 mM nucleotide concentration. The addition of cAMP significantly stimulates fluid secretion to 3.1 nl/min over the next hour. The advantage of this experimental protocol is the use of each tubule as its own control which makes the data suitable for paired t tests.

Storing small fluid volumes: The analysis of secreted fluid yields information about the types of fluid secreted under control and experimental conditions. In the case of *Aedes aegypti* the analyses are done on nanoliter and subnanoliter fluid volumes because of the small volumes of fluid secreted. These small fluid volumes must be kept under oil, as evaporation alters volume and the concentrations of dissolved solutes. Samples may also evaporate under oil, especially at high temperatures or in oil which has not been water equilibrated. For best results, the analyses should be done as soon as possible. If samples must be stored, they should be stored under water-equilibrated oil at subzero temperatures. For more detail see Bonventre et al.[5]

[5] J. V. Bonventre, K. Blouch, and C. Lechene, in "X-Ray Microscopy in Biology" (M. A. Mazat, ed.), p. 307. Univ. Park Press, Baltimore, Maryland, 1981.

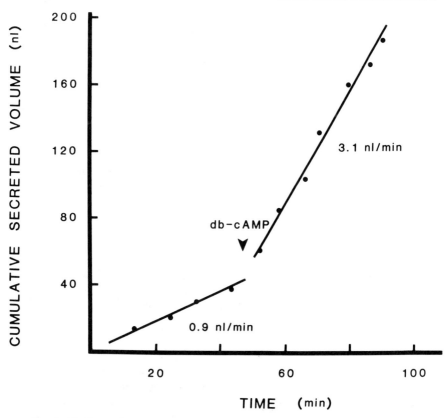

FIG. 3. Fluid secretion rates measured with the Ramsay method. Cumulative volume of secreted fluid is plotted against time. The slopes are fluid secretion rates under control conditions and in the presence of dibutyryl-cAMP (1 mM). [From J. C. Williams and K. W. Beyenbach, *J. Comp. Physiol.* **149**, 511 (1983).]

Osmotic pressure measurements: Osmotic pressures are measured by the freezing point depression against known standards using the Clifton nanoliter osmometer. In our laboratory we use pure NaCl osmotic standards ranging from 150 to 390 mOsm in increments of 30 mOsm. The use of the same volume (0.3 nl) of standards and unknowns improves the precision and accuracy of the measurements. In general, the fluid secreted by Malpighian tubules is isosmotic to the peritubular medium[6] whereas the

---

[6] J. E. Phillips, *Am. J. Physiol.* **241**, R241 (1981).

fluid excreted from the rectum may be isosmotic or hypoosmotic to the hemolymph.[7]

Electron probe analysis of secreted fluid: Bonventre et al.,[5] Quinton,[8] and Roinel[9] give detailed accounts of the elemental analysis of biological fluids by X-ray spectroscopy. In our laboratory we use a modified version of the method of Roinel.[9] Briefly, samples of secreted fluid (~ 100 pl) are deposited under light paraffin oil on the polished surface of a beryllium block (9 × 8 × 7 mm). After droplets of (1) known Na^+, K^+, and Cl^- concentrations in Ringer (standards), (2) peritubular Ringer, and (3) unknown secreted fluid samples have been deposited, the oil is washed off by immersing the block in spectrograde chloroform. The block is then washed twice again, each time in fresh spectrograde chloroform. The procedure also dehydrates the samples, leaving the solutes in each sample behind in 80- to 100-μm spots of uneven appearance and large crystal size. A light puff of breath on the cooled block briefly wets the surface again, hydrating and dehydrating the samples to yield spots of uniform appearance and small crystal size. The blocks are then stored in a desiccator under vacuum until examination under the electron probe.

We use a JEOL superprobe 733 (Japanese Electron Optical Laboratory, Peabody, MA) with KeVex SiLi crystals. The instrument can be used to scan the energy spectrum of the X rays emitted from the spots (energy-dispersive spectroscopy, EDS) for the qualitative analysis of the kinds of elements present in the samples, or to count the X rays emitted at a particular wavelength (wavelength-dispersive spectroscopy, WDS) for the quantitative analysis of the Na^+, K^+, or Cl^- concentrations in the samples. All X-ray counts are corrected automatically for beam current (50 nA at 15 keV accelerating voltage) and background noise with the aid of a Trachor–Northern automation system (TN-2000, Middleton, WI) and a DEC PDP-11 computer.

Table I shows the Na^+, K^+, and Cl^- concentrations in the hemolymph and in fluid secreted by isolated Malpighian tubules of female yellow fever mosquitoes. The electron probe analysis shows that (1) insect hemolymph is not unlike vertebrate plasma in the concentrations of Na^+, K^+, and Cl^-, (2) the fluid secreted spontaneously by isolated Malpighian tubules contains nearly equimolar quantities of Na^+ and K^+, and (3) cAMP, which stimulates the rate of fluid secretion (Fig. 3), does so by stimulating the secretion of NaCl. As will be shown below, cAMP and the mosquito natriuretic factor (MNF, an M_r 2000 peptide) stimulate the principal cells

---

[7] J. C. Williams, Jr., H. H. Hagedorn, and K. W. Beyenbach, *J. Comp. Physiol.* **153**, 257 (1983).
[8] P. M. Quinton, *Micron* **9**, 57 (1978).
[9] N. Roinel, *J. Microsc. (Paris)* **22**, 261 (1975).

TABLE I
CONCENTRATION OF Na^+, K^+, AND Cl^- IN *Aedes* HEMOLYMPH AND FLUID SECRETED BY ISOLATED MALPIGHIAN TUBULES UNDER CONTROL CONDITIONS AND STIMULATION WITH cAMP (RAMSAY METHOD)[a]

	Mosquito hemolymph	Secreted fluid	
		Control	cAMP
Na^+	96 ± 7 (10)	94 ± 5 (18)	178 ± 7 (7)[b]
K^+	6.5 ± 1.0 (10)	91 ± 8 (18)	17 ± 1 (7)[b]
Cl^-	135 ± 19 (10)	161 ± 5 (16)	185 ± 4 (5)

[a] Numbers in parentheses indicate the number of experiments. [From J. C. Williams, Jr., and K. W. Beyenbach, *J. Comp. Physiol.* **149**, 511 (1983).]
[b] Significantly different from control, paired t test, $p < 0.05$.

of Malpighian tubules to secrete an NaCl-rich fluid. In the intact insect the initial diuresis which commences during feeding on blood is an NaCl diuresis.[7] We therefore believe that MNF is released at this time to eliminate the unwanted serum fraction of the blood meal. In isolated Malpighian tubules the specific increase of NaCl secretion in nearly isosmotic proportion with water leads to the dilution and consequently the low concentration of K^+ in secreted fluid; Net K^+ secretion rates are not affected by cAMP whereas net Na^+ secretion rates are increased six- to sevenfold (Tables I and II; Fig. 3).

Analysis of organic solutes in secreted fluid: The collection and analysis of secreted fluid in the Ramsay assay also provide the opportunity to examine the mechanism and regulation of organic solute transport. However, this area of research has not been widely explored, presumably for lack of biochemical micromethods that are necessary to measure pico- and femtomolar quantities of organic solutes.[2] Radioimmunoassay (RIA) techniques have considerable promise in the study of organic solute secretion by Malpighian tubules. For example, Rafaeli *et al.*[10] have used RIA methods to detect and measure cAMP concentrations in secreted fluid. Moreover, the cAMP RIA has found application as a bioassay in the search for diuretic factors.

In Vitro Microperfusion of Malpighian Tubules. In general, the Burg method of *in vitro* microperfusion of epithelial tubules (see [12], this volume) allows Ussing-type experiments to be conducted on short seg-

[10] A. Rafaeli, M. Pines, P. S. Stern, and S. W. Applebaum, *Gen. Comp. Endocrinol.* **54**, 35 (1984).

TABLE II
Stimulation of Na^+ Secretion in Isolated Malpighian Tubules of the Yellow Fever Mosquito[a]

Measurement	Control	Dibutyryl-cAMP
Electrophysiological data from isolated perfused Malpighian tubules (method of Burg)[b]		
Transepithelial voltage (mV)	30.1 ± 7.3 (7)	64.1 ± 13.3 (7)[d]
Transepithelial resistance (kΩ·cm)	14.9 ± 2.6 (7)	9.9 ± 1.9 (7)[d]
Basolateral membrane voltage (mV)	−77.1 ± 6.3 (7)	−23.8 ± 2.0 (7)[d]
Apical membrane voltage (mV)	−107.2 ± 7.2 (7)	−87.9 ± 14.8 (7)[d]
Fractional resistance, basolateral membrane	0.63 ± 0.11 (7)	0.33 ± 0.08 (7)[d]
Data from secreting Malpighian tubules (method of Ramsay)[c]		
Fluid secretion rate (nl/min)	0.65 ± 0.05 (18)	2.17 ± 0.17 (18)[d]
Na^+ secretion rate (pmol/min)	69 ± 7 (18)	454 ± 43 (7)[d]
K^+ secretion rate (pmol/min)	68 ± 8 (18)	43 ± 6 (7)
Cl^- secretion rate (pmol/min)	118 ± 10 (16)	451 ± 50 (5)[d]

[a] Numbers in parentheses indicate the number of experiments.
[b] From D. H. Petzel, M. M. Berg, and K. W. Beyenbach, *Am. J. Physiol.* **253**, R701, 1987.
[c] From J. C. Williams, Jr., and K. W. Beyenbach, *J. Comp. Physiol.* **149**, 511 (1983).
[d] $p < 0.001$.

ments of small tubular epithelia such as renal tubules, sweat ducts, salivary ducts, and Malpighian tubules. Prior to Burg's advance the functions of these tubular epithelia were inferred from so-called model tissues, namely epithelial sheets like the gall bladder, urinary bladder, or frog skin which could be mounted in Ussing chambers[11] for measurements of transepithelial solute and water flows.

As in Ussing chamber experiments the advantage of the Burg method lies in the experimental control of the solutions bathing the peritubular (basolateral membrane) and lumenal (apical membrane) side of the tubule. This is important in studies of transepithelial transport, namely, in defining the transepithelial and transmembrane electrochemical potentials of the transported solutes.

Because of the small biomass of renal tubules measurements of transepithelial or transmembrane fluxes are difficult, time consuming, tedious, and not always possible (for lack of a suitable isotope). For this reason electrophysiological methods of studying transepithelial transport have

---

[11] H. H. Ussing and E. E. Windhager, *Acta Physiol. Scand.* **61**, 484 (1964).

been fruitful. Provided that transepithelial transport is at least partly electrogenic and/or conductive, the electrophysiological approach has two advantages: (1) on-line readout of the data: the results of experimental treatments may be observed with almost instantaneous time resolution; (2) high resolution: currents as low as 10^{-12} A can be measured accurately, which is several orders of magnitudes lower than currents measured via isotope fluxes.

In our laboratory we have successfully used the *in vitro* microperfusion method together with electrophysiological methods to (1) elucidate the epithelial transport mechanism of Na^+ secretion in Malpighian tubules, and (2) isolate the mosquito natriuretic factor (probably a hormone) using the transepithelial voltage of the isolated perfused tubule as a rapid, preliminary bioassay for diuretic activity.

Electrophysiological methods for investigating Na^+ secretion: Secretory segments of Malpighian tubules (0.5–1 mm long) are isolated and perfused *in vitro* from right to left as shown in Fig. 4. The tubule is cannulated with a double-barreled perfusion pipet. Through one barrel of the pipet, the perfusion channel, current ($I_{T,0}$; 500–1000 nA) is injected into the tubule lumen for measurements of the transepithelial resistance (R_T), and

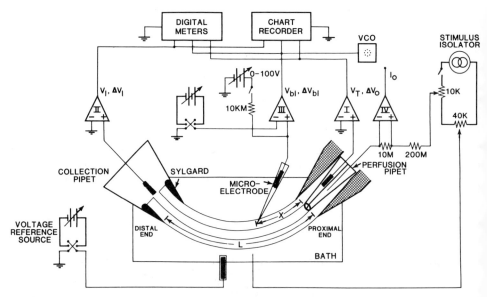

FIG. 4. Diagram of isolated renal tubules prepared for the measurement of transepithelial and transmembrane electrical properties. VCO, voltage control oscillator. [From K. W. Beyenbach and E. Frömter, *Am. J. Physiol.* **248**, F282 (1985).]

the fractional resistances of the basolateral (fR_{bl}) and apical (fR_a) membranes. Through the other barrel the transepithelial voltage ($V_{T,0}$) is measured. At distance x from the mouth of the perfusion pipet, a principal cell of the tubule is punctured with conventional microelectrode which measures the basolateral membrane potential (V_{bl}). The distal end of the perfused segment is aspirated into a holding pipet and sealed electrically from the peritubular bath with the uncured elastomer (Sylgard 124, Corning, NY). The transepithelial voltage at the distal end of the tubule ($V_{T,1}$) is measured via an electrode immersed in the fluid exiting from the tubule lumen. This voltage measurement is essential in the cable analysis of the transepithelial and membrane fractional resistances. Accordingly, the transepithelial resistance is

$$R_T = [(\Delta V_{T,0}\lambda)/(\Delta I_{T,0})] \tanh (L/\lambda) \tag{2}$$

where λ is the length constant of the tubule and L is the length of the perfused tubule segment exposed to the bath. Since the transepithelial voltage is the sum of the voltages across the basolateral and apical membrane, the voltage deflections at the site of the microelectrode impalement are

$$\Delta V_{T,x} = \Delta V_{bl,x} + \Delta V_{a,x} \tag{3}$$

From this the fractional resistances of the basolateral and apical membranes can be derived as the voltage divider ratio,

$$\Delta V_{bl,x}/\Delta V_{T,x} = R_{bl}/(R_a + R_{bl}) = fR_{bl} \tag{4}$$

$$\Delta V_{a,x}/\Delta V_{T,x} = R_a/(R_a + R_{bl}) = fR_a \tag{5}$$

The voltage deflection across the tubule wall at the site of the microelectrode impalement ($\Delta V_{T,x}$) is defined as

$$\Delta V_{T,x} = \Delta V_{T,0} \frac{\cosh [(L-x)/\lambda]}{\cosh (L/\lambda)} \tag{6}$$

In the above equations, the Malpighian tubule is treated as a single cable of constant cross-sectional area and uniform electrical property. For more detail the interested reader is referred to the chapter by Greger (see [16], this volume) and the treatments by Helman.[12]

Using the *in vitro* microperfusion techniques of Burg together with transepithelial and transmembrane electrophysiological measurements in Malpighian tubules, we have obtained the results summarized in Table II. Under control conditions the transepithelial voltage is lumen positive; the transepithelial resistance is moderate. The basolateral membrane with a

[12] S. I. Helman, *Yale J. Biol. Med.* **45**, 339 (1972).

voltage of 77 mV has a higher resistance than the apical membrane. In the presence of cAMP, the transepithelial voltage hyperpolarizes consistent with the stimulation of active transport of a cation, namely Na^+. At the same time, the transepithelial resistance and the resistance of the basolateral membrane fall significantly, indicating a major effect of cAMP on the basolateral membrane. Since these electrical effects of cAMP are dependent upon the presence of Na^+ in the bath,[13,14] the data indicate that cAMP increases the basolateral membrane Na^+ conductance as part of the mechanism for stimulating transepithelial Na^+ secretion.[14]

Voltage response of the isolated perfused Malpighian tubule as a rapid bioassay for diuretic hormones: In previous searches for diuretic hormones the Ramsay method has been used as the key bioassay.[15] In brief, several Malpighian tubules are isolated and prepared for *in vitro* secretion studies as shown in Fig. 2. Some tubules (control) are incubated in plain Ringer solutions; others are incubated in Ringer containing suspected diuretic factors or hormones. Alternatively, a single Malpighian tubule is used initially as its own control, and the bathing Ringer is then changed to include the suspected secretagogue. We found this experimental approach to be very time consuming, requiring between 1 and 2 hr to evaluate the diuretic potency of a single test substance. Moreover, the use of the Ramsay method as a bioassay becomes very labor intensive when a large number of assays must be run, as for instance in attempts to isolate and purify diuretic molecules from the insect. A typical high-pressure liquid chromatography (HPLC) column run generates over 100 fractions, most of which (or all of which) do not contain diuretic activity. However, one would spend weeks to evaluate the diuretic potency of these unknown HPLC fractions. Hence the need for a rapid bioassay for diuretic potency was immediately apparent when our laboratory first expressed an interest in searching for the diuretic hormone in *Aedes aegypti*.

We found the transepithelial voltage of the *in vitro* perfused Malpighian tubule to be a reliable first indicator of diuretic potency. We reasoned that agents which stimulate fluid secretion must affect the transepithelial voltage of the tubule since changes of fluid secretion are mediated via changes in electrolyte transport and its associated changes in tubule electrophysiology. For example, cAMP stimulates fluid secretion by stimulating NaCl secretion with its associated hyperpolarization of the transepithelial voltage (Table II). The advantage of the transepithelial voltage assay lies in the

---

[13] J. C. Williams and K. W. Beyenbach, *J. Comp. Physiol.* **154**, 301 (1984).

[14] D. B. Sawyer and K. W. Beyenbach, *Am. J. Physiol.* **248**, R339 (1985).

[15] S. H. P. Maddrell, *in* "Neurohormonal Techniques in Insects" (T. A. Miller, ed.), p. 81, Springer-Verlag, New York, 1980.

large number of samples that can be screened for potential diuretic activity in a short period of time. For example, as described below, we use only one Malpighian tubule to test the diuretic potency of more than 100 HPLC fractions in a matter of 2 hr.

The tubule is isolated and perfused *in vitro* with Ringer at rates between 2 and 5 nl/min. Higher perfusion rates damage the epithelium. The tubule is also superfused with Ringer, i.e., the bath (~ 500-μl volume) is perfused at a rate between 1 and 2 ml/min. Bath inflow is by gravity flow; outflow is by suction. The tubule is connected to measure transepithelial voltage only. Test substances (suspected secretagogues, tissue extracts, HPLC fractions) are injected into the bath inflow line, one at a time. If the substance has diuretic potency, it can be expected to affect the transepithelial voltage. Accordingly the transepithelial voltage will change as the substance enters the bath making contact with the tubule, and the transepithelial voltage will return to control values as the substance is washed out of the bath (continuous bath flow). Figure 5 illustrates the results of such a bioassay. From a total of 120 HPLC fractions of an aqueous extract of 12,000 mosquito heads, 3 groups of HPLC fractions affect the transepithelial voltage. Factor I elutes in fractions 44 and 45, factor II elutes in fractions 48, 49, and 50, and factor III elutes in fraction 58.[16]

Two caveats are in order: (1) Changes in electrolyte and fluid transport must not always be accompanied by electrophysiological changes. Some epithelial transport systems are "electrically silent"; hence changes in solute and water transport may not always be paralleled by changes in epithelial voltage or resistance. However, in every Malpighian tubule examined to date, transepithelial transport of electrolytes is electrogenic, giving rise to transepithelial voltages. (2) Not every voltage response of the tubule must indicate a diuretic effect of the test substance. The transepithelial voltage of an epithelium is variably influenced by a host of variables, some related, others unrelated to epithelial transport. For this reason the voltage assay for diuretic activity described above is a preliminary assay only. Once biologically active substances have been identified in the voltage assay, it is necessary to consult the Ramsay assay for the true test of diuretic potency.

Following this two-step assay, first the effects on transepithelial voltage, and then the effects of the voltage-activity substances on fluid secretion, we learn that factor I does not have diuretic activity, and that factors II and III have diuretic activity (Fig. 5).

In conjunction with (1) more detailed electrophysiological studies described above, (2) electron-probe analysis of secreted fluid,[16] and (3) RIA measurements of intracellular cAMP concentrations in Malpighian tu-

---

[16] D. H. Petzel, H. H. Hagedorn, and K. W. Beyenbach, *Am. J. Physiol.* **249**, R379 (1985).

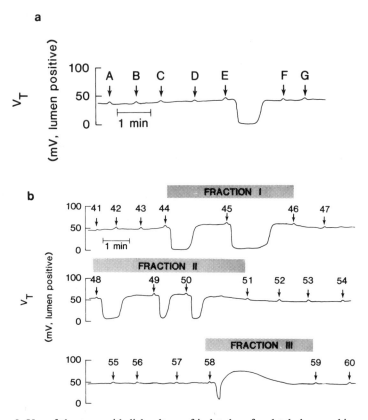

FIG. 5. Use of the transepithelial voltage of isolated perfused tubules as a bioassay for diuretic hormones. (a) The voltage response of a single Malpighian tubule to HPLC pools containing between 20 and 30 HPLC fractions each. Only pool E has an effect on transepithelial voltage. (b) The effect on transepithelial voltage of each HPLC fraction of pool E is shown. Three fractions, each isolated by inactive fractions, are identified. Fractions II and III are peptides (M_r 1800–2400) which stimulate fluid secretion. Because of their selective stimulation of NaCl secretion via increased active transport of Na^+ we have called fractions II and III the mosquito natriuretic factors (MNF). [From D. H. Petzel, H. H. Hagedorn, and K. W. Beyenbach, *Am. J. Physiol.* **249,** R379 (1985).]

bules,[17] we find that factor III stimulates NaCl and fluid secretion by the same mechanism cAMP stimulates NaCl and fluid secretion (Table II). Factor III also increases intracellular cAMP activity in Malpighian tubule epithelial cells.[17] On the basis of this evidence we have called factor III the

---

[17] D. H. Petzel, M. M. Berg, and K. W. Beyenbach, *Am. J. Physiol.* **253,** R701 (1987).

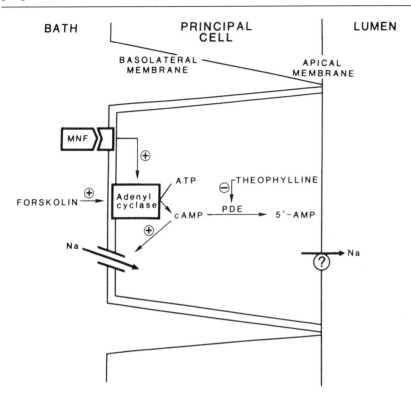

FIG. 6. Mechanism of action of mosquito natriuretic factor (MNF) in primary cells of mosquito Malpighian tubules. The peptides lead to an increase of intracellular cAMP concentration, which increases the Na^+ conductance of the basolateral membrane. These effects favor the entry of Na^+ into the cell. The mechanism of Na^+ extrusion across the apical membrane is largely unknown. [From D. H. Petzel, M. M. Berg, and K. W. Beyenbach, *Am. J. Physiol.* **253**, R701 (1987).]

mosquito natriuretic factor (MNF). Like cAMP it selectively stimulates NaCl and fluid secretion by increasing the Na^+ conductance of the basolateral membrane (Fig. 6). It is a peptide of $M_r \sim 2000$; it is most likely the natriuretic, diuretic hormone in the mosquito, but its hormone status needs to be confirmed in the intact insect.[18]

Patch-Clamp Studies of Malpighian Tubule Ion Channels. With the advent of patch-clamp techniques, the opportunity is given to study epithelial transport at the level of the basolateral and apical membranes of epithelial cells. Before the basolateral membrane can be approached by the

[18] D. H. Petzel, H. H. Hagedorn, and K. W. Beyenbach, *Am. J. Physiol.* **250**, R328 (1986).

patch electrode, the basement membrane of the epithelial cells must be removed. This may be accomplished by incubating isolated Malpighian tubules in Ringer solution containing elastase.[19] However, we find that elastase treatment, which dissociates the epithelial cells from one another, also causes cell membranes to form blebs (unpublished observations). How these blebs affect the physiology of the cell and its membranes is unknown.

With some practice it is possible to split a segment of a Malpighian tubule down its long axis. This is done by holding one end of the tubule with two fine forceps and by gently tearing along the long axis of the tubule. The tubule tears along the borders of the principal cells. "Unzipped" segments usually consist of two to five principal cells, serially attached by their lateral borders. These segments curl backward, prominently exposing the brush-bordered apical membrane surface. In a similar way, single principal cells may be isolated. Single cells or split tubule segments may then be draped across a droplet of uncured sylgard. In this way the apical membrane of principal cells is displayed upward and is accessible to patch electrodes. We have found no difficulties in forming gigaohm seals with apical membrane patches. Among the many current activities in these membrane patches we have identified a Cl^- channel with moderate channel conductance and slow channel kinetics (unpublished observations).

Tubular Epithelia of Elasmobranchs

Unique Tubular Functions

Marine sharks are slightly hyperosmotic to sea water by virtue of the high concentrations of urea and tetramethyl ammonium oxide (TMAO) in their extracellular fluid. Extracellular NaCl concentrations are significantly less than sea-water NaCl concentrations. Consequently marine sharks are subjected to the dual threats of (1) osmotic water loading, like fresh-water fish, and (2) NaCl loading, like marine fish.

The renal tubules of elasmobranchs are unique in at least two ways. First, recent studies have shown a highly organized anatomical arrangement that includes counter-current flow systems.[20] This is unusual since the kidneys of other fish do not show an orderly anatomical organization. The significance of this is unknown. Second, the tubules must conserve urea and TMAO if high plasma concentrations of these compounds are to be maintained.[21] How the elasmobranch kidney accomplishes this is unknown.

[19] G. Levinson and T. J. Bradley, *Tissue Cell* **16**, 367 (1984).
[20] H. Hentschel and M. Elger, *Verb. Otsch. Ges.* **1982**, 263 (1982).
[21] B. Schmidt-Nielsen and L. Babinowitz, *Science* **146**, 1587 (1964).

Next to the renal tubules, the tubules of the rectal gland contribute importantly to the elimination of environmental salt and water loads. The tubules secrete an isosmotic fluid which contains NaCl at approximately twice the plasma concentration.

Sources of Marine Elasmobranchs

Sharks and rays are the most studied species of elasmobranch. They are best studied at marine stations, because they do not do well in captivity. Even at marine stations they do not survive captivity for much longer than 2 weeks. The Mount Desert Island Biological Laboratory in Salisbury Cove, Maine, is perhaps best known for work on the mechanisms of osmotic and ionic regulation in sharks. It is at this laboratory where Shannon first undertook renal clearance studies in shark,[22] and where others since then have worked out some of the details of epithelial transport in the rectal gland and the kidney.

Isolation and Perfusion of Rectal Gland Tubules

The procedure for the dogfish shark, *Squalus acanthias,* is described. After extirpation of the rectal gland it is dacapsulated by blunt dissection. Thin slices (<0.3 mm thick) are cut perpendicular to the long axis of the gland and placed in shark Ringer solution. Tubules are isolated by freehand dissection by tearing the slice apart, beginning at the periphery and progressing toward the central lumen of the gland. Rectal gland tubules start as blind tubules at the periphery and soon merge with other tubules so that only unbranched pieces of tubules are short (0.3–1.3 mm in length). The tubules are teased away at their junctions, opening the tubule lumen at this end. The blind end of the tubule is then frayed open with dissection needles. The dissection is carried out at 10–12°.

Rectal gland tubules have been perfused *in vitro* since 1983. The first perfusion revealed lumen-negative transepithelial voltages (reflecting Cl⁻ secretion) that are stimulated by cAMP, vasoactive intestinal peptide (VIP), and adenosine, and inhibited by furosemide.[23] Using *in vitro* microperfusion techniques of tubules together with conventional and ion-selective microelectrodes Greger and colleagues[24-26] have worked out the details and the stoichiometry of NaCl secretion in isolated rectal gland tubules.

[22] J. A. Shannon, *J. Cell. Comp. Physiol.* **5**, 301 (1934).
[23] J. N. Forrest, Jr., F. Wang, and K. W. Beyenbach, *J. Clin. Invest.* **72**, 1163 (1983).
[24] R. Greger and E. Schlatter, *Pfluegers Arch.* **402**, 63 (1984).
[25] R. Greger and E. Schlatter, *Pfluegers Arch.* **402**, 364 (1984).
[26] R. Greger, E. Schlatter, F. Wang, and J. N. Forrest, Jr., *Pfluegers Arch.* **402**, 376 (1984).

Isolation and Perfusion of Renal Tubules

To this date *in vitro* microperfusions of shark renal tubules have been conducted in our laboratory only.[27,28] In particular, we have studied the mechanism of fluid secretion in proximal tubules isolated from the caudal portion of the shark kidney. The isolated kidney is stored in Ringer solution on ice. Thin slices (1-3 mm) of the caudal kidney are cut perpendicular to the long axis of the kidney and transferred to dissection dishes filled with shark Ringer solution (~10°). Proximal tubules are dissected freehand with fine dissection needles and forceps (Dumoxel #5, A. Dumont and Fils, Switzerland) while viewing the tissue under magnification ($\times 40 - \times 100$). In our studies we have isolated the large, brush-bordered proximal tubule, i.e., the proximal segment II according to the nomenclature of Hentschel and Elger,[20] or tubule segment V according to the nomenclature of Ghouse et al.[29] The tubules have a large expanded tubule lumen (due to fluid secretion) with an average inner diameter of 47 μm. The outer diameter is approximately 117 μm. The lumenal border has a brush border as well as eyelash-like cilia. Isolated segments of the proximal tubule II can be several millimeters long. However, we found a total length of 1-1.5 mm to be most suitable for the electrophysiological studies described below.

Electrophysiological Evidence for Cl^- Secretion in Shark Proximal Tubules

Isolated shark proximal tubules are prepared for *in vitro* microperfusion and cable analysis as shown in Fig. 4. In addition, an epithelial cell of the tubule is impaled with a conventional microelectrode for the measurement of membrane voltages and fractional resistances. Lumenal Cl^- substitutions (isosmotic replacement with gluconate) are made in the absence and presence of cAMP. The results of a typical experiment (Fig. 7) show that cAMP significantly increases the apical membrane Cl^- conductance. This is documented by (1) the depolarization of the intracellular voltage ($V_{bl,x}$), (2) the lumen-negative transepithelial voltage ($V_{T,0}$), and (3) the decrease of the transepithelial resistance (R_T) and the fractional resistance of the apical membrane (f_{R_a}) when cAMP is first added to the bath. In the presence of cAMP a large Cl^- diffusion potential is also observed across the apical membrane when the lumenal Cl^- concentration is reduced 10-fold. This apical membrane response is not observed in the absence of cAMP (Fig. 7). These data are internally consistent with cAMP increasing the

---

[27] K. W. Beyenbach and E. Frömter, *Am. J. Physiol.* **248**, F282 (1985).
[28] D. B. Sawyer and K. W. Beyenbach, *Am. J. Physiol.* **249**, F884 (1985).
[29] A. M. Ghouse, B. Parsa, J. W. Boylan, and J. C. Brennan, *Bull. Mt. Desert Isl. Biol. Lab.* **8**, 22 (1968).

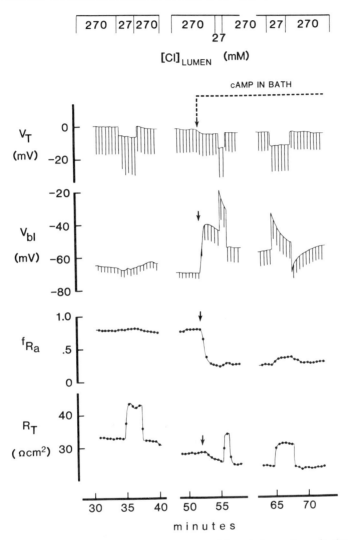

FIG. 7. Cyclic AMP increases the Cl⁻ conductance of the apical membrane in shark renal proximal tubules. The tubule is studied by the methods depicted in Fig. 4. Under control conditions (perfusion with symmetrical Ringer solution) the fractional resistance of the apical membrane f_{R_a} is high and the apical membrane does not respond to 10-fold changes of the lumenal Cl⁻ concentration (no change in intracellular voltage, $V_{bl,x}$). In the presence of cAMP the fractional resistance of the apical membrane falls and the apical membrane now responds to lumenal Cl⁻ substitutions. [From K. W. Beyenbach and E. Frömter, *Am. J. Physiol.* **248**, F282 (1985).]

apical membrane Cl⁻ conductance. On the basis of (1) other electrophysiological experiments[27] and (2) electron probe analysis of secreted fluid in nonperfused tubules[28] (method described previously), we established secondary active transport of Cl⁻ as the driving force for fluid secretion. Hence, marine sharks appear to employ the same epithelial transport mechanism to secrete NaCl and H_2O from the kidney as well as the rectal gland.

Teleost Renal Tubules

Unique Tubular Functions

As the most diverse class of vertebrates, the bony fish (teleosts) can be expected to exhibit a diverse pattern of renal functions. Indeed, renal functions range from glomerular to totally aglomerular strategies with its associated consequences on tubular functions. In aglomerular kidneys tubular transport mechanisms alone must be responsible for the renal contributions to extracellular fluid homeostasis. However, to date renal tubules of aglomerular fish have not been studied *in vitro,* and the tubular mechanism of aglomerular urine formation remains an unwritten chapter in the renal physiology of vertebrates.[30]

Not all glomeruli are filtering at all times in fish. Hence questions arise about tubular functions in the case of nonfiltering nephrons. According to our observations in sea water- and fresh water-adapted fish, the tubules may still continue to function albeit in a secretory aglomerular way. As will be shown later we have observed net NaCl and fluid secretion in nonperfused proximal tubules isolated from glomerular kidneys of sea water- and fresh water-adapted fish. The physiological role of NaCl and fluid secretion in glomerular proximal tubules is unknown.[31]

In proximal tubules of marine fish the epithelial transport system for Mg^{2+} and SO_4^{2-} secretion are particularly well expressed. Urine/plasma Mg^{2+} concentration ratios in excess of 100 stem in part from the secretory transport mechanisms in the proximal tubule. No other vertebrate tissue expresses a Mg^{2+} transport so clearly as renal proximal tubules of marine fish.[32] Similarly, transport systems serving the tubular secretion of organic solutes are prominently developed in fish proximal tubules and have been studied in a number of laboratories. Finally, euryhaline fish are able to dwell in fresh-water and sea-water environments. Their kidneys cope with environmental water loading and salt depletion when the fish is in fresh

[30] K. W. Beyenbach, *Renal Physiol.* **8,** 222 (1985).
[31] K. W. Beyenbach, *Am. J. Physiol.* **250,** R753 (1986).
[32] W. H. Cliff, D. B. Sawyer, and K. W. Beyenbach, *Am. J. Physiol.* **250,** R616 (1986).

water; their kidneys cope with salt loading and water depletion when the fish is in sea water. The tubular functions, changing from elimination to conservation in the two environments, the transport mechanisms involved, and the regulation of these functions have yet to be explored using modern experimental techniques.

Fish Sources

Marine stations, fish hatcheries, and aquaria are perhaps the best sites for studying freshly caught or well-maintained fish. Improvements in the shipping of live animals by the United States Postal Service or commercial couriers are such that the study of fish is no longer restricted to the sites of their natural occurrence. Several marine stations (most notably the Woods Hole Oceanographic Institute on the east coast) will ship a number of small atlantic fish species caught off the coast of Massachusetts. Similarly, marine stations on the west coast are suppliers of Pacific fish species. For studies of fresh-water fish, federal, state, and private fish hatcheries are ready suppliers of fish important to commerce and sports fishing. Not to be forgotten are local pet stores for tropical and other fish species.

Isolation of Renal Tubules

The kidney of fish consist of hematopoeitic, lymphoid, interrenal, chromaffin, and renal tissues. In general, renal vascular and tubular tissues become more dominant proceeding from the head kidney (opisthonephros) to the caudad kidney (metanephros). For this reason tubules from the caudad kidney are usually selected for *in vitro* studies. Still, the blood supply to the caudal kidney is so abundant (renal arterial and venous portal) that the renal tubules cannot be seen until most of the blood has been removed from the tissue. This is done in our laboratory by cutting under Ringer solution $3 \times 3 \times 3$-mm sections of the caudad kidney using a razor blade. The sections are then gently teased open without causing excessive damage to the tubules. The teased piece of tissue is then held firmly by a forcep and rapidly moved back and forth through the Ringer bath. This action removes most of the blood and the interrenal elements from the tissue, leaving a clear mesh of renal tubules. When illuminated from the side (fiber optic light source) and viewed from above against a black background, undamaged tubules are uniformly transparent. "White spots" indicate tubule damage, either crushing or tearing wounds.

Proximal and distal tubules are identified on the basis of their general geometry and appearance (presence or lack of a lumenal brush border) and connection to glomeruli or junctions to other tubules. Rigorous criteria for the identification of specific tubular segments are not available because (1) the microanatomy of renal tubules has not been described in most fish species, and (2) fish kidneys lack a clear tubular and vascular organization which could serve as reference points in the isolation of specific tubule segments. However, with little experience it is not difficult to distinguish proximal tubules from distal tubules. Once the isolated tubule segment is connected for perfusion *in vitro,* further examination of the tubule under higher magnification ($\times 100 - \times 400$) as well as physiological indicators (transepithelial voltage and resistance) may help identify the particular segment under study. Although it is still difficult to distinguish segments I, II, and III of the proximal tubules, we have no difficulties in consistently selecting for study one type of proximal tubule.[32-34] All isolation procedures are carried out at the *in vivo* ambient temperature of the fish.

In Vitro Studies of Fish Renal Tubules

The Teased Tubules Method. This method has been used to study the uptake of radiolabeled solutes by a mesh of renal tubules. Since the mesh contains all tubule segments along the nephron, the method has been applied in uptake studies where a single tubule segment, i.e., the proximal tubule, makes up the bulk of the renal parenchyma. This is true of flounder kidneys, where the proximal tubule comprises more than 80% of the entire nephron. Forster,[35] Kinter,[36] and Kleinzeller and McAvoy[37] have used the teased tubule method to study the renal cellular uptake of Phenol Red and sugars, respectively.

From thin slices of the caudad flounder kidney, $3 \times 3 \times 3$-mm pieces of tissue are cut with a razor blade. Each piece of tissue is then gently teased under Ringer solution with dissecting needles as described above. The mesh of tubules is then stored in Ringer solution on ice until use. Tissue viability is assessed by the tubular uptake of Chlorphenol Red at room temperature from saline containing dye at a concentration of 25 μM.

In the typical uptake experiment, three to six pieces of tubule mesh are incubated in a volume of appropriate saline containing the labeled substance of interest and other additions (competitive inhibitors, blockers, etc.). Incubation periods last from 30 to 90 min at the *in vivo* ambient tissue of the fish. At the end of the incubation the tubule mesh is removed and plotted with filter paper (Whatman #541 W & R Balston Ltd., England). The individual pieces (5–15 mg) are then weighed on a microbal-

---

[33] K. W. Beyenbach, D. H. Petzel, and W. H. Cliff, *Am. J. Physiol.* **250,** 608 (1986).
[34] K. W. Beyenbach, *Nature (London)* **299,** 54 (1982).
[35] R. P. Forster, *Science* **108,** 65 (1948).
[36] W. B. Kinter, *Am. J. Physiol.* **211,** 1152 (1966).
[37] A. Kleinzeller and E. M. McAvoy, *J. Gen. Physiol.* **62,** 169 (1973).

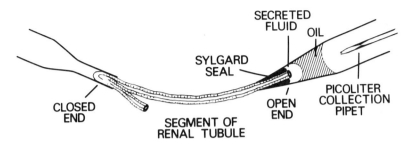

FIG. 8. Preparation of isolated renal tubule for the study of net fluid secretion. One end of the tubule is crimped closed; the other end is left open so that the rate of fluid secretion and the composition of secreted fluid can be measured. [From K. W. Beyenbach, *Nature (London)* **299**, 54 (1982).]

ance (Cahn, Paramount, CA) and extracted to measure the quantity of isotope taken up. Corrections are made for the extracellular fluid volume using inulin as extracellular volume marker. Uptake rates are normalized to tissue weight. The results are usually interpreted to reflect the uptake mechanism across the basolateral membrane of renal tubules.

The Nonperfused Single Tubule Method. There are two versions of this method. In the method introduced by Grantham et al.,[38] using mammalian proximal tubules one end of the tubule is crimped closed and the other end is cannulated. The cannulating pipet has a constant bore and is filled with Ringer solution containing a single oil droplet. In the event the tubule reabsorbs fluid, the oil meniscus moves toward the lumen; in the case of net secretion, the oil meniscus moves away from the tubule lumen. Grantham's study showed that in rabbit pars recta (straight portion of the proximal tubule) net reabsorption takes place under control conditions, but net secretion occurs when the peritubular medium contains organic acids.[39]

In the other method, practiced in our laboratory, one end of the renal tubule is crimped closed, and the other end is aspirated into a holding pipet and sealed from the bath with uncured Sylgard (Fig. 8). This method allows the direct measurement of fluid secretion rates and the collection of secreted fluid for electron probe analysis. Accordingly, we have observed fluid secretion in proximal tubules isolated from the kidneys of the shark (see above), the winter flounder,[34] and the killifish.[40]

The rates of fluid secretion in nonperfused fish proximal tubules fluctu-

---

[38] J. J. Grantham, R. C. Irwin, P. B. Qualizza, D. R. Tucker, and F. C. Whittier, *J. Clin. Invest.* **52**, 2441 (1973).
[39] J. J. Grantham, *Physiol. Rev.* **56**, 248 (1976).
[40] W. H. Cliff and K. W. Beyenbach, *Am. J. Physiol.* **254**, R154 (1988).

TABLE III
FLUID SECRETION RATES ($\dot{V}_s$) IN NON-PERFUSED RENAL PROXIMAL TUBULES[a]

	$\dot{V}_s$ [pl/(min·mm tubule length)]	
	Control	dbAMP (2 mM)
Pseudopleuonectes americanus		
Flounder proximal tubule (marine)[b]	36.6 ± 4.2 (53)	Not measured
Fundulus heteroclitus		
Killifish (marine)[c]	49.0 ± 8.7 (16)	Not measured
Killifish (fresh-water adapted)[d]	56 ± 17 (6)	115 ± 27 (6)[e]

[a] Numbers in parentheses indicate the number of experiments.
[b] From K. W. Beyenbach, D. H. Petzel, and W. H. Cliff, *Am. J. Physiol.* **250,** 608 (1986).
[c] From K. W. Beyenbach, *Am. J. Physiol.* **250,** R753 (1986).
[d] From W. H. Cliff and K. W. Beyenbach, *Am. J. Physiol.* **254,** R154 (1988).
[e] $p < 0.05$, paired t test.

ate in the picoliter range (Table III). Such low rates of transepithelial volume flow cannot be detected using the customary method of measuring transepithelial volume flows, where the tubule is perfused *in vitro* with a volume marker (see the section, Reptilian Renal Tubules). The resolution of the volume marker method has a lower detection limit at about 100 pl/(min · mm length) when the tubules are perfused at the customary rate of 5–10 nl/min. Hence transepithelial volume flows less than 100 pl/(min · mm) are not detected. This is perhaps the reason why fluid secretion has not been observed in other vertebrate renal tubules.[30,33] Hence one of the advantages of working with nonperfused renal tubules as shown in Fig. 8 is the detection of subnanoliter transepithelial volume flows and the collection of primary secreted fluid.

The compositional analysis (electron probe) of fluid secreted by fish proximal tubules reveals that the dominant secreted electrolytes are Na^+ and Cl^- (Table IV). The tubules also secrete significant quantities of Mg^{2+} and sulfur (probably SO_4^{2-}, but the major osmotic solutes are Na^+ and Cl^-. This is surprising, since it has long been thought that fluid secretion in proximal tubules of marine fish is driven solely by the secretion of divalent ions.[41] However, the direct examination of fluid secretion by the methods described above (Fig. 8) revealed NaCl secretion to be the driving force for

[41] L. J. Renfro, *Am. J. Physiol.* **238,** F92 (1980).

TABLE IV
COMPOSITION OF FLUID SECRETED *in Vitro* BY FISH PROXIMAL TUBULES[a]

(mM)

Species	Na$^+$	Cl$^-$	Mg^{2+}	S	K$^+$	P
Winter flounder (marine)[b]	152 ± 7 (22)	155 ± 4 (22)[d]	26 ± 4 (22)[d]	10 ± 1 (22)[d]	Not measured	Not measured
Killifish (marine)[b]	129 ± 5 (16)[d]	159 ± 3 (16)	29 ± 3 (16)[d]	11 ± 2 (16)[d]	Not measured	Not measured
Killifish (fresh water-adapted)[c]	138 ± 9 (16)[d]	160 ± 6 (16)	19 ± 4 (16)[d]	8 ± 3 (16)[d]	1.1 ± 0.2 (16)[d]	6.4 ± 1.3 (16)[d]

[a] Numbers in parentheses indicate the number of experiments.
[b] From K. W. Beyenbach, *Am. J. Physiol.* **250,** R753 (1986).
[c] From W. H. Cliff and K. W. Beyenbach, *Am. J. Physiol.* **254,** R154 (1988).
[d] $p < 0.05$, paired *t* test, significantly different from peritubular Ringer solution.

fluid secretion.[31,33,34] Indeed, the removal of Mg^{2+} and SO_4^{2-} from the peritubular bath abolishes the secretion of these two ions, but it does not eliminate fluid secretion.[34] Under these conditions, the dominant electrolytes in secreted fluid are again Na^+ and Cl^-, and only the removal of Na^+ or Cl^- from the peritubular medium brings fluid secretion to a halt.[33,34] As will be shown later, there appear to be two different transport systems which secrete NaCl into the tubule lumen.

In Vitro Microperfusion of Fish Renal Tubules. To date only three laboratories have perfused fish renal tubules *in vitro*. The laboratory of Burg was first to examine the secretion of organic solutes in isolated perfused proximal tubules of the flounder.[42] The laboratory of Nishimura and colleagues[43] has studied Cl^- transport in isolated distal tubules of the fresh-water trout. The laboratory of Beyenbach has studied NaCl, $MgSO_4$, and fluid secretion in isolated proximal tubules of the winter flounder and the killifish.[30-34]

Need for low perfusion rates: In measurements of transepithelial solute and water flows, low perfusion rates are advised. Burg ([12], this volume) originally perfused mammalian proximal tubules at rates between 5 and 10 nl/min, which is appropriate for mammalian proximal tubules and commensurate with mammalian single-nephron glomerular filtration rates. However, nanoliters perfusion rates are too high in studies of fish proximal tubules since they do not allow significant changes in the solute and water content of the perfusate to take place. Two examples are cited to illustrate this point.

Detection of small transepithelial volume flow: In the first example[33] the attempt is made to measure transepithelial volume flows in flounder proximal tubules with [^{14}C]inulin or [^{14}C]polyethylene glycol as volume marker (see [12], this volume). The tubules are perfused at rates between 5 and 10 nl/min. In no perfusion experiment does the isotope activity of the collected fluid differ significantly from that in the perfusate. On average (17 tubules, 139 measurements), there is a slight decrease in the activity of the volume marker, but the calculated transepithelial volume flow rates (0.13 ± 0.007 nl in a secretory direction) do not reach statistical significance for two reasons: the variability of the data and, more importantly, the perfusion of the lumen at too high a rate.[33] We know that flounder proximal tubules secrete fluid at rates between 10 to 60 pl/(min · mm), when the tubules are not perfused at all (Fig. 8, Table III). Obviously such low secretion rates are inadequate to significantly dilute the volume marker in the perfusate to indicate that secretion of fluid has taken place.

[42] M. B. Burg and P. W. Weller, *Am. J. Physiol.* **217**, 1053 (1969).
[43] H. Nishimura, M. Imai, and M. Ogawa, *Am. J. Physiol.* **244**, F247 (1983).

Detection of low-capacity solute transport systems: In the second example,[32] in our study of transepithelial Mg^{2+} secretion in flounder proximal tubules, we are compelled to perfuse the tubules at very low perfusion rates, i.e., at rates between 0.1 and 0.6 nl/min in order to measure increased Mg^{2+} concentrations in the collected fluid. Again we knew from nonperfused tubules that Mg^{2+} is secreted into the tubule lumen. When the peritubular Mg^{2+} concentration was 1 mM, Mg^{2+} concentrations in secreted fluid are between 20 and 30 mM (Table IV). However, in spite of the high concentrations against which Mg^{2+} can be secreted into the tubule lumen, the capacity of the Mg^{2+} transport system is low; the transport maximum, J_{max}, is 1.5 pmol min^{-1} mm^{-1} tubule length,[32] so that perfusion rates normally used in microperfusion experiments (5–10 nl/min) do not allow Mg^{2+} concentrations to build up in the collectate. This is illustrated in Fig. 9, which shows the Mg^{2+} concentration in the collectate falling as the perfusion rate increases. Given an Mg^{2+} transport maximum

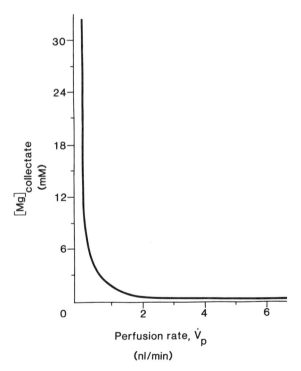

FIG. 9. The need for low *in vitro* perfusion rates in order to study low-capacity transport systems is illustrated.

of 1.5 pmol min^{-1} mm^{-1}, Mg^{2+} concentrations in the collectate will not rise significantly unless the perfusion rate is kept below 1 nl/min.

The need for subnanoliter perfusion rates may extend beyond the study of transport in fish renal tubules. It is conceivable that high perfusion rates in the study of other renal tubules, in fish or other vertebrates, have masked other low-capacity transport systems.

Electrophysiological studies of in vitro perfused fish renal tubules: Killifish proximal tubules: After our initial summer studies of fluid secretion in flounder renal proximal tubules at the Mount Desert Island Biological Laboratory,[33,34] we were interested in working on fish proximal tubules in our home laboratory year round. Large fish are difficult to transport inland; small fish can be shipped in small containers via the United States Postal Service (see Fish Sources, above). Accordingly, we acquire sea-water killifish from the Woods Hole Oceanographic Institute and maintain them in sea-water or fresh-water aquaria in our laboratory.

As shown in Tables III and IV proximal tubules isolated from sea water- and fresh water-adapted killifish secrete NaCl and fluid spontaneously. Moreover, it is the first segment of the proximal tubule which is capable of secreting fluid. Fluid secretion is driven by NaCl secretion, as in flounder and shark proximal tubules. In addition fluid secretion in killifish proximal tubules is stimulated by cAMP (Table III) and agents which increase endogenous intracellular cAMP concentrations (theophylline and forskolin).[40]

Probing the mechanisms of transepithelial NaCl secretion with the electrophysiological methods described above (Fig. 4) we seem to have identified two apparently separate transport systems. The first is inhibited by amiloride and SITS (4-acetamido-4'-isothiocyano-2,2'-disulfonic stilbene) and resistant to furosemide, the second is stimulated by cAMP and inhibited by furosemide. These conclusions obtain from studies of proximal tubules isolated from fresh water-adapted killifish. As shown in Table III, proximal tubules isolated from fresh water-adapted killifish secrete fluid spontaneously at rates averaging 56 pl/(min · mm tubule length).

Spontaneous rates of fluid secretion are not inhibited by furosemide or bumetanide; these inhibitors of the Na$^+$–K$^+$–2Cl$^-$ symporter also fail to affect the electrophysiology of the isolated perfused tubule. However, SITS, presumably an inhibitor or Cl$^-$/HCO$_3^-$ exchange, reduces the fractional resistance of the basolateral membrane and hyperpolarizes this membrane. Amiloride, an inhibitor of Na$^+$/H$^+$ exchange at the concentration used (10^{-4} M), increases the fractional resistance and the total transepithelial resistance while depolarizing the basolateral membrane voltage. Both SITS and amiloride also reduce the rates of spontaneous transepithelial fluid secretion. Accordingly, we propose that spontaneous rates of fluid secretion are supported in part by the entry of Na^+ and Cl^- across the basolateral membrane by, respectively, amiloride- and SITS-sensitive transport pathways. At the present time we do not know the mechanisms by which Cl^- leaves the cell across the apical membrane; nor do we know the fate of Na^+ which has entered the cell. As a first hypothesis we speculate that the SITS cotransport system elevates intracellular Cl^- to electrochemical potentials greater than those in the extracellular fluids. Chloride ion is then free to diffuse out of the cell across the apical membrane provided this membrane offers some Cl^- conductance. Sodium ion, which has entered the cell via the Na^+/H^+ exchange, may be returned to the bath via the ouabain-sensitive Na^+/K^+-ATPase.

Table III also shows that cAMP stimulates the rate of fluid secretion. Since the concentration of NaCl in secreted fluid does not change (Table IV) the stimulation of NaCl transport is indicated. Endogenous intracellular cAMP concentrations which have been raised by forskolin or theophylline apparently have the same effect. Detailed electrophysiological studies of the cAMP effects have revealed two major actions of cAMP: (1) cAMP increases the Cl^- conductance of the apical membrane as it does in shark proximal tubules (Fig. 7), and (2) cAMP induces a marked bumetanide sensitivity of the basolateral membrane. In cAMP-stimulated tubules bumetanide inhibits fluid secretion, but it is of interest that only the cAMP-stimulated component of fluid secretion is inhibited. Spontaneous rates of fluid secretion are not affected. Hence it appears that cAMP invokes a transport mechanism of NaCl secretion that is separate from the mechanism of spontaneous fluid secretion. Further experiments need to be conducted to confirm this hypothesis.

Trout distal tubules: Nishimura et al.[43] have perfused segments of the distal tubule isolated from the fresh-water trout, *Salmon gairdneri.* The tubules are perfused by the method of Burg ([12], this volume). Measurements of the transepithelial voltage reveal lumen-positive potentials of 18 mV on average. Measurements of transepithelial water flows using inulin or iothalamate as volume marker show that the distal tubule has a low water permeability. Measurements of transepithelial isotope fluxes of Cl^- revealed the absorptive flux to be 62% greater than the secretory flux. The authors conclude that the distal tubule dilutes the tubular fluid, as it does in other vertebrate distal tubules.

On the question of proximal NaCl and fluid secretion in glomerular nephrons: Secretion of NaCl by secondary active transport of Cl^- (energy dependent) does not make good physiological sense in glomerular renal tubules which already filter NaCl into the tubule lumen. The rationale for the physiological benefit of having two separate routes for delivering NaCl and fluid into the tubule lumen is not clear. Smith might have argued that proximal NaCl and fluid secretion is a preadaption in marine fish for life without glomeruli. Modern species of marine fish are entirely aglomerular, relying on secretory mechanisms in the kidney. Proximal NaCl and fluid secretion might also be related to glomerular intermittency, where it may continue to generate tubular fluid and in this way maintain a basic tubular function in the event the parent glomerulus has stopped filtering. In this context it is of interest that we observe NaCl and fluid secretion not in every proximal tubule, but only in a small subpopulation of renal proximal tubules. Environmental factors do not seem to have a significant role since we observe NaCl and fluid secretion in proximal tubules isolated from fresh water- and sea water-adapted fishs. Clearly, the physiological role of NaCl and fluid secretion will remain obscure until a great deal more is learned about the regulatory mechanisms of tubular transport and glomerular filtration in the kidneys of fish.

Amphibian Renal Tubules

Unique Tubular Functions

Amphibian renal tubules have a central position in renal physiology. The reason for this is not necessarily the uniqueness of their functions but rather the ready availability of amphibians worldwide, the low cost and ease of maintenance and, most importantly, the large size of their renal tubules. The large epithelial cells have made the amphibian renal tubule the experimental model of choice in electrophysiological studies of renal cellular physiology. Every tubule segment along the length of the amphibian nephron has been studied *in situ* in the doubly perfused kidney. Often the first definitive experiments have been conducted in amphibian renal tubules with considerable impact on our understanding of vertebrate renal physiology.

It would appear that *in vivo* microperfusion and *in situ* micropuncture of amphibian renal tubules have nearly preempted the *in vitro* microperfusion techniques of Burg from making many significant advances in this vertebrate kidney. However, very sophisticated experimental methods have been added to *in vitro* microperfusion methods, thereby making further significant advances in renal physiology.

Amphibian Sources

Three popular amphibian species of renal physiologists are the salamander, *Necturus ambystoma,* and *Amphiuma.* They are available from a number of sources, including NASCO (Fort Atkinson, WI), Mogul Ed

(Oshkosh, WI), Connecticut Valley Biological Company (Southampton, MA), Carolina Biological Supply, and Charles Sullivan or Amphiuma of North America (both in Nashville, TN).

Isolation and in Vitro Microperfusion of Amphibian Renal Tubules

Our present understanding of tubular transport across amphibian renal tubules rests largely on *in vivo* microperfusion and *in situ* microperfusion experiments. The isolation of specific tubule segments and the methods for *in vitro* microperfusion are not uniquely different from those of other nonmammalian renal tubules, and they will not be repeated here. Instead, we will briefly review the new analytical methods that have been added to *in vitro* microperfusion techniques and used with good success in *in vitro* perfused amphibian tubules.

No other studies of *in vitro* microperfused tubules exemplify better the advantages of electrophysiology than the *in vitro* studies in amphibian renal tubules.

Amphibian renal tubules are the only renal tubules in which the lumen has been cannulated with electrodes for transepithelial voltage clamp experiments and measurements of the short-circuit current.[44,45]

The impalement of the epithelial cells of *in vitro* microperfused renal tubules with conventional microelectrodes was first performed in salamander proximal tubules by Sackin.[46] This was accomplished about 7 years ago. Today, nearly every *in vitro* perfused tubular segment from amphibian and other vertebrate kidneys has been successfully punctured with conventional microelectrodes, yielding important information about membrane voltages and conductances.

An extension of the intracellular recording techniques described here led others to measure intracellular ion activities with ion-selective microelectrodes. And again it was the proximal tubule of the salamander which allowed the regulation of intracellular pH to be studied by Boron and Boulpaep[47,48] in isolated perfused renal tubules.

Sackin has further exploited the advantages of both electrophysiological approaches and isolated proximal tubules of the salamander to probe the nature of the paracellular pathway and its role in transepithelial volume flow.[49,50]

[44] K. R. Spring and C. V. Pazanelli, *J. Gen. Physiol.* **60**, 181 (1972).
[45] R. Delaney and L. C. Stoner, *J. Membr. Biol.* **64**, 45 (1982).
[46] H. Sackin and E. L. Boulpaep, *Am. J. Physiol.* **241**, F39 (1981).
[47] W. F. Boron and E. L. Boulpaep, *J. Gen. Physiol.* **81**, 19 (1983).
[48] W. F. Boron and E. L. Boulpaep, *J. Gen. Physiol.* **81**, 53 (1983).
[49] H. Sackin, *Am. J. Physiol.* **251**, F319 (1986).
[50] H. Sackin, *Am. J. Physiol.* **251**, F334 (1986).

Detailed studies of intracellular volume regulation in isolated perfused renal tubules were conducted by Guggino, Giebisch, and co-workers.[51,52] By adding the video-optical techniques of Welling et al.[53] and Strange and Spring[54] and intracellular microelectrode techniques to the methods of *in vitro* microperfusion of amphibian renal tubules they studied the relationships between basolateral membrane ion transport and intracellular volume.

These few examples illustrate the fact that amphibian renal tubules are the preparation of choice in the development of pioneering experimental techniques in the study of isolated renal tubules.

Reptilian Renal Tubules

Among the reptiles, only the renal tubules of garter snakes *(Thamnophis spp.)* have been isolated and perfused *in vitro*. Fluid absorption and the transport of organic anions, organic cations, and glucose have been studied in proximal tubules, and fluid absorption and transepithelial electrical properties have been studied in distal tubules.

Unique Tubular Functions

Most reptiles inhabit desiccating environments. Most reptiles are land animals, but there are also aquatic, marine, and fresh-water species. Hence their kidneys have the ability to maintain extracellular fluid homeostasis in the face of water deprivation and water loading.

Renal functions in reptiles are unique in several ways. Reptiles excrete uric acid as nitrogenous waste. Accordingly, reptilian renal tubules are the tubules of choice in investigations of the cellular mechanisms of urate excretion. The excretion of uric acid eliminates the obligatory water loss associated with the excretion of urea in other vertebrate kidneys. In addition, ureteral urine empties into the cloaco where NaCl and fluid absorption continues. In this way nearly 100% of the renal water may be returned to the animal under conditions of water deprivation.

In comparison with the kidneys of fish and amphibians, the vascular and tubular organization is highly developed in reptilian kidneys. The significance of the highly organized structural arrangements, including counter-current flows through loops, is unknown.

---

[51] W. B. Guggino, H. Oberleithner, J. Giebisch, and L. M. Amgel, *J. Gen. Physiol.* **86**, 81 (1985).
[52] W. B. Guggino, *Am. J. Physiol.* **250**, F430 (1986).
[53] L. W. Welling, D. J. Welling, and T. J. Ochs, *Am. J. Physiol.* **245**, F12 (1983).
[54] K. Strange and K. R. Spring, *Kidney Int.* **30**, 192 (1986).

Some snakes hibernate during the winter months. The glomerular and tubular functions in these hibernating poikilotherms have not been studied in detail and remain an intriguing question from the perspective of extracellular fluid homeostasis and glomerular and tubular physiologies.

The kidneys of reptiles are sexually dimorphic. In male kidneys the most distal segment of the nephron is the so-called sexual segment. Its outer diameter is approximately twice as large as any other tubule segment. The cells of the sexual segment are filled with secretory granules which give the tubule a white, opaque, glandular appearance. The secretions of this segment are thought to form a copulatory plug in the female after mating.

Reptile Sources

Reptiles are commercially available from a variety of sources. These are the large biological supply houses such as Wards, Carolina Biology, and Connecticut Valley. We obtain species of the common garter snake from Nasco (Fort Atkinson, WI) Midwest Reptile and Animal Sales (Fort Wayne, IN), or Mogul Ed (Oshkosh, WI).

Isolation of Reptilian Renal Tubules

Snake Proximal Tubules. General approach to perfusion: The basic principles of perfusion are the same as those outlined in the next chapter. The animals are killed by decapitation and the kidneys quickly removed and placed in oxygenated Ringer solution on ice (see below, Composition of Ringer solutions). The kidney tissue remains in excellent condition in Ringer solution on ice for the entire day. The kidneys are elongated organs consisting of a series of lobules arranged along the ureter and renal portal vein on the medial border. A single lobule is removed from a kidney and transferred to another dish of oxygenated Ringer at room temperature ($25 \pm 2°$) for dissection. The capsule is removed from the lobule and masses of tubules are teased from it with fine forceps (usually Dumont 7B, Switzerland). These masses of tubules are then transferred to the stage of a stereomicroscope and the individual tubules are teased free with fine needles (made from jeweler's brooches or insect pins) under a magnification of about $\times 21 - \times 50$. Most teasing is performed at a very low magnification ($\times 25$) with some checking of cell borders at higher magnification ($\times 50$). The proximal tubule can easily be identified beginning at the glomerulus (a short, thin, ciliated neck segment actually is interposed between the glomerulus and the true proximal nephron) and extending to a long, thin, ciliated intermediate segment. If young adult garter snakes are chosen (optimum age indicated by a body with maximum diameter of about 0.5 in.), the connective tissue between the tubules is very flimsy and the tubules can be easily separated. Older animals have much tougher connective tissue between the tubules and separation is difficult without damage to the tubule cells. The average length of the entire proximal tubule varies depending on its position in the lobule (longest tubules near the center: shortest tubules near the caudad and cephalad ends) but average about 4.5 mm (4.5 ± 1.62 mm; mean ± SD from 31 tubules).[55] In general the length of the segments teased out for perfusion averages 1.5 mm but may vary from about 1.0 to 2.00 mm.[55,56] Segments shorter than 1.0 mm are difficult to use for transport studies because so much of each segment (300–400 μm) is taken up in the holding pipets.

Segments of the reptilian proximal tubule: The snake proximal tubule can be divided into two major segments based on the transepithelial transport of *p*-aminohippurate (PAH).[56] One segment extends from the narrow neck segment below the glomerulus through a series of convolutions, culminating in the first major bend in the proximal tubule. This segment, referred to as the proximal-proximal segment, is not capable of transporting PAH from the peritubular side to the lumen against a significant concentration gradient. The second portion of the proximal tubule extends from this bend to the beginning of the thin intermediate segment. Thus it comprises the remainder of the proximal tubule and is termed the distal-proximal tubule. This segment is capable of transporting PAH from the peritubular side to the lumen against a significant concentration gradient. This distal-proximal segment, which is generally slightly easier to obtain in longer lengths than the proximal-proximal portion, has been used most extensively for the studies of organic anion and cation transport. Once isolated, each tubule segment is transferred in a drop of Ringer to the perfusion chamber and perfused as described in the next chapter. The average outside diameter during perfusion is about 55 μm and the average inside diameter, about 27 μm. Perfusion is normally performed at 25 ± 2°, the same temperature at which the animals themselves are maintained.

Composition of Ringer solutions: The standard Ringer used for the dissection of tubules and for the perfusate and bathing medium for proximal tubules contains the following (in mM): NaCl, 126.0; KCl, 3.0; NaHCO$_3$, 24.0; NaH$_2$PO$_4$, 10; MgSO$_4$, 1.2, and CaCl, 1.8 (Table V).[55] In addition, under control conditions, the Ringer solution for the outside bathing medium contains 4 g/100 ml neutral dextran (M_r 40,000 ± 3000) to approximate the plasma protein concentration.

For studies of fluid absorption or the transport of organic anions in which the removal of sodium is desired, the bicarbonate buffer is replaced

---

[55] W. H. Dantzler, *Am. J. Physiol.* **224,** 445 (1973).
[56] W. H. Dantzler, *Am. J. Physiol.* **226,** 534 (1974).

TABLE V
COMPOSITION OF RINGER SOLUTIONS

Ringer solutions	NaHCO$_3$	Tris-Cl	HEPES	NaCl	KCl	NaH$_2$PO$_4$	Na$_2$HPO$_4$	K$_2$HPO$_4$	MgSO$_4$	CaCl	CaSO$_4$	Choline chloride	LiCl
Standard bicarbonate[a]	24			126	3	1			1.2	1.8			
Tris[b]		24		128	1				1.2	1.8			
Phosphate[b]				150	3	0.4	1.6		1.2	1.8			
HEPES[c]			24	126	3	1			1.2	1.8			
Sodium free (choline)[b]		24						1	1.2	1.8		128	
Sodium free (lithium)[b]		24						1	1.2	1.8			128
Tris, sodium free,[b,e] low chloride (sucrose)		24			1			1	1.2	1.8			
Tris, sodium free,[b,f] low chloride (lactose)		24			1			1	1.2	1.8			
HEPES, sodium free,[c,g] low chloride (sucrose)			20					2.5	1.2	1.8			
Chloride free[b,h] (CH$_3$SO$_4$)	24								1.2		1.8		
Chloride free[d,i] (isethionate)	24								1.2		1.8		
Chloride free[d,j] (thiocyanate)	24								1.2		1.8		
Sodium free[b,k] [tetramethyl ± ammonium (TMA)]		24			1			1	1.2		1.8		

[a] W. H. Dantzler, *Am. J. Physiol.* **224**, 445 (1973).
[b] W. H. Dantzler and S. K. Bentley, *Am. J. Physiol.* **234**, F68 (1978).
[c] W. H. Dantzler and O. H. Brokl, *Am. J. Physiol.* **250**, F407 (1986).
[d] W. H. Dantzler and S. K. Bentley, *Am. J. Physiol.* **241**, F632 (1981).

Ringer solutions were made isosmotic with 216 mM sucrose[e]; 250 mM sucrose[e]; 126 mM sodium methyl sulfate[h]; 126 mM sodium isethionate and 3 mM potassium isethionate[i]; 126 mM sodium thiocyanate and 3 mM potassium thiocyanate[j]; 128 mM tetramethylammonium chloride.[k]

with Tris–Cl and sodium chloride is replaced with the chloride salt of an organic cation or another inorganic cation (e.g., lithium).[57] For studies of organic cation transport in which the removal of Na^+ is desired, the bicarbonate buffer is replaced with HEPES and sodium chloride is replaced with sucrose (Table V).[58,59] Although this last replacement undoubtedly alters the ionic strength of the solution, it is the best substitute possible when organic cation or other inorganic cation substitutes for sodium cannot be used.

When the effects of chloride replacements on organic anion or cation transport are examined, $CaCl_2$ is replaced with $CaSO_4$ and NaCl and KCl are replaced with the sodium and potassium salts of the isethionate, methyl sulfate, or thiocyanate (Table V).[57,60] When the effect of changes in the potassium concentration on fluid absorption or organic anion transport is studied, the sodium concentration is adjusted appropriately to maintain the osmolality constant.[61–63]

The pH of the standard solution is 7.6, which is appropriate for these animals at a temperature of 25°. Alterations in pH are best effected by titrating the N-2-hydroxyethylpiperazine-N'-2-ethanesulfonic acid (HEPES) buffer with NaOH and then adjusting the NaCl concentration appropriately to maintain the sodium concentration constant. However, simple phosphate-buffered solutions can be used, although the pH is less stable.

The osmolarity of all these solutions is approximately 280 mOsm. This can be routinely checked for each new solution with any standard osmometer involving the vapor pressure or freezing point depression approach. The bicarbonate-buffered solutions are bubbled with a 95% O_2–5% CO_2 gas mixture and the Tris-, HEPES-, phosphate-buffered solutions are bubbled with 100% O_2 or compressed air.[57,64] Compressed air is sufficient to maintain a high oxygen tension in the media.

Snake Distal Tubules. The snake distal tubule in the common garter snake (the only reptilian distal tubule studied to date) is best identified in male snakes by its attachment to the sexual segment.[65] As described above, after removal of the renal capsule, tubules are teased from the tubules with

---

[57] W. H. Dantzler and S. K. Bentley, *Am. J. Physiol.* **234**, F68 (1978).
[58] W. H. Dantzler and O. H. Brokl, *Am. J. Physiol.* **250**, F407 (1986).
[59] C. T. Hawk and W. H. Dantzler, *Am. J. Physiol.* **246**, F476 (1984).
[60] W. H. Dantzler and S. K. Bentley, *Am. J. Physiol.* **241**, F632 (1981).
[61] W. H. Dantzler, *Am. J. Physiol.* **227**, 1361 (1974).
[62] W. H. Dantzler and S, K. Bentley, *Am. J. Physiol.* **229**, 191 (1975).
[63] H. W. Randle and W. H. Dantzler, *Am. J. Physiol.* **255**, 1206 (1973).
[64] W. H. Dantzler and S. K. Bentley, *Am. J. Physiol.* **230**, 256 (1976).
[65] K. W. Beyenbach and W. H. Dantzler, *Am. J. Physiol.* **234**, F238 (1978).

forceps. In male snakes, the large, white, opaque sexual segment stands out against the small, translucent tubules of the remaining nephron. The distal tubule forms a sharp transition with the sexual segment, which marks the most distal portion of the distal tubule. The tubule is severed here and at a site 2–3 mm upstream.

In Vitro Microperfusion of Reptilian Proximal Tubules

Fluid Absorption. Net fluid absorption (J_v, nl min^{-1} mm^{-1} tubule length) by proximal tubules is studied in the standard manner with a volume marker in the perfusate (see the next chapter). The volume markers that have been used successfully with these proximal tubules are [*methoxy*-^{3}H]inulin, [*carboxyl*-^{14}C]inulin, ^{3}H- or ^{14}C-labeled polyethylene glycol (PEG), sodium ferrocyanide [$Na_4Fe(CN)_6 \cdot 10H_2O$], and potassium ferrocyanide [$K_4Fe(CN)_6 \cdot 10H_2O$]. [*methoxy*-^{3}H]Inulin must be dialyzed before use for 24 hr at room temperature with dialysis tubing of M_r 3500 cut-off (Arthur H. Thomas Co., Philadelphia, PA) to remove any free tritium and low-molecular-weight fragments.[66] Polyethylene glycol gives the same fluid absorption values as inulin, does not deteriorate with storage, and does not require dialysis before use. Therefore, it appears to be the volume marker of choice under most circumstances.[60] The particular radioactive label used (^{3}H or ^{14}C) depends on the label of other substances whose transport is studied simultaneously. When only fluid absorption is being determined, then the particular label used does not matter. Sodium or potassium ferrocyanide, whose concentration in the perfusate is measured as the concentration of iron by electron microprobe analysis, also gives fluid absorption values identical with those obtained with inulin or PEG.[59] It is more difficult and time consuming to measure the iron in sodium ferrocyanide by electron microprobe analysis than to determine radioactivity. However, if volume absorption must be determined simultaneously with the fluxes of substances with low radioactivity, then the use of ferrocyanide permits the determination of both the fluxes and the volume absorption with much greater accuracy than with dual isotope counting.

Regardless of the choice of volume marker, the fluid absorption is always determined as the difference between the initial perfusion rate (determined from the change in concentration of volume marker from the initial perfusate of the collected fluid) and the collection rate. The result is generally divided by the length of the perfused tubule to give a value in nanoliters minute^{-1} millimeter^{-1} tubule length. The change in the concentration of the volume marker is always small. However, the standard perfusion rate in these tubules is maintained low (about 4 nl min^{-1})[57] to

---

[66] D. W. Barfuss and W. H. Dantzler, *Am. J. Physiol.* **231**, 1716 (1976).

mimic the low normal filtration rate[67] and, thus, the change in concentration of the volume marker is much greater than in mammalian tubules, which are perfused at more rapid rates.

Probing the mechanism of fluid absorption: The process of net fluid absorption, studied by the above methods, in these isolated, perfused reptilian renal tubules is particularly interesting. With the standard, sodium-containing, bicarbonate-buffered Ringer solution as both perfusate and bathing medium and dextran in the bathing medium, J_v averages about 0.90 nl min^{-1} mm^{-1} in both the proximal-proximal and the distal-proximal segments.[57] This rate appears to be reasonable for the tubules *in vivo*. Moreover, in the presence of sodium-containing perfusate and bathing medium, fluid absorption is not dependent on the buffer used (bicarbonate, phosphate, or Tris) in the lumenal or peritubular fluid.[57] It is, however, reduced by about 18–25% by the removal of colloid from the peritubular fluid.[57]

Of greatest interest with regard to the mechanism of fluid absorption in these reptilian tubules are the effects of substitutions for sodium, chloride, or both in the solutions bathing the tubules (Table V).[57] When sodium in the perfusate is replaced with choline, net fluid absorption virtually ceases. When sodium in the bathing medium is also replaced with choline, so that both solutions are identical (except for the dextran in the bathing medium), net absorption returns to the control rate. The results are the same when sodium is replaced with tetraethylammonium, when sodium and the equivalent amount of chloride are replaced with sucrose, and when chloride alone is replaced with methyl sulfate. However, when sodium in the perfusate or in both the perfusate and the bathing medium is replaced with lithium, net fluid absorption does not change from the control rate. The fluid absorption at control rates, regardless of the composition of the perfusate and bathing medium, is isosmotic (within the limits of the measurements) and can be at least partly inhibited by cold and cyanide.[57] Even with sodium present, however, it cannot be inhibited by ouabain or other cardiac glycosides or by the removal of potassium from the bathing medium,[68] and it apparently is not dependent on the nature of the buffer (bicarbonate, phosphate, or Tris) used.[57]

Morphometric analyses of perfused proximal tubules: Recently, cell volumes and cell surface areas have been measured by ultrastructural morphometric methods in tubules perfused under control conditions and in tubules perfused with the sodium in both the perfusate and bathing medium replaced with choline (Table V).[69] Following measurements of

[67] J. Bordley III and A. N. Richards, *J. Biol. Chem.* **101**, 193 (1933).
[68] W. H. Dantzler and S. K. Bentley, *Renal Physiol.* **1**, 268 (1978).
[69] W. H. Dantzler, O. H. Brokl, R. B. Nagle, D. J. Welling, and L. W. Welling, *Am. J. Physiol.* **251**, F150 (1986).

fluid absorption as described above, each tubule is fixed with half-strength Karnovsky's fixitive[70] while still attached to the perfusion pipets and while perfusion pressure is maintained. After standard embedding and sectioning procedures, transverse sections are examined in a transmission electron microscope. Morphometric analyses are made by well-described computer-assisted operations[71] and manual point- or intercept-counting operations.[72] These initial studies show that, during the 20- to 25-min perfusion in the absence of sodium, significant morphological changes occur while net fluid absorption stays at the control level. The volume of the cells doubles and the volume of the intercellular spaces nearly quintuples. The areas of the lateral and apical cell membranes approximately double, but their surface densities remain constant. Therefore, the larger cells in the absence of sodium have proportionately enlarged surface areas so that the volume-to-surface ratio remains constant. These morphological changes occur simultaneously with the maintenance of control fluid absorption and may play a permissive role in such maintenance in the absence of sodium.

Reflection Coefficients. The transepithelial reflection coefficients for experimental solutes, such as those substituted for sodium and chloride in the fluid absorption experiments discussed above, are determined by comparing the increment in net fluid absorption produced by an osmotic gradient of one of these solutes (J_v, solute) with that produced by an apparently equal osmotic gradient of raffinose (J_v, raffinose) (assuming that raffinose has a reflection coefficient of 1.00).[57,63] Thus

$$r \text{ solute} = \Delta J_v \text{ solute}/\Delta J_v \text{ raffinose} \tag{7}$$

In practice this means adding equal osmotic concentrations of either the experimental solute or raffinose to the bath and observing the net fluid absorption compared to the control situation. In these experiments three solutions are used as the bathing medium: (1) standard bicarbonate-buffered Ringer (solution 1, Table V); standard bicarbonate-buffered Ringer plus 100 mOsm raffinose, and (3) standard Ringer plus 100 mOsm of the substance for which the reflection coefficient is being determined. The osmolarities are checked with a freezing point or vapor pressure osmometer and the amount of solute adjusted appropriately. No dextran is present in the bath. The perfusate is always standard bicarbonate-buffered Ringer.

First, three 50-min control collections are taken to determine the basic net fluid absorption rate. Second, the bath is changed to one of the other solutions and three more 5-min collections are taken. Third, the bath is changed to the other experimental solution and three more 50-min collec-

---

[70] M. J. Karnovsky, *J. Cell Biol.* **27**, 137A (1965).
[71] J. N. Simone, L. W. Welling, and D. J. Welling, *Lab. Invest.* **41**, 334 (1979).
[72] E. R. Weibel, "Stereological Methods." Academic Press, New York, 1979.

tions are taken. Finally, the control solution is restored to the bath and the last three 5-min collections are made. The order in which the experimental solutions are used is randomized to avoid systemic bias. Mean values for each group of the three 5-min collections are obtained. The control value for J_v is the mean of the series obtained at the beginning and end of the experiments. The reflection coefficient is then determined from the increments in J_v produced by each experimental solution as described above.

At best, the measurements of reflection coefficients are only approximations of the true effective osmotic pressures exerted by these solutes. The primary difficulty with the measurements is that osmotic and ionic equilibration tends to occur across the epithelium during the measurements of J_v. In order to minimize such equilibration, only short segments (about 1.0 mm in total length) are perfused at rapid rates (about 15 nl min^{-1}) and the perfusion rate in a given tubule is held as constant as possible throughout the control and experimental periods. This often requires changes in the perfusion pressure because the cell volume and, thus, the lumen diameter change with the changes in osmolarity of the bathing medium. Because of these osmotic effects, the tubules are easily damaged during the course of the measurements. Therefore, it is essential that the final set of control measurements be approximately the same as the first set of control measurements (within about 10% of the initial value). If this is not the case, the entire experiment must be discarded. Although these measurements do not give a true measure of the effective osmotic pressure of a given solute, they do give a reasonable indication of the relative reflection coefficients for solutes studied in the same experiment. Table VI lists some measurements of reflection coefficients for a number of solutes across the proximal-proximal and distal-proximal segments of snake renal tubules.[57] These were all determined in the same set of experiments.

Hydraulic Conductivity. The osmotic conductivity (L_p, in ml cm^{-2} sec^{-1} atm^{-1}) of proximal tubules to water also can be estimated from the increment in net fluid absorption (ΔJ_v) produced by increasing the osmolarity of the bathing medium 100 mOsm with raffinose. It is calculated from the following relationship.[73]

$$L_p = \Delta J_v / \Delta \Pi \tag{8}$$

in which $\Delta \pi$ is the mean osmotic pressure difference between the bath and the perfusate. For these experiments, we assume that raffinose has a reflection coefficient of 1.0. The protocol is the same as that described previously for the determination of the reflection coefficients except that only the increment in J_v produced by the addition of raffinose to the bath is

[73] J. P. Kokko, *J. Clin. Invest.* **49,** 1838 (1970).

TABLE VI
REFLECTION COEFFICIENTS[a]

Substance	Proximal-proximal tubule	Distal-proximal tubule
Na$^+$	0.53 ± 0.057 (5)	0.61 ± 0.051 (9)
Choline chloride	0.77 ± 0.025 (5)	0.75 ± 0.027 (5)
Tetramethylammonium chloride	0.75 ± 0.025 (4)	0.76 ± 0.033 (5)
LiCl	0.87 ± 0.020 (5)	0.87 ± 0.046 (6)
NaCH$_3$SO$_4$	0.86 ± 0.050 (5)	0.73 ± 0.044 (6)
Sucrose	0.98 ± 0.021 (5)	0.97 ± 0.039 (6)

[a] Numbers in parentheses indicate number of experiments. [From W. H. Dantzler and S. K. Bentley, *Am. J. Physiol.* **234**, F68 (1978).]

determined. The same precautions used for the measurements of the reflection coefficients are required. The values obtained to date (1.29 × 10^{-5} ml cm^{-2} sec^{-1} atm^{-1} for the proximal-proximal segment and 0.71 × 10^{-5} ml cm^{-2} sec^{-1} atm^{-1} for the distal-proximal segment)[57] are about one-fifth to one-half those obtained in a similar manner for mammalian proximal tubules.[74,75] Although these values for the osmotic hydraulic conductivity may be lower than those for the hydrostatic hydraulic conductivity and probably underestimate the true osmotic hydraulic conductivity, they still indicate that perfusion pressure could not account for the observed fluid absorption in the absence of sodium.

Transepithelial Voltage. The transepithelial electrical potential is evaluated in these tubules in a manner identical to that described for other, isolated, perfused tubules (see [16], this volume). Under control circumstances with standard bicarbonate-buffered Ringer (solution 1, Table V) as the perfusate and standard bicarbonate-buffered Ringer with dextran as the bathing medium, the transepithelial potential averages about −0.50 mV, lumen negative.[60] Detailed studies have not been made on the origin of this potential, but it is not enhanced by the presence of small amounts of organic substrates in the perfusate. Also, these tubules are almost totally insensitive to ouabain and no effect of this substance on this small transepithelial potential is observed.

Transepithelial Organic Solute Transport: Experimental Methods. Transepithelial organic solute fluxes: The transport of organic anions, organic cations, and glucose has been and continues to be the subject of study with these tubules. The transepithelial transport is studied with flux

[74] J. P. Kokko, M. B. Burg, and J. Orloff, *J. Clin. Invest.* **50**, 69 (1971).
[75] J. P. Kokko and C. C. Tisher, *Kidney Int.* **10**, 64 (1976).

measurements, the techniques for which are described elsewhere (see [13], this volume). However, several points need to be kept in mind for these flux studies. For organic anions, such as p-aminohippurate, which undergoes net transport from the bathing medium to the lumen (net secretion) and virtually no movement from the lumen to the basic medium, the net flux is simply studied by measurement of the unidirectional flux from the bath to the lumen.[56] This is also the case for uric acid, although there is a substantial passive reabsorptive flux for this substance.[55] For other substances, such as organic cations, for which there are substantial saturable fluxes in both the bath-to-lumen and the lumen-to-bath directions, the transepithelial transport is best evaluated by determinations of the unidirectional fluxes.[58] Sometimes the direction of the net flux is obvious from the unidirectional fluxes. However, in other cases, when the unidirectional fluxes are similar in magnitude and there is considerable variability among tubules, it may be necessary to measure the net flux directly.[58] This can be done by using the same solution containing the substance being studied for both the perfusate and the bathing medium. An appropriate volume marker must also be added to the perfusate to correct for the volume absorption and to permit determination of the net flux.

The effects of various possible inhibitors or of alterations in the concentrations of constituents of the media on the fluxes can be determined. In these studies, each tubule is always used as its own control so that statistics can be performed by paired analysis. The protocol usually involves three control periods, three experimental periods in which the bathing medium, perfusate, or both are changed, and three more control periods. The final control periods are important in determining the reversibility of any effects.

Determination of intracellular organic solute concentrations: Extraction of organic solute: One of the most important measurements in these studies of the transport of organic molecules is the determination of the concentration of the substance being studied in the cell water immediately following a flux measurement. This requires the rapid removal of the tubule from the pipets and extraction of the transported substance with minimal contamination by perfusate and bathing medium or with corrections for such contamination.[55] In practice, the tubule is pulled free from the pipets with a glass needle or fine steel forceps and transferred through the oil layer covering the bathing medium into 10 μl or trichloroacetic acid (TCA) under mineral oil for extraction of the compound. This transfer can be accomplished in under 2 sec. Extraction is ordinarily performed for 1 hr, although it is usually complete within a matter of minutes. Although such acid extraction may alter the structure of a given compound, this is generally not important because radioactively labeled compounds are used and the amount present in the tissue is simply determined from the total radioactivity (which is extracted) and the known specific activity of the compound. The movement of the tubule through the oil layer usually removes the bathing medium that would normally be transferred with it. However, in cases in which it is absolutely essential that a correction be made for even the smallest amount of transferred bathing medium (for example, following a bath-to-lumen flux measurement in which the bath contains a high concentration of the labeled substance), an appropriate labeled extracellular volume marker (inulin, PEG, mannitol, etc.) is added to the bathing medium and the transferred volume is determined from the radioactivity of the marker extracted from the tubule and the radioactivity per unit volume of the medium. When a radioactively labeled volume marker is present in the perfusate, it is generally not possible to have such a marker in the bathing medium. In these cases, it is necessary to determine the volume of extracellular bathing medium transferred with a series of nonperfused tubules handled in the same manner as the perfused tubules and apply the average value obtained to the perfusate tubules.

For these measurements of the concentration in the cell water, the radioactivity in the lumen can be eliminated by ending the perfusion with mineral oil so that following the last collection, at the time the tubule is removed from the pipets, the tubule is completely filled with oil.[76,77] In practice, this is difficult because it requires prefilling the perfusion pipet with oil behind the estimated required amount of perfusate (rapid changes from perfusate to oil, as from one perfusate to another, are not possible because of the viscosity of oil). It is generally simpler to calculate the amount of experimental substance transferred in the lumen from the amount of the volume marker extracted and the ratio of the volume marker to the experimental substance in the perfusate.[66] The total amount of volume marker extracted is divided by this ratio to determine the amount of experimental substance in the lumen at the time of extraction. This amount is subtracted from the total amount of experimental substance extracted from the tubule. Generally, less than 10% of the total radioactivity extracted from a tubule is attributable to material transferred in the lumen or with the bathing medium.[58,66]

Determination of the epithelial cell water content: Following the extraction in TCA, each tubule is further extracted in chloroform, dried, and weighed on a quartz-fiber ultramicrobalance.[78] For these tubules, the dry weight is multiplied by 1.249 to correct for weight loss during the TCA and

---

[76] M. B. Burg and P. F. Weller, *Am. J. Physiol.* **217**, 1053 (1969).
[77] B. M. Tune, M. B. Burg, and C. S. Patlak, *Am. J. Physiol.* **217**, 1057 (1969).
[78] S. L. Bonting and B. R. Mayron, *Microchem. J.* **5**, 31 (1961).

chloroform extractions.[55] This factor was established by weighing a series of tubules before and after the extraction procedure.[55] The water content of these tubules is determined by the method of Burg et al.[79] by incubating them for 1 hr in Ringer containing tritiated water. Each tubule is then incubated for an additional 30 sec in Ringer containing a ^{14}C-labeled volume marker (inulin or PEG) under mineral oil. The tubule is then pulled from this drop with a fine glass needle through the mineral oil and into 10 μl of 0.75 N HNO$_3$. After 1 hr of extraction, the tubule is removed, washed in chloroform, dried, and weighed on a quartz-fiber ultramicrobalance as described above. Appropriate corrections are made for weight losses from the extractions in HNO$_3$ and chloroform, also as described above. The radioactivity in the HNO$_3$ extract is then determined. The measurements of the tubule water content are corrected individually for any extracellular water by the ^{14}C-labeled volume marker determinations. Although HNO$_3$ extraction was used in the initial determinations of cell water, TCA extraction works as well. In control Ringer, the water content of these tubules is about 78%.[55] This average volume is used in determining the concentrations of various substances in the cell water of the tubules under control conditions. This water content can be determined in a similar manner with major changes in the Ringer composition. However, it appears to vary no more than 2% for most solutions, a value within the error of the determination (W. H. Dantzler, unpublished observations).

Estimates of membrane permeability: Apparent permeability of the lumenal membrane: For uncharged or negatively charged organic substances that undergo net secretion from the bath to the lumen with no carrier-mediated transport back into the cells from the lumen and that develop a higher concentration in the cells than in the bath or lumen (for example, uric acid and PAH), an apparent permeability of the lumenal membrane (P_L, cm sec^{-1}) to the substance can be calculated from the net bath-to-lumen flux and the concentration difference across the lumenal membrane.[55,77] For this purpose Eq. (9) is used.

$$P_L = J \text{(substance)}/\{A_L[\bar{C}_T - (C_L/2)]\} \tag{9}$$

In this equation, J (substance) is the net secretion of a substance from bath to lumen, A_L is the surface area of the lumenal membrane per unit length, $\bar{C}_T$ is the mean concentration of the substance in the tissue water, and C_L is the concentration of the substance in the collected tubule fluid. Since none of the substance is present in the initial perfusate and since the concentration of the substance in the collected fluid is the highest concentration achieved, $C_L/2$ is used as an approximation of the mean concentra-

---

[79] M. Burg, J. Grantham, M. Abramow, and J. Orloff, *Am. J. Physiol.* **210**, 1293 (1966).

tion of the substance in the tubule lumen.[55] Although the true surface area of the lumenal membrane cannot be determined with certainty in perfused proximal tubules (because of the brush border), the area of the membrane simply based on the mean lumenal diameter during perfusion is used to facilitate comparisons among tubules and under various sets of circumstances. It must be emphasized that the permeability calculated in this manner is an "apparent" permeability only and that movement across the membrane, although down the electrochemical gradient, may well be carrier mediated.

Apparent permeability of the peritubular membrane: For uncharged or negatively charged organic substances that undergo net secretion from the bath to the lumen and that develop a higher concentration in the cell water than in the bath or lumen (for example, uric acid or PAH), an apparent permeability of the peritubular membrane (P_p, cm sec^{-1}) to the substance can be determined from the efflux of the substance across the peritubular membrane of tubules with oil-filled lumens.[61] This approach was devised to limit efflux to movement from the cells across the peritubular membrane only. In these experiments, tubules are teased from fresh tissue and measured as described above. The peritubular membrane is calculated from these measurements on the assumption that the tubule is a right cylinder. Each tubule is then attached to the perfusion pipet and the lumen is filled with mineral oil. The tubule segments (usually four to six) are then incubated together in the appropriate Ringer with the radioactively labeled substrate whose transport is being studied until a steady state is reached. The organic substrates studied to date have low lipid solubilities and do not enter the oil in the tubule lumen, but this should be verified for each new substance.

After the tubules have reached a steady state (traditionally a 60-min period is allowed, but less time is often enough), they are transferred on a glass needle to 0.5 ml of the appropriate Ringer that is free of labeled substrate and washed for 15 sec to remove any labeled substrate on the outside of the tubules. Experiments in which tubules were incubated with a labeled volume marker indicate that this is a sufficient wash to eliminate the volume marker and, thus, presumably all the extracellular fluid.

The tubules are then transferred on a glass needle through a series of 0.5-ml baths of the appropriate Ringer free of substrate. They are vigorously stirred in each efflux bath. The entire 0.5 ml of each efflux bath is counted to determine the amount of labeled substrate lost from the cells. The time during which the tubules are in each bath varies from about 15 sec for the early efflux periods to about 3 min for the final two efflux periods. The efflux periods are kept as short as is consistent with obtaining enough labeled substrate for adequate counting. The total time for all efflux periods is about 15–17 min for each experiment. After the tubules are removed from the final efflux bath, they are extracted in TCA to remove the remaining labeled substrate, further extracted in chloroform to remove the oil, dried, and weighed as described above. The concentration of labeled substrate at the start of each efflux experiment is then determined from the total radioactivity in all efflux baths and the radioactivity extracted from the tubules at the end of the experiment. The concentration of labeled substrate in the tubule cell water at the start of each succeeding efflux period is determined by subtracting the labeled substrate lost in the preceding efflux period. In order to determine the mean concentration of labeled substrate in the cell water during each efflux period, the concentration at the beginning of each efflux period is plotted against time on semilog paper. For the substrates studied to date, the concentration decreases in a linear fashion with time until 10–15% of the initial concentration is reached. This indicates that a single component is involved in about 90% of the efflux. The data used in calculating the peritubular membrane permeability are all taken from the linear component. The mean cell water concentration during each efflux period is determined at the midpoint of each period from the line on the semilog plot connecting the concentrations at the beginning and end of the period.

The efflux from the cells to the bath is proportional to the concentrations in the cells and the constant of proportionality is the permeability coefficient. The efflux of labeled substrate is plotted against the mean cell water concentration for each period (determined as described above) on linear graph paper and the permeability determined from the slope of the curve. As in the case of the permeability of the lumenal membrane, it must be stressed that the permeability of the peritubular membrane is "apparent" only and that the movement of the substrate may be carrier mediated.

Efflux coefficients: The efflux across the peritubular membrane can also be studied in a similar manner for substances that are transported out of the cells against an electrochemical gradient (for example, N-methylnicotinamide, NMN; W. H. Dantzler, unpublished observations). The purpose in these cases is not to determine a permeability coefficient but simply to study the transport across a single membrane. The process, however, is almost identical to that described above for the determination of the peritubular membrane permeability. Usually in these cases only two 1-min efflux collections are made. Because the concentration in the cells in these cases is below that necessary to saturate the efflux transport and because small changes in the intracellular concentration would be expected to influence the efflux, the efflux during each period is standardized by dividing the mean intracellular concentration during that period. Although this gives a value with the units of a permeability coefficient (cm sec^{-1}), it is not a true permeability because transport out of the cells is against an electrochemical gradient. It is simply termed an efflux coefficient.

An efflux coefficient across the lumenal membrane can be obtained in an analogous manner to that across the peritubular membrane for substances that are transported from the cells to the lumen against an electrochemical gradient (again, for example, NMN; W. H. Dantzler, unpublished observations). In these cases, efflux across the lumenal membrane is measured in tubules covered with mineral oil. The technique was devised to limit efflux to movement from the cells across the lumenal membrane only and to permit the use of each tubule as its own control for efflux into experimental media. Initially, each tubule is prepared for perfusion with control Ringer as bathing medium and control Ringer containing radioactively labeled substrate and a radioactively labeled volume marker as the perfusate. The area of the lumenal membrane is calculated as described above. Once perfusion is established, the bathing medium is replaced with mineral oil and perfusion pressure is reduced to just sufficient to keep the lumen open to the control diameter. A period of 30 min is allowed for the labeled substrate to attain a steady state distribution between the lumen and the cells. Less time is probably required for most substrates and this period can be varied as necessary. At the end of this time, the perfusate is changed to control Ringer free of labeled substrate and labeled volume marker, the perfusion rate is increased to about 30 nl min^{-1}, and a series of 2-min collections of perfusate are made to measure efflux from the cells. Any collections that contain any labeled volume marker, and, thus, any initial perfusate are discarded. This should only apply to the first collection at this perfusion rate. Following five additional collections, the perfusate can be changed to an experimental one and five more collections made. Timed controls indicate that no change occurs with time alone during these 10 collections. The rapid perfusion rate is maintained throughout to prevent accumulation of labeled substrate in the lumen. At the end of the last collection, the tubule is harvested and the concentration of labeled substrate in the cell water is determined as described above. The concentration of labeled substrate in the cell water at the start of each efflux period is calculated from the labeled substrate in the collected fluid and the residual labeled substrate in the cell water at the end of the experiment. An efflux coefficient for each period is determined as for the efflux across the peritubular membrane and mean values for the five control periods and the five experimental periods are used as control and experimental values, respectively, for that tubule.

Finally it should be noted that for any substance for which the concentration in the cells is less than that in bathing medium or perfusate, an apparent permeability can be determined across either membrane in the direction from the bath to the cell or from the lumen to the cell if the net flux in those directions and the concentration difference across the membranes are determined. These are all determined as described above. The equation used is basically the same as that for the apparent permeability of the lumenal membrane discussed above [Eq. (9)] with the appropriate membrane area (determined as discussed above), the appropriate net flux of substrate, and the appropriate mean substrate concentrations in the lumen or the bath and the cells. Of course, since the flux is in the direction into the cells, the concentration difference is reversed from that shown in Eq. (9). This approach has been most useful in the study of glucose transport with snake tubules.

Transport of Anionic, Cationic, and Neutral Organic Solutes across Reptilian Proximal Tubules. Cellular mechanisms of organic anion transport: Net secretion of PAH: With the techniques described above, the transport of some organic anions (PAH, urate, and lactate), organic cations [tetraethylammonium (TEA) and NMN], and glucose has been studied in the isolated, perfused snake tubules. Net PAH transport against a concentration gradient from bath to lumen occurs only in the distal portion of the proximal tubule.[56] This transepithelial secretory transport process is saturable with an apparent K_m of about 0.01 mM and a V_{max} of about 350 fmol min^{-1} mm^{-1}.[80] During the transport from bath to lumen against a concentration gradient, the concentration of PAH in the cell water is greater than that in either the bath or the lumen.[56] Since there is no evidence of binding of PAH within the cells, these data are compatible with transport into the cells against a concentration gradient at the peritubular side. If it is assumed that the inside of the cells is electrically negative compared to the bath, and the PAH is transported as an anion, then transport into the cells at the peritubular side is against an electrical gradient as well. p-Aminohippurate can then move down an electrochemical gradient from the cells to the lumen, and the apparent permeability of the lumenal membrane to PAH (3.5 × 10^{-5} cm sec^{-1}) is about seven times the apparent permeability of the peritubular membrane (0.5 × 10^{-5} cm sec^{-1}).[56,61] These values are consistent with the concept of active transport of PAH into the cells at the peritubular side and some form of passive movement (probably carrier mediated) into the lumen.

The transepithelial permeability to PAH calculated from these independently measured lumenal and peritubular membrane permeabilities (0.44 × 10^{-5} cm sec^{-1}) is very close to that determined directly from PAH

---

[80] W. H. Dantzler, *in* "Amino Acid Transport and Uric Acid Transport" (S. Silbernagl, F. Lang, and R. Greger, eds.), pp. 169–180. Thieme, Stuttgart, Federal Republic of Germany, 1976.

efflux from the lumen (0.67×10^{-5} cm sec^{-1}), suggesting that transepithelial efflux crosses the cells.[80] In fact, the rate of transepithelial efflux of PAH predicted from the membrane permeabilities is virtually identical to that actually measured.[80] These data, showing only a low transepithelial permeability and only a small transcellular lumen-to-bath efflux, are consistent with the observation that net PAH secretion shows almost no dependence on perfusion rate.

The relationship of inorganic ions to the transport step for PAH at the peritubular side of the cells has also been studied. The removal of potassium from the bathing medium reversibly inhibits the active transport step for PAH into the cells at the peritubular side without altering the apparent passive permeability of the peritubular membrane.[61] Although the dependence of this transport step on potassium is apparent, the mechanism involved in the dependence is not. Sodium is also necessary for PAH transport. Removal of sodium from the bathing medium reversibly depresses net PAH secretion. Apparently, this involves both a depression of the active transport step into the cells at the peritubular membrane and an increase in the passive permeability of that membrane. These effects are not a result of an increase in the cytosolic concentration of calcium.[81] The relationship of sodium to PAH transport does not appear to involve a simple cotransport step. However, the inhibitory effects of SITS on the transport step for PAH into the cells at the peritubular side[82] suggests that this step may involve countertransport for other anions whose movement might be influenced by sodium. Finally, studies on the effects of a low calcium concentration using lanthanum, and verapamil suggest that the entry of calcium into the cells, but probably not the cytosolic concentration of calcium, is important for the transport step at the peritubular membrane and for the maintenance of the normal passive permeability of that membrane to PAH.[83]

Although movement of PAH from the tubule cells to the lumen during the process of net secretion is down an electrochemical gradient and could occur by simple passive diffusion, the data suggest that this is a mediated process. The movement of labeled PAH across the lumenal membrane is inhibited and the apparent permeability of this membrane to PAH is reduced by the presence of unlabeled PAH, Phenol Red, probenecid, and SITS.[82,84] This mediated transport step from the cells to the lumen for PAH (and, presumably, other organic acids that share this system) is not in-

---

[81] W. H. Dantzler and O. H. Brokl, *Am. J. Physiol.* **246,** F175 (1984).
[82] W. H. Dantzler and S. K. Bentley, *Am. J. Physiol.* **238,** F16 (1980).
[83] W. H. Dantzler and O. H. Brokl, *Am. J. Physiol.* **246,** F188 (1984).
[84] W. H. Dantzler and S. K. Bentley, *Am. J. Physiol.* **236,** F379 (1979).

fluenced by the presence or absence of sodium or potassium in the lumen.[61,64] Moreover, it is not dependent on chloride in the lumen, but it may be dependent on the presence of a major anion to which the membrane is highly permeable and may involve anion exchange.[60] It also may be dependent on calcium entry into the cells.[81,83] Finally, inhibition of this lumenal exit step appears to reduce uptake of PAH at the peritubular membrane in a secondary fashion, suggesting that there may be some feedback coupling between the two transport systems.[60,82,84]

Net secretion of urate: Net urate transport from the bath to the lumen occurs against a concentration gradient throughout the proximal tubule, but not in the distal tubule.[55,80] It appears that net urate secretion, like that of PAH and other organic anions, occurs by a process of transport into the cells against an electrochemical gradient at the peritubular side and movement from the cells down an electrochemical gradient at the lumenal side.[55] There are important differences from the transport process for PAH and similar organic anions. First, the apparent passive permeability of the lumenal membrane to urate (0.75×10^{-5} cm sec^{-1}) is much lower than that of the peritubular membrane (3.10×10^{-5} cm sec^{-1}), suggesting an inefficient system if urate transported into the cells is to move easily into the lumen and not back into the bath.[80] Second, the transport step into the cells on the peritubular side appears to be dependent in part on the presence of glomerular filtrate (or an equivalent artificial perfusate) moving through the lumen. Third, net urate secretion varies directly with perfusion rate, suggesting significant back diffusion into the bath at low perfusion rates.[55] Indeed, significant efflux from lumen to bath that varies with the perfusion rate can be demonstrated.[55] Moreover, the transepithelial permeability determined directly from this efflux (about 2.4×10^{-5} cm sec^{-1}) is about four times that (0.60×10^{-5} cm sec^{-1}) calculated from the independently measured values for the lumenal and peritubular membranes given above.[80] This observation suggests that much of the apparent passive backflux occurs between the cells.[80] The relatively high passive permeability of the peritubular membrane, the apparent dependence of the transport step at this membrane on the presence of fluid in the lumen, and the apparent large passive backflux between the cells all may function to reduce the accumulation of urate in the cells or lumens of nephrons that are not filtering. Fourth, saturation of the transepithelial urate transport process occurs at much higher bath concentrations than saturation of the PAH transport process. The K_m for urate transport (about 0.15 mM) is about 15 times that obtained for PAH transport under similar circumstances.[80] Fifth, in contrast to PAH transport, net urate secretion is unaffected when all the sodium in the bathing medium and perfusate is replaced with choline.[63] Thus, the transport of urate into the cells at the peritubular side has neither a direct nor an indirect dependence on sodium. However, as in the case of PAH transport, removal of potassium from the bathing medium reversibly inhibits net urate secretion.[63] The active transport step for urate into the cells at the peritubular side appears to be completely inhibited in the absence of potassium.[63] Sixth, the movement of urate from the cells to the lumen during the net secretory process is not readily inhibited, and there is no evidence that it is mediated in any fashion.[84,85] It could occur by simple passive diffusion. Finally, there is additional evidence that urate is transported by a separate pathway from that for PAH or other organic anions. As noted above, urate is secreted throughout the proximal tubule whereas PAH is secreted only in the distal portion of the proximal tubule. Moreover, urate secretion is not inhibited by high concentrations of PAH and PAH secretion is not inhibited by high concentrations of urate.[86]

Net reabsorption of lactate: The endogenous organic acid lactate is freely filtered at the glomerulus and undergoes net absorption. In these isolated tubules, the unidirectional flux of lactate from the lumen to the bath is greater than the unidirectional flux from the bath to the lumen, the flux ratio being about 3.0.[87] The unidirectional fluxes and flux ratios are essentially the same in both proximal and distal portions of the proximal tubule. Of particular interest is the observation that the lactate transported from the lumen to the bath is not metabolized. Since these experiments were performed so that the concentration gradients during the measurements of the unidirectional fluxes were identical and so that no exchange diffusion could have occurred (since the transepithelial voltage is too small to have accounted for the flux ratio, and since solvent drag could not have affected the fluxes), the absorptive flux must involve some sort of energy-requiring transport process.[87] In this regard, the unidirectional flux from the lumen to the bath is reduced reversibly by about 75% when all the sodium in the perfusate and bath is replaced by choline or tetramethylammonium, but the unidirectional flux from the bath to the lumen is reduced by only 25%.[88] It seems likely that lactate is absorbed by an active sodium-dependent process deriving its energy from the electrochemical gradient for sodium.[88]

Although lactate transported from the lumen to the bath is not metabolized, lactate transported into the cells of nonperfused tubules from the peritubular side is significantly metabolished.[87,89] These data suggest that

[85] S. K. Mukherjee and W. H. Dantzler, *Pfluegers Arch.* **403**, 35 (1985).
[86] W. H. Dantzler, *Handb. Exp. Pharmacol.* **51**, 185 (1978).
[87] P. H. Brand and R. S. Stanbury, *Am. J. Physiol.* **238**, F218 (1980).
[88] P. H. Brand and R. S. Stanbury, *Am. J. Physiol.* **240**, F388 (1981).
[89] P. H. Brand and R. S. Stanbury, *Am. J. Physiol.* **238**, F296 (1980).

absorbed lactate and metabolized lactate are taken up by the tubule cells at opposite membranes and that the pools of absorbed and metabolized lactate are separate. They support the idea that renal absorptive transport conserves substrate for the entire organism whereas peritubular uptake supplies nutrients for the tubule cells.

Mechanism of organic cation transport: The transport of the organic cations TEA and NMN has also been studied with these tubules.[58,59,73] For TEA[59] the unidirectional lumen-to-bath and the unidirectional bath-to-lumen fluxes both exhibit saturation kinetics. However, the bath-to-lumen flux also exhibits an apparent diffusive component whereas the lumen-to-bath flux does not.

Net secretion of TEA: Transport into the cells across both the lumenal and peritubular membranes is apparently against an electrochemical gradient and is inhibited by cyanide. The K_m for the lumen-to-bath flux (about 6.0 μM) is about one-third the K_m for the bath-to-lumen flux (about 20 μM), indicating that the lumenal transporter has a greater affinity for TEA than does the peritubular transporter. However, the V_{max} for the bath-to-lumen flux (about 150 fmol min^{-1} mm^{-1}) is about six times the V_{max} for the lumen to bath flux (about 27 fmol min^{-1} mm^{-1}), indicating a greater capacity for the peritubular than the lumenal transporter and apparently accounting for net tubular secretion of TEA. The fluxes are inhibited only by high concentrations of NMN. The lumen-to-bath flux but not the bath-to-lumen flux is significantly reduced (by about 40%) by the replacement of sodium in bathing medium and perfusate with sucrose, indicating some form of partial sodium dependency of the lumenal transporter for TEA. All the data indicate that there is active (either primary or secondary) TEA transport into the cells at both the lumenal and peritubular membranes but net transepithelial transport in the bath-to-lumen direction.[59]

Net reabsorption of NMN: Although all organic cations are generally considered to share the same transport system in renal tubules, the system for NMN clearly differs from that for TEA in these isolated perfused snake renal tubules.[58] For NMN, the unidirectional lumen-to-bath and bath-to-lumen fluxes both saturate. The K_m for the lumen-to-bath flux (about 60 μM) is slightly lower than the K_m for the bath-to-lumen flux (about 86 μM). The V_{max} for the lumen-to-bath flux is also slightly higher (about 491 fmol min^{-1} mm^{-1}) than the V_{max} for the bath-to-lumen flux (about 483 fmol min^{-1} mm^{-1}). However, neither of these differences is statistically significant. Nevertheless, direct net flux measurements indicate that net absorption always occurs and that the rate of such absorption is equivalent to the difference between the independently measured unidirectional fluxes. Thus, for NMN two relatively large unidirectional fluxes differ only slightly, resulting in a small net absorptive flux, whereas for TEA two much smaller unidirectional fluxes differ markedly, resulting in a substantial net secretory flux.[58,59] Moreover, the NMN transport is not inhibited even by very high concentrations of TEA.[58]

Transport of NMN into the cells across both the lumenal and peritubular membranes during the flux measurements, in contrast to the transport of TEA, is down an electrochemical gradient.[58] Inhibition of NMN transport with NMN analogs (which primarily block entry into the cells) suggests that the transport step into the cells across the lumenal membrane during the lumen-to-bath flux is more specific than is the transport step into the cells across the peritubular membrane during the bath-to-lumen flux.[58] Transport out of the cells across both the lumenal and peritubular membranes during flux measurements is apparently against an electrochemical gradient.[58] Measurements of the efflux across the individual cell membranes, as described above, indicate that transport across the peritubular membrane, but not the lumenal membrane, can be driven by countertransport or by proton–organic cation exchange (W. H. Dantzler, unpublished observations).

Mechanism of transepithelial transport of neutral organic solute: reabsorption of glucose: Glucose transport by these isolated, perfused renal tubules shows some differences from that in other vertebrate renal tubules.[66] As in other vertebrates, net saturable absorption occurs in the proximal tubule. However, the maximum rate of net absorption is twice as great in the distal portion of the proximal tubule as in the proximal portion,[66] exactly the opposite of the situation in mammalian tubules.[90] The apparent permeability of the lumenal membrane in the lumen-to-cell direction (about 10×10^{-5} cm sec^{-1}) is about 15 times the permeability in the cell-to-lumen direction (about 0.65×10^{-5} cm sec^{-1}). The apparent permeability of the lumenal membrane in the lumen-to-cell direction is reduced by about 50% when sodium is removed from the lumen, although this is probably not a maximum reduction because sodium could still enter the lumen from the bath at this time. Moreover, the apparent permeability of the lumenal membrane in the lumen-to-cell direction is reduced to essentially identical with the apparent permeability in the cell-to-lumen direction when phlorizin is added to the perfusate. However, the glucose concentration in the cells is always below the concentration in the lumen during the process of maximum absorption. This low intracellular concentration of transported glucose cannot be explained by metabolism. The low intracellular glucose concentration, the high control value for the permeability of the lumenal membrane in the lumen-to-cell direction, and the

[90] B. M. Tune and M. B. Burg, *Am. J. Physiol.* **220**, 87 (1971).

effects of phlorizin and a sodium-free perfusate on this permeability all suggest that glucose enters the cells across the lumenal membrane by a saturable, phlorizin-sensitive, sodium dependent form of facilitated diffusion during maximum absorption.[66] Whether or not a form of secondary active transport, as suggested for many other species,[91] also functions at the lumenal membrane of this species when the glucose concentration in the lumen is low cannot be determined from these data.

Although the movement of glucose out of the cells across the peritubular membrane is considered to be a passive process in many species, it clearly appears to be mediated in these isolated, perfused snake tubules.[66] The concentration of glucose in the cells of tubules perfused with equal concentrations of glucose in the perfusate and bathing medium is only about 60% of these concentrations during maximum absorption.[66] This observation alone suggests that glucose is actively transported out of the cells on the peritubular side. This possibility is further supported by the observations that the glucose concentration in the cells tends to decrease further compared with that in the bath when entry from the lumenal side is blocked with phlorizin and to rise to equal that in the bath when the tubules spontaneously stop transporting glucose.[66] Finally, the apparent permeability of the peritubular membrane to glucose, determined in the bath-to-cell direction (about 0.46×10^{-5} cm sec^{-1}), is too low to permit glucose absorption across this membrane at the observed rates by simple passive diffusion.[66]

In Vitro Microperfusion of Reptilian Distal Tubules

Amiloride-Sensitive Na$^+$ Reabsorption. Electrophysiological evidence for active reabsorptive Na$^+$ transport: When perfused *in vitro,* snake distal tubules generate lumen-negative transepithelial voltages which are dependent upon the presence of Na in the lumen perfusate.[65] The transepithelial resistance is approximately 79 $\Omega \cdot$ cm^2, which ranks this tubule among the moderately tighter epithelia. The perfusion of the lumen with amiloride (10^{-5} M) abolishes all transepithelial potentials while doubling the transepithelial resistance.[92] Amiloride is a well-known inhibitor of Na$^+$ transport across tight (aldosterone-sensitive) epithelia. Hence, the transepithelial responses of the tubule to amiloride and to Na$^+$ concentrations in the tubule lumen support the conclusion of active Na$^+$ reabsorption in this tubule segment.

---

[91] W. H. Dantzler, *in* "Renal Transport of Organic Substances: Proceedings in Life Sciences" (R. Greger, F. Lang, and S. Silbernagl, eds.), pp. 290–308. Springer-Verlag, Berlin, 1981.

[92] K. W. Beyenbach, B. M. Koeppen, W. H. Dantzler, and S. I. Helman, *Am. J. Physiol.* **239**, F412 (1980).

Self-inhibition of Na^+ transport. It appears that the mechanism of Na^+ transport in the distal tubule is poised to reabsorb Na^+ from low concentrations.[92] Transepithelial voltage and resistance are only in a steady state when the Na^+ concentration in the lumen is less than 30 mM. The perfusion with higher Na^+ concentrations initially yields results consistent with the stimulation of Na^+ reabsorption. This stimulatory response lasts only 5–10 min. Thereafter, the transepithelial voltage (V_T) decays toward zero, the transepithelial resistance (R_T) increases, and the epithelial cells of the tubule swell.[65,92] The calculated short-circuit current, I_{sc},

$$I_{sc} = V_T/R_T \tag{10}$$

reflecting Na^+ reabsorption falls in parallel with the decrease of the Na^+ conductance of the apical membrane.[92] These results indicate that the perfusion of the distal tubule lumen with Na^+ concentrations greater than 30 mM (but still less than normal Ringer concentrations) reduces the Na^+ permeability of the apical membrane. The intracellular mechanism responsible for the Na^+-induced inhibition of Na^+ entry across the apical membrane is unknown.

The Distal Tubule as Diluting Segment. Transepithelial water permeabilities: In studies of transepithelial water flow, the snake distal tubule is found to be virtually impermeable to transepithelial osmotic water flow. The hydraulic conductivity measures 1.2×10^{-7} cm^3 sec^{-1} cm^{-2} atm^{-1} and is not affected by vasopressin. Investigating the symmetry of the tubule response to hypertonic solutions presented from the lumenal or peritubular side reveals the apical membrane and also the tight junctions, as the water-impermeable barriers of the epithelium (Fig. 10). No change in the epithelial cell volume is observed with lumenal hypertonicity. In contrast, the epithelial cells shrink markedly when the peritubular bath becomes hyperosmotic.

Attempts to measure the water permeability of the basolateral membrane: As shown in Fig. 10, the exposure of the tubule to hyperosmotic fluids in the bath causes the cells to shrink, indicating the water permeability of the peritubular membrane. The loss of cell water to the bath reduces the height of cells with the effect that the lumen diameter increases (Fig. 10). The lumen diameter (D) is inversely proportional to the core resistance (R_c), a parameter in the cable equations

$$R_c = 4r/\Pi D^2 \tag{11}$$

where r is the resistivity of the fluid (Ringer) in the tubule lumen. Accordingly, changes in the lumen diameter, reflecting changes in cell volume, can be followed electrically with nearly instantaneous time resolution, and used to measure the water permeability of the peritubular membrane. Though attractive in theory, the experiment cannot be performed because

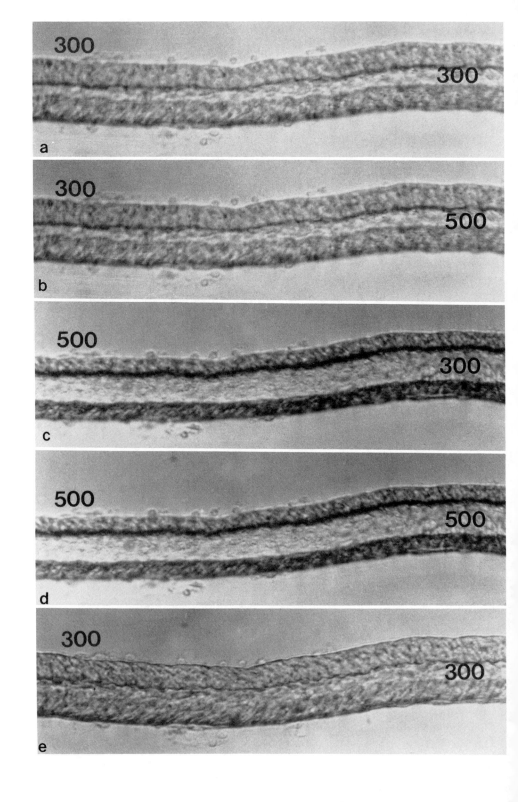

the time constant of the bath change is not significantly different from the time constant of the tubule response.[93] However, with the use of a smaller bath chamber and an improved laminar flow system, as described by Strange and Spring,[54] it should be possible to perform this experiment. Nevertheless, our preliminary studies of water flow across the snake distal tubule established the water-permeable and -impermeable barriers at the apical and basolateral membranes, respectively. In view of Na^+ reabsorption across a water-impermeable epithelium, the snake distal tubule is a diluting segment that may be useful *in vivo* when the animal needs to excrete hypotonic water loads.

Bird Renal Tubules

Unique Tubular Functions

Birds and mammals are the only vertebrate classes which produce urine more concentrated than the plasma through countercurrent multiplication in the loop of Henle. In addition, birds excrete uric acid, which further reduces the need for obligatory water loss in renal excretion. Undoubtedly, urinary concentration mechanisms and uricotelism both contribute significantly toward reducing flight loads and therefore toward improved metabolic efficiencies while airborne. It appears that birds have developed their unique renal strategies by borrowing from reptiles and mammalian kidneys. The renal cortex of bird kidneys is composed of reptilian-type nephrons and the renal medulla is composed of mammalian-type nephrons.[94,95]

The only segments of avian renal tubules perfused *in vitro* are the early distal tubule of the reptilian-type nephrons and the thick ascending limb of the mammalian-type nephrons from Japanese quail *(Coturnix coturnix)*. These studies have been performed by Nishimura and colleagues.[96-98]

[93] K. W. Beyenbach, *Am. J. Physiol.* **246**, F290 (1984).
[94] E. J. Braun and W. H. Dantzler, *Am. J. Physiol.* **222**, 617 (1972).
[95] E. J. Braun, *Comp. Biochem. Physiol.* **71**, 8 (1982).
[96] H. Nishimura, T. Miwa, and J. R. Bailey, *J. Exp. Zool.* **232**, 697 (1984).
[97] H. Nishimura, M. Imai, and M. Ogawa, *Am. J. Physiol.* **250**, R333 (1986).
[98] T. Miwa and H. Nishimura, *Am. J. Physiol.* **250**, R341 (1986).

FIG. 10. Water-permeable and -impermeable barriers of snake renal distal tubule. The osmolality of the lumenal perfusate and peritubular Ringer solution are indicated (300 or 500 mOsm). Peritubular hyperosmolality reduces the volume of the epithelial cells; lumenal hyperosmolality has no effect on cell volume. [From K. W. Beyenbach, *Am. J. Physiol.* **246**, F290 (1984).]

Bird Sources

In the United States the most studied avian species is the chicken because of its commercial value. There are a large number of suppliers. Consult local feed stores, agricultural colleges, and university departments of poultry for the best local sources. Nishimura and collaborators obtain their Japanese quail (or bobwhite quail) from E. Q. S. Manufacturing Company (Savannah, GA). Brown and Dantzler obtain species of European starlings by trapping them in the Arizonan desert.

Basic Methods for Isolation and Perfusion of Tubules

The basic principles of perfusion are those described by Burg ([12], this volume). The techniques and the types of measurements are those already described in this chapter. The only differences concern the isolation and maintenance of the tissue and these are described here.[97,98] The experiments are performed on 3-to-6-week-old Japanese quails. In animals older than 6 weeks, the connective tissue combined with the fragility of the renal tubules makes it impossible to prepare segments of adequate length and condition for perfusion (H. Nishimura, personal communication; W. H. Dantzler, unpublished observations). The animals are killed by decapitation. The kidneys are quickly removed, cut into blocks 4-5 mm thick, and placed in chilled avian Ringer containing 1% bovine serum albumin (fraction V; Sigma, St. Louis, MO). The albumin is added to keep the tissue fragments from sticking to the dissection instruments. The basic avian Ringer contains (in mM): NaCl, 115.0; $NaHCO_3$, 25.0; KCl, 5.0; sodium acetate, 10.0; Na_2HPO_4, 2.0; NaH_2PO_4, 0.5; $CaCl_2$, 1.9; $MgCl_2$, 1.0; D-glucose, 8.3; L-alanine, 5.0. It is bubbled with 95% O_2, 5% CO_2 and the pH is maintained at 7.4. Several intact medullary cones are isolated and transferred to fresh Ringer for further dissection.

Isolation of Thick Ascending Limbs. The outer cortical area is removed and the tubular structures in the medullary cones are carefully separated with fine forceps and needles to isolate the thick ascending limbs. In the avian kidney, the descending thin limb always thickens before the bend of the loop and the entire ascending limb is of uniform diameter and wall thickness from the tip of the loop to the base of the cone. Thus, if one can trace upward the thick limb from the bend of the loop, one can readily identify the thick ascending limb. However, as Nishimura *et al.*[97] note, within the basal region of the medullary cone it is difficult to differentiate between the thick descending limb and the thick ascending limb because the thick descending limb has considerable length in this region and the transition to the thin limb is not easy to recognize. Moreover, the thin descending limb is too fragile to permit dissection of the descending limb in a retrograde fashion. Therefore, in dissecting thick ascending limbs and differentiating them from thick descending limbs, it is best to trace them from the bend of the loop.

Isolation of Reptilian-Type Nephrons. The reptilian-type nephrons are located in the outer, cortical regions of the avian kidney. The tubules are less tightly bound together than the portions of the mammalian-type nephrons in the medullary cones. The intermediate segment of the reptilian-type nephrons (analogous to the thin limb of Henle's loop in the mammalian-type nephrons) is very short and the transition to the early distal tubule is impossible to define clearly under a dissecting microscope. Therefore, Nishimura et al.[97] defined the portion of the distal tubule near the glomerulus as the early distal segment. The parent glomerulus generally remains firmly attached to the dissected segment.

As noted above, the basic perfusion method and the methods for the measurements of transepithelial potentials and water and ion fluxes are essentially the same as those previously described in this chapter. The perfusate and bathing medium are identical avian Ringer except that the bathing medium contains 1% bovine serum albumin. The osmolality of the avian Ringer is about 312 mOsm and the difference in osmolality between the solution with and without albumin is less than 2 mOsm. Studies to date have been performed at 37°, although the avian body temperature is about 41°, to slow deterioration of the tissue *in vitro* and to reduce evaporation from the bathing.[97]

Transepithelial Potential. The transepithelial potential in the thick ascending limb of Henle's loop of the mammalian-type nephrons averages about +9.0 mV, lumen positive.[97] This voltage decreases with increasing perfusion pressure and is depressed in a reversible manner by the addition of furosemide to the lumen, sodium cyanide to the bathing medium, or ouabain to the bathing medium. Removal of sodium or chloride from the perfusate and bathing medium reduces the transepithelial voltage to a value that is not statistically different from zero.[97]

The distal tubule of the reptilian-type nephrons demonstrates a varying transepithelial voltage depending on the location with regard to the parent glomerulus.[97] The transepithelial potential is about −3.0 mV, lumen negative, at 300 μm or more distal to the attachment to the parent glomerulus. This potential approaches zero and then becomes lumen positive as the perfused area approaches the attached parent glomerulus. It averages about +4.0 mV, lumen positive, where the parent glomerulus is in close contact with the tubule. These observations suggest that the thick ascending limb of Henle's loop of the avian mammalian-type nephrons resembles the thick ascending limb of true mammalian nephrons. They also suggest that the earliest distal segment of the avian reptilian-type nephrons is similar and probably functions as a diluting segment, as in amphibians.

Transport of Water and Chloride. Measurements of water and chloride transport across the thick ascending limbs of the avian mammalian-type nephrons further support their similarity to the thick ascending limbs of true mammalian nephrons.[98] Net transepithelial water movement (J_v) is essentially zero (-0.01 ± 0.02 nl min^{-1} mm^{-1}) when this segment is perfused and bathed with isosmotic solutions. It increases only slightly (to about 0.24 nl min^{-1} mm^{-1}) when an osmotic gradient of 120 mOsm (bath > lumen) is imposed. Thus the L_p is extremely low (about 12×10^{-9} cm^2 sec^{-1} atm^{-1}) and cannot account for significant water movement. With the osmotic gradient present the addition of arginine vasotocin (the natural antidiuretic hormone in birds) even in extremely high concentrations has absolutely no effect on J_v or L_p. The lumen-to-bath chloride flux (about 370 pmol min^{-1} mm^{-1}) is substantially greater than the bath-to-lumen flux (about 100 pmol min^{-1} mm^{-1}) and is not influenced by arginine vasotocin, further supporting the idea that this portion of the tubule functions as a diluting segment and contributes to the generation of an osmotic gradient in the medullary cones.

Acknowledgments

The work of the authors has been supported by grants from the National Science Foundation and the National Institutes of Health. Portions of this chapter were written by W. H. Dantzler during tenure as an Alexander von Humboldt United States Senior Scientist Award at the University of Würzburg, West Germany.

[12] Microdissection of Kidney Tubule Segments

By P. A. WRIGHT, M. B. BURG, and M. A. KNEPPER

Introduction

The mammalian nephron consists of approximately 20 distinct renal tubule segments. Microdissection of fresh kidney tissue and isolation of individual tubule segments is a valuable experimental technique to determine the metabolic characteristics of these renal tubule segments. Some of the segments contribute relatively little to the total tissue mass, but play important functional roles. The metabolic characteristics of these segments can be assessed through microdissection and application of the appropriate biochemical techniques to the single tubules, e.g., measurement of the enzyme activities and metabolite concentrations. Other standard biochemical preparations such as tissue slices, mixed tubule suspensions, cell suspensions, and subcellular and mitochondrial preparations cannot always adequately address the heterogeneous nature of renal tissue.

In some studies, tubule segments have been dissected from lyophilized tissue as described by Lowry and Passonneau[1]; however, the process of lyophilization itself may be detrimental to the function of some enzymes.[2] In this chapter, we will present the method of microdissection of fresh (unfrozen) kidney tissue. We will describe the dissection of the rat kidney and the reader is referred to Burg *et al.*[3] and Morel *et al.*[4] for a detailed description of rabbit kidney dissections. Finally, we will discuss the dissection of kidneys pretreated with collagenase because collagenase aids the dissection of tubule segments while preserving cellular function. Tubules pretreated with collagenase, however, are not satisfactory for isolated tubule perfusion studies.[3]

Dissecting Apparatus

To discriminate the different renal tubule segments, it is necessary to use a high-quality stereomicroscope. In our laboratory we use Wild M-8 dissection microscopes (Switzerland) with direct illumination from below the dissecting dish with diffuse light. Direct illumination is particularly critical for inner medullary dissections since the margins of individual collecting ducts and thin limbs are nearly impossible to discern with indirect or dark-field illumination. Direct illumination permits visualization of thin limbs that adhere to the medullary collecting duct surface and must be stripped off using dissecting forceps.

Tubules can be dissected with dissecting forceps (Dumont No. 5, Switzerland), dissecting needles, or a combination of both. The tubules tend to stick to rough surfaces, so it is necessary to create a smooth surface on the tips of the forceps or needle with abrasive paper (0.3–30 μm; Thomas Scientific, Philadelphia, PA).

Collagenase Perfusion

The rat is killed by decapitation and the left kidney is isolated by ligating the aorta at two sites: just above the aortic bifurcation and between

[1] O. H. Lowry and J. V. Passonneau, "A Flexible System of Enzymatic Analysis," p. 121. Academic Press, New York, 1972.
[2] L. C. Garg, M. A. Knepper, and M. B. Burg, *Am. J. Physiol.* **240,** F536 (1981).
[3] M. B. Burg, J. Grantham, M. Abramow, and J. Orloff, *Am. J. Physiol.* **210,** 1293 (1966).
[4] F. Morel, D. Chabardes, and M. Imbert, *Kidney Int.* **9,** 264 (1976).

the origin of the left and right renal arteries. A small incision is made in the aorta below the left kidney and the tip of a heat-tapered cannula (PE-90; Clay Adams, Parsippany, NJ) is advanced toward the kidney and secured with surgical silk (#000). Prior to perfusion, the vena cava is severed in order to facilitate the flow of fluid out of the kidney. The kidney is perfused initially with 10 ml of ice-cold bicarbonate-free dissecting solution [composition (mM): NaCl, 135; Na_2HPO_4, 1; Na_2SO_4, 1.2; $MgSO_4$, 1.2; KCl, 5; $CaCl_2$, 2; glucose, 5.5; HEPES, 5; pH 7.4], followed by 10 ml of dissecting solution containing collagenase (type I, 300 U/mg, Sigma Chemical Co., St. Louis, MO) and bovine serum albumin (0.1% BSA, fraction V, Boehringer Mannheim Biochemicals, Indianapolis, IN). The dissection solution is bubbled with 100% O_2 prior to the addition of collagenase and BSA.

After perfusion, the left kidney is excised and a coronal section is made which contains the entire corticomedullary axis. This section is trimmed to yield a pyramid along the corticomedullary axis with the papillary tip as the apex. The pyramid can be cut into various sections: the cortex, the outer stripe of the outer medulla, the inner stripe of the outer medulla, and the inner medulla (Fig. 1). Individual sections are then transferred into test tubes containing 1 ml of the same collagenase solution as used to perfuse the kidneys. The test tubes are incubated for 20 min (outer stripe–outer medulla), 30 min (cortex, inner stripe–outer medulla), and 40–60 min (inner medulla) at 37° in a shaking water bath or a block heater and orbital shaker. The samples are suffused with 100% O_2 during the incubation. The incubated tissue is rinsed with collagenase-free dissection medium, and stored on ice until dissection of the tubules. In general, tubules should be dissected immediately, but some tubules are more metabolically active

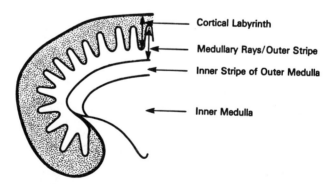

FIG. 1. Regional architecture of the mammalian kidney. Diagram depicts a hemisected unipapillate kidney such as found in rat or rabbit. See Table I for the location of renal tubule segments in each of the regional sections.

than others and should be given priority. For instance, for enzyme assays it is advisable to isolate and transfer proximal tubules within 30 min, whereas inner medullary tubules are viable for at least 90 min on ice.

Dissection

Cortex

Tubules are dissected in the bicarbonate-free dissecting solution at 17°. Table I summarizes the tubule segments found in each of the regional sections. In the cortex the following segments can be identified and dissected: glomerulus, segments 1 and 2 of the proximal convoluted tubule (S_1, S_2), the cortical thick ascending limb (CTAL), the distal convoluted tubule (DCT), the cortical connecting tubule (CNT), and the cortical collecting duct (CCD). Glomeruli are easy to identify because of their small

TABLE I
LOCATION OF RENAL TUBULE SEGMENTS

Major region	Minor region	Tubule segment	Abbreviation
1. Cortex	Cortical labyrinth	Glomeruli	GLM
		Early proximal convoluted	PCT (S_1)
		Late proximal convoluted	PCT (S_2)
		Connecting tubule	CNT
	Medullary rays	Glomeruli	GLM
		Early proximal straight	PST (S_2)
		Cortical thick ascending limb	CTAL
		Cortical collecting duct	CCD
2. Outer medulla	Outer stripe	Late proximal straight	PST (S_3)
		Medullary thick ascending limb, outer stripe	MTAL-OS
		Outer medullary collecting duct, outer stripe	OMCD-OS
	Inner stripe	Medullary thick ascending limb, inner stripe	MTAL-IS
		Outer medullary collecting duct	OMCD-IS
		Descending thin limb of short loops (short descending limb)	SDL
		Descending thin limb of long loops, upper part (long descending limb)	LDL_u
3. Inner medulla		Inner medullary collecting duct	IMCD
		Descending thin limb of long loops (long descending limb), upper part	LDL_u
		Descending thin limb of long limbs (long descending limb), lower part	LDL_l
		Thin ascending limb	tAL

ball-like shape. Proximal convoluted tubules are located in the cortical labyrinth, have relatively large diameters (40–50 μm), are dark in color, and have no visible lumen. The S_1 segment is identified by its attachment to a glomerulus and does not exceed 1 mm in length. The S_2 segment of the proximal tubule begins about 1 mm after the glomerulus, is convoluted for most of its length, but gradually transforms into a straight segment before descending into the outer medulla (see the next section). The CTAL and CCD segments run parallel in the medullary rays of the cortical slice. The CTAL is straight and narrow, with a clear lumen, and single cells cannot be visualized under high magnification. The CCD does not have a defined lumen, but rather a bumpy "cobblestone" appearance due to single cells that can be visualized under high magnification. The CCD is unbranched throughout most of its length, but is formed from the junction of two smaller tubules (termed "mutual collecting tubules") in the superficial cortex. The DCT, found in the cortical labyrinth, is similar in appearance to the CTAL, but is slightly thicker and has a distinctive convolution in the region where it contacts the vascular pole of the glomerulus. The CNT is located in the vicinity of the arcuate arteries at the corticomedullary junction and the interlobular arteries in the cortical labyrinth. The CNT has a similar appearance to the CCD, but is darker in color, with a more irregular surface, and branches of equal thickness.

Outer Medulla

The following tubules can be dissected from the outer medulla: segment 3 of the proximal tubules (S_3), the medullary thick ascending limbs (MTAL), the outer medullary collecting ducts (OMCD), the upper portion of the long descending loop of Henle (LDL_u), and the short descending loop of Henle (SDL) (Table I). The S_3 proximal straight tubule is located in the outer stripe of the outer medulla. The S_3 segment can be confused with the late, straight S_2 segment; however, the S_3 tubule is identified by its rapid transformation into the descending limb of the loop of Henle. The S_3 segment has a similar appearance to the cortical proximal tubule, and although it is mostly straight, it is not unusual to find bends. The MTAL and OMCD lie in parallel in the outer medulla and are usually dissected from the inner stripe. They are easily distinguished and maintain the same general features described above for the CTAL and CCD. There are two distinct thin descending limbs in the inner stripe of the outer medulla, the descending limbs of the short loops (SDL) and the upper portion of the descending limbs of the long loops (LDL_u). The SDL is very straight and very narrow. The LDL_u is much thicker, often two-thirds to three-quarters as thick as the thick ascending limbs. The LDL_u generally takes a wavy course through the inner stripe.

Inner Medulla

In the inner medulla the following tubules can be dissected: the inner medullary collecting ducts (IMCD), the lower portion of the long descending limb of Henle (LDL_l), and the thin ascending limb of the loop of Henle (tAL) (Table I). The thin limbs and IMCD are oriented in parallel. The IMCD is highly branched and relatively long segments can be stripped from a piece of tissue by starting at the papillary tip and dissecting upward. The IMCD surface should be cleaned of adhering thin limbs when possible.

Transfer

The length of each tubule (0.4–1.4 mm) is first measured with an ocular micrometer. Prior to tubule transfer, tubules are "washed" by dragging them across the dissecting dish with the tips of the forceps to a free area. The individual tubules are transferred from the dissecting dish to separate assay tubes by attaching them to small glass beads (0.5-mm diameter, Thomas Scientific Co.) A single bead is placed on top of the tubule with clean forceps and, in most cases, the tubule will stick spontaneously to the glass bead on contact. The glass bead is drawn by negative pressure onto the end of a disposable yellow plastic tip of an automatic pipettor (0–100 μl, Gilson, France), set to 5 μl. The beads are slightly larger than the lumen of the pipet tip, and therefore the bead remains on the outside of the pipet tip until it is safely transferred to a test tube. This transfer method is more reliable than fluid-phase transfer because the beads are visible with the naked eye and there is less fluid carryover.

With most enzyme assays, it is necessary to permeabilize the cells. Tubules (attached to glass beads) are typically placed in a hypoosmotic solution (25 μl of 1 mg/ml BSA in distilled water) and frozen on dry ice. This permeabilization step has been shown to be effective for various enzyme assays, including aldose reductase,[5] Na^+,K^+-ATPase,[6,7] and glutamate dehydrogenase[8] and phosphate-dependent glutaminase (P. A. Wright and M. A. Knepper, unpublished observations, 1990). For certain enzymes, it may be necessary to add small quantities of stabilizing agents to the hypoosmotic medium.[1] Studies show that the presence of BSA is critical to enzyme activity, although the exact concentration is not critical and can be varied between 0.5 and 2 mg/ml.[5]

---

[5] J. M. Sands, Y. Terada, L. M. Bernard, and M. A. Knepper, *Am. J. Physiol.* **256**, F569 (1989).

[6] A. Doucet, A. I. Katz, and F. Morel, *Am. J. Physiol.* **237**, F105 (1979).

[7] Y. Terada and M. A. Knepper, *Am. J. Physiol.* **256**, F218 (1989).

[8] P. A. Wright and M. A. Knepper, *Am. J. Physiol.* (in press) 1990.

[13] Measurement of Transmural Water Flow in Isolated Perfused Tubule Segments

By JAMES C. WILLIAMS, JR., and JAMES A. SCHAFER

The measurement of water, or volume, flow across renal tubules is required in various segments of the nephron in a variety of studies in which the quantification of water flow is the primary objective, but it is also required for studies in which *solute movement* is the focus. The accurate measurement of volume flow is critical to the measurement of net fluxes of solute across the epithelium because of its potential effect on solute flux via solvent drag, and also because of its effect on the average concentration of the lumenal solute, as discussed in Chapter [20] of this volume. The methods described below can be used to measure volume flux in any segment of the nephron, or in any perfused tubular structure. This description presumes that the reader is familiar with the basics of the isolated, perfused tubule system as detailed Wright *et al.* ([12], this volume) and by Chonko *et al.*[1]

The basic technique involves the determination of the difference between the rate at which volume enters the perfused segment (the perfusion rate, $\dot{V}_0$) and the rate at which fluid leaves the segment and enters the collection pipet (the collection rate, $\dot{V}_L$). It is also necessary to normalize this flux measurement to some index of tubule mass or surface area. Presumably the transport rate will be proportional to the membrane surface area involved, which in turn may be proportional to some readily measurable parameter such as the length of the perfused segment. Thus the volume flux is often divided by the length of the tubule segment, and the tacit assumption is made that the surface area of the epithelium per length of tubule is constant among segments from the same part of the nephron. This is probably not a bad assumption, and does avoid the difficulties of assessing the true surface area of the segment, a value that can be obtained only by careful morphometric analysis of electron micrographs of the segment in question.[2,3] Thus, the rate of volume flow in a perfused segment can be written as

$$J_v = (\dot{V}_0 - \dot{V}_L)/L \tag{1}$$

where L is the length of the perfused segment.

The units of J_v are commonly nanoliters minute^{-1} millimiter^{-1}. The

---

[1] A. M. Chonko, J. M. Irish III, and D. J. Welling, *Methods Pharmacol.* **4B**, 221 (1978).
[2] L. W. Welling and D. J. Welling, *Kidney Int.* **8**, 343 (1975).
[3] D. J. Welling and L. W. Welling, *Fed. Proc. Fed. Am. Soc. Exp. Biol.* **38**, 121 (1979).

use of nanoliters in this measurement suggests the first difficulty in accurately determining J_v: the precise measurement of small volumes. In the section below (Measurement of Collection Rate), we describe pipet systems that allow such precise measurement of small volumes, which is required for determination of both the rates of perfusion and collection.

Ideally, one should be able to fix the rate of perfusion of an isolated tubule to a known value by driving the perfusion with a precision syringe pump. In practice, this does not provide the required accuracy. At least one source of imprecision in this method is that the rate of fluid delivery by the pump can easily be greater than the actual rate of perfusion. In cannulating the tubule segment with glass pipets, the tubule is pulled by suction into a position to seal between the outside of the perfusion pipet and the inside of the holding pipet. Although this seal must be tight enough to cause most of the fluid delivered through the perfusion pipet to enter the tubule lumen, some of the delivered fluid can leak between the perfusion pipet and the tubule back into the holding pipet, so that the volume of fluid delivered by the perfusion pipet is not necessarily the volume perfused through the tubule. For this reason the rate of perfusion is usually determined by including a volume marker in the perfusate, as described below (see section "Measurement of Perfusion Rate"). Also, most laboratories use a static pressure head (e.g., a fluid column in a length of tubing attached to the perfusion pipet) to drive perfusion, a system that is much less expensive than a syringe pump and that also provides the experimenter with some knowledge of the maximum intralumenal pressure to which the tubule is exposed.

The rate of collection of the fluid emerging from the perfused tubule (the collectate) can be determined by collecting the fluid in calibrated pipets over timed intervals. The precision of this measurement depends on the proper construction of the pipet holding the collection end of the tubule as well as that of the pipets used to measure the volumes. The details of methods for measuring the collection rate are described in the next section.

Measurement of Collection Rate

If one wishes to measure accurately and precisely the rate at which fluid emerges from a perfused segment, one must first have the end of the tubule sealed into the mouth of a holding pipet so that none of the perfused fluid escapes into the bath and no bath fluid can enter the holding pipet to mix with the collectate. One must also make sure that no evaporation of the collectate occurs within the holding pipet. The seal between the glass micropipet and the tubule is produced by using a holding pipet with a mouth diameter closely matching the outer diameter of the perfused segment, and the seal is enhanced by using Sylgard liquid resin as described below. In order to prevent evaporation, the collectate accumulates in the holding pipet under a layer of a nonpolar solvent which is less dense than water. (We normally use *n*-decane or light mineral oil.) The fluid exiting the tubule is allowed to collect in the holding pipet and is removed at regular intervals by inserting a sampling pipet through the holding pipet and aspirating the collected fluid until the meniscus of the solution returns to some reference point, leaving as small a residual volume in the holding pipet as possible. In order for the meniscus to be well defined and for the solution collection to proceed smoothly, it is necessary that the inner bore of the collecting pipet be clean. Therefore, we carefully clean these pipets between uses by aspirating volumes of water, chloroform, and acetone through them under vacuum.

Figure 1 shows two types of holding pipets that have been used in our laboratory, as well as two types of sampling pipets that are used for removing the collectate from the holding pipet and determining its volume. The holding pipets and the fixed chamber volumetric pipet shown in Fig. 1 are constructed using a microforge as described by Chonko *et al.*[1] The holding pipets are made from standard flint glass tubing 0.084-in. o.d. and 0.064-in. i.d., while the fixed chamber volumetric pipets are usually made from 0.020-in. o.d., 0.014-in. i.d. glass (both obtainable from Drummond Scientific Co., Broomall, PA). In both cases other capillary sizes could be used according to the volume to be collected and any particular demands of the situation. The constant-bore volumetric pipet is made from custom-pulled glass capillaries (also available from Drummond Scientific Co.), and details of its construction are provided below.

The holding pipet shown in the upper part of Fig. 1 is the one used most frequently in our laboratory. It consists of a simple taper of the glass shaped to the desired mouth diameter. Sylgard resin (Sylgard 184 silicone elastomer; Dow Corning Corp., Midland, MI) is polymerized in the tip of the pipet by applying a small amount of the liquid resin mixed with its curing agent to the tip, blowing excess resin out of the tip, and warming the tip with the filament of the microforge while continuing to blow air through the tip to keep it open. The polymerized Sylgard at the tip provides an attachment site for unpolymerized resin, which does not stick well to glass. The usual strategy is to make several of these pipets with mouth diameters spanning the range of tubule diameter expected. After a tubule segment has been cannulated and is perfusing, the outer diameter of the distal end of the segment is measured using a reticle in the ocular of the microscope, and a holding pipet is selected from the assortment to match the tubule. A small amount of unpolymerized Sylgard resin is then applied to the tip of the pipet, and the pipet is maneuvered into a position so that the end of the tubule segment can be aspirated into the mouth of the pipet. If the proper

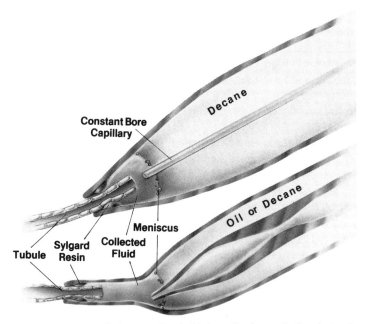

FIG. 1. Configurations of pipets used to hold the collecting end of an isolated perfused tubule and to collect fluid samples. The outer pipet holds the tubule and the Sylgard resin provides a seal that isolates the collecting fluid from the bathing medium, even electrically. In both systems the fluid collects under a layer of decane or light mineral oil. At periodic intervals the sampling pipet is used to remove the accumulated fluid until the meniscus reaches some reference point close to the holding pipet tip, so that the residual volume left in the pipet is minimal (<5-10 nl). *Upper:* Conventional holding pipet with a constant-bore sampling pipet, described in Figs. 2 and 3. *Lower:* A modified holding pipet in combination with a fixed volume collection pipet. The tip of this holding pipet has a smaller inside diameter so that the tubule lumen is slightly constricted, but only at a point, which provides a firm grip on the tubule segment. In this example, the tip of the holding pipet is angled so that a long narrow segment will lie parallel to the bottom of the perfusion chamber, allowing the tubule and the meniscus in the pipet to be in the same focal plane. The sampling pipet has a fixed volume in the chamber defined from the tip to the constriction. The chamber of this pipet and some of the barrel above is filled initially with light mineral oil and fluid is constantly taken up into the pipet by suction so that the meniscus in the holding pipet is kept at the reference level. The time required for the fluid in the sampling pipet to reach the middle of the constriction is recorded. When the pipet is filled the fluid sample is withdrawn for analysis.

amount of Sylgard has been applied to the mouth of the holding pipet (usually an amount just sufficient to completely occlude the pipet tip), the end of the tubule can be pulled into the pipet accompanied by a small volume of bathing medium, and the Sylgard will then flow around the tubule to form a good seal between the inside of the pipet and the bath.

The style of holding pipet shown in the lower part of Fig. 1 has a tip that contacts the tubule segment over a smaller area than does the pipet discussed above. It is made by pulling a straight shaft that is 60 to 100 μm in diameter, and sectioning the end to leave a flat opening that is then "fire polished" with the filament of the microforge until the opening shrinks to the required size. The extended nose of this pipet eliminates the possibility of the end of the tubule coming in contact with the volumetric sampling pipet, and allows for a more precise location of the meniscus. With this style of pipet it is also possible to rinse the inside of the pipet with water and acetone while the tubule is attached, because the flow of collectate out of the tubule into the neck of the pipet prevents these cleaning fluids from coming in contact with the tubule. A disadvantage of this pipet style is the difficulty one has in measuring the diameter of the opening at the tip. The thick glass at the tip distorts the image of the opening as seen in the light microscope when the pipet is viewed from the side, so that good measurements require that the tip be viewed end on, preferably under water or oil to reduce reflection off the glass. The holding pipet shown in the bottom half of Fig. 1 also has a bend in the final taper with an angle of 60°. This modification was made by Delon Barfuss in our laboratory in order to allow the final taper to be positioned horizontally and thus parallel to the bottom of the perfusion chamber, allowing the tubule and the meniscus in the pipet to be in the same focal plane. The angled nose of the pipet is also an advantage when it is necessary to bring a condenser lens close to a short tubule without interference from the holding pipets.[4] Obviously, it is possible to combine features of each holding pipet style in manufacturing pipets for specialized needs.

Also shown in Fig. 1 are two styles of volumetric pipets: a constant-bore pipet in the upper part of the figure and a fixed volume pipet in the lower. With the constant-bore pipet, the rate of collectate formation is determined by collecting fluid over a precisely measured time interval and then calculating its volume by measuring the length of the fluid column in the constant-bore capillary. With a fixed volume pipet, the capacity of the pipet is determined by prior calibration as described below and the collection rate is calculated from the time required to fill it precisely to that calibrated volume.

The constant-bore pipet is made from custom-manufactured glass capillary, the inner diameter of which is typically 30 to 50 μm, giving the capillary a volume of 1 to 2 nl/mm length. The relation between length and diameter of constant-bore tubing is shown in Fig. 2, and the details of the holder constructed in order to support the fine and fragile capillary are shown in Fig. 3. The quality of the constant-bore capillary can be assessed

---

[4] K. L. Kirk, P. D. Bell, D. W. Barfuss, and M. Ribadeneira, *Am. J. Physiol.* **248**, F890 (1985).

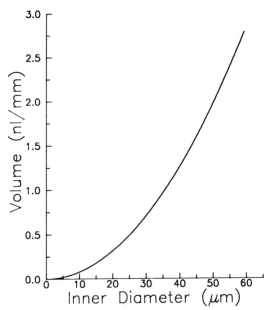

FIG. 2. Relationship between the inside diameter of a constant-bore capillary and its volume per millimeter of length. This relationship can be used to select the size of capillary needed for a particular application.

by drawing a short column of fluid (1 to 2 cm) into the pipet and measuring the length of this fluid column at various positions along the length of the capillary; if the bore of the capillary is constant, the measurement of the length of the fluid column should be the same regardless of where the column is placed along the capillary. The constant-bore pipet is calibrated by drawing up a fluid containing radiolabeled volume marker (such as tritiated methoxyinulin) as described in detail below. The capillary is then

FIG. 3. Construction of a constant-bore sampling pipet. A 15-cm length of the constant-bore capillary extends from one end of the assembly to the other. The outer glass and handle glass are used to support the fine capillary tubing and to facilitate the attachment of a silastic tube to the handle end for suction when the sample is aspirated, or for pressure when it is ejected. An air-tight seal is made between the capillary and the outer glass with dental wax, and between the handle glass and outer glass with fast-setting epoxy cement. The epoxy bead is also used to form a tight seal between the handle glass and the silastic tubing when the latter is attached. Typical glass sizes: constant-bore capillary, 40-μm i.d., 120-μm o.d.; outer glass, 0.0140-in. i.d., 0.020-in. o.d., 14 cm long; handle glass, 0.027-in. i.d., 0.047-in. o.d., 3–4 cm long.

laid next to a millimeter scale and the length of the fluid column in the pipet is measured under a dissecting microscope. (We find that transillumination of the pipet with diffuse light passed through an opalized glass filter allows us to see the bore of the capillary more clearly and reduces reflections.) The fluid is then ejected into a vial for liquid scintillation counting. The activity per volume in the calibrating fluid is determined by serial dilution of larger volumes to obtain a value for the counts per minute per nanoliter (cpm/nl) in the fluid that can then be used to calculate the pipet volume. Several lengths of a fluid column within the pipet should be used for the calibration to ensure a realistic assessment of the variability present in a given pipet.

Decane is used to bound the fluid column in the constant-bore pipet in order to prevent evaporative losses from the collected fluid. We have found that the low viscosity of decane helps to eliminate problems with the pipet clogging, which was experienced when more viscous oils were used with this type of pipet. We also stain the decane with a small amount of Oil Red O (Sigma Chemical Co.; St. Louis, MO) in order to make the decane column and the interface between the aqueous solution and the decane easily observable even to the naked eye. A small amount of decane stained with the Oil Red O is drawn into the tip of the pipet (about 0.5 cm) both before and after the fluid volume to be measured. These columns of red-stained decane make the ends of the fluid column more visible for measurement of its length and also prevent breakup of the fluid column as it is ejected from the pipet, the trailing decane acting to sweep any droplets of aqueous fluid out of the pipet. Following each use of the pipet, it is cleaned by successively aspirating filtered water, chloroform, and acetone through the bore.

The fixed volume pipet is constructed on the microforge. The volume of the pipet is chosen by the rate of perfusion expected for an experiment and the time desired between collections. A typical pipet volume for an experiment in which the rate of volume absorption is to be measured might be 60 nl. Thus, at a perfusion rate of about 12 nl/min, collections would be required every 5 min. For pipets of this volume, we find that the shape of the fixed volume chamber is approximately cylindrical with tapers at the ends of the cylinder to the tip and constriction. Thus, to estimate the dimensions of a pipet required for a given volume, one can model the chamber as a cylinder, as shown in Fig. 4A, where a nomogram is provided for easy estimation of the diameter and length that can be used for a given volume. For smaller volumes, the shape of the chamber becomes dominated by the tapers at the ends of the chamber, so that a double-cone model of the chamber shape is more appropriate, as in Fig. 4B. Figure 4C gives a nomogram for subnanoliter chamber volumes, which are used for special purposes such as measuring rates of fluid secretion in certain renal

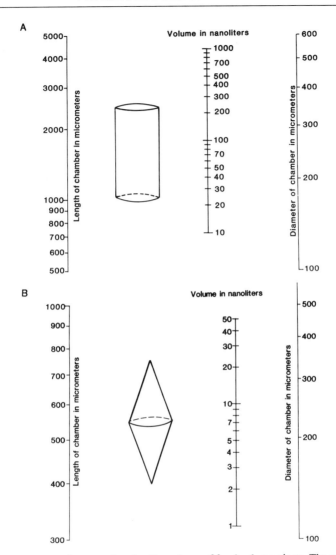

FIG. 4. Nomograms for approximating the volume of fixed volume pipets. The volume of a pipet can be estimated by laying a ruler between the length of the chamber on the left and its inside diameter on the right. However, it is more useful to choose the volume desired before construction and then use the nomogram to determine the combination of length and inside diameter that will produce the desired volume. *Nomogram A:* For chambers that are dominated by a relatively long cylindrical region. *Nomogram B:* For small-volume pipets in which the chamber shape approximates a double cone. *Nomogram C:* For subnanoliter pipets used in removing known volumes from a collection sample.

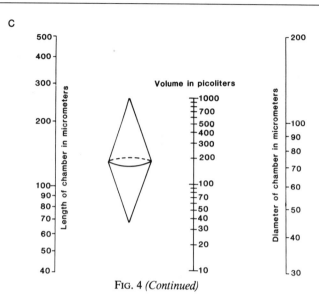

FIG. 4 *(Continued)*

tubules,[5] or when measuring small samples of the collected fluid for subsequent ultramicroanalytical procedures, e.g., for electron probe liquid droplet analysis to determine the elemental composition of the collected fluid.[6]

The fixed volume pipet is calibrated using a solution containing a radiolabeled volume marker, much as described above for the constant-bore pipet. The pipet is not used unless the coefficient of variation of a series of volume determinations is less than 1%. We find that silanization of the pipet (e.g., with Prosil-28, Curtin Matheson Scientific, or a similar compound) improves sample handling. After each sample is ejected into scintillation fluid and the pipet rinsed with the fluid in the vial, the pipet is rinsed with water, acetone, chloroform, and acetone again and dried by attaching the pipet to a vacuum line. We have used light mineral oil to bound the sample in these pipets, rather than decane, and have had no problems with clogging or visibility of the oil–aqueous meniscus.

An advantage of this pipet over the constant-bore pipet is that the length-measuring step is avoided. A disadvantage is that the collection must be removed at the precise time when the pipet has been filled by the collected fluid to the calibrated volume. If this time is missed, one can pull the pipet tip back into the oil layer and pull up a column of oil. The pipet

[5] K. W. Beyenbach, *Nature (London)* **299**, 54 (1982).
[6] J. V. Bonventre, K. Blouch, and C. Lechene, in "X-Ray Microscopy in Biology" (M. A. Mazat, ed.), pp. 307–366. Univ. Park Press, Baltimore, Maryland, 1981.

tip is then placed back in the collecting fluid and a second volume is collected. The time recorded is that required to fill 2 vol of the pipet and both volumes are combined for subsequent analysis. However, this procedure requires that the pipet be sufficiently clean that no droplets from the first volume remain in the chamber to displace volume that should be filled by the second volume of fluid. With the constant-bore pipet, the time of collection is not as critical, and the collectate need only be removed often enough that one collection does not exceed the capacity of the constant-bore pipet. Also, with the constant-bore pipet, one need only focus on the tip of the pipet to ensure that the meniscus of the collectate is left at the proper location following each collection. With the fixed volume pipet, one must focus on the tip of the pipet as well as at the constriction, with the time of collection being the point at which the meniscus within the pipet reaches the center of the constriction while the meniscus of the collectate is at its proper location. Overall, however, we cannot suggest that one of these pipets is better than the other for measurements of volume absorption, as each has been used successfully for a number of years in this laboratory.

Measurement of Perfusion Rate

As stated above, the rate of perfusion of an isolated tubule is measured from the rate of delivery of a volume marker to the collection pipet. If we know the concentration of the volume marker in the perfusate (in cpm/nl), and can measure its rate of appearance in the collectate (in cpm/min), it is easy to calculate the rate of perfusion (in nl/min) as the rate of collection of the volume marker divided by the concentration of the marker in the perfusate. Given this, what kind of volume marker is appropriate for an experiment with isolated tubules? First, the volume marker must not leak across the epithelium of the isolated tubule to any significant extent, it must not bind to the tubule or to the glass pipets, and it must not be either transported or metabolized by the tubule cells in any way. Second, one must be able to use the volume marker in low enough concentrations in the perfusate that it does not significantly affect the perfusate osmolality or concentration, and yet still be able to measure its concentration accurately. Finally, the volume marker must not be too expensive; because it is included in all perfusates for every experiment, the cost of the volume marker can quickly become prohibitive.

Methods such as the microinulin determination used in micropuncture have been developed to measure the concentrations of volume markers chemically. More recently, the electron probe has been used to measure raffinose concentrations (as elemental carbon) in fluid samples collected from individual proximal tubules *in vivo*,[7] and it could also be applied to similar samples from isolated perfused tubule segments. Raffinose has also been measured using an ultramicrofluorometric method for the measurement of volume absorption in isolated, perfused tubules.[8] Raffinose can also be measured using a redox electrode utilizing the galactose oxidase enzyme.[9] The advantage of this method is that the electrode can be positioned within the holding pipet immediately adjacent to the fluid exiting the end of the tubule. Because the ratio of the emergent raffinose concentration to that in the perfusate should be directly proportional to the volume absorption rate, this provides a continuous measurement of the transepithelial volume flow. However, all direct chemical determinations of a volume marker at present have two drawbacks. First, they usually require volume marker concentrations in excess of 1 mM in order to achieve accuracy, and this puts constraints on the perfusate composition. Second, it is difficult to determine the volume marker concentration with sufficient accuracy to measure the volume flow. Usually the investigator is trying to measure volume flow rates of the order of 0.5–1.0 nl/min with an accuracy of at least ±0.1 nl/min for a tubule length of 1.0 mm perfused at 10 nl/min or greater. Thus the volume marker concentration must be measured with a precision of ±1%, which may not be achievable with ultramicrochemical methods. For this reason it is generally easiest to use a radioactively labeled volume marker that has a very high specific activity so that its concentration in the perfusate can be much less than 1.0 mM.

In earlier reports of isolated tubule perfusion, ^{125}I-labeled albumin was frequently used as a volume marker. It is our experience that albumin and other proteins tend to bind to the glass pipets used to sample the collectate and produce highly unreliable results. Therefore, we have used instead tritiated methoxyinulin as a volume marker for many years with good success. It is a highly water-soluble compound that is large enough to be poorly permeant across renal epithelia and is nontoxic. It is relatively inexpensive, but does require dialysis every 2 weeks or so (as described below) in order to remove low-molecular-weight fragments that appear to develop with storage. ^{14}C-labeled inulin, polyethylene glycol, and cyanocobalamin also work well as volume markers, but are considerably more expensive than tritiated methoxyinulin.

We dialyze [^{3}H]methoxyinulin (New England Nuclear Corp., Boston, MA; specific activity, 100–500 mCi/g) in our laboratory every 2 weeks, combining any old lots with new lots of the compound as available. For a new lot, 5 mCi of the compound is dissolved in 5 ml of deionized water and placed in a length of dialysis tubing (Spectrapor 3; Spectrum Medical

[7] R. R. Warner and C. Lechene, *J. Gen. Physiol.* **79,** 709 (1982).
[8] J. L. Garvin, M. B. Burg, and M. A. Knepper, *Am. J. Physiol.* **249,** F785 (1985).
[9] J. Geibel, H. Völkl, and F. Lang, *Pfluegers Arch.* **400,** 388 (1984).

Industries, Inc., Los Angeles, CA; molecular weight cutoff, 3500, 11.5-mm diameter). (We rinse the dialysis tubing with deionized water continuously overnight before use to remove preservatives.) The length of tubing is sealed at each end and placed in a stirred flask with 1 liter of deionized water. The water is exchanged four to five times over the course of 2–3 days. We have found that in the first few hours of dialysis there is a relatively rapid loss of ^3H to the bathing solution, but this loss declines to a low constant rate after about 24 hr (unpublished observations). Thus we suspect that most of the low-molecular-weight contaminants are lost in these first few hours of dialysis and that the later, slow rate of loss is due to the penetration of full-size methoxyinulin through a minor population of larger pores in the dialysis tubing. After dialysis, the methoxyinulin solution is removed from the dialysis tubing and stored in the refrigerator in a vial. For each experiment we add methoxyinulin to the perfusate by evaporating an amount of the solution containing 50–100 μCi under a stream of nitrogen. The residue is then dissolved in 0.5 to 2 ml of the desired perfusate solution.

The precise determination of the activity of the radiolabeled volume marker in the perfusate is important to the accuracy of the volume absorption measurement. This is best done by taking samples for liquid scintillation counting with the same pipet used to measure the collectate, thus reducing the inherent error that would be introduced by the use of another pipet. Samples of perfusate can be taken from the bulk perfusate that is mixed for introduction into the perfusion pipet; with this method, one must be careful that no evaporation occurs in the bulk perfusate between the times the sample is taken and the perfusate is injected into the pipet system, and that no dilution of the volume marker occurs during the injection into the perfusion pipet from fluid already present there. Another method for sampling the perfusate is to place the tip of the perfusion pipet under oil and take samples of perfusate directly from the tip of the perfusion pipet. This second method requires that the constant-bore or fixed volume pipet be mounted temporarily in a micromanipulator to collect the perfusate sample, but does ensure that the perfusate samples taken are of the same fluid that will perfuse the tubule.

An important test of the assumptions inherent in the use of a volume marker is the check of leakage of the volume marker from the perfusate into the bath during an experiment. The rate of leakage of a volume marker, or the lumen-to-bath flux of the marker, is best measured by collecting all volume exiting the bath chamber in a flow-through bath system (described in [20], this volume) and counting the activity of the radiolabel in this bath fluid. The rate of leakage acceptable for an experiment depends somewhat on the segment being perfused. Generally, the rates of leakage of label from methoxyinulin are greater in segments of the proximal nephron than in more distal segments, and this probably reflects real differences in epithelial permeability to medium- and large-sized molecules. The rate of leakage of a volume marker is commonly reported as a rate of volume loss, reflecting the idea that leakage of volume marker is evidence of sites of bulk leakage of perfusate from lumen to bath. One can use the leakage rate as one criterion for acceptance of data from an experiment, requiring, for instance, that the rate of leakage be less than 1% of the perfusion rate. However, this criterion is arbitrary; for a given combination of species, tubule segment and volume marker, the rate of apparent bulk leakage for an acceptable experiment may well be different from 1%.

It is difficult to predict how a leakage of perfusate may affect a measurement of volume absorption because the answer depends on the nature of the leak. If leakage occurs entirely through a break in the epithelium near the cannulated end of the tubule, the leak will have no effect on the validity of the rate of volume absorption calculated from the marker appearing in the collectate, because the amount of volume marker in the collectate will still accurately reflect the rate of perfusion of most of the length of the tubule. However, if the leak occurs further along the length of the tubule, either through a single break in the epithelium or all along the length of the perfused segment, it will result in an underestimate of the rate of perfusion and thus in an underestimate of the rate of volume absorption. Therefore, although volume absorption data from a perfused tubule with a high rate of leakage are not to be trusted, one can be sure that the calculated rate of volume absorption is a minimum estimate of the actual rate.

Normalization of Water Flows

The rate of volume absorption in an isolated perfused tubule segment is usually quantified in terms of some index of the "amount" of the transporting epithelium. The easiest way in which to normalize the flow is in terms of the length of the perfused segment, as discussed in the Introduction. However, one could also normalize it in terms of either the actual or apparent surface area of the lumenal membrane. The apparent surface area of the lumenal membrane is calculated merely as the surface area of a right circular cylinder that would precisely fill the lumen of the perfused segment. Thus it would be given as πdL, where d is the inside diameter of the tubule, and L is the length of the tubule segment that is being perfused, both measured by an ocular micrometer during the experiment.

However, due to the complexity of the lumenal surface this calculation underestimates the true surface area, especially in the case of proximal tubule segments. In some cases, morphometric studies have been con-

TABLE I
SURFACE AREA ESTIMATES FOR SELECTED RABBIT NEPHRON SEGMENTS[a]

Nephron Segment	Inside diameter (μm)	Apparent lumenal surface area (μm^2/mm)	Morphometric lumenal surface area (μm^2/mm)
Proximal convoluted[b]	25.3	79.5×10^3	2900×10^3
Proximal straight[b]	26.2	82.3×10^3	1200×10^3
Medullary thick ascending[c]	15.0	47.1×10^3	80×10^3
Cortical collecting[d]	25.0	78.5×10^3	91×10^3

[a] Lumenal diameters were obtained from ocular micrometer measurements in isolated perfused segments *in vitro* and from measurements in sections prepared for electron microscopy. Morphometric lumenal membrane area was determined in the latter preparation according to the methods described in the references below. The apparent lumenal membrane was calculated as πdL (see text) where d is the lumenal diameter and L is a length of 1 mm or 1000 μm.
[b] From L. W. Welling and D. J. Welling, *Kidney Int.* **8**, 343 (1975).
[c] From L. W. Welling, D. J. Welling, and J. J. Hill, *Kidney Int.* **13**, 144 (1978).
[d] From L. W. Welling, A. P. Evan, and D. J. Welling, *Kidney Int.* **20**, 211 (1981).

ducted in order to determine the true surface area of various nephron segments. Results of such morphometric measurements are compared with the apparent lumenal surface area in Table I.

Errors in the Measurement of Volume Absorption

Of the three measurements that are combined for a measurement of the rate of volume absorption (collection rate, concentration of marker in collectate, and length or surface area of the tubule), errors in determining the concentration of volume marker in the collectate can have the biggest impact on the error in the volume absorption measurement. It is also the measurement of the concentration of the volume marker in the collectate that is most affected by the rate of perfusion in an isolated tubule. This is shown by rewriting the equation for the rate of volume absorption, using the measured quantities:

$$J_v = \dot{V}_L(R - 1)/L \qquad (2)$$

where R is the ratio of the concentration of the volume marker in the collectate to that in the perfusate, or the relative increase in the volume marker concentration in the collectate over that in the perfusate. (For samples taken with fixed volume pipets, R is equal to the counting rate for the sample of collectate divided by the mean counting rate of samples of the perfusate.)

By inspection of this equation, we can see that an error in the measurement of the length of the tubule will introduce a fixed, systematic error into the volume absorption calculation. However, this fixed error will be the same for all collection periods within an experiment, provided that the perfused length remains constant, so that comparison of rates of volume absorption between experimental periods in a single tubule is virtually unaffected by an error in the measurement of the tubule length. This statement is equally true if one uses the apparent lumenal surface area of the perfused segment, or even the morphometrically determined surface areas (Table I) to give the rate of volume absorption relative to the approximate surface area of actual lumenal membrane.

Also by inspection of Eq. (2), one can see that errors in the measurement of the collection rate will be directly reflected in the error in the measurement of volume absorption. That is, if the rate of collection is overestimated by 1%, the rate of volume absorption calculated from this collection period will also be overestimated by 1%, the other parameters being correct. If the errors in measuring the collection rate from period to period during an experiment are systematic—as might happen, for instance, if the volumetric pipet were inaccurately calibrated—then comparisons of rates of volume absorption between periods are largely unaffected. However, one also expects some random error in the measurement of collection rate from period to period during an experiment. This random error will introduce "noise" in the system and reduce the ability to detect a small difference in the rate of volume absorption between two experimental periods. The random error introduced by errors in determining collection rate are rarely serious and, more importantly, are not likely to be affected by the rate of perfusion, unlike the measurement of the concentration of the volume marker in the collectate.

The measurement of volume absorption is critically dependent on the existence of a detectable difference in the concentration of the volume marker between the perfusate and collectate. For a given rate of volume absorption, a slow rate of perfusion will result in a higher concentration of the volume marker in the collectate than will a more rapid rate of perfusion, so that the concentration difference between perfusate and collectate

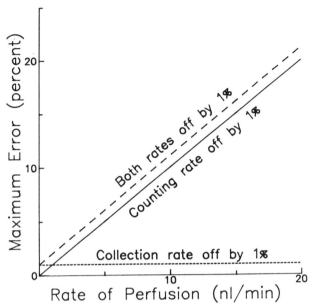

FIG. 5. Effect of parameter errors on the maximum total error involved in the determination of a net volume flow. The maximum error involved in the volume flow determination is plotted as a function of the perfusion rate. The effect of 1% errors in the collection rate, or the volume marker counts, or both are shown.

will be easier to measure at slow rates of perfusion. *This means that measurement of volume absorption at slow rates of perfusion is inherently more accurate than at more rapid rates.* A graphical expression of this phenomenon is shown in Fig. 5, where the maximum error in the calculation of the rate of volume absorption is plotted against the rate of perfusion for three conditions. The lower line shows that a 1% random error in the determination of the collection rate yields an equal error in the rate of volume absorption, and this error is unaffected by the rate of perfusion. (This assumes, of course, that one can measure the rate of collection with the same precision at both high and low perfusion rates, which is reasonable.) The middle line shows the maximum error expected in the measured rate of volume absorption if there is a 1% random error in the determination of the amount of volume marker in the collectate. Such an error in this determination could occur, for example, as the result of random error in counting the radioactivity in the sample. Because an error of this type will be more significant when the concentrations of the volume marker in the perfusate and collectate are very close, the error in the volume absorption determination increases with the rate of perfusion. Combining 1% errors in both the collection rate and volume marker concentration yields a maximum-error bound not much greater than that for the concentration error alone.

Therefore, it is very important when measuring volume absorption to determine the concentration (or activity) of the volume marker with the best precision possible. This is the basis of the requirement stated above that the radiolabel on the volume marker must have a high specific activity, which will allow accurate measurements of the amount of volume marker in a sample. The error analysis also points out the difficulty in measuring volume absorption at high rates of perfusion, a phenomenon that is visible in the scatter of data points in experiments in which the effect of perfusion rate on volume absorption is examined.[10,11]

Alternative Methods of Measuring Transepithelial Volume Flow Rates

Although most investigators measure transepithelial volume flow from the difference between measured perfusion and collection rates, two other, more direct methods have been described which have advantages in certain applications. These are the "crimped end" method and the direct, quantitative collection of absorbate. In the crimped end method described by Grantham and colleagues,[12,13] the tubule is mounted and perfused from one end in the standard fashion; however, the opposite end of the tubule is then drawn into a holding pipet in such a way that the lumen is completely occluded, thus stopping the perfusion. If net volume efflux from the lumen occurs, the lost volume is replaced by fluid from the perfusion pipet, which is kept at a positive pressure to maintain a patent tubule lumen. Therefore, the rate at which fluid flows from the perfusion pipet to the lumen is equal to the rate of volume absorption, and it can be quantified by adding a highly permeant radioactive marker such as ^3H_2O to the perfusion solution. The rate of ^3H_2O appearance in the bathing solution can then be used to calculate the rate of volume absorption from the known dpm/nl of ^3H_2O in the perfusate.[12] Alternatively, the rate of the volume loss or gain in the perfusion pipet can be measured using a narrow-bore perfusion pipet and following the rate of movement of a droplet of a nonpolar solvent such as naphtha in the pipet.[13] The primary problem with either method of measuring volume movement in the crimped end preparation is that the various solutes normally present in the perfusate would not be expected to be absorbed from the lumen at the same rates. Thus preferentially ab-

---

[10] J. A. Schafer, C. S. Patlak, S. L. Troutman, and T. E. Andreoli, *Am. J. Physiol.* **234**, F340 (1978).

[11] J. C. Williams, Jr., and J. A. Schafer, *Am. J. Physiol.* (in press) (1987).

[12] L. W. Welling and J. J. Grantham, *J. Clin. Invest.* **51**, 1063 (1972).

[13] J. J. Grantham, P. B. Qualizza, and R. L. Irwin, *Am. J. Physiol.* **226**, 191 (1974).

sorbed solutes would tend to develop lower concentrations in the lumen and less permanent solutes higher concentrations as volume absorption occurred.[14] For this reason, it is not possible to know the exact composition of the lumenal compartment while the transepithelial volume movement is being measured.

More recently, it has been found that proximal tubules continue to absorb fluid at normal rates even when immersed in light mineral oil. Under these conditions, absorbate forms on the peritubular surface and it can be quantitatively sampled at regular intervals for the determination of its rate of formation and composition.[15,16] This method gives a very accurate measurement of the rate of volume absorption, as well as of any leakage of tubular fluid from the lumen. The drawbacks of this method are that the composition of the peritubular compartment cannot be independently controlled—it is determined by the composition of the primary absorbate—and the technique can be applied only to those tubule segments that exhibit spontaneous volume absorption.

Measurement of Osmotic Water Permeability

In experiments that measure the water permeability of tubular epithelia, the accurate measurement of volume absorption is especially critical. This permeability is calculated from the water flow that occurs in response to a difference in osmotic pressure across the epithelium. The result is expressed as a water flow permeability parameter, P_f (in units of cm/sec), which is linearly related to the hydraulic conductivity of the epithelium.[17] Although one could drive the water flow with a difference in hydrostatic pressure across the epithelium, it is usually not possible to produce high-enough transmural pressures to drive measurable rates of water flow without compromising the integrity of the epithelium or causing the tubule to detach from the pipet systems. In addition, driving volume flow with hydrostatic pressure at normal rates of perfusion can cause an increase in the osmolality of the perfusate that can counter much of the driving force of the pressure.[11] Therefore, an osmotic gradient is used for driving water flow for the measurement of osmotic water permeability. In these experiments the tubule is perfused and bathed with fluids of differing osmolality and the rate of volume transport driven by this osmotic gradient is measured. The isolated tubule preparation is a good choice for this method

---

[14] J. C. Williams, Jr., and J. A. Schafer, (unpublished observations using a mathematical model).
[15] D. W. Barfuss and J. A. Schafer, *Am. J. Physiol.* **241**, F597 (1981).
[16] D. W. Barfuss and J. A. Schafer, *Am. J. Physiol.* **247**, F117 (1984).
[17] J. A. Schafer and T. E. Andreoli, *Membr. Transp. Biol.* **4B**, 473 (1978).

because unstirred layers in the perfusate or bath are minimal,[18] but there is a problem in measuring the average osmolality gradient that is present along the length of the tubule. As water moves across the tubule wall in response to the imposed osmotic difference, the osmolality of the fluid in the lumen of the tubule is altered by the water flow, thus dissipating the osmotic gradient across the tubule wall. So long as this dissipation of the osmotic difference occurs exclusively by the movement of water, and not solute, across the wall of the tubule, one can calculate the osmotic water permeability from the bath-to-perfusate osmotic difference and the rate of volume absorption measured in the perfused segment[19] (and rearranged from Al-Zahid et al.[20]):

$$P_f = \frac{\dot{V}_0 C_o}{A \overline{V}_w} \left[\frac{C_o(1-R)}{R C_o^2 C_b} + \frac{1}{C_b^2} \ln \left(\frac{(RC_o - C_b)}{R(C_o - C_b)} \right) \right] \quad (3)$$

where C_o and C_b are, respectively, the osmolalities of the initial perfusate solution and the bathing solution, $\dot{V}_0$ is the perfusion rate, R is the ratio of volume marker concentration in the collectate to that in the initial perfusate, A is the apparent surface area of the lumen, and $\overline{V}_w$ is the partial molal volume of water (~ 18 cm^3/mol).

This equation has been found to work well for determinations of the osmotic water permeability in segments of the cortical collecting tubule,[20,21] in which the rates of solute movement across the tubule wall are negligible in comparison with the rates of water movement that can be driven by osmotic differences. The equation is not, however, appropriate for use in the proximal tubule of the nephron, in which solute permeabilities are very high. If one wishes to determine the osmotic water permeability in the proximal tubule using osmotic gradients of a permeant solute (such as NaCl), some other approach must be used. One solution to this problem is to measure the rate of volume absorption as well as the osmolality of the collectate in a tubule perfused and bathed with solutions of differing osmolality. With these measurements, one knows the osmolality difference at the beginning and end of the tubule. If it can be demonstrated that a significant fraction of the initial osmolality difference is still present at the collecting end of the perfused segment an approximation of the average osmotic difference can be obtained as the arithmetic or geometric mean of the osmolality differences at the perfusion and collection ends.

Another solution to the problem of measuring the osmotic water permeability in the face of substantial solute movement is to look at the

---

[18] C. A. Berry, *Am. J. Physiol.* **249**, 729 (1985).
[19] R. DuBois, A. Verniory, and M. Abramow, *Kidney Int.* **10**, 478 (1976).
[20] G. Al-Zahid, J. A. Schafer, S. L. Troutman, and T. E. Andreoli, *J. Membr. Biol.* **31**, 103 (1977).
[21] M. C. Reif, S. L. Troutman, and J. A. Schafer, *Kidney Int.* **26**, 725 (1984).

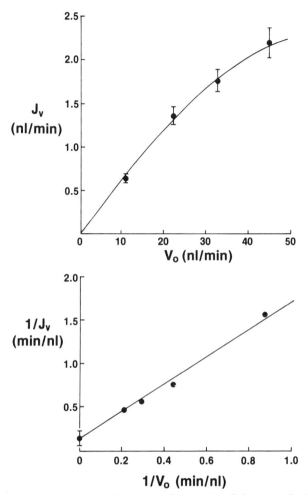

Fig. 6. Relation between osmotic volume flow and perfusion rate. In this series of experiments proximal convoluted tubules with an average length of 0.86 mm were perfused at 21° with a solution having an osmolality of 270 mOsm/kg H_2O, while the bathing solution osmolality was 290 mOsm/kg H_2O. *Upper:* In the absence of active volume absorption at room temperature, the net osmotic water flow (J_v, not normalized for tubule length) rises as the perfusion rate increases due to the fact that a longer length of the tubule were exposed to a greater osmolality difference as the perfusion rate increases. *Lower:* The same data are plotted as inverse functions. The y intercept is the inverse of the maximum flow rate that would obtain at an infinite perfusion rate when the whole length of the tubule is exposed to the osmolality difference between the initial perfusate and the bathing solution. (Data from Schafer et al.[10])

dependence of the rate of volume absorption driven by the osmotic difference on the rate of perfusion of the tubule segment. It can be reasoned that at an infinitely high rate of perfusion, the osmotic difference along the length of the tubule would be constant and equal to that at the perfusion end. This would produce the maximum rate of volume flow possible for the given bath-to-perfusate osmotic difference. The asymptotic approach of the rate of volume absorption to this maximum value as the rate of perfusion is increased, is shown from some actual experiments in the upper panel of Fig. 6. Note that there is a unique relationship between the rate of volume absorption and the rate of perfusion for a value of osmotic water permeability. Thus, one can use an experimental curve of the rate of volume absorption plotted against the rate of perfusion and match it to these theoretical curves.[10,22] Alternatively, the data can be transformed by plotting the inverse of the volume flow as a function of the inverse of the perfusion rate, as shown in the bottom panel of Fig. 6. With such a plot, one can extrapolate a linear fit to intercept the ordinate, giving a measure of the rate of volume absorption expected at an infinite rate of perfusion. This method has been applied to the proximal tubule, and was used to demonstrate the very high osmotic water permeability in this segment.[10,22] Nevertheless, there are definite problems with this approach. First, there is no reason *a priori* to assume a linear relation for the inverse plot. Second, even assuming a linear relation as in Fig. 6, the scatter of the data, especially from the highest perfusion rates, causes the estimate of the y intercept to have a large associated uncertainty. Thus, often the best estimate of the water flow permeability can be derived from the water flow observed at the highest practical perfusion rate, assuming that the whole perfused length is exposed to the same osmolality difference as that present at the perfusion end. Obviously, this approximation would be a least lower bound to the true water permeability. Unfortunately, this is presently the best available approach to estimating water permeability in nephron segments such as the proximal tubule where both water and solute permeabilities are very high.

Acknowledgments

We would like to thank the technicians and coinvestigators who have worked with them in the development of many of the methods described in this chapter, in particular Ms. Susan L. Troutman and Dr. Delon W. Barfuss. Support for many of the studies referenced from our laboratory was provided by NIH Research Grant 1-RO1-DK25519.

---

[22] J. A. Schafer, S. L. Troutman, M. L. Watkins, and T. E. Andreoli, *Kidney Int.* **20**, 588 (1981).

[14] Identification and Study of Specific Cell Types in Isolated Nephron Segments Using Fluorescent Dyes

By George J. Schwartz and Qais Al-Awqati

Fluorescent dyes allow noninvasive assessment of the properties of a number of cellular compartments, such as the cytoplasm, mitochondria, and vesicular organelles. An advantage of these agents is that a number of functions can be simultaneously studied using dyes with different spectral properties. In tissues with different cell types a careful choice of a fluorescent dye can facilitate the study of a function of only one of these cell types. For instance, in a tissue where the cells have widely different rates of endocytosis, short application of a fluorescent macromolecule will permit the study of the cell type with a high rate of endocytosis without the need for cell fractionation. In fact, even a rare cell in a heterogeneous tissue can be identified and studied individually using dyes that preferentially partition into some compartment of that cell type. A further advantage of fluorescent dyes is that they can be derivatized to change their compartmental localization without sacrificing the spectral property that is physiologically useful. For instance, fluoresceins have pH-sensitive excitation spectra but are impermeant. Esterfication of the dyes will increase their permeability and allow them to accumulate in compartments with a high esterase activity such as the cytoplasm. Alternatively they may be coupled to macromolecules which will permit study of the pH of endocytic and lysosomal compartments. They can also be attached to ligands, allowing the study of pH in the environment of the ligand receptor.

All of these studies assume that the dyes do not alter cell functions, but where tested that seems to be a reasonable assumption. Newer dyes with higher quantum yields have also improved the signal-to-autofluorescence ratio such that the concentration needed to produce adequate signals is lower than previously required. Further, their spectral properties inside the cell can be independently analyzed to allow quantitation of the function being studied even when the intracellular spectrum is different from that in solution.

We have used fluorescent dyes to study the function of individually identified cells in heterogeneous epithelia of the kidney. Each nephron is axially differentiated into 10 or more segments, each with its characteristic structure and function. Some of these segments have multiple cell types as studied by morphological methods. While these methods are useful in cataloging the different morphologies, they are performed on fixed tissues. Hence, the conclusions of such studies regarding the physiology of individual cell types merely correlate changes in morphology with changes in the state of the animal or tissue, rather than providing direct evidence for involvement of an individual cell type with a specified function. This is a particular problem in the kidney, which adapts readily to changes in the animal's environment and diet. After the function of individual cells is established, morphological methods provide powerful tools for examination of the cellular events that underlie the function of these cells. In addition, they can reveal the polar distribution of specific molecules that give that cell its characteristic function. Transport of ions and water is one of the fundamental properties of epithelial cells, and a comparison of the location of the transport protein and its functional presence in the apical or basolateral membrane is one of the major aims of epithelial physiology. Below we describe methods that we have used to identify specific cell types as well as to study the function of individual membrane domains using fluorescent dyes in isolated perfused nephron segments. In particular we have studied the function of the collecting tubule, a heterogeneous epithelium composed of two cell types, the principal cell and the intercalated cell. The latter, accounting for only a third of the total number of cells, is specialized for H^+/HCO_3^- transport.

Isolation and Perfusion of Defined Nephron Segments

Tubule segments were isolated by freehand dissection of rabbit kidneys using sharpened forceps. Cortical collecting ducts (1–2.5 mm), isolated under a dissecting microscope from medullary rays in the inner cortex, extended superficially into the midcortex. Outer medullary collecting tubules were isolated from medullary rays deep to the corticomedullary junction. Proximal straight tubules were easily recognized by their shiny appearance and were dissected out of the cortex. Other segments can be dissected using methods described in the chapter by Wright *et al.*, ([12], this volume).

Fluorescence Microscopy

Fluorescence studies were performed on a Nikon Diaphot inverted epifluorescence equipped with a 75-W xenon lamp and a Farrand microspectrofluorometer with a photometer attached to the front camera port (Farrand Optical, Valhalla, NY). A 50% neutral density filter was used to reduce the intensity of the fluorescent beam. The green fluorescence of fluoresceins was observed using a Nikon B filter cassette (excitation at 420–485 nm; dichroic mirror at 510 nm; emission >520 nm). Rhodamine fluorescence was observed using a Nikon G filter cassette (excitation at 535–550 nm; dichroic mirror at 580 nm; emission >580 nm) that was mounted on the same unit, so that red and green fluorescence could be observed sequentially in the same area without perceptibility moving the field. A red cut-off filter placed in the B filter cassette prevented the appearance of any rhodamine emission in the fluorescein fluorescence. Photomicrographs were taken with a Nikon FG 35-mm camera attached to the camera port and Kodak Ektachrome or Tmax 400 film.

Endocytosis[1,2]

To test for endocytosis we added fluorescein isothiocyanate (FITC) coupled to an M_r 70,000 dextran (Sigma Chemical Co., St. Louis, MO) to the perfusate in a concentration of 10 mg/ml. We have also derivatized proteins such as bovine serum albumin (BSA) with fluorescein or rhodamine isothiocyanate and used them as markers for endocytosis. The fluorescent molecules were dialyzed for 24 to 48 hr against the appropriate buffers to remove unreacted fluorophores. Alternatively, these could be removed by centrifugation chromatography on Sephadex G-50 columns. Uptake of these macromolecules occurs only by endocytosis. After 20–60 min, the fluorescent macromolecule was removed and the tubule was perfused with an acidic (pH 5) fluorophore-free solution for 10 min. This maneuver served to wash away fluorescent material that was bound to the lumenal surface. Then the segment was perfused with the initial solution (pH 7.4 and free of fluorophore) and examined under light and fluorescence microscopy.

In the proximal straight tubule all cells showed vigorous endocytosis. We were able to study the pH of the endocytic vesicle using excitation ratio fluorometry as described below (Measurement of Cell pH). We found that these vesicles were acidified by a proton pump to a pH of approximately 6.0. In the cortical collecting tubule, only about 1% of the cells endocytosed the markers, while in the outer medullary collecting tubule 25% of the cells internalized these macromolecules. Counterstaining with other dyes showed that these were intercalated cells. We found that endocytosis was enhanced when the fluorophores were perfused while the bathing medium was changed to one with a low pCO_2.

Figure 1A shows a proximal tubule after it was perfused with FITC–dextran. Note the punctuate fluorescence beneath the brush border of all

---

[1] G. J. Schwartz and Q. Al-Awqati, *J. Clin. Invest.* **75,** 1638 (1985).
[2] G. J. Schwartz, L. M. Satlin, and J. E. Bergmannn, *Am. J. Physiol.* **255,** F1003 (1988).

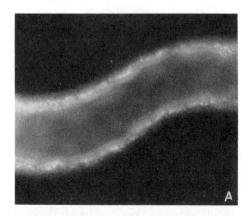
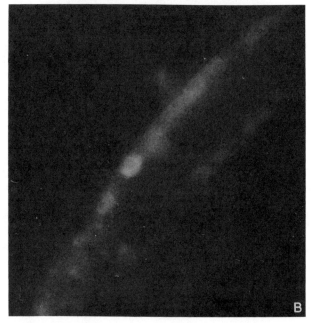

FIG. 1. (A) Endocytosis in a proximal straight tubule. (From Schwartz and Al-Awqati.[1])
(B) Endocytosis in a cortical collecting tubule. (From Satlin and Schwartz.[14])

cells. In Fig. 1B, a cortical collecting tubule was perfused with rhodamine-BSA. Note that only two cells in this tubule have internalized the marker.

Exocytosis[1,3]

To show that CO_2 causes fusion of the vesicles with the lumenal membrane (exocytosis) we loaded the endocytotic vesicles at 37° using a solution that contained NaCl (145 mM), K_2HPO_4 (2.5 mM), $MgSO_4$ (1.2 mM), $CaCl_2$ (2.0 mM), sodium lactate (4.0 mM), sodium citrate (1.0 mM), L-alanine (6.0 mM), and glucose (5.5 mM) at pH 7.4 and 1–10 mg/ml of fluorescent dextran in the absence of CO_2. We then exposed both lumenal and basolateral surfaces of the tubule segment to 6% CO_2 substituting 25 mM $NaHCO_3$ for an equal amount of NaCl. We also performed experiments where the cortical collecting tubules were loaded with fluorescent dextran at an ambient pCO_2 of 37 mmHg at pH 7.1. Then we increased the pCO_2 of perfusate and bath isohydrically to 114 mmHg. We tested for fusion in these two types of experiments by measuring the fluorescence intensity at 520 nm while exciting at 460 nm those cells that internalized the dextran. A reduction in fluorescence would imply loss of fluorescent dextran, presumably by exocytosis.

We examined the same cells photometrically during sequential 10- to 15-min periods, first in 100% O_2 and then in 94% O_2–6% CO_2. We focused the photometer on a single cell and took readings at 460 nm excitation before and after increasing the pCO_2. Although this wavelength is not the isosbestic point of fluorescein, it shows a small change in intensity with a change of pH, making it useful for the measurement of total dye content.

We found that increasing the pCO_2 did not reduce the pH of endocytic compartments, which averaged 5.7 in the cortical collecting duct, 6.0 in the medullary collecting tubule, and 5.95 in the proximal straight tubules.[1] The pH of these compartments increased on addition of a permeant base such as NH_4Cl, or proton ionophores such as nigericin and CCCP (carbonyl cyanide-p-chloromethoxyphenylhydrazone). These results suggest that the low pH of these compartments was generated by a proton pump. Increasing the pCO_2 rapidly reduced the total fluorescence in all three segments examined by 25–50%. That the total fluorescence did not decline while the signal was measured at the same initial ambient pCO_2 for similar periods of time implies that the fluorescent dextran was secreted from the tubules. Further, pretreatment of the tubules with 0.5 mM colchicine abolished the decrease of fluorescence in response to increases in pCO_2, indicating it is due to exocytotic fusion of vesicles with the lumenal membrane.

[3] G. J. Schwartz, J. Barasch, and Q. Al-Awqati, *Nature (London)* **318**, 368 (1985).

Density and Distribution of Acid Intracellular Compartments

Proton secretion in intercalated cells occurs via an H^+-ATPase, a protein that is present in the apical plasma membrane as well as being packaged within acidic cytoplasmic vesicles.[3,4] The presence of these vesicles in the collecting duct was demonstrated using acridine orange, a weak base that accumulates within acidic compartments. The peak emission of acridine orange is concentration dependent, being green at low concentration and undergoing a red shift at high concentration.[5] Cortical and outer medullary collecting ducts were incubated in 10 μM acridine orange for 3 min, rinsed, and examined under epifluorescence excitation. We found that most of the vesicle-rich cells in the collecting tubule possessed discrete apical densities of orange (acidic) vesicles and basally located nuclei. In the outer medullary collecting duct a second pattern of vesicular staining, characterized by a diffuse distribution of vesicles among the apical and lateral cell regions, was found frequently. Basally located vesicles were rarely found in the collecting duct. In the proximal tubule all cells showed a high density of acid vesicles throughout the cell.

To confirm that H^+ pumps acidified the vesicles and resulted in the accumulation of acridine orange therein, we incubated three cortical collecting tubules isolated from adult rabbits in acridine orange and 100 μM CCCP (a protonophore). Within 5 min none of the tubules demonstrated orange staining of vesicles, indicating that acridine orange was not trapped within vesicles in the absence of pH gradients.

While the density of acid vesicles can be readily assessed using this method, the location of these vesicles in individual cells is less easily determined. This is likely due to the fact that there is much light scattering in this dye and, without confocal microscopy, it is difficult to obtain optical sections of the cell that would permit resolution of apical or basolateral location of the vesicles. Frequently, even when using routine fluorescence microscopy, one can find an optical section that will demonstrate a specific localization of acid vesicles to one or another region of the cell.[3]

Figure 2 shows a cortical collecting tubule taken from a neonatal rabbit. Note the presence of apical vesicles (arrows) that accumulate acridine orange in only a few cells of the epithelium, presumably the intercalated cells.

Mitochondrial Stain

Intercalated cells were further functionally identified using 3,3'-dipentyloxacarbocyanine [Di-O-C_5(3); Molecular Probes, Junction City, OR], a

---

[4] D. Brown, S. Hirsch, and S. Gluck, *Nature (London)* **331**, 622 (1988).
[5] S. Gluck, C. Cannon, and Q. Al-Awqati, *Proc. Natl. Acad. Sci. U.S.A.* **79**, 4327 (1982).

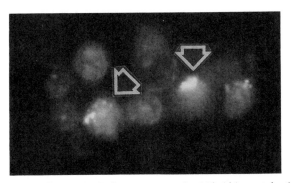

FIG. 2. Cortical collecting tubule from a neonatal rabbit kidney stained with acridine orange. (From Satlin and Schwartz.[8])

permeant fluorescent cation[6] that selectively localizes to intracellular sites of high negative potential such as mitochondria.[7] Cortical and medullary collecting ducts were incubated at 38° for ~3 min in a medium containing 10^{-8} M Di-O-C_5(3). Intercalated (mitochondria-rich) cells in the adult kidney were characterized by a reticular, perinuclear, mitochondria-associated staining pattern under epifluorescence illumination. Using this protocol it was possible to stain the intercalated cells only[3]; the principal cells did not give a well-defined stain (Fig. 3).

To confirm that the preferential uptake of Di-O-C_5(3) in intercalated cells reflected the presence of a large mitochondrial transmembrane potential, we tested the effects of various agents known to disrupt the functional state of the mitochondria.[8] Included among the agents were the K^+ ionophore valinomycin (20 μM) and the protonophore CCCP (100 μM), both of which dissipate the electrochemical gradient across the mitochondrial membrane, and sodium azide (5 mM), an inhibitor of electron transport. Cortical collecting ducts were transferred to a chamber treated with poly-L-lysine, warmed to 38°, and suffused with 100% O_2. Standard medium (pH 7.4) was routinely used as the bathing solution except in those studies involving valinomycin, which were conducted in a high-K^+ buffer. After a 10-min incubation, Di-O-C_5(3) was added to the bath to yield a final concentration of 10^{-8} M. Three minutes later, observations were made and the inhibitor was added. The effect of each inhibitor on the mitochondrial retention of Di-O-C_5(3) in the cortical collecting tubule was assessed under epifluorescence illumination in the continuous presence of the cyanine dye.

[6] R. L. Cohen, K. A. Muirhed, J. E. Gill, A. S. Waggoner, and P. K. Horan, *Nature (London)* **290**, 593 (1981).
[7] L. V. Johnson, M. L. Walsh, B. J. Bockus, and L. G. Chen, *J. Cell Biol.* **88**, 526 (1981).
[8] L. M. Satlin and G. J. Schwartz, *Am. J. Physiol.* **253**, F622 (1987).

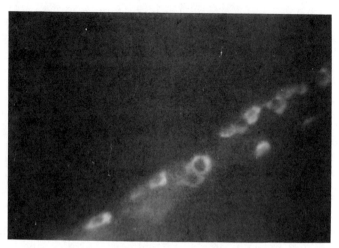

FIG. 3. Staining of a cortical collecting tubule with a cationic carbocyanine, Di-O-C$_5$(3). (From Schwartz et al.[3])

Prestaining of the collecting tubule with Di-O-C$_5$(3) followed by exposure to either CCCP or valinomycin led to release of the fluorescent cation from the mitochondria, resulting in a diffuse cytoplasmic fluorescence and loss of selective mitochondrial fluorescence within 15 min. Intercalated cells prestained with Di-O-C$_5$(3) and then exposed to sodium azide slowly (15–30 min) lost their mitochondria-associated fluorescence as metabolism was inhibited; control tubules retained their fluorescence during this time period.

Peanut Lectin Binding

Le Hir et al.[9] found that intercalated cells of the rabbit collecting tubule bind peanut lectin to their apical surface. In the rabbit kidney only these cells and those of the thin descending limb bind peanut lectin. In the kidney of other species peanut lectin binding is less specific.[10,11] Peanut lectin can be readily coupled to fluorescein or rhodamine (as the isothiocyanates), and is now commercially available. Perfusion of the tubule with 5–10 µg/ml fluorescent lectin at room temperature for 5 min followed by a wash is usually sufficient to give a good signal. A crescentic cap is readily

[9] M. Le Hir, B. Kaissling, B. M. Koeppen, and J. B. Wade, *Am. J. Physiol.* **242,** C117 (1982).
[10] T. Farraggiana, F. Malchiodi, A. Prado, and J. Churg, *J. Histochem. Cytochem.* **30,** 451 (1982).
[11] H. Hamada, *Acta Histochem. Cytochem.* **16,** 189 (1983).

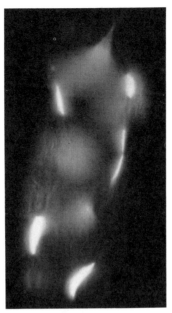

FIG. 4. Staining of a cortical collecting tubule with rhodamine–peanut lectin. (From Schwartz et al.[2])

observed on the apical surface of intercalated cells in optical sections when the cells are viewed from the side (Fig. 4). In *en face* views, the whole apical surface is seen to be stained.

Because peanut lectin is a fluorescent macromolecule it could be internalized by endocytosis. To distinguish between surface binding and endocytosis, we reduced the ambient temperature or fixed the cells with 3% paraformaldehyde. Neither maneuver changed the pattern of staining in the cortical collecting tubule.

Measurement of Cell pH

Cell pH was measured by excitation ratio fluorometry using permeant derivatives of fluorescein. The excitation spectrum of fluorescein is pH sensitive.[12] At acid pH the fluorescence intensity at 520–535 nm when excited at 490–500 nm is reduced, but when excited at 450–460 nm it remains reasonably constant. Hence the ratio of the fluorescence intensity at the two exciting wavelengths represents a measure of pH in the environment of the dye. To measure cell pH in individually identified cells we

[12] S. Ohkuma and B. Poole, *Proc. Natl. Acad. Sci. U.S.A.* **75**, 3327 (1978).

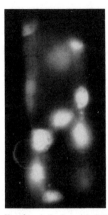

FIG. 5. Staining of a cortical collecting tubule with BCECF for 5 min. Note the selective staining of only some cells.

perfused the tubules with 5,6-dicarboxyfluorescein (6-CF) or the tetracarboxy derivative, 2′, 7′-bis-(2-carboxyethyl)-5-(and -6) carboxy fluorescein (BCECF) (Molecular Probes) at $10-20 \,\mu M$ concentrations for 10–20 min. While all cells of the proximal tubule stained uniformly, those in the collecting duct showed heterogeneous staining (Fig. 5). The intercalated cells selectively concentrated this dye in their cytoplasm after only brief exposures, making it a method that can be used to study these cells in the intact epithelium without interference from other cell types. The preferential loading is likely to be due to their high concentrations of cytoplasmic carbonic anhydrase, a potent esterase which can rapidly hydrolyze these dyes to their impermeant fluorescent states.[13]

After removal of the esterified compounds, equilibration for at least 10 min was allowed to minimize the effects on subsequent pH_i measurements of any cellular acidification caused by the release of acetic acid moieties. We then excited a cell sequentially at 490, 450, and 490 nm using bandpass interference filters (Ditric Optics, Hudson, MA) placed on a slider in front of the fluorescent beam. We exposed it to a 10-μm spot of fluorescent light for only the time required to perform the excitation ratio measurements (3–5 sec) to minimize photobleaching. The emission signal was recorded at 535 nm. If the second 490-nm excitation signal was not within 10% of the initial 490-nm excitation measurement, the reading was discarded and a different cell was examined. The scale for the photometer measurements was the most sensitive (0.01 μA), yielding average readings at 490 nm of $\sim 0.01 \,\mu$A. Background fluorescence readings in the bathing medium and cellular autofluorescence were consistently $<0.0002 \,\mu$A.

[13] J. van Adelsberg and Q. Al-Awqati, *J. Cell Biol.* **102**, 1638 (1986).

At the end of the experiment an intracellular calibration was performed. The tubule was equilibrated for 5–10 min in each of two to three different potassium and phosphate buffers of known pH to which 20 μM nigericin was added. Nigericin is a K^+-H^+ exchange ionophore that clamps the cell pH to that of the bathing solutions. The calibration curve, which was obtained with approximately the same intensity as the readings during the various maneuvers, bracketed the observed ratio values. The linearity of the slope over the pH range 6.6–7.6 allowed us to solve for unique pH values using standard fluorescence units per pH unit.

The sensitivity and ease of these studies can now be improved using a Silicon-intensifier Target (SIT) camera with recording and storage of the images on video tape.[14] Another advantage of this method is that more than one cell in a field can be studied, thereby reducing the number of animals that need to be used for each experimental maneuver.

Identification of Different Cells Using Response of pH_i to Various Maneuvers[3,14]

Measurement of cell pH in epithelia can be used to localize a variety of acid–base transporters to the apical or basolateral membrane. By changing the composition of the bathing medium separately from the lumenal medium, one can infer the location of Na^+/H^+ and Cl^-/HCO_3^- exchangers, or the presence of H^+/HCO_3^- pumps and permeabilities. For instance, we used this method to demonstrate that there are two types of intercalated cells: one has a Cl^-/HCO_3^- exchanger on the apical membrane while the other has it on the basolateral membrane. Peanut lectin (PNA)-labeled cells secrete HCO_3^- via apical Cl^-/HCO_3^- exchangers. We identified these cells by double-fluorescence labeling using rhodamine–PNA (Fig. 6B) and BCECF (Fig. 6A). We measured the change in pH_i in response to the removal of lumenal Cl^- in individually identified PNA-labeled cells that stained brightly with BCECF or 6-CF. Tubules were studied initially with Cl^-- and HCO_3^--containing solutions in lumen and bath. The lumenal solution was then replaced with Cl^--free perfusate containing (in millimoles): sodium gluconate (110), $NaHCO_3$ (25), K_2HPO_4 (2.5), calcium acetate (4.0–8.0), $MgSo_4$ (1.2), sodium lactate (4.0), trisodium citrate (1.0), alanine (6.0), and glucose (5.5) at pH 7.4 in 94% O_2/6% CO_2. Use of 4 or 8 mM calcium did not affect the magnitude of the change in pH_i in response to the removal of Cl^-. After 5–10 min, excitation ratio fluorometry was again repeated. Then the original solution was restored to the lumen and, after 5–10 min, postexperimental measurements were

---

[14] L. M. Satlin and G. J. Schwartz, *J. Cell Biol.* **109**, 1279 (1989).

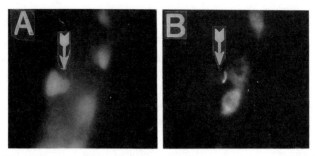

FIG. 6. Simultaneous staining of a cortical collecting tubule with BCECF (A) and rhodamine–peanut lectin (B). Note the ready identification of cells that stain for both dyes (arrows). (From Schwartz et al.[2])

performed; all excitation signals were recorded on videotape. The cell pH alkalinized in response to removal of lumenal Cl^-, suggesting that there exists an apical Cl^-/HCO_3^- exchanger. Further, when Cl^- was removed from the bathing solution, the cell pH of these identified cells decreased further, demonstrating the presence of an apical Cl^-/HCO_3^- exchanger (in series with a basolateral Cl^- permeability).

Conclusions

These methods have allowed us to study the physiological and cell biological characteristics of individually identified cells in an intact epithelium. For instance, we were able to study endocytosis and exocytosis in different cell types and to show that these processes play an important role in regulating transepithelial proton transport. Further, we examined the density, intracellular location, and intensity of acidifications of these endosomes in different cell types and observed that these findings correlated with the function of the cells. We also were able to study the regulation of cell pH in identified cell types, finding that acid–base transporting cells had an intracellular pH that was higher than that of their immediate neighbors, which raises questions regarding cell-to-cell communication among cells of an epithelium. These measurements of cell pH were used to identify and characterize the polar distribution of Cl^-/HCO_3^- exchangers, thereby allowing us to discover that there were two functionally distinct types of intercalated cells: one secretes acid and the other secretes HCO_3^- into the lumen.

The power of these methods is especially apparent in an epithelium with multiple cell types provided that each cell type can be labeled by a fluorescent dye in a characteristic manner. While our studies concentrated on pH-sensitive dyes, intracellular calcium can be studied simultaneously.[13] Such studies could be extended to other heterogeneous organs or cultured cell systems. Obvious possibilities include studies of ductal and acinar cells in exocrine glands, endothelial, neural, and muscle cells in skeletal muscle, different cell types in various regions of the brain such as cerebellum and hippocampus, and nonclonal established cell lines and primary cultures isolated from various organs.

[15] Functional Morphology of Kidney Tubules and Cells *in Situ*

By BRIGITTE KAISSLING and MICHEL LE HIR

Histologic and fine structural techniques applied to the study of the kidney proved to be important tools in formulating various concepts of kidney function. For instance, models for the urinary concentration mechanism in mammals[1-4] would not have been possible without the detailed knowledge of the structural organization of the renal medulla. More recently, the concept that structural modulation of renal epithelia plays an important role in the regulation of transepithelial electrolyte transport processes[5-9] has been established on the basis of structural study of isolated renal tubules[10,11] and more extensively by analysis of renal tubules in their natural environment *in situ*.[12-31] Such investigations made evident that the

[1] J. L. Stephenson, *Kidney Int.* **2**, 85 (1972).
[2] J. P. Kokko and F. C. Rector, *Kidney Int.* **2**, 214 (1972).
[3] C. C. Tisher, *Am. J. Physiol.* **220**, 1100 (1971).
[4] W. Kriz, *Am. J. Physiol.* **241**, R3 (1981).
[5] B. Kaissling, *Am. J. Physiol.* **243**, F211 (1982).
[6] B. Kaissling, *Fed. Proc., Fed. Am. Soc. Exp. Biol.* **44**, 2710 (1985).
[7] K. M. Madsen and C. C. Tisher, *Fed. Proc., Fed. Am. Soc. Exp. Biol.* **44**, 2704 (1985).
[8] B. A. Stanton, *Fed. Proc., Fed. Am. Soc. Exp. Biol.* **44**, 2717 (1985).
[9] J. B. Wade, *Curr. Top. Membr. Transp.* **13**, 123 (1980).
[10] B. Kaissling, B. M. Koeppen, J. B. Wade, and M. Le Hir, *Acta Anat.* **111**, 72a (1981).
[11] J. B. Wade, R. G. O'Neil, J. L. Pryor, and E. L. Boulpaep, *J. Cell Biol.* **81**, 439 (1979).
[12] L. Bankir, C. Fischer, S. Fischer, K. Jukkala, H.-C. Specht, and W. Kriz, *Pfluegers Arch.* **412**, 42 (1988).
[13] N. Bouby, L. Bankir, M. M. Trinh-Trang-Tan, W. W. Minuth, and W. Kriz, *Kidney Int.* **28**, 456 (1985).
[14] W. L. Clapp, K. M. Madsen, J. W. Verlander, and C. C. Tisher, *Kidney Int.* **31**, 1080 (1987).

extent of some structural changes correlates well with the magnitude of some functional responses.[32]

The advantage of the study of renal epithelia *in situ,* in comparison to that of isolated tubules, is the possibility of discrimination of different cell types within one segment, of sampling great numbers of nephron segments within the same kidney, of simultaneous assessment of internephron heterogeneity, as well as of observation of the specific tubular environment.

In this chapter we will give a brief overview of the correlation of certain structures with the function of electrolyte transport in renal epithelia and outline some morphological techniques for the qualitative and quantitative evaluation of renal tubules *in situ,* with their strengths and limitations. A few points which are particularly important in the application of these techniques to renal tissue will be explained in more detail. In an appendix we will provide a short overview of histotopographical relationships within the kidney zones.

Structural Basis for Transepithelial Electrolyte Transport

Compartments

The single-layered epithelia of the renal tubules function as selective barriers between two extracellular compartments with different physico-

---

[15] J. Dorup, *J. Ultrastruct. Res.* **92,** 101 (1985).
[16] G. P. Hansen, C. C. Tisher, and R. R. Robinson, *Kidney Int.* **17,** 326 (1980).
[17] D. Hirsch, M. Kashgarian, E. L. Boulpaep, and J. P. Hayslett, *Kidney Int.* **26,** 798 (1984).
[18] B. Kaissling, S. Bachmann, and W. Kriz, *Am. J. Physiol.* **248,** F374 (1985).
[19] B. Kaissling and M. Le Hir, *Cell Tissue Res.* **224,** 469 (1982).
[20] B. Kaissling and B. Stanton, *Am. J. Physiol.* **255,** F1256 (1988).
[21] M. Kashgarian, T. Ardito, D. J. Hirsch, and J. P. Hayslett, *Am. J. Physiol.* **253,** F8 (1987).
[22] N. Koechlin, J. M. Elalouf, B. Kaissling, N. Roinel, and C. de Rouffignac, *Am. J. Physiol.* **256,** F814 (1989).
[23] K. M. Madsen and C. C. Tisher, *Am. J. Physiol.* **245,** F670 (1983).
[24] K. M. Madsen and C. C. Tisher, *Lab. Invest.* **51,** 268 (1984).
[25] K. M. Madsen and C. C. Tisher, *Am. J. Physiol.* **250,** F1 (1986).
[26] A. Rastegar, D. Biemesderfer, M. Kashgarian, and J. P. Hayslett, *Kidney Int.* **18,** 293 (1980).
[27] B. A. Stanton, D. Biemesderfer, J. B. Wade, and G. Giebisch, *Kidney Int.* **19,** 36 (1981).
[28] B. A. Stanton, A. Janzen, G. Klein-Robbenhaar, R. Defronzo, G. Giebisch, and J. Wade, *J. Clin. Invest.* **75,** 1327 (1985).
[29] D. L. Stetson, J. B. Wade, and G. Giebisch, *Kidney Int.* **17,** 45 (1980).
[30] J. W. Verlander, K. M. Madsen, P. S. Low, K. P. Allen, and C. C. Tisher, *Am. J. Physiol.* **255,** F115 (1988).
[31] R. K. Zalups, B. A. Stanton, J. B. Wade, and G. Giebisch, *Kidney Int.* **27,** 636 (1985).
[32] B. A. Stanton and B. Kaissling, *Am. J. Physiol.* **255,** F1269 (1988).

chemical compositions. With their apical pole the epithelial cells are in contact with the lumenal compartment, which contains the tubular fluid. The base of the cells is situated on a basement membrane which is continuous with the interstitial ("serosal") compartment. The individual cells are joined to a continuous epithelial sheet by junctional complexes. These separate the lumenal compartment from the lateral intercellular spaces. The latter communicate through the "basal slits" with the renal interstitium.

Transport Pathways

Movement of solutes from one compartment to the other can proceed either across the junctional complexes, via the paracellular pathway, or across the cells, via the transcellular pathway. The two transport routes are not strictly independent of each other. They interfere in the lateral intercellular spaces.

Paracellular Pathway. The junctional complexes encircle the epithelial cells like belts and consist of the tight junction (zonula occludens) and the intermediate junction (zonula adhaerens). In addition, patches of desmosomes (maculae adhaerentes) may also be present. Gap junctions are not specific for epithelia. Among the renal tubular cells only proximal tubule cells are connected by gap junctions. It is the tight junction which functions as an effective barrier between the lumenal compartment and the lateral intercellular spaces. The selective permeability qualities of the tight junction for solutes depend on its structural and physicochemical properties and are specific for each nephron segment.

Tight Junction Structure. The tight junction seals adjacent cells together by close apposition of their outer membrane leaflets. The structure of the contacts of the two opposing membranes can be judged best in replicas of the interior face of membranes split by freeze-fracturing. On these the contacts appear as "strands," composed of "particles." It is still unclear whether they are constituted by proteins[33,34] or lipids.[35,36] The density of the particles within the strands and their chemical and physical qualities (such as electrical charges), the lateral continuity of the strands, their number, and their configuration (parallel or netlike) all determine the

[33] L. A. Staehelin, *J. Cell Sci.* **13**, 763 (1973).
[34] B. R. Stevenson, J. D. Siliciano, M. S. Mooseker, and D. A. Goodenough, *J. Cell Biol.* **103**, 755 (1986).
[35] B. Kachar and T. S. Reese, *Nature (London)* **296**, 464 (1982).
[36] P. Pinto Da Silva and B. Kachar, *Cell (Cambridge, Mass.)* **28**, 441 (1982).

transjunctional resistance.[37–41] One additional important parameter is the total length of the tight junction[37,42,43]; cells with elaborate lumenal outline, such as S_1 cells of proximal tubules, have long tight junctional belts; those with polygonal outline, such as collecting duct cells, have short ones. Generally, epithelia with tight junctional belts of great total length, of shallow depth, and with discontinuous strands of low particle density are characteristic for leaky, low-resistance epithelia, whereas tight junctional belts with short total length, continuous parallel strands, and a high density of particles are found in tight, high-resistance epithelia. The efficiency of the paracellular pathway is modulated by interference with the transcellular pathway.

Investigations of the paracellular pathway exclusively by morphological techniques can give only rough estimations of its qualities. In particular, they give no information on the permeability of the tight junction to water or specific solutes.

Changes in the morphological appearance of the tight junctional strands in mammalian renal epithelia in association with defined physiological changes have not been observed. Under pathological conditions disruption of the junctional complexes may occur.[44]

Transcellular Pathway. The tight junctions topographically delimit the lumenal from the basolateral cell membrane domains. The polar distribution of a variety of membrane components, such as enzymes, channel proteins, and receptors, is the fundamental prerequisite for transcellular vectorial transports of solutes across epithelia. For example, in renal tubule cells the enzyme Na^+,K^+-ATPase, which is the motor for active transcellular sodium and sodium-related transports, is confined exclusively to the lateral and to infolded basal cell membrane domains[45,46]; a given species of carrier proteins is generally localized in either the lumenal (e.g., carrier for sodium/glucose cotransport) or in the basolateral membrane domain (e.g., carrier for sodium-independent glucose diffusion).

Active transports, i.e., transcellular movement against electrochemical gradients, are directly or indirectly linked to the consumption of ATP. It is,

[37] P. Claude and D. A. Goodenough, *J. Cell Biol.* **58,** 390 (1973).
[38] A. Martinez-Palomo and D. Erlij, *Proc. Natl. Acad. Sci. U.S.A.* **72,** 4487 (1975).
[39] C. Pricam, F. Humbert, A. Perrelet, and L. Orci, *Lab. Invest.* **30,** 286 (1974).
[40] A. Schiller and R. Taugner, *Cell Tissue Res.* **223,** 603 (1982).
[41] A. Schiller, W. G. Forssmann, and R. Taugner, *Cell Tissue Res.* **212,** 395 (1980).
[42] D. R. DiBona and G. R. Mills, *Fed. Proc., Fed. Am. Soc. Exp. Biol.* **38,** 134 (1979).
[43] K. Kuhn and E. Reale, *Cell Tissue Res.* **160,** 193 (1975).
[44] R. E. Bulger, *Lab. Invest.* **30,** 136 (1974).
[45] S. A. Ernst, *J. Cell Biol.* **66,** 586 (1975).
[46] M. Kashgarian, D. Biemesderfer, M. Caplan, and B. Forbush, *Kidney Int.* **28,** 899 (1985).

thus, not surprising that most cell types with high active transport rates also display large amounts of mitochondria.

Forms of Cell Membrane Amplification. In most cells of the renal tubules the surface area of the basolateral membrane domain is much greater than the projection area of the cells on the tubular basement membrane (basal planar area). In some cell types (S_1 and S_2 of proximal tubule, thick ascending limb cells, distal convoluted tubule cells) the surface area, almost exclusively, of the lateral cell membrane is amplified. This is achieved by the formation of large, lamella-like lateral cell processes. These processes interdigitate intimately with those of the neighboring cells.[47-49] The intercellular space confined between the lateral cell membranes of adjacent cell processes is of uniform narrow width (~20 nm) and converges toward the tight junctional belt. This structural arrangement suggests extensive interference between the trans- and paracellular transport routes.

In other cell types (collecting duct cells, S_3 of proximal tubules) the surface areas of lateral cell membranes are increased only relatively little by formation of small lateral folds and microvilli.[50,51] In such cell types infoldings of the cell membrane within the basal cell portion may substantially contribute to the amplification of the basolateral surface area. Often the extracellular space between the basal infoldings is not directly continuous with the lateral intercellular spaces. Both spaces can be very variable in width. Because of the structural arrangement in this type of epithelium direct interference between trans- and paracellular pathways seems to be less evident than in epithelia with predominantly lateral membrane amplification.

Basically, the surface area of the lumenal cell membrane domain reveals three modes of amplification (with respect to the lumenal planar area, the plane at the level of the tight junctional belt): formation of brush border, of stubby microvilli, and of microfolds.

The lumenal surface of proximal tubule cells is homogeneously covered by narrowly arranged microvilli which are all of uniform length and shape. They form the so-called brush border. The collected surface area of brush border microvilli is similar to (rabbit) or even larger (S_3 segment in rat) than the surface area of the basolateral cell membranes.[48]

---

[47] C. C. Tisher and K. M. Madsen, *in* "The Kidney" (B. M. Brenner and F. C. Rector, eds.), Vol. 1, 3rd ed., p. 3. Saunders, Philadelphia, Pennsylvania, 1987.
[48] L. W. Welling and D. J. Welling, *Kidney Int.* **8,** 343 (1975).
[49] L. W. Welling and D. J. Welling, *Kidney Int.* **9,** 385 (1976).
[50] L. W. Welling, A. P. Evan, and D. J. Welling, *Kidney Int.* **20,** 211 (1981).
[51] L. W. Welling, A. P. Evan, D. J. Welling, and V. H. Gattone, *Kidney Int.* **23,** 358 (1983).

In all segments downstream of the proximal tubule the lumenal surface area of the cells (except intercalated cells; see below) is amplified by the formation of more or less numerous, rather stubby microvilli.[52,53] Their shape and their distribution on the lumenal cell area are generally inhomogeneous. Usually their density is greatest in the peripheral cell regions along the junctional belt, and decreases toward the central cell regions above the nucleus. Their collected surface area is generally severalfold smaller than the surface area of basolateral cell membranes.

The microfold formation found in some cell types, for instance in intercalated cells, seems to be associated with rapid modulation of the lumenal surface area involving specific membrane recycling systems.

Correlation of Structure with Transport

The differences among the various cell types of the nephron with respect to the surface area of their *basolateral cell membranes* are obvious. The surface area is quantitatively correlated with the activity of Na^+, K^+-ATPase and the sodium transport rate of the cells. The interdependence of these three parameters has been demonstrated by three types of experimental approach: first, the correlation between the enzyme activity and the transport rate for sodium, both calculated per unit tubular length, has been proved by numerous studies on isolated tubules, mostly those of rabbits[54-58]; second, the proportionality of enzyme activity, measured per unit protein or dry weight of tissue, and the surface area of basolateral cell membranes assessed per unit tissue volume (as surface density, S_V) by morphometric methods, has been shown in distal tubular segments of rabbits[19,59] and in proximal tubules of rats[60]; and third, the direct correlation between transport rate and membrane surface area, both per unit tubular length, was suggested by combined structural and functional studies in rat kidneys[27,28,32] and in isolated tubules from rabbits.[11]

Taken together, these studies give evidence that *differences* in the surface area of basolateral membranes among nephron segments correspond

---

[52] B. Kaissling and W. Kriz, *Adv. Anat. Embryol. Cell Biol.* **56**, 1 (1979).
[53] W. Kriz and B. Kaissling, in "The Kidney: Physiology and Pathophysiology" (D. W. Seldin and G. Giebisch, eds.), p. 265. Raven, New York, 1985.
[54] A. Doucet and A. I. Katz, *Am. J. Physiol.* **238**, F380 (1980).
[55] C. E. Ganote, F. F. Grantham, H. L. Moses, M. B. Burg, and J. Orloff, *J. Cell Biol.* **36**, 355 (1968).
[56] S. K. Mujais, *Am. J. Physiol.* **254**, F845 (1988).
[57] R. G. O'Neil and R. W. Hayhurst, *J. Membr. Biol.* **85**, 169 (1985).
[58] C. S. Wingo, *Am. J. Physiol* **253**, F1136 (1987).
[59] M. Le Hir, B. Kaissling, and U. C. Dubach, *Cell Tissue Res.* **224**, 493 (1982).
[60] W. Pfaller, *Adv. Anat. Embryol. Cell Biol.* **70**, 1 (1982).

to parallel differences in their enzyme activity and sodium transport rate, and that *changes* in the surface area of basolateral cell membranes within a segment correlate with the changes in Na^+,K^+-ATPase activity and sodium transport rate. Thus, based on the surface area of basolateral cell membranes of renal tubules some conclusions can be drawn on their capacity for transcellular electrolyte transport *in vivo*.

Direct quantitative correlation of *lumenal surface area*, specific membrane components, and sodium transport rate has not been described. The increases in the lumenal surface area which have been observed in association with chronic increases in sodium absorption were mostly smaller than those in the surface area of the basolateral membrane in the same cells.[11,18,26,27,29] In contrast, lumenal microfold formation and modulation of the lumenal membrane area by membrane recycling, occurring in intercalated cells (IC cells) and collecting duct cells (CD cells), seem to be quantitatively related to the regulation of the transport of other solutes, such as protons and water, respectively.

Recycling of Membrane. IC cells: The lumenal surface pattern of IC cells is variable. Intercalated cells with extensive formation of microfolds can be found in the same tubular cross-section together with IC cells with an almost smooth lumenal surface. By immunocytochemical techniques it has been demonstrated that some membrane domains (membrane of specific vesicles in the apical cytoplasm, parts of the lumenal or of the infolded basal and lateral cell membrane) of IC cells possess a proton ATPase. It is assumed that proton secretion is promoted by the rapid insertion of the specific vesicles into the lumenal membrane, associated consequently with decreased vesicular membrane area and increased lumenal surface area.[15,24] Proton reabsorption and bicarbonate secretion seem to be associated with endocytotic uptake of lumenal membrane domains into the cell, which are then stored in vesicular form within the cell.[7,61,62] Thereby the lumenal surface area decreases whereas the vesicular surface area increases.

The proton ATPase of IC cells seems to be associated with 10-nm specific particles—studs—on the cytoplasmic membrane face,[63] which can be taken advantage of as a "marker" for morphological localization of the enzyme. The occurrence of studs seems to coincide with "rod"-shaped *intra*membrane particles which can be demonstrated by freeze–fracture techniques.[29] The functional significance and biochemical nature of rod-shaped particles are not yet clear.

[61] D. Brown, *Am. J. Physiol.* **256**, F1 (1989).
[62] G. J. Schwartz, J. Barasch, and Q. Al-Awqati, *Nature (London)* **318**, 368 (1985).
[63] D. Brown, S. Gluck, and J. Hartwig, *J. Cell Biol.* **105**, 1637 (1987).

In the basolateral cell membrane of some proton-secreting IC cells an anion exchanger protein (band 3 protein for HCO_3^-/Cl^- exchange) has been revealed by immunocytochemical methods.[64] Recycling of this protein has not yet been observed.

CD cells: In the typical CD cells, lumenal membrane recycling is involved in antidiuretic hormone (ADH)-controlled transcellular water transport. Immediately beneath the lumenal membrane of these cells elongated vesicles are present, in the limiting membrane for which freeze-fracture techniques have demonstrated aggregates of intramembranous particles[65-67]—"aggrephores"—which seem to contain water channels. On stimulation with ADH the vesicles fuse with the apical cell membrane and deliver the water-permeable proteins to the limiting cell membrane. The vesicles are retrieved from the membrane after withdrawal of ADH. Under some functional circumstances, the ADH-induced increase in the frequency of particle aggregates in the lumenal membrane correlates roughly with the ADH-induced increase in water permeability.[68]

Mitochondrial Density and Transport Rate. The direct correlation between mitochondrial density, mitochondrial enzyme activity, and sodium transport rate is less evident than that between the surface of basolateral cell membranes, Na^+,K^+-ATPase, and sodium transport rate. Mitochondrial energy is not used exclusively for transport processes and other energy sources exist (glycolysis). However, in most renal cell types observed changes in mitochondrial density and mitochondrial enzyme activity were roughly of the same magnitude as those in the basolateral cell membrane area.

Time Course of Structural Changes. Increases in the surface area of basolateral cell membrane and in the density of mitochondria represent slow adaptive responses of the cells since they require *de novo* synthesis of large amounts of membrane materials. In cells of distal tubules and collecting ducts significant quantifiable increases in basolateral cell membrane area have been observed not earlier than 3 days after stimulation of their transport activity *in vivo* by treatment with mineralocorticoid hormones[21,27,28,69]; or with various diuretic drugs.[18,20] No further increases were observed beyond 7 days of treatment.[20,69] In the given models the changes were associated with hypertrophy and hyperplasia of the respective cells.

[64] D. Drenckhahn, K. Schluter, D. P. Allen, and V. Bennett, *Science* **230**, 1287 (1985).
[65] R. M. Hays, N. Franki, and G. Ding, *Kidney Int.* **31**, 530 (1987).
[66] J. B. Wade, *Fed. Proc., Fed. Am. Soc. Exp. Biol.* **44**, 2687 (1985).
[67] J. B. Wade, D. L. Stetson, and S. A. Lewis, *Ann. N.Y. Acad. Sci.* **372**, 106 (1981).
[68] H. W. Harris, Jr., J. B. Wade, and J. S. Handler, *J. Clin. Invest.* **78**, 703 (1986).
[69] M. J. Field, B. A. Stanton, and G. H. Giebisch, *J. Clin. Invest.* **74**, 1792 (1984).

Decreases in the surface area of basolateral membrane, of mitochondrial density, and of transport rate have been less extensively documented. In collecting duct cells unequivocal adaptive decreases in the surface area of basolateral membrane could be detected about 3 days after reduction of their cellular transport activities,[28] in distal convoluted tubule cells (DCT cells) after at least 10 days or more.[12,22]

Increases in the surface area of lumenal membrane may follow similar time courses as those of basolateral membranes if they involve *de novo* synthesis of membrane proteins. However, modulation in lumenal membrane area which involves membrane recycling occurs much faster, within minutes after a specific stimulus.[70]

Strategies for Investigation of Renal Epithelia *in Situ*

From the above-outlined correlations between some cellular structures of renal tubules and solute transport rate it follows that qualitative and quantitative evaluation of membrane areas, and with some restriction mitochondrial and vesicular densities in renal tubular segments, may be a useful tool to estimate the influence of various conditions *in vivo* on the transport capacity of renal segments and cell types.

Microscopic Approaches

The following light and electron microscopic techniques are the most suitable and most commonly used for the morphological investigation of renal tissue.

Light microscopy on 1μm sections (magnifications $<\times 1250$) gives rapid orientation on the quality of the tissue fixation and allows a first estimation of structural differences among animals. The optical resolution on semithin sections is very high. Cell types and to a certain extent also cell organelles are unambiguously distinguished. Examination of semithin sections strengthens the validity of the electron microscopic observations as the semithin sections can display larger areas than ultrathin sections. Large semithin sections are very convenient for the quantitative assessment of the distribution of renal cell types and segments within the different renal zones.

Routine transmission electron microscopy (TEM) has the broadest application in the structural investigation of kidneys. It allows in the same section screening at low EM magnifications ($\times 1000-2000$), overviews of entire cells with their membranes and organelles at intermediate magnifi-

---

[70] L. M. Satlin and G. S. Schwartz, *J. Cell Biol.* **109**, 1279 (1989).

cations (×2000–12,000), and detailed analysis of cell organelles at high magnifications (>×12,000).

Together, light microscopic study of semithin sections and conventional transmission electron microscopy constitute the basis for the quantitative evaluation of renal tissue (see the section, Quantitative Evaluation of Renal Epithelia, below).

Light and transmission electron microscopy can be combined with immunocytochemical methods for the localization of specific proteins. Assessment of differences in density within membranes or organelles is possible if particulate markers, such as colloidal gold, are used as labels for antibodies.

Freeze–fracture methods are of particular importance in the study of tight junction and membrane structure. With this technique the tissue, frozen at <−214° in melting nitrogen, is fractured. Thereby cellular membranes are split and their interior faces are exposed. They can be observed in thin replicas made of the fractured surfaces. Thus, estimation of intramembranous particle density (intramembranous particles probably correspond to integral membrane proteins) within the smooth membrane matrix is possible. In combination with rapid freezing and freeze–drying methods the fine structure of true membrane surfaces and of some components of the cytoskeleton can be observed in detail. However, these methods are difficult to apply to the systematic study of large tissue areas and of selected cell types in the kidney since the fracture plane through the tissue block generally cannot be precisely and voluntarily predetermined, and recognition of cell types may be difficult.

Scanning electron microscopy is applied on critical point-dried tissue blocks. Scanning electron microscopy can give overviews of large tissue areas at low magnifications (<×500) and reveals nicely the three-dimensional pattern of cellular surfaces at higher magnifications (<×20,000). Heterogeneity with respect to different surface patterns of cells within a tubule can be particularly well assessed. For instance, the proton-secreting IC cells with their extensive formation of narrow, very complex microfolds are easily recognized among the other cell types of the segment.[47,71] However, study of intracellular structures is not routinely possible with this technique.

Processing of the Tissue for Microscopic Investigation

Fixation of Renal Tissue. Good tissue preservation is the basic requirement for all qualitative and quantitative morphologic investigations. The fixed tissue structure should be as closely equivalent to the structure *in vivo*

---

[71] A. LeFurgey and C. C. Tisher, *J. Membr. Biol.* **61**, 13 (1981).

as possible. The structural preservation of renal tissue is particularly problematic: first, the tubules collapse and the renal cells swell after even the shortest interruption of blood circulation and oxygen supply; second, in the renal cortex high osmotic gradients exist between the lumenal and peritubular compartments of some segments, e.g., of distal tubules; third, the osmotic concentrations are different among the zones, and even within the zones osmotic gradients may be very pronounced.

The technique of retrograde vascular perfusion of the fixative solution without prior rinsing of the vasculature with buffers[72] can cope relatively well with all three problems. The greatest advantage of this technique is the almost instantaneous exchange of blood against the fixative without any prior interruption of the renal blood flow. In kidneys fixed by this method the tubular lumena are open. Functionally induced changes of the structure are well preserved. The validity of this method is apparent when micrographs of the fixed tubules are compared with the structure of living tubules observed *in vitro* by optical microscopic methods, using differential interference contrast and optical sectioning.[73,74,74a] The fixed tissue can be studied by all the techniques mentioned above.

Fixation Procedure. The abdominal aorta of the anesthetized animal must be exposed by careful surgery and clamped immediately distal to the renal arteries. From the iliac bifurcation, the aorta is cannulated with a flexible tube [e.g., by intramedic polyethylene tubing (Clay Adams and Company, Parsippany, NJ) with the largest possible internal diameter] that extends to the clamp and is secured. The tube must be filled with a small volume (about 0.1 ml in rats) of saline containing heparin in order to prevent direct contact between the fixative and the blood during insertion of the tube and, thus, to prevent its obstruction by fixed or coagulated blood. The tube is connected over a valve, equipped with a stopcock, to the container with the fixative solution. All tubings have to be absolutely free of air bubbles. Then a large incision is made into the vena cava at the level of the renal veins in order to allow free outflow of the blood and the fixative. At this time, even in small animals, relatively important blood samples can be drawn if necessary. After the incision into the vein the aortic clamp is released, and the stopcock must be opened immediately to allow flow of the fixative.

Since we omit prior washout of the blood with buffers or saline the fixative solution must be perfused by high pressure. For all animals we use a pressure of 0.3 bar in the fixative container. With this high pressure the flow in the renal vasculature is so fast that the blood is pushed out of the

---

[72] B. Kaissling, *Cell Tissue Res.* **212**, 475 (1980).
[73] K. L. Kirk, J. A. Schafer, and D. R. DiBona, *J. Membr. Biol.* **79**, 65 (1984).
[74] K. Strange and K. R. Spring, *J. Membr. Biol.* **96**, 27 (1987).
[74a] H. Völkl and F. Lang, *Pfluegers Arch.-Eur. J. Physiol.* **412**, 1 (1988).

vessels before it becomes solid by fixation and thus obstructive. The fixative is distributed in the kidney cortex within less than a second from onset of the fixative flow, as evidenced by the color change of the kidneys. As the fixative is filtered in the glomeruli the epithelia are fixed from the vascular side as well as from the tubular lumen. The fixative solution is also excreted into the urinary bladder.

The flow of the fixative is stopped after exactly 2 min from its start. The fixative process will continue even without further flow since the vascular bed and the tubular lumena are filled with the solution. The kidneys can be removed from the body, cleaned of all adherent tissue, weighed, and stored in vials with the same fixative solution at 4° until further processing (possible for at least a year). In any case the total fixation duration should be not less than 5 min.

If enzyme or immunohistochemical studies are planned, the following modification is advantageous. The kidneys remain *in situ* after shutting off the fixative flow and exactly 5 min after its start the fixation is stopped by washout of the fixative. This is carried out by perfusion for 10 min with adequate buffer solutions at hydrostatic pressure (~100 cm). Then the kidneys are removed and stored in the buffer solution. If the latter contains sodium azide (~0.05%) the tissue can be maintained at 4° for several months without any noticeable structural deterioration.

All organs which get their blood supply from the point proximal to the tip of the cannula will be perfused regularly and fixed (except lungs) with this technique. In animals larger than rats it may be advantageous to clamp the aorta above the renal arteries *after* onset of the perfusion in order to reduce the amount of required fixative solution.

Fixation Solution. Fixative: Aldehydes are the most commonly used fixatives. Glutaraldehyde (GA) seems to induce an irreversible cross-linking of proteins and to react faster than paraformaldehyde (PFA). The latter seems to penetrate the cells faster than GA. Therefore, both aldehydes are usually combined in the fixative solution. Mixtures of 1.5% GA and 1.5% PFA (or with higher aldehyde concentrations) in buffer solutions (see next paragraph) have proved to give very good structural preservation of the tissue. Yet, since GA strongly inhibits enzyme activity and the expression of antigenic epitopes of proteins, it is advantageous to reduce the GA concentration in the fixative solution if the tissue is to be used for enzyme or immunocytochemical investigations. With solutions containing 2.5% PFA and 0.1% GA, perfused by the above-described technique, the structure of the cells is almost as good as with higher GA concentrations, and many enzymatic activities (e.g., various ATPases) and epitopes are preserved; with GA concentrations below 0.1%, even if combined with 4% PFA, the structural preservation is not as good.

In any case the aldehyde solutions have to be freshly prepared since aldehydes tend to cross-link in older solutions and to lose their good fixation qualities. Glutaraldehyde is commercially obtained as concentrated stock solution (e.g., EM grade, 8, 10, 25, and 50% from EMS, Washington, PA). A concentrated PFA solution (e.g., 25 or 10%) is made by dissolving the powder (e.g., Merck, Darmstadt, FRG) in water by gently heating to about 60° and carefully adding a few drops of 1 M NaOH until the solution clears.

Vehicle: We dilute the concentrated aldehyde solutions with a mixture of 6 vol of 0.1 M cacodylate buffer, pH 7.4, adjusted with sucrose to 300 mOsm (this buffer is used also to wash out the fixative), with 4 vol of a polyhydroxyethyl starch solution (M_r 200,000 in 0.9% saline; Pentaspan 10%, McGraw Laboratory Division, American Hospital Supply Corp., Evanston, IL; HÄS 10%, Fresenius, Bad Homburg, BRD) in order to increase the oncotic pressure of the solution. To that we add 3 mM MgCl$_2$ and 2 mM picric acid. The latter improves membrane contrast and stains the solution yellow. This is of practical advantage since occasionally unperfused regions in the kidney can be well recognized because of the lack of yellow stain and can be discarded from further tissue processing. We perfuse the fixation solution at room temperature.

Osmolality: The total osmolality of the fixation solution is far above physiological values. Yet the experience of numerous laboratories has shown that osmolality due to the aldehydes is negligible, since their reflection coefficient is low, provided they have been freshly prepared.

Some investigators recommend adjusting the osmolality of the vehicle solution to the expected approximate osmolality of the kidney zone to be studied.[75] When fixing without prior washout of the blood we realized that with 2.5% PFA and 0.1% GA in the 300-mOsm vehicle solution all kidney zones are almost equally well fixed and do not show osmotic swelling or shrinkage. According to our experience the quality of the fixation is influenced more by the rapidity of the initial perfusion flow and fluid exchange (in addition to the pressure, both depend on the diameter of the tubing in the aorta and on the large incision into the vena cava) than by the vehicle buffer osmolality.

Processing of the Fixed Tissue for Sectioning. Light and transmission electron microscopic investigations are carried out on thin sections (1 μm to 20 nm) of tissue. Such sections can be cut only after adequate "hardening" of the tissue. This can be achieved by more or less complete dehydration of small tissue pieces over graded series of alcohols followed by impregnation and embedding into fluid resins (epoxy, acrylic) which

---

[75] S.-O. Bohman, *J. Ultrastruct. Res.* **47,** 329 (1974).

harden by polymerization, or by freezing the tissue. Dehydration and embedding into resins is generally accompanied by some degree of tissue shrinkage. This is, however, negligible in comparison with shrinkage occurring with conventional embedding in paraffins for light microscopy. But it is important to use the same methods for tissue processing in all tissue samples.

Embedding into Conventional Epoxy Resin. The tissue blocks must be small enough to allow impregnation and polymerization of the resin throughout the entire block. Yet, larger blocks are advantageous in the qualitative and quantitative analysis of the tissue. For embedding into conventional epoxy resin (e.g., from EMS, Washington, PA) we cut thin (<500 μm) but large ($\sim 3 \times 3$ mm in rat kidneys) slices with sharp razor blades. Prior to complete dehydration all remnants of the fixative solution must be washed out with the osmotically equilibrated buffer and the slices are postfixed with OsO_4 (e.g., 1% in water for 60 min at 4°). The resin-impregnated slices are placed flat into the covers of Beem capsules (Better Equipment for Electron Microscopy, Inc., Bronx, NY), which have had their tips cut off, and the capsules are filled with the resin. Polymerization is induced by a chemical agent within the given resin mixture, generally at temperatures around 60° (epoxy). The blocks must stay at this temperature for several days (depending on the mixture used) in order to be completely polymerized. Thereafter, semithin (~ 1 μm) and ultrathin sections (100–20 nm) can be cut with ultramicrotomes, using glass and diamond knives, respectively. The sections are studied by light or electron microscopy after adequate staining procedures.

Embedding at Low Temperatures. Some resins (e.g., Lowicryl; Chemische Werke LOWI GmbH, Waldkreiburg, FRG; LR-White; The London Resin Company, Ltd., Basingstoke, Hampshire, England) do not need complete dehydration, as they are water soluble to some small extent and they polymerize with UV light at lower temperatures ($<-35°$).[76-78] They are used especially in immunocytochemical studies. Embedding in these resins requires the preparation of very small tissue blocks (<1 mm^3), which should not be treated with osmium, because UV light will not be able to penetrate a block which, thus, will not be polymerized throughout. The tissue blocks are placed into translucent, O_2-impermeable gelatin capsules, and exposed at low temperatures to indirect UV light for about 24 hr and to direct UV light at room temperature for another 2 days. Low-tempera-

[76] E. Carlemalm, R. M. Garavito, and W. Villiger, *J. Microsc. (Oxford)* **126**, 77 (1982).
[77] L. G. Altman, B. S. Schneider, and D. S. Papermeister, *J. Histochem. Cytochem.* **32**, 1217 (1984).
[78] G. T. Simon, J. A. Thomas, K. A. Chorneyko, and E. Carlemalm, *J. Electron Microsc. Tech.* **6**, 317 (1987).

ture embedding needs special devices; the unpolymerized resins (Lowicryl) are highly allergenic and sectioning is more delicate than that of conventional epoxy resins.

Freezing. Another method for "hardening," avoiding dehydration of the tissue by alcohols or solvents and its embedding into resins, is freezing of fixed pieces of tissue which have been cryoprotected by impregnation with 2.3 M sucrose or other adequate media.[79] For freezing we use liquefied propane, cooled with liquid nitrogen. Semithin and ultrathin sections can be cut with glass or cryodiamond knives at temperatures ranging between -70 and $-120°$ with a special cryoultramicrotome. Whereas semithin sectioning of rather large tissue blocks ($>2 \times 2$ mm) is relatively easy, ultrathin sectioning of frozen tissue needs great skill and considerable practical experience. The usable section areas are generally too small to provide overviews on several cells, but in small areas the technique can yield high-resolution images.

Basic Immunocytochemical Procedures. Immunocytochemical techniques are used to locate a specific protein in the tissue. The tissue is exposed to the primary antibody, which is directed against the protein in question. The tissue structures to which the primary antibody has bound are revealed by incubation with a secondary, labeled antibody (commercially available; e.g., labeled with colloidal gold or peroxidase for light and electron microscopy; with fluorescent dyes for fluorescence microscopy). The secondary antibody is directed against immunoglobulins of the animal in which the primary, specific antibody has been raised. Instead of the secondary antibody, labeled protein A can be used in many cases.

In some instances binding of specific lectins to membranes seems to give some information on the physiological state of a cell.[70,80] The procedures for lectin histochemistry are very similar to those of immunohistochemistry.[81,82]

Basically, immunostaining of the tissue can be carried out on tissue slices before embedding (preembedding methods), on tissue sections after embedding (postembedding methods), and/or on semithin and ultrathin sections of frozen unembedded tissue.

In preembedding techniques the entire procedure of immunostaining is made on slices (10 to 100 μm; made with a vibratome, a tissue chopper, or a cryostat) of the adequately fixed unembedded tissue, which thereafter are embedded into epoxy resins in the same way as for routine electron

---

[79] K. T. Tokuyasu, *Histochem. J.* **21**, 163 (1989).
[80] D. Brown, J. Roth, and L. Orci, *Am. J. Physiol.* **248**, C348 (1985).
[81] J. Roth and D. J. Taatjes, *Eur. J. Cell Biol.* **39**, 449 (1985).
[82] J. Roth, *J. Histochem. Cytochem.* **31**, 987 (1983).

microscopy.[83] The relatively easy handling, the good tissue preservation, and the good reaction intensity at those sites to which the antibodies can access are advantages of this method. Yet, the poor penetration of the antibody molecules into the tissue and cells constitutes a serious limiting factor of this technique.

In postembedding techniques immunostaining is carried out on ultrathin or semithin sections of embedded tissue.[83] In these sections intra- and extracellular antigens should be, in principle, accessible to the antibodies. However, the dehydration steps before embedding in resins as well as the high peaks of temperature ($>80°$) occurring during polymerization of epoxy resins may alter the configuration of the proteins and therefore may destroy many epitopes. The polymerized resin may also limit the access of the antibodies to the antigenic sites in the section. Therefore treatment of the sections with caustic agents is often necessary before immunostaining.[84] The binding of antibodies to their antigens in such sections is generally weak.

In tissue, embedded in Lowicryls and LR-White, antigenic sites are better preserved than in tissue embedded in conventional epoxy resin. By ultrathin sectioning of the tissue embedded in the resin is apparently "broken off" from interfaces and, thus, the antigens are much more accessible to the antibodies.[85]

In semithin and ultrathin sections of frozen tissue the retention of epitopes is much superior and access of antibodies to them is better than in any embedded tissue. With fluorescent dyes [e.g., fluorescein isothiocyanate (FITC) from e.g., Dakopatts, Glostrup, Denmark] used as label on semithin sections, very high resolution and precise assignment of an antigen to basolateral, vesicular, or lumenal membranes and often even to Golgi apparatus and to other subcellular organelles are possible. The fading of the FITC which normally occurs during observation of the sections can be substantially inhibited by adding 2.5 g 1,4-diazabicyclo[2.2.2]octane (DABCO; Sigma, St. Louis, MO) to 100 ml of the glycerin gel for covering the slides.[86] This allows conservation of the slides (at 4° in the dark) for several months and their repeated and extensive observation without significant loss of the fluorescent signal.

Peroxidase techniques [peroxidase–antiperoxidase (PAP); avidin–biotin complex (ABC)] can detect low antigen concentrations, but the

[83] D. Kerjaschki, H. Savada, and M. G. Farquhar, *Kidney Int.* **30**, 229 (1986).
[84] M. H. Maxwell, *J. Microsc. (Oxford)* **112**, 253 (1968).
[85] E. Kellenberger, W. Dürrenberger, E. Villiger, E. Carlemalm, and M. Wurtz, *J. Histochem. Cytochem.* **35**, 957 (1987).
[86] G. D. Johnson, R. S. Davidson, K. C. McNamee, G. Russel, D. Goodwin, and D. J. Holborow, *J. Immunol. Med.* **55**, 231 (1982).

localization of the antigen is less precise than with fluorescent dyes due to the diffusion of the chromogen.

The silver-enhanced gold-staining technique for light microscopy gives good contrast and has very high sensitivity.[87] With epipolarization filters even very low concentrations of an antigen can be detected. Yet, due to the size of the gold-silver complexes, their assignment to subcellular localizations is less precise.

Applied on ultrathin frozen sections, the technique with colloidal gold (using 3- to 5-nm gold particles without silver enhancement; from Aurion (Wageningen, Netherlands) allows the most precise localization of antigens since no diffusion of the label occurs. Quantification of the labeling intensity is possible.[88] Yet, as mentioned above, the technique of ultrathin sectioning is delicate.

Detailed methods for tissue embedding, freezing, sectioning, and immunocytochemistry are given in standard treatises on immunocytochemical[89-91] and EM methods.[92-95]

Quantitative Evaluation of Renal Epithelia

Morphological differences among renal cell types as well as their morphological modulation, associated with changes in their transport capacity, are often evident by simple light and electron microscopic study. However, their quantitative evaluation represents a valuable or even necessary complement because it gives objective data and it becomes possible to assess correlations between morphological and functional data.

Basic Parameters in Stereology

Standard planimetric and stereological techniques allow the estimation of lengths, of surfaces, and of volumes of structures and bodies in three-di-

---

[87] J. M. Lucoq and J. Roth, *Tech. Immunocytochem.* **3**, 203 (1985).
[88] M. Bendayan, A. Nanci, and F. W. K. Kan, *J. Histochem. Cytochem.* **35**, 983 (1987).
[89] J. M. Pollack and S. van Noorden, "Immunocytochemistry." Wright, Bristol, England, 1986.
[90] G. R. Bullock and P. Petrusz (eds.), *Tech. Immunocytochem.* **3**, 203 (1985).
[91] J. A. Bourne, "Handbook of Immunoperoxidase Staining Methods." Dako Corp., Santa Barbara, California, 1983.
[92] M. A. Hayat, "Correlative Microscopy in Biology: Instrumentation and Methods." Academic Press, Orlando, Florida, 1987.
[93] M. A. Hayat, "Fixation for Electron Microscopy." Academic Press, New York, 1981.
[94] M. A. Hayat, "Principles and Techniques of Electron Microscopy: Biological Applications," third ed. MacMillan Press, 1989.
[95] M. A. Hayat, "Positive Staining for Electron Microscopy." Van Nostrand-Reinhold, New York, 1975.

mensional complex tissues from a two-dimensional plane through the tissue. Such a plane is displayed by sections, cut across the tissue, although they themselves have a definite thickness. Correction factors for the section thickness have to be used only in cases in which the diameter of the structure in question is close to or less than the section thickness (e.g., some types of vesicles, microvilli).[15] The n-dimensional structure of the tissue is represented on the sectional plane by its $(n - 1)$ dimensional image: lines are seen as points, surfaces as lines and bodies as areas.[60,96,97] Usually the structures in question within a tissue are assessed as "densities," meaning they are related to a unit volume of the tissue.

The volume density relates the (collected) volume V of structures (e.g., of vesicles or mitochondria) to a unit volume V of the tissue. It is expressed as V_V in percentage units and is directly proportional to the corresponding ratio of surface areas, A_A, displayed on the tissue section.

The surface density indicates the surface area of a structure within a unit of tissue volume. It is expressed as surface area S per volume V, S_V (in $\mu m^2/\mu m^3$); e.g., the surface area S of basolateral cell membranes, S_{blm}, in a given volume V of an epithelium, V_{epi}, is S_V (blm/epi). In the section the surface density is assessed as the so-called boundary length B of the structure per area A, B_A (in $\mu m/\mu m^2$); e.g., B_A (blm/epi). The B_A is proportional to the S_V by multiplication with the factor r/π, $B_A \times 4/\pi = S_V$. The factor $4/\pi$ corrects for the fact that the intersected membranes are not parallel to the sectional plane.[96]

The surface area of membranes is often related also to the surface area of the basal or lumenal projection of the cell, the so-called planar area. This ratio is called *amplification* and constitutes a dimensionless factor. For example, the ratio of the surface area S of the basolateral cell membrane, S_{blm}, to the surface area S of the tubular basement membrane, S_{bm}, is expressed as S_S (blm/bm). It is assessed by measuring on the micrograph the corresponding length of the basolateral (or lumenal) membrane and the length of the basement membrane (or the shortest distance between the tight junctions).

Densities can be correlated well with biochemical data (e.g., enzymatic activities) calculated per tissue volume or weight. Differences in densities may be due to changes of the structure in question or of the volume confining the structure. In the case of correlation of structural data of renal tubules with transport data, assessed per tubular length, it seems reasonable to relate the surface areas S and volumes V of the respective structures to a

[96] E. R. Weibel, "Stereological Methods: Practical Methods for Morphometry," Vol. 1. Academic Press, New York, 1979.

[97] E. R. Weibel and R. P. Bolender, *in* "Principles and Techniques of Electron Microscopy" (M. A. Hayat, ed.), Vol. 3. Van Nostrand-Reinhold, New York, 1973.

given tubular length L, and to express them, for example, as S_L (blm/tub) (in $\mu m^2/\mu m$) and V_L (mito/tub) (in $\mu m^3/\mu m$), respectively. These ratios are the product of the epithelial volume per given tubular length, V_L (epi/tub), and of the respective density (S_V, V_V). The V_L (epi/tub) can be derived from the epithelial area of a tubular cross-section, multiplied with the unit of length (i.e., section thickness).

In tubules with cellular heterogeneity the proportion of each cell type can be calculated as the fractional volume, which corresponds to a volume density. Thus, it is the collected volume of the given cell type in percentage units of the total epithelial volume of the tubular sample. The latter can be composed by a collection of respective tubular profiles. The sample size can be delimited, e.g., by a given total number of nuclear profiles.

The collected volume of a given segment (V_{seg}) within a unit volume of the corresponding zone (V_z) is the fractional volume of the segment V_V (seg/z) in percentage units. This can be related to the absolute, total volume of the zone.

Techniques of Measurement

All stereologic or planimetric measurements can be made on enlarged micrographs (electron or light microscopic) or on direct projections of the microscopic images. When using a semiautomatic image analyzer (e.g., MOP-Videoplan, Kontron/Zeiss) the structures can be measured by tracing or outlining them on a magnetic tablet with a sensor pen. The data are directly calculated and displayed by the connected computer.

Another possibility (realizable without expensive equipment) is to overlay the micrographs with a transparent grid with points and lattices of lines and to count test points (P_T) or intersections (I_i). The ratio of test points falling onto the structure in question, e.g., mitochondria, to the total number of points falling onto the reference area, e.g., epithelium, corresponds to the ratio of the surface areas in the section, A_A (mito/epi), and thus to the respective volume density. The size of the area in question (test area, A_T) is calculated by the equation $A_T = P_T \times d^2$, where d is the distance between the points and lines, respectively. The distance d has to be chosen according to the size of the structure in question. Generally, it must be at least as large as the diameter of the smallest units to be measured in order that no more than one test point falls onto the same structure.

Because of the anisotropic orientation of the membranes in renal cells test grids composed by connected semicircles, as developed by Merz,[98] are used generally in the estimation of the boundary length of membranes. With this grid system the boundary length B of, e.g., basolateral mem-

[98] W. A. Merz, *Mikroskopie* **22**, 132 (1967).

branes, is calculated according to the equation $B_{blm} = I_i \times d$, where I_i is the number of intersections of the membranes with the curvilinear test grid and d is the diameter of the semicircles.

Both the semiautomatic and the point-counting techniques yield similar results and can be applied in the estimation of relative and absolute parameters, as S_V, S_A, and V_V, and in that of V_L and planimetry of volumes of zones, respectively, at the electron and light microscopic level. Detailed practical advice and the necessary mathematical background for the stereologic techniques are given in Weibel.[60,96,97]

It should be kept in mind that all structural data, derived from measurements in the fixed and embedded tissue, are only equivalent to their real value *in vivo*. The exact magnitude of shrinkage or swelling of the tissue, occurring by the tissue processing, cannot be determined. Therefore it is important that strictly identical methods for tissue preparation be used for all animals within an experiment. Only in that case the observed qualitative and quantitative structural differences in the tissue may be relevant and reflect true differences *in vivo* in approximately the same order of magnitude.

Sampling Procedures

Prior to sampling, the population of cells to be analyzed must be precisely defined: first, with regard to the cell type [e.g., S_1 of the proximal tubule, CD cells of the cortical collecting duct (CCD) and the outer medullary collecting duct (OMCD)]; second, with regard to the location of the cell within the segment, as some segments, e.g., the thick ascending limb (TAL) and connecting tubule (CNT), display pronounced intrasegmental, so-called "axial" heterogeneity. That means that the cells at the beginning and at the end of the segment are quantitatively differently equipped with respect to their transport machinery; and third, with regard to the nephron population, since the structural equipment for transport of cells in corresponding segments differs among nephron populations, e.g., superficial nephrons and juxtamedullary nephrons.

For recognition of the cell type it is generally necessary that the cells display the lumenal and basal cell membrane on the section. For definition of functionally corresponding locations of the cells within a segment it is preferable to use landmarks within the tissue rather than a given distance from a border of a zone. In the different tissue blocks the sectional plane may not always be perpendicular to the border. Furthermore, the dimensions of the zone may have changed under the specific experimental conditions. Thus, in differently treated animals cells located at an equal distance from the given border may be situated, nevertheless, in functionally different renal compartments (e.g., in one case in the upper cortical half, in the other in the deep cortex). Therefore it is advantageous to use specific histotopographical relationships as landmarks; e.g., the upper portion of the cortical collecting duct (CCD) may be defined by the vicinity of exclusively S_2 segments of the proximal tubule, the lower portion of the CCD by the exclusive juxtaposition of S_3 segments; e.g., the fused portion (arcade) of the connecting tubule may be recognized by its situation immediately adjacent to cortical radial vessels.

In order to be able to determine the precise location of cells the sections (semithin and ultrathin) must be rather large. They should confine at least one and, if possible (e.g., in rat and mouse), both borders of a zone, and at least two medullary rays with the surrounding labyrinth in the cortex or corresponding areas in other zones.

The essential condition for objective data is the randomness of sampling of the structures to be analyzed. The cells or structures defined as above will constitute the population within which the sampling has to be made. Randomness of sampling is given, if within this strictly defined population *each* tubular profile or cell, displayed on the section and fulfilling the specific criteria, is sampled up to a total number that has been fixed in advance. If several sections must be screened they should be cut from different tissue blocks or be at least 100 μm distant from each other, in order to exclude multiple analysis of the same structure. It is advantageous to use always the same screening procedure on the sections, e.g., to start always in the same corner of the sections and to move in a meandering way through them.

Tubular parameters as V_L (epi/tub) and fractional volumes of cells within segments can be measured on semithin sections, since these allow unequivocal distinction of segments and cell types and can display large enough areas. Profiles of tubules will rarely be cut exactly perpendicular to their longitudinal axis. In order to limit overestimation of the cross-sectional area (or circumference), criteria for the acceptability of profiles for analysis must be given. Some researchers accept only profiles, in which the ratio of the longest to the shortest diameter does not exceed 1.2; others accept ratios up to 1.5. It may be necessary to use sections from several differently oriented tissue blocks for the collection of a sufficient number of profiles, in particular if the tubules are all arranged in parallel (e.g., tubules in the inner stripe and inner medulla).

The evaluation of cellular parameters, such as membrane areas [e.g., S_L (blm/tub)], and vesicular and mitochondrial densities [e.g. V_L(mito/tub), etc.], is made on *ultrathin sections*. If not enough entire cross-sectional tubular profiles are available on the ultrathin sections the respective data can be obtained in two steps: first, the respective densities of the cellular parameters [e.g., S_V(blm/epi)] are measured in corresponding fragments of tubular profiles on the ultrathin sections; second, semithin sections, made from the same tissue block as the ultrathin sections, are used in order to collect a sufficiently great number of complete cross-sectional profiles for calculation of the V_L(epi/tub). Multiplication of the respective means constitutes the mean value of S_L or S_V for the given cellular parameter for the animal.

For determination of fractional volumes of segments within a given zone [V_V(seg/z)] it is most essential that both borders of the zone are confined within the sectional area. Each analyzed section should be from a different tissue block.

The determination of the volume of the kidney zones needs serial sectioning through one entire kidney. For this purpose one of the two perfusion-fixed kidneys of the animal can be embedded in paraffin according to conventional histological techniques. Serial sections of 7- to 10-μm thickness are cut, in parallel to the hilar face of the kidney. In this plane the exact delimitation of the zones on the sections is possible whereas in other planes the borders of the zones appear blurred. Sectioning is made beginning from the convex face of the kidney. The very first section, then each fiftieth or sixtieth section (valid for the size of a rat kidney) is stained, e.g., by the hematoxylin eosin technique, and on adequate enlargements the borders of the zones (and of pelvic space, if present) are delineated.[13,18] Their surface areas can be estimated according to standard planimetric procedures, as discussed above. The volumes of the zones are calculated as the sums of the areas measured on the sections, multiplied by the section thickness and the distance between the sections. The total kidney parenchymal volume is the sum of the volumes minus pelvic space. However, as the tissue shrinkage is much greater with paraffin embedding (> 50%) than with resin embedding (< 10%), again only the relative data can be used. These data can be related either to the weight of the kidneys or to the absolute kidney volume. The latter can be determined, for example, by immersion of the fixed kidney in water.

Sample Size for Statistical Analysis. Variation of individual measurements of a given cellular parameter within one animal and among animals of the same experimental group is rather large. However, most investigators agree that analysis of 10 to 15 micrographs for cellular parameters and a similar number for tubular parameters is sufficient to yield reliable data for 1 animal (e.g., Ref. 27). The individual measurements of the parameters are averaged and their mean is considered as the value for the animal. Statistical analysis is made with these means. The sample size n for the statistical analysis of the data is given by the number of animals.[27] In studies with more than two experimental groups the means calculated for each animal within the groups are usually submitted to an analysis of variance before testing for significance by adequate statistical methods.[99]

Appendix: Histotopographical Criteria for Recognition of Tubular Segments

Cortex (C)

The cortex comprises the compartments of the labyrinth (L) and the medullary rays (MR).[100] Within the labyrinth are localized the cortical radial vessels (including lymphatics), nerves, the afferent and efferent arterioles, the renal corpuscles, and the convoluted nephron portions. Within the medullary rays the straight nephron portions are grouped together with the cortical collecting ducts. All tubules are surrounded by peritubular capillaries and interstitial fibroblasts.

Proximal Tubules (S_1, S_2, S_3). S_1 segments of the proximal tubule are most unequivocally identified at the urinary pole of the renal corpuscles and in their immediate vicinity. S_2 segments can most easily be distinguished in the upper half of the medullary rays, but they are situated in addition in the cortical labyrinth. S_3 segments are found in the central parts of the medullary rays in the lower cortical half and within the outer stripe.

Cortical Thick Ascending Limbs (CTALs). These are present within the medullary rays and extend a short distance beyond the macula densa at the vascular pole of renal corpuscles, situated within the cortical labyrinth. In that location TAL cells can be distinguished from the distal convoluted tubule because of the central position of their nucleus.

Distal Convoluted Tubules (DCTs). These are present exclusively in the cortical labyrinth. They begin in the vicinity of their corresponding glomerulus. The nucleus of a DCT cell is situated immediately beneath the lumenal cell membrane. The apical face of the nucleus is often flattened.

Connecting Tubules (CNTs). Connecting tubules have a heterogeneous cell population, comprising CNT and IC cells. Connecting tubules of superficial nephrons (unbranched CNTs) are found in the superficial cortex. Generally, they do not contact the renal capsule. The fused portions of the CNTs (arcades) from midcortical and deep nephrons are located in parallel and in close vicinity to the cortical radial vessels and consequently they are often grouped around afferent arterioles.

[99] G. W. Snedecor and W. G. Cochran, "Statistical Methods." Iowa State Univ. Press, Ames, 1972.
[100] W. Kriz, L. Bankir *et al., Kidney Int.* **33**, 1 (1988).

Collecting ducts are composed of the collecting duct cells proper, the CD cells, and of IC cells. In most animals the latter are usually lacking in the last two-thirds of the inner medullary collecting duct (IMCD).

Cortical Collecting Ducts (CCDs). The beginning of the CCDs (often denominated as "initial collecting ducts," ICTs) is situated in the cortical labyrinth. Initial collecting ducts may touch the renal capsule before they turn and enter the medullary ray. As "medullary ray collecting ducts" they run straightly toward the medulla.

Outer Medulla (OM)

The arcuate vessels are at the border between the cortex and the outer medulla. They are easily recognized by their large patent lumena in perfusion-fixed kidneys and facilitate the delimitation of the cortex from the outer stripe of the outer medulla.

Outer Stripe (OS). The outer stripe comprises the S_3 segments of proximal tubules, TALs, and outer medullary collecting ducts (OMCDs), surrounded by capillary-like wide venous vessels, ascending from the inner stripe and inner zone, and a few true capillaries. Interstitial fibroblasts are sparse.

The S_3 segments of juxtamedullary nephrons often have a very tortuous course. Together with their corresponding TALs, they are situated the most distant from the collecting ducts.

Inner Stripe (IS). The inner stripe comprises two distinct compartments: the vascular bundles with the descending and ascending vasa recta and the interbundle compartments with peritubular capillaries, interstitial fibroblasts, and the tubular structures [descending thin limbs (DTLs), medullary thick ascending limbs of Henle's loop (MTALs), and OMCDs].

In some species (e.g., rat, mouse, psammomys) the descending limbs of short loops (sDTLs) of Henle are closely associated with or fully integrated into the bundles, whereas the descending thin limbs of long loops (lDTLs) are among the CDs and TALs in the interbundle region. In these species the TALs belonging to the short loops are found in the neighborhood of CDs and the TALs belonging to long loops are in the vicinity of the vascular bundles! Thick ascending limbs of exclusively long loops can be found just at the border between the inner stripe and the inner zone. The TAL reveals important axial changes along its course through the inner and outer stripe of the medulla.

Inner Medulla (IM)

The inner medulla is sharply demarcated from the inner stripe of the outer medulla by its distinctly lighter color.

No specific compartmentation with respect to tubular and vascular structures is apparent. The inner zone comprises the descending thin limbs of long loops (lDTLs) and their ascending thin limbs (ATLs) in addition to the inner medullary collecting ducts (IMCDs), running in parallel with descending and ascending vasa recta. Lipid-laden interstitial cells may be rather prominent between the structures of this zone.

[16] An Electrophysiological Approach to the Study of Isolated Perfused Tubules

By RAINER GREGER

Introduction

The isolated perfused tubule preparation was originally designed to mimic the elegant experimental studies of the perfused squid axon. In the late 1960s and 1970s the method of *in vitro* perfusion was almost exclusively applied to renal tubule segments.[1] The analytical approach focused on flux measurements using chemical determinations or radioisotopes.[2] Only in the last 8–10 years has the method of *in vitro* perfusion of isolated tubules also been applied to other tubular structures, such as the glandular ducts of pancreas[3] sweat glands,[4,5] Malpighi tubules,[6] insect gut segments,[7] and tubules of the shark rectal gland.[8] Furthermore, electrophysiological techniques have been refined for such preparations, and today the spectrum includes transepithelial, as well as transmembrane, potential and resistance measurements, determinations of intralumenal and intracellular ionic activities using ion-selective microelectrodes, and the patch-clamp analysis of ionic channels in the basolateral and lumenal cell membranes. These electrophysiological measurements can now be combined with (1) net flux and unidirectional ion flux determinations, (2) optical methods for the determination of cell volume, fluorimetric methods for the assessment

---

[1] M. Burg, J. Grantham, M. Abramow, and J. Orloff, *Am. J. Physiol.* **210,** 1293 (1966).
[2] J. A. Schafer and T. E. Andreoli, *Membr. Transp. Biol.* **4B,** 473 (1979).
[3] I. Novak and R. Greger, *Pfluegers Arch.* **411,** 58 (1988).
[4] K. Sato, *Pfluegers Arch.* **407** (Suppl. 2), S100 (1986).
[5] P. M. Quinton, *Pfluegers Arch.* **391,** 309 (1981).
[6] F. Hevert, *Verh. Dtsch. Zool. Ges.* **77,** 105 (1984).
[7] K. Strange and J. E. Phillips, *J. Membr. Biol.* **83,** 25 (1985).
[8] J. N. Forrest, Jr., F. Wang, and K. W. Beyenbach, *J. Clin. Invest.* **72,** 1163 (1983).

of intracellular proton activity[9] and for that of cytosolic calcium,[10] and (3) metabolic measurements of substrate utilization[11] and O_2 consumption. Considering this wide spectrum of methods to study intact perfused tubules, with easy and rapid access to both extracellular compartments, it is understandable that many new concepts and models for epithelial transport have been suggested. Much more is yet to come, as the method is now applied increasingly to many tubular structures and it even can be used for bioassays in pharmacology. When compared to other approaches to study epithelial transport, e.g., the measurement in the intact organ or organism, or the other extreme, the flux analysis in membrane vesicles or liposomes, the isolated perfused tubule combines the advantages of both approaches. It is an intact biological preparation, like in the intact organism, and yet it permits an analysis of cellular mechanisms such as the vesicle preparation. Moreover, disadvantages of the other two alternative methods are avoided. The *in vitro* perfusion is simplified, when compared to the intact organ, as the two extracellular compartments can be defined and easily modified. Also, the *in vitro* perfused tubule does not suffer from the disadvantage of isolated membrane vesicles where the preparations and the conditions are highly artificial. In the following sections I shall briefly list and describe the different levels of electrophysiological analysis of the *in vitro* perfused tubule. Details already described in recent articles will be omitted, and I will focus on the most recent achievements.

Transepithelial Measurements

Transepithelial ionic transport may generate a transepithelial voltage. This voltage can be caused by active transport. If this is the case, it will persist even in strictly symmetric solutions on the two epithelial cell sides, and it will be blocked by metabolic inhibitors and inhibitors of ionic transport. This voltage was coined "active transport potential" by Frömter,[12] a technical term which is both precise and also sufficiently vague, as to emphasize the black box character of this voltage. If the two solutions bathing the epithelium differ in composition, water fluxes or ionic fluxes will superimpose additional voltages on top of the "active transport voltage." The quantitative analysis of the magnitude of these additional streaming and diffusion voltages will allow characterization of the permeability properties of the intact epithelium.

[9] J. Geibel, G. Giebisch, and W. Boron, *Am. J. Physiol.* (in press) (1990).
[10] R. Nitschke, U. Fröbe, and R. Greger, *Proc. Forefront Symp. Nephrol., 3rd* (1989).
[11] S. P. Soltoff, *Annu. Rev. Physiol.* **48,** 9 (1986).
[12] E. Frömter, *J. Physiol. (London)* **288,** 1 (1979).

The technical aspect of these voltage measurements has been described explicitly by Burg et al.[13] and many others (for reviews, see Greger[14] and Ullrich and Greger[15]). As shown in Fig. 1, the perfusate and the collected fluid are connected via salt agar bridges, or via flowing KCl electrodes and KCl to Ag/AgCl or Hg/Hg$_2$Cl$_2$ half-cells to the high-impedance ($> 10^{10}$ Ω) input of two electrometers. The reference of both instruments is connected to the bath, again using the same components (salt agar bridge or KCl electrode:KCl:Ag/AgCl or Hg/Hg$_2$Cl$_2$). With this approach the transepithelial voltage of the lumenal solution is measured with respect to the bath solution at the perfusion and on the collection end of the tubule segment. It is generally agreed to express this voltage such that the bath is the point of reference. In the absence of the tubule the circuit is checked for symmetry and the recorded voltage should be close to zero. With the tubule in place the voltages recorded at the two ends of the tubule should be very similar. If this is not the case, the rate of tubule perfusion may be too small, thus leading to a variation of the composition of the perfusate as it flows along the tubule. To avoid this pitfall the rate of lumenal perfusion should be increased. Obviously, the adequate rate will depend on the preparation. For renal tubules a rate in excess of 10 nl/min should be sufficient. For glandular ducts the rate may be even lower. Another and probably more frequent cause for discrepant voltage measurements of the perfusion and the collection sites is due to the fact that inadequate sealing of either the collection or perfusion pipet results in short circuiting of the voltage. Sylgard 184, a silicon liquid with good isolation properties, is placed around the tubule to avoid this problem (cf. Fig. 1).

Transepithelial Electrical Resistance

The measured value of the transepithelial resistance of an isolated perfused tubule segment contains the lumped information about the parallel ionic pathways of the epithelium, i.e., the paracellular pathway and the parallel transcellular pathway. The latter resistance is composed of the serial arrangement of the resistances of the lumenal and of the basolateral cell membrane. The electrical resistance of the cytosol is probably small when compared to that of the cell membranes. It is obvious that the complex nature of the transepithelial resistance precludes any straightforward interpretation of its value or its alteration by experimental maneu-

[13] M. B. Burg, L. Isaacson, J. Grantham, and J. Orloff, *Am. J. Physiol.* **215**, 788 (1968).
[14] R. Greger, *Mol. Physiol.* **8**, 11 (1985).
[15] K. J. Ullrich and R. Greger, in "The Kidney: Physiology and Pathophysiology" (D. W. Seldin and G. Giebisch, eds.), p. 427. Raven, New York, 1985.

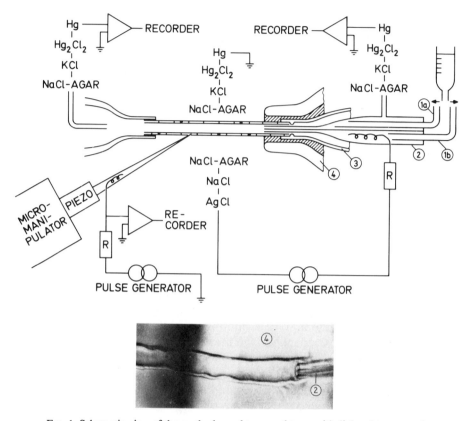

FIG. 1. Schematic view of the methods used to record transepithelial and transmembrane voltages, as well as the transepithelial resistance and the voltage divider ratio in isolated perfused tubule segments. The tubule segment is perfused from right to left through a dual-channel perfusion pipet (1a, 1b). The active channel (2) is determined by the position of the three-way valve. The composition of the perfusate in the two channels of the perfusion pipet can be varied through two fluid-exchange pipets (1a and 1b). The tubule segment is held by the holding pipet (3) and by the collection pipet. For optimal insulation an outer Sylgard pipet (4) is used. The transepithelial voltage is measured at the perfusion and the collection side of the tubule segment via NaCl–agar bridges and calomel half-cells and referenced to the grounded bath. The transepithelial resistance is determined by the injection of rectangular current pulses, produced by a current generator and defined in their magnitude by a resistor (R). The current pulses are injected through channel 2b of the perfusion pipet. The current loop is closed by a separate NaCl–agar bridge and by a stimulation isolation unit, such that the current loop is off ground. Individual cells are impaled through the basolateral membrane, using microelectrodes mounted on a piezo stepping element. The position of the microelectrode is controlled by a micromanipulator. The voltage across the basolateral cell membrane is recorded via an Ag/AgCl half-cell, and is referenced to the grounded bath. The input resistance of the microelectrode is monitored by the injection of current pulses through a second current injection loop. *Inset:* Photograph of an isolated, *in vitro* perfused rabbit cortical thick ascending limb segment ($\times 400$). The inner diameter of the tubule corresponds to 15 μm. Note the dual-channel perfusion pipet (2). The diaphragm separating the two channels is clearly visible. Also note the light reflex from the Sylgard in the opening of the Sylgard pipet (4). The Sylgard snugly covers the perfused end of the tubule segment.

vers. Consequently, statements like "leaky" or "tight" epithelium on the basis of absolute resistance values ($\Omega \cdot cm^2$ or $\Omega \cdot cm$), should be avoided until it is clear what contribution is made by the transcellular versus the paracellular route of ion movement. Nevertheless, the magnitude of the transepithelial resistance (R_{te}) is useful in calculating the equivalent short-circuit current (I_{sc}) from the transepithelial "active transport voltage" V_{te}:

$$I_{sc} = \frac{V_{te}}{R_{te}}$$

Obviously, I_{sc} rather than V_{te}, is directly related (also see section The Equivalent Short-Circuit Current) to the active ion flux.[16] In addition, any changes in the value of the transepithelial resistance produced by inhibitors whose mode of action is familiar are very helpful in the analysis of the mechanism of ionic transport. For instance, transcellular resistance can be increased drastically by blocking the ionic channels present in such membranes: K^+ channels may be blocked by Ba^{2+}, tetraethylamine (TEA^+), quinine, lidocaine, adenosine trisphosphate, Ca^{2+} antagonists, etc. (cf. Petersen,[17] Greger and Gögelein,[18] and Bleich et al.[19]); Na^+ channels may be blocked by amiloride (cf. O'Neil and Boulpaep[20] and Palmer and Frindt[21]); Cl^- channels may be blocked by diphenylamine-2-carboxylate (Di Stefano et al.[22]) and 5-nitro-3(2-phenylpropylamino)benzoate (Wangemann et al.[23]; see also the chapter on chloride channel blockers by Greger, [45], this volume).

The technical aspect of resistance measurements in isolated perfused tubules has been dealt with explicitly in our recent reviews.[15,24,25] As shown in Fig. 1, a current pulse is injected into the tubule lumen, and the corresponding voltage deflection is measured at the perfusion and at the collection ends of the tubule. The magnitude of the input resistance, i.e., the ratio of the voltage deflection at the perfusion end over the injected current, the voltage attenuation along the tubule, and the length of the tubule are used to calculate the transepithelial resistance. The cable equa-

---

[16] R. Greger, Physiol. Rev. **65**, 760 (1985).
[17] O. H. Petersen, Am. J. Physiol. **251**, G1 (1986).
[18] R. Greger and H. Gögelein, Kidney Int. **31**, 1055 (1987).
[19] M. Bleich, E. Schlatter, and R. Greger, Pfluegers Arch. **415**, 449 (1990).
[20] R. G. O'Neil and E. L. Boulpaep, J. Membr. Biol. **50**, 365 (1979).
[21] L. G. Palmer and G. Frindt, Proc. Natl. Acad. Sci. U.S.A. **83**, 2767 (1986).
[22] A. Di Stefano, M. Wittner, E. Schlatter, H. J. Lang, H. Englert, and R. Greger, Pfluegers Arch. **405**, (Suppl. 1), S95 (1985).
[23] P. Wangemann, M. Wittner, A. Di Stefano, H. C. Englert, H. J. Lang, E. Schlatter, and R. Greger, Pfluegers Arch. **407** (Suppl. 2), S128 (1986).
[24] R. Greger, Pfluegers Arch. **390**, 30 (1981).
[25] R. Greger and E. Schlatter, Pfluegers Arch. **396**, 315 (1983).

tions employed for this calculation have been reported repeatedly.[24,25,26] Also, the problem requiring that the current not be injected directly into the voltage-recording perfusion pipet has been discussed explicitly in previous reports.[14,24,25] To solve this problem, we have developed a dual-channel perfusion pipet: The voltage is recorded through one channel and the current is injected through the second channel (cf. Fig. 1).

The Equivalent Short-Circuit Current

As stated above, the value of the equivalent short-circuit current (I_{sc}) can be calculated from the transepithelial voltage (V_{te}) and the transepithelial resistance (R_{te}) simply by applying Ohm's law. The relevance of this parameter is not easily understood. Consider a situation where the voltage clamping was technically feasible in an isolated perfused tubule, with an axial current electrode placed in the lumen, a corresponding current electrode in the bath, and with the transepithelial voltage measured as discussed above. (In fact, a similar approach has been used in the large renal tubule of *Necturus* by Spring and Paganelli,[27] but it is hardly feasible in small tubule segments of mammalian kidney.) In an epithelium which transports ions electrogenically, clamping the voltage to zero would lead to a directly measured short-circuit current. This short-circuit current corresponds to that calculated from V_{te} and R_{te} under so-called "open-circuit conditions." However, the two values may not be necessarily identical, since the state of the epithelium under short-circuit conditions is different from that under open-circuit conditions. Both currents, the calculated one and the one directly measured (if voltage clamping was feasible), do, however, relate to the electrogenic ion transfer mechanism, i.e., to the flux of the actively transported ionic species. In the case where the epithelium is fairly simple and possesses only one predominant active ion flux, the equivalent short-circuit current determination is an easy way to measure the magnitude of this flux electrically.[16] A more direct way to measure the flux of actively transported ions would seem to be conventional flux analyses by tracers or chemical determinations. However, these are technically more demanding and have a very poor time resolution. Furthermore, they require a certain tubule length and fairly low perfusion rates. At these low rates, the experimental conditions (ionic composition and transepithelial voltage) are not constant along the length of the tubule. When the epithelium under study is a black box, both methods of flux determination, i.e., chemical and electrophysiological, can be used to complement each

[26] E. L. Boulpaep and G. Giebisch, *Methods Pharmacol.* **4B**, 165 (1978).
[27] K. R. Spring and C. V. Paganelli, *J. Gen. Physiol.* **60**, 181 (1972).

other as in the paradigmatic studies of Ussing and Zerhan.[28] Recently we have been able to combine flux studies, including electron microprobe analysis of nanoliter samples, with transepithelial electrical measurements in the thick ascending limb of Henle's loop,[29] and we were able to prove that the measured I_{sc} matched reasonably well with the reabsorptive rate of Cl$^-$.

Intracellular Voltage Measurements

Measurements of intracellular voltages in small cells and even in isolated perfused tubules, which are usually covered by a rather rigid basement membrane, have been made possible by technical refinements in microelectrode production, yielding microelectrodes with extremely fine tips of some 70-nm diameter[30] and making available piezo electric steppers. The initial attempts gave fairly low voltages[31] or the rate of success was extremely poor.[25] Today, these measurements are a standard procedure, the reproducibility of the measurements is satisfactory (± a few millivolts), and they are stable for periods exceeding 1 hr. When current pulsing for the determination of the transepithelial resistance is used, as discussed above, and shown in Fig. 1, the intracellular voltage measurement also contains information about the resistances of the cell membranes. The transepithelial voltage deflection at the site of the cellular impalement (ΔV_{te}) is composed of a voltage drop across the lumenal (ΔV_l) and one across the basolateral membrane (ΔV_{bl}). The magnitude of the voltage drops corresponds largely to the magnitude of the membrane resistances. Thus, from the ratio of the voltage deflection recorded on the impaled cell membrane (ΔV_{bl}) over that calculated for the respective other membrane ($\Delta V_{te} - \Delta V_{bl}$), the so-called voltage divider ratio (α) is determined as

$$\alpha = \frac{\Delta V_{bl}}{\Delta V_{te} - \Delta V_{bl}}$$

From the above it may be concluded that α should also be equal to the ratio of basolateral membrane resistance divided by that of the lumenal membrane. This is true only with some reservations, which have been

---

[28] H.-H. Ussing and I. Zerhan, *Acta Physiol. Scand.* **23**, 110 (1951).

[29] M. Wittner, A. Di Stefano, P. Wangemann, R. Nitschke, R. Greger, C. Bailly, C. Amiel, N. Roinel, and C. de Rouffignac, *Pfluegers Arch.* **412**, 516 (1988).

[30] R. Greger, H. Oberleithner, E. Schlatter, A. C. Cassola, and C. Weidtke, *Pfluegers Arch.* **399**, 29 (1983).

[31] B. Biagi, T. Kubota, M. Sohtell, and G. Giebisch, *Am. J. Physiol.* **240**, F200 (1981).

discussed in detail elsewhere.[16,18,32] Even if α cannot be taken as a quantitative estimate of the resistance ratio, its variation with experimental maneuvers is useful, because it permits qualitative conclusions about relative changes in membrane resistances.

Intracellular voltage measurements serve several purposes. They are useful in characterizing the channels, electrogenic pumps, and carriers responsible for the voltage. Apart from the control voltage measurements, the changes in membrane voltage, caused by a variation of the ionic composition in the lumenal or bath perfusates or addition of transport inhibitors of known mode of action, can be used to construct models of ionic transport. Details of this kind of approach have been discussed previously.[16,18,26] It may suffice to emphasize here that the isolated perfused tubule is especially suited for this type of analysis, since the composition of the perfusates can be varied with ease and, above all, very rapidly. In the case that the epithelium under study does permit equating of α and R_{bl}/R_l (see above) it is possible to quantitatively analyze the absolute values of the membrane resistances and the respective contributions of the individual ionic species.

The ability to obtain "good" cellular impalements is not consistent even in the many laboratories in which this technique has now become a routine procedure. Nevertheless, much of the mystery of the early days has gone, and with the new technical advances the success rate has increased. One important advance is the use of filament borosilicate glass for the preparation of the microelectrodes; the filament allows easy filling with KCl solutions. Several companies now provide reproducible batches of glass. We have adopted the most simple procedure. We do not boil the glass nor do we incubate it in any cleaning or conditioning solution. We simply pull the glass and fill the electrodes (usually with 1 mol/liter KCl) directly prior to their usage. The pulling procedure is very critical if the tips need to be small. Several commercially available pullers serve the purpose. In my experience, the shape of the heating filament is very critical. Also, I try not to change the settings of the puller once they have been optimized. If several kinds of electrodes are manufactured in the same laboratory, it is advisable to use one puller for each kind of electrode rather than "playing" with the "ideal settings" all day. To obtain cellular impalements we visualize the microelectrode optically ($\times 400$, differential interference contrast) and place it to approach the tubule with an angle of some 30° with respect to the horizontal and the tubule axes. First, we indent the cell membrane with the microelectrode which is placed in the micromanipulator (shown schematically in Fig. 1). Then we activate a piezo stepper to drive the

---

[32] D. I. Cook and E. Frömter, *Pfluegers Arch.* **403**, 388 (1985).

microelectrode for approximately 0.5–1 μm with a time constant of 10–50 msec. Other workers still recommend tapping the table for this step. The rate of successful impalements will depend on the preparation. It may be as high as 30% of the trials or as low as <1%, as we have experienced in a recent study in macula densa cell segments.[33] The judgment to call an impalement successful is a matter of some controversy. Nevertheless, most workers would agree that certain minimum requirements must be fulfilled. These requirements are summarized as follows:

1. Tip potential: The microelectrode must not have a large tip potential since the magnitude of this potential will vary when the electrode is placed intracellularly. We accept up to 5 mV.

2. Voltage deflection: On impalement the voltage should jump abruptly, usually to a more negative value of about −70 mV.

3. Input resistance: To document the microelectrode resistance, most investigators inject small current pulses into the electrode (cf. Fig. 1) and calculate the input resistance from the observed voltage deflection. Upon impalement, the microelectrode resistance should not increase significantly since the input resistance of the epithelia studied thus far using this technique is only in the 10-MΩ range, or below, while the input resistance of the microelectrodes is between 100 and 200 MΩ. Increases in microelectrode resistance during impalement in excess of 30% most likely indicate partial clogging of the tip; and there is a danger that with time the tip potential becomes more negative and thus contributes largely to the recorded voltage.

4. Stability: Among the criteria discussed here, the stability of the voltage is the most controversial and ambiguous criterion. Some investigators request stability of ±1 mV in the control phase. Others accept less stringent limits. In addition, most investigators accept impalements which collapse directly after the impalement and subsequently recover the initial spike value. This recovery is called "sealing" of the microelectrode tip, and the physical concept assumes that the microelectrode tip initially punched a hole which seals around the electrode tip thereafter. This concept may be correct, at least if one is able to show that the recovery of the voltage is paralleled by an increase in the input resistance of the cell. However, recovery may also be observed without any appreciable increase in the input resistance of the microelectrode. In that case, we attribute the recovery of the membrane voltage to cell-to-cell communication whereby surrounding cells assist the impaled cell to deal with the extra load of Na^+ and Cl^- entering the K^+ escaping through the leak created by the impalement.

[33] E. Schlatter, M. Salomonsson, A. E. G. Persson, and R. Greger, *Pfluegers Arch.* **414**, 286 (1989).

Our view is supported by straightforward observations. For example, when a microelectrode is placed in a thick ascending limb cell which showed no voltage collapse and if furosemide is added to the lumenal fluid, we observe a hyperpolarized membrane voltage. Here we can assume that the uptake of Na^+ via the lumenal cell membrane is very small or close to zero, since the only relevant uptake mechanism for Na^+ ($Na^+/2Cl^-/K^+$ carrier) is blocked.[34] If we now block the exit mechanism for Na^+, namely, the Na^+,K^+-ATPase by ouabain, no change in the voltage occurs. If we repeat this experiment in a cell which shows some initial collapse in membrane voltage, but which thereafter recovers to the normal voltage, ouabain induces depolarization because Na^+ will enter into the cell via the impalement leak, even in the presence of furosemide. It is worth noting in this context that individual epithelial cells in culture, i.e., cells prior to confluency, do only very rarely permit stable voltage measurements. The reason may be that these individual cells are not assisted in the "fight" against the impalement damage by their neighborhood cells (own unpublished observation).

5. *Voltage Recovery:* When the microelectrode is removed, the voltage must collapse abruptly to a value very close (about ± 1 mV) to the baseline prior to the impalement. Moreover, the microelectrode input resistance should stay almost the same (cf. above).

To sum up criteria for successful impalements, it is probably fair to conclude that individual impalement data should be preferentially published as original recordings. This has the advantage that the reader is able to judge directly what criteria are applicable. Similarly, the time course of voltage changes, observed after some pertubation, is best seen in original records. This time course, if the pertubation was fast enough, will contain information about the rates or even allow for a distinction between primary and secondary events going on in the cell.

Intracellular Ionic Activities

What has been described above about intracellular voltage measurements is also applicable for measurements with ion-selective microelectrodes. Additional problems arise from the fact that the tips of ion-selective microelectrodes are usually larger, since the conductance of the exchange resin as a filling solution is very low when compared to KCl. The resistance of the ion-selective microelectrode may be very high. The resulting poor time resolution (time constant = electrode resistance · capacitance) may be

---

[34] R. Greger and E. Schlatter, *Klin. Wochenschr.* **61**, 1019 (1983).

improved with shielding and capacitance compensation techniques[35] or one may even increase the tip diameter such as is the case for Ca^{2+}-selective microelectrodes. Double-barreled microelectrodes generate additional problems, such as the coupling of the nonselective channel and the ion-selective channel. Several studies in the isolated perfused tubule have now utilized double-barreled microelectrodes.[35-40] Details on the methodology should be taken from these reports. Problems of coupling persist and fairly large microelectrode tips are required even with the new types of glass, like the two-filament "family type" glass, in which a small tube, used for the nonselective channel, is melted onto a larger tube, used for the exchange resin. General information on the manufacture of ion-selective microelectrodes is contained in recent symposia reports[41] or monographs.[42]

Patch-Clamp Analysis in the Isolated Tubule

The application of the patch-clamp technique[43] to the isolated tubule is a recent achievement. The aim of the technique is to form a high-resistance seal ($> 10^9$ Ω) with a patch pipet on the lumenal (apical) or the basolateral cell membrane. The general approach designed by ourselves is shown in Fig. 2. The lumenal membrane can be approached by cutting a window into the tubule[21,44,45] or by inserting a microelectrode into the open end of a perfused tubule (cf. Fig. 2 and Refs. 19 and 46-49). The lumenal membrane generates high-resistance seals easily without any pretreatment. This is especially surprising for the lumenal membrane of the proximal renal tubule, since this membrane contains microvilli ("brush border" membrane), many of which must fold into the mouth of the patch pipet and still

[35] K. Yoshitomi and E. Frömter, *Pfluegers Arch.* **402**, 300 (1984).
[36] B. Biagi, M. Sohtell, and G. Giebisch, *Am. J. Physiol.* **241**, F677 (1981).
[37] R. Greger and E. Schlatter, *Pfluegers Arch.* **402**, 63 (1984).
[38] R. Greger and E. Schlatter, *Pfluegers Arch.* **402**, 364 (1984).
[39] R. Greger, E. Schlatter, F. Wang, and J. N. Forrest, Jr., *Pfluegers Arch.* **402**, 376 (1984).
[40] S. Sasaki, T. Shiigai, and J. Takeuchi, *Am. J. Physiol.* **249**, F417 (1985).
[41] R. Greger and E. Schlatter, in "Ion Measurements in Physiology and Medicine" (D. Kessler, K. Harrison, and J. Höper, eds.), p. 301. Springer-Verlag, New York, 1985.
[42] R. C. Thomas, "Ion-Sensitive Intracellular Microelectrodes: How to Make and Use Them." Academic Press, New York, 1978.
[43] B. Sakmann and E. Neher, "Single-Channel Recording:" Plenum, New York, 1983.
[44] B. M. Koeppen, K. W. Beyenbach, and S. I. Helman, *Am. J. Physiol.* **247**, F380 (1984).
[45] M. Hunter, A. Lopes, E. Boulpaep, and G. Giebisch, *Proc. Natl. Acad. Sci. U.S.A.* **81**, 4237 (1984).
[46] H. Gögelein and R. Greger, *Pfluegers Arch.* **401**, 424 (1984).
[47] H. Gögelein and R. Greger, *Pfluegers Arch.* **406**, 198 (1986).
[48] R. Greger, H. Gögelein, and E. Schlatter, *Pfluegers Arch.* **409**, 100 (1987).
[49] E. Schlatter, M. Bleich, and R. Greger, *Proc. Kongr. Ges. Nephrol., 20th* (1989).

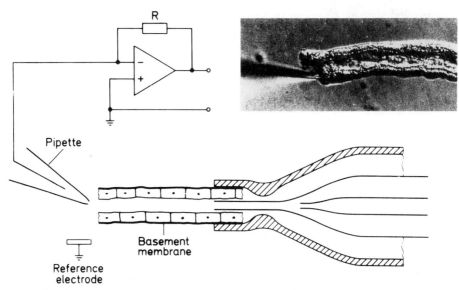

Fig. 2. Patch-clamp analysis of an *in vitro* perfused tubule. The tubule segment is perfused with a pipet system similar to that shown in Fig. 1. If necessary the free end of the tubule is immobilized by additional suction pipets attached to this end (not shown in this figure). The patch pipet is mounted to a micromanipulator and can be inserted into the open lumen or pressed against the lateral membrane. *Inset:* A patch-clamp experiment in a rabbit late proximal tubule (×400). The inner diameter of the tubule is 20 μm. The patch pipet is pressed against the basolateral membrane and, after a gigaohm seal with this plasma membrane is achieved, a cell-attached recording can be obtained. (After Gögelein and Greger.[46])

permit the membrane–glass interaction and the formation of a gigaseal. The access to the basolateral membrane is not as straightforward since the basement membrane prevents any electrically tight contact of the patch pipet and the cell membrane. Several methods to overcome this problem have been developed. The basement membrane can be digested by collagenase treatment.[46,50] The success of this treatment depends on the purity (source!) of the collagenase. The optimal conditions are defined according to the rate of success. Another approach is to mechanically dissect the tubule. In one such method, the nonperfused tubule end is dissected so that a clean and smooth surface of lateral membrane is exposed (cf. Fig. 2 and Refs. 46, 48, and 51). The patch pipet then can be pressed against the lateral membrane and a gigaseal formed. One possible problem with this approach may come from the fact that the tight junction area of the end of

[50] L. Parent, J. Cardinal, and R. Sauve, *Am. J. Physiol.* **254,** F105 (1988).
[51] R. Greger, M. Bleich, and E. Schlatter, *Renal Physiol. Biochem.* **13,** 37 (1990).

the tubule is interrupted, and that the accessible lateral membrane may adopt the permeability characteristics of the lumenal membrane. We do not think that this is the case in the preparations which we have examined, since the type of channels found in lateral membrane patches was always different from that found in the lumenal membrane.[47,48,51,52] Another method for dissection was developed recently.[53] The basement membrane is removed mechanically by dissection, and a "window" is produced through which the basolateral membrane is accessible.

To date, patch-clamp analysis has been applied to several nephron segments[19,21,44-46,49,50,53,54] and to glandular tubules.[48,52,55,56] Experiments performed were of the cell-attached and cell-excised type.

In one recent study, a new concept for cell-attached patches was introduced.[47] It was shown that in some experiments with a reasonably high seal resistance ($> 10^{10}$ Ω), a cytosolic negative voltage was recorded in zero-current clamp mode. This voltage was as high as -70 to -80 mV and collapsed entirely when the patch was pulled off the cell. The surprising aspect of this finding was that the recorded voltage corresponded qualitatively and even quantitatively to the membrane voltage as recorded by an inserted microelectrode. Our interpretation of these findings is that what is being measured as a seal conductance of the patch pipet reflects mostly nonselective pathways in the membrane area of the patch and not the leak pathway between glass and membrane. In fact, it is reasonable to postulate that the resistance of this latter barrier is much larger than 10 GΩ. To date it is not clear which pathways in the patch are responsible for the voltage transmission and whether these pathways are also present in untouched cell membrane. Nevertheless, this high negative voltage is very helpful for determining whether the patch is still on the cell or whether it has been ripped off unintentionally.

There are two advantages of the combination of the patch-clamp technique with the *in vitro* perfused tubule preparation: (1) The epithelium is intact and unaltered. This aspect may be of importance since we do not know for certain whether the preparation of isolated epithelial cells (cell separation, digestion, etc.) leaves unaltered the functional characteristics of the channel proteins. Similar reservation may also be applicable to cultured epithelial cells. In fact, in a recent patch-clamp study in isolated and cultured thick ascending limb cells[57] a K^+ channel was reported which does

[52] H. Gögelein, R. Greger, and E. Schlatter, *Pfluegers Arch.* **409**, 107 (1987).
[53] K. Kawahara, G. Giebisch, and M. Hunter, *Am. J. Physiol.* **253**, F448 (1987).
[54] J. Taniguchi, K. Yoshitomi, and M. Imai, *Am. J. Physiol.* **256**, F246 (1989).
[55] R. Greger, E. Schlatter, and H Gögelein, *Pfluegers Arch.* **409**, 114 (1987).
[56] H. Gögelein, E. Schlatter, and R. Greger, *Pfluegers Arch.* **409**, 122 (1987).
[57] S. E. Guggino, W. B. Guggino, N. Green, and B. Sacktor, *Am. J. Physiol.* **252**, C121 (1987).

not match at all with the K^+ channel found in isolated *in vitro* perfused thick ascending limb segments.[19] (2) Information about the electrical properties of the isolated perfused tubule obtained by patch-clamp analysis can be compared to information derived from intracellular voltage measurements. Consequently, the channel properties found with the patch-clamp technique supplement otherwise well-defined models. This aspect of data evaluation is of importance, since up to now no standards are available which could tell whether a certain channel property, as revealed by patch-clamp analysis, is "normal" or artifactual.

Conclusions

The isolated perfused tubule is a preparation offering certain unique advantages: (1) The segments and cell types under study are well selected by dissection; (2) Both compartments, lumen and peritubular space, are easily accessible and the experimental conditions can be freely defined; (3) Several techniques can be employed in this preparation: flux analysis, quantitative morphology, intracellular ion activities, and electrical properties. Each of these methodological approaches supplements the others. The flux analyses were performed shortly after the invention of the technique some 20 years ago. The optical and the electrical methods are very young and still developing. The most recent addition involves cytosolic Ca^{2+} measurements by the Fura-2 method in isolated *in vitro* perfused tubules.[10] This chapter describes the methods of transepithelial and transmembrane voltage and current measurements which can now be performed routinely. The field is still rapidly growing and new combinations of refined methodology, such as the patch-clamp technique combined with the *in vitro* perfusion technique, will help to describe the properties of tubules in which the ability to transport has been left intact.

Acknowledgments

I gratefully acknowledge the secretarial assistance by Mrs. E. Viereck. Also, I am indebted to Dr. I. Novak for reviewing this chapter. Work from my laboratory, cited in this chapter, has been supported by Deutsche Forschungsgemeinschaft (Gr 480/8-10).

[17] Hormonal Receptors in the Isolated Tubule

By FRANÇOIS MOREL and DANIEL BUTLEN

Introduction

It is now well established that many hormones contribute to the adjustment of kidney functions by controlling specific cell properties in their respective target structures along the nephron.[1] Different experimental approaches have been used in an attempt to determine the sites in the kidney where the receptors specific to each of these hormones are located.

As for other tissues, a first approach consisted of binding studies of radiolabeled hormones to acellular preparations obtained from tissue homogenates with further characterization of the corresponding receptors according to the usual kinetic and pharmacological criteria.[2-6] However, this kind of approach allows only a rough estimation of receptor distribution throughout the kidney (for example, cortex versus outer medulla or papilla) due to the highly heterogeneous cell composition of the different kidney regions.

Other approaches were therefore developed in order to overcome the major limitation inherent in kidney tissue cell organization. These investigations include the following:

1. *Autoradiographic studies,* performed on tissue slices after appropri-

---

[1] F. Morel and A. Doucet, *Physiol. Rev.* **66,** 377 (1986).
[2] J. Bockaert, C. Roy, R. Rajerison, and S. Jard, *J. Biol. Chem.* **248,** 5922 (1973).
[3] S. J. Marx, C. H. Woodward, J. D. Aurbach, H. P. Glossmann, and H. T. Keutmann, *J. Biol. Chem.* **248,** 4797 (1973).
[4] F. P. Di Bella, T. P. Dousa, S. S. Miller, and C. D. Arnaud, *Proc. Natl. Acad. Sci. U.S.A.* **73,** 723 (1974).
[5] J. M. Schmitz, R. M. Graham, A. Sagalowsky, and W. A. Pettinger, *J. Pharmacol. Exp. Ther.* **219,** 400 (1981).
[6] M. Milavec-Krizman, J. P. Evenou, H. Wagner, R. Berthold, and A. P. Stoll, *Biochem. Pharmacol.* **34,** 3951 (1985).
[7] M. J. Osborne, B. Droz, P. Meyer, and F. Morel, *Kidney Int.* **8,** 245 (1975).
[8] C. Bianchi, J. Gutkowska, G. Thibault, R. Garcia, J. Genest, and M. Cantin, *Am. J. Physiol.* **251,** F594 (1986).
[9] C. Koseki, Y. Hayashi, S. Torikai, M. Furuya, N. Ohnuma, and M. Imai, *Am. J. Physiol.* **250,** F210 (1986).
[10] M. E. Stoeckel, M. J. Freund-Mercier, and J. M. Palacios, *J. Endocrinol.* **113,** 179 (1987).
[11] R. J. Summers, *Fed. Proc., Fed. Am. Soc. Exp. Biol.* **43,** 2917 (1984).
[12] D. P. Healy, P. A. Münzel, and P. A. Insel, *Circ. Res.* **57,** 278 (1985).
[13] M. F. Rouleau, H. Warshawsky, and D. Goltzman, *Endocrinology (Baltimore)* **118,** 919 (1986).

ate *in vivo* or *in vitro* application of the radiolabeled hormone,[7-14] have been performed. These studies make it possible to detect hormone-binding sites at a resolution scale which depends on tissue slice thickness and on the nature and energy of the radiations emitted by the isotope used. In principle, tritiated or ^{125}I-labeled iodinated ligands should allow the localization of hormonal receptors over cell membranes and cytoplasmic or nuclear areas by using electron microscopy. Unfortunately, with high-resolution microscopy and ultrathin sections, the number of silver grains detected per cell is vanishingly small, whereas with thick sections in light microscopy, the resolution level is rather poor. In addition, the biochemical criteria required for analyzing receptor properties are difficult to apply to this type of approach.

2. *Immunocytochemical studies* using antibodies directed against hormonal receptors, associated with antibody visualization with a second enzyme-coupled antibody, have also been considered.[15] However, the number of specific antibodies against hormone receptors is still very limited. If the antibody used is directed against the hormone itself, the hormone molecules bound to their tissue receptors can be detected[16-19] so long as the epitope recognized by the antibody is exposed and accessible on the hormone–receptor complex and the hormone does not dissociate during the experimental procedure.

3. *Kidney tissue fractionation* by sequential sieving or density gradient centrifugation after digestion of interstitial tissue with proteolytic enzymes leads to preparations enriched in defined kidney structures. Depending on the conditions used, glomeruli, kidney microvessels, proximal tubules, and, more recently, medullary thick ascending limbs and even collecting tubules were obtained in sufficient amounts[20,21] to analyze hormone–receptor interactions[9,22-29] hormone-regulated enzyme activities,[20,30-34] or hormone-induced intracellular Ca^{2+} mobilization[34] by using conventional biochemical approaches. Of course, the reliability of the data is limited by

[14] A. Milton, M. Hellfritzsch, and E. I. Christensen, *Diabetes Res.* **7,** 189 (1988).
[15] M. Shimonaka, T. Saheki, H. Hagiwara, Y. Hagiwara, H. Sono, and S. Hirose, *Am. J. Physiol.* **253,** F1058 (1987).
[16] R. Ravid, D. F. Swaab, and C. W. Pool, *J. Endocrinol.* **105,** 133 (1985).
[17] M. Naruse, K. Nitta, T. Sanaka, K. Naruse, H. Demura, T. Inagami, K. Shizume, and N. Sugino, *Acta Endocrinol. (Copenhagen)* **119,** 235 (1988).
[18] G. L. Andersson, A. Skottner, and E. Jennische, *Acta Endocrinol. (Copenhagen)* **119,** 557 (1988).
[19] M. Luo, R. Faure, Y. A. Tong, and J. H. Dussault, *Acta Endocrinol. (Copenhagen)* **120,** 451 (1989).
[20] D. Schlondorff, *Kidney Int.* **30,** 201 (1986).
[21] F. A. Gesek, D. W. Wolff, and J. W. Strandhoy, *Am. J. Physiol.* **253,** F358 (1987).
[22] J. D. Sraer, J. Sraer, R. Ardaillou, and D. Mimoune, *Kidney Int.* **6,** 241 (1974).

the cell homogeneity of the preparations. Moreover, many kidney cell types have not been isolated by tissue fractionation until recently (for instance, thin limbs, distal tubules, and connecting tubules).

4. *Kidney tissue microdissection* allows one to obtain clean and well-defined pieces of tubules from nearly all nephron segments, in particular after tissue pretreatment with collagenase.[20,35] Such microdissected fragments of tubule may be used *in vitro* for many kinds of physiological and biochemical studies, including biological responses to hormones in microperfused tubules,[36] enzyme activation by hormones,[1,20,35,37-43] and detection of hormonal receptors either by direct binding measurements[44-58] or by autoradiographic studies.[9,59-62] As the amount of tissue obtained by microdissection is exceedingly low, even when several homologous fragments of tubule are pooled, appropriate micromethods had to be developed for each kind of study.

[23] A. De Léan, P. Vinay, and M. Cantin, *FEBS Lett.* **193**, 239 (1985).
[24] E. Meezan and P. Freychet, *Mol. Pharmacol.* **16**, 1095 (1979).
[25] D. J. Pillion, J. F. Haskell, and E. Meezan, *Am. J. Physiol.* **255**, E504 (1988).
[26] A. Chaudhari and M. A. Kirschenbaum, *Prostoglandins Leukotrienes Med.* **20**, 55 (1985).
[27] P. R. Sundaresan, M. Barac-Nieto, G. W. Stambo, and S. L. Kelvie, *J. Cardiovasc. Pharmacol.* **13**, 16 (1989).
[28] R. A. Felder, M. Blecher, G. M. Eisner, and P. A. Jose, *Am. J. Physiol.* **246**, F557 (1984).
[29] S. McArdle, L. C. Garg, and F. T. Crews, *J. Pharmacol. Exp. Ther.* **248**, 12 (1989).
[30] G. Wirthensohn and W. G. Guder, *Physiol. Rev.* **66**, 469 (1986)
[31] J. A. Shayman, R. J. Auchus, and A. R. Morrison, *Biochim. Biophys. Acta* **888**, 171 (1986).
[32] K. A. Hruska, D. Moskowitz, P. Esbrit, R. Civitelli, S. Westbook, and M. Huskey, *J. Clin. Invest.* **79**, 230 (1987).
[33] S. McArdle and L. C. Garg, *J. Pharmacol. Exp. Ther.* **248**, 682 (1989).
[34] M. A. Burnatowska-Hledin and W. S. Spielman, *J. Clin. Invest.* **83**, 84 (1989).
[35] F. Morel, D. Chabardès, and M. Imbert-Teboul, *Methods Pharmacol.* **4B**, 297 (1978).
[36] K. J. Ullrich and R. Greger, in "The Kidney: Physiology and Pathophysiology" (D. W. Seldin and G. Giebisch, eds.), Vol. 1, p. 427. Raven, New York, 1985.
[37] F. Morel, D. Chabardès, and M. Imbert, *Kidney Int.* **9**, 264 (1976).
[38] M. G. Brunette, D. Chabardès, M. Imbert-Teboul, A. Clique, M. Montégut, and F. Morel, *Kidney Int.* **15**, 357 (1979).
[39] F. Morel, D. Chabardès, and M. Imbert-Teboul, *Curr. Top. Membr. Transp.* **13**, 415 (1980).
[40] F. Morel, *Am. J. Physiol.* **240**, F159 (1981).
[41] M. Imbert-Teboul, D. Chabardès, A. Clique, M. Montégut, and F. Morel, *Am. J. Physiol.* **247**, F316 (1984).
[42] N. Murayama, B. T. Ruggles, S. M. Gapstur, J. L. Werness, and T. P. Dousa, *J. Clin Invest.* **76**, 474 (1985).
[43] N. M. Griffiths, D. Chabardès, M. Imbert-Teboul, S. Siaume-Perez, F. Morel, and N. I. Simmons, *Pfluegers Arch.* **412**, 363 (1988).
[44] A. Doucet and A. I. Katz, *Am J. Physiol.* **241**, F605 (1981).
[45] S. M. K. Lee, M. A. Chekal, and A. I. Katz, *Am. J. Physiol.* **244**, F504 (1983).
[46] K. Tomita and J. J. Pisano, *Am. J. Physiol.* **246**, F732 (1984).

This chapter describes in detail two types of microtechniques which were applied to microdissected tubules *in vitro* in order to establish the presence of hormonal receptors in well-defined nephron segments and to analyze directly or indirectly their properties.

General Aspects Related to Receptor Studies Using Microdissected Nephron Segments

The amount of total protein contained per millimeter of tubular length ranges between about 20 ng (thin segments of Henle's loop) and 250 ng (proximal tubules), depending on the nephron segments and animal species.[35,63] Consequently, protein contents cannot be measured if experimental samples contain only one or a few pieces of microdissected tubules. Therefore, the total length of tubule per sample is determined (from photographic pictures) and length is used as a reference.[35] If necessary, data can be expressed per milligram protein from published tables giving the total protein contents per millimeter of tubule for the different nephron segments.[35,63]

In purified membrane fractions from various homogeneous tissues, the total number of receptors to a given hormone generally falls in a range from 0.1 to 1 pmol binding sites/mg membrane protein. Assuming that similar values also apply to kidney nephrons, one may calculate that 1 mm of tubule would contain about 2 to 20 attomoles (amol) of receptors (1

---

[47] R. A. Felder, M. Blecher, P. L. Calcagno, and P. A. Jose, *Am. J. Physiol.* **247**, F499 (1984).
[48] E. Kusano, R. Nakamura, Y. Asano, and M. Imai, *Tohoku J. Exp. Med.* **142**, 275 (1984).
[49] R. Nakamura, D. S. Emmanouel, and A. I. Katz, *J. Clin. Invest.* **72**, 388 (1983).
[50] D. Butlen, S. Vadrot, S. Roseau, and F. Morel, *Pfluegers Arch.* **412**, 604 (1988).
[51] H. Endou, A. Miyanoshita, and K. Y. Jung, *Proc. Jpn. Acad.* **65**, 21 (1989).
[52] D. Butlen and F. Morel, *Pfuegers Arch.* **404**, 348 (1985).
[53] S. K. Mujais, S. Kauffman, and A. I. Katz, *J. Clin. Invest.* **77**, 315 (1986).
[54] D. Butlen, M. Mistaoui, and F. Morel, *Pfluegers Arch.* **408**, 356 (1987).
[55] J. K. Kim, S. N. Summer, J. Durr, and R. W. Schrier, *Kidney Int.* **35**, 799 (1989).
[56] B. Semmekrot, S. Roseau, G. Vassent, and D. Butlen, *Mol. Cell. Endocrinol.* **68**, 35 (1990).
[57] A. I. Katz, in "Modern Techniques of Ion Transport" (B. M. Brenner and J. H. Stein, eds.), Vol. 15, p. 149. Churchill Livingstone, New York, 1987.
[58] K. L. Kirk, *Am. J. Physiol.* **255**, C622 (1988).
[59] N. Farman, A. Vandewalle, and J. P. Bonvalet, in "Biochemistry of Kidney Function" (F. Morel, ed.), Vol. 21, p. 285. Elsevier, Amsterdam, 1982.
[60] N. Farman, P. Pradelles, and J. P. Bonvalet, *Am. J. Physiol.* **249**, F923 (1985).
[61] A. Gnionsahe, M. Claire, N. Koechlin, J. P. Bonvalet, and N. Farman, *Am. J. Physiol.* **257**, R87 (1989).
[62] C. Manillier, N. Farman, J. P. Bonjour, and J. P. Bonvalet, *Am. J. Physiol.* **248**, F296 (1985).
[63] B. D. Ross and W. G. Guder, in "Metabolic Compartmentation" (H. Sies, ed.), p. 363. Academic Press, New York, 1982.

amol = 10^{-18} mol). If a tritiated hormone with a specific radioactivity of 1.85 TBeq/mmol is used, the maximal radioactivity specifically bound to receptors is expected to be about 0.2 to 2 cpm/mm of tubule. Obviously, when tritiated ligands only are available, it is necessary to pool a great number of homologous pieces of nephron segments for each individual determination and to increase the duration of radioactivity measurements in order to have enough signal.[44-48] However, if an ^{125}I-labeled iodinated hormone with a specific radioactivity of 74 TBeq/mmol is available, the corresponding saturating values are about 8 to 80 cpm/mm of tubule; in this case, a few pieces of tubules per sample are sufficient to obtain accurate results.[49-56]

For hormonal receptors functionally coupled to a membrane-bound enzyme such as adenylate cyclase, the signal amplification resulting from enzyme activation is sufficient to observe the response elicited *in vitro* by a hormone in a single piece of tubule by measuring the rate of cyclic AMP production under appropriate conditions.[35] We do not present here the microassay of adenylate cyclase activity in permeabilized tubules, since it has been extensively described and discussed earlier.[35,37-43,55] This assay allowed localization, along the nephrons of different mammalian species, of the sites of action where various hormones increase adenylate cyclase activity through a coupling mechanism involving the stimulatory α_s subunit of the G transducing protein.[40] However, the inhibitory effects exerted on adenylate cyclase by other agonists (whose receptors are negatively coupled to the enzyme via the inhibitory α_i subunit of the G protein) could not be analyzed using this microassay, because this inhibition is hardly elicited in permeabilized cells. In contrast, such inhibitory effects are reproducibly observed in intact living cells by measuring the amounts of cyclic AMP generated from endogenous ATP. Intracellular cyclic AMP contents, as well as cyclic GMP contents, are currently measured using radioimmunoassays.[55,64-73]

[64] R. M. Edwards, B. A. Jackson, and T. P. Dousa, *Am. J. Physiol.* **240,** F311 (1981).
[65] S. Torikai, M. S. Wang, K. L. Klein, and K. Kurokawa, *Kidney Int.* **20,** 649 (1981).
[66] D. Chabardès, M. Montégut, M. Imbert-Teboul, and F. Morel, *Mol. Cell. Endocrinol.* **37,** 263 (1984).
[67] S. Umemura, D. D. Smyth, and W. A. Pettinger, *Am. J. Physiol.* **250,** F103 (1986).
[68] D. Chabardès, C. Brick-Ghannam, M. Montégut, and S. Siaume-Perez, *Am. J. Physiol.* **255,** F43 (1988).
[69] K. Takaichi and K. Kurokawa, *J. Clin. Invest.* **82,** 1437 (1988).
[70] R. M. Edwards and M. Gellai, *J. Pharmacol. Exp. Ther.* **244,** 526 (1988).
[71] I. Dublineau, J. M. Elalouf, P. Pradelles, and C. de Rouffignac, *Am. J. Physiol.* **256,** F656 (1989).
[72] D. Chabardès, M. Mongtégut, M. Mistaoui, D. Butlen, and F. Morel, *Pfluegers Arch.* **408,** 366 (1987).
[73] H. Nonoguchi, M. A. Knepper, and V. C. Manganiello, *J. Clin. Invest.* **79,** 500 (1987).

In the following sections, we summarize first the microdissection method for isolating well-localized nephron portions from collagenase-treated kidney tissue. Then, we report micromethods for measuring cyclic AMP and cyclic GMP contents in single pieces of living tubules incubated *in vitro*. Finally, we describe the conditions allowing the direct characterization of hormonal receptors and specific binding sites for other nonhormonal drugs in microdissected tubules by using either tritiated or ^{125}I-labeled iodinated ligands.

Experimental results from our laboratory will serve as examples illustrating the possibilities and limitations of the micromethods.

Microdissection of Nephron Segments

Microdissection of glomeruli and tubules from rat, mouse, rabbit, and human kidneys is performed as reported earlier.[35,74-76] The kidney is perfused via a catheter introduced in the abdominal aorta of the rat or the mouse, in the renal artery of the rabbit, or in a branch of the renal artery of the human kidney with 4 ml (rat, mouse) or 20 ml (rabbit, human) of a chilled modified Hanks' solution containing either (1) HEPES–NaOH, 20 mM, pH 7.4; NaCl, 137 mM; KCl, 5 mM; MgSO$_4$, 0.8 mM; Na$_2$HPO$_4$, 0.33 mM; NaH$_2$PO$_4$, 0.44 mM; NaHCO$_3$, 4 mM; MgCl$_2$, 1 mM; CaCl$_2$, 1 mM (used as an activator of collagenase); glucose, 5 mM; lactic acid, 3 mM; sodium acetate, 10 mM; bovine serum albumin (BSA), 0.1%; and collagenase from *Clostridium histolyticum* (CLS, 146–196 U/mg, from Worthington Biochemical Corp., Freehold, NJ), 0.25% (rat, mouse) or 0.10% (rabbit, human) in perfusion medium A used for ligand-binding assay experiments); or (2) HEPES–NaOH, 20 mM, pH 7.5; NaCl, 120 mM; KCl, 5 mM; NaH$_2$PO$_4$, 4 mM; MgSO$_4$, 1 mM; NaHCO$_3$, 4 mM; CaCl$_2$, 1.5 mM; glucose, 5 mM; alanine, 1 mM; lactic acid, 10 mM; pyruvic acid, 1 mM; BSA, 0.1%; dextran (M_r, 40,000 to maintain colloid osmotic pressure), 3%; and collagenase, 0.4%, in perfusion medium B used for cyclic nucleotide radioimmunoassay experiments. Due to large variations in collagenase activity from one batch to another, it is advisable to test the action of new batches of collagenase on the kidney before routine use.

[74] M. Imbert-Teboul, D. Chabardès, M. Montégut, A. Clique, and F. Morel, *Endocrinology (Baltimore)* **102**, 1254 (1978).
[75] D. Chabardès, M. Imbert-Teboul, M. Gagnan-Brunette, and F. Morel, in "Biochemical Nephrology" (W. G. Guder and U. Schmidt, eds.), p. 447. Huber, Bern, Switzerland, 1978.
[76] D. Chabardès, M. Gagnan-Brunette, M. Imbert-Teboul, O. Gontcharevskaia, M. Montégut, A. Clique, and F. Morel, *J. Clin. Invest.* **65**, 439 (1980).

Thin pyramids are cut along the corticopapillary axis of the kidney; the tissue is then incubated for 15-20 min at 30° (rat, mouse), 15-30 min at 35° (rabbit), or 60-90 min at 35° (human) in an aerated 0.07% collagenase solution A or B. After careful rinsing, the tissue is microdissected by hand with the help of thin steel needles under stereomicroscopic observation in ice-cold microdissection medium C or D (of the same composition as A or B medium, respectively, except for the omission of collagenase and the addition of appropriate chemicals for cyclic nucleotide assays or hormonal binding assays where required).

Superficial glomeruli are microdissected from the outer cortex and juxtamedullary glomeruli from deep cortex close to the corticomedullary transition.

The proximal convoluted tubule (PCT) may be dissected over its full length; its early portion is recognized by the attached glomerulus. The cortical portion of the pars recta (PR) is distinguished from the other structures present in the medullary rays by its larger diameter. The terminal portion of the pars recta is identified in the outer stripe of the outer medulla by its continuity with the thin descending limb of Henle's loop.

With regard to the thin segments of Henle's loop, it is important to mention that the thin descending limb (TDL) and the thin ascending limb (TAL) are difficult to distinguish from each other by appearance only after microdissection. Therefore, TDLs must be microdissected from their attachment with PR in the outer medulla, whereas TALs are dissected in the inner medulla and are identified by their attachment with the medullary thick ascending limbs.

Medullary thick ascending limbs (MALs) are easily distinguishable from the adjacent structures by their "bright" appearance and a diameter that is narrow compared to those of PR and medullary collecting tubules, but large compared to TDLs. Cortical thick ascending limbs (CALs) are also easy to recognize from their straight shape, narrow diameter (specially in the rabbit), and "bright" appearance, as well as from the presence of the macula densa (MD).

The so-called distal convoluted tubule (DCT), i.e., the nephron portion included between the macula densa and the first branching with another tubule, is the most heterogeneous part of the nephron as judged from stereomicroscopic observation as well as from biochemical and physiological properties.[37]

In the rabbit, up to four different successive portions can be observed and isolated along the DCT. The first portion (DCT_a) actually corresponds to the CAL end part extending over a short and variable distance beyond the MD; DCT_a may be absent in some nephrons. The second portion is of "bright" appearance (DCT_b); it has a larger diameter than the CAL, a tortuous shape, and a relatively constant length (about 0.4 mm as an average value). The third portion exhibits a "granular" aspect (DCT_g), a larger diameter, and a variable length (from 0.2 to 0.8 mm); in most nephrons, DCT_g is connected to a portion of the cortical collecting tubule of similar "granular" appearance, the CCT_g or arcade; DCT_g and CCT_g form together the "connecting tubule."[37] Finally, a fourth DCT portion of narrow diameter and variable length may be present in DCTs localized in the very superficial cortex. This segment has a "light" appearance (DCT_l) and represents the initial portion of the light cortical collecting tubule (CCT_l) to which it is always branched.

The cortical collecting tubule (CCT) includes, in the rabbit, two different segments which are likely to be of different embryonic origin,[37] namely, a granular segment (CCT_g) forming an arcade to which several DCT_gs are branched (connecting tubule), and a straight segment of light appearance (CCT_l) running in the medullary rays from the kidney surface down to the medulla. Then, collecting tubules give outer medullary collecting tubules (MCTs) followed by inner medullary collecting tubules (IMCTs).

In the rat, mouse, and human kidneys, the bright, granular, and light portions of the DCT are difficult to isolate, because the transitions between successive DCT portions are rather progressive in these species. Moreover, all CCT portions are of the light type.

The required nephron segments are dissected over an appropriate length and separated from the adjacent structures according to the morphological criteria described above. Each piece of tissue is aspirated with a small volume of microdissection medium using a polyethylene catheter (coated with BSA so that the tubules do not stick to the wall) and transferred onto a hollow bacteriological glass slide. In order to limit the area covered by the sample over the glass, the siliconized slides are pretreated by evaporating to dryness a 2-μl droplet of 0.1% BSA solution deposited in the center of the hollow glass slide. After tightly covering the samples with another Vaseline-coated glass slide (to obtain a water-tight seal), the tubules are photographed for length determination and, if necessary for binding experiments, may be stored overnight at 4° until use.

Except where otherwise indicated, assays are further performed on bacteriological glass slides.

Cyclic Nucleotide Radioimmunoassays

The generation of cyclic AMP from endogenous ATP and of cyclic GMP from endogenous GTP requires the use of well-preserved living segments of tubule. In other words, this method implies that intracellular

ATP or GTP contents are not limiting factors for endogenous cyclic nucleotide production and that the nucleotides formed during the incubation are not degraded by intracellular phosphodiesterases. Consequently, the tubules have to be incubated in the presence of bioenergetic substrates and cyclic nucleotide-dependent phosphodiesterase inhibitors. Cyclic AMP and cyclic GMP accumulated in well-defined nephron segments are measured by radioimmunoassays using their appropriate kits such as, for example, those produced by Institut Pasteur Production (Paris, France).

Experimental Procedures

Microdissected nephron segments are first rinsed at 4° in medium E, i.e., solution D complemented with bacitracin, 0.1% (as proteolytic enzyme inhibitor), and 3-isobutyl-1-methylxanthine (IBMX), 1 mM, or Ro 20-1724, 50 μM (as phosphodiesterase inhibitors).

The experiment itself begins by preincubating all samples in 2 μl of medium E for 10 min at 30°. Then the reaction is initiated by adding 2 μl of solution E containing the required hormones or drugs. The incubation is carried out for 4 min at 35°. The reaction is stopped by quickly transferring the pieces of tissue together with 1 μl of incubate into a Durham glass tube containing 10 μl of formic acid in absolute ethanol (5%, v/v). After an overnight evaporation to dryness at 40°, 40 μl of potassium phosphate buffer, 50 mM, pH 6.2, is added and the samples are frozen and kept at $-20°$ until radioimmunoassay. In each experiment, several samples containing no piece of tissue are run in the same way and used as experimental blanks to prepare the zero samples of the standard curves (see Preliminary Control Experiments, below).

Samples containing increasing amounts of unlabeled cyclic nucleotides are prepared for the standard curves (from 1.2 to 156 fmol cyclic AMP/tube or from 0.3 to 80 fmol cyclic GMP/tube) and stored frozen at $-20°$ until use.

Prior to radioimmunoassay, the cyclic nucleotides contained in biological samples and in samples of the standard curves are acetylated by adding 1 μl of a mixture of acetic anhydride and triethylamine (1/2, v/v) in order to enhance the sensitivity of the assay.[77] Tracer amounts of cyclic [^{125}I]-AMP or cyclic [^{125}I]-GMP (about 80 Beq) and the corresponding specific antibody in potassium phosphate buffer (50 mM, pH 6.2) are added to all samples (final volume, 80 μl). The samples are then incubated for 24 hr at 4°. The complex antigen–antibody is precipitated by adding 50 μl bovine γ-globulin (0.5% in potassium phosphate buffer, 50 mM, pH 6.2) and, 30

---

[77] J. F. Harper and G. Brooker, *Cyclic Nucleotide Res.* **1**, 207 (1975).

min later, 250 μl polyethylene glycol (16% in potassium phosphate buffer, 50 mM, pH 6.8), followed immediately by a centrifugation at 250 g for 20 min at 4°. Supernatants are discarded and the pellets counted by γ-spectrometry.

Typical standard curves used for the assays of cyclic AMP and cyclic GMP are given in Fig. 1.

Comments

Preliminary Control Experiments. It has been shown that (1) standard curves obtained with cyclic nucleotide dilutions delivered in Durham tubes pretreated with formic acid–ethanol solution and evaporated to dryness are not different from the usual standard curves[66]; (2) the salts and biochemical compounds contained in incubate E do not modify significantly the displacement curves. Experimental blanks (measured in the absence of antibody) and the experimental zero of the standard curve, B_0 (measured in the absence of unlabeled antigen), are the same as blanks and B_0 found under standard conditions. When a difference is observed between experimental B_0 and standard B_0, it never exceeds the equivalent of 2 fmol for cyclic AMP and 0.3 fmol for cyclic GMP[72]; (3) the addition of known

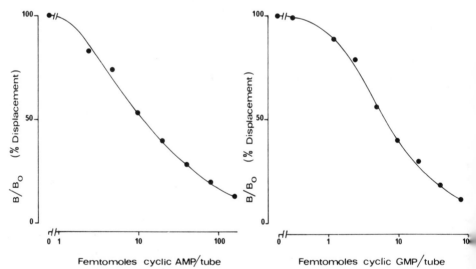

FIG. 1. Standard displacement curves for cyclic AMP and cyclic GMP radioimmunoassays. The inhibition of cyclic [^{125}I]AMP (left) and cyclic [^{125}I]GMP (right) binding to their corresponding specific antibody (B) by increasing amounts of unlabeled cyclic nucleotides is expressed as a percentage of the radioactivity bound in the absence of nonradioactive nucleotide (B_0). Each point is the mean of four determinations.

amounts of nucleotide to control samples incubated and treated as usual leads to an average recovery amounting to 88 ± 7% of the expected value; the recovery is independent of the presence or absence of tubule in these controls.[66]

Cyclic nucleotides generated by the biological samples are assayed after a 4-min incubation at 35° because control experiments showed that the rate of increase in cyclic AMP and/or cyclic GMP content is roughly constant up to 6 min and decreases progressively for longer incubation times.[66,72] This 4-min duration was also chosen to reduce the scatter of data inherent in very short incubation times.

Sensitivity of Radioimmunoassays. Cyclic nucleotide contents measured in tissue samples are taken into account only when the displacement obtained falls between 20 and 80% on the titration curves. The sensitivity limits of the methods allow accurate and reproducible measurement of 2.0 to 80 fmol of cyclic AMP and 1.0 to 40 fmol of cyclic GMP (Fig. 1). Under these conditions, basal cyclic nucleotide production can be detected by pooling at least three glomeruli (1.7 ± 0.2 and 0.35 ± 0.03 fmol of cyclic AMP and cyclic GMP formed/4 min/glomerulus, respectively); however, production levels are too low to be detected, even in the presence of IBMX, in all other nephron segments tested.[66,72]

On the other hand, cyclic GMP production induced in glomeruli by 0.1 μM atriopeptin II can be measured even in the absence of phosphodiesterase inhibitor: 4.3 ± 0.7 versus 9.4 ± 0.7 fmol cyclic GMP/4 min/glomerulus for samples incubated in the absence or the presence of 1 mM IBMX, respectively.[72]

Reliability of the Microtechniques. The validity of the microassay employed for cyclic AMP is supported by the following pieces of evidence: (1) a close linearity of cyclic AMP contents as a function of tubular length in MCT samples stimulated by 1 nM arginine vasopressin (AVP) ($r = 0.92$)[66]; (2) induction by 10 μM forskolin of greater cyclic AMP accumulations in CCT and MCT than that induced by 1 μM AVP (CCT: 372 ± 42 versus 142 ± 17 and MCT: 327 ± 29 versus 155 ± 10 fmol cyclic AMP/ 4 min/mm, respectively),[66] an observation demonstrating that intracellular ATP content is not a limiting factor for endogenous cyclic AMP production under the conditions used. In addition, no detectable amount of cyclic AMP is found in the incubate of collecting tubules stimulated either by vasopressin or norepinephrine, indicating that all the formed nucleotide remains contained in epithelial cells[66]; and (3) the dose dependency and saturability of AVP-stimulated cyclic AMP production by MCT samples (Fig. 2). Furthermore, vasopressin concentrations leading to threshold, half-maximal, and maximal cyclic AMP accumulations in living collecting tubules are lower than those inducing the corresponding adenylate cyclase

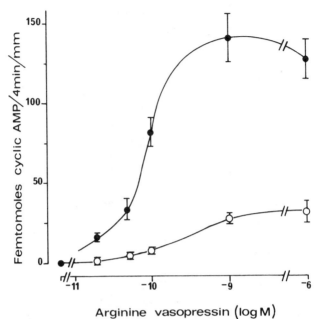

FIG. 2. Inhibition by norepinephrine of the arginine vasopressin-induced cyclic AMP accumulation in isolated outer medullary collecting tubules (MCT). Single pieces of MCT microdissected from rat kidney were incubated for 4 min at 35° with 1 mM IBMX, 10 μM propanolol, and the indicated concentrations of arginine vasopressin, in the absence (●—●) or in the presence of 10 μM norepinephrine (○—○). Values are means ± SE calculated from 8 to 32 replicate samples prepared in 2–8 different experiments. (Adapted from Chabardès et al.[66])

responses in permeabilized collecting tubules.[66,74] This observation indicates that the sensitivity of the hormone transduction mechanisms is better preserved in living than in permeabilized tubules.

The method also allows the pharmacological identification of adrenoceptors of the α_2 type in MCT segments. Thus, a marked and reproducible inhibition of the vasopressin-induced cyclic AMP accumulation is obtained in the presence of norepinephrine and propanolol (Fig. 2). This inhibition by norepinephrine is not altered by prazosin, but it is suppressed by phentolamine or yohimbine.[66]

In regard to cyclic GMP, the accuracy of the method used was assessed by experiments showing that (1) cyclic GMP production by glomeruli stimulated with 1 nM atriopeptin II increases linearly throughout the wide range of glomeruli number tested ($r = 0.96$)[72] and (2) that atriopeptin II induces dose-dependent and saturable cyclic GMP accumulation by glomeruli above basal production (Fig. 3).

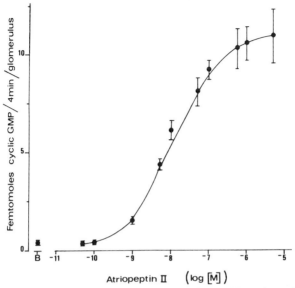

FIG. 3. Dose-dependent cyclic GMP accumulation induced by atriopeptin II in rat glomeruli. Assays were performed by incubating glomeruli for 4 min at 35° in the presence of 1 mM IBMX and the indicated concentrations of atriopeptin II. The number of glomeruli per sample was four for basal determination, three for low atriopeptin II concentrations, including 1 nM, and two for higher concentrations. Values are means ± SE of 6–21 determinations performed in 2 different experiments. B, Basal cyclic GMP production. (Redrawn from Chabardès et al.[72])

Hormone and Drug-Binding Microassays

The usual radioligand-binding techniques used for the characterization of hormonal receptors in acellular preparations (i.e., incubation of fractions in the presence of labeled ligand followed by separation of bound radioactivity from free ligand) cannot be applied to microdissected nephron segments without great caution due to the very low amounts of radioactivity expected to be bound. Thus, as already indicated, the use of tritiated ligands requires pooling a great number of homologous pieces of tubule, whereas a single or a few nephron segments are sufficient for binding studies using ^{125}I-labeled iodinated hormone. Consequently, the experimental procedure convenient for an accurate and rapid separation of bound from free radioactivity depends on the nature of the labeled ligand and on the number of tubules used. Generally, filtration techniques are employed for tritiated ligand assays whereas extensive washings of the samples are performed in binding experiments using ^{125}I-labeled iodinated ligands (see Table I). In addition, the use of nephron segments requires

TABLE I
SEPARATION OF [^{125}I]GLUCAGON BOUND TO RAT
MEDULLARY THICK ASCENDING LIMBS FROM
FREE LABELED HORMONE[a,b]

	Counts per hour
Total radioactivity per sample	1.95×10^6
Background	896 ± 42
Blank	940 ± 43
Total binding per millimeter (after blank subtraction)	2645 ± 128
Nonspecific binding per millimeter (after blank subtraction)	101 ± 89

[a] Values reprinted from Butlen and Morel.[52]
[b] Samples of medullary thick ascending limbs (MALs, about 1 mm long each) were incubated for 60 min at 30° in a 3-μl droplet of glucagon-binding assay medium C containing 7.5 nM [^{125}I]glucagon in the absence (total binding) or in the presence (nonspecific binding) of 5 μM unlabeled glucagon. Radioactivity bound to tubules was separated from free ligand by repeated washings of the segments and blanks were determined and subtracted, as indicated under *Experimental Procedures*.

determination of optimal incubation conditions which prevent ligand degradation or metabolism by tissue samples. For instance, ^{125}I-labeled α-rat atrial natriuretic peptide 1–28 (α-[^{125}I]rANP) binding can be performed only at low temperature (4°) and in the presence of proteolytic enzyme inhibitors[54] because high ANP degrading enzyme activities are present at higher temperatures and result in a marked decrease in binding (Fig. 4).

Finally, it is worth noting that this binding microtechnique can also be applied to the study of the binding properties of nonhormonal labeled ligands as, for example, the distribution of benzodiazepine- and ouabain-binding sites along the nephron. In both cases, the huge number of binding sites present in renal membranes allows the use of tritiated ligands and samples containing a single piece of nephron segment.[78,79]

[78] D. Butlen, *FEBS Lett.* **169**, 138 (1984).
[79] G. El Mernissi and A. Doucet, *Am. J. Physiol.* **247**, F158 (1984).

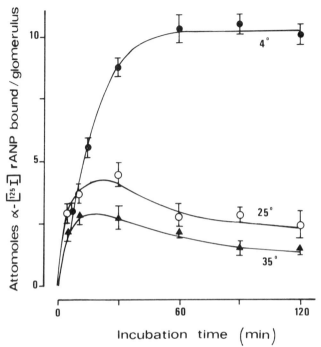

FIG. 4. Effect of temperature on the kinetics of α-[^{125}I]rANP binding to rat glomeruli. Glomeruli were incubated at the indicated temperatures in the presence of 0.50 nM α-[^{125}I]rANP. Each point is the mean value ± SE of five determinations performed using five glomeruli per sample, corrected for the corresponding nonspecific binding measured in the presence of 5 μM unlabeled atriopeptin II. (Redrawn from Butlen et al.[54])

Experimental Procedures

Table II summarizes the binding conditions adopted in different studies using either [^{125}I]glucagon (CNTS, Orsay, France; 74 TBeq/mmol), ^{125}I-labeled α-rat atrial natriuretic peptide 1–28 (α-[^{125}I]rANP, Amersham, Buckinghamshire, England; 74 TBeq/mmol), or [^{3}H]aldosterone (NEN, Boston, MA; 2.96 TBeq/mmol).

For [^{125}I]glucagon- and α-[^{125}I] rANP-binding assays, the binding reaction is stopped by adding to the sample 200 μl of chilled microdissection medium C. The pieces of tissue are sucked out and washed twice in 200 μl of the same cold medium. A 0.5-μl droplet containing the pieces of tissue is aspirated from the last rinse, whereas an equivalent volume of this rinse is used as a blank reference. After drying, all samples are counted for 60 min in a γ spectrometer.

TABLE II
STANDARD ASSAY CONDITIONS FOR BINDING MEASUREMENTS OF [^{125}I]GLUCAGON, α-[^{125}I]rANP, AND [^{3}H]ALDOSTERONE TO ISOLATED NEPHRON SEGMENTS

Condition	[^{125}I]Glucagon[a]	α-[^{125}I]rANP[b]	[^{3}H]Aldosterone[c]
Final volume, duration, and temperature	3 μl, 60 min at 30°	3 μl, 120 min at 4°	1 μl, 60 min at 25°
Tissue samples	One to four pieces of tubule (0.5–2 mm)	Four to six glomeruli	Ten to 40 pieces of tubule (5–20 mm)
Buffer	HEPES–NaOH, 20 mM, pH 7.4	HEPES–NaOH, 20 mM, pH 7.4	Tris–HCl, 10 mM, pH 7.4
Modified Hanks solution[d]	+	+	+
Proteolytic enzyme inhibitors	Bacitracin, 0.1%	Bacitracin, 0.1% + aprotinin, 15 μM	—
Other compounds	—	Tyrosine, 10 μM, + NaI, 10 μM	—
Labeled ligand	0.1 to 0.8 nM	0.1 to 5.3 nM	0.4 to 90 nM
Nonspecific binding[e]	5 μM unlabeled glucagon	5 μM unlabeled atriopeptin II	2000-fold excess of unlabeled aldosterone

[a] From Butlen and Morel.[52]
[b] From Butlen et al.[54]
[c] From Doucet and Katz.[44]
[d] The modified Hanks' solution has the same composition as the microdissection medium C.
[e] The unlabeled ligand is added only for non-specific binding determinations.

TABLE III
STANDARD ASSAY CONDITIONS FOR BINDING MEASUREMENTS OF [^{3}H]PK 11195 AND
[^{3}H]OUABAIN TO ISOLATED NEPHRON SEGMENTS

Condition	[^{3}H]PK 11195[a]	[^{3}H]Ouabain[b]
Final volume, duration, and temperature	3 μl, 120 min at 4°	1 μl, 30–45 min at 37°
Tissue samples	One to two pieces of tubule (0.5–1 mm)	One piece of tubule (0.5 mm)
Buffer	Tris-HCl, 20mM, pH 7.4	Tris-HCl, 10mM, pH 7.4
Salts	Hanks' salts[c] + CaCl$_2$, 1 mM	MgSO$_4$, 3 mM, + NaH$_2$PO$_4$, 3 mM, + NaVO$_3$, 1 mM
Other compounds	—	Sucrose, 250 mM
Labeled ligand	0.5–20 nM	0.5–10 μM
Nonspecific binding[d]	5 μM unlabeled PK 11195	100-fold excess of unlabeled ouabain

[a] From Butlen.[78]
[b] From El Mernissi and Doucet.[79]
[c] See the saline composition of the medium C used for the microdissection of the kidney tissue.
[d] The unlabeled ligand is added only for nonspecific binding determinations.

For [^{3}H]aldosterone-binding assays, in which a large number of pieces of tubules are pooled, the reaction is stopped by cooling the samples at 4° and radioactivity bound to tubules is separated by rapid filtration through Millipore filters (HAWP 1300, 0.45 μm) followed by filter washing with 1 ml of cold saline buffer under vacuum. The filters are counted for 50 min in a liquid scintillation spectrometer.

Table III depicts the assay conditions adopted in binding measurements of two drugs to isolated tubules: [^{3}H]ouabain (NEN, 0.74 TBeq/mmol) and a marker of peripheral benzodiazepine-binding sites,[80] the tritiated 1-(2-chlorophenyl)-N-methyl-N-(1-methylpropyl)-3-isoquinoline carboxamide ([^{3}H]PK 11195; Amersham, 1.48 TBeq/mmol).

For [^{3}H]PK 11195-binding assays, the radioactivity bound to tubules is separated from the free labeled ligand by filtration of samples through Whatman GF/C glass filters. Filters are washed five times with 5 ml of Tris-HCl, 20 mM, pH 7.4 and MgCl$_2$, 1 mM, at 4°. Radioactivity retained on the filter is counted for 50 min by liquid scintillation spectrometry.

[^{3}H]Ouabain-binding assays are performed under conditions where Na$^+$, K$^+$-ATPase is blocked in its phosphorylated conformation by vana-

---

[80] G. Le Fur, M. L. Perrier, N. Vaucher, F. Imbault, A. Flamier, J. Benavides, A. Uzan, C. Renault, M. C. Dubroeucq, and C. Guérémy, *Life Sci.* **32**, 1839 (1983).

date (an inhibitor with a higher affinity for the phosphorylation site than ATP or phosphate).[79] Individual tubules are transferred with microdissection medium onto small disks of dry BSA, deposited in the center of pieces of aluminum foil (5 × 5 mm) previously placed in the cavities of a chilled aluminum plaque. The microdissection solution is then aspirated and replaced by 1 μl of chilled incubation medium. Incubation is started by immersing the plaque in a water bath at 37°. After 30–45 min, the incubation is stopped by putting the plaque back on ice. Tubules are rinsed four times with 2 μl of microdissection medium, and the plaque is left for 60 min at 4° (during this postincubation step at 4°, nonspecific binding decreases by more than 95% whereas specific binding remains constant).[79] Then tubules are rinsed four times again, and the pieces of aluminum foil containing the tubules are transferred into counting vials for radioactivity measurements of the samples by liquid scintillation spectrometry.

In each experiment, several blank samples without tubule are treated under similar conditions.

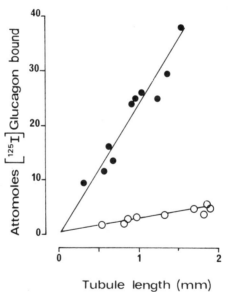

FIG. 5. Effect of tubule length on total and nonspecific binding of [^{125}I]glucagon to medullary thick ascending limbs (MALs) microdissected from rat kidney. Tubules were incubated for 60 min at 30° with 7.5 nM [^{125}I]glucagon in the absence (total binding, ●—●) or in the presence (nonspecific binding, ○—○) of 5 μM unlabeled glucagon. The equations of the linear regression lines computed from data are the following:(●), $y = 25.0x - 0.3$ ($r = 0.96$, $p < 0.01$); (○), $y = 2.6x - 0.1$ ($r = 0.93$, $p < 0.01$). (Redrawn from Butlen and Morel.[52])

For each concentration of radioligand used, total binding is measured in the presence of radioligand alone and nonspecific binding in the presence of the radioligand plus a large excess of unlabeled ligand added together at the beginning of the binding reaction.

The results are calculated as follows: ligand specifically bound to tubule (mol/mm) = $(x - y)$/SRA, where x is the total radioactivity bound per millimeter after blank subtraction, y is the nonspecific radioactivity bound per millimeter after blank subtraction, and SRA is the specific radioactivity of the radioligand used (counts/mol).

When competition for binding of structural analogs is studied in pharmacological experiments, nephron fragments are incubated in the presence of a constant amount of labeled ligand (usually equal to the concentration leading to half-maximal occupancy of the specific binding sites) and increasing amounts of the corresponding unlabeled ligand or unlabeled structural analogs.

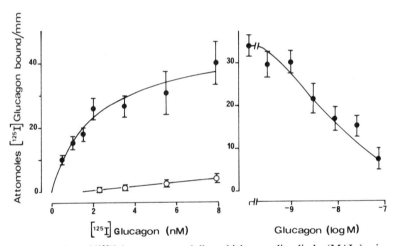

FIG. 6. Binding of [^{125}I]glucagon to medullary thick ascending limbs (MALs) microdissected from rat kidney. *Left:* Dose dependency of specific (●—●) and nonspecific (○—○) binding of [^{125}I]glucagon to MALs. Values are means ± SE of four to six determinations. The transformation of the dose-dependent specific binding curve in Scatchard coordinates gives a straight line corresponding to the following equation: [RH]/[H] = (-0.43×10^{-3})[RH] + (20.4×10^{-3}) ($r = 0.95, p < 0.01$), where [RH] is expressed as attomoles·microliter^{-1}·millimeter^{-1} and [H], free hormone concentration as attomoles·microliter^{-1}. *Right:* Dose-dependent inhibition of [^{125}I]glucagon binding to MALs by unlabeled glucagon; MALs were incubated in the presence of 5.1 nM [^{125}I]glucagon and of the indicated concentrations of unlabeled glucagon. Values (means ± SE of four or five determinations) were corrected for nonspecific binding measured in the presence of 5 μM unlabeled glucagon. (Redrawn from Butlen and Morel.[52])

Comments

Various prerequisites must be fulfilled in order to identify specific binding sites with true hormonal receptors, as illustrated for glucagon and atrial natriuretic peptide in Figs. 5–7: (1) linearity between binding activities and tubular length (Fig. 5) or glomeruli number[54,56]; (2) dose-dependent and saturable specific binding as a function of labeled hormone concentration (Fig. 6, left); (3) dose-dependent inhibition of binding by increasing amounts of the corresponding unlabeled ligand or structural analogs exhibiting agonist or antagonist properties (Fig. 6, right, and Fig.

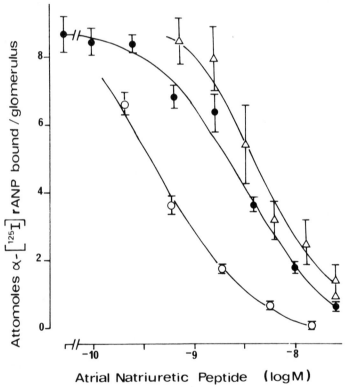

FIG. 7. Dose-dependent inhibition of α-[^{125}I]rANP binding to rat glomeruli by unlabeled atrial natriuretic peptide analogs. Glomeruli were incubated for 2 hr at 4° in the presence of 0.54 nM α-[^{125}I]rANP and the indicated connections of unlabeled α-rANP 1–28 (●), rANP 3–28 (○), or atriopeptin II (△). Values (means ± SE of four to six determinations) were corrected for nonspecific binding measured in the presence of 5 μM atriopeptin II. (Redrawn from Butlen et al.[54])

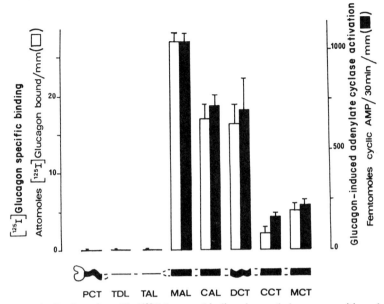

FIG. 8. Distribution of specific [^{125}I]glucagon-binding sites and glucagon-sensitive adenylate cyclase along the rat nephron. [^{125}I]Glucagon-binding capacity (□): Tubules were incubated in the presence of 7.5 nM [^{125}I]glucagon. Values (means ± SE of 10–50 determinations) were corrected for nonspecific binding measured in the presence of 5 μM unlabeled glucagon. Maximal adenylate cyclase activations induced by 1 μM glucagon (■): Values (means ± SE of 10–50 determinations) were corrected for the corresponding basal enzyme activities. PCT, proximal convoluted tubule; TDL, thin descending limb; TAL, thin ascending limb; MAL, medullary thick ascending limb; CAL, cortical thick ascending limb; DCT, distal convoluted tubule) CCT, cortical collecting tubule; MCT, outer medullary collecting tubule. (Adapted from Butlen and Morel.[52])

7); and (4) no impaired binding by unrelated hormones or biologically inactive analogs.[44-56]

Furthermore, the specific [^{125}I]glucagon-binding sites observed along the rat nephron most likely correspond to the physiological glucagon receptors triggering adenylate cyclase activation, because specific [^{125}I] glucagon-binding sites are detected only in the nephron segments which also contain a glucagon-sensitive adenylate cyclase activity (Fig. 8), i.e., the thick ascending limb of Henle's loop, the distal tubule, and the cortical and outer medullary portions of the collecting tubule. The maximal specific binding level measured in each of these nephron portions varies in parallel with maximal adenylate cyclase response to glucagon (Fig. 8), and the apparent dissociation constants for glucagon binding (hormone concentrations leading to half-maximal binding) measured in MAL and MCT segments, respectively, are similar to glucagon concentrations leading to half-maximal adenylate cyclase activations.[52,81]

The [^{3}H]aldosterone-binding microassay described here does not allow one to distinguish between cytoplasmic and nuclear receptors. Therefore, the [^{3}H]aldosterone-binding sites identified in the collecting tubule of the rabbit nephron (about 11 amol/mm) most likely correspond to the sum of cytosolic and nuclear receptors.[44] These tubular binding sites exhibit the following recognition pattern for various steroids: aldosterone > deoxycorticosterone > spironolactone > dexamethasone > > 5 α-dihydroxytestosterone = progesterone = 17βestradiol, as expected of mineralocorticoid receptors. This conclusion is supported by the observation that adrenalectomy decreases Na$^+$, K$^+$-ATPase activity in collecting tubules and that this activity is restored by physiological doses of aldosterone.[82]

The mapping of peripheral benzodiazepine-binding sites along the rat nephron is performed using [^{3}H]PK 11195, a chemical drug which is a potent marker of peripheral benzodiazepine "receptors," though chemically unrelated to benzodiazepines.[80] Huge levels of high-affinity PK 11195-binding sites revealing a sequence of stereospecificity (PK 11195 = Ro 5-4864 > > clonazepam) are identified in glomeruli (0.3 fmol/glomerulus), in thick ascending limbs, and in collecting tubules (1.1–1.8 fmol/mm).[78] This observation leads to as yet unanswered questions regarding the possible functional or physiological significance of peripheral benzodiazepine-binding sites in the kidney.[83]

The number of Na$^+$,K$^+$-ATPase units and the molecular activity of Na$^+$,K$^+$-ATPase along the nephron are determined by measuring in parallel [^{3}H]ouabain-binding sites and Na$^+$,K$^+$-ATPase activity in single nephron segments microdissected from rabbit kidney.[79] A close linear correlation is obtained in the different nephron segments between the number of specific [^{3}H]ouabain-binding sites (ranging from 3 fmol/mm in the MCT to 31 fmol/mm in the DCT$_b$) and the maximal enzyme activity. The specific activity per ATPase unit is similar in all nephron segments (2000 ATP molecules hydrolyzed per minute per ouabain-binding site at 37°).[79] In addition, measurements of dissociation constants for ouabain binding (K_D) and inhibition constants of ATPase activity by ouabain (K_I) in the successive nephron segments reveal a parallel decrease in K_D and K_I values from the proximal tubule to the collecting tubule.[84]

[81] C. Bailly, M. Imbert-Teboul, D. Chabardès, A. Hus-Citharel, M. Montégut, A. Clique, and F. Morel, *Proc. Natl. Acad. Sci. U.S.A.* **77**, 3422 (1980).
[82] G. El Mernissi and A. Doucet, *Pfluegers Arch.* **402**, 258 (1984).
[83] D. S. Lukeman and D. D. Fanestil, *J. Pharmacol. Exp. Ther.* **241**, 950 (1987).
[84] A. Doucet and C. Barlet, *J. Biol. Chem.* **261**, 993 (1986).

It should be mentioned in conclusion that similar binding micromethods can be applied to investigate the action sites of other drugs and certain diuretics which are currently employed for pharmacological studies of kidney functions.[85-88]

Acknowledgments

The authors are indebted to Mrs. Violette Biausque and to Mrs. Sylvie Siaume-Perez for their help in preparing the manuscript.

[85] W. N. Suki, B. J. Stinebaugh, J. P. Frommer, and G. Eknoyan, in "The Kidney: Physiology and Physiopathology" (D. W. Seldin and G. Giebisch, eds.), Vol. 2, p. 2127. Raven, New York, 1985.
[86] E. Giesen-Crouse, P. Frandeleur, M. Schmidt, J. Schwartz, and J. L. Imbs, *J. Hypertension* **3**, S211 (1985).
[87] P. Barbry, C. Frelin, P. Vigne, E. J. Cragoe, Jr., and M. Lazdunski, *Biochem. Biophys. Res. Commun.* **135**, 25 (1986).
[88] P. Vigne, M. Lazdunski, and C. Frelin, *Eur. J. Pharmacol.* **160**, 295 (1989).

[18] Metabolism of Isolated Kidney Tubule Segments[1]

By GABRIELE WIRTHENSOHN and WALTER G. GUDER

Introduction

In the last century morphologists had already observed that the kidney consists of different structural and functional units. Each kidney contains about 10^6 nephrons with glomeruli producing the primary filtrate and the tubule forming the final urine in many subsequent segments. To study the physiological and biochemical functions of the different nephron segments (Fig. 1) special techniques have been developed. Micropuncture and microperfusion were used to study tubular transport. Biochemical functions (enzymes, substrates, metabolic pathways) were studied mainly in whole kidney homogenates and tissue slices until the late 1950s, when more special techniques were developed which allowed biochemical studies at the defined segment level.

In this chapter we present a short review on the isolation and analysis of defined nephron segments, giving some examples of the determination of enzyme activities and metabolic pathways in them. The advantages and disadvantages of the different techniques are also discussed.

[1] In memory of Helen B. Burch.

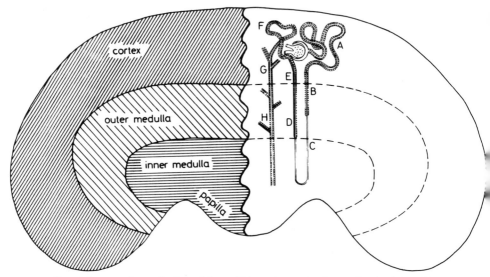

FIG. 1. Structural organization of the rat kidney. A, proximal convoluted tubule (PCT); B, proximal straight tubule (pars recta, PR); C, thin descending limb of Henle's loop (TDL); D, medullary thick ascending limb of Henle's loop (MAL); E, cortical thick ascending limb of Henle's loop (CAL); F, distal convoluted tubule (DCT); G, cortical collecting tubule (CCT); H, medullary collecting tubule (MCT).

Isolation Procedures for Defined Segments

Tubule Suspensions

Cortical Tubules. In 1962, Burg and Orloff[1a] isolated tubule suspensions from rabbit renal cortex with the aid of collagenase. The suspensions exhibited morphologic and metabolic properties of proximal tubules. This method was later extended to rat,[2-4] dog,[5] mouse,[6] and human[7] kidney. This preparation was morphologically demonstrated to contain more than 90% proximal tubules.[8] An additional Percoll gradient step allows further

---

[1a] M. B. Burg and J. Orloff, *Am. J. Physiol.* **203**, 327 (1962).
[2] N. Nagata and H. Rasmussen, *Proc. Natl. Acad. Sci. U.S.A.* **65**, 368 (1970).
[3] W. G. Guder, W. Wiesner, B. Stukowski, and O. Wieland, *Hoppe-Seyler's Z. Physiol. Chem.* **352**, 1319 (1971).
[4] W. G. Guder, *Biochim. Biophys. Acta* **584**, 507 (1979).
[5] G. Baverel, M. Bonnard, E.d'Armagnac de Castanet, and M. Pellet, *Kidney Int.* **14**, 567 (1978).
[6] G. Wirthensohn and W. G. Guder, *Pfluegers Arch.* **404**, 94 (1985).
[7] G. Baverel, M. Bonnard, and M. Pellet, *FEBS Lett.* **101**, 282 (1979).
[8] W. Pfaller, W. G. Guder, G. Gstraunthaler, P. Kotanko, J. Jehart, and S. Pürschel, *Biochim. Biophys. Acta* **805**, 152 (1984).

purification[9] and separation of a small fraction of pure distal convoluted tubule segments.[10]

This preparation can be applied to all metabolic studies of function which have been found to be preferentially, if not exclusively, localized in the proximal tubule. Renal gluconeogenesis, amino acid, lipid, and ketone body metabolism, as well as oxygen consumption and hormonal regulation and degradation exhibited optimal rates when studied in this model.[11] In addition this preparation can be used to study intracellular compartmentation[8,12] as well as membrane composition of proximal tubules.[13]

Medullary Tubules. More recently a similar technique has been applied to prepare tubules from the outer and inner medulla.[14,15]

Outer medullary tubule suspensions consist mainly of thick ascending limb segments, contaminated with other tubule types and cellular debris which can be removed by density gradient centrifugation on Percoll leading to a suspension of 96% thick ascending limbs.[14] When studying oxygen consumption, substrate metabolism, and hormone action, they establish properties of the mitochondria-rich cells of the ascending limb of Henle's loop.

Inner medullary tubule suspensions consist mainly of collecting tubule segments with some thin limbs and can be used as a model for studying metabolism of the renal inner medulla.[15]

Isolated Cells

With the aid of calcium chelators [ethylenediaminetetraacetic acid (EDTA), citrate], tight junctions connecting tubule cells can be weakened to obtain isolated cell suspensions. These were prepared from rabbit renal cortex[16] and outer medulla[17] and further purified by free-flow electrophoresis.[18] This procedure led to proximal and distal tubule cells of high purity, whose metabolic functions, however, were not so well maintained compared to tubule suspensions.[19] Further purification of distal cell types can

---

[9] D. W. Scholer and I. S. Edelman, *Am. J. Physiol.* **237**, F350 (1979).
[10] P. Vinay, A. Gougoux, and G. Lemieux, *Am. J. Physiol.* **241**, F403 (1981).
[11] G. Wirthensohn and W. G. Guder, *Physiol. Rev.* **66**, 469 (1986).
[12] W. G. Guder and S. Pürschel, *Int. J. Biochem.* **12**, 63 (1980).
[13] C. LeGrimellec, M.-C. Giocondi, B. Carriere, S. Carriere, and J. Cardinal, *Am. J. Physiol.* **242**, F246 (1982).
[14] M. E. Chamberlin, A. LeFurgey, and L. J. Mandel, *Am. J. Physiol.* **247**, F955 (1985).
[15] G. Wirthensohn, F. X. Beck, and W. G. Guder, *Pfluegers Arch.* **409**, 411 (1987).
[16] P. Poujeol and A. Vandewalle, *Am. J. Physiol.* **249**, F74 (1985).
[17] J. Eveloff, W. Haase, and R. Kinne, *J. Cell Biol.* **87**, 672 (1980).
[18] A. Vandewalle, B. Köpfer-Hobelsberger, and H.-G. Heidrich, *J. Cell Biol.* **92**, 505 (1982).
[19] H.-G. Heidrich, A. Vandewalle, B. Köpfer-Hobelsberger, and W. G. Guder, in "Cell Function and Differentiation" (A. Evangelopoulos, ed.). Liss, New York, 1982.

be obtained by immunoabsorption and lectin absorption techniques[20-22] and can be used as a basis for metabolic studies and cell culture techniques.

Microdissection Procedures

At the beginning of this century methods were developed to isolate single nephrons.[23] This maceration technique was done by treatment of the tissue with 75% hydrochloric acid. Although proteins were denatured, well-preserved epithelial structures could be obtained. To isolate discrete nephron segments for the study of enzyme activities or metabolic pathways, however, two different techniques have been applied and have become well established in the past 20 years.

Microdissection from Lyophilized Tissue Sections. This technique, initially developed by Linderstrom–Lang *et al.*[24] for metabolic studies in microbial cells, was extended and amended by O. Lowry and associates[25] for microdissection of freeze–dried tissue sections. This technique is well suited to the isolation and identification of single tubular structures from freeze–dried renal sections.[26]

In this procedure fresh kidney tissue is quickly cut into small cones of 2-mm thickness, mounted on a tissue holder, and plunged into liquid nitrogen to maintain structural and biochemical integrity and reduce the development of ice crystals.[27] The tissue sample is transferred to a cryostat and sections of 16-μm thickness are cut with a microtome. This procedure preserves the full activity of most enzymes (for exceptions, see Pros and Cons of the Microdissection Procedures, below). Single segments are dissected using an alternate periodic acid–Schiff base-stained section as a guide.[26] Single tubular structures obtained from these tissue sections are weighed on a quartz fiber balance calibrated with quinine crystals.[26] Dry weights of about 1–50 ng can be quantified with this method.

[20] M. Le Hir and U. C. Dubach, *Histochemistry* **74,** 521 (1982).
[21] W. S. Spielman, W. K. Sonnenberg, M. L. Allen, L. J. Arend, K. Gerozissis, and W. L. Smith, *Am. J. Physiol.* **251,** F348 (1986).
[22] A. Vandewalle, M. Tauc, F. Cluzeaud, P. Ronco, F. Chatelet, P. Verroust, and P. Poujeol, *Am. J. Physiol.* **250,** F386 (1986).
[23] K. Peter, "Untersuchungen über Bau und Entwicklung der Niere." Gustav Fischer Verlag, Jena, German Democratic Republic, 1909.
[24] K. Linderstrom-Lang, H. Holter and A. S. Ohlson, C.R. Lab. Carlsberg, Ser. Chim. **21,** 315 (1938).
[25] O. H. Lowry, *J. Histochem. Cytochem.* **1,** 420 (1953).
[26] U. Schmidt and U. C. Dubach, *Prog. Histochem. Cytochem.* **2,** 185 (1971).
[27] U. Schmidt, U. C. Dubach, W. G. Guder, B. Funk, and K. Paris, *in* "Biochemical Aspects of Kidney Function" (S. Angielski and U. C. Dubach, eds.), p. 22. Huber, Bern, Switzerland, 1975.

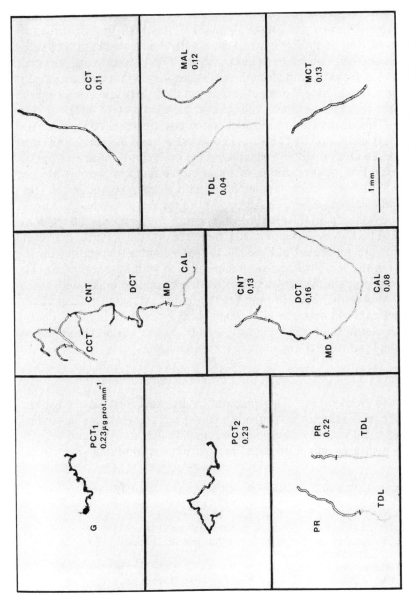

FIG. 2. Isolated structures of rabbit nephron. G, Glomerulus; MD, macula densa; for other abbreviations see Fig. 1. Protein content of dissected structures was calculated from tubular length, using the data of Vandewalle et al.[35]

The amount of tissue needed depends on the sensitivity of the assay procedure and the activity/concentration of the measured enzyme/substrate in the respective segment.

Microdissection from Fresh Tissue. This method, first described by Burg and Orloff in 1962,[1a] is based on the free-hand preparation of single nephron segments from fresh renal tissue. The procedure is, however, limited to kidneys of animals with easily accessible connective tissue, as is found in rabbit kidney. To separate different medullary and distal nephron structures in other animals, collagenase pretreatment of the kidney is needed. This disintegration of renal connective tissue by collagenase was extensively examined by Morel and co-workers[28] and has been applied by many different groups[29-33] to overcome the problem of intranephron heterogeneity. For the preparation of single-tubule samples, kidneys are perfused with isotonic saline supplemented with glucose, bovine serum albumin, and collagenase.[28,34,35] After ligation of the renal vein and the ureter the same solution is injected under pressure until rupture of the renal capsule. The kidney is sliced along the corticomedullary axis and small pyramids are incubated at 35–37° for 20–40 min (depending on the animal) in the same collagenase buffer bubbled with a stream of air. The small tissue pyramids are washed with collagenase-free buffer and microdissection is carried out with fine needles under a stereomicroscope. Each dissected nephron segment is photographed for confirmation of identity and measurement of tubular length. Figure 2 shows some representative photographs of tubular segments of rabbit nephron.

Advantages and Limitations of the Various Isolation Procedures

Tubule Suspensions. This preparation has the great advantage that many identical samples of relatively well-defined segments can be studied. Routine methods may be applied for metabolite, oxygen, and enzyme activity measurements. Compared to isolated cells and tubule-derived cell

---

[28] M. Imbert, D. Chabardès, M. Montégut, A. Clique, and F. Morel, *Pfluegers Arch.* **354,** 213 (1975).

[29] A. Doucet, A. I. Katz, and F. Morel, *Am. J. Physiol.* **237,** F105 (1979).

[30] H. Endou, H. Nonoguchi, J. Nakada, Y. Takehara, and H. Yamada, in "Kidney Metabolism and Function" (R. Dzurik, B. Lichardus, and W. G. Guder, eds.) p. 26. Nijhoff, Boston, Massachusetts, 1985.

[31] R. M. Edwards, B. A. Jackson, and T. P. Dousa, *Am. J. Physiol.* **238,** F269 (1980).

[32] N. Farman, A. Vandewalle, and J.-P. Bonvalet, *Am. J. Physiol.* **244,** F325 (1983).

[33] G. Wirthensohn, A. Vandewalle, and W. G. Guder, *Kidney Int.* **21,** 877 (1982).

[34] F. Morel, *Am. J. Physiol.* **240,** F159 (1981).

[35] A. Vandewalle, G. Wirthensohn, H.-G. Heidrich, and W. G. Guder, *Am. J. Physiol.* **240,** F492 (1981).

cultures a high degree of viability and differentiation is maintained. On the other hand, several limitations have to be considered when using these preparations. Besides effects of collagenase, preparations are always contaminated with minor fractions of other cell types. This can be especially disturbing if isotopes are applied to study metabolic pathways. Likewise minor contaminations can lead to major errors if the studied enzyme is very low in the main cell type.[36] In addition, metabolite levels may change during preparation and therefore must be compared with the *in vivo* levels.

Pros and Cons of the Microdissection Procedures. A great advantage of the "Lowry" technique is the low amount of tissue needed for preparation (<10 mg fresh weight). Moreover, the rapid freezing of the tissue in liquid nitrogen preserves the *in vivo* conditions. This is of special importance for the study of human tissue, which is available only as biopsy or after nephrectomy. The technique allows storage of tissue in a deep-frozen state over several months to years before dissection is performed. In the hands of a well-trained person this microdissection procedure provides a high degree of flexibility and accuracy.

Limitations, on the other hand, arise mainly from the fact that this isolation technique does not provide full integrity of tubular cells. For this reason metabolic pathways and hormone actions cannot be studied. Moreover several enzymes, such as pyruvate carboxylase[37] and adenylate cyclase (W. G. Guder, unpublished observations), lose their activity on lyophilization and/or freezing. In contrast to fresh tissue dissection the former method does not allow the dissection of different segments in one tubule, offering the possibility of studying intranephron heterogeneity. Furthermore, only fresh tissue dissection is suitable for studying the sites of hormone actions[6,28,34,38] and metabolic pathways such as triacylglycerol[39] and glucose[40] synthesis as well as CO_2,[41,42] NH_3,[43,44] and lactate[45] formation.

---

[36] W. G. Guder and A. Rupprecht, *Eur. J. Biochem.* **52**, 283 (1975).
[37] H. B. Burch, R. G. Narins, C. Chu, S. Fagioli, S. Choi, W. McCarthy, and O. H. Lowry, *Am. J. Physiol.* **235**, F246 (1978).
[38] F. Morel and A. Doucet, *Physiol. Rev.* **66**, 377 (1986).
[39] W. G. Guder and G. Wirthensohn, in "Biochemistry of Kidney Functions" (F. Morel, ed.), p. 95. Elsevier, Amsterdam, 1982.
[40] A. Maleque, H. Endou, C. Koseki, and F. Sakai, *FEBS Lett.* **166**, 154 (1980).
[41] F. LeBouffant, A. Hus-Citharel, and F. Morel, *Pfluegers Arch.* **401**, 346 (1984).
[42] F. LeBouffant, A. Hus-Citharel, and F. Morel, in "Biochemistry of Kidney Functions" (F. Morel, ed.), p. 363. Elsevier, Amsterdam, 1982.
[43] H. Nonoguchi, S. Uchida, T. Shiigai, and H. Endou, *Pfluegers Arch.* **403**, 229 (1985).
[44] D. W. Good and M. B. Burg, *J. Clin. Invest.* **73**, 602 (1984).
[45] S. Bagnasco, D. Good, R. Balaban, and M. Burg, *Am. J. Physiol.* **248**, F522 (1985).

The disadvantage of fresh tissue dissection lies in the use of collagenase, which may alter tubular membrane permeabilities and affect enzyme activities.[33,35,46] In addition the material is of limited use in the measurement of metabolites since the preparation and incubation procedure may change their concentrations. If regulation *in vitro* is, however, to be studied at the single-nephron level, freshly dissected tubules are adequate.

Microanalytical Procedures

To study metabolite concentrations, enzyme activities, hormone actions, or metabolic pathways in single-nephron segments obtained by the microdissection procedures described, sensitive microassays are needed which allow the detection of specific products/metabolites in the picomolar range. This can be achieved by either of the following techniques, which have been successfully applied to renal tissue. Considerable increase in sensitivity can be obtained by decreasing assay volumes to the nanoliter range. This tiny volume is pipetted into small holes (4×3 mm) of a Teflon rack sealed with a drop of paraffin oil (oil well) to prevent evaporation.[47] Reactions in a few microliters can be performed between two glass slides with opposed depressions, an arrangement allowing radiochemical[41,42] and cycling[30] procedures to be performed. Table I summarizes enzymes, substrates, and pathways which have been successfully analyzed with one of the following techniques.

Enzymatic Cycling

Enzymatic cycling analyses were performed with methods based on specific enyzmatic reactions which result in oxidation or reduction of pyridine nucleotides.[47] Amplification of signals is obtained by use of enzymatic cycling of nucleotides.[48] The following examples illustrate the principles of enzymatic cycling as it is used to study enzyme activities or renal metabolites.

Citrate Synthase[49]. The tissue to be studied is added to a reagent that carries out the following reaction:

$$\text{Oxaloacetate} + \text{acetyl} - \text{CoA} \rightarrow \text{citrate} + \text{CoA}$$

---

[46] G. Wirthensohn, A. Vandewalle, and W. G. Guder, *Biochem. J.* **198**, 534 (1981).
[47] O. H. Lowry and J. V. Passonneau, "A Flexible System of Enzymatic Analysis," Academic Press, New York, 1972.
[48] O. H. Lowry, *Mol. Cell. Biochem.* **32**, 135 (1980).
[49] H. B. Burch, *in* "Biochemistry of Kidney Functions" (F. Morel, ed.), p. 297. Elsevier, Amsterdam, 1982.

TABLE I
ENZYMES, SUBSTRATES, AND PATHWAYS MEASURED IN SINGLE ISOLATED NEPHRON SEGMENTS[a]

Methods	Enzymes	Substrate	Pathways
Enzymatic cycling	Mg^{2+}- and Na$^+$,K$^+$-ATPase (1)	Adenine nucleotides (10, 12)	Gluconeogenesis (15)
	Phosphofructokinase (2)	Guanine nucleotides (10)	Ammoniagenesis (16)
	Pyruvate kinase (2)	Aspartate (13)	
	Fructose-1,6-bisphosphatase (3, 4)	Glutamate (13)	
	Phosphoenolpyruvate carboxykinase (3)	Glutamine (13)	
	Glucose-6-phosphatase (3)	Glutathione (13)	
	Phosphate-dependent glutaminase (5)	Fructose 1-phosphate (10)	
	Aspartate aminotransferase (6)	Glucose 6-phosphate (10)	
	Alanine aminotransferase (6)	Glycerol phosphate (10)	
	Branched chain amino acid aminotransferase (7)	6-Phosphogluconate (10)	
	D-Amino-acid oxidase (6)	Glycogen (14)	
	Malic enzyme (8)	Glucose (14)	
	cGMP phosphodiesterase (9)		
	Guanylate kinase (9)		
	Fructose-1-phosphate aldolase (10)		
	Fructose-1,6-bisphosphate aldolase (10)		
	5-Nucleotidase (9)		
	Glutamine synthetase (11)		
	Fructokinase (10)		
	Glycerol kinase (12)		
	3-Oxoacid transferase (8)		
	Adenylate kinase (9)		
	Glycerol-3-phosphate dehydrogenase (10)		
	Citrate synthase (8)		
	Fumarase (8)		
	3-Hydroxyacyl-CoA dehydrogenase (8)		
	cAMP phosphodiesterase (9)		
Fluorometry	Lactate dehydrogenase (17)		Lactate formation (21)
	Isocitrate dehydrogenase (17)		Ammoniagenesis (22)
	Malate dehydrogenase (17)		
	Glucose-6-phosphate dehydrogenase (17)		
	Glutamate dehydrogenase (17)		

(continued)

TABLE I (continued)

Methods	Enzymes	Substrate	Pathways
Photometry	Acid phosphatase (18)		
	Alkaline phosphatase (19)		
	N-Acetyl-β-D-glucosaminidase (18, 19)		
	β-Galactosidase (18)		
	3-Hydroxyacyl-CoA dehydrogenase (20)		
	Cytochrome P-450 (23)		
	Succinate dehydrogenase (24)		
	Leucine aminopeptidase (25)		
	γ-Glutamyl transpeptidase (26)		
	Glutaminase (26)		
	Glutathione-S-transferase (27)		
	Glutamate dehydrogenase (26)		
	α-Glutamylcysteine synthetase (28)		
	2-Oxoglutarate dehydrogenase (29)		
Radiochemical	Phosphoenolpyruvate carboxykinase (30)		CO_2 formation (43)
	Hexokinase (30)		Triacylglycerol synthesis (44)
	Glycerol kinase (31)		Betaine formation (45)
	Choline kinase (32)		Vitamin D hydroxylation (46)
	3-Oxoacid-CoA transferase (33)		Phospholipid turnover (47)
	Adenylate cyclase (34)		
	Na^+,K^+-ATPase (35)		
	Ca^{2+}-ATPase (36)		
	H^+-ATPase (38)		
	Anion ATPase (39)		
	Corticosteroid dehydrogenase (40)		
	Protein kinase (cAMP) (41)		
	Phosphodiesterase (42)		
Luminescence	Glucose-6-phosphate dehydrogenase (48)		
	3-Hydroxybutyrate dehydrogenase (48)		
Radioimmunoassay	Kallikrein (kininogenin) (49)	cAMP (51)	
	Kininases (50)		

Catalase inhibition	D-Amino-acid oxidase (52)
	α-Hydroxyacid oxidase (52)
Histochemical staining	Succinate dehydrogenase (53)
	Na$^+$,K$^+$-ATPase (54)

[a] References: (1) U. Schmidt and M. Horster, *Methods Pharmacol.* **4B**, 259 (1978); (2) H. Schmid, A. Mall, M. Scholz, and U. Schmidt, *Hoppe-Seyler's Z. Physiol. Chem.* **361**, 819 (1980); (3) H. B. Burch, R. G. Narins, C. Chu, S. Fagiolo, S. Choi, W. McCarthy, and O. H. Lowry, *Am. J. Physiol.* **235**, F246 (1978); (4) H. Schmid, M. Scholz, A. Mall, U. Schmidt, W. G. Guder, and U. C. Dubach, in "Biochemical Nephrology" (W. G. Guder and U. Schmidt, eds.), p. 282. Huber, Bern, Switzerland, 1978; (5) N. P. Curthoys and O. H. Lowry, *J. Biol. Chem.* **248**, 162 (1973); (6) A. W. K. Chan, S. G. Parry, H. B. Burch, S. Fagioli, T. R. Alvey, and O. H. Lowry, *J. Histochem. Cytochem.* **27**, 751 (1979); (7) H. B. Burch, N. Cambon, and O. H. Lowry, *Kidney Int.* **28**, 114, 1985; (8) H. B. Burch, T. E. Bross, C. A. Brooks, B. R. Cole, and O. H. Lowry, *J. Histochem. Cytochem.* **32**, 731 (1984); (9) B. R. Cole, A. E. Hays, J. G. Boylan, H. B. Burch, and O. H. Lowry, *Am. J. Physiol.* **243**, F349 (1982); (10) H. B. Burch, S. Choi, C. N. Dence, T. R. Alvey, B. R. Cole, and O. H. Lowry, *J. Biol. Chem.* **255**, 8239 (1980); (11) H. B. Burch, S. Choi, W. Z. McCarthy, P. Y. Wong, and O. H. Lowry, *Biochem. Biophys. Res. Commun.* **82**, 498 (1978); (12) H. B. Burch, A. E. Hays, M. D. McCreary, B. R. Cole, M.-Y. Chi, C. N. Dence, and O. H. Lowry, *J. Biol. Chem.* **257**, 3676 (1982); (13) J. E. Brehe, A. W. K. Chan, T. R. Alvey, and H. B. Burch, *Am. J. Physiol.* **231**, 1536 (1976); (14) P. Needleman, J. V. Passonneau, and O. H. Lowry, *Am. J. Physiol.* **215**, 655 (1968); (15) A. Maleque, H. Endou, C. Koseki, and F. Sakai, *FEBS Lett.* **116**, 154 (1980); (16) H. Nonoguchi, S. Uchida, T. Shiigai, and H. Endou, *Pfluegers Arch.* **403**, 229 (1985); (17) U. Schmidt and U. C. Dubach, *Prog. Histochem. Cytochem.* **2**, 185 (1971); (18) M. Le Hir and U. C. Dubach, in W. G. Guder, *Int. J. Biochem.* **12**, 41 (1980); (19) H. Schmid, A. Mall, and H. Bockhorn, *J. Clin. Chem. Clin. Biochem.* **24**, 961 (1986); (20) M. Le Hir and U. C. Dubach, in "Biochemistry of Kidney Functions" (F. Morel, ed.), p. 87. Elsevier, Amsterdam, 1982; (21) S. Bagnasco, D. Good, R. Balaban, and M. Burg, *Am. J. Physiol.* **248**, F522 (1985); (22) D. W. Good and M. B. Burg, *J. Clin. Invest.* **73**, 602 (1984); (23) H. Endou, C. Koseki, S. Hasamura, K. Kakuno, K. Hajo, and F. Sakai, in "Biochemistry of Kidney Functions" (F. Morel, ed.), p. 319 Elsevier, Amsterdam, 1982; (24) B. Höhmann, Habilitationsschrift, Heidelberg, 1973; (25) J. Sudo, *Folia Pharmacol. Jpn.* **78**, 27 (1981); (26) H. Shimada, H. Endou, and F. Sakai, *Jpn. J. Pharmacol.* **32**, 121 (1982); (27) L. G. Fine, E. J. Goldstein, and J. M. Arias, *Kidney Int.* **8**, 474 (1975); (28) H. Heinle, A. Wendel, and U. Schmidt, *FEBS Lett.* **73**, 220 (1977); (29) M. Le Hir and U. C. Dubach, *Pfluegers Arch.* **395**, 239 (1982); (30) A. Vandewalle, G. Wirthensohn, H.-G. Heidrich, and W. G. Guder, *Am. J. Physiol.* **240**, F492 (1981); (31) G. Wirthensohn, A. Vandewalle, and W. G. Guder, *Biochem. J.* **198**, 534 (1981); (32) G. Wirthensohn, A. Vandewalle, and W. G. Guder, *Kidney Int.* **21**, 877 (1982); (33) W. G. Guder, S. Pürschel, and G. Wirthensohn, *Hoppe Seyler's Z. Physiol. Chem.* **364**, 1727 (1983); (34) M. Imbert, D. Chabardès, M. Montégut, A. Clique, and F. Morel, *Pfluegers Arch.* **354**, 213 (1975); (35) A. I. Katz, A. Doucet, and F. Morel, *Am. J. Physiol.* **237**, F114 (1979); (36) A. Doucet and A. I. Katz, *Am. J. Physiol.* **242**, F346 (1982); (37) A. Ait-Mohamed, S. Marsy, C. Barlet, C. Khadouri, and A. Doucet, *J. Biol. Chem.* **261**, 12526 (1986); (38) L. E. Garg and N. Narang, *Can. J. Physiol. Parmacol.* **63**, 1291 (1985); (39) M. Ben Abdelkhalek, J. Lee, C. Barlet, and A. Doucet, *J. Membr. Biol.* **89**, 225 (1986); (40) W. Schulz, H. Siebe, and K. Hierholzer, in "Molecular Nephrology" (Z. Kovacevic and W. G. Guder, eds.), p. 361. de Gruyter, Berlin, 1987; (41) R. M. Edwards, B. A. Jackson, and T. P. Dousa, *Am. J. Physiol.* **238**, F269 (1980); (42) B. A. Jackson, R. M. Edwards, and T. P. Dousa, *Kidney Int.* **18**, 512 (1980); (43) F. LeBouffant, A. Hus-Citharel, and F. Morel, *Pfluegers Arch.* **401**, 346 (1984); (44) W. G. Guder and G. Wirthensohn, in "Biochemistry of Kidney Functions" (F. Morel, ed.), p. 95. Elsevier, Amsterdam, 1982; (45) G. Wirthensohn and W. G. Guder, in "Biochemistry of Kidney Functions" (F. Morel, ed.), p. 119. Elsevier, Amsterdam, 1982; (46) H. Kawashima, S. Torikai, and K. Kurokawa, *Proc. Natl. Acad. Sci. U.S.A.* **78**, 1199 (1981); (47) G. Wirthensohn and W. G. Guder, *Pfluegers Arch.* **404**, 94 (1985); (48) W. G. Guder, S. Pürschel, A. Vandewalle, and G. Wirthensohn, *J. Clin. Chem. Clin. Biochem.* **22**, 129 (1984); (49) W. G. Guder, J. Hallbach, G. Wirthensohn, R. Linke, E. Fink, and W. Müller-Esterl, in "Molecular Nephrology" (Z. Kovacevic and W. G. Guder, eds.), p. 377. de Gruyter, Berlin, 1987; (50) I. Marchetti, S. Roseau, and F. Alhenc-Gelas, *Kidney Int.* **31**, 744 (1987); (51) D. Chabardès, M. Montégut, M. Imbert-Teboul, and F. Morel, *Mol. Cell. Endocrinol.* **37**, 263 (1984); (52) M. Le Hir and U. C. Dubach, *FEBS Lett.* **127**, 250 (1981); (53) H. Schmid, *Basic Appl. Histochem.* **28**, 221 (1984); (54) S. A. Ernst, *J. Cell Biol.* **66**, 586 (1975).

After oxaloacetate is destroyed by addition of NaOH and heating, a solution is added containing reagents which catalyze the following reaction:

$$\text{Citrate} \longrightarrow \text{acetate} + \text{oxaloacetate} \underset{\text{NADH}}{\overset{\text{malate}}{\rightleftarrows}} \text{NAD}^+$$

Excess NADH is destroyed by hydrochloric acid and the following cycle is carried out for 60 min:

$$\text{Malate} \underset{\text{malate dehydrogenase}}{\overset{\text{NAD}^+}{\rightleftarrows}} \text{oxaloacetate} \quad \text{NADH} \underset{\text{alcohol dehydrogenase}}{\overset{\text{ethanol}}{\rightleftarrows}} \text{acetaldehyde}$$

The cycling enzymes are malate dehydrogenase and alcohol dehydrogenase. The cycle is stopped and malate measured in the indicator reaction with the malate dehydrogenase and aspartate aminotransferase reagent:

$$\text{Malate} \underset{\text{NAD}^+}{\overset{\text{oxaloacetate}}{\rightleftarrows}} \text{NADH} \underset{}{\overset{\text{glutamate}\;\;\text{aspartate}}{\rightleftarrows}} \text{2-oxoglutarate}$$

The fluorescence of NADH is then measured.

Similar cycles can be obtained with $NADP^+/NADPH$, which is reported to be more sensitive[26] and can provide amplification rates of 60,000/hr.[49]

ATP[49]. The protocol for the measurement of ATP uses an amplification by the $NADP^+/NADPH$ cycle.

Tissue enzyme activities and tissue NADPH are destroyed by heating and subsequent acidification of the sample. Then the following reaction is carried out:

$$\text{Glucose} + \text{ATP} \rightarrow \text{glucose 6-phosphate} + \text{ADP}$$
$$\text{Glucose 6-phosphate} + \text{NADP}^+ \rightarrow \text{6-phosphogluconate} + \text{NADPH}$$

The excess $NADP^+$ is destroyed with NaOH and an aliquot of the sample is added to the cycling reagent for 60 min according to the reaction

$$\text{NH}_4^+ + \text{2-oxoglutarate} \underset{}{\overset{\text{Glutamate}\;\;\text{NADP}^+}{\rightleftarrows}} \text{NADPH} \underset{}{\overset{\text{glucose 6-phosphate}}{\rightleftarrows}} \text{6-phosphogluconate}$$

In the last step 6-phosphogluconate is measured with excess $NADP^+$ and 6-phosphogluconate dehydrogenase by determination of fluorescence of NADPH. As can be seen from Table I, a wide variety of enzymes and substrates have been determined in defined nephron segments with enzymatic cycling.

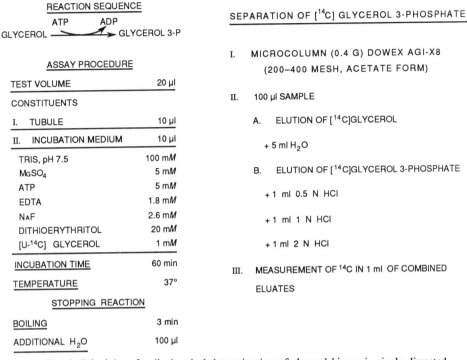

FIG. 3. Principles of radiochemical determination of glycerol kinase in single dissected nephron segments.

Fluorometry

Enzymes with relatively high activities can be measured in pooled samples without enzymatic cycling by direct fluorometry of NADH and NADPH (Table I). Moreover, the method introduced by Mead et. al.[50] allows the determination of hydrolases by fluorimetric measurement of liberated methylumbelliferon at 365/450 nm.[51]

Photometry

Only a few enzymes have been quantitated by direct photometry, using indicator reactions with high absorption coefficients[52] (see Table I).

[50] J. A. R. Mead, J. N. Smith, and R. T. Williams, *Biochem. J.* **61**, 569 (1955).
[51] M. Le Hir and U. C. Dubach, *Histochemistry* **63**, 245 (1979).
[52] B. Höhmann, R. Zwiebel, R. Yamagata, and R. Kinne, *Pfluegers Arch.* **312**, 110 (1969).

Radiochemical Procedures

High sensitivity has been obtained with the use of radiochemical methods. Thus the elegant studies of Morel and co-workers[34,38] on the localization of hormone-dependent adenylate-cyclase along the nephron were carried out by measuring [^{32}P]cAMP formed during incubation of single nephron segments with [α-^{32}P]ATP.[28] The high sensitivity of the assay was achieved by high specific activity of the precursor and a small incubation volume.

Furthermore, labeled substrates have been applied successfully to determine enzyme activities,[33,35,46,53-60] triacylglycerol synthesis,[39] and CO_2[41,42] production from different renal substrates in defined nephron segments.

The method shown in Fig. 3 is based on the conversion of the substrate glycerol catalyzed by glycerol kinase.[46,61] Substrate and product are separated by ion-exchange chromatography on small columns. The principle of this assay can be used for the study of a variety of enzyme activity measurements.[33,46,53]

Luminometric Methods

Certain molecules can release energy as light emission. This process is called chemiluminescence.[62] The rate of light intensity depends on the rate of the oxidation reaction and thus gives a direct indication of the concentration of the reacting molecule(s).

More recently bioluminescence techniques catalyzed by luciferases found in bacteria, fish, fungi, etc., have been applied to the measurement of picomolar quantities of ATP and NAD(P)H.[60,63]

---

[53] A. I. Katz, A. Doucet, and F. Morel, *Am. J. Physiol.* **237**, F114 (1979).
[54] A. Doucet and A. I. Katz, *Am. J. Physiol.* **242**, F346 (1982).
[55] A. Ait-Mohamed, S. Marsy, C. Barlet, C. Khadouri, and A. Doucet, *J. Biol. Chem.* **261**, 12526 (1986).
[56] L. E. Garg and N. Narang, *Can. J.Physiol. Pharmacol.* **63**, 1291 (1985).
[57] M. Ben Abdelkhalek, J. Lee, C. Barlet, and A. Doucet, *J. Membr. Biol.* **89**, 225 (1986).
[58] W. Schulz, H. Siebe, and K. Hierholzer, *in* "Molecular Nephrology" (Z. Kovacevic and W. G. Guder, eds.), p. 361. de Gruyter, Berlin, 1987.
[59] B. A. Jackson, R. M. Edwards, and T. P. Dousa, *Kidney Int.* **18**, 512 (1980).
[60] W. G. Guder, S. Pürschel, and G. Wirthensohn, *Hoppe-Seyler's Z. Physiol. Chem.* **364**, 1727 (1983).
[61] E. A. Newsholme, J. Robinson, and K. Taylor, *Biochim. Biophys. Acta* **132**, 338 (1967).
[62] M. A. DeLuca (ed.). this series, Vol. 57.
[63] W. G. Guder, S. Pürschel, A. Vandewalle, and G. Wirthensohn, *J. Clin. Chem. Clin. Biochem.* **22**, 129 (1984).

A. Tubule segments lyophilized in 5 μl Hanks' stored at -80°
B. Incubation
   5 μl tubule (H₂O added) | 3HOB + NAD⁺ ⟶ AcAc + NADH |
   45 μl reagent containing
   + D,L-3HOB   10 mM   time: 15 min
   NAD⁺   5 mM   temp.: 25°
   Dithiothreitol   1 mM   stop: 3-min boiling
   Tris, pH 8.2   40 mM   Addition of 50 μl 100 mM
                                     Phosphate, pH 6.8

C. Detection of NADH
   Reaction
   1. NADH + FMN —FMN Red.→ NAD⁺ + FMNH₂

   2. FMNH₂ + Myristinaldehyde —Luciferase→ FMN + Myristinol + light

   50 μl sample; + 60 μl Luciferase reagent (Boehringer, Mannheim);
   30-sec delay, 30-sec assay in Luminometer SAI 3000

FIG. 4. Principles of bioluminescence determination of 3-hydroxybutyrate dehydrogenase in single dissected nephron segments. 3HOB, 3-Hydroxybutanoate; AcAc, acetoacetate; FMN, flavinmononucleotide.

Reduced flavin mononucleotide-dependent luciferase from *Photobacterium fisherii* has been used to measure NAD(P)⁺-dependent enzymes in isolated nephron structures of rabbit,[60] human,[60] and mouse and rat[63] kidney. Figure 4 shows an example of the different steps and reagents for determination of 3-hydroxybutyrate dehydrogenase.

Radioimmunoassay

The radioimmunoassay technique was applied to measure kininogenin (kallikrein) activity in rabbit[64] and mouse[65] nephron using dog plasma kininogen as substrate. The bradykinin radioimmunoassay was performed using rabbit antibodies[64] in microtiter vessels.[65] The same technique can be applied to measure the activity of kininases[66] and kininogen.[67]

[64] J. Marchetti, M. Imbert-Teboul, F. Alhenc-Gelas, J. Allegrini, J. Menard, and F. Morel, *Pfluegers Arch.* **401**, 27 (1984).
[65] W. G. Guder, J. Hallbach, G. Wirthensohn, R. Linke, E. Fink, and W. Müller-Esterl, in "Molecular Nephrology" (Z. Kovacevic and W. G. Guder, eds.), p. 377. de Gruyter, Berlin, 1987.
[66] J. Marchetti, S. Roseau, and F. Alhenc-Gelas, *Kidney Int.* **31**, 744 (1987).
[67] J. Hallbach, G. Adams, G. Wirthensohn, and W. G. Guder, *Biol. Chem. Hoppe-Seyler* **368**, 1151 (1987).

Enzyme Inhibition

Peroxisomal enzymes lead to the production of H_2O_2 during oxidation of their resepctive substrates. Le Hir and Dubach[68] developed a microassay based on the inhibition of exogenous catalase by 3-amino-1,2,4-triazole in the presence of a source of H_2O_2. The percentage catalase inhibited in a given time is proportional to the activity of the oxidase.

Histochemical Staining

Only a few enzymes were quantitated by light or electron microscopic staining procedures. They correlated well with results obtained with other techniques when measuring succinate dehydrogenase[69] and Na^+,K^+-ATPase.[70]

[68] M. Le Hir and U. C. Dubach, *FEBS Lett.* **127**, 250 (1981).
[69] H. Schmid, *Basic Appl. Histochem.* **28**, 221 (1984).
[70] S. A. Ernst, *J. Cell Biol.* **66**, 586 (1975).

[19] Endocytosis and Lysosomal Hydrolysis of Proteins in Proximal Tubules

By THOMAS MAACK and C. HYUNG PARK

Introduction

Proteins and polypeptides are extensively endocytosed by proximal tubular cells of the kidney in all animal species tested to date. This process is of fundamental physiological importance for the plasma homeostasis of several circulating low-molecular-weight proteins (LMWP) and polypeptide hormones. Indeed, a great fraction of the metabolic clearance rate of these substances takes place in the kidney by a process that involves filtration with subsequent tubular uptake from the lumenal side by an adsorptive endocytic process followed by lysosomal hydrolysis of the absorbed protein.[1-3] Even for larger proteins, such as albumin, whose filtration is markedly hindered by the normal glomerular barrier, the low filtered loads are almost completely absorbed by the proximal tubular

[1] T. Maack, *Kidney Int.* **16**, 251 (1979).
[2] T. Maack, C. H. Park, and M. J. F. Camargo, *in* "Physiology and Pathology of Electrolyte Metabolism" (D. Seldin and G. Giebish, eds.), p. 1773. Raven, New York, 1984.
[3] D. A. Wall and T. Maack, *Am. J. Physiol.* **248**, C12 (1985).

epithelium such that minimal amounts are excreted into the urine.[3] When glomerular permselectivity is increased by disease processes, one of the main consequences is a major increase in the filtration of albumin with consequent albuminuria, increased absorption rates of albumin, and eventually deposition of protein absorption droplets within tubular cells.[1,3]

Synopsis of the Process of Tubular Absorption and Metabolism of Proteins

Some of the overall properties and mechanisms of tubular uptake and metabolism of proteins are very briefly summarized in this section. The reader is referred to review articles for inclusive description and literature references on the subject.[1-3]

Endocytic uptake of filtered proteins takes place mostly in proximal tubular cells. Positively charged groups of proteins bind to negatively charged lumenal cell surfaces of proximal tubular cells, resulting in the eventual concentration of protein in clathrin-coated regions of the cell membrane which are localized at the bottom of the microvilli. Net charge of the protein molecule as well as the particular morphological localization of the endocytic regions at the bottom of long and narrowly spaced microvilli, rather than specific receptors, are the major determinants of the selectivity for tubular uptake of proteins. Endocytosis of proteins is a saturable process with very high capacity and relatively low affinity compared to normal filtered loads of protein. Proteins of same net charge (cationic or anionic) but not opposite charge compete for the absorption process. In addition to charge, size, conformation, and possibly nature of the cationic moiety (lysine or arginine) of the protein are the main determinants of the uptake kinetics.

Macromolecules that do not bind to the cell membrane, such as neutral dextran or inulin, are absorbed exclusively by fluid-phase endocytosis and may, therefore, be used as markers for this process (see below). Similarly, horseradish peroxidase, because of its ease of visualization by histochemical techniques, has been traditionally used as a marker for fluid-phase endocytosis although unquestionably there is some binding of this protein to cell membranes. For most proteins, absorption by fluid-phase endocytosis is but a small fraction of its transport by adsorptive endocytosis. In most instances, uptake of proteins from the peritubular side is quantitatively negligible compared to its endocytosis by the lumenal side of proximal tubular cells.

Invagination of the endocytic regions of the cell membrane containing the adsorbed proteins results in their transport to an endosomal compartment. Eventually endosomes fuse with lysosomes and the absorbed protein is hydrolyzed to completion within these cell organelles. The resulting amino acids are released across basolateral membranes back to the circulation, minimal amounts being released across the lumenal membrane to the urine. As in other endocytic processes the mechanisms of intracellular transport and fusion of endosomes with lysosomes are still very poorly understood. In most instances lysosomal hydrolysis is the final fate of absorbed proteins. Similarly to the uptake process, lysosomal hydrolysis is a process of high capacity and, therefore, absorbed proteins do not easily accumulate within proximal tubular cells. The rate of hydrolysis depends on the nature of the protein and, in turn, on the resistance of the protein to lysosomal acid hydrolases. At steady state, lysosomal hydrolysis must balance endocytic uptake; if not, there would be an infinite accumulation of protein within the cells. Morphologically detectable accumulation of endogenous proteins within renal tubular cells occurs in pathological conditions in which there is a very large excess of filtered load of proteins (e.g., albumin in the nephrotic syndrome, Bence Jones protein in multiple myeloma, lysozyme in myelocytic leukemia) or in conditions in which acidification of the intralysosomal milieu is impaired (e.g., anoxia, administration of lysosomotropic weak bases).

Methodological Approaches

Several methodological approaches may be used to study endocytosis of proteins, polypeptides, and macromolecules by proximal tubular cells. These approaches encompass (1) morphological techniques including histochemical, immunological, or autoradiographic visualization of the absorbed protein with the optical[4] or electron microscope,[5] (2) differential centrifugation techniques to isolate organelles involved in the intracellular transport and metabolism of absorbed proteins,[6-8] (3) clearance techniques in intact animals to study uptake rates of filtered low-molecular-weight proteins and polypeptides,[9] (4) clearance and biochemical techniques in isolated perfused rat kidney preparations to determine both endocytic uptake and lysosomal hydrolysis of absorbed proteins,[10-12] (5) micropunc-

[4] W. Straus, *J. Cell. Biol.* **21,** 295 (1964).
[5] A. B. Maunsbach, *J. Ultrastruct. Res.* **15,** 197 (1966).
[6] A. B. Maunsbach, this series. Vol. 31, p. 330.
[7] J. Hjelle, J.-P. Morin, and A. Trouet, *Kidney Int.* **20,** 71 (1981).
[8] F. Bode, K. Baumann, and R. Kinne, *Biochim. Biophys. Acta* **433,** 299 (1976).
[9] J. F. Harrison and A. D. Barnes, *Clin. Sci.* **38,** 533 (1970).
[10] T. Maack, *Kidney Int.* **30,** 142 (1986).
[11] M. J. F. Camargo, B. E. Sumpio, and T. Maack, *Am. J. Physiol.* **247,** F656 (1984).
[12] V. Johnson and T. Maack, *Am. J. Physiol.* **233,** F185 (1977).

ture and microperfusion of proximal tubules in the rat kidney *in vivo* (physiological[13,14] and morphological[5,15] techniques), and (6) microperfusion of isolated segments of proximal tubules of the rabbit using physiological, biochemical, and morphological techniques.[16-22]

The choice of techniques depends essentially on the experimental question. To study in detail the intracellular pathways of tubular absorption and metabolism of proteins from a qualitative or semiquantitative point of view, morphological and differential centrifugation techniques using either whole animal, isolated perfused rat kidney, or microperfusion techniques have provided essential information on the organelles, enzymes, and cell processes involved in endocytosis and metabolism of filtered proteins. To determine the quantitative aspects, kinetics, and pathways of tubular absorption of LMWP and polypeptides, clearance techniques in intact animals and in the isolated perfused rat kidney have been shown to provide very useful information. In this regard, studies in filtering and nonfiltering isolated perfused kidney preparations provided the broader insights on the overall process of endocytic uptake and lysosomal metabolism of these substances. More direct information on cellular mechanisms of endocytosis can be obtained by microperfusion of proximal tubules *in vivo* or in isolated nephron segments. For proteins and macromolecules that are poorly filtered by the kidney, the latter technique is mandatory since errors in the determination of glomerular sieving coefficients of these substances are of such magnitude as to make the results obtained in whole organ experiments practically uninterpretable from a quantitative point of view.[1,2]

In the present chapter we will consider methods of determining and analyzing endocytosis and metabolism of proteins in isolated perfused proximal tubules of the rabbit. For a detailed description of the other techniques the reader is directed to the references cited above.

---

[13] M. A. Cortney, L. L. Sawin, and D. D. Weiss, *J. Clin. Invest.* **49**, 1 (1970).
[14] C. Cojocel, M. Franzen-Sieveking, C. Beckmann, and K. Baumann, *Pfluegers Arch.* **390**, 211 (1981).
[15] E. I. Christensen, *Kidney Int.* **10**, 301 (1976).
[16] J. E. Bourdeau, F. A. Carone, and C. E. Ganote, *J. Cell Biol.* **54**, 382 (1971).
[17] J. E. Bourdeau, E. R. Y. Chen, and F. A. Carone, *Am. J. Physiol.* **225**, 1399 (1973).
[18] C. H. Park and T. Maack, *J. Clin. Invest.* **73**, 767 (1984).
[19] C. H. Park, *Am. J. Physiol.* **255**, F520 (1988).
[20] D. R. Peterson, F. A. Carone, S. Oparil, and E. I. Christensen, *Am. J. Physiol.* **242**, F112 (1982).
[21] L. Larsson, W. L. Clapp III, C. H. Park, and C. G. Tisher, *Am. J. Physiol.* **253**, F95 (1987).
[22] K. M. Madsen and C. H. Park, *Am. J. Physiol.* **252**, F1290 (1987).

Endocytic Uptake and Lysosomal Hydrolysis of Proteins by Isolated Perfused Proximal Convoluted Tubules of the Rabbit

Procedure for Perfusion of Proximal Convoluted Tubules

The general procedures and instrumentation for perfusion of proximal convoluted tubules are similar to those described by Burg et al.,[23] and are described in detail in other chapters of this book. For studies on protein uptake, three pipets are used on the perfusion side: a holding pipet into which one end of the tubule is aspirated, a perfusion pipet which is centered inside the holding pipet and filled with the tubule perfusate, and an exchange pipet advanced as far as possible into the perfusion pipet to exchange the tubule perfusate during the experiment. In some experiments the exchange pipet may be dispensable. The perfusion pipet is connected to a gravity reservoir whose height determines the lumenal perfusion rate. Unless the influence of lumenal flow rate on protein absorption is to be studied, a perfusion rate of 10–15 nl/min is recommended for proximal convoluted tubules. Two pipets are used on the collection side: a holding pipet into which the other end of the tubule is aspirated and a constant-bore collection pipet (i.d. = 200 μm). The constant-bore pipet may be further calibrated using a solution containing a known concentration of a radioisotope (e.g., [^{14}C]inulin). The holding pipet is filled with colored mineral oil for better visualization of the collected fluid.

Siliconization and Protein Coating of Glass Micropipet

For studies on protein uptake in isolated nephron segments it is essential to known precisely the concentration of the protein in the delivered perfusion fluid as well as the concentration of the protein in collected tubular fluid. A major problem in this regard is binding of protein to the perfusion and collection micropipets. This is particularly true when low concentrations of protein, similar to those which normally reach the tubular fluid *in vivo,* are microperfused. Siliconization alone does not totally preclude this problem. In our studies on albumin absorption in isolated perfused proximal convoluted tubules of the rabbit, in addition to siliconization it was necessary to apply a protein coat to the collection pipets.[18] All glass pipets used in the microperfusion system are first siliconized with a 5% solution of Siliclad. This solution is aspirated into the pipets several times for approximately 30 min. Then, the siliconized pipets are washed with distilled water and dried at 100° for 30 min. Since in the siliconized perfusion pipet the volume of perfusion fluid greatly exceeds the internal

[23] M. B. Burg, J. Grantham, M. Abramow, and J. Orloff, *Am. J. Physiol.* **210**, 1293 (1966).

surface area of the micropipet, binding of albumin to the glass does not substantially decrease the concentration of this protein in the solution. Nevertheless at the lowest concentrations used (4.6 µg/ml), the concentration of albumin in fluid delivered from the pipet may be as much as 10% lower than that in the original perfusion fluid. Therefore it is essential to measure the concentration of the protein in the delivered perfusion fluid. This is accomplished at the end of the experiment by drastically increasing perfusion rate to 200–500 nl/min and determining the concentration of the protein in the fluid collected with the constant-bore pipet in the collection side. Independent determination of the concentration of protein in the tubule perfusate can be performed by delivering the perfusion fluid from the perfusion pipet directly into an oil droplet either before or after the experiment. The perfusion fluid is then collected by the constant-bore micropipets from the oil droplet.

More troublesome is the protein binding to the collection micropipet because of the low fluid volume/glass surface area ratio of the constant-bore micropipet. For example, when collection pipets were only siliconized we recovered only 3.5 ± 1% of albumin from a solution containing 4.6 µg/ml albumin.[18] Only at concentrations of albumin greater than 1 mg/ml does binding of this protein to collection micropipets become negligible. This problem can be avoided by precoating the collection side pipets with albumin or another protein. For this purpose the siliconized collection side micropipets are equilibrated with a solution containing 10 mg/ml albumin or 2 mg/ml insulin for 30 min. Then, the pipets are washed and dried, as in the procedure for siliconization. With this method more than 95% of albumin at a concentration of 4.6 µg/ml is recovered from the constant-bore micropipet. The binding of protein to glass micropipets is so strong that the same protein-coated micropipets may be reused several times, even after extensive washing with distilled water.

Labeling of Proteins

We prefer to label proteins with tritium using the reductive methylation procedure of Tack et al.[24] or with ^{125}I, using an immobilized preparation of lactoperoxidase and glucose oxidase.[25] Using the reductive methylation procedure we labeled albumin with tritium to a specific activity of approximately 200 Ci/mmol and using Enzymobeads (Bio-Rad Laboratories, Richmond, CA), we labeled albumin with ^{125}I to a specific activity of 40

---

[24] F. F. Tack, J. Dean, D. Eliat, P. E. Lorenz, and A. N. Schechter, *J. Biol. Chem.* **285**, 8842 (1980).
[25] Bio-Rad Laboratories, "Enzymobeads," Tech. Bull. 1071. Bio-Rad Lab., Rockville Center, New York.

Ci/mmol.[18] The fundamental criterion for choosing a particular label and labeling procedure is that the labeled protein should conserve as close as possible the physicochemical characteristics of the native protein since endocytosis of proteins by proximal tubular cells is dependent on these characteristics. In this sense, the reductive methylation procedure is the most reliable. For albumin, the reductive methylation procedure is carried out at 4° in 0.32 ml of a solution containing crystalline bovine albumin, HCHO, and ^{3}HNaBH$_4$ (60 Ci/mmol) at the following molar ratios: HCHO/lysine residues of albumin, 2.4; ^{3}HNaBH$_4$/HCHO, 0.2. Labeled albumin is separated from unreacted material by using disposable G25 Sephadex columns preequilibrated and eluted with phosphate buffer. The labeled albumin is extensively dialyzed for 3 days against a 100-fold excess phosphate buffer in the cold room with repeated changes of the buffer solution to remove the remaining free label. The dialyzed labeled albumin is aliquoted and stored at $-70°$. Once thawed, the labeled protein should not be refrozen and reused since repeated freezing and thawing denaturates protein molecules. In our hands, the labeled albumin is stable for at least 1 year. Analysis of the labeled albumin both by gel chromatography and isoelectric gel focusing shows essentially that a single labeled protein is obtained by this procedure.[18] In the case of albumin, mild procedures for labeling this protein with ^{125}I are also adequate since experiments in isolated perfused proximal convoluted tubules of the rabbit showed that the nature of the label (^{3}H or ^{125}I) did not influence tubular absorption or metabolism of albumin.[18]

Composition of Lumenal Perfusion and Bathing Solutions

The tubule perfusate and bath solutions have identical concentration of the following salts and organic solutes (in millimolar units): NaCl (105), NaHCO$_3$ (25), sodium acetate (10), NaH$_2$PO$_4$ (0.92), KCl (4.84), KH$_2$PO$_4$ (0.6), CaCl$_2$ (1.97), MgSO$_4$ (0.81), D-glucose (8.0), and L-alanine (5.0). The pH of the solutions is 7.4 when equilibrated with 95% O$_2$–5% CO$_2$. In addition, the bath solution contains 6.5 g/dl of fraction V bovine albumin. Labeled protein and amounts of unlabeled protein to make up a desired concentration in lumenal fluid are added to the tubule perfusate. Since concentrations of labeled protein are measured in nanoliter samples of collected tubular fluid, enough radioactivity counts must be added to the perfusion solution to obtain accurate results. We obtained good radioactivity counting resolution by adding to the tubule perfusate 55,000 cpm/μl of [^{3}H$_3$C]albumin (70 μCi/ml, with a counting efficiency of 35%). The tubule perfusate containing the labeled and unlabeled protein is extensively dialyzed against a solution with the same composition as the bath solution to remove label not associated to the protein and any excess of phosphate or other ions and substances used in the labeling procedure. After dialysis, [*carboxy*-^{14}C]inulin, a marker of fluid reabsorption, leakiness of the preparation, and fluid-phase endocytosis (see below) is added to the tubule perfusate to a concentration of 5–20 μCi/ml. Several modifications of the perfusion and/or bath solutions (e.g., addition of inhibitors, removal of ions, addition of competing proteins) may be performed to investigate particular phenomena which may influence endocytosis and lysosomal hydrolysis of proteins. In all instances, however, the osmolality of the perfusion fluid and bath solution should be close to 290 mOsmol/kg H_2O and within 3 mOsm/kg H_2O of each other.

General Protocol

S_1 or other segments of proximal tubules, measuring 0.6–2.0 mm in length are dissected and perfused at 37° at perfusion rates of 10–15 nl/min. A perfusate without inulin or protein may be used for the equilibration period (see below). The bath solution is constantly gassed with 95% O_2–5% CO_2. To determine protein uptake, after 30–40 min of perfusion (equilibration period) the tubule perfusate is changed to a perfusion solution containing inulin, the labeled protein, and the desired concentration of unlabeled protein. Three to seven 10-min collection periods may be performed in which the tubular fluid is collected by the constant-bore collection pipet and the entire bathing solution is collected and exchanged with several washes of fresh bath solution prewarmed to 37°. At the end of all sampling periods the rate of perfusion is drastically increased to 200–500 nl/min to preclude any significant tubular uptake of protein or fluid reabsorption. Triplicate samples of this rapidly perfused tubular fluid are collected and determination of inulin and protein in the collected tubular fluid is taken as the concentration of these substances delivered from the perfusion pipet. After this step the perfusate is again changed to the control perfusate and the tubules are thoroughly perfused with this solution to remove any protein or inulin remaining in the tubular lumen. The bath solution is also exchanged several times to eliminate contamination from the peritubular side. The perfusion micropipet is retracted, and the tubule length is measured with the aid of an eye piece micrometer. Then, the tubule is aspirated into the holding pipet of the perfusion side and delivered in a vial containing 0.1 N HNO_3. Extraction in this acid is carried out overnight and the radioactivity is determined by standard procedures.

Figure 1 illustrates results of experiments using the above protocol with [$^3H_3$C]albumin. Panel A shows the disappearance of radioactivity from lumenal perfusate, accumulation of radioactivity in tubule cells, and ap-

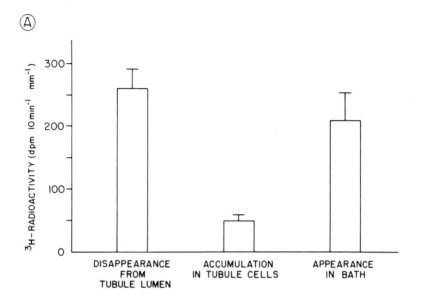
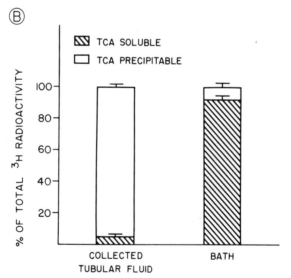

FIG. 1. Fate of absorbed [^3H_3C]albumin in isolated perfused proximal convoluted tubules of the rabbit. Six tubules were perfused with an albumin concentration of 0.03 mg/ml for 121 ± 4 min. Efflux of radioactivity from the tubule lumen and appearance of radioactivity in the bath solution were determined over the entire span of perfusion. At the end of the experiment, the radioactivity remaining in the tubules was determined (see text). (A) The rate of efflux of ^3H radioactivity from the lumen, the rate of accumulation of radioactivity in tubule cells, and in the bath solution were determined in each tubule. Results (mean ± SE)

pearance of radioactivity in the bathing solution. Panel B depicts the trichloroacetic acid (TCA)-precipitable and soluble radioactivity in collected tubular fluid and bath solution.

Criteria for Adequacy of Experiment

Artifactual results may be obtained if the tubules are leaky, either because the seals in the holding pipets at the perfusion and collection sides are imperfect or because the tubule has been damaged during the dissection procedure. Inulin appearance in the bath solution is the fundamental test for leaks. In a good preparation the appearance of inulin should be negligible and results should be discarded if the amount of inulin appearing in the bath solution is more than 5% of the perfused load of this marker. In our studies on albumin endocytosis, the inulin leak was $1.6 \pm 0.4\%$ of perfused load of [^{14}C]inulin.[18]

In nonleaky tubules, the viability of the preparation can be tested by measuring fluid reabsorption with inulin. In proximal convoluted tubules, fluid reabsorption averages approximately 1 nl min^{-1}/mm tubule length, and tubules with less than half of this average value should be discarded, except when the influence of fluid reabsorption on protein uptake is being determined.[18] Normally proximal tubules have a small lumenal negative potential which disappears in poor preparations. Therefore, it is useful, albeit not essential, to measure transtubular electrical potential differences. The methodology of determining this parameter is described by Burg and Orloff,[26] by Friedman *et al.*,[27] and in other chapters of this volume.

Recovery of perfused ^{3}H or ^{125}I radioactivity must be complete, i.e., the cumulative amount of radioactivity in collected tubular fluid plus cumulative radioactivity in the bath solution plus radioactivity in the proximal

[26] M. B. Burg and J. Orloff, *Am. J. Physiol.* **219**, 1714 (1970).
[27] P. A. Friedman, J. F. Figueiredo, T. Maack, and E. E. Windhager, *Am. J. Physiol.* **240**, F558 (1981).

---

are expressed as disintegrations per minute per 10 min per millimeter of tubule length to normalize for differences in perfusion time and tubule length. The amount of radioactivity accumulated in tubule cells is relatively small compared to that appearing in the bath solution. Note that the sum of radioactivity in tubule cells and bath solution nearly equals the disappearance of radioactivity from the lumen, indicating adequate recovery in these experiments (see text). (B) Corresponding values for TCA-soluble and TCA-precipitable radioactivity in collected tubular fluid and bath solution. Note that the bulk of the radioactivity in collected tubular fluid is TCA precipitable, indicating lack of significant lumenal hydrolysis of albumin, and that the bulk of radioactivity in the bath solution is TCA soluble, indicating that absorbed albumin is hydrolyzed within proximal tubular cells and the metabolic products are released to the bath solution. (Reproduced from Park and Maack[18] by permission.)

tubule at the end of perfusion must equal the total amount of microperfused ^{3}H or ^{125}I radioactivity, within the experimental errors of the measurements. Panel A of Fig. 1. shows that in the experiments with [^{3}H$_3$C]albumin the sum of radioactivity accumulated in kidney tissue plus the radioactivity appearing in the bath is practically identical to the radioactivity that disappeared from the lumenal perfusate, indicating a near complete recovery of radioactivity.

For studies on endocytic uptake it is essential to verify the integrity of the labeled protein in the lumenal fluid since some proteins and polypeptides may be partly hydrolyzed by brush border proteases. This is best accomplished by analyzing the nature of ^{3}H or ^{125}I radioactivity in the collected tubular fluid with the TCA-precipitation method or with chromatographic methods (see section Determination of Intact Protein and Metabolites).

Finally, in determining endocytic uptake of proteins it is important to assure that steady state conditions prevail during the course of the experiments, otherwise the results are not quantitatively interpretable. For this purpose, at least two sequential determinations with the same concentration of protein must be performed and the results should be the same within the experimental errors of the determinations.

Measurement and Calculation of Fluid and Protein Absorption

Fluid reabsorption (J_v, in nl min^{-1}/mm tubule length) is determined by the equation $J_v = V_c[(C/P)_{\text{inulin}} - 1]L$, where V_c is the collection rate in nanoliters per minute, $(C/P)_{\text{inulin}}$ is the ratio of [^{14}C] inulin (in cpm) per nanoliter of collected fluid (C) and perfusate standard (P), and L is the length of the tubule in millimeters.

Protein absorption (J_{pr}, in ng min^{-1}/mm tubule length) is determined by the equation $J_{pr} = (V_p P_{pr} - V_c C_{pr})/L$ where V_c is the fluid collection rate (in nl/min), V_p is the fluid perfusion rate (in nl/min) determined by $V_p = V_c(C/P)_{\text{inulin}}$, and P_{pr} and C_{pr} are the protein concentrations in tubule perfusate and collected fluid, respectively. Protein concentrations may be calculated from their specific activities, namely the ratio of counts per minute per milliliter and counts per minute per milligram protein in perfusion fluid.

Table I provides the values of J_v and J_{albumin} and several of the intermediate parameters described in the above equations in a series of six tubule perfusions with 0.03 mg/ml of [^{3}H$_3$C]albumin.

Figure 2 shows a complete endocytic uptake curve for albumin by proximal convoluted tubules of the rabbit at albumin concentrations ranging from near physiological (0.0012 mg/ml) to very high (10 mg/ml) concentrations in tubular fluid.

TABLE I
[^3H_3C]ALBUMIN ABSORPTION AND FLUID REABSORPTION IN ISOLATED PERFUSED PROXIMAL CONVOLUTED TUBULES OF THE RABBIT[a]

Experiments	V_p (nl/min)	P_{inulin} (dpm/nl)	C_{inulin} (dpm/nl)	$(C/P)_{inulin}$ (dpm/nl)	$P_{albumin}$ (mg/ml)	$C_{albumin}$ (mg/ml)	$(C/P)_{albumin}$ (mg/ml)	$J_{albumin}$ (ng/min · mm)	J_v (nl/min · mm)
1	11.3	9.0	10.4	1.15	0.030	0.031	1.04	0.0261	1.29
2	11.1	9.1	10.2	1.12	0.030	0.030	1.00	0.0323	1.11
3	13.3	9.0	9.4	1.04	0.030	0.029	0.97	0.0270	0.56
4	11.9	11.8	14.3	1.22	0.030	0.031	1.03	0.0250	1.04
5	11.1	11.8	12.6	1.07	0.030	0.031	1.03	0.0147	0.72
6	10.1	11.9	12.5	1.05	0.030	0.031	1.02	0.0134	0.74
Mean	11.5	10.4	11.6	1.11	0.030	0.0305	1.02	0.0231	1.91
± SE	± 0.4	± 0.6	± 0.8	± 0.03		± 0.0003	± 0.01	± 0.0030	± 0.11

[a] Proximal convoluted tubules of the rabbit were perfused as described in the text. V_p, Perfusion rate; P_{inulin} and C_{inulin}, [^{14}C]inulin concentrations in tubule perfusate and collected tubular fluid, respectively; $(C/P)_{inulin}$, collected tubular fluid/tubule perfusate concentration ratio of [^{14}C]inulin; $P_{albumin}$ and $C_{albumin}$, concentrations of albumin in tubule perfusate and collected tubular fluid, respectively; $(C/P)_{albumin}$, collected tubular fluid/tubule perfusate concentration ratio of albumin; $J_{albumin}$, absorption rate of albumin; J_v, fluid reabsorption rate. Albumin concentrations in tubule perfusate and collected tubular fluid were calculated from the specific activity of [^3H_3C]albumin (see text). Reproduced from Park and Maack[18] by permission.

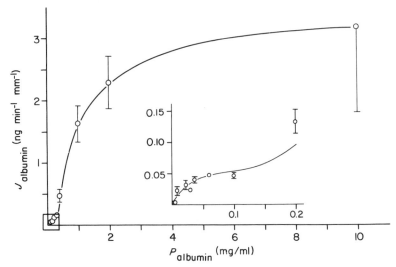

FIG. 2. Kinetics of albumin absorption in isolated proximal convoluted tubules of the rabbit. Albumin absorption rates ($J_{albumin}$) are plotted against tubule perfusate concentrations of albumin ($P_{albumin}$). Each of 57 tubules was perfused with a fixed $P_{albumin}$ ranging from 0.0012 to 10 mg/ml. Values are the mean ± SE of at least three tubules. The absorption curve has at least two components: (1) an overall high capacity–low affinity system which saturates at 3.7 ng min^{-1} mm^{-1} tubule length and has an apparent K_m of 1.2 mg/ml; (2) a low-capacity system (inset), which saturates at 0.0064 ng min^{-1} mm^{-1} tubule length, the near-physiological filtered loads of albumin in mammals, and has an apparent K_m of 0.0031 mg/ml. (Reproduced from Park and Maack[18] by permission.)

Determination of Fluid-Phase Endocytosis

Inulin does not bind to and does not permeate cell membranes. Consequently, inulin gains access to the cell interior exclusively by fluid-phase endocytosis and can be used to determine quantitatively this process in isolated perfused nephron segments. Experiments are performed as described above, using a higher concentration of inulin in the tubule perfusate (above 15 μCi/ml). This higher concentration is necessary to obtain adequate resolution of radioactivity counts since only a very small fraction of the perfused [^{14}C]inulin accumulates in tubular cells (see below). Tubules are perfused for approximately 60 min, and perfused fluid and bath solution are collected as described above to determine the rate of fluid reabsorption, the inulin load, and the tubular leakiness to inulin. At the end of perfusion the tubule is treated as described above and the radioactivity accumulated in tubular cells is measured. In a published series of experiments ($n = 4$), at a fluid perfusion rate of 11.5 nl/min, 9.7 ± 1.7 cpm/mm tubule length was detected after 60 min of perfusion with a perfusate containing 17 μCi/ml of [^{14}C]inulin.[18] This represents an uptake rate of less than 0.1% of the perfused load in 1 hr, or a fluid endocytosis rate of approximately 10 pl min^{-1} mm^{-1} tubule length. This value is less than 1/1000 of the perfused load of fluid and less than 1/100 of fluid reabsorption and albumin uptake by the tubules. The rate of fluid endocytosis was not altered by the presence of albumin in the tubule perfusate.[18] Consequently, fluid-phase endocytosis plays a negligible role in fluid reabsorption and albumin uptake in proximal convoluted tubules of the rabbit. In view of its small value, fluid-phase endocytosis is a difficult parameter to measure with precision. Nevertheless it provides two important points of information: (1) the degree by which tubular cell uptake of a particular protein is due to fluid-phase and absorptive endocytosis, and (2) the effect, if any, of the protein itself on the rate of fluid-phase endocytosis, which is also a measure of the rate of formation of endocytic vesicles.

Determination of Intact Protein and Metabolites

In most instances TCA precipitability is sufficient to establish whether the label in collected tubular fluid and bath solution remains associated with the protein or is in the form of metabolic products. Collected tubular fluid (100 to 200 nl) is dissolved in 0.1 ml of a solution with the same composition as the bath solution. Then the mixture is precipitated with an equivolume of ice-cold 10% TCA, and centrifuged in the cold (2000 rpm) for 20 min. The supernatant is removed and the precipitate is resuspended with 0.1 ml 10% TCA and recentrifuged. The same procedure is used with 0.1 ml of the experimental bath solution. Radioactivity in combined supernatants and precipitates is determined by standard procedures.

Panel B of Fig. 1 show the results of this method as applied to collected tubular fluid and bath solution of tubules perfused with [^{3}H$_{3}$C]albumin. As can be seen, practically all radioactivity in collected tubular fluid is TCA precipitable (intact labeled albumin) whereas the opposite is true in the bath solution. This indicates that there is negligible lumenal hydrolysis of albumin and that albumin absorbed by the tubular cells is metabolized and the resulting amino acids are delivered to the bath solution.

For more precise assessment of the nature of the label in collected tubular fluid and bath solution, and for the characterization of the radioactivity in tubule cells, chromatographic procedures may be employed. The particular chromatographic procedure to be used depends on the protein to be tested and therefore will not be described here. The reader is directed to Park[19] for an example of size-exclusion chromatographic analysis of radioactivity in collected tubular fluid, bath solution, and tubule using Sephadex G-100.

Summary

The techniques and basic protocols described above can be readily reproduced by investigators with experience in perfusion of isolated nephron segments. They can be modified and adapted by the investigator to address specific issues. In particular, isolated perfused nephron segments have also been successfully used for elucidation of biochemical and morphological aspects of endoyctosis and lysosomal hydrolysis of macromolecules, proteins, and polypeptides. The reader is directed to the references cited under Methodological Approaches in this chapter for a description of these techniques. Although studies on the endocytic uptake and metabolism of proteins and polypeptides using isolated perfused nephron segments have made significant inroads in our understanding of these fascinating and important biological processes, much remains to be learned. Hopefully, future uses of the technique will further advance our knowledge in this field.

Acknowledgment

Supported by the National Institutes of Health Research Grant DK-14241.

[20] Flux Measurements in Isolated Perfused Tubules

By JAMES A. SCHAFER and JAMES C. WILLIAMS, JR.

Solute fluxes are measured in isolated perfused tubule segments in much the same way that they are in any epithelium. The only additional considerations occur because of the small volume of the lumenal compartment. This imposes difficulties both in measuring solute or radioactive label concentrations in the fluid samples and in ensuring that the composition of the lumenal compartment is known and remains relatively constant along the length of the perfused segment.

General Considerations

Two types of fluxes can be measured: a net solute flux or a unidirectional flux. The net flux represents the total mass transport of a substance across the epithelium. The net flux is, therefore, the difference between the two unidirectional fluxes across the epithelium: the lumen-to-bath (absorptive) flux and the bath-to-lumen (secretory) flux. Because of the relatively much larger volume of the bathing solution compartment compared to the mass of the tissue and the lumenal compartment, it is impractical if not impossible to measure net fluxes from changes in the amount of the solute in the bathing solution. Thus the net flux is measured from the rate of appearance or disappearance of the solute in question from the lumen. The solute concentration is usually measured by some ultramicro-chemical method in samples of the perfusion solution and collectate (i.e., the fluid which exits the lumen at the collection side) in order to calculate the net flux. Alternatively, it is possible to determine the net solute movement using a radioactive label to measure solute concentration as described below.

If the solute is present on only one side of the tubular epithelium (lumenal perfusate or bathing solution), then the net flux measured chemically or isotopically must be identical to the unidirectional flux. Alternatively, one can add a radioactive label with the solute on one side of the membrane (cis side) and measure its rate of appearance on the opposite side (trans side). This can be done either when the solute concentration on the trans side is zero, or when it is present at a finite concentration. Which of the two methods is chosen depends on the demands of the particular experimental situation as well as on the questions being asked.

In the measurement of fluxes using radioactive labels it is essential to ascertain that the label being measured is associated with the solute of interest and not some metabolite thereof. Obviously, this is not a problem for individual elements such as ^{22}Na. However, considerable caution must be exercised in interpreting radioactive fluxes of metabolizable organic solutes. For example, when D[^{14}C] glucose is added to the solution bathing a proximal convoluted tubule in order to measure the bath-to-lumen flux, we have found using two-dimensional thin-layer chromatography of the collected fluid that only about 30% of the ^{14}C counts entering the lumen were associated with D-glucose and the rest of the counts appeared to be associated with lactate and pyruvate.[1] Therefore, in cases such as these one must use some ultramicro-analytical technique to determine the form of the radioactive label on both sides of the epithelium, and within the cell if intracellular levels of the isotope are being measured. We have used thin-layer chromatographic procedures for quantifying the labeled metabolites of not only glucose, but also amino acids.[1-3]

It is usually desirable to arrange the experimental conditions so that there is no net transepithelial volume flow during the measurement of a solute flux. If volume flow is present then solvent drag can add to or subtract from the solute flux. The magnitude of the solvent drag component of the solute flux can be approximated as $(1 - \sigma)CJ_v$, where σ is the

---

[1] D. W. Barfuss and J. A. Schafer, *Am. J. Physiol.* **240**, F332 (1981).
[2] D. W. Barfuss and J. A. Schafer, *Am. J. Physiol.* **236**, F149 (1979).
[3] J. A. Schafer and M. L. Watkins, *Pfluegers Arch.* **401**, 143 (1984).

reflection coefficient of the solute, C is the arithmetic mean of the solute concentration in the perfusate and bathing solutions, and J_v is the rate of volume flow. It can be seen that if J_v is zero or if $\sigma = 1.0$ there will be no solvent drag component of the solute flux to complicate its interpretation. Thus it is usually necessary to monitor J_v unless one can be sure that the reflection coefficient of the solute in question is 1.0. The latter assumption is reasonable in the case of all but the smallest solutes in mammalian distal nephron segments, but solute reflection coefficients less than 1.0 have been proposed for Na^+ and Cl^- and other solutes in the proximal nephron.[4-6]

Methods for Measuring Net Fluxes

Net fluxes are measured as the rate of disappearance or appearance of the solute in the tubule lumen or, in other words, as the difference in the rate at which the solute is perfused and the rate at which it is collected at the opposite end of the tubule. The net flux (J_{net}; in pmol min^{-1} mm^{-1}) is then calculated as

$$J_{net} = (\dot{V}_o C_o - \dot{V}_L C_L)/L \tag{1}$$

where $\dot{V}_o$ and $\dot{V}_L$ are, respectively, the perfusion and collection rates in nanoliters per minute; C_o and C_L are, respectively, the solute concentrations in the initial perfusate and collected fluid; and L is the length of the tubule segment. (As discussed in [13], this volume, the solute flux can be normalized alternatively in terms of the apparent or morphometrically measured lumenal membrane surface area.) In order to perform calculations using Eq. (1), it is necessary to know both the perfusion and collection rates as well as the solute concentrations entering and leaving the tubule segment.

In order to determine the solute concentration in the collected fluid, which usually is no more than 100 nl in volume, it is necessary to use ultramicro-analytical techniques appropriate to the solute in question. The solute concentration entering the segment can usually be assumed to be the same as that in the bulk perfusate solution, but it is desirable to confirm this by applying the same ultramicromethods to samples of the perfusate solution, preferably sampled from the tip of the perfusion pipet after the tubule segment has been removed using the same sampling pipet and handling procedures as for the collected fluid.

There are many methods for quantifying amounts of various solutes in

---

[4] E. Frömter, G. Rumrich, and K. Ulrich, *Pfluegers Arch.* **343**, 189 (1973).
[5] J. A. Schafer, C. S. Patlak, and T. E. Andreoli, *J. Gen. Physiol.* **66**, 445 (1975).
[6] J. A. Schafer and T. E. Andreoli, *Membr. Transp. Biol.* **4B**, 473 (1978).

small sample volumes.[7] One of the most accurate is the electrometric determination of Cl⁻ concentration according to the method of Ramsay *et al.*[8] Instruments for this purpose are now available from at least two sources (World Precision Instruments, Inc., New Haven, CT; and Idea Computers, Inc., Richmond, TX). Various cation concentrations can be determined by ultramicro-flame photometry,[9] with a helium glow photometer,[10] or with atomic absorption spectroscopy.[11] Vurek[12] has developed an ultramicro colorimeter that has been used for the measurement of picomolar quantities of urea, creatinine, Ca^{2+}, Mg^{2+}, phosphate, Cl^-, Na^+, and K^+.[13,14] A similar instrument has been constructed for fluorimetric analyses in which NADH is produced or consumed including those for urea, lactate, ammonium, raffinose, and CO_2.[14,15] Both methods can be used with standard analysis kits developed for the automated analysis of many substances in clinical laboratories. Electron probe microanalysis of freeze-dried picoliter fluid samples for elements of atomic number 6 or greater is also in use.[17,18] However, the details of all of these specialized techniques, important as they are for measurement of net fluxes, are beyond the scope of this chapter.

As an alternative approach to the determination of solute concentrations in the ultramicrovolumes, radioactive labeling techniques can be used. However, when measuring net fluxes it is essential that the specific activity (e.g., DPM/μmol) of the label be precisely the same in the perfusate and bathing solution. The only way to ensure that this condition is met is to use the same solution containing the radioactive label to both perfuse and bathe the tubule segment. This approach has obvious limitations, especially when there are requirements for differing concentrations of other solutes such as protein in the two solutions.

Examination of Eq. (1) indicates the importance of all five parameters in the determination of the net flux. Even if it can be assumed or directly determined that there is no net volume flow, so that $\dot{V}_L = \dot{V}_o$, the net flux is still very dependent on the rate of perfusion. Rearranging Eq. (1) for a

---

[7] R. Greger, F. Lang, F. G. Knox, and C. Lechene, *Methods Pharmacol.* **4B**, 105 (1978).
[8] J. A. Ramsay, R. H. J. Brown, and P. C. Croghan, *J. Exp. Biol.* **32**, 822 (1955).
[9] G. Malnic, R. M. Klose, and G. Giebisch, *Am. J. Physiol.* **206**, 674 (1964).
[10] G. G. Vurek and R. L. Bowman, *Science* **149**, 448 (1965).
[11] D. W. Good and F. S. Wright, *Am. J. Physiol.* **236**, F192 (1979).
[12] G. G. Vurek, *Anal. Biochem.* **114**, 288 (1981).
[13] G. G. Vurek and M. A. Knepper, *Kidney Int.* **21**, 656 (1982).
[14] Y. Terada and M. A. Knepper, *Am. J. Physiol.* **257**, F893 (1989).
[15] D. W. Good and G. G. Vurek, *Anal. Biochem.* **130**, 199 (1983).
[16] E. A. Mroz, R. J. Roman, and C. Lechene, *Kidney Int.* **21**, 524 (1982).
[17] C. Lechene, in "Microprobe Analysis as Applied to Cells and Tissues" (T. Hall, P. Echlin, and R. Kaufmann, eds.), p. 351. Academic Press, New York, 1974.
[18] R. R. Warner and C. Lechene, *J. Gen. Physiol.* **79**, 709 (1982).

condition of no net volume flow, one obtains

$$J_{net} = \dot{V}_o[(C_o - C_L)/L] \qquad (2)$$

It can be seen from Eq. (2) that the net flux will be directly proportional to the perfusion rate for any given difference in solute concentration. However, the perfusion rate has its greatest influence on the accuracy of the net solute flux measurement because of its effect on the solute concentration difference that is measured between the perfusate and collected fluid $(C_o - C_L)$. It can be appreciated intuitively that the higher the perfusion rate, the smaller this concentration difference will be for any given J_{net}.

Figure 1 illustrates this relationship graphically for one particular example. In the figure we assume that the true net flux for the tubule segment in question is 100 pmol min^{-1} mm^{-1} (which is a fairly large flux). The net solute concentration difference between the perfused and collected fluid that would have to be measured is then plotted as a function of the perfusion rate. It can be seen that, depending on the error involved in

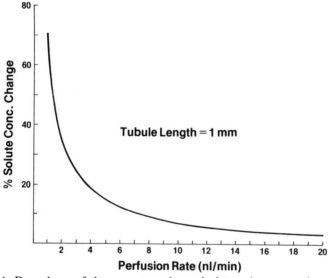

FIG. 1. Dependence of the percentage change in lumenal concentration of a solute $[(C_o - C_L)/C_o]$ on the perfusion rate. In order to illustrate this relationship, we assume that a 1.0-mm nephron segment is reabsorbing a solute, present in the initial perfusate at a concentration of 140 mM (e.g., Na$^+$), at a rate of 100 pmol min^{-1} mm^{-1}. The resulting difference in the solute concentration between the initial perfusate and collectate is then calculated as $J_{net}(L/\dot{V}_o)$.

determining C_o and C_L, it might be necessary to perfuse the segment at quite slow rates in order to measure this concentration difference accurately. Also, the perfusion rate would have to be measured as accurately as possible using the volume marker method described in [13], this volume. These factors must be considered carefully in designing experiments to measure net fluxes, and depending upon the results obtained, the length of the tubule segment and the perfusion rate may have to be altered to optimize the ability to measure the solute concentration difference. Obviously, these constraints would be more severe for solutes with lower net fluxes.

Measurement of Unidirectional Lumen-to-Bath Fluxes

The unidirectional lumen-to-bath flux of a solute can be measured in the same manner as the net flux of the solute if the solute is present in the perfusate but not the bathing solution. However, it is more common to measure one unidirectional flux in the presence of the oppositely directed flux, and in this situation the use of a radioactive label is essential. When measuring the lumen-to-bath flux, labeled solute is added to the perfusate solution at tracer concentrations of 25–100 μCi/ml in order to achieve sufficient disintegrations per minute in samples of either the perfusate or the bathing solution. Because the volume marker, usually [^{3}H]methoxyinulin, is also added to the perfusate at 25–50 μCi/ml, high tracer concentrations are also necessitated by the double-label counting procedures. In every case the *final concentration* of the solute whose flux is to be measured must be calculated by taking into account the total amount of solute added with the tracer. If very low concentrations of the solute are to be used, it is important to use a tracer with a high specific activity, and in many such cases no "cold" solute is added above that already present in the tracer preparation.

In measuring a lumen-to-bath flux it is also important to recognize that the specific activity of the tracer will change along the perfused length of the tubule. Because of the relatively small volume of fluid that is perfused through the lumen per minute, entry of cold solute from the bathing solution will tend to lower the specific activity of labeled solute from its initial level in the perfusate. This problem is especially apparent in those cases where there is net secretion of the solute so that the bath-to-lumen flux exceeds the lumen-to-bath flux. In this situation one must either estimate the average lumenal specific activity, or perfuse more rapidly to maintain a nearly constant lumenal specific activity. The lumen-to-bath flux can then be calculated either from the rate at which the label disappears from the lumenal perfusate, or from the rate at which the label appears in the bathing solution, remembering that both must be corrected for possible metabolism of the solute.

Measurement of the Lumenal Disappearance of Tracer

Measurement of a unidirectional lumen-to-bath flux (J_{lb}; in pmol min^{-1} mm^{-1}) involves essentially the same procedures used in the measurement of a net flux with the exception that one determines the rate of perfusion and collection of the radioactive label. Thus the flux is given as

$$J_{lb} = (\dot{V}_o C_o^* - \dot{V}_L C_L^*)/(S^*L) \qquad (3)$$

where C_o^* and C_L^* are, respectively, the tracer concentrations (dpm/nl) in the initial perfusate and the collected fluid, and S^* is the specific activity of the tracer in the lumen (in dpm/pmol).

If it is known that the bath-to-lumen flux is much less than the lumen-to-bath flux, then S^* can be approximated as the specific activity in the original perfusate solution. However, if there is a significant backflux, then some approximation of the average lumenal specific activity must be made. If the total solute concentration in the collected fluid can be measured conveniently, and the calculated specific activity of the tracer in the collected fluid is within 50–80% of that in the initial perfusate, it may be sufficient to approximate the true average lumenal specific activity as the arithmetic mean of the perfused and collected specific activities.[1] However, it is usually not easy to measure the collected solute concentration chemically, and one usually wants to have a constant specific activity along the perfused length of the tubule, especially when a saturable transport mechanism is involved. For these reasons, changes in the lumenal specific activity are often circumvented experimentally by using a rapid perfusion rate. However, the use of a rapid perfusion rate means that there will be little difference in the collected and perfused tracer concentrations for the same reasons discussed in connection with the measurement of a net flux (see Fig. 1 and the related text). The solution to this apparent dilemma is to measure the lumen-to-bath flux from the rate of appearance of the tracer in the bathing solution rather than from its rate of loss from the lumen. In this way, the experimenter can still measure the flux while maintaining a high perfusion rate and thus a known and nearly constant lumenal specific activity. However, in either case it is necessary to do the enabling calculations of the expected change in lumenal specific activity at the perfusion rate used and, therefore, bath-to-lumen fluxes of the solute in question should always be examined by additional experiments.

Measurement of Tracer Appearance in the Bathing Solution

The lumen-to-bath flux can be easily calculated from the rate of appearance of the tracer (corrected for metabolism) in the bathing solution. Because of the large volume of the bathing solution relative to the tubule, the specific activity of the tracer will remain essentially unchanged at zero, especially if the bathing solution is continuously exchanged as described below. In other words, there should be no complications from backflux of tracer accumulating in the bathing solution. Under these conditions the lumen-to-bath flux can be calculated as

$$J_{lb} = \text{dpm}_b/(S^*Lt) \qquad (4)$$

where dpm_b is the disintegrations per minute of the labeled solute delivered to the bathing solution during the time period t. It is often desirable to correct the total count rate in the bathing solution for counts that may appear artifactually through bulk fluid leakage from the lumen to the bathing solution. This can be done if one assumes that all volume marker appearing in the bathing solution represents such a bulk leakage, as estimated by the rate of appearance of the volume marker in the bath. Then the disintegrations per minute of solute entering the bathing solution by entrainment in this leak flow will be equal to the leak flow rate times the solute tracer concentration in the perfusate. Thus the corrected dpm_b would be given as

$$\text{dpm}_b \text{ (corrected)} = \text{dpm}_b - (\text{dpm}_{vm}/C_{vm}^*)C_o^* \qquad (5)$$

where dpm_{vm} is the disintegrations per minute of volume marker in the bathing solution, C_{vm}^* is the tracer concentration (dpm/nl) of volume marker in the perfusate, and C_o^{**} is the solute tracer concentration (dpm/nl) in the perfusate.

The determination of the total counts of both the solute tracer and the volume marker appearing in the bathing solution over time t requires either that the total bathing solution volume exposed to the tubule be counted, or that its volume be measured accurately and an aliquot counted. We have found it most convenient to maintain a continuous bath flow through our perfusion chamber using the apparatus shown schematically in Fig. 2. The roller pump is adjusted to provide a bathing solution flow rate of 0.3 ml/min from a reservoir maintained at the desired temperature and equilibrated with an appropriate gas mixture. A constant bath volume is maintained in the perfusion chamber by setting a stainless steel suction tube at a fixed height. The aspirated bathing solution is collected in scintillation vials that are removed and replaced at intervals (t) of 10 min.

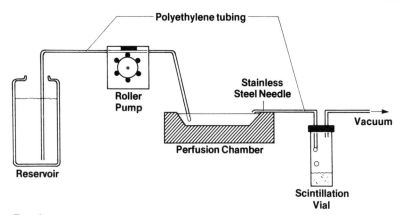

FIG. 2. Arrangement for constant perfusion and collection of bathing solution. The bathing solution is pumped from a reservoir, where it is maintained at the desired temperature and in equilibrium with the desired gas mixture, into the Lucite perfusion chamber at a rate of 0.3 ml/min. The bathing solution is aspirated continuously from the opposite end of the chamber by means of a suction needle, which is positioned to maintain a given volume of bathing solution in the perfusion chamber. The aspirated bathing solution is collected in scintillation vials that are exchanged at regular intervals.

Thus each bathing solution sample is 3.0 ml in volume, and can be prepared easily for liquid scintillation counting by the addition of an aqueous fluor cocktail such as Aquasol-2 (NEN Research Products; Boston, MA). However, if this scintillation fluid is different from the one used to count the perfusate and collectate samples, the difference in quenching must be taken into consideration so that all counts are converted into true disintegrations per minute.

We have found that this method of complete collection of the bathing solution samples is particularly advantageous for obtaining very reliable measurements of both high and low lumen-to-bath fluxes and it allows one to maintain a sufficiently rapid perfusion rate to ensure a constant specific activity of the lumenal solute. In many cases when there is a sufficient change in the lumenal solute tracer concentration ($C_o^* - C_L^*$) without a significant change in its specific activity, one can also use the rate of lumenal disappearance of the tracer as a confirmation of the lumen-to-bath flux determined by bath appearance. We perform both calculations when reducing raw data from experiments, but we have found that there is less variability among replicate samples using the rate of bath appearance. This is expected because the latter calculation depends only on the measurement of total counts appearing in the bathing solution, whereas the former depends on accurate determination of the tracer concentrations in the perfusate and collectate, as well as of the perfusion and collection rates. We have also found that the bath collection method also gives us a very accurate measurement the rate of volume marker leakage. For these reasons we now use this system for all experiments in which such measurements are to be made.

The rate of tracer appearance in the bathing solution can also be used to calculate the lumen-to-bath flux under some circumstances when the lumenal perfusate specific activity is changing along the tubule length. If the opposing unidirectional fluxes of a solute across the epithelium are high but nearly equal, the net flux approximates zero. Thus the solute concentration in the lumen will remain nearly constant and will approximate the perfused concentration C_o if there is little transepithelial volume movement (i.e., if $\dot{V}_o \approx \dot{V}_L$). However, due to the large unidirectional fluxes, the specific activity of the solute in the lumen will fall rapidly along the perfused length. *Under these specific conditions,* if N_o disintegrations per minute of tracer are perfused in a time period t_o, and N_b disintegrations per minute appear in the bathing solution over time period t_b, the lumen-to-bath flux may be calculated as[19]

$$J_{lb} = (-\dot{V}_o C_o/L) \ln[1 - (N_b t_o)/(N_o t_b)] \qquad (6)$$

Measurement of Unidirectional Bath-to-Lumen Fluxes

Measurement of bath-to-lumen tracer fluxes involves the same general considerations as lumen-to-bath fluxes but, because of the large volume of the bathing solution, the tracer concentration in this compartment remains constant. On the other hand, the flux of label is directed into a much smaller volume, the lumenal perfusate, and thus the flux of the labeled solute into the lumen raises its concentration in that compartment so that it may approach that of the bathing solution. For this reason, one usually tries to perfuse the tubule segment at rates sufficiently high to prevent the collected tracer concentration from exceeding 5% of that in the bathing solution. If the solute in question is actively transported in the lumen-to-bath direction, then the collected tracer concentration must be kept even lower.

Also, because of the large volume of the bathing solution, the only practical way of measuring the tracer movement is from its appearance in the collectate. If there is no tracer in the initial perfusate, then all counts appearing in the collectate must have derived from the bathing solution. The primary difficulty in conducting these flux experiments is the expense

---

[19] M. B. Burg and N. Green, *Am. J. Physiol.* **224**, 659 (1973).

involved in adding radioactively labeled solute to the larger volume of bathing solution at a sufficiently high concentration (usually 100–200 µCi/ml) that the counts appearing in the collectate can be determined accurately. Obviously, it would be monetarily impossible to perfuse the bathing solution continuously if it contained label at this concentration. For this reason, we use a very small volume of bathing solution (150–250 µl) which is exchanged only when we wish to change the solution composition.

Special problems with the use of a small bath volume include the prevention of evaporation and therefore concentration of the bathing solution, keeping the bathing solution well mixed, and the maintenance of equilibrium between the bath and the desired gas mixture. In order to circumvent these problems, we use a modified perfusion chamber that has a small trough in the bottom, as shown in Fig. 3. The trough is the area in which the tubule is perfused and contains the small volume of bathing solution. This trough is then covered with a layer of about 1 ml of silicone fluid (dimethylpolysiloxane 200 fluid; Accumetrics, Elizabethtown, KY) which prevents evaporation but has a relatively high permeability to gases. We usually stain the silicone fluid with Oil Red O so that the interface between it and the aqueous bath can be readily observed. In order to keep the bathing solution well mixed, a small glass tube is placed in the trough

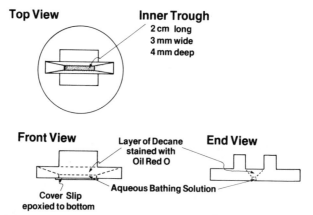

FIG. 3. Diagram of perfusion chamber used for bath-to-lumen tracer fluxes. The chamber is constructed of Lucite with a narrow and shallow trough milled in the bottom. The bottom surface of this trough is covered by a glass coverslip through which the tubule segment can be observed. The dimensions of the trough are chosen so as to give a total bathing solution volume on the order of 250 µl. The bathing solution is then covered by a layer of silicone fluid, which is placed in the upper portion of the chamber.

and a column of fluid is continuously drawn up and down in the tube by a reciprocating pump mechanism.

Appropriate gas partial pressures are maintained in the bathing solution by superfusing the gas mixture over the silicone fluid. When working with HCO_3^- buffer solutions we find that it is necessary to superfuse CO_2 at higher partial pressures in order to maintain the pH at 7.4. We regulate the flow of an 80% O_2/20% CO_2 gas mixture with a precision flow regulator and gauge, while measuring the solution pH with a miniature combination electrode placed in the chamber trough and with the silicone fluid layer in place. We find that the gas flow rate required to maintain the bath pH in the range 7.35–7.45 is not critical and can be easily reproduced from day to day. However, if close control of the pH is required for the experiment, the miniature electrode can be left in the trough and the pH continuously monitored and adjusted as required. The primary difficulty in using the small volume perfusion chamber is imposed by the narrowness and straight walls of the trough which limit the ability to maneuver the perfusion and collection pipet assemblies.

The bath-to-lumen flux (J_{bl}; in pmol min^{-1} mm^{-1}) can then be measured from the rate of appearance of the tracer in the collectate (the fluid exiting from the tubule lumen into the holding pipet):

$$J_{bl} = \text{dpm}_L/(S^*tL) \tag{7}$$

where dpm$_L$ is the total disintegrations per minute appearing in the fluid removed from the holding pipet over the collection time interval t, and S^* is the specific activity of the tracer in the bathing solution (in dpm/pmol). S^* is calculated from the solute concentration added to the bathing solution and the measured tracer concentration (in dpm/nl). The tracer concentration is constantly measured during the course of an experiment by removing volumetric samples of the bathing solution (e.g., 2 µl at 10-min intervals) and counting them. The constancy of the tracer concentration is a good criterion of the absence of any evaporation from the bath.

Because of the high solute tracer concentration in the bathing solution, it is impossible to measure the leak of a volume marker (labeled by a second isotope such as [^{3}H]methoxyinulin) from the lumen into the bathing solution. If it is important to ascertain the leak rate, one can add the volume marker to the bathing solution and monitor its rate of appearance in the collectate. This procedure precludes the simultaneous measurement of the perfusion rate, but the calculation of the bath-to-lumen flux is not dependent on this parameter. Furthermore, if the experimental conditions are arranged to prevent any transepithelial volume flow, then the perfusion rate will be equal to the collection rate, which is monitored in the usual fashion by the rate of volume sampling from the holding pipet.

If the volume marker is added to the bathing solution, it is also possible to use its rate of appearance in the collectate to estimate the rate of bulk leak and thus to correct the disintegrations per minute of the solute tracer for such leakage. One uses the same approach as discussed in the case of lumen-to-bath fluxes above [Eq. (5)], except that the volume marker concentration and solute tracer concentrations are referenced to the bathing solution rather than to the perfusate.

Measurement of Intracellular Solute Concentration

Although ion-selective electrodes can be used to measure the intracellular activity of some ions, such techniques are not available for the determination of cell content of organic ions or nonelectrolytes. Nevertheless, it is possible to measure the intracellular concentration of many solutes by extraction from the isolated tubules.

Theoretically, if one could measure the intracellular solute concentration, it would be possible to measure solute fluxes across the individual membranes of the perfused tubule segment. In practical terms, however, it is possible to obtain only one measurement of the intracellular solute concentration in each experiment. Because this measurement involves removal and extraction of the tubule, it will always be the concentration at the end of the experiment. Thus it is not feasible to obtain continuous measurements of intracellular concentration, nor to obtain paired concentration values under different experimental conditions within any given experiment.

Nevertheless, the single measurement of intracellular concentration in an experiment can provide important information about the transport characteristics of the individual cell membranes. For example, during lumen-to-bath flux measurements or D-glucose or L-amino acids in proximal tubule segments it is found that the intracellular solute concentration exceeds that in the lumen in spite of the fact that the bathing solution concentration is zero.[1-3] For these nonelectrolytes this indicates that the solute must be actively transported into the cell across the lumenal membrane. In proximal straight tubule segments it is also found that during a bath-to-lumen flux measurement the intracellular glycine concentration exceeds that in the bathing solution, demonstrating active transport into the cell across the basolateral membrane.[20]

Because of the minute quantities of solute that would be accumulated in such a small intracellular volume (less than 0.8 nl/mm tubule length for mammalian nephron segments), these measurements can be made conve-

[20] D. W. Barfuss, J. M. Mays, and J. A. Schafer, *Am. J. Physiol.* **238**, F324 (1980).

niently only by using a radioactive tracer. Furthermore, the counts extracted from the tubule must be corrected for metabolism by chemical analysis as for other samples (see above).

In order to obtain a valid measurement of the intracellular solute concentration, it is necessary to detach the tubule from the pipet assemblies and transfer it as rapidly as possible into an extraction medium with minimal contamination from adhering extracellular fluid containing the tracer. This is accomplished by detaching the segment from the collection-side holding pipet while maintaining perfusion. The tubule is then snatched from the perfusion-side pipets using a glass needle with a hook at the tip or with fine forceps. The tubule is pulled rapidly from the bathing solution through a covering layer of light mineral oil, which serves to strip peritubular fluid off the segment, and it is placed immediately into an extraction medium of 3% trichloroacetic acid. After 30–60 min the extract is removed and counted for the solute tracer. It is necessary to correct the disintegrations per minute obtained for contamination by tracer in any adhering lumenal or peritubular fluid. This is done by adding the volume marker to the same solution as the solute tracer, and calculating the intracellular concentration as

$$C_i = [\text{dpm}_e - (\text{dpm}_{vm} C_s^*/C_{vm}^*)]/(S^* V_i) \tag{8}$$

where C_i is the intracellular solute concentration, dpm_e is the total extracted solute tracer disintegrations per minute, dpm_{vm} is the disintegrations per minute of the volume marker tracer in the extract, C_{vm}^* is the tracer concentration of the volume marker in the extracellular solution (in dpm/nl), C_s^* is the solute tracer concentration in the extracellular solution (in dpm/nl), S^* is the specific activity of the solute tracer (in dpm/pmol), and V_i is the intracellular volume (in nl).

The intracellular volume can be determined as the difference in the weight of the tubule segment before and after desiccation minus the extracellular volume. However, determinations of the small tubule weights can be accomplished only by custom-designed balances such as the quartz fiber fishpole method.[21] We have found that it is equally precise to determine the intracellular volume from the optically measured lumenal radius (r_i) and the outside radius of the tubule (r_o), assuming that approximately 70% of this volume is nonaqueous. Thus the intracellular volume is given as $0.7\pi L(r_o^2 - r_i^2)$.[2,21] Because of the errors in measuring these radii accurately, the total cumulative error in the volume determination is estimated to be ±30%. Nevertheless, for most solutes that are actively transported the resulting uncertainty in the actual intracellular solute concentration is acceptable.

[21] M. B. Burg and P. F. Weller, *Am. J. Physiol.* **217**, 1053 (1969).

The intracellular concentration of a solute during the measurement of a steady state unidirectional flux can also be used to estimate the relative permeabilities of the lumenal and basolateral membranes. Assuming that no significant paracellular flux occurs, in the steady state the fluxes across these two membranes must be equal to each other and to the transepithelial unidirectional flux. The measurement of the intracellular solute concentration at the end of the experiment allows one to assess the solute concentration gradient across each individual membrane. Thus, at least for a nonelectrolyte, the apparent solute permeability (see below) of each membrane would be given by the quotient of the unidirectional flux and the concentration difference across the membrane.

Calculation of Permeability Coefficients

The presence of a net flux of a solute in the absence of a transepithelial electrochemical potential gradient, or the demonstration of a ratio of unidirectional fluxes different from that predicted by the Ussing flux ratio test,[22] can be used as evidence for active absorption or secretion of the solute in question. But the flux measurements themselves do not provide direct information about the nature of the transport processes involved. Usually one wishes to know how the flux responds to differences in the electrochemical driving force, that is, to assess the apparent permeability characteristics of the epithelium. In the case of a purely diffusional flux, one would expect to find that the flux is a linear function of the driving force, whereas in the case of a facilitated diffusion process, either active or passive, the flux would be a saturable function of the solute concentration. For this reason, both net and unidirectional fluxes are usually measured at varying concentrations.

Nonelectrolyte Permeabilities

For a nonelectrolyte, the only relevant driving force in the absence of net volume flow is the concentration difference. Thus in the case of a unidirectional flux we can define an apparent permeability (sometimes referred to as a flux coefficient) as the ratio of the flux to the concentration of the solute on the cis side of the epithelium, i.e., J_{lb}/C_l or $-J_{bl}/C_b$ (by convention the secretory flux is taken as negative), but the resulting units (pmol min^{-1} mm^{-1} mM^{-1}) are clumsy. Therefore, it is customary to convert the calculated permeability to more conventional units of μm per second by taking into account the apparent lumenal surface area per

[22] S. G. Schultz, "Basic Principles of Membrane Transport." Cambridge Univ. Press, Cambridge, England, 1980.

millimeter length of the tubule segment, πd, where d is the inside diameter of the tubule in micrometers. With unit conversions the permeabilities are calculated as 16.67 $(J_{lb}/\pi d C_l)$ or -16.67 $(J_{bl}/\pi d C_b)$. If it can be demonstrated that the flux is a linear function of the concentration over a reasonable concentration range, then it can be presumed, although not proved, that the flux involves free diffusional permeation.

Apparent permeability coefficients can also be directly calculated from the tracer concentration in the perfused and collected tubular fluid without the intermediate calculation of the actual flux. If the reflection coefficient (σ) of the solute is known, or can be presumed to be zero, the apparent permeability (k_{lb}; in μm/sec) can be calculated in the presence of a net volume absorption as[23]

$$k_{lb} = [(\dot{V}_o - \dot{V}_L)/A]\{[\ln(C_o^*/C_L^*)/\ln(\dot{V}_o/\dot{V}_L)] + [(1 + \sigma)/2]\} \quad (9)$$

where A is the apparent lumenal surface area ($\pi d L$). When there is no net volume flow ($\dot{V}_o = \dot{V}_L$), the apparent permeability can be calculated as[23]

$$k_{lb} = (\dot{V}_o/A) \ln(C_o^*/C_L^*) \quad (10)$$

Because the tracer specific activity in the bathing solution remains constant during the measurement of a bath-to-lumen flux, the permeability in this direction can be calculated directly from the quotient of the flux and concentration as discussed above. However, if there is net volume flow, J_{lb} must be corrected by the solvent drag contribution to the net flux as given by $J_v(1 - \sigma)C$, where J_v is the transepithelial volume flow and C is the average of the lumenal and bathing solution solute concentrations.

In cases where the bath-to-lumen flux is sufficiently high that the collectate tracer concentration becomes appreciable, the apparent bath-to-lumen permeability can still be calculated if the apparent permeability in the lumen-to-bath direction (k_{lb}) is known and is a linear function of the lumenal solute concentration. In this case one uses the relationship[24]

$$k_{bl} = (C_L/C_b^*)\{k_{lb}/[1 - \exp(-Ak_{lb}/\dot{V})]\} \quad (11)$$

where $\dot{V}$ is the arithmetic mean of the perfusion and collection rates, presuming these are similar.

Permeabilities of Ions

In the case of ions one must consider both electrical and concentration driving forces in the calculation of apparent permeability coefficients. The

[23] J. A. Schafer and T. E. Andreoli, *J. Clin. Invest.* **51**, 1264 (1972).
[24] M. Imai and J. P. Kokko, *J. Clin. Invest.* **53**, 393 (1974).

most useful formulation of the relationship between a flux and its electrochemical potential driving force comes from the Goldman–Hodgkin–Katz equation, which assumes a constant voltage profile across the diffusion barrier.[22] In the case of an epithelial cell layer this assumption is obviously incorrect if the flux is proceeding transcellularly. However, in this case one would not expect the process to involve significant free diffusional movement across both of the cell membranes in any case. The equation is of most use when calculating the permeability of the paracellular pathway for the backflux of an ion, when it can be reasonably assumed that the junctional complex constitutes a single rate-limiting barrier to the ion movement. In this case the permeability (μm/sec) can be calculated from the unidirectional flux (in pmol min^{-1} mm^{-1}) as[22]

$$k_{bl} = (16.67 J_{bl}/\pi d C_b)\{[1 - \exp(zFV_e/RT)]/(zFV_e/RT)\} \qquad (12)$$

where d is the lumenal diameter (in μm), C_b is the concentration of the ion in the bathing solution, V_e is the transepithelial voltage (lumen with respect to bathing solution), R is the gas constant, T is the absolute temperature, z is the valence, and F is the Faraday constant. In the case of a lumen-to-bath flux the permeability can be calculated as

$$k_{lb} = (16.67 J_{lb}/\pi d C_l)\{[1 - \exp(-zFV_e/RT)]/(-zFV_e/RT)\} \qquad (13)$$

where C_l is the mean lumenal ion concentration. In either case one must establish that the unidirectional flux is a linear function of the electrochemical potential driving force before concluding that the flux measured is in fact due to free diffusional movement and thus that one is calculating a true permeability coefficient.

Acknowledgments

The authors would like to thank the technicians and coinvestigators who have worked with them in the development of many of the methods described in this chapter, in particular Ms. Susan L. Troutman and Dr. Delon W. Barfuss. Support for many of the studies referenced from the authors' laboratory was provided by NIH Research Grant 1-RO1-DK25519.

[21] Measurements of Volume and Shape Changes in Isolated Tubules

By WILLIAM B. GUGGINO, DIANE MARKAKIS, and L. MARIO AMZEL

Introduction

Maintenance of cell size and shape is a universal function of all cells. Volume regulatory mechanisms function at constant external osmolality to maintain renal cells at an optimum cell volume in normal transport conditions. Once the external osmotic environment changes, volume will be altered because cells are permeable to water. For example, when exposed to abrupt increases or decreases in extracellular osmolality, cells shrink or swell initially. The initial response is transient because regulatory mechanisms activate to restore cells back to their optimum volume by inducing a change in intracellular solute and water content.[1]

To study how a cell regulates volume, it is necessary to have available techniques to measure cell volume following a perturbation in extracellular osmolality. Although relatively easy to measure in individual cells such as red cells, determining the volume of renal cells poses unique challenges. First, because the kidney contains many nephrons, each having several different segments, the particular portion of the nephron must be isolated from the kidney and perfused *in vitro* before volume regulation can be studied. Second, because renal cells are polar, possessing both apical and basolateral membranes, it is important to have access to both membranes to study the unique role that each cell membrane plays in volume regulation. Third, because the plasma membrane surface of renal cells is rich in morphological structures such as basal infoldings, lateral interdigitations, and apical brush borders, special procedures must be developed to measure cell volume accurately. Fourth, because most renal cells are highly permeable to water, at least across their basolateral cell membranes, exposures of the cells to anisotonic solutions must be very rapid. Finally, because renal cells regulate volume quickly the techniques for measuring cell volume must have excellent time resolution in order to detect them.

Described here are techniques designed specifically for measuring cell volume in isolated and perfused renal tubules using a combination of video and optical techniques. The methods allow access to both apical and basolateral solutions and provide excellent time resolution.

---

[1] C. Montrose-Razifadeh and W. B. Guggino, *Annu. Rev. Physiol.* (in press) (1990).

Tubule Preparation

The nephron segment to be studied is isolated and perfused according to the methods outlined in a previous chapter. Briefly, tubules are placed within a perfusion chamber and mounted on a concentric perfusion pipet system. On the perfusion side is an outer holding pipet and an inner perfusion pipet. Within the inner perfusion pipet is another exchange pipet for rapidly exchanging lumenal solutions. The collection side has an outer holding pipet containing at the pipet tip a small drop of Sylgard (Dow Corning Corp., Midland, MI), a dimethylpolysiloxane fluid. This fluid is available in viscosities between 100 and 1000 centistoles (cs). These two viscosities can be mixed to obtain a consistency which will hold the tubule in the collecting pipet without collapsing the lumen. Because the dimensions of a single portion of the tubule will be measured during changes in cell volume, it will be important that the tubule be held firmly between perfusion and collection sides. The lumenal and bath perfusion systems should be designed to allow for fast fluid exchange rates (< 1 sec).

Rapid changes in the bath solution tend to cause the tubule to shift position in the bath. If the microscope is focused on a cell, a small shift in position would make it difficult to keep the cell in focus. This problem can be minimized with the use of a four-way valve perfusion system. One leg is connected to the chamber and one to a waste container. This arrangement will facilitate solution changes between normal and anisotonic solutions while both are continuously flowing (one to the bath chamber and the other to the waste container).

It may be important to change both apical and basolateral solutions simultaneously. Because of inherent differences in lumenal and bath perfusion systems, it is necessary to determine the time from the initial switching of the valve to the time when the solution actually contacts the tubule by placing a dye in one of the perfusion solutions.

Optical and Video Apparatus

The tubule is viewed through an inverted microscope modified to include differential interference optics (DIC). Although DIC is commercially available for most inverted microscopes, it may not be possible to adapt the commercial DIC for use with the isolated perfused tubule preparation. If this is the case, a DIC system which works well with isolated and perfused tubules can be constructed as follows. The light source is polarized with a glass polarizing filter placed in the condenser housing. The condenser lens is removed and replaced by a revolving microscope nosepiece with at least two objective sockets. A $\times 32$ long-working distance objective (E. Leitz, Inc., Rockleigh, NJ) is placed in the first socket. In the back of this objective is placed a DIC prism (474571, Carl Zeiss, Inc., New York, NY). The long working distance objective is important to allow access to the chamber by both perfusion and collection pipets while illuminating the tubule with a narrow beam of light. A ×10 objective is placed in the second socket. The ×10 lens is used to position the pipets and while perfusing tubules. For DIC the ×32 objective is used as a condenser.

The objective used will depend on the tubule studied. Basically, the objective should be of high enough power to allow the tubule to occupy the whole field of view. The aim is to measure both cellular and outer tubular dimensions, simultaneously. For some mammalian tubules, such as the thin descending limb of Henle's loop, a ×100 oil immersion objective will be necessary. A second DIC prism is placed in the back of the objective (also from Carl Zeiss, Inc.) and aligned in the same direction as the first. The image is analyzed by a second polarizer positioned until complete extinction of the polarized light is obtained. This optical system has a narrow depth of field, making it possible to focus clearly on specific levels within the tubule.

The tubule image is viewed by a television camera. The television signal must be mixed with a time signal from a signal generator in order to determine rates of change in cell volume. The time signal generator should have a time resolution of 0.1 sec. Tubule image and time are recorded either on video tape or disk. Images of tubules are analyzed in stop-frame mode of the video recorder. Tubule dimensions (see below for details) can be measured either by hand directly from the video screen, or by using a video position analyzer (For-A Corporation of America, Boston, MA), or by a computer using the appropriate video frame grabbing hardware.

Level of Focus

To assess cell volume, the level of focus of the microscope is adjusted at approximately the center of the tubule lumen. This will give a side view of the cells along a portion of the tubule (see Figs. 1, 2, and 3). Cell images should be recorded only if the midpoint of the cell is in focus, if both apical and basolateral surfaces can be seen clearly, and if the outer tubule diameter can be measured accurately. All tubule images should be obtained near the perfusion pipet to obtain maximum stability of the preparation.

Overall Strategy for Measuring Cell Volume

The strategy for measuring volume will vary among the different nephron segments, depending on the shape of the apical surface of the cells. The approach is to consider the tubule as having two cylinders; an exterior cylinder and an interior cylinder. The whole tubule encompasses the external cylinder, whereas the internal cylinder is the tubule lumen. Hence, volume of the external cylinder minus the internal cylinder is the tubule cell volume. A length of tubule is chosen which may be a cell length, if it is possible to determine cell borders accurately, or simply a convenient length of tubule.

The concept is to take one optical section of the tubule by establishing the level of focus of the microscope in the middle of the tubule lumen. Using dimensions of the tubule and the cell layer, the volume of cells within an annulus of the tubule is determined. As is true of most nephron segments, changes in cell volume occur primarily via changes in cell height. Because the cells are tethered by the basement membrane, changes in cell length are minimal. This, however, should be verified for each segment studied.

The methods described in the following sections ignore membrane amplifications and treat the cells as if they have a smooth surface. Because the methods approximate the true cell volume, all experimental data should be reported as percentage of control to minimize problems with estimation errors.

Simple Double-Cylinder Model

If changes in cell shape following changes in medium osmolality are uniform along the tubule, such as occurs in the proximal tubule (see Fig. 1), a simple double-cylinder model can be used to estimate cell volume.[2]

A portion of the tubule is selected with the level of focus adjusted at the center of the lumen; the volume (V) of an annulus of length, L, is determined according to the equation

$$V = \pi L[(R_o)^2 - (R_o - H)^2] \qquad (1)$$

where R_o is the outer tubule radius and H is the thickness of the cell layer, including the thickness of the brush border measured at several points along a portion of the tubule and averaged. V should be normalized per length of tubule by dividing V/L.

In order to make sure that Eq. (1) is an accurate representation of the cell outline, the cross-sectional area of the cellular layer is measured with a planimeter or by a computer and compared with the cross-sectional area (A_s) calculated from

$$A_s = LH \qquad (2)$$

[2] A. G. Lopes and W. B. Guggino, *J. Membr. Biol.* **97**, 117 (1987).

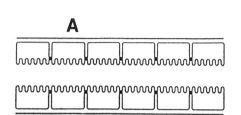 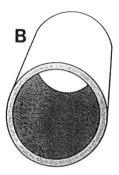

Fig. 1. Simple double-cylinder model. (A) A drawing of a length of proximal tubule viewed from the side with the center of focus at the center of the tubule lumen. Although the cells possess an apical brush border, the lumen surface is uniform, making it possible to represent the tubule as a simple double cylinder as shown in (B).

Complex Double-Cylinder Models

Two basic approaches can be taken to estimate cell volume when the lumenal surface of tubule cells does not have a uniform shape across the whole cell surface (see Figs. 2 and 3). One approach uses a double-cylinder model but recognizes that the inner radius varies around the circumference and in the axial direction. An estimate is made of the inner tubule radius which is used to calculate the inner lumenal volume. Total cell volume is the difference between outer tubule volume, calculated from the volume of a simple cylinder, and the estimated lumenal volume. The second approach is to divide the epithelial cell layer into two portions. The volume of one portion, considered the base layer, is estimated using a simple double-cylinder model as described above for the proximal tubule. A second function is used to estimate only that portion of cell layer which protrudes above the base layer. The overall cell volume is the sum of the two cell portions.

Two examples of nonuniform shape changes during cell swelling which illustrate these two approaches are shown in Figs. 2 and 3. Although the

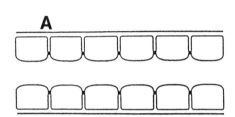 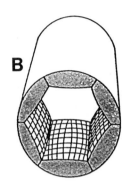

FIG. 2. Complex double-cylinder model of the diluting segment. (A) A drawing of the side view of the *Amphiuma* diluting segment with the level of focus as in Fig. 1. Note that the apical surface of the cells bulges into the lumen at the center of each cell. The shape of the bulge can be represented by a parabolic function. The corresponding shape of the tubule representing a parabolic function is shown in (B).

treatments outlined are designed specifically for the diluting segment of the *Amphiuma* kidney[3] and thin descending limb of Henle's loop,[4] similar techniques can be applied to other nephron segments which behave in a similar fashion.

Cells of the Diluting Segment. Cells of the diluting segment swell by bulging into the lumen in such a way that a large central portion of the lumenal surface of the cell extends into the lumen beyond the level of the tight junction. Junctions between cells are clearly visible, allowing for an accurate determination of cell length. With the center of focus in the center of the tubule lumen the following parameters can be measured: the outer tubule diameter (R_o), the minimum cell height (h_o) at the junction between cells (values measured at each junction are averaged), the maximum cell height (h_m) at the midpoint of the cell, and the cell length (L), which is taken as the distance between successive points at which h_o is measured.

[3] W. G. Guggino, H. Oberleithner, and G. Giebisch, *J. Gen. Physiol.* **86**, 31 (1985).
[4] A. G. Lopes, L. M. Amzel, D. Markakis, and W. B. Guggino, *Proc. Natl. Acad. Sci. U.S.A.* **85**, 2873–2877 (1988).

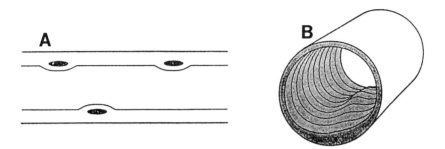

FIG. 3. Complex double-cylinder model of the thin descending limb of Henle's loop. (A) A drawing of the side view of the tubule with the level of focus as in Fig. 2A. Note that the tubule is uniform except in the region of the nucleus. The shape of the nuclear bulge can be represented by a Gaussian function. The corresponding shape of an annulus of the tubule is shown in (B).

A parabolic function of the form

$$f(x) = h_o + (4/L^2)(h_m - h_o)(L - x)x \tag{3}$$

is a reasonable approximation of the true profile of a cell along the tubule axis. The longitudinal (viewed at a plane that contains the tubule axis) cross-sectional area (A_s) of each cell outline is then obtained as

$$A_s = \int_0^L f(x)dx = L[(1/3)h_o + (2/3)h_m] \tag{4}$$

Longitudinal cross-sectional areas of cell outlines can be measured with a planimeter or by computer and compared to Eq. (4) to determine the validity of the approach.

The volume of the cells for length L of the tubule can be calculated as follows:

$$V = \pi \int_0^L [R_o^2 - R_i^2(x)]dx \tag{5}$$

where R_i is the inner radius and $R_i(x) = R_o - f(x)$. Replacing $f(x)$ by Eq.

(3) in Eq. (5) and integrating one obtains

$$V = \pi L[(R_o^2 - R_i^2) + (4/3)R_i(h_m - h_o) - (8/15)(h_m - h_o)^2] \quad (6)$$

with $R_i = R_o - h_o$. Equation (6) gives a good approximation to the volume if the values of h_o and h_m represent averages around the circumference of the tubule.

A more precise description requires that one consider the variations around the circumference of the thickness h at a given position x along the tubule [$h(x)$]. This requires the introduction of the dependence of h on an angular variable Θ. With this additional dependence, integration of the equation for the cross-sectional area indicates that a better approximation for the values of h_m and h_o in Eq. (6) is given by

$$h_o = 0.7 h_m(0) + 0.3 h_o(0) \quad (7)$$

and

$$h_m = 0.7 h_m(L/2) + 0.3 h_o(L/2) \quad (8)$$

The equation uses the experimental values for h around the tubule circumference at the beginning [$h_m(0), h_o(0)$] and at the middle [$h_m(L/2), h_o(L/2)$] of a cell. Values of h_o and h_m obtained with Eqs. (7) and (8) can be used in Eq. (6) to give a good approximation of the volume.

If the maximum and minimum h along the axial direction and along the circumference are the same, the formula for the volume obtained by the double integration is quite compact. The formula can be simplified without much loss of accuracy to give

$$V = \pi L[R_o^2 - (R_o - \langle h \rangle)^2] \quad (9)$$

where $\langle h \rangle$ can be calculated as $0.49 h_m + 0.51 h_o$.

Cells of the Thin Descending Limb of Henle's Loop (TDL). The TDL cells provide a different challenge to the measurement of cell volume. These cells are uniformly thin except in the nuclear region, which bulges into the lumen (Fig. 3). In contrast to the diluting segment, the junctions between cells are difficult to detect with light microscopy. Choosing a portion of the tubule where the midpoint of the nuclear region is in focus, the following parameters can be measured: R_o, R_i, the width at half-height (WHH), h_o, here defined as the height of the cell layer in the nonnuclear region of the tubule, and h_m, the maximum cell height measured in the nuclear region. The "excess height" on the nuclear region, $f(x)$, is then defined as $h_m - h(x)$, with a maximum value $a = h_m - h_o$. The width of the nuclear region can be measured at $f(x) = a/2$.

The Gaussian function

$$f(x) = a \exp(-x^2/2\sigma^2) \quad (10)$$

is used to represent the outline of the "excess height" of a cell in the nuclear region where a is the maximum excess height and σ is derived from the width at half-height (WHH):

$$\sigma = (WHH/2)/[(2 \ln 2)^{1/2}] \qquad (11)$$

The parameters h_m, h_o (and therefore $a = h_m - h_o$), and WHH can be easily measured from an outline of the side view of a cell.

The cross-sectional area (A_s) of the side view of the nuclear region was calculated by integrating the Gaussian function

$$A_s = a \int_{-L/2}^{L/2} \exp(-x^2/2\sigma^2) dx \qquad (12)$$

The upper and lower limits of the integral are determined empirically. The value $L/2 = 2\sigma$ gives the best approximation to the measured area and cell length. The area is calculated as follows:

$$A_s = \sqrt{2\pi}\sigma a (0.9545) \qquad (13)$$

where the value 0.9545 is the tabulated value of the Gaussian integral for $L/2 = 2\sigma$.

Assuming the cell is symmetric, the volume of the nuclear region (V_n) can be approximated by integrating the bivariate Gaussian function of x and y where $y = r_i \Theta$ (r_i is the tubule radius in the nonnuclear region of the tubule and Θ is the angular tubule variable in radians) with $\sigma_x = \sigma_y$, that is,

$$f(x,y) = a \exp[-(x^2 + y^2)/2\sigma^2] \qquad (14)$$

Using this function one gets

$$V_n = 2\pi a \sigma^2 (0.9545)^2 \qquad (15)$$

To obtain the volume of the nonnuclear region the volume of an annulus (V_a) of length L can be calculated as follows:

$$V_a = \pi L R_o^2 - \pi L R_i^2 \qquad (16)$$

where R_o and R_i ($R_o = R_i + h_o$) are the outer and inner tubular radii, respectively. The total volume, V_t, is then obtained as $V_n + V_a$.

[22] Transport in Isolated Cells from Defined Nephron Segments

By ROLF K. H. KINNE

Introduction

During recent years studies on defined isolated nephron segments *in vitro* have contributed significantly to our knowledge of transport properties, metabolic pathways,[1] and hormone receptors[2] in the various regions of the nephron. These segments have been obtained either by microdissection from sometimes collagenase-treated kidney slices or, when larger amounts were required, by sedimentation procedures after dissociation of the tissue into single tubules. The microdissected tubules have also been used as starting material for primary cell cultures or for the establishment of continuously growing cell lines (for reviews, see Refs. 3–5). Since almost all nephron segments, however, contain several types of cells which differ in their morphology and (probably) also in their functional characteristics, transport properties found in a defined nephron segment cannot unequivocally be attributed to a distinct cell type. Thus methods were developed for the isolation of homogeneous cell populations from defined nephron segments and such preparations were used to define the transport properties of these cells. For transport studies the isolated cells offer additional advantages over the use of isolated tubules. Paracellular transport processes are absent in these preparations; therefore transport parameters found in such cells represent only cellular uptake or transcellular transport. In addition, in contrast to isolated tubular segments with a possibly collapsed lumen, the lumenal and contralumenal sides of the epithelial cells are exposed to the incubation medium and diffusional barriers are very small. Hence supply of oxygen and substrates is optimal and the composition of the medium bathing the lumenal and contralumenal cell membrane is well defined.

In this chapter studies are summarized in which transport-related parameters, such as ion-dependent energy metabolism, ion contents, and ion fluxes, have been measured in isolated renal cells and have been used to

---

[1] U. Schmidt and U. C. Dubach, *Pfluegers Arch.* **306**, 219 (1969).
[2] F. Morel, D. Chabardès, and M. Imbert, *Kidney Int.* **9**, 264 (1976).
[3] D. Schlöndorff, *Kidney Int.* **30**, 201 (1986).
[4] D. M. Scott, *Differentiation (Berlin)* **36**, 35 (1987).
[5] G. J. A. Gstraunthaler, *Renal Physiol. Biochem.* **11**, 1 (1988).

identify transport systems and their regulation. First the methods used to isolate the cells are described, then criteria to identify isolated renal cells are discussed. A description of various methods to measure transport-related parameters follows. Finally the results obtained for the different tubular cells with regard to their transport properties are summarized and discussed in view of conclusions derived from other methodological approaches.

Isolation of Renal Cells: General Considerations

In order to obtain a homogeneous cell population from the kidney the following steps are usually involved. First the kidney is freed from nonrenal cells and proteins such as red blood cells and plasma by perfusing the kidney after it has been removed from the animal. Sometimes this initial perfusion is followed by an additional perfusion in which specific agents are instilled into the renal tissue to facilitate the second major step in cell isolation, the dissociation into tubular segments and—subsequently or simultaneously—into single cells. The third main step is separation of the cells, which is achieved either on the basis of their difference in density, in surface charge, or expression of cell-specific proteins recognized by antibodies or lectins. These three main steps in the isolation procedure differ from species to species and must be adapted to the region of the kidney in which the tubular cells are located.

For example, the rabbit kidney has turned out to be well suited for cell isolation because the cells are relatively easily removed from their surroundings and the yield and viability of the cell preparations are high. With regard to the different areas of the kidney the renal cortex is readily dissociated and cortical cells can be obtained under mild experimental conditions. The renal medulla and papilla usually require prolonged enzymatic treatment before cells can be separated from the interstitium and from each other. In the following, procedures for the isolation of cells from the renal cortex, the renal outer medulla, and the renal papilla are described.

Isolation of Renal Cortical Cells (Proximal Tubule, Cortical Segments of the Distal Tubule)

Perfusion Solutions and Perfusion Conditions

The perfusion solutions and the perfusion conditions that have been employed by various authors are summarized in Table I. For initial perfu-

TABLE I
ISOLATION OF PROXIMAL TUBULE CELLS: INITIAL PERFUSION SOLUTIONS[a]

	Isotonic (Poujeol and Vandewalle[6])	Isotonic (Nord et al.[7])	Isotonic (Heidrich and Dew[8,9])	Hypertonic (Soltoff and Mandel[10,11])
		Earle's balanced salt solutions		
Salt				
NaCl	115 mM	137 mM	116.4 mM	115 mM
NaHCO$_3$	25 mM		26.2 mM	25 mM
NaH$_2$PO$_4$	4 mM	0.3 mM	1.15M	2 mM
KH$_2$PO$_4$		0.5 mM		
KCl	5 mM	3 mM	5.3 mM	5 mM
MgSO$_4$	0.1 mM	0.4 mM	0.8 mM	1 mM
CaCl$_2$	0.25 mM	2 mM	1.7 mM	1 mM
MgCl$_2$		0.5 mM		
Substrates				
Glutamine	4 mM	4 mM	2.56 mM	
Lactate				4 mM

Alanine		1 mM	1 mM
Glucose		6 mM	5 mM
Citrate			
Mannitol			25 mM
pH	7.2 (gassed with 5% CO_2/95% O_2)	7.4 (buffered with 5 mM Tris-Cl and 6 mM Tris base)	7.4 (gassed with 95% O_2/5% CO_2)
		1 g/liter	
		0.84 g/liter	
		Oxygenated before use	
Colloids		0.5% BSA	17.11 g sucrose, 10% BSA
			6% dextran
Temperature	37°	4°	25°
Perfusion rate	15 ml/min for 2–3 min		15 ml/min for 5 min (pulsating peristaltic pump)
			45–50 ml/min for 5–10 min
			37°
Additional treatments		2 ml 0.5% iron oxide in initial perfusion solution to load glomerula with iron	10-min treatment with collagenase (0.8 mg/ml initial perfusion solution)

[a] Additional perfusion: 25 mM NaCl, 25 mM NaHCO$_3$, 4 mM NaH$_2$PO$_4$, 1 mM MgSO$_4$, 1 mM lactate, 4 mM alanine, 5 mM glucose, 1 mM EDTA, pH 7.3 (gassed with 95% O$_2$/5% CO$_2$).

sion usually a modified balanced salt solution is used.[6-11] In addition, in order to obtain short segments with an open lumen, mannitol is used during the perfusion.[11] Mannitol is filtered by the glomerula into the lumen of the proximal tubule and as a nonreabsorbed solute prevents the collapse of the tubules. All perfusion solutions contain some metabolic substrates such as glutamine and glucose. The inclusion of such substrates seems to be advisable when the perfusion is performed at 37° and when well-oxygenated perfusion solutions are used. The combination of substrates, oxygen, and 37° is an attempt to maintain renal function close to normal during the 2- to 5-min period required for perfusion. This attempt probably has only limited success, as studies with isolated perfused kidneys indicate. Irregular perfusion patterns and the low oxygen capacity of the perfusion solutions are severe limitations to this approach.[12-14] Another possibility is to stop the function of the renal cells as fast as possible and to reduce ischemic tissue damage. This can be achieved, for example, by chilling the kidney immediately after removal on aluminum plates and by perfusing the organ with an ice-cold solution which does not necessarily have to contain substrates or oxygen. Perfusion with specifically balanced solutions or inhibitors of electrolyte transport can also maintain tissue integrity. Furthermore, colloidal substances such as dextran or bovine serum albumin[7-11] can be included. In the perfused kidney these substances have to some extent a protective effect in maintaining the normal function of the kidney; their usefulness during the initial perfusion is not unequivocally established.

An additional perfusion is sometimes used subsequently to initiate the second phase of cell separation, the dissociation. Thereby chelating agents such as EDTA or citric acid are employed. These agents can also be included initially, as done by Heidrich and Dew,[8,9] who observed that the use of enzymes in cell dissociation can make cell surfaces extremely adhesive, precluding proper subsequent separation as, for example, by free-flow electrophoresis. Another purpose for an additional perfusion can be to specifically change the properties of the glomerula. Perfusing the kidney with iron oxide that is trapped in the glomerula makes a removal of the latter by a magnet possible.[10]

---

[6] P. Poujeol and A. Vandewalle, *Am. J. Physiol.* **249,** F74 (1985).
[7] E. P. Nord, D. Goldfarb, N. Mikhail, P. Moradeshagi, A. Hafezi, S. Vaystub, E. J. Cragoe, Jr., and L. G. Fine, *Am. J. Physiol.* **250,** F539 (1986).
[8] H.-G. Heidrich and M. E. Dew, *Curr. Probl. Clin. Biochem.* **6,** 109 (1976).
[9] H.-G. Heidrich and M. E. Dew, *J. Cell Biol.* **74,** 780 (1977).
[10] S. P. Soltoff and L. J. Mandel, *J. Gen. Physiol.* **84,** 601 (1984).
[11] S. P. Soltoff and L. J. Mandel, *J. Gen. Physiol.* **84,** 623 (1984).

TABLE II
Isolation of Renal Cortical Cells: Dissociation Methods

Vandewalle et al.[15]	Heidrich and Dew[8,9]	Garcia–Perez and Smith[16]
Incubate slices for 20 min at 25° in EDTA containing Ca^{2+}-free BBS under a stream of 95% O_2/5% CO_2 Mix slices with BBS^a + 0.25% $CaCl_2$ Pass through tissue press	Press through a fine tissue Press on 40-mesh nylon gauze Stir through gauze with 100 ml (for 3–4 g cortex) 1:1 Earle's solution (without citrate) and electrophoresis medium (11 mM triethanolamine, 11 mM acetic acid, 2.5 mM $MgCl_2$, 5 mM $CaCl_2$, 5 mM glucose, 285 mM sucrose, 5 mg% BSA, pH 7.5 with 1 M KOH)	Combine 5 g minced kidney cortex + 24 ml 0.1% collagenase (CLS II, Worthington) in Krebs buffer, pH 7.2 Incubate 40 min at 37° in plastic culture tube under water-saturated 10% CO_2 atmosphere Agitate by siphoning five times up and down through a widebore 10-ml pipet Continue incubation for 20 min at 37°

^{a} BBS, = bicarbonate-buffered solution consisting of (in mM) NaCl, 115; $NaHCO_3$, 25; NaH_2PO_4, 4; $MgSO_4$, 1; lactate, 4; alanine, 1; KCl, 5; glucose, 5; 95% O_2 – 5% CO_2; pH 7.3.

Dissociation Methods

In order to release tubule segments from the interstitium and to dissociate the epithelial cells from one another and the basement membrane, either substances which chelate calcium or enzymes acting on collagen and mucopolysaccharides are used[15–17] (see Table II). The use of chelators is advisable when the surface properties of the cells must be preserved for subsequent cell separation on the basis of surface charge[8,9] or presence of antigens[18–20] or lectins.[17,21,22] Enzymes can be applied when the separation is achieved on the basis of size and density.[6,15] Dissociation also always

[12] H.-J. Schurek and W. Kriz, *Lab. Invest.* **53**, 145 (1985).
[13] M. Brezis, S. Rosen, K. Spokes, P. Silva, and F. H. Epstein, *Am. J. Pathol.* **116**, 327 (1984).
[14] F. H. Epstein, M. Brezis, P. Silva, and S. Rosen, *Mol. Physiol.* **8**, 525 (1985).
[15] A. Vandewalle, M. Tauc, F. Cluzeaud, P. Ronco, F. Chatelet, P. Verroust, and P. Poujeol, *Am. J. Physiol.* **250**, F386 (1986).
[16] A. Garcia-Perez and W. L. Smith, *Am. J. Physiol.* **244**, C211 (1983).
[17] P. Ronco, M. Geniteau, P. Poujeol, C. Melcion, P. Verroust, and A. Vandewalle, *Am. J. Physiol.* **250**, C506 (1986).
[18] C. Koseki, Y. Yamaguchi, M. Furusawa, and H. Endou, *Kidney Int.* **33**, 543 (1988).
[19] M. A. Burnatowska-Hledin and W. S. Spielman, *Am. J. Physiol.* **254**, F907 (1988).
[20] G. Fejes-Tóth and A. Náray-Fejes-Tóth, *Am. J. Physiol.* **256**, F742 (1989).
[21] M. L. Allen, A. Nakao, W. K. Sonnenburg, M. Burnatowska-Hledin, W. S. Spielman and W. L. Smith, *Am. J. Physiol.* **255**, F704 (1988).
[22] C. Grupp, J. Steffgen, D. Cohen, S. Langhans, and H. E. Franz, *Kidney Int.* **37**, 1163 (1990).

involves mechanical disruption by mincing the tissue, passing it through a tissue press, stirring through gauze, or suctioning through pipets with defined bore diameter. The last two procedures are difficult to describe in detail in the publications. The specifics of the methods and information on potential pitfalls should be obtained directly from the authors using the particular method.

Dissociation usually takes place at room temperature to accelerate the action of chelators or enzymes. Care has to be taken that under these conditions the cells are optimally supplied with substrates and oxygen and that mechanical stress on the cells is minimized by using mild mechanical forces and preventing foaming of the tissue suspension. At the air–water interface cells can be easily destroyed. The majority of authors also uses plastic dishes for dissociation or siliconized glassware[23] in order to prevent mechanical rupture at and adsorption to the surface of the dish. Another potential source of cell damage is the harsh rims of Pasteur pipet tips. Smoothing by a quick passage through the flame of a Bunsen burner alleviates this problem.

In order to minimize cell damage the dissociation period should also be kept as short as possible; frequent analysis of the suspension by phase-contrast microscopy gives a good indication of whether the tissue is sufficiently fragmented to proceed to the cell separation step. The dissociation period for renal cortex is usually rather short; an enzymatic treatment, for example, requires only 20–30 min in comparison to 2 hr required for inner medulla.

Separation of Cells

The most commonly used method to separate tubular cells from the other major contaminant, the glomerula, involves sieving the dissociated cell suspension through meshes of decreasing pore size (see Table III). In some instances the glomerula are removed by a weak magnetic field after preloading the glomerula with iron oxide (see above). After filtration the cell suspension is composed (to about 80 to 90%) of proximal tubule cells; the remaining cells are derived from the distal tubule. A further separation of proximal tubule from distal tubule cells can be achieved by density gradient centrifugation in Percoll gradients,[23] by free-flow electrophoresis,[8,9] by immunoadsorption,[17] or by lectin-coated[22] or antibody-coated magnetic beads.[21] The latter procedure has so far predominantly been used to provide starting material for cell culture, but has recently been scaled up to yield sufficient cells for biochemical analysis and for transport studies.[18,20]

---

[23] P. Vinay, A. Gougoux, and G. Lemieux, *Am. J. Physiol.* **241**, F403 (1981).

Isolation of Medullary Cells

General Remarks

The isolation procedures described below have been used for the isolation of both outer medullary thick ascending limb of Henle's loop[24] (TALH) and collecting duct cells[25-28] and inner medullary (papillary) collecting duct cells.[29,30] In designing the isolation methods it turned out that thick ascending limb cells were particularly vulnerable when exposed to hypoxic conditions. This vulnerability was documented in studies on isolated perfused kidneys where hypoxia induced damage of TALH cells, especially when these cells were forced to perform reabsorptive work.[12-14] It is, therefore, essential to reduce cell function after removal of the kidney from the animal as effectively as possible and to pay special attention to optimum conditions for cell function during the lengthy dissociation steps.

It also must be stressed that extremely careful (but speedy) dissection of the starting material is a prerequisite for the final purity of the cells. Thus, for the isolation of TALH cells[24] the renal cortex must be trimmed off very carefully to avoid contamination with proximal tubule cells. With regard to the papillary collecting ducts, care must be taken not to include even small amounts of red outer medulla. The TALH cells present in the outer medulla have very high transport and metabolic activities and, therefore, even a small contamination of the final preparation can significantly affect the results.[31]

Perfusion Solutions and Conditions

The perfusion solutions and perfusion conditions are summarized in Table IV. As in the cortex an isotonic modified salt solution containing some substrates and protective colloids is used. The perfusion of the whole kidney with dissociating enzymes is necessary to expose the tissue as completely as possible to the enzyme action. An effective perfusion is also necessary to remove blood components which are highly concentrated in the red outer medulla.

---

[24] J. Eveloff, W. Haase, and R. Kinne, *J. Cell Biol.* **87**, 672 (1980).
[25] M. L. Zeidel, J. L. Seifter, S. Lear, B. M. Brenner, and P. Silva, *Am. J. Physiol.* **251**, F379 (1986).
[26] M. L. Zeidel, P. Silva, and J. L. Seifter, *J. Clin. Invest.* **77**, 1682 (1986).
[27] M. L. Zeidel, P. Silva, and J. L. Seifter, *J. Clin. Invest.* **77**, 113 (1986).
[28] M. L. Zeidel, P. Silva, B. M. Brenner, and J. L. Seifter, *Am. J. Physiol.* **252**, F551 (1987).
[29] J. B. Stokes, C. Grupp, and R. K. H. Kinne, *Am. J. Physiol.* **253**, F251 (1987).
[30] D. L. Dworzack and J. J. Grantham, *Kidney Int.* **8**, 191 (1975).
[31] C. Grupp, I. Pavenstädt-Grupp, R. W. Grunewald, C. Bevan, J. B. Stokes III, and R. K. H. Kinne, *Kidney Int.* **36**, 201 (1989).

TABLE III
ISOLATION OF PROXIMAL TUBULE CELLS: SEPARATION METHODS

Vandewalle et al.[15]	Poujeol and Vandewalle[6]	Nord et al.[7]	Heidrich and Dew[8,9]	Garcia-Perez and Smith[16]
Centrifuge for 5 min at 75 g	Sieve through 40-μm mesh nylon gauze	Homogenize in Wheaton Dounce homogenizer (four strokes) at 4°	Filter through 20-μm mesh nylon gauze at 4°	Centrifuge suspension at 1000 g for 5–10 min
Rotate for 3 min in rotatory evaporator	Centrifuge for 5 min at 75 g	Resuspend homogenate in 400 ml modified Hanks' solution at 25° [137 mM NaCl, 3 mM KCl, 2 mM CaCl$_2$, 0.5 mM MgCl$_2$, 0.4 mM MgSO$_4$, 0.5 mM KH$_2$PO$_4$, 0.3 mM Na$_2$HPO$_4$, 5 mM D-glucose, 4 mM L-lactate, 1 mM alanine, 0.2% (w/v) BSA]	Centrifuge 5 min at 150 g	pellet + 10 ml 0.2% saline for 30 sec + 10 ml of 1.6% saline for 30 sec. } hypotonic lysis to remove red blood cells
Sieve through 20-mesh nylon gauze	Rotate for 3 rotatory evaporator		Resuspend in 40 ml electrophoresis buffer (for composition see Table II)	Filter through several Gelman stainless steel meshes (pore size, 0.25 μm)
Centrifuge for 5 min at 75 g	Centrifuge for 5 min at 75 g		Resuspend pellet in 20 ml medium	Spin filtrate at 800 g for 10 min
Sieve through 100-μm nylon mesh		Sieve through two nylon mesh screens: upper mesh (250-μm) traps larger elements of the homogenate; lower mesh (80-μm) traps glomeruli and proximal tubules	Rotate for 2 min in rotatory evaporator	Resuspend pellet in 10% BSA/PBS, pH 7.2
Centrifuge for 5 min at 75 g			Filter through glass wool (Pasteur pipet) (retains cell aggregates and pieces of tubules)	Centrifuge for 10 min at 800 g to collect cells and remove cell debris
Pour resuspended pellet in BBS + Ca^{2+}		Scrape glomeruli and proximal tubules off the 80-μm mesh with spatula, suspend in 40 ml Hanks' solution	Spin 5 min at 150 g	Immunoadsorption follows
Sieve through 40-mesh nylon gauze			Suspend pellet in 1 ml electrophoresis medium (~40 × 10^6 cells/ml)	
		Extract ion-laden glomeruli with a magnetic stirring bar	Subject to free-flow electrophoresis	

Centrifuge enriched tubule suspensions for 2 min at 500 g

Resuspend pellet in hypotonic buffer (75 mmol/liter): 5 mM KCl, 1 mM Na$_2$HPO$_4$, 5 mM D-glucose, 1 mM lactate, 1 mM alanine, 26 mM NaHCO$_3$, 0.2% BSA, pH 7.35–7.40

Agitate suspension in a shaker bath at 280 cpm for 60–90 sec to release cells from tubular basement membranes

Resuspend in modified Hanks' solution

Filter sequentially through 80- and 20-μm mesh and a column of glass wool packed in a 10-ml pipet to remove tubule fragments

Repeat passage through 20-μm mesh two times

Centrifuge for 2 min at 500 g

Resuspend cells in buffer

TABLE IV
ISOLATION OF MEDULLARY CELLS: INITIAL PERFUSION SOLUTIONS[a]

Eveloff et al.[24]	Zeidel et al.[25,26]
	Joklik's buffer
Salts	
NaCl (111.1 mM)	Phenolsulfonphthalein (0.03 mM)
	Asparagine (0.8 mM)
NaHCO$_3$ (23.8 mM)	Glycine (4.0 mM)
	Histidine (0.15 mM)
NaH$_2$PO$_4$ (9.6 mM)	Leucine (0.4 mM)
	Lysine (0.8 mM)
	Phenylalanine (0.1 mM)
	Proline (0.3 mM)
KCl (5.4 mM)	Tryptophan (0.2 mM)
	Tyrosine (0.3 mM)
	Valine (0.4 mM)
	Panthothenate (0.002 mM)
	Choline (0.01 mM)
	Folic acid (0.002 mM)
CaCl$_2$ (1.1 mM or 0.2 mM)	Inositol (0.01 mM)
	Nicotinamide (0.01 mM)
	Pyridoxal (0.01 mM)
	Riboflavin (0.0003 mM)
	Thiamin (0.003 mM)
	Penicillin G (75 U/ml)
	Streptomycin (50 mg/ml)

Substrates
  Glucose (11.1 mM)
Specific additions
  Collagenase (0.2%)
  Hyaluronidase (0.25%)
pH
  7.4 (gassed with 5% CO$_2$/95% O$_2$)
Colloids
  Fetal calf serum (10%)
Temperature
  4°
Perfusion rate
  50 ml in ~4 min

[a] Additional perfusion: 150 ml Joklik's buffer + 0.2% collagenase and 0.25% hyaluronidase for isolation of TALH cells and outer medullary collecting duct cells; for papillary collecting duct cells, hyaluronidase is omitted.

Dissociation Methods

As depicted in Table V, the dissociation of the medullary cells requires a combination of enzymatic treatment and mechanical disruption of the tissue. In addition, long periods of incubation are required to disrupt the connective tissue surrounding the tubules. The dissociation usually occurs in two steps: first tubular segments[32] are released, then single cells are obtained.[24] Both tubular segments and dissociated cells can be used as starting material for further purification.[24,32] In some instances the cells cannot be dissociated completely and clumps of cells remain which resist all mechanical and enzymatic treatment. This has been observed, for example, for rat papillary collecting duct cells.[29]

Separation of Cells

Since the various cells present in the outer and inner medulla differ in their size and their nucleus-to-cytoplasm ratio, density gradients are the preferred tool for separation (see Table VI). The density gradients consist of osmotically almost inert materials, such as Ficoll or Nycodenz,[33,34] dissolved in a balanced saline solution.[24-26]

Another approach to cell separation takes advantage of the resistance of certain renal cells, such as the collecting duct cells, to hypotonic lysis.[35] By exposing a mixed cell population of papillary origin to a medium with low osmolality cells other than collecting duct cells are lysed. The intact collecting duct cells can consequently be separated easily from the debris of the other cells. Cells isolated by this procedure have been used mainly as starting material for primary cell cultures; the viability of these cells and their purity have, however, not been studied systematically. Also, immunodissection has been employed to isolate mitochondria-rich cells from other medullary collecting tubules.[19]

Characterization of Isolated Cells

Morphology

Isolated cells can be identified by their specific morphology, specific enzyme content, or presence of specific surface markers. The morphologi-

---

[32] M. E. Chamberlin, A. LeFurgey, and L. J. Mandel, *Am. J. Physiol.* **247,** F955 (1984).
[33] D. Rickwood, T. Ford, and J. Graham, *Anal. Biochem.* **123,** 23 (1982).
[34] T. C. Ford and D. Rickwood, *Anal. Biochem.* **124,** 293 (1982).
[35] F. C. Grenier and W. L. Smith, *Prostaglandins* **16,** 759 (1978).

TABLE V
ISOLATION OF MEDULLARY CELLS: DISSOCIATION METHODS

Rabbit kidney: TALH cells and outer medullary collecting duct cells (Eveloff et al.[24])	Rabbit kidney: inner medullary collecting duct cells (Zeidel et al.[25,26])	Rat kidney: papillary collecting duct cells (Stokes et al.[29])	Rat kidney: papillary collecting duct cells (Dworzack and Grantham[30])	Dog kidney: papillary collecting duct cells (Grenier and Smith[35])
Excise outer medulla taking care to remove all cortical tissue	Excise inner medulla	Mince the carefully dissected papilla (avoid contamination with red outer medulla)	Mince dissected papilla	Mince dissected papilla (0.5 g) with razor blades
Mince medulla	Mince medulla	Incubate for 45 min (12 papilla/10 ml) at 37° in 15 ml plastic centrifuge tube	Place in 10 ml incubation medium in a 300-ml siliconized centrifuge tube	Place in 3 ml Krebs–Ringer buffer, pH 7.5, containing trypsin (0.5 mg/ml) at room temperature
Incubate in Joklik's buffer: [containing 0.2% collagenase, 0.25% hyaluronidase, and 10% fetal calf serum (FCS)]: 40 ml for two kidneys, 1 hr at 37°, pH 7.4, gassed with 95% O_2/5% CO_2	Incubate in 0.2% collagenase in Joklik's buffer for 2 hr at 37° pH 7.4, gassed with 95% O_2/5% CO_2	Incubation medium: NaCl (118 mM), HEPES (16 mM), Na-HEPES (17 mM), glucose (14 mM), KCl (3.2 mM), $CaCl_2$ (2.5 mM), $MgSO_4$ (1.8 mM), KH_2PO_4 (1.8 mM), pH 7.4, with 0.2% collagenase, 0.2% hyaluronidase, gassed continuously with room air	Incubation medium: NaCl (150 mM), KCl (5 mM), NaH_2PO_4 (1.2 mM), $NaHCO_3$ (25 mM), $CaCl_2$ (1 mM), $MgSO_4$ (1.2mM), glucose (5 mM), sodium acetate (10 mM), urea (300 mM), calf serum (5%), collagenase (2 mg/ml), gently agitated by bubbling with 5% CO_2/95% O_2	Draw tissue up and down through a disposable pipet (3-mm bore) 15–20 times for 5 min and filter through stainless steel mesh (0.25-mm² pore size)
Mechanically dissociate clumps by suction (10 times) through a wide-mouth syringe	Mechanically dissociate clumps by suction (10 times) through a wide-mouth syringe	After 45 min add DNase to 0.001%	Incubate for 60 min at 23–26°	Remove residue from filter
Sediment dissociated tubules for 3 min at 50 g	Layer cell suspension on 16% Ficoll cushion (see text)		Spin down tubular fragments	Resuspend in 3 ml Krebs–Ringer and completely disperse tissue by repeated suction through a disposal pipet (1-mm bore) (15–20 times/min), filter through stainless steel mesh
Wash tubules once with Joklik's buffer + FCS				

- Transfer in 50 ml Joklik's buffer + FCS and 0.25% trypsin
- Incubate 20 min at 22°, gassed with 95% O_2/5% CO_2
- Sediment tubules by centrifugation (25 sec at 750 g)
- Resuspend sediment in 20 ml Joklik's buffer + FCS and 0.25% trypsin
- Centrifuge supernatant for 5 min at 500 g to collect cells
- Repeat the above seven times
- Store cells at 4° in Joklik's buffer + FCS until further use
- Combine cells, centrifuge for 5 min at 500 g, resuspend in 20 ml Joklik's buffer + FCS, filter through Teflon mesh (36-μm grid size)

- Continue incubation for 45 min
- Aspirate cells through a Pasteur pipet 10–12 times every 15 min to dissociate cell clumps

- Resuspend in fresh collagenase medium
- Incubate additional 60 min

TABLE VI
ISOLATION OF MEDULLARY CELLS: SEPARATION METHODS

TALH cells and outer medullary collecting duct cells		Papillary collecting duct cells		
Eveloff et al.[24]	Chamberlin et al.[32]	Rabbit (Zeidel et al.[25,26])	Rat (Stokes et al.[29])	Rabbit (Dworzack and Grantham[30])
Prepare Ficoll-400 in 137 mM NaCl, 5 mM KCl, 0.33 mM Na$_2$HPO$_4$, 0.44 mM KH$_2$PO$_4$, 1.mM CaCl, 1 mM MgCl$_2$, 0.8 mM MgSO$_4$, 29 mM Tris-HCl, pH 7.4 (10 ml 30.7% cushion, 60 ml 30.7% → 2.6%; tube, length 16 cm, i.d. mm)	Prepare isotonic 55% Percoll solution in 115 mM NaCl, 25 mM NaHCO$_3$, 2 mM NaH$_2$PO$_4$, 5 mM KCl, 35 mM mannitol, with no Ca^{2+} or Mg^{2+}, gassed with 95% O$_2$/5% CO$_2$	Layer mixed papillary cell suspension on 16% Ficoll gradient prepared in Joklik's medium (12)	Centrifuge cell suspension for 2 min at 28 g	separate by differential sedimentation at 1 g
	Combine 3 ml Percoll solution with 0.4–0.6 ml tubule suspension	Centrifuge for 45 min at 2300 g; inner medullary collecting duct cells are located at the top of the 16% Ficoll layer	Discard supernatant	Dilute filtrate 1:2 with H$_2$O and allow to stand for 4–5 min
			Resuspend in collagenase- and hyaluronidase-free buffer (see text); add 0.001% DNase	add 3–4 ml Krebs–Ringer solution
			Repeat this procedure twice	Collect cells by centrifugation at 500 g for 2 min

Place 10 ml of cells in Joklik's + FCS on continuous Ficoll gradient	Place in 4-ml centrifuge tube and, spin for 30 min at 20,000 g at 4°
Spin 40 min at 2100 g_{max} at 4°	Obtain,three regions: 1.015–1.030 g/ml (cell debris) 1.045–1.060 g/ml (undefined tubules) 1.063–1.070 g/ml (TALH tubules)
Collect cells in 4-ml fractions with a wide-mouth pipet	
Dilute samples 10 times with Joklik's buffer	
Sediment cells for 5 min at 4000 g_{max}	Dilute fractions in salt solution given above + 10 mM D-glucose, 0.8% dextran, 1 mM MgSO$_4$, 1 mM CaCl$_2$, adjust mannitol concentration to 30 mM

cal identification of renal cells released from their tubular structure and from their neighboring landmarks is sometimes difficult, therefore the features summarized in Table VII have proved useful in distinguishing various cells.[24] Additional features can be found in reviews on the morphology of the kidney.[36] Figure 1 gives representative examples for the morphological appearance of some typical isolated cells. Preliminary morphological identification can also be performed at a light microscopic level using phase contrast or Nomarsky optics. The definite identification is, however, possible only by electron microscopy. A particular feature of proximal tubule cells is the extensive enlargement of the lumenal surface by numerous long microvilli. These surface structures can easily be identified by scanning electron microscopy.[8,9]

Cellular Components and Cell Surface Markers

An important and feasible tool for cell identification is the presence of specific components inside the cell or at the cell surface (see Table VIII). These components usually comprise cytosolic or membrane-bound enzymes[37] and lectin-[38] or antibody-binding sites.[39,40] Thus distal tubule cells can be identified and purified by the use of cell-specific monoclonal antibodies,[17] the thick ascending limb of Henle's loop cell carries Tamm–Horsfall protein[41] and some of the intercalated cells in the distal tubule and collecting duct stain positive with an antibody against the band 3 protein of the erythrocyte.[42] Selective lectin binding can be used to estimate the number of papillary collecting duct cells in rat and rabbit *(Dolichos biflorus)* or some of the intercalated cells *(Arochis hypogaea)*.[43-47]

---

[36] B. Kaissling and W. Kriz, *Adv. Anat. Embryol. Cell Biol.* **56,** 1 (1979).
[37] A. I. Katz, A. Doucet, and F. Morel, *Am. J. Physiol.* **237,** F114 (1979).
[38] D. Brown, J. Roth, and L. Orci, *Am. J. Physiol.* **248,** C348 (1985).
[39] V. L. Schuster, S. M. Bonsib, and M. L. Jennings, *Am. J. Physiol.* **251,** C347 (1986).
[40] M. LeHir, B. Kaissling, B. M. Koeppen, and J. B. Wade, *Am. J. Physiol.* **242,** C117 (1982).
[41] E. A. Schenk, R. H. Schwartz, and R. A. Lewis, *Lab. Invest.* **25,** 92 (1981).
[42] D. Drenckhahn, K. Schlüter, D. P. Allen, and V. Bennett, *Science* **230,** 1287 (1985).
[43] M. Le Hir and U. C. Dubach, *Histochemistry* **74,** 521 (1982).
[44] M. Le Hir and U. C. Dubach, *Histochemistry* **74,** 531 (1982).
[45] M. Le Hir, B. Kaissling, B. M. Koeppen, and J. B. Wade, *Am. J. Physiol.* **242,** C117 (1982).
[46] R. G. O'Neil and R. A. Hayhurst, *Am. J. Physiol.* **248,** F449 (1985).
[47] H. Holthöfer, *J. Histochem. Cytochem.* **31,** 531 (1983).

TABLE VII
STRUCTURAL MARKERS OF IDENTIFICATION OF SINGLE ISOLATED CELLS FROM THE RABBIT RENAL MEDULLA[a]

Cells	Plasma membrane	Nucleus	Mitochondria	Cytoplasmic inclusions
Proximal tubule (pars recta)	Long microvilli, thick basolateral membrane infoldings	Round, elongated, often indented	Small, ovoid, dark	Numerous vesicles (mean diameter ~0.7 μm; electron dense material)
TALH	A few short microvillus projections, basolateral membrane is highly infolded, appearing as vesicles	Elongated with indentations	Numerous, large, dark, densely packed, filling up most of cell, mostly round, larger than in proximal and collecting tubule cells, cristae visible	Vesicles mostly smaller than mitochondria
Collecting duct (light or principal cells)	A few short microvilli, no infoldings	Large, round	Dark, small, less numerous than in proximal tubule	Small, light vesicles (mean diameter ~0.3 μm)
Thin loop, interstitial endothelial cells	Irregular surface with microvillous-like extrusions	Irregular shape	Very few	Some vesicles; cytoplasma is only a small ring around nucleus

[a] Adapted from Eveloff et al.[24]

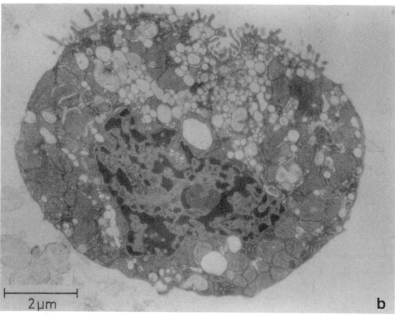

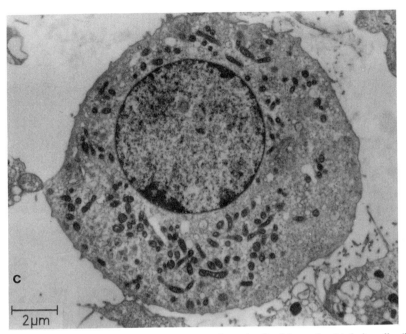

FIG. 1. Morphological appearance of isolated renal cells. (a) proximal tubule cells. (b) medullary thick ascending loop of Henle cell. (c) inner medullary collecting duct cell. Fig. 1a reprinted from Heidrich and Dew[9] and Fig. 1b and c from Eveloff et al.[24] by permission.

Transport-Related Parameters in Isolated Cells

Identification of Sodium Uptake Systems by Determination of Metabolic Activity

Since in epithelial cells, as in other cells, each sodium ion entering the cell is ultimately removed from the cell by active transport via Na^+, K^+-ATPase,[48-50] sodium movement through the cell is combined with an increased metabolic activity. The most frequently used indicators of this activity is O_2 consumption. The degree of commitment of the cell to transport and various transport pathways can be identified by ion replacement experiments and the use of inhibitors that interact specifically with different sodium pathways.

[48] S. I. Harris, R. S. Balaban, and L. J. Mandel, *Science* **208**, 1148 (1980).
[49] S. I. Harris, R. S. Balaban, L. Barrett, and L. J. Mandel, *J. Biol. Chem.* **256**, 10319 (1981).
[50] R. Kinne, *Miner. Electrolyte Metab.* **9**, 270 (1983).

TABLE VIII
Markers for Renal Cells

Proximal tubule cells	Cells of the thick ascending limb of Henle's loop	Distal tubule cells	Intercalated cells	Collecting duct cells
Enzymes: Alkaline phosphatase, aminopeptidase M, γ-Glutamyltranspeptidase Phosphoenolpyruvate-carboxykinase[54]	Calcitonin-sensitive adenylate cyclase (rabbit)[2] Na$^+$,K$^+$-ATPase[1,37]			Arginine/vasopressin-sensitive adenylate cyclase[2] Hexokinase (and other glycolytic enzymes)[38,54]
Antigens:	Tamm–Horsfall protein[41]		Band III protein[42]	
Lectin-binding sites[38]: Monoclonal antibodies	Monoclonal antibodies	Various monoclonal antibodies	Peanut lectin	*Dolichos biflorus* (rat)[29]

A summary of the results obtained from O_2 consumption studies on isolated cells (or highly purified tubules) is given in Table IX. The degree of commitment of a particular renal cell to sodium transport can be estimated from the degree of inhibition of respiration by removal of sodium or by addition of ouabain, the specific inhibitor of the Na^+, K^+-ATPase[51]. The data obtained so far show that ouabain has its strongest action on the cells in the thick ascending limb of Henle's loop. About 50% of the energy production in these cells is dedicated to sodium transport. In cells derived from cortical proximal tubules the inhibition averages about 30%. In collecting duct cells of the rabbit, the inner medullary or papillary collecting duct cells show an inhibition of O_2 uptake by about 25%, whereas the inhibition in rat papillary cells is even smaller. The pathways used by sodium to enter the cells can be defined from such energetic studies as follows. In rat proximal tubule, anion replacement does not affect sodium entry-related respiration, suggesting different pathways for sodium and chloride. In the thick ascending limb sodium uptake (and therefore O_2 consumption) is dependent on the presence of chloride and is inhibited by low concentrations of loop diuretics such as furosemide or bumetanide.[52] This sodium uptake is mediated by a $Na^+-K^+-2Cl^-$ cotransport system that has also been extensively characterized in other physiological and biochemical studies. There seems to be a (small) additional sodium entry because ouabain has a small but significant effect in the presence of bumetanide or in the absence of chloride. This entry step seems to be insensitive to low concentrations of amiloride and its nature remains to be determined. In the rabbit and rat papillary cells an amiloride-sensitive O_2 consumption can be demonstrated, suggesting the presence of an amiloride-sensitive sodium channel in these cells.[53] Thus, the findings in the intact epithelium and in the isolated cells basically compare very favorably.

It should be emphasized in this context that CO_2 production has also been used as a parameter to determine sodium-dependent energy expenditure.[54] Such data are in very good agreement with those obtained in O_2 consumption measurements.

In cells in which the majority of energy production proceeds via anaerobic pathways, such as glycolysis, lactate production under conditions of inhibited mitochondrial respiration seems to be a more suitable indicator of metabolic activity. In rat papillary cells, for example, O_2 consumption is

[51] R. S. Balaban, S. P. Soltoff, J. M. Storey, and L. J. Mandel, *Am. J. Physiol.* **238**, F50 (1980).
[52] J. Eveloff, E. Bayerdörffer, P. Silva, and R. Kinne, *Pfluegers Arch.* **389**, 263 (1981).
[53] M. L. Zeidel, D. Kikeri, P. Silva, M. Burrowes, and B. M. Brenner, *J. Clin. Invest.* **82**, 1067 (1988).
[54] R. K. H. Kinne (ed.), "Renal Biochemistry: Cells, Membranes, and Molecules." Elsevier, Amsterdam, 1985.

TABLE IX
SODIUM TRANSPORT-RELATED O$_2$ CONSUMPTION

	Proximal tubule cells	TALH cells[a]	Cortical distal cells	Outer medullary collecting duct cells	Papillary collecting duct cells
Inhibition by ouabain	Glucose, lactate, alanine[49] (25%)[b] Glucose, lactate, alanine[51] (45%) Acetate + glucose[52] (27%)	Acetate + glucose[24] (52%) Glucose[32] (43%)	Lactate + alanine[15] (31%) Lactate + alanine + succinate[15] (38%)	Pyruvate + acetate[25] (9%)	Pyruvate + acetate[25,28] (26.7%; rabbit) Glucose (10%; Sprague–Dawley rats) Glucose (15%; Wistar rats)
Inhibition by amiloride		Acetate + glucose (0%)		Pyruvate + acetate[25] (1.6%)	Pyruvate + acetate[25] (22%; rabbit) Glucose[51] (19%; Wistar rats) Glucose[51] (5%; Sprague–Dawley rats)
Stimulation by addition of K$^+$	Succinate + glucose[48] (31%)				
Inhibition by loop diuretics	Acetate + glucose[52] (19%)	Acetate + glucose[24] (50%) Glucose[23] (43%)			Glucose[51] (5%; Sprague–Dawley rats) Glucose[51] (5%; Wistar rats)
Inhibition by Cl$^-$ removal	Acetate + glucose[52] (7%)	Acetate + glucose[24] (50%) Glucose[32] (43%)			
Inhibition by Na$^+$ removal	Acetate + glucose[52] (32%)	Acetate + glucose[24] (50%)			

[a] TALH, Thick ascending limb of Henle's loop.
[b] Numbers in parentheses indicate percentage of cell's energy production dedicated to sodium transport.

not significantly affected by furosemide whereas a significant and large reduction of lactate production by furosemide can be observed.[31]

Measurements of Ion Fluxes

Transport of ions across the plasma membrane can be detected directly by a variety of techniques. First, the cell ion content can be maintained at its physiological level and fluxes can be determined with the use of radioisotopes. Second, the cells can be depleted of particular ions, and the reuptake of solutes is investigated by measuring the change in the intracellular concentration with time. For both techniques it is necessary to separate at a certain point isolated cells and incubation medium and to correct for extracellular fluid adhering to the isolated cells. Separation of cells from the surrounding medium can be achieved quite effectively by centrifugation through a silicon oil layer[7] or by a rapid filtration technique.[55] In the latter method cells are retained on a filter kept under light suction and the extracellular fluid is removed by repeated washing. A modification of the latter technique is to pellet cells onto a suitable filter support and to rinse this support by several insertions into buffer solutions. The silicon layer technique can be adapted to cells of different density and is theoretically the method of choice because no washing steps are involved during which the cellular ion content might be altered. Sometimes, however, this method is not suitable because the cells of interest aggregate and thus do not penetrate the silicon layer or include too much extracellular medium. It is also possible that the cells remain as cell clumps, which raises the same problems as stated above. The filtration and sedimentation methods are relatively easy; however, loss of intracellular ions can occur during the washing steps. A quite effective washing solution which prevents ion leakage is an ice-cold buffered ammonium acetate solution. Ammonium acetate apparently blocks K^+ channels and a lowered temperature reduces ion fluxes quite effectively. Another possibility is to use known transport inhibitors such as amiloride, bumetanide, barium, and 9-carboxyanthracen, during removal of the extracellular medium.[56]

Correction for extracellular space is routinely done by using substances that do not enter cells readily,[6,7,55] such as radiolabeled inulin, polyethyleneglycol, or sucrose. However, attention must be paid to several potential problems with these substances. First, inulin might contain traces of small breakdown products which can enter the cell, thus leading to an overestimation of the extracellular water content. Underestimation can occur when the high-molecular-weight marker does not penetrate all com-

[55] I. Pavenstädt-Grupp, C. Grupp, and R. K. H. Kinne, *Pfluegers Arch.* **413**, 378 (1989).
[56] I. Pavenstädt-Grupp, C. T. Grupp, H. E. Franz, and R. K. H. Kinne, in "Microbeam Analysis–1989", (P. E. Russel, ed.), p. 107. San Francisco Press, Inc., San Francisco, 1989.

partments of the cell which represent extracellular space. Such sometimes inaccessible spaces can be, for example, extensive basolateral infoldings of epithelial cells as in the thick ascending limb of Henle's loop.

The ion content of the cell sample is usually determined by atomic absorption spectroscopy or when isotopes have been used by liquid scintillation techniques. The values are then corrected for the extracellular ion content by using the inulin space. In most instances the water content of the cells is also determined by measuring the differences between the wet weight and the dry weight of the sample or from the difference between the total water content measured by tritiated water and the inulin space. Values for intracellular water are about $2-3$ μl/mg cell protein or $2-3 \times 10^{-12}$ liter/cell. If the cell suspension consists of single nonaggregated cells their volume can also be estimated with a Coulter counter.[57]

Use of Indicators

No separation of the cells from the extracellular medium is necessary when intra- and extracellular ions can be distinguished with indicators which are restricted either to the intracellular or the extracellular compartment. Indicators such as 6-carboxyfluorescein acetate enter the cell, are intracellularly deesterified, and exhibit a pH-dependent fluorescence which can be monitored in single cells by a microfluorometer or in cell suspensions.[26,27,58] Because 6-carboxyfluorescein acetate can also be hydrolyzed extracellularly and 6-carboxyfluorescein tends to leak slowly but appreciably out of the cells, a correction for changes in extracellular fluorescence must be made for cell suspensions by addition of acid to the outside compartment. Similarly, the intracellular fluorescence signal must be corrected for changes in intracellular concentration by measuring the fluorescence of the indicator at a wavelength where no pH dependence of the signal is found. Specific equipment has been developed for such procedures; "emission wavelength ratios" or "excitation wavelength ratios" can be employed. There are now indicators available for intracellular pH, Ca^{2+}, Na^+, and Cl^-.[59-67] Another, naturally abundant, intracellular pH indicator is inorganic phosphate. Depending on its protonation it possesses a differ-

[57] C. Knoblauch, M. H. Montrose, and H. Murer, *Am. J. Physiol.* **256,** C252 (1989).
[58] J. R. Chaillet, K. Amsler, and W. F. Boron, *Proc. Natl. Acad. Sci. U.S.A.* **83,** 522 (1986).
[59] W. R. Jacobs and L. J. Mandel, *J. Membr. Biol.* **97,** 53 (1987).
[60] S. Olsnes, T. I. Tønnessen, J. Ludt, and K. Sandvig, *Biochemistry* **26,** 2778 (1987).
[61] W. V. Breuer, E. Mack, and A. Rothstein, *Pfluegers Arch.* **411,** 450 (1988).
[62] R. R. Ratan, M. L. Shelanski, and F. R. Maxfield, *Proc. Natl. Acad. Sci. U.S.A.* **83,** 5136 (1986).
[63] P. Fong, N. P. Illsley, J. H. Widdicombe, and A. S. Verkman, *J. Membr. Biol.* **104,** 233 (1988).

ent electromagnetic shift in nuclear magnetic resonance (NMR) investigations. The shift of the intracellular phosphate thus is pH dependent and has been used to study intracellular pH regulation. Nuclear magnetic resonance has also been used for the determination of intracellular sodium content and transmembranal sodium fluxes. In this instance an extracellular "shift reagent" allows the distinction between the intra- and extracellular compartment.[68-71]

As a further method electron probe microanalysis has been applied recently to the study of ion fluxes in epithelial cells.[55,72,73] In this method, after removal of the extracellular medium, single cells or cell clusters are quick frozen and freeze dried below the critical point for water in order to prevent loss of ions from the cells. The cells are then analyzed in a scanning electron microscope equipped with an energy-dispersive X-ray detector. Depending on its atomic number each element emits X rays of a particular energy and can thus be discriminated and determined quantitatively. The advantage of this method is that several physiologically interesting ions, such as sodium, potassium, and chloride, can be detected simultaneously in the same cells. In addition, the number of cells required for the analysis is minute, using isolated collecting duct cells which are obtained in the form of clusters of 20-30 cells; 10 clusters or 300 cells are sufficient to

---

[64] J. A. Thomas, in "Optical Methods in Cell Physiology," p. 311. Wiley (Interscience), New York, 1986.

[65] H. E. Edwards, J. K. Thomas, G. R. Burleson, and C. F. Kulpa, *Biochim. Biophys. Acta* **448**, 451 (1976).

[66] P. A. Negulescu, A. Harootunian, A. Minta, R. Y. Tsien, and T. E. Machen, *J. Gen. Physiol.* **92**, 26a (1988).

[67] G. Grynkiewicz, M. Poenie, and R. Y. Tsien, *J. Biol. Chem.* **260**, 3440 (1985).

[68] S. R. Gullans, M. J. Avison, T. Ogino, G. Giebisch, and R. G. Shulman, *Am. J. Physiol.* **249**, F160 (1985).

[69] Y. Boulanger, P. Vinay, and M. Boulanger, *Am. J. Physiol.* **253**, F904 (1987).

[70] A. M. Kumar, R. K. Gupta, and A. Spitzer, *Proc. Meet. Am. Soc. Nephrol., 18th* p. 207A (1985).

[71] A. W. H. Jans, R. Willem, E. R. Kellenbach, and R. K. H. Kinne, *Magn. Reson. Med.* **7**, 292 (1988).

[72] J. L. Eveloff and J. Calamia, *Am. J. Physiol.* **250**, F176 (1986).

[73] R. C. Harris, J. L. Seifter, and C. Lechene, *Am. J. Physiol.* **251**, C815 (1986).

[74] J. G. Haggerty, E. J. Cragoe, Jr., C. W. Slayman, and E. A. Adelberg, *Biochem. Biophys. Res. Commun.* **127**, 759 (1985).

[75] H. Cantiello, I. Thompson, and C. Rabito, *Fed. Proc.* **43**, 448 (1984).

[76] A. W. H. Jans, K. Amsler, B. Griewel, and R. H. K. Kinne, *Biochim. Biophys. Acta* **927**, 203 (1987).

[77] A. W. H. Jans, E. S. Krijnen, J. Luig, and R. H. K. Kinne, *Biochim. Biophys. Acta* **931**, 326 (1987).

TABLE X
TRANSPORT SYSTEMS IN ISOLATED RENAL CELLS

Cell type	Method	Separation of intra- and extracellular compartment	Transport system	Reference
Proximal tubule	^{22}Na uptake DMO (5,5-dimethyloxazoli-dine-2,4-dione) distribution	Centrifugation through silicon oil mixture	Na$^+$/H$^+$ exchanger	Nord et al.[7]
	D-[methyl-^{14}C] glucose uptake		Na$^+$-D-glucose cotransporter	Nord et al.[7]
	[^{32}P]Phosphate uptake		Na$^+$-phosphate cotransporter	Nord et al.[7]
Proximal tubule	[^{32}P]Phosphate uptake	Sedimentation 1 min in microfuge	Na$^+$-phosphate cotransporter	Poujeol and Vandewalle[6]
	[^{3}H]Glucose uptake		Na$^+$-D-glucose cotransporter	Poujeol and Vandewalle[6]
LLC-PK$_1$ cells	6-Carboxyfluorescein	Microfluorometry of single cells	Na$^+$/H$^+$ exchanger Cl$^-$/HCO$_3^-$ exchanger	Chaillet et al.[58] Chaillet et al.[58]
	^{22}Na uptake		Na$^+$/H$^+$ exchanger	Haggerly et al.[74]

Cell/Tissue	Method	Detection	Transporter	Reference
LLC-PK$_1$ cells	^{31}P NMR	Different resonances of intra- and extracellular phosphate	Na$^+$/H$^+$ exchanger Cl$^-$/HCO$_3^-$ exchanger	Cantiello et al.[75] Jans et al.[76,77]
LLC-PK$_1$ cells	^{22}Na NMR	Extracellular shift reagent [Dy(PPP$_i$)]$_2^{7-}$	Na$^+$/H$^+$ exchanger Na$^+$-D-glucose cotransporter	Jans et al.[71] Jans et al.[71]
Thick ascending limb	^{22}Na uptake ^{86}Rb uptake	Rapid filtration method	Na$^+$-K$^+$-Cl$^-$ cotransporter	Eveloff and Calamia[72]
	O$_2$ consumption	Not necessary	Na$^+$-K$^+$-Cl$^-$ cotransporter	Chamberlin et al.[32] Eveloff et al.[24]
Medullary collecting duct	6-Carboxyfluorescein	Fluorescence measurement in cell suspension	H$^+$-ATPase Cl$^-$/HCO$_3^-$ exchanger	Zeidel et al.[26,28] Zeidel et al.[26,28]
Papillary collecting duct	Electron probe X-ray microanalysis	Sedimentation	Amiloride-sensitive Na$^+$ channel Na$^+$-K$^+$-Cl$^-$ cotransporter	Pavenstädt et al.[55]
	Lactate production	Not necessary	Amiloride-sensitive Na$^+$ channel	Grupp et al.[31] Zeidel et al.[25,28]
	O$_2$ consumption	Not necessary	Amiloride-sensitive Na$^+$ channel	Zeidel et al.[25,28]

obtain reproducible and consistent results. The drawback of the method, which has been employed previously with some modifications in whole epithelia, kidney sections, and primary cultures of renal cells, is to obtain intracellular concentrations since the determination of the water content in such a small number of cells has not yet been possible. To date, therefore, usually the phosphorus content of the cells, representing mainly the phosphorus in the nucleus, is used as the point of reference.

Transport Systems in Isolated Renal Cells

In Table X some studies are compiled in which isolated renal cells have been used to demonstrate the presence of transport systems, to study their regulation, or to study the maintenance of the intracellular milieu. The table also includes some work on cultured cells derived from renal tissue in order to broaden the spectrum of methods available to study transport processes in isolated cells. It is noteworthy that, compared to studies using intact tubules, all transport systems found in the tubules have also been demonstrated in the isolated cells. In several instances studies with isolated cells were the first studies to demonstrate a transport function or its regulation in a renal cell. This holds, for example, for the H^+-ATPase in the medullary collecting duct,[27] for the stimulation of the Na^+-K^+-Cl^- cotransport system by cell shrinkage in the thick ascending limb of Henle's loop,[72] and for the action of the atrial natriuretic factor and cGMP on the amiloride-sensitive sodium channel in rabbit papillary collecting ducts.[25,28] Thus, studies on isolated cells are complementary to other approaches to renal physiology at higher levels of organization (intact kidney, microdissected tubule) as well as at a lower level (plasma membrane vesicles and isolated transport proteins).

Concluding Remarks

The successful separation of homogeneous populations of renal cells represents an additional step to resolve the complexity of the renal tubule with its longitudinal and spatial inhomogeneity. Although separation methods for only a few cell types have been established it can be foreseen that the availability of cell-specific antibodies in combination with immortalization techniques for differentiated epithelial cells[78] will increase the specificity of selection and the availability of a great number of renal cells which currently are not accessible to studies at the cellular level.

Such studies can be criticized due to the fact that epithelial cells are separated from their nearest neighbors and thereby their properties may

[78] D. M. Scott, C. MacDonald, H. Brzeski, and R. Kinne, *Exp. Cell Res.* **166**, 391 (1986).

change. The evidence available to date indicates, however, that this seems not to be the case for the transport properties which have been investigated. Nevertheless, it must always be borne in mind that channels and transport systems might be modified by the isolation procedure or by the experimental procedure used to measure transport in isolated cells. This uncertainty is, however, compensated for by the fact that studies with intact single cells provide, under well-defined conditions, information on the regulation of intracellular parameters, the activity of various transport systems, and about the mutual interaction and coupling of such transport processes. Thus transport studies with isolated cells represent the first step into a higher order of organization where information about transport systems obtained from purified systems or in plasma membrane vesicles is integrated into the minimum living unit of the body, the cell. Understanding of these cellular processes is indispensable for the understanding of the function of organs such as the kidney.

Acknowledgment

The author wishes to thank Mrs. Daniela Mägdefessel for skillful secretarial work.

[23] Primary Culture of Isolated Tubule Cells of Defined Segmental Origin

By MICHAEL F. HORSTER and MASAYOSHI SONE

Introduction

Primary cultures of cells derived from defined nephron segments of the mammalian kidney, in the past decade, have become a most valuable tool with which to study segmental renal cell functions *in vitro*. There are obvious advantages to this technique. The epithelial cell population is homogeneous, e.g., cells from the proximal convoluted tubule and those from the proximal straight tubule can be cultured separately. The segmental cell population may be cultured as a polarized monolayer on solute- and fluid-permeable supports, thereby rendering the epithelium accessible from its apical or its basolateral side for studies on bidirectional vectorial solute transport and electrophysiological analysis. The defined epithelial monolayer can be exposed to asymmetric media to elucidate the distribution of cytoplasmic membrane transporters and hormone receptors. Primary homogeneous cultures of defined nephron cells contribute to distinguishing between processes of epithelial growth and those of segment-specific epithelial differentiation, e.g., by selective hormonal supplementation of the media.

Most of the disadvantages of primary cultures of defined nephronal origin have, by now, been minimized. The low cell number of primary cultures has been balanced by the refinement of analytical techniques, e.g., the analysis of intracellular free calcium by fluorescent dyes. The loss of segment-specific transport characteristics with time in culture has been successfully limited by improvements in cell isolation procedures, in the substrata, and in media composition.

The present state of primary culture of segmental nephron populations or subpopulations (e.g., intercalated cells) will be outlined in two major sections: Culture Techniques and Segmental Epithelial Transport Functions in Culture. Many of the primary culture techniques are not only segment specific but also particularly designed for the type of question to be asked; obviously, the data only testify to the quality and specificity of any culture technique.

Culture Techniques

Cell Isolation for Culture

Proximal Convoluted Tubule. To isolate cells from the proximal convoluted tubule (PCT),[1] three major procedures have been applied successfully: (1) Enzymatic disaggregation of cortical tissue and subsequent separation of nephron epithelia from glomeruli, (2) purification of cortical epithelial fragments by discontinuous density gradient centrifugation, and (3) microdissection of individual PCT segments.

Enzymatic disaggregation[4] of cortical tissue is initiated by perfusion of the kidney with phosphate-buffered saline (PBS) and subsequent perfusion with a 0.5% solution of iron oxide which is trapped in glomeruli. Cortical tissue is minced, gently homogenized, and the homogenate is passed through two mesh screen filters (253 and 83 μm); tubules and glomeruli on top of the 83-μm mesh are resuspended and glomeruli are separated by gentle stirring with a magnetic stir bar. The resulting suspension of proxi-

---

[1] Abbreviations for nephron segments follow the standard nomenclature for structures of the kidney,[2] and cell culture terminology corresponds to that suggested by the Tissue Culture Association.[3]
[2] W. Kriz and L. Bankir (eds.), *Kidney Intern.* **33,** 1 (1988).
[3] W. I. Schaeffer, *In Vitro Cell. Dev. Biol.* **20,** 19 (1984).
[4] D. S. Chung, N. Alavi, D. Livingstone, S. Hiller, and M. Taub, *J. Cell Biol.* **95,** 118 (1982).

mal (and other cortical) tubule fragments is seeded into culture dishes. This protocol has been utilized in studies of rabbit PCT cultures[5-10] and in the mouse kidney.[11]

Monoclonal antibodies to renal microvillous membrane proteins have been prepared to isolate a pure population of rat proximal tubule cells,[12,13] thereby preventing the nonproximal contamination inherent in the disaggregation protocol. This technique of "immunodissection" of nephron populations of defined segmental origin had been introduced by Garcia-Perez and Smith[14] to isolate cortical collecting tubule cells.

Purification of cortical tubule fragments[15] by gradient centrifugation has been applied to the cortex of canine kidney[16,17] and rabbit kidney.[18-20] Suspensions of cortical tubule fragments are prepared by different protocols; the tissue pellet is resuspended in a 50% Percoll solution and centrifuged (12,200 g; 30 min), whereby a gradient of density between 1.05 and 1.35 is established which separates the cortical tissue suspension into four distinct bands, of which band IV represents a more than 98% purified PCT suspension. Minor modifications of this principal preparation have been introduced.[18]

Microdissection of individual cortical nephron segments has been applied to rabbit PCT[21-24] and proximal straight tubule (PST),[23,24] to human PCT,[25] and to rabbit distal convoluted (DCT) tubule.[26] Proximal convo-

[5] L. M. Sakhrani, B. Badie-Dezfooly, W. Trizna, N. Mikhail, A. G. Lowe, M. Taub, and L. G. Fine, *Am. J. Physiol.* **246,** F757 (1984).
[6] L. M. Sakhrani, N. Tessitore, and S. G. Massry, *Am. J. Physiol.* **249,** F346 (1985).
[7] L. G. Fine, B. Badie-Dezfooly, A. G. Lowe, A. Hamzeh, J. Wells, and S. Salehmoghaddam, *Proc. Natl. Acad. Sci. U.S.A.* **82,** 1736 (1985).
[8] J. Norman, B. Badie-Dezfooly, E. P. Nord, I. Kurtz, J. Schlosser, A. Chaudhari, and L. G. Fine, *Am. J. Physiol.* **253,** F299 (1987).
[9] N. Alavi, R. A. Spangler, and C. Y. Jung, *Biochem. Biophys. Acta* **899,** 9 (1987).
[10] A. Aboolian and E. P. Nord, *Am. J. Physiol.* **255,** F486 (1988).
[11] S. S. Blumenthal, D. L. Lewand, M. A. Buday, N. S. Mandel, G. S. Mandel, and J. G. Kleinman, *Am. J. Physiol.* **257,** C419 (1989).
[12] R. C. Stanton, D. L. Mendrick, H. G. Rennke, and J. L. Seifter, *Am. J. Physiol.* **251,** C780 (1986).
[13] R. C. Stanton and J. L. Seifter, *Am. J. Physiol.* **253,** C267 (1988).
[14] A. Garcia-Perez and W. L. Smith, *Am. J. Physiol.* **244,** C211 (1983).
[15] P. Vinay, A. Gougoux, and G. Lemieux, *Am. J. Physiol.* **241,** F403 (1981).
[16] M. S. Goligorsky, D. N. Menton, and K. A. Hruska, *J. Membr. Biol.* **92,** 151 (1986).
[17] K. A. Hruska, M. Goligorsky, J. Scoble, M. Tsutsumi, S. Westbrook, and D. Moskowitz, *Am. J. Physiol.* **251,** F188 (1986).
[18] E. Bello-Reuss and M. R. Weber, *Am. J. Physiol.* **251,** F490 (1986).
[19] K. G. Dickman and L. J. Mandel, *Am. J. Physiol.* **257,** C333 (1989).
[20] M. J. Tang, K. R. Suresh, and R. L. Tannen, *Am. J. Physiol.* **256,** C532 (1989).
[21] M. Horster, *Pfluegers Arch.* **382,** 209 (1979).

luted tubule segments are microdissected from thin cortical slices using fine forceps or needles and transferred by pipet or tweezers into the matrix of a culture dish. Outgrowth of epithelial cells[21] occurs from the open ends within 24 hr (Fig. 1). This protocol is applied when low cell numbers are not disadvantageous[22] and when pure preparations of proximal or distal tubule subsegments are essential.[24,26]

Thick Ascending Limb of Loop of Henle. Primary cultures of the medullary thick ascending limb (MTAL) have been initiated by microdissection of MTAL segments from the outer medulla of rabbit kidney[21,27,28] and from mouse kidney.[29] Also, linear density gradient centrifugation[30] and centrifugal elutriation[31] have been utilized to obtain MTAL cells for culture.

The method of indirect immunoaffinity ("immunodissection") has been extended to MTAL cells.[32] This most promising approach yields large and highly purified populations of medullary (MTAL) and of cortical thick ascending limb (CTAL) cells. The method takes advantage of the fact that the Tamm–Horsfall glycoprotein of loop cells is immunoreactive with anti-goat immunoglobulin G; the latter is applied to the culture dishes to separate the adherent MTAL or CTAL cells from other cortical and medullary epithelial cells.

A cell line from microdissected MTAL (MAL-1) has been established and propagated,[28] of which a clone (MAL-1/A3) serves for the analysis of cytoplasmic membrane channels by electrophysiological techniques.[33–35]

---

[22] J. Merot, M. Didet, B. Gachot, S. Le Maout, M. Tauc, P. Poujeol, L. Othmani, and M. Gastineau, *Pfluegers Arch.* **413**, 51 (1988).

[23] M. Suzuki, A. Capparelli, O. D. Jo, Y. Kawaguchi, Y. Ogura, T. Miyahara, and N. Yanagawa, *Kidney Intern.* **34**, 268 (1988).

[24] M. Suzuki, Y. Kawaguchi, S. Kurihara, and T. Miyahara, *Am. J. Physiol.* **257**, F724 (1989).

[25] P. D. Wilson, M. A. Dillingham, R. Breckon, and R. J. Anderson, *Am. J. Physiol.* **248**, F436 (1985).

[26] J. Merot, M. Bidet, B. Gachot, S. Le Maout, N. Koechlin, M. Tauc, and P. Poujeol, *Am. J. Physiol.* **257**, F288 (1989).

[27] M. Burg, N. Green, S. Sohraby, R. Steele, and J. Handler, *Am. J. Physiol.* **242**, C229 (1982).

[28] N. Green, A. Algren, J. Hoyer, T. Triche, and M. Burg, *Am. J. Physiol.* **249**, C97 (1985).

[29] J. D. Valentich and M. F. Stokols, *Am. J. Physiol.* **251**, C312 (1986).

[30] D. M. Scott, *Differentiation* **36**, 35 (1987).

[31] E. D. Drugge, M. A. Carroll, and J. C. McGiff, *Am. J. Physiol.* **256**, C1070 (1989).

[32] M. L. Allen, A. Nakao, W. K. Sonnenburg, M. Burnatowska-Hledin, W. S. Spielman, and W. L. Smith, *Am. J. Physiol.* **255**, F704 (1988).

[33] S. E. Guggino, W. B. Guggino, N. Green, and B. Sacktor, *Am. J. Physiol.* **252**, C121 (1987).

[34] S. E. Guggino, W. B. Guggino, N. Green, and B. Sacktor, *Am. J. Physiol.* **252**, C128 (1987).

[35] M. Cornejo, S. E. Guggino, and W. B. Guggino, *J. Membr. Biol.* **110**, 49 (1989).

DISSECTION

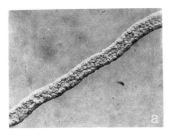

no protease

EXPLANTATION

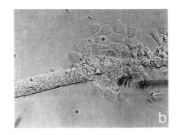

attachment factors

GROWTH

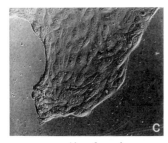

growth factors
-migration, proliferation-

DIFFERENTIATION

instructional factors
-epithelial properties-

FIG. 1. Principal steps in primary culture of cells derived from microdissected nephron segments. (a) Dissection without protease. (b) Substratum (attachment factors) for initial adherence and migration of cells. (c) Growth factors in medium induce cell proliferation. (d) Instructional factors in medium (e.g., steroid hormones) regulate the expression of segmental differentiated characteristics. Example shows cortical collecting duct (CCD) culture. [From M. F. Horster, J. Fabritius, and M. Schmolke, *Pfluegers Arch.* **405**, S158 (1985), by permission].

Medullary thick ascending limb cells have been transfected[30] with the plasmid pSV2-neo DNA plus DNA from the early region of Simian virus 40 (SV40); these cells have been maintained in culture for about 2 years.

Collecting Tubule. Isolation procedures for cortical collecting duct (CCD) and for inner medullary collecting duct (IMCD) cells differ substantially.

CCD: Cultures of cortical collecting duct cells have been initiated either by microdissection of segments from the rabbit kidney[21,36-38] and human kidney,[25] or by the method of immunoaffinity (immunodissection). The isolation by immunodissection of a pure CCD cell population from mixed cortical cell suspensions has been achieved either by selective immunoadsorption of canine CCD cells to monoclonal antibodies against Madin–Darby canine kidney (MDCK) cells[14] or of rabbit CCD cells to monoclonal antibodies against CCD cells.[39-42]

Intercalated cells (IC) of the rabbit CCD have been isolated by solid-phase immunoabsorption with a monoclonal antibody[43] or by adsorption to peanut lectin-coated plates.[44]

Cortical collecting duct cells, obtained by immunodissection, have been transfected with an adenovirus 12–SV40 hybrid; the continuous cell line (RCCT-28A) retains differentiated structural and functional properties.[45]

Explants of S-shaped bodies and of collecting duct anlagen from the nephrogenic zone of the neonatal rabbit kidney[46] yield pure populations of principal cells in primary culture.[47]

IMCD: The most widely applied technique for isolation of inner medullary collecting duct cells[48] takes advantage of the differential sensitivity of inner medullary cells to hypotonic treatment. The inner medulla is excised and finely minced; the fragments are incubated in collagenase (0.1%; 1 to 2 hr), filtered, and the suspension is exposed to a hypotonic solution (about 100 mOsm/kg) to disrupt cells other than those of the IMCD.

[36] P. D. Wilson and M. F. Horster, *Am. J. Physiol.* **244**, C166 (1983).
[37] B. S. Dixon, R. Breckon, C. Burke, and R. J. Anderson, *Am. J. Physiol.* **254**, C183 (1988).
[38] M. F. Horster, M. Schmolke, and R. Gleich, *Miner. Electrolyte Metab.* **15**, 137 (1989).
[39] W. S. Spielman, W. K. Sonnenberg, M. L. Allen, L. J. Arend, K. Gerozissis, and W. L. Smith, *Am. J. Physiol.* **251**, F348 (1986).
[40] G. Fejes-Tóth and A. Naray-Fejes-Tóth, *Am. J. Physiol.* **253**, F1302 (1987).
[41] A. Naray-Fejes-Tóth, O. A. Carretero, and G. Fejes-Tóth, *Hypertension* **11**, 392 (1988).
[42] L. J. Arend, M. Burnatowska-Hledin, and W. S. Spielman, *Am. J. Physiol.* **255**, C581 (1988).
[43] D. B. Light, E. M. Schwiebert, G. Fejes-Tóth, A. Naray-Fejes-Tóth, K. H. Karlson, F. V. McCann, and B. A. Stanton, *Am. J. Physiol.* **258**, F273 (1990).
[44] J. van Adelsberg, J. D. Edwards, D. Herzlinger, C. Cannon, M. Rater, and Q. Al-Awqati, *Am. J. Physiol.* **256**, C1004 (1989).
[45] L. J. Arend, J. S. Handler, J. S. Rhim, F. Gusovsky, and W. S. Spielman, *Am. J. Physiol.* **256**, F1067 (1989).
[46] W. W. Minuth and W. Kriz, *Cell Tissue Res.* **224**, 335 (1982).
[47] P. Gross, W. W. Minuth, W. Kriz, and E. Frömter, *Pfluegers Arch.* **406**, 380 (1986).
[48] F. C. Grenier, T. E. Rollins, and W. L. Smith, *Am. J. Physiol.* **241**, F94 (1981).

Whereas in the initial studies, and in a few others, IMCD cells were isolated from the rabbit kidney,[49,50] most of the work in cultured IMCD cells isolated by hypotonic treatment has utilized cells derived from the rat inner medulla.[51-67]

Cultured epithelial cells from the rabbit kidney papilla have been passaged and cloned to establish a cell line[68] which is being utilized to study the regulation of osmolytes, such as sorbitol.[69]

Requirements for Growth and Differentiation in Culture

Basal Nutrient Media and Media Supplements. For PCT cultures, the basal growth medium introduced in the first study on disaggregated rabbit cortical tubule tissue in culture[4] is a mixture (1/1, v/v) of Dulbecco's minimum essential medium (DMEM) and Ham's F-12, buffered with N-2-hydroxyethylpiperazine-N'-2-ethanesulfonic acid (HEPES) and sodium bicarbonate. This basal nutrient medium was supplemented with

---

[49] J. A. Shayman, K. A. Hruska, and A. R. Morrison, *Biochem. Biophys. Res. Commun.* **134**, 299 (1986).
[50] J. A. Shayman, R. J. Auchus, and A. R. Morrison, *Biochem. Biophys. Acta* **888**, 171 (1986).
[51] M. Sato and M. J. Dunn, *Am. J. Physiol.* **247**, F423 (1984).
[52] S. Ishikawa, T. Saito, and T. Kuzuya, *J. Endocrinol.* **107**, 15 (1985).
[53] S. Ishikawa, T. Saito, K. Okada, T. Kuzuya, K. Kangawa, and H. Matsuo, *Biophys. Biochem. Res. Commun.* **130**, 1147 (1985).
[54] R. G. Appel and M. J. Dunn, *Hypertension* **10**, 107 (1987).
[55] J. G. Kleinman, S. S. Blumenthal, J. H. Wiessner, K. L. Reetz, D. L. Lewand, N. S. Mandel, G. S. Mandel, J. C. Garancis, and E. J. Gragoe, Jr., *J. Clin. Invest.* **80**, 1660 (1987).
[56] S. Ishikawa, T. Saito, and T. Kuzuya, *J. Endocrinol.* **113**, 199 (1987).
[57] S. M. Wall, S. Muallem, and J. A. Kraut, *Am. J. Physiol.* **253**, F889 (1987).
[58] S. Ishikawa, T. Saito, and T. Kuzuya, *Kidney Intern.* **33**, 536 (1988).
[59] S. Ishikawa, K. Okada, and T. Saito, *Endocrinology* **123**, 1376 (1988).
[60] D. B. Light, F. V. McCann, T. M. Keller, and B. A. Stanton, *Am. J. Physiol.* **255**, F278 (1988).
[61] M. Konieczkowski and M. J. Dunn, *J. Cell. Physiol.* **135**, 235 (1988).
[62] I. Teitelbaum, A. Strasheim, and T. Berl, *Kidney Intern.* **35**, 647 (1989).
[63] R. C. Harris, *Am. J. Physiol.* **256**, F1117 (1989).
[64] I. N. Slotki, J. H. Schwartz, and E. A. Alexander, *Am. J. Physiol.* **257**, F210 (1989).
[65] L. P. Brion, J. H. Schwartz, H. M. Lachman, B. J. Zavilowitz, and G. J. Schwartz, *Am. J. Physiol.* **257**, F486 (1989).
[66] D. B. Light, E. M. Schwiebert, K. H. Karlson, and B. A. Stanton, *Science* **243**, 383 (1989).
[67] D. B. Light, D. A. Ausiello, and B. A. Stanton, *J. Clin. Invest.* **84**, 352 (1989).
[68] S. Uchida, N. Green, H. Coon, T. Triche, S. Mims, and M. B. Burg, *Am. J. Physiol.* **253**, C230 (1987).
[69] A. W. Siebens and K. R. Spring, *Am. J. Physiol.* **257**, F937 (1989).

bovine insulin (5 μg/ml), human transferrin (5 μg/ml), and hydrocortisone (5×10^{-8} M). In canine PCT cultures,[16,17] prostaglandin E_1 (25 ng/ml) and triiodothyronine (5×10^{-12} M) were added. In some preparations, dexamethasone (5×10^{-8} M) was added to the basal DMEM–Ham's F-12 medium[13,22,26] instead of hydrocortisone. Epidermal growth factor (EGF, 10 ng/ml) has been added to some cultures derived from microdissected segments.[21,22] Serum, either fetal calf serum (FCS; 10%, v/v)[24] or fetal bovine serum (FBS; 10%, v/v)[70] has been added in some PCT studies to otherwise defined media.

For TAL cultures, different growth media and supplements are being used. The basal growth medium,[27–29,33–35] in general, is Coon's Ham's F-12–Leibovitz L15, but DMEM–Ham's F-12[31] and DMEM[30] have also been shown to support TAL growth in culture.

The supplements to Coon's Ham's F-12–Leibovitz L15, in addition to transferrin, insulin, hydrocortisone, triiodothyronine, prostaglandin E_1, as in PCT cultures, are dibutyryl-cAMP (0.5×10^{-3} M), putrescine (1.6×10^{-6} M), spermine (3.0×10^{-8} M), spermidine (2.5×10^{-7} M), and selenium (1.0×10^{-8} M). Proliferation of cells from microdissected TAL segments confirmed earlier reports in which defined, hormonally supplemented media had been shown to support growth of epithelial sheets from nephron fragments.[21] The minimum serum requirement for TAL cultures appears to be 1% (v/v).

In CCD cultures, hormone-supplemented basal growth media are used in studies on cells derived from microdissected segments, whereas cultures initiated from immunodissected (high-density seeding) CCD cells require less elaborate growth conditions. Basal growth media are RPMI 1640[21,36,40,41,71] or RPMI 1640/Ham's F-12[38] or DMEM.[14,39,42]

Hormonal supplements, as in supplemented growth media for PCT cells, are insulin, transferrin, PGE_1, dexamethasone, and triiodothyronine.[21,25,38,72] Serum (FCS; 10%, v/v) is added in nonsupplemented CCD growth media.[39,42]

Inner medullary collecting duct cultures, in general, are grown by now in the basal growth medium DMEM–Ham's F-12 (1/1, v/v), with insulin, transferrin, hydrocortisone, and triiodothyronine added[51] as in CCD hormone-supplemented media. Serum (1%, v/v) is supplemented by some groups,[52,62] and high serum concentrations have been used to culture IMCD cells on glass coverslips or to establish cell lines from IMCD.[68]

---

[70] R. C. Harris, J. L. Seifter, and C. Lechene, *Am. J. Physiol.* **251**, C815 (1986).

[71] B. S. Dixon, R. Breckon, J. Fortune, E. Sutherland, F. R. Simon, and R. J. Anderson, *Am. J. Physiol.* **257**, F808 (1989).

[72] M. G. Currie, B. R. Cole, K. DeSchryver-Kecskemeti, S. Holmberg, and P. Needleman, *Am. J. Physiol.* **244**, F724 (1983).

These lines (GRB-PAP1; PAP-HT25) are being studied in a completely defined, serum-free medium.[69]

Substrata and Support of Monolayer Cultures. The term substratum indicates biological material placed between cells and support before seeding, to serve as some substitute for extracellular matrix. Support and substratum may be identical.

In nephron segmental cell culture, the type of support depends on the type of study. Transcellular vectorial transport and transepithelial electrophysiological work require a permeable support (e.g., collagen membrane; low-pore-size filters), whereas apical uptake studies and metabolic work are done in cell monolayers grown on solid support (e.g., plastic or glass).

In PCT cultures, plastic dishes and multiwell dishes of different sizes are used. They may be coated with collagen[25,26]; PCT cell growth on cover glass requires either collagen[24] or poly-L-lysine coating.[70] The permeable support is a collagen membrane fit into an Ussing-type chamber,[18] or commercially available filter membranes[16,17,22,73] which are covered with collagen as substratum. Collagen gels on solid supports, as introduced for TAL passages,[29] have also been applied to subculture PCT cells.[23]

In TAL cultures, the permeable collagen membrane support has been introduced by Burg *et al.*[27] to measure transepithelial voltage. The support is a polycarbonate cup with a 1-mm hole in its bottom which is covered with a thin preformed collagen sheet. The collagen is preincubated to adsorb some serum protein. Sheep amnion has been denuded of its epithelial cells and arranged in an embroidery hoop-type cup. Cells grown on this natural substratum as support were passaged alternately between amnion- and collagen-coated plastic dishes; this strategy led to the establishment of the first cell lines (GRB-MAL1; GRB-MAL2) derived from a defined nephron segment, the MTAL.[28]

Passage of MTAL cultures within their substratum, a hydrated collagen gel, has been used to establish cell lines (MTAL-IC; MTAL-IP) from mouse kidney.[29] Monolayers within the hydrated collagen gel receive nutrients from apical and basolateral cell sides; the gels are cut into small pieces and transferred for subculture to freshly prepared host collagen gels, i.e., intercellular junctions are not disrupted during passage.

Once established, the cell line, e.g., MAL1 clone A3,[28] is maintained and investigated on collagen-coated plastic dishes.[33,35] Transfected cell lines derived from MTAL[30] have been similarly maintained in culture flasks.

Cultures of cortical collecting tubule cells (CCD) require conditions for

[73] J. G. Blackburn, D. J. Hazen-Martin, C. J. Detrisac, and D. A. Sens, *Kidney Intern.* **33**, 508 (1988).

attachment and proliferation on the substratum which differ from those in cultures of inner medullary collecting tubule cells (IMCD).

CCD: Attachment and outgrowth from microdissected segments were first shown to occur on a collagen substratum covered with rabbit plasma or with a mixture of rabbit plasma and chick embryo extract.[21] The support was a plastic dish with a gas-permeable bottom. This arrangement was used in studies on growth kinetics and enzyme expression in response to hormonal supplements in rabbit and human CCD.[25,36] Collagen-coated glass slides are convenient for enzyme studies on small cell populations grown from microdissected CCD.[37,71]

In vectorial transport studies, CCD cells are cultured either on collagen-coated filter membranes[40] or on a collagen membrane[38] glued to a Lucite disk which is inserted into an Ussing chamber (Fig. 2). At high seeding densities (e.g., when isolated by immunodissection), CCD cells attach and proliferate on microcarriers[40] and on plastic dishes,[39,41,42] as is the case for high-density seeding of transfected CCD cells.[45]

A natural substratum, the capsula fibrosa of the kidney, serves in studies on principal cells derived in culture from explanted S-shaped bodies and collecting duct Anlagen.[46]

Inner medullary collecting duct cultures, ever since the first report,[48] are being grown to confluence on solid plastic support without substratum. For vectorial ion transport of IMCD, cells are cultured as monolayers on coated filter membranes.[48,60]

Segmental Epithelial Transport Functions in Culture

Solute Transport and Metabolism

Sodium ion-dependent hexose transport has been measured in cultured PCT cells.[4-6,9,12,17,20] Proximal convoluted tubule cultures accumulate α-methylglycoside against a concentration gradient in the presence of Na^+; phlorizin (0.1 mM), but not phloretin (0.1 mM), inhibits the energy-dependent uptake.[4] The kinetics of this uptake are consistent with a single saturable system; oxidation of glucose to CO_2 indicates the presence of aerobic metabolism and of hexose monophosphate shunt in cultured PCT cells.[5] The saturation kinetics show that K_M, but not V_{max}, is dependent on the Na^+ concentration of the medium; ouabain (10^{-3} M) inhibits (by 60%) the steady state α-methylglucose accumulation.[9]

Sodium ion H^+ antiport activity has been associated with growth expression in rabbit and rat PCT cultures.[7,12,70] Quiescent PCT cultures, when exposed to growth stimuli (e.g., insulin), show an increase of amiloride-sensitive Na^+ uptake, of Na^+-dependent H^+ efflux, and of ouabain-sensitive Rb^+ uptake, indicating that stimulation of the Na^+/H^+ antiporter

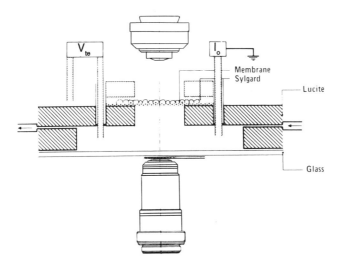

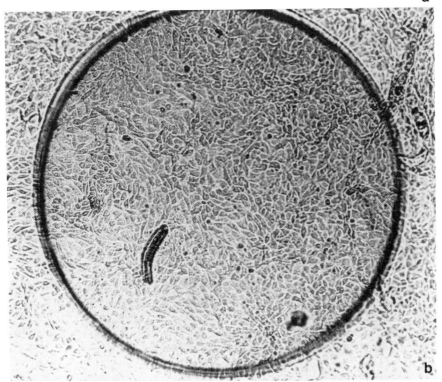

FIG. 2. Example of arrangement for transepithelial measurement of electrical and ion transport parameters on the stage of an inverted microscope. (a) Scheme of Ussing-type chamber. Cells were cultured on a collagen-coated permeable membrane. (b) Microscopic appearance of the confluent monolayer of CCD cells on membrane covering the center hole of a Lucite disk. (From M. F. Horster, M. Schmolke, and R. Gleich,[38] by permission).

is an early event in cellular hypertrophy.[7] Proximal convoluted tubule cells, isolated by immunodissection, exhibit the amiloride-sensitive Na^+/H^+ antiporter activity.[12] The Na^+/H^+ antiporter mediates a major fraction of Na^+ entry; cellular acidification increases the antiporter activity and it activates, via coupling through the elevated intracellular Na^+ content, the activity of the primary active Na^+/K^+ antiporter.[70]

In cultured IMCD cells, the amiloride-sensitive Na^+/H^+ antiporter has been demonstrated.[57]

Uptake of $^{22}Na^+$ into IMCD cells is stimulated by increased extracellular pH, whereas the efflux is inhibited, consistent with the presence of an Na^+/H^+ antiporter in this ultimate nephron segment.[61]

Hydrogen ion secretion by IMCD cells, obtained from mature rat kidney, has been followed during postmitotic development in culture.[65] While quiescent cells in domes and in the center of growing cell clusters showed active proton transport, but rarely activity of the Na^+/H^+ antiporter, the dividing cells at the periphery of the clusters showed only Na^+/H^+ antiporter activity. While proton pump activity developed, these cells acquired corresponding properties, such as carbonic anhydrase activity. Mitotic activity, therefore, and the expression of a differentiated active ion transport, appear to be irreconcilable events.

The presence of these two modes of acid extrusion, via the proton ATPase and the Na^+/H^+ antiporter, had been demonstrated by cellular pH measurements in IMCD cells grown on glass slides,[55] indicating that both H^+ transport systems may contribute to cell pH regulation and to urinary acidification. Hydrogen ion secretion in quiescent IMCD cells is stimulated by an increase of the intracellular calcium concentration.[64] The resulting changes of the intracellular pH are independent of Na^+ and metabolic energy, but they are dependent upon extracellular calcium. The calcium-dependent cell alkalinization was mimicked by cell depolarization, suggesting a regulatory contribution of membrane ionic channel activity.

Vectorial ion transport in cultured CCD cells has generated impressive transepithelial solute gradients ($Na^+, K^+, Cl^-, PO_4^{2-}$, pH), and a transepithelial osmotic gradient of 14.3 mOsm/liter.[40]

Secretion of HCO_3^- has been demonstrated across morphologically differentiated monolayers of intercalated (IC) cells obtained from the cortical distal nephron.[44] Transepithelial HCO_3^- transport is Cl^- dependent and mediated by an apical antiporter.

Uptake of the major osmolyte sorbitol has been examined[69] in a cell line (GRB-PAP-HT25) derived from papillary epithelial cells in hyperosmotic media.[68] Sorbitol efflux, as measured by sorbitol tracer uptake, was demonstrated to be nonsaturable and independent of other osmolyte permeabilities; the sorbitol permeability increase after a decrease in medium osmolality was not influenced by several species of transport inhibitors or by ion substitutions, suggesting a rather unusual and novel mechanism of organic solute membrane transport in IMCD cells.

Oxygen consumption and ATP content have been measured in PCT cells which had been cultured either in stationary dishes, shaking flasks, or in dish cultures transferred to shaking flasks.[19] While oxygen consumption was fully maintained in 24-hr shaking cultures, dish cultures, within 6 hr, showed a continuous decline in oxygen consumption accompanied by a stimulation of lactate production, and they exhibited a fall in ATP and K^+ content. The hypoxic changes of these dish cultures were reversible on transfer into shaking conditions, indicating that primary PCT cultures on solid plastic substrata rapidly change from oxidative metabolism to glycolysis. By contrast, PCT cells grown to confluency (day 6) showed increased glycolytic metabolism with enhanced glucose uptake and lactate production, and decreased oxygen consumption rates with elevated ATP levels.[20] In CCD cells cultured on filters, similarly, the amount of basolateral glucose decreased while lactate accumulated in the basolateral medium.[40]

Intracellular Signal Systems

Intracellular calcium concentrations have been assessed and altered in cultured PCT cells.[6,8,10,16,24,70]

Increased cytosolic calcium inhibits Na^+-dependent phosphate uptake and amiloride-sensitive Na^+ uptake in rabbit PCT cultures.[6] Parathyroid hormone (PTH) increases the intracellular free calcium concentration in filter cultures of canine PCT cells.[16,17] The rise of calcium induced by the hormone was concentration dependent, it was not mimicked by adenosine 3′, 5′-cyclic monophosphate (cAMP), and it was dependent on extracellular calcium.[17] Similarly, angiotensin II, when added to rabbit PCT cultures at high concentrations (5×10^{-8} to 5×10^{-6} M), led to a fast, concentration-dependent, and transient rise in cytosolic free calcium concentration.[8]

Bradykinin, by contrast, induces a transient rise of intracellular calcium in PCT cultures which is independent of the calcium concentration in the medium, and which is nonresponsive to subsequent activation by angiotensin II,[10] suggesting that both peptide hormones release calcium from a common intracellular pool. Intracellular calcium, when monitored in single cultured cells of PCT and of PST, responds differentially to PTH and to calcitonin (CT). While PTH elicits a transient rise of calcium in PCT, calcium in PST is activated by CT.[24] The effects are specific for the segment, and they are dependent upon the presence of extracellular calcium. Phosphate uptake, correspondingly, is inhibited by PTH in cultured PST cells, but not in PCT cells.[23]

PTH effects on cyclic AMP in PCT cells have become established criteria for segment-specific and viable PCT cultures.[4,17,18,20,22–26]

In CCD cells in primary culture, adenosine analogs to the adenosine receptors (A_1 and A_2) activate a pertussis toxin-sensitive mobilization of intracellular calcium which is independent of the extracellular calcium concentration.[42] This transient increase in cytosolic free calcium by adenosine analogs is probably mediated through stimulation of the phosphoinositide turnover.[45]

Intracellular calcium in cultured IMCD cells is rapidly mobilized by bradykinin.[49] The rise was independent of extracellular calcium, supporting the role of calcium as second messenger in IMCD cells. Consistent with this effect is the finding that the peptide hormone induces hydrolysis of phosphatidylinositol 4,5-biphosphate in IMCD cells.[50] Arginine vasopressin (AVP), similarly, has been demonstrated to utilize the intracellular calcium messenger system.[59] AVP increased both the cellular free calcium concentration (as measured by the dye fluorescence in cells grown on glass slips) and the cyclic AMP production. This AVP-induced elevation of cellular free calcium, via the V_2 receptor, depends both on the cellular pool and on extracellular calcium.

Medullary thick ascending limb cells, when transfected with early region SV40 genes,[30] expressed differentiated transport functions such as the apical $Na^+/K^+/Cl^-$ symport system, the basolateral Na^+,K^+-ATPase antiport system, and a Ba^+-sensitive K^+ channel in the apical cell membrane.

Na^+,K^+-ATPase activity and its expression under hormonal influence in cultured CCD cells was demonstrated by direct measurements of the enzyme.[38] Sodium pump activity increased during the selective substitution of triiodothyronine, or dexamethasone, or aldosterone in physiological concentrations, indicating the selective and differential hormonal dependence of the Na^+/K^+ antiporter activity.

The regulation of the cyclic AMP system has been investigated in cultured CCD cells by direct measurements of the formation of cAMP.[14,37,39,41] Increased formation of cAMP was demonstrated in canine CCD cultures[14] and in rabbit CCD cultures[39] in response to AVP, PGE_2, and isoproterenol, but not to calcitonin. Different activators of protein kinase C, in a systematic study on cultured CCD,[37] have been shown to inhibit the AVP-stimulated adenylate cyclase activity by distinctly different mechanisms.

Protein kinase C is activated by bradykinin via phospholipase C, and it inhibits AVP-stimulated cAMP production at the level of the hormone receptor or the coupling of the receptor to G_s.[71]

AVP in immunodissected CCD cells in culture causes a dose-dependent increase in cAMP production which is not altered by the simultaneous addition of atrial natriuretic factor (ANF). The ANF- induced stimulation of cyclic guanosine 3′,5′-monophosphate (cGMP), likewise, is not influenced by AVP, indicating that interactions of ANF and AVP on the effector system (e.g., osmotic water permeability) are likely to occur distal to the cyclic nucleotide formation.[41]

Inner medullary collecting duct cells in culture have become a primary target in the study of the cAMP second messenger system in renal epithelia.[48,51-54,56,58,59,62,63] Inner medullary collecting duct cells synthesized cAMP in response to AVP, oxytocin, and PTH in culture, while kinins (e.g., bradykinin) but not AVP caused the formation of PGE_2, suggesting direct kinin effects on prostaglandin release.[48] Analysis of the interactions of AVP, prostaglandins, and cAMP in IMCD cells[51] indicated that the increased PGE_2 production after AVP treatment of IMCD cells may be related to effects of prostaglandins at a site distal to cAMP synthesis within the messenger system. Calmodulin was shown to regulate, in IMCD cells, the basal adenylate cyclase activity, as well as the AVP-, PGE_2-, and forskolin-activated enzyme activity.[52]

Atrial natriuretic factor inhibited AVP- and forskolin-stimulated cAMP levels, and it increased cellular cGMP levels.[53] The response of cellular cGMP to ANF in cultured IMCD cells from prehypertensive salt-sensitive (Dahl) and salt-resistant rats[54] suggested that these cells may differ in their guanylate cyclase activity, which may be related to the differing natriuretic response of these two strains. The response of cAMP to AVP or forskolin in IMCD cells is dependent on the presence of potassium in the medium.[56] Somatostatin attenuated the cellular cAMP production in response to AVP, glucagon, and cholera toxin, while pertussis toxin treatment of IMCD cells abolished the inhibitory effects of somatostatin on cAMP production,[58] suggesting that the inhibitory effect of somatostatin is mediated through the G_i protein.

AVP, in addition to its stimulatory effect on second messenger production, was shown in IMCD cells to concomitantly increase the cellular free calcium concentration. The calcium mobilization may be mediated through the V_2 receptor of AVP.[59]

Specific and high-affinity receptors for epidermal growth factor (EGF) have been demonstrated in 4- to 7-day cultures of rat IMCD.[63] Epidermal growth factor stimulated [^{3}H] thymidine incorporation, cAMP accumulation, and PGE_2 production, suggesting modulatory functions of the peptide in the terminal nephron segment. The generation of the second messenger cAMP in IMCD is further controlled by adrenergic hormones.[62] While isoproterenol increased basal cAMP production but had no further effect on the AVP-stimulated level, norepinephrine decreased AVP-stimulated cAMP generation.

Electrophysiology

Electrophysiological studies have been performed in PCT cells cultured to confluence on permeable substrata.[16,18,22,73] Transepithelial voltages (apical side negative) and resistances across PCT monolayers were -2.6 mV[18] and 7 Ω-cm^2, -0.28 mV[16] and 95 Ω-cm^2, -0.13 mV[22] and 37 Ω-cm^2, and -2.0 mV[73] and 310 Ω-cm^2 in human PCT. Basolateral membrane potential difference was -51 mV,[18] and apical voltages were -48.4 mV[18] and -61.5 mV.[22] In distal bright convoluted tubules (DCT$_b$), transepithelial voltage was -3.1 mV in 20-day cultures and increased to -22.3 mV in 30-day cultures.[26] The voltage was amiloride and calcitonin sensitive, suggesting calcitonin-modulated apical Na$^+$ channels in DCT$_b$.

Transepithelial voltage in cultured TAL cells has been demonstrated in primary MTAL cultures[27,31] and in MTAL cell lines derived by passage from primary cultures.[28,29] The voltage in primary MTAL cultures was apical side positive (9.9 mV) and was sensitive to furosemide but not to amiloride in the apical medium, suggesting the presence of *in vivo* characteristics, such as the Na$^+$/K$^+$/Cl$^-$ cotransport system and the Na$^+$,K$^+$-activated ATPase.[27]

When grown on plastic support, MTAL cells expressed ouabain-sensitive "domes," reflecting vectorial solute transport, and a transepithelial voltage of 4.7 mV across monolayers grown on a filter support.[31] In two cell lines derived from MTAL[28] by passages on amnion, transepithelial voltage in one (GRB-MAL1) never differed from zero, whereas significant voltages (apical side positive) appeared in passages four and five of the other (GRB-MAL2).

In a mouse kidney MTAL cell line (M-MTAL-1C), normal transport functions of the medullary thick ascending limb were not expressed; instead, the line exhibited electrogenic Na$^+$ absorption which was inhibited by amiloride.[29] However, after transplantation of the line in diffusion chambers into allogeneic mice, transport properties characteristic for MTAL were reexpressed.

Transepithelial voltage in cultured CCD was first measured in canine CCD cells obtained by immunodissection.[14] The low voltage of rabbit principal cell cultures was subsequently shown to increase with aldosterone in the medium while transepithelial electrical resistance remained unchanged.[47]

Transepithelial voltage in cultured CCD has been induced by hormones.[38] Triiodothyronine, dexamethasone, or aldosterone selectively and differentially induced the postmitotic appearance of the voltage (12.7 to 28.4 mV) in CCD cells grown on collagen membranes. Transepithelial electrical resistance across monolayers of intercalated cells (IC), isolated by the peanut-lectin adsorption technique,[44] was 595 Ω-cm^2. Electrically tight IMCD monolayers on filter membranes apparently are difficult to establish.[48]

Ion channel characteristics of the apical cytoplasmic membrane of cultured PCT cells have been evaluated.[22,74] Patch-clamp analysis revealed a nonselective cationic channel with a conductance of 13 pS which, in excised patches, had a conductance of 20 to 28 pS and required a high calcium concentration on its cytoplasmic side.[22] The apical membrane of cultured PCT cells is further characterized by two types of K$^+$ channels: A large-conductance, calcium- and voltage-activated K$^+$ channel, similar to the maxichannels of other epithelia, and a K$^+$ channel with a smaller conductance blocked by apamin.[74] The first evaluation of ion channel characteristics of MTAL cells demonstrated an apical K$^+$ conductance and sensitivity to loop diuretics in the cell line GBR-MAL1 clone A3; the most often found channel of the apical cytoplasmic membrane is a Ca^{2+}-activated, maxi-K$^+$ channel.[33] The apical cell membrane voltage (-46 mV) of this MTAL clone is furosemide sensitive. Analysis of the influence of several K$^+$ channel blockers on the Ca^{2+}-activated K$^+$ channel (channel conductance 127 pS) suggested that the apical K$^+$ channels of these MTAL cells are homologous to those found in muscle cells.[34] Elevated intracellular H$^+$ concentration was demonstrated to reduce the open probability of the Ca^{2+}-activated K$^+$ channel by interfering with the Ca^{2+} binding to the channel gate.[3]

Intercalated cells, isolated by immunoadsorption and grown on collagen-coated cover slips, have been studied by the patch-clamp technique.[43] In the excised inside-out patches of the apical membrane, a large-conductance (303 pS) anion channel was found; the channel activity was inhibited by chloride channel blockers, but it was not altered by changes of the cytoplasmic calcium or proton concentrations.

Ion channels in cultured IMCD cells have been analyzed by the same technique.[60,66,67] In the apical membrane, a nonselective and amiloride-sensitive cation channel (28 pS) was found that was insensitive to membrane voltage and pH; the channel has been postulated to mediate electrogenic Na$^+$ absorption.[60] Both the atrial natriuretic peptide (ANP) and its second messenger cGMP inhibited the channel activity in cell-attached patches and in inside-out patches, respectively.[66] Further characterization demonstrated that an α subunit of the G$_i$ protein directly activates the cation channel in IMCD cells.[67]

---

[74] J. Merot, M. Bidet, S. La Maout, M. Tauc, and P. Poujeol, *Biochem. Biophys. Acta* **987**, 134 (1989).

Conclusions

Techniques for primary segmental epithelial culture and the expression of segmental nephron functions in culture, as presented, lend strong support to the notion that this "reductionistic" approach to the kidney has, by now, been instituted. The tacit assumption that "primaries" of segmental nephron populations are difficult to set up and "dedifferentiate" rapidly can no longer be regarded as justified. One may envision that cell lines derived from defined cultured nephron populations, in contrast to renal cell lines of nondefined origin, maintain the genotypic stability and phenotypic expression of their *in vivo* segment.

In the final evaluation of this issue, without exposing improper subjectivity, the recent remarkable work on intracellular signaling systems and cyptoplasmic membrane ion channel activity in primary segmental cultures justifies the opinion that the interdependence of particular properties in a given nephron cell type may be expressed best in primary cultures.

[24] Tissue Culture of Established Renal Cell Lines

By N. L. SIMMONS

Introduction

The existence of several established epithelial cell lines derived from whole kidneys, which retain significant histotypic features, has ensured their popularity as general rather than specific models for studies of epithelial polarity, ion transport, and hormonal regulation. The most frequently used cell lines are MDCK (derived from a cocker spaniel kidney by Madin and Darby[1] and LLCPK$_1$ (derived from a hog kidney by Hull *et al.*[2]). There is continued interest in defining renal cell lines that may display useful properties, e.g., PK$_1$ and JTC cells. Established cell lines derived initially from human renal carcinomas (Cur, Caki-1) have received relatively little attention.[3]

An important caveat to the use of all established cell lines is to be aware of the possibility of phenotypic heterogeneity, not least from the origin of whole kidney enzymatic digests of most of the existing cell lines, but also from the existence of selection pressures arising from unique culture con-

---

[1] C. R. Gausch, W. L. Hard, and T. F. Smith, *Proc. Soc. Exp. Biol. Med.* **122,** 931 (1966).
[2] R. N. Hull, W. R. Cherry, and G. W. Weaver, *In Vitro* **12,** 670 (1976).
[3] N. F. Nelson, W. Cieplak, S. C. Dacus, and M. D. Prager, *J. Cell. Physiol.* **126,** 435 (1986).

ditions in individual laboratories. The use of clonal cell populations, with recloning to ensure homogeneity and stability of the phenotype of interest, is to be recommended. Even with cloned cell populations the observed phenotype may not be truly representative of any particular nephron segment or renal cell type and, indeed, the responsiveness to renal acting hormones in both the MDCK[4] and LLCPK$_1$ cell lines[5] is neither typical of cortical collecting tubules nor proximal tubule, respectively.

In this chapter I will describe techniques and procedures for the routine culture and generation of epithelial layers used in our laboratory for MDCK cells.

Stock Cultures

Stock cultures of MDCK cells are grown in 75-cm^2 glass (or plastic) tissue culture flasks in medium consisting of minimum essential Eagle's medium (MEM) supplemented with 2 mM glutamine, nonessential amino acids, and 5% (v/v) each of fetal calf serum and donor horse serum at 37° with kanamycin antibiotic (1 μg/cm^3), in an air/5% CO$_2$ atmosphere. Antibiotic is usually included, but the use of antibiotic-free medium is preferable, if only to promote good aseptic practices. The exact concentration of fetal serum used is varied, but the minimum quantity is used that supports adequate growth. It is also a good practice to heat inactivate serum (complement) at 60° for 20 min. A defined serum-free medium has been formulated for the MDCK cell line by Taub et al.[6] and contains Dulbecco's modified Eagle's medium: Ham's F12 [50% (v/v) each], 5 μg/cm^3 insulin, 5 μg/cm^3 transferrin, 50 nM hydrocortisone, 5 pM triiodothyronine, 0.1 μM prostaglandin E$_1$, 10 nM selenium dioxide, and 1.1 g/liter NaHCO$_3$/10 mM N-2-hydroxyethylpiperazine-N'-2-ethanesulfonic acid (HEPES) buffer. The general utility of this medium for the differing phenotypes displayed has not been established and our experience is that the growth rate declines after five to six passages. Cultures are seeded at a split ratio of 1:10. From a 75-cm^2 flask between $5-7.2 \times 10^7$ cells may be harvested in 7 days with a single medium change at day 3. Tissue culture flasks are placed on top of a thick glass plate to ensure minimal temperature variation in the constant-temperature cabinet/hot room[7] since similar patterns to those observed in LLCPK$_1$ cells were observed in MDCK

---

[4] K. E. Meier and P. A. Insel, in "Tissue Culture of Epithelial Cells" (M. Taub, ed.), p. 145. Plenum, New York, 1985.
[5] A. Wohlwend, D. B. Vassalli, and L. Orci, Am. J. Physiol. 250, C682 (1986).
[6] M. Taub, L. Chuman, M. H. Saier, and G. Sato, Proc. Natl. Acad. Sci. U.S.A. 76, 3338 (1979).
[7] E. M. Adler, L. J. Fluk, J. M. Mullin, and A. Kleinzeller, Science 217, 851 (1982).

cultures. The number of serial passages is limited (typically 15) and frequent recourse is made to the frozen stocks (see Freezing of Cell Stocks, below). Since the importance of "bottom feeding" in the maintenance of epithelial characteristics is now generally recognized for some cell lines,[8] the use of impermeant supports could be questioned. Since significant transjunctional permeability of nonelectrolytes in leaky epithelia has been documented, this may not be a serious problem in some cultures. If a serious problem is envisaged by the use of impermeant supports, restriction of seeding densities and growth times to ensure subconfluent conditions should be considered.

Formation of a single-cell suspension for serial passage of stock tissue culture flasks is made by washing the cell monolayer three times in a $Ca^{2+}-Mg^{2+}$-free Dulbecco's salt solution, followed by incubation in 5 cm^3 of 0.25% trypsin$-$2 mM ethylenediaminetetraacetic acid (EDTA) in $Ca^{2+}-Mg^{2+}$-free Earle's basal salt solution at 37°, until the complete detachment of the cell monolayer is achieved. With MDCK cells particular importance is attached to this step, as incomplete harvesting of the cells would result in an inherent selection pressure. A single-cell suspension is obtained by gentle syringing through a wide-bore needle (2-mm i.d.). Trypsin is neutralized by the addition of complete growth medium and cell density determined directly by electronic cell counting and sizing of an aliquot of the cell suspension diluted in Isoton (Coulter Electronics Ltd., Luton, England).[9]

We presently buy all tissue culture stocks from Gibco (Paisley, Scotland) though in the past we have obtained supplies from Flow laboratories (Rickmansworth, England). The dependence of the researcher on the high quality of sterile products produced commercially does not need to be emphasized.

Preparation of Cell Clones

I have already stressed the importance of the use of cell clones that express stable phenotypic properties. Table I demonstrates the measured phenotype of two extensively used and characterized clonal cell lines of MDCK. Also, the properties of MDCK cells from different sources are shown. The following procedure, presently used in our laboratory for cloning MDCK cells, is from a method outlined by K. Soderberg at European Molecular Biology Laboratory (Heidelberg, West Germany). A monodisperse cell suspension is made from a logarithmically growing

---

[8] J. S. Handler, A. S. Preston, and R. E. Steele, *Fed. Proc., Fed. Am. Soc. Exp. Biol.* **43**, 2221 (1984).

[9] L. Boardman, M. Huett, J. F. Lamb, J. P. Newton, and J. Polson, *J. Physiol. (London)* **241**, 771 (1974).

TABLE I
VARIATION IN THE STABLE PHENOTYPIC CHARACTERISTICS BETWEEN DIFFERENT MDCK CELL STRAINS AND CELL CLONES[a]

MDCK cell	Diuretic-sensitive K$^+$ transport [nmol/(10^6 cells·min)]	Transepithelial resistance (Ω-cm^2)	Adenylate cyclase (-fold stimulation over basal)
Strain 1[b]	8.8 + 2.6	2500–7000	2.85 + 0.24
Strain 2[b]	ND[c]	100 (cation)[d]	1.07 + 0.04
Clone DL$^e_{17}$	ND[c]	100 (cation)[d]	1.03 + 0.08
Clone CL8$^e_{1b}$	3.9 + 0.7	2500–4000	1.50 + 0.05
MDCKN[f]	1.8 + 0.4	1500–2500	1.80 + 0.03

[a] With respect to some of the measured parameters described in the text. Diuretic-sensitive transport is that portion of the ouabain-insensitive flux that is inhibited by 0.1 mM loop diuretic (furosemide, bumetanide). Transepithelial resistance was determined from epithelial layers grown on uncoated Millipore filters which were then clamped into Ussing chambers. Isoprenaline (100 μM)-stimulated adenylate cyclase activity was determined in disrupted cells in the presence of IBMX.
[b] From G. Barker and N. L. Simmons, *Q. J. Exp. Physiol.* **66**, 61 (1981).
[c] ND, Not detectable.
[d] Cation refers to the existence of a cation-selective paracellular pathway.
[e] From E. L. Rugg and N. L. Simmons, *Q. J. Exp. Physiol.* **69**, 405 (1984).
[f] From J. A. McRoberts, C. Trong-Trang, and M. H. Saier, *J. Biol. Chem.* **258**, 12320 (1983).

stock culture, as outlined above. Typically 1000 cells are then plated directly onto a 10-cm diameter Petri dish (Nunc plastic, Gibco, Paisley, Scotland). The growth medium is enriched by the use of 20% fetal calf serum. Plates are then incubated for approximately 14 days until single-cell colonies are visible (2–3 mm in diameter). The plates are then washed in Ca^{2+}–Mg^{2+}-free Earle's salt solution and sterile cloning cylinders (5 mm long, 7-mm o.d, 6-mm i.d.) are placed around isolated colonies by sealing them to the culture dish with autoclaved high-viscosity vacuum grease. The individual colony is then trypsinized as above and the resulting cell suspension transferred to 1 well of a 12-well plate for further expansion. A second cloning ensures the clonal nature of the cell culture. Time-lapse photomicroscopy of subconfluent cultures has demonstrated significant cellular motility and cell aggregation. An appropriate cell density for formation of single-cell colonies is therefore critical.

Freezing of Cell Stocks

All cell stocks are maintained frozen above liquid nitrogen. Concentrated cell suspensions (to 1 × 10^6 cells/cm^3) are made by centrifugation (250 g) and resuspension in an appropriate volume of complete culture medium supplemented with 10% (v/v) dimethylsulfoxide (DMSO) as cryoprotectant. Freezing of cultures should be performed so that there is a controlled initial reduction (no greater than $-5°/\text{min}$) to obviate the formation of large ice crystals. We achieve this by placing the vials (5 cm³ total capacity), wrapped in cotton wool, in an expanded polystyrene carton and placing this in a $-90°$ cabinet overnight. Final equilibration to liquid nitrogen temperature may then be achieved rapidly. On thawing, vials are directly incubated in a shaking water bath held at 37°. It should be noted that different epithelial parameters have recently been reported for a single batch of MDCK cells held in aliquots, frozen, thawed, and recultured.[10]

Detection of Mycoplasma Contamination

Mycoplasma is a common contaminant of established cell lines. Routine detection of such contamination is detected by DNA fluorochromes such as bisbenzimide (Hoechst 33258, Aldrich Chemical Co. Ltd., Gillingham, England), which allows the detection of mycoplasma.[11] Hoe 33258 (5 mg) is dissolved in 100 cm³ of Ca^{2+}–Mg^{2+}-free phosphate-buffered saline (PBS) and stored in the dark at 4°. Cell monolayers seeded directly onto glass coverslips [or Millipore (Bedford, MA) filter supports] are rinsed in PBS and then fixed in 3:1 (v/v) methanol/acetic acid for 10 min and then air dried. The cell layers are then stained with a 1:100 dilution of the stock Hoe 33258 for 10 min, washed in distilled water, and wet mounted for observation under incident light epifluorescence (UV excitation, LP 420 barrier filters; Zeiss, Oberkochen, FRG). Elimination of mycoplasma from established cell lines can be attempted using cell cloning or repeated incubation with agents such as BMcycline (Cat. No. 799050; Boehringer-Mannheim, Mannheim, FRG).

Measurements of Transmembrane Solute Transport

Measurements of plasma membrane solute transport or receptor density may be made directly using radiotracers in cells cultivated on small 3-cm dishes or in 6-, 12-, or 96-well dishes, depending on particular requirements imposed by the specific activities of the solute of interest and the capacity or density of membrane transport sites or receptor density present on individual cultures. Specific attention should be focused on the cellular location of the system of interest, as access to the basolateral surfaces may be considerably restricted in confluent cultures. Even in subconfluent cultures autoradiography has demonstrated the restriction of cellular uptake of certain amino acids to cells on the periphery of cell

---

[10] R. F. Husted, M. J. Welsh, and J. B. Stokes, *Am. J. Physiol.* **250**, C214 (1986).
[11] T. R. Chen, *Exp. Cell Res.* **104**, 255 (1977).

islands in LLCPK$_1$ cells.[12] In any case the density may be an important consideration, if high cellular densities are required for the expression of transport function. An alternative approach would be to ensure disruption of the apical tight junction using incubations in Ca^{2+}–Mg^{2+}-free medium containing EDTA,[13] or to perform the experiments on cells in suspension.[12] Adequate controls must then be performed to demonstrate that these procedures do not interfere with the membrane transport of interest. For MDCK cells where K$^+$ transport (unidirectional influx) is measured using the K$^+$ congener ^{86}Rb under equilibrium exchange conditions we aim to produce subconfluent layers after 3 days of culture. A seeding density of 5×10^4 cells/cm^2 is used to realize a final density for experiments of $1.5-2.0 \times 10^5$ cells/cm^2.

All experimental incubations are performed in nonsterile conditions on a bench incubator made of an aluminum trough supported on a Perspex plate through which thermostated water is forced by a flow heater (Churchill Instrument Co. Ltd., Uxbridge, England). This necessitates the use of phosphate, Tris, or HEPES-buffered experimental medium. The usual composition of the experimental buffer is 137 mM NaCl, 5.4 mM KCl, 2.8 mM CaCl$_2$, 1.2 mM MgSO$_4$, 0.3 mM NaH$_2$PO$_4$, 0.3 mM KH$_2$PO$_4$, 14 mM Tris, 12 mM HCl, pH 7.4, at 37°. A preincubation step of 20 min usually precedes flux determinations, due to a transiently elevated flux upon transfer from the growth medium to the experimental medium. Glucose (5 mM) and donor horse serum [1% (v/v) or lower] are the only additions required to maintain cell viability for prolonged periods (up to 3 hr). The flux measurement is initiated by the addition of the prewarmed radiotracer-containing solution (3.7 kBq/cm^3 ^{86}Rb; Amersham International, Amersham, England), and terminated by its rapid removal by suction via a water-driven pump and by repeated washing (five or six times, ~30 sec each time) in ice-cold tracer-free wash medium (usually the incubation medium). With cultured cells on plastic such washes achieve extremely rapid temperature reductions (~100 msec). Artifactual loss of isotope is thus low, but may be reduced by inclusion of transport inhibitors in the "stopping solution," e.g., HgCl$_2$ for facilitated sugar transport. The adequacy of the wash protocol must be checked by the efficient clearance of an extracellular marker such as [^{14}C]- or [^{3}H]inulin. The total time for measurement of undirectional uptake requires that the flux is undertaken in the linear portion of the uptake curve. For ^{86}Rb a 5- or 10-min influx period is usual. For other ions, e.g., ^{22}Na, extremely rapid influx determinations must be used (10 sec); this is not easily performed since high

[12] F. V. Sepulveda and J. D. Pearson, *J. Cell. Physiol.* **118**, 211 (1984).
[13] K. Matlin, D. F. Bainton, M. Pesonen, D. Louvard, N. Genty, K. Simons, *J. Cell Biol.* **97**, 627 (1983).

isotope-specific activities need to be used. It should be noted that determinations of influx of Na^+ in MDCK cells in some studies are made under conditions of net uptake in Na^+-depleted cells.[14] Flux determinations are normalized to cell number/plate (most usual), protein, or DNA content using an aliquot of the extracted well. If flux is normalized by cell number determinations extraction of isotope is made with a trypsin–2 mM EDTA solution to form a single-cell suspension as outlined above. An important control that is always performed is to test for nonspecific uptake of radiotracer on plastic Petri dishes in the absence of seeded cells. This is especially important for binding determinations.

Unidirectional efflux is performed using a single-plate technique with continual replacement (every 3 min) of the external medium. This necessitates the preequilibration of the intracellular fluid with ^{86}Rb by incubation in experimental medium containing 37 kBq/cm^3 ^{86}Rb for 3 hr. Rate coefficients for ^{86}Rb loss can be calculated from semilog plots of ^{86}Rb loss against time.

Measurements of net ion movements and of total ionic composition are made by washing the cell layers in ice-cold wash medium (usually isotonic choline chloride for Na^+ and K^+ determinations) and by final extraction in double glass-distilled water for 3–4 hr. Sodium ion and K^+ are directly determined using flame photometry.

Determination of cell volumes in monolayer cultures is most conveniently measured using the technique of Kletzein et al.[15] Cell layers are incubated to equilibrium (60 min) with 37 kBq/cm^3 of the nonmetabolizable sugar 3-O-methyl-D-[^{14}C]glucose as an intracellular space marker. It is important to check that no uphill transport occurs (as would occur in LLCPK$_1$ cells) and that there is no intracellular binding. Washing and extraction is carried out as described for K^+ fluxes.

If the demonstration of the epithelial polarity of transport or binding is required, influx, efflux, and binding determinations are possible on epithelial layers grown upon permeant supports, using techniques identical in principle to those described. The filter support in this case will impose a serious "unstirred layer" for measurements across the basal surfaces.

Preparation of Epithelial Layers on Permeant Supports

Uncoated Millipore (MF) Supports

Epithelial layers of MDCK cells have been successfully generated with pore sizes up to 2 μm. Generally the larger the pore size the slower the

---

[14] M. J. Rindler, M. Taub, and M. H. Saier, *J. Cell Biol.* **254,** 11431 (1979).

[15] R. F. Kletzein, M. W. Pariza, J. E. Becker, and V. R. Potter, *Anal. Biochem.* **69,** 537 (1975).

attainment of a confluent layer. We have not successfully generated epithelial layers in larger pore sizes without coating the support with collagen. Seeding is carried out by directly applying the cell suspension at high cell densities (2×10^5 cells/cm^2 of permeable support) onto Millipore filters (2.5-cm diameter) (boiled in 100 cm^3 distilled water to remove the wetting agent) and held in presterilized Swinnex holders (Millipore Ltd., Harrow, England). The holder is incubated at 37° in an air/5% CO_2 atmosphere for up to 1 hr to allow cell attachment. After this time the filter is removed from the holder, the layer is washed free of nonadherent cells, and the layer is finally floated in complete medium to which insulin is invariably added. After 3 days confluent epithelial layers are formed and we usually perform our experiments at this time. However, the medium may be changed every 3 days and the epithelial layers maintained. For routine experiments involving measurement of epithelial parameters we favor this method, as the epithelial layers may then be clamped directly into Ussing chambers in which edge effects are reduced in a conventional way. It should be stressed that edge effects exist even with epithelia grown in mini-Marbrook-type chambers, as epithelial cells are unlikely to form functional junctions with the vertical walls of the chamber. The morphology of MDCK cells certainly varies toward the outer edge of the cup under such conditions.

Collagen-Coated Millipore Filters

Collagen (5 mg/cm^3) [made from rat tails or purchased from Sigma (St. Louis, MO)] in 1% acetic acid is poured directly onto 2.5-cm Millipore filter disks contained in a 10-cm Petri dish and spread to give an even coverage. Gelation is achieved by exposure to ammonia vapor from 1 to 2 cm^3 of ammonia solution. The ammonia solution is removed and the filters allowed to dry. The collagen gel is then cross-linked by immersion in 4% glutaraldehyde solution for 60 min. Repeated washing with MEM ensures removal of any residual glutaraldehyde. Cell seeding is achieved by placing the collagen-coated filters in small (3 cm) dishes and proceeding as above.

Assessment of Confluency on Millipore Substrates

Assessment of confluency in cell layers grown on Millipore filters is made by a conventional fixation (2% glutaraldehyde, staining with hemotoxylin) and dehydration schedule. Filters in absolute alcohol are then cleared in a 50:50 mixture of absolute alcohol and toluene, and finally in toluene. The filters are air dried and mounted on microscope slides in Canada balsam, which, since it has the same refractive index as the filter, renders it translucent. The epithelial layer may then be visualized directly.

Growth of Epithelial Layers on Collagen-Coated Nylon Mesh

Collagen gel may be used to form a translucent substrate in a nylon mesh support, as initially described by Cereijido et al.[16] Disks of nylon mesh (2.5 cm; Simonyl N118, Henry Simon Textiles, Stockport Cheshire, England) are first cleaned by sonication in acetone (5 min), boiling in distilled water (10 min), and then washing in an ascending series of alcohols before being dried. They are then dipped in acetic acid–collagen solution and processed as are Millipore filters. The strength of the collagen gel was often sufficient such that the glutaraldehyde cross-linking step could be omitted. The omission of the glutaraldehyde fixation step facilitated microelectrode penetration through the collagen matrix. Final sterilization of the collagen/nylon substrates was achieved by short-wave UV irradiation (30 min). Storage of both collagen-covered Millipore filters and nylon/collagen matrices was possible in MEM for up to 1 month at 4°.

An alternative translucent substrate based on chitosan has been developed and shown to be perfectly adequate for the formation of MDCK epithelia.[22]

Growth in Marbrook-Type Chambers

A variety of arrangements have been adopted to allow direct culture of epithelial layers in which separate access to each epithelial surface is achieved. Our initial method was to clamp Millipore filters directly into commercially available (Hendley Engineering Co., Loughton, Essex, England) nylon mini-Marbrook chambers.[17] These chambers consist of two portions to which the filter directly clips. They may be sterilized preassembled and have small feet to allow direct access of medium to the basal surfaces of the developing epithelium. Seeding is achieved by direct addition of the concentrated cell suspension to the chamber outlined above, except that fresh growth medium is finally added to both the upper chamber (1.5 cm^3) and the 3-cm Petri (3 cm^3) dish in which the mini-Marbrook chamber is placed. During growth the upper chamber often becomes acidified before the lower chamber, indicating functional (H$^+$) transport.

The great advantage of the mini-Marbrook arrangement is for longer term experiments in which epithelial cultures can be monitored continually using salt bridges cleaned by alcohol wipes, and for experiments in which access to both epithelial surfaces is required for collection of secreted products, for biochemical labeling experiments, or for transepithelial cell migration studies.

---

[16] M. Cereijido, E. S. Robbins, W. J. Dolan, C. A. Rotunno, and D. D. Sabatini, *J. Cell Biol.* **77**, 853 (1978).

[17] J. C. W. Richardson and N. L. Simmons, *FEBS Lett.* **105**, 201 (1979).

More recently several different designs have been introduced encompassing similar design features to those existing in the commercial mini-Marbrook chambers.[18] In addition, different sizes of presterilized chambers consisting of membrane filter glued to a polystyrene body are now available [e.g., Millicell-HA; a transparent membrane filter (Biopore, Millicell-CM) allows direct microscopy of growing epithelial layers. Such transparent membranes need to be precoated with a biological matrix such as collagen; our own experience is that lower epithelial resistance values are achieved with Millicell-CM compared with Millicell-HA.] A novel adaptation of this idea was developed by the Simons group in Heidelburg.[19] The Millipore millititer HA 96-well plate for microfiltration can be used to produce a large number of identical epithelial layers suitable for rapid screening of the effects of agents such as monoclonal antibodies (or other bioactive reagents) on epithelial junction formation. The impermeable backing is first stripped from the 96-well plate, and the sterile packaging bubble can be used as the bottom chamber. Cell seeding is achieved, exactly as described above. Our experience is that entirely adequate epithelial layers may be generated, though the electrical resistance is of a somewhat lower magnitude.

Measurement of Transepithelial Electrical Properties (Resistance, Short-Circuit Current, Dilution, and Biionic Potential Differences (pd)

A detailed and basic discussion of this topic, including construction, calibration, and use of reversible electrodes and salt bridges, is given in an excellent review by Watlington *et al.*[20] As previously mentioned, epithelial layers may be mounted directly using large areas and silicone seals to reduce edge effects,[21] or whole mini-Marbrook chambers may be clamped directly into specially adapted chambers. For the purpose of the present chapter it is worth stressing the importance of the correct geometry needed with respect to the measuring and current-passing Ringer bridges, when applied to the p.d. measurement of the electrical resistance of "leaky" epithelial layers, and of the necessity of making an appropriate correction for the series resistance afforded by the solution resistor present between the potential-sensing electrodes. This point is also of practical importance

---

[18] R. E. Steele, A. S. Preston, J. P. Johnson, and J. S. Handler, *Am. J. Physiol.* **251**, C136 (1986).
[19] D. Gumbiner and K. Simons, *J. Cell Biol.* **102**, 457 (1986).
[20] C. O. Watlington, J. C. Smith, and E. G. Huf, *Exp. Physiol. Biochem.* **3**, 49 (1970).
[21] N. L. Simmons, C. D. A. Brown, and E. L. Rugg, *Fed. Proc., Fed. Am. Soc. Exp. Biol.* **43**, 2225 (1984).

when short circuiting the epithelial layer. In standard Ussing chambers made of Perspex, drilled guide holes for the potential and current-passing agar bridges are present, ensuring that no change in the distance between the potential-sensing electrodes occurs on chamber assembly/disassembly, and that the current-passing bridges are placed sufficiently distant from the epithelial layer to ensure an even current density (the rule of thumb is two times the radius of the epithelial layer). For accurate determination of epithelial properties, it is simply not sufficient to dip the agar bridges in the bathing solutions, though this may be adequate for rapid screening.

Often the DC electrical resistance of the epithelium might not be the most convenient or reliable method for determining the functional presence of an epithelial layer. If a significant paracellular pathway for ions is present with ionic mobilities differing from free solution, a more convenient method is to elicit a dilution or biionic pd. For MDCK epithelial layers this can be achieved by replacing the basal bathing solution for one in which the NaCl is replaced in an equimolar fashion by choline chloride.[16] The resulting concentration gradient for Na^+ and the cationic permselectivity of the paracellular pathway will render the basal bathing solution up to 45 mV electropositive with respect to the apical bathing solution. Replacement of the basal bathing solution is recommended as the epithelial layer is less subject to damage from solution turbulence under these conditions.

Measurement of Bidirectional Transepithelial Solute Fluxes

Simultaneous measurement of bidirectional transepithelial fluxes using ^3H-or ^{14}C-labeled solutes or ^{22}Na and ^{24}Na is recommended whenever possible in order to eliminate time-dependent variation and between-monolayer variation. Isotopes are added to both bathing solutions (volume = 8 cm^3) and after an initial 30-min flux period, 1-cm^3 samples of each chamber are taken. Two further 1-cm^3 samples at 60 and 90 min are taken and fluxes averaged over these two 30-min periods. Flux is calculated from the rate of tracer appearance on the contralateral side, to which radiotracer was added. Isotope equilibration should not exceed 1–2%, and the specific activities used should allow increments of 5×10^2 Bq/cm^3 in the contralateral chamber.

---

[22] P. Popowicz, J. Kurzyca, B. Dolinska, and J. Popowicz, *Biomed. Biochim. Acta* **44**, 1329 (1985)

[25] Giant MDCK Cells: A Novel Expression System

By H. Oberleithner, A. Schwab, H.-J. Westphale, B. Schuricht, B. Püschel, and H. Koepsell

Introduction

Over the past years a large number of cDNA sequences have been reported. Most of them are characterized by their homology to DNA sequences or to proteins which have been described earlier. In many cases the functional relevance of the cloned cDNAs has been established by injecting the corresponding cRNAs into oocytes from *Xenopus laevis* and demonstrating functional properties which are related to the injected cRNA. Although the amphibian oocytes are considered to be a valuable expression system they cannot be employed in all cases since they contain many endogenous cell functions that may be either expressed or dormant.[1] The number of endogenous transport systems detected in oocytes from *Xenopus laevis* has become obvious in the past few years.[2] It has to be kept in mind that the expression of a function after injection of cRNA does not necessarily mean that the expressed protein is sufficient to perform the induced function.[1] It has been demonstrated, for example, that the cRNA coding for the β subunit of the Na^+,K^+-ATPase is able to increase Na^+,K^+-ATPase activity in oocytes from *Xenopus laevis*.[3] Furthermore, if a function to be investigated after injection of cRNA is not detected in an expression system, it does not necessarily mean that the expression system does not contain any components of the investigated functional protein. Thus, it has to be defined whether by cRNA injection (1) the absence of a functional protein, (2) the absence of a protein subunit, (3) a mutation in the functional protein, or (4) the lack of a posttranslational modification is supplemented. For this reason it is desirable to do cRNA injections and to study expression in a number of different cell lines which have been characterized in detail. Although there is no doubt as to the special merits of *Xenopus* oocytes as living "test tubes"[1,4] microinjection of mRNA into intact mammalian cells offers several distinct advantages. As pointed out

---

[1] H. Soreg, *CRC Crit. Rev. Biochem.* **18,** 199 (1985).
[2] W. M. Weber, W. Schwarz, and H. Passow, *J. Membr. Biol.* **111,** 93 (1989).
[3] J. R. Emanuel, J. Schulz, X.-M. Zhou, R. B. Kent, D. Housman, L. Cantley, and R. Levenson, *J. Biol. Chem.* **263,** 7726 (1988).
[4] J. B. Gurdon, C. D. Lane, H. R. Woodland, and G. Marbaix, *Nature (London)* **233,** 177 (1971).

recently,[5] it provides greatly enhanced sensitivity in detecting translation products from small amounts of mRNA and can also be applied to almost every cell type. The recipient cell kept in culture can be subjected to a variety of manipulations (e.g., specific hormone treatment; induction or inhibition of mitosis) and its response after translation of the injected mRNA can be studied directly. Molecular mechanisms of differential translation of mRNA molecules and differential protein turnover in determining eucaryotic gene regulation[6] may be most adequately studied in mRNA-injected somatic cells in culture. If a mutant is used as a recipient, the microinjection of wild-type mRNA and transient restoration of the mutant phenotype can serve as a powerful technique for characterizing the complementing mRNA and the nature of the mutation.[5] Finally, genetic diseases such as cystic fibrosis or essential hypertension, which are supposed to be the result of dysregulation of plasma ionic channels and transport carriers, can be studied. Whether such cellular malfunctions are due to the expression of defective regulatory proteins may be elucidated in experiments in which mRNA from abnormal tissue is injected into normal cells in culture.

In order to facilitate the injection process and to allow the application of microelectrode techniques we fused single MDCK cells to giant cells.[7] MDCK cells, originally derived from dog kidney,[8] represent a stable epithelial cell line deficient in Na^+-dependent glucose transport.[9] In this chapter we describe a technique by which an epithelial cell line can be used for expression experiments.

Basically, cells are harvested from a subconfluent monolayer and fused to form giant cells. After a 24-hr resting period giant cells are separated from single cells and seeded on appropriate supports. At this stage a portion of enriched giant cells can be stored in the freezer (-50 to $-70°$). Twenty-four hours after plating giant cells are injected and again incubated. Another 24 hr later the cells are used for electrical measurements. In this four-step procedure (fusion, separation, injection, analysis) only viable cells enter the last stage in which the analysis is performed.

---

[5] P.-F. Lin, D. B. Brown, P. Murphy, M. Yamaizumi, and F. H. Ruddle, this series, Vol. 151, p. 371.

[6] J. E. Darnell, Jr., *Nature (London)* **297**, 365 (1982).

[7] U. Kersting, H. Joha, W. Steigner, B. Gassner, G. Gstraunthaler, W. Pfaller, and H. Oberleithner, *J. Membr. Biol.* **111**, 37 (1989).

[8] S. H. Madin and N. B. Darby, Jr., *Proc. Soc. Exp. Biol. Med.* **98**, 574 (1958).

[9] M. J. Rindler, L. M. Chuman, L. Shaffer, and M. H. Saier, Jr., *J. Cell Biol.* **81**, 635 (1979).

MDCK Cell Fusion

In a recent paper the various steps of cell-to-cell fusion were described in detail.[7] In principle, any type of cultured cells can be fused by this procedure. Depending on the cell line some modifications of the procedure may be required. Cells in the mitotic phase will give the best yield when they have been treated with proteases. Fused MDCK cells vary in size from 20 to 300 μm in diameter. Usually, cells with diameters larger than 100 μm die within 24 hr. The fusion protocol described previously[7] yields cells in the range of 20 to 100 μm. The fused cells contain 2 to 50 cell nuclei. This number depends on the number of individual cells fused to the giant cell and on the frequency of nuclear fusion. When the fusion procedure is completed the cells are seeded into a Petri dish and incubated for 24 hr (37°) in culture medium. Over this period of time viable single and fused cells firmly attach to the bottom of the dishes while damaged cells remain in the supernatant. This is considered an important step of selection that helps to select viable cells for injection experiments.

Preparation of Giant Cells for Microinjections

For expression experiments the giant cells have to be identified first, then injected individually, and, finally, 24 hr after microinjection, reidentified in order to perform the electrical analysis. To keep trace of individual giant cells, the cells must be seeded at very low density *(vide infra)*. Since after fusion only some 10% of the cells meet the requirements for microinjection (e.g., cell size) the fused cells must be enriched. This was done under sterile conditions as follows (Fig. 1): 24 hr after fusion the MDCK cells are trypsinized; they then are suspended in 3 ml of culture medium and filtered by carefully pipetting the suspension onto a fine nylon mesh (nyboldt PA − 30/13, Schweizerische Seidengaze AG, Zürich, Switzerland; pore diameter 30 μm) which is fixed between two Plexiglas rings. Single cells retained in the sieve are washed out with 3 ml of culture medium. Then the sieve is immersed in medium in another culture dish and shaken to remove the giant cells from the mesh. After this procedure there are approximately 200 cells/Petri dish. Cell size is taken as the criterion to differentiate between single and fused MDCK cells. Thus, a cell diameter larger than 27 μm is assumed to indicate fused cells. The medium filtered through the nylon mesh contained $13 \pm 1\%$ ($n = 6$) fused cells whereas $69 \pm 6\%$ ($n = 8$) fused cells were found in the medium in which the mesh was rinsed after filtration. To enable the identification of individual single fused cells a minor fraction of the suspension of filtered giant cells is

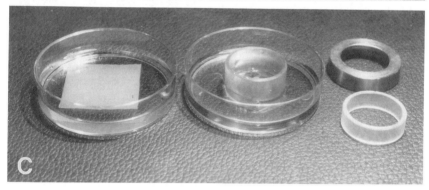

FIG. 1. Procedure of giant cell separation. (A) From left to right: plastic ring I, nylon mesh (3 × 3 cm), plastic ring II, heavy metal ring. (B) Sterile giant cell separation in Petri dish. (C) Giant cells harvested from the filter (left dish), single cells recovered from the filtered fluid (right dish).

transferred onto a small dish which is made by gluing a plastic ring (25-mm i.d., 1-mm height) onto a spherical microscope glass coverslip (No. 2 thickness, 30-mm diameter). To localize fused cells on the coverslip five small rings (1 mm in diameter, arranged assymmetrically in the center) were engraved by a diamond cutter on the back. When 100 cells are transferred onto the coverslip about 5 giant cells attach in the circular area of each ring. This procedure enables the experimenter to identify the individual cells and to facilitate the localization of injected cells. Now the coverslips are put into appropriate culture dishes and the cells are again incubated for 24 hr. Thereafter the viable cells are firmly attached to the glass surface. In order to improve cell attachment glass coverslips can be coated with poly(lysine) [poly(L-lysine), 0.1 g/liter; Serva, Heidelberg, FRG]. Cells are virtually spherical when not attached to solid surfaces (Fig. 2). A few hours of incubation leads to cell attachment and the cells flatten. Hours to days later cells are only a few micrometers in height, but up to 500 μm in diameter. Nuclei are clearly visible.

Preparation of mRNA from Rat Intestine

Solutions

Guanidine solution A: 6 M guanidine hydrochloride, 100 mM 2-mercaptoethanol, 20 mM sodium citrate, pH 7.0.
Guanidine solution B: 6 M guanidine hydrochloride, 250 mM 2-mercaptoethanol, 100 mM sodium acetate, pH 5.2
Extraction buffer: 0.5% sodium dodecyl sulfate (SDS), 100 mM NaCl, 50 mM sodium acetate, pH 5.2, 5 mM ethylenediaminetetraacetic acid (EDTA).

Directly after slaughter the jejunum of female Wistar rats is perfused with ice-cold Ringer solution and everted with a glass rod. Thereafter, the jejunum is rinsed three times with ice-cold Ringer solution and the mucosa is removed by scraping with the edge of a microscopic slide. RNA is extracted in the presence of guanidine hydrochloride[10]; remaining proteins are removed by phenol/chloroform extraction[11] and the mRNA is isolated by affinity chromatography on oligo(dT) cellulose.[12] Briefly, 2-g portions of

---

[10] J. M. Chirgwin, A. E. Przybyla, R. J. MacDonald, and W. J. Rutter, *Biochemistry* **18**, 5294 (1979).
[11] W. Braell and H. F. Lodish, *Cell (Cambridge, Mass.)* **28**, 23 (1982).
[12] H. Aviv and P. Leder, *Proc. Natl. Acad. Sci. U.S.A.* **69**, 1408 (1972).

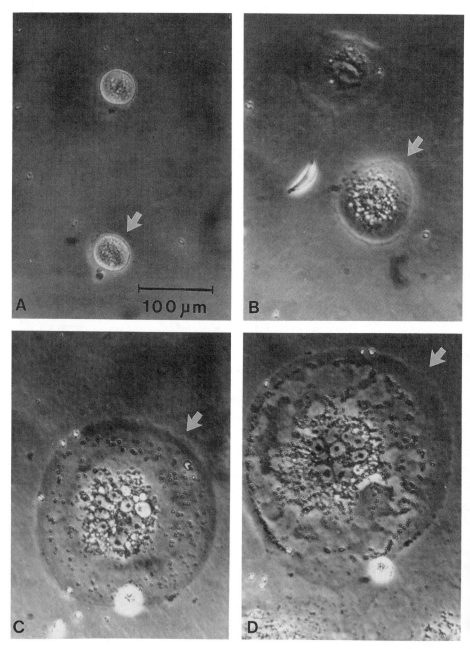

Fig. 2. Time course of giant cell attachment after the fusion process. (A) Two fused MDCK cells a few minutes after fusion. (B) The same cells 3 hr after fusion, (C) 18 hr after fusion, and (D) 42 hr after fusion. Arrows indicate the same cell. Please note that magnification is identical from (A) to (D).

scraped mucosa are added to 20 ml of guanidine solution A and homogenized for 30 sec (0°). The suspension is then centrifuged 10 min (4°) at 11,000 g. The RNA in the supernatant is precipitated by incubation at $-20°$ overnight after addition of 0.025 vol of 1 M acetic acid and 0.75 vol of ethanol. The precipitated RNA is collected by centrifugation (15 min, 13,000 g, 4°), resuspended in a quarter of the original volume of guanidine solution B, and, after 0.5 vol of ethanol has been added, precipitated by overnight incubation at $-20°$. After another centrifugation, resuspension in guanidine solution B, and precipitation the RNA is collected by centrifugation, washed twice with 80% ethanol, and finally dried briefly under vaccuum. RNA from 8 g of intestinal mucosa is dissolved in 3.5 ml of 0.1% (w/v) SDS. Short incubations (about 5 min) at 65° and very careful homogenation will help get all material in solution. The volume is increased by addition of 3.5 ml of extraction buffer. To remove proteins, 7 ml phenol/chloroform/isoamylalcohol (25/24/1, v/v/v) is added and the suspension is mixed (5 min, 22°) and centrifuged for 10 min (4°) at 4000 g. The obtained aqueous phase is extracted another time as described above and saved. To collect residual RNA, the phenol phases are mixed with 3 ml of 0.1% (w/v) SDS plus 3 ml of extraction buffer and centrifuged once more. The combined aqueous phases are extracted three times with an equal volume of buffer-saturated diethyl ether to remove phenol traces. Finally, after addition of 0.1 vol of 3 M sodium acetate, pH 5.2, and 2.2 vol of ethanol, the mRNA is precipitated by an overnight incubation at $-20°$. The final purification of mRNA by affinity chromatography on oligo (dT) cellulose and the analysis of the purified mRNA by electrophoresis on 1% (w/v) agarose gels containing 2.2 M formaldehyde are performed exactly as described by Maniatis et al.[13] Ethidium bromide staining of the gel shows that the final mRNA is apparently free of RNases and contains only minor contamination of 28S and 18S RNA.

Giant Cell Microinjection

Preparation of Attached Cells for Microinjection

Cell injections are feasible only in virtually spherical cells which have a diameter larger than 40 μm. Thus, fused cells attached to the coverslip are treated with EDTA-buffered Ca^{2+}-free Ringer solution containing 0.05% trypsin. After addition of this solution (37°) the appearance of the cells

[13] T. Maniatis, E. F. Fritsch, and J. Sambrook, "Molecular Cloning: A Laboratory Manual." Cold Spring Harbor Lab., Cold Spring Harbor, New York, 1982.

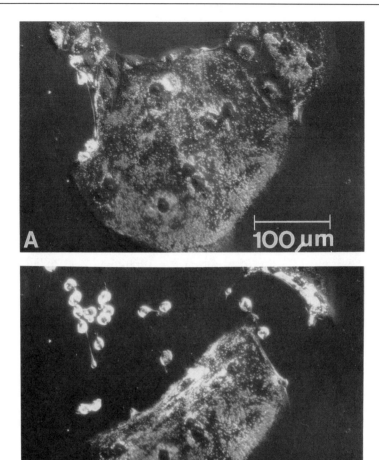

FIG. 3. Time course of partial giant cell detachment by trysinization and mRNA microinjection. (A) Giant cell 48 hr after fusion. (B) The same cell 3 min after application of Ca^{2+}-free EDTA-buffered Ringer solution containing trypsin. (C) The same cell 5 min later. (D) The same cell 7 min later; now trysinization is stopped and normal Ringer solution is superfused. The tip of the injection pipet is gently pressed against the plasma membrane. (E) mRNA is injected by one to three pressure pulses (0.1 sec duration) until cell volume is increased by about 10%. (F) The pipet is withdrawn; the cell remains intact.

changes rapidly from flat to spherical (Fig. 3). When after a few minutes a spherical cell shape is approached trypsinization is stopped and complete detachment of cells is avoided by superfusing regular Ringer solution. Now a giant cell is identified within one of the five rings by means of an inverted microscope (IM 35, Zeiss, Oberkochen, FRG).

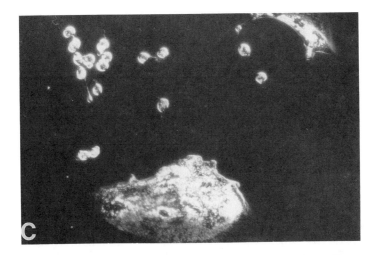

Fig. 3. *(Continued)*

Preparation of Micropipets

Micropipets, as used for patch-clamp experiments, are pulled by the two-stage procedure described by Hamill *et al.*[14] We use borosilicate glass tubings with internal fibers (o.d. 1.5 mm, i.d. 0.85 mm; Hilgenberg GmbH, Malsfeld, FRG) and pull short-shank micropipets by means of a two-stage puller (PP-83, Narishige, Tokyo, Japan). The diameter of the pipet tips

---

[14] O. P. Hamill, A. Marty, E. Neher, B. Sakmann, and F. J. Sigworth, *Pfluegers Arch.* **391,** 85 (1981).

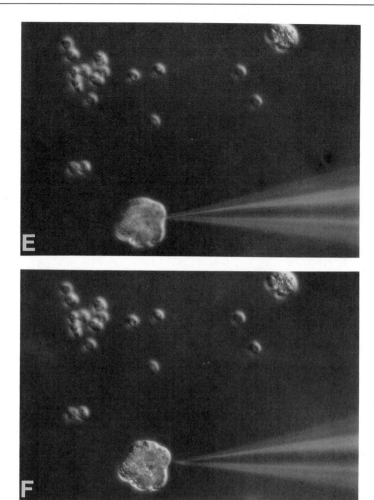

Fig. 3. *(Continued)*

(about 1 μm) and the short shank guarantee undisturbed pressure-controlled fluid delivery through the pipet into the cell.

Filling the Micropipets with mRNA

Poly(A^+) mRNA has been stored frozen in small fractions [5 μg/100 μl in 100 mM N-2-hydroxyethylpiperazine-N'-2-ethanesulfonic acid (HEPES) buffer, pH 7.4] in plastic vials. The microinjection pipets are loaded with the mRNA from the back. This is performed via a hand-pulled glass tubing filled with a few microliters of mRNA and then connected to a glass syringe. After the filling procedure the micropipet is mounted on a pipet holder carried by a micromanipulator. We use a remote-controlled motor-driven micromanipulator (Zeiss, Oberkochen, FRG) firmly attached to the stage of an inverted microscope (IM 35, Zeiss).

Pressure-Induced Microinjection

We use a home-made fluid injection device in order to control the mRNA delivery through the pipet tip. This microinjection system is based on electronically controlled pneumatic valves that can be precisely opened over periods of 0.1 to 10 sec to allow a preset pressure to be maintained on the pipet injection system. The injection pressure can be varied between 100 and 4000 mbar. A continuous and constant "holding pressure" (between 10 and 100 mbar) rests on the pipet fluid over the entire length of the experiment to keep the pipet tip open. A "cleaning pressure" (up to 10,000 mbar) can be intermittently applied to free the pipet tip from cell material. All functions are remote controlled, facilitating the microinjection procedure considerably. Using the patch-clamp pipets with 1-μm tip diameters we apply about 20 to 50 mbar constant holding pressure (to prevent plugs at the tip and fluid running backward into the shank), inject a giant MDCK cell by one to five injection pulses (injection pressure 800–1200 mbar, each pulse 0.1 sec), and use the cleaning pressure (about 5000 mbar) before and after each injection. As documented in Fig. 3 we press the pipet tip gently onto the plasma membrane but usually do not penetrate. Applying the injection pressure leads to the perforation of the membrane patch located at the pipet tip. This minimizes cell damage and rarely leads to cell death. The amount injected depends critically on the tip's resistance, which in turn is determined by its geometry, but also by the viscosity of the fluid to be injected and by some variable intrinsic properties of the cell itself. With some expertise we are able to estimate the amount of fluid delivered into the cell by the instantaneous cell volume increase induced by the injection. We usually increase the cell volume by about 10%. Since the cells regulate their volume the increase of cell volume is transient and can be observed only during the injection period. Based on these assumptions we have calculated an mRNA delivery of 1 pg/giant MDCK cell. In the experiments performed so far the relationship between the amount of mRNA injected and the magnitude of the electrical signal obtained was not systematically analyzed. Our pragmatic approach was to inject as much mRNA as possible without obviously damaging the giant cells. If a cell is damaged by the microinjection it usually becomes detached from the solid support. With some expertise damage occurs only in less than 10% of the injected cells. Within 1 hr about 20 cell injections, which are usually done with one to three pipets, can be performed.

After injection the cells are documented in terms of their location and the amount of injected mRNA so that the injected cells can be identified individually after 24 hr. After injection the Ringer solution is exchanged by culture medium (plus antibiotics) and the injected cells are incubated for expression.

Signal Analysis

About 24 hr after mRNA injection the injected cells are studied by electrophysiological techniques. Intracellular cell membrane and specific plasma membrane resistances of fused MDCK cells are measured according to techniques published recently.[15] Shortly, after trypsinization (see above and Fig. 3), cells are constantly superfused with HEPES-buffered Ringer solution and inspected with an inverted microscope. An mRNA-injected cell is identified as described above and gently penetrated by two microelectrodes (filled with 0.1 M KCl). One microelectrode is used to inject negative current pulses (0.5×10^{-9} A, 200-msec duration) while the other is used to monitor the cell membrane potential (V_m). The cell input resistance (R) is calculated according to Ohm's law: $R = \Delta V_m/I$, where ΔV_m is the cell membrane potential deflection in response to the injected current pulse (I). Then R is related to the cell surface (F): $(\Omega \cdot cm^2) = R (\Omega) \cdot F (cm^2)$. Since the fused cells are approximately spherical, F is calculated as the surface of a sphere ($F = \pi r^2$) where r (measured optically) is the radius of the sphere. In order to test for the rheogenic sodium-glucose cotransporter induced by the mRNA injection we superfused the cell under study with Ringer solution with (5 mM) or without glucose in a rapid sequence. The glucose-induced depolarization (ΔV_m) was recorded and used to calculate the transmembrane current (I_g) carried by the sodium-glucose cotransporter. We again applied Ohm's law: $I_g = \Delta V_m/R$. In this equation ΔV_m is the glucose-induced membrane depolarization (mV) and R is the membrane resistance times area ($\Omega \cdot cm^2$).

Table I shows the data obtained in 10 mRNA-injected cells. In contrast to noninjected cells that show absolutely no electrical response to glucose we observed a clear-cut electrical signal in the mRNA-injected cells. The glucose-induced current is explained by the rheogenic entry of Na$^+$ and glucose via a specific cotransport system.[16]

---

[15] H. Oberleithner, U. Kersting, S. Silbernagl, W. Steigner, and U. Vogel, *J. Membr. Biol.*, **111**, 49 (1989).

[16] E. Frömter, *Pfluegers Arch.* **393**, 179 (1982).

TABLE I
ELECTRICAL PROPERTIES OF mRNA-INJECTED
GIANT MDCK CELLS

Cell diameter (10^{-4} cm):	60 ± 7
Cell surface (10^{-5} cm^2):	11 ± 1
Cell membrane potential (mV):	-50 ± 3
Glucose-induced depolarization (mV):	16 ± 2
Cell input resistance (10^6 Ω):	30 ± 3
Resistance related to surface ($\Omega \cdot$cm^2):	2900 ± 350
Glucose-induced current (10^{-6} A$\cdot$cm^{-2}):	4.3 ± 0.7
Number of observations:	10

Evaluation of the Model and the Experimental Yield

With some experience, of 100 microinjections about 60% of the cells survive for at least 24 hr. About 50% of the cells that have survived the injection procedure can be identified 24 hr later. The electrical analysis (impalement with two microelectrodes, etc.) kills another 30% of the successfully identified mRNA-injected cells. Thus, out of 100 injection trials about 10 fused MDCK cells reach the final stage and are successfully analyzed by microelectrode techniques.

We assume that with more experience and with some further modifications of the experimental procedure the number of successful measurements per injected cells can be further increased. Compared to the *Xenopus* oocyte system[1] the giant MDCK cells are supposed to have several advantages that compensate for the disadvantages of this system. The disadvantages are the requirement of cell fusion and the small size of the giant cells compared to the oocytes. The main advantages are (1) that the experimenter can choose a cell line which has been characterized in terms of functional properties and presence and characteristics of the protein(s) under investigation and (2) that the injected cells can be maintained in culture for long periods. Since with the described method or a modification thereof any cell line in culture can be used to express and to study cellular functions which lead to an alteration of the membrane potential, it is considered an important method for future experiments.

Acknowledgments

The authors thank F.-J. Sauer and J. Bendisch for the construction of the pneumatic microinjection system and Mrs. I. Ramoz and Mrs. I. Schönberger for typing the manuscript. The study is supported by "Deutsche Forschungsgemeinschaft," SFb 176-A6.

[26] Isolation of Lumenal and Contralumenal Plasma Membrane Vesicles from Kidney

By EVAMARIA KINNE-SAFFRAN and ROLF K. H. KINNE

Introduction

One of the most striking morphological and functional features of epithelial cells is their polarity. This polarity extends from a specific arrangement of intracellular organelles and cytoskeletal elements to a specific organization and lipid and protein composition of the two areas of the plasma membranes[1-3] covering the opposing faces of epithelial cells. The elucidation of this polarity has for the last two decades fascinated biochemists and physiologists; their studies have yielded a more or less complete picture of the status quo of the polarity. This picture was crucial for the description of transcellular transport as a series of consecutive steps of transport reactions whose characteristics achieve the vectorial movement of solutes not only across the plasma membranes but also across cells in epithelial cell layers. More recently polarity of epithelial cells has attracted the interest of cell biologists who are studying in the most general sense differentiation of cells, and, in particular, the establishment of cell membrane polarity by sorting of specific membrane components to one cell side or the other. Epithelial cells have thus become a very useful model for studies on the basic mechanism of cell membrane targeting and the control thereof.[4,5]

Isolation of membranes derived from a defined pole of the epithelial cell therefore often constitutes an essential step in the investigation of membrane components, membrane properties, and their regulation. This chapter is intended to provide some guidelines to isolate lumenal and contralumenal membrane fractions from renal epithelia. Experience has shown that the isolation of lumenal membranes is usually easier to achieve, probably because this cell membrane represents that cell pole which has undergone the highest degree of differentiation. Thus a separation of this membrane from the contralumenal membrane and from intracellular membranes can in most instances be accomplished based on the relatively simple principle of "differential precipitation."[6] The isolation of contralu-

[1] P. R. Dragstein, R. Blumenthal, and J. S. Handler, *Nature (London)* **292**, 718 (1981).
[2] J. S. Rodman, L. Seidman, and M. G. Farquhar, *J. Cell Biol.* **102**, 77 (1986).
[3] K. Simons and S. D. Fuller, *Annu. Rev. Cell Biol.* **1**, 243 (1986).
[4] K. Simons and G. van Meer, *Biochemistry* **27**, 6197 (1986).
[5] K. Simons, *Kidney Int.* **23**, S201 (1987).
[6] A. G. Booth and A. J. Kenny, *Biochem. J.* **142**, 575 (1974).

menal membranes is still difficult, requires a higher degree of sophistication, and is therefore prone to higher variability. The scope of this chapter is limited to technical aspects of vesicle isolation; functional, biochemical, and physiological aspects of the approaches used to study epithelial transport employing vesicles can be found in some recent reviews.[7-13]

Purification of Lumenal Membranes from Proximal Tubule

Lumenal membranes are usually membranes which contain a high amount of glycoproteins or glycolipids; carbohydrate chains contain sialic acid residues as their terminal component and are, therefore, negatively charged at neutral pH. This high surface charge density forms the basis for the purification of these membranes by differential precipitation employing divalent cations such as magnesium or calcium in a low ionic strength medium. In the following, flow diagrams will be presented which describe two procedures that have been applied successfully to isolate brush borders from rat kidney cortex and hog kidney cortex. Some modifications are necessary for other species, such as higher divalent cation concentrations for nonmammalian kidneys[14] or a third precipitation step.[15]

Method 1: Purification of Brush Border Membranes from Rat Kidney Cortex (Isolation by Differential Precipitation)[16]

Rat kidney cortex slices (outer layer of cortex)
  One-half millimeter thick, 3 g from 10 rats
  Homogenize as a 10% suspension (w/w) for 2 min at 4° in a buffer containing 10 mM mannitol, 2 mM Tris-HCl, pH 7.1, with a blender (Waring, ESGE, Polytron)
  Filter through cheesecloth. Under mixing, add CaCl$_2$ (prepared in distilled water) to a final concentration of 10 mM. Keep on ice. After 15 min dilute 1:1 with mannitol buffer (4°) containing 10 mM CaCl$_2$

---

[7] E. Kinne-Saffran and R. Kinne, *Ann. N.Y. Acad. Sci.* **341**, 48 (1980).
[8] J. E. Lever, *CRC Crit. Rev. Biochem.* **7**, 1987 (1980).
[9] H. Murer and R. Kinne, *J. Membr. Biol.* **55**, 81 (1980).
[10] H. Murer, J. Biber, P. Gmaj, and B. Stieger, *Mol. Physiol.* **6**, 55 (1984).
[11] R. K. H. Kinne, *Comp. Biochem. Physiol.* **90A**, 721 (1988).
[12] H. Murer and P. Gmaj, *Kidney Int.* **30**, 171 (1986).
[13] R. D. Mamelok, D. F. Groth, and S. B. Prusiner, *Biochemistry* **19**, 2367 (1980).
[14] J. Eveloff, M. Field, R. Kinne, and H. Murer, *J. Comp. Physiol.* **135**, 175 (1980).
[15] N. A. Wolff and R. Kinne, *J. Membr. Biol.* **102**, 131 (1988).
[16] C. Evers, W. Haase, H. Murer, and R. Kinne, *Membr. Biochem.* **1**, 203 (1978).

> Centrifuge at 500 g for 12 min
> Remove supernatant as completely as possible
> Discard pellet
> Centrifuge supernatant at 15,000 g for 12 min

Crude brush border membrane fraction

> Resuspend pellet in 15 ml mannitol buffer with a Potter–Elvehjem homogenizer (10 strokes at 1000 rpm), clearance 0.1 mm. Add $CaCl_2$ to a final concentration of 10 mM. Keep on ice. After 15 min dilute 1:1 with mannitol buffer containing 10 mM $CaCl_2$
> Centrifuge at 750 g for 12 min
> Discard pellet
> Centrifuge supernatant at 15,000 g for 12 min

Purified brush border membranes

> Resuspend in 60 ml of buffer as required for further experiments (for transport studies use 100 mM mannitol, 20 mM Tris–HEPES, pH 7.4) with a glass–Teflon homogenizer (10 strokes at 1000 rpm). Centrifuge at 48,000 g for 20 min
> Homogenize pellet in 0.6 ml buffer (sterilized by filtration through a 0.2-μm membrane filter) as required for further experiments by sucking the suspension 10 times through a steel needle (26 gauge) into a 1 ml plastic syringe

Brush border membrane vesicles

All centrifugation and homogenization steps should be performed at 4°.

There are several aspects which have to be considered at the various steps. These are reviewed below.

Step 1: Tissue Selection and Pretreatment. If the starting material is not homogeneous, it is important to enrich the cell type of interest. In the kidney, dissection of the various kidney regions is necessary. Thin cortical slices (~3–5 mm thick) prepared from the outer layer contain mainly S_2 segments (or pars convoluta) of the superficial proximal tubule. The outer medulla contains S_3 segments (or pars recta) of the proximal tubule. Since it has been shown that these two segments differ in their transport properties[17-26] careful separation is necessary. In species such as the

---

[17] R. Kinne, in "Renal Biochemistry: Cells, Membranes, Molecules" (R. K. H. Kinne, ed.), p. 99. Elsevier, Amsterdam, 1985.
[18] R. J. Turner and A. Moran, *Am. J. Physiol.* **242**, F406 (1982).
[19] R. J. Turner and A. Moran, *J. Membr. Biol.* **70**, 37 (1982).
[20] K. E. Jørgensen and M. I. Sheikh, *Biochem. J.* **223**, 803 (1984).
[21] K. E. Jørgensen and M. I. Sheikh, *Biochem. Biophys. Acta* **814**, 23 (1985).
[22] K. E. Jørgensen and M. I. Sheikh, *Biochim. Biophys. Acta* **860**, 632 (1986).

flounder, in which the kidney contains hematopoietic tissue, tubular cells and hematopoietic cells have to be separated. In general, the tissues used in this preparation procedure should be devoid of red blood cells, since erythrocyte ghosts tend to copurify with lumenal membranes.[6] This can be achieved, for example, by blotting the tissue slices briefly on dry filter paper. In most instances it has turned out to be advantageous to freeze the tissue at $-70°$ for at least 24 hr before initiating the experiment. Purity and yield of the membranes increase markedly if frozen instead of freshly obtained tissue is used. The homogenization step can be carried out in almost any blender of the appropriate volume. Small membrane vesicles can also be obtained by nitrogen cavitation.[27] As long as the ratio between tissue wet weight and homogenization buffer is kept at 1/10 the initial amount of tissue can be varied from 0.3 to 200 g.

Addition of Divalent Cations. The concentration of divalent cations and the kind of cation must be chosen for the studies to be done after the separation. Some tissues, especially from lower vertebrates, require up to 30 mM calcium to obtain tight membrane vesicles.[28,29] Membrane-bound phospholipases may be activated by the addition of calcium; this activation leads to the formation of lysophospholipids, which increase membrane permeability for small cations such as protons.[30] Furthermore, divalent cations are trapped inside the vesicles and tightly bound to the former cytoplasmic face of the membrane. About 50 nmol Ca^{2+}/mg protein has been found in hog kidney brush border membranes isolated using 10 mM calcium for the differential precipitation.[31] Thus for studies on the calcium dependence of membrane processes the use of magnesium and EGTA is recommended.[32]

It also should be noted that the most reproducible manner for calcium addition is to add a defined volume of the tissue homogenate to a beaker which already contains the appropriate amount of 1 M calcium chloride

---

[23] U. Kragh-Hansen, H. Røigaard-Petersen, C. Jacobsen, and M. I. Sheikh, *Biochem. J.* **220**, 15 (1984).
[24] H. Røigaard-Petersen and M. I. Sheikh, *Biochem. J.* **220**, 25 (1984).
[25] H. Røigaard-Petersen, C. Jacobsen, and M. I. Sheikh, *Am J. Physiol.* **253**, F15 (1987).
[26] H. Røigaard-Petersen, C. Jacobsen, and M. I. Sheikh, *Am. J. Physiol.* **254**, F628 (1988).
[27] J. E. Lever, *CRC Crit. Rev. Biochem.* **7**, 187 (1980).
[28] J. Eveloff, R. Kinne, E. Kinne-Saffran, H. Murer, P. Silva, F. H. Epstein, J. Stoff, and W. B. Kinter, *Pfluegers Arch.* **378**, 87 (1978).
[29] N. A. Wolff, R. Kinne, B. Elger, and L. Goldstein, *J. Comp. Physiol. B.* **157**, 573 (1987).
[30] I. Sabolić and G. Burckhardt, *Biochim. Biophys. Acta* **772**, 140 (1984).
[31] S. M. Grassl, E. Heinz, and R. Kinne, *Biochim. Biophys. Acta* **736**, 178 (1983).
[32] J.-T. Lin, Z.-J. Xu, C. Lovelace, E. E. Windhager, and E. Heinz, *Am. J. Physiol.* **257**, F126 (1989).

solution. Rapid swirling distributes the calcium evenly in the suspension. Pipetting a concentrated calcium solution on top of the (often foaming) homogenate is not as reproducible; addition of solid calcium salts leads to high local concentrations which affect the selectivity of precipitation.

Preparation of the Crude Brush Border Membrane Fraction. In some tissue homogenates the lumenal membrane vesicles are so small that a large fraction of them is lost in the supernatant at this centrifugation step. The speed of centrifugation and the centrifugation time must be adjusted accordingly.[29] The recovery of smaller fragments is also the aim of the additional centrifugation step described by Vannier et al. (second method, see below).[33] However, since in the latter procedure mitochondrial and lumenal membranes cosediment, the subsequent alkaline treatment is required.[34] This treatment introduces further cations into the vesicles which might interfere with subsequent investigations.

Purified Brush Border Membranes. In some tissues the purity of the brush border membrane fraction obtained after two steps of calcium precipitation is not satisfactory with regard to the contamination with lysosomes and basolateral plasma membranes. In this instance a third precipitation step can be added or other means of separation, such as density gradient centrifugation,[35,36] must be employed.

Final Steps of Suspension. The aims of the final steps are the suspension of the membrane vesicles in the desired buffer and the obtainment of a homogeneous suspension of vesicles. Because for transport experiments highly concentrated vesicle suspensions are used, careful homogenization using a fine needle and a syringe is important for good reproducibility. The inclusion of air into the syringe during the homogenization should be avoided since it leads to loss of membrane material and potentially to disruption of membrane vesicles.

Method 2: Purification of Brush Border Membranes on a Preparative Scale by Differential Precipitation and Alkaline Treatment (Modified after Vannier et al.[33] and Meldolesi et al.[34])

Slices

↓ Homogenize hog kidney cortex slices (3–5 mm thick, 200 g) at 4° in 1200 ml buffer (10 mM mannitol, 2 mM Tris, pH 7.1, adjusted with

---

[33] C. Vannier, D. Louvard, S. Maroux, and P. Desnuelle, *Biochim. Biophys. Acta* **455**, 185 (1976).
[34] J. Meldolesi, J. D. Jamieson, and G. E. Palade, *J. Cell Biol.* **49**, 109 (1971).
[35] L. P. Karniski and P. S. Aronson, *Am. J. Physiol.* **253**, F515 (1987).
[36] R. Kinne and G. Sachs, in "Physiology of Membrane Disorders" (T. E. Andreoli, J. F. Hoffman, D. D. Fanestil, and S. G. Schultz, eds.), p. 83. Plenum, New York, 1986.

    HCl) in a Waring blender (full speed) twice, 1 min each time, with a 2-min interval
    Filter through cheesecloth
    Add CaCl$_2$ (final concentration, 10 mM)
    Keep on ice for 15 min with occasional stirring
    Centrifuge at 1500 g for 12 min
    Discard pellet, centrifuge supernatant at 30,000 g for 3 hr

Crude brush border membrane fraction

    Suspend pellet in 150 ml salt solution (170 mM NaCl, 2 mM Tris, pH 7.1, adjusted with HCl) by homogenizing in a Waring blender (full speed for 20 sec). Add 350 ml 200 mM Tris/HCl, pH 7.8, and mix
    Centrifuge at 10,000 g for 10 min
    Discard pellet
    Centrifuge supernatant at 105,000 g for 1 hr, discard supernatant

Purified brush border membranes

    ↓ Continue as described in Method 1 to obtain vesicles

The yield of membranes is about 10% if the first method is employed and about 27–30% if the second is used. The purity of the two final membrane fractions is comparable. The brush border membrane vesicles are almost exclusively oriented right side out[37]; their intravesicular space as determined by D-glucose uptake is about 2–4 μl/mg protein. Typical transport properties have been reviewed recently.[7,10-12]

Isolation of Contralumenal (Basolateral) Membrane Vesicles from Rat Kidney Cortex

Two methods have been successfully applied to enrich basolateral plasma membrane with acceptably low contamination by lumenal membranes. In both methods a crude plasma membrane fraction is first prepared in which basolateral membranes are enriched compared to brush border membranes and intracellular organelles. Then, either the small difference in apparent density or electrophoretic mobility is employed for the final purification. Small differences in density are effective in shallow regions of Percoll gradients[13,38,39]; the different electrophoretic mobilities become evident during free-flow electrophoresis.[40-42]

---

[37] W. Haase, A. Schäfer, H. Murer, and R. Kinne, *Biochem. J.* **172**, 57 (1978).
[38] V. Scalera, Y.-H. Huang, B. Hildman, and H. Murer, *Membr. Biochem.* **4**, 49 (1981).
[39] K. Inui, T. Okano, M. Takano, S. Kitazawa, and R. Hori, *Biochim. Biophys. Acta* **647**, 150 (1981).

Method 3: Purification of Basolateral Plasma Membranes from Rat Kidney Cortex Using Percoll Density Gradient Centrifugation[38]

Slices
> One-half millimeter thick, 1.5 g from five rats
> Homogenize at 4° in 35 ml sucrose buffer (250 mM sucrose, 10 mM triethanolamine/HCl, pH 7.6, and 0.1 mM phenylmethylsulfonylfluoride) with 20 strokes in a Potter–Elvehjem homogenizer at 1200 rpm (clearance = 0.1 cm)
> Dilute 1:2 with sucrose buffer
> Centrifuge at 2500 g for 15 min
> Discard pellet
> Centrifuge at 20,500 g for 20 min
> Remove supernatant carefully, add 5 ml of sucrose buffer on top of the pellet, suspend fluffy layer of pellet by swirling and scraping
> ↓ Discard supernatant and residual pellet

Fluffy layer of second pellet
> Add 20 ml sucrose buffer
> Homogenize in glass–Teflon homogenizer with 20 strokes at 1200 rpm
> ↓

Crude plasma membranes
> Add Percoll [8% (v/volume of membrane suspension)]
> Centrifuge at 48,000 g for 30 min, switch off automatic brake, recover fractions between 13 and 17 ml of the gradient from the top
> Dilute 1:10 with sucrose-free buffer (10 mM triethanolamine/HCl, pH 7.4, 0.1 mM phenylmethylsulfonylfluoride)
> ↓ Centrifuge at 48,000 g for 30 min

Purified basolateral membranes
> ↓ Continue as described in Method 1 to obtain vesicles

In basolateral membrane fractions obtained after Percoll density gradient centrifugation, Na$^+$,K$^+$-ATPase, a marker enzyme for basolateral membranes, is usually enriched 20-fold and brush border membranes are enriched 2- to 3-fold.[38] The yield is about 5% of the enzyme activity. The following points deserve further description. The homogenization step is

---

[40] H. G. Heidrich, R. Kinne, E. Kinne-Saffran, and K. Hannig, *J. Cell Biol.* **54**, 232 (1972).
[41] R. A. Reynolds, H. Wald, P. D. McNamara, and S. Segal, *Biochim. Biophys. Acta* **601**, 92 (1980).
[42] M. S. Medow, K. S. Roth, K. Ginkinger, and S. Segal, *Biochem. J.* **214**, 209 (1983).

critical with regard to purity and yield of the final membrane fraction. The aim of the homogenization is to obtain relatively large sheets of basolateral plasma membranes, which can be enriched in the crude membrane fraction and then separated on the Percoll gradient. Clearance of the pestle, speed of rotation, as well as forces applied during the vertical movement of the homogenization tube have to be relatively gentle and should be standardized as much as possible. Different batches of Percoll differ slightly in the density profile of their gradients established during the centrifugation. Colored density beads should, therefore, be used to determine whether the gradient can effectively separate blue and orange marker beads. For optimal membrane separation the blue markers ($d = 1.038$) are found between 19 and 25 ml and the orange marker ($d = 1.043$) at 31–32 ml.

Removal of Percoll from the fractions requires additional centrifugation. For the upper 18 ml of the gradient a 1:10 dilution in sucrose-free buffer followed by centrifugation at 48,000 g for 30 min yields tightly packed pellets of Percoll on top of which the membranes can be retrieved as a fluffy layer.

In the following a combination of differential centrifugation and free-flow electrophoresis is depicted. This method is a modification of the electrophoretic separation described below[40] in that a plasma membrane fraction is used which is already highly enriched in basolateral membranes compared to brush border membranes.

Method 4: Purification of Basolateral Plasma Membranes from Rat Kidney Cortex Using Free-Flow Electrophoresis[39]

Rat kidney cortex slices

　One-half millimeter thick, ~6 g wet from 20 rats

　Homogenize in 30 ml sucrose buffer (250 mM sucrose, 10 mM triethanolamine/HCl adjusted to pH 7.4 with NaOH) with eight strokes in a Potter–Elvehjem homogenizer at 300 rpm (clearance = 0.025 cm)

　Centrifuge at 1475 g for 10 min

　Discard supernatant

　Resuspend pellet in 2 M sucrose (1 ml/g starting tissue)

　Homogenize with three strokes in a Potter–Elvehjem homogenizer at 1000 rpm

　Centrifuge at 13,300 g for 10 min

　Discard pellet

　Dilute supernatant to isotonicity by the addition of 7 vol of distilled water

> Homogenize with three strokes in a Potter–Elvehjem homogenizer
> ↓ Centrifuge at 33,000 g for 15 min

Upper fluffy layer of pellet
> Resuspend in half of the supernatant
> Homogenize with three strokes in a Potter–Elvehjem homogenizer
> Centrifuge at 33,000 g for 15 min
> Resuspend upper fluffy layer in sucrose buffer (1 ml/g starting tissue)
> Homogenize with three strokes in a Potter–Elvehjem homogenizer
> Centrifuge at 33,000 g for 15 min
> Suspend upper fluffy layer of pellet in electrophoresis buffer (250 mM
> ↓ sucrose, 8.5 mM triethanolamine titrated with acetic acid to pH 7.4)

Crude basolateral membrane fraction
> Adjust protein concentration to 3.5–5.5 mg protein/ml
> Homogenize with five strokes in tight-fitting Potter–Elvehjem homogenizer at 1000 rpm
> ↓ Inject into free-flow electrophoresis chamber

Free-flow electrophoresis
> Adjust chamber buffer flow to 2 ml/hr/fraction, 875 V, chamber temperature 4.5°
> Determine distribution of protein and marker enzymes
> Combine fractions with highest Na$^+$,K$^+$-ATPase activity, centrifuge at 33,000 g for 20 min
> ↓ Resuspend pellet in vesicle buffer (see Method 1)

Basolateral plasma membranes

This method has a yield of 25%, the final membrane fraction exhibits a 16-fold enrichment of Na$^+$,K$^+$-ATPase, and the specific activity of marker enzymes for brush border enzymes is lower than in the starting material.[41]

The aforementioned method can also be used to isolate brush border membranes and basolateral membranes simultaneously in high purity from rat renal cortex.[40] For this purpose a slightly different procedure is used to prepare the crude plasma membrane fraction that is to be subjected to free-flow electrophoresis. This crude plasma membrane fraction contains both lumenal and contralumenal membranes but has low mitochondrial, lysosomal, and endoplasmic contamination.

This method was the first to be described for the preparation of highly purified brush border membranes and basolateral membranes.[40] It played a pivotal role in the discovery of a variety of transport systems and enzymes in renal membranes and the establishment of the polarity of the proximal tubule cell with regard to transport systems and enzymes.[17]

Method 5: Simultaneous Purification of Lumenal and Contralumenal Membranes by Free-Flow Electrophoresis[40]

Rat kidney cortex slices
- One-half millimeter thick, 3 g from 10 rats
- Homogenize in 10 ml sucrose buffer (250 mM sucrose, 10 mM triethanolamine/HCl, pH 7.6, adjusted with NaOH) in loose-fitting Potter–Elvehjem homogenizer, 10 strokes by hand and 3 strokes at 300 rpm
- Centrifuge at 700 g for 10 min
- Discard pellet
- Centrifuge supernatant at 700 g for 10 min
- Discard pellet
- Centrifuge supernatant at 16,000 g for 20 min

Upper fluffy layer of pellet
- Resuspend in 5 ml of supernatant by swirling and scraping
- Add residual supernatant
- Homogenize gently in loose-fitting Potter–Elvehjem homogenizer with three strokes at 800 rpm
- Centrifuge at 16,000 g for 20 min
- Resuspend upper fluffy layer of pellet in 10 ml sucrose buffer, homogenize gently with three strokes, add 10 ml of sucrose buffer
- Centrifuge at 16,000 g for 20 min
- Resuspend upper fluffy layer of pellet in electrophoresis buffer (250 mM sucrose, 8.5 mM triethanolamine titrated with acetic acid to pH 7.4)

Crude plasma membrane fraction
- D. free-flow electrophoresis as described in Method 4

Brush border membranes
Basolateral membranes

Free-flow electrophoresis is also sometimes used to further purify basolateral membranes enriched by Percoll gradient centrifugation.[43]

The degree of vesiculation in isolated basolateral membranes is usually lower than in brush border membrane fractions, resulting in smaller intravesicular spaces (2 μl/mg protein). The orientation is random; some preparations contain a slightly higher amount of inside-out vesicles. Further transport properties have been reviewed recently.[10,19]

---

[43] B. Hagenbuch and H. Murer, *Pfluegers Arch.* **407**, 149 (1986).

Purity of Lumenal and Contralumenal Membrane Preparation

Purity of lumenal and contralumenal membrane preparations is usually established by the use of marker enzymes assumed to be localized exclusively in one or the other membrane. Despite the fact that targeting of membrane proteins into the basolateral plasma membrane may also include some incorporation of enzymes destined for the lumenal membrane,[4,5] the assumption of an almost exclusive localization of marker enzymes is an appropriate approximation for the purpose of purity determinations under physiological conditions. In pathophysiological situations—such as short ischemia, for example—the distribution of marker enzymes changes due to lateral diffusion of the membrane proteins across the tight junctions.[44] Enrichment of marker enzymes therefore partly loses its value as a criterion of purity. It is noteworthy that differential precipitation of brush border membranes leads also to an enrichment of lysosomes and peroxisomes in the final membrane preparation,[16] a fact that has to be considered when studying functions that are exerted both by microvilli and lysosomes. The basolateral membranes isolated by Percoll density gradient centrifugation also show an enrichment of lumenal membranes.[38]

It might be of interest to consider the latter contamination on a quantitative basis. The highest enrichment achieved for basolateral membranes from renal cortex is about 20-fold, indicating that in the starting material about 5% of the protein constitutes basolateral plasma membranes. For brush border membranes an enrichment of about 15-fold is the average, thus about 7% of the cell protein represents brush border membranes. A threefold enrichment of the marker enzyme alkaline phosphatase in a basolateral membrane fraction therefore suggests that 21% of the protein in this fraction is derived from lumenal membranes. Taking into account the twofold higher degree of vesiculation of the brush border membranes, about 42% of the overall intravesicular space could represent intravesicular space surrounded by lumenal membranes. This very high degree of "functional" cross-contamination relative to the minor contamination derived from enzymatic measurements demonstrates that any enrichment of lumenal marker enzymes in contralumenal membrane fractions should be avoided. Otherwise wrong conclusions on the properties and distribution of transport systems can result. An interesting example for such functional cross-contamination are studies on the sodium dependence of phosphate transport in basolateral plasma membranes.[43]

---

[44] B. A. Molitoris, C. A. Hoilien, R. Dahl, D. J. Ahnen, P. D. Wilson, and J. Kim, *J. Membr. Biol.* **106**, 233 (1988).

The methods described above can also be used to isolate brush border membranes from S_3 segments (or pars recta) of proximal tubules. For this purpose the cortex and outer medulla are carefully dissected and the latter is used as starting material.[18-26]

Isolation of Plasma Membrane Vesicles from Medullary Thick Ascending Limb

The renal outer medulla contains in its inner stripe, among other cells,[45] the thick ascending limb of Henle's loop (TALH). This tubular segment has a very high Na^+,K^+-ATPase concentration in the contralumenal membrane and its lumenal membrane is rich in Na^+-K^+-$2Cl^-$ cotransporter and K^+ channel activity.[46] The membrane vesicles isolated from this region of the kidney have, therefore, been used so far to study mainly the molecular mechanism of these three transport elements.

Lumenal Plasma Membranes

Lumenal plasma membranes derived from TALH cells can be identified by functional criteria, such as Na^+-K^+-$2Cl^-$ cotransport activity[47] or capacity to bind with high affinity and specificity to inhibitors of this transport system, such as bumetanide.[48] Since only about 20% of the starting material represent TALH cells,[45] cell separation[49,50] has to precede some of the isolation methods described below.[51]

Method 6: Enrichment of Lumenal Plasma Membranes from TALH Cells by Differential Centrifugation

Isolated TALH cells[47,49]

> Prepare 24 mg protein from two cell preparations using six rabbits; store at $-70°$ in 4 ml sucrose buffer (250 mM sucrose, 10 mM triethanolamine, pH 7.6, adjusted with HNO_3)
>
> Homogenize at 4° in a total volume of 15 ml sucrose buffer upon thawing with 40 strokes in a tight-fitting Potter–Elvehjem homogenizer at 1000 rpm

---

[45] B. Kaissling and W. Kriz, *Adv. Anat. Embryol. Cell Biol.* **56**, 1 (1979).
[46] R. Greger, *Physiol. Rev.* **65**, 760 (1985).
[47] J. Eveloff and R. Kinne, *J. Membr. Biol.* **72**, 173 (1983).
[48] B. Forbush III and H. C. Palfrey, *J. Biol. Chem.* **258**, 11787 (1983).
[49] J. Eveloff, W. Haase, and R. Kinne, *J. Cell Biol.* **87**, 672 (1980).
[50] M. E. Chamberlin, A. LeFurgey, and L. J. Mandel, *Am. J. Physiol.* **247**, F955 (1984).
[51] R. Kinne, this series.

Cell Homogenate
(see *Tissue homogenate* below)
Outer strip of outer medulla[52]

    Obtain 10 g wet wt from four or five rabbits
    Mince in 5 ml sucrose buffer (see above)
    Homogenize at 4° in a total of 60 ml buffer with 15 strokes in a loose-fitting Dounce (glass–glass) homogenizer. Filter homogenate through two layers of cheesecloth

Tissue Homogenate

    Centrifuge at 700 g for 10 min, discard pellet
    Centrifuge supernatant at 16,000 g for 20 min
    Save supernatant
    Resuspend upper white fluffy layer of pellet in 12–15 ml sucrose buffer by homogenizing with 10 strokes in a loose-fitting Potter–Elvehjem homogenizer
    Centrifuge at 16,000 g for 20 min
    Repeat twice
    Centrifuge combined four supernatants at 100,000 g for 60 min
    Resuspend pellet in small volume of vesicle buffer [100 mM sucrose, 1 mmol Mg(NO$_3$)$_2$, 20 mM triethanolamine adjusted to pH 7.4 with H$_2$SO$_4$] by repeated passage through a 26-gauge needle

Crude lumenal membrane fraction

The yield of membrane protein is about 1.5 mg if isolated TALH cells are used and 25 mg if renal tissue is employed. The transport properties of the two membrane fractions are similar; bumetanide (10^{-4} M) inhibits chloride-dependent sodium uptake by about 50%.[47,52] No enrichment of Na$^+$,K$^+$-ATPase is observed, indicating that the membrane fraction is not contaminated with basolateral membranes. As expected from the anatomical arrangement, membranes derived directly from renal outer medulla contain appreciable amounts of brush border membranes. This cross-contamination is avoided when isolated TALH cells are used. The intravesicular volume of the membrane preparations is about 2 μl/mg membrane protein. Further transport properties of these vesicles are described in more detail in some recent publications.[47,52–54]

---

[52] R. Kinne, E. Kinne-Saffran, B. Schölermann, and H. Schütz, *Pfluegers Arch.* **407**, S168 (1986).

[53] B. König, S. Ricapito, and R. Kinne, *Pfluegers Arch.* **399**, 173 (1983).

[54] R. Kinne, E. Kinne-Saffran, H. Schütz, and B. Schölermann, *J. Membr. Biol.* **94**, 279 (1986).

Another method to enrich lumenal plasma membranes from TALH cells, one employing density gradient centrifugation in addition to differential centrifugation, is compiled below.

Method 7: Preparation of Lumenal Membranes from Red Outer Medulla Using Metrizamide Density Gradient Centrifugation[52-56]

Red Outer Medulla
> Prepare 0.6 g/rabbit kidney
> Homogenize at 4° in 10 ml sucrose buffer (250 mM sucrose, 50 mM KCl, 2 mM $MgCl_2$, 1 mM EGTA, 10 mM 4-morpholinopropanesulfonic acid (MOPS), pH 7.2, adjusted with triethanolamine) in a tight-fitting Potter–Elvehjem homogenizer with five strokes at 1000 rpm
> Centrifuge at 6000 g for 15 min
> Save supernatant
> Resuspend pellet in original volume of sucrose buffer
> Rehomogenize with five strokes in Potter–Elvehjem homogenizer
> Centrifuge at 6000 g for 15 min
> Discard pellet
> Combine supernatants
> Centrifuge at 48,000 g for 30 min
> Resuspend pellet in the same buffer as above except for 25 mM imidazole/acetic acid instead of MOPS buffer

Crude microsomes
> Adjust protein concentration to 3–4 mg/ml
> Layer 1 ml of crude microsomes on continuous metrizamide gradient [5–15% (w/v), on top of a 1-ml cushion of 30% metrizamide]
> Centrifuge for 16 hr at 106,000 g
> Collect the fraction between 3 and 5 ml
> Dilute in 25 mM imidazole acetate, 1 mM EDTA (pH 7.4)
> Centrifuge at 140,000 g for 90 min
> Resuspend in a small volume of aforementioned buffer by repeatedly drawing and expelling the suspension with a micropipet

Crude lumenal membrane fraction

In the metrizamide gradient lumenal and contralumenal membranes are

---

[55] D. A. Klaerke, S. J. D. Karlish, and P. L. Jørgensen, *J. Membr. Biol.* **95**, 105 (1987).
[56] C. Burnham, S. J. D. Karlish, and P. L. Jørgensen, *Biochim. Biophys. Acta* **821**, 461 (1985).

only partially separated. The "lumenal" membrane fraction shows Na^+-K^+-$2Cl^-$ cotransporter and K^+ channel activity, these activities have been reconstituted, and the regulation of the K^+ channel has been studied in some detail.[55,56] Furosemide inhibition of barium-insensitive ^{86}Rb uptake is about 30%. Basolateral membranes of the thick ascending limb cells are characterized by a very high Na^+,K^+-ATPase activity.[57,58] This marker enzyme has been successfully employed in isolating membrane fractions highly enriched in this enzyme and thus predominantly derived from these cells.

Method 8: Purification of Inside-Out Oriented Basolateral Plasma Membrane Vesicles from Outer Medulla Using Hypaque Gradient Centrifugation[59]

Outer medulla
> Dissect tissue from transverse slices of pig kidney using scissors
> Add 5 ml sucrose/histidine buffer (250 mM sucrose, 30 mM histidine, pH 7.2) per 1 g tissue
> Homogenize with five strokes in a very loose-fitting Potter–Elvehjem homogenizer (clearance = 0.05 cm) at 1000 rpm
> Filter through cheesecloth
> Homogenize in a Potter–Elvehjem homogenizer (0.01-cm clearance) with 10 strokes at 1000 rpm
> Centrifuge at 7500 g for 15 min
> Discard pellet
> Centrifuge supernatant for 30 min at 48,000 g
> Resuspend light upper portion of the pellet in sucrose/histidine buffer

Crude microsomes
> Layer 1 ml of membrane suspended in sucrose/histidine buffer (20 mg protein/ml) on 20 ml continuous Hypaque gradient. Hypaque gradients are prepared from 8% (w/v) Hypaque (prepared by dilution of 60% Hypaque in sucrose/histidine buffer) and 20% Hypaque (w/v) (prepared in 25 mM imidazole)
> [alternatively a step gradient (16%/20%/23%; 10/10/3 ml) can be used]
> Centrifuge for 2 hr at 102,000 g

[57] J. L. F. Shaver and C. Stirling, *J. Cell Biol.* **278**, (1978).
[58] U. Schmidt and U. C. Dubach, *Pfluegers Arch.* **306**, 219 (1969).
[59] K. N. Dzhandzhugazyan and P. L. Jørgensen, *Biochim. Biophys. Acta* **817**, 165 (1985).

Collect fractions between 13 and 17 ml from the top or membranes at 0/16% interface
Dilute 1 : 10 with sucrose/histidine buffer
Centrifuge for 1 hr at 100,000 g
Resuspend pellet in desired volume of buffer

Right-side-out basolateral plasma membrane vesicles

This method yields membranes with a specific activity of Na^+,K^+-ATPase of about 10 μmol of P_i/(mg protein · min). The membranes are all oriented right side out since almost all of the Na^+,K^+-ATPase is cryptic and can be determined only after opening the membrane vesicles with detergents. Similar preparations can be obtained using linear metrizamide gradients.[59] Such vesicles have been employed so far mainly in studies on the mechanism of action of Na^+,K^+-ATPase; other transport properties of these vesicles are only poorly characterized.

Isolation of Plasma Membrane Vesicles from the Inner Medulla (Renal Papilla)

The renal inner medulla, similar to the outer medulla, contains a variety of different tubular segments such as the thin descending limb of Henle's loop, the papillary collecting ducts, a substantial number of capillaries, as well as a significant number of interstitial cells. Plasma membranes derived from this region of the kidney can, therefore, be expected to show a large degree of heterogeneity unless either defined cell populations[60] or a sophisticated sequence of several separation procedures is employed. In addition, there are only a few enzymatic markers which identify specific membranes derived from a defined cell type. The cell that has attracted most interest hitherto is the collecting duct, on the one hand because it is the site of action of antidiuretic hormone, on the other hand because of its capability to strongly acidify the urine.

In the following sections separation procedures are described which yield purified basolateral plasma membranes of the collecting duct, identified by their vasopressin-sensitive adenylate cyclase activity, and plasma membrane fractions, which probably are derived from the lumenal surface or reserve vesicles thereof (as judged by their proton pump and cAMP-dependent protein kinase activity). The final separation in both procedures occurs via free-flow electrophoresis; the second procedure includes a density gradient which enriches lumenal or contralumenal membranes before the final purification step.

[60] J. B. Stokes, C. Grupp, and R. K. H. Kinne, *Am. J. Physiol.* **253,** F251 (1987).

Method 9: Simultaneous Purification of Lumenal and Contralumenal Membranes from Bovine Papilla by Free-Flow Electrophoresis[61]

White papilla from bovine kidney
> Mince 6 g wet wt (from two kidneys) in a small volume of ST buffer (250 mM sucrose, 10 mM triethanolamine/HCl, pH 7.6)
> Dilute to 30 ml with ST buffer
> Homogenize 10-ml aliquots in a Dounce (glass–glass) homogenizer (five strokes by hand with a loose-fitting pestle)
> Combine homogenates and filter through two layers of cheesecloth
> Homogenize filtrate in a Dounce glass–glass homogenizer (15 strokes by hand with a tight-fitting pestle)
> Centrifuge for 10 min at 700 g
> Discard pellet
> Centrifuge supernatant for 10 min at 10,000 g
> Save supernatant
> Rehomogenize pellet (15 strokes by hand with a tight-fitting pestle) in 5 ml ST buffer; add 10 ml ST buffer
> Centrifuge for 10 min at 10,000 g
> Discard pellet
> Combine supernatants
> ↓ Centrifuge for 1 hr at 100,000 g

Crude plasma membrane fraction
> Suspend pellet in 4 ml electrophoresis buffer (280 mM sucrose, 8.5 mM acetic acid, 8.5 mM triethanolamine, pH 7.4, adjusted with 2 M NaOH)
> Homogenize by 10 strokes in a tight-fitting homogenizer
> Dilute with 10 ml electrophoresis buffer
> Centrifuge for 10 min at 3000 g
> Save supernatant
> Repeat twice
> ↓ Subject supernatant to free-flow electrophoresis

Free-flow electrophoresis
> Arrange electrophoresis conditions as in Method 4
> ↓ Combine fractions according to specific activity of Ca^{2+}-ATPase or

---

[61] I. L. Schwartz, L. J. Shlatz, E. Kinne-Saffran, and R. Kinne, *Proc. Natl. Acad. Sci. U.S.A.* **71**, 2595 (1974).

vasopressin-sensitive adenylate cyclase (contralumenal membranes) or nonmitochondrial HCO_3^--stimulated ATPase (lumenal membranes)

Dilute with 0.1 M Tris-HCl buffer, pH 7.4

Centrifuge for 1 hr at 100,000 g

Suspend in desired buffer (see Method 1)

Enriched lumenal and contralumenal membranes

Method 10: Isolation of Lumenal and Contralumenal Membranes from Canine Inner Medulla[62]

Dissected papilla from mongrel dogs

Homogenize in ST buffer (250 mM sucrose, 20 mM Tris-HCl, pH 7.4) as a 10% (w/v) suspension by two 2- to 5-sec bursts at full speed in a Waring blender and five strokes of a loose-fitting Teflon–glass homogenizer at 2000 rpm

Filter homogenate through a double layer of cheesecloth

Centrifuge for 10 min at 600 g

Discard pellet

Centrifuge supernatant for 20 min at 10,000 g

Save pellet (P_{10})

Crude contralumenal plasma membranes

Centrifuge supernatant for 1 hr at 100,000 g

Save pellet (P_{100})

Crude lumenal plasma membranes

Further purification of contralumenal membranes

Resuspend P_{10} in 5 ml ST buffer

Homogenize by five strokes in a loose-fitting Dounce homogenizer

Load on top of a linear sucrose gradient [34–45% (w/w), dissolved in 20 mM Tris-HCl, pH 7.4]

Centrifuge for 7.5 hr at 100,000 g

Collect fractions between 36 and 41% sucrose

Dilute to 8% sucrose with 20 mM Tris-HCl, pH 7.4

Sediment membranes by centrifugation for 1 hr at 100,000 g

Save pellet for free-flow electrophoresis as detailed in Method 4

---

[62] R. Iyengar, D. S. Mailman, and G. Sachs, *Am. J. Physiol.* **234**, F247 (1978).

Further purification of lumenal membranes
- Resuspend P_{100} in 5 ml ST buffer
- Homogenize by five strokes in a loose-fitting Dounce homogenizer
- Load on top of an exponential gradient prepared by mixing 7.5% Ficoll in ST buffer and 43% (w/w) sucrose dissolved in 20 mM Tris-HCl, pH 7.4
- Centrifuge for 45 hr at 100,000 g
- Collect second protein peak at density 1.15
- Dilute fractions to 8% sucrose with 20 mM Tris-HCl buffer, pH 7.4
- Sediment membranes by centrifugation for 1 hr at 100,000 g
- Save pellet for free-flow electrophoresis as detailed in Method 4

Purified lumenal plasma membranes

The lumenal membranes obtained by the first method show a 10-fold enrichment for the anion-stimulated ATPase; the method including density gradient centrifugation prior to free-flow electrophoresis yields membranes with a 26-fold enrichment of the anion-stimulated ATPase. Both membrane fractions show considerable intrinsic cAMP-dependent protein kinase activity. The basolateral membrane fractions are enriched 16- to 20-fold in Ca^{2+}-ATPase or vasopressin-sensitive adenylate cyclase, respectively. The lumenal membrane vesicles were found to be osmotically reactive[63] and to possess a proton pump which can generate quite high proton gradients across the membrane.[64]

Some technical aspects of the isolation procedures presented above require some further comment. Homogenization of papillary tissue is difficult because of its high content in connective tissue. Thorough mincing into small pieces of tissue is therefore essential prior to using the loose-fitting Dounce homogenizer. Filtration through cheesecloth removes connective tissue which would impede the final homogenization in the tight-fitting Dounce homogenizer. During free-flow electrophoresis the membranes show a high tendency to aggregate; occasional suction of the membrane suspension through a 26-gauge needle alleviates this problem.

In view of the intense exchange between the lumenal membrane and endosomes in the collecting duct, purified endosomes can also be used to study specific functions of the lumenal membrane.[65]

[63] R. Kinne and I. L. Schwartz, in "Disturbances in Body Fluid Osmolality" (T. E. Andreoli, J. Grantham, F. C. Rector, eds.), p. 37. *Am. Physiol. Soc.,* Baltimore, 1977.

[64] E. Kinne-Saffran and R. Kinne, in "Hydrogen Ion Transport in Epithelia" (J. G. Forte, D. G. Warnock, and F. C. Rector, Jr., eds.), p. 247. Wiley, New York, 1984.

[65] A. S. Verkman, W. I. Lencer, D. Brown, and D. A. Ausiello, *Nature (London)* **333,** 268 (1988).

Concluding Remarks

Despite the apparently straightforward methods given above the isolation of lumenal and contralumenal vesicles from defined renal tubular segments can by no means be regarded as routine. There are problems with regard to the purity of the membrane fractions and with regard to their unambiguous assignment to a particular cell type and a particular cell surface. The extent to which these factors may interfere with experiments must be considered in every instance.[66] Generally speaking there is no problem obtaining membrane vesicles; the tendency of renal plasma membranes to vesiculate spontaneously is quite strong and transport studies can be performed relatively easily. But here again a cautious approach to techniques, results, and conclusions is advisable.[12]

Acknowledgment

The author wishes to thank Mrs. Daniela Mägdefessel for skillful secretarial work.

[66] E. Kinne-Saffran and R. Kinne, this series, Vol. 172, p. 3.

[27] Transport Studies by Optical Methods

By G. SACHS, E. RABON, and S. J. D. KARLISH

Introduction

The introduction of vesicle technology in the early 1960s by Kaback demonstrated that it was possible to retain transport properties of membranes after homogenization of cells and fractionation.[1] Various types of experimental design, involving the use of ionophores coupled with tracers or radioactive weak bases to measure potential or pH gradients, were applied to bacterial, mitochondrial, or chloroplast vesicles.[2] In the 1970s optical probes gradually replaced radioactive methods in many vesicle and then intact cell applications. This chapter will describe probes of pH and potential frequently utilized for vesicles and cells in the current literature.[3]

[1] H. R. Kaback and E. R. Stadtman, *Proc. Natl. Acad. Sci. U.S.A.* **55,** 920 (1966).
[2] P. Mitchell, *Biol. Rev.* **41,** 455 (1966).
[3] B. Chance, *MTP Int. Rev. Sci.: Biochem.* **3,** 1 (1975).

Optical probes of Ca^{2+}, largely confined to intact cell studies, have been described elsewhere.[4]

Principles of pH Gradient Measurements in Cells and Vesicles

The ability to show a differential of pH depends on the ability to obtain a measurement that reflects the pH in the medium or in the internal space of the membrane-bound compartment. Prior to the development of chemical probes, pH electrode measurements of medium pH or pH microelectrode measurements of intracellular pH gave useful information.[2] Naturally, measurement of internal pH was limited by the size of the electrode relative to the size of the compartment of interest. This allowed the use of pH microelectrodes for measurements of pH_i in large cells, but not in small epithelial cells nor in membrane vesicles.

For such purposes various radiochemical methods were developed, based on the distribution of weak bases or weak acids. In the 1960s the most frequently used probe for cells was the weak acid 5,5-dimethyloxazolidine-2,4-dione (DMO),[5] whereas for vesicles, weak bases such as methylamine[6] were more useful since intravesicular acidification was usually measured. The major difference between cells and vesicles is simplicity. There are several compartments of different pH in cells whereas in vesicles, although heterogeneous in size, there is usually a predominance of one functional class. The use of radioactive chemical probes, weak bases such as methylamine, benzylamine, or aminopyrine,[7] and weak acids such as DMO or salicylate depends on the relative impermeability of the unprotonated or deprotonated (charged) species relative to the neutral species. The choice of a suitable weak base is in part determined by the pH gradient expected. As shown below, if the pK_a of the weak base is more than 1 log unit different from the pH of the compartment, the ratio of the concentration of the weak base translates directly into the ratio of the hydrogen ion concentration gradient. For intact cells containing more than one acidic compartment, the more acidic pH can be estimated by selection of a weak base probe with a pK_a between that of the two compartments. One example is the use of aminopyrine ($pK_a = 5.0$) in parietal cells.[7]

The pH gradient, $pH_0 - pH_i$, is derived using the following equations

---

[4] R. Y. Tsien, *Annu. Rev. Biophys. Bioeng.* **12**, 91 (1983).
[5] T. C. Butler, *Adv. Tracer Methodol.* **2**, 189 (1965).
[6] S. Ramos and H. R. Kaback, *Biochemistry* **16**, 854 (1977).
[7] T. Berglindh, *Biochim. Biophys. Acta* **464**, 217 (1977).

for the case of a weak base:

$$[B]_i/[B]_o = R \quad \text{(distribution ratio)} \tag{1}$$

where B_i is the base accumulated in cells or vesicles and B_o is the base in the medium.

$$R = [1 + 10^{(pK_a - pH_i)}]/[1 + 10^{(pK_a - pH_o)}] \tag{2}$$

Taking logarithms and assuming $pK_a - pH_o > 1$,

$$pH_o - pH_i = \log[B]_i - \log[B]_o \tag{3}$$

gives less than a 10% error.

This calculation is sufficient to estimate the pH gradient in a single membrane-bound compartment, provided there is a uniform response in all the compartments. The measurement of the ratio of distribution of radioactive compound depends on an accurate estimate of the medium/vesicle volume ratio. This is obtained by distribution of a passively distributed substrate such as [^{14}C]sucrose. In cells, it is necessary to find the difference in pH across the plasma membrane and also between cytosol and several other membrane-bound compartments, as well an estimate of the relative volume of the compartment relative to the volume of the total cellular space. In general, several cellular compartments such as lysosomes or secretory granules are acidic, but some, such as mitochondria, are relatively alkaline. The simultaneous measurement of weak base and weak acid distribution allows a more accurate determination of the compartmental pH in cells.[8]

The measurement of distribution of isotopic probes depends on separation of the cells or vesicles from the medium, without disturbance of the pH gradient. This is achieved by centrifugation, filtration, or column chromatography. This technique is limited in sensitivity by the volume of the membrane compartment that traps the radioactive probe and the completeness of separation. Time resolution is limited by the rapidity of separation of the radioactive compartment from the medium. Radiochemical methods have been replaced largely by optical methods, which are generally more rapid and sensitive, albeit less quantitative.

Optical Methods

Optical probes or dyes have the advantage that it is not necessary to separate cells or vesicles from the medium not is it necessary to have a large

---

[8] P. Geck and E. Heinz, *Ann. N.Y. Acad. Sci.* **341,** 57 (1980).

volume of cells or vesicles relative to the medium. Two types of pH probes have been developed, each with its particular area of use. One type depends on trapping of the dye in the compartment of interest, the other is akin to the radiochemical probe in that redistribution of the dye is monitored. These will be discussed separately, since probe entrapment has been applied almost exclusively to the determination of cytosolic, secondary lysosomal, or endocytic vacuole pH,[9] while dye redistribution has been applied to determination of pH gradients in vesicles.

Fluorescein Methods

Fluorescein and its derivative, biscarboxyethyl fluorescein (BCECF), are fluorescent dyes which change in quantum yield as a function of pH. The principle of using this dye as an indicator of intracellular pH depends on the derivatization of the free carboxy groups to acetate or acetomethoxy esters to reduce fluorescence and allow penetration across biological membranes. Nonspecific cytoplasmic esterases then hydrolyze the ester, regenerating a relatively membrane-impermeable, fluorescent pH reporter. The BCECF dye has a complex spectral response to changes of pH, so that excitation at two wavelengths (one isosbestic, the other varying in quantum yield as a function of pH) and measurement of emission at a single wavelength allows pH calibration dye environment independent of the degree of cell loading.[4] It is also possible to obtain pH measurements with this type of dye at a single excitation wavelength or, with excitation at 480 nm, to obtain pH measurements of single cells viewed under a microscope without specialized UV optics.[10]

The esterases that liberate the free dye intracellularly appear to be cytosolic, so that measurement of pH response of the probe determines cytosolic pH. The following describes both single and dual wavelength excitation methods used for loading parietal cells and measuring intracytosolic pH changes as a function of stimulation.

Single-Wavelength Method. Cells are loaded with 2 μM acetomethoxybiscarboxyethyl carboxyfluorescein (BCECF-AM) by incubation at 37° for 20 min. Dye loading can be followed fluorimetrically since the BCECF-AM is nonfluorescent and converted, following esterase action, to the fluorescent BCECF. The cells are washed twice with dye-free medium and resuspended at a concentration of 5×10^8 cells/ml. About 10^7 cells is added to a stirred fluorimeter cuvette and maintained at 37°. The excitation wavelength is 505 nm and the emission wavelength is 530 nm. At the end of the experiment, the cells are lysed with digitonin and the fluores-

[9] P. D. Yamashiro, S. R. Fluss, and F. R. Maxfield, *J. Cell Biol.* **97**, 929 (1983).
[10] A. M. Paradiso, R. Y. Tsien, and T. E. Machen, *Nature (London)* **325**, 447 (1987).

cence signal of the dye is compared when the medium pH is varied by small additions of HCl. A pH electrode is used to measure the medium pH. This provides a calibration curve which is usually corrected by a constant factor of 0.15 to compensate for a 5-nm red shift of the excitation spectrum of the intracellular dye.[4] When nigericin is used to set pH_i equal to pH_o (see Dual-Wavelength Method, below) the correction factor is unnecessary. Single-cell measurements, with single-wavelength excitation, are uncorrected for loss of responding probe by leakage from the cells.

Dual-Wavelength Method. Cells are loaded and suspended in medium in a cuvette as above. The emission fluorescence is monitored at 530 nm, with excitation at two wavelengths, 450 and 500 nm. Fluorescence is highly pH sensitive at 500 nm and insensitive to pH at 450 nm. For this measurement a double-beam excitation fluorimeter is particularly convenient; however, a single-beam instrument can also be used since the need for 450-nm excitation is less stringent and values can be obtained by interpolation of data obtained with excitation at either side of the excitation at 500 nm. Calibration is carried out using known changes in intracellular pH where pH_i is made equal to pH_o using the exchange ionophore, nigericin. The pH range is usually from 6.3 to 7.8 to bracket the vast majority of expected changes of pH_i. The ratio of fluorescence emission at 530 nm is obtained with excitation at 450 and 500 nm at each pH_o ($=pH_i$). This ratio is insensitive to loss of dye by photobleaching or dye leakage, since fluorescence generated at 450-nm excitation compensates for such artifacts.[10] This compensation by the dual-excitation method makes it the method of choice for single-cell or imaging analysis of pH_i in a microscope system. In both methods, dye leakage should be estimated. The easiest method is to allow the cells to settle in the fluorimeter, with the stirrer off, and measure the free fluorescence. In the time course of a usual experiment, in healthy cells, less than 5% of the dye leaks. For a difference of 1 pH unit between cell and medium (medium greater than cell pH) a 5% error would overestimate pH_i by no more than 0.05 pH units.

Fluorescein Dextran. This insoluble form of fluorescein has been used to measure pH_i in endocytosed material in intact cells and in vesicles prepared from cells that have endocytosed the dye.[9,11] The principle of calibration is as above, using single-wavelength fluorescence.

Weak Base Optical Probes

There are a variety of optical probes that are useful for measurements of changes of intravesicular pH. These are protonatable amines that dis-

---

[11] A. S. Verkman, *Biophys. J.* **57**, 18a (1990).

tribute across a membrane according to the pH gradient as shown in Eqs. (1–3). When concentrated in the vesicular space, the dyes stack and fluorescence is quenched. Dyes such as 9-aminoacridine and acridine orange have been used extensively for optical measurements of pH gradients in vesicles.[12,13] Both absorbance and optical methods can be used with acridine orange whereas only fluorescent methods are applicable with 9-aminoacridine.

Acridine Orange. This is a fluorescent dye that undergoes a spectral shift upon stacking. A dual-beam spectrophotometer is used to measure a difference in absorbance at an isosbestic point, 446 nm, and at the wavelength of maximal absorbance, 496 nm. Fluorescence is measured with the excitation wavelength at 490 nm and the emission wavelength set at 530 nm. Acridine orange is useful in microscopic studies where image analysis is performed because at higher concentrations the emission wavelength undergoes a marked red shift to wavelength >600 nm. Intracellular acidic compartments are visualized by the change in emission from green to orange to red.[14]

Vesicle studies: For absorbance measurements, acridine orange is used at a concentration of 5–10 μM in vesicle suspensions containing 0.02 to 0.1 mg protein/ml of medium. In purified gastric vesicles, this provides an effective intravesicular volume of 0.044 to 0.22 μl/ml. It is usually important to preincubate the protein and the dye to ensure that steady state has been reached. Certain types of cuvettes absorb acridine orange, giving a baseline shift. The use of plastic or quartz cuvettes minimizes this effect, as does the use of high ionic strength medium (150 mM). At the end of the experiment an exchange ionophore is added to dissipate the pH gradient so as to obtain the level of absorbance at a zero pH gradient value.

Calibration of the dye signal can be done in two ways. In the first, the vesicles can be preequilibrated in media of different pH between pH 5.5 and pH 9. At each pH, enough base is added to produce a constant change of 1 pH unit. this is the constant pH pulse method. Alternatively, the vesicles are equilibrated at a low pH, for example 5.5, and base is added progressively to increase external pH. The absorbance change is measured and the pH of the external medium is monitored on each addition. In both cases the absorbance difference is plotted as a function of the pH gradient. Again, ionophore is added to dissipate the gradient to establish the baseline.[15]

[12] S. Schuldiner, H. Rottenberg, and M. Avron, *Eur. J. Biochem.* **25**, 64 (1972).
[13] P. Dell'Antone, R. Colonna, and G. F. Azzone, *Eur. J. Biochem.* **24**, 553 (1972).
[14] D. R. Dibona, S. Ito, T. Berglindh, and G. Sachs, *Proc. Natl. Acad. Sci. U.S.A.* **76**, 6689 (1979).
[15] E. Rabon, H. Chang, and G. Sachs *Biochemistry* **17**, 3345 (1978).

Fluorescence measurements utilize a lower concentration of dye (0.1–1 μM) as well as a lower relative quantity of vesicles (0.002 to 0.05 mg protein/ml medium). It is possible to calibrate fluorescent measurements of the pH gradient by generation of known pH gradients to produce a standard for the fluorescent shift or by measuring the intravesicular volume by an independent technique (e.g., [^{14}C]sucrose or mannitol space) and applying Eq. (4):

$$\text{Dye}_i/\text{dye}_o = [\text{H}^+]_i/[\text{H}^+]_o = QV/[(100 - Q)v] \quad (4)$$

where Q is the percentage quench, V is the medium volume, and v is the volume of the intravesicular space.[15]

There are some artifactual interactions of anions with acridine orange. For example, the lipophilic anion, SCN$^-$, sometimes used to induce a shunt conductance in membrane vesicles, interacts directly with the dye. Direct dye/anion interaction can give misleading results.

Cell studies: It is not possible to use the accumulation and quench of acridine orange in intact cells to measure pH gradients. This is presumably due to binding of the dye intracellularly, producing optical responses that are unrelated to pH gradients. However, using the red shift of the dye, when confirmed by reagents that collapse acid gradients (e.g., ionophores or NH$_4^+$), it is possible to determine the presence and location of acid spaces.[14]

9-Aminoacridine. This is also a fluorescent probe of acid accumulation in acidic spaces and is used like acridine orange.[12,15] However, since there is no spectral shift, this dye can be used only fluorimetrically. The excitation wavelength is 398 nm and the emission wavelength is 430 nm. Again, upon accumulation, the dye stacks and fluorescence is quenched. Calibration of the pH gradient is done by the methods described for acridine orange. The quantum yield of this dye is less than that of acridine orange, necessitating a range of 5 to 10 μM dye for the fluorimetric experiments.

It is also possible to use the accumulation of this dye in intact cells. It is not possible to measure the subcellular location of quenching, but increased fluorescence is obtained at low dye concentrations with accumulation into acidic spaces. In parietal cells, acidic spaces were monitored by using 0.1 μM dye in the perfusion medium bathing cells and using an epifluorescence microscope equipped with a photodiode array spectrophotometer. Changes of cytosolic pH were monitored simultaneously using BCECF emission at 530 nm.[16]

---

[16] S. Muallem, D. Blissard, E. Cragoe, and G. Sachs, *J. Biol. Chem.* **263**, 14703 (1988).

Measurement of Potential in Cells and Vesicles

Three methods for measurement of intracellular potentials involve microelectrodes, radioactive ion distribution, and optically active ions. For vesicles, only the latter two techniques are feasible.

The distribution of radioactive lipid-permeable cations or anions was initially used to determine potentials in mitochondria and vesicles. The cations most frequently used were $^{86}Rb^+$ in the presence of an ionophore such as valinomycin (which sets the cationic transference number to unity) or Tl^+ (which was able to permeate the plasma membrane via lipid as well as K^+ pathways). [^{14}C]SCN was the most frequently used anion. Subsequently, cations such as [^3H]triphenylmethylphosphonium (TPMP) or [^3H]tetraphenylphosphonium (TPP) were synthesized to avoid the use of ionophores.[17] These methods, as for the radioactive probes of pH gradients, depend on the separation of cells or vesicles from the medium and calibration requires an accurate knowledge of the relative volume of the cell or vesicle suspension. It has become more convenient to use optical probes of potential gradients. In general, these methods have been applied to vesicles more frequently than to cells.

Cationic Dyes

The most frequently used of the cationic dyes are the carbocyanine dyes. Two of these will be described. In both cases, it is possible to use either the absorbance difference between the isosbestic wavelength and absorbance maximum or fluorescent quench to monitor dye uptake. The fluorescence is usually the more sensitive method, requiring less dye and less relative intravesicular volume. These dyes are typically used to monitor potentials that are net negative with respect to medium, but have on occasion been used to monitor positive potentials.[18]

The calibration of the dye response depends on the formation of known outward monovalent cation gradients (either K^+, Li^+, or H^+) in the presence of the appropriate ionophore [valinomycin for K^+; N,N'-diheptyl-N,N'-diethylether 5,5-dimethyl-3,7-dioxanone (AS701) for Li^+; and either TCS or FCCP for H^+]

The potential difference (E_{mv}), vesicle interior negative, is then given by the Nernst potential:

$$E_{mv} = RT/F \ln([C^+]_i/[C^+]_o) = 58 \log([C^+]_i/[C^+]_o) \qquad (5)$$

[17] S. Ramos, S. Shuldiner, and H. R. Kaback, *Proc. Natl. Acad. Sci. U.S.A.* **73**, 1892 (1976).
[18] A. Waggoner, *J. Membr. Biol.* **27**, 317 (1974).

The dye response is thus calibrated against the calculated potential.

Diethyl Oxocarbodicyanine (DOCC). For this dye, the absorbance difference between 598 and 636 nm is measured as a function of generation of a potential difference. Dye (4 μM) is added to a vesicle suspension at a relative volume of 0.15 μl internal space/ml of medium. The difference in absorbance is followed as a function of time. The baseline is determined by abolishing the potential using the simultaneous addition of valinomycin (if K$^+$ is present) and TCS. The coupled action of these two conductive ionophores collapses the potential across the vesicle membrane.

Calibration of this signal can be carried out using known pH gradients. The vesicles are preequilibrated at pH 6, 6.5, or 7 and then the external pH is increased in increments from pH 8 to pH 9.5. At each pH change the absorbance is measured for 20 sec and then the protonophore, TCS, is added at 1 μM. The change of medium pH, verified with a pH electrode, and the change in absorbance due to TCS, are plotted as a function of $58(pH_i - pH_o) = E_m$.[15]

DiS-C$_3$(5) (3,3'-Dipropylthiadicarbocyanine Iodide). This is the dye used most frequently for negative intracellular or intravesicular potentials. In the fluorescence mode, transfer of the dye from membrane to aqueous domain or accumulation results in quenching of the fluorescence, proportional to the transmembrane potential. The excitation wavelength is 640 nm and the emission wavelength is 670 nm. The dye is used at a concentration of 1 μM and the relative vesicle volume is 0.1–0.3 μ/ml medium.

For calibration, the vesicles are preequilibrated in high KCl solution (~150 mM) and diluted about 20-fold in the fluorimeter cuvette containing 1 μM dye and no K$^+$. After a 20-sec equilibration, valinomycin is added to provide the initial value of the potential difference. After the initial excursion, K$^+$ in the form of the salt of an impermeant anion (sulfate, gluconate, or cyclamate) is added in minimal volume to the cuvette to progressively decrease the K$^+$ gradient. The difference in fluorescence is noted and the change in fluorescence is then plotted as a function of the calculated potential difference. The stability of the vesicular cation gradient in the presence of dye and valinomycin should be independently verified. It is also useful to compare the changes obtained with the addition of TMA$^+$ salts to ensure that the effect is K$^+$ and valinomycin specific.

With intact cells, if the plasma membrane potential is desired, uptake into mitochondria must be excluded. In some cells (e.g., pancreatic acinar cells, red blood cells) special precautions are not necessary and the [K$^+$]$_i$ can be obtained by the null point method (i.e., the K$^+$ concentration in the medium where no change in fluorescence is observed, in the presence of valinomycin). Again, this method of $[K^+]_i$ determination must be independently verified. The value of the plasma membrane potential is obtained as above, by calibration with different medium K^+ concentrations in the presence of valinomycin. The assumption is that $E_m = M_{K^+}$ in the presence of the ionophore.[19]

However, when cells have a significant mitochondrial content (about 10% of cell volume or greater) the contribution of the large negative intramitochondrial potential obscures the plasma membrane potential. Here it is necessary, as has been done for ascites cells, for example, to obliterate the mitochondrial potential with a cocktail composed of a protonophore, a mitochondrial ATPase inhibitor, and an inhibitor of oxidation.[20] These conditions result in nonphysiological conditions so that few studies are now done in such cells with these dyes.

Anionic Dyes

The dyes that have been used most frequently are of the oxonol series. Of the several that have been tried, only oxonol VI does not interact with ionophores such as valinomycin.[21] Oxonol VI is readily calibrated by standard methods and the use of this dye in reconstituted Na^+,K^+-ATPase vesicles will be described here using the absorbance difference method, rather than the change in fluorescence.[22]

The difference spectrum of this dye is obtained at an absorbance of 590 nm and an absorbance maximum at 628 nm. Vesicles were suspended in transport medium with continuous stirring at 20° and oxonol VI added to give a final concentration of $0.2-2\ \mu M$. the absorbance difference was monitored for about 5 min before a change in potential was initiated with the addition of ATP and other suitable ligands.

Calibration of this anionic dye was achieved by inducing positive intravesicular potentials using pH gradients with the protonophore, FCCP, or with Li^+, using the Li^+ ionophore, AS701. For the first method FCCP at $5\ \mu M$ was first added to the vesicles with the dye. Aliquots of H_2SO_4 were then added to alter medium pH (monitored with a pH electrode) and the rapid absorbance changes (less than 1 sec) measured. Blanks are run to monitor the effect of pH on dye absorbance. A pH of greater than 5.5 is suitable for calibration. Below that pH the effect of H^+ on dye absorbance is too great. The Li^+ method allows an extended calibration of the poten-

---

[19] P. J. Sims, A. S. Waggoner, C. H. Wang, and J. F. Hoffman, *Biochemistry* **13**, 3315 (1974).
[20] P. C. Laris, D. P. Bahr, and R. R. Chaffee, *Biochim. Biophys. Acta* **376**, 415 (1975).
[21] A. Waggoner, *Enzymes Biol. Membr.* **3**, 313 (1985).
[22] R. Goldshleger, Y. Shahak, and S. J. D. Karlish, *J. Membr. Biol.* **113**, 139–154 (1990).

tial. This also has the advantage of avoiding the use of K^+ and valinomycin, which may not be suitable when K^+ transport is being monitored. Vesicles are suspended in varying LiCl solutions (1–50 mM) maintained isotonic with Tris-HCl to a total concentration of 300 mM; 10 μM ionophore, AS701, in 25 μM histidine, pH 7.0, is added and the maximum absorbance change noted. This ionophore cannot be used in the presence of Na^+ since there is significant carriage of this cation.[23]

An interesting application of the rate of dissipation of this dye signal, the measurement of membrane resistance, has been described which can be applied with all the other potential dye probes. The technique used in the case of the Na^+,K^+-ATPase was to stop pump activity (e.g., with vanadate) and measure the decay of the signal. If the decay can be described by a single exponential, the rate constant, k, can be obtained. The resistance of the membrane (Ω-cm^2) is calculated as

$$R_m = (1/k)C_m \tag{6}$$

where C_m is the capacitance of the membrane (~ 1 μF cm^{-2}).[24]

[23] A. Shanser, D. Samuel, and R. Korenstein, *J. Am. Chem. Soc.* **105**, 3815 (1983).
[24] H.-J. Apell, and B. Bersch, *Biochim. Biophys. Acta* **903**, 480 (1987).

[28] Stoichiometry of Coupled Transport Systems in Vesicles

By R. James Turner

Secondary active transport systems enable cells or organelles to drive the flux of a transported substrate, S, against its electrochemical gradient by coupling the flux of S to that of another solute (referred to here as an "activator"), A, which is moving down its electrochemical gradient. The stoichiometry of a such a coupled transport event is of considerable interest since it plays a major role in the determination of both the concentrating capacity of the cell or organelle for S and the energetic cost of the transport process.[1]

In this chapter I describe the methods which are currently used to determine the stoichiometries of co- and countertransport systems in membrane vesicle preparations. Generally speaking these methods are also applicable to intact cell or organelle preparations, the primary advantage of

[1] R. J. Turner, *Ann. N.Y. Acad. Sci.* **456**, 10 (1985).

vesicles being that it is easier to control and modify the intravesicular and extravesicular media.

In order to fully appreciate the principles and problems involved in the experimental methods decribed here it is important to understand the thermodynamic basis of coupled transport events. This topic is reviewed briefly below. A more detailed discussion is given in a previous publication.[1]

Thermodynamics of Coupled Transport

Basic Principles

Consider first, for simplicity, a cell or organelle whose outer membrane contains a cotransport system for S and A, where S is an electroneutral molecule and A is an ion with one positive charge (the general case of a charged substrate and multiple activators is treated at the end of this section). Assume also that there are no other primary or secondary active transport systems for S in the membrane. On thermodynamic grounds it can be shown that the ratio of the intracellular to extracellular substrate concentrations, $[S_I]/[S_O]$, must obey the following inequality,

$$\ln([S_I]/[S_O]) \leq n_A[\ln([A_O]/[A_I]) + F\Delta\psi/RT] \qquad (1a)$$

or equivalently,

$$[S_I]/[S_O] \leq [([A_O]/[A_I]) \exp(F\Delta\psi/RT)]^{n_A} \qquad (1b)$$

where F, R, and T have their usual thermodynamic interpretations, $\Delta\psi = \psi_O - \psi_I$ is the transmembrane electrical potential, and n_A is the $A:S$ *coupling stoichiometry* (the number of moles of A translocated per mole of S *via the transporter*).

The biological significance of Eq. (1) can be best appreciated by considering the situation where $\ln([A_O]/[A_I]) + R\Delta\psi/RT$ is maintained at some fixed (positive) value by other systems in the cell (e.g., Na^+, K^+-ATPase), and where the distribution of S across the membrane has reached a steady state. Equation (1) may then be thought of as a thermodynamic constraint on the steady state concentration gradient of S resulting from its coupled transport with A. Thus the equality in Eq. (1) gives the maximum thermodynamically allowed concentrating capacity of the system for S. In this case the substrate and activator electrochemical gradients will be at thermodynamic equilibrium via the coupled transporter and the net carrier-mediated flux of both will be zero. The extent to which any real system approaches this thermodynamic limit depends on the magnitude of its relevant leak pathways. These may be divided into two distinct classes:

those that occur via the coupled transporter (*internal* leaks) and those which do not (*external* leaks). External leaks include passive diffusion and any mediated transport pathways for the substrate across the membrane in question. Internal leaks arise when the flux of activator and substrate via the carrier is not tightly coupled, that is, when activator translocation without substrate or substrate translocation with a variable number of activator ions (possibly zero) can occur via the transporter. The presence of internal leaks adds a significant complication to stoichiometric determinations since in this case the coupling stoichiometry may vary with experimental conditions. There have, in fact, been several reports in the literature of transporters that apparently display this type of variability.[2]

In a system with internal or external leak pathways the flux via the coupled transporter will reach a steady state determined by kinetic as well as thermodynamic factors. This kinetic steady state will necessarily result in a value of $[S_I]/[S_O]$ that is less than that predicted by thermodynamic equilibrium.[1]

Multiple Activators and Charge Stoichiometry

The more general form of Eq. (1) for multiple activators and charged substrates is given by

$$\ln([S_I]/[S_O]) \leq \sum_A n_A \ln([A_O]/[A_I]) + qF\Delta\psi/RT \qquad (2)$$

where the sum is over all activators, that is, over all solutes co- or countertransported with S via the carrier. Here n_A is positive if A is cotransported with S and negative if A is countertransported with S. The *charge stoichiometry, q* (the number of moles of charge translocated per mole of S *via the transporter*), is given by

$$q = Z_s + \sum_A n_A Z_A$$

where Z_S and Z_A are the charges on S and A, respectively.

Flux Measurements: The Rapid Filtration Technique

A variety of methods have been used to monitor transport or transport-related events in vesicle preparations, but the most commonly applied and generally applicable procedure for studying coupled transport systems is to measure the flux of radiolabeled ligands using the "rapid filtration" technique. In this chapter I discuss mainly experiments employing this tech-

---

[2] A. A. Eddy, *Biochem. Soc. Trans.* **8**, 271 (1980).

nique; however, stoichiometric determinations may equally well be carried out using other methods of flux determination.

The basic procedure for the rapid filtration technique as carried out in our laboratory is the following. Vesicles (10–50 µl, 1–3 mg protein/ml) are combined with an incubation medium (10–100 µl) containing radioactively labeled ligands (10–100 µCi/ml) and other constitutents as required in a 12 × 75 disposable borosilicate glass test tube. After an appropriate incubation time, 1.5 ml of an ice-cold stop solution is added and the resulting mixture is removed with a Pasteur pipet and applied to a Millipore filter (HAWP, 0.45 µm; Bedford, MA) under light suction. The filter is then quickly washed with a further 4.5 ml of stop solution, placed in an appropriate scintillation cocktail, and counted for radioactivity along with samples of the incubation medium and appropriate standards. Efflux studies can be carried out in the same way using vesicles preloaded with labeled substrate. The actual volumes, protein concentrations, and isotopic activities employed in a given experiment will depend on the transporter under study as well as on tissue availability and the cost of radiolabeled ligands.

The apparatus we employ consists of a fritted glass filter holder (Millipore XX10 025 02) connected via flexible tubing to a collection bottle and then to a vacuum pump. We typically use Eppendorf series 4700 fixed volume pipets to dispense vesicles and incubation media; however, any pipet with comparable accuracy and reproducibility (both <1%) is acceptable. If incubation times of <5 sec are desired we use the rapid sampling device described by Turner and Moran.[3] A rapid sampling device is also available commercially from Inovativ Labor AG (Adliswil, Switzerland). We have occasionally carried out flux measurements at incubation times of <2 sec by hand with the aid of a metronome.

Owing to the quantitative nature of stoichiometric determinations considerable care is required in the design and performance of experiments in order to avoid the introduction of artifactual or systematic errors. The importance of choosing an appropriate stop solution to prevent isotope efflux or influx during the stopping and washing procedure and the necessity of measuring initial rates of transport have been stressed earlier.[4] Also, when working with coupled transport processes that are electrogenic, variations in membrane potential with experimental conditions must be avoided.[3-5] In our vesicle preparations we typically clamp membrane potentials at the diffusion potential of a permeant anion such as nitrate or thiocyanate or at the potassium diffusion potential using the potassium

---

[3] R. J. Turner and A. Moran, *Am. J. Physiol.* **242,** F406 (1982).
[4] R. J. Turner, *J. Membr. Biol.* **76,** 1 (1983).
[5] U. Hopfer, *Am. J. Physiol.* **254,** F89 (1978).

ionophore valinomycin (2.5–12.5 μg/mg membrane protein). Further discussion of this problem and methods for testing the effectiveness of the above voltage clamping procedures are given in previous publications.[3,4,6,7]

Methods of Measuring Stiochiometry

Four methods have been devised for measuring the stoichiometry of coupled transport systems. Some experimentation may be required to determine the range of experimental conditions (if any) under which a given method may be successfully performed on a particular carrier. For this reason it is advisable to have accumulated a good experimental data base on the system under study before attempting stoichiometric measurements. Useful information includes the time course of uptake, the magnitude of specific vs nonspecific uptake, the degree of substrate binding and metabolism by the vesicles, and the dependence of uptake on substrate and activator concentrations (the latter is actually a stoichiometric determination—see Method 2, below). It is also important to ensure that all activator species have been identified; a number of secondary active transporters have recently been shown to be coupled to multiple activators (see, e.g., Refs. 6–8).

All four methods of stoichiometric determination are equally applicable to cotransport and countertransport systems. In every case the roles of substrate and activator are interchangeable. The use of each method for the determination of both coupling and charge stoichiometries is discussed at appropriate points below.

Method 1 (Direct Method)

The direct method for determining the A:S coupling stoichiometry of a secondary active transport system consists of simultaneously measuring and comparing substrate and activator fluxes via the transporter. There are two ways in which this might be done: (1) by measuring the total transporter-related substrate and activator fluxes (e.g., by measuring fluxes in the presence and absence of a specific inhibitor of the transporter), or (2) by measuring the activator-dependent substrate flux and the substrate-dependent activator flux. For tightly coupled transporters both procedures should give the same result. In the presence of significant internal leak pathways, however, only the former procedure will give the correct coupling stoichiometry since the latter procedure does not take into account

[6] Y. Fukuhara and R. J. Turner, *Am. J. Physiol.* **248,** F869 (1985).
[7] R. J. Turner, J. N. George, and B. J. Baum, *J. Membr. Biol.* **94,** 143 (1986).
[8] R. J. Turner, *J. Biol. Chem.* **261,** 16060 (1986).

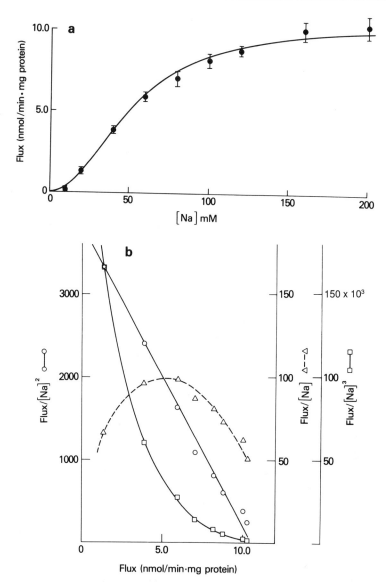

FIG. 1. Initial rate of sodium-dependent succinate uptake into renal outer cortical brush border membrane vesicles measured as a function of sodium concentration. Vesicles were prepared in THM buffer (10 mM HEPES plus 100 mM mannitol buffered with Tris to pH 7.5) containing 600 mM mannitol and 100 mM KSCN. The (isoosmotic) incubation media were THM buffer containing 100 mM KSCN and sufficient [^{14}C]succinate and NaCl to give final extravesicular concentrations of 0.2 and 0–200 mM, respectively. Choline replaced sodium isoosmotically to obtain the various sodium concentrations studied. Mem-

uncoupled fluxes via the transporter. The direct method is conceptually the most straightforward and in some ways the most satisfying method of measuring coupling stoichiometries.

An electrophysiological version of the direct method has been used by several authors to measure charge stoichiometry in intact epithelia[9,10]; however, no attempts to measure substrate flux and charge translocation directly in cell suspensions or vesicle preparations appear to have been attempted.

Method 2 (Activation Method)

The activation method for determining the A:S stoichiometry of a transport system consists of measuring substrate flux as a function of activator concentration. As discussed below the interpretation of activation method data is considerably more model dependent than the other methods described here. It should also be noted that the activation method is sensitive both to activators that are actually transported with the substrate (energetic activation) and to activators that simply affect substrate flux without being themselves transported (catalytic activation), for example, by binding at a modifier site on the carrier. Thus, whereas the other methods for determining stoichiometries discussed here are sensitive only to transported activators, the activation method provides a measure of the total (catalytic plus energetic) activator involvement in the transport event. For this reason I refer here to the result of an activation method experiment as the A:S stoichiometry rather than the A:S *coupling* stoichiometry. Note that only transported (energetic) activators are relevant to the *thermodynamics* of coupled transport [Eqs. (1) and (2)].

A hyperbolic relationship between activator-dependent substrate flux and activator concentration provides good evidence for a 1:1 activator:substrate stoichiometry since this behavior is expected for a

[9] E. Frömter, *Pfluegers Arch.* **393**, 179 (1982).
[10] D. S. Misfeldt and M. J. Sanders, *J. Membr. Biol.* **70**, 191 (1982).

---

brane potentials were clamped at zero by the presence of 100 mM KSCN in all solutions and by the addition of 12.5 μg valinomycin/mg protein to the vesicles. (a) A plot of uptake vs sodium concentration. The line drawn through the data points is a fit to Eq. (3) obtained using the nonlinear regression routine NONLIN (Systat, Inc., Evanston, IL). The fit is given by J_{max} = 10.6 ± 1.2 nmol/min/mg protein, $K_{0.5}$ = 54.2 ± 9.3 mM, and n = 2.1 ± 0.4. (b) Plots of flux/[Na]^n vs flux for n = 1 (△), 2 (○), and 3 (□), illustrating the way in which data can be fit to Eq. (3) by trial and error. A more quantitative procedure involving linear regression analysis is described in the text. [Redrawn from Y. Fukuhara and R. J. Turner, *Am. J. Physiol.* **245**, F374 (1983).]

broad class of 1:1-type models.[1,11] In this case a plot of flux/[A] vs flux will be linear. In contrast, a sigmoidal dependence of flux on activator concentration as shown in Fig. 1a for sodium-dependent succinate transport in renal outer cortical brush border membrane vesicles indicates the involvement of multiple activator ions per substrate translocation event. The most common method for interpreting such data is based on analysis in terms of the Hill equation[12]:

$$J_S = J_{max}[A]^n/(K_{0.5}^n + [A]^n) \tag{3}$$

Here J_S is the substrate flux, n is the A:S stoichiometry (the "Hill coefficient"), and J_{max} and $K_{0.5}$ are constants. It can be shown that a tightly coupled transporter with activator:substrate stoichiometry n and strong cooperativity between activator-binding sites will obey such a flux equation. [Note that for $n = 1$, Eq. (3) reduces to the hyperbolic relationship expected for a 1:1 system.]

In order to get the most information from the activation method it is important to obtain flux data over the widest possible range of activator concentrations. Fluxes at low activator concentrations are particularly important since these often define the sigmoidal portion of the activation curve which identifies and characterizes the involvement of multiple activator ions. Two procedures for analyzing activation method data, based on the Hill equation, are given below.

Procedure 1. In this procedure one simply fits the experimental data to Eq. (3) using a least-squares regression or manual method. The line drawn through the data points in Fig. 1a is a nonlinear least-squares fit obtained using a commercially available nonlinear regression computer program (see figure caption). In carrying out such a fit the points are typically weighted in proportion to the reciprocals of their variances.

In our laboratory we frequently employ an alternative to nonlinear least-squares analysis in fitting activation data to Eq. (3). This method takes advantage of the fact that Eq. (3) predicts that a plot of $J_S/[A]^n$ vs J_S will be linear. Thus one can search manually for a value of n which results in such a linear plot, as we have done in Fig. 1b, or one can carry out a similar search more quantitatively with the aid of a linear least-squares regression routine. This latter procedure is used routinely in our laboratory.[6] We have modified a linear regression program so that it fits a straight line to a $J_S/[A]^n$ vs J_S plot over a range of values of n. The points are weighted in proportion to the reciprocals of the squares of their relative

---

[11] R. J. Turner, *Biochim. Biophys. Acta* **649**, 269 (1981).
[12] I. H. Segel, "Enzyme Kinetics." Wiley, New York, 1975.

experimental errors. The best value of n is taken to be the one that minimizes the sum of the squares of the deviations of the experimental points from the calculated regression line. The statistical errors on J_{max} and $K_{0.5}$ are then given directly from the linear least-squares fit and the statistical error on n can be determined using an F test. This procedure was found to give results virtually identical to nonlinear least-squares analysis while requiring only a simpler, more readily available, linear least-squares regression routine.

Owing to the rather restrictive assumptions made in deriving the Hill equation (see above) it is unlikely that any real transporter will obey it well. Thus one should expect nonintegral Hill coefficients from the above fitting procedures and/or systematic deviations from linearity on $J_S/[A]^n$ vs J_S plots. In general, the Hill coefficient derived from fitting activation method data to Eq. (3) should be regarded as a lower limit on the A:S stoichiometry.[1]

Procedure 2. This procedure is recommended when J_{max} can be accurately determined from the experimental data, i.e., when J_S can be measured at activator concentrations which are sufficiently high that saturation is assured. In this case a Hill plot of $\log[J_S/(J_{max} - J_S)]$ vs $\log[A]$ can be made. For data fitting Eq. (3) such a plot will be linear with slope n. Hill plots for sodium-dependent sulfate transport in renal brush border membrane vesicles at pH 8.0 and pH 6.0 are illustrated in Fig. 2. The linearity of the Hill plot for pH 8.0 with slope 1.02 provides good evidence for a 1:1 sodium:sulfate stoichiometry at this pH while the break in the Hill plot for pH 6.0 may be interpreted as the result of a shift in the sodium:sulfate

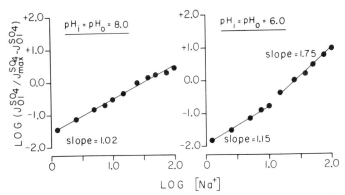

FIG. 2. Hill plots of the initial rate of sodium-dependent sulfate uptake ($J_{OI}^{SO_4}$) into renal brush border membrane vesicles measured as a function of sodium concentration. Measurements were made at pH 8.0 equilibrium (left) and pH 6.0 equilibrium (right). J_{max} was estimated by extrapolation of a $J_S/[Na]$ vs J_S plot. (From Ahearn and Murer.[13])

stoichiometry from approximately 1 : 1 to approximately 2 : 1 with increasing sodium concentration under these experimental conditions.[13]

Given our present (meager) understanding of the kinetics of coupled transport systems a Hill plot is unquestionably the best method for extracting the maximal amount of information from activation method data. I would, however, give two notes of caution. First, because the assumptions made in deriving Eq. (3) are rather restrictive, and because the Hill transformation itself is rather complex, it is not entirely clear what relationship the slope of the Hill plot will bear to the activator : substrate stoichiometry for more realistic flux equations; thus such plots should be interpreted with some caution. This point is discussed in more detail in an earlier publication.[1] Second, considerable care should be taken in the determination of J_{max} since a small change in this constant can cause a large shift in the Hill plot, especially at high activator concentrations. Indeed it is advisable to test the effects of experimental uncertainties in J_{max} on the shape of the Hill plot to be sure that any conclusions reached are not overly dependent on the choice of this parameter.

The complex way in which $\Delta\psi$ enters into the kinetic equations of most coupled transport models makes this method very difficult to apply to the determination of charge stoichiometries. To my knowledge no attempts to carry out determinations of this type have been made.

Method 3 (Steady State Method)

The steady state method for determining the coupling and charge stoichiometries of a secondary active transport system consists of measuring the steady state substrate gradient generated by the transporter in the presence of known (measured or applied) activator electrochemical gradients. *If the transporter is operating at or close to thermodynamic equilibrium* Eq. (2) becomes (the case of a single activator is treated for simplicity)

$$\ln([S_I]/[S_O]) = n_A \ln([A_O]/[A_I]) + qF\Delta\psi/RT \qquad (4)$$

Thus, for example, n_A can be calculated from the slope of a plot of $\ln([S_I]/[S_O])$ vs $\ln([A_O]/[A_I])$, at constant $\Delta\psi$, and q can be determined in a similar fashion from a plot of $\ln([S_I]/[S_O])$ vs $F\Delta\psi/RT$ at constant activator chemical gradient. Alternatively, if $[S_I]/[S_O]$, $[A_O]/[A_I]$, and $\Delta\psi$ are measured under two or more independent experimental conditions, n_A and q can be calculated from the simultaneous equations resulting from substitution of these numerical values in Eq. (4). Steady state method calculations are particularly simple if measurements can be carried out in the presence

[13] G. A. Ahearn and H. Murer, *J. Membr. Biol.* **78**, 177 (1984).

of an activator gradient alone ($\Delta\psi = 0$) or an electrical gradient alone $[\ln([A_O]/[A_I])] = 0]$.

The steady state method is typically applied to whole cells or cell fragments containing an intact activator gradient or membrane potential gradient-generating system. In this case, the relevant gradient can be maintained at a constant value for sufficient time that a steady state substrate gradient can be established and measured.

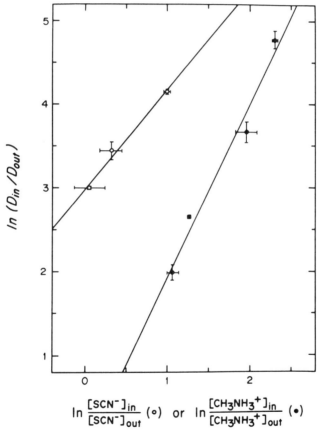

FIG. 3. Steady state dopamine (D) uptake into chromaffin granule ghosts measured as a function of membrane potential (O) and pH gradient (●). $\Delta\psi$ and pH were varied by adding ATP and different amounts of FCCP to ghosts suspended in sucrose and KCl media, respectively (see text). The steady state gradients of radiolabeled dopamine, SCN^- and $CH_3NH_3^+$ were measured after 60 min of incubation at 25°. The slopes of the lines drawn through the data points are 1.18 ± 0.14 (O) and 2.06 ± 0.24 (●). (From Knoth et al.[15])

An example of the determination of the coupling and charge stoichiometries of the H^+/dopamine countertransporter in chromaffin granule ghosts is shown in Fig. 3. In these experiments Knoth et al.[14,15] were able to generate stable pH and potential gradients independently of one another using an H^+-ATPase which is also present in the granule membrane. Briefly, they found that in the absence of permeant ions the H^+-ATPase generated a membrane potential but, owing to the high buffering capacity of the ghosts, no pH gradient. The magnitude of the resulting membrane potential could be varied by the addition of different amounts of the H^+ ionophore FCCP. On the other hand, when the permeant anion Cl^- was present in the ghost preparation it could follow the actively pumped protons into the ghosts, dissipating the membrane potential and lowering the internal pH. The magnitude of the resulting pH gradient could likewise be varied by adding the uncoupler FCCP. In these experiments the transmembrane potential and pH gradients were monitored using the distribution ratios of radiolabeled thiocyanate and methylamine, respectively.[14] The slopes of the lines in Fig. 3 indicate an H^+ : dopamine coupling stoichiometry of 2 : 1 and a charge stoichiometry of 1 : 1. These results are consistent with the exchange of two H^+ and one dopamine$^+$ by the transporter.

As stressed above the stoichiometry derived from the steady state method will be correct only when the transporter is operating at or near thermodynamic equilibrium since it is only then that the $\leq$ sign in Eq. (2) becomes an equality [Eq. (4)]. In the presence of significant internal or external leaks S_I/S_O may deviate significantly from the value predicted by thermodynamic equilibrium and the steady state method will yield an underestimate of the true stoichiometry. The inability of this method to take into account external leak pathways is a potentially serious shortcoming which has received some attention in the literature.[1,4,16] The presence of contaminating cells or vesicular material can also lead to errors in measurements of relevant substrate and activator gradients, since this material may behave quite differently from the cells or vesicles containing the system under study.

Stoichiometries obtained from applications of the steady state method to preparations that cannot maintain activator or electrical gradients over time must be interpreted with some care, since in such preparations the sustained steady state substrate gradient assumed in Eq. (4) never occurs.[1]

---

[14] J. Knoth, M. Zallakian, and D. Njus, *Biochemistry* **20**, 6625 (1981).
[15] J. Knoth, M. Zallakian, and D. Njus, *Fed. Proc., Fed. Am. Soc. Exp. Biol.* **41**, 2742 (1982).
[16] I. R. Booth, W. J. Mitchell, and W. A. Hamilton, *Biochem. J.* **182**, 687 (1979).

Method 4 (Static Head Method)

The static head method[17-20] of determining coupling and charge stoichiometries provides a means of circumventing some of the practical difficulties and limitations of the steady state method. Like the steady state method the static head method is based on Eq. (4), which gives the condition for thermodynamic equilibrium for a tightly coupled transporter. At thermodynamic equilibrium there is no net flux of substrate or activator via such a transporter because the thermodynamic driving forces of the substrate and activator electrochemical graidents are balanced. The principle of the static head method is similar to the steady state method in that one attempts to determine experimental conditions where Eq. (4) holds. Here, however, this determination is based directly on measurements of transport rates via the carrier rather than on steady state distribution ratios produced by the experimental system as a whole. The method is most easily understood by considering a concrete example.

A static head experiment for sodium-dependent D-glucose transport in renal outer medullary brush border membrane vesicles is shown in Fig. 4. In this experiment vesicles were preloaded with known concentrations of sodium and labeled glucose and then diluted 1:5 into appropriate glucose-free media, thus establishing an intravesicular-to-extravesicular glucose gradient of 6:1. The glucose retained in these vesicles was then measured as a function of time at various extracellular sodium concentrations. A control run carried out in the absence of sodium was used to measure efflux via unrelated sodium-independent pathways (leaks and/or contaminants). The static head condition (no net substrate flux via the sodium-dependent transporter) is characterized by that external sodium concentration that causes the test run to superimpose on the control. Because membrane potentials were clamped at zero in this experiment (by KSCN and valinomycin, see Figure 4 caption), Eq. (4) reduces to

$$\ln([S_I]/[S_O]) = n_A \ln([A_O]/[A_I]) \tag{5}$$

The values of n indicated on the figure are the sodium:glucose coupling stoichiometries that would be predicted by Eq. (5) were that extravesicular sodium concentration to result in static head conditions. Figure 4 indicates that $n_A \simeq 1.8$ for this system.

---

[17] R. J. Turner and A. Moran, *J. Membr. Biol.* **67**, 73 (1982).
[18] Y. Fukuhara and R. J. Turner, *Biochim. Biophys. Acta* **770**, 73 (1984).
[19] R. J. Turner and A. Moran, *J. Membr. Biol.* **70**, 37 (1982).
[20] J. L. Kinsella and P. S. Aronson, *Biochim. Biophys. Acta* **689**, 161 (1982).

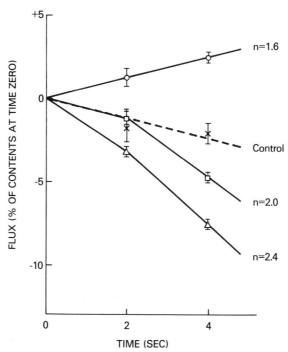

FIG. 4. The results of a static head experiment designed to measure the sodium:glucose coupling stoichiometry for sodium dependent D-glucose transport in renal outer medullary (late proximal tubule) brush border membrane vesicles. Vesicles were prepared in THM buffer containing 100 mM KSCN, 0.25 mM D-[^{14}C]glucose and L-[^{3}H]glucose, 12.5 μg valinomycin/mg protein, and 20 mM NaCl plus 50 mM choline chloride. At time zero the vesicles were diluted 1:5 into incubation media made up of THM buffer containing 100 mM KSCN, sufficient NaCl to give final extravesicular concentrations of 62 mM (O), 49 mM (□), or 42 mM (△), and sufficient choline chloride to produce a solution isoosmotic with the intravesicular medium. The control run (×) was carried out in the absence of sodium at 70 mM choline chloride equilibrium. The stereospecific efflux (or influx) of D-glucose has been expressed as a percentage of the total (equilibrium) intravesicular glucose at time zero. (From Turner and Moran.[19])

A static head charge stoichiometry determination for the outer medullary sodium-dependent D-glucose transporter is shown in Fig. 5. Here membrane potentials were generated by a potassium gradient in the presence of valinomycin. Thus $\Delta\psi$ is given to a good approximation by the potassium diffusion potential, that is, $\Delta\psi = (RT/F)\ln(K_I/K_O)$. Since no sodium gradients were present in this experiment Eq. (4) reduces to

$$\ln(S_I/S_O) = qF\Delta\psi/RT = q\ln(K_I/K_O) \qquad (6)$$

The values of q indicated on the figure are the charge stoichiometries that would be predicted by Eq. (6) were that potassium diffusion potential to result in static head conditions. Figure 5 indicates that $q = 2$ for this system.

Since, in contrast to the steady state method, the static head procedure is based directly on the determination of transport rates, activator gradients and potentials need only be maintained by the experimental system over times long enough to measure fluxes. This makes this method ideal for vesicle preparations that often can neither generate electrochemical gradients nor maintain them over long time intervals. Also, difficulties arising from the presence of external leaks and contaminants in the preparation

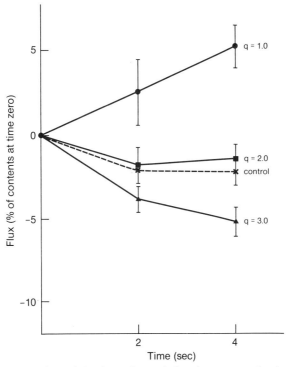

FIG. 5. The results of a static head experiment designed to measure the charge stoichiometry of sodim-dependent D-glucose transport in renal outer medullary brush border membrane vesicles. Vesicles were prepared in 120 mM K$_2$SO$_4$, 20 mM Na$_2$SO$_4$, 1 mM labeled glucose and valinomycin (12.5 μg/mg protein), then diluted (1:5) into isoosmotic media such that the final extravesicular concentrations of Na$_2$SO$_4$ and labeled glucose were 20 and 0.25 mM, respectively, and the final extravesicular concentrations of K$_2$SO$_4$ were 30 mM (●), 60 mM (■), and 75 mM (▲). N-Methyl-D-glucamine sulfate was added to the incubation medium to maintain isoosmolarity with the intravesicular solution. In the control run (×) 20 mM Na$_2$SO$_4$ was replaced by 60 mM mannitol in all media. (From Turner.[1])

can be taken into account explicitly in the static head method by an appropriate control run (see Figs. 4 and 5). As in the steady state method, internal leaks will result in an underestimate of coupling and charge stoichiometries by the static head procedure.

Concluding Remarks

In conclusion, it is worth reemphasizing that stoichiometric determinations must be tailored to the coupled transport system under investigation and to the experimental system in which it is found. When characterizing a new transporter in our laboratory we typically carry out an activation method experiment first. This experiment provides us not only with a stoichiometric determination but also with information concerning the activator concentration dependence of the transporter. We then attempt the direct method and the steady state method since these are often technically easier experiments than the static head method. Since most of the vesicle systems we work with are not capable of maintaining electrochemical gradients over time our steady state experiments cannot be interpreted quantitatively[1]; however, they can provide qualitative evidence for energetic coupling by demonstrating concentrative uptake of substrate due to activator or electrical gradients.[8,21] Finally, we attempt the static head method. Although this method is often technically difficult, it provides a direct thermodynamic demonstration of coupled transport and is free from many of the problems associated with other methods of measuring stoichiometry.

[21] Y. Fukuhara and R. J. Turner, *Am. J. Physiol.* **245**, F374 (1983).

[29] Phosphate Transport in Established Renal Epithelial Cell Lines

By J. BIBER, K. MALMSTRÖM, S. RESHKIN, and H. MURER

Introduction

Phosphate (P_i) is reabsorbed from the lumenal fluid of the renal proximal tubule via a secondary active transport mechanism involving sodium/phosphate cotransport across the microvillar brush border membrane. Numerous properties of this transport system, such as its kinetic characteristics and functional changes in response to hormones (e.g., parathyroid hormone), acid/base changes, and nutritional stimuli (e.g., phosphate diet), have been described.[1-3]

For the elucidation of the regulatory mechanisms underlying the functional changes of the Na/P_i-cotransport system, isolated renal cortex brush border membrane vesicles (see also Chapter 26 of this volume for isolation procedures) have been used extensively. In particular, isolated renal cortex brush border membrane vesicles have been used to test for protein phosphorylation[4,5] and ADP ribosylation[6-8] as possible post-translational regulatory events. This kind of approach, however, has turned out to be unsuccessful, mainly because of the lack of (yet unknown) cellular factors in the isolated membranes and because the isolated membrane vesicles are closed and of a right-side-out orientation.[9] This makes it difficult if not impossible to study posttranslational modifications of the Na/P_i cotransporter at its cytosolic domain.

In the past few years, cultured renal cells have opened new possibilities to explore the regulatory events underlying the functional changes of the Na/P_i-cotransport system. This chapter will describe the basic methodology for measuring P_i transport in established renal epithelial cells and in apical membrane vesicles prepared thereof. It is by these methods that it has become possible to describe the basic mechanisms involved in the regulation of the Na/P_i-cotransport system by parathyroid hormone and by the concentration of extracellular phosphate. So far such studies have been performed with the following renal cell lines (see Table I): OK cells, derived from an adult American opossum[10]; LLC-PK_1 cells, derived from pig kidney[11]; JTC-12 cells, derived from a monkey kidney[12]; and MDCK cells, derived from dog kidney.[13]

Similarly, regulation of the Na/P_i-cotransport system can be studied in primary cultures of proximal tubular epithelial cells, as demonstrated by

---

[1] J.-P. Bonjour and J. Caverzasio, *Rev. Physiol. Biochem. Pharmacol.* **100**, 162 (1984).
[2] P. Gmaj and H. Murer, *Physiol. Rev.* **66**, 36 (1986).
[3] C. L. Mizgala and G. A. Quamme, *Physiol. Rev.* **65**, 431 (1985).
[4] J. Biber, K. Malmström, V. Scalera, and H. Murer, *Pfluegers Arch.* **398**, 221 (1983).
[5] M. R. Hammerman and K. A. Hruska, *J. Biol. Chem.* **257**, 992 (1982).
[6] M. R. Hammerman, *Am. J. Physiol.* **251**, F385 (1986).
[7] S. A. Kempson and N. P. Curthoys, *Am. J. Physiol.* **245**, C449 (1983).
[8] P. Gmaj, J. Biber, S. Angielski, G. Stange, and H. Murer, *Pfluegers Arch.* **400**, 60 (1984).
[9] W. Haase, A. Schaefer, H. Murer, and R. Kline, *Biochem. J.* **172**, 57 (1978).
[10] H. Koyama, C. Goodpasture, M. M. Miller, R. L. Teplitz, and A. D. Riggs, *In Vitro* **14**, 239 (1978).
[11] R. N. Hull, W. R. Cherry, and G. W. Weaver, *In Vitro* **12**, 670 (1976).
[12] T. Takaoka, H. Katsuta, M. Endo, K. Sato, and H. Okumura, *Jpn. J. Exp. Med.* **32**, 351 (1962).
[13] C. R. Gaush, W. C. Hard, and T. F. Smith, *Proc. Soc. Exp. Biol. Med.* **122**, 931 (1966).

TABLE I
BASIC CHARACTERISTICS OF Na/P_i COTRANSPORT IN ESTABLISHED RENAL EPITHELIAL CELL LINES[a]

Cell line[b]	Kinetic parameters[c]		Na/P_i transport regulated by
	~K_m for P_i (μM)	V_{max}^d	
OK	86 ± 8	10.3 ± 1.8; 37° (1)	Parathyroid hormone (1, 2)
	94 ± 7	2.1 ± 0.2; 37° (3)	and P_i deprivation (4)
OK (filter)			
Apical	340 ± 50	1.1 ± 0.04; 37° (13)	
Basolateral	5000	2.3 (13)	
LLC-PK_1	95 ± 15	0.83 ± 0.14; 22° (5)	P_i deprivation
	125	1.56; 37°C (9)	
	196 ± 33	0.82 ± 0.07; 37° (10)	
JTC-12	120	0.152; 37° (11)	P_i deprivation (12)
MDCK	374 ± 48	0.53 ± 0.05; 37° (10)	P_i deprivation (10)

[a] The data are taken from the references given in paranthesees: (1) K. Malmström and H. Murer, *Am. J. Physiol* **251**, C23 (1986); (2) J. A. Cole, S. L. Eber, R. E. Poelling, P. K. Thorne, and L. R. Forte, *Am. J. Physiol.* **253**, E221 (1987); (3) J. Caverzasio, R. Rizzoli, and J.-P. Bonjour, *J. Biol. Chem.* **261**, 3233 (1986); (4) J. Biber, J. Forgo, and H. Murer, *Am. J. Physiol.* **255**, C155 (1988); (5) J. Biber, C. D. A. Brown, and H. Murer, *Biochim. Biophys. Acta* **735**, 325 (1983); (6) J. Caverzasio, C. D. A. Brown, J. Biber, J.-P. Bonjour, and H. Murer, *Am. J. Physiol.* **248**, F122 (1985); (7) J. Biber and H. Murer, *Am. J. Physiol.* **249**, C430 (1985); (8) C. A. Rabito, *Am. J. Physiol.* **245**, F22 (1983); (9) L. Noronha-Blob, C. Filburn, and B. Sacktor, *Arch. Biochem. Biophys.* **234**, 265 (1984); (10) B. Escoubet, K. Djabali, and C. Amiel, *Am. J. Physiol.* **256**, C322 (1989); (11) Y. Takuwa and E. Ogata, *Biochem. J.* **230**, 715 (1985); (12) Y. Takuwa, Y. Takeachi, and E. Ogata, *Clin. Sci.* **71**, 307 (1986); (13) S. Reshkin, J. Forgo, and H. Murer, *Pfluegers Arch.*, in press (1990).
[b] If not indicated otherwise, cells have been grown to confluency in Petri dishes.
[c] V_{max} is given in nanomoles P_i per milligram per minute at 22 or 37° as indicated.
[d] All parameters were determined at an extracellular pH of 7.2 to 7.4.

various laboratories.[14–17] The methodological background for cultivating and using primary cell cultures of proximal tubular cells will, however, not be described here.

For the successful use of an established renal epithelial cell line some premises must be considered.[18,19]

[14] M. A. Wagar, J. Seto, S. D. Chung, S. Hiller-Grohol, and M. Taub, *J. Cell. Physiol.* **124**, 411 (1985).
[15] L. Noronha-Blob and B. Sacktor, *J. Biol. Chem.* **261**, 2164 (1986).
[16] Y. Kinoshita, M. Fukase, A. Miyauchi, M. Takenaka, M. Nakada, and T. Fujita, *Endocrinology (Baltimore)* **119**, 1954 (1986).
[17] G. Friedlander and C. Amiel, *J. Biol. Chem.* **264**, 3935 (1989).
[18] J. S. Handler, *Kidney Int.* **30**, 208 (1986).
[19] L. M. Sakhrani and L. G. Fine, *Miner. Electrolyte Metab.* **9**, 276 (1983).

1. Morphologically, cells must be polarized (apical pole being morphologically separated from the basolateral pole by tight junctions). As a consequence, an asymmetric distribution of various enzymatic and transport activities should be observed.

2. From (1) it follows that vectorial transport of water and other solutes should be observed. Very often vectorial solute transport into the space between the plastic surface and the basolateral membranes leads to the formation of the so-called domes.

3. Identification of metabolic and transport functions and hormonal responses typical for specific nephron segments[20,21] helps to classify the respective cell line as a model system for studying the cellular functions of a specific nephron segment. It should be mentioned here that no established renal cell line exists so far that exactly resembles a certain nephron segment

Transport Studies with intact Cell Monolayers

Comments

1. Since any activity of cells in culture might vary depending on the "cell age" and/or on the state of confluency, it is important to establish first the exact culture conditions (e.g., days in culture) which will allow one to investigate a particular question. To illustrate that the rate of Na/P_i cotransport is not necessarily proportional to the state of confluency, initial Na/P_i cotransport in the cell lines LLC-PK$_1$ and OK is shown (Fig. 1) as a function of days in culture.

2. The time dependency of the uptake of a solute should be established as well. Because phosphate is metabolized rapidly within the cell, transport measurements should be performed within the linear range of uptake and in as short a time as possible. As illustrated in Fig. 2, the uptake of P_i into OK cells grown in Petri dishes is linear for at least 6 min. In order to completely exclude any possible contribution by metabolic activities, the basic observations, such as apparent K_m and V_{max} values, as obtained with intact cell nonolayers, should be confirmed with isolated plasma membrane (apical or basolateral) vesicles.

3. Transport is best measured at 37°. Yet results obtained at 25° are qualitatively similar.

4. When transport experiments are performed with cells grown as a monolayer on Petri dishes, uptake of the added solute occurs preferentially

---

[20] F. Morel and A. Doucet, *Physiol. Rev.* **66**, 377 (1986).
[21] G. Wirthenson and W. G. Guder, *Physiol. Rev.* **66**, 469 (1986).

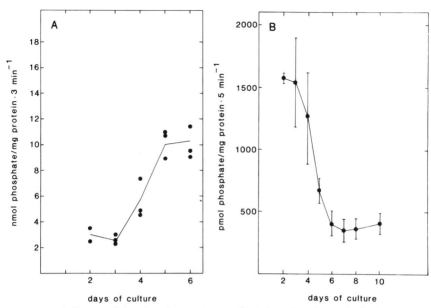

FIG. 1. Influence of days in culture on net sodium-dependent P_i uptake in OK (A) and LLC-PK$_1$ cells (B). At day 0 the cells were seeded as described in Refs. 25 and 26. In both cases confluency was reached after day 5. Media were changed every second day.

through the apical membrane. In order to measure uptake of a solute through both the apical and the basolateral membranes, cells are best grown on filters, such that the growth medium has access to both sides of the cell.[22] With the appropriate controls for tightness, it will be possible to measure transport through the apical or basolateral membrane separately. This approach has been used to study polarized Na/P_i cotransport in LLC-PK$_1$[23] and OK cells.[24]

Experimental Procedures

Cells Grown in Petri Dishes. Cells are grown to confluency in plastic culture dishes using standard protocols.[25-26] We routinely used dishes (35-mm diameter) containing at confluency approximately 0.5 mg of total cellular protein.

[22] R. E. Steele, A. S. Preston, J. P. Johnson, and J. S. Handler, *Am. J. Physiol.* **251**, C136 (1986).
[23] C. A. Rabito, *Am. J. Physiol.* **245**, F22 (1983).
[24] S. Reshkin and H. Murer, *Pfluegers Arch.*, in press (1990).
[25] K. Malmström and H. Murer, *Am. J. Physiol.* **251**, C23 (1986).
[26] J. Biber, C. D. A. Brown, and H. Murer, *Biochim. Biophys. Acta* **735**, 325 (1983).

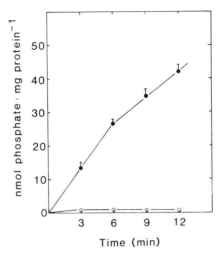

FIG. 2. Time course of P_i uptake into OK cells grown in Petri dishes. P_i transport was determined either in the presence of NaCl (●) or N-methyl-D-glucamine (MGA) (○). (From Malmström and Murer.[25])

The growth medium is aspirated off and the cell monolayer is quickly rinsed twice with incubation buffer lacking the transport solute (in mM): 137 NaCl (replaced by 137 N-methyl-D-glucamine-HCl for sodium-independent transport), 5.4 KCl, 2.8 CaCl$_2$, 1.2 MgSO$_4$, and 10 HEPES/Tris, pH 7.4. Transport is initiated by adding 1 ml of the above medium containing 0.1 mM KH$_2$^{32}PO$_4$ (0.5 to 1.0 μCi/ml). The Petri dish is agitated slowly (at the desired temperature) and at the given time point the radioactive medium is aspirated off and the monolayer is washed four times with 2 ml of an ice-cold stop solution (in mM): 137 NaCl, 14 Tris-HCl, pH 7.4.

To measure total radioactivity taken up, the cells are solubilized with 1 ml of 0.5% Triton X-100. Homogeneous solubilization is normally achieved within 30 min, such that 100 to 300 μl of the homogenate can be used directly for scintillation counting. Total protein is determined by a modified Lowry procedure.[27]

The washing procedure is controlled (blank value) by adding 1 ml of the isotope-containing buffer, which afterward is aspirated off immediately. As an example, the background for ^{32}P obtained after four washes is of the order of 200 to 500 cpm (compared to ~15,000 cpm of total uptake) and cannot be reduced further by more extensive washing.

Cells Grown on Permeant Filter Supports. For this purpose Millicell-CM filter inserts (12-mm diameter, 0.45-μm pore size; Millipore, Bedford,

[27] J. R. Dulley and P. A. Grieve, *Anal. Biochem.* **64**, 176 (1975).

MA) are used. If cells will not grow on bare filters, filters must be covered with a layer of collagen. To grow OK cells on Millicell-CM filters, rat tail collagen (R-type, 2 mg/ml; Serva, Heidelberg, FRG) is diluted four-fold in sterile ethanol:water (1:1) and 50 μl of this solution is added under sterile conditions, such that the total area is equally covered. Collagen-coated filter inserts are placed in a 24-well culture plate and permitted to dry overnight. The next day, 300 μl of a suspension of trypsinized cells ($\sim 5 \times 10^5$ cells/ml) is added to each filter insert. Place the inserts in the incubator for 2 hr. After this period, the medium is aspirated off and 500 μl of fresh medium is added to each side of the insert. If necessary, the medium should be changed every second day. Prior to the experiment the monolayer is checked at the light microscopic level for intactness.

Tests for monolayer permeability: There are a number of ways to measure the tightness of the monolayer during its development and after confluency. Transepithelial diffusion can be estimated by measuring the movement across the monolayer of large nontransported/nonmetabolized compounds, such as inulin.[28] On the other hand, transepithelial movement of ^{22}Na can be measured with bare filters and filters containing an intact monolayer.[29] Usually substantial movement of radioactivity occurs within the first 5 min through bare filters, which is equilibrated in about 30 min. In contrast, when an intact monolayer has been formed only about 5% of the radioactivity can be found in the transcompartment after 30 min. If the cell monolayer is incubated for 30 min in a calcium-free medium (containing in addition 1 mM EGTA) prior to the addition of the radioactivity, evidence that the impermeability results from the formation of tight junctions can be obtained. If high enough, the determination of transepithelial electrical resistance would be another way to monitor monolayer integrity.

Transport studies: With some modifications, to facilitate the use of the filter inserts, the procedure and solutions are essentially as described for the transport studies with cells grown in plastic Petri dishes (see above). Just prior to uptake, the filter inserts are removed from the growth medium and gently rinsed on both sides twice in sodium-free uptake solution. Then uptake solutions (but without the substrate) are added to the appropriate filter insert compartment (500 μl to each side). Uptake is initiated by mixing 50 μl of the same uptake solution containing the 11-fold amount of the desired substrate to one or the other compartment.

Uptake is stopped and unspecifically bound radioactivity is effectively removed from the relatively high plastic surface area by rapid aspiration of

[28] M. J. Caplan, H. C. Anderson, G. E. Palade, and J. D. Jamieson, *Cell (Cambridge, Mass.)* **46,** 623 (1986).

[29] J. G. Haggerty, N. Agarwal, R. F. Reilly, E. A. Adelberg, and C. W. Slayman, *Proc. Natl. Acad. Sci. U.S.A.* **85,** 6797 (1988).

the uptake solution and careful rinsing of the filter and filter insert in an ice-cold isotonic solution containing a high concentration (up to 100 mM) of the substrate. Total radioactivity incorporated into the monolayer is assessed by counting the whole filter insert which has been mixed with scintillation cocktail.

Non-specific binding (blanks) is assessed by measuring zero time uptake by starting uptake and immediately aspirating off the uptake solution. Total counts in the uptake solution are measured for each filter by removing 10 μl of the complete uptake solution.

Isolation of Apical Membrane Vesicles

Comments

1. For successful isolation of apical membrane vesicles it is necessary to start with a large quantity of cells ($\sim 2 \times 10^8$). Usually, sufficient membrane vesicles (~ 0.5 mg of protein) can be obtained when starting with cells cultivated either in one or two roller bottles (each 840 cm^3) or in one or two large Petri dishes (each 625 cm^2). Alternatively, cells might also be cultivated on microcarrier beads such as Cytodex (Pharmacia; Piscataway, NJ) or Biosilon (Nunc).

2. Characterization of the isolated apical membranes is best performed by the determination of enzymatic activities for which the cellular localization is well established. For example, in the established cell lines LLC-PK$_1$ and MDCK the enzyme alkaline phosphatase (EC 3.1.3.1) and leucine aminopeptidase (EC 3.4.11.2) have been localized by immunohistochemistry in the apical membrane[30,31] whereas the Na$^+$K$^+$-ATPase (EC 3.6.1.3) is exclusively localized in the basolateral membrane.[32-34]

For the isolation of transport-competent apical membrane vesicles from the cell lines OK and LLC-PK$_1$ the following two procedures have been used successfully.[25,35] The enzymatic characterization of these preparations is illustrated in Table II. Other very similar procedures have been

---

[30] C. A. Rabito, J. I. Kreisberg, and D. Wight, *J. Biol. Chem.* **259**, 574 (1984).
[31] D. Louvard, *Proc. Natl. Acad. Sci. U.S.A.* **77**, 4132 (1980).
[32] J. W. Mills, A. D. C. McKnight, J. H. Dayer, and D. A. Ausiello, *Am. J. Physiol.* **236**, C157 (1979).
[33] J. F. Laub, P. Odgen, and N. L. Simmons, *Biochim. Biophys. Acta* **644**, 333 (1981).
[34] C. A. Rabito and R. Tchao, *Am. J. Physiol.* **238**, C43 (1980).
[35] C. D. A. Brown, M. Bodmer, J. Biber, and H. Murer, *Biochim. Biophys. Acta* **769**, 471 (1984).

TABLE II
CHARACTERIZATION OF APICAL MEMBRANES ISOLATED FROM LLC-PK$_1$ AND OK CELLS

Enzyme	Specific activities[a]		Enrichment	Yield (% of homogenate)
	Homogenate	Apical membranes		
Alkaline phosphatase (EC 3.1.3.1)				
LLC-PK$_1$	318 ± 38	2087 ± 25	7.2 ± 0.5	38 ± 7
OK	Not detectable			
Leucine aminopeptidase (EC 3.4.11.2)				
LLC-PK$_1$	16.4 ± 3.0	93 ± 11	6.1 ± 0.6	31 ± 7
OK	28.0 ± 1.5	152 ± 42	5.4 ± 1.2	5.6 ± 1
γ-Glutamyl transferase (EC 2.3.2.2)				
LLC-PK$_1$	Not determined			
OK	22 ± 6	175 ± 6	7.8 ± 1.2	19 ± 6
Na$^+$,K$^+$-ATPase (EC 3.6.1.3)				
LLC-PK$_1$	13.8 ± 3.3	7.4 ± 1.9	0.6 ± 0.1	5.4 ± 3.0
OK	67 ± 25	125 ± 58	1.9 ± 0.6	4.3 ± 1.9

[a] In micromoles per minute per milligram protein. Values are taken from Refs. 25 and 35.

described by other laboratories.[36-38] Typically, the specific activities of the apical marker enzymes are enriched (compared to the homogenate) by a factor of 6 to 10. For reasons not yet clear, the separate isolation of the basolateral membrane from the presently discussed established renal cell lines has not been described so far.

Experimental Procedures

Isolation of Apical Membrane Vesicles from $LLC-PK_1$ Cells. $LLC-PK_1$ are grown in Spinner flasks on microcarrier beads[35] (e.g., Cytodex; Pharmacia) which are prepared according to the manufacturer's instructions. Confluent cell monolayers on microcarrier beads are first separated from the growth medium by centrifugation at 1000 g for 2 min. Cells are then separated from the microcarriers by resuspending the beads in a hypotonic medium (30 mM mannitol, 20 mM HEPES/Tris, pH 8.0 at 4°) and vortexing for 1 min. To ensure maximum recovery a mild sonication step (bath sonicator) might be included here. The cellular debris is separated from the microcarriers by filtration through a 100-μm nylon mesh which is washed with the above hypotonic medium. The filtrate is centrifuged at 31,000 g for 15 min in a Sorvall SS-34 rotor. The pellet is then resuspended in 300 mM mannitol, 20 mM HEPES/Tris, pH 7.4 and homogenized with a Polytron homogenizer (Typ PT-10; Kinematica, Switzerland) at 4° for 4 min (setting 4). $MgCl_2$ is added to the homogenate up to 10 mM. After 30 min (on ice) the homogenate is centrifuged at 2400 for 15 min (SS-34 rotor). The resulting supernatant is further centrifuged at 30,000 g for 45 min. The final pellet is resuspended in the buffer of choice by repeated syringing through a fine (22-gauge) needle.

When grown in large Petri dishes or in roller bottles cells are harvested by scraping with a rubber policeman directly into 300 mM mannitol, 20 mM HEPES/Tris, pH 7.4. The subsequent steps are the same as described above.

Isolation of Apical Membranes from OK Cells. In contrast to the $LLC-PK_1$ cells, OK cells must be homogenized differently in order to use the Ca^{2+}/Mg^{2+}-precipitation technique for the isolation of the apical membranes. Confluent cell monolayers (grown in large Petri dishes as described in Ref. 25) are rinsed once with ice-cold 300 mM mannitol, 5 mM EGTA, 12 mM Tris-HCl (pH 7.1) and then scraped off using a rubber policeman into approximately 30 ml of the mannitol/ETGA buffer. Cells are then

---

[36] J. E. Lever, B. G. Kennedy, and R. Vasan, *Arch. Biochem. Biophys.* **234**, 330 (1984).
[37] A. Moran, J. S. Handler, and R. J. Turner, *Am. J. Physiol.* **243**, C293 (1982).
[38] J. E. Lever, *J. Biol. Chem.* **257**, 8680 (1982).

harvested by a centrifugation (400 g for 5 min), resuspended in 7 ml of the mannitol/ETGA buffer, and diluted with 13 ml of H_2O. Disruption of the cells is performed by a nitrogen cavitation bomb (Parr Instruments) which is equilibrated at a pressure of 500 to 750 psi for 30 min. To the resulting homogenate $MgCl_2$ is added up to 10 mM. After 30 min (on ice), the homogenate is centrifuged at 2400 g for 10 min (Sorvall SS-34) and the supernatant is further centrifuged at 30,000 g for 30 min. The final pellet is washed once more with a buffer appropriate for the subsequent transport experiments.

Transport Studies. Phosphate transport into apical membranes isolated from the cultivated cells is performed by the filtration technique (using cellulose acetate filters of 0.45-μm pore size) which has been described extensively in this volume ([28]). Typical results obtained with apical membrane vesicles isolated from LLC-PK_1 and OK cells are illustrated in Fig. 3. In contrast to whole cells, sodium-independent transport of phosphate into the isolated membranes is determined by replacing sodium with potassium.

Specific Problems Isolated to ^{32}P Isotopes. For a correct calculation of the transport rate of P_i into the vesicles one has to correct by a blank value (see also [28], this volume) which corrects for nonspecific absorption of ^{32}P to the filters. However, this blank value can vary significantly depending on the batch of ^{32}P delivered. To avoid this problem—which is not completely understood—we recommend filtering the final incubation medium through a 0.22-μm filter prior to the transport experiment. Furthermore, the stock solution of ^{32}P often is contaminated by pyrophosphate.[39]

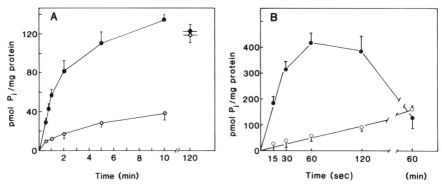

FIG. 3. Time course of phosphate transport into apical membrane vesicles prepared from LLC-PK_1 (A) and OK cells (B). The uptake of 0.1 mM phosphate was determined by applying an inwardly directed gradient of either 100 mM NaCl (●) or 100 mM KCl (○). (Data from Malmström and Murer[25] and Brown et al.[35])

As pyrophosphate formation is reduced in acid media, it is recommended that the isotope be kept in 0.02 N HCl if not purchased as such already.

Acknowledgment

The skillfull technical assistence of J. Forgo is gratefully acknowledged. This work has been supported by the Swiss National Funds (SNF Grant No. 3.881.085 and 3.881.185).

[39] R. J. Kessler, D. A. Vaughn, and D. D. Fenestil, *Anal. Biochem.* **158**, 117 (1986).

[30] ATP-Driven Proton Transport in Vesicles from the Kidney Cortex

By IVAN SABOLIĆ and GERHARD BURCKHARDT

Introduction

A variety of organelles in eukaryotic cells belong to the vacuolar system, which is involved in endocytosis, transport and degradation of internalized macromolecules, as well as in biosynthesis, processing, storage and release of intracellular molecules. The vacuolar system includes clathrin-coated vesicles, endocytotic or pinocytotic vesicles, lysosomes, Golgi membranes, endoplasmic reticulum, and secretory vesicles. Recent studies have demonstrated that all these structures have a low internal pH, which is generated by ATP-driven proton pumps (H^+ pumps, H^+-ATPases; for a review, see Ref. 1). The characteristics of these proton pumps are similar in all organelles investigated so far. They exhibit a preference for ATP, electrogenicity, stimulation by permeant anions, insensitivity to vanadate, ouabain, and oligomycin, and inhibition by the carboxyl group reagent N,N'-dicyclohexylcarbodiimide (DCCD) and the sulfhydryl group reagent N'-ethylmaleimide (NEM). These characteristics indicate that vacuolar proton ATPases belong to a separate class of ATPases which are different from the oligomycin-inhibitable mitochondrial and bacterial F_0F_1-type H^+-ATPase and the vanadate-inhibitable E_1E_2-type H^+-ATPase present in gastric mucosa, yeast, and fungi.

Membrane vesicles containing a proton pump with characteristics similar to those of the vacuolar H^+-ATPase have been isolated from bovine

[1] I. Mellman, R. Fuchs, and A. Helenius, *Annu. Rev. Biochem.* **55**, 663 (1986).

renal medulla[2] and rat renal proximal tubules.[3-5] These vesicles are localized underneath the lumenal membrane of proximal tubule cells and of intercalated cells in collecting ducts. It has been suggested that, under appropriate stimulus, these vesicles fuse with the lumenal membrane, leading to an increased, ATP-driven proton secretion into the tubular lumen.[2,6] In turn, proton secretion can be decreased by retrieval of proton pumps from the lumenal membrane and their storage in endocytotic vesicles.

In this chapter we describe an improved preparation of endocytotic vesicles from rat renal cortex. This method yields vesicles which are more enriched in proton pump activity than those prepared by a method previously published.[4,5] The preparation of endocytotic vesicles takes about 6 hr and does not require an ultracentrifuge. Therefore, endocytotic vesicles from kidney can serve as an excellent model to study the characteristics of the vacuolar proton pump (H^+-ATPase).

Experimental Procedure

Animals

For the experiments described here we prepared endocytotic vesicles from the kidney cortex of Wistar rats. The same results can be obtained with vesicles prepared from an equivalent amount of tissue of mouse, rabbit, or pig kidney.

Equipment Used in the Experiments

  1. Motor-driven glass/Teflon homogenizers (loose fitting and tight fitting)
  2. A Sorvall RC 5B centrifuge equipped with an SS-34 rotor (DuPont de Nemours, Bad Nauheim, FRG)
  3. A spectrofluorophotometer Shimadzu RF 510 with a chart recorder
  4. A Shimadzu UV-300 double-beam/dual wavelength spectrophotometer with a chart recorder (Shimadzu, Kyoto, Japan)

[2] S. Gluck and Q. Al-Awquati, *J. Clin. Invest.* **73,** 1704 (1984).
[3] R. W. Gurich and D. G. Warnock, *Am. J. Physiol.* **251,** F702 (1986).
[4] I. Sabolić, W. Haase, and G. Burckhardt, *Am. J. Physiol.* **248,** F835 (1985).
[5] I. Sabolić and G. Burckhardt, *Am. J. Physiol.* **250,** F817 (1986).
[6] G. J. Schwartz and Q. Al-Awquati, *J. Clin. Invest.* **75,** 1638 (1985).

Buffers

All buffers were made in deionized water with chemicals of reagent grade. The buffers were filtered through Sartorius filters (type 11307, pore size 0.2 μm) before use.

Buffers and Solutions for Isolation of Endocytotic Vesicles

Homogenizing buffer: 300 mM mannitol, 12 mM HEPES/Tris, pH 7.4

Potassium buffer: 300 mM mannitol, 100 mM potassium gluconate, 5 mM MgSO$_4$, 5 mM HEPES/Tris, pH 7.0

Percoll: Undiluted solution (Pharmacia Fine Chemicals, Uppsala, Sweden)

Buffers and Solutions for the Measurement of Proton Pump Activity

Gluconate buffer: 300 mM mannitol, 100 mM potassium gluconate, 5 mM MgSO$_4$, 6 μM acridine orange (Eastman Kodak, Rochester, NY), 5 mM HEPES/Tris, pH 7.0

Chloride buffer: 300 mM mannitol, 100 mM KCl, 5 mM MgSO$_4$, 6 μM acridine orange, 5 mM HEPES/Tris, pH 7.0

0.5 M ATP: The stock solution of 0.5 M ATP (disodium salt, Boehringer–Mannheim, Mannheim, FRG) was prepared in water and neutralized with Tris to pH 7.0. This solution can be kept frozen ($-20°$) for several weeks

Buffers and Solutions for Measurement of Proton ATPase Activity

ATPase buffer A: 330 mM mannitol, 110 mM potassium gluconate, 5.5 mM MgSO$_4$, 2.2 mM ouabain, 1.1 mM levamisole {L(−)-2,3,5,6-tetrahydro-6-phenylimidazo[2,1−b]thiazole hydrochloride; Sigma, St. Louis, MO}, 5.5 μg/ml oligomycin (Serva, Heidelberg, FRG), 55 mM HEPES/Tris, pH 7.0

ATPase buffer B: The same as above except that 110 mM potassium gluconate is replaced by 110 mM KCl

Substrate/enzyme buffer: 5.5 mM phosphoenolpyruvate (PEP; 3-cyclohexylammonium salt), 55 mM ATP (disodium salt), 2.42 mM NADH, 105 U/ml pyruvate kinase (PK), 195 U/ml lactic acid dehydrogenase (LDH, both from rabbit muscle, in glycerol solution), 5 mM HEPES/Tris, pH 7.0. All compounds were from Boehringer–Mannheim, FRG.

Homogenizing buffer and potassium buffer can be kept frozen ($-20°$) for 2 weeks. The same holds for gluconate and chloride buffers as well as for ATPase buffers A and B. Substrate/enzyme buffer can be made in two steps. The buffer containing ATP and PEP can be prepared and kept frozen ($-20°$) for at least 2 weeks. The appropriate amounts of NADH, PK, and LDH are added immediately before use.

Other Solutions

Stock solutions of NEM (200 mM, Serva, Heidelberg, FRG) and vanadate (200 mM Na$_2$HVO$_4$, Merck, Darmstadt, FRG) were prepared in water. The following stock solutions were made in ethanol: valinomycin (1 mM), CCCP (carbonyl cyanide-p-chloromethoxyphenylhydrazone, 2 mM) (both from Boehringer–Mannheim, FRG), DCCD (200 mM, Calbiochem, Los Angeles, CA), and diethylstilbestrol (DES) (10 mM, Serva, Heidelberg, FRG).

Assays

Protein was measured by the method of Bradford[7] using bovine serum albumin as a standard. The activities of leucine arylamidase (EC 3.4.11.2) and acid phosphatase (EC 3.1.3.2) were determined by using commercial kits (Merckotest 3359 and 3305, respectively). Na$^+$,K$^+$-ATPase (EC 3.6.1.3) activity was measured by a coupled optical test according to Berner and Kinne.[8] The activities of succinate-cytochrome c oxidoreductase (EC 1.3.99.1) and KCN-resistant NADH oxidoreductase (EC 1.6.99.2) were determined as described by Fleischer and Fleischer[9] and by Sottocasa *et al.*,[10] respectively.

Measurement of the Proton Pump Activity

To visualize the formation and dissipation of transmembrane proton gradients, we used the ΔpH-sensitive fluorescent dye acridine orange.[4,5] The weak base acridine orange accumulates in vesicles whenever intravesicular pH is acidic relative to the pH of the extravesicular buffer. The fluorescence of the accumulated dye is quenched. Thereby, the degree of fluorescence quenching (or decrease of absorption) is a measure for the magnitude of the ΔpH across the vesicular membrane.

The assay mixture contained 2.0 ml of gluconate or chloride buffer, appropriate ionophores and inhibitors added from stock solutions or an

---

[7] M. M. Bradford, *Anal. Biochem.* **72**, 248 (1976).
[8] W. Berner and R. Kinne, *Pfluegers Arch.* **361**, 269 (1976).
[9] S. Fleischer and B. Fleischer, this series, Vol. 10, p. 406.
[10] G. L. Sottocasa, L. Kuylenstierna, L. Ernster, and A. Bergstrand, *J. Cell Biol.* **32**, 415 (1967).

equivalent amount of solvent in control experiments (not more than 1% of the total volume), and vesicles. Amount of vesicles and concentrations of ionophores and inhibitors are indicated in the figure legends. The reaction was started by addition of a stock solution of ATP (final concentration 1.5 mM). The fluorescence was continuously monitored at room temperature (excitation 493 nm; emission 525 nm). During the measurement, the samples were stirred.

Measurement of Proton ATPase Activity

The ATPase activity was measured by a coupled optical assay at 37°. Fifty micrograms of vesicle protein was added into 2.0 ml of ATPase buffer A or B and preincubated at 37° for 10 min in the absence (controls containing solvent) or presence of various ionophores and inhibitors added from stocks (the concentrations of which are indicated in the corresponding figure legends below). The measurement of ATPase activity was started by addition of 0.2 ml substrate/enzyme buffer. The decrease in NADH absorbance was monitored continuously at 340 nm. During the measurement, the samples were constantly stirred.

Isolation of Endocytotic Vesicles

All steps in the preparation were performed in the cold, i.e., on ice and in refrigerated centrifuges. The flow scheme of the isolation procedure is shown in Fig. 1. In a typical experiment, three male Wistar rats were killed by cervical dislocation. The kidneys were taken out and immediately immersed in an ice-cold Ringer's solution. After removing the capsule, cortical slices (~0.3-mm thickness) were cut off by a razor blade. The pooled tissue (2 to 2.5 g wet) was put into 35 ml of homogenizing buffer and homogenized with 20 strokes in a loose-fitting glass/Teflon Potter homogenizer (1200 rpm) (Braun, Melsungen, FRG). Therefore the choice of homogenizer is critical. Homogenization by a tight-fitting glass/Teflon Potter or by blenders leads to a severely reduced yield in endocytotic vesicles. After homogenization, another 35 ml of homogenizing buffer was added and the suspension was centrifuged at 2500 g for 15 min. The pellet (P_1) was discarded and the supernatant (S_1) was centrifuged at 20,000 g for 20 min. Most of the resulting supernatant (S_2) was decanted and saved. The rest of the supernatant (about 2 ml) was used to disperse the fluffy upper part of the pellet (P_{2a}) by careful swirling of the tube. Therefore care was taken not to disturb the hard, yellowish brown mitochondrial pellet (P_{2b}). Supernatant S_2 and the dispersed fluffy pellet P_{2a} were combined and centrifuged at 48,000 g for 30 min. The resulting supernatant (S_3) was

HOMOGENATE (Kidney cortex from 3 rats = ca. 2.5g of tissue - immerse in 35 ml homogenizing buffer, homogenize with loose-fitting glass/Teflon Potter with 20 strokes at 1,200 rpm. Add another 35 ml of homogenizing buffer and mix well)

Centrifuge at 4,600 rpm for 15 min

Discard pellet (P_1)

SUPERNATANT (S_1)

Centrifuge at 13,000 rpm for 20 min

Discard dark and hard yellowish pellet (P_{2b})

SUPERNATNANT (S_2) + FLUFFY PELLET (P_{2a})

Centrifuge at 20,000 rpm for 30 min

Discard supernatant by sucking off (S_3)

PELLET (P_3) (Disperse P_3 with 30 ml homogenizing buffer with a tight-fitting glass/Teflon Potter by 10 strokes at 1,200 rpm. Add 6.1g undiluted Percoll to 32g suspension of homogenate (16% wt/wt Percoll). Mix well)

Centrifuge at 20,000 rpm for 30 min

THE LAST 5 ml FROM THE GRADIENT (P_4)
(Disperse with 30 ml potassium buffer. Leave on ice for 30 min)

Centrifuge at 20,000 rpm for 30 min

Discard completely supernatant (S_5) by suction

PELLET (P_5) (Resuspend P_5 with 1 ml potassium buffer and transfer into an Eppendorf reaction tube)

Centrifuge at 4,600 rpm for 15 min

Remove supernatant (S_6) by a syringe with a needle

PELLET (P_6) (This pellet represents endocytotic vesicles. Resuspend pellet in 50 µl of potassium buffer)

FIG. 1. Flow scheme of preparation of rat renal cortical endocytotic vesicles, using a Sorvall RC 5B centrifuge equipped with an SS-34 rotor.

siphoned off and discarded. The pellet (P_3) contained crude plasma membranes and endocytotic vesicles.

Endocytotic vesicles were separated from other membranes on a Percoll density gradient. For this purpose, fraction P_3 was resuspended in 30 ml of homogenizing buffer by 10 strokes (1200 rpm) in a tight-fitting glass/Teflon Potter homogenizer. Undiluted Percoll (6.1 g) was added to 32 g of vesicle suspension [16% (w/w) Percoll]. The suspension was well mixed and centrifuged at 48,000 g for 30 min. For analytical purposes, the resulting self-orienting gradient was fractionated from top to bottom by pumping a 60% sucrose solution onto the bottom of the centrifuge tube via a steel cannula. As shown in Fig. 2, each milliliter fraction of the gradient was tested for protein concentration ($A_{280\,nm}$), activity of the marker enzyme for basolateral membranes, Na^+,K^+-ATPase (closed triangles), activity of a marker enzyme for lumenal membranes, leucine arylamidase (open circles), and ATP-driven proton pump activity in the presence of chloride as a marker for endocytotic vesicles (closed circles). As demonstrated in our previous publication,[4] proton pump-containing vesicles at the bottom of the Percoll gradient can be filled with the fluid-phase markers horserad-

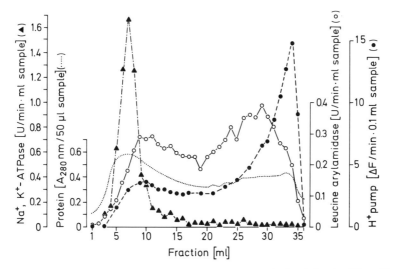

FIG. 2. Distribution of protein (·····), marker enzymes for brush border (○) and basolateral membranes (▲), and ATP-driven, Cl^--stimulated proton pump activity (●) along a Percoll gradient. Protein and enzyme activities were measured as described in the text. Proton pump activity (H^+ pump) was measured by adding a 0.1-ml aliquot to 2.0 ml acridine orange-containing chloride buffer with 0 or 1.5 mM ATP. The initial rates of acridine orange quenching in the presence of ATP were corrected for quenchings observed in the absence of ATP. One unit (U) corresponds to 1 μmol hydrolyzed Na^+,K^+-ATPase or 1 μmol L-leucine-p-nitroanilide (leucine arylamidase).

ish peroxidase and fluorescein isothiocyanate-labeled dextran administered intravenously *in vitro,* proving the identity of these vesicles with endocytotic vesicles.

Proton pump activity was present in all fractions from the gradient, with a sharp peak on the bottom. The peak of proton pump is separated from the peaks of basolateral and brush border membranes. However, it partially overlaps with the activity of leucine arylamidase. For preparative purposes, the last 5 ml of the gradient, which contained the highest activities of ATP-driven proton pump and relatively low leucine arylamidase activities, was pooled (P_4) and diluted with 30 ml of cold potassium buffer. The sample was kept on ice for 30 min and then centrifuged at 48,000 g for 30 min. The clear supernatant (S_5) was completely removed by suction. The fluffy white membrane pellet (P_5), which overlayed the hard glassy pellet of Percoll, was transferred into an Eppendorf tube and dispersed by vigorous vortexing in 1 ml of potassium buffer. The sample was centrifuged (by using a tube adapter) at 2500 g for 15 min. The slightly opaque supernatant (S_6), which displays a high activity of proton pump, but is heavily contaminated with brush border membranes, was removed completely by careful suction using a syringe with a needle. The white–yellow pellet on the bottom of the tube (P_6), which contains endocytotic vesicles, was dispersed in 50 μl of potassium buffer and immediately transferred into a new Eppendorf tube. Care was taken not to resuspend membranes adherent to the walls of Eppendorf tube since this material contains a high amount of brush border membranes. The protein concentration in the final vesicle preparation was around 10 mg/ml in a total volume of about 60 μl.

We would like to emphasize that all centrifugation steps following membrane separation on the Percoll gradient must be performed at physiological ionic strength. The usage of buffers of low ionic strength does not allow one to pellet endocytotic vesicles tightly and to separate them appropriately from contaminating brush border membranes.

The isolated endosomes can be kept frozen in liquid nitrogen for at least 2 weeks without visible loss of proton pump or ATPase activity. The deposition of vesicles at 4° overnight or longer, however, results in a time-dependent drop in ATPase activity and a significant increase in proton and potassium conductances in the membranes.[4]

Characterization of Renal Endocytotic Vesicles

Measurement of the Proton Pump

In Fig. 3 we show the quenching of acridine orange fluorescence after addition of ATP to a homogenate of the renal cortex (upper part) and to

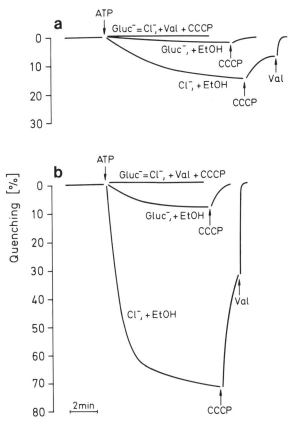

FIG. 3. ATP-driven acridine orange quenchings in (a) homogenate [150 μg protein] and (b) isolated endocytotic vesicles [50 μg protein]. The indicated amounts of sample protein were added to acridine orange-containing gluconate (gluc⁻) or chloride buffer (Cl⁻), which in final concentrations contained either 0.5% ethanol (EtOH) or 5 μM valinomycin plus 10 μM CCCP (Val + CCCP). The reaction was started by addition of ATP (final concentration 1.5 mM). At the indicated time (arrow), proton gradients were dissipated by addition of CCCP (10 μM) and valinomycin (5 μM).

isolated endocytotic vesicles (lower part). In both homogenate and endocytotic vesicles fluorescence decrease is accelerated in the presence of the permeant anion chloride (Cl⁻, + EtOH) as compared to the weakly permeant anion gluconate (Gluc⁻, + EtOH). This result indicates the presence of an electrogenic ATP-driven proton pump that requires the uptake of a chloride ion as a charge compensation for each pumped proton.[5] This proton pump generates a transmembrane ΔpH (inside acidic) which can be dissipated by the addition of an uncoupler, CCCP. This dissipation is faster when CCCP is added together with valinomycin. In this case each proton flowing out of the vesicles through CCCP can be electrically compensated for by the influx of a potassium ion through valinomycin. In the absence of valinomycin, potassium conductance of the vesicle membrane is too small to allow for a rapid potassium movement.[5] If valinomycin and CCCP were present from the very beginning of the measurement (Fig. 3, Gluc$^-$ = Cl$^-$, + Val + CCCP), no ΔpH developed in the presence of ATP. Similarly, in the absence of ATP no ΔpH was created (data not shown).

From Fig. 3 it is also evident that the rate of fluorescence decrease and the final degree of quenching are much greater in isolated endocytotic vesicles. This difference is observed although the curves for homogenate were obtained with a threefold greater amount of protein (150 μg/assay as compared to 50 μg/assay for purified endocytotic vesicles). Consequently, endocytotic vesicles contain more ATP-driven proton pumps per milligram protein than the homogenate. To estimate the degree of purification of endocytotic vesicles we determined the initial rates of ATP-driven, Cl$^-$-stimulated acidifications by drawing the tangent to the initial part of the fluorescence recordings with renal homogenate and isolated vesicles. The initial rates were expressed as fluorescence changes per minute per milligram protein [ΔF/min · mg protein]. The initial rates in the presence of ATP were corrected for fluorescence changes recorded in the absence of ATP (baseline drift). The ratio of initial rates observed with endocytotic vesicles over those recorded with homogenates yields the enrichment factor (cf. Table II). This calculation is meaningful only if the initial fluorescence changes are proportional to the amount of protein within the test assay. We found proportionality for protein amounts up to 0.25 mg/assay with homogenates and 0.15 mg/assay with purified endocytotic vesicles (data not shown).

Recovery of Protein and Proton Pump in Vesicle Preparations

In Table I is summarized the yield of protein and of ATP-driven, Cl$^-$-stimulated proton pump activity during the most important purification steps. The yield of the protein in the final vesicle preparations was only 0.13%, whereas the yield of the proton pump was 5.6%. The total recovery of the proton pump activity during the preparation ranged between 90 and 105% (data not shown), thus showing no artificial activation or inactivation of proton pump activity.

Sometimes, the limited yield of endocytotic vesicles makes it necessary to start the preparation with more tissue than shown in Fig. 1. This can be achieved with the same method, but the ratio of tissue (in grams wet weight) over homogenizing buffer (in milliliters) must be kept equal to that described in Fig. 1. The same holds for all other steps in the isolation

TABLE I
RECOVERY OF PROTEIN AND ATP-DRIVEN H⁺ PUMP ACTIVITY DURING PREPARATION OF ENDOCYTOTIC VESICLES

Fraction	Protein		ATP-driven H$^+$ pump		
	Total (mg)	Yield (%)	Total activity (ΔF/total protein)[a]	Yield (%)	Specific activity (ΔF/mg protein)
Homogenate	524 ± 15.4[b]	100	7790 ± 825	100	15.0 ± 1.83
S_1	255 ± 11.6	48.9 ± 2.71	6525 ± 1010	82.4 ± 5.6	25.9 ± 4.27
$S_2 + P_{2a}$	216 ± 8.7	41.1 ± 0.99	3710 ± 509	47.2 ± 2.5	17.6 ± 3.02
P_3	35 ± 2.5	6.7 ± 0.34	2708 ± 301	34.8 ± 1.4	80.3 ± 14.1
P_4	4.3 ± 0.24	0.83 ± 0.03	892 ± 101	11.5 ± 0.9	198.5 ± 15.8
P_5	1.65 ± 0.08	0.32 ± 0.02	970 ± 117	12.4 ± 0.2	601 ± 33.2
P_6	0.67 ± 0.05	0.13 ± 0.01	440 ± 52	5.6 ± 0.2	649 ± 33.3

[a] ΔF, Acridine orange fluorescence change.
[b] Shown are means ± SEM from four experiments.

procedure. Moreover, overloading of the Percoll gradient with more material than indicated must be avoided as this causes poorer resolution on the gradient and bigger contamination of endosomes with brush border membranes.

Enzymatic Characteristics of Vesicle Preparations

Table II shows the specific activities of the proton pump and of marker enzymes for cellular membranes in homogenate and isolated endocytotic vesicles and their respective enrichment factors. As determined from the initial rate of acidification the vesicles were enriched in ATP-driven, Cl$^-$-stimulated proton pump activity greater than 40-fold. A small enrichment of about 1.5-fold was found for the marker enzyme for lumenal membranes, leucine arylamidase, and for the lysosomal marker, acid phosphatase. No enrichment was found in the activities of marker enzymes for contralumenal membranes, mitochondria, and endoplasmic reticulum. This pattern of enrichment factors indicates that our preparations of endocytotic vesicles are not significantly contaminated with respective cellular membranes. Table II also indicates that endocytotic vesicles possess no other marker enzyme besides the ATP-driven proton pump.

Effect of Inhibitors

As indicated by the curves in Fig. 4, ATP-driven Cl$^-$-stimulated proton pump in renal cortical endocytotic vesicles is not significantly inhibited by

TABLE II
PROTON PUMP AND ENZYME ACTIVITIES IN HOMOGENATE AND ENDOCYTOTIC VESICLES

Enzyme	n	Homogenate	Endocytotic vesicles	Enrichment factor
Proton pump (endocytotic vesicles)[a]	10[b]	16.5 ± 2.13	715 ± 124.4	44.3 ± 5.64
Leucine arylamidase (brush border membranes)[c]	8	106.5 ± 10.3	167.3 ± 28.6	1.53 ± 0.14
Na,$^+$ K$^+$-ATPase (basolateral membranes)[c]	8	151.9 ± 32.6	65.2 ± 20.8	0.42 ± 0.11
Acid phosphatase (lysosomes)[d]	6	82.0 ± 2.19	121.6 ± 7.65	1.49 ± 0.10
Succinate cytochrome c oxidoreductase (mitochondria)[c]	6	213.9 ± 20.6	243.8 ± 31.3	1.17 ± 0.15
KCN-resistant NADH oxidoreductase (endoplasmic reticulum)[c]	6	198.1 ± 12.1	226.8 ± 20.9	1.18 ± 0.17

[a] ΔF/per minute · milligram protein.
[b] Shown are means ± SEM for the number of experiments indicated by n.
[c] Micromoles per minute · milligram protein.
[d] Micromoles per 30 min · milligram protein.

5 μg/ml oligomycin. This result clearly shows that the proton pump in endocytotic vesicles is not identical to the mitochondrial H$^+$-ATPase which is blocked by <5 μg/ml oligomycin. Furthermore, the lacking effect of oligomycin on the acridine orange signals in Fig. 4 excludes any significant contamination of our preparation with mitochondria or submito-

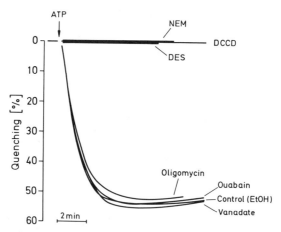

FIG. 4. Effect of various inhibitors on ATP-driven, Cl$^-$-stimulated proton pump in isolated endocytotic vesicles. Vesicles (25 μg protein) were suspended in chloride buffers which contained 1% ethanol (EtOH) or the indicated inhibitors (defined in text) and preincubated at 37° for 10 min. The reaction was then started by addition of ATP (final concentration 1.5 mM). The concentrations of inhibitors were the following: ouabain, 2 mM; vanadate, 1 mM; oligomycin, 5 μg/ml; NEM, 1 mM; DCCD, 1 mM; and DES, 50 μM.

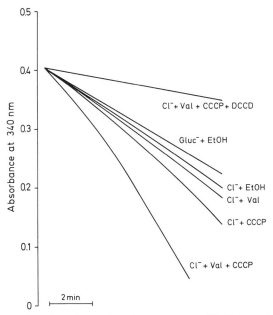

FIG. 5. ATPase activity in endocytotic vesicles as measured by the coupled optical assay. Vesicles (50 μg protein) were diluted in ATPase buffer A (Gluc⁻) or B (Cl⁻) which contained either 1% ethanol (EtOH) or 5 μM valinomycin (Val) or 10 μM CCCP or both (Val + CCCP), or, in addition to both ionophores, 1 mM DCCD (Val + CCCP + DCCD). Before starting the reaction with substrate/enzyme buffer (see text), vesicles were preincubated with the indicated ionophores and inhibitors at 37° for 10 min.

chondrial particles. Also ouabain, an inhibitor of Na^+,K^+-ATPase, does not inhibit proton pumping as well as vanadate, an inhibitor of various ATPases which are phosphorylated during the catalytic cycle.[11] However, the pump is completely inhibited by N'-ethylmaleimide (NEM), N,N'-dicyclohexylcarbodiimide (DCCD), and diethylstilbestrol (DES). The strong sensitivity to NEM also distinguishes the proton pump in endocytotic vesicles from that in mitochondria, as the mitochondrial pump is insensitive to NEM.[2]

Proton ATPase Activity in Endocytotic Vesicles

The measurement of ATPase activity in endocytotic vesicles by using a coupled optical assay is shown in Fig. 5. Endocytotic vesicles exhibit an ATPase activity in the absence of permeant anions (Gluc⁻ + EtOH). The

[11] S. O'Neal, D. B. Rhoads, and E. Racker, *Biochem. Biophys. Res. Commun.* **89**, 845 (1979).

activity is slightly stimulated by chloride (Cl^- + EtOH) and further by valinomycin in the presence of potassium (Cl^- + Val). As shown previously,[12] chloride and valinomycin plus potassium dissipate the inside positive membrane potential created by the electrogenic proton pump and thus release a constraint on the pump activity. A greater stimulation was observed with the uncoupler CCCP (Cl^- + CCCP). The highest activity was recorded when both membrane potential and proton gradient were completely abolished (Cl^- + Val + CCCP). Stimulation of ATPase activity by compounds which dissipate membrane potential, proton gradients, or both directly indicates the presence of an electrogenic H^+-ATPase in endocytotic vesicles. The (Val + CCCP)-stimulated ATPase activity was strongly inhibited by DCCD (Cl^- + Val + CCCP + DCCD). Similar inhibitions can be observed with 1 mM NEM and 1 mM DES (not shown). However, we have found that NEM and DES partially interact with the optical assay, leading to an overestimation of the inhibition of ATPase activity.

The maximal stimulation of ATPase activity by valinomycin plus CCCP was used to localize H^+-ATPase-containing vesicles along the Percoll gradient. ATPase activity was measured in each fraction from the gradient in the presence of oligomycin with or without valinomycin plus CCCP. As shown in Fig. 6A, (Val + CCCP)-stimulated ATPase activity is distributed along the Percoll gradient similarly to the ATP-driven, Cl^--stimulated proton pump measured by acridine orange quenching (compare hatched area in Fig. 6A with the distribution of the H^+ pump in Fig. 2). This result proves the coexistence of proton pump and H^+-ATPase in the same vesicles.

The former experiments showed that the proton pump (H^+-ATPase) in endocytotic vesicles is inhibited by DCCD. However, the distribution of DCCD-sensitive ATPase along a Percoll gradient indicates that only a small part of DCCD-sensitive ATPase is present at the bottom of the gradient, where the peak of proton pump (H^+-ATPase) was found. The majority of DCCD-sensitive ATPase activity is found at the top of the gradient and copurifies with Na^+,K^+-ATPase (compare Fig. 6B with Fig. 2). A similar distribution was also observed for the NEM- and DES-sensitive ATPases. For these measurements, the signals had to be corrected for the nonspecific interaction of NEM and DES with the optical assay (data not shown). Such corrections, however, are not recommended for practical use as the interaction of inhibitors with the optical assay depends on both concentration of inhibitors and concentration of PK and LDH in the assay mixture. We rather recommend the measurement of liberated phosphate, whenever N-ethylmaleimide or diethylstilbestrol are applied.[13]

[12] G. Burckhardt, B. Moewes, and I. Sabolić, *J. Clin. Chem. Clin. Biochem.* **24,** 682 (1986).
[13] I. Sabolić and G. Burckhardt, *Biochim. Biophys. Acta* **937,** 398 (1988).

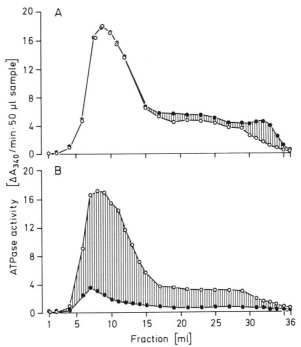

FIG. 6. Distribution of valinomycin plus CCCP-stimulated ATPase (A) and DCCD-sensitive ATPase (B) along a Percoll gradient. Fifty microliters of the fraction sample was diluted in 2.0 ml ATPase buffer B which contained 1% ethanol (A and B; ○) or 5 μM valinomycin plus 10 μM CCCP (A; ●) or 1 mM DCCD (B: ●) and 1.5 mM ATP. (Val + CCCP)-stimulated ATPase and DCCD-sensitive ATPase are shown by the hatched areas in (A) and (B), respectively. Before starting the reaction, the assay mixtures were preincubated with the indicated ionophores and inhibitors for 10 min at 37°.

Comments

We would like to emphasize again that the way of tissue homogenization is of critical importance for the success of the preparation. More rigorous homogenization procedures using blenders lead in our hands to a decreased yield in endocytotic vesicles and to a loss in proton pump activity. Recently, endocytotic vesicles have been prepared from the rabbit kidney cortex using a blender for homogenization and a different centrifugation protocol, including an overnight spin at 100,000 g in a sucrose density gradient.[3] The endocytotic vesicles prepared by this method differed from ours by having a smaller density and by possessing an Na^+/H^+ exchanger in addition to the ATP-driven H^+ pump. The degrees of purification cannot be compared at present, as Gurich and Warnock[3] measured the enrichment of membrane-bound horseradish peroxidase (enrichment

2.6-fold in isolated vesicles over the starting homogenate[3]), whereas we determined the enrichment of ATP-driven proton uptake (>40-fold with respect to the homogenate).

As shown in our experiments, rat renal cortical endosomes have no endogenous specific marker enzyme. Therefore, the proton pump is the only parameter which can be used throughout the experimental procedure to check the purity of the vesicles. Alternatively, one can use (Val + CCCP)-stimulated ATPase activity as an indicator of the proton ATPase in these vesicles. However, determination of the initial rates of ATP-driven, Cl^--stimulated intravesicular proton uptake measured by acridine orange quenching (cf. Fig. 3, Cl^-, + EtOH) is much easier and more sensitive.

Although the proton pump (H^+-ATPase) in isolated endocytotic vesicles is strongly inhibited by DCCD, NEM, and DES, neither of these inhibitors can be used as a specific marker for the pump. Experiments showing the distribution of inhibitor-sensitive ATPases indicate that much higher activities of DCCD-, NEM-, and DES-sensitive ATPases are present in fractions of the Percoll gradient that do not contain endocytotic vesicles with an ATP-driven proton pump.

[31] Aldosterone

By Diana Marver

Introduction

The principal and most well-documented renal target sites for the steroid aldosterone are the cortical and outer medullary collecting tubules (CCT/oMCT).[1-3] In the former, the primary role of mineralocorticoids is to enhance Na^+ reabsorption and K^+ secretion, while in the latter, aldosterone stimulates H^+ secretion. There is also indirect evidence to suggest that the medullary thick ascending limb of Henle (mTALH),[4,5] the distal convoluted tubule (DCT),[6,7] the connecting segment (CNT),[8] and the

---

[1] J. B. Gross, M. Imai, and J. P. Kokko, *J. Clin. Invest.* **55**, 1284 (1975).
[2] G. J. Schwartz and M. B. Burg, *Am. J. Physiol.* **235**, F576 (1978).
[3] D. Stone, D. Seldin, J. Kokko, and H. Jacobson, *J. Clin. Invest.* **72**, 77 (1983).
[4] B. A. Stanton, *J. Clin. Invest.* **78**, 1612 (1986).
[5] J. Work and R. L. Jamison, *J. Clin. Invest.* **80**, 1160 (1987).
[6] N. Farman, A. Vandewalle, and J. P. Bonvalet, *Am. J. Physiol.* **242**, F69 (1982).
[7] D. H. Ellison, H. Velazquez, and F. S. Wright, *J. Clin. Invest.* **83**, 113 (1989).
[8] L. C. Garg, M. A. Knepper, and M. B. Burg, *Am. J. Physiol.* **240**, F536 (1981).

papillary collecting duct (PCD)[9] may also be aldosterone sensitive, although the exact role played by mineralocorticoids at these segments, and indeed whether these represent classical or receptor-mediated target sites, is less clear. Of the naturally occurring steroids, desoxycorticosterone and corticosterone have appreciable mineralocorticoid activity. Since corticosterone is in many species the circulating glucocorticoid, most experiments defining aldosterone actions employ adrenalectomized animals or cultured distal tubular epithelia and contrast the cellular effects of either replacement aldosterone or dexamethasone. These are often complemented by controls given either steroid + spirolactones to block the mineralocorticoid (type I) receptor or RU 486, RU 38486, or ZK 98,734 to block the glucocorticoid (type II) receptor. This is necessary due to the sequence similarity of these proteins[10] and therefore the ability of high doses of aldosterone to bind to type II sites and high doses of dexamethasone to bind to type I sites. As both mineralocorticoids and glucocorticoids have actions on Na^+, K^+, and H^+ handling by the kidney,[11] a type I versus type II receptor-mediated process is best defined by first determining which of these two receptors is involved rather than quantitating the net effect of a given steroid on Na^+, K^+, or acid excretion rates.

Thus the following sections first deal with the determination of the kinetics and steroid specificity of the mineralocorticoid receptor. Examples will be given for a macroassay using renal cortical slices and a microassay utilizing isolated CCTs. Clearance studies, classically carried out in adrenalectomized rats, will also be described in detail. Finally, the biochemical assay of two enzymes frequently associated with aldosterone action, namely Na^+,K^+-ATPase and citrate synthase, will be defined at the ultramicro- and microlevel. State-of-the-art issues and techniques regarding evaluation of aldosterone action on target tissues will also be discussed in the concluding section.

Receptors

The renal mineralocorticoid receptor is defined by its high affinity and selectivity for the steroid aldosterone and synthetic congeners such as 9α-fluorocortisol. The K_d of aldosterone for this corticosteroid type I site ranges from about 0.1 to 3 nM, depending on temperature, species, and

[9] E. Higashihara, N. W. Carter, L. Pucacco, and J. P. Kokko, *Am. J. Physiol.* **246**, F725 (1984).
[10] J. L. Arriza, C. Weinberger, G. Cerelli, T. M. Glaser, B. L. Handelin, D. E. Housman, and R. M. Evans, *Science* **237**, 268 (1987).
[11] D. Marver, in "Handbook of Renal Physiology" (E. E. Windhager, ed.). Waverly, Baltimore, Maryland, in press.

TABLE I
TYPICAL EQUILIBRIUM DISSOCIATION CONSTANTS DETERMINED FOR [^{3}H]ALDOSTERONE BINDING TO MINERALOCORTICOID (TYPE I) AND GLUCOCORTICOID (TYPE II) RENAL RECEPTORS

Tissue source	Renal preparation	Temperature (°C)	K_d (nM)		Ref.
			Type I	Type II	
Rat	Slices	4	0.09	41	12
	Cytosol	4	3.9	53	13
	Slices	37	0.5	25	14
Rabbit	CCTs[a]	25	2.0	ND[b]	15
	Slices	37	3.0	25	16
Human	Slices	25	2.0	ND[b]	17

[a] CCTs, Cortical collecting tubules.
[b] ND, Not determined.

assay conditions. A selection of reported values appears in Table I.[12-17] If assays are corrected for nonspecific labeling, then Scatchard analyses of [^{3}H]aldosterone binding as a function of free aldosterone concentration will yield two specific binding sites, the lower affinity site equivalent to the type II classical glucocorticoid receptor. The K_d of aldosterone for type II binding sites ranges from 5 to 50 nM[18] and that of dexamethasone for the type I site from 17 to 25 nM (Table II).[19-21] On average, it has been estimated that dexamethasone has about one-fiftieth the potency of aldosterone for the type I site, while aldosterone has roughly one-fifth the potency of dexamethasone for the type II site.[14,19]

A third corticosteroid binding site [type III, or corticosteroid binding globulin (CBG)-like] has been described in kidney that has a high affinity for desoxycorticosterone and corticosterone, and low affinity for aldoster-

[12] D. G. Warnock and I. S. Edelman, *Mol. Cell. Endocrinol.* **12**, 221 (1978).
[13] M.-E. Rafestin-Oblin, A. Michaud, M. Claire, and P. Corvol, *J. Steroid Biochem.* **8**, 19 (1977).
[14] J. W. Funder, D. Feldman, and I. S. Edelman, *Endocrinology (Baltimore)* **92**, 994 (1973).
[15] A. Doucet and A. I. Katz, *Am. J. Physiol.* **241**, F605 (1981).
[16] D. Marver, *Endocrinology (Baltimore)* **106**, 611 (1980).
[17] P. Fuller and J. W. Funder, *Kidney Int.* **10**, 154 (1976).
[18] D. Marver, in "Biochemical Actions of Hormones" (G. Litwack, ed.). p. 385. Academic Press, Orlando, Florida, 1985.
[19] J. W. Funder, D. Feldman, and I. S. Edelman, *Endocrinology (Baltimore)* **92**, 1005 (1973).
[20] M. C. Lan, B. Graham, F. C. Bartter, and J. D. Baxter, *J. Clin. Endocrinol. Metab.* **54**, 332 (1982).
[21] N. Farman, A. Vandewalle, and J. P. Bonvalet, *Am. J. Physiol.* **244**, F325 (1983).

TABLE II
TYPICAL EQUILIBRIUM DISSOCIATION CONSTANTS DETERMINED FOR
[^{3}H]DEXAMETHASONE BINDING TO GLUCOCORTICOID (TYPE II) AND
MINERALOCORTICOID (TYPE I) RENAL RECEPTORS

Tissue source	Renal preparation	Temperature (°C)	K_d (nM)		Ref.
			Type I	Type II	
Rat	Slices	25	25	5	14,19
	Slices	37	17	6	20
Rabbit	Cytosol	30	ND[a]	15	21
Human	Slices	4	ND[a]	40	17

[a] ND, Not determined.

one and particularly dexamethasone.[22-24] Because there is at present much interest in the role of this site and its relationship to aldosterone action in target tissue, it will be discussed in some detail. The concentration of the type III binding site is highest in the rat CCT[23,24] and, in the original studies, appeared to be a receptor based on two properties. First, labeled sites were localized to both the cytoplasm and nucleus, the latter being the presumed target of corticosteroid receptors.[22,23] Second, when renal cytosol was labeled with 52 nM [^{3}H]corticosterone in the presence of 100× aldosterone and 100× dexamethasone (the K_d of corticosterone for type III site = 3 nM at 25°), the specifically labeled protein gave a characteristic corticosteroid–receptor 8S complex on low salt density gradients.[22] More recently, there has been the suggestion that this site may not be a receptor, but a microsomal enzyme, and specifically an 11β-hydroxysteroid dehydrogenase (11β-OHSD) activity that converts compounds such as corticosterone (compound B) to 11-dehydro-B and cortisol to cortisone. The products happen to have much less affinity for the type I and type II receptors than the parent compounds. Thus, it has been postulated that the physiological role of the renal enzyme is to ensure that local concentrations of corticosterone or cortisol remain low so as to disallow appreciable

[22] D. Feldman, J. W. Funder, and I. S. Edelman, *Endocrinology (Baltimore)* **92**, 1429 (1973).
[23] J. M. Strum, D. Feldman, B. Taggart, D. Marver, and I. S. Edelman, *Endocrinology (Baltimore)* **97**, 505 (1975).
[24] S. M. K. Lee, M. A. Chekal, and A. I. Katz, *Am. J. Physiol.* **244**, F504 (1983).
[25] P. M. Stewart, J. E. T. Corrie, C. H. L. Shackleton, and C. R. W. Edwards, *J. Clin. Invest.* **82**, 340 (1988).
[26] C. R. W. Edwards, D. Burt, M. A. McIntyre, E. R. DeKloet, P. M. Stewart, L. Brett, W. S. Sutanto, and C. Monder, *Lancet* **2**, 986 (1988).
[27] J. W. Funder, P. T. Pearce, R. Smith, and A. I. Smith, *Science* **242**, 583 (1988).

binding to, and activation of, the mineralocorticoid receptor.[25-28] A glance at the relative affinities of compound B for type I/II/III receptors or binding sites in Table III[29-37] indicates that circulating 10–20 nM concentrations of free corticosterone would occupy appreciable fractions of all three corticosteroid-binding sites, including the mineralocorticoid receptor. Lan et al.[20] were among the first investigators to recognize that if receptor assays accurately monitored affinity constants, then during peak diurnal corticosterone secretion rates, the mineralocorticoid receptor would be dominated by the concentration of this steroid rather than by aldosterone. The presence of renal 11β-OHSD may be the reason why this potential problem doesn't exist in vivo.

11β-Hydroxysteroid dehydrogenase is actually a composite of two activities, an 11-reductase activity which predominates in the kidney and an 11-oxidase activity which predominates in the liver.[38,39] It is only one of several corticosterone-metabolizing enzymes in the kidney, and is most easily demonstrable in female as opposed to male rats.[40] However, from the data given in Table III, it would appear unlikely that the type III site is equivalent to 11β-OHSD since the concentration of compound B required to half-maximally stimulate enzyme activity is about 1000× greater than that required to half-maximally occupy the type III site (best described by the data for CCTs). Also note that cortisol (compound F) does not bind to the type III site in CCTs, yet it is a preferred ligand, along with corticosterone, for 11β-OHSD. This means we do not as yet understand the nature and role of the type III binding site in the CCT. In keeping with the concept that the type III site and 11β-OHSD are separate entities, Edwards et al.[26] localized 11β-OHSD activity only to the proximal tubule and vasa recta of the rat by immunocytochemistry, although a cortical distal tubule fraction

---

[28] C. R. W. Edwards, D. Burt, and P. M. Stewart, J. Steroid Biochem. **32**, 213 (1989).
[29] D. Feldman, J. W. Funder, and I. S. Edelman, Am. J. Med. **53**, 545 (1972).
[30] Z. S. Krozowski and J. W. Funder, Proc. Natl. Acad. Sci. U.S.A. **80**, 6056 (1983).
[31] M. Jobin and F. Perrin, Can. J. Biochem. **52**, 101 (1974).
[32] P. L. Ballard, in "Glucocorticoid Hormone Action" (J. D. Baxter and G. G. Rousseau, eds.), p. 25. Springer-Verlag, Berlin, 1979.
[33] I. E. Bush, S. A. Hunter, and R. A. Meigs, Biochem. J. **107**, 239 (1968).
[34] D. R. Koerner, Biochim. Biophys. Acta **176**, 377 (1969).
[35] A. Ulmann, J. Menard, and P. Corvol, Endocrinology (Baltimore) **97**, 46 (1975).
[36] P. M. Stewart, R. Valentino, A. M. Wallace, D. Burt, C. H. L. Shackleton, and C. R. W. Edwards, Lancet **2**, 821 (1987).
[37] D. Armanini, I. Karbowiak, Z. Krozowski, J. W. Funder, and W. R. Adam, Endocrinology (Baltimore) **111**, 1683 (1982).
[38] V. Lakskmi and C. Monder, Endocrinology (Baltimore) **116**, 552 (1985).
[39] J. S. Jenkins, J. Endocrinol. **34**, 51 (1966).
[40] H. Siebe, D. Tsiakiras, and K. Hierholzer, Pfluegers Arch. **400**, 372 (1984).

TABLE III
A Comparison of the Apparent K_d or K_m of Compound B, Glycyrrhetinic Acid, or Carbenoxolone for Various Corticosteroid Receptors, Corticosteroid-Binding Globulin (CBG) or the Enzyme 11β-Hydroxysteroid Dehydrogenase (11β-OHSD)

Ligand	Source	Preparation	Temperature (°C)	Site	K_d/K_m	Steroid selectivity[a]	Ref.
Compound B	Rat	Slice	37	Type I	25 nM	Aldo > DOC > B > Dex	15, 29, 30
	Rat	Slice	25	Type II	12 nM	Dex > B > DOC = Aldo	29
	Rat	Slice	25	Type III	3 nM	B > F > DOC > Prog > Aldo > Dex	22
	Rat	CCTs	25	Type III	25 nM	B > Prog > DOC > Aldo >>> Dex > F	24
	Rat	—	37	CBG	67 nM		31, 32
	Rat	MC[b]	20	11β-OHSD	29 μM	B > F > DOC > Prog >> Aldo >>> Dex	27, 33, 34
Glycyrrhetinic acid	Rat	Slice	25	Type I	2 μM	B > F >>> Aldo	35
	Rat	—	25	CBG	No aff.[c]		35
	Human	In vivo	37	11β-OHSD	<2 μM[d]		36
Carbenoxolone	Rat	Cytosol	22	Type I	~8 μM		37
	Rat	Cytosol	22	Type II	~23 μM		37
	Rat	In vivo	37	11β-OHSD	<8 μM[d]		27

[a] Aldo, Aldosterone; DOC, deoxycorticosterone; B, corticosterone; Dex, dexamethasone; F, cortisol; Prog, progesterone.
[b] MC, liver microsomes.
[c] No aff., No appreciable affinity at all concentrations tested.
[d] Affinity judged to be greater than value indicated by K_d shown.

separated on Ficoll gradients appeared as active as the cortical proximal tubule fraction with respect to metabolizing cortisol to cortisone. In order to reconcile these conflicting data, they suggested that either this enzyme acts as a paracrine rather than autocrine modulator, ensuring that blood bathing the distal nephron or filtrate delivered to the distal nephron is much reduced in corticosterone or cortisol, or that a unique 11β-OHSD activity resides in the distal nephron, undetected by this particular antibody raised against the hepatic microsomal enzyme.

Table III also includes values for the apparent affinity of glycyrrhetinic acid and carbenoxolone (18β-glycyrrhetinic acid hemisuccinate), a derivative of the former compound, for the type I receptor. It has been well recognized for years that licorice addicts can present with the same clinical features as patients with hyperaldosteronism. The flavoring in licorice is obtained from the *Glycyrrhiza labra* plant. Glycyrrhizic acid contained within the plant extract is metabolized *in vivo* to the active compound glycyrrhetinic acid, which has a modest affinity for the mineralocorticoid receptor. Thus it was originally felt that the salt-retaining properties of glycyrrhetinic acid and its derivatives were due to their affinity for the type I site. However, it appeared that they were more potent *in vivo* than their type I site affinities would predict. Furthermore, their potency was dependent on an intact adrenal gland.[41,42] Since the plasma half-life of [^{3}H]cortisol was more than doubled in normal volunteers given licorice over 1 week, and there was a significant rise in the ratio of urinary allotetrahydrocortisol plus tetrahydrocortisol to tetrahydrocortisone, it was suggested that glycyrrhetinic acid was also an efficient inhibitor of 11β-OHSD.[28] In support of the critical role of this enzyme in suppressing the potency of cortisol/corticosterone relative to the type I site, a patient with a congenital deficiency of this enzyme exhibited a marked antinatriuretic and kaliuretic response to hydrocortisone, in contrast to a natriuresis following dexamethasone.[25] Furthermore the patient presented with hypokalemia, hypertension, and a suppressed renin–angiotensin axis. Finally, although endogenous plasma cortisol concentrations were normal, tracer [^{3}H]cortisol had a prolonged half-life of 131 min compared to values in normal volunteers of 40 min. Thus the presence of this corticosteroid-metabolizing enzyme may play an important role in defining occupancy of the mineralocorticoid receptor and its responsiveness to a series of corticosteroids *in vivo* or in tissue preparations.

Several other pieces of information or possibilities should be considered. Aldosterone is also metabolized by the kidney to less active deriva-

[41] J. G. G. Borst, S. P. Ten Holt, L. A. De Vries, and J. A. Molhuysen, *Lancet* **2**, 657 (1953).
[42] W. I. Card, W. Mitchell, J. A. Strong, N. R. W. Taylor, S. L. Tompsett, and J. M. G. Wilson, *Lancet* **2**, 663 (1953).

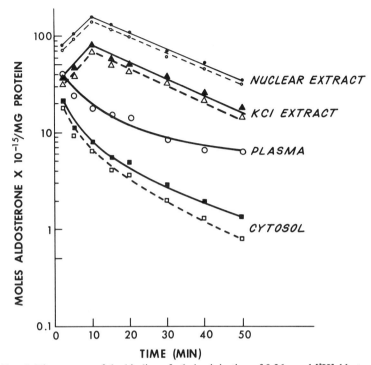

FIG. 1. Time course of the binding of a bolus injection of 0.26 nmol [^{3}H]aldosterone to renal cytosolic receptors (■), and receptors associated with nuclei (●, ▲), compared to corresponding plasma concentrations (○). Dashed lines represent specific binding and solid lines total binding. (From Marver et al.[44] by permission of the publisher.)

tives via a 5α-reductase activity.[43] The kidney ranks second only to the liver in its ability to metabolize steroids. Further, if either an 11β-OHSD or 5α-reductase activity is associated directly with corticosteroid target tissues such as the CCT and is important to the control of steroid action, it is not clear as yet if they prevent binding of active steroid to receptors, or if they are important to turning off a steroid signal. For instance, Fig. 1 shows the time course of binding of a tracer intravenous dose of [^{3}H]aldosterone to specific renal cytoplasmic receptors, as well as receptors associated with nuclear binding sites.[44] Since aldosterone does not bind appreciably to plasma proteins, the half-life of the unmetabolized steroid is reported to be on the order of 20–30 min. Notice that binding to receptors in the cytosol fell with a $t_{1/2}$ similar to the fall in plasma [^{3}H]aldosterone concentrations.

[43] M. McDermott, S. Latif, and D. J. Morris, J. Steroid Biochem. 19, 1205 (1983).
[44] D. Marver, D. Goodman, and I. S. Edelman, Kidney Int. 1, 210 (1972).

In contrast there was a rise in nuclear associated complexes over the first 10 min followed thereafter by a steady decline. Nuclear retention time, however, was not markedly extended relative to plasma concentrations or the concentration of [^{3}H]aldosterone bound to cytoplasmic complexes. On the other hand, the half-life of circulating glucocorticoids is longer than that of aldosterone due to stabilization by CBG. Second, inhibition of 11β-OHSD apparently extends their half-lives. If these glucocorticoids could now occupy the mineralocorticoid receptor, the half-life of the biological response would most likely be significantly enhanced relative to aldosterone actions. Thus a steroid-metabolizing activity close to the site of action of steroid may be important to determining prompt cessation of receptor activation of transcription, to allow the type I site to be sensitive to not only the rapid rise in plasma aldosterone, but also to the rapid fall.

Given this introduction, the discussion above has practical implications when assessing the steroid specificity of type I receptors. First, the reader is advised to consider that any contaminating CBG may significantly reduce free corticosterone levels due to the high concentration of this protein relative to the receptor. Thus a 10-fold concentration of added unlabeled compound B in competition experiments with [^{3}H]aldosterone may represent a substantially lower addition due to sequestration of compound B by CBG. Such an artifact would underestimate the affinity of corticosterone for any of the corticosteroid-binding sites in such an assay. This artifact can be diminished by perfusing kidneys well before use and by ensuring that the volume-to-tissue ratio remains high. Some investigators who use cytosol to monitor type I sites also pretreat their preparations with hydroxylapatite to remove CBG.[30] Since the preference is to use intact tissue, a good control is to simply add a known amount of [^{3}H]corticosterone to parallel well-perfused kidney slices or isolated tubule segments and determine the concentration of free labeled steroid following the assay.

Another significant problem is obviously the metabolism of compound B during the assay. This is most easily minimized by carrying out assays at ice temperature. Although carbenoxolone or glycyrrhetinic acid could be added as inhibitors of 11β-OHSD, a report of their successful use for competition analyses *in vitro* has not been forthcoming as yet, and results may be compromised by their modest affinity for the type I receptor. Furthermore, 11β-OHSD is obviously not the only corticosterone/cortisol-metabolizing activity in kidney.[40] Some laboratories, however, are evaluating the identity and proportion of [^{3}H]corticosteroid metabolites generated during assay by HPLC[45] as a means of determining the extent of steroid metabolism during assay.

[45] R. L. Duncan, W. M. Grogan, L. B. Kramer, and C. O. Watlington, *Am. J. Physiol.* **255**, F736 (1988).

The Cytoplasmic Mineralocorticoid Receptor Assay

Anyone about to embark on an assay of corticosteroid receptor kinetics by Scatchard analyses[46] should first read three papers that deal with binding artifacts and proper interpretation of the Scatchard plot.[47-49] If at all possible, a computer program such as LIGAND[50] should be used to aid in characterization of the labeled sites. To assay mineralocorticoid receptors, generally high specific activity [^{3}H]aldosterone concentrations of 5×10^{-11} to 10^{-7} M are made, $\pm 10^{-5}$ M unlabeled aldosterone to determine nonspecific binding. To reduce binding of [^{3}H]aldosterone to the type II site, incubations can also include about 50 nM RU 26899.[51,52] In a pilot study, choose three concentrations of [^{3}H]aldosterone within each log (such as 1, 2.5, and 5×10^{-10} M) and carry out the binding assays. For the repeat analysis, try to adjust the exact concentrations used so that a minimum of five to seven points are available to determine any one slope on Scatchard plots of bound/free (B/F) vs B. If there appears to be more than one binding site involved, then sufficient concentrations should be used such that these five to seven points do not rely on values close to or at the transition zone between slopes. Furthermore, these minimum five to seven points should contain concentrations at or near the K_d, with several more above and below that value. The investigator must also be prepared to show that equilibrium binding has been achieved at the time and temperature used in order to use the Scatchard analysis. At 4°, this can vary from 2 to 4 hr to overnight.

Either adrenalectomized rat or rabbit renal slices or cytosol may be used. (If isolated nephron segments such as the cortical collecting tubule are to be incubated, then the rabbit may be preferred, since in this animal nephron segments can be hand dissected without the aid of collagenase.) If using the rat, a convenient and well-tolerated anesthetic is chloral hydrate. It should be prepared fresh each day (3.6 g% in saline) and given at a dose of 1 ml/100 g body wt. This is also a good anesthetic agent to use for rat adrenalectomies. The kidneys are perfused via the abdominal aorta with about 30–50 ml of room-temperature saline or 0.25 M sucrose. Kidney slices can be made with a Stadie–Riggs tissue slicer, but are best made by a commercial slicer set for 275-μm-thick sections. Some investigators further

---

[46] G. Scatchard, *Ann. N.Y. Acad. Sci.* **51**, 660 (1949).
[47] G. C. Chamness and W. L. McGuire, *Steroids* **26**, 538 (1975).
[48] K.-J. Chang, S. Jacobs, and P. Cuatrecasas, *Biochim. Biophys. Acta* **406**, 294 (1975).
[49] H. E. Rosenthal, *Anal. Biochem.* **20**, 525 (1967).
[50] P. J. Munson and D. Rodbard, *Anal. Biochem.* **107**, 220 (1980).
[51] M. E. Rafestin-Oblin, M. Lombes, P. Lustenberger, P. Blanchardie, A. Michaud, G. Cornu, and M. Claire, *J. Steroid Biochem.* **25**, 527 (1986).
[52] C. E. Gomez-Sanchez and E. P. Gomez-Sanchez, *Endocrinology (Baltimore)* **113**, 1004 (1983).

mince these slices by hand for better oxygenation of the tissue during incubation with steroid. The slices or minces are rinsed with copious amounts of ice-cold saline or 0.25 M sucrose, and then well drained. Optimize the volume-to-tissue ratio, and attempt to place about the same amount of tissue in each incubation tube or flask. Add tissue to an incubation medium already containing the steroids at the appropriate concentrations. In the past we have used an incubation medium containing (in mM): NaCl, 135; KH_2PO_4, 5; $MgCl_2$, 0.5; $CaCl_2$, 1.0; and Tris base, 5 (pH 7.4 final). Gently mix the fractions during incubation. At the end of the incubation, remove the tissue (minces are best filtered onto nylon cloth, 132 mesh), and retain the residual radioactive medium to determine free [^{3}H]aldosterone concentrations. Do not rely on a calculated free concentration. Homogenize the tissue in 1-2 ml of ice-cold 0.25 M sucrose, 10 mM Tris-HCl buffer, 10 mM sodium molybdate, 1.5 mM ethylenediaminetetraacetic acid (EDTA), pH 7.4, using a Teflon-glass homogenizer. Homogenates are briefly centrifuged in a refrigerated centrifuge at 600 g for 10 min followed by centrifugation of resultant supernatants at 10,000 g for 10 min, and then at 34,000 g for 10 min. Bound steroid in the final supernatant can be separated by at least one of five methods; charcoal-dextran, DE-52 cellulose columns, DE-52/DE-81 filters, G-50 Sephadex columns, or hydroxylapatite.[12,30,53] Specific binding is corrected for protein. If cytoplasmic or cytosolic incubations are preferred (as opposed to slices), then well-perfused kidneys are first homogenized as above, except that a 100,000 g centrifugation (for 1 hr) of the 10,000 g supernatant is required to prepare cytosol. If the reader wishes to avoid molybdate (see nuclear or chromatin-binding assays below), then 20-25% glycerol can be substituted to protect protein receptors from degradation.[13,54] Again the ratio of medium to tissue should remain high to avoid artifacts.[12,48]

It has become increasingly clear that problems with CBG contamination and paracrine metabolism of steroid *in vitro* can be minimized by utilizing isolated nephron structures (such as the CCT) for binding assays. Dissection of the same in rabbit can be accomplished with the anatomical guide written by Kaissling and Kriz.[55] Dissections (performed in Petri dishes resting on blocks of ice) should be completed within 1 hr of sacrifice, and normally 12-16 CCTs (1-2 mm) can be dissected per person per hour. Although the use of collagenase with rabbit kidneys can increase the number of tubules harvested per unit time, this can compromise the preparation due to contaminating protease activity. Thus collagenase

---

[53] D. Marver, *Endocrinology (Baltimore)* **106**, 611 (1980).
[54] G. E. Swaneck, L. L. H. Chu, and I. S. Edelman, *J. Biol. Chem.* **245**, 5382 (1970).
[55] B. Kaissling and W. Kriz, *Adv. Anat. Embryol. Cell Biol.* **56**, 1 (1979).

should be avoided unless absolutely necessary; it is better to have more than one person dissecting any given kidney. If collagenase is used, the investigator may find that DNA leaking from damaged tubules will make them "sticky" and difficult to remove once aspirated into a collection pipet. The dissection medium contains (in mM): NaCl, 115; KCl, 5; NaHCO$_3$, 25; NaH$_2$PO$_4$, 2.3; sodium acetate, 10; MgSO$_4$, 1; glucose, 8; alanine, 5; and CaCl$_2$, 1.8 (pH 7.4 after bubbling with 95% O$_2$–5% CO$_2$). For type I site analyses using [^{3}H]aldosterone with a specific activity of 80 Ci/mmol, it has been possible to carry out Scatchard plots using between 5- and 20-mm total length of CCT per assay.[15] Data are most easily normalized to millimeters of tubule length, determined with an eyepiece micrometer. Total volume of the assay is 5 μl. The tubules are usually harvested in about 2–3 μl of the dissection medium and added to an individual well. Steroid mixes are ultimately added to bring the total volume up to 5 μl, and the assay plate transferred from ice temperatures to room temperature or to a dish of sand equilibrated at the temperature of choice. Generally tubules do better if the temperature does not exceed 28 or 30°. The other option is to carry out the binding assay at ice temperatures. A convenient way to transfer tubules from the dissecting dish to the assay plate is to affix about a 3-in length of PE-50 tubing to the end of a 25- or 50-μl Hamilton Co. syringe (Reno, NV) having a cemented needle and a repeating Hamilton dispensor attached. Mark off a 2- to 3-μl volume on the tip of the PE-50 tubing with a black marking pen. Aspirate your tubules into the PE-50 tubing and eject the volume into the well. The most pupular assay plates for incubations utilizing 1- to 5-μl total volumes are the covered plastic microtest plates (60 × 5-μl wells, Sarstedt Co., #82.602, Princeton, NJ) that have been lined around the walls with moistened filter paper to reduce evaporation during assay. When assays are performed utilizing smaller volumes, water-saturated mineral oil should be used to cover the volume to prevent evaporation. Following the assay, the plate is returned to ice temperature, the tubules are again aspirated, and the tissue collected on a filter apparatus following ejection of the sample (such as a Millipore filter, 0.45-μm pore diameter).[15] To ensure efficient transfer of tissue, both the well and the tubing can be rinsed with another 5 μl of media. Samples harvested on the filters are washed with ice-cold incubation medium (about 1 ml) and then counted for radioactivity. Controls include incubations with steroid and without tissue. Counting times have to be sufficiently long to make accurate judgments about the receptor kinetics. Since the tubules fall to the bottom of the well during the assay, it is also possible to make a direct measurement of free [^{3}H]aldosterone by removing a measured microvolume from the medium above the tubules at the end of the assay, just before transferring the tissue to the filters.

Nuclear Binding Assays

In order to direct gene transcription, corticosteroid receptor complexes have to undergo an activation step that greatly enhances their affinity for key chromatin-binding sites. Thus at times it is important to evaluate the kinetics of binding of ^{3}H-labeled steroid–receptor complexes to another template, such as isolated target tissue nuclei or, better still, isolated chromatin. For instance, although [^{3}H]aldosterone-labeled type I receptors have a high affinity for low-capacity (or specific) nuclear binding sites, [^{3}H]SC 26304-labeled type I receptors do not.[56] This suggests that the spirolactones antagonize aldosterone by competing with this active steroid for cytoplasmic receptors. Once bound, the resultant [^{3}H]spirolactone complex is incapable of undergoing transformation or activation to a state that can be retained by chromatin acceptor sites. Our previous work has suggested that at least part of this chromatin acceptor site for activated complexes is the major groove of DNA since the major groove-intercalating agents proflavine sulfate and ethidium bromide have significant inhibitory effects on the binding of [^{3}H]aldosterone-labeled receptor complexes to chromatin, in contrast to the more minor effects of the minor groove intercalator actinomycin D and the minor groove reporter, netropsin.[57]

Figures 2 and 3 illustrate the mechanism of action of spirolactones such as SC 26304 and the importance of evaluating the affinity of labeled type I receptors for nuclear or chromatin-binding sites. In the top panel of Fig. 2, adrenalectomized rats were injected with 0.3 µg of aldosterone plus 0 to 600 µg SC 26304 (hatched bars) and urinary K^+/Na^+ ratios were determined and compared to diluent injected controls (open bars). Bottom panels represent specific binding of [^{3}H]aldosterone to either receptors within the cytoplasm or bound to nuclei after injection of steroids *in vivo*. The figure shows that specific binding of [^{3}H]aldosterone fell as a function of the dose of added unlabeled SC 26304 as did the physiological response to aldosterone, indexed by the urinary K^+/Na^+ ratios. The fact that no dose of SC 26304 produced a significant rise in the urinary K^+/Na^+ ratio compared to diluent controls suggests that this mineralocorticoid antagonist has minimal agonist activity under the conditions of the assay. It also appeared that [^{3}H]SC 26304-labeled cytoplasmic receptors did not bind to nuclei. When adrenalectomized rats were injected with either [^{3}H]aldosterone or [^{3}H]SC 26304 *in vivo*, both groups of animals had significant labeling of cytoplasmic receptors at 2 min postinjection that fell with time postinjection (see Fig. 1); however, only rats injected with [^{3}H]aldosterone

---

[56] D. Marver, J. Stewart, J. W. Funder, D. Feldman, and I. S. Edelman, *Proc. Natl. Acad. Sci. U.S.A.* **71**, 1431 (1974).

[57] I. S. Edelman and D. Marver, *J. Steroid Biochem.* **12**, 219 (1980).

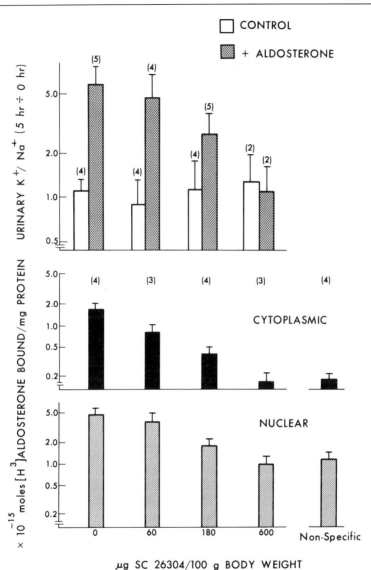

FIG. 2. A comparison of the dose-dependent antagonist effect of spirolactone SC 26304 on an aldosterone-mediated increase in the urinary K^+/Na^+ ratio and on the binding of [^{3}H]aldosterone to renal cytoplasmic and nuclear bound receptor complexes. (From Marver et al.[56] by permission of the publisher.)

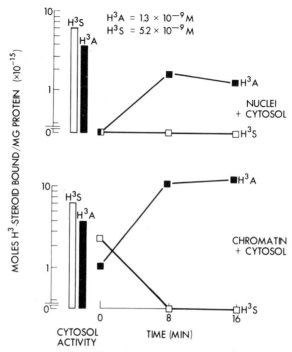

FIG. 3. A comparison of the ability of renal cytoplasmic receptors prelabeled with either [³H]aldosterone (³HA) or [³H]SC 26304 (³HS) to be retained by renal nuclear or isolated chromatin acceptor sites at 25°. (From Marver et al.[56] by permission of the publisher.)

had significant nuclear labeling at either 2 or 10 min postinjection.[56] Figure 3 supports this finding and illustrates an *in vitro* transfer reaction in which renal cytosol prelabeled with either [³H]aldosterone or [³H]SC 26304 was mixed with purified renal nuclei or chromatin fractions. Following incubation at 25°, only the [³H]aldosterone–receptor complexes were retained by high-affinity acceptor sites within the nuclei, or associated with chromatin.

If the reader wishes to evaluate binding to nuclei or chromatin, some slight modifications have to be made in the homogenization buffer given above. Kidneys are homogenized in 0.25 M sucrose–3 mM $CaCl_2$ and filtered through nylon cloth, 132 mesh, to remove large clumps of tissue. These kidneys may already have been labeled with [³H]aldosterone *in vivo,* or may be the source of subcellular fractions for *in vitro* transfer assays. Following homogenization, the 600 g × 10 min pellet is retained, and the supernatant further centrifuged to isolate either a cytoplasmic or cytosolic fraction, as above. The 600 g pellet is thoroughly resuspended in a small volume of homogenizing buffer and layered on top of 30 ml of 2.2 M

sucrose – 3 mM CaCl$_2$ in an ultracentrifuge tube, and the interface gently disturbed. The preparation is then centrifuged in a swinging bucket rotor at 65,000 g for 1 hr. After discarding the dense sucrose, the pellet is resuspended in a small volume of homogenizing medium to allow dispersion; the nuclei can be reharvested when needed by centrifuging the preparation at 600 g.

Whether labeling of nuclei occurs *in vivo* or *in vitro,* two nuclear extracts are obtained following exposure to [^{3}H]aldosterone. The first involves a Tris extract, in which loosely bound receptor complexes are released by osmotically shocking the nuclei by suspension in 4–5 ml of 0.1 M Tris-HCl, 3 mM CaCl$_2$ (pH 7.4) for 10 min on ice. The osmotically shocked nuclei are harvested after centrifugation at 19,000 g for 10 min in a refrigerated centrifuge, and the supernatant treated so as to remove bound vs free steroid. The pellet is then extracted with 0.4 M KCl for 10 min and centrifuged as above to remove tightly bound radioactive receptor complexes.

Purified nuclei are also used to isolate chromatin by the method of Mainwaring and Peterken.[58] All other extraction methods and incubation methods utilizing chromatin were the same as those described for nuclei, except that chromatin requires centrifugation at 10,000 g for 10 min to pellet the insoluble material before, and immediately after, the assay. *In vitro* transfer assays between labeled cytosol and unlabeled nuclei or chromatin must be carried out at greater than ice temperature (usually 25°) to allow for activation of the receptor complexes. Molybdate cannot be used to transfer reactions between isolated cytoplasm and isolated nuclei or chromatin because it also prevents activation of the receptor. If the reader wishes to stabilize the receptor during incubation, then cytosol can be diluted with glycerol to a final concentration of 20–25%. The presence of steroid also stabilizes receptors.

Glycerol Density Gradients

Corticosteroid receptors bound to agonists and antagonists have characteristic equilibrium sedimentation profiles in low and high salt gradients. Because of the lability of type I sites, receptors labeled *in vitro* are added to linear gradients of 10–30% spectroquality glycerol in a final concentration of 0.1 M Tris-HCl (pH 7.2–7.4), 1.5 mM EDTA, 1–10 nM [^{3}H]aldosterone or 1–10 nM [^{3}H]aldosterone plus a 100-fold concentration of unlabeled steroid (low salt gradients) or the same milieu containing 0.4 M KCl (high salt gradients). Parallel gradients might substitute [^{3}H]SC 26304 for [^{3}H]aldosterone. Gradient tubes are centrifuged for 40 hr (4°) at 283,000 g

[58] W. I. P. Mainwaring and B. M. Peterken, *Biochem. J.* **125**, 285 (1971).

in a swinging bucket rotor. Gradient fractions are then individually treated to remove bound vs free steroid. Agonists bound to cytoplasmic receptors characteristically form an 8–9S complex on low salt gradients and a 4.5S complex on high salt gradients. Antagonists such as [^{3}H]SC 26304 form only a 4S receptor complex on low salt gradients and a 3S receptor complex on high salt gradients.[56] The nuclear Tris osmotic shock complex sediments at 3S irrespective of salt concentration in the gradient. In contrast, the 0.4 M KCl extract sediments at 4S in high salt and pellets to the bottom of the tube in low salt.[54] It is not clear if the latter finding is due to other nuclear or ribonuclear proteins extracted with 0.4 M KCl. Although the exact meaning of these transitional states is still debated, glucocorticoid type II sites bound to agonists or antagonists behave in a similar manner.[59] Such analyses are often helpful, since CBG sediments at about 4S on low salt gradients, and thus can be clearly dissociated from 8S [^{3}H]corticosterone-labeled receptor complexes.

Clearance Studies

Clearance studies are usually carried out in adrenalectomized rats rather than rabbits for two practical reasons. The first concerns cost considerations, and the second concerns the difficulty in obtaining complete pre- and post-steroid urine samples in the female rabbit, the sex which has been most widely used in this species. Anatomical considerations make it extraordinarily difficult to catheterize the female, and it is likely that during a 6-hr assay period, there will be no spontaneous voiding of urine. The rabbit is also resistant to any humane coercion to void. Therefore it is the male rat which is normally used for both clearance studies and companion receptor assays. Male rather than female rats are used because some years ago Morris et al.[60] determined that the antinatriuretic and kaliuretic response to aldosterone in the adrenalectomized rat was greater in males than females.

Animals are used 5–7 days postsurgery, and maintained in the interim on normal rat chow and saline drinking water. To enhance the sensitivity of the urinary Na^+/K^+ index, animals are placed on a low potassium diet the night before the clearance assay.[61] On the morning of the study a baseline urine sample is collected by inducing the rat to micturate into a test tube. This is most easily accomplished by exposing the rat to a brief

---

[59] N. Kaiser, R. J. Milholland, R. W. Turnell, and F. Rosen, *Biochim. Biophys. Acta* **125**, 285 (1971).

[60] D. J. Morris, J. S. Berek, and R. P. Davis, *Endocrinology (Baltimore)* **92**, 989 (1973).

[61] G. M. Fimognari, D. D. Fanestil, and I. S. Edelman, *Am. J. Physiol.* **213**, 954 (1967).

whiff of ether or gently pulling up on or gently pinching the tail near its base. This does not cause any discomfort to the animal but is usually very effective with regard to obtaining a urine sample. Rats are then placed in individual metabolism cages with fine mesh screens at the base to prevent loss of any stool into the collected urine. Animals are provided water to drink for the duration of the assay. The substitution of water for saline for this brief period ensures that saline is not accidentally added to the voided urine sample. If antagonists are to be injected along with agonists, they are usually administered at $t = -30$ min, followed by the agonist at $t = 0$ hr. Stock aldosterone or dexamethasone solution in 100% ethanol can be maintained at refrigerator temperatures at $2-5 \times 10^{-3}$ M, and diluted just before assay into sterile saline. About 1 μg/100 g body wt of aldosterone should give a maximal effect on urinary Na^+/K^+ excretion rates.

The spirolactones are generally solubilized just before injection. Due to the low water solubility of some of these compounds and the high doses required to block the action of aldosterone, they may have to be given as a suspension intraperitoneally (ip). We have consistently used SC 26304, available from the G.D. Searle Company (Chicago, IL), for investigational purposes. It is made up in a stock of 4 mg/ml ethanol:propylene glycol:saline in the ratio 2:5:5 by volume. The agonist and antagonist properties of each spirolactone vary and are species specific.[62,63] Thus it is necessary to include a group of rats that obtains the same dose of spirolactone plus only the aldosterone diluent. If injection volumes are small, it also may be necessary to give 3–5 ml of sterile saline ip/100 g body wt to maintain good urine volumes during the period of the assay or to give glucocorticoid-maintenance doses of dexamethasone to the animals the day before and the morning of the assay (about 1.5 μg bid/100 g body wt ip). Once the steroids are injected, the investigator can then follow one of two courses. Either each rat is placed into a metabolism cage and a 0- to 5-hr complete urine collection harvested, or this time is split such that the 0- to 1-hr or 0- to 2-hr urine fraction is first collected and assayed separately. Usually this is ultimately discarded due to the fact that there is a lag time of about 1–2 hr before there is a significant effect of aldosterone on urinary Na^+ and K^+ excretion rates. Therefore the sample of interest is that obtained between 1 and 5 hr or 2 and 6 hr postagonist. Urinary Na^+ and K^+ (U_{Na^+} and U_{K^+}) are determined by flame photometer, and creatinine (U_{Cr}) by the alkaline picric acid method[64]; records of urinary volumes (V)

---

[62] C. Sakauye and D. Feldman, *Am. J. Physiol.* **231**, 93 (1976).

[63] J. W. Funder, D. Feldman, E. Highland, and I. S. Edelman, *Biochem. Pharmacol.* **23**, 1493 (1974).

[64] R. L. Haden, *J. Biol. Chem.* **56**, 469 (1923).

per unit time are maintained. Statistical evaluations have included the simple urinary ratio of Na^+/K^+, $U_{Na^+}(V)$ or $U_{K^+}(V)$ (in mEq/hr), or U_{Na^+} or U_{K^+} divided by urinary excretion of creatinine.[65,66] The latter represents a crude correction for possible changes in glomerular filtration rate (GFR) that can influence K^+ excretion. A superior protocol adds blood samples obtained during the assay to allow for the estimation of the fractional excretion (FE) of both Na^+ and K^+. This requires determination of plasma creatinine, Na^+, and K^+ (P_{Cr}, P_{Na^+}, and P_{K^+}). It is important that the blood not be hemolyzed. $FE_{Na^+}(\%) = (U_{Na^+}/U_{Cr})(P_{Cr}/P_{Na^+})$; FE_{K^+} is determined in a similar manner.

An even more accurate determination of changes in the fractional excretion of sodium or potassium can be obtained by anesthetizing the rats and immobilizing them on a heated table, maintained at body temperature. Cannulas can be placed into a jugular vein, a femoral artery, and the bladder for the infusion of steroids (or inulin or saline to replace urinary losses), sampling of blood, and the collection of urine. Either plasma and urinary creatinine can be monitored for determination of GFR, or inulin can be injected and plasma and urine concentrations of this marker monitored by a cold inulin method.[67] Figure 4 presents data from Horisberger and Diezi[65] that were obtained by the latter method. In this case the investigators injected 0.5 μg dexamethasone/100 g body wt to help maintain good urinary flows and to improve the general well being of the animals before undergoing anesthesia. Aldosterone was given as a bolus of 0.1 μg/100 g body wt, followed by an infusion of 0.1 μg/100 g body wt/hr for the duration of the study. Inulins were used to determine the FE_{Na^+} and GFR.[65] What can be appreciated is the significant fall in FE_{Na^+} with time and the significant rise in $U_{K^+}(V)$. FE_{K^+} was not plotted for these studies. As it happens the purpose of this protocol was to show that actinomycin D inhibits both the antinatriuretic and kaliuretic effects of aldosterone. Thus, these groups are also plotted in this figure.

Biochemical Assays

A cartoon of a CCT principal cell involved in aldosterone-mediated Na^+ reabsorption is shown in Fig. 5. The rate-limiting step in the action of aldosterone on Na^+ transport is thought to depend on an induced protein that increases the lumenal membrane Na^+ permeability. Whether the protein(s) constitute the Na^+ channel per se, or regulate the open probability

[65] J.-D. Horisberger and J. Diezi, *Am. J. Physiol.* **246**, F201 (1984).
[66] T. J. Campen, D. A. Vaughn, and D. D. Fanestil, *Pfluegers Arch.* **399**, 93 (1983).
[67] H. E. Harrisson, *Proc. Soc. Exp. Biol. Med.* **49**, 111 (1942).

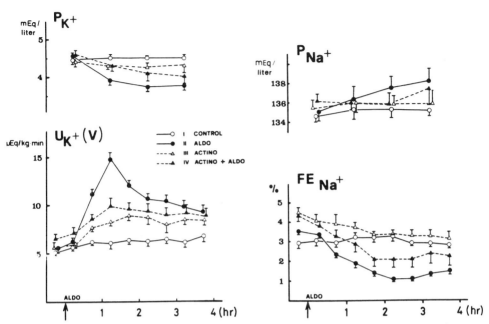

FIG. 4. The influence of intravenous aldosterone (ALDO) on the urinary excretion of K^+ [$U_{K^+}(V)$], the fractional excretion of Na^+ (FE_{Na^+}), and plasma K^+ (P_{K^+}) and Na^+ (P_{Na^+}) as a function of time. Some animals in this study also received actinomycin D (ACTINO). (From Horisberger and Diezi[65] by permission of the authors and the publisher.)

of the channel, or the insertion of sublemmal pools of Na^+ channels, is not known. Amiloride, however, is a well-recognized inhibitor of flux through these channels. In addition to increasing lumenal membrane Na^+ permeability, work of some years ago in toad bladder showed that aldosterone enhances the apparent V_{max} activity of several key Krebs cycle enzymes, including citrate synthase.[68] Since there are relatively few maneuvers that lead to a rise in cellular V_{max} citrate synthase activity, aldosterone-dependent elevations in activity have been used as marker for aldosterone-sensitive segments along the nephron.[69] Investigators have suggested that the rise in activity of key Krebs cycle enzymes allows the capacity of cellular ATP synthesis to be increased in concert with energy demands. Aldosterone also appears to increase both the synthesis and expression of the Na^+ pump subunits in target epithelia such as *Xenopus laevis* kidney A6 cells or

[68] E. Kirsten, R. Kirsten, A. Leaf, and G. W. G. Sharp, *Pfluegers Arch.* **300**, 213 (1968).
[69] D. Marver and M. J. Schwartz, *Proc. Natl. Acad. Sci. U.S.A.* **77**, 3672 (1980).

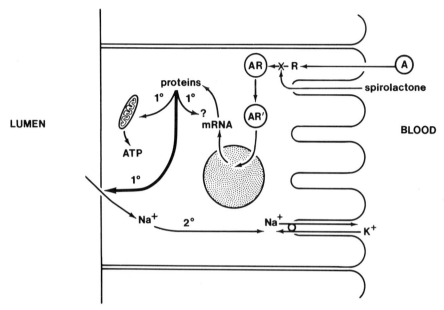

FIG. 5. A schematic of a CCT principal cell in which aldosterone (A) binds to a cytoplasmic receptor (R) and this complex undergoes transformation (AR → AR'), followed by the induction of several proteins. One of these proteins regulates lumenal membrane Na^+ permeability in some manner, while others may be key mitochondrial enzymes such as citrate synthase. Aldosterone also increases Na^+,K^+-ATPase activity in the adrenalectomized rabbit by an amiloride-sensitive mechanism. This suggests that Na^+ may play some permissive role in the processing or activation of this enzyme. The net effect is enhanced Na^+ reabsorption and K^+ secretion.

the rabbit CCT.[70,71] When steriod is given to adrenalectomized rabbits, the increase in V_{max} Na^+,K^+-ATPase activity occurs within 3 hr after 10 μg/kg body wt aldosterone, but not after the administration of the same dose of dexamethasone. Coadministration of the Na^+ channel inhibitor, amiloride, prevents the rise in Na^+,K^+-ATPase activity, but apparently does not block the rise in the abundance of Na^+,K^+-ATPase subunits,[71,72] nor does it block the rise in citrate synthase activity observed following aldosterone administration (Fig. 6). The aggregate amiloride studies have been interpreted as meaning one of two things: some hypothesize that the processing and insertion of Na^+ pumps is somehow dependent on Na^+ entry, and is

[70] F. Verrey, J. P. Kraehenbuhl, and B. C. Rossier, *Mol. Endocrinol.* **3**, 1369 (1989).
[71] K. J. Petty, J. P. Kokko, and D. Marver, *J. Clin. Invest.* **68**, 1514 (1981).
[72] K. Geering, M. Girardet, C. Bron, J. P. Kraehenbuhl, and B. C. Rossier, *J. Biol. Chem.* **257**, 10338 (1982).

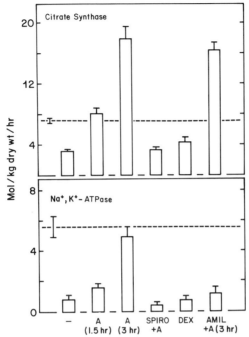

FIG. 6. Citrate synthase and Na^+,K^+-ATPase assays in CCTs from adrenalectomized rabbits injected *in vivo* with either aldosterone (A), spirolactone SC 26304 plus aldosterone (SPIRO + A), dexamethasone (DEX), or amiloride plus aldosterone (AMIL + A). Animals were sacrificed at either 1.5 or 3 hr after steroid injection. The dashed line in each panel represents values in normal rabbit CCTs. (From Marver[11] by permission of Waverly Press.)

therefore secondary to the action of aldosterone on lumenal membrane Na^+ permeability. Others suggest that amiloride acts at another cellular site that is involved in posttranscriptional processing of Na^+,K^+-ATPase (but not citrate cynthase) and that the rise in Na^+,K^+-ATPase activity does not require the permissive action of enhanced Na^+ entry following aldosterone. This remains to be resolved.

Two methods will be described below for Na^+,K^+-ATPase (ultramicro- and microassays) and one for citrate synthase activity (ultramicroassay). The ultramicromethods are about 50–100× more sensitive than the micromethods. They are also about the same order of magnitude more difficult. The ultramicromethod has the advantage that enzyme activities can be assayed along the length of a single CCT, and both ouabain-sensitive and ouabain-insensitive ATPase activity can be determined on the same segment. Furthermore, activity is normalized to tubule dry weight, a more sensitive index than millimeters of tubule length. The sensitivity of the ultramicroassays is such that the investigator is limited only by techniques available to obtain and transfer small amounts of tissue. It has been possible, for instance, to monitor Na^+,K^+-ATPase activity in a single plaque of macula densa cells.[73] The micromethods, on the other hand, usually require 0.5–2 mm of tubule for each assay well. Both types of assay, however, rely on at least one coupling reaction ultimately leading to the production of a fluorescent product.

Before initiating an ultramicro-enzyme assay, the investigator should read the book by Lowry and Passonneau,[74] which not only describes the various enzymatic recycling assays available, but details methods such as how to make hand-blown glass nanoliter transfer pipets, quartz fiber balances for weighing tubules, and various tools for transferring dried tubules before assay, and where to order the vacuum tubes to freeze–dry tissue samples.

For the ultramicroassay, individual tubule segments are rinsed briefly in distilled water immediately following dissection, placed on a glass slide, and the slide in turn placed on a block of dry ice. Once frozen, the slide is placed into a holder which fits within a vacuum drying assembly made by Ace Glass, Inc. (Vineland, NJ). The holder is also part of the drying assembly furnished by the company. The base of the vacuum flask containing the holder is placed into a ethylene glycol–methanol mixture maintained at -50 to $-60°$ and the top attached to a freeze–drying apparatus. It generally takes 48–72 hr to completely freeze–dry the tissue samples. Once dried, they are maintained under vacuum at $-80°$ until assay. On the day of the assay, the flasks are brought to room temperature and the dried segments cut sequentially by hand into approximately 50- to 100-μm lengths under the microscope. Each length is transferred to the tip of a quartz fiber balance on a hair (cemented to a holder with dental wax) from a camel's hair brush. Static electricity is reduced by the use of a small radium-impregnated foil, available commercially (see Ref. 74). It may also be necessary to increase the humidity in the room.

The displacement of the quartz fiber balance is directly proportional to weight of the tubule. These balances are calibrated beforehand by noting the displacement of the fiber by small crystals of quinine hydrobromide, subsequently added to 1 ml of 0.1 N sulfuric acid and fluorescence recorded against quinine hydrobromide standards of known concentrations. Figure 7 illustrates the quartz fiber balance, while Fig. 8 indicates the

[73] J. Schnermann and D. Marver, *Pfluegers Arch.* **407**, 82 (1986).
[74] O. H. Lowry and J. V. Passonneau, "A Flexible System of Enzymatic Analysis." Academic Press, New York, 1972.

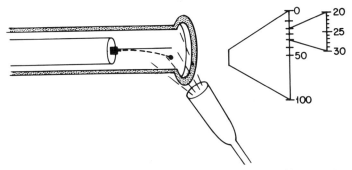

FIG. 7. A schematic of a quartz fiber balance used to weigh pieces of freeze-dried tubules. The displacement of the fiber is proportional to the weight of the tubule (in nanograms).

sensitivity of one of our balances (0.186 ng/20 μm). An individual piece of CCT used for a single assay weighs about 10–25 ng.

Ultramicro-Na$^+$,K$^+$-ATPase Assay

The steps of the ultramicro-Na$^+$,K$^+$-ATPase assay are schematically given in Fig. 9. In brief, the assay is as follows: tissue samples, phosphate standards, or media blanks are transferred into a droplet of medium within a well of a Teflon plate, and the sample immediately covered with water-equilibrated mineral oil. The reaction is initiated by the addition of an Na$^+$, K$^+$, and Mg^{2+}·ATP-containing solution and stopped after incubation at 37° by heating the plate for 4 min on a multiblock heater at 100°.

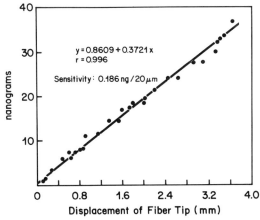

FIG. 8. Demonstration of the sensitivity of the quartz fiber balance.

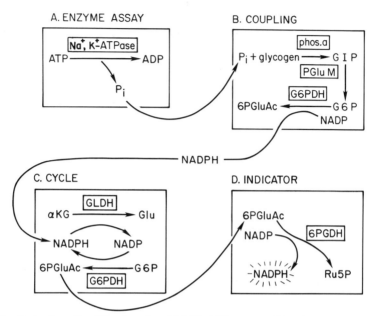

FIG. 9. A schematic of the ultramicro-Na$^+$,K$^+$-ATPase assay. Phosphate generated in step A in the presence of fragments of tubules and excess Na$^+$, K$^+$, and Mg^{2+}·ATP is ultimately coupled to the fluorescent product, NADPH, in step D. phos a, Phosphorylase a; GIP, glucose-1-phosphate; PGluM, phosphoglucomutase; 6PGluAc, 6-phosphogluconolactone; G6PDH, glucose-6-phosphate dehydrogenase; αKG, α-ketoglutarate; GLDH, glutamate dehydrogenase; 6PGDH, 6-phosphogluconate dehydrogenase; Ru5P, ribulose 5-phosphate. For details, see text.

The plates are efficiently maintained at 37° during the ATPase assay by placing them in a dish containing sand equilibrated at this temperature. The final concentrations of constituents of the medium during the ATPase reaction are (in mM): Tris-HCl (pH 7.4), 100; NaCl, 53; KCl, 5; MgCl$_2$, 3; vanadate-free Na$_2$ATP, 3; EDTA, 0.1; and 0.1% bovine serum albumin (BSA) (±1.5 mM ouabain).[71] Due to the sensitivity of this method to phosphate, glassware for reagents must not be cleaned in phosphate-containing detergents. The total volume of this part of the assay (Fig. 9A) is about 300–500 nl, depending upon available nanoliter pipets. The inorganic phosphate generated in the first reaction is then coupled to NADPH equivalents in the presence of NADP$^+$, 5'-AMP, MgCl$_2$, glycogen, glucose-6-phosphate dehydrogenase (G6PDH), phosphoglucomutase (PGluM), and phosphorylase a (phos. a) (Fig. 9B). Glycogen (0.8% stock in 20 mM NaOH) is activated before use by heating at 60° for 10 min. Phosphorylase a is first washed and then activated before use. To wash phosphorylase *a*, approximately 2.8 mg is placed in 1 ml of ice-cold 50 mM low-fluorescence blank imidazole buffer, pH 6.9. The mixture is kept on ice for 30 min and then centrifuged at 12,000 *g* for 8 min. This is repeated three times. The final pellet is resuspended in 0.5 ml of an activation buffer containing (in mM): imidazole buffer, 50; dithiothreitol, 0.5; 5′-AMP, 0.1; EDTA, 1; and 0.02% BSA, pH 6.9, and incubated 1 hr at 37°. The activated enzyme is useable for several days if kept in the refrigerator. The enzymes G6PDH and PGluM are also washed several times by repeated precipitation with ammonium sulfate solutions. Approximately 3800 nl of a mixture (B) containing these enzymes, substrates, and cofactors is added to part A. The concentrations of the reactants in B (in mM) are as follows: imidazole buffer, 50; NADP$^+$, 0.03; 5′-AMP, 0.01; MgCl$_2$, 0.5; and EDTA, 1; also included are glycogen, 0.08%; BSA, 0.05%; G6PDH, 3 μg/ml; PGluM, 9 μg/ml; and phosphorylase *a*, 100 μg/ml (pH 6.9). This coupling step is so critical that it is necessary to pretest the reaction mix before addition to the microassay. This is done by the addition of 8 nmol of phosphate to 450 μl of B mix, and determining both the length of time it takes to complete the reaction and the extent of conversion of phosphate to NADPH. A conversion of 80% or more is considered acceptable. If less, phosphorylase *a* has to be reactivated, or a new batch of enzyme prepared as above. Generally the reaction is complete in 30–60 min at 37°. When the mix is added to the sample wells and incubated for the same length of time, residual NADP$^+$ is ultimately destroyed by the addition of approximately 3800 nl of 0.17 N NaOH and heating for 30 min at 80°. The NADPH, stable in alkali, is then used in a cycling reaction (Fig.

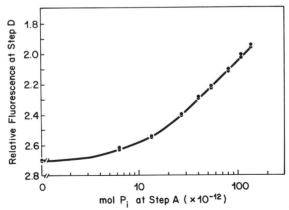

Fig. 10. A standard curve for the ultramicro-Na$^+$,K$^+$-ATPase assay relating phosphate entered in step A with the rise in fluorescence in step D.

9C) containing α-ketoglutarate (αKG), glucose 6-phosphate (G6P), ammonium acetate, glutamate dehydrogenase (GLDH), and G6PDH. To accomplish this, approximately 280 nl of each sample is added to the bottom of a 1.5-ml snap-cap Eppendorf tube maintained on ice. Then 30 μl of mix C is added containing (in mM): Tris-HCl, 100; αKG, 5; G6P, 1; ammonium acetate, 20; and 5'-ADP, 0.3 (pH 8.0). The solution also has 0.02% BSA, 150 μg/ml GLDH, and 35 μg/ml G6PDH. Because of the sensitivity of this reaction to ammonium sulfate, both the GLDH and G6PDH aliquots have to be centrifuged to remove ammonium sulfate before addition to the mix. After cycling for about 40 min at 30°, the reaction is stopped by incubating the tubes in a boiling water bath for 4 min. Then one of the products of the cycle, 6-phosphogluconolactone, is coupled to the generation of NADPH in the presence of NADP$^+$ and 6-phosphogluconate dehydrogenase (6PGDH) (Fig. 9D). This is done by the addition of 1 ml of a mix containing (in mM): Tris-HCl, 20; EDTA, 0.1; ammonium acetate, 60; MgCl$_2$, 0.003; NADP$^+$, 0.1, and 0.05% BSA plus 1 μg/ml 6PGDH (pH 8.0). After 30 min at 37°, the fluorescent product NADPH in each sample is contrasted with fluorescence observed in phosphate standards added into the initial reaction, once the samples have cooled to room temperature. Figures 10 and 11 provide a typical standard curve for P$_i$ and the linearity of activity with regard to time of incubation in part A, and tubule weight.

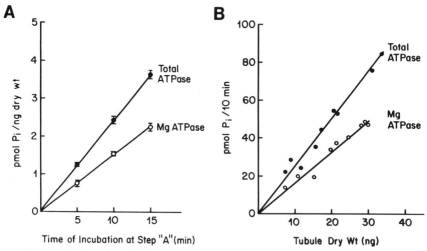

FIG. 11. The linearity of ATP hydrolysis in CCTs as a function of time of incubation in step A of the ultramicroassay (A) and as a function of tubule dry weight (B). (From Petty et al.[71] by permission of the publisher.)

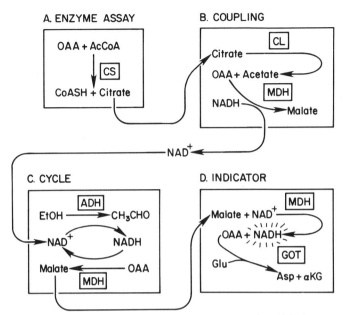

FIG. 12. A schematic of the ultramicro-citrate synthase assay, in which citrate generated in the presence of tubules and excess substrate (step A) is ultimately coupled to the fluorescence of NADH in step D. OAA, oxaloacetate; AcCoA, acetyl-CoA; SC, citrate synthase; CoASH, coenzyme A; CL, citrate lyase; MDH, malate dehydrogenase; ADH, alcohol dehydrogenase; GOT, glutamate–oxaloacetate transaminase. For further details, see Fig. 13. (From Marver and Schwartz[69] by permission of the publisher.)

Micro-Na$^+$-ATPase Assay

Tubule lengths are measured with an eyepiece micrometer. Approximately 1–2 mm is used per assay. The sample is aspirated with 2.5 μl of dissecting solution (using a Hamilton syringe fitted with PE50 tubing as described above) and transferred to the very bottom of a 1.5-ml snap-cap Eppendorf centrifuge tube containing 2.5 μl distilled water. The tubes are rapidly frozen in dry ice–acetone and thawed twice. The tubes are placed in an ice bath and 5 μl of a mix added, such that the final concentrations of the reactants are (in mM) as follows: NaCl, 60; choline chloride, 60; KCl, 30; low-fluorescence blank imidazole buffer, pH 7.4, 50; MgCl$_2$, 5; vanadate-free Na$_2$ATP, 5; ascorbic acid, 1; EDTA, 1; phosphoenolpyruvic acid, 10; NADH, 1; and 2 U/ml pyruvate kinase plus 1.4 U/ml lactic dehydrogenase (±1.5 mM ouabain). Some tubes contain reagent blanks or 0–4000 pmol Na$_2$ADP. Standard curves are carried out in duplicate, using either reagent with or without ouabain. Following addition of the mix, tubes are placed in a 37° water bath for 15–30 min. The reaction is stopped by the

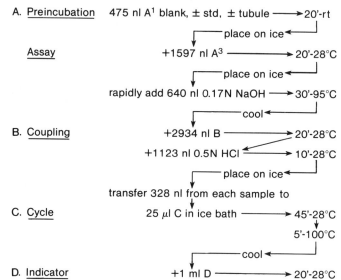

FIG. 13. Details of the ultramicro-citrate synthase assay. Compositions of solutions added are as follow: A^1, 0.6 M KCl, 50 mM 2-amino-2-methyl-1,3-propanediol buffer at pH 8.8, 0.5% BSA; A^3, 50 mM Tris-HCl at pH 7.8, 1.55 mM oxaloacetate, 1.04 mM acetyl-CoA, 0.05% BSA; B, 200 mM Tris-HCl at pH 7.0, 80 μM $ZnCl_2$, 100 μM NADH, 4 μg bacterial citrate lyase/ml, 1 μg malate dehydrogenase/ml; 0.02% BSA; C, 100 mM Tris-HCl (pH 8.0), 300 mM ethanol, 2 mM 2-mercaptoethanol, 2 mM oxaloacetate, alcohol dehydrogenase, and malate dehydrogenase; D, 25 mM 2-amino-2-methyl-1,3-propanediol buffer (pH 9.9), 10 mM glutamate, 200 μM NAD^+, 2 μg malate dehydrogenase/ml, and 2 μg glutamate-oxaloacetate transaminase/ml. (From Marver and Schwartz[69] by permission of the publisher.)

addition of 7.5 μl of 1 N HCl, and the samples placed back into ice. The Eppendorf tubes are centrifuged in a Eppendorf microfuge for 10 sec to ensure that the droplet of acid is coalesced with the droplet containing the tubule or standard. After allowing the tubes to sit 20 min at room temperature to destroy excess NADH, 270 μl of 6 N NaOH is added to each tube while vortexing the sample, and the tubes placed in a 60° water bath for 20 min (in the dark). Then 710 μl of ice-cold distilled water is added. When the samples are at room temperature, they are transferred to cuvettes and read on a fluorometer (340 μm excitation, 480 emission) under reduced lighting conditions.[75]

Ultramicro-Citrate Synthase Assay

Nephron segments are freeze-dried, cut, and weighed as indicated above for the ultramicroassay of Na^+,K^+-ATPase. A schematic for the

[75] R. G. O'Neil and W. P. Dubinsky, *Am. J. Physiol.* **247**, C314 (1984).

assay of this enzyme is shown in Fig. 12, and the actual volumes, temperatures, and constituents are shown in Fig. 13. For this assay, only the enzyme malate dehydrogenase requires washing by precipitation several times with ammonium sulfate. Finally, Fig. 14 shows a typical standard curve for citrate, and the linearity of product formation as a function of time of incubation and tubule weight.[69]

[³H] Ouabain-Binding Analyses

These studies are best done in intact rabbit nephron segments, due to the low affinity of ouabain for rat Na^+,K^+-ATPase. Under B_{max} conditions, the amount of specifically bound ouabain per millimeter tubule length should count the number of Na^+ pumps, since only one ouabain binds per pump. In combination with V_{max} Na^+,K^+-ATPase ATP hydrolysis measurements or V_{max} $^{86}Rb^+$ uptake measurements in intact tubules, the maximal turnover per pump can be calculated. Tubules are incubated in $1-5$ μl of medium containing 0.25 M sucrose, 3 mM $NaSO_4$, 3 mM Na_2HPO_4, 1 mM sodium orthovanadate, 10 mM Tris-HCl (pH 7.4), and 10^{-5} M [³H]ouabain $\pm$ 1 mM unlabeled ouabain (10^{-6} to 10^{-5} M both saturate the high-affinity site and completely inhibits Na^+,K^+-ATPase in the rabbit CCT). Following 45 min at 30° (or that amount of time required to achieve binding equilibrium), the tubules are cooled, rinsed with medium without ouabain, and incubated for 1 hr at 0°.[76] This reduces the extent of nonspecific binding of [³H]ouabain. Samples are counted for radioactivity to determine the apparent number of Na^+ pumps.

Conclusion

This chapter has outlined various receptor assays, an *in vivo* clearance assay, and biochemical assays for Na^+,K^+-ATPase and citrate synthase. With the availability of cDNA probes to monitor message levels for the mineralocorticoid receptor and for Na^+,K^+-ATPase, it should be possible to significantly advance our understanding of the regulation of gene expression by the mineralocorticoid receptor in the future. Also, ongoing studies regarding the role of steroid-metabolizing enzymes should aid our understanding of the sensitivity of the type I receptor to circulating aldosterone as opposed to circulating glucocorticoid. Furthermore, recent developments suggest that we will soon know more about the subcellular transduction pathways that are involved in control of both lumenal membrane Na^+ permeability and the processing of Na^+,K^+-ATPase in aldosterone target tissues.

[76] A. Doucet and C. Barlett, J. Biol. Chem. **261**, 993 (1986).

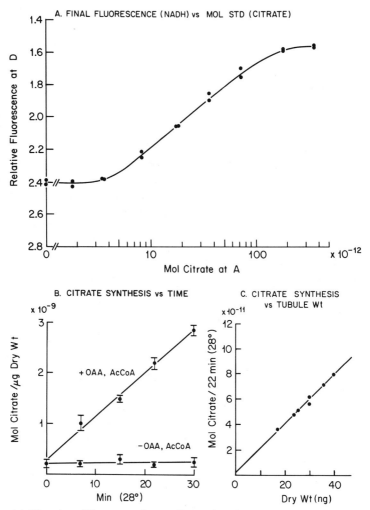

FIG. 14. Linearity of fluoresence in step D as a function of citrate in step A (panel A), linearity of citrate synthesis as a function of incubation time with pieces of CCT segments (panel B), and linearity of citrate synthesis as a function of nanograms tubule added in step A of the reaction (panel C). (From Marver and Schwartz[69] by permission of the publisher.)

Acknowledgments

The author wishes to thank Dr. Isidore S. Edelman, who has been an innovative leader as well as a catalyst in the field of aldosterone research through the continuing work of his many trainees.

[32] The Cellular Action of Antidiuretic Hormone

By DENNIS BROWN, ALAN VERKMAN, KARL SKORECKI, and DENNIS AUSIELLO

Introduction

Binding of the antidiuretic hormone, vasopressin, to receptors on the basolateral plasma membrane of kidney collecting duct principal cells stimulates a cascade of events that results in a rapid and dramatic increase in the water permeability of the apical plasma membrane of these cells. This increase in permeability allows the developing urine within the tubule lumen to equilibrate osmotically with the hypertonic interstitium of the renal medulla, and represents an important mechanism for water conservation by terrestrial vertebrates. The importance of this regulatory action of vasopressin can be seen in mutant strains of rats and mice that either do not produce vasopressin (*hypothalamic* diabetes insipidus—Brattleboro rats)[1] or do not respond to vasopressin (mice with *nephrogenic* diabetes insipidus).[2] These animals produce an amount of urine equivalent to their body weight each day, and must make up for this enormous fluid loss by constantly drinking. Both central and nephrogenic forms of diabetes insipidus occur in man.

A great deal of effort has been expended to understand how vasopressin causes a plasma membrane that is unusually impermeable to water to become water permeable within a relatively short period of time. This effort has resulted in the development of a model for the action of vasopressin that involves the movement of putative "water channels" to and from the apical plasma membrane during hormonal stimulation and hormone withdrawal. This recycling hypothesis has been the subject of a number of reviews,[3-7] and is schematically illustrated in Fig. 1. Much of our knowledge concerning vasopressin action was obtained using vasopressin-sensitive amphibian epithelia—the frog and toad urinary bladder and epidermis—but in this chapter we will concentrate mainly on work carried out directly on the kidney collecting duct, which has been the main focus

---

[1] H. Valtin, *Ann. N.Y. Acad. Sci.* **394**, 1 (1982).
[2] D. Naik and H. Valtin, *Am. J. Physiol.* **217**, 1183 (1969).
[3] J. B. Wade, D. L. Stetson, and S. A. Lewis, *Ann. N.Y. Acad. Sci.* **372**, 106 (1981).
[4] R. M. Hays, *Am. J. Physiol.* **245**, C289 (1983).
[5] J. S. Handler, *Am. J. Physiol.* **235**, F375 (1988).
[6] D. Brown, *Am. J. Physiol.* **256**, F1 (1989).
[7] A. S. Verkman, *Am. J. Physiol.* **257**, C837 (1989).

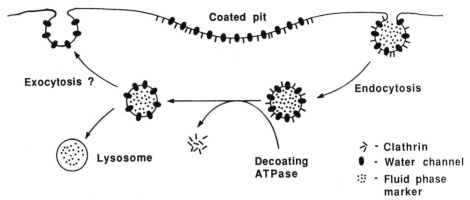

FIG. 1. Diagram of proposed pathway of water channel recycling in collecting duct principal cells. Intramembranous particles that are believed to represent water channels are concentrated in clathrin-coated pits at the cell surface and are endocytosed in coated vesicles. These vesicles are rapidly decoated, and while their fluid-phase content may eventually reach multivesicular lysosomes, the water channels themselves may escape degradation and be recycled back to the apical membrane. The nature of the vesicles carrying water channels to the membrane has not been elucidated in the collecting duct, nor have sites of fusion of these vesicles at the apical membrane been described following vasopressin administration in this tissue. (From Brown[6] by permission.)

of studies in our laboratories. Where applicable, work using amphibian models will also be described.

Investigations on the mechanism of action of vasopressin have used a particularly wide variety of techniques, from target analysis studies on hormone–receptor interaction at the basolateral plasma membrane, to morphological studies on vasopressin-induced rearrangement of apical plasma membrane components of target cells. Recent work by our laboratories has made use of biophysical techniques to study directly the water permeability of fluorescently labeled endosomes from target cells, and fluorescent techniques are now being applied to measure the water permeability of isolated perfused collecting ducts with rapid time resolution. The aim of this chapter is to describe in more detail the techniques that have been used in our studies on vasopressin action, and to describe briefly the information that has been gained by this multifaceted approach to a cell biological problem.

Morphological Studies

Conventional Freeze–Fracture

Freeze–fracture has made an invaluable contribution to the study of membrane structure because in these replicas, the interior, hydrophobic portion of biological membranes is revealed.[8] This appears as a smooth matrix in which intramembranous particles (IMPs) are embedded. Intramembranous particles are believed to represent integral membrane proteins, and reconstitution experiments have shown that many proteins can produce IMPs when added to otherwise smooth, pure lipid liposomes.[9-11] Freeze-fracturing involves freezing a tissue specimen at low temperature (about $-150°$), fracturing the frozen specimen with a cooled microtome blade, and shadow casting the frozen, fractured surface with platinum and carbon.[12] Both the fracturing and shadowing processes are carried out at high vacuum. Typically, the tissue to be examined is fixed with glutaraldehyde (often 2%), chopped into smaller pieces and infiltrated with 30% glycerol in buffer for 1–2 hr to protect the tissue against damage caused by ice crystal formation during freezing. Once infiltrated, pieces of tissue are placed on copper or gold supports and frozen in Freon 22 cooled with liquid nitrogen. They are then placed in a freeze-fracture device such as a Cressington CFE40 (Cressington Scientific Instruments, Watford, England) and fractured at about $-110°$, at a vacuum of better than 10^{-6} torr. The fracturing can be performed either using a cooled microtome blade, or in a specialized double-replica device that snaps open and splits the specimen into two complementary parts. The fractured surface of the tissue is replicated by first evaporating a 2-nm layer of platinum (at 45° to the surface) to give contrast, followed by a 20-nm layer of carbon (at 90° to the surface) to give mechanical strength to the replica. The underlying tissue is removed by floating the replica on a drop of sodium hypochlorite bleach, and the replica can be cleaned further with a chloroform/methanol mixture and, finally, distilled water. Replicas are picked up from the surface of water onto copper microscope grids.

Freeze-fracture electron microscopy has revealed that specialized membrane domains known as IMP clusters (or aggregates) appear on the apical plasma membrane of sensitive cells (Fig. 2) in parallel with the increased transepithelial flow of water induced by vasopressin.[13-16] In the toad bladder and the toad epidermis, a correlation has been shown between

---

[8] D. Branton, *Proc. Natl. Acad. Sci. U.S.A.* **55**, 1048 (1966).
[9] D. W. Pumplin and D. M. Fambrough, *J. Cell Biol.* **97**, 1214 (1983).
[10] W. J. Vail, D. Papahadjopoulous, and M. A. Moscarello, *Biochim. Biophys. Acta* **345**, 463 (1974).
[11] J. Yu and D. Branton, *Proc. Natl. Acad. Sci. U.S.A.* **73**, 3891 (1976).
[12] H. Moor, K. Muhlethaler, H. Waldner, and A. J. Frey-Wyssling, *J. Biophys. Biochem. Cytol.* **10**, 1 (1961).
[13] W. A. Kachadorian, J. B. Wade, and V. A. DiScala, *Science* **190**, 67 (1975).
[14] M. C. Harmanci, P. Stern, W. A. Kachadorian, H. Valtin, and V. A. DiScala, *Am. J. Physiol.* **235**, F440 (1978).
[15] D. Brown, A. Grosso, and R. C. DeSousa, *Am. J. Physiol.* **245**, C334 (1983).
[16] J. Chevalier, J. Bourguet, and J. S. Hugon, *Cell Tissue Res.* **152**, 129 (1974).

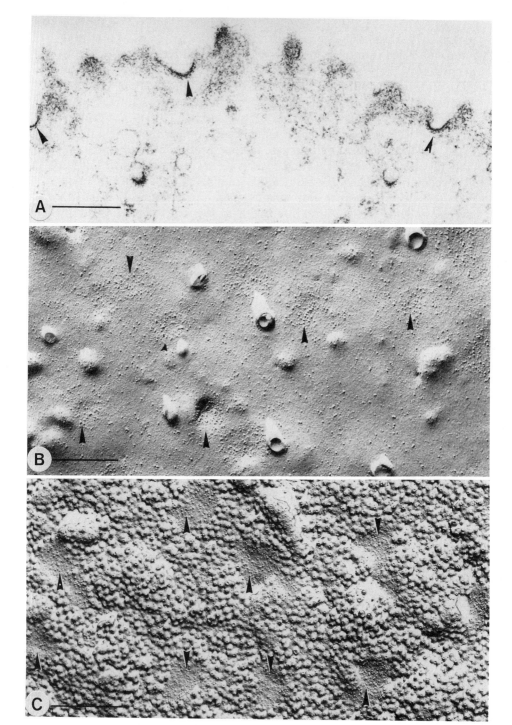

the magnitude of the transepithelial water flow and the frequency of IMP clusters on the apical plasma membrane.[15,17] In the collecting duct, IMP clusters (Fig. 2B) are absent from principal cells of mice with nephrogenic diabetes insipidus[18] and, in Brattleboro rats, the frequency of apical membrane clusters is directly related to the vasopressin-induced increase in urine osmolality upon hormone infusion.[19] This evidence is believed to indicate that the IMP clusters contain water channels and are, therefore, water-permeable patches in the apical plasma membrane. Similar IMP clusters are also found on the limiting membranes of cytoplasmic vesicles in amphibian bladder[20,21] and epidermis, and it was proposed that these vesicles act as a reservoir of water channels that can be rapidly inserted into the apical plasma membrane, by exocytosis, upon vasopressin stimulation. Vesicle fusion events at the apical membrane increase sharply in the presence of vasopressin, and IMP clusters are frequently associated with sites of vesicle fusion.[22] Following vasopressin withdrawal, the apical plasma membrane returns to an impermeable state and this correlates with a decrease in the number of IMP clusters and an increase in endocytosis of apical membrane patches.[23,24] Using fluorescent tracers, we have shown directly that water channels are endocytosed during vasopressin-induced

---

[17] W. A. Kachadorian, S. D. Levine, Wade, J. B., V. A. DiScala, and R. M. Hays, *J. Clin. Invest.* **59**, 576 (1977).

[18] D. Brown, G. I. Shields, H. Valtin, J. F. Morris, and L. Orci, *Am. J. Physiol.* **249**, F582 (1985).

[19] M. C. Harmanci, P. Stern, W. A. Kachadorian, H. Valtin, and V. A. DiScala, *Am. J. Physiol.* **239**, F560 (1980).

[20] F. Humbert, R. Montesano, A. Grosso, R. C. DeSousa, and L. Orci, *Experientia* **33**, 1364 (1977).

[21] J. B. Wade, *J. Membr. Biol.* **40** (Spec. Issue), 281 (1978).

[22] J. Muller, W. A. Kachadorian, and V. A. DiScala, *J. Cell Biol.* **85**, 83 (1980).

[23] R. A. Coleman, H. W. Harris, and J. B. Wade, *J. Histochem. Cytochem.* **35**, 1405 (1987).

[24] H. W. Harris, J. B. Wade, and J. S. Handler, *J. Clin. Invest.* **78**, 703 (1986).

FIG. 2. This plate shows lumenal plasma membranes of collecting duct principal cells from Brattleboro rats treated with vasopressin for four consecutive days. (A) A thin section; (B and C) freeze-fracture replicas. Vasopressin treatment induces the appearance of IMP clusters (arrowheads) on these membranes, in parallel with an increase in urine osmolality. When vasopressin-treated kidneys are exposed to filipin before fracture, the typical filipin–sterol complexes are abundant in the apical membrane, but are absent from the IMP clusters [arrowheads in (C)]. This property can be used to identify the IMP clusters in thin sections, since most of the apical plasma membrane is disrupted by filipin (A), while the regions corresponding to IMP clusters retain a trilaminar structure. In (A), three such regions are visible (arrowheads) and they all correspond to coated pits. In this way, it was shown that vasopressin-induced IMP clusters are located in coated pits on the cell surface. Bars = 0.5 μm.

membrane recycling in kidney collecting ducts and toad urinary bladders.[25,26]

Freeze–Fracture after Filipin Treatment of Tissue

The polyene antibiotic, filipin, has been widely used as a morphological probe for the detection of cholesterol in biological membranes.[27,28] Exposure of tissues to filipin, even after chemical fixation, results in the formation of filipin–sterol complexes (FSC) in the membrane that are easily visible both in freeze–fracture replicas (Fig. 2C) and in conventional thin sections. In both artificial and natural membranes treated with filipin, there is a close correlation between the number of FSC detected morphologically and the cholesterol content of the membrane.[29–31] The pattern of filipin labeling of several cell types and subcellular compartments has, therefore, generally been assumed to reflect differences in the cholesterol content between different membrane domains. However, recruitment of bimolecular filipin–sterol complexes into larger, morphologically detectable FSC complexes (about 20–30 nm in diameter) is dependent on the ability of these smaller complexes to migrate laterally and coalesce in the plane of the membrane. The appearance of detectable, large FSC can be inhibited in membranes by factors that prevent lateral mobility, such as low temperature, cytoskeletal constraints, or a high local density of IMPs.[32,33] In addition, some membrane-associated proteins, including clathrin, can also inhibit complex formation.[34] This property of clathrin was of importance in the discovery that ADH-induced IMP aggregates, presumed to be water channels, are located in clathrin-coated pits (Fig. 2) on the surface of collecting duct principal cells.[35] For this purpose, rat kidneys were fixed in glutaraldehyde (2.5% in phosphate-buffered saline or 0.1 M sodium cacodylate buffer) and were then cut into thin, 20- to 30-μm

---

[25] A. S. Verkman, W. I. Lencer, D. Brown, and D. A. Ausiello, *Nature (London)* **333**, 268 (1988).
[26] L.-B. Shi and A. S. Verkman, *J. Gen. Physiol.* **94**, 1101 (1989).
[27] P. M. Elias, D. S. Friend, and J. Goerke, *J. Histochem. Cytochem.* **27**, 1247 (1979).
[28] N. J. Severs and H. Robenek, *Biochim. Biophys. Acta* **737**, 373 (1983).
[29] A. J. Verkleij, B. De Kruijff, W. J. Gerritsen, R. A. Demel, L. L. M. Van Deenen, and P. H. J. Ververgaert, *Biochim. Biophys. Acta* **291**, 577 (1973).
[30] T. W. Tillack and S. C. Kinskey, *Biochim. Biophys. Acta* **323**, 43 (1973).
[31] D. S. Friend and E. L. Bearer, *Histochem. J.* **13**, 535 (1981).
[32] D. Brown, R. Montesano, and L. Orci, *J. Histochem. Cytochem.* **30**, 702 (1982).
[33] C. A. Feltkamp and A. W. M. Van der Waarden, *Exp. Cell Res.* **140**, 289 (1982).
[34] R. Montesano, A. Perrelet, P. Vassalli, and L. Orci, *Proc. Natl. Acad. Sci. U.S.A.* **76**, 6391 (1979).
[35] D. Brown and L. Orci, *Nature (London)* **302**, 253 (1983).

slices with a vibratome. These slices were incubated in 150 μM filipin in sodium cacodylate buffer overnight at room temperature, in the dark, and were then prepared for freeze-fracture as described above.

When filipin was applied to collecting ducts of Brattleboro rat kidneys that had been treated with vasopressin, small pits and protuberances typical of FSC were present throughout most of the membrane, but were specifically absent from the vasopressin-induced clusters (Fig. 2C). This absence of FSC enabled these membrane domains to be identified in thin sections of filipin-treated tissue, since these regions retained a normal, trilaminar appearance while the rest of the membrane was disrupted by the presence of the FSC (Fig. 2A). In this way, IMP clusters were found to be located in clathrin-coated pits, which are involved in endocytotic events in other cell types. Quantitative studies revealed a 10-fold increase in the number of coated pits in thin sections of vasopressin-treated principal cells, in parallel with the appearance of IMP clusters seen by freeze-fracture of vasopressin-treated tissue.[35]

Thus, the application of freeze-fracture and thin-sectioning techniques allowed the detection of a morphological endpoint of vasopressin action, i.e., the appearance of IMP clusters on apical plasma membranes in parallel with the hormonally induced increase in water permeability. The fact that, in the collecting duct, these IMP clusters are located in coated pits implies that membrane segments containing water channels are constantly recycled during vasopressin action, a prediction that was confirmed in subsequent experiments using horseradish peroxidase and fluorescent molecules as markers of endocytosis in the kidney.[25,36]

Tracer Studies Using Horseradish Peroxidase and Fluorescent Probes

Horseradish Peroxidase. At the ultrastructural level, horseradish peroxidase (HRP) can be used as a tracer for endocytosis. Its use was first described in the kidney by Strauss,[37] and then at the electron microscope level by Graham and Karnovsky.[38] Because of our observation that vasopressin increases the number of coated pits in principal cells, we injected HRP into the jugular vein of Brattleboro rats, treated and not treated with vasopressin, and quantified the number of HRP-labeled endocytotic vesicles in collecting duct principal cells (Fig. 3) under these conditions.[36] Kidneys were fixed by perfusion with 2% glutaraldehyde 15 min after injection of HRP, and slices of kidney were incubated with 1 mg/ml diaminobenzidine and 0.03% H_2O_2 for 10 min to reveal sites of HRP

---

[36] D. Brown, P. Weyer, and L. Orci, *Eur. J. Cell Biol.* **46,** 336 (1988).
[37] W. Strauss, *J. Cell Biol.* **21,** 295 (1964).
[38] R. C. J. Graham and M. J. Karnovsky, *J. Histochem. Cytochem.* **14,** 291 (1966).

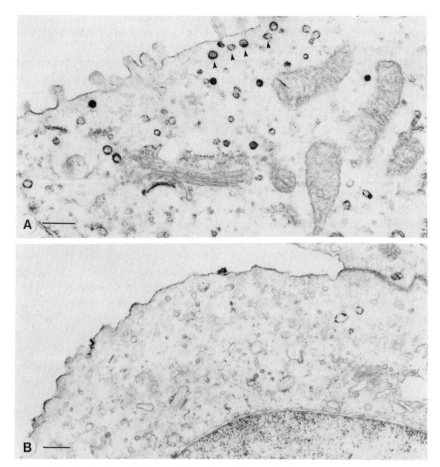

FIG. 3. Thin sections of the apical region of two principal cells, one from a normal Long-Evans rat (A) and one from a vasopressin-deficient Brattleboro rat (B). Both animals were injected with horseradish peroxidase for 15 min prior to fixation of the kidneys and subsequent incubation to reveal the peroxidase reaction product. In the normal animal, many apical vesicles are labeled with HRP reaction product, showing an active endocytotic uptake of this tracer by this cell. These vesicles are believed to be involved in the endocytosis and subsequent recycling of water channels between the cytoplasm and the plasma membrane. In the Brattleboro rat, vesicular recycling is considerably reduced, as shown by the paucity of HRP-labeled vesicles in the apical cytoplams. Bars = 0.5 μm. (From Brown et al.[36] by permission.)

accumulation in vesicles. After a 1-hr incubation with 1% osmium tetroxide, tissues were dehydrated in graded ethanol and embedded in LX-112 epoxy resin as for conventional electron microscopy.

In this way, we showed that vasopressin induces a five- to sixfold increase in the number of HRP-labeled vesicles in the apical cytoplasm of principal cells from Brattleboro rats.[36] Ultrastructural examination showed that this uptake occurred primarily via clathrin-coated pits, although due to the rapid decoating of vesicles once they are free in the cytoplasm[39] the bulk of the HRP-labeled vesicles in the cytoplasm are smooth (uncoated) vesicles. Studies on isolated, perfused cortical collecting ducts from rabbit also demonstrated the involvement of coated vesicles in endocytosis, and showed a marked enhancement of HRP uptake after vasopressin withdrawal,[40] similar to previous findings in the toad urinary bladder.[3,24]

Fluorescent Probes of Endocytosis. In the same way that HRP can be used as an ultrastructural tracer of endocytosis, fluorescent probes can be used as tracers for light microscopy. When fluorescein dextrans are applied to the surface of epithelial cells, they are internalized by the cells and are trapped inside vesicles that are in the process of endocytosis during the time the marker is present (Fig. 4). Fluorescein isothiocyanate (FITC)-dextran has been used in the toad urinary bladder as a marker of endocytosis, and it was reported that endocytosis is markedly increased during vasopressin withdrawal in the presence of a transepithelial osmotic gradient, i.e., a dilute Ringer's solution on the mucosal surface, and a more concentrated Ringer's solution on the serosal surface.[24] We have applied fluorescent tracers to the study of vasopressin-induced recycling of water channels in collecting duct, and have developed a methodology that allows FITC-dextran to be used as a probe for the morphological and functional study of endosomes in the kidney.[41] Briefly, FITC-dextran [0.5-1 ml of a 25 mg/ml solution in phosphate-buffered saline (PBS)/100 g body wt] is injected into the jugular vein of Brattleboro rats, treated or untreated with vasopressin. For functional studies, a crude microsomal fraction of kidney is prepared (see next section) and functional measurements of water permeability, proton pumping, and nonelectrolyte permeability can be made on the vesicles in which FITC-dextran is entrapped. For morphological studies, the kidneys are perfused with paraformaldehyde-lysine-periodate

---

[39] J. E. Rothman and S. L. Schmidt, *Cell (Cambridge, Mass.)* **46**, 5 (1986).
[40] K. Strange, M. C. Willingham, J. S. Handler, and H. W. Harris, Jr., *J. Membr. Biol.* **103**, 17 (1988).
[41] W. I. Lencer, P. Weyer, A. S. Verkman, D. A. Ausiello, and D. Brown, *Am. J. Physiol.* **258**, C309 (1989).

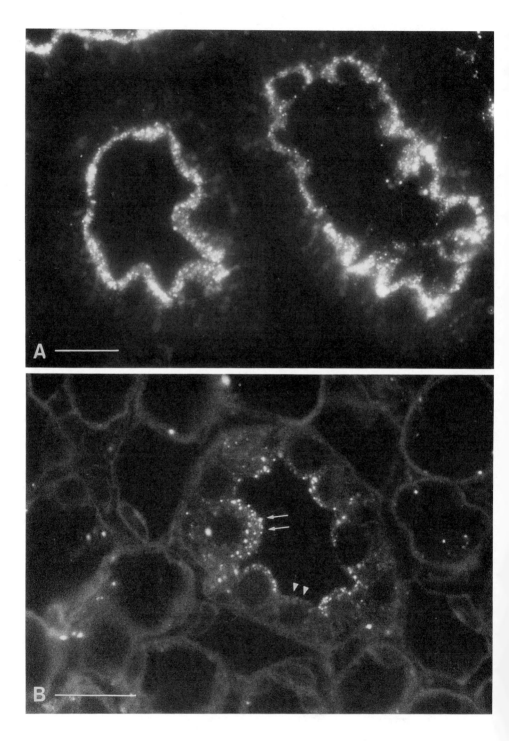

fixative, immersed in 2.3 M sucrose, and frozen in Freon 22 cooled with liquid nitrogen. Semithin frozen sections are cut on an ultracryomicrotome, and the internalized FITC-dextran can be visualized by epifluorescence microscopy. These procedures can be carried out at different time points after injection of the FITC-dextran.

Combining these procedures enabled us to show that endosomes containing water channels are derived from collecting tubule principal cells, and are concentrated at the apical pole of these cells in the distal papilla of the kidney. In addition, we also showed that these same vesicles do not contain functional proton pumps,[42] a characteristic that distinguishes these specialized endosomes from endosomes in many other cell types which acidify their lumenal content by means of a vacuolar proton-pumping ATPase on their limiting membrane. In different kidney regions, other cells also internalize FITC-dextran, such as all cells in the proximal tubule (Fig. 4A) and intercalated cells in the inner stripe and the proximal papilla (Fig. 4B).

Optical Measurement of Water Permeability in Vesicles and Tubules

Identification of Water Channels in Membranes

The direct measurement of water permeability across isolated membrane vesicles and intact epithelia provides important functional information about the presence of water channels which, when combined with data from morphological studies, provides considerable insight into the cell biology of vasopressin action. The important biophysical parameters which describe membrane water permeability are as follows: (1) the osmotic

[42] W. I. Lencer, D. Brown, M. A. Arnaout, D. A. Ausiello, and A. S. Verkman, J. Cell Biol. **107**, 810 (1989).

FIG. 4. Localization of FITC-labeled endosomes in semithin frozen sections of renal cortex and papilla from rats infused with FITC-dextran. There is a bright labeling of apically located vesicles in epithelial cells. The strongest fluorescence is seen in proximal convoluted tubules (A), where it is concentrated in a band at the base of the brush border. FITC-dextran does not appear to label endocytic vesicles in the vicinity of basolateral membranes. In the proximal region of the papilla, collecting ducts contain mainly principal cells, but intercalated cells account for about 10% of the epithelial cell population. Both cell types endocytose lumenal FITC-dextran into apical vesicles, but the internalization is greater in intercalated cells (arrows). Note that the uptake by principal cells is heterogeneous; some principal cells appear to take up little if any fluorescent probe (arrowheads) while in adjacent cells, a marked accumulation of labeled vesicles is visible beneath the apical plasma membrane. Bars = 20 μm.

water permeability coefficient (P_f), (2) the diffusional water permeability coefficient (P_d), (3) the activation energies (E_a) for P_f and P_d, and (4) the reflection coefficients for small solute molecules. Based on studies in the human erythrocyte and artificial membranes containing amphotericin and gramicidin water channels, several characteristics of water channels have been defined empirically. Membranes containing water pores or channels generally have high P_f (>0.01 cm/sec), $P_f/P_d > 1$, low E_a (<6 kcal/mol), and, if the pores are physically large, solute reflection coefficients < 1.[42a] In addition, water transport in biological membranes containing water channels is usually inhibited by mercurial sulfhydryl reagents.

Measurement of Water Permeability in Isolated Vesicles

New methods have been developed to measure the water permeability characteristics of isolated membrane vesicles. For measurement of P_f, the time course of vesicle volume change is measured in response to rapid application of a transmembrane osmotic gradient.

Two optical methods have been used to measure the instantaneous vesicle volume, as illustrated in Fig. 5. The amount of light absorbed or scattered by a vesicle suspension is a strong function of vesicle volume.[43] In Fig. 5 (top), brush border membrane vesicles from rabbit proximal convoluted tubule were subjected to a 65 mOsm inwardly directed gradient of the impermeant solute sucrose. Mixing of the vesicle suspension with the hyperosmotic buffer was accomplished in <1 msec by use of a stopped flow apparatus. The inward osmotic gradient results in outward osmotic water movement, decreased vesicle volume, and increased light scattering. P_f can be calculated from the time course of light scattering, the measured relationship between light scattering and vesicle volume, and the vesicle surface-to-volume ratio. Incubation of vesicles with $HgCl_2$ caused a marked inhibition of water transport. By use of the light-scattering method, it was shown that apical and basolateral vesicles from proximal tubules contain water channels,[44] and that granules from toad urinary bladder have extremely low water permeability, and are thus not involved in the vasopressin hydroosmotic response.[45]

The light absorbance or scattering method can be used effectively only when the sample is biochemically homogeneous because signals arising from both desired and contaminating membranes are measured together. There are potential difficulties in the interpretation of scattering or absorb-

---

[42a] A. Finkelstein, "Distinguished Lecture Series of the Society of General Physiologists, vol. 4." Wiley Interscience, New York, p. 153 (1987).
[43] A. S. Verkman, J. A. Dix, and J. L. Seifter, *Am. J. Physiol.* **248**, F650 (1985).
[44] M. M. Meyer and A. S. Verkman, *J. Membr. Biol.* **96**, 107 (1987).
[45] A. S. Verkman and S. K. Masur, *J. Membr. Biol.* **104**, 241 (1988).

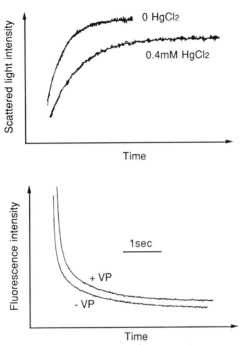

FIG. 5. Measurement of vesicle water permeability (osmotic water transport) in vesicles by light-scattering (top) and fluorescence-quenching techniques (bottom). The two curves at the top of the figure (modified from Meyer and Verkman[44]) show the time course of osmotic water transport in isolated brush border membrane vesicles from rabbit kidney cortex, measured by light scattering. Vesicles were subjected to a 65 mM inwardly directed sucrose gradient in a stopped flow apparatus. The time course of increased scattered light intensity corresponds to vesicle shrinkage due to water efflux. When 0.4 mM HgCl$_2$ was incubated with the vesicles for 10 min prior to the experiment, a marked inhibition of water transport was seen ($-$HgCl$_2$, $P_f = 0.0073$ cm/sec; $+$ HgCl$_2$, $P_f = 0.0038$ cm/sec). The bottom curves (modified from Verkman et al.[25]) show the time course of osmotic water transport in endocytic vesicles measured by the fluorescence-quenching technique. Vesicles were isolated from rat kidney cortex, 15 min after intravenous infusion of 6-carboxyfluorescein into Brattleboro rats, with or without vasopressin (VP). Vesicles were exposed to a 100 mOsm inwardly directed sucrose gradient in a stopped flow apparatus. The time course of decreasing fluorescence reflects vesicle shrinkage due to water efflux, which results in self-quenching of the fluorophore trapped within the vesicles. Vasopressin treatment of the animals has no effect on the water permeability characteristics of the cortical endosome population, which are derived mainly from proximal tubules.

ance data because the intensity of scattered or transmitted light depends on nonvolume factors, including solution refractive index, vesicle aggregation, and vesicle movement in the light beam. To eliminate these difficulties, a fluorescence quenching method was developed to measure vesicle water and solute permeabilities.[46] Membrane-impermeable fluorophores that undergo concentration-dependent self-quenching were incorporated into vesicles. In response to an inwardly directed gradient of an impermeant solute, osmotic water efflux causes vesicle shrinkage, an increase in the intravesicular fluorophore concentration, and an instantaneous decrease in vesicle volume (Fig. 5, bottom). The fluorescence signal is insensitive to refractive index, aggregation, and movement artifacts. Importantly, it is possible to measure water and solute permeability in a selected population of a nonpurified vesicle mixture by labeling vesicles selectively with an impermeant fluorescent marker. This approach has been used successfully to identify and characterize water channels in endocytic vesicles from kidney proximal tubule,[47] kidney collecting tubule,[25] and toad urinary bladder.[26]

It was shown that vasopressin induces the formation of a population of endosomes in kidney collecting tubule and toad urinary bladder that contain functional water channels. In contrast, the water permeability of cortical endosomes, derived mainly from proximal tubules, was unaffected by vasopressin (Fig. 5, bottom). By definition, P_d is measurable from the transport of labeled water in the absence of an osmotic gradient. Water can be labeled magnetically (^{1}H NMR), isotopically (D_2O), or radioactively (^3H_2O). P_d is very difficult to measure because of the rapid rates for water exchange; typical exchange times are 5–10 msec for small cells and liposomes, and 0.1–2 msec for membrane vesicles. ^{1}H NMR has been used to measure P_d in kidney proximal tubule cells.[44,48] Recently, a fluorescence method to measure P_d was developed based on the sensitivity of the impermeant fluorophore aminonaphthalene trisulfonic acid (ANTS) to solution H_2O/D_2O content.[49] Vesicles containing ANTS in an H_2O buffer were mixed with an isosmotic D_2O buffer in a stopped flow apparatus. P_d was calculated from the time course of increasing fluorescence as H_2O/D_2O exchange occurs. By simultaneous measurement of ANTS fluorescence and vesicle light scattering or transmittance, P_f and P_d were measured in a single experiment. E_a for osmotic or diffusional water permeability is measurable from the temperature dependence of P_f or P_d by the Arrhenius relation. Measurement of solute reflection coefficients has

[46] P. Y. Chen, D. Pearce, and A. S. Verkman, *Biochemistry* **27**, 5713 (1988).
[47] R. Ye, L. B. Shi, W. I. Lencer, and A. S. Verkman, *J. Gen. Physiol.* **93**, 885 (1989).
[48] A. S. Verkman and K. Wong, *Biophys. J.* **51**, 717 (1987).
[49] R. Ye and A. S. Verkman, *Biochemistry* **28**, 824 (1989).

been difficult because of rapid solute transport rates and refractive index effects in the light-scattering method. The fluorescence-quenching method has been used to measure reflection coefficients from the initial transmembrane volume movement in response to a series of inward solute gradients.[46] Recently, the NaCl reflection coefficient was measured in apical and basolateral vesicles from proximal tubule by the solvent drag of Cl^- as measured by the fluorescence of the entrapped Cl^- indicator 6-methoxy-n-(3-sulfopropyl) quinolinium (SPQ).[50]

Measurement of Water Permeability in Intact Kidney Tubules

Measurement of transepithelial P_f and P_d in the kidney collecting tubule is required for definition of the biophysics of the vasopressin-sensitive water channel and the vasopressin signaling mechanism. P_f is measured classically from the dissipation of an osmotic gradient along an isolated perfused kidney tubule. For example, tubules are perfused through the lumen with a 300 mOsm buffer containing an impermeant marker ([^{3}H]inulin) and bathed in a 400 mOsm buffer. P_f is calculated from the ratio of [^{3}H]inulin counts at the end vs the beginning of the tubule, the tubule length and diameter, and the flow rate at the beginning of the tubule. P_d is measured from the dissipation of an ^3H_2O gradient along the tubule. Tubules are perfused with buffer containing ^3H_2O and bathed in an isosmotic buffer. P_d is calculated from the ratio of ^3H_2O counts, tubule geometry, and lumen flow. Because timed fluid collections are required for measurement of P_f and P_d, the time resolution was limited to one time point per one or more minutes.

Recently new fluorescence methods were developed to measure P_f and P_d with 1-sec time resolution in the isolated perfused kidney collecting tubule.[51,52] For measurement of P_f, the [^{3}H]inulin was replaced by a fluorescent marker having volume-dependent fluorescence. The fluorescence along the length of the tubule was measured continuously and used to infer lumen osmolality, and thus P_f. For measurement of P_d, the ^3H_2O was replaced by ANTS in a D_2O buffer and the tubule was bathed in an isosmotic H_2O buffer. P_d was determined from the fluorescence along the length of the tubule. These methods were used to measure the second-to-second response of tubule water permeability to addition and removal of vasopressin.

With addition of saturating doses of vasopressin, P_f increases from 15 to 250×10^{-4} cm/sec at 37°. Interestingly, there is a lag period of 20–22 sec

[50] D. Pearce and A. S. Verkman, *Biophys. J.* **55**, 1251 (1989).
[51] M. Kuwahara, C. A. Berry, and A. S. Verkman, *Biophys. J.* **54**, 595 (1988).
[52] M. Kuwahara and A. S. Verkman, *Biophys. J.* **54**, 587 (1988).

in which P_f does not increase, followed by a rise in P_f to its maximal value over 5–10 min.[53] The lag time decreased to 11 sec when vasopressin was replaced by the permeable cAMP analog 8-Br-cAMP. From the analysis of the kinetics of the turn-on of P_f following addition of different concentrations of 8-Br-cAMP, it was concluded that the cAMP-dependent step which regulates steady state water permeability was not the rate-limiting step in the vasopressin signaling mechanism. In another study of the turn-on and turn-off of P_f in response to vasopressin addition, it was found that P_f turned on fastest in the presence of a lumen-to-bath osmotic gradient compared to no gradient or a bath-to-lumen gradient.[54] In response to vasopressin removal, P_f turned off fastest in the presence of a bath-to-lumen gradient. These results were interpreted in terms of a model in which transcellular water flow had a direct influence on the targeting of intracellular vesicles containing water channels.

Target Analysis Studies of Vasopressin-Sensitive Adenylate Cyclase

A fundamental problem in biology is the delineation of mechanisms of activation of membrane-bound receptor and enzyme systems as they occur in the lipid bilayer. This is particularly true for adenylate cyclase. Results obtained from solubilized and purified components do not necessarily reflect events *in situ*, since the techniques involved, of necessity, result in the disruption of biologically important subunit assembly. In contrast, the technique of radiation inactivation is one of a limited number of approaches that can be applied to probe the interaction of proteins in the membrane during the process of enzyme activation without the disruption of subunit assembly.

In the case of adenylate cyclase, studies in detergent-solubilized systems had suggested that dissociation of the α from the β/γ components of the stimulatory GTP-binding protein G_s occurred as a crucial step in the activation process.[55] However, corroboration of this observation in the intact membrane was not available. Accordingly, we turned to the technique of radiation inactivation and analysis of the results by target theory to search for subunit dissociation in the intact membrane.

[53] M. Kuwahara and A. S. Verkman, *J. Membr. Biol.* **110**, 57 (1989).
[54] M. Kuwahara and A. S. Verkman, *Kidney Int.* **35**, 188 (1989).
[55] A. G. Gilman, *J. Clin. Invest.* **73**, 1 (1984).

Target Theory

The technique of radiation inactivation has been used as a tool to explore the size and functional structure of enzyme systems.[56] According to target theory, discussed in more detail in a recent review in this series,[57] the probability that a molecule of interest (whose biological activity is being measured) will survive one or more destructive radiation "hits" is given by a Poisson distribution. Accordingly, biological activity decreases as an exponential function of radiation dose, and the slope of exponential loss of function with radiation dose, is related to the radiation-sensitive target size (molecular mass) of the molecule of interest.[58] While the majority of enzyme systems studied give single exponential inactivation curves with excellent agreement between the molecular mass determined by target analysis and by classical biochemical techniques,[56] application of this approach to complex multisubunit receptor and enzyme systems is more complex.[59] A number of factors may affect the activity-dose relation, including subunit stoichiometry, relative molecular mass and activity of subunits, the occurrence of transfer of radiation energy among subunits, the extent of equilibration at the time of enzyme assay, and multihit requirement for inactivation. The inactivation curve shapes for various permutations of these factors that transitions in activity-dose relations between the inactive and activated state of an enzyme could yield information about the nature of the enzyme activation process itself.[60,61] In particular, a specific curve shape transition was expected for an enzyme activation process which required, as a crucial or rate-limiting step, the dissociation of an inactive multimer to one or more active subunits. The expected curve shape transition in this case if from a concave downward ln(activity) vs dose relation in the multimeric (inactive) state of the enzyme to a linear ln(activity) vs dose relation in the dissociated (active) state (Fig. 6). Furthermore, the limiting slope for the concave downward curve in the inactive state and the actual slope for the linear ln(activity) vs dose relation in the active state are expected to coincide with the molecular mass of the species contributing to overall final enzyme activity. In turn, details of the

---

[56] E. S. Kempner and W. Schlegel, *Anal. Biochem.* **92**, 2 (1979).
[57] E. S. Kempner and S. Fleischer, this series, Vol. 172, p. 410.
[58] G. R. Kepner and R. L. Macey, *Biochim. Biophys. Acta* **163**, 188 (1968).
[59] A. S. Verkman, K. L. Skorecki, and D. A. Ausiello, *Am. J. Physiol.* **250**, C103 (1986).
[60] A. S. Verkman, D. A. Ausiello, C. Y. Jung, and K. L. Skorecki, *Am. J. Physiol.* **250**, C115 (1986).
[61] A. S. Verkman, K. L. Skorecki, and D. A. Ausiello, *Proc. Natl. Acad. Sci. U.S.A.* **81**, 150 (1984).

FIG. 6. Expected pattern of radiation inactivation curve transition for a dissociative activation process.

initial curve shape (concavity) in the inactive state are determined by the relative molecular masses and stoichiometries of the inactive and active subunits contributing to the multimer.

With the background of this theory, the pattern of curve shape transition between the unstimulated and vasopressin-stimulated states of adenylate cyclase in a pig kidney epithelial cell line (LLC-PK$_1$) was investigated.

Experimental Procedures

Radiation Inactivation. The experimental procedures used have been previously described in detail.[60,61] LLC-PK$_1$ cell cultures were treated or untreated with various doses of lysine–vasopressin, frozen by immersing in liquid nitrogen, and radiated under flowing liquid nitrogen using a Van de Graaf generator.

Enzyme Assay. Following radiation, adenylate cyclase activity was assayed in a 10,000-*g* cell pellet by measuring the rate of conversion of [^{32}P]ATP in [^{32}P]cAMP as previously described.[62] Adenylate cyclase activity was expressed as picomoles of cAMP produced per milligram of protein per 20 min. γ-Glutamyl transpeptidase was also measured in each membrane particulate preparation using a kit supplied by Sigma (St. Louis, MO).

Data Analysis. Adenylate cyclase activity at each radiation dose was expressed as a fraction of corresponding activity in the nonradiated sample. Curves were then plotted for ln(fractional activity) vs radiation dose

---

[62] K. L. Skorecki, A. S. Verkman, C. Y. Jung, and D. A. Ausiello, *Am. J. Physiol.* **250**, C115 (1986).

for each experimental condition. For linear relations, a line was fitted by a weighted least-squares procedure; the molecular mass (M) was then determined from the slope using the relation $M = (640 \times 1.03) \times$ slope, where the slope is expressed in megaradiation units^{-1}. For nonlinear ln(fractional activity) vs dose relaitons, curves were fitted by weighted nonlinear least-squares fits to given enzyme activation models to be tested.

Interpretation of Results

The most striking result for adenylate cyclase activity was the transition in the activity–dose relation between the basal and vasopressin-stimulated states. The basal activity–dose relation was concave downward, whereas the curve shape for vasopressin-stimulated activity was linear (Fig. 7; compare with theoretical curves in Fig. 6). The limiting slope for the basal curve shape and the slope for the vasopressin-stimulated linear relation were parallel and corresponded to a molecular weight of approximately 170 kDa. This curve shape transition is consistent with a dissociative enzyme activation process, in which the inactive state of the enzyme is a multimer and in which the fully activated state of the enzyme requires dissociation of one of the subunits. In the case of adenylate cyclase, the most straightforward interpretation is that this step represents dissociation of the α_s subunit of the stimulatory GTP-binding protein G_s. The final molecular weight of the active species (170 kDa) corresponds to the sum of the molecular weight of this dissociated α_s subunit (45 kDa) and the molecular weight of the catalytic subunit of adenylate cyclase, now known to be 120 kDa.[63] It is important to note that target theory prerdicts that the pattern of curve shape transition for a multistep activation process will be determined by that step which yields the moiety whose concentration limits overall final activity.[59,61] Furthermore, the molecular weight derived from slopes of the activity–dose relations reflects the components which must come together for the activation process. In the case of adenylate cyclase, this constitutes α_s plus the catalytic unit.

The same experimental approach could be applied to activators of adenylate cyclase other than vasopressin in this system. These include activators which bypass the receptor entirely and act at the level of G_s (guanylyl imidodiphosphate) or at the level of a catalytic unit (forskolin). When this was done, similar patterns of curve shape transition from concave downward to linear activity–dose relations were observed for vasopressin-sensitive adenylate cyclase in the LLC-PK$_1$ cell membranes.

[63] J. Krupinski, F. Coussen, H. A. Bakalyar, W. J. Tang, P. O. Feinstein, K. Orth, O. Slaughter, R. R. Reed, and A. G. Gilman, *Science* **244**, 1558 (1989).

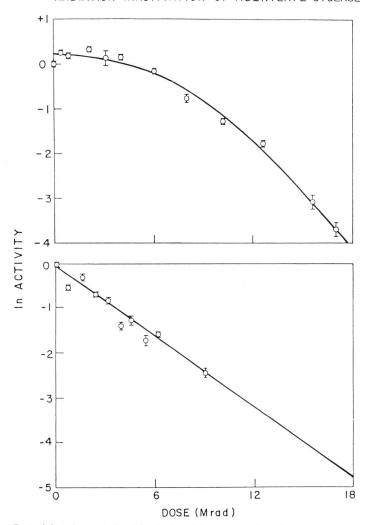

FIG. 7. Activity–dose relationship for basal (upper panel) and vasopressin-stimulated (lower panel) adenylate cyclase activity. Note the similarity between these curve shapes and those in Fig. 6, which are the theoretical curves expected for a dissociative activation process. (From Verkman et al.[59])

Taken together the radiation inactivation approach yielded a remarkably consistent pattern of results indicating that subunit dissociation occurs as part of the vasopressin-sensitive adenylate cyclase activation process in the intact membrane. There was a dramatic ligand-induced transition in activity–dose curve shapes consistent with a dissociative process. Furthermore, the limiting slope for the basal and hormone-activated curves yielded molecular weights corresponding to the sum of the molecular weights of the catalytic unit with the dissociated α_s subunit. In addition, results not shown here indicated that half-maximal curve shape transition occurred at a ligand concentration which produced half-maximal adenylate cyclase activation and that the state of subunit assembly prior to radiation did not affect the pattern of curve shape transition, indicating absence of energy transfer among subunits.[59] This finding is consistent with previous reports indicating a lack of energy transfer among noncovalently linked subunits.[64] This constellation of findings (dissociative pattern of curve shape transition, expected molecular weight for the active moiety, congruence of hormone concentration for enzyme activation and curve shape transition, internal validation of lack of energy transfer among subunits) in these experiments, done without detergent solubilization of enzyme components, lends strong support to the applicability of the subunit dissociation model to the intact membrane.

Conclusion

Over the last several years, we have combined a variety of experimental disciplines to examine the cellular mode of action of the antidiuretic hormone, vasopressin. The combination of morphology and biophysics has proved especially valuable for the investigation of water channel recycling at the apical membrane, while target analysis has provided novel information about initial steps in the vasopressin-induced stimulation of adenylate cyclase at the basolateral plasma membrane of vasopressin-sensitive epithelial cells. In addition to the battery of techniques that have already been applied to the study of vasopressin action, resolution of major remaining questions will require the application of additional methodologies in the fields of protein chemistry and molecular biology. In particular, the identification and molecular cloning of the vasopressin-sensitive water channel is an active area of research in different laboratories. This is likely to be a difficult endeavor due in part to the lack of a highly specific inhibitor of bulk water flow across membranes. Nevertheless, the availability of a sensitive functional assay for membrane water permeability as described above will be a key factor in examining the properties of any putative channel molecule after reconstitution into lipid bilayers or after expression of the molecule in an appropriate cell system.

[64] E. S. Kempner and J. H. Miller, *Science* **222**, 585 (1983).

[33] Purification and Reconstitution of Epithelial Chloride Channels

By DONALD W. LANDRY, MYLES A. AKABAS, CHRISTOPHER REDHEAD, and QAIS AL-AWQATI

Chloride channels are found in most cells, in which they exist in the plasma membrane as well as in intracellular membranes. Electrophysiological studies of epithelia have shown that these channels are located in the apical plasma membrane in some cells and in the basolateral membrane of others.[1] Because the intracellular activity of Cl^- is above electrochemical equilibrium in epithelial cells and in smooth muscle, Cl^- channels play a dominant role in the control of the membrane potential and the transepithelial transport of ions. In intracellular membranes such as the Golgi, endosomes, and lysosomes, these channels are present in parallel to the H^+-ATPase, an electrogenic proton pump that serves to acidify the contents of the vesicles. The Cl^- channel plays an important role in regulation of the degree of acidity of these organelles by shunting the membrane potential generated by the ATPase and allowing maximal acidification.[2] Chloride channels are also regulated by a variety of second messengers such as protein kinase A, protein kinase C, and intracellular Ca^{2+}.[3] Several important diseases are associated with abnormalities of this channel. In myotonia, there appears to be a defect in the muscle Cl^- conductance. In cholera and secretory diarrhea, cyclic AMP opens the epithelial Cl^- channel, leading to massive diarrhea. In cystic fibrosis, neither protein kinase A or C can open this channel.

With the development of the patch-clamp method it became apparent that Cl^- channels exhibit a wide variety of electrophysiological behaviors. Single-channel conductances vary from 1 pS to several hundred picoSiemens; some channels show marked rectification while others have linear current–voltage relationships, and most channels do not show marked halide selectivity. The following questions then arise: What is the molecular basis of this diversity of electrophysiological behaviors? Are all Cl^- channels derived from a single protein or are there multiple proteins? Are

---

[1] E. Schlatter and R. Greger, *Pfluegers Arch.* **405**, 367 (1985); R. L. Shoemaker *et al.*, *Biochim. Biophys. Acta* **858**, 235 (1986); M. J. Welsh and C. M. Liedtke, *Nature (London)* **322**, 467 (1986).
[2] J. Glickman *et al.*, *J. Cell Biol.* **97**, 1303 (1983).
[3] R. A. Schoumacher *et al.*, *Nature (London)* **330**, 752 (1987); M. Li *et al.*, *Science* **244**, 1353 (1989); T.-C. Hwang *et al.*, *Science* **244**, 1351 (1989).

all proteins derived from a single gene or from a related family of genes? Finally, what is the molecular basis of these diseases? To answer such questions requires purification of the protein and development of reagents, such as antibodies and oligonucleotide probes. The initial stages of development of these reagents are described below.

Strategy for Purification of the Chloride Channel

To purify a protein requires a functional assay and for a channel the obvious function is ion transport. However, membrane proteins must be solubilized for purification and to assay from this condition requires either reconstitution or the development of a ligand through which to construct a binding assay. A high-affinity ligand would also allow the use of affinity chromotography. Regardless of whether one uses classical or affinity chromatography, reconstitution of the purified proteins and demonstration of the transport function remains the *sine qua non* of channel purification. Because it is expected that during purification large losses will be incurred, one must also search for starting materials that are highly enriched in Cl^- channel activity and also readily available in large amounts. This usually restricts one to the use of solid organs such as liver, kidney, and brain.

Unlike the case of other channels that have been purified, such as the nicotinic acetylcholine receptor, the voltage-gated Na^+ and Ca^{2+} channels, and the γ-aminobutyric acid (GABA) receptor, no useful ligands were available for the epithelial Cl^- channel. Using compounds that were known to inhibit Cl^- transport, e.g., ethacrynic acid, bumetanide, anthranilic acids, and stilbene disulfonates, we started by screening these compounds and their derivatives using an assay of voltage-sensitive $^{36}Cl^-$ uptake into bovine renal cortex microsomes.[4] Approximately 100 compounds were screened and 15 examined in more detail. The indanyloxyacetic acids (IAA) constituted an active class of compounds and the most potent of these, IAA-94, inhibited Cl^- transport with a K_i of 1 μM. [^{3}H]IAA-94 was initially generated by acid-catalyzed isotope exchange to a specific activity of 0.6 Ci/μmol. An equilibrium binding assay was developed in which the bound was separated from the free ligand by filtration on glass fiber filters after 18 hr of incubation. The ligand bound to the kidney microsomes with a K_d of 0.6 μM and for a panel of inhibitors the rank order of potency for $^{36}Cl^-$ transport inhibition correlated with [^{3}H]IAA-94 displacement, all suggesting that IAA-94 binds to the channel. These results were described in detail elsewhere.[4]

[4] D. W. Landry, *et al., J. Gen. Physiol.* **90,** 779 (1987).

We describe our methods for (1) preparation of bovine kidney membranes, (2) an improved radiolabeling of IAA-94, (3) solubilization of the Cl⁻ channel while maintaining the ligand-binding function, (4) synthesis of IAA affinity columns, (5) affinity purification of Cl⁻ channels from kidney cortex microsomes, and (6) reconstitution of Cl⁻ channels into proteoliposomes.

Preparation of Kidney Membranes[4]

Adult cow kidney, freshly harvested and cooled on ice, is dissected for superficial cortex slices; usually 85 g is used as starting material. The slices are minced, mixed with 450 ml iced homogenization buffer (250 mM sucrose, 5 mM Tris-HCl, 1 mM dithiothreitol, 1 mM EGTA, pH 8.3), and homogenized with a precooled Waring blender set to high speed for 30 sec, off for 15 sec, then again on for 30 sec. All buffers are kept at 4°. The homogenate is spun at 6000 g for 15 min, and the supernatant is spun at 32,000 g for 1 hr. The resulting pellet consists of a brown lower layer and a fluffy white upper layer which is separated with gentle swirling for 5 sec. The fluffy pellet (40 ml) is diluted with 40 ml of buffer A (250 mM sucrose/10 mM imidazole/pH 7) and spun at 40,000 g for 1 hr. The pellet is resuspended in 80 ml of 1 M potassiumthiocyanate (KSCN) and after 30 min spun at 40,000 g for 1 hr. The pellet is resuspended in 20 ml of buffer A containing seven protease inhibitors [50 μg/ml antipain, 2 μM leupeptin, 10 μM pepstatin A, 100 μM (p-amidinophenyl)methanesulfonyl fluoride, 10 μM iodoacetamide, 100 μM N-tosyl-L-phenylalamine chloromethyl ketone, and 1 mM EDTA] and stored at −70°

[³H]IAA-94

IAA-94 is tritiated at the C_α to the carboxylic acid group. This proton exchanges, but only under forcing conditions. Thus, IAA-94 in dry tetrahydrofuran be silylated with trimethylsilylchloride (1.1 Eq) and quantitatively deprotonated with lithium cyclohexylisopropylamide (1.1 Eq) at −78° (Fig. 1). The lithium amide is formed *in situ* from the corresponding amine and n-butyllithium. The silyl ester enolate is quenched with high specific acitivity tritiated water, which also regenerates the carboxylic acid. Acid–base extraction followed by high-pressure liquid chromatography (HPLC) on a Spherisorb 5μ octyl column with CH_3OH/H_2O/acetic acid (125:80:1) elution yields [³H]IAA-94 with specific activity of 12.4 Ci/mmol. Using acid-catalyzed exchange with [³H]trifluoroacetic acid at 120°, a specific activity of only 0.6 Ci/mmol may be obtained.

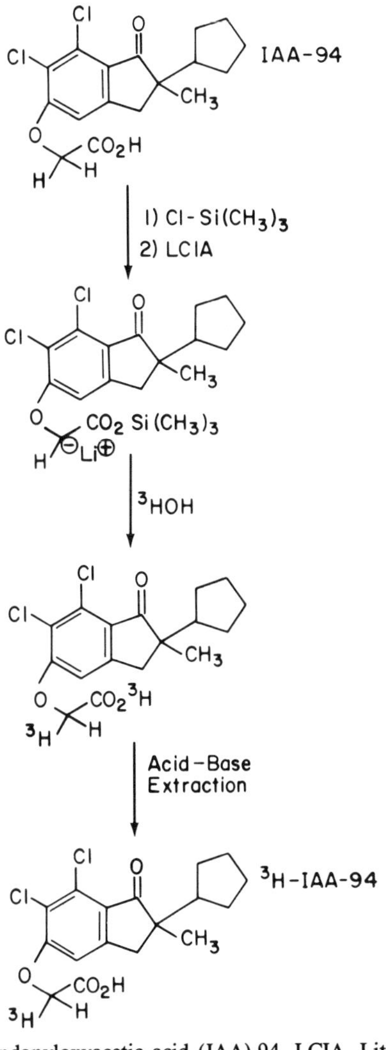

FIG. 1. Tritiation of indanyloxyacetic acid (IAA)-94. LCIA, Lithium cyclohexylisopropylamide.

Solubilization of the Chloride Channel

In preparation for chromatographic purification, a solubilization system was sought that would maintain the capacity of the channel for binding to [^{3}H]IAA-94. In this binding assay, aliquots of vesicles or solubilized vesicles (150 μl) are incubated at 4° with 0.1 μM [^{3}H]IAA-94. After

1.5 hr bound ligand is separated from free by rapid gel filtration through a 0.5 × 3 cm G-50 Sephadex column and the eluate counted for total binding in a liquid scintillation counter (Beckman Instruments, Inc., Fullerton, CA). Nonspecific binding is defined as that occurring in the presence of 100 μM IAA-94 and the difference between total and nonspecific binding yields specific [^{3}H]IAA-94 binding. A number of detergents were screened by this method, but N-octylglucoside alone was effective. Bovine kidney cortex microsomes at a final protein concentration of 10 mg/ml are solubilized by dropwise addition of a freshly prepared 10% solution of N-octylglucoside. The optimum solubilizing concentration is 1.4%. After 90 min at 4° the sample is spun at 100,000 g; analysis of the supernatant shows that 60% of the protein and 20% of the binding sites are solubilized. Although a minority of the sites are solubilized, the degree of solubilization is stable from day to day. Addition of 10% glycerin maintains the number of solubilized binding sites even after freezing at −70° or incubation at 4° for 24 hr. The IC$_{50}$ for ligand displacement by IAA-94 is approximately 2 μM in the solubilized vesicles, similar to that in the intact vesicles.

Synthesis of an Indanyloxyacetic Affinity Column[5]

IAA-92, an indanyloxyacetic acid similar in structure and potency to IAA-94, was chosen as the basis of an affinity ligand because of the availability of derivatives, IAA-21 and IAA-23, suitable for coupling (Fig. 2). IAA-21 is reacted through the primary amino group, with either cyanogen bromide (CNBr)-activated Sepharose or N-hydroxysuccinimide-activated CH Sepharose (Fig. 3). IAA-23 reacts only with the cyanogen bromide-activated support. Each of the ligand–resin combinations is effective and further results are from IAA-23 on CNBr-activated Sepharose. Dry CNBr-activated Sepharose 4B is reswollen and washed on a sintered glass funnel with 100 vol of ice-cold 1 mM HCl. After 15 min the resin is washed with 3 vol of 0.1 M NaHCO$_3$/0.5 M NaCl/pH 9. IAA-23 dissolved in 2 ml 0.1 M NaHCO$_3$/0.5 M NaCl/pH 9 is added to the swollen resin and the mixture is agitated. At a ratio of 1.5 μmol ligand to 1.0 ml swollen resin, coupling efficiency is 97%, as demonstrated by depletion of the ligand from the supernatant in serial UV spectrograms (IAA-23 UV$_{max}$ = 268 nm). After 48 hr the resin is washed with alternating 0.1 M Tris/pH 8.5 and 0.1 M sodium acetate/pH 4.5 and stored in 250 mM sucrose/10 mM imidazole/pH 7 with 0.02% NaN$_3$. Before use, the resin is washed with 250 mM sucrose/10 mM imidazole/10% glycerin/0.6% n-octylglucoside/pH 6.0.

The IAA-23 affinity resin (1 ml) is incubated with 6 mg of solubilized

---

[5] D. W. Landry, et al., Science **224**, 1469 (1989).

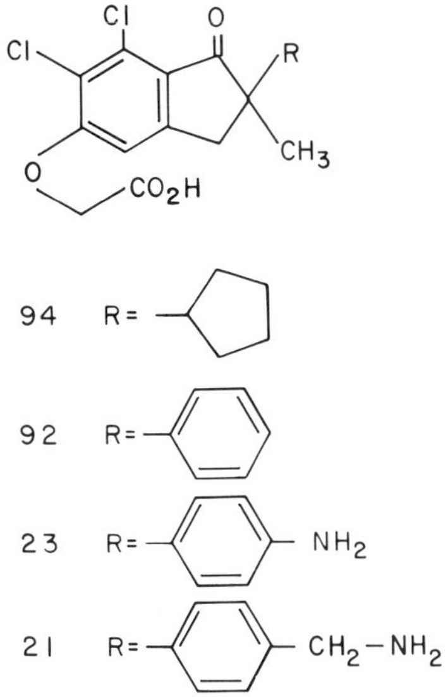

FIG. 2. Structures of IAA ligands for affinity chromatography.

bovine kidney cortex microsomes and agitated for 18 hr at 4° for 18 hr. The supernatant is depleted of 55% of the [^{3}H]IAA-94-binding sites, but less than 10% of protein, compared to a control incubation without resin. The extent of specific depletion is not improved by increasing the ratio of resin to protein. Decreasing the ligand density to 0.3 μmol/ml swollen resin decreases the specific depletion and increasing the density to 6 μmol/ml resin only increases nonspecific protein binding.

Affinity Purification of the Chloride Channel[5]

IAA-23 resin (1 ml) is incubated with solubilized bovine kidney membrane vesicles (1.5 ml). The mixture is agitated for 18 hr at 4° and then transferred to a column (0.5 × 10 cm). The column is washed at a rate of 0.3 ml/min with 35 ml 250 mM sucrose/10 mM imidazole/10% glycerin/ 0.6% n-octylglucoside/100 μM benzoic acid/1% ethanol/pH 6 and then eluted with 7 ml of a solution containing 100 μM IAA-94 in place of 100 μM benzoic acid, but otherwise identical. Benzoic acid, which does not affect IAA-94 binding, is added to the wash to minimize the perturbation due to 100 μM IAA-94 in the eluant. The seven protease inhibitors added to the vesicles at the time of their preparation are also present during solubilization and affinity chromatography. Extensive washing of the resin before elution is necessary to remove nonspecifically bound proteins. The last wash fraction and the specific eluate are precipitated by the addition of

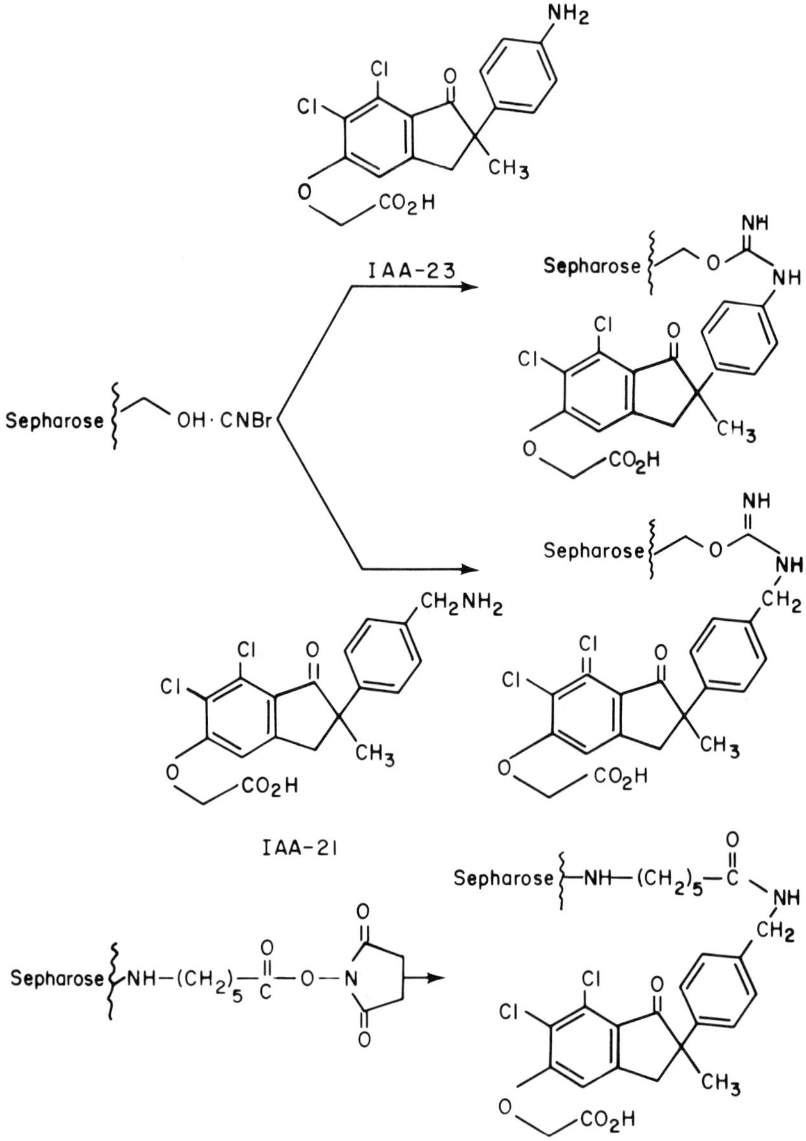

FIG. 3. Synthesis of IAA-21 and IAA-23 Sepharose affinity columns.

3–5 vol of acetone and stored at $-20°$ for 18 hr. The precipitate is mixed with sample buffer containing 15 mM dithiothreitol and the proteins are separated by SDS gel electrophoresis. A silver-stained gel is shown in Fig. 4, demonstrating the presence of four proteins of M_r 97,000, 64,000, 40,000, and 27,000. The yield is difficult to estimate, but from 6 mg of protein

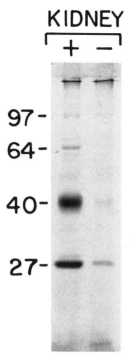

FIG. 4. Electrophoresis of purified Cl⁻ channel proteins from kidney. The IAA-23 affinity resin is mixed with solubilized membranes and after 18 hr transferred to a column. The resin is washed and then eluted with a solution containing 100 μM IAA-94. The eluate (+) and last wash fraction (−) are precipitated in acetone and chromatographed on SDS-PAGE. Proteins are detected by silver staining of the gel.

loaded on the affinity column, approximately 1 μg is eluted based on the intensity of silver staining on the SDS gel.

Reconstitution[5]

The above procedures result in purification of drug-binding proteins. To demonstrate that one or all of these proteins are indeed the Cl⁻ channel requires incorporation of the proteins into artificial lipid bilayers and demonstration of Cl⁻ channel activity. We describe below two kinds of reconstitution assays that we have used, one involving vesicle flux and the other a planar lipid bilayer technique. The purified proteins are first incorporated into phospholipid vesicles by a detergent dialysis procedure.[6] To

---

[6] M. Kasahara and P. C. Hinkle, *Proc. Natl. Acad. Sci. U.S.A.* **73**, 396 (1976).

form the vesicles, affinity column eluate is concentrated to 1 ml, added to asolectin and N-octylglucoside, vortexed, and placed in a dialysis tube. The addition of other reagents and the composition of the dialysate varies depending on the specific reconstitution assay.

Coreconstitution of Chloride Channels with Bacteriorhodopsin

To demonstrate that the purified proteins represented Cl^- channels we measured voltage-dependent $^{36}Cl^-$ uptake into liposomes. Initially, we found that the uptake of the tracer into liposomes was very high and represented a significant problem in terms of signal to noise ratio. It had been previously demonstrated that phospholipid vesicles have a high neutral Cl^- transport activity.[7] To counteract this problem, we coreconstituted the purified proteins with bacteriorhodopsin, a light-activated electrogenic proton pump. Previous studies had shown that bacteriorhodopsin reconstitutes asymmetrically and that bacteriorhodopsin vesicles, upon exposure to light, generate a membrane potential and a pH gradient oriented positive outside and acid inside, respectively.[8] Media with high buffering capacity minimize the pH component of the electrochemical gradient, and the membrane potential component can be used to drive $^{36}Cl^-$ uptake. The advantage of this coreconstitution is that one is able to start the reaction instantaneously by switching on the light, which should generate a membrane potential immediately. We use this assay to reconstitute crude solubilized vesicles as well as the affinity-purified proteins.

Thus, to 0.8 ml of concentrated affinity-purified protein, 0.2 ml of 1 mg/ml bacteriorhodopsin and 10 mg asolectin are added, all in a final concentration of 1.5% n-octylglucoside. The mixture is dialyzed against 100 mM KCl/10 mM Tris/pH 8.0 (2 × 1 liter) for 1 day and for a further day against 50 mM piperazine-N,N'-bis-2-ethanesulfonic acid (PIPES) titrated to pH 7.0 with Tris base. Following dialysis, the proteoliposomes are frozen to $-80°$ until used. Immediately before use the vesicles are thawed to room temperature and sonicated for 25 sec in a Branson 2200 bath sonicator. The same method is used to reconstitute unpurified Cl^- channel proteins with 0.8 ml of kidney vesicles containing 6 mg protein/ml.

For measurement of tracer uptake, 2.3 μCi/ml of $^{36}Cl^-$ is added to thawed and sonicated liposomes in a glass fluorimetric cuvette. The vesicles are exposed to light from a Kodak 650H slide projector. At desired times, samples (80–120 μl) are removed and eluted through 8-cm gluco-

---

[7] Y. Toyoshima and T. E. Thompson, *Biochemistry* **14**, 1525 (1975).
[8] D. Oesterhelt and W. Stoeckenius, *Proc. Natl. Acad. Sci. U.S.A.* **70**, 2853 (1973).

nate-loaded anion-exchange columns (IRN-78, Rohm and Haas Co., Philadelphia, PA) with 1 ml of 250 mM sucrose/10 mM imidazole/pH 7.0 to remove extravesicular $^{36}Cl^-$. The uptake in the dark is measured from identical aliquots of each preparation and should be linear with time. This uptake is subtracted from the light-dependent flux.

Reconstitution of the purified proteins into bacteriorhodopsin vesicles demonstrates a light-induced $^{36}Cl^-$ uptake (Fig. 5, closed squares). This uptake is increased when a higher amount of purified protein was reconstituted (Fig. 5, closed squares). Both of these uptakes are substantially reduced by pretreatment of the vesicles with valinomycin to collapse any membrane potential generated. The uptake in the presence of valinomycin (Fig. 5B, open symbols) is equal to the uptake in liposomes reconstituted with bacteriorhodopsin, but no purified Cl⁻ channels (not shown). Comparison of the magnitude of the $^{36}Cl^-$ uptake of the crude and purified vesicles suggests that we have purified the proteins by at least a factor of a 1000, assuming that there is about 1 μg of reconstituted purified proteins.

Reconstitution of Chloride Channel Proteins into Planar Bilayers

The concentrated purified channel proteins are added to 10 mg of asolectin and 9 mg of N-octylglucoside, vortexed, placed in a dialysis tube (Spectra-por, M_r 14,000 cutoff), and dialyzed against 1 liter of 10 mM KCl/700 mM sucrose/10 mM HEPES titrated to pH 7.0 with KOH for 15 hr. The dialysate is changed and dialysis continued for an additional 3 hr. Vesicles are stored on ice until use.

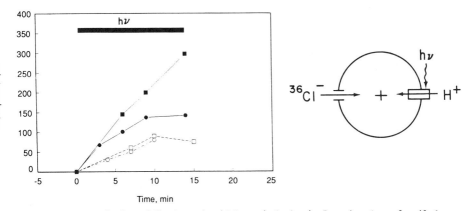

FIG. 5. Reconstitution of Cl⁻ channels with bacteriorhodopsin. Less than 1 μg of purified protein (open or closed circles) or a 10-fold higher amount (open or closed squares) is reconstituted. Valinomycin is added before the light was turned on (open symbols).

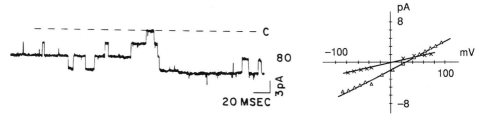

FIG. 6. Single-channel recording (left) and $I-V$ relations (right) of purified Cl⁻ channels reconstituted into planar lipid bilayers. The trace shows the channel in 350 mM KCl, in which the single-channel conductance is 70 pS; 26 pS in symmetrical 150 mM KCl. A downward deflection represents channel opening. C represents closed state. There are either three identical channels in the record (or one channel with two subconductance states). Holding potential (in millivolts) is indicated to the right of the trace. $I-V$ relation for this channel was obtained in the presence of a 150 to 10 mM KCl gradient (▲) or with symmetrical 150 mM KCl (✕).

Planar bilayers (4% asolectin in decane) are formed by the brush technique in a 100-μm hole[9] in a Teflon partition. Vesicles are squirted at the membrane through a micropipet positioned 20–50 μm from the planar bilayer.[10] The cis (vesicle-containing) chamber has a buffer of either 350 mM KCl/10 mM CaCl$_2$/10 mM HEPES/pH 7.0 with KOH or 150 mM KCl/400 mM urea/20 mM hemicalcium gluconate/10 mM HEPES/pH 7.0 with KOH. The trans side contains a similar buffer with either 100 mM KCl or 10 mM KCl but without urea. The single-channel currents are amplified with a home-made current-to-voltage converter and amplifier and recorded on a PCM-video tape recorder (Indec Systems, Inc., Sunnyvale, CA). The data are digitized and analyzed using interactive programs on a laboratory computer system (Indec Systems). Records are filtered at 300 Hz prior to digitization through an 8-pole Bessel filter (Frequency Devices). Potentials given are those in the cis chamber relative to virtual ground in the trans chamber.

We found only anion channels and one of them is shown in Fig. 6, where one vesicle appears to have incorporated three identical channels with a linear $I-V$ relationship and a single-channel conductance of 30 pS. We cannot exclude the possibility that this tracing represents a single 90-pS channel with two subconductance states of 30 and 60 pS. Other channels with different electrophysiological characteristics were also seen.

[9] A. Finkelstein, this series, Vol. 32, p. 387.
[10] M. H. Akabas et al., J. Cell Biol. **98**, 1063 (1984).

Concluding Comments

We have applied the above methods to successfully purify and reconstitute Cl⁻ channels from a variety of sources, including apical membranes from bovine tracheal mucosa, sarcolemma from rabbit striated muscle, and bovine thyroid. We observed electrophysiologically distinct channels from each source. That apparently different Cl⁻ channels from these diverse sources would bind to an IAA affinity column was unexpected. Since a positively charged analog to the IAAs also inhibits Cl⁻ transport, IAA compounds do not merely compete for a Cl⁻-binding site. This suggests a fundamental structural similarity among the various Cl⁻ channels purified by IAA affinity chromatography. We have scaled our procedures, obtaining adequate quantities of purified proteins for N-terminal sequencing and the generation of antibodies. By means of these reagents, we are investigating Cl⁻ channel structure.

[34] Reconstitution and Fractionation of Renal Brush Border Transport Proteins

By HERMANN KOEPSELL and STEFAN SEIBICKE

Introduction

It has been known for many years that several Na⁺-coupled transport systems exist in brush border membranes of small intestine and renal proximal tubules, which are responsible for the reabsorption of D-glucose, amino acids, and anions.[1-3] The Na⁺-D-glucose cotransporter is the best investigated Na⁺-coupled transport system from renal and intestinal brush border membranes, and reconstitution and purification experiments have been nearly exclusively performed with this transporter.[4-17] Since the

---

[1] S. Silbernagl, E. C. Foulkes, and P. Deetjen, *Rev. Physiol. Biochem. Pharmacol.* **74,** 105 (1975).
[2] K. J. Ullrich, *Annu. Rev. Physiol.* **41,** 181 (1979).
[3] H. Murer and G. Burckhardt, *Rev. Physiol. Biochem. Pharmacol.* **96,** 1 (1983).
[4] R. K. Crane, P. Malathi, and H. Preiser, *Biochem. Biophys. Res. Commun.* **71,** 1010 (1976).
[5] R. K. Crane, P. Malathi, and H. Preiser, *FEBS Lett.* **67,** 214 (1976).
[6] J. T. Lin, M. E. M. Da Cruz, S. Riedel, and R. Kinne, *Biochim. Biophys. Acta* **640,** 43 (1981).
[7] W. B. Im, K. Y. Ling, and R. G. Faust, *J. Membr. Biol.* **65,** 131 (1982).

Na$^+$-D-glucose cotransporter can be considered as a model for Na$^+$-coupled transport proteins, we will mainly describe procedures which have been developed from experiments in which the activity of this transporter was measured. In contrast to all other Na$^+$-coupled transporters a high-affinity competitive inhibitor (phlorizin) is available for the Na$^+$-D-glucose cotransporter so that the function of the substrate-binding site can be controlled by binding measurements. Although many attempts have been reported to purify and to reconstitute the solubilized Na$^+$-D-glucose cotransporter,[6-17] in none of these cases was Na$^+$ gradient-dependent D-glucose uptake (following protein fractionation and reconstitution) higher than in intact brush border membrane vesicles.[18]

By affinity-labeling experiments with the side group-specific reagents fluorescein isothiocyanate and N-acetylimidazole, Wright and co-workers showed that in intestinal brush border membranes polypeptides with M_r values of 149,000, 97,000, 75,000, 39,000, and 29,000 contain Na$^+$-binding sites,[19] that the M_r 97,000 polypeptide contains a proline-binding site,[20] and that a D-glucose-binding site of the Na$^+$-D-glucose cotransporter is present on the M_r 75,000 polypeptide.[21] Recently the cDNA of an M_r 73,080 polypeptide has been cloned, which is supposed to be identical with the labeled M_r 75,000 polypeptide of the intestinal Na$^+$-D-glucose cotransporter.[22] When mRNA derived from this clone was injected into oocytes from *Xenopus laevis,* Na$^+$-D-glucose cotransport into the oocytes was drastically increased. Moreover, it was reported that after transfection of cos-7 cells with this clone Na$^+$-dependent D-glucose transport could be measured.[23] The oocytes contain an endogenous Na$^+$-D-glucose cotrans-

---

[8] H. Koepsell, H. Menuhr, I. Ducis, and T. F. Wissmüller, *J. Biol. Chem.* **258,** 1888 (1983).
[9] K.-Y. Ling and R. G. Faust, *Int. J. Biochem.* **17,** 365 (1985).
[10] P. Malathi and H. Preiser, *Biochim. Biophys. Acta* **735,** 314 (1983).
[11] A. Kano-Kameyama and T. Hoshi, *Jpn. J. Physiol.* **33,** 955 (1983).
[12] J.-T. Lin, K. Schwarc, and A. Stroh, *Biochim. Biophys. Acta* **774,** 254 (1984).
[13] J. C. Poirée, M. Starita-Geribaldi, and P. Sudaka, *Biochim. Biophys. Acta* **858,** 83 (1986).
[14] H. Koepsell, K. Korn, D. Ferguson, H. Menuhr, D. Ollig, and W. Haase, *J. Biol. Chem.* **259,** 6548 (1984).
[15] M. Starita-Geribaldi, J.-C. Poiree, and P. Sudaka, *Anal. Biochem.* **165,** 406 (1987).
[16] J.-S. R. Wu and J. E. Lever, *Biochemistry* **26,** 5958 (1987).
[17] M. Silverman and P. Speight, *J. Biol. Chem.* **261,** 13820 (1986).
[18] H. Koepsell, *Rev. Physiol. Biochem. Pharmacol.* **104,** 65 (1986).
[19] B. E. Peerce and E. M. Wright, *J. Biol. Chem.* **260,** 6026 (1985).
[20] E. M. Wright and B. E. Peerce, *J. Biol. Chem.* **259,** 14993 (1984).
[21] B. E. Peerce and E. M. Wright, *J. Biol. Chem.* **259,** 14105 (1984).
[22] M. A. Hediger, M. J. Coady, T. S. Ikeda, and E. M. Wright, *Nature (London)* **330,** 379 (1987).
[23] B. Birnir, H.-S. Lee, M. A. Hediger, and E. M. Wright, *FASEB J.* abstr. 1948 (1989).

porter,[24,25] and the apparent absence of Na^+-D-glucose cotransport in nontransfected cos-7 cells, which are derived from simian kidney, may be due to a number of different things such as, e.g., a point mutation in the transporter, absence of the transporter, absence of one subunit of the transporter, or a defect in a posttranslational modification of the transporter. Thus, since the concentration of the active transporter in the cRNA-injected oocytes and in the transfected cos-7 cells may be increased after expression of one subunit of the transporter[26,27] it cannot be determined whether the M_r 75,000 polypeptide is a subunit or the only polypeptide component of the intestinal Na^+-D-glucose cotransporter.

Employing the D-glucose analog 10-N-(bromoacetyl)amino-1-decyl-β-D-glucopyranoside (BADG), which covalently binds, we have shown that polypeptides with molecular weights and respective isoelectric points of 75,000 and pH 5.4 and 75,000 and pH 6.9 are components of porcine renal Na^+-D-glucose cotransporters which contain D-glucose-binding sites.[28] Furthermore, evidence has been obtained that a polypeptide with a molecular weight and isoelectric point of 47,000 and pH 5.4 is also part of porcine renal Na^+-D-glucose cotransporters. This polypeptide has also been labeled with BADG and reacts with monoclonal antibodies, which cross-react with the M_r 75,000 polypeptide and inhibit Na^+-D-glucose cotransport and Na^+-dependent phlorizin binding in porcine kidney.[29] The monoclonal antibodies directed against the M_r 75,000 and 47,000 polypeptides of the porcine renal Na^+-D-glucose cotransporter cross-react with M_r 75,000 and 47,000 polypeptides from rat kidney and rat intestine.[30,31] Some of these antibodies bind to brush border membranes and subapical vesicles of renal proximal tubules[30] and to intestinal brush border membranes[31] and stimulate Na^+-dependent phlorizin binding and/or Na^+-D-glucose cotransport in brush border membrane from rat intestine.[31] Thus, the Na^+-D-glucose cotransporters from kidney and intestine appear to be homologous but not identical. The functional unit of the renal and intestinal Na^+-D-glucose cotransporter has not been defined. Radiation inactivation experiments support the view that the Na^+-D-glucose cotransporter

---

[24] H. Aoshima, H. Ishii, and M. Anan, *Mol. Brain Res.* **2**, 263 (1987).
[25] W.-M. Weber, W. Schwarz, and H. Passow, *J. Membr. Biol.* **111**, 93 (1989).
[26] H. Soreq, *CRC Crit. Rev. Biochem.* **18**, 199 (1985).
[27] J. R. Emanuel, J. Schulz, X.-M. Zhou, R. B. Kent, D. Housman, L. Cantley, and R. Levenson, *J. Biol. Chem.* **263**, 7726 (1988).
[28] M. Neeb, U. Kunz, and H. Koepsell, *J. Biol. Chem.* **262**, 10718 (1987).
[29] H. Koepsell, K. Korn, A. Raszeja-Specht, S. Bernotat-Danielowski, and D. Ollig, *J. Biol. Chem.* **263**, 18419 (1988).
[30] W. Haase and H. Koepsell, *Eur. J. Cell Biol.* **48**, 360 (1989).
[31] W. Haase, K. Heitmann, W. Friese, D. Ollig, and H. Koepsell, *Eur. J. Cell Biol.* (in press).

is on oligomeric protein and that the functional unit for phlorizin binding is smaller than that for D-glucose transport. In these studies the functional molecular mass for Na^+-D-glucose cotransport has been determined to be 345,000, whereas for the functional unit binding phlorizin a molecular mass of 230,000 or 110,000 has been reported.[32-34] The role of the M_r 75,000 and 47,000 polypeptides during Na^+-D-glucose cotransport is not known. Both polypeptides could be subunits of the transporter, but it is also possible that the M_r 75,000 polypeptide is a precursor, or the M_r 47,000 polypeptide a degradation product of the transporter. Thus, purification and reconstitution experiments are required to find out whether the M_r 75,000 and/or the M_r 47,000 polypeptide is engaged in Na^+-D-glucose cotransport.

This chapter describes procedures for (1) isolating renal brush border vesicles, (2) solubilizing and precipitating Na^+ cotransport proteins without denaturation, (3) renaturating the Na^+-dependent high-affinity phlorizin-binding activity after solubilization, (4) reconstituting Na^+ cotransport proteins into large proteoliposomes, and (5) partially purifying the renal Na^+-D-glucose cotransporter.

Reagents which are employed for the procedures are as described below:

TM buffer: 2 mM Tris-HCl, pH 7.1, 10 mM D-mannitol
TRA buffer: 10 mM triethanolamine-HCl, pH 7.4, 150 mM NaCl, 5 mM ethylenediaminetetraacetic acid (EDTA), 10% (v/v) glycerol
$CaCl_2$, 1 M, prepared in distilled water
KC buffer: 20 mM imidazole cyclamate, pH 7.4, 0.1 mM magnesium cyclamate, 100 mM potassium cyclamate

Isolation of Renal Brush Border Vesicles (Modification of a Method of Booth and Kenny[35])

Procedure

The renal cortex is dissected from pig kidneys which are obtained directly after slaughter and transported in ice-cold Ringer solution. The preparation of membrane vesicles is performed at 8°. A volume of 900 ml of TM buffer is added per 100 g kidney cortex and the tissue is minced for

---

[32] J.-T. Lin, K. Szwarc, R. Kinne, and C. Y. Jung, *Biochim. Biophys. Acta* **777**, 201 (1984).
[33] M. Takahashi, P. Malathi, H. Preiser, and C. Y. Jung, *J. Biol. Chem.* **260**, 10551 (1985).
[34] R. J. Turner and E. S. Kempner, *J. Biol. Chem.* **257**, 10794 (1982).
[35] A. G. Booth and A. J. Kenny, *Biochem. J.* **142**, 575 (1974).

5 min in a household mixer. Then 10 ml of 1 M CaCl$_2$ is added. After a 15-min incubation the mixture is centrifuged for 12 min at 1600 g. The pellet is discarded and the supernatant is centrifuged 12 min at 17,500 g. The pellet obtained after this centrifugation step is resuspended in 100 ml TM buffer and homogenized by 10 strokes with a tightly fitting motor-driven glass–Teflon homogenizer. Then 1 ml of 1 M CaCl$_2$ is added and the mixture is allowed to stand for 15 min. Thereafter the mixture is centrifuged for 12 min at 3500 g. The supernatant obtained after centrifugation is then centrifuged for 12 min at 20,000 g. The obtained sediment is suspended in 10 ml of TRA buffer and centrifuged for 20 min at 48,000 g. The sediment after this final centrifugation is suspended in 5 ml TRA buffer, homogenized by five strokes with a glass–Teflon homogenizer, and then frozen in portions of 1 ml in liquid nitrogen.

For transport measurements in intact membrane vesicles,[8,14] 1 ml of vesicles is thawed by a 5-min incubation at 37°, diluted with 20 ml KC buffer (22°), and spun down by a 20-min centrifugation at 48,000 g (22°). The pellet is washed twofold by suspending in 20 ml KC buffer and centrifuging for 20 min at 48,000 g (22°).

Activities of Na$^+$-Coupled Transporters in Intact Membrane Vesicles

In the membrane vesicles the specific activity of the brush border marker enzyme alkaline phosphatase was more than 20-fold enriched compared to the homogenate of renal cortex. Sodium ion cotransport was demonstrated for D-glucose, L-glutamic acid, L-alanine, L-proline, phosphate, sulfate, L-lactate, and succinate. D-Glucose and L-glutamate uptake was measured at 37° at different substrate concentrations at equilibrium (100 mM K$^+$) and in the presence of initial concentration differences [89 mM Na$^+$ (out > in) and 89 mM K$^+$ (in > out)] and the Na$^+$/K$^+$ gradient-dependent uptake rates were calculated. The V_{max} values for Na$^+$/K$^+$ gradient-dependent uptake of D-glucose and L-glutamate were 1.1 and 0.14 nmol/(mg protein · sec), respectively. Measuring Na$^+$-dependent high-affinity phlorizin binding in the membrane vesicles,[8] K_D values between 0.3 and 0.5 μM were obtained. The number of high-affinity phlorizin-binding sites in the presence of Na$^+$ was between 0.09 and 0.12 nmol/mg protein.

Solubilization of Brush Border Membrane Proteins and Precipitation of Na$^+$ Cotransport Proteins

Solubilization of Na$^+$ cotransport proteins has been performed with Triton X-100, cholate, deoxycholate and octylglucoside.[4-15] In these ex-

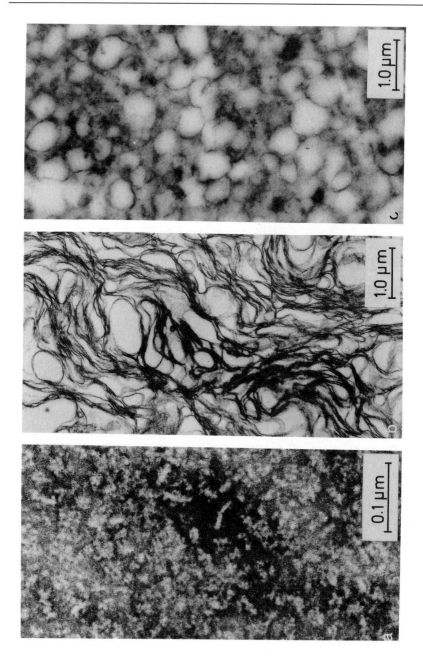

periments irreversible inactivation of Na^+ cotransport proteins could be avoided if the ratio of detergent to membrane protein and/or membrane lipid was sufficiently low and if Na^+ and D-glucose were present during solubilization. A solubilization procedure will be described below after which highly active Na^+ cotransport of D-glucose, L-alanine, L-glutamic acid, succinate, lactate, and proline could be measured after reconstitution.[8,14] During this procedure brush border membranes are dissociated into mixed micelles of detergent, lipids, and proteins. Since the protein-to-detergent ratio is high it is assumed that an irreversible dissociation of the oligomeric transport proteins is avoided.

Procedure

Eight to 10 1-ml portions of frozen membrane vesicles, which contain 250 mg of membrane protein in total, are thawed for 5 min at 37°. The membrane vesicles are combined and the concentration of protein is adjusted to 25 mg/ml by addition of TRA buffer. Ten milliliters of these membrane vesicles is added to 12.5 ml ice-cold TRA buffer containing 2 M D-glucose and stirred for 15 min in ice. Then, during continuous stirring in ice, 2.5 ml of TRA buffer containing 4% (w/v) sodium deoxycholate is added dropwise. Thereafter the mixture is allowed to stand for 1 hr at 8°. The deoxycholate and protein concentration is reduced to 0.2% (w/v) and 5 mg/ml of protein, respectively, by adding 25 ml of TRA buffer containing 1 M D-glucose (8°). The mixture is then centrifuged for 1 hr at 200,000 g (8°) and the clear supernatant is removed. In Fig. 1a, an electron micrograph of this supernatant, which was negatively stained, is shown. Mixed micelles with diameters between 50 and 500 Å can be seen. To the supernatant 115 ml of TRA buffer containing 1 M D-glucose and 0.057% (w/v) sodium deoxycholate is added and the mixture is dialyzed for 3 hr or for 16 hr (8°) against 5 liters of TRA buffer containing 0.1% (w/v) sodium deoxycholate. A fraction of the mixed micelles in the supernatant first aggregate (see Fig. 1b, 3-hr dialysis) and form large proteoliposomes after

FIG. 1. Morphological analysis of solubilization and reassociation of proteins and lipids from brush border membranes. Micrographs of (a) brush border membranes which were solubilized with deoxycholate, (b) aggregated mixed micelles which were formed after a 3-hr dialysis, and (c) large proteoliposomes which were formed after a 16-hr dialysis are shown. The samples were prepared as described in the text. The solubilized membranes in (a) were fixed with glutaraldehyde, applied to grids which were coated with polyvinylformal and carbon, washed with distilled water, and negatively stained with ammonium molybdate. The samples in (b) and (c) were fixed with glutaraldehyde, embedded, sectioned, and evaluated by electron microscopy as described previously.[14]

long dialysis (see Fig. 1c, 16-hr dialysis). This insoluble material after dialysis is isolated by 1 hr of centrifugation at 200,000 g. For reconstitution experiments and for measurements of high-affinity phlorizin binding or D-glucose uptake the pellets are suspended in 5 ml KC buffer (8°), centrifuged for 30 min at 200,000 g (8°), and then washed twice by suspending in 5 ml of KC buffer and centrifuging for 30 min at 200,000 g (8°). The final pellet is suspended with KC buffer at a concentration of protein greater than 5 mg/ml.

Comments

In the sediments obtained after dialysis some brush border membrane polypeptides were enriched whereas others were reduced.[8] After the 3- or 16-hr dialysis identical patterns of polypeptides were found in sediments. Moreover, about the same number of Na^+-dependent high-affinity phlorizin-binding sites was found in both types of sediment. Furthermore, Na^+ cotransport of D-glucose, L-glutamate, L-alanine, L-lactate, and DL-methylsuccinate could be reconstituted from both sediments by freezing and thawing. From the 16-hr dialysis sediment Na^+ cotransport could be directly reconstituted by freezing and thawing. From the 3-hr dialysis sediment significant Na^+ cotransport could be reconstituted only if nontransporting proteoliposomes were formed first by solubilization and detergent removal and if these were frozen and thawed in a second step.

Renaturation of Phlorizin Binding after Solubilization of the Na^+-D-Glucose Cotransporter

In the presence of Na^+ the K_D for phlorizin binding to membrane vesicles is about 1000-fold lower than the K_i for the competitive inhibition of phlorizin binding by D-glucose.[36-40] Since we found that high-affinity phlorizin binding in the presence of Na^+ is also competitively inhibited by phloretin, the aglucon of phlorizin,[41] it can be assumed that phlorizin binds not only at the D-glucose-binding site of the Na^+-D-glucose cotransporter but also at a closely related hydrophobic protein domain. Thus, high-affin-

[36] M. Silverman and J. Black, *Biochim, Biophys. Acta* **394**, 10 (1975).
[37] R. J. Turner and M. Silverman, *Biochim. Biophys. Acta* **507**, 305 (1978).
[38] R. J. Turner and M. Silverman, *J. Membr. Biol.* **58**, 43 (1981).
[39] H. Koepsell, H. Menuhr, T. F. Wissmüller, I. Ducis, and W. Haase, *Ann. N.Y. Acad. Sci.* **358**, 267 (1980).
[40] H. Koepsell and A. Madrala, *Top. Mol. Pharmacol.* **4**, 169 (1987).
[41] H. Koepsell, M. Neeb, K. Korn, U. Kunz, and E. Zoubek, in "Molecular Nephrology: Biochemical Aspects of Kidney Function" (Z. Kovačević and W. G. Guder, eds.), p. 13. de Gruyter, Berlin, 1987.

ity phlorizin binding is supposed to be highly dependent on the tertiary structure of the Na^+-D-glucose cotransporter. Actually it was found that Na^+-dependent high-affinity phlorizin binding cannot be measured when brush border membranes are solubilized.[39,42] However, Na^+-dependent high-affinity phlorizin binding can be recovered if the detergent concentration is reduced and membrane proteins precipitate.[8,17,39,42] In contrast to a recent report, where less than 10% of the Na^+-dependent phlorizin-binding sites was recovered after solubilization,[17] quantitative recovery can be achieved if after solubilization with deoxycholate or octylglucoside the Na^+-D-glucose cotransporter is precipitated with polyethylene glycol 4000 and washed as described below.

Procedure

Reactivation of phlorizin-binding sites by precipitation with polyethylene glycol 4000 can be obtained with brush border membrane proteins, which have been solubilized in TRA buffer with (1) deoxycholate (see above) or with (2) 0.75–1.0% octylglucoside (in the presence of 0.25 or 1 M D-glucose). The concentration of the solubilized membrane proteins should be between 0.5 to 5 mg of protein/ml. If the protein concentration is lower than 1 mg/ml, 1 mg of phosphatidylserine/ml should be added to increase the recovery of phlorizin-binding sites during precipitation.

In the following steps the temperature should be carefully kept at 6°. Five hundred microliters of solution containing the solubilized proteins is mixed with 500 μl of KC buffer containing 40% (w/v) of polyethylene glycol 4000 and the mixture is allowed to stand for 15 min. Then the precipitated proteins are spun down by centrifugation for 15 min at 50,000 g. The pellet is washed twice by diluting with 10 ml of KC buffer and by a 15-min centrifugation at 200,000 g.

Reconstitution of Na^+ Cotransport after Solubilization of Transport Proteins

To measure Na^+ gradient-dependent uptake after reconstitution of Na^+ cotransport proteins into proteoliposomes an inwardly directed Na^+ gradient must persist for the time of measurement. Since long measuring times are required to detect low transport activities it is favorable to reconstitute the transporters into large proteoliposomes which have a low passive permeability for Na^+. In such proteoliposomes an accumulation of cotransported substrates above the equilibrium concentration may be demonstrated in the presence of an inwardly directed Na^+ gradient.

[42] H. Koepsell, Z. Jakubowski, and A. Madrala, *Mol. Physiol.* **8**, 77 (1985).

Large proteoliposomes with a mean diameter around 0.5 μm can be formed by different procedures: (1) long dialysis of brush border membranes, which have been solubilized with deoxycholate in the presence of 1 M D-glucose, against TRA buffer containing 0.1% (w/v) deoxycholate (see Fig. 1c); (2) removal of deoxycholate or octylglucoside from solubilized brush border membranes on a Sephadex G-50 column[14]; (3) dialysis of brush border membrane proteins, which have been solubilized with octylglucoside, against detergent-free buffer (unpublished observations); and (4) a modified freeze–thaw reconstitution procedure[8,14,43] which will be described in detail. In all four types of proteoliposomes highly active Na^+-dependent high-affinity phlorizin binding was found. However, only in the proteoliposomes formed by freezing and thawing was significant Na^+ gradient-dependent uptake of D-glucose measured. The proteoliposomes which were formed by freezing and thawing had the lowest passive permeability for Na^+ and D-glucose. However, the difference in passive permeability does not explain the missing transport in the other types of proteoliposomes since, also after a short incubation of 10 sec, Na^+ gradient-dependent D-glucose uptake was not found in these proteoliposomes. If we accept the hypothesis that the Na^+-D-glucose cotransporter, and possibly also other Na^+ cotransport proteins, are oligomeric proteins, one can imagine that polypeptide components of the transporters, which may have been dissociated during solubilization, may reassociate during freezing and thawing. This may happen because during decrease of temperature some types of phospholipids aggregate below their transition temperature and membrane proteins may be concentrated in fluid membrane areas.[44–47]

The method of freeze–thaw reconstitution consists, first, of the formation of nontransporting monolamellar proteoliposomes from solubilized transport proteins. These can be prepared by different methods of detergent removal (see Step I,A,B,C below). This step is required to observe transport after freezing and thawing. Presumably transporter subunits are properly aligned in lipid membranes during the detergent removal. In a second step of the reconstitution procedure liposomes are formed from extraneous lipids, which will be added to the nontransporting monolamellar proteoliposomes during freeze–thaw reconstitution. These liposomes

[43] I. Ducis and H. Koepsell, *Biochim. Biophys. Acta* **730,** 119 (1983).
[44] D. Papahadjopoulos, K. Jacobson, S. Nir, and T. Isac, *Biochim. Biophys. Acta* **311,** 330 (1973).
[45] E. J. McMurchie and J. K. Raison, *Biochim. Biophys. Acta* **554,** 364 (1979).
[46] T. Lookman, D. A. Pink, E. W. Grundke, M. J. Zuckermann, and F. deVerteuil, *Biochemistry* **21,** 5593 (1982).
[47] H. G. Kapitza, D. A. Rüppel, H.-J. Galla, and E. Sackmann, *Biophys. J.* **45,** 577 (1984).

may be large and mainly multilamellar and can be formed by shaking (see Step II,A below and Fig. 2a), or may be smaller and mainly monolamellar and can be formed by shaking, sonication, and differential centrifugation (see Step II,B and Fig. 2c). Third, to form large transporting proteoliposomes by freezing and thawing (See Step III and Fig. 2b and d), the nontransporting proteoliposomes are mixed with the liposomes from extraneous lipids and the mixture is heated up, frozen, and thawed. During this step reassociation of transporter subunits may occur. The thawed sample is then warmed to 37°, spun down, homogenized, and protein–lipid aggregates and multilamellar proteoliposomes are removed by a low-speed centrifugation.

Procedure for Freeze–Thaw Reconstitution

Step I: Formation of Nontransporting Proteoliposomes

  A. By a 16-hr dialysis of brush border membranes, which have been solubilized with sodium deoxycholate in the presence of 1 M D-glucose against TRA buffer containing 0.1% deoxycholate: See above (Fig. 1c)
  B. By removal of detergent from membrane proteins, which have been solubilized with deoxycholate or octylglucoside during gel chromatography: Two milliliters of TRA buffer containing 1 M D-glucose, 0.4% deoxycholate (or 1% octylglucoside), plus at least 10 mg of protein and 3 mg of endogenous phospholipids is applied to a 50-cm Sephadex G-50 column with an internal diameter of 1.6 cm. The columns are equilibrated with KC buffer and run at a flow rate of 0.3 ml/min. The formed proteoliposomes appear in the void volume and are sedimented by a 1-hr centrifugation at 200,000 g (8°). The sediment is suspended in KC buffer at a protein concentration greater than 5 mg/ml
  C. By removal of octylglucoside from solubilized membrane proteins during dialysis: This method can be used with small amounts of protein. At minimum 1 mg of membrane proteins, which have been solubilized in 1 ml of TRA buffer containing 250 mM D-glucose, 0.75% (w/v) octylglucoside, 1 mg of phosphatidylcholine, and 1 mg of phosphatidylserine, is dialyzed for 12 hr against TRA buffer containing 250 mM D-glucose (8°). Thereafter the proteoliposomes are sedimented by a 30-min centrifugation at 200,000 g and washed twice with KC buffer by a 30-min centrifugation at 200,000 g (8°). The final pellet is suspended in KC buffer at a concentration of protein greater than 5 mg/ml

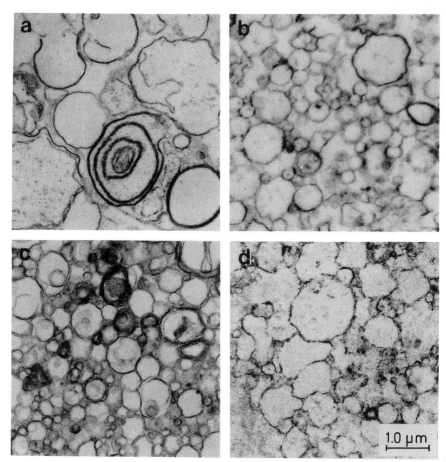

FIG. 2. Morphological analysis of some steps during the formation of transporting proteoliposomes by freezing and thawing. In (a) large, mainly multilamellar cholesterol/phosphatidylserine liposomes are shown which were formed as described in the text (Step II,A). These liposomes were mixed with nontransporting proteoliposomes which were formed by detergent removal (see Step I,B). In (b) the final transport-active proteoliposomes which were formed by freezing, thawing, and centrifugation of this mixture (see Step III) are shown. In (c) smaller, mainly monolamellar cholesterol/phosphatidylserine/phoshatidylcholine liposomes (see Step II,B) are presented. These were added to nontransporting proteoliposomes, which were formed as described in the text (see Step I,C). In (d) the final transport-active proteoliposomes obtained from this mixture after freezing, thawing, and centrifugation (see Step III) are shown. The samples were prepared for electron microscopy as in Fig. 1b and c.

To reduce the vesicle size the nontransporting proteoliposomes, which have been formed by Steps I,A and I,B, can be sonicated for 30 sec with a tip sonicator (Labsonic 1510, Braun Melsungen AG, probe diameter 4 mm, energy setting 100 W, test tube diameter 1 cm, sample volume 2 ml). This step is not obligatory; however, it may lead to higher uptake rates after freeze–thaw reconstitution.

Step II: Formation of Liposomes to Be Added During Freeze–Thaw Reconstitution

A. To form large, mainly multilamellar cholesterol/phosphatidylserine liposomes: Two milligrams of cholesterol in chloroform/methanol (1/1, v/v) and 4 mg of phosphatidylserine from bovine brain in chloroform/methanol (1/1, v/v) are added to a 25-ml round-bottom flask and dried under a stream of nitrogen. The flask should be rotated during drying. Then some small glass beads and 1 ml of KC buffer are added and the flask is shaken for 3 hr under nitrogen at 22°

B. To form mainly monolamellar cholesterol/phosphatidylserine/phosphatidylcholine liposomes: Solutions of chloroform/methanol (1/1, v/v) containing 50 mg of cholesterol, 50 mg of phosphatidylserine from bovine brain, plus 50 mg of phosphatidylcholine from egg yolk are added to a 250-ml round-bottom flask and dried by rotation under nitrogen to a thin lipid film. Then 24 ml of KC buffer is added and the mixture is shaken for 3 hr under nitrogen at 22°. Thereafter the suspension is heated to 41° and then sonicated in 3-ml portions for 30 sec with a tip sonicator (probe diameter 4 mm, test tube diameter 1 cm, energy setting 100 W). Lipid aggregates and multilamellar liposomes are removed by centrifugation for 15 min at 10,000 g (25°). Liposomes are separated from the supernatant by a 15-min centrifugation at 100,000 g (25°). The pellet is suspended in 650 μl of KC buffer at 25° (Fig. 2c)

Step III: Formation of Transporting Large Proteoliposomes by Freezing, Thawing, and Centrifugation.

To reconstitute functional active transport systems from nontransporting proteoliposomes, which have been formed by Steps I,A or I,B, 1 ml of these types of nontransporting proteoliposome is mixed with 1 ml of multilamellar cholesterol/phosphatidylserine liposomes, which have been formed by Step II,A. On the other hand, to reconstitute brush border membrane proteins, which have been incorporated into nontransporting proteoliposome by Step I,C, 50 to 200 μl of these proteoliposomes is mixed with the same volume of mainly monolamellar cholesterol/phosphatidylserine/phosphatidylcholine liposomes, which have been formed by Step

II,B. The mixtures of the nontransporting proteoliposomes and of the liposomes formed from exogenous lipids are incubated for 15 min at 41° and frozen in liquid nitrogen. For transport measurements the frozen samples are thawed for 5 min in a shaking water bath at 37°. The samples are then diluted with KC buffer (37°) and centrifuged for 15 min at 150,000 g. The pellets are resuspended in KC buffer (37°) and multilamellar proteoliposomes and protein–lipid aggregates are removed by a 30-sec centrifugation at 8000 g (Fig. 2b and d).

Comments

By the described freeze–thaw procedure highly active Na^+ cotransport of D-glucose, L-alanine, L-proline, L-glutamic acid, lactate, and succinate could be reconstituted.[14] In Fig. 3 time courses of D-glucose and L-glutamate uptake are shown in proteoliposomes which were obtained after freezing and thawing of a mixture of type I,B nontransporting proteoliposomes and mainly multilamellar cholesterol/phosphatidylserine liposomes (type II,A). The uptake measurements were performed at K^+ equilibrium or in the presence of an inwardly directed Na^+ plus an outwardly directed K^+ gradient. This type of proteoliposome is very impermeable to Na^+ and D-glucose. From Fig. 3 it can be seen that even after a 1-hr incubation, a Na^+ gradient-dependent increase of D-glucose concentration in the proteoliposomes was observed. However, proteoliposomes formed by freeze–thaw reconstitution of a mixture of type I,C nontransporting porteoliposomes and mainly monolamellar cholesterol/phosphatidylcholine/phosphatidylserine liposomes (type II,B) had a much higher passive permeability for Na^+ and D-glucose. With the electron microscope it can be seen that both types of transporting proteoliposomes are mainly monolamellar (compare Fig. 2b and d). Their mean diameter was 0.50 ± 0.37 μm as revealed from measurements on electron micrographs. It was found that low passive permeability for Na^+ and D-glucose in freeze–thaw reconstituted proteoliposomes was correlated with the addition of cholesterol.[43] Addition of phosphatidylserine during reconstitution was required to obtain highly active Na^+-D-glucose cotransport. However, following reconstitution, identical Na^+ gradient-dependent uptake rates of L-glutamate and L-alanine were measured if during freeze–thaw reconstitution either cholesterol/phosphatidylserine or cholesterol/phosphatidylcholine liposomes were added.

Partial Purification of the Na^+-D-Glucose Cotransporter

Many attempts have been made in our group as in other laboratories[6-18,42] to purify the Na^+-D-glucose cotransporter by fractionation of

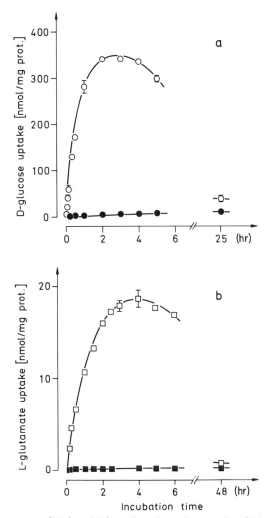

FIG. 3. Time courses of Na^+ and K^+ gradient-dependent uptake of (a) D-glucose and (b) L-glutamate. D-Glucose and L-glutamate uptake was measured in proteoliposomes, which were formed by freeze–thaw reconstitution in the same way as those presented in Fig. 2b. The reconstituted proteoliposomes containing 100 mM K^+ were preincubated without (b) or with 2×10^{-8} M valinomycin. Then they were incubated with 0.2 mM radioactively labeled D-glucose or 6 μM radioactively labeled L-glutamate at K^+ equilibrium (closed symbols) or in the presence of concentration differences of 89 mM Na^+ (out > in) and 89 mM K^+ (in > out) (open symbols). Uptake was measured as described previously.[8,14]

solubilized membrane polypeptides and by demonstrating Na^+-dependent high-affinity phlorizin binding and/or Na^+-D-glucose cotransport during and after fractionation. To fractionate the membrane proteins in the presence of detergents the following procedures were employed: (1) affinity chromatography on a phlorizin polymer,[6] (2) affinity chromatography with concanavalin A,[9,10] (3) chromatography with hydroxylapatite,[7,9] (4) gel filtration,[11,13,17] (5) chromatofocusing,[11,12,17] (6) gel electrophoresis,[15] (7) ion-exchange chromatography,[16] and (8) density gradient centrifugation.[42] Unfortunately these experiments were hampered by the fact that phlorizin binding and/or D-glucose transport was partially or totally inactiviated during separation. It proved to be extremely difficult to find conditions which allow the separation of the Na^+-D-glucose cotransporter and the detection of phlorizin binding or transport activity in protein fractions obtained during separation. Thus, in these experiments Na^+-D-glucose cotransport after reconstitution could not be significantly enriched over that measured in intact brush border membrane vesicles[6-14,18,42] and Na^+-dependent high-affinity phlorizin binding related to the amount of membrane protein could not be enriched more than 10-fold over that measured in intact brush border membranes.[8,17,18,39,42,48] Furthermore, the yields of active transport protein were low.

We were able to optimize conditions (presence of Na^+, D-glucose, and phosphatidylserine) under which the Na^+-D-glucose cotransporter remained active for more than 48 hr after solubilization of brush border membranes with deoxycholate or octylglucoside. However, employing the same buffers and detergent concentrations and adding phosphatidylserine and/or other phospholipids more than 95% of Na^+-D-glucose cotransport activity and phlorizin binding was irreversibly inactivated when the solubilized membrane proteins were separated by gel chromatography, ion-exchange chromatography, or density gradient centrifugation. Residual transport or phlorizin-binding activity was found only in fractions which still contained a relatively complex polypeptide composition. Since these and other observations[29,32-34,49] suggest that the Na^+-D-glucose cotransporter is an oligomeric protein we (1) tried to purify the transporter without dissociating the presumed polypeptide components and (2) we are currently attempting to purify transporter components which have been identified by affinity labeling and with monoclonal antibodies[28,29] in order to later attempt the reassociation of the functional transporter.

[48] In a recent report[16] the authors claim about a 50-fold enrichment of Na^+-dependent binding sites when they reactivate their protein with an excess of phospholipids. However, the significance of these data cannot be evaluated, since saturation of the analyzed phlorizin binding was not unequivocally demonstrated (see Fig. 2 in Ref. 16).

[49] H. Koepsell, G. Fritzsch, K. Korn, and A. Madrala, *J. Membr. Biol.* **114**, 113 (1990).

As starting material for the purification of the functional transporter complex and of transporter components, aggregated mixed micelles were used which were prepared from deoxycholate-solubilized brush border membranes as described above (Fig. 1b and c and Fig. 4a and b). In the aggregated mixed micelles the number of Na^+-dependent high-affinity phlorizin-binding sites is about four times higher than in intact membrane vesicles and they contain concentrations of Na^+ cotransporters for lactate, succinate, L-alanine, and L-glutamate which are at least as high as in the intact brush border membrane. A further purification of the native Na^+-D-glucose cotransporter was achieved if from the aggregated mixed micelles extraneous proteins were removed by incubation with a low concentration of deoxycholate and with alkaline pH.[50] In the removed proteins no phlorizin binding was detected whereas in the remaining mixed micelles the number of high-affinity phlorizin-binding sites was increased about eightfold over that measured in intact membrane vesicles (600 to 1000 pmol phlorizin/mg of protein). The transporter was partially inactivated since the total recovery of phlorizin-binding sites was only 5–7%. After reconstitution of the partially purified sample Na^+-coupled transport of D-glucose was detected.[40] In addition Na^+ gradient-dependent uptake of L-alanine and L-glutamate was found. By this procedure, in which about 0.2 to 0.4 mg protein of the partially purified transporter was obtained from 100 mg of total brush border membrane protein, most membrane polypeptides with molecular weights larger than 80,000 and smaller than 40,000 were removed (Fig. 4c). A polypeptide with a molecular weight of about 75,000 was constantly enriched and differing amounts of polypeptides with molecular weights of 60,000 and 47,000 were obtained in different preparations (see Ref. 50 and Fig. 4c). The polypeptides with the apparent molecular weights of 75,000, 60,000, and 47,000 are supposed to be components and splitting products of the porcine renal Na^+-D-glucose cotransporter since they are labeled with the covalently binding D-glucose analog BADG[28] and react with monoclonal antibodies directed against the transporter.[29] Since the M_r 60,000 polypeptide observed in porcine kidney was not observed in rat or rabbit kidney when protease inhibitors were added directly after slaughter, the M_r 60,000 polypeptide observed in pig is supposed to be a proteolytic splitting product of the M_r 75,000 polypeptide. After purification more than 60% of the protein in the sample was represented by the polypeptide components of the Na^+-D-glucose cotransporter with the apparent molecular weights of 75,000, 60,000, and 47,000.

To obtain larger amounts of polypeptide components of the renal Na^+-D-glucose cotransporter, brush border membranes were solubilized

---

[50] H. Koepsell, M. Neeb, and A. Madrala, *INSERM Symp.* **26**, 117 (1986).

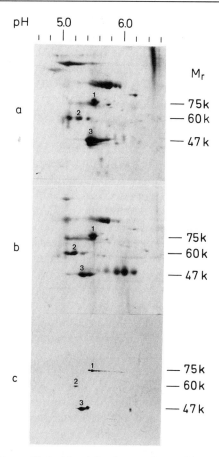

FIG. 4. Protein patterns of intact brush border membranes (a), aggregated mixed micelles (b), and the partially purified Na^+-D-glucose cotransporter (c). The preparation of brush border membranes (a) and of aggregated mixed micelles (b), which were formed from solubilized brush border membranes by a 3-hr dialysis against a buffer containing 0.1% deoxycholate, is described in the text. A preparation of partially purified Na^+-D-glucose cotransporter (c) was obtained by deoxycholate and alkaline treatment of aggregated mixed micelles (see text). Proteins were first separated by isoelectric focusing in rod gels and then by SDS gel electrophoresis in slab gels.[14] Staining of the proteins was performed with silver. Polypeptides which have been labeled with covalently binding D-glucose analogs[28] are indicated by the numbers 1–3. Affinity labeling of polypeptides 1 and 3 could be protected by D-glucose or phlorizin. Several monoclonal antibodies directed against the Na^+-D-glucose cotransporter cross-react with polypeptides 1, 2, and 3.[29] Since solubilized brush border membrane proteins do not enter two-dimensional gels quantitatively the purity of fractionated proteins is often overestimated in two-dimensional gels.

with decaethylene glycol lauryl ether ($C_{12}E_{10}$) and separated by ion-exchange chromatography (Fig. 5). The obtained protein fractions were separated by SDS gel electrophoresis, blotted to nitrocellulose membranes, and analyzed with monoclonal antibodies which react with the polypeptide components of the renal Na^+-D-glucose cotransporter.[29,30] This procedure copurifies polypeptide components of the Na^+-D-glucose cotransporter with the apparent molecular weights of 75,000 and 60,000 (see Fig. 5, fractions 9 and 10). Starting from 100 mg of total brush border membrane protein about 2 to 4 mg protein can be separated, which consists of about 90% of these polypeptides. This purification procedure needs 1 day and can be scaled up easily. After further optimization and/or after adding a further purification step it should be possible to obtain milligram amounts of purified polypeptide components of the transporter.

Procedure to Remove Extraneous Proteins from Aggregated Mixed Micelles

Brush border membranes are solubilized with 0.4% deoxycholate in the presence of 1 M D-glucose, the nonsolubilized membranes are removed by a 1-hr centrifugation at 200,000 g, and the supernatant is dialyzed for 3 hr against TRA buffer (8°) containing 0.1% deoxycholate as described above. The formed aggregated mixed micelles (Fig. 1b) are sedimented by a 1-hr centrifugation at 200,000 g. The sediment containing between 25 and 30 mg of protein is suspended in 2 ml of TRA buffer (8°) and homogenized by five strokes with a glass-Teflon homogenizer. To remove proteins from the aggregated mixed micelles by incubation with deoxycholate solutions (1) TRA buffer containing 2 M D-glucose, (2) TRA buffer containing protease inhibitors and, finally, (3) TRA buffer containing 4% sodium deoxycholate are added so that the indicated final concentrations of protein (5 mg/ml), deoxycholate (0.2%, w/v), D-glucose (1 M), phenylmethylsulfonyl fluoride (PMSF) (0.1 mM), leupeptin (10 μg/ml), benzamidine (1 mM), and aprotinin (20 μg/ml) are obtained. After a 1-hr incubation at 8° the sample is centrifuged for 1 h at 200,000 g. The sediment is suspended in 1 ml of TRA buffer. Then 4 ml of 300 mM D-glucose in water, pH 7.4 (8°), is added and the mixture is centrifuged for 15 min at 300,000 g (8°). The pellet is suspended in 500 μl of an aqueous solution of 300 mM D-glucose, pH 7.4 (0°), and 2 ml of an ice-cold aqueous solution of 300 mM D-glucose, which has been adjusted to pH 12 with NaOH, is added. After homogenization by five strokes with a glass-Teflon homogenizer the mixture is incubated for 10 min on ice. Thereafter it is centrifuged for 15 min at 300,000 g and the pellet is suspended in KC buffer and analyzed for Na^+-dependent high-affinity phlorizin binding. For

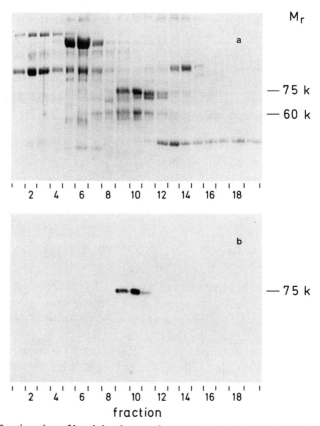

FIG. 5. Fractionation of brush border membrane proteins by ion-exchange chromatography. Solubilized brush border membrane proteins were applied to a DEAE-column and the proteins were eluted by an NaCl gradient (see text). The protein pattern of the different protein fractions which was analyzed by SDS-polyacrylamide gel electrophoresis, Western blotting, and staining, is shown in (a). (b) An autoradiogramm which was obtained after immunoreaction of the blotted protein fractions with a monoclonal antibody against the Na^+-D-glucose cotransporter (R4A6).[29] This antibody reacts nearly exclusively with the M_r 75,000 polypeptide component of the transporter. Also, the M_r 60,000 polypeptide, which copurifies with the M_r 75,000 polypeptide (see fractions 9 and 10), is considered to be a component or degradation product of the porcine renal Na^+-D-glucose cotransporter, since several monoclonal antibodies against the porcine renal Na^+-D-glucose cotransporter[29] cross-react with the M_r 75,000 and 60,000 polypeptides.

reconstitution of Na^+-D-glucose cotransport nontransporting proteoliposomes are formed by dialysis after solubilization with octylglucoside (see above, Step I,C), cholesterol/phosphatidylcholine/phosphatidylserine liposomes are added (see above, Step II,B) and freeze–thaw reconstitution was performed (see above, Step III).

Partial Purification of Polypeptide Components of the Renal Na^+-D-Glucose Cotransporter by Ion-Exchange Chromatography

Brush border membranes containing 5 mg of protein are solubilized by gently stirring (1 hr, room temperature) in 1 ml of 10 mM NaH_2PO_4/Na_2HPO_4, pH 7.4, containing 2 M D-glucose, 2% (w/v), decaethylene glycol lauryl ether ($C_{12}E_{10}$), 20 µg aprotinin, and 0.25 mM diisopropylfluorophosphate. Unsolubilized material is removed by ultracentrifugation (1 hr, 200,000 g, 22°), and the supernatant containing about 3.5 mg of protein is applied to a DEAE-column (i.d. 1.6 cm, height 2.5 cm) which is equilibrated with running buffer [room temperature, 10 mM NaH_2PO_4/Na_2HPO_4, pH 7.4, 1% (w/v) $C_{12}E_{10}$, 1 M D-glucose, 1 µg/ml aprotinin, 0.1 mM diisopropylfluorophosphate]. After washing with 6 vol of running buffer the bound proteins are eluted with a linear gradient of NaCl in running buffer. To analyze the protein fractions for immunoreactivity with monoclonal antibodies against the Na^+-D-glucose cotransporter, the protein fractions obtained from the column are dialyzed extensively against 0.1% (w/v) of SDS in water. Then they are lyophilized and suspended in a small volume of sample buffer for SDS polyacrylamide gel electrophoresis. After gel electrophoresis and Western blotting the polypeptide components of the Na^+-D-glucose cotransporter are detected with our monoclonal antibodies R4A6 and R1C4, which react with the Na^+-D-glucose cotransporter.[29]

Open Questions

Concerning the structure of the Na^+-D-glucose cotransporter and of other Na^+-coupled cotransport proteins, many questions remain open. The polypeptide composition of the different transport systems is not known although components of the Na^+-D-glucose cotransporter, a component of the Na^+-proline cotransporter, and, furthermore, some polypeptides containing Na^+-binding sites have been identified.[19-21,28,29] In addition, a component of the intestinal Na^+-D-glucose cotransporter from rabbit and from man has been cloned[22,51] which shows some homology with a presumed polypeptide component of a human renal Na^+-D-glucose cotransporter.[52] For the Na^+-D-glucose cotransporter an oligomeric protein has to be assumed since radiation inactivation experiments showed that the functional molecular weight for Na^+-D-glucose cotransport is larger than 300,000.[32,33] It has to be elucidated whether the transporter is composed of one or of different types of polypeptides. In the first case the transporter

[51] E. Turk, M. A. Hediger, and E. M. Wright, *FASEB J.* abstr. 1950 (1989).

may be a tetramer of M_r 75,000 polypeptides since an M_r 75,000 polypeptide has been identified as a component of the Na^+-D-glucose cotransporter in intestine and in kidney,[19,21,28,29] and may be identical to the recently cloned polypeptides.[22,51] An M_r 47,000 polypeptide has been shown to be also related to Na^+-D-glucose cotransport, since this polypeptide was also labeled with covalently binding D-glucose analogs[28] and cross-reacts with monoclonal antibodies which alter the function of the renal and intestinal Na^+-D-glucose cotransporter and bind to the M_r 75,000 polypeptide component of the transporter.[29,30,31] Since proteolytic digestion cannot be completely excluded it cannot be decided whether the functional transporter is exclusively composed of M_r 75,000 polypeptides or whether it is composed of partially homologous polypeptide subunits with the molecular weights of 75,000 and 47,000.

The structure of the Na^+-D-glucose cotransporter from intestine and kidney appears to be similar but not identical since some of our monoclonal antibodies, which inhibit Na^+-D-glucose-cotransport and react with M_r 75,000 and 47,000 polypeptides in pig kidney, bind to M_r 75,000 and 47,000 polypeptides in rat kidney and rat intestine and have differential effects on Na^+-D-glucose cotransport and phlorizin binding in kidney and intestine.[29-31]

Recently we obtained data which suggest that different Na^+ cotransport systems contain homologous amino acid sequences and are functionally linked to each other. Thus, employing monoclonal antibodies against components of the Na^+-D-glucose cotransporter which alter Na^+ gradient-dependent D-glucose transport and/or Na^+-dependent phlorizin binding, it was found that antibody binding to native brush border membrane proteins was not only altered by D-glucose (and not by D-mannose) but also by lactate of by L-alanine (and not be D-alanine). Since the effects of D-glucose and lactate or of D-glucose and L-alanine were not additive they cannot be explained by cross-reactivity of the antibodies with separate Na^+ cotransporters.[29] Functional coupling of different cotransporters was assumed since with one monoclonal antibody which inhibits Na^+-D-glucose cotransport, effects of L-alanine or of L-lactate on the antibody binding were observed only if D-glucose was also present during the antigen–antibody reaction. Functional coupling of Na^+ cotransport systems could also be demonstrated by affinity-labeling experiments. Thus, we recently observed that covalent labeling of the renal M_r 75,000 component of the Na^+-D-glucose cotransporter with 10-N-(bromoacetyl)amino-1-decyl-β-D-glucopyranoside, which is reduced in the presence of D-glucose,[28] was increased if 1 mM L-lactate was present.[53] Notwithstanding that there are different ways to explain functional coupling it may be speculated that the different Na^+ cotransporters, which are supposed to be oligomeric proteins, contain partially identical subunits.[54] Another possibility is that the Na$^+$ cotransporters are composed of highly homologous polypeptides with molecular weights around 75,000 and that the active transporters are di- or oligomers of M_r 75,000 polypeptides, which may be specific for one or two transport systems.

To finally elucidate the compositions of the Na$^+$ cotransporters the labeled components have to be isolated under native conditions or have to be produced by genetic experiments in quantitative amounts. They will then have to be reconstituted separately and after reassociation of different combinations of polypeptides. Then Na$^+$ cotransport will need to be measured with different substrates. In spite of the progress obtained in the last years this challenging task seems to be difficult. However, this important step is required to understand the mechanism of Na$^+$ cotransport in kidney and intestine.

---

[52] M. A. Hediger, E. Turk, A. M. Pajor, T. K. Mohandas, and E. M. Wright, *FASEB J.* abstr. 1951 (1989).
[53] H. Koepsell, unpublished observations.
[54] D. Zilberstein, I. J. Ophir, E. Padan, and S. Schuldiner, *J. Biol. Chem.* **257**, 3692 (1982).

Section II

Stimulus Secretion Coupling in Epithelia

[35] Receptor Identification

By RAKESH VINAYEK and JERRY D. GARDNER

Receptors play a key role in regulating cellular processes by enabling the cell to interact with the outside environment. A receptor is a molecular entity on the target tissue which, when occupied by an agonist, initiates a biological response. A receptor consists of two essential elements—a regulatory unit or binding site that recognizes and binds a specific ligand and an effector unit or transducer that transduces the agonist–receptor interaction into a biological response through a cascade of events initiated by the activation of an effector system such as adenylate cyclase or phospholipase C.

Whereas peptides and catecholamines interact with receptors localized on the cell surface, steroids and thyroxine interact with receptors present intracellularly. In the present chapter, we will focus on the methods of receptor identification for receptors localized on the cell surface.

Biological Activity

Materials

The following are required: Agonist, antagonist, and an *in vitro* or *in vivo* biological system, e.g., dispersed cells (*in vitro* model) or isolated organ preparation (*in vivo* model). A suitable assay system for measuring biological activity should be easy to prepare, give an easily measurable biological response, give reproducible results, and preferably permit multiple representative samples to be taken at appropriate times.

Methods

Prepare the assay system and measure the response to an agonist. The response is usually measured using varying concentrations of an agonist with fixed incubation times. Kinetic studies are usually not helpful in receptor identification.

Table I lists the various tissues in which measurement of a biological response has indicated the presence of a particular type of receptor.[1-44]

---

[1] J. D. Gardner and R. T. Jensen, *in* "Physiology of the Gastrointestinal Tract" (L. R. Johnson, ed.), 2nd ed., p. 1109. Raven, New York, 1987.

[2] A. H. Soll, *Handb. Physiol. Sect. 6: Gastrointestinal Syst.* **2**, 193 (1989).

TABLE I
DIFFERENT TYPES OF TISSUES TESTED TO IDENTIFY DIFFERENT CLASSES OF RECEPTORS[a]

Class of receptors	Tissue preparation[b]
Muscarinic cholinergic	Pancreatic acinar cells,[1] gastric parietal cells,[2] gastric chief cells,[3-6] gastric fundic mucosal cells,[7,8] small and large intestinal epithelium and smooth muscle cells,[9,10] lower esophageal sphincter muscle,[11-13] gastric smooth muscle cells,[14] gall bladder smooth muscle cells[14]
CCK	Pancreatic acinar cells,[1] gastric smooth muscle cells,[15] gall bladder smooth muscle and sphincter of Oddi,[16] gastric chief cells,[3,4,5] gastric parietal cells,[2] and gastric fundic somatostatin cells,[2,17] lower esophageal sphincter muscle[14,16,18]
Gastrin	Pancreatic acinar cells,[19] gastric chief cells,[3,4,15] parietal cells[2] and gastric fundic mucosal cells,[2] gastric smooth muscle cells,[14,16,20] lower esophageal sphincter muscle[18]
Bombesin	Pancreatic acinar cells,[1] antral gastrin cells,[2] gastric smooth muscle cells,[16] small intestinal smooth muscle cells,[16,21] gall bladder smooth muscle cells,[16,22] lower esophageal smooth muscle[16,23]
Substance P	Pancreatic acinar cells,[1] gastric smooth muscle cells,[16] gall bladder smooth muscle cells,[14,16,22] small intestinal smooth muscle cells[14,24,25]
VIP preferring	Pancreatic acinar cells,[1] gastric smooth muscle cells,[14,16] small intestinal smooth muscle cells,[16] hepatocytes,[26] small and large intestinal epithelium[27]
Secretin preferring	Pancreatic acinar cells,[1] gastric and fundic mucosal glands,[28,29] gastric chief cells[3]
Histamine	Gastric parietal cells,[2] gastric chief cells[3] and gastric fundic mucosal cells,[2] small intestinal smooth muscle cells[16]
Cholera toxin	Pancreatic acinar cells,[1,30] small intestinal epithelium[31]
β-Adrenergic	Gastric smooth muscle cells,[16,32,33] gastric fundic somatostatin and gastric mast cells[2]
Opiate	Gastric smooth muscle cells,[14,16] gall bladder smooth muscle cells,[16,34] small intestinal epithelium,[27] small intestinal smooth muscle cells[16,35]
Somatostatin	Pancreatic acinar cells,[36-39] gastric chief cells,[3] gastric parietal cells,[2,40] antral gastrin cells[2,41]
EGF	Gastric parietal cells,[2] gastric fundic mucosal cells,[2] small intestinal epithelium,[27] hepatocytes[26]
CGRP	Pancreatic acinar cells,[1,42] gastric smooth cells,[16] gastric parietal cells[43]
Motlin	Gastric smooth muscle cells[16,44]

[a] CCK, Cholecystokinin; VIP, vasoactive intestinal peptide; EGF, epidermal growth factor; CGRP, calcitonin gene-related peptide.
[b] Review articles are cited whenever possible to minimize the number of citations. Moreover, references are not intended to be all inclusive; instead, representative examples have been selected.

An agonist interacts with its receptors on the target tissue and causes the final response through a series of intermediate responses between receptor interaction and the final response of the target tissue. If one assumes that the agonist (A) interacts reversibly with its receptor (R) and forms an agonist–receptor complex (AR) that produces the final response through a series (n) of intermediate responses, the following reaction sequence can be written:

$$A + R \underset{K_2}{\overset{K_1}{\rightleftarrows}} AR \xrightarrow{(n)} \text{final response}$$

where K_1 and K_2 are the rate constants for association and dissociation of the agonist with its receptor, respectively. The initial interaction between the agonist and its receptor is analogous to the interaction of a substrate with an enzyme and the magnitude of the response can be analyzed using an equation that is similar to that for enzymatic product formation:

$$RS = RS_{max} [A]/(K_d + [A])$$

---

[3] S. J. Hersey, in "Physiology of the Gastrointestinal Tract" (L. R. Johnson, ed.), 2nd ed., p. 947. Raven, New York, 1987.

[4] D. K. Kasbekar, R. T. Jensen, and J. D. Gardner, *Am. J. Physiol.* **244,** G392 (1983).

[5] H. R. Koelz, S. J. Hersey, G. Sachs, and C. S. Chew, *Am. J. Physiol.* **243,** G218 (1982).

[6] V. E. Sutliff, S. Rattan, J. D. Gardner, and R. T. Jensen, *Am. J. Physiol.* **257,** G226 (1989).

[7] R. Hammer, *Scand. J. Gastroenterol.* **17** (Suppl. 65), 5 (1980).

[8] M. G. Herawi, G. Lambrecht, E. Mustschler, U. Moser, and A. Pfeiffer, *Gastroenterology* **94,** 630 (1988).

[9] X.-Y. Tein, R. Wahawisan, L. J. Wallace, and T. S. Gaginella, *Life Sci.* **36,** 1949 (1985).

[10] R. Wahawisan, L. J. Wallace, and T. S. Gaginella, *J. Pharm. Pharmacol.* **38,** 150 (1986).

[11] R. K. Goyal and S. Rattan, *Gastroenterology* **74,** 598 (1978).

[12] R. Gilbert, S. Rattan, and R. K. Goyal, *J. Pharmacol. Exp. Ther.* **230,** 284 (1984).

[13] S. Rattan and R. K. Goyal, *Trends Pharmacol. Sci., Suppl. p. 78 (1984).*

[14] G. M. Makhlouf, in "Physiology of the Gastrointestinal Tract" (L. R. Johnson, ed.), 2nd ed. p. 555. Raven, New York, 1987.

[15] J. A. Cherner, V. E. Sutliff, D. M. Grybowski, R. T. Jensen, and J. D. Gardner, *Am. J. Physiol.* **254,** G151 (1988).

[16] G. M. Makhlouf and J. R. Grider, *Handb. Physiol. Sect. 6: Gastrointestinal Syst.* **2,** 281 (1989).

[17] A. H. Soll, T. Yamada, J. Park, and L. P. Thomas, *Am. J. Physiol.* **247,** G558 (1984).

[18] S. Rattan and R. K. Goyal, *Gastroenterology* **90,** 94 (1986).

[19] D.-H. Yu, M. Noguchi, Z.-C. Zhou, M. L. Villanueva, J. D. Gardner, and R. T. Jensen, *Am. J. Physiol.* **253,** G793 (1987).

[20] D. Menozzi, J. D. Gardner, R. T. Jensen, and P. N. Maton, *Am. J. Physiol.* **257,** G73 (1989).

[21] R. Micheletti, J. R. Grider, and G. M. Makhlouf, *Regul. Pept.* **21,** 219 (1988).

[22] C. Severi, J. R. Grider, and G. M. Makhlouf, *J. Pharmacol. Exp. Ther.* **245,** 195 (1988).

[23] T. Von-Schrenck, P. Heinz-Erian, T. Moran, S. A. Mantey, J. D. Gardner, and R. T. Jensen, *Am. J. Physiol.* **256,** G747 (1989).

where RS is the response and RS_{max} is the maximal response. A is the agonist concentration and K_d (K_2/K_1) is the apparent dissociation constant for the agonist–receptor complex. This equation describes a simple rectangular hyperbola. There is no response at [A] = 0. When [A] = K_d the response is half-maximal and the maximal response is obtained asymptotically as [A] increases above K_d. A sigmoid dose response is obtained if the RS is plotted as a function of log[A]. An example of a sigmoid dose–response curve is illustrated by vasoactive intestinal peptide (VIP)-stimulated amylase secretion from dispersed pancreatic acini. As shown in Fig. 1, with increasing concentrations of VIP amylase secretion increases, becomes maximal, and then remains constant.

The stoichiometric pattern of the biological response for some agonists may vary from the pattern described above for VIP. For example, cholecystokinin (CCK)-8-stimulated amylase secretion from pancreatic acini gives a biphasic dose–response curve. As shown in Fig. 2, with increasing concentrations of CCK-8 enzyme secretion increases, becomes maximal, and then decreases. Broad biphasic dose–response curves such as that

---

[24] J.-C. Souquet, J. R. Grider, K. N. Bitar, and G. M. Makhlouf, *Am. J. Physiol.* **249**, G533 (1985).

[25] Deleted in proof.

[26] G. Rosselin, *Handb. Physiol. Sect. 6: Gastrointestinal Syst.* **2**, 245 (1989).

[27] M. Laburthe and B. Amiranoff, *Handb. Physiol. Sect. 6: Gastrointestinal Syst.* **2**, 215 (1989).

[28] C. Gespach, D. Bataille, N. Vauclin, L. Moroder, E. Wünsch, and G. Rosselin, *Peptides* **7** (Suppl. 1), 155 (1986).

[29] W. Bawab, C. Gesbach, J.-C. Marie, E. Chastre, and G. Rosselin, *Life Sci.* **42**, 791 (1988).

[30] J. D. Gardner and A. J. Rottman, *Biochim. Biophys. Acta* **585**, 250 (1979).

[31] J. Holmgren, I. Lonnroth, and L. Svennerholm, *Infect. Immun.* **8**, 208 (1973).

[32] T. Honeyman, P. Merriam, and F. S. Fay, *Mol. Pharmacol.* **14**, 86 (1977).

[33] C. R. Scheid, T. W. Honeyman, and F. S. Fay, *Nature (London)* **277**, 32 (1979).

[34] C. Severi, J. R. Grider, and G. M. Makhlouf, *Life Sci.* **42**, 2373 (1988).

[35] K. N. Bitar and G. M. Makhlouf, *Life Sci.* **37**, 1545 (1985).

[36] J. P. Esteve, C. Susini, N. Vaysse, H. Antoniotti, E. Wunsch, G. Berthon, and A. Ribet, *Am. J. Physiol.* **247**, G62 (1984).

[37] T. Matozaki, C. Sakamoto, M. Nagao, and S. Baba, *J. Biol. Chem.* **261**, 1414 (1986).

[38] T. Matozaki, C. Sakamoto, M. Nagao, and S. Baba, *Horm. Metab. Res.* **20**, 141 (1988).

[39] N. Vigurie, N. Tahiri-Jouti, J.-P. Esteve, P. Clerc, C. Logsdon, M. Svoboda, C. Susini, N. Vaysse, and A. Ribet, *Am. J. Physiol.* **255**, G113 (1988).

[40] J. Park, T. Chiba, and T. Yamada, *J. Biol. Chem.* **262**, 14190 (1987).

[41] R. F. Harty, D. G. Malco, and J. E. McGuigan, *Gastroenterology* **81**, 707 (1981).

[42] Z.-C. Zhou, M. L. Villanueva, M. Noguchi, S. W. Jones, J. D. Gardner, and R. T. Jensen, *Am. J. Physiol.* **251**, G391 (1986).

[43] Y. Umeda and T. Okada, *Biochem. Biophys. Res. Commun.* **146**, 430 (1987).

[44] D. S. Louie and C. Owyang, *Am. J. Physiol.* **254**, G210 (1988).

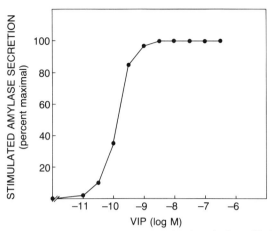

Fig. 1. A sigmoid dose–response curve for vasoactive intestinal peptide (VIP)-stimulated amylase secretion from dispersed acini from guinea pig pancreas. Acini were incubated with VIP for 30 min at 37°. Stimulation of amylase secretion is expressed as percentage maximal stimulation. (Data from Gardner and Jensen.[1])

illustrated in Fig. 2 raise the possibility of the existence of multiple subclasses of receptors for the agonist.[1]

Differences in the relative potencies of different agonists in producing a biological response can help to identify different classes of receptors. For example, as illustrated in Fig. 3 the difference in the relative potencies of CCK-8 and gastrin in terms of stimulating amylase secretion from dispersed pancreatic acini and causing contraction of gastric smooth muscle cells is helpful in distinguishing CCK receptors on pancreatic acini[19] from gastrin receptors on gastric smooth muscle cells.[20] In pancreatic acini, CCK-8 is 1000-fold more potent than gastrin in terms of stimulating amylase secretion (Fig. 3, left). In contrast, in gastric smooth muscle cells, CCK-8 is equipotent with gastrin in terms of causing contraction of smooth muscle cells (Fig. 3, right).

As illustrated in Fig. 4, occupation of a receptor by an agonist usually activates one of two functionally distinct biochemical pathways in the target cell. One pathway involves receptor occupation, activation of adenylate cyclase, increased cellular cyclic AMP, activation of cyclic AMP-dependent protein kinase (protein kinase A), phosphorylation of one or more cellular proteins, and then, after a series of unknown steps, the final response. The other pathway includes receptor occupation, activation of phospholipase C, formation of diacylglycerol and inositol phosphates, mobilization of cellular calcium, activation of protein kinase C, phosphoryla-

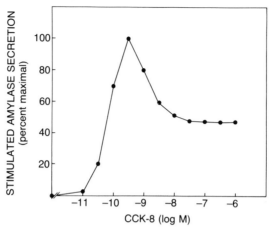

FIG. 2. A biphasic dose-response curve for cholecystokinin (CCK)-8 stimulated amylase secretion from dispersed acini from guinea pig pancreas. Acini were incubated with CCK-8 for 30 min at 37°. Stimulation of amylase secretion is expressed as percentage maximal stimulation. (Data from Gardner and Jensen.[1])

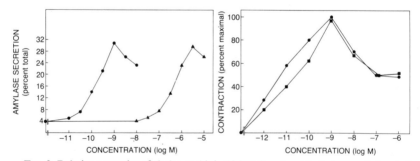

FIG. 3. Relative potencies of cholecystokinin (CCK)-8 and gastrin in terms of stimulating amylase secretion from pancreatic acini (left) and causing contraction of gastric smooth muscle cells (right). *Left:* Dose-response curves for CCK-8- (●) and gastrin (▲)-stimulated amylase secretion in dispersed acini from guinea pig pancreas. Acini were incubated at 37° for 30 min with the indicated concentrations of CCK-8 or gastrin. Amylase secretion is expressed as the percentage of amylase activity in the acini at the beginning of the incubation that was released into the incubation medium during the incubation (i.e., percentage total). *Right:* Dose-response curves for CCK-8 (●) and gastrin (■)-stimulated contraction of dispersed smooth muscle cells from guinea pig stomach. Cells were incubate at 31° for 30 sec with the indicated concentrations of CCK-8 or gastrin and results are expressed as a percentage of the decrease in cell length obtained with 1 nM CCK-8. (Results from Yu et al.[19] and Menozzi et al.[20])

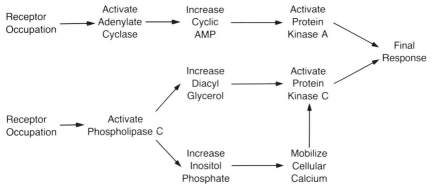

FIG. 4. Two pathways mediating action of an agonist.

tion of one or more cellular proteins, and then, after a series of unknown steps, the final response.

Usually the initial steps in these two pathways are functionally distinct. That is, agonists that cause release of cellular calcium usually do not increase cyclic AMP or alter the increase in cyclic AMP caused by other agonists. Similarly, agonists that increase cellular cyclic AMP usually do not alter calcium mobilization or alter the increase in calcium mobilization caused by other agonists.

Although the two different pathways have initial steps that are functionally distinct, in some systems, such as pancreatic acinar cells,[45] gastric chief cells,[46] or gastric parietal cells,[47] these two pathways interact synergistically at some presently unknown step, resulting in potentiation of enzyme secretion. That is, the increase in response caused by an agonist that increases cyclic AMP in combination with an agonist that causes calcium mobilization is substantially greater than the sum of the effect of each agonist acting alone. Although the biochemical basis of potentiation is not known, this phenomenon serves to amplify the signal generated in response to occupation of a particular class of receptors by an agonist. In other systems, such as gastrointestinal smooth muscle,[24] these two functionally distinct pathways counteract each other in that activation of the phospholipase C pathway causes contraction, whereas activation of the adenylate cyclase pathway causes relaxation.

[45] M. J. Collen, V. E. Sutliff, G. Z. Pan, and J. D. Gardner, *Am. J. Physiol.* **242**, G423 (1982).
[46] J.-P. Raufman, D. K. Kasbekar, R. T. Jensen, and J. D. Gardner, *Am. J. Physiol.* **245**, G525 (1983).
[47] A. H. Soll, *Gastroenterology* **83**, 216 (1982).

If the presence of one agonist potentiates the response of the system to a second agonist, the two agonists must act through two functionally distinct classes of receptors. For example, in dispersed pancreatic acini[45] or dispersed chief cells,[46] potentiation of enzyme secretion occurs when carbachol or CCK-8 is combined with secretion of VIP.[45,46]

Measurement of changes in mediator or second messenger (such as changes in cyclic AMP or calcium) in response to the action of an agonist on the target tissue can be helpful in identifying different classes of receptors on the target tissue. If the interaction of two agonists with the target tissue results in activation of two different pathways, then those agonists must act through two different classes of receptors. For example, VIP and CCK, which interact with different classes of receptors on pancreatic acini, stimulate amylase secretion by increasing two different second messengers (i.e., by increasing cyclic AMP and mobilizing cellular calcium, respectively).[45] The converse, however, is not true, i.e., if two agonists cause a change in the same second messenger on interaction with the target tissue, then those agonists do not necessarily act through the same classes of receptors. For example, carbachol and CCK, which stimulate amylase secretion by interacting with two different classes of receptors on pancreatic acini, do so by increasing the same second messenger (i.e., mobilizing cellular calcium.)[45]

Dose-response curves for an agonist plotted on a logarithmic scale, first without and then with a competitive antagonist, are parallel and reach the same maximal response. The competitive antagonist causes a parallel rightward shift of the dose-response curve for the agonist. Figure 5 (left) illustrates the effect of a CCK receptor antagonist, proglumide analog 10 (CR 1409), on CCK-8-stimulated amylase secretion from dispersed pancreatic acini.[48] As shown in Fig. 5 (left), increasing concentrations of the antagonist cause a parallel rightward shift of the dose-response curve for the ability of CCK-8 to stimulate amylase secretion with no change in the maximal secretion.[48] A Schild plot[49] of these results gives a straight line with a slope that is not significantly different from unity (Fig. 5, right). The value of K_i (concentration of antagonist required to occupy 50% of the receptors with no agonist present) can be calculated from the Schild plot using the equation $y = ax + K_i$, where a is the slope and x is the concentration of the antagonist. The value of K_i is the intercept of the straight line on the abscissa.

---

[48] R. T. Jensen, D.-H. Yu, and J. D. Gardner, in "Cholecystokinin Antagonists" (R. Y. Wang, ed.), p. 53. Liss, New York, 1988.
[49] H. O. Schild, Br. J. Pharmacol. **4**, 277 (1949).

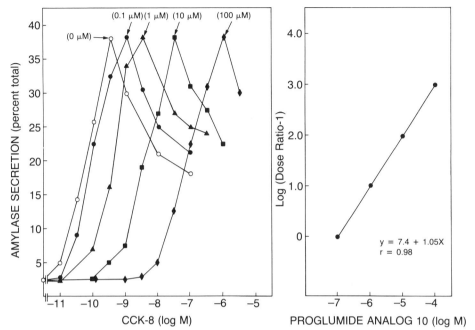

FIG. 5. Effect of various concentrations of proglumide analog 10 (CR 1409) on the dose–response curve for the action of CCK-8 on amylase secretion. Acini were incubated with the concentrations of proglumide analog 10 given in parentheses plus the indicated concentrations of CCK-8. Amylase secretion was measured during a 30-min incubation at 37° and is expressed as the percentage of amylase activity in the acini at the beginning of the incubation that was released into the incubation medium during the incubation (i.e., percentage total). The right panel illustrates results from the left panel plotted in the form described by Schild.[49] Dose-ratio is the ratio of the concentrations of CCK-8 required to give half-maximal stimulation of amylase secretion in the presence of a given concentration of proglumide analog 10 to the concentration of CCK-8 required to give half-maximal stimulation of amylase secretion without proglumide analog 10. The best fit to the line ($y = ax + b$) and correlation coefficient (r) were calculated by least-squares analysis. (Results from Jensen et al.[48])

Receptor antagonists can be used to identify and characterize various classes of receptors. For example, if an antagonist interacts with a single class of receptors, the antagonist will inhibit the biological activity of all agonists that interact with the same class of receptors as the antagonist but will not alter the action of agonists that interact with the other class of receptors. For example, in pancreatic acini (D-Phe^6)Bn(6–13)NH$_2$ is an antagonist that inhibits the action of agonists that interact with bombesin receptors [e.g., bombesin, gastrin-releasing peptide (GRP), and neuromedin C] but does not alter the action of agonists that interact with other classes of receptors.[50] Moreover, the antagonist will give the same value of K_i for each agonist that interacts with the same class of receptors as the antagonist.

If a receptor antagonist gives different values of K_i with each of several different agonists, one can conclude that each agonist interacts with a different class of receptors. One can also conclude that the antagonist interacts with several different classes of receptors. An example of this sort of phenomenon is illustrated by the action of (D-Pro^4,D-Trp^{7,9,10})substance P-4-11 on the stimulation of pancreatic enzyme secretion caused by substance P and bombesin.[51] The value of K_i obtained with bombesin (13.3 μM) is significantly greater than the value of K_i obtained with substance P (2.8 μM). On the other hand, the finding that an antagonist gives the same value of K_i with two different agonists does not necessarily indicate that these two different agonists interact with the same class of receptors. For example, the value of K_i on pancreatic acini for (D-Pro^4,D-Trp^{7,9,10})substance P-4-11 for inhibiting the action of substance P (2.8 μM) is not significantly different from the value of K_i for the antagonist for inhibiting the action of CCK-8 (3.7 μM). Receptor-binding studies show clearly, however, that substance P interacts with substance P receptors, that CCK-8 interacts with CCK receptors, and that the antagonist has the same apparent affinity for both classes of receptors.[51]

If several different receptor antagonists interact with more than one class of receptors, one may be able to use the relative potencies of the antagonists to distinguish the different classes of receptors. For example, the benzodiazepine derivatives, L-364,718 and L-365,260, each interacts with pancreatic CCK receptors and with gastrin receptors on gastric smooth muscle cells.[52] L-364,718 is more potent than L-365,260 in inhibiting the action of CCK on pancreatic enzyme secretion (Fig. 6, left) whereas L-365,260 is more potent than L-364,718 at inhibiting gastrin-induced contraction of gastric smooth muscle cells (Fig. 6, right).

An antagonist may also help to distinguish different subclasses of receptors by having different actions at the different subclasses. For example, in rat pancreatic acini, CCK-JMV-180, an analog of the C-terminal heptapeptide of CCK, distinguishes between high-affinity CCK receptors and low-affinity CCK receptors by acting as an agonist at the high-affinity CCK

---

[50] L. H. Wang, D. H. Coy, J. E. Taylor, N.-Y. Jiang, S. H. Kim, J.-P. Morean, S. C. Huang, S. A. Mantey, H. Frucht, J. D. Gardner, and R. T. Jensen, *Biochemistry* (in press).

[51] L. Zhang, S. Mantey, R. T. Jensen, and J. D. Gardner, *Biochim. Biophys. Acta* **972**, 37 (1988).

[52] S. C. Huang, L. Zhang, H.-C. Chiang, S. A. Wank, P. N. Maton, J. D. Gardner, and R. T. Jensen, *Am. J. Physiol.* **257**, G169 (1989).

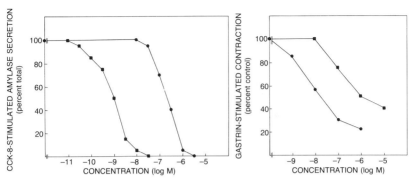

FIG. 6. Relative potencies for L-364,718 and L-365,260 to inhibit CCK-8-stimulated amylase secretion (left) or gastrin-stimulated contraction of gastric smooth muscle cells (right). *Left:* Comparison of the abilities of L-365,260 (●) to inhibit CCK-8-stimulated amylase secretion from guinea pig pancreatic acini. Acini were incubated at 37° for 30 min with no additions or 0.1 nM CCK-8 alone or plus the indicated concentrations of L-364,718 or L-365,260. Results are expressed as the percentage of stimulation caused by 0.1 nM CCK-8 alone (i.e., percentage control). *Right:* Comparison of the abilities of L-364,718 (■) and L-365,260 (●) to inhibit gastrin-stimulated contraction of guinea pig gastric smooth muscle cells. Dispersed gastric smooth muscle cells were incubated alone or with 0.1 nM gastrin with or without the indicated concentrations of L-364,718 and L-365,260. Results are expressed as the percentage of the contraction caused by 0.1 nM gastrin (percentage control). (Results from Huang et al.[52])

receptors and as an antagonist at the low-affinity CCK receptors.[53] As described earlier in this chapter, the dose–response curve for CCK-8-stimulated amylase secretion is biphasic in that as the concentration of CCK-8 increases enzyme secretion increases, reaches a maximum, and then decreases. In contrast, with increasing concentrations of CCK-JMV-180, amylase secretion increases, reaches the same maximum as that obtained with CCK-8, but does not decrease with supramaximal concentrations.[53] Moreover, supramaximal concentrations of CCK-JMV-180 inhibit the downstroke of the dose–response curve for CCK-8-stimulated enzyme secretion.[53]

Binding of Radiolabeled Ligands

Materials

The following are required: Radiolabeled ligand, agonist, antagonist, and tissue preparation e.g., dispersed cells or membranes, tissue sections,

[53] H. A. Stark, C. M. Sharp, V. E. Sutliff, J. Martinez, R. T. Jensen, and J. D. Gardner, *Biochim. Biophys. Acta* **1010**, 145 (1989).

or slices. A suitable tissue preparation for measuring binding of a radiolabeled ligand should be easy to prepare, give measurable binding, and permit multiple representative samples to be taken at appropriate times. The procedure for tissue preparation should not alter or degrade receptors of interest. A radiolabeled ligand used for binding studies should be of sufficiently high specific activity to allow binding to be measured, have a half-life sufficiently long so that the radiolabeled ligand can be used over a period of 2–3 months without a significant loss of specific activity, and have little or no degradation by the tissue preparation.

Methods

Incubate the radiolabeled ligand with a particular tissue preparation and measure the amount of radioactivity associated with the tissue at an appropriate time and temperature. Measurement of binding of radiolabeled ligand requires a technique that will separate radioactivity associated with the tissue (bound radioactivity) from that present in the incubation medium (free radioactivity). The three techniques used most frequently have been (1) washing by alternate centrifugation and resuspension whereby tissue-associated radioactivity is sedimented, (2) washing by filtration whereby the tissue-associated radioactivity is retained on the filter, and (3) washing by immersing slides with radioactivity bound to mounted tissue sections using an appropriate wash solution to wash away the free radioactivity.

The choice of a particular technique is usually determined by the nature of the tissue preparation and the characteristics of the radiolabeled ligand. For example, a membrane preparation may not sediment in a sufficiently short period of time to allow one to use centrifugation to separate bound from free radioactivity. A radiolabeled ligand may bind to the filter and thereby prevent one from obtaining a meaningful estimate of the amount of radioactivity bound to the tissue preparation. Irrespective of the technique used, the extent to which the technique influences the values obtained must be kept in mind and potential contributions of such artifacts as binding of a radiolabeled ligand to the filter should be assessed by performing appropriate control experiments. With the availability of appropriate radiolabeled ligands of high specific activity, it has become possible to measure directly the binding function of a receptor. Table II lists various radiolabeled ligands used in binding studies to identify different classes of receptors in various epithelial tissues.[1,2,19,30,36–39,42,53–74]

[54] L. Larose, Y. Dumont, J. Asselin, J. Morisset, and G. G. Poirer, *Eur. J. Pharmacol.* **76**, 247 (1981).
[55] J.-P. Dehaye, J. Winand, P. Poloczek, and J. Christophe, *J. Biol. Chem.* **259**, 294 (1984).

TABLE II
RADIOLABELED LIGANDS USED IN BINDING STUDIES TO IDENTIFY
DIFFERENT CLASSES OF RECEPTORS[a]

Class of receptors	Radiolabeled ligands used for binding studies
Muscarinic cholinergic	[^{3}H]QNB,[1,53] [^{3}H]NMS[1,54]
CCK	[^{125}I]CCK,[55] [^{125}I]CCK-8[56] [^{3}H]L-364.718[1,57,58]
Gastrin	[^{125}I]Gastrin,[19] [Leu^{15}-^{125}I]gastrin-17[59]
Bombesin	[Tyr^4-^{125}I]Bombesin[1,60] [^{125}I]GRP[61]
Substance P	[^{125}I]Physalemin,[62,63] [^{125}I]substance P[63]
VIP preferring	[^{125}I]VIP[1,64]
Secretin preferring	[^{125}I]secretin,[1,65] [^{125}I]VIP[1,64]
Cholera toxin	[^{125}I]Cholera toxin[1,30]
β-Adrenergic	l-[^{3}H]DHA,[66] dl-[^{125}I]HYP[67] dl-[^{3}H]propanolol[68]
Opiate	[^{3}H]DAGO,[69] [^{3}H]naloxone[70,71] [^{3}H]DADLE[70,71] [^{3}H]Dynorphin,[71,72] [^{3}H]u-70,593[73]
Somatostatin	[Tyr^1-^{125}I]Somatostatin[37,38] [Tyr^{11}-^{125}I]Somatostatin[36,39]
EGF	[^{125}I]EGF[2,74a]
CGRP	[^{125}I]CGRP[1,42]

[a] QNB, Quinuclidinyl benzilate; NMS, N-methyl-scopolamine; CCK, cholecystokinin; GRP, gastin releasing peptide; VIP, vasoactive intestinal peptide; DHA, dihydroalprenolol; HYP, hydroxybenzylpindolol; DAGO, (D-Ala,[2] NMe-Phe,[4] Glyol)enkephalin; DADLE, (2D-alanine-5D-leucine)enkephalin; EGF, epidermal growth factor; CGRP, calcitonin gene-related peptide.

[b] Review articles are cited whenever possible to minimize the number of citations. Moreover, references are not intended to be all inclusive; instead, representative examples have been selected.

---

[56] R. T. Jensen, G. F. Lemp, and J. D. Gardner, *J. Biol. Chem.* **257**, 5554 (1982).
[57] S. A. Wank, R. T. Jensen, and J. D. Gardner, *Am. J. Physiol.* **255**, G106 (1988).
[58] R. S. L. Chang, V. J. Lotti, and T. B. Chen, *Biochem. Pharmacol.* **36**, 1709 (1987).
[59] R. S. L. Chang, V. J. Lotti, T. B. Chen, and K. A. Kunkel, *Mol. Pharmacol.* **30**, 212 (1987).
[60] A. H. Soll, D. A. Amirian, L. Thomas, and A. Ayalon, *Gastroenterology* **82**, 1184A (1982).

Binding of a radiolabeled ligand is usually reversible and temperature dependent, usually reaches a steady state after 30–60 min at 37°, and is limited to a finite number of receptors. Binding of a radiolabeled ligand is a function of its concentration. A plot of the amount of bound radioactivity versus the concentration of radiolabeled ligand shows two components (Fig. 7). One is a saturable component and reflects binding of the radiolabeled ligand to a finite number of binding sites. The other component is nonsaturable and may reflect "trapping" of the radiolabeled ligand between cells or membranes, or adsorption of the radiolabeled ligand to the walls of the test tube or to the filter. Experimentally, total binding of the radiolabeled ligand is determined by measuring the amount of radioactivity that is bound when the tissue preparation in incubated with the radiolabeled ligand. Nonsaturable binding is determined by measuring the amount of radioactivity bound when the incubation medium contains a large excess of nonradioactive ligand. Saturable binding is calculated as total binding minus nonsaturable binding. As shown in Fig. 7, the higher the concentration of radiolabeled ligand, the greater the contribution of the nonsaturable component to total binding. In general, one wants to maximize total binding and minimize the fraction of nonsaturable binding; therefore, one should use a concentration of radiolabeled ligand that gives easily detected total binding and a value of nonsaturable that is less than 30% of the total binding. In some systems, the nonsaturable component

---

[61] R. T. Jensen, T. Moody, C. Pert, J. E. Rivier, and J. D. Gardner, *Proc. Natl. Acad. Sci. U.S.A.* **75**, 6139 (1978).

[62] M. Nakumura, M. Oda, K. Kaneko, Y. Akaiwa, N. Tsukada, H. Koniatsu, and M. Tsuchiya, *Gastroenterology* **94**, 968 (1988).

[63] R. T. Jensen and J. D. Gardner, *Proc. Natl. Acad. Sci. U.S.A.* **76**, 5679 (1979).

[64] R. T. Jensen, S. W. Jones, Y.-A. Lu, J.-C. Xu, K. Folkers, and J. D. Gardner, *Biochim. Biophys. Acta* **804**, 181 (1984).

[65] B. M. Bissonnette, M. J. Collen, H. Adachi, R. T. Jensen, and J. D. Gardner, *Am. J. Physiol.* **246**, G710 (1984).

[66] R. T. Jensen, C. G. Charlton, H. Adachi, S. W. Jones, T. L. O'Donohue, and J. D. Gardner, *Am. J. Physiol.* **245**, G186 (1983).

[67] R. J. Lefkowitz, C. Mukherjee, M. Coverstone, and M. G. Caron, *Biochem. Biophys. Res. Commun.* **60**, 703 (1974).

[68] G. D. Aurbach, S. A. Fedak, C. J. Woodward, J. S. Palmer, D. Hauser, and T. Troxler, *Science* **186**, 1223 (1974).

[69] A. Levitzki, D. Atlas, and M. L. Steer, *Proc. Natl. Acad. Sci. U.S.A.* **71**, 2773 (1974).

[70] B. K. Handa, A. L. Lane, J. A. H. Lord, B. A. Morgan, M. J. Rance, and C. F. C. Smith, *Eur. J. Pharmacol.* **70**, 531 (1981).

[71] J. M. Hiller, L. M. Angel, and E. J. Simon, *Science* **214**, 468 (1981).

[72] J. M. Hiller, L. M. Angel, and E. J. Simon, *Mol. Pharmacol.* **25**, 249 (1984).

[73] R. R. Goodman and S. H. Snyder, *Life Sci.* **31**, 1291 (1982).

[74] R. A. Lahiti, P. F. Von Voightlander, and C. Barsuhn, *Life Sci.* **31**, 2257 (1982).

[74a] N. Gallo-Payet and J. S. Hugon, *Endocrinology* **116**, 194 (1985).

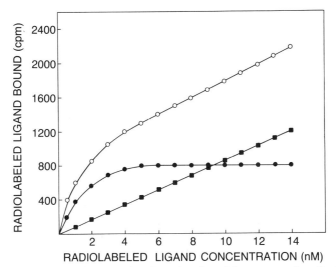

FIG. 7. Binding of a radiolabeled ligand as a function of the concentration of radiolabeled ligand. The tissue preparation was incubated with indicated concentrations of radiolabeled ligand and the amount of bound radiolabeled ligand was determined. The curve describing total binding (○) can be resolved into two components: a nonsaturable component (■) and a saturable component (●). Total binding is the amount of radioactivity that is bound when the tissue preparation is incubated with the radiolabeled ligand alone. Nonsaturable binding is determined by measuring the amount of radioactivity bound when the incubation medium also contains a large excess of nonradioactive ligand. Saturable binding is calculated as total binding minus nonsaturable binding. (Results from Gardner.[75])

may contribute such a large fraction of the total binding that even with the lowest possible concentration of the radiolabeled ligand at its maximum specific activity, one cannot measure the saturable component with any meaningful precision.

In most instances binding of a ligand to its receptor reflects a bimolecular reaction and can be described by the following equation:

$$[L] + [R] \underset{K_2}{\overset{K_1}{\rightleftarrows}} [LR] \tag{1}$$

where [L] is the concentration of the free ligand, [R] is the concentration of unoccupied receptors, [LR] is the concentration of ligand–receptor binding site complex, and K_1 and K_2 are the rate coefficients for association and dissociation, respectively.

The stoichiometry of the binding reaction can be described by the equation

$$[LR] = RT[L]/[L] + K_d \tag{2}$$

[75] J. D. Gardner, *Gastroenterology* **76**, 202 (1979).

where RT is the total concentration of receptors and K_d is the apparent dissociation constant for the binding reaction (K_2/K_1). This equation illustrates that if binding can be described by a simple bimolecular reaction, the parameters of the reaction, RT and K_d, can be calculated by plotting [LR] measured at various concentrations of [L]. A plot of [LR] vs [L] will have the form of a rectangular hyperbola. It may be relatively difficult to determine the values of K_d and RT directly from the graph, particularly if one is unable to use sufficiently high concentrations of ligand to produce maximal binding. To circumvent this difficulty, Eq. (2) can be transformed into various linearized forms (e.g., a Scatchard plot[75a]) to facilitate graphic evaluation of the values for K_d and RT.

Because binding of the radiolabeled ligand is limited to a finite number of receptors on the cell surface, one can monitor the interaction of a particular agent with the receptors by measuring that agent's ability to compete with the radiolabeled ligand for occupation of binding sites and by so doing inhibit binding of radioactivity. In general, those ligands that inhibit binding of given radiolabeled ligand interact with the same class of receptors whereas those ligands that do not inhibit binding of the radiolabeled ligand do not interact with the same class of receptors as the radiolabeled ligand. The affinity of a ligand for its receptor is reflected by the range of concentrations over which it inhibits binding of the radiolabeled ligand. The higher the affinity of the ligand for the receptor, the lower the concentration of the ligand that inhibits binding of radioactivity.

In experiments in which binding of radiolabeled ligand to its receptor is measured using incubations containing radiolabeled ligand alone as well as radiolabeled ligand plus unlabeled ligand and the amount of radioactivity bound is expressed as a fraction of the amount bound with no added unlabeled agent, Eq. (2) can be expanded to

$$R^* = (K_d^2[L] + K_d^1 K_d^2)/(K_d^2[L] + K_d^1[A] + K_d^1 K_d^2) \qquad (3)$$

where [L] is the concentration and K_2^1 is the dissociation constant for the radiolabeled ligand, [A] is the concentration and K_d^2 is the dissociation constant of the unlabeled ligand, and R^* is the amount of radiolabeled ligand bound expressed as a fraction of radioactivity bound with no added unlabeled agent.

When an appropriate radiolabeled ligand is available for measuring binding to a given class of receptors, one can measure the degree of inhibition of binding of a radiolabeled ligand by a possibly large number of unlabeled ligands. Assuming that all the agents are competing for the same class of receptors, one estimates the affinity (K_d) of the radiolabeled ligand

---

[75a] G. Scatchard, *Ann. N.Y. Acad. Sci.* **51**, 660 (1949).

from a plot of inhibition of the binding by the same agent (ligand) as the radiolabeled ligand then calculates the affinities (values of K_d) for other agents, using Eq. (3). The values of K_d and RT can also be determined using a nonlinear, least-squares, curve-fitting computer program (e.g., LIGAND[76]). Programs such as this calculate the binding parameters efficiently and allow convenient formulation and comparison of variety of models so that the best model may be chosen on a sound, statistically meaningful basis. Such programs are particularly useful for analyzing binding with multiple classes or subtypes of receptors in the same tissue preparation.

Differences in the relative potencies of various agonists in terms of inhibiting binding of a given radiolabeled ligand to its receptors (different values of K_d) may help to distinguish different classes of receptors. For example, pharmacological studies based on the relative potencies of various CCK-related peptides (e.g., CCK-8, gastrin, and CCK-4) to interact with CCK receptors on pancreatic and other tissues have provided evidence that different classes of CCK receptors exist in these tissues.

1. Pancreatic CCK receptors (also "CCK-A" receptors) are present on pancreatic acinar cells,[1,56,77-79] gall bladder muscle,[16,80-83] rat anterior pituitary cells,[84] inhibitory neurons of the lower esophageal sphincter,[18] human gastric leiomyosarcomas,[85] and the area postrema and other discrete areas of the brain.[86] Differences in the relative potencies of various CCK-related peptides in terms of interaction with this class of CCK receptors is illustrated by differences in their abilities to inhibit binding of [^{125}I]CCK-8 to rat pancreatic membranes.[79] As shown in Fig. 8, top, gastrin is 1000-fold less potent and CCK-4 is 30,000-fold less potent than CCK-8 in terms of inhibiting binding of [^{125}I]CCK-8.[79] Moreover, interaction of this class of CCK receptors with CCK is critically dependent on the presence of sulfate

[76] P. J. Munson and D. Rodbard, *Anal. Biochem.* **107**, 220 (1980).
[77] R. T. Jensen, G. F. Lemp, and J. D. Gardner, *Proc. Natl. Acad. Sci. U.S.A.* **77**, 2079 (1980).
[78] H. Sankaran, I. D. Goldfine, C. W. Deveney, K.-Y. Wong, and J. A. Williams, *J. Biol. Chem.* **255**, 1849 (1980).
[79] R. B. Innis and S. H. Snyder, *Proc. Natl. Acad. Sci. U.S.A.* **77**, 6917 (1980).
[80] J. R. Chowdhury, J. M. Berkowitz, M. Praissman, and J. W. Fara, *Experientia* **32**, 1173 (1976).
[81] B. Rubin, S. L. Engel, A. M. Drungis, M. Dzelzkalns, E. V. Grigas, M. H. Waugh, and E. Yiacas, *J. Pharm. Sci.* **58**, 955 (1969).
[82] M. J. Shaw, E. M. Hadar, and L. J. Miller, *J. Biol. Chem.* **262**, 14313 (1987).
[83] R. W. Steigerwalt, I. D. Goldfine, and J. A. Williams, *Am. J. Physiol.* **247**, G709 (1984).
[84] T. Reisine and R. T. Jensen, *J. Pharmacol. Exp. Ther.* **236**, 621 (1986).
[85] L. J. Miller, *Am. J. Physiol.* **247**, G402 (1984).
[86] T. Moran, P. Robinson, M. Goldrich, and P. P. McHugh, *Brain Res.* **362**, 175 (1986).

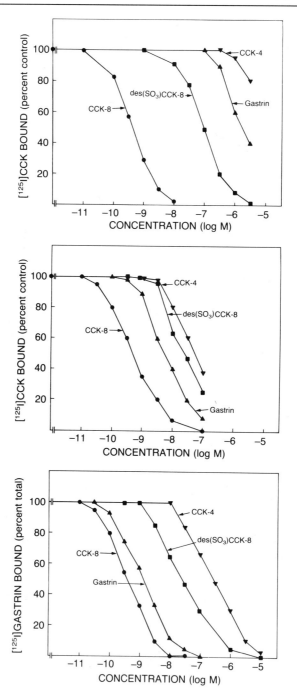

in the seventh position from the COOH terminus. That is, as shown in Fig. 8 (top), desulfated CCK-8 [des(SO_3)CCK-8] is 300-fold less potent than CCK-8 in terms of inhibiting binding of [^{125}I]CCK-8 to rat pancreatic membranes.[79]

2. Central cortical CCK receptors (also called "CCK-B" receptors) are located in the cerebral cortex and widely distributed in other areas of the central nervous system.[79,87,88] The relative potencies of various CCK-related peptides (CCK-8, gastrin, and CCK-4) for this class of CCK receptors is different from those described for CCK-A receptors. As shown in Fig. 8 (middle), gastrin is only 10-fold less potent and CCK-4 100-fold less potent than CCK-8 in terms of inhibiting binding of [^{125}I]CCK-8 to membranes from guinea pig cerebral cortex.[79] Moreover, interaction of this class of CCK receptors with CCK is dependent on the presence of sulfate to a lesser degree than is the case for CCK-A receptors. That is, as shown in Fig. 8 (middle), des(SO_3)CCK-8 is 50-fold less potent than CCK-8 in terms of inhibiting binding of [^{125}I]CCK-8 to membranes from guinea pig cerebral cortex[79] whereas des(SO_3)CCK-8 is 300-fold less potent than CCK-8 in terms of inhibiting binding of [^{125}I]CCK-8 to pancreatic membranes (Fig. 8, top).

3. Another class of CCK receptors (also called "gastrin receptors") is found in parietal cells,[2,60] gastric glands,[2,88] gastric smooth muscle,[20] lower esophageal sphincter muscle,[18] pancreatic acinar cells,[19] fundic somatostatin-releasing cells,[89] and certain colonic tumor cell lines.[90] Gastrin re-

---

[87] S. E. Hays, M. C. Beinfeld, R. T. Jensen, F. I. C. Goodwin, and S. M. Paul, *Neuropeptides* **1**, 53 (1980).
[88] A. Saito, H. Sankaran, I. D. Goldfine, and J. A. Williams, *Science* **208**, 1155 (1980).
[88a] R. S. L. Chang, V. J. Lotti, M. E. Keegan, and K. A. Kunkel, *Biochem. Biophys. Res. Commun.* **134**, 895 (1986).
[89] A. H. Soll, D. A. Amirian, L. P. Thomas, J. Park, J. D. Elashoff, M. A. Beaven, and T. Yamada, *Am. J. Physiol.* **247**, G715 (1984).

FIG. 8. Relative abilities of CCK-8 and structurally related peptides to inhibit binding of [^{125}I]CCK to rat pancreatic membranes (top) or guinea pig brain membranes (middle) and to inhibit binding of [^{125}I]gastrin to guinea pig pancreatic acini (bottom). *Top and middle:* Membranes were incubated for 30 min at 37° with 50–100 pM [^{125}I]CCK plus the indicated concentrations of peptides. Saturable binding of [^{125}I]CCK is expressed as the percentage of radioactivity bound in the absence of added nonradioactive peptides. *Bottom:* Ability of CCK-8 and related peptides to inhibit binding of [^{125}I]gastrin to guinea pig pancreatic acini. Acini were incubated for 60 min at 37° with 130 pM [^{125}I]gastrin plus the indicated concentrations of peptides. Saturable binding of [^{125}I]gastrin is expressed as the percentage of the radioactivity bound in the absence of added nonradioactive peptides. (Results from Yu *et al.*[19] and Innis and Snyder[79]. Values in the upper and middle panels redrawn from data by Innis and Synder[79] by permission of the authors.)

ceptors and CCK-B receptors may actually be the same, as both classes of CCK receptors have the same affinity for gastrin and, compared to CCK-A receptors, are dependent to a lesser degree on the presence of sulfate in the seventh position from the COOH terminus for full potency of CCK. As shown in Fig. 8 (bottom), interaction with this class of CCK receptors is characterized by CCK-8 being equipotent or one- to threefold more potent than gastrin in terms of inhibiting binding of [^{125}I]gastrin in pancreatic acini.[19] Moreover, compared to CCK-A receptors, des(SO$_3$)CCK-8 is 70-fold less potent and CCK-4 is 1000-fold less potent than CCK-8 or gastrin in terms of inhibiting binding of [^{125}I]gastrin to pancreatic acini.[19]

In pancreatic acini, dose–response curves for inhibition of binding of [^{125}I]CCK-8 by CCK-related peptides are broad in that the maximally effective concentration of a given peptide is at least 1000 times greater than the threshold concentration (Fig. 9).[1,53,57] The broad dose–response curve for inhibition of binding of [^{125}I]CCK-8 can be accounted for by the fact that pancreatic acini possess two subclasses of receptors for CCK.[1,53] Analysis of the curve for inhibition of binding of [^{125}I]CCK-8 by CCK-8 using a nonlinear, least-squares, curve-fitting program (LIGAND[76]) indicates that pancreatic acini possess two subclasses of CCK receptors. One subclass of receptors has a high affinity and low capacity for CCK-8 and the other subclass has a low affinity and high capacity for CCK-8.[1,53]

Another example of subclasses of receptors on the target tissue is illustrated by VIP receptors on pancreatic acini. Analysis of the ability of VIP to inhibit binding of [^{125}I]VIP has indicated that pancreatic acini possess two subclasses of VIP receptors.[1,91,92] One subclass has a high affinity and low capacity for VIP and the other subclass has a low affinity and high capacity for VIP.[1,91,92]

Thus, the differences in the relative affinities (different values of K_d) of an agonist for subclasses of its receptor can be helpful in distinguishing different subclasses of the receptor in the same tissue.

As mentioned previously, a receptor antagonist can be used to distinguish different classes of receptors by determining the value of K_i of the antagonist for a class/classes of receptors. K_i, which is the concentration of the antagonist required to occupy 50% of the receptors with no agonist present, can be calculated from the formula $K_i = [I](K_d/K_d^1)$ where $[I]$ is the concentration of the antagonist and K_d and K_d^1 are the apparent dissocia-

---

[90] P. Singh, B. Rae-Venter, C. M. Townsend, T. Khalil, and J. C. Thompson, *Am. J. Physiol.* **249,** G761 (1985).
[91] J. P. Christophe, T. P. Conlon, and J. D. Gardner, *J. Biol. Chem.* **251,** 4629 (1976).
[92] P. Robberecht, T. P. Conlon, and J. D. Gardner, *J. Biol. Chem.* **251,** 4635 (1976).

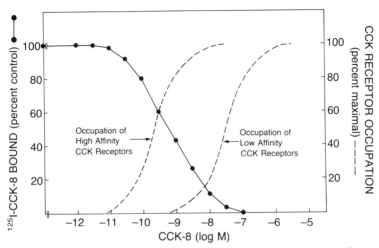

FIG. 9. Ability of CCK-8 to inhibit binding of [^{125}I]CCK-8 to dispersed acini from guinea pig pancreas and the calculated abilities of CCK-8 to occupy high-affinity CCK receptors and low-affinity CCK receptors. Acini were incubated for 60 min at 37° with 50 pM [^{125}I]CCK-8 alone or plus indicated concentrations of CCK-8. Saturable binding of [^{125}I]CCK-8 is expressed as the percentage of radioactivity bound in the absence of nonradioactive CCK-8. The dashed lines representing occupation of high-affinity CCK receptors and low-affinity CCK receptors by CCK-8 were calculated from the values of K_d for CCK-8 for high-affinity CCK receptors and low-affinity CCK receptors obtained by analysis of the binding data for the curve for inhibition of binding of [^{125}I]CCK-8 using the nonlinear, least-squares, curve-fitting program, "LIGAND."[96] (Results from Wank et al.[57])

tion constants for binding of the radiolabeled ligand in the absence and presence of the antagonist, respectively.

An antagonist will give the same value of K_i for each agonist that interacts with the same class of receptors as the antagonist. For example, in pancreatic acini (D-Pro^2, D-Trp^{7,9})substance P is an antagonist that inhibits binding of [^{125}I]substance P and [^{125}I]physalaemin, each of which interacts with substance P receptors.[93] The value of K_i for the antagonist for inhibiting binding of [^{125}I]substance P (1.36 μM) is not significantly different from the value of K_i for the antagonist for inhibiting binding of [^{125}I]physalaemin (1.54 μM).[93] On the other hand, if a receptor antagonist gives different values of K_i for the agonists, each agonist must interact with a different class of receptors. Moreover, this finding indicates that the antagonist interacts with more than one class of receptors. For example, the benzodiazipine derivative, L-364,718, interacts with CCK receptors

---

[93] R. T. Jensen, S. W. Jones, Y.-A. Lu, J.-C. Xu, K. Folkers, and J. D. Gardner, Biochim. Biophys. Acta **804**, 181 (1984).

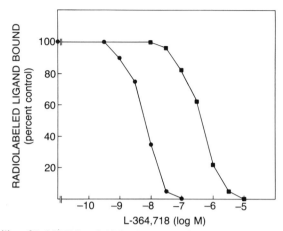

FIG. 10. Ability of L-364,718 to inhibit binding of [^{125}I]CCK-8 (●) or [^{125}I]gastrin (■) to dispersed acini from guinea pig pancreas. Acini were incubated for 60 min at 37° with 50 pM [^{125}I]CCK-8 plus 100 nM gastrin or 100 pM [^{125}I]gastrin with or without the indicated concentrations of L-364,718. Gastrin (100 nM) was added with [^{125}I]CCK-8 to prevent binding of the tracer to gastrin receptors to measure only binding to CCK receptors. Results are expressed as the percentage of saturable binding without added L-364,718 (i.e., percentage control). (Results from Huang et al.[52])

and gastrin receptors and in pancreatic acini inhibits binding of [^{125}I]CCK-8 and binding of [^{125}I]gastrin (Fig. 10).[52] However, the relative potency of the antagonist in inhibiting binding of [^{125}I]CCK-8 differs from the relative potency of the antagonist in terms of inhibiting binding of [^{125}I]gastrin. As shown in Fig. 10, L-364,718 is more potent at inhibiting binding of [^{125}I]CCK-8 than at inhibiting binding of [^{125}I]gastrin to pancreatic acini. Moreover, the value of K_i for the antagonist for inhibiting binding of each radiolabeled ligand differs. For example, the value of K_i for L-364-718 for inhibiting binding of [^{125}I]CCK-8 (4 nM) is significantly different from the value of K_i for the antagonist inhibiting binding of [^{125}I]gastrin (500 nM). Thus, the difference in the relative potency of L-364,718 for CCK receptors and gastrin receptors (with different values of K_i for the antagonist for each class of receptors) helps to distinguish CCK receptors from gastrin receptors.[52]

For most gastrointestinal peptides, antagonists do not help to distinguish between different subclasses of receptors. For example, both L-364,718 and L-365,260 interact with both high-affinity and low-affinity pancreatic CCK receptors and each antagonist has the same affinity for both subclasses of CCK receptors (Table III).[52]

TABLE III
AFFINITIES OF L-364,718 AND L-365,260 FOR
TWO SUBCLASSES OF CCK RECEPTORS ON
GUINEA PIG PANCREATIC ACINI[a]

Agent added	Inhibition of binding of [^{125}I]CCK-8 (K_i in nM)	
	High affinity	Low affinity
L-364,718	4.0	3.0
L-365,260	580	560

[a] Values are means ± SE. Acini were incubated at 37° for 60 min with 50 pM [^{125}I]CCK-8 with or without various concentrations of L-364,718 or L-365,260. Gastrin (100 nM) was added with [^{125}I]CCK-8 to prevent binding of the tracer to gastrin receptors. Dose–inhibition curves were analyzed using a nonlinear least-squares curve-fitting program (LIGAND).[76] [^{125}I]CCK-8 binding was fitted with a two-site model with one class of receptors having high affinity for CCK-8 and the other class of receptors having low affinity for CCK-8. (Results from Huang et al.[52])

Solubilization of Receptors

Receptor solubilization allows one to analyze the physicochemical and biochemical properties of the receptor and may allow one to purify the receptor to homogeneity to determine at least a portion of its amino acid sequence. The active state of solubilized receptor is demonstrated by its ability to bind a radiolabeled ligand.

Materials

The following are required: Detergent, radiolabeled ligand, and an appropriate tissue preparation.

Methods

Plasma membranes are prepared from the tissue of interest and incubated in a solubilizing solution containing an appropriate concentration of a suitable detergent. The solubilization is carried out under continuous agitation for an appropriate period of time at an appropriate pH and temperature. The incubation mixture is then centrifuged at 100,000 g for 1 hr and the supernatant containing the solubilized receptor is removed for measurement of binding of a radiolabeled ligand.

Several factors influence the binding of a radiolabeled ligand to its solubilized receptor: Type and concentration of detergent, type of buffer, pH and ionic strength, assay temperature and duration. From a practical point of view, the starting point in the development of an assay may be the use of phosphate buffer, pH 7 – 8, and temperature of 25° to determine the suitable type and concentration of detergent that will allow for successful solubilization and detection of maximal saturable binding of a radiolabeled ligand to the receptor. Recent studies of several receptors, including the insulin receptor, the opiate receptor, and the angiotensin II receptor, indicate that successful solubilization occurs with detergents with high critical micelle concentration (CMC) such as 3-[(3-cholamidopropyl)dimethylammonio]-1-propane sulfonate (CHAPS) and octylglucoside at a concentration below their CMC.[94-96] In most cases, the concentration of a detergent above its CMC invariably results in a loss of functional activity of the receptor. Once a suitable type and concentration of a detergent is determined for successful solubilization of the receptor, "fine tuning" of buffer type, pH, temperature, as well as duration of assay are frequently necessary to optimize the assay conditions.

The proper separation of free ligand from ligand bound to solubilized receptors is of paramount importance in determining saturable binding to the solubilized receptor. Several separation techniques are available to achieve the desired separation of free ligand and ligand bound to the solubilized receptor; these include gel filtration, equilibrium dialysis, use of nonspecific agents like polyethylene glycol and ammonium sulfate to precipitate the ligand – receptor complex, use of specific agents like antibodies that are specific for the solubilized receptor, or incorporation of the solubilized receptor – ligand complex in phospholipid vesicles and subsequent separation of the vesicles from the incubation medium by centrifugation.[97]

Characterization of solubilized receptors usually involves (1) measurement of binding of a radiolabeled ligand to its solubilized receptor, (2) evaluation of other receptor functions using receptor reconstitution studies, and (3) evaluation of the physicochemical properties of the solubilized receptor.

[94] R. J. Gould, B. H. Ginsberg, and A. A. Spector, *Biochemistry* **20**, 6776 (1981).
[95] A. M. Capponi, M. A. Birabean, and M. B. Vallotton, *J. Recept. Res.* **3**, 289 (1983).
[96] W. F. Simmonds, G. Koski, R. A. Streaty, L. M. Hjelmeland, and W. A. Klee, *Proc. Natl. Acad. Sci. U.S.A.* **77**, 4623 (1980).
[97] M. F. El-Refai, *in* "Receptor Biochemistry and Methodology" (J. C. Venter and L. C. Harrison, eds.), Vol. 1, p. 99. Liss, New York, 1984.

Evaluation of the ability of the solubilized receptor to bind a radiolabeled ligand is one of the methods to determine if the receptor has been solubilized in an active state. Solubilization may result in alteration in the binding characteristics of a receptor and one may have to test several different radiolabeled ligands to select the one that gives maximal saturable binding to the receptor. The change in binding characteristics of a receptor may reflect a change in the affinity of the receptor for its ligand or a change in the number of receptors or both. For example, after solubilization, the affinity of solubilized CCK receptors for CCK receptor agonists is reduced 50–100 times compared to that of membrane-bound CCK receptors.[58,98] On the other hand, the affinity of solubilized CCK receptors for CCK receptor antagonists is similar to that of membrane-bound CCK receptors.[58,98] Since the affinity of CCK receptors for antagonists remains unaltered after solubilization, a radiolabeled high-affinity CCK receptor antagonist such as [^{3}H]L-364,718 offers distinct advantages over [^{125}I]CCK in assaying solubilized CCK receptors during purification studies. The reduction in both number and affinity of solubilized receptors for its agonist is illustrated by prolactin receptors on human mammary glands. Solubilization of prolactin receptors with Triton X-100 results in a reduction of the number of receptors and in the affinity of the receptors for prolactin compared to membrane-bound receptors.[99]

Solubilized receptors have been used in receptor reconstitution studies to study receptor functions in a defined environment. For example, the asialoglycoprotein receptor has been successfully solubilized and reconstituted into phospholipid vesicles.[100] The reconstituted receptor displays the same ligand specificity, pH sensitivity, and calcium dependence as the native receptor in intact cell membranes.

Receptor solubilization also enables one to determine the size and chemical structure of the receptor. The most optimal size for all membrane proteins in detergent solution is the smallest that is achievable with retention of receptor function. The usual methods for assessing the size of a solubilized receptor are gel filtration and gel electrophoresis.[101]

Solubilized receptors purified to homogeneity have been used to determine amino-terminal sequence of the receptor protein by microsequence analysis. For example, this method has been successfully used to determine the amino-terminal sequence of the four polypeptides of the nicotinic

---

[98] J. Szecowka, I. D. Goldfine, and J. A. Williams, *Regul. Pept.* **10,** 71 (1985).
[99] R. P. Shiu and H. G. Friesen, *J. Biol. Chem.* **249,** 7902 (1974).
[100] R. D. Klausner, K. Bridges, H. Tsunoo, R. Blumenthal, J. N. Weinstein, and G. Ashwell, *Proc. Natl. Acad. Sci. U.S.A.* **77,** 5087 (1980).
[101] A. Davis, in "Receptor Biochemistry and Methodology" (J. C. Venter and L. C. Harrison, eds.), Vol. 2, p. 161. Liss, New York, 1984.

acetylcholine receptor, thus allowing identification of complementary DNA (cDNA) clones coding for these subunits.[102]

Covalent Cross-Linking

Covalent cross-linking has proved to be a useful tool in identifying cell surface receptors. Covalent cross-linking involves joining two molecules by using a chemical reactive agent (an affinity label) or a photoactivatable agent (a photoaffinity label). An affinity label bears a reactive group that can form a suitable covalent bond with a suitably oriented reactive group on molecules with which it interacts. A photoaffinity label yields a highly reactive intermediate, usually a carbene or nitrene, on photolysis that forms a covalent bond with molecules with which it interacts. Since solubilization of a receptor may result in a loss of the affinity of the receptor for a radiolabeled ligand, this problem can be circumvented by covalent cross-linking of the radiolabeled ligand to its receptor prior to solubilization. The resultant radiolabeled receptor complex can then be solubilized using an appropriate detergent.

The most frequently used affinity cross-linking agents in receptor identification are homobifunctional agents that rely on NH_2-reactive N-hydroxysuccinimide ester [e.g., disuccinimidyl suberate (DSS)], or heterobifunctional agents that rely on the SH- and NH_2-reactive N-hydroxysuccinimide ester [e.g., m-maleimidobenzoic acid N-hydroxysuccinimide ester (MBS)]. In photoaffinity labeling, cross-linking relies on the photoactivatable aryl azide groups [an example of a photoactivatable analog of CCK is ($Nle^{28,31}$, $6NO_2$-Trp^{30})CCK-26-33].

Materials

The following are required: Cross-linking reagent, radiolabeled ligand or photoreactive ligand, and an appropriate tissue preparation.

Methods

In affinity cross-linking, plasma membranes from an appropriate tissue are first incubated with an appropriate radiolabeled ligand to allow binding to occur. The cross-linking reaction is then carried out by incubating the ligand–receptor complex with an appropriate cross-linking reagent at an appropriate temperature for an appropriate period of time. The reaction is terminated by adding an appropriate quench solution containing free

---

[102] B. M. Conti-Tronconi, M. W. Hunkapiller, J. M. Lindstrom, and M. A. Raftery, *Proc. Natl. Acad. Sci. U.S.A.* **79**, 1488 (1982).

amines, such as Tris. The membranes are then pelleted by centrifugation at 10,000 g for 15 min.

In photoaffinity labeling, membranes are first incubated with an appropriate photoactivatable radiolabeled ligand to allow binding to occur. Cross-linking of the radiolabeled ligand receptor complex is then achieved by irradiating the membranes with an ultraviolet (UV) light source (photolysis) for an appropriate period of time and at appropriate temperature.

The resulting covalently linked radiolabeled ligand–receptor complex obtained by either method is then solubilized using an appropriate detergent. The solubilized complex is then subjected to gel filtration or gel electrophoresis and autoradiography to assess the properties of the receptor, as discussed earlier in the section, Solubilization of Receptors.

The data obtained by cross-linking of radiolabeled ligand to binding proteins reflects only the molecular proximity of reactive groups on both the ligand and its receptor with in the ligand–receptor complex. A change in the chemical specificity of the cross-linking reagent may give a different labeling pattern of the receptor. For example, studies performed by covalently cross-linking [^{125}I]CCK to its binding sites on plasma membranes from pancreatic acinar cells with a homobifunctional agent (DSS) indicate that the minimal structure of the CCK receptor is a protein with an M_r of 70,000–80,000 and joined by disulfide bonds to a protein with an M_r of 40,000–55,000.[103-106] In similar studies with the heterobifunctional agent, m-maleimidobenzoic acid N-hydroxysuccinimide ester (MBS), [^{125}I]CCK was cross-linked to form proteins with M_r values of 47,000, 80,000–95,000, 30,000–140,000 and > 200,000.[107]

In addition to the nature of the cross-linking agent, the chemical composition of the labeled peptide can also influence the pattern and mobility of labeled proteins. For example, cross-linking studies using DSS or UV irradiation of labeled membranes from guinea pig pancreas by photoactivatable radioiodinated CCK-39 detected components with M_r values of 57,000, 76,000–89,000, and 130,000–150,000, respectively.[108] On the other hand, cross-linking studies using an iodinated analog of the C-terminal region of CCK ("short probe" of CCK, and several monofunctional photoactivatable short probes of CCK) each identified the same proteins with an M_r of 85,000–95,000.[108-111]

[103] S. A. Rosenzweig, L. J. Miller, and J. D. Jamieson, *J. Cell Biol.* **96,** 1288 (1983).
[104] C. Sakamoto, I. D. Goldfine, and J. A. Williams, *J. Biol. Chem.* **258,** 12707 (1983).
[105] C. Sakamoto, J. A. Williams, K. Y. Wong, and I. D. Goldfine, *FEBS Lett.* **15,** 63 (1983).
[106] M. Svoboda, M. Lambert, J. Furnelle, and J. Christophe, *Regul. Pept.* **4,** 163 (1982).
[107] L. D. Madison, S. A. Rosenzweig, and J. D. Jamieson, *J. Biol. Chem.* **259,** 14818 (1984).
[108] A. Zahidi, D. Fourmy, J.-M. Darbon, L. Pradayrol, J.-L. Scemama, and A. Ribet, *Regul. Pept.* **15,** 25 (1986).

From the above-mentioned studies, it appears that the pattern and mobility of labeled proteins may be influenced by the chemical composition of the cross-linking reagent as well as by the chemical composition of the labeled peptide used; therefore, caution must be exercised when analyzing native receptor structures on the basis of affinity-labeling data alone.

Receptor Identification: Protein and Amino Acid Sequencing

Purification of a protein is an essential step if one has to determine its amino acid sequence by traditional methods. Receptor proteins are, however, large, poorly soluble in aqueous media, and partially buried in the lipid environment of the cell membrane. Hence, the first step in purifying a receptor protein involves solubilization of the membrane receptor as described in the section, Solubilization of Receptors. When multiple sources of tissues are available for receptor purification, their specific binding capacities may be used as a guide to choosing the tissue most likely to provide the best yield of receptor.

Once the tissue has been selected and solubilization of the receptor achieved, a choice of purification strategy must be made. To achieve an adequate yield of pure receptor protein, it is essential to use the least possible number of successive fractionation steps and ensure that each step be efficient in terms of percentage recovery of receptor protein. Because of the low abundance of receptor protein, schemes based purely on series of physicochemical fractionations such as gel filtration, ion-exchange chromatography, and high-performance liquid chromatography[112] are not likely to provide an adequate yield of purified protein. At least one or two highly selective steps based on specific properties of the receptor are usually required (e.g., affinity chromatography using immobilized ligand,[113] receptor antibody[114] or lectin[115]).

Because of the low abundance and large size of receptor proteins, traditional protein sequencing methods and recently introduced microsequencing methods are demanding and time-consuming processes, and no receptor protein has been sequenced using this approach. Fortunately,

---

[109] R. K. Pearson and L. J. Miller, *J. Biol. Chem.* **262,** 869 (1987).
[110] R. K. Pearson, L. J. Miller, S. P. Powers, and E. M. Hadar, *Pancreas* **2,** 79 (1987).
[111] U. G. Klueppelberg, S. P. Powers, and L. J. Miller, *Gastroenterology* **94,** 230A (1989).
[112] G. Zubay, in "Biochemistry" (G. Zubay, ed.), 2nd ed., p. 100. Macmillan, New York, 1988.
[113] J. Szecowka, G. Hallden, I. D. Goldfine, and J. A. Williams, *Regul. Pept.* **24,** 215 (1989).
[114] L. C. Harrison and A. Itin, *J. Biol. Chem.* **255,** 12066 (1980).
[115] L. T. Duong, E. M. Hadar, L. J. Miller, and G. P. Vlasuk, *J. Biol. Chem.* **264,** 17990 (1989).

however, recombinant DNA technology has markedly advanced the field of protein sequencing and as discussed in the next section, several receptor proteins have been sequenced using recombinant DNA technology.

Gene Sequencing

During the past few years, a number of membrane-bound receptors have been successfully approached with respect to the determination of their likely primary structure based on DNA sequence isolated from screening cDNA or genomic libraries constructed from tissues that synthesize the receptor protein of interest. Gene sequencing of the receptor protein may allow identification of various subtypes of the receptor. For example, the cloning of cDNA and genes encoding muscarinic receptors has established that there are at least five distinct subtypes of muscarinic receptors, each encoded by a separate gene.[116-121] From the primary structure of the receptor, likely structural and functional domains of the receptor can be ascribed or predicted. For example, the primary structure of the epidermal growth factor (EGF) receptor appears to be divided into four structural/functional domains[122]: (1) EGF-binding domain, (2) transmembrane domain, (3) protein kinase C domain, and (4) C-terminal component (autophosphorylation domain). The structural and/or functional homology between two receptor proteins may indicate an evolutionary relationship between them, i.e., these proteins may have evolved from a common ancestor and may belong to a same family. For example, analysis of the amino acid sequence of the mammalian β-adrenergic receptor indicates significant amino acid homology with bovine rhodopsin.[123] The sequence homology between the two receptors parallels similarities in their functions, i.e., both rhodopsin and β-adrenergic receptors are involved in

[116] T. Kubo, K. Fukuda, A. Mikami, A. Maeda, H. Takakeshi, M. Mishina, T. Haga, K. Haga, A. Ichiyama, K. Kangawa, M. Kojima, H. Matsuo, T. Hirose, and S. Numa, *Nature (London)* **323**, 411 (1986).

[117] T. Kubo, A. Maeda, K. Sugimoto, I. Akiba, A. Mikami, H. Takahashi, T. Haga, K. Haga, A. Ichiyama, K. Kangawa, H. Matsuo, T. Hirose, and S. Numa, *FEBS Lett.* **209**, 367 (1987).

[118] E. G. Peralta, J. W. Winslow, G. L. Peterson, D. H. Smith, A. Ashkenazi, J. Ramachandran, M. Schimerlik, and D. Capon, *Science* **236**, 600 (1987).

[119] T. I. Bonner, N. J. Buckley, A. L. Young, and M. R. Brann, *Science* **237**, 527 (1987).

[120] T. I. Bonner, A. C. Young, M. R. Brann, and N. J. Buckley, *Neuron* **1**, 403 (1988).

[121] E. G. Peralta, A. Ashkenazi, J. W. Winslow, D. H. Smith, J. Ramchandran, and D. J. Capon, *EMBO J.* **6**, 3923 (1987).

[122] T. Hunter, *Nature (London)* **311**, 414 (1984).

[123] R. A. F. Dixon, B. K. Kobilka, D. J. Strader, J. L. Benovic, H. G. Dohlman, T. Frielle, M. A. Bolanowski, C. D. Bennett, E. Rands, R. E. Diehl, R. A. Mumford, E. E. Slater, I. S. Sigal, M. G. Caron, R. J. Lefkowitz, and C. Strader, *Nature (London)* **321**, 75 (1986).

signal transduction that involves interaction with the guanine nucleotide regulatory proteins, transducins and G_s, respectively.

The DNA sequences encoding a specific receptor protein can be obtained from three sources: (1) chromosomal DNA, which contain essentially all the genetic information; (2) mRNA isolated from the tissue of interest; or (3) chemical synthesis of oligonucleotides if the partial amino acid sequence is known. Short synthetic oligonucleotides corresponding to partial sequences in the protein of interest can be used to isolate the entire gene from a collection of genomic or mRNA sequences.

In principle, any gene of interest may be isolated if either a nucleic acid or antibody probe is available. Colony or plaque hybridization with nucleic acid probe is the method of choice to screen genomic DNA or cDNA library to find a copy of gene of interest. However, for receptor genes that are expressed at low levels, it may be difficult to prepare sufficiently pure mRNA from which a nucleic acid probe may be derived. Alternatively, nucleic acid probes may be chemically synthesized to match a known amino acid sequence in the protein. Presently, as little as 5 pmol of pure polypeptide can be used to determine an amino acid sequence by microsequence analysis from which an oligonucleotide probe may be designed. It is necessary to determine only enough of the amino acid sequence to allow the chemical synthesis of two or three different sets of oligonucleotides that would hybridize to the receptor mRNA or cDNA. Using this approach several receptors have been cloned and their complete amino acid sequence determined. These include the nicotinic and muscarinic acetylcholine receptor,[102,116-121] the transferrin receptor,[124] the asialoglycoprotein receptor,[125] the EGF receptor[126-128] the low-density lipoprotein (LDL) receptor,[129] and, more recently, the insulin receptor.[130]

If oligonucleotide probes are not available, it may still be possible to identify receptor-specific clones by taking advantage of *in vitro* translation

---

[124] A. McClelland, L. C. Kuhn, and F. H. Ruddle, *Cell (Cambridge, Mass.)* **39,** 267 (1984).

[125] K. Drickamer, J. F. Mamon, G. Binns, and J. O. Leung, *J. Biol. Chem.* **259,** 770 (1984).

[126] A. Ullrich, L. Coussens, J. S. Hayflick, T. J. Dull, A. Gray, A. W. Tam, J. Lee, Y. Yarden, T. A. Libermann, J. Schlessinger, J. Downward, E. L. V. Mayes, N. Whittle, M. D. Waterfield, and P. H. Seeburg, *Nature (London)* **309,** 418 (1984).

[127] C. R. Lin, W. S. Chen, W. Kruiger, L. S. Stolarsky, W. Weber, R. M. Evans, I. M. Verma, G. N. Gill, and M. G. Rosenfeld, *Science* **224,** 843 (1984).

[128] Y.-L. Xu, S. Ishic, A. J. L. Clark, M. Sullivan, R. K. Wilson, D. P. Ma, B. A. Roe, G. T. Merlino, and I. Pastan, *Nature (London)* **309,** 800 (1984).

[129] D. W. Russell, W. J. Schneider, T. Yamamoto, K. L. Luskey, M. S. Brown, and J. L. Goldstein, *Cell (Cambridge Mass.)* **37,** 577 (1984).

[130] A. Ullrich, J. R. Bell, E. Y. Chen, R. Herrera, L. M. Petruzzelli, T. J. Dull, A. Gray, L. Coussens, Y.-C. Liao, M. Tsubokawa, A. Mason, P. H. Seeburg, C. Grunfeld, O. M. Rosen, and J. Ramchandran, *Nature (London)* **313,** 756 (1985).

or expression systems that generate a receptor protein recognizable by its ability to bind to its specific ligand or to an antibody known to be receptor specific. These procedures still depend on obtaining a sufficiently large cDNA library statistically likely to contain receptor clones. Expression of a receptor protein can be done in a cell-free translation system[131,132] or in hosts such as bacteria (e.g., *Escherichia coli*),[133] oocytes (e.g., *Xenopus* oocytes),[134] or mammalian cells (e.g., COS cells).[135–137]

Methods available using cell-free translation systems are the "hybridization arrest of translation system"[131] and the positive hybridization selection system.[132] There are some advantages to these methods: e.g., the clone in the cDNA bank need not be full length provided it is long enough to hybridize strongly to a significant part of the mRNA. While most cell-free systems will not correctly synthesize and process membrane receptors to a functional state, cellular systems such as the *Xenopus* oocyte are able to do so.[138,139] Molecules of cDNA can be inserted into vectors that form their expression in host cells. Such plasmids or phages are called "expression vectors." Using this approach several receptors have been cloned and their amino acid sequence determined. These include the γ-aminobutyric acid (GABA) receptor,[140] substance P receptor,[141] and substance K receptor.[134] It seems that this technique will prove useful in cloning other receptors from various epithelial tissues in the near future.

---

[131] B. M. Paterson, B. E. Roberts, and E. L. Kuff, *Proc. Natl. Acad. Sci. U.S.A.* **74**, 4370 (1977).

[132] R. T. Ricciardi, J. S. Miller, and B. E. Roberts, *Proc. Natl. Acad. Sci. U.S.A.* **76**, 4927 (1979).

[133] D. L. Kaufman and A. J. Tobin, in "Receptor Biochemistry and Methodology" (J. C. Venter and L. C. Harrison, eds.), Vol. 3, p. 241. Liss, New York, 1984.

[134] Y. Masu, K. Nakayama, H. Tamaki, Y. Harada, M. Kuno, and S. Nakanishi, *Nature (London)* **329**, 836 (1987).

[135] B. Seed and A. Aruffo, *Proc. Natl. Acad. Sci. U.S.A.* **84**, 3365 (1987).

[136] A. Aruffo and B. Seed, *Proc. Natl. Acad. Sci. U.S.A.* **84**, 8573 (1987).

[137] K. Yamasaki, T. Taga, Y. Hirata, H. Yawata, Y. Kawanishi, B. Seed, T. Taniguchi, T. Hirano, and T. Kishimoto, *Science* **241**, 825.

[138] C. D. Lane, *Curr. Top. Dev. Biol.* **18**, 89 (1983).

[139] J. A. Williams, D. J. McChesney, M. C. Calayag, V. R. Lingappa, and C. D. Logsdon, *Proc. Natl. Acad. Sci. U.S.A.* **85**, 4939 (1988).

[140] P. R. Schofield, M. G. Darlison, N. Fujita, D. R. Burt, F. A. Stephenson, H. Rodriguez, L. C. Rhee, J. Ramachandran, V. Reale, T. A. Glencorse, P. H. Seeburg, and E. A. Barnard, *Nature (London)* **328**, 221 (1987).

[141] Y. Yokota, Y. Sasai, K. Tanaka, T. Fujiwara, K. Tsuchida, R. Shigemoto, A. Kakizuka, H. Ohkubo, and S. Nakanishi, *J. Biol. Chem.* **264**, 17649 (1989).

[36] cAMP Technologies, Functional Correlates in Gastric Parietal Cells

By CATHERINE S. CHEW

Introduction

Although methods for measurement of cyclic nucleotides have been available for over 20 years, the ability to correlate functional activity in a specific epithelial cell type with altered cAMP metabolism has only recently become possible. The major deterrent to such studies has been the inability to isolate a single cell type from a heterogeneous population while preserving cell viability and hormonal responses similar to those observed *in vivo*. Several recent review articles describe excellent overviews of the progress made in the past few years toward in attainment of these goals.[1-3]

In this chapter methods for isolation and purification of one epithelial cell type, the gastric parietal cell, are described in detail. The methodology has been optimized for the study of the relationship between hormone/paracrine-induced increases in parietal cell HCl secretion and changes in cAMP metabolism and cAMP-related activities in these cells. However, with a few modifications, most of the techniques can be readily transferred to studies of other secretory cell types. In all such studies the most important criteria include the careful correlation of secretory response with biochemical measurements and characterization of all measured responses with respect to dose, time, and specificity of antagonists.

Overview of Cell Isolation Techniques

For initial studies it is more practical to use partially purified cells which can be obtained in much larger quantities than highly enriched cells. Collagenase digestion of the gastric mucosa, a procedure that was originally described by Berglindh and Öbrink,[4] produces small groups of cells or gastric glands which contain approximately 50% parietal and 50% chief, or

---

[1] A. H. Soll and T. Berglindh, *in* "Physiology of the Gastrointestinal Tract" (L. R. Johnson, ed.), p. 883. Raven, New York, 1987.
[2] J. G. Forte and A. Soll, *in* "Handbook of Physiology–the Gastrointestinal System III" (J. G. Forte, ed.), p. 207. Oxford, 1989.
[3] C. S. Chew, *in* "Handbook of Physiology—The Gastrointestinal System III" (J. G. Forte, ed.), p. 255. Oxford, 1989.
[4] T. Berglindh and K. J. Öbrink, *Acta Physiol. Scand.* **96**, 150 (1976).

pepsinogen-secreting cells. Glands are well characterized with respect to hormone responsiveness, and parietal cells in glands appear to function much as parietal cells function and respond in vivo.[4-7]

Once cellular viability and responsiveness to hormones have been established and characterized, more rigorous purification procedures may be initiated. The enrichment technique we presently use is a modification of several earlier methodologies.[4,7-9] As with gland isolation, the procedure includes high-pressure, retrograde perfusion of the gastric mucosa in situ, which removes red blood cells and effectively loosens the gastric mucosa from underlying connective tissue and smooth muscle. In order to produce single cells in place of glands, a brief pronase digestion is used prior to collagenase. When pronase is used, the incubation time is kept as brief as possible, typically not more than 20 min, and the subsequent collagenase digestion is limited to a maximum of 30 min to avoid cell receptor damage. When collagenase alone is used, the tissue should be digested into glands within 30–45 min.

Parietal cells are separated from other cell types by isosmotic gradient centrifugation followed by centrifugal elutriation. The density gradient separation is performed first because this step requires only 10–15 min. With elutriation each run takes approximately 15 min and only $1-5 \times 10^7$ cells can be separated during a run with an average enrichment of 50–65%. Parietal cells are enriched 70–85% on the gradient depending on the size of the parietal cell fraction taken from the gradient. Final purification of enriched parietal cell fractions (>95%, Fig. 1) can then be accomplished within a few minutes by elutriation. Parietal cells isolated on the density gradient are contaminated with surface epithelial cells that have a smaller diameter. These contaminating cells are readily removed from the parietal cell fraction by centrifugal elutriation, which separates cells according to size as well as density. In our hands neither separation when performed alone yields as pure a preparation as the two combined.[10] Advantages of the combined purification techniques include (1) the ability to isolate cells at near 100% purity, (2) avoidance of the use of the calcium or magnesium chelators (EGTA/EDTA) which may damage cell membranes, (3) preservation of responses to known secretory stimulants, and (4) relatively rapid cell isolation and enrichment.

---

[5] T. Berglindh, H. Helander, and K. J. Öbrink, *Acta Physiol. Scand.* **97**, 401 (1976).
[6] C. S. Chew, G. Sachs, S. J. Hersey, and T. Berglindh, *Am. J. Physiol.* **238**, G312 (1980).
[7] C. S. Chew and S. J. Hersey, *Am. J. Physiol.* **242**, G504 (1982).
[8] A. H. Soll, *J. Clin. Invest.* **61**, 381 (1978).
[9] T. Berglindh, *Fed. Proc., Fed. Am. Soc. Exp. Biol.* **440**, 1203 (1985).
[10] C. S. Chew and M. R. Brown, *Biochim. Biophys. Acta* **888**, 116 (1986).

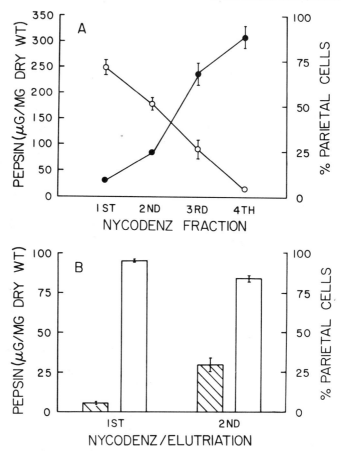

FIG. 1. Parietal cell enrichment using Nycodenz gradients and Nycodenz plus elutriation. (A) Following pronase/collagenase digestion of gastric mucosa, cells were layered on gradients and centrifuged. The number of parietal cells present in each fraction was determined by differential cell count. Pepsin, which is present in chief cells, was measured in each fraction and expressed in terms of cellular dry weight. (B) Centrifugal elutriation of the top two fractions from gradients in (A) (means ± SEMs, $n = 5-8$). In more recent experiments, parietal cell enrichment of the top fraction has been improved to $98 \pm 1\%$ ($n = 20$). Increased purity is obtained by gradually increasing and decreasing centrifuge speeds and careful removal of the top fraction from the gradient. (From Chew and Brown.[10])

The major problem we and others have encountered in isolation of hormonally responsive cells and glands arises in the initial tissue digestion step, which requires the use of pronase for cells and crude collagenase for glands and cells. Pronase should be used for only a short time—20 min or less. The tissue is only slightly digested at the time the enzyme is removed from the incubation medium. With collagenase we have found considerable variation in the ability of different lots obtained from both Worthington (Freehold, NJ) and Sigma Chemical Company (St. Louis, MO) to digest the gastric mucosa and in the quality of hormone responsiveness obtained after tissue digestion. Careful attention to this problem is essential for successful cell and gland isolation. We purchase small quantities of several different lots of collagenase and test each for speed of digestion and preservation of hormone responsiveness. As a general rule a particular lot is deemed acceptable if the tissue is digested by 0.5–1 mg collagenase/ml within 30–45 min and a secretory response to the weakest agonist, which in rabbit parietal cells is gastrin, is retained. We have successfully used several different types of collagenase from *Clostridium histolyticum,* including Sigma types I, II, IV, and V and Worthington types II and III. There appears to be little correlation between collagenase activity and speed of digestion. Pure collagenase is almost devoid of activity. With few exceptions, the most successful lots we have used contained, in units per milligram dry weight, 150–450 collagenase, 0.6–0.8 clostripain, 100–300 caseinase, and 0.04–0.1 tryptic activity. Unfortunately, collagenase with such characteristics is not always available. Collagenase is stored desiccated at $-20°$ in 100-mg amounts to avoid loss of activity due to repeated freezing and thawing. Pronase is kept in the refrigerator. Our experience with pronase is limited to that produced by Calbiochem (*Streptomyce griseus,* Cat. No. 53702; Los Angeles, CA). This pronase has been highly reliable in our hands and does not appear to cause detectable receptor damage when used judiciously.

Cell Isolation Procedure

Solutions.

1. Phosphate-buffered saline: 149.6 mM NaCl (biological grade), 3 mM K$_2$HPO$_4$, 0.64 mM NaH$_2$PO$_4$. The solution is stable several weeks at room temperature
2. Pronase/collagenase digestion medium: 130 mM NaCl, 12 mM N-2-hydroxyethylpiperazine-N'-2-ethanesulfonic acid (HEPES), 3 mM NaHPO$_4$, 3 mM Na$_2$HPO$_4$, 3 mM KHPO$_4$, 2 mM MgSO$_4$, 1 mM CaCl$_2$, 0.01 mg/ml Phenol Red, 10 mM glucose, 1 mg/ml bovine serum albumin (BSA, fraction V), 0.5–1 mg/ml collagenase or 0.5 mg/ml pronase. Glucose, BSA, and enzymes are added immediately before use, pH adjusted to 7.4 with 1 N NaOH. The basic solution is stable for 7–10 days when refrigerated
3. Cellular incubation medium: 114.4 mM NaCl, 5.4 mM KCl,

5 mM Na$_2$HPO$_4$, 1 mM NaH$_2$PO$_4$, 1.2 mM MgSO$_4$, 1 mM CaCl$_2$, 10 mM HEPES, 0.01 mg/ml Phenol Red, 10 mM glucose, 1 mM pyruvate, 2 mg/ml BSA, 0.5 mM dithiothreitol (DTT). Glucose, pyruvate, BSA, DTT, and HEPES are added immediately before use and pH adjusted to 7.4 with 1 N NaOH. Inclusion of DTT is optional. Low concentrations of DTT have been found to enhance hormonal responses[7]; however, concentrations much above 0.5 mM may be inhibitory. Fetal calf serum potently suppresses cellular responses (unpublished observations)
4. Cellular dilution medium for gradient separation: Add 9 mg/ml BSA to 25 ml cellular incubation medium
5. Nycodenz (Accurate Chemical Co. Westbury, NY, sterile isotonic solution, density 1.15) supplemented with 10 mg/ml BSA, 0.5 mM DTT, 2.4 mM KCl, 1.2 mM MgSO$_4$, 15 mM HEPES. Bovine serum albumin and DTT reduce cell clumping. The pH is adjusted to 7.4 with 0.1 M Tris
6. Nycodenz diluent: 132 mM NaCl, 5.4 mM MgSO$_4$, 0.5 mM DTT, 15 mM HEPES, 10 mg/ml BSA. The pH is adjusted to 7.4 with 0.1 M Tris. The density of this solution is approximately 1.006

Gradient Preparation

Nycodenz is diluted 2:1, 1:1, and 1:2 with diluent. Two milliliters of undiluted Nycodenz is placed in each 15 × 100 mm tube, then 2 ml each of 2:1, 1:1, and 1:2 dilutions are carefully layered over the undiluted Nycodenz. The respective densities of each solution should be approximately 1.139, 1.095, 1.073, and 1.049. Gradients may be made the day before and stored horizontally at 4° until use. With horizontal storage, linear gradients are formed. We have achieved good results with both step and linear gradients.

Perfusion of Mucosa

The perfusion system is arranged so that one end of the tubing from the perfusion pump head (Masterflex pump, controller, #7018 pump head; Cole Parmer, Chicago, IL) is attached to a PE 240 cannula by a three-way stopcock. The other end of the pump tubing is placed in 1 liter of warm, oxygenated, phosphate-buffered saline immediately after cannulation of the animal. The saline is gassed continuously with 100% oxygen and maintained at 37° until use.

Care should be taken not to stress the animal. Anesthetic can be administered to unrestrained animals using a syringe/cannula. (The syringe/cannula is made by carefully breaking off the metal portion of a 25-gauge needle from its shaft and inserting it into a 12-in. length of PE 50 tubing. The other end of the tubing is attached to a 5-ml syringe with an intact 25-gauge needle.) The ear is shaved with a single-edge razor blade using a hemostat as a holder. The needle is gently inserted into the animal's outer ear vein, bevel up. Successful venipuncture is determined by drawing back on the syringe plunger and observing blood moving freely into the cannula. The animal is then anesthetized by gradually administering the anesthetic while observing blink reflexes. Anesthesia should be just sufficient to block reflexes. Rabbits are extremely sensitive to anesthetics and rapid administration often results in death.

Once the animal is anesthetized, the abdomen is rapidly opened by midline incision and the abdominal aorta located. The pump tubing and attached cannula are filled with warm, oxygenated saline. No air bubbles should be trapped in the tubing. The PE 240 cannula is placed in the aorta as distally as possible using standard cannulation procedures and secured by surgical ties. Heparin (1000 U) is injected via the stopcock into the cannula, which is then flushed with saline. Approximately 30 sec later the cannula is opened and the animal exsanguinated through the stopcock. Clamps are then rapidly placed on the small intestine immediately distal to the stomach and on the thoracic aorta just above the diaphragm to block movement of the perfusate from the gastric mucosa. A small section of the liver is excised to allow an exit for the perfusate. Throughout the surgical procedure, care should be taken to keep all exposed organs damp, using a wash bottle filled with warm saline.

Perfusion is begun by gradually increasing the pump speed to approximately 200 ml/min. A good perfusion will cause complete blanching of the stomach. After perfusion with 800–900 ml of saline, the stomach is rapidly excised and the contents rinsed away with saline. The surface of the mucosa is wiped gently with paper towels moistened in saline to remove surface mucous cells. The mucosa is rinsed again in saline, then gently scraped onto a 6 × 6 in. section of Plexiglas with a central depression. The tissue is minced with curved surgical scissors, then rinsed twice with saline. The saline is poured off and the tissue transferred into a 250-ml round-bottomed flask which contains the collagenase or pronase mixture (see below). Prewarmed medium (50 ml) should be added to the collagenase or pronase immediately before use to avoid loss of enzymatic activity. All steps following removal of the stomach should be performed as rapidly as possible to prevent tissue anoxia.

Gland Isolation

For isolation of glands, the minced mucosa is digested in a 250-ml round-bottomed flask in a 37° water bath with collagenase digestion medium for approximately 30 min. Constant rapid stirring with a large egg-shaped stirring bar and magnetic stirrer is essential for good tissue dissociation. Completeness of digestion is confirmed by periodically sampling and microscopically examining the contents of the flask to determine precisely when connective tissue surrounding the glands has been digested away. For all types of digestion, the tissue is oxygenated by dimpling the surface of the medium with a steady stream of oxygen. If necessary, 1 N NaOH is added to maintain the pH at 7.3–7.4.

Once glands are produced, the medium is diluted with 25–30 ml incubation medium and filtered through nylon mesh to remove undigested tissue and debris. The glands are rinsed three times by settling for 10 min, aspirating the supernatant, and gently resuspending the glands with a fire-polished Pasteur pipet. The initial dilution for functional measurements is 1 ml settled glands/16–18 ml medium, which is equivalent to 1.5–2 mg glandular dry wt/ml.

Cell Purification

To produce single cells, the minced mucosa is digested 20 min in pronase digestion medium, rinsed three times by low-speed centrifugation in medium without proteolytic enzymes, then digested with four-fifths the amount of collagenase used for gland isolation. During collagenase digestion, the tissue should be triturated every 5 min to speed dissociation. Near 100% free cells are normally produced after 20–30 min with collagenase. Following collagenase digestion, cells are rinsed twice in incubation medium by brief centrifugation, then resuspended in cellular dilution medium (0.1–0.15 ml packed cells/ml of medium). Two milliliters of resuspended cells is carefully pipetted onto a 15 × 100 mm gradient tube which contains 8 ml of the preformed gradient. Gradients are centrifuged at room temperature in a swinging bucket rotor at 1000 g for 8 min. Parietal cells are enriched at the top of the gradient. Usually there are two parietal cell fractions—one in a distinct band immediately below the dilution medium at the top of the gradient and a second, more diffuse band below the first. A large band of mixed cells and mucus occupies the center of the gradient with the lowermost band typically containing >85% chief or pepsinogen-secreting cells. If one plans to isolate only parietal cells, the bottom layer of Nycodenz can be omitted and cells can be diluted to 0.05 ml packed cells/ml of medium with 4 ml of cells used/gradient tube. This greatly reduces the cost of the gradients and reduces cell clumping. Cells are carefully removed from the gradient with a fire-polished pipet, diluted in cellular incubation medium, and centrifuged at 1000–2000 g for 60 sec. The cell pellet is immediately resuspended in cellular incubation medium at a dilution of 0.1–0.2 ml packed cells/5 ml medium (total cell number = $4-5 \times 10^6$).

For >98% parietal cells, only the top fraction is subjected to elutriation using a Beckman (Fullerton, CA) J2 21 centrifuge equipped with a Beckman JE 6B elutriator rotor and standard chamber. For less demanding applications (90–95% enrichment), the top two fractions can be elutriated. The total yield from both fractions is approximately 0.5–1.0 ml packed cells. Cells are elutriated by injecting 5-ml aliquots of cells diluted 0.1/5 into the elutriator chamber at a flow rate of 20 ml/min. The rotor speed is maintained at 1950 rpm throughout the run. Flow rate is increased to 40 ml/min, then to 60 ml/min. Two to three 50-ml fractions are collected at each flow rate. Parietal cells are enriched in the 60 ml/min fractions. At the end of the run, parietal cell fractions are pooled and examined microscopically for purity. Parietal cells are easily recognized because they have a large diameter (18 μm on average), are granular because they are mitochondria rich, and stain pink when a drop of Polysciences (Warrington, PA) multiple stain solution is added (other cells stain blue or purple). Trypan Blue exclusion is typically near 100%. The initial dilution for functional measurements is 0.1 ml packed cells/20 ml medium (1–1.5 mg dry wt/ml where 1 mg dry wt is equivalent to approximately 10^6 cells).

Functional Measurements

Respiration

Measurement of cellular oxygen consumption is a very reliable technique for assessing viability and hormone responsiveness. Increased parietal cell respiration has been shown to be correlated with increased HCl secretion both *in vivo* and *in vitro*. Since parietal cells have many mitochondria, it is relatively easy to perform such measurements using standard manometric techniques.[5-8] We have obtained excellent results with a Gilson respirometer Middleton, WI. The temperature of the respirometer is set to 37° with air as the gas phase. Glands or parietal cells are diluted to a final concentration of 1.5–2 mg dry wt/ml with incubation medium. Two-milliliter aliquots are added to 25-ml respirometer flasks and readings taken every 10 min after an initial 20-min temperature equilibration. Results are expressed as microliters O_2 consumed per milligram cellular dry weight.

Amino[^{14}C]pyrine Accumulation

In vivo, parietal cells secrete HCl in quantities sufficient to lower the pH of gastric juice to 1 or lower. HCl is also secreted *in vitro* by the intact gastric mucosa and isolated parietal cells. With amphibian mucosae mounted in Ussing chambers at room temperature, changes in pH can be measured with a pH electrode. Such direct measurements are more difficult with mammalian mucosae, in which problems with anoxia arise because the tissue is relatively thick and the temperature must be maintained at 37°. With isolated parietal cells anoxia is not a problem, but HCl secretion cannot be measured directly because HCl secreted by the apical membrane into the incubation medium is immediately neutralized by HCO_3^- secreted by the basolateral membrane. Estimates of HCl secretion can be obtained, however, through the use of weak bases such as aminopyrine, which has a pK of 5.[4] At pH values above 5, aminopyrine passes freely across cell membranes. When protonated, aminopyrine is no longer membrane permeable. Thus, aminopyrine will be trapped in any cellular compartment when the pH of that compartment drops below 5. Within limits, the lower the pH, the more aminopyrine will be trapped. Gastric glands possess a central lumen that becomes sealed off during the isolation procedure. In single parietal cells the intracellular canaliculi, which face the lumen *in vivo,* also become sealed off from the extracellular medium. It appears that aminopyrine is trapped in the lumen of glands and in intracellular canaliculi of parietal cells.

Measurements of aminopyrine accumulation are made by incubating 1–2 mg dry wt/ml parietal cells or glands suspended in 2 ml incubation medium in capped 25-ml Erlenmeyer flasks. It is not necessary to oxygenate the cells. Amino[^{14}C]pyrine (0.1 μCi/ml, specific activity 80–90 mCi/mmol; New England Nuclear, Boston, MA) is added to the cells before they are pipetted into the flasks to assure uniform distribution of the label. After preincubation for 30–45 min in a metabolic shaker (37°, 50–70 oscillations/min), test agents are added at timed intervals. Samples are taken for measurement of amino[^{14}C]pyrine uptake by withdrawing 1 ml of well-mixed cell suspension from each flask. The samples are then placed in a 1.5-ml microfuge tube, centrifuged 10 sec, and supernatants withdrawn. The surface of the cell pellet is rapidly washed by gently pipetting 1 ml of medium into the tube, then withdrawing the medium. The wash medium is discarded. Pellets are dissolved in 200 μl concentrated nitric acid (50°, 10 min), cooled to room temperature, then diluted with 300 μl H_2O. Fifty-microliter aliquots of supernatants and 100 μl of dissolved pellets are counted in a liquid scintillation counter with dpm correction. Aminopyrine accumulation is expressed as the ratio of intracellular vs extracellular amino[^{14}C]pyrine. Nonspecific trapping of label is determined by incubating cells with 10 mM sodium thiocyanate (SCN), which completely abolishes parietal cell HCl secretion. Extracellular water trapped in the pellet has been found to be 2× the dry weight in glands and 1× the dry weight in parietal cells when inulin is used as a marker.[4,7] The aminopyrine accumulation ratio in cells is calculated as follows:

$$\frac{[(\text{Cellular dpm}/100\ \mu l) - (\text{SCN dpm}/100\ \mu l)] \times 5}{\dfrac{\text{cellular dry wt/ml}}{\text{supernatant dpm}/\mu l}}$$

For glands the cellular dry weight is multiplied by 2 to correct for the difference in trapped extracellular water.

When carefully controlled, measurement of amino[^{14}C]pyrine accumulation appears to be a reliable measure of parietal acid secretory activity. Under certain conditions, however, this measurement will yield artifactual results. One should remember, for example, that aminopyrine is a weak base and any agent that has similar properties can compete with aminopyrine for protons. Changes in incubation media which cause glands to dissociate or changes which cause cell swelling or shrinkage may nonspecifically affect aminopyrine accumulation. Also, certain agents such as cAMP analogs and phosphodiesterase inhibitors can cause cells to clump so severely that anoxia may result. The best approach is to observe the physical appearance of cells during the incubation and to correlate changes in aminopyrine accumulation with cellular respiration whenever possible. If the observed changes are closely correlated, it is less likely that they are artifactual.

Cyclic Nucleotide-Related Measurements

cAMP Accumulation and Activation of Adenylate Cyclase

Now that cAMP radioimmunoassay kits and cAMP-binding proteins are commercially available, it is not difficult to measure cAMP in cellular extracts. Cells or glands are diluted and incubated as for measurement of amino[^{14}C]pyrine accumulation. Accurate determination of cellular cAMP content depends on rapid fixation and elimination of factors which may interfere with the assay. One of the most reliable methods of cell fixation is addition of 50% trichloroacetic acid TCA (reagent grade) to a final concentration of 5%. After centrifugation of the sample to remove precipitated protein, TCA is extracted from the supernatant with 0.7 × 5 cm AG 50W-x4 columns by placing 1 ml of sample on the column, allowing the sample to enter the column, then rinsing with 3 ml H$_2$O. cAMP is eluted with an additional 4–5 ml H$_2$O.[6] The columns may be reused many times after regeneration by rinsing with 4 ml 1 N NaOH, 20 ml H$_2$O, then 4 ml 1 N HCl. The deionized water used with the columns should be high-pressure liquid chromatography (HPLC) grade for optimal results. Immediately before use columns are rinsed with 20 ml H_2O. Recovery of cAMP is monitored by inclusion of 1000–2000 dpm [^{3}H]cAMP/100 μl final extract volume.

Samples are concentrated by evaporation or lyophilization, then resuspended in assay buffer. If the less sensitive Gilman protein-binding assay is used, the sample should be resuspended at a concentration of 100 mg dry wt equivalent/50 μl assay buffer and 50-μl aliquots assayed. For radioimmunoassay (RIA) samples must be diluted 10-fold or more, depending on assay sensitivity. A good control for detection of interfering factors is to incubate cellular extracts with and without a known amount of exogenously added cAMP overnight at 30° with and without phosphodiesterase (3′,5′-cyclic nucleotide, beef heart, 0.2×10^{-4} U/μl) in 50 mM Tris, pH 8.0, 6 mM $MgCl_2$, 5 mM 2-mercaptoethanol. The reaction is stopped by boiling for 2 min and the samples assayed in the cAMP assay. Prolonged treatment with phosphodiesterase should destroy all authentic cAMP.

Adenylate cyclase activity is measured by briefly sonicating 1–2 mg dry wt/ml cells (5–10 sec, 18 W, Branson model 185 sonifier/microprobe, Farmingdale, NY) in ice-cold 2 mM Tris, pH 8.5, 1 mM DTT. For optimal hormonal responses, cells should be permeable to Trypan Blue but not totally disrupted. Aliquots (100 μl) of the crude sonicate are added to assay tubes containing (final concentration in 0.5 ml) 50 mM Tris, pH 7.8, 10 mM $MgCl_2$, 5 mM KCl, 1 mM DTT, 1 mM 3-isobutyl-1-methyxanthine (IBMX), 1 mM ATP, 0.1% BSA, and an ATP-regenerating system of 10 mM creatine phosphate and 35 U/ml creatine phosphokinase. Inclusion of 0.1 mM GTP in the incubation medium will increase basal cyclase activity and enhance the response to most hormones.

Reactions are linear for at least 30 min but are typically allowed to run for 10 min at 37°, then terminated by acidification to pH 4.5 with 5 μl 1 N HCl to decrease the pH to 4.5 and heating at 100° for 1 min. After centrifugation at 10,000 g to remove precipitated protein, aliquots of supernatants are assayed for cAMP in either the protein-binding assay or RIA. The amount of supernatant assayed depends on the amount of cAMP produced and must be empirically determined. For measurement of basal adenylate cyclase activity, 50-μl aliquots of supernatant yield assay values of 4–8 pmol cAMP with blanks of 0.1–0.3 pmol. With this technique, we have found cAMP production to be linear for at least 30 min. Enzyme activity may be expressed as picomoles cAMP per milligram homogenate dry weight or picomoles cAMP per milligram homogenate protein.

Assay of cAMP-Dependent Protein Kinase(s)

The assay of cAMP-dependent protein kinase in crude homogenates is difficult due to the presence of endogenous inhibitors, cAMP-independent protein kinases, and type I and type II isozymes, as well as binding of free catalytic subunit to particulate fractions. High salt concentrations (>0.1 M) cause the type I but not the type II isozyme to dissociate. Omission of salt results in binding of the free catalytic subunit to particulate fractions and reassociation of the catalytic and regulatory portions of the type II isozyme. When cells are stimulated with hormones that activate cAMP-dependent protein kinase, cellular disruption may release cAMP from compartments that were not accessible to the protein kinase of interest. Also, dilution of the homogenate may cause cAMP to dissociate from the regulatory subunit of cAMP-dependent protein kinase.[11]

cAMP-Dependent Protein Kinase Isozyme Identification

The first step should be to determine the isozyme type(s) present in the cell of interest. This can be accomplished by photoaffinity labeling of isozyme subunits in crude homogenates or cell fractions with 8-azido-[^{32}P]cAMP, separation with SDS-PAGE, and detection by autoradiography. The isozymes may also be separated on DEAE-Sephacel columns or on Pharmacia Mono Q HR 5/5 columns (Piscataway, NJ). In this case fractions may be assayed for cAMP-dependent protein kinase activity as well as photoaffinity labeling of regulatory subunits.[12]

For photoaffinity labeling or column purification, intact cells are sonicated in 10–20 mM bis-Tris-propane, pH 6.8, 10% glycerol, 0.1 M NaCl, 1 mM DTT, 10 mM benzamidine, 0.5 mM phenylmethylsulfonyl fluoride (PMSF), 5 mM EDTA, 1 mM EGTA, 1 mM IBMX, 1 μg/ml each leupeptin and pepstatin. Sonicates may be fractionated by differential centrifugation and pellets dissolved in sonication medium plus 0.5% Triton X-100 or the entire homogenate may be treated with Triton X-100. Photoaffinity labeling is performed by a modification of the method of Walter et al.[13] in a total assay volume of 100 μl. For assay of homogenates or column fractions, 20 μl labeling medium containing 250 mM 2[N-morpholino]ethane sulfonic acid (MES), pH 6.2, 1 mM EGTA, 1 mM DTT, 1 mM IBMX, 50 mM MgCl$_2$, and 5 μM 8-azido [^{32}P]cAMP (specific activity 20,000 cpm/pmol) is added to assay tubes followed by 80 μl homogenate diluted to 100–150 μg protein/80 μl or 80 μl undiluted column fraction. After incubation in the dark on ice for 30–60 min to allow the label to bind isozyme subunits, photoaffinity labeling is achieved by exposure to UV light (Mineralight model UVG-5, 10-cm distance, Ultraviolet Products, San Gabriel, CA) for 15 min. Reactions are terminated by addition of 20 μl 10% sodium dodecyl sulfate (SDS), 15% glycerol, 250 mM Tris, pH 7.8,

[11] D. A. Flockhart and J. D. Corbin, *CRC Crit. Rev. Biochem.* **12**, 133 (1982).
[12] C. S. Chew, *J. Biol. Chem.* **260**, 7540 (1985).
[13] U. Walter, I. Uno, A. Y. C. Liu, and P. Greengard, *J. Biol. Chem.* **252**, 6494 (1977).

0.05% Bromphenol Blue, and heating at 50° for 5 min. After cooling, add 20 μl of 0.35 M DTT and freeze at $-20°$.

Samples (25–30 μl) are separated on discontinuous SDS-PAGE gels (0.75-mm thick, 4.5% stacking gel and 8% running gel) at 18°, 30 mA/gel, according to Laemmli.[14] If unlabeled molecular weight standards are used, gels are stained in 0.025% (w/v) Coomassie Brilliant Blue R 250, 25% (v/v) isopropanol, 10% (v/v) glacial acetic acid for 1 h with shaking, then destained overnight in 7% methanol, 5% glacial acetic acid. To dry, soak gels in 7% methanol, 5% glacial acetic acid, 1% glycerol for 15–20 min, place on cellophane membrane backing which has been presoaked in deionized H_2O, and cover with Saran Wrap. To prevent gels from cracking during drying, take care not to stretch them and be sure no air bubbles are trapped in any of the layers. Gels are dried under vacuum for $1-1\frac{1}{2}$ hr with heat and 20–30 min without heat, then placed in Kodak X-ray exposure holders with Du Pont Cronex intensifying screens. Saran Wrap should be left on gels to prevent fogging of X-ray film. The film (Kodak Xomat AR) is taped to the gel and the cassette placed under a heavy object to assure good contact between film and gel and left at -20 to $-80°$ for 2–10 days. Film may be developed manually using Kodak GBX developer and rapid fixer as per manufacturer's instructions. Labeled regulatory subunits are identified according to molecular weight (R_I, 48,000–49,000; R_{II}, 51,000–57,000).[13,15,16] If more accurate identification is required, samples should be subjected to two-dimensional electrophoresis and identified by pI and molecular weight. A detailed description of this technique is given in the section, cAMP-Dependent Protein Phosphorylation in Intact Parietal Cells.

Detection of Hormonal Activation of cAMP-Dependent Protein Kinase(s)

Once isozyme types have been identified, cell disruption and assay conditions can be adjusted appropriately. Cells are incubated as for measurement of amino[^{14}C]pyrine accumulation. Aliquots (1 ml) are withdrawn, placed in 1.5-ml microfuge tubes, and rapidly pelleted (1-sec burst) in a microfuge. The supernatant is rapidly withdrawn and pelleted cells resuspended in 1 ml ice-cold medium containing 20 mM bis-Tris-propane, pH 6.8, 10 mM benzamidine, 0.1 mM PMSF, 5 mM EDTA, 1 mM EGTA, 1 mM IBMX, 1 mM DTT, 10% glycerol. If both isozyme types are

---

[14] U. K. Laemmli, *Nature (London)* **227**, 680 (1970).
[15] J. Erlichman, D. Sarkar, N. Fleischer, and C. S. Rubin, *J. Biol. Chem.* **255**, 8179 (1980).
[16] T. Jahnsen, S. M. Lohmann, U. Walter, L. Hedin, and J. S. Richards, *J. Biol. Chem.* **260**, 15980 (1985).

present, 0.1 M NaCl should be included to prevent subunit reassociation. If only the type II isozyme is present, the NaCl concentration may be increased to 0.25 M. This concentration of NaCl is not recommended for the type I isozyme because it dissociates in high salt solutions.[12] For assay of total cAMP-dependent protein kinase activity, cells can be solubilized by including 0.5% Triton X-100 in the disruption medium. For cell fractionation studies, cells are sonicated on ice for 4 × 5 sec (18 W, microprobe) in medium without Triton X-100, centrifuged at appropriate speeds, and pellets resuspended in medium containing 0.5% Triton X-100.

Cell fractions should be assayed as soon as they have been prepared to prevent dissociation/reassociation of isozyme subunits. With parietal cell homogenates we have found that homogenate protein concentration in assay tubes is optimal at approximately 100 μg/ml. With this amount of protein, cAMP-dependent protein kinase activity is linear for at least 30 min. When protein concentrations are increased much above 100 μg/ml, cAMP-dependent protein kinase activity is reduced and the assay becomes nonlinear within 1–5 min. Protein kinases that are cAMP-independent contaminate cell extracts. These kinases may substantially affect the apparent ratio of cAMP-dependent protein kinase activity (substrate phosphorylation with no added cAMP/substrate phosphorylation + exogenous cAMP). Under some conditions, the Walsh cAMP-dependent protein kinase inhibitor (PKI) may be used to control for cAMP-independent protein kinase activity. However, crude PKI may cause an anomalous increase in basal protein kinase activity in parietal cell homogenates. This effect appears to be localized in a particulate fraction and is independent of cAMP (Fig. 2). It has not been determined whether PKI or a contaminant activates a protein kinase in this fraction or whether a contaminant is utilized as a substrate. One should be aware, however, that crude PKI may stimulate phosphorylation in cellular homogenates. Hence, use of PKI may not accurately predict the relative amounts of cAMP-dependent and -independent protein kinases.

The cAMP-dependent protein kinase assay, which is a modification of the method of Corbin and Reimann,[17] is initiated by addition of 20 μl cell extract (5–10 μg protein) to a 10 × 75 mm tube containing (final concentration in a total volume of 75 μl) 50 mM HEPES, pH 6.8, 15 mM MgCl$_2$, 1 mM DTT, 0.1 mM [^{32}P]ATP (specific activity, 500–1000 cpm/pmol) with 1 mg/ml histone (Sigma, type IIS) as substrate. Total cAMP-dependent activity is determined by adding 2 μM cAMP. When used, PKI is added to achieve a final concentration of 100 μg/ml. After a 10- to 15-min incubation at 30°, reactions are terminated by pipetting 50-μl aliquots

---

[17] J. D. Corbin and E. M. Reimann, this series, Vol. 38, p. 287.

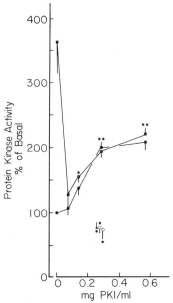

FIG. 2. Stimulatory and inhibitory effects of cAMP-dependent protein kinase inhibitor on histone kinase activity in a 100,000 g particulate fraction from rabbit parietal cells. Upper and lower lines, protein kinase activity $\pm \mu M$ cAMP; open circle, open square, and solid triangle represent, respectively, 900 g particulate, 14,000 g particulate, and 100,000 g supernatant + cAMP. PKI, Protein kinase inhibitor. Values are means ± SEMs, $n = 5$. (From Chew.[12])

from each tube onto 2.3-cm filter disks (Whatman grade 3, labeled with a #2 pencil) and rapidly dropping the disks into cold 10% TCA, 6 mM sodium pyrophosphate (10–20 ml/filter). The filters are gently agitated for 30 min, then rinsed twice more at room temperature in fresh TCA over a period of 10–12 hr. After a brief rinse in acetone, filters are dried, placed in 10 ml counting cocktail (as Ecoscint, National Diagnostics), and counted in a liquid scintillation counter. Protein kinase activity is expressed as picomoles ^{32}P transferred from ATP to substrate per milligram extract protein · minute.

cAMP-Dependent Protein Phosphorylation in Intact Parietal Cells

Agonists that increase parietal cell cAMP content have recently been shown to increase phosphorylation of several parietal cell proteins.[18-20]

[18] C. S. Chew and M. R. Brown, *Am. J. Physiol.* **253**, G823 (1987).
[19] I. M. Modlin, M. Oddsdottir, T. E. Adrian, M. J. Zdon, K. A. Zucker, and J. R. Goldenring, *J. Surg. Res.* **42**, 348 (1987).
[20] T. Urushidani, D. K. Hanzel, and J. G. Forte, *Biochim. Biophys. Acta* **930**, 209 (1987).

There are a large number of phosphoproteins in unstimulated parietal cells, hence it is difficult to demonstrate increased phosphorylation upon stimulation. In our experience, two-dimensional analysis is essential for detection of changing phosphoproteins in whole cell extracts. Since phosphoproteins may move from one cellular compartment to another following stimulation, it is recommended that initial studies be performed with whole cell extracts to avoid potential artifacts. We have not found wide pH range isoelectric focusing (IEF) gels to produce sufficient resolution of low-abundance phosphoproteins because they are obscured by high background associated with more abundant unchanging phosphoproteins. Hence, narrow pH range IEF gels are recommended. The following protocol, which we adapted from several different published procedures,[14,21-24] has been found to produce reliable two-dimensional data. We use electrophoresis-grade chemicals whenever possible. Where applicable preferred suppliers are indicated.

Cellular Incubation Media and Rinse Buffer

1. Phosphate-free medium, low BSA: 122.4 M NaCl, 5.4 mM KCl, 1.2 mM MgSO$_4$, 1 mM CaCl$_2$, 10 mM glucose, 1 mM pyruvate, 1 mg/ml BSA, 0.5 mM DTT, 0.01 mg/ml Phenol Red, 20 mM HEPES, pH 7.4. Solution is made the same way as cellular incubation medium for cell isolation
2. Phosphate-free medium, no BSA: As in (1), except BSA is omitted
3. Phosphate-buffered saline: As described for cellular isolation procedure

Stop Reagent and DNase/RNase Solution

1. Stop reagent: 0.3% SDS, 1% 2-mercaptoethanol (SDS/BME) in deionized water (DIW). Store at $-70°$ in 0.5- to 1-ml aliquots
2. DNase/RNase (10×): Mix together 952 μl of 1.5 M Tris-HCl, 48 μl of 1.5 M Tris base, 1.5 mg RNase (Worthington), 3.0 mg DNase (Worthington DPFF), 150 μl 1 M MgCl$_2$, and bring to a final volume of 3 ml with DIW. Store at $-70°$ in 50-μl aliquots

IEF Reagents

1. Reducing/solubilizing buffer (for pH 5-7 IEF gels): Mix together 5.71 g urea, 0.15 g DTT (Bio-Rad, Richmond, CA), 200 μl pH 5-7

[21] J. W. Pollard, in "Methods in Molecular Biology" (J. M. Walker, ed.), Vol. 1, p. 81. Humana, Clifton, New Jersey, 1984.
[22] J. E. Celis and R. Bravo, "Two-Dimensional Gel Electrophoresis of Proteins," Academic Press, Orlando, Florida, 1984.
[23] J. I. Garrels, *J. Biol. Chem.* **254,** 7961 (1979).
[24] P. H. O'Farrell, *J. Biol. Chem.* **250,** 4007 (1975).

ampholines (40%; LKB, Piscataway, NJ), 50 µl pH 3.5–9.5 ampholines (40%, LKB), and bring to a final volume of 10 ml with 2% 3-[(3-cholamidopropyl)dimethylammonio]-1-propane sulfonate (CHAPS). Dissolve slowly. Do not heat. Store at $-70°$ in 0.5-ml aliquots

2. Acrylamide/DATD stock: Dissolve 7.99 g acrylamide (Bio-Rad) and 1.41 g N,N'-diallyltartardiamide (DATD) in DIW, final volume 25 ml, and store at $-20°$
3. IEF gel stock (for pH 5–7 IEF gels): Mix together 5.42 g urea, 1.33 ml acrylamide/DATD stock, 2.0 ml 10% CHAPS, 400 µl pH 5–7 ampholines, 100 µl pH 3.5–9.5 ampholines, and dilute to a final volume of 9.5 ml with DIW. Store at $-70°$ in 3-ml aliquots
4. Riboflavin/TEMED (25 ml): Dissolve 1 mg riboflavin and 200 µl N,N,N',N'-tetramethylethylenediamine (TEMED, Bio-Rad) in DIW, final volume 25 ml, and store in a dark bottle at 4°
5. Fast green overlay solution: Mix together 100 µl reducing/solubilizing buffer (#1), 100 µl DIW, 5 µl 1% Fast Green. Store at $-20°$
6. Lower electrode solution: 0.06 M H_2SO_4, 3 liters
7. Upper electrode solution: 0.04 M NaOH, 1 liter
8. IEF gel equilibration buffer: Dissolve 6.0 g SDS (Bio-Rad), 0.02 g Bromophenol Blue (Bio-Rad), 1.54 g DTT, 25 ml 0.5 M Tris-HCl, pH 6.8, in a final volume of 200 ml with DIW. Stable for several weeks at room temperature
9. Agarose (1%): Add 0.1 g low electroendosmosis agarose (LKB) to 10 ml DIW. Store at 4°. Heat to 50° to dissolve immediately before use

SDS-PAGE Reagents and Standards

1. Tank buffer: Dissolve 3 g Tris (Bio-Rad), 14.4 g glycine (Bio-Rad), 1 g SDS in 1 liter final volume DIW. Total volume depends on type of apparatus used. We generally make up 5 liters for the Hoeffer SE600 and 15 liters for the Bio-Rad Protean II. Store at 4°. May reuse two or three times. After third use discard, as gels will not run as well with continued reuse. At room temperature, pH should be approximately 8.6 and is not adjusted
2. Acrylamide (29.2%), 0.8% Bis stock: Dissolve 29.2 g acrylamide (99.9%, Bio-Rad), 0.8 g Bis [N,N'-methylenebis(acrylamide)], (Bio-Rad) in DIW, final volume, 100 ml. Store in dark bottle at 4°. TOXIC
3. Running gel buffer (4×): Combine 1.5 M Tris and 0.008 M EDTA. Adjust to pH 8.9 with concentrated HCl and store at 4°

4. Stacking gel buffer (4×): Combine 0.5 M Tris and 0.008 M EDTA. Adjust to pH 6.8 with concentrated HCl and store at 4°
5. SDS (10%): Dissolve 5 g SDS in DIW, final volume 50 ml. Stable at room temperature
6. Ammonium persulfate (10%): Dissolve 0.1 g ammonium persulfate (Bio-Rad) in 10 ml DIW immediately before use
7. SDS standards: Seal bottoms of six or seven IEF gel tubes (7.5 × 150 mm, 1 mm i.d.) with several layers of Parafilm. Combine 0.45 ml IEF equilibration buffer [IEF reagent (8) above], 0.005 g agarose (LKB), and 50 μl Bio-Rad high- or low-molecular-weight standard in a 12 × 75 mm tube. Warm to approximately 50° to dissolve agarose. Quickly transfer to IEF tubes using a 1-ml syringe with an 18-mm cannula. After gels are cool, extrude onto Parafilm and store frozen in a tightly sealed container until use.

Gel Preparation

First dimension IEF gels are poured a few hours before use. The bottoms of gel tubes (7.5 × 130 mm, 1 mm i.d.) are sealed with three layers of Parafilm. Gels are placed in a tube rack. For 12–14 gels 10 cm in length, 3 ml IEF gel stock is degassed for 5 min, mixed with 6 μl 10% ammonium persulfate and 120 μl TEMED (Bio-Rad), and quickly placed in tubes using a 5-ml syringe with an 18-cm cannula. The tip of the cannula is kept below the surface of the gel solution in tubes as tubes are filled to avoid introduction of air bubbles. Tubes are gently tapped to remove any air bubbles trapped in the bottom. Gels are overlayed with 20 μl water-saturated isopropanol and polymerized by placing gels approximately 6 in. from a strong fluorescent light for $1\frac{1}{2}$–2 hr. Before loading samples, the isopropanol is removed and gels are rinsed three times with reducing/solubilizing buffer.

The lower halves (running gels) of second dimension gels are most conveniently poured the afternoon before gels are to be run. Depending on the gel-casting apparatus used and the width of the gel plates (15 vs 20 cm), approximately 25–30 ml of gel solution is required per 1.5-mm thick, 11.5-cm height running gel. For 100 ml of 12% running gel solution, the following components are filtered through Whatman filter paper into a small vacuum flask: 40 ml of 29.2% acrylamide, 0.8% Bis stock, 25 ml of running gel buffer, 1 ml 10% SDS, 0.5 ml 10% ammonium persulfate, 33.5 ml DIW. The solution is degassed for 5 min under vacuum. Immediately before gel is poured, add 50 μl TEMED. Swirl briefly to mix and rapidly pipet 25–30 ml of gel solution into gel sandwich prepared as per manufacturer's instructions. Immediately overlay gels with 2–3 ml water-saturated isopropanol. Allow to polymerize 1 hr. Remove isopropanol and replace with running gel buffer (4× diluted with DIW). Cover and let stand at room temperature overnight. If gels are stored at 4°, it is essential that they be at room temperature when upper stacking gels are poured because otherwise polymerization will be poor. Poor off buffer, rinse two or three times with DIW, and invert to drain just before stacking gels are poured.

Stacking gels are poured approximately 1 hr before use. For stacking gels, allow approximately 7 ml/gel for the Hoeffer apparatus and 11 ml/gel for the Bio-Rad Protean II 20-mm gels. For each 10 ml of stacking gel solution, filter through Whatman filter paper into a vacuum flask: 1.6 ml acrylamide/Bis stock, 2.5 ml stacking gel buffer, 0.1 ml 10% SDS, 5.7 ml DIW, and 70 μl 10% ammonium persulfate. Degas 5 min. Add 7 μl TEMED. Swirl to mix and pipet immediately onto running gel to within 2–3 mm of the top of the plate. Better polymerization is obtained if the tops of running gels are briefly rinsed with a small amount of the stacking gel mixture before pouring the stacking gel. Gently overlay gel with water-saturated isopropanol and allow to polymerize 30–45 min. Decant isopropanol and rinse with DIW two or three times. Invert to drain. Use immediately.

Cellular Incubation and Processing

Highly enriched parietal cells are diluted 0.1/10 with phosphate-free, low-BSA medium and incubated in a shaking water bath for 1 hr at 37° with 0.5–1 mCi/ml carrier-free ^{32}P. The ^{32}P is dried via a stream of N_2 gas in the flask that is to be used for cellular incubation. After 1 hr cells are diluted to a final concentration of 0.1/20 (~1 mg dry wt/ml) with warm phosphate- and BSA-free medium. The incubation is continued for 30 min, then 0.5–1 ml of cells is aliquoted into 1.5-ml microfuge tubes. Following agonist/antagonist additions and incubation for specified amounts of time, reactions are terminated by brief centrifugation (~1 sec) in an Eppendorf microfuge. The radioactive supernatant is removed and placed in a radioactive waste container. Cells are rinsed by resuspension in ice-cold PBS and brief centrifugation. The PBS is carefully withdrawn from the tube with a glass disposable pipet and discarded in radioactive waste. Cell pellets are quickly resuspended in 100 μl hot SDS/BME and heated for 3 min at 100° in a dry heat block. Dissolved pellets are placed on ice and incubated with 10 μl 10× DNase/RNase for 30 min on ice, then rapidly frozen and lyophilized.

Lyophilized residues are dissolved in 100 μl reducing/solubilizing buffer and centrifuged 30 sec in a microfuge to remove any undissolved particulate material. Twenty to 30 μl of the clear supernatant is loaded with a 50-μl Hamilton syringe onto an IEF gel under a 3-μl layer of Fast Green overlay solution. Samples are overlayed with 1 μl of carbamylytes (Pharmacia, pI range 4.8–6.7), which serve as internal IEF standards. Gels are placed in a Hoeffer IEF unit containing 2.5 liters of lower electrode solution. The upper chamber is filled with 800 ml upper electrode solution and gels filled with this solution by submerging and inserting the tip of the Hamilton syringe into the top of each gel to remove air bubbles. Gels are focused for 13–14 hr, 400 V, 18°, then 1 hr longer at 1000 V. Gels are loosened from the tube walls by injecting a small amount of 1% glycerol between the upper and lower ends of the gels and tube walls, then quickly extruded into 5 ml IEF equilibration buffer using a 50-ml syringe connected to a short length of plastic tubing. After a 10-min equilibration, gels are drained by pouring onto nylon mesh, placed on a clear, thin plastic sheet, carefully straightened using a glass rod to avoid breakage, and gently teased onto the SDS-PAGE gel with a thin weighing spatula, taking care not to trap air bubbles between the two gels. Frozen SDS-PAGE standards are cut into 1-cm sections and low- and high-molecular-weight sections placed at either end of the IEF gel. Standards and the IEF gel is sealed onto the SDS-PAGE gel with a small amount of warm 1% agarose. Second dimension gels are routinely run at 200 V constant voltage, 18°, 4–5 hr (tracking dye 1 cm from bottom of gel). Gels are fixed, stained with Coomassie Blue R, destained, and dried as described in the section, Assay of cAMP-Dependent Protein Kinase(s). Silver stain is not recommended as it quenches ^{32}P emission in gels. Dried gels are covered with fresh Saran Wrap to prevent film fogging and placed in X-ray cassettes with Kodak XAR-5 film for 15–24 hr. After the initial exposure, additional exposures are obtained, depending on the amount of radioactivity present in gels. We routinely make three exposures at 15–24 hr, 48 hr, and 3 days. This allows us to detect low and high concentrations of radiolabeled proteins and circumvents problems with film linearity. Autoradiographs may be quantitated manually using a densitometer or by cutting out the spots of interest and counting in a liquid scintillation counter. Global quantitation and analysis can be achieved via computer-based image analysis using commercially available software and hardware such as that offered by Kodak (Bio Image) or with a radioanalytic imaging system such as the AMBIS. A detailed discussion of methodology associated with computer-based techniques can be found in the chapter by J. Garrels in Ref. 22.

Correlation of Cellular Function with cAMP Metabolism

Whenever possible, functional responses should be correlated with cAMP-related measurements within the same cell population. This approach establishes that each preparation possesses appropriate hormonal responsiveness and allows direct correlation of biochemical and functional responses. A wide range of agonist/antagonist concentrations should be tested in order to establish whether or not the observed effects are of potential physiological relevance. Supramaximal concentrations of test agents may alter cellular activity but the observed effects may have little or nothing to do with a specific receptor-mediated response. With agonists direct comparisons of time courses and EC_{50} values for functional and biochemical responses provide useful information. For antagonists, experiments should be designed to determine (1) whether the inhibition is competitive or noncompetitive, (2) whether the effect is reversible, and (3) the time course of inhibition. Determination of pA_2 values using standard pharmacologic approaches is an excellent technique for establishing whether the same receptor subtype is involved in both types of response (Fig. 3).

With the development of more sophisticated cell isolation and functional measurement techniques, it is becoming increasingly apparent that many hormonal control mechanisms cannot be explained by invoking cAMP as the sole mediator of a response. Two recent examples that point to multiple control mechanisms include the observations that glucagon and histamine, which, respectively, elevate cAMP in hepatocytes and parietal cells, also elevate intracellular Ca^{2+} in these cells.[10] Thus, a single agonist may initiate multiple cellular responses. These responses are probably compartmentalized within the cell and may be controlled by multiple hormones. Future investigations will focus on attempts to develop more

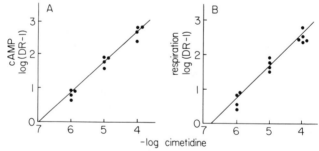

FIG. 3. Determination of pA_2 values for cimetidine antagonism of histamine-stimulated increased in cAMP and respiration in rabbit gastric glands. Dose ratios (DR) were determined by generating a series of histamine dose-response curves with indicated concentrations of cimetidine, then calculating the ratio of the EC_{50} for agonist (histamine) plus antagonist (cimetidine) to the EC_{50} for agonist alone. The pA_2 (intercept) is defined as the negative log of the antagonist concentration that produces a DR of 2. The larger the pA_2, the more potent the antagonist. Values are from five separate experiments. (Data from Chew et al.[6])

specific inhibitors of various cellular enzymes, as cAMP-dependent and cAMP-independent protein kinases, and introduction of these inhibitors into living cells. Experiments designed to identify the substrates of these enzymes are also of key importance. Recent developments in digitized video image analysis and confocal microscopy hold great promise as tools for determining whether or not hormonally induced increases in cAMP and activation of cAMP-dependent protein kinase isozymes are indeed compartmentalized. Such investigations will be greatly aided by the development of fluorescent probes for cAMP that respond to changes in concentration as do the recently developed fluorescent calcium probes.[25]

[25] G. Grynkiewicz, M. Poenie, and R. Tsien, *J. Biol. Chem.* **260**, 3440 (1985).

[37] Stimulus–Secretion Coupling: General Models and Specific Aspects in Epithelial Cells

By HOWARD RASMUSSEN

General Models of Stimulus–Response Coupling

In spite of a great diversity of intracellular messengers, of cellular organization, and of cellular response, a few central motifs underlie our present views of stimulus–response coupling whether the response is the secretion of a steroid hormone such as aldosterone from the nonpolar adrenal glomerulosa cell, or the secretion of Na^+ and Cl^- from the polar cells of the epithelium lining the small intestine. The two messenger systems with clearly established roles are the cAMP messenger system and the Ca^{2+} messenger system. Much of what we know about the operation of these systems has been learned from studies of stimulus–response coupling in nonpolar cells. The first part of this chapter will be concerned with our present views of these general models of stimulus–response coupling. Many fewer details have been defined in the process of stimulus–response coupling in polar cells. Nonetheless, the same molecular components have been identified in such polar cells. As a consequence, the models, developed from an analysis of stimulus–response coupling in nonpolar cells, have been applied to stimulus–response coupling in polar cells. However, such applications have been done largely without due consideration being given to the unique features which may dictate a different organizational arrangement in polar cells. The second part of this chapter will consider some of the issues which need to be resolved before a general model(s) of stimulus–response coupling in polar cells can be elaborated.

General Models

The overall organization of the two major messenger systems is remarkably similar, but there are some important differences. Furthermore, the Ca^{2+} and cAMP messenger systems do not regulate distinctly different types of cellular processes, but regulate many of the same processes, often interacting in one of several ways to regulate a particular cellular response. Hence, a separate discussion of the organization of each system provides a basis for understanding their main features but artificially separates their roles, whereas, in fact, Ca^{2+} and cAMP commonly function together as synarchic messengers. Thus, a discussion of the organization of each system will be followed by a discussion of the interrelations, one with another, and with the production and actions of eicosanoids.

The cAMP Messenger System. The classic model of cAMP action proposed that cAMP serves as a second or intracellular messenger in the action of a variety of peptide hormones or first messengers. As eventually formulated, the model encompassed the following sequence of events: (1) hormone–receptor interaction (at the plasma membrane), (2) activation of adenylate cyclase, (3) increase in cAMP concentration in the cell cytosol, (4) activation of specific cAMP-dependent protein kinase(s), and (5) cellular response as a consequence of changes in the contents of specific phosphoproteins.[1]

As originally developed, this model implied that the response was unidirectional and temporally constant. Furthermore, the off signals — phosphodiesterase to hydrolyze the second messenger, cAMP, and phosphoprotein phosphatase to catalyze the dephosphorylation of the phosphoproteins — were considered to function largely as means of terminating response when first messenger concentration fell. However, it is now apparent that the cAMP messenger system is much more elegant than the original model implied. First, at the level of the plasma membrane receptor(s), there are two classes of receptors — stimulatory receptors which are linked to the catalytic unit of the adenylate cyclase via a specific class of guanine nucleotide regulatory proteins, G_s or N_s; and inhibitory receptors which are linked to the cyclase via another class of G proteins, G_i. A dual control of cyclase activity is a common feature of many cells.[2,3] This dual control provides a means by which cAMP production rate and cAMP concentration can be precisely regulated. Second, there are several different classes of phosphodiesterases, both free in the cytosol and bound to mem-

---

[1] G. A. Robison, R. W. Butcher, and E. W. Sutherland, "Cyclic AMP." Academic Press, New York, 1971.
[2] E. M. Ross and A. G. Gilman, *Annu. Rev. Biochem.* **49**, 553 (1980).
[3] M. Rodbell, *Nature (London)* **284**, 17 (1980).

branes, including the plasma membrane, which are activated in response to first messenger action.[4]

A characteristic feature of many cells in which cyclase is activated by a first messenger is that cAMP content rises initially, then falls to a lower level and remains at this lower level during the sustained phase of the response. Yet, when cAMP production rate is measured, it remains high. In other words, the subsequent fall in cAMP content is not due to a transient activation of cyclase. There is a sustained activation of cyclase associated with a more slowly developing activation of phosphodiesterase, i.e., during sustained activation of the cell by first messenger the most characteristic feature of the cAMP messenger system is a sustained increase in rate of cAMP turnover.[5] Whether this turnover is a means of conveying information from cell surface to cell interior remains to be determined (see below, the discussion of a similar phenomenon in the Ca^{2+} messenger system). Part of the change in cAMP metabolism during the sustained phase of the response, and part of the explanation for a high rate of cAMP production associated with a minimal elevation of cAMP concentration, is an increased rate of cAMP efflux from the cell. The regulatory significance of this change is also not known.

In addition to activating protein kinases, intracellular cAMP also inhibits the activity of phosphoprotein phosphatases. This is achieved in two ways.[6,7] In the first case, certain small specific intracellular proteins (inhibitor proteins) are phosphorylated by the cAMP-dependent protein kinases, after which they bind with high affinity to the phosphatases but are poor substrates for their action. Hence, they effectively block the dephosphorylation of other phosphoprotein products of the cAMP-dependent protein kinases.

In order to understand the second mechanism of phosphatase inhibition, it is necessary to describe the mechanism by which cAMP acts on protein kinases.[8] The inactive forms of these kinases exist as tetramers of two regulatory (R) and two catalytic (C) subunits [$R_2C_2 + 4cAMP \rightleftarrows 2R(cAMP)_2 + 2C$]. The free catalytic units are active kinases. Until recently the R(cAMP) subunits were considered inert from a regulatory point of view. However, the R(cAMP) of type II protein kinase can serve as a substrate for C and become P–R(cAMP), in which form it is a specific inhibitor of a certain class of phosphoprotein phosphatase.[7] This feature of

---

[4] R. W. Butcher, *Adv. Cyclic Nucleotide Res. Protein Phosphorylation Res.* **16**, 1 (1984).
[5] G. B. Sala, K. Hayashi, K. J. Catt, and M. L. Dufau, *J. Biol. Chem.* **254**, 3861 (1979).
[6] P. Cohen, *Curr. Top. Cell. Regul.* **14**, 117 (1978).
[7] B. S. Khatra, R. Printz, C. E. Colb, and J. D. Corbin, *Biochem. Biophys. Res. Commun.* **5**, 99 (1972).
[8] E. G. Krebs, *Curr. Top. Cell. Regul.* **5**, 99 (1972).

the molecular interactions in the cAMP messenger system strongly suggests that when the first messenger activates a cell by activating adenylate cyclase, response is determined by the combined activation of cyclase and inhibition of phosphatase activity. This dual control of the content of specific phosphoproteins may provide an explanation of why the cAMP concentration is not as high during the sustained phase of cellular response as it is during the initial phase of the response, and yet the response is cAMP dependent throughout. If so, regulation of phosphatase activity is of equal importance to that of kinase activity in determining cellular response. At present, our knowledge of the regulation of phosphoprotein phosphatase activity is rudimentary.

The Ca^{2+} Messenger System. The model of Ca^{2+} messenger system function developed to account for agonist-induced control of smooth muscle contraction was operationally quite similar to the classic second messenger model of cAMP action: hormone–receptor interaction leads to a rise in the intracellular $[Ca^{2+}]_c$, the second messenger, which in turn activates the calmodulin-dependent enzyme, myosin light chain kinase. [Calmodulin (CaM) is a nearly universal intracellular Ca^{2+} receptor protein.] The resulting phosphorylation of myosin light chains (MLC $\rightleftarrows$ MLC · P) leads to a contractile response.[9] Sustained response requires a sustained elevation of Ca^{2+} and/or MLC · P. However, neither of these changes persist during a sustained cellular response.[10,11] Rather, there is a temporal sequence of regulatory events.[12] Our current view of the organization of the calcium messenger system involves a two-branch model in which the flow of information via each branch is responsible for a specific temporal component of the response.[13,14]

Hormone–receptor (or neurotransmitter–receptor) interaction leads to the activation of a specific membrane-associated phospholipase C.[15–17] Receptor is coupled to phospholipase via a specific G protein as other receptors are coupled to adenylate cyclase. Activation of the phospholipase C leads to the hydrolysis of phosphatidylinositol 4,5-bisphosphate resulting in the production of two messengers: a water-soluble inositol phosphate,

[9] R. S. Adelstein and E. Eisenberg, *Annu. Rev. Biochem.* **49**, 921 (1980).
[10] J. P. Morgan and K. G. Morgan, *J. Physiol. (London)* **351**, 155 (1984).
[11] P. J. Silver and J. T. Stull, *Mol. Pharmacol.* **25**, 267 (1984).
[12] S. Park and H. Rasmussen, *Proc. Natl. Acad. Sci. U.S.A.* **82**, 8835 (1985).
[13] H. Rasmussen and P. Q. Barrett, *Physiol. Rev.* **64**, 938 (1984).
[14] H. Rasmussen, *N. Engl. J. Med.* **314**, 1094, 1164 (1986).
[15] Y. Nishizuka, *Philos. Trans. R. Soc. London,* **B 302**, 101 (1983).
[16] M. J. Berridge, *Biochem. J.* **220**, 345 (1984).
[17] R. F. Irvine, E. E. Anggard, A. J. Letcher, and C. P. Downes, *Biochem. J.* **229**, 505 (1985).

1,4,5-inositol trisphosphate (1,4,5-InsP$_3$) and the lipid-soluble diacylglycerol (rich in arachidonic acid).

There is a transient rise in the concentration of 1,4,5-InsP$_3$ followed by a fall to a value slightly above the basal value, even though it seems likely that 1,4,5-InsP$_3$ production remains high. The fall in 1,4,5-InsP$_3$ is due to its rapid conversion first to 1,3,4,5-inositol tetrabisphosphate, and then to 1,3,4-InsP$_3$. Only 1,4,5-InsP$_3$ has a defined messenger function. This function is to induce the release of a pool of intracellular Ca^{2+}, thought to be located in the endoplasmic reticulum. This leads to an increase in [Ca^{2+}]$_c$ and the activation of CaM-dependent enzymes, including CaM-dependent protein kinases and the Ca^{2+} pump of the plasma membrane. As a consequence of the transient nature of the rise in 1,4,5-InsP$_3$ concentration and of the activation of the plasma membrane Ca^{2+} pump, the [Ca^{2+}]$_c$ rises only transiently, and the falls close to its basal value during the sustained phase of cellular response. Nonetheless, this transient increase in [Ca^{2+}]$_c$ (lasting 2–4 min) is sufficient to activate CaM-dependent protein kinases. As a result, a specific subset of cellular proteins become phosphorylated (Pr$_a$ ⇌ Pr$_a$ · P).[18] As a consequence, cellular response is initiated. However, the duration of this phase of the response is considerably longer (15–20 min) than the duration of the signal, the [Ca^{2+}]$_c$ transient, (2–4 min) needed to generate it. Two possible explanations for this difference in duration are possible: (1) either (as in the cAMP messenger system) phosphoprotein phosphatase activity is inhibited in response to agonist along with activation of kinase; and/or (2) the kinase itself displays a unique pattern of response. There is at present no evidence for the former mechanism, but there is for the latter. The type II CaM-dependent protein kinase undergoes autophosphorylation which converts it from Ca^{2+}–CaM-activated form to a Ca^{2+}- and CaM-insensitive one. This form is an active protein kinase.[19] Hence, the half-life of an active form of kinase is considerably longer than that of the Ca^{2+} transient which brings about its initial activation.

The increase in diacylglycerol content of the plasma membrane, along with the transient increase in [Ca^{2+}]$_c$, lead to the conversion of the enzyme, protein kinase C or C kinase, from its Ca^{2+}-insensitive to its plasma membrane-associated, Ca^{2+}-sensitive form. Additionally, in many cells, hormone–receptor interaction leads to an increase in Ca^{2+} influx rate.[13] The cell compensates by increasing efflux rate so that during the sustained phase of the response, the key regulatory events are an increase in Ca^{2+}

---

[18] P. Q. Barrett, I. Kojima, K. Kojima, K. Zawalich, and H. Rasmussen, *Biochem. J.*, **238**, 893 (1986).

[19] S. G. Miller and M. B. Kennedy, *Cell (Cambridge, Mass.)* **44**, 861 (1986).

cycling across the plasma membrane and a Ca^{2+}-sensitive C kinase associated with this membrane.[14] The activity of the enzyme is controlled by the rate of Ca^{2+} cycling. Whether the enzyme reads this cycling as a flux, or as a change in concentration of Ca^{2+} in a specific submembrane domain, $[Ca^{2+}]_{sm}$, is not known. We have adopted the convention of $[Ca^{2+}]_{sm}$ for the present. In any case, as a consequence of C kinase activity, a second subset of cellular proteins, Pr_b, become phosphorylated ($Pr_b \rightleftarrows Pr_b \cdot P$).[18] These mediate the sustained phase of cellular response.

In this model, Ca^{2+} serves as a messenger during each phase of cellular response, but its subcellular locus of action and its molecular target differs. During the initial phase, a change in $[Ca^{2+}]_c$ is the message and CaM-dependent enzymes are the targets; during the sustained phase, a change in $[Ca^{2+}]_{sm}$ is the message and C kinase the target.

A comparison of the two messenger systems indicates that even though there is some organizational similarity, there are apparently key differences. In particular, the model of the cAMP system emphasizes the importance of the simultaneous control of both protein kinase and phosphoprotein phosphatase activities in determining the contents of various phosphoproteins, but indicates no temporal sequence of cellular events. The model of the Ca^{2+} system, on the other hand, emphasizes the importance of a temporal sequence and says little of phosphoprotein phosphatase. However, these dissimilarities are due more to incomplete evidence that to established data. It is more than likely that the control of phosphoprotein phosphatase activities will be as important a feature of the Ca^{2+} messenger system as it is in the cAMP system. It is already known that the inhibitor protein, which is a substrate for cAMP-dependent protein kinase, is also a substrate for the Ca^{2+}-CaM-dependent protein kinase, phosphorylase b kinase. Conversely, it is likely that a temporal sequence of events will be found to characterize the operation of the cAMP as well as the Ca^{2+} messenger system.

Interrelationships between Messenger Systems. As discussed above, these two major messenger systems share similar organizational motifs, but appear to be rather self-contained systems. Such a representation is useful from a didactic point of view, but is a distortion of the biological reality. An overriding fact of cell regulation is that these two messenger systems interact at nearly every level, and usually serve in an integrated way to control cellular response: Ca^{2+} and cAMP nearly always serve as synarchic messengers in the regulation of cellular response.[20]

To fully appreciate the importance of such synarchic regulation, one must understand that the particular physiological response of a differen-

[20] H. Rasmussen, "Calcium and cAMP as Synarchic Messengers." Wiley, New York, 1981.

tiated cell type, i.e., aldosterone secretion from an adrenal glomerulosa cell or contraction of a vascular smooth muscle cell, is controlled by multiple extracellular messenger inputs which may and often are both neural and humoral. These multiple inputs may be either stimulatory or inhibitory, and may involve activation or inhibition of either one or the other messenger system. An important feature of this type of organization is that a single type of receptor is coupled to only one type of transducing system, e.g., β-receptors for catecholamines are linked exclusively to the stimulatory branch of adenylate cyclase activation, α_1-catecholamine receptors to a stimulation of the phospholipase C in the Ca^{2+} messenger system, and α_2-receptors to an inhibition of adenylate cyclase.

A particularly striking feature of synarchic regulation is the differing patterns of interaction between the two messenger systems. One can recognize at least five patterns: (1) coordinate, in which a particular extracellular messenger simultaneously stimulates adenylate cyclase via one receptor type, and phosphatidylinositol 4,5-bisphosphate hydrolysis via another; (2) hierarchical, in which a low concentration of a particular messenger activates, for example, phosphatidyl inositol (PI) hydrolysis and a higher concentration activates adenylate cyclase; (3) redundant, in which one extracellular messenger activates the cyclase system and another the PI system, with either alone inducing the cellular response; (4) sequential, in which activation of one system, e.g., cyclase, leads to the activation of the second; and (5) antagonistic, in which activation of one system, cyclase, blocks the activation of the second. This classification is a minimal one in that it does not include all those instances in which negative inputs via the cyclase system alter response to activation of the Ca^{2+} messenger system. Additionally, in the actual control of a given cellular response, there are usually multiple interactions between events in the two systems so the control of a given response often displays one of the above patterns as a major motif and one or more additional patterns as minor motifs. Even so, this classification is useful in calling attention to the varied operational modes by which these messenger systems operate, and thereby emphasizing the great inherent plasticity in this type of organization. From a practical point of view, what this nearly universal association between the operations of these two systems means is that during studies aimed at defining the mechanism by which a particular cellular response is controlled, it is necessary to elucidate the roles of both systems, rather than one or the other.

Integration of Messenger Systems in a Cellular Response. Great progress has been made at the cellular level in defining the nature of the interactions between Ca^{2+} and cAMP. A summary of this information is provided in Table I. Noteworthy is the fact that there is not a complete

TABLE I
Interaction between Ca^{2+} and cAMP Messenger System[a]

1. Ca^{2+} regulates cAMP metabolism:
   a. Ca^{2+} activates phosphodiesterase
   b. Ca^{2+} activates adenylate cyclase
2. cAMP regulates PI turnover:
   cAMP inhibits PIP_2 hydrolysis
3. cAMP regulates Ca^{2+} metabolism:
   a. cAMP activates voltage-dependent Ca^{2+} channels in plasma membrane
   b. cAMP activates Ca^{2+}-ATPase in plasma membrane
   c. cAMP activates C^{2+}-ATPase in membrane of sarcoplasmic reticulum
4. cAMP regulates sensitivity of response elements to activation by Ca^{2+}:
   a. cAMP-dependent phosphorylation of Ca^{2+}-CaM-dependent enzyme increases its sensitivity to activation by Ca^{2+} — positive sensitivity modulation
   b. cAMP-dependent phosphorylation of Ca^{2+}-CaM-dependent enzyme decreases its sensitivity to activation by Ca^{2+} — negative sensitivity modulation
5. cAMP and Ca^{2+} act at same cellular site:
   cAMP- and Ca^{2+}-CaM-dependent protein kinases phosphorylate same substrate protein (often at separate molecular loci)

[a] PI, phosphatidylinositol; PIP_2, phosphatidylinositol 4,5-bisphosphate; CaM, calmodulin.

symmetry of interrelationships. Thus, even though cAMP-dependent phosphorylation of Ca^{2+}-regulated enzymes is a common mechanism by which cAMP alters response to a given amplitude (concentration) of Ca^{2+} message, the converse has yet to be described. It is possible that this is simply a failure of investigators to detect such a relationship yet, or it may be related to the evolution of these messenger systems. In any case, in terms of the present discussion of stimulus–response coupling in epithelial cells, cataloguing the numerous types Ca^{2+}–cAMP interactions does not provide an appropriate background for discussing the organization of these messenger systems in polar cells. A more appropriate introduction is provided by discussing a specific example of Ca^{2+}–cAMP interactions in a particular nonpolar target cell responding to a specific stimulus: K^+-mediated aldosterone secretion from the adrenal glomerulosa cell.

A point emphasized in the discussion of the organization of the Ca^{2+} messenger system was that the cellular locus and molecular site of Ca^{2+} action differed in the initial and sustained phases of cellular response. In this view, during the sustained phase of cellular response, the cycling of Ca^{2+} across the plasma membrane is the message and the C kinase associated with the plasma membrane is the molecular target for this message. However, cells may employ Ca^{2+} as a messenger in mediating a sustained response, but not utilize the C kinase branch in doing so. An example is the K^+-dependent stimulation of aldosterone secretion from the adrenal glomerulosa cell.[21,22] This cell type is particularly sensitive to changes in the extracellular K^+ concentration. An increase in $[K^+]$ from 3.5 to 7.0 mM is sufficient to induce a fourfold increase in aldosterone production rate. This change in K^+ concentration does not activate the hydrolysis of phosphatidyl 4,5-bisphosphate nor activate C kinase. The rise in K^+ concentration leads to a fourfold increase in Ca^{2+} influx rate, which leads to an initial gain in total cell calcium that is soon compensated for by a fourfold increase in Ca^{2+} efflux rate, so that during the sustained phase of cellular response the total cell calcium remains constant. The sudden increase in Ca^{2+} influx rate induced by the rise in K^+ concentration leads to a rise in $[Ca^{2+}]_c$, but this, too, is transient so that $[Ca^{2+}]_c$ during the sustained phase of the response is only slightly above its original basal value. Nonetheless, aldosterone production rate continues at a high level. However, this physiological response is critically dependent on a second effect of K^+. The increase in K^+ concentration leads to a Ca^{2+}-dependent (presumably mediated via calmodulin) activation of adenylate cyclase. The resulting rise in $[cAMP]_c$ concentration is transient, but the cAMP production rate remains high. Thus, the sustained phase of the response depends upon the synarchic messengers, Ca^{2+} and cAMP. The interesting feature of this sustained response is that neither $[Ca^{2+}]_c$ nor $[cAMP]_c$ is significantly greater than its respective basal value (that seen in the unstimulated cell), but their respective turnover and production rates are both three- to fourfold higher in the stimulated as compared to the control cell. Somehow the rates of cycling of these messengers, and/or the effects of this cycling on the local concentration of the respective messenger in a subcellular domain at or just beneath the plasma membrane, are the messengers: a situation analogous to the cases in which the C kinase branch provides the second order transducing event at the plasma membrane essential for sustained cellular response. Hence, information flow from cell surface to cell interior is by a pathway other than simply a diffusion of messenger.

[21] I. Kojima, K. Kojima, and H. Rasmussen, *Biochem J.* **228**, 69 (1985).
[22] P. J. Hyatt, J. F. Tait, and S. A. S. Tait, *Proc. R. Soc. London,* **B 227**, 21 (1986).

Autocrine and Paracrine Inputs. A particularly important feature of many tissue responses is the involvement of paracrine and/or autocrine agents in the control of cellular response. This is particularly true in such epithelial tissues as the intestinal mucosa. It is not possible to consider all of these agents in detail, but a discussion of one class, the eicosanoids, and their relationships to the Ca^{2+} and cAMP messenger systems provides a basis for understanding their role.[14] The eicosanoids are a class of lipid molecules derived from the oxygenation of arachidonic acid.[23,24] There are two major pathways for their synthesis: the lipoxygenase pathway, giving rise to 5-hydroperoxyeicosatetraenoic acid (5-HETE) and leukotrienes; and the cyclooxygenase pathway, giving rise to prostacyclin, prostaglandins, and thromboxanes. Each class of these compounds can serve as a mediator of some aspect of cellular response. A major unresolved issue is whether these agents act only via surface receptors (as do endocrine and neurocrine agents) or whether they can act directly on intracellular receptors. There is at present convincing evidence for the existence of surface receptors for these agents, but not for the existence of intracellular receptor.

Under basal conditions in most cells, the concentration of free arachidonic acid is very low, and hence the rates of production of the various eicosanoids are low. The major substrates from which arachidonic acid is released upon activation of a cell are linked to the Ca^{2+} messenger system and are catalyzed by a class of enzymes known as phospholipase of the A_2 class. The diacylglycerol (DG) generated when an agonist activates a cell via the Ca^{2+} messenger system is rich in arachidonic acid and can serve as a source of free arachidonic acid (AA) for eicosanoid synthesis. There are several pathways by which the liberation of AA from DG can occur, but the major one appears to be the sequential conversion of DG to phosphatidic acid (DG + ATP $\rightleftarrows$ PA + ADP) catalyzed by DG kinase, and then the action of PA-specific phospholipase A_2 (PLA_2) leading to lysophosphatidic acid (LPA) and AA (PA $\rightleftarrows$ LPA + AA). The other source of arachidonic acid is from other membrane lipids, in particular phosphatidylcholine (PC). The AA is derived directly from the PC as a result of the action of a general Ca^{2+}-sensitive phospholipase A_2 [PC $\rightleftarrows$ lysophosphatidylcholine (LPC) + AA]. The relative importance of each of these pathways in particular cells is not yet known, but it is clear that in certain cases (platelets) the two pathways can be activated separately.

The platelet release reaction can serve as a model of how eicosanoid synthesis and action are related to both the Ca^{2+} and cAMP messenger

[23] B. Samuelson, *Science* **220**, 568 (1983).
[24] B. Samuelson, M. Goldyne, E. Granström, M. Hamburg, S. Hammarström, and C. Malmsten, *Annu. Rev. Biochem.* **47**, 997 (1978).

systems.[25] Thrombin is a potent activator of the platelet release reaction. When thrombin binds to its receptor, the hydrolysis of phosphatidylinositol 4,5-bisphosphate (PIP_2) occurs, leading to the production of 1,4,5-$InsP_3$ and DG. The DG is rich in arachidonic acid, and serves both to activate C kinase (see above) and as a source for arachidonic acid. Additional AA is derived from the PLA_2-mediated hydrolysis of PC. The AA is rapidly metabolized via the cyclooxygenase pathway to thromboxane A_2 (TXA_2), prostaglandin E_2) (PGE_2) and prostacylin (PGI_2). The TXA_2 acts as a positive feedback regulator, via a specific surface receptor, to activate PIP_2 hydrolysis. The PGE_2 and PGI_2 act as negative feedback regulators by binding to specific surface receptors and causing the activation of adenylate cyclase. The subsequent rise in $[cAMP]_c$ inhibits the thrombin- and/or TXA_2-mediated response (via the Ca^{2+} messenger system) by (1) causing a resequestration of Ca^{2+} and a fall in $[Ca^{2+}]_c$, (2) causing a change in the sensitivity of Ca^{2+}-CaM-dependent enzymes, e.g., myosin light chain kinase, to activation by Ca^{2+}, and (3) by directly inhibiting the activity of the phospholipase C responsible for PIP_2 hydrolysis. A specific secretory product of the release reaction is ADP. This also serves as a positive feedback modifier which acts via a surface receptor to bring about an inhibition of adenylate cyclase, a fall in $[cAMP]_c$, and thereby allowing thrombin and/or TXA_2 to exert their effects. Hence, a complex mixture of positive and negative signals acting via receptors linked either to the Ca^{2+} or the cAMP messenger systems determines whether a thrombin-induced platelet aggregation reaction will progress or stop.

Special Problems in Epithelia Cells

The majority of the studies from which these general models of stimulus–response coupling have been developed have been carried out in nonpolar cells. The obvious reason for this bias is that a variety of isolated nonpolar cell types from various organs can be obtained in high yield and appear to be metabolically stable for hours, thus allowing one to carry out detailed studies of stimulus–response coupling in isolated cells. Much less satisfactory preparations of polar epithelial cells have been generally available until quite recently. Nonetheless, polar cells do possess all the molecular components of both the cAMP and Ca^{2+} messenger systems. Hence, the models of stimulus–response coupling, derived from the studies in nonpolar cells, have been applied to similar events in polar cells.[26] However, one

---

[25] M. B. Feinstein, in "Calcium and Cell Physiology" (D. Marme, ed.). Springer-Verlag, Berlin, 1985.
[26] H. Rasmussen, I. Kojima, W. Apfeldorf, and P. Barrett, Kidney Int. **29**, (1986).

striking difference exists between these two classes of cells. In nonpolar cells, the receptors for a particular hormone are more or less uniformly distributed over the entire cell surface. Such is not the case in polar cells: hormone receptors appear to be confined to the surface of the basolateral plasma membrane, and appear not to be present on the brush border or lumenal cell membrane. Yet there is considerable evidence that specific hormones produce changes in transport functions at the lumenal membrane. This means there is a considerable distance between the site of intracellular messenger generation on the basolateral membranes and presumed messenger action on the lumenal membrane.

Spatial Separation of Messenger Generation and Action

This separation between the site of messenger synthesis and action poses a particularly interesting problem. One line of evidence has been taken to signify that the solution to the problem is simple and straightforward: messenger concentration rises and remains elevated throughout the duration of agonist action, and as a consequence messenger diffuses from its site of synthesis to its site of action. The major evidence in support of this view is the demonstration that hormone-sensitive adenylate cyclase is localized largely to the basolateral membrane of renal cells, and the cAMP-dependent protein kinase on the brush border or lumenal membrane, i.e., the message generator and receiver are located at appropriate sites within the cell.[27,28] The difficulties with this proposal are, first, that the cell possesses a complement of enzymes, phosphodiesterases, which hydrolyze the messenger and which the messenger is sure to encounter on its journey from one side of the cell to another. Second, if the model of K^+ action discussed above is valid, it implies that in nonpolar cells a change in $[cAMP]_c$ may not be the major messenger during a sustained response, but a change in rate of cAMP turnover may be the messenger. The same is likely to apply to polar cells. Finally, insufficient attention has been paid to the fact that cAMP-dependent protein kinases can be activated either by a change in cAMP concentration or by a change in their sensitivity to activation by cAMP. Hence, a cAMP-dependent kinase located on the brush border may alter its activity in response to a signal other than a change in [cAMP], a signal that would change its sensitivity to activation by cAMP.

The problem is even more acute in the case of the Ca^{2+} messenger

[27] R. Kinne, L. J. Shlatz, E. Kinne-Safran, and I. L. Schwartz, *J. Membr. Biol.* **24**, 145 (1975).
[28] L. J. Shlatz, I. L. Schwartz, E. Kinne-Safran, and R. Kinne, *J. Membr. Biol.* **24**, 131 (1985).

system. It has become increasingly apparent that when a variety of nonpolar cells are activated by their particular agonists, the rise in $[Ca^{2+}]_c$ is transient so that during the sustained phase of the response $[Ca^{2+}]_c$ is at or close to its basal value.[14] Nevertheless, cell response continues. Given the major intracellular Ca^{2+}-buffering systems existing in all cells, it is very unlikely that a physiologically appropriate increase in Ca^{2+} influx rate across the basolateral membranes of a polar cell will lead to an increase in $[Ca^{2+}]$ in the domain of the lumenal membrane. Something else must occur. In the case of nonpolar cells, we have seen that during prolonged cellular response an increase in Ca^{2+} cycling across the plasma membrane can function in at least two different ways to mediate response: either by linking the Ca^{2+} cycling to an activated C kinase, or coupling Ca^{2+} cycling with adenylate cyclase activation and increase cAMP turnover. In neither of these cases is it yet clear how the information being generated in the domain of the plasma membrane mediates the sustained phase of cellular response. Nonetheless, one might anticipate similar modes of information generation at the level of the basolateral membrane of the polar cell. In this regard, it is of particular interest that phorbol ester activators of protein kinase C, when added to the serosal surface of strips of rat small intestine, stimulated Cl^- transport and hence NaCl secretion under conditions wherein one would not expect a change in the intracellular free Ca^{2+} concentration.[29,30] Hence, in polar intestinal mucosa cells, activation of C kinase can induce a cell response without altering intracellular Ca^{2+} concentration.

It is possible to prepare quite pure preparations of both brush border and basolateral membranes from the mucosal cells of the small intestine or the proximal renal tubule. Analysis of the enzyme composition of brush border membranes from proximal tubule reveals a very high content of protein kinase C.[31] It is clear that this enzyme is not there simply because of an adventitious association, but is a normal component of this structure. Its presence in such quantities is presumptive evidence for its playing a key role in regulating transport processes across this membrane. On the other hand, its location at this site poses a problem in terms of how its activity is controlled. The current model of its activation is one in which hormone–receptor interaction leads to the local intramembranous generation of DG which serves to activate C kinase. Since, in polar cells, the generation of DG is presumably limited to the sites in the basolateral membrane linked

[29] J. D. Fondacaro and L. S. Henderson, *Am. J. Physiol.* **249**, G422 (1985).
[30] C. S. Weikel, J. J. Sando, and R. L. Guerrant, *J. Clin. Invest.* **76**, 2430 (1985).
[31] P. Q. Barrett, K. Zawalich, and H. Rasmussen, *Biochem. Biophys. Res. Commun.* **128**, 494 (1985).

to the hormone receptor, it is not at all evident how this messenger could bring about the activation of the C kinase in the lumenal membrane. Neither diffusion of DG across the cell, nor its lateral diffusion in the plane of the membrane, are likely. The lumenal localization of C kinase challenges our present views of stimulus–response coupling. Control of its activity must be by a mechanism other than that thought to operate in nonpolar cells. The flow of information from one cell surface to another in a polar cell may be conveyed by the cytoskeleton and/or intracellular matrix, by global ionic changes, or by a protein kinase cascade. These possibilities remain to be explored.

Ca^{2+} as Messenger in Cells Capable of Transcellular Ca^{2+} Transport

A number of epithelial cells carry out the transcellular transport of Ca^{2+}, yet appear to employ the Ca^{2+} messenger system to couple stimulus to response. The usual model of this transcellular transport is one in which Ca^{2+} is transported into the cell across the lumenal membrane down its concentration gradient, enters the bulk cytosol, diffuses across the cell, and is then actively transported out of the cell across the basolateral membrane either by a Ca^{2+} pump and/or Na^+/Ca^{2+} exchange. On the other hand, it is now quite clear that very small changes in $[Ca^{2+}]_c$ are sufficient to activate Ca^{2+}-dependent enzymes, and that the $[Ca^{2+}]_c$ is extremely well regulated. The problem of maintaining cellular Ca^{2+} homeostasis and avoiding the inadvertent activation of Ca^{2+}-dependent enzymes must be particularly acute in the duodenal mucosal cell. This cell carries out the transcellular transport of Ca^{2+}, but is exposed to a highly variable Ca^{2+} load. Hence, if Ca^{2+} entry and transcellular diffusion of Ca^{2+} via the cytosol was the mechanism underlying transcellular transport, one would anticipate periodic changes in $[Ca^{2+}]_c$ as the load of Ca^{2+} to be transported varied. The present evidence suggests that this does not occur because the Ca^{2+} to be transported across the cell does not diffuse freely but is packaged in membrane vesicles which form at the lumenal surface, traverse the cell, fuse with the basolateral membrane, and discharge their Ca^{2+} load extracellularly.[32] In other words, there is a compartmentation of the transported Ca^{2+}. It does not normally enter the cytosol to perturb cellular Ca^{2+} homeostasis and the operation of the Ca^{2+} messenger system. It is not clear whether such transcellular vesicular transport moves Ca^{2+} across other epithelial cells such as those in the proximal and distal renal tubules.

[32] D. K. Kreutter and H. Rasmussen, *in* "Regulation of Intestinal Electrolyte Transport" (M. Donowitz and G. N. G. Sharp, eds.). Liss, New York, 1984.

The Use of Isolated Brush Border Membranes in the Study of Stimulus-Response Coupling

A major tool in the study of the subcellular basis of transcellular transport across epithelial cells has been the isolated brush border membrane vesicle. The Na^+-dependent entry of glucose, phosphate, etc., has been characterized with such preparations. Given this success, it was only natural to assume that this tool would also be of value in defining the molecular details of messenger action in such cells. However, only limited success has been achieved. The aim is to be able to alter, for example, the state of phosphorylation of specific proteins in the brush border membrane vesicle and correlate such changes with changes in the rate of transport of specific ions or metabolites. The problem again is one of polarity. As usually prepared, brush border membrane vesicles are predominantly right side out so that the external face of the membrane is accessible and the internal face, or cytosolic face, is inaccessible. It is on this latter face that kinases are presumed to catalyze the phosphorylation of specific proteins and thereby change transport rates. Yet, added kinases and their specific activators cannot reach their target if simply added to the medium containing intact vesicles. Reversible permeabilization of the vesicles with incorporation of appropriate kinases, ATP, and activators into the internal compartment is the obvious solution. However, to date it has not been possible to reversibly permeabilize these vesicles in such a way that the resealed vesicles have the same transport rates as they display before permeabilization.

There are other factors which must be considered in the relationship between phosphorylation events in brush border membranes and the regulation of transcellular transport. In current models of either parathyroid hormone (PTH) action on renal tubular phosphate reabsorption, or arginine vasopressin (AVP) action on tubular H_2O reabsorption, hormone–receptor interaction on the basolateral membranes leads to the activation of adenylate cyclase, and thus to a rise in the cellular cAMP content leading to the phosphorylation of specific brush border membrane proteins by the activation of intrinsic cAMP-dependent protein kinase. However, in the case of each of these two hormones there is now considerable evidence that each mediates a major part of its effect by altering the turnover of membrane units between the brush border membrane and an intracellular membrane pool possessing either phosphate transporters, in the case of PTH-sensitive proximal renal tubules, or H_2O-permeable units in the case of AVP-sensitive collecting tubules.[33,34] Since the molecular basis underly-

---

[33] P. Gmaj and H. Murer, *Physiol. Rev.* **66**, 36 (1986).
[34] J. Wade, D. L. Stetson, and S. A. Lewis, *Ann. N.Y. Acad. Sci.* **372**, 106 (1981).

ing the shift of membrane units to and from the intracellular to the brush border membranes is not known, it is not clear at which site the putative phosphorylation events occur which are responsible for the shifts of membrane units from one pool to another. It is quite possible the specific phosphorylation of brush border membrane proteins is not the target of second messenger action. Furthermore, if membrane insertion and removal are the major means of regulating transport processes at the level of the brush border membrane, then studies attempting to correlate changes in protein phosphorylation induced in isolated brush border membrane vesicles with changes in rates of transport of specific ions or metabolites are doomed to fail.

Acknowledgments

Supported by grants from the National Institutes of Health (AM 19813), the Muscular Dystrophy Association, and the National Dairy Council.

[38] Metabolism and Function of Phosphatidylinositol-Derived Arachidonic Acid

By LOWELL E. HOKIN

Introduction

Arachidonate metabolites, i.e., prostaglandins, leukotrienes, and thromboxanes, which are collectively referred to as eicosanoids, are potent regulators of various physiological responses. Since mammalian cells do not store eicosanoids and since unstimulated cells contain very little, if any, unesterified arachidonate, the levels of these compounds are primarily regulated by the availability of free arachidonate, which must be liberated mainly from esterified lipids.[1] It appears that phosphatidylinositol (PI) in all mammalian tissues is rich in arachidonate at the 2-position of glycerol.[2] This has been demonstrated in a variety of tissues, e.g., pancreas[3,4] and platelets.[5] The polyphosphoinositides have also been shown to contain a

[1] S. Bergstrom, M. Danielsson, and B. Samuelsson, *Biochim. Biophys. Acta* **90**, 207 (1964).
[2] R. S. Rana and L. E. Hokin, *Physiol. Rev.* **70**, 115 (1990).
[3] R. W. Keenen and L. E. Hokin, *Biochim. Biophys. Acta* **84**, 458 (1964).
[4] M. W. Banschbach, R. L. Geison, and M. Hokin-Neaverson, *Biochim. Biophys. Acta* **663**, 34 (1981).
[5] V. G. Mahadevappa and B. J. Holub, *J. Biol. Chem.* **258**, 5337 (1983).

large proportion of arachidonate at the 2-position of glycerol.[6] Besides phosphoinositides, phosphatidylcholine (PC) and phosphatidylethanolamine (PE) are also esterified with arachidonate to varying degrees. Irvine[7] has discussed in detail the various mechanisms by which arachidonate levels are regulated in mammalian cells. More recent reviews on the subject, including second messenger functions of released arachidonate, have recently appeared.[2,8]

Increased release of arachidonate and subsequent synthesis of eicosanoids appear to be related to the agonist-induced phosphoinositide responses in several tissues.[9,10] This is supported by the observation that in many cell types and tissues, activation of the same receptors that control phosphoinositide breakdown also results in the liberation of arachidonate and/or eicosanoids. Although it is now well established that arachidonate is released from phospholipids in response to certain stimuli, it is still not known for most tissues what proportion of the total arachidonate released is derived from phosphoinositides and/or other phospholipids; its mechanism of release also remains unknown.

Mechanisms of Arachidonate Liberation

Depending on the agonist and the type of tissue involved, several mechanisms for arachidonate release appear to exist. Two mechanisms have been investigated in some detail. One of these involves sequential actions of phospholipase C and diacylglycerol (DAG) and monoacylglycerol (MAG) lipase, while the other involves direct action of phospholipase A_2 on phospholipids.

Phospholipase C–DAG–Lipase Pathway. This mechanism for arachidonate release was first proposed by Majerus and Rittenhouse (also known as Rittenhouse–Simmons) and collaborators and extended by Dixon and Hokin. In platelets[10,11] and pancreatic minilobules,[9] there is agonist-evoked liberation of free arachidonate and, at least in pancreas, this appears to proceed almost entirely via sequential DAG and MAG lipase action on DAG liberated from the phosphoinositides (Table I). Since DAG lipase preferentially cleaves the fatty acyl group at the sn-1 position, arachidonate liberation requires the action of MAG lipase on the sn-2 position of MAG. In several cells, e.g., erythrocytes,[12] DAG lipase has been shown to be

---

[6] R. Baker and W. Thompson, *Biochim. Biophys. Acta* **270**, 489 (1972).
[7] R. F. Irvine, *Biochem. J.* **204**, 3 (1982).
[8] A. A. Abdel-Latif, *Life Sci.* **45**, 757 (1989).
[9] J. F. Dixon and L. E. Hokin, *J. Biol. Chem.* **259**, 14418 (1984).
[10] P. W. Majerus, E. J. Neufeld, and D. B. Wilson, *Cell (Cambridge, Mass.)* **37**, 701 (1984).
[11] S. E. Rittenshouse, *Cell Calcium* **3**, 311 (1982).
[12] R. H. Mitchell and R. Coleman, *Biochem. J.* **124**, 49P (1971).

present in the plasma membrane. In pancreatic minilobules, several observations support the phospholipase C–DAG–lipase pathway for release of arachidonate[9]:

1. Stimulation of phosphoinositide breakdown with the secretogogue, caerulein, was associated with an increase in the steady state level of 1-stearoyl-2-arachidonoyl-*sn*-glycerol and a release of substantial amounts [up to 50% of the total PI breakdown (see below)] of stearate, arachidonate, and glycerol (Tables I–III). Unlike the situation in platelets, there was no appreciable accumulation of MAG.

2. With [^{3}H]glycerol as precursor, the accumulation of phosphatidic acid (PA), DAG, triacylglycerol (TAG), and glycerol agreed closely with the loss of PI, indicating that there was no contribution of other pathways leading to glycerol. However, with the fatty acids as precursors, the accumulation of PA, DAG, TAG, and arachidonate or stearate slightly exceeded the loss of label from PI, suggesting that there was some contribution of fatty acids by the action of phospholipases A_1 and A_2 on PI (see Tables I to III). However, stimulation of pancreatic minilobules with either caerulein or carbachol did not give rise to increased amounts of lysophosphatidylinositol (LPI),[9] glycerophosphorylinositol, or glycerophosphorylinositol mono- or bisphosphate[13] products of phospholipases A_1 and A_2 action on phosphoinositides.

3. The DAG lipase inhibitor, RHC 80267,[14] reduced the secretogogue-stimulated liberation of free stearate, arachidonate, and glycerol and elevated the steady state level of 1-stearoyl-2-arachidonoyl-*sn*-glycerol. RHC 80267 had no effect on phospholipids. A similar mechanism of arachidonate release has also been suggested in thyrotropin-stimulated thyroid lobes[15] and mast cells.[16]

Phospholipase A_2 Pathway. Diacylphospholipids: In platelets[17] and neutrophils,[18] the presence of phospholipase A_2 specific for PA has been described, and it has been implicated in the release of arachidonate from PA produced during PI turnover[19] (however, see Neufeld and Majerus[20]). There appear to be multiple pathways for release of arachidonate in platelets. An additional pathway to that described above is arachidonate release in platelets via a phospholipase A_2-mediated breakdown of PC, PE, and PI.[21–24]

[13] M. C. Sekar, J. F. Dixon, and L. E. Hokin, *J. Biol. Chem.* **262**, 340 (1987).
[14] C. A. Sutherland and D. Amin, *J. Biol. Chem.* **257**, 14006 (1982).
[15] S. Levasseur, M. Kostelek, and G. Burke, *Prostaglandins* **27**, 673 (1984).
[16] Y. Okano, K. Yamada, K. Yano, and Y. Nozawa, *Biochem. Biophys. Res. Commun.* **145**, 1267 (1987).
[17] M. M. Billah, E. G. Lapetina, and P. Cuatrecasas, *J. Biol. Chem.* **256**, 5399 (1981).

TABLE I
Changes in Radioactivity in Lipids on Caerulein Stimulation of [^{14}C]Arachidonate-Prelabeled Pancreatic Minilobules in the Absence and Presence of RHC 80267[a]

	Radioactivity in [^{14}C]arachidonate-prelabeled lipids (dpm/100 mg)			
Lipid	−Caerulein	+Caerulein	Increment	p value
−RHC 80267				
Total	585,000 ± 12,700	571,000 ± 6,840		NS[b]
PI[c]	52,000 ± 7,830	25,000 ± 548	−27,000	≪ 0.001
PA	1,190 ± 104	9,870 ± 360	+8,680	≪ 0.001
DAG	4,740 ± 159	5,870 ± 204	+1,130	0.002
TAG	89,100 ± 1,350	104,000 ± 811	+14,900	<0.001
Arachidonate	23,300 ± 417	36,000 ± 789	+12,700	<0.001
PC	273,000 ± 7,830	267,000 ± 7,590		NS
PS	7,080 ± 11	6,670 ± 142	−410	0.0075
PE	45,200 ± 1,630	45,900 ± 1,030		NS
MAG	372 ± 51	443 ± 56		NS
+100 μM RHC 80267				
Total	575,000 ± 37,943	573,000 ± 23,510		NS
PI	52,600 ± 10,160	29,000 ± 854	−23,600	≪ 0.001
PA	1,400 ± 43	9,630 ± 342	+8,230	≪ 0.001
DAG	8,740 ± 530	13,000 ± 176	+4,260	<0.001
TAG	90,900 ± 1,100	101,000 ± 1,950	+10,100	0.0045
Arachidonate	24,000 ± 900	31,100 ± 1,700	+6,500	0.004
PC	271,000 ± 10,160	259,000 ± 13,195		NS
PS	8,340 ± 221	7,890 ± 190		NS
PE	47,200 ± 418	46,300 ± 400		NS
MAG	550 ± 54	648 ± 162		NS

[a] Minilobules were prepared and prelabeled as described. The prelabeling medium contained 0.4 μCi of [^{14}C]arachidonate/ml. Two-milliliter samples of minilobules were incubated at 37° for 15 min in the presence or absence of 100 μM RHC 80267. This was followed by incubation for 30 min without and with caerulein. Lipids were extracted as described. The average radioactivity in the Bligh and Dyer extract was 576,000 dpm/100 mg. The values are means of triplicate incubations and are expressed as the mean ± SD. (Reproduced by permission of the *Journal of Biological Chemistry*.[9])

[b] NS, Not significant.

[c] Abbreviations defined in text.

---

[18] E. G. Lapetina, M. M. Billah, and P. Cuatrecasas, *J. Biol. Chem.* **255**, 966 (1980).
[19] E. G. Lapetina, *J. Biol. Chem.* **257**, 7314 (1982).
[20] E. J. Neufeld and P. W. Majerus, *J. Biol. Chem.* **258**, 2461 (1983).
[21] M. M. Billah, E. G. Lapetina, and P. Cuatrecasas, *J. Biol. Chem.* **255**, 10227 (1980).
[22] V. G. Mahadevappa and B. J. Holub, *J. Biol. Chem.* **259**, 9369 (1984).
[23] H. Van Den Bosch, *Biochim. Biophys. Acta* **604**, 191 (1980).
[24] M. Waite, *Handb. Lipid Res.* **5** (1987).

TABLE II
CHANGES IN RADIOACTIVITY IN LIPIDS ON CAERULEIN STIMULATION OF
[³H]GLYCEROL-PRELABELED PANCREATIC MINILOBULES IN THE ABSENCE AND
PRESENCE OF RHC 80267[a]

Lipid	Radioactivity in [¹⁴C]glycerol-prelabeled lipids (dpm/100 mg)			p value
	−Caerulein	+Caerulein	Increment	
−RHC 80267				
Total	619,000 ± 17,100	614,000 ± 39,000		
PI[b]	162,000 ± 2,940	48,800 ± 1,410	−113,200	≪0.001
PA	2,200 ± 188	20,300 ± 1,780	+18,100	<0.001
DAG	10,800 ± 1,240	25,400 ± 935	+14,600	<0.001
TAG	71,200 ± 2,300	103,000 ± 3,460	+31,800	<0.001
Glycerol	45,700 ± 2,300	77,400 ± 2,820	+31,700	<0.001
PC	122,000 ± 4,730	129,000 ± 2,200		NS[c]
PS	2,140 ± 480	1,930 ± 206		NS
PE	80,500 ± 3,630	93,000 ± 2,800	+12,500	0.008
MAG	1,260 ± 164	2,260 ± 259	+1,000	0.005
+100 μM RHC 80267				
Total	661,000 ± 31,500	593,000 ± 18,900		
PI	159,000 ± 3,700	53,300 ± 1,440	−105,700	≪0.001
PA	3,740 ± 602	20,700 ± 478	+16,960	≪0.001
DAG	33,800 ± 690	62,400 ± 469	+28,600	≪0.001
TAG	78,500 ± 1,610	92,500 ± 1,380	+14,000	<0.001
Glycerol	27,500 ± 860	45,200 ± 1,320	+17,700	≪0.001
PC	126,000 ± 4,060	129,000 ± 3,140		NS
PS	2,520 ± 308	2,280 ± 355		NS
PE	80,900 ± 6,450	96,900 ± 6,690	+16,000	0.04
MAG	2,590 ± 308	4,360 ± 285	+1,770	0.0025

[a] Minilobules were prepared and treated exactly as described in Table I except that 25 μCi of [³H]glycerol was substituted for [¹⁴C]arachidonate. The average radioactivity in the Bligh and Dyer extract was 609,000 dpm/100 mg. (Reproduced by permission of the *Journal of Biological Chemistry*.[9])
[b] Abbreviations defined in text.
[c] NS, Not significant.

Alkylacyl phospholipids: Another source of arachidonate which so far has received little attention is ether phospholipids. In some tissues, such as, for example, rabbit platelets,[25] alkylacylglycerophosphorylcholine appears to be a significant source of metabolizable arachidonate. In a recent study

[25] M. Chignard, J. P. Le Couedic, E. Coeffier, and J. Benveniste, *Biochem. Biophys. Res. Commun.* **124**, 637 (1984).

TABLE III
EFFECT OF CAERULEIN ON THE RADIOACTIVITY IN LYSOPHOSPHATIDYLINOSITOL AND
OTHER LIPIDS IN PANCREATIC MINILOBULES PREPARED FROM PANCREATA
PRELABELED WITH [³H]GLYCEROL[a]

Lipid	Radioactivity in [¹⁴C]glycerol-prelabeled lipids (dpm/100 mg)			p value
	−Caerulein	+Caerulein	Increment	
2 min				
Total	824,830 ± 22,971	823,810 ± 38,280		NS[b]
PI[c]	212,000 ± 2,690	188,000 ± 3,160	−24,000	<0.001
LPI	1,430 ± 198	1,158 ± 63		NS
PA	13,800 ± 189	26,200 ± 1,500	+12,400	≪0.001
DAG	15,300 ± 390	14,600 ± 446		NS
TAG	162,300 ± 1,090	167,000 ± 1,047	+4,700	0.0055
Glycerol	37,100 ± 1,420	46,100 ± 3,200	+9,000	0.035
MAG	1,970 ± 202	4,350 ± 440	+2,380	0.0015
5 min				
Total	819,630 ± 42,080	808,930 ± 16,640		NS
PI	211,000 ± 5,370	159,000 ± 2,740	−52,000	<0.001
LPI	1,690 ± 495	1,420 ± 107		NS
PA	9,270 ± 185	27,200 ± 2,560	+17,930	<0.001
DAG	12,900 ± 284	15,900 ± 695	+3,000	0.003
TAG	137,800 ± 7,490	157,000 ± 771	+19,200	<0.011
Glycerol	50,400 ± 1,800	67,100 ± 695	+16,700	<0.001
MAG	1,760 ± 164	3,370 ± 282	+1,610	<0.001

[a] Twelve pancreata were prelabeled as described. Five microcuries of [³H]glycerol was present in each prelabeling vessel. Minilobules were prepared, and the incubation was carried out as described. The extraction and chromatography were modified for LPI, as described. The water/methanol phase from the Bligh and Dyer extract was concentrated in a Speed Vac and chromatographed in two systems for separation of glycerol, as described. Radioactivity was normalized to an average total radioactivity of 826,000 dpm/100 mg in the Bligh and Dyer extract. (Reproduced by permission of the *Journal of Biological Chemistry*.[9])
[b] Abbreviations defined in text.
[c] NS, Not significant.

with rat platelets,[26] stimulation with thrombin was shown to cause a loss of arachidonate from diacyl-*sn*-glycero-3-phosphoinositol (PI) and alkylacyl- and diacyl-*sn*-glycero-3-phosphocholine. Further work is required to establish metabolic pathways for ether phospholipids in other tissues.

[26] O. Colard, M. Breton, and G. Bereziat, *Biochem. J.* **233,** 691 (1986).

Ca^{2+} Dependency of Phospholipase A_2 and Its Relationship to the Phosphoinositide Cascade

The affinity of phospholipase A_2 for Ca^{2+} is much lower (higher K_d) than that of phospholipase C,[24] suggesting that phospholipase A_2 action on phospholipids is likely to be stimulated by agonist-evoked rises in Ca^{2+}. Thus, arachidonate release could conceivably respond to the transient rises in Ca^{2+} evoked by inositol 1,4,5-trisphosphate [$I(1,4,5)P_3$]. The possibility that a rise in intracellular Ca^{2+} may be important in activating phospholipase A_2 is suggested by the observation that Ca^{2+} ionophores are potent activators of phospholipase A_2 in cells.[27] A caveat is that the $I(1,4,5)P_3$-induced rise in cytosolic Ca^{2+} in most tissues appears to be far below the high micromolar or low millimolar concentrations required to activate phospholipase A_2 in cell-free preparations.

Involvement of Protein Kinase C in Arachidonate Release

In several cell types,[2] treatment with phorbol myristoyl acetate (PMA) stimulates release of arachidonate and/or eicosanoids. These observations support a potential role for protein kinase C (PKC) in arachidonate release. In platelets, PMA or oleoyl acetyl glycerol (OAG) alone did not stimulate arachidonate release but potentiated agonist-induced[28] and Ca^{2+} ionophore-induced[29,30] release of arachidonate. This observation suggests that arachidonate release may depend on a cooperative interaction between phosphoinositide-mediated activation of PKC and the $I(1,4,5)P_3$-mediated rise in intracellular Ca^{2+}.

Role of Released Arachidonate

A detailed discussion of various functions of eicosanoids is beyond the scope of this chapter. Only some relatively recent developments are emphasized here.

Arachidonate Metabolites as Mitogens. Arachidonate or its metabolites have been suggested to play an undefined role in cell replication.[31-33]

---

[27] S. G. Laychock and J. W. Putney, Jr., in "Cellular Regulation of Secretion and Release" (P. M. Conn, ed.), p. 53. Academic Press, New York, 1982.
[28] S. Krishnamurthi, S. Joseph, and V. V. Kakkar, *Biochim. Biophys. Acta* **927**, 429 (1987).
[29] S. P. Halenda and A. G. Rehm, *Biochem. J.* **248**, 471 (1987).
[30] S. P. Halenda, G. B. Zavoico, and M. B. Feinstein, *J. Biol. Chem.* **260**, 12484 (1985).
[31] R. M. Burch, A Luini, D. E. Mais, D. Corda, J. Y. Vanderhoek, L. D. Kohn, and J. Axelrod, *J. Biol. Chem.* **261**, 11236 (1986).
[32] R. Gartner, W. Greil, R. Demharter, and K. Horn, *Mol. Cell. Endocrinol.* **42**, 145 (1985).
[33] A. J. R. Habenicht, J. A. Glomset, M. Goerig, R. Gronwald, J. Grulich, U. Loth, and G. Schettler, *J. Biol. Chem.* **260**, 1370 (1985).

Hydroxyeicosatetraenoic acids (HETEs) have been shown to act as mitogens in several types of cells, such as fibroblasts[34] and epidermal cells.[35]

Possible Involvement of Arachidonate and/or Its Metabolites in Phospholipase C Action. A number of investigators have reported that arachidonate and/or its metabolites are capable of activating phospholipase C in intact cells[36,37] and in cell-free preparations.[27,38] However, the biochemical mechanism of this activation remains to be established. In placental cells, stimulation of *myo*-[^{3}H]inositol release by arachidonate from labeled cells was insensitive to lipoxygenase or cyclooxygenase inhibitors,[37] indicating that the effect was mediated either directly by arachidonate or indirectly by metabolites that were insensitive to the inhibitors utilized in this study. Since arachidonate metabolites are usually released from the cell, they can have physiologically important effects on the same cell or on neighboring cells that may possess the appropriate receptors. In this respect, the lipoxygenase product, leukotriene B_4, appears to be a good candidate for the amplification of the phospholipase C signal since it is exported from cells and has been shown to act as an agonist, leading to receptor-mediated activation of phospholipase C[39,40] and Ca^{2+} mobilization.[41] On the other hand, prostacyclin[42] and prostaglandin E_2[43] appear to be negative modulators since these agents have been shown to inhibit phosphoinositide breakdown or generation of inositol phosphates. It is possible that the inhibition of phosphoinositide hydrolysis by prostaglandin E_2 is mediated through the generation of cAMP since both dibutyryl-cAMP and forskolin inhibit phosphoinositide breakdown.[43]

Activation of Ca^{2+} Channels by Arachidonate. Recent observations suggest that arachidonate may act as a messenger to open channels in the plasma membrane. Two types of ATP-, Mg^{2+}-, and Ca^{2+}-insensitive, potassium-selective channels were activated by intracellular arachidonic acid in neonatal rat atrial cells[44] and in smooth muscle cells.[45] Certain other fatty acids or PC were also active.

[34] W. Dodge and M. Thomas, *Biochem. Biophys. Res. Commun.* **131**, 731 (1985).
[35] C. C. Chan, L. Duhamel, and A. Ford-Hutchinson, *J. Invest. Dermatol.* **85**, 333 (1985).
[36] W. Siess, F. L. Siegel, and E. G. Lapetina, *J. Biol. Chem.* **258**, 11236 (1983).
[37] P. Zeitler and S. Handwerger, *Mol. Pharmacol.* **28**, 549 (1985).
[38] R. F. Irvine, A. J. Letcher, and R. M. C. Dawson, *Biochem. J.* **178**, 497 (1979).
[39] T. Anderson, W. Schlegel, A. Monod, K. H. Krause, O. Stendahl, and D. P. Lew, *Biochem. J.* **240**, 333 (1986).
[40] S. Mong, G. Chi-Rosso, J. Miller, K. Hoffman, K. A. Razgaitis, P. Bender, and S. T. Crooke, *Mol. Pharmacol.* **30**, 235 (1986).
[41] D. W. Goldman, L. A. Gilford, D. M. Olson, and E. J. Goetzl, *J. Immunol.* **135**, 525 (1985).
[42] E. G. Lapetina, *Biochem. Biophys. Res. Commun.* **120**, 37 (1984).
[43] T. Takenawa, J. Ishitoya, and Y. Nagai, *J. Biol. Chem.* **261**, 1092 (1986).
[44] D. Kim and D. E. Clapham, *Science* **244**, 1174 (1989).
[45] R. W. Ordway, J. V. Walsh, Jr., and J. J. Singer, *Science* **244**, 1176 (1989).

Arachidonate and Ca^{2+} Mobilization. There have been several reports that arachidonate activates intracellular mobilization of Ca^{2+}. Kolesnick *et al.*[46] showed that 3 μM arachidonate added exogenously to cloned pituitary GH_3 cells stimulated $^{45}Ca^{2+}$ efflux and prolactin secretion. Eicosatetraenoic acid and indomethacin, inhibitors of the lipoxygenase and cyclooxygenase pathways for arachidonate metabolism, respectively, did not inhibit either of these effects, suggesting a direct effect of arachidonate on $^{45}Ca^{2+}$ efflux and prolactin secretion. The arachidonate effect was subsequentially further characterized.[47] At concentrations below 5 μM, arachidonate stimulated the exchange of intracellular for extracellular Ca^{2+}; above 5 mM, arachidonate decreased membrane-bound Ca^{2+}. Arachidonate also decreased the total cellular Ca^{2+} content without increasing cytosolic Ca^{2+}. These findings could be explained by a mechanism in which arachidonate stimulates Ca^{2+} extrusion from and depletes Ca^{2+} stores within GH_3 cells.

Arachidonate itself has been shown to mobilize Ca^{2+} from isolated organelles, such as mitochondria[48] and sarcoplasmic reticulum,[49] and it has been suggested that arachidonate may be an intracellular mediator for Ca^{2+} mobilization from mitochondria in stimulated hepatocytes.[50] More recently, utilizing pancreatic islets[51,52] and liver microsomes,[53] arachidonate was shown to release Ca^{2+} from a pool that was distinct from the $I(1,4,5)P_3$-sensitive pool. In rabbit platelets, arachidonate metabolites, especially lipooxygenase products, have been suggested to play a role in Ca^{2+} uptake.[54]

Activation of PKC. Arachidonate metabolism may play a key role in PKC activation (for reviews, see Refs. 2 and 55). Arachidonate-derived oxygenation products, particularly lipoxin A (5,6,15-L-trihydroxy-7,9,11,13-eicosatetraenoic acid), are potent intracellular activators of PKC and may also be involved in the modulation of substrate specificity of this enzyme.[56] Type I PKC is relatively insensitive to DAG but is highly sensitive to arachidonate.[55]

[46] R. N. Kolesnick, I. Musacchio, C. Thaw, and M. C. Gershengorn, *Am. J. Physiol.* **246**, E458 (1984).
[47] R. N. Kolesnick and M. C. Gershengorn, *J. Biol. Chem.* **260**, 707 (1985).
[48] I. Roman, P. Gmaj, C. Noweicka, and S. Angielski, *Eur. J. Biochem.* **102**, 615 (1979).
[49] A. M. Cheah, *Biochim. Biophys. Acta* **648**, 113 (1981).
[50] J. A. Whiting and G. J. Barritt, *Biochem. J.* **206**, 121 (1982).
[51] S. A. Metz, B. Draznin, K. E. Sussman, and J. W. Leitner, *Biochem. Biophys. Res. Commun.* **142**, 251 (1987).
[52] B. A. Wolf, J. Turk, W. R. Sherman, and M. L. McDaniel, *J. Biol. Chem.* **261**, 3501 (1986).
[53] K. M. Chan and J. Turk, *Biochim. Biophys. Acta* **928**, 186 (1987).
[54] T. Lee, B. Malone, and F. Snyder, *Arch. Biochem. Biophys.* **223**, 33 (1983).
[55] Y. Nishizuka, *Biofactors* **1**, 17 (1988).
[56] A. Hansson, C. N. Serhan, J. Haeggstrom, M. Ingelman-Sundberg, and B. Samuelsson, *Biochem. Biophys. Res. Commun.* **134**, 1215 (1986).

Possible Compartmentalization of Arachidonate

In the exocrine pancreas, the elevation in PI-derived arachidonate on stimulation of enzyme secretion with several agonists is several orders of magnitude greater than the stimulated formation of prostaglandin E_2 (PGE_2) and $PGF_{2\alpha}$.[9,57,58] The difference between the amount of arachidonate released and prostaglandin formed may be due, at least in part, to cyclooxygenase being the rate-limiting reaction. Also, it could be due in part to increased arachidonate in compartments which are inaccessible to prostaglandin-synthesizing enzymes. Such a possibility is supported by the following observations. First, in human platelets, arachidonate released from phospholipids is preferentially utilized by cyclooxygenase rather than by lipooxygenase.[59] Second, in neutrophils, while zymosan and ionophore both liberate arachidonate, only the arachidonate released by ionophore is converted to HETEs.[60]

Balance Sheet Analysis of the Disposition of Moieties in PI on Stimulation of PI Breakdown

Prelabeling of Whole Pancreata

This is carried out as described previously for prelabeling with [^{14}C]arachidonate.[9] Pancreata are carefully dissected from 30-g white male adult Swiss ICR mice immediately after killing by cervical dislocation. The pancreata are incubated separately for 30 min at 37° in stoppered 25-ml Erlenmeyer flasks containing 2 ml of Krebs–Henseleit bicarbonate saline, 0.2% glucose, 2×10^{-4} M carbamylcholine (CCh), and varying concentrations of one of the following: [^{14}C]arachidonate, 50 mCi/mmol; [^{14}C]stearate, 50 mCi/mmol; myo-[^{3}H]inositol, 15.8 Ci/mmol; [^{3}H]glycerol, 500 mCi/mmol; and [^{32}P]orthophosphate, "carrier free." All radioactive compounds are obtained from New England Nuclear (Boston, MA). Fatty acid solutions are dried under N_2, taken up in dimethyl sulfoxide, and added to the medium (final dimethyl sulfoxide concentration, 0.25%). The pancreata are then incubated for 30 min as above but without CCh and isotope and with 10^{-4} M atropine sulfate and 1% fatty acid-free bovine serum albumin (BSA) (quenching medium).

---

[57] M. W. Banschbach and M. Hokin-Neaverson, *FEBS Lett.* **117**, 131 (1980).
[58] H. Bauduin, N. Galand, and J. M. Boeynaems, *Prostaglandins* **22**, 35 (1981).
[59] L. Sautebin, D. Caruso, G. Galli, and R. Paoletti, *FEBS Lett.* **157**, 173 (1983).
[60] C. E. Walsh, B. M. Waite, M. J. Thomas, and L. R. Dechatelet, *J. Biol. Chem.* **256**, 7228 (1981).

Preparation and Incubation of Minilobules

Twelve pancreata, freshly excised or prelabeled as described above, are dissociated into minilobules by collagenase treatment by the method of Amsterdam et al.[61] with the following modifications. The collagenase concentration is reduced to 80 units/ml. Seven shearing cycles are used.

Prelabeling of minilobules is similar to that of the whole pancreata as described above. The volumes of the prelabeling and quenching media are 20 ml for minilobules from one preparation. After quenching, the minilobules are suspended in 20 ml of incubation medium (Krebs–Henseleit bicarbonate saline containing 0.2% glucose, 0.2 mg/ml of BSA, amino acids, and 0.1 mg of soybean trypsin inhibitor). Large pieces are allowed to settle out for 15 sec, and the more homogeneous supernatant material is removed. Aliquots of this material are taken while stirring with a magnetic stirrer. In experiments where the DAG lipase inhibitor, RHC 80267, is used, separate batches are incubated with and without the drug for 15 min. Fresh medium of the same composition is used for the final incubation. The final incubation is carried out with one of two different volumes, depending on the incubation time. For incubation times of 5 min or less, an 0.8-ml volume is used so that the incubation can be stopped rapidly by addition of the initial chloroform/methanol/HCl extraction medium. For longer incubation times, 0.8-ml aliquots are diluted with incubation medium to 2.0 ml. For these longer incubation times, the reaction is stopped by cooling in ice, and the minilobules are separated from the medium so that the extraction can be carried out on the usual scale. The lipid levels remain constant during work-up. The final incubation is in incubation medium without and with 0.5 μg/ml of caerulein, which is a decapeptide analog of cholecystokinin/pancreozymin, the physiological hormone which stimulates enzyme secretion in the pancreas. Incubations are carried out in stoppered vessels gassed with 5% CO_2 in O_2 with shaking. Identical samples are incubated in triplicate. Experiments were carried out at least three times, giving a minimum of nine estimations per lipid per condition.[9] All glassware used in the preparation and work-up of minilobules is siliconized with "Surfasil" (Pierce Chemical Co., Rockford, IL).

Extraction

The method of Bligh and Dyer[62] is used to extract phospholipids and neutral lipids. Tissue sample and medium with a combined volume of 0.8 ml is homogenized with 2 ml of methanol, 1 ml of chloroform (with a

---

[61] A. Amsterdam, T. E. Solomon, and J. D. Jamieson, *Methods Cell Biol.* **20**, 362 (1978).
[62] E. G. Bligh and W. J. Dyer, *Can. J. Biochem. Physiol.* **37**, 911 (1959).

trace of butylated hydroxytoluene), and 0.032 ml of concentrated HCl. One milliliter of chloroform and 1 ml of medium are then added in sequence, with mixing. When the radioactivity in LPI is measured, $2\ M$ KCl in medium is substituted for medium.[63] The phases are separated by centrifugation (3000 g for 5 min). The chloroform phase is saved, and the water/methanol and solid phases are washed once with 2 ml of chloroform. The chloroform extracts are combined.

Chromatography

Phospholipids are analyzed by the method of Portoukalian *et al.*[64] Merck (Darmstadt, FRG) silica gel 60-precoated thin-layer chromatography (TLC) plates are activated for 30 min at 110°. The samples are applied in a line 2 cm long and 2 mm wide and chromatographed in two dimensions (first dimension, tetrahydrofuran/acetone/methanol/water, 50:20:40:8; second dimension, chloroform/acetone/methanol/acetic acid/water, 50:20:10:15:5). Phosphatidylserine, 5 µg, and 20 µg of PA are added as carrier. When the radioactivity in LPI is measured, 100 µg of carrier is added to minilobules after incubation and 100 µg is added prior to chromatography. Also, a different developing system is used (first dimension, chloroform/methanol/acetic acid/water, 75:45:12:3[63]; second dimension, tetrahydrofuran/acetone/methanol/water, 50:20:40:8).[64]

The neutral lipids are chromatographed in one dimension on silica gel 60 TLC plates in hexane/diethyl ether/acetic acid, 65:35:4.[9] Thirty micrograms of monolein, diolein, triolein, and arachidonate (or stearate) are each added as carrier. All neutral lipid extracts are run in duplicate.

Iodine-stained spots were initially scraped into 10 ml of Aquasol (New England Nuclear Corp., Boston, MA) and counted. It has been shown that with ^{14}C-labeled lipids, recovery of radioactivity with this scintillation fluid is essentially quantitative, except for PC, which gives 70–85% recovery[65,66]; addition of 1 ml of water to the scintillation vial prior to adding the 10 ml of Aquasol gives quantitative recovery of PC radioactivity.[66] When we added water, the recovery of radioactivity in PC was approximately 15% higher; this did not affect the results with this nonresponsive lipid.[9] The recoveries in radioactivity in the other lipids are not affected by the addition of water. With ^{3}H-labeled lipids, the recovery of radioactivity is not quantitative with Aquasol alone. For the [^{3}H]glycerol and *myo*-[^{3}H]inositol experiments, the lipids on the TLC chromatograms are eluted in Aqua-

---

[63] M. M. Billah and E. G. Lapetina, *J. Biol. Chem.* **257**, 5196 (1982).
[64] J. Portoukalian, R. Meister, and G. Zwingelstein, *J. Chromatogr.* **152**, 569 (1978).
[65] D. Kretchevsky and S. Malhotra, *J. Chromatogr.* **52**, 498 (1970).
[66] R. A. Webb and D. F. Mettrick, *J. Chromatogr.* **67**, 75 (1972).

sol/water (9:1; water added first), which has been shown to give quantitative elution of ^{3}H-labeled lipids from silica gel.[66]

Aliquots for total radioactivity are taken after the chloroform extract has been taken to dryness and resuspended in a known volume of chloroform. Total radioactivity is determined for all samples. Values for each lipid are normalized to the average total dpm as follows:

$$\text{Corrected dpm} = (\text{dpm in spot}) \left(\frac{\text{average dpm from all lipid extracts}}{\text{dpm in individual lipid extracts}} \right)$$

It was found that the greatest reproducibility in triplicate samples in a single experiment was obtained by this procedure.[9] The validity of this procedure was confirmed by our observation of no significant effects of any experimental condition on the recovery of total radioactivity in the appropriate Bligh and Dyer phases (see Tables I to III). With this method of normalization and with minilobules prepared from a single batch of pooled pancreata and prelabeled after preparation, the standard deviations of replicate samples are quite low as compared to whole pancreata[67] (see Tables I to III).

The wet weight of minilobules is determined by measuring the total lipid–phosphate in weighed whole pancreas and in the minilobules added to the incubation vessel. Radioactivities are adjusted to 100 mg (wet weight) of minilobules.

Phosphatidylinositol phosphorus is determined as follows. The PI spots are scraped from the TLC plates, and phosphorus is analyzed by the method of Bartlett.[68] The silica gel is removed by centrifugation before the colors are read in the spectrophotometer.

Typical Data

Increases in Labeling of PA, DAG, Free Arachidonate, Free Stearate, and TAG on Stimulation of PI Breakdown in [^{14}C]Arachidonate- or [^{14}C]Stearate-Prelabeled Pancreata

In order to prelabel PI in mouse pancreas, it is preferable to stimulate with a cholinergic drug, followed by cholinergic blockade with atropine.[9] On cholinergic stimulation, PA reaches maximum labeling, but PI does not. On cholinergic blockade with atropine, radioactivity in PA drops to near basal levels, and the radioactivity in PI reaches maximal levels due to

---

[67] P. J. Marshall, D. E. Boatman, and L. E. Hokin, *J. Biol. Chem.* **256**, 844 (1981).
[68] G. R. Bartlett, *J. Biol. Chem.* **234**, 466 (1959).

the conversion of the PA to PI. A second stimulation must be carried out with an atropine-insensitive agonist, such as caerulein. We found that this prelabeling procedure was necessary to demonstrate maximum breakdown of prelabeled PI (J. F. Dixon and L. E. Hokin, unpublished observations). Under these conditions, the caerulein-stimulated decrement in [^{14}C]-arachidonate-prelabeled PI was essentially the same as that of chemically measured PI.

Table I shows the effects of caerulein on the radioactivity in phospholipids and neutral lipids prelabeled with [^{14}C]arachidonate. The drop in [^{14}C]PI was accompanied by statistically significant increases in [^{14}C]PA, [^{14}C]DAG, [^{14}C]arachidonate, and [^{14}C]TAG. In some experiments, there were statistically significant decreases in [^{14}C]phosphatidylserine (PS) and increases in [^{14}C]PE, which were less than 10%. The radioactivity in PC in unstimulated and stimulated minilobules generally agreed within 5% or less.

The increased level of [^{14}C]arachidonoyl-DAG on stimulation with caerulein supports a phospholipase C-catalyzed breakdown of PI in the acinar pancreas. The increase in arachidonoyl-TAG is presumably due to synthesis from DAG and arachidonate following the release of both as a consequence of PI breakdown. Forty-seven percent of the caerulein-induced decrement in [^{14}C]PI could be accounted for as free [^{14}C]arachidonate. This increased level of [^{14}C]arachidonate suggests further breakdown of DAG by DAG lipase (see below).

Results similar to those seen with [^{14}C]arachidonate-prelabeled lipids were seen with [^{14}C]stearate-prelabeled lipids.[9] Namely, there was a fall in radioactivity in [^{14}C]stearate-prelabeled PI and a rise in ^{14}C in PA, DAG, TAG, and free stearate.

Increases in Labeling in PA, DAG, TAG, and Free Glycerol on Stimulation of [^{3}H]Glycerol-Prelabeled PI Breakdown

Table II shows the caerulein-induced changes in lipids prelabeled with [^{3}H]glycerol. The loss of PI (70%) was similar to the loss of PI prelabeled with radioactive fatty acids. Highly significant increases in radioactivity were seen in PA, DAG, TAG, and free glycerol. The caerulein-induced release of free glycerol was 28% of the PI breakdown. This further confirms that an appreciable amount of the DAG moiety is not conserved in the phosphoinositide cycle. A significant increase in [^{14}C]glycerol-prelabeled MAG was also seen on caerulein stimulation, although this increase accounted for less than 1% of the PI breakdown. There was a small but significant increase in [^{3}H]glycerol-labeled PE on stimulation. Other lipids did not change on caerulein stimulation.

Inhibition of the Conversion of DAG to Arachidonate, Stearate, and Glycerol by the Specific Lipase Inhibitor, RHC 80267

Sutherland and Amin[14] have shown specific inhibition of DAG lipase by RHC 80267 at concentrations in the micromolar range. If the caerulein-induced increases in [^{14}C]arachidonate and [^{3}H]glycerol seen in Tables I and II were due to their release from DAG by DAG lipase, it would be predicted that in the presence of the lipase inhibitor the caerulein-stimulated formation of fatty acids and glycerol would be inhibited and more DAG would accumulate. This was, in fact, found to be the case (Tables I and II). To make possible statistical analysis of these effects, three separate experiments were performed for each label. The caerulein-induced increments in [^{14}C]arachidonoyl-, [^{14}C]stearoyl-, and [^{14}C]glycerol-DAG were increased by 39% ($p = 0.01$), 67% ($p = 0.0095$), and 116% ($p < 0.001$), respectively, in the presence of the inhibitor. The basal levels of [^{14}C]arachidonoyl- and [^{14}C]glycerol-DAG were also almost twice as great in the presence of RHC 80267 as in its absence (Tables I and II), suggesting that DAG is breaking down in the absence of caerulein. The increments in free [^{14}C]arachidonate and [^{3}H]glycerol on stimulation were reduced by 42% ($p = 0.033$) and 39% ($p = 0.0025$) in the presence of RHC 80267. The radioactivity in TAG in the unstimulated minilobules was not affected by the inhibitor, suggesting that RHC 80267 did not inhibit TAG lipase to any extent (Tables I and II). However, there were inhibitions by RHC 80267 of the caerulein-induced increments in [^{14}C]arachidonoyl-and [^{3}H]glycerol-TAG by 59% ($p = 0.002$) and 56% ($p = 0.025$), respectively. This is consistent with the view that the caerulein-induced increments in [^{14}C]arachidonoyl- and [^{3}H]glycerol-TAG were at least partly due to the caerulein-induced increases in radioactive fatty acids, which were reduced by RHC 80267.

Radioactivity in LPI on Stimulation of PI Breakdown in Minilobules Prelabeled with [^{3}H]Glycerol

If the breakdown of PI were due in part or entirely to a stimulation of deacylation by a phospholipase A_2, one would expect to see an increase in radioactivity in LPI on stimulation with caerulein. Table III shows the levels of radioactivity in LPI and other relevant lipids on incubation without and with caerulein after prelabeling with [^{3}H]glycerol. When prelabeled with [^{3}H]glycerol, the radioactivity in PI was reduced on stimulation with caerulein at 2 and 5 min. Short incubation times were chosen to increase the chance of trapping radioactive LPI (maximum breakdown of PI in minilobules requires 30 min). At 5 min, the radioactivity in LPI was 0.38% of that in PI. There was no increase in radioactivity in LPI on stimulation with caerulein. With *myo*-[³H]inositol prelabeling, there was an increase in radioactivity in the methanol/water phase after removal of the chloroform phase of the Bligh and Dyer extract, and this increase was equal to the loss of radioactivity in PI or the total lipid extract.[9] This would indicate that essentially all of the radioactivity lost by PI would be accounted for by *myo*-inositol or *myo*-inositol phosphates. Essentially all of the caerulein-induced decrement in the chloroform extract could be accounted for by the decrement in PI.[9] This indicates that decrements in the polyphosphoinositides were a small fraction of the decrement in the chloroform extract. This was directly confirmed by counting the radioactivities in the polyphosphoinositides (J. F. Dixon and L. E. Hokin, unpublished observations). There was very little [³H]glycerol-labeled LPC and no increase on stimulation (data not shown), suggesting no phospholipase A_2 action on PC.

Acknowledgments

Wish to thank Karen Wipperfurth and Barbara Bollig for their assistance in the preparation of the manuscript. The work reported here was supported in part by a grant from the National Institutes of Health, Number HL-16318.

[39] Measurement of Intracellular Free Calcium to Investigate Receptor-Mediated Calcium Signaling

By CARL A. HANSEN, JONATHAN R. MONCK, and JOHN R. WILLIAMSON

Introduction

Change in the cytosolic free Ca^{2+} concentration functions as an important intracellular signaling mechanism whereby hormones and growth factors regulate many different cellular processes such as secretion, metabolism, neurotransmitter release, cell growth, and differentiation.[1-3] Receptor-mediated Ca^{2+} signaling is initiated by the generation of D-*myo*-inositol 1,4,5-trisphosphate [Ins(1,4,5)P_3], which is formed from the hydrolysis of phosphatidylinositol 4,5-bisphosphate in the plasma membrane through receptor-mediated activation of phospholipase C. The binding of

---

[1] M. J. Berridge and R. F. Irvine, *Nature (London)* **341**, 197 (1989).
[2] S. P. Soltoff and L. C. Cantley, *Annu. Rev. Physiol.* **50**, 207 (1988).
[3] J. R. Williamson and J. R. Monck, *Annu. Rev. Physiol.* **51**, 107 (1989).

Ins(1,4,5)P_3 to specific receptors located in the membrane of the internal Ca^{2+} store opens a Ca^{2+} channel and thereby causes a rapid increase in the cytosolic free Ca^{2+} concentration. This increase is transient because of an activation of Ca^{2+} efflux mechanisms in the plasma membrane, notably Ca^{2+}-ATPase. In addition, receptor occupancy is associated with an enhanced influx of Ca^{2+}, which maintains the cytosolic free Ca^{2+} above resting levels for the duration of the agonist response.[3] Despite a general acceptance of the overall sequence of events responsible for increasing the cytosolic free Ca^{2+}, details of the mechanisms regulating each step remain unresolved. Recent studies suggest the presence of an intricate network of positive and negative controls that provide continuous regulation of the responses elicited by receptor activation. These mechanisms account for the diversity of cellular Ca^{2+} responses seen in different tissues, as well as the marked heterogeneity of responses seen in single cells within a population, including an oscillatory behavior of the Ca^{2+} signal. An elucidation of the detailed molecular mechanisms underlying these regulatory processes requires that changes in the cytosolic Ca^{2+} concentration be reliably measured. The first section of this chapter describes the use of tetracarboxylate-derived fluorescent Ca^{2+} indicators to measure changes in the cytosolic free Ca^{2+} in single cells and suspensions of cells. This is followed by a discussion of several approaches that can be used in conjunction with these indicators to study the biochemical mechanisms regulating the Ca^{2+} signal.

Measurement of Ca^{2+} with Fluorescent Indicators

Historically, a number of techniques have been used to measure the changes that occur in the cytosolic free Ca^{2+} concentration during cell stimulation.[4] These include null-point titration, Ca^{2+}-selective microelectrodes, and aequorin. However, these techniques are limited by the necessity for disruption of the cell membrane. The development of a family of fluorescent Ca^{2+} indicators, initially quin2, then Fura-2, Indo-1 and Fluo-3, which have the property that they can be loaded into cells in an esterified form with subsequent cleavage of the ester group to release the indicator into the cytosol, has allowed the direct measurement of Ca^{2+} changes in intact cells.[5]

The essential property of these dyes is that they bind Ca^{2+} with a K_d in the physiological range and that the Ca^{2+}-bound and free forms of the indicators have different fluorescence properties. Thus, when Fura-2 is excited at 340 nm, a change from the free form to the Ca^{2+}-bound form of

---

[4] R. Y. Tsien, *Annu. Rev. Biophys. Bioeng.* **12**, 91 (1983).
[5] R. Y. Tsien, *Methods Cell Biol.* **30**, 127 (1989).

the indicator results in an increase in fluorescence. If the fluorescence of the Ca^{2+}-free form and the Ca^{2+}-bound form are known, the cytosolic free Ca^{2+} concentration can be calculated from the measured fluorescence. An additional important feature of Fura-2 and Indo-1 is that the excitation or emission spectra, respectively, shift upon binding Ca^{2+}. This allows ratiometric measurements of Ca^{2+} in single cells and has the intrinsic advantage that the fluorescence signal is independent of the dye concentration, thereby making the measurements relatively insensitive to dye leakage and changes of cell shape and volume.

Procedures for Loading Cells with Indicator

An extremely valuable property of these indicators is that they can be loaded into cells as their acetoxymethyl esters. The cell membrane is relatively permeable to these esters, which diffuse into the cytosol where they are cleaved by endogenous esterases to release the indicator form of the molecule. Since the permeant form is removed, the concentration gradient of the ester is maintained and more ester enters the cell, allowing accumulation of the indicator. Typically a suspension of cells is incubated for 5–30 min with 1–10 μM Fura-2 or Indo-1 ester, giving a final concentration in the cell of 50–500 μM. The exact concentration and incubation period required vary widely between cell type and must be determined empirically. Loading can be enhanced in some cells (e.g., hepatocytes, PC12 cells) by using tissue culture media such as Leibovitz-15. The active factor is unknown but could be a vitamin acting as a cofactor to the esterase. Pluronic acid has also been used to aid dispersal of the Fura-2-tetraacetoxymethyl ester (Fura-2/AM).[6] An alternative method of loading is microinjection of Fura-2-free acid directly into the cell. This method avoids problems of incomplete hydrolysis of the esters, but is limited primarily to studies with single cells.

Instrumentation Required for Fluorescence Measurements

The equipment required to make measurements of Ca^{2+} using fluorescent probes requires an illumination system, a sample holder, and a detection system. The illumination system usually uses a mercury or xenon arc lamp, both of which have suitable spectral characteristics. The wavelengths can be selected with monochromators or, less expensively, with interference filters. The rapidly increasing number of fluorescent indicators for Ca^{2+}, Mg^{2+}, Na^+, and pH allows the same instrument to be used for measurement of a variety of intracellular ions. Neutral density filters may

[6] M. Poenie, J. Alderton, R. Steinhardt, and R. Y. Tsien, *Science* **233**, 886 (1986).

be necessary to reduce light levels since the high-intensity output can cause problems of dye photobleaching and phototoxicity, particularly with the mercury lamp if 360 nm is chosen as an excitation wavelength. For ratiometric measurements, some mechanism for changing the excitation or emission wavelength is needed. A chopper mirror or spinning wheel may be used to change the excitation wavelength, as required for Fura-2, or a dichroic beamsplitter to select the two emission wavelengths for Indo-1. The sample holder can be a cuvette box for measurement of cells in suspension or on monolayers, or, alternatively, a cell chamber is mounted on an epifluorescent microscope for single-cell measurements. Detection devices include a photomultiplier tube, an intensified television camera (SIT or ISIT) or cooled charge-coupled device (CCD) camera attached to a camera port on the microscope. Recent technological advances in computer array processing hardware and image analysis techniques have made it possible to make Ca^{2+} measurements at high spatial and temporal resolution within single cells.

Cell Preparations

Cell Suspension and Monolayers. Much of our basic understanding of the Ca^{2+} signal has been derived from studies using cells either in bulk suspension or attached to coverslips placed in a cuvette. This approach is operationally the most straightforward, requires the least investment in hardware, and is generally adequate to answer basic questions relating to Ca^{2+} signaling, such as dose–response relationships for Ca^{2+}-mobilizing agonists, whether the source of the Ca^{2+} is derived from internal stores or extracellular Ca^{2+} entry, and whether there is cross-talk between a Ca^{2+}-mobilizing agonist and other receptor-mediated events. The data obtained can be correlated with other biochemical and physiological information measured in the same cell preparation. The major limitation of these preparations, however, is that the measured signal originates from a large number of cells and, thus, represents an averaged response of the cell population. Studies on single cells have revealed a large degree of heterogeneity in the Ca^{2+} transients elicited by an agonist, including oscillations of the cytosolic free Ca^{2+}. Hence, the Ca^{2+} signaling pattern observed from a cell population is not necessarily indicative of the responses occurring in individual cells. Likewise, the average biochemical and physiological measurements in these bulk cell preparations may not be representative of the events occurring at the single-cell level. Population studies in certain cell preparations, however, may reflect single-cell events. Confluent monolayers of bovine pulmonary artery endothelial cells have been shown to respond to histamine with synchronous oscillations in the cytosolic free Ca^{2+} concentration, suggesting that all cells in the monolayer behave similarly.[7] The presence of gap junctions between cells in monolayers of other cell types suggests that similar synchronously responding preparations may be obtainable. Use of such cell preparations should allow Ca^{2+} changes to be compared kinetically with changes in other metabolites. For example, by measuring the level of $Ins(1,4,5)P_3$ in these monolayers at the peak and the base of the Ca^{2+} spike, an assessment can be made of the hypothesis that cyclic generation of $Ins(1,4,5)P_3$ underlies the mechanism of receptor-stimulated Ca^{2+} oscillations.

Single Cells. The heterogeneity of responses observed in single cells includes differences in the time between exposure to agonist and the onset of the response (latency period), the peak Ca^{2+} levels reached, and the duration of the plateau phase of elevated Ca^{2+}.[8] Also, there is a propensity of a proportion of the cells to respond with repetitive spikes in the cytosolic free Ca^{2+} (oscillations), although with a pattern that can be highly variable. Furthermore, in certain cell populations, only a portion of the cells respond to a particular agonist. Hence, in order to assess accurately potential differences in the mechanism of Ca^{2+} signaling by different agonists or during specific experimental manipulations, control experiments must be performed on the same cell. By using an inverted microscope with an open cell chamber, several perfusion pipets (i.d. ~ 10 μm) can be placed near the surface of an individual cell. The cell can then be perfused with solutions containing different mixtures of agonists and antagonists. Likewise, the composition of the medium surrounding a particular cell, e.g., Ca^{2+}-free medium, can be manipulated. When the perfusion pipet is turned off, the original medium is restored to the cell surface and the applied compounds are effectively removed from the cell by diffusion into the bulk medium. This is illustrated in Fig. 1, which shows the changes in the cytosolic free Ca^{2+} of a single Fura-2-loaded hepatocyte incubated in a medium containing 1.3 mM Ca^{2+} and stimulated successively for 2, 15, 30, and 60 sec with a maximal concentration of phenylephrine (2.5 μM). Each pulse of hormone was delivered to the cell at 3-min intervals. This approach can yield considerable information concerning the mechanisms responsible for generating the overall Ca^{2+} response with a particular agonist. Thus, the peak height was independent of time of agonist exposure, suggesting that sufficient $Ins(1,4,5)P_3$ was produced within the first 2 sec to elicit a maximal release of Ca^{2+} from the internal Ca^{2+} stores. However, a distinct plateau phase of intracellular Ca^{2+} was observed only with phenylephrine exposure times greater than 15 sec, suggesting that Ca^{2+} entry into the cell occurred subsequent to intracellular Ca^{2+} mobilization. Finally, removal of phenylephrine caused a prompt return of the cytosolic free Ca^{2+} from the plateau

---

[7] S. O. Sage, D. J. Adams, and C. van Breeman, *J. Biol. Chem.* **264**, 6 (1989).
[8] J. R. Monck, E. E. Reynolds, A. P. Thomas, and J. R. Williamson, *J. Biol. Chem.* **263**, 4569 (1988).

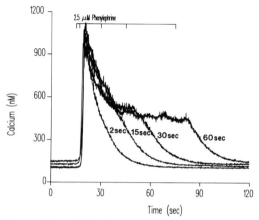

FIG. 1. Calcium ion transients induced by repeated application of 2.5 μM phenylephrine for increasing periods of time to a single Fura-2-loaded hepatocyte.

level to the prestimulated basal level, indicating that there is a close coupling between receptor occupation and the mechanism responsible for Ca^{2+} entry.

Single-cell Ca^{2+} measurements can also be combined with microinjection techniques to introduce nonpermeable compounds into the cell. Techniques successfully used involve pressure microinjection, which rapidly introduces a bolus of compound into the cell, and whole cell patch-clamp dialysis,[9] in which a constant level of experimental compound can be maintained within the cytosol compartment. Both these approaches allow simultaneous monitoring of changes in ion currents or membrane potential.

Calibration for Estimation of Ca^{2+} Concentrations

The following discussion of calibration procedures focuses on the indicator Fura-2, but is applicable to other members of this family of Ca^{2+} indicators. The Ca^{2+}-bound form of the indicator is excited at 340 nm, the Ca^{2+}-free form at 380 nm, and the fluorescent emissions are measured using a 480 to 580-nm bandpass filter. Calibration approaches fall into two categories; single-wavelength and ratiometric methods. The principles underlying the two methods are essentially the same.[10] The formulas used to allow calculation of the free Ca^{2+} from the fluorescence changes entail the following assumptions:

1. The Ca^{2+} indicator exists in only two fluorescent forms, i.e.,

$$c_T = c_f + c_b \tag{1}$$

where c_f, c_b, and c_T are the Ca^{2+}-free, Ca^{2+}-bound, and total concentrations of indicator, respectively.

2. For a mixture of two forms of indicator, the total fluorescent intensity (F) is described by Eq. (2):

$$F = S_f c_f + S_b c_b \tag{2}$$

where S_f and S_b are the proportionality constants for the fluorescence of Ca^{2+}-free and Ca^{2+}-bound forms of the indicator, respectively.

3. Equation (3) for 1:1 complexation describes the binding of Ca^{2+} to the indicator, i.e.,

$$c_b = c_f [Ca^{2+}]/K_d \tag{3}$$

where K_d is the dissociation constant for Ca^{2+} binding to Fura-2 (224 nM at 37°, 1 mM $MgCl_2$, 120 mM KCl, and 20 mM NaCl[10]).

For single-wavelength calibration, the standard Eq. (4) is derived based on the above three assumptions[11]:

$$[Ca^{2+}] = K_d(F - F_{min})/(F_{max} - F) \tag{4}$$

Calculation of the Ca^{2+} concentration requires that the following parameters are known: the K_d of the Ca^{2+}-indicator complex, the fluorescence under saturating Ca^{2+} conditions ($F_{max} = S_b c_T$ and $c_b = c_T$), and the fluorescence under zero Ca^{2+} conditions ($F_{min} = S_f c_T$ and $c_f = c_T$).

For radiometric measurements, assumption (2) requires two equations, one for each wavelength, and Eq. (5) is then derived[10]

$$[Ca^{2+}] = K_d[(R - R_{min})/(R_{max} - R)]\beta \tag{5}$$

where R is the ratio of fluorescence at two suitable wavelengths (e.g., F^{340}/F^{380}) and β is the fluorescence at 380-nm excitation under Ca^{2+}-free conditions divided by the fluorescence under Ca^{2+}-saturating conditions. For Fura-2 using the 340/380 ratio, $\beta = S_f^{380}/S_b^{380}$. The Ca^{2+} concentration can be calculated if K_d, R_{min}, R_{max}, and β are known. The last three parameters can be estimated in an analagous manner to the single-wavelength method by measuring the fluorescence at both excitation wavelengths, at zero, and at saturating Ca^{2+} concentrations.

For the measurement of indicator fluorescence at infinite and zero Ca^{2+} concentrations, the intracellular Fura-2 must be exposed to media of defined Ca^{2+} concentration. One method involves the disruption of plasma membrane with detergent to release the Fura-2 into the medium.[11] Detergents used include digitonin and saponin, which selectively permeabilize the plasma membrane by complexing cholesterol. Less specific detergents such as Triton X-100 also permeabilize intracellular membranes and may cause the release of indicator trapped in intracellular compartments. A second method uses Ca^{2+} ionophores (ionomycin or Br-A23187) to permeabilize the plasma membrane to Ca^{2+}, which allows equilibration of intracellular and extracellular Ca^{2+} concentrations.[12] A23187 is unsuitable since it is fluorescent with ultraviolet excitation. The ionophore method has the advantage that the indicator remains in an environment closer to that of the cytosol. The ionophore method must be used for monolayer and single-cell studies, since disruption of the cell alters the indicator concentration in the light path. A saturating concentration of Ca^{2+} is readily achieved by exposing the indicator to the Ca^{2+} concentration in the extracellular medium (typically 1–2 mM). A "zero" Ca^{2+} concentration can be attained by chelating the Ca^{2+} ions with an excess of EGTA. The Ca^{2+} chelation can be aided by alkalinization of the medium to pH 8.3 with Tris base, as this procedure decreases the K_d of EGTA for Ca^{2+}. An alternate method for obtaining F_{min} uses Mn^{2+}.[12] The fluorescence in the presence of Mn^{2+} does not provide F_{min} directly, since Mn^{2+} quenches both the Ca^{2+}-sensitive and -insensitive components of the Fura-2 fluorescence by different amounts that are instrument dependent and must be determined empirically. By definition, F_{min} is the Ca^{2+}-insensitive fluorescence.

Before calculation of the ratio, it is important to correct F^{340} and F^{380}, as well as their F_{min} and F_{max} counterparts, for all fluorescence values not associated with Fura-2 in the cytosol [i.e., assumption (1)]. Therefore, background fluorescence and autofluorescence must be subtracted from the fluorescence at each wavelength. Autofluorescence for the purposes used here represents the contribution of endogenous fluorochromes in the cell [e.g., NAD(P)H and FADH] and background light in the system (e.g., fluorescence of glass in the light path or dark currents of the measuring device). Provided that these are constant during the experiment no corrections need be made for the single-wavelength method as the contributions to F, F_{min}, and F_{max} cancel out. However, problems arise when the autofluorescence changes during the experiment. For example, when liver cells are stimulated by hormones, NAD(P)H levels increase with a consequent increase in total fluorescence emission.[13] These changes must be corrected

---

[11] R. Y. Tsien, T. Pozzan, and T. J. Rink, *Nature (London)* **295**, 68 (1982).
[12] T. R. Hesketh, G. A. Smith, J. P. Moore, M. V. Taylor, and J. C. Metcalfe, *J. Biol. Chem.* **258**, 4876 (1983).

for either by subtracting values obtained from control experiments in the absence of indicator or by using separate emission channels for the nucleotides and indicator and correcting for cross-over between channels.[13] These corrections are relatively minor with Fura-2 and Indo-1, which have a higher fluorescence yield than quin2. For ratiometric measurements, the autofluorescence must be subtracted before calculation of the ratio.

A factor commonly encountered with cell suspensions is the presence of extracellular indicator. At normal extracellular Ca^{2+} concentrations, extracellular indicator is saturated and gives an artificially high estimate of the intracellular Ca^{2+} concentration. The F_{min} and F_{max} values must be corrected by subtracting the contribution from the extracellular indicator. There are two methods for obtaining the correction factors. According to the first, EGTA is added to the medium prior to addition of ionophore or detergent. This gives a rapid decrease in fluorescence due to removal of Ca^{2+} from extracellular indicator, followed by a slower decrease as cellular Ca^{2+} is depleted. The initial fall of fluorescence after EGTA addition must also be subtracted from the cellular fluorescence (F) before values are used in the calibration equation. The calibration proceeds by adding detergent or ionophore to determine the F_{min} followed by adding excess Ca^{2+} to determine the F_{max}. The second method uses a low concentration of Mn^{2+} (typically 50–100 μM) to quench extracellular indicator and the Mn^{2+} is subsequently removed by the addition of the heavy metal chelator diethylenetriaminepentaacetic acid. As with the EGTA method, the initial decrease in fluorescence is used to correct F, F_{max}, and F_{min}. However, corrections must also be made for the ability of Mn^{2+} to quench the fluorescence of the Ca^{2+}-insensitive forms of the indicator.

For ratiometric estimation of Ca^{2+} with Fura-2 or Indo-1, a similar protocol is followed to determine the fluorescence of the indicators at saturating and zero Ca^{2+} concentrations, allowing calculation of R_{max} and R_{min}, respectively. It is important to note that corrections for extracellular indicator and autofluorescence must be made prior to calculation of the ratio. Likewise, for the Mn^{2+} method to be employed, the corrections for Mn^{2+} quenching of Ca^{2+}-insensitive and -sensitive fluorescence at both wavelengths must be made before calculation of the Ca^{2+} values, as described by Bjornsson et al.[14]

---

[13] J. R. Williamson, C. A. Hansen, A. Verhoeven, K. E. Coll, R. A. Johanson, M. T. Williamson, and C. Filburn, in "Cell Calcium and the Control of Membrane Transport" (L. J. Mandel and D. C. Eaton, eds.), pp. 94–116. Rockefeller Univ. Press, New York, 1986.

[14] O. G. Bjornsson, J. R. Monck, and J. R. Williamson, Eur. J. Biochem. 186, 395 (1989).

Potential Problems

Several potential problems can arise which result in an inaccurate calculation of cytosolic free Ca^{2+} concentration from the Fura-2 fluorescence signals, particularly if single-wavelength calibration is employed. The concentration of Fura-2 being measured may not remain constant during the experiment. This will generate changes in the measured light intensity independent of changes in Ca^{2+} concentration. There can be changes in the morphology of the cells, such as contraction, which move the cell out of the region within the focal zone of the microscope. Contraction artifacts can often be overcome by using the ratiometric calibration.[15] Leakage of Fura-2 from the cell can occur during incubation. This can be quite large at 37°, but if the experiments are performed at lower temperatures (25-30°) leakage is considerably reduced and is not a problem for short-term experiments. Probenecid, an inhibitor of anion transport, has been shown to reduce the rate of dye leakage when added to the loading and incubation media.[16] In any case, the ratiometric calibration procedure can correct for changes in cell Fura-2 content, provided that autofluorescence and extracellular indicator are not problems, as is usually the case for single-cell measurements. Another problem is photobleaching. Becker and Fay[17] recently showed that photobleaching of Fura-2 yields a product that is fluorescent so that, contrary to previous practice, it cannot be assumed that the ratiometric calibration will correct for photobleaching. To circumvent this problem, the experiments must be performed under conditions that minimize exposure to light. Fortunately, the sensitivity of the common measuring devices (photomultiplier, SIT camera, etc.) are such that an adequate signal can be detected at low illumination intensities that cause minimal photobleaching over short incubation periods.

Another problem can occur if the assumption that Fura-2 exists only in two forms, i.e., Ca^{2+} free and Ca^{2+} bound, is violated. Scanlon *et al.*[18] have recently demonstrated the presence of an additional species of Fura-2 that is Ca^{2+} insensitive. We have observed a similar problem under certain loading conditions (e.g., using high Fura-2 concentrations and short incubation times). If cells are incubated for 15-20 min in distilled water, most of their fluorescence is lost as the Fura-2 leaks out. The cells remain attached to the coverslip and retain a residual fluorescence that is not autofluorescence and is Ca^{2+} insensitive. This fluorescence, which has been seen in hepatocytes, GH_3 cells, and in cardiac ventricular myocytes, either

---

[15] L. Barcenas-Ruiz and W. G. Wier, *Circ. Res.* **61,** 148 (1987).
[16] P. M. McDonough and D. C. Button, *Cell Calcium* **10,** 171 (1989).
[17] P. L. Becker and F. S. Fay, *Am. J. Physiol.* **253,** C613 (1987).
[18] M. Scanlon, D. A. Williams, and F. S. Fay, *J. Biol. Chem.* **262,** 6308 (1987).

corresponds to Fura-2 that is in a subcellular compartment[19] or is a Ca^{2+}-insensitive form of Fura-2 that remains tightly bound to cell membranes or proteins.[20] Scanlon et al.[18] suggested that such a species could result from the partial hydrolysis of the fura-2/AM. The consequence is that only a small fractional change in Fura-2 fluorescence is observed, leading to an underestimation of the calculated free Ca^{2+}. Fortunately, in many cell types these problems are not important or can be overcome by optimizing loading conditions. However, it is clear that for each experimental system an evaluation of the properties of the indicator within the cell is required before a meaningful estimation of cytosolic Ca^{2+} concentration can be made.

Investigation of Biochemical Mechanisms Regulating Ca^{2+} Signaling in Fura-2-Loaded Cells

The complexity of the signal transduction process involved in Ca^{2+} signaling has made it difficult to study directly the biochemical mechanisms regulating Ca^{2+} fluxes. When Ca^{2+}-mobilizing agonists bind to their receptors, not only is Ins(1,4,5)P_3 formed and Ca^{2+} signaling initiated, but many other biochemical events are activated. These include (1) activation of G proteins in the plasma membrane that may interact with target proteins other than phospholipase C, (2) formation of diacylglycerol and activation of protein kinase C, which phosphorylates a number of cellular proteins, and (3) activation of other protein kinases and phosphatases. Calcium itself activates specific protein kinases and Ca^{2+}-dependent proteins. A specific Ca^{2+}-binding protein is thought to be responsible for feedback inhibition of Ins(1,4,5)P_3 binding to its internal receptor.[21] Certain Ca^{2+}-mobilizing hormones, such as growth factors, are tyrosine kinases,[2] while other agonists, such as glucagon, increase both Ca^{2+} and cAMP levels.[22] Furthermore, Ins(1,4,5)P_3 is rapidly metabolized to a variety of other inositol polyphosphates, whose functions are not yet known. Generation of multiple intracellular signals thus provides many possibilities for either positive or negative interactions.

Two basic approaches have been utilized to isolate specific aspects of the Ca^{2+} signaling process and, hence, assess their relative importance for cell function. The most widely used has been to perturb one component of

---

[19] A. Margaroli, D. Miliani, J. Meldolesi, and T. Pozzan, J. Cell Biol. **105**, 2145 (1987).
[20] S. M. Baylor and S. Hollingworth, J. Physiol. (London) **403**, 151 (1988).
[21] P. F. Worley, J. M. Baraban, S. Suppattapnoe, V. S. Wilson, and S. H. Snyder, J. Biol. Chem. **262**, 12132 (1987).
[22] F. D. Sistare, R. A. Picking, and R. C. Haynes, J. Biol. Chem. **260**, 12744 (1985).

the signaling mechanism (e.g., by pretreatment of cells with phorbol esters or pertussis toxin) followed by observing its effect on an agonist-induced Ca^{2+} transient. Alternatively, the membrane-associated signal transduction events can be bypassed by directly introducing known intracellular messenger molecules into the cytosol and monitoring their effects on the Ca^{2+} transient relative to those induced by the agonist. Both approaches rely on the specificity of perturbing agents used. The following section discusses several approaches that have been used to isolate and subsequently study specific components of the Ca^{2+} signal.

Cations that Substitute for Ca^{2+}

In general, receptor-mediated Ca^{2+} entry involves opening of relatively nonselective divalent cation channels. Hence, certain divalent cations have proved to be useful tools to study Ca^{2+} channels. Electrophysiologists have long used Ba^{2+} and Sr^{2+} to substitute for Ca^{2+} in measuring Ca^{2+} channel conductances. In many cells, plasma membrane permeability to Mn^{2+} and Co^{2+} is stimulated by Ca^{2+}-mobilizing agonists through activation of the same cation channels responsible for receptor-mediated Ca^{2+} entry. These cations also bind to Fura-2 and consequently affect its fluorescence. Manganese ions and Co^{2+} quench Fura-2 fluorescence to a few percent of the Ca^{2+}-saturated Fura-2 fluorescence, whereas Ba^{2+} and Sr^{2+} interact with Fura-2 in a manner similar to Ca^{2+}.[23] Since the entrance of these cations into the cell can be monitored by Fura-2, they can be used to study the mechanism of receptor-mediated cation influx independently of changes in the cytosolic free Ca^{2+} concentration.

The most widely used of these cations has been Mn^{2+}.[24] Cells are placed in nominally Ca^{2+}-free medium and 1 mM $MnCl_2$ is added to the extracellular medium. After a basal rate of Fura-2 quenching is established, an agonist is added. An enhanced rate of Fura-2 quenching indicates agonist stimulation of cation influx. Unfortunately, it is difficult to convert the rate of quenching to a rate of Ca^{2+} influx because the affinities of the cations for Fura-2 are quite different from the binding of Ca^{2+} to Fura-2. Furthermore, the agonist stimulates release of internal Ca^{2+}, which competes with the entering cation for Fura-2. This approach provides a qualitative characterization of cation permeability in unstimulated and stimulated cells. Application of this approach has demonstrated that platelets, neutrophils, hepatocytes, and endothelial cells possess agonist-stimulated Mn^{2+} entry, wheras parotid and lacrimal acinar cells do not.[23,24] These data suggest that

[23] C. Kwan and J. W. Putney, *J. Biol. Chem.* **265**, 678 (1990).
[24] T. Hallem, R. Jacob, and J. Merritt, *Biochem. J.* **255**, 179 (1988).

distinct Ca^{2+} channels may be activated in different cell types by receptor stimulation.

A recent study suggests that Ba^{2+} and Sr^{2+} may be extremely useful in the study of Ca^{2+} signaling.[23] Both of these cations not only permeate Ca^{2+} channels activated by receptor stimulation, but they also appear to elicit Ca^{2+}-activated events. Interestingly, it appears that in lacrimal acinar cells Sr^{2+}, but not Ba^{2+}, is capable of refilling $Ins(1,4,5)P_3$-sensitive internal Ca^{2+} stores. Hence, preactivation of cells in medium containing Sr^{2+} in place of Ca^{2+} may provide a useful system for the study of agonist-stimulated cation influx in the absence of internal Ca^{2+} release.

Selective Inactivation of Ca^{2+} Entry and the $Ins(1,4,5)P_3$-Sensitive Ca^{2+} Stores

Since receptor-mediated Ca^{2+} signaling often involves both release of internal Ca^{2+} stores and stimulation of Ca^{2+} entry, experimental manipulations in which these two components are selectively inhibited need to be used. Elimination of extracellular Ca^{2+} entry can be achieved relatively easily by incubation of cells in a nominally Ca^{2+}-free extracellular medium ($[Ca^{2+}] \sim 10 \mu M$). If cells rapidly lose their intracellular Ca^{2+}, EGTA must be added just prior to addition of the agonist. However, the buffering capacity of the extracellular medium must be adequate to prevent a change in pH by the H^+ released from EGTA. Alternatively, receptor-stimulated Ca^{2+} entry can usually be blocked by high extracellular concentrations of La^{3+} or Ni^{2+} (5–10 mM).

Depletion of the internal Ca^{2+} stores has been more difficult to achieve. Two compounds have recently become available that should allow development of a stable system in which the internal Ca^{2+} stores remain depleted: 2,5-di-(*tert*-butyl)-1,4-benzohydroquinone (tBuBHQ), which inhibits microsomal Ca^{2+} sequestration,[25] and thapsigargin, a tumor-promoting sesquiterpene lactone.[26] Both of these compounds cause a rapid rise in the cytosolic free Ca^{2+} concentration, which remains elevated above initial levels for approximately 20 min, and the $Ins(1,4,5)P_3$-sensitive Ca^{2+} store becomes depleted.

These compounds have been used to examine several unresolved aspects of the Ca^{2+} signal. Putney has proposed that Ca^{2+} entry is linked to emptying of the internal $Ins(1,4,5)P$-sensitive Ca^{2+} store.[27] In its most simple form, the capacitative Ca^{2+} entry model does not require any signal

---

[25] G. Kass, S. Duddy, G. Moore, and S. Orrenius, *J. Biol. Chem.* **264**, 15192 (1989).

[26] H. Takemura, A. R. Hughes, O. Thastrup, and J. W. Putney, *J. Biol. Chem.* **264**, 12266 (1989).

[27] J. W. Putney, *Cell Calcium* **7**, 1 (1986).

other than emptying of the internal store to stimulate extracellular Ca^{2+} entry. Addition of 2 μM thapsigargin to parotid acinar cells was indeed found to stimulate Ca^{2+} entry, presumably without generation of any agonist-mediated signal other than depletion of the internal Ca^{2+} store. In hepatocytes, however, addition of 10 μM tBuBHQ emptied the internal Ca^{2+} store, but did not stimulate Ca^{2+} entry,[25] suggesting that the capacitative Ca^{2+} entry mechanisms do not operate in these cells. These studies further support the concept that different mechanisms for agonist-stimulated Ca^{2+} entry exist in different cell types.

Alternatively, the Ca^{2+}-mobilizing properties of these compounds can be used to increase the cytosolic free Ca^{2+} concentration, thereby permitting the effects of agonists on sequestration of the released Ca^{2+} to be examined. In tBuBHQ-treated hepatocytes, addition of vasopressin, angiotensin II, and phenylephrine resulted in the rapid decrease in the cytosolic free Ca^{2+} concentration.[25] Interestingly, addition of phorbol myristoyl acetate (PMA) did not lower the cytosolic Ca^{2+}. These data suggest that other signals are generated upon receptor activation which affect mechanisms responsible for removal of the cytosolic Ca^{2+} in addition to activation of protein kinase C.

Microinjection Techniques to Introduce Nonpermeable Compounds

Use of techniques for microinjection in conjunction with monitoring cytosolic Ca^{2+} concentration in Fura-2 loaded single cells provides the possibility of examining directly the function of potential intracellular messenger molecules in the absence of activation of membrane signal transduction events. Use of this approach is illustrated in Fig. 2 for microinjection of $Ins(1,4,5)P_3$ into hepatocytes. Pressure microinjection is utilized and the amount of $Ins(1,4,5)P_3$ injected is altered by varying the duration of the 40-psi pressure pulse as noted on the top of the figure. Microinjection of $Ins(1,4,5)P_3$ into the hepatocyte elicited a Ca^{2+} spike, the height of which was dependent on the amount of $Ins(1,4,5)P_3$ injected. The latency period was less than 200 msec, showing that the longer latency periods observed following addition of an agonist must result in steps involved in the activation of phospholipase C, perhaps in the generation of a threshold level of $Ins(1,4,5)P_3$.[28]

A limitation of pressure microinjection, which becomes important when metabolism occurs, is that the compound is introduced into the cell as a bolus. $Ins(1,4,5)P_3$ is rapidly metabolized to $Ins(1,4)P_2$ by 5-phospha-

[28] J. R. Monck, R. E. Williamson, I. Rogulja, S. J. Fluharty, and J. R. Williamson, *J. Neurochem.* **54**, 278 (1990).

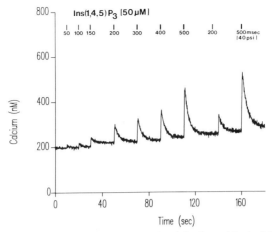

FIG. 2. Calcium ion transients induced by microinjection of Ins(1,4,5)P$_3$ into Fura-2-loaded hepatocytes. The micropipet contained 100 μM Ins(1,4,5)P$_3$ in 125 mM KCl. The volume injected was varied by changing the duration of the pressure pulse.

tases and to Ins(1,3,4,5)P$_4$ by Ins(1,4,5)P$_3$ 3-kinase, which accounts for the transient nature of the Ca^{2+} response in Fig. 2. The phosphorothioate derivative of Ins(1,4,5)P$_3$, thioIP$_3$, is not hydrolyzed, is not metabolized to Ins(1,3,4,5)P$_4$, and is only slightly less effective than Ins(1,4,5)P$_3$ in eliciting Ca^{2+} mobilization.[29] Figure 3 illustrates that when thioIP$_3$ was pressure injected into a hepatocyte using a 200-msec pulse at 30 psi, the cytosolic free Ca^{2+} remained elevated when the cells were incubated in normal Ca^{2+} medium (top trace). Under these conditions, the Ins(1,4,5)P$_3$-sensitive Ca^{2+} store rapidly becomes depleted since the Ca^{2+} channels remain open due to the continued presence of thioIP$_3$. An effect of Ins(1,4,5)P$_3$ on Ca^{2+} flux across the plasma membrane was also observed after injection of thioIP$_3$, as revealed by the removal of the extracellular Ca^{2+} in the environment of the cell by perfusing excess EGTA onto the cell at the peak of the Ca^{2+} response. Under these conditions, the cytosolic free Ca^{2+} gradually declined as the Ca^{2+} was transported out of the cell without being compensated by a simultaneous Ca^{2+} influx. This experiment illustrates that the combined approach of altering both the intracellular and extracellular environment of the cell, together with Ca^{2+} measurements, allows Ca^{2+} influx to be investigated separately from internal Ca^{2+} mobilization. The conclusion reached from these and similar experiments is that Ins(1,4,5)P$_3$

[29] C. W. Taylor, M. J. Berridge, A. M. Cooke, and B. V. Potter, *Biochem. J.* **259**, 645 (1989).

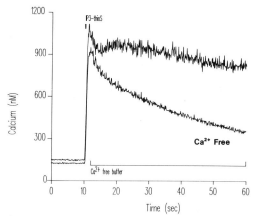

FIG. 3. Calcium ion transients induced by microinjection of thioIP$_3$ into a Fura-2-loaded hepatocyte exposed to normal (1.3 mM) and nominally Ca^{2+}-free external medium. The micropipet contained 50 μM thioIP$_3$ in 125 mM KCl.

has a dual second messenger role, at least in hepatocytes, of stimulating Ca^{2+} entry across the plasma membrane as well as Ca^{2+} release from intracellular stores.[30]

Microinjection techniques can obviously be used to introduce any compound of interest into the cells. Injection of specific enzymes such as phospholipase C[31] and antibodies to inositol lipids[32] and other inositol phosphates[9] have been used to study various aspects of the signal transduction process. As more reagents become available, microinjection techniques promise to offer a powerful approach toward unraveling the molecular mechanisms responsible for receptor-mediated Ca^{2+} signaling.

Acknowledgment

Supported by National Institutes of Health Grants (DK-15120, HL-14461) and an American Heart Association Grant-in-aid to C.A.H.

---

[30] C. A. Hansen, I. R. Siemems, and J. R. Williamson, in "Advances in Second Messenger and Phosphoprotein Research" vol 24. (Y. Nishizuka, M. Endo and C. Tanaka, eds.), pp. 128–133, Raven, New York, 1990.

[31] M. R. Smith, S. Ryu, P. Suh, S. Rhee, and H. Kung, *Proc. Natl. Acad. Sci. U.S.A.* **86**, 3659 (1989).

[32] K. Fukami, K. Matsuoko, O. Nakanishi, A. Yamakawa, S. Kawai, and T. Takenawa, *Proc. Natl. Acad. Sci. U.S.A.* **85**, 9057 (1988).

[40] Two-Stage Analysis of Radiolabeled Inositol Phosphate Isomers

By K. A. WREGGETT, D. J. LANDER, and R. F. IRVINE

Introduction

Many studies on inositol phosphate (InsP) formation in receptor-stimulated tissues prelabeled with [^{3}H]inositol either measure total radiolabeled inositol phosphates as an overall estimate of phosphoinositidase C (EC 3.1.4.11) activity[1] or, because inositol triphosphate (InsP$_3$) is a well-established second messenger derived from inositides, they measure InsP$_1$, InsP$_2$, and InsP$_3$ separately.[2] With the more recent discovery of Ins(1,3,4,5)P$_4$ and its potential second messenger role,[3] this InsP is also sometimes assayed separately by an adaptation[4] of the original Dowex anion-exchange method.[2]

For simple questions concerning the degree of activation of phosphoinositidase C, such methods are entirely adequate. Recent advances, however, have indicated that the metabolism and function of inositol phosphates are much more complicated than had been thought (for review, see Ref. 3). This has two immediate practical implications. First, the inositol phosphate fractions eluted from anion-exchange columns are mixtures of isomers. For example, the InsP$_3$ fraction is predominantly a mixture of Ins(1,3,4)P$_3$ and Ins(1,4,5)P$_3$,[5] the latter of which is biologically the active species,[3] so that any attempt to relate InsP$_3$ (total) to any aspect of cell biology, such as mobilization of intracellular Ca^{2+}, will be flawed. Second, such a complexity of inositol phosphate metabolism must exist for a reason, and this is most likely to be, at least in part, because of the function of Ins(1,3,4,5)P$_4$.[3] In order to develop a complete picture of inositol phosphate metabolism, analysis of the production of each of the various inositol polyphosphate isomers must be performed. There are several different ways of doing this (e.g., Refs. 5–9), all of which involve the analytical tool

[1] M. J. Berridge, C. P. Downes, and M. R. Hanley, *Biochem. J.* **206**, 587 (1982).
[2] M. J. Berridge, R. M. C. Dawson, C. P. Downes, J. P. Heslop, and R. F. Irvine, *Biochem. J.* **212**, 473 (1983).
[3] R. F. Irvine, R. M. Moor, W. K. Pollock, P. M. Smith, and K. A. Wreggett, *Philos. Trans. R. Soc. London B* **320**, 281 (1988).
[4] I. R. Batty, S. R. Nahorski, and R. F. Irvine, *Biochem. J.* **232**, 211 (1985).
[5] R. F. Irvine, E. E. Änggård, A. J. Letcher, and C. P. Downes, *Biochem. J.* **229**, 505 (1985).
[6] N. M. Dean and J. D. Moyer, *Biochem. J.* **242**, 361 (1987).
[7] J.-C. Sulpice, P. Gascard, E. Journet, F. Renda, D. Renard, J. Poggioli, and F. Giraud, *Anal. Biochem.* **179**, 90 (1989).

of high-pressure liquid chromatography (HPLC). However, all of these published methods share the major drawback of being expensive either in time or equipment: on-line detection of radioactivity means that automatic injection and complex gradients (involving time-consuming programs) can be used, but this equipment is costly to buy and run, and can detect only high (and thus expensive) amounts of radioactivity; on the other hand, while collection of fractions overcomes these drawbacks, it is labor intensive, with sometimes hundreds of fractions to be collected for each sample. Either way, the prospect of a typical simple experiment with, for example, controls and 3 stimulation conditions in quadruplicate (= 16 samples), is daunting.

For our detailed analysis of radiolabeled inositol phosphate isomers of cultured cells[10] we have therefore adopted a different strategy—that of two-stage processing. In this approach, each sample is initially analyzed by simple ion-exchange resins such as either Dowex or Am-Preps in the formate form. A portion of each eluted fraction phosphate is counted for radioactivity, and from this we have an initial assessment of the experimental results (i.e., whether sufficient radioactivity has been incorporated and whether the stimuli have worked in a quantitatively satisfactory manner). Then the remaining fractions are directly analyzed by isocratic anion-exchange HPLC using an automatic injector (optional) to separate the isomers. It is possible to separate $InsP_1$, $InsP_2$, and $InsP_3$ isomers in this way, but in practice it is not always necessary. For example, at an early time of stimulation, if there is little accumulation of $InsP_1$ and $InsP_2$, then only $InsP_3$ isomers need to be analyzed, whereas in some experimental systems only the $InsP_1$ accumulates enough radioactivity for analysis, and in that case, since they approximately define the two routes of metabolism of $Ins(1,4,5)P_3$, differentiating $Ins(1)P$ from $Ins(4)P$ can be very informative.

First Stage of Analysis: Ion-Exchange Resins

The most widely used method of separating inositol phosphates is by Dowex anion exchange, using the formate form of the exchange resin[1,2,4] (for a discussion and exact description, see Ref. 11). An alternative is to use a prepackaged anion-exchange resin,[12] which has the two-fold advantage of

---

[8] J. L. Meek, *Proc. Natl. Acad. Sci. U.S.A.* **85,** 4162 (1986).
[9] G. W. Mayr, *Biochem. J.* **254,** 585 (1988).
[10] W. K. Pollock, K. A. Wreggett, and R. F. Irvine, *Biochem. J.* **256,** 371 (1988).
[11] R. F. Irvine, *in* "Phosphoinositides and Receptor Mechanisms" (J. W. Putney, Jr., ed.), p. 89. Liss, New York, 1986.
[12] K. A. Wreggett and R. F. Irvine, *Biochem. J.* **245,** 655 (1987).

lower ionic strength eluates (making desalting easier) and, more importantly, of speed, especially with few (less than 20) samples. The original prepackaged columns used were QMA SEP-PAKS (Waters, Milford, MA), but since publication of that method,[12] the manufacturers have changed the specifications of the packing material and these are no longer suitable. Here we describe the use of an alternative material, Am-Preps (Amersham, Cardiff, England), and also the introduction of a means of processing 10 samples at a time. Using this method, a typical 20-sample experiment can be processed into $InsP_{1-4}$ in about 20 min.

Figure 1A shows the elution profiles of inositol phosphate standards using a gradient of ammonium formate with formic acid, and Fig. 1B shows the elution of these standards using a stepwise gradient as would be used in practice. The solutions are all made by dilution with water of a 1.0 M ammonium formate, 0.1 M formic acid stock solution; this is important, as it keeps the pH constant (compare the standard Dowex eluates,[2] which use varying ammonium formate concentrations with constant 0.1 M formic acid). The only exception is the first elution after water, and that is the 50 mM ammonium formate buffered with NaOH to a pH of 8.0. This effectively elutes glycerophosphoinositol (GroPIns), but not $InsP_1$; we have found that Am-Preps do not tolerate the tetraborate-containing eluants usually used[2,12] for GroPIns.

As with Dowex, a thorough wash with water is advisable before beginning elution of a tissue-derived sample because in most such samples there will be a great excess of radiolabeled free inositol to be removed. The Am-Preps are converted from the chloride to the formate form by pumping 10 ml of 3.0 M ammonium formate/0.1 M formic acid followed by water. Am-Preps do not store well after usage, and we recommend new ones for each experiment. Multiple elution can be used in a single experiment (i.e., recycling and using again for a different sample), but some batches of Am-Preps are reliable (>95% separation of $InsP_3$ and $InsP_4$) only for a single sample and in general multiple elution is not advisable. We routinely include a few thousand disintegrations per minute of $[^{32}P]Ins(1,4,5)P_3$ (prepared as below) as an internal standard to monitor the reliability and performance of the anion exchanger. This can be added to the cellular lysate before its transfer from culture wells in order to estimate sample recovery.

A principal drawback of using these prepackaged columns is that if samples are processed one at a time, much of the time saved by the more rapid elution[12] (10 min/sample) is lost. Figure 2 shows a way around this disadvantage, and that is to run 10 Am-Preps concurrently. This set-up employs a Watson-Marlow 10-channel peristaltic pump (model 5025, Falmouth, Cornwall, England). We set the pump to deliver about 0.1

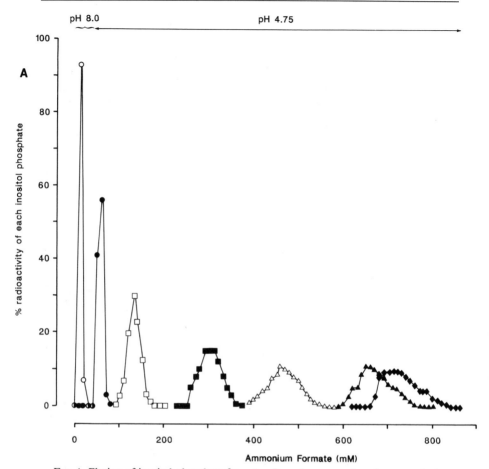

FIG. 1. Elution of inositol phosphate from Am-Preps by ammonium formate solutions. (A) Gradient elution: Two milliliters of each solution was passed through Am-Preps loaded with inositol phosphates as described in the text. Note that the solutions from 10 to 40 mM were buffered to pH 8.0. GroPIns, (○); Ins(3)P, (●); Ins(1,4)P_2, (□); Ins(1,4,5)P_3, (■); Ins(1,3,4,5)P_4 (△); Ins(1,3,4,5,6)P_5, (▲); InsP_6, (◆). (B) Isocratic batch elution: In these experiments, 10 ml of each solution shown here passed through the Am-Preps. Symbols are as in (A).

ml/sec (and find the error to be ± 1% by volume). We also have incorporated a suitable electronic timer device to deliver a constant volume of each eluant.

The samples are loaded by a 10-channel manifold, and elution can either use the same manifold (the eluting solutions are placed in a glass trough) or the lines to the Am-Preps can be switched by T-splitters and

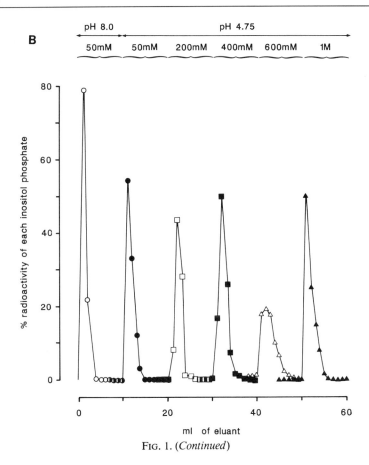

FIG. 1. (*Continued*)

cord clamps to a bunched arrangement so that beakers can be used. Note that the Am-Preps are held by syringes; the advantage of this is that bubbles can easily be removed if necessary. Scintillation vials are used to collect fractions. Using this set-up, 10 samples can be run as quickly as 1, so that a complete cycle of loading, washing, elution of GroPIns and $InsP_{1-6}$ can be achieved in about 10 min.

Second Stage of Analysis: Isocratic Ion-Exchange High-Pressure Liquid Chromatography (HPLC)

Up to a third of each of the various InsP fractions (usually 6 ml total from Am-Preps or 10 ml from Dowex[1,2]) is used to determine eluted

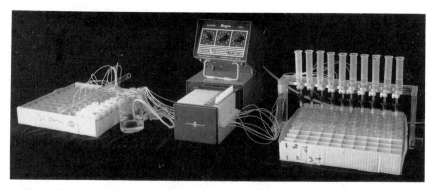

Fig. 2. Experimental set-up for processing 10 Am-Preps at the same time. On the left, a 10-channel manifold is loading 10 samples. Next to that is a beaker with water for washing through the Am-Preps after sample loading, which can then be filled with the various eluting solutions in turn. The pumping is switched to this beaker by freeing a tube clamp and transferring the clamping to the manifold. In the centre is a Watson–Marlow 10-channel peristaltic pump, controlled by an automated timer (behind). The 10 tubes are then linked (on the right) to 10 T-piece joints to which are attached, above, syringes which can be brought into use to remove air bubbles from the Am-Preps should they occur. Below, the Am-Preps are linked through syringe adaptors (Amersham).

radioactivity. As discussed above, this is to ensure that the experiment is worth the investment of HPLC time. The remainder, as required, is then injected onto a Partisil SAX-10 column (25 × 0.65 cm, Technicol, Ltd., Stockport, England) running isocratically with sodium phosphate buffered to pH 3.8 with phosphoric acid.[13] The presence of ammonium formate in the fractions from Am-Preps or Dowex (except for $InsP_1$, see below) does not interfere with the resolution of the various isomers.

For example, Fig. 3 shows a separation of InsP isomers by injection of Dowex fractions onto an isocratically running Partisil SAX-10 column. Figure 3A shows that $InsP_1$ isomers, Ins(1)P and Ins(4)P (prepared as in Ref. 13), are resolved when 4 ml (injected as two sequential 2-ml aliquots using a 2-ml loading loop) of 0.05 M ammonium formate/0.025 M formic acid (= the solution in which InsP elutes from Dowex,[2] diluted fivefold with water) is loaded onto the column running with water. After loading of the last aliquot, the eluting buffer is then switched to 0.04 M NaH_2PO_4. The collection window is initially determined by prior trial runs of an AMP standard (which is then included in each sample to monitor column fidelity) and can be confined to only 5 min around AMP, with 0.25 min/fraction, i.e., 20 fractions/sample; if the HPLC column is not near the end of its life, the elution time of AMP will vary by no more than 1 fraction over consecutive runs. In practice, therefore, 2 ml of the InsP fraction from

[13] K. A. Wreggett and R. F. Irvine, *Biochem. J.* **262,** 997 (1989).

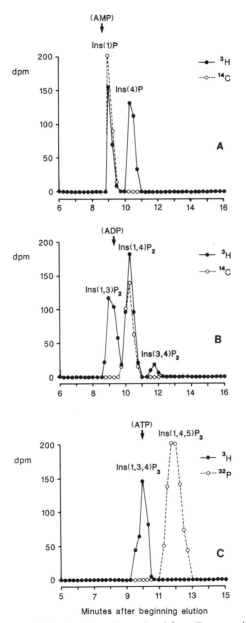

FIG. 3. Isocratic HPLC of inositol phosphates eluted from Dowex anion-exchange resins. (A) InsPs, (B) InsP$_2$s, (C) InsP$_3$s. For details of injection an elution protocols, see the text. Positions of nucleotide markers are shown by arrows. Each run contained an internal standard of an inositol phosphate: (A) [^{14}C]Ins(3)P, (B) [^{14}C]Ins(1,4)P$_2$, (C) [^{32}P]Ins(1,4,5)P$_3$. The HPLC column was a Partisil SAX-10 packed by Technicol, Limited (Stockport, England).

a Dowex column is diluted by 4 vol of water, and is injected either into a 4-ml loading loop, or as 2×2 ml aliquots with water running through the column. The water is then switched to 0.04 M NaH$_2$PO$_4$ and then during a suitable time window, bracketing AMP, fractions are collected.

Figure 3 (B and C) shows similar isomer separations for InsP$_2$ and InsP$_3$ isomers. Here the Dowex eluates do not have to be diluted before injection. InsP$_2$ fractions (2×2 ml in 0.1 M formic acid, 0.4 M ammonium formate) are injected with the column running at 0.1 M NaH$_2$PO$_4$ and eluted at 0.24 M NaH$_2$PO$_4$; ADP is used as a marker to judge the collection window. Note that using this elution regime, Ins(4,5)P$_2$ elutes after Ins(3,4)P$_2$, and Ins(1,5)P$_2$ elutes between Ins(1,4)P$_2$ and Ins(3,4)P$_2$ (unpublished observations).

Isomers of InsP$_3$ (Fig. 3C), with ATP as the internal marker, were loaded as 3×2 ml injections (in 0.1 M formic acid, 1.0 M ammonium formate) with the column running at 0.35 M NaH$_2$PO$_4$, switched to 0.55 M NaH$_2$PO$_4$. Not shown in the figure, but shown in independent runs, we found that Ins(3,4,5)P$_3$ elutes just after (not separated completely from) Ins(1,4,5)P$_3$, and Ins(2,4,5)P$_3$ elutes after Ins(1,4,5)P$_3$ with baseline separation. In practice, in acutely labeled cells with few intermediates accumulating from the synthesis of Ins(1,3,4,5,6)P$_5$ and InsP$_6$,[3] the only likely isomers to be analyzed are Ins(1)P, Ins(4)P, Ins(1,3)P$_2$, Ins(1,4)P$_2$, Ins(3,4)P$_2$, Ins(1,3,4)P$_3$, and Ins(1,4,5)P$_3$, and these are all illustrated in Fig. 3.

It is important to emphasize two points here. First, the loading and eluting conditions described here are for experiments in which the amounts of radioactivity in inositol phosphates are very low, so that maximal loading of Dowex eluates (4–6 ml) is required. If only 2 ml or less is to be loaded with InsP$_1$, InsP$_2$, or InsP$_3$, the samples can be loaded with the column already running in the respective eluting conditions (0.04, 0.24, and 0.55 M NaH$_2$PO$_4$, pH 3.8, respectively), i.e., no switching of solutions is required. Second, it is essential, before leaving a column to inject automatically with a short collection window, to check with the relevant adenosine nucleotide standard that the column is running reproducibly with a good retention time.

Discussion

The advantages of two-stage processing of samples have been discussed in the Introduction, and here we have described an alternative to Dowex columns for the first stage, and a universal (to follow either Am-Preps or Dowex) second stage which is quick, amenable to automation, and which minimizes the number of fractions to be collected. We should note that

InsP$_3$ is often eluted from Dowex in 0.8 M ammonium formate/0.1 M formic acid, and then InsP$_4$ in 1.0 M ammonium formate/0.1 M formic acid, and although we have not explored this here, we see no reason in principle why InsP$_4$ isomers should not subsequently be separated from this eluate by HPLC on Partisil SAX or WAX columns (see, e.g., Refs. 14 and 15). This isomer separation has been shown to delineate more exactly the large increase in amounts of Ins(1,3,4,5)P$_4$ which is generated by stimulation of cells in which high unchanged levels of Ins(3,4,5,6)P$_4$ masked that increase almost completely.[15]

Preparation of Internal Standards

Although adenine nucleotides are very useful as indicators of column behavior and for identifying collection windows, they are not reliable as markers for identification of particular inositol phosphate isomers. In our original method[5] we found that in about 99 out of 100 runs, ATP coelutes with Ins(1,3,4)P$_3$; but in some columns we have found instances where the separation of Ins(1,3,4)P$_3$ and Ins(1,4,5)P$_3$ is excellent, but both of these elute several minutes after ATP. Needless to say, although this is extremely rare, one could lose InsP$_3$ isomers this way. It is in fact an inescapable principle of chromatography that there is nothing as good as an internal standard of one of the compounds that you are actually trying to separate. For this reason, we always include in our sample for InsP$_3$ analysis an internal standard of [^{32}P]Ins(1,4,5)P$_3$. This can be bought from commercial sources, but, with a half-life of only 14.2 days, is expensive to keep in stock. We describe here a simple method for its preparation from [γ-^{32}P]ATP; the latter is readily available and, as specific activity is not paramount, we can invariably scrounge "expired" ATP from other laboratories for no cost.

The method for preparation is to make an extract from brain enriched in 1-phosphatidylinositol-4-phosphate kinase [PtdIns(4)P kinase] [and free of back reactions such as those catalyzed by PtdIns(4,5)P$_2$ phosphomonoesterase and 1-PtdIns(4,5)P$_2$ phosphodiesterase] and phosphorylate PtdIns(4)P with [γ-^{32}P]ATP. The PtdIns [5-^{32}P]PtdIns(4,5)P$_2$ is then degraded by sequential removal of the fatty acids and the glycerol moiety to yield [5-^{32}P]Ins(1,4,5)P$_3$.

[14] T. Balla, A. J. Baukal, G. Guillemette, R. O. Mogan, and K. J. Catt, *Proc. Natl. Acad. Sci. U.S.A.* **83**, 9323 (1986).
[15] L. R. Stephens, P. T. Hawkins, C. J. Barker, and C. P. Downes, *Biochem. J.* **253**, 721 (1988).

PtdInsP Kinase Fraction

This is prepared by a modification of the method of Smith.[16]

Requirements.

HPLC ion-exchange column (BioGel TSK-DEASE-5-PW, 75 × 7.5 mm)

Buffer A: 250 mM sucrose, 50 mM ammonium acetate, 50 mM Tris/acetate, 10 mM Mg^{2+} acetate

Buffer B: 250 mM sucrose, 1 M ammonium acetate, 50 mM Tris/acetate, 10 mM Mg^{2+} acetate

The pH of buffers A and B may be adjusted to 7.4 with acetic acid, and both buffers should be filtered and degassed.

Rat brain (cerebrum and cerebellum from a 200-g rat) homogenized in 5 ml buffer A

Procedure

1. Centrifuge at 100,000 g for 60 min at 4°.
2. Decant supernatant and store on ice.
3. Equilibrate the DEAE HPLC column at 0.5 ml/min for 60 min with buffer A.
4. Load the brain supernatant onto the DEAE column using a 5-ml loop, as 2 × 2.6-ml aliquots. As the second batch is loaded the program is initiated as follows (flow rate = 1 ml/min; arrows indicate sampling times):

Time (min)	A (%)	B (%)
Zero	100	0
14	100	0
→ 14	75	25
22	75	25
→ 22	55	45
30	55	45
→ 30	0	100
32	0	100
35	100	0 (linear decrease over 3 min)

5. A 4-ml "window" is collected from 26 to 30 min. This eluate, containing the PtdIns(4)P kinase activity, is stored in aliquots, frozen, or used directly. This whole procedure takes 2 hr or less, and fresh enzyme is most active, though it keeps well frozen.

[16] P. M. Smith, Ph.D. thesis. Univ. of Cambridge, Cambridge, England, 1989.

If the column performance deteriorates after a number of preparations it can be cleared (usually) with 100 ml containing 50 mg pepsin, 2.92 g NaCl, 0.57 ml acetic acid. Also, 0.2 M NaOH can be passed through as an alternative (we find a combination of one after the other is most efficacious).

The preparation of ^{32}P-labeled PtdIns(4,5)P$_2$ is then as follows.

6. Combine

200 nmol PtdIns(4)P
2 μmol pig liver phosphatidylethanolamine

in chloroform and dry down by blowing N$_2$ over them as a mixture. The lipid is then suspended by adding 0.5 ml water and sonicating the tube in a sonicating bath.

7. Add to this

50 μl 1 M Tris/acetate, pH 7.4
100 μl 200 mM Mg^{2+} acetate
500 μl DEAE PtdIns(4)P kinase fraction
  5 μl 10 mM ATP
  5 μl [^{32}P]ATP (50 μCi)
140 μl H$_2$O

Incubate for 4 hr at room temperature, then add 200 nmol PtdIns(4,5)P$_2$ as a carrier.

8. Add

5 ml chloroform/methanol (1:2, v/v)
20 μl concentrated HCl

Let stand for 5 min, then add

1.42 ml chloroform
1.67 ml 1M HCl

9. Centrifuge, remove upper phase, and wash the lower phase (Whirlimix, centrifuge) six times with chloroform/methanol/1 M HCl (3/48/47, v/v).

10. Take to dryness, and incubate for 1 hr at 59° with 3 ml of Clarke and Dawson's[17] deacylating reagent. Process exactly as in Clarke and Dawson,[17] using volumes in proportion to 3 ml of deacylating reagent. The final 1 ml of aqueous solution of deacylated phospholipids then has the glycerol moieties removed (based on Ref. 18) by adding 1 ml 50 mM sodium periodate. Incubate at room temperature for 30 min. Add 150 μl

[17] N. Clarke and R. M. C. Dawson, *Biochem. J.* **195**, 301 (1981).
[18] D. M. Brown and J. C. Stewart, *Biochim. Biophys. Acta* **125**, 413 (1966).

10% (w/v) ethylene glycol, wait 15 min, then add 0.5 ml fresh 1% (w/v) 1 : 1 dimethylhydrozine buffered to pH 4 with formic acid.[18] After 2 hr, pour this solution through 30 ml of Dowex-50 cation-exchange resin in the H^+ form, adjust the pH of the solution which passes through to 6 with KOH.

The $[^{32}P]Ins(1,4,5)P_3$ is not entirely pure at this stage, and for use as a substrate for $Ins(1,4,5)P_3$-5-phosphates or $Ins(1,4,5)P_3$-3-kinase it should be purified by HPLC. As an internal marker for HPLC it is suitable without further purification.

Acknowledgments

We are grateful to Andy Letcher for help with some of the standards and HPLC, to Nullin Divecha for refinements in the $[^{32}P]Ins(1,4,5)P_3$ preparation, and to Susan Porter for preparing the manuscript.

Section III
Pharmacological Agents in Epithelial Transport

[41] Pharmacological Agents of Gastric Acid Secretion: Receptor Antagonists and Pump Inhibitors

By BJÖRN WALLMARK and JAN FRYKLUND

Introduction

One important aspect of peptic ulcer therapy and of other acid-related disorders is reduction of gastric acid. This can be achieved by surgery, by agents that neutralize gastric acid, and by agents that inhibit gastric acid secretion. Pharmacological inhibition of gastric acid secretion includes several different classes of agents, such as anticholinergics, histamine H_2 receptor antagonists, and inhibitors of the gastric proton pump, the H^+,K^+-ATPase. The discovery that this enzyme constitutes the proton pump of the stomach[1-3] has formed the basis for a new pharmacological approach to the inhibition of gastric acid secretion: the development of H^+,K^+-ATPase inhibitors. This class of compounds falls into two main groups: the acid-activated sulfhydryl group-directed inhibitors and the reversible K^+-site antagonists. The former group is best characterized by omeprazole, while the latter class of drugs may be best represented by SCH 28080 (see Fig. 1).

The inhibitory effect of the omeprazole class of H^+,K^+-ATPase inhibitors shown in Fig. 1, depends on (1) their weak base properties, which allow them to accumulate in the acidic compartments of the parietal cell, (2) their acid-catalyzed conversion to form an inhibitor of the H^+,K^+-ATPase (for omeprazole, this is the sulfenamide derivative denoted by I in Fig. 1), and (3) their subsequent reaction with lumenally accessible sulfhydryl groups of the H^+,K^+-ATPase (Fig. 1). By this mechanism the mother compound is rearranged in acid to form a new SH-directed compound, and can therefore be described as an acid-activated prodrug.[4-6]

The K^+-site antagonists differ from the omeprazole-type drugs in their mechanism of action. They reversibly inhibit the H^+,K^+-ATPase by com-

---

[1] A. L. Ganser and J. G. Forte, *Biochim. Biophys. Acta* **307**, 169 (1973).
[2] G. Sachs, H. H. Chang, E. Rabon, R. Schachmann, H. Lewin, and G. Saccomani *J. Biol. Chem.* **251**, 7690 (1976).
[3] B. Wallmark, H. Larsson, and L. Humble, *J. Biol. Chem.* **260**, 13681 (1985).
[4] B. Wallmark, B.-M. Jaresten, H. Larsson, B. Ryberg, A. Brändström, and E. Fellenius, *Am. J. Physiol.* **245**, G64 (1983).
[5] B. Wallmark, A. Brändström, and H. Larsson, *Biochim. Biophys. Acta* **778**, 549 (1984).
[6] P. Lindberg, P. Nordberg, T. Alminger, A. Brändström, and B. Wallmark, *J. Med. Chem.* **29**, 1327 (1986).

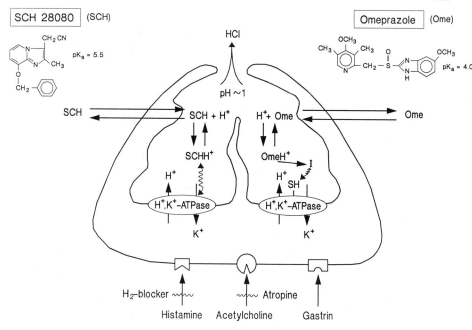

FIG. 1. Inhibition of gastric H^+,K^+-ATPase by omeprazole and SCH 28080.

petitively interacting with the extracellular K^+-activation site of the enzyme. The best studied compound of this class is SCH 28080 (Fig. 1).[7-10]

In this chapter we will describe the inhibitory characteristics in some representative experimental models commonly used for the evaluation of inhibitors of acid secretion.

Isolated Gastric Mucosa

This model is an organ preparation and allows studies of acid secretion without the interference from blood and nerve factors. Acid secretion by the intact gastric mucosa is readily measured by titration of the secreted acid. The guinea pig preparation is particularly suitable, since the gastric mucosa is thin, and therefore easily oxygenated, and it is highly sensitive to histamine.

---

[7] C. K. Scott and E. Sundell, *Eur. J. Pharmacol.* **112**, 268 (1985).
[8] W. Beil, I. Hackbarth, and K.-F. Sewing, *Br. J. Pharmacol.* **88**, 19 (1986).
[9] D. J. Keeling, S. M. Laing, and J. Senn-Bilfinger, *Biochem. Pharmacol.* **37**, 2231 (1988).
[10] B. Wallmark, C. Briving, J. Fryklund, K. Munson, R. Jackson, J. Mendlein, E. Rabon, and G. Sachs, *J. Biol. Chem.* **262**, 2077 (1987).

Preparation and Responsiveness to Different Secretagogues

Male guinea pigs weighing 300–500 g can be used. The animals are starved overnight but allowed free access to water. After cervical dislocation, the stomach is removed, opened, and washed in a medium containing (in mM): Na$^+$, 140.9; K$^+$, 5.8; Ca^{2+}, 2.6; Cl$^-$, 125.8; SO$_4^{2-}$, 1.2; HCO$_3^-$, 24.9; HPO$_4^{2-}$, 1.2; glucose, 25; at pH 7.4. The solution is gassed with 95% O$_2$ and 5% CO$_2$. Two pieces of the fundus (one for control purposes) of each stomach are excised and the muscular layer (serosal side) removed through careful dissection. During the preparation the mucosa is repeatedly moistened with the medium. The pieces of stripped mucosa are mounted on plastic funnels with the secretory side (mucosal side), 2 cm^2, facing inward. The mucosa is fixed to the funnel by tying a ligature around the neck of the funnel. The funnel is then immersed in an organ water bath containing 40 ml of the above medium. The bottom of the bath is provided with inlets for gassing (95% O$_2$, 5% CO$_2$) and a catheter for drug administration or exchange of serosal solution. Three to 5 ml of mucosal solution is added. The solution is identical to the serosal medium with the exception that NaHCO$_3$ and K$_2$HPO$_4$ are replaced by equimolar concentrations of NaCl and KCl. The unbuffered mucosal solution is gassed with 100% O$_2$. The mucosal and serosal solutions are adjusted to the same level and the organ bath temperature is set at 36°. Acid secretion into the unbuffered medium inside the funnel is measured by continuous titration (end point pH 3.8) with 0.025 M NaOH, using a Radiometer pH stat (pHM 62, TTT 60), and autoburette (ABU 80).[11,12] (Copenhagen, Denmark).

About 30 min after mounting, the isolated gastric mucosa spontaneously secretes acid at a rate of 2–3 μmol H$^+$/(cm$^2 \cdot$ hr). Increasing concentrations of histamine are added from the serosal side until the maximal stimulatory concentration of 2×10^{-5} M is reached (EC$_{50}$ = 6×10^{-6} M), which increases acid secretion to two or three times the basal rate.[12] The steady state levels for each histamine concentration are attained after about 20 min.

The response to the intracellular activator N^6,O^2-dibutyryladenosine-3′,5′-cyclic adenosine monophosphate (dbcAMP) is both low and slow. Pretreatment with the phosphodiesterase inhibitor 3-isobutyl-1-methylxanthine (IMX) at a concentration of 10^{-5} M enhances the response to 1 mM dbcAMP over a 4-hr period to about the same magnitude as for histamine.[12,13] The response to carbacholine at a maximal effective con-

---

[11] S.-E. Sjöstrand, B. Ryberg, and L. Olbe, *Naunyn-Schmiedebergs Arch. Pharmakol.* **330,** 37 (1977).
[12] H. Larsson, E. Carlsson, and G. Sundell, *Digestion* **29,** 12 (1984).
[13] S.-E. Sjöstrand, B. Ryberg, and L. Olbe, *Acta Physiol. Scand., Spec. Suppl.* p. 181 (1977).

centration of 2×10^{-7} M is about 50% of the stimulatory effect of histamine. Pentagastrin has no stimulatory effect on the preparation.

Effect of Cimetidine and Omeprazole

Cimetidine, at 10^{-5} M, displaces the histamine concentration response curve to the right in a parallel manner, resulting in a K_B value of $2.4 \pm 0.28 \times 10^{-6}$ M. Omeprazole depresses the maximal response to histamine in a concentration-dependent manner (50% inhibition at $5 \times 10^{-7} M$) without any displacement of the histamine concentration response curve.[12]

In isolated mucosa stimulated with dbcAMP, cimetidine at 10^{-5} M was without effect, whereas omeprazole, at a concentration of 10^{-6} M, inhibited acid secretion by 65% (Fig. 2).[12] This indicates that omeprazole inhibits acid secretion in the parietal cell at a point beyond the histamine H_2 receptor-mediated production of cAMP.

Since the proton pump catalyzes the exchange of cytosolic H^+ for lumenal K^+, inhibition of the pump will lead not only to a decreased proton secretion but also to a decreased K^+ absorption. Accordingly, omeprazole, in contrast to cimetidine, induces an increase in the K^+ outflow rate in parallel with inhibition of stimulated acid secretion.[4]

Isolation of Gastric Glands

The gastric gland can be isolated through enzyme digestion of the surrounding lamina propria. Isolation of rabbit gastric glands[14] is an easy and reproducible method which demands a minimum of equipment for the isolation procedure. From one rabbit stomach enough material for several hundred incubations is obtained. The parietal cells constitute 33 to 36% of the volume of the rabbit gastric gland. Mucous neck cells, chief cells, and some endocrine cells have also been identified.[14,15] Gastric glands can also be prepared from other species, including man.[16] Since acid secretion in the isolated gland preparation cannot be directly titrated, the acid formed has to be estimated by other methods, such as accumulation of permeable weak bases or by quantification of respiration.

In order to prepare isolated gastric glands, a rabbit is anesthetized by an intravenous injection of mebumal (30 mg/ml) via the marginal ear vein. The abdomen is opened with a midline incision and a ligature is loosely

---

[14] T. Berglindh and K. J. Öbrink, *Acta Physiol. Scand.* **96**, 401 (1976).
[15] J. Fryklund, H. F. Helander, B. Elander, and B. Wallmark, *Am. J. Physiol.* **254**, G399 (1988).
[16] E. Fellenius, B. Elander, B. Wallmark, U. Haglund, H. F. Helander, and L. Olbe, *Clin. Sci.* **64**, 423 (1983).

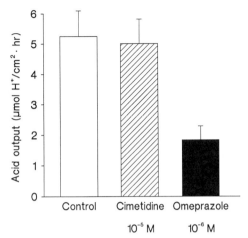

FIG. 2. Effects of omeprazole and cimetidine on acid secretion in isolated guinea pig gastric mucosa. Acid secretion was stimulated with 1 mM N^6,2′-O-dibutyryladenosine 3′,5′-cyclic monophosphate (dbcAMP) and 10 μM 3-isobutyl-1-methylxanthine (IMX). Values are mean ± SE of six to eight preparations. (Data from Larsson et al.[12])

placed around the intestine. The abdominal aorta is then cannulated in retrograde direction with a polythene catheter (PP 280, inner diameter 2.15 μm) and 5.0 ml of a heparin solution (250 IE/ml) is subsequently injected. After 10 sec the heparin syringe is removed and the rabbit is bled through the cannula (~ 50 ml). The ligature around the intestine (mesenteric artery) is tightened and the perfusion with oxygenated phosphate-buffered saline (PBS) containing (in mM): NaCl, 149.6; $K_2HPO_4 \cdot 3H_2O$, 3.0; $NaH_2PO_4 \cdot H_2O$, 0.64 (at pH 7.4 and 37°) is performed under high pressure (80 kPa). The chest is quickly opened and the thoracic aorta is clamped just above the diaphragm. This directs most of the PBS solution through the gastric blood vessels. When the liver starts to stiffen an incision is made in one of the lobes to allow free outflow of the perfusate. When the stomach is exsanguinated and shows signs of edema, after about 500 ml PBS or less, it is rapidly removed and the antrum is cut off. The stomach is opened, emptied, and rinsed quickly, first under 37° tap water and then in 37° PBS, and finally gently blotted with filter paper. The mucosa is easily separated from the underlying submucosal layer by carefully scraping with a pincette. The mucosa is transferred to a small beaker containing 10 ml PBS and cut into 2- to 3-mm pieces with a pair of scissors. Thereafter the mucosa is washed repeatedly in PBS and transferred into a 200-ml flask containing 50 ml collagenase solution (in mM): NaCl, 130; $NaHCO_3$, 12.0; $NaH_2PO_4 \cdot H_2O$, 3.0; Na_2HPO_4, 3.0; $K_2HPO_4 \cdot 3H_2O$, 3.0; $MgSO_4 \cdot 7H_2O$, 2.0; $CaCl_2 \cdot 2H_2O$, 1.0; indomethacin, 0.01; N^α-tosyl-L-lysine chloromethyl ketone (clostripain and trypsin inhibitor), 0.1; the solution also contains albumin (1.0 mg/ml), glucose (2.0 mg/ml), and, finally, 12 units (25–50 mg) collagenase (type IA; Sigma, St. Louis, MO). The flask is gassed with 100% oxygen, sealed and incubated at 37° for 30–60 min while being gently stirred with a magnetic bar. The digestion is complete when no crude material is detected in the flask, and when free single glands (200–800 μm) can be seen under a light microscope. The suspension is then filtered through a nylon cloth (120 mesh) into a beaker. The glands are washed three times in the incubation medium (in mM): NaCl, 132.4; KCl, 5.4; $NaH_2PO_4 \cdot H_2O$, 1.0; Na_2HPO_4, 50; $MgSO_4 \cdot 7H_2O$, 1.2; $CaCl_2 \cdot 2H_2O$, 1.0, to which albumin (2 mg/ml) and glucose (2 mg/ml) are added. After the last wash the glands are allowed to settle properly, and the volume of the loose pellet is estimated with a pipet. Addition of five parts of the incubation buffer results in a concentration of about 10 mg gland dry wt/ml buffer.

Determination of Acid Formation in Gastric Glands

Acid formation in gastric glands is assessed by measurement of oxygen consumption,[17] glucose oxidation,[18] or by accumulation of dimethylamine [^{14}C]aminopyrine.[17]

Oxygen Consumption. In order to supply the proton pump with the necessary amount of ATP, the parietal cell has an extremely high oxidative capacity, which is reflected morphologically in the high mitochondria content (30–40% of the cytoplasmic volume). Accordingly, a close correlation between acid secretion and oxygen consumption has been demonstrated in several experimental models.[19,20] The high oxidative rate during acid secretion can also be measured by using a ^{14}C-labeled substrate, i.e., glucose.[18]

The oxygen consumption is measured with a differential respirometer (Gilson, Middletown, WI, USA). The respirometer consists of a water bath with holders for 14–20 Warburg flasks. One milliliter of the above gland suspension is added to 2 ml of incubation medium containing test agents. One hundred microliters of 20% KOH solution is applied to a folded filter paper and is used as CO_2 absorber. The mixture is allowed to equilibrate with the reference chamber for 30 min at 37°. Oxygen consumption is then recorded at 10-min intervals for 60 min. A flask without glands is run as a blank. A correction for ambient temperature and pressure is made accord-

[17] T. Berglindh, H. F. Helander, and K. J. Öbrink, *Acta Physiol. Scand.* **97**, 401 (1976).
[18] J. Fryklund, K. Gedda, D. Scott, G. Sachs, and B. Wallmark, *Am. J. Physiol.* **258**, G719 (1990).
[19] W. H. Bannister, *J. Physiol. (London)* **177**, 429 (1965).
[20] F. G. Moody, *Am. J. Physiol.* **215**, 127 (1968).

ing to Eq. (1):

$$K = 273(PP_w)/760T \qquad (1)$$

where P is barometric pressure (in mmHg), P_w is vapor pressure of water (in mmHg) at temperature T, and T is temperature at the micrometer (in degrees K).

Oxygen consumption is calculated by multiplying K with the micrometer value. After correction for the blank, oxygen uptake is expressed as microliters O_2 per milligram gland dry weight · hour or converted to nanomoles O_2 per milligram gland dry weight · hour.

The gland dry weight is calculated from the mean weight of eight separate pipettings of the gland suspension used.

Glucose Oxidation. One microcurie of [^{14}C]glucose (uniformly labeled) is added to the incubation mixture. After a 30-min equilibration an incubation vessel is withdrawn and analyzed for $^{14}CO_2$ production from [^{14}C]glucose during this period. In order to release the dissolved carbon dioxide in the medium and to stop the reaction 1.5 ml of 2 M perchloric acid is added at the end of the incubation period, and the flasks are incubated for an additional 30 min. Medium not containing glands is incubated in parallel for the calculation of spontaneous oxidation of [^{14}C]glucose. The $^{14}CO_2$ trapped on the filters is quantified by liquid scintillation counting. The carbon dioxide released is calculated from Eq. (2)

$$\text{Nanomoles } CO_2 = \frac{\text{trapped (dpm} - \text{dpm}_{\text{blank}}) \times 6}{\text{specific activity of [}^{14}C\text{]glucose (dpm/nmol)}} \qquad (2)$$

and expressed as nanomoles CO_2 per milligram gland dry weight · hour. Alternatively, glucose oxidation can be measured in capped glass scintillation vials, where the CO_2 trapper is placed in a micro-test tube inside the vial.[18]

[^{14}C]Aminopyrine (AP) Accumulation. The acid produced by the parietal cell is secreted into the secretory canaliculi or into the gland lumen. The total amount of acid sequestered in the gland is indirectly quantified by the use of a weak base such as AP.[17] The base form is freely permeable across the membranes and becomes much less permeable when protonated. The degree of accumulation of AP is determined by the acidity and the volume of the acidic compartment. The distribution ratio of AP between the gland and the incubation medium thus becomes a measure of the amount of acid present in the gland.

[^{14}C]Aminopyrine (0.2 μCi), test substances, and incubation medium to 1.2 ml are put in a 3-ml conical scintillation tube (Lumax 5125). The incubation is started by adding 0.3 ml of the above gland suspension. To avoid aggregation of the glands, the tubes are capped and incubated in a horizontal position in an orbital shaking water bath. After a 30- or 60-min incubation (see below), samples are centrifuged at room temperature in the same tubes at 2000 g for 2 min. Two hundred microliters of the supernatant is transferred to a new 3-ml tube. The remaining supernatant is immediately siphoned off and discarded. The gland pellet is digested in 250 μl 0.5 M NaOH at 60° for 1–2 hr, or at room temperature overnight. After addition of 3 ml of standard cocktail to the supernatant and the dissolved pellet and vigorous shaking, the samples are counted in a liquid scintillator counter. The extent of AP accumulation is calculated according to Eq. (3) as the distribution ratio of AP between the intraglandular water (IGW) and the incubation medium after correction of supernatant contamination of the pellet. The IGW can be measured by radiolabeled inulin but is usually approximated as 2× gland dry weight.[17,18]

$$\text{AP ratio} = \frac{[\text{dpm pellet}/(2 \times \text{gland dry wt})] - (\text{dpm supernatant}/200)K}{\text{dpm supernatant}/200} \quad (3)$$

where K is wet weight − dry weight − IGW.

To estimate gland weights, 10 tubes are dried at 100° for 1 hr, allowed to cool, and then weighed. The same amount of glands as above is pipetted into the tubes, which are then centrifuged, the supernatants siphoned off, and the wet weights recorded. The glands are then dried at 100° for 1 hr and the dry weights recorded.

Responsiveness to Different Stimuli in the Gastric Gland Preparation

The gastric gland preparation responds to histamine and carbacholine but not to gastrin (Fig. 3). Acid secretion can also be evoked by dbcAMP.

AP Accumulation. The unstimulated AP ratio, often referred to as the basal level, is about 10% of the maximally stimulated level. The addition of histamine stimulates AP uptake, with an EC_{50} value of about 3 μM. The maximal AP ratio is about 150 and occurs at 3×10^{-5} M following a 60-min incubation period. The response to carbacholine is more variable and transient. The EC_{50} value is reached at about 2 μM. The maximal AP ratio is about 50 at 10^{-5} M following a 30-min incubation period. Addition of dbcAMP to the incubation medium increases the basal AP ratio to about 250 at a concentration of 1 mM. The EC_{50} value is about 0.2 mM.[18]

Oxygen Consumption and Glucose Oxidation. The basal consumption of O_2 is about 300 nmol O_2/(mg·hr) [8–10 μl/(mg·hr)] and the unstimulated glucose oxidation measured under the same conditions is about 200 nmol CO_2/(mg·hr). Histamine, carbacholine, and dbcAMP stimulate O_2

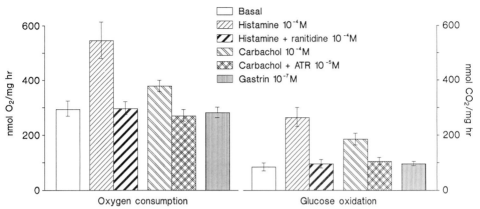

FIG. 3. Stimulation and inhibition of respiration in gastric glands. Oxygen consumption and glucose oxidation were measured simultaneously in a respirometer. Values are mean ± SE of four preparations. (Data from Fryklund et al.[18])

uptake and glucose oxidation in a parallel manner with EC_{50} values similar to those found in AP-uptake studies. The maximal responses were reached at the same concentrations as when the AP technique was used (Fig. 3).[18]

Responsiveness to Different Inhibitors

The basal AP uptake can be inhibited by H^+,K^+-ATPase inhibitors such as omeprazole and SCH 28080, but not by H_2 antagonists.[4,10] The pump inhibitors do not interfere with basal respiration, indicating that they do not interact with basal mitochondrial respiration.

Both omeprazole and SCH 28080 concentration-dependently inhibit dbcAMP-stimulated AP uptake and oxygen consumption in parallel in the gastric gland preparation (Fig. 4). Thus, inhibition of the proton pump reduces the acid accumulation, which is related to the stimulated component of oxygen consumption.

Inhibition of acid accumulation, stimulated by different secretagogues, in the gland preparation can serve as a model for differentiating the inhibitory patterns for receptor antagonists and proton pump inhibitors. In glands stimulated by histamine and the intracellular messenger dbcAMP, omeprazole and SCH 28080 concentration-dependently inhibited both agonists with the same IC_{50} value (omeprazole, 0.5 μM; SCH 28080, 0.2 μM). Cimetidine, on the other hand, was effective only against histamine stimulation, in agreement with its histamine H_2 receptor-blocking properties ($IC_{50} = 10\ \mu M$) (Fig. 5).

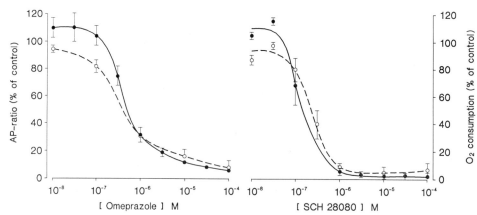

FIG. 4. Effect of omeprazole and SCH 28080 on dbcAMP-stimulated [^{14}C]aminopyrine (AP) uptake (●) and oxygen consumption (○) in isolated gastric glands. Values are mean ± SE of three or four preparations. [Data in part from J. Fryklund and B. Wallmark, *J. Pharmacol. Exp. Ther.* **236**, 248 (1986).]

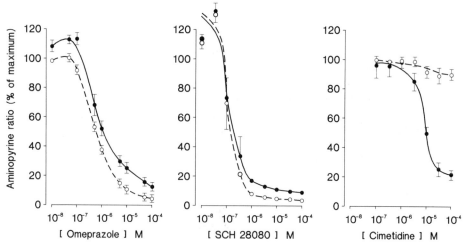

FIG. 5. Inhibitory action of omeprazole, SCH 28080, and cimetidine on histamine-A (●) and dbcAMP-stimulated (○) [^{14}C]aminopyrine uptake in isolated gastric glands. Values are mean ± SE of three or four preparations. (Data from Wallmark et al.[4,10])

Permeable Glands: Preparation and Effect of Different Inhibitors

The parietal cell membrane can be made permeable to large molecules, such as ATP, by exposure to digitonin.[21] This results in a preparation in which acid secretion can be driven by addition of exogenous ATP, even in the presence of oligomycin, suggesting that this substrate is the only energy source needed for acid formation in the gland preparation.

Omeprazole concentration-dependently inhibited the ATP-stimulated acid accumulation measured with AP in the permeable gland preparation. The IC_{50} value of 0.27 μM was close to that found in intact glands. Cimetidine was without significant effect even at a concentration of 100 μM.[4] The inhibitory effect of omeprazole in this model and the lack of inhibitory effect on unstimulated oxygen uptake are evidence that the compound interacts directly with the proton pump. The permeabilized gland system also enables H^+,K^+-ATPase activity to be measured *in situ*.[22]

Effect of Neutralization on the Inhibitory Action of Omeprazole and SCH 28080

Both omeprazole and SCH 28080 are weak bases. Omeprazole has a pK_a of 4 and that of SCH 28080 is 5.5. In the stimulated gastric gland, the pH of the parietal cell secretory canaliculi can be calculated by the AP technique to be about 1.5. This low pH will result in protonation and therefore in formation of a net positive charge of the inhibitors, when present in the acidic secretory canaliculi. Since the charged species have a much more limited permeability than the unprotonated ones, the protonated species will accumulate in these acidic compartments.

The acidic space of the gastric gland can be buffered by using a weak base such as imidazole ($pK_a = 7$). In the gland preparation it has been shown at 3 mM that this compound totally abolishes the AP uptake without having any effect on stimulated oxygen consumption.[23] Thus, imidazole neutralizes the gland without exerting any effect on the proton pump activity. Accordingly, imidazole neutralization results in a shift of the concentration response curve for SCH 28080 to the right,[10] owing to decreased concentration of the protonated compound at its inhibitory lumenal K^+ site of the H^+,K^+-ATPase. Neutralization of the glands prior to the addition of omeprazole results in a near-complete loss of its inhibitory

---

[21] D. H. Malinowska, H. R. Koelz, S. J. Hersey, and G. Sachs, *Proc. Natl. Acad. Sci. U.S.A.* **78**, 5908 (1981).
[22] S. J. Hersey, L. Steiner, S. M. Matheravidather, and G. Sachs, *Am. J. Physiol.* **254**, G856 (1988).
[23] J. Fryklund, K. Gedda, and B. Wallmark, *Biochem. Pharmacol.* **37**, 2543 (1988).

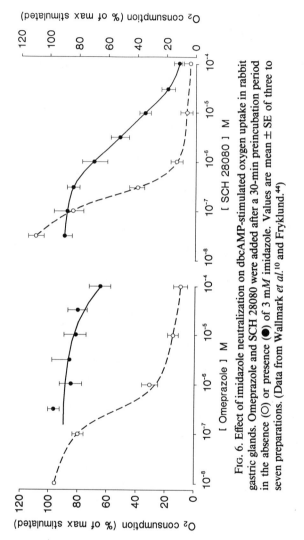

FIG. 6. Effect of imidazole neutralization on dbcAMP-stimulated oxygen uptake in rabbit gastric glands. Omeprazole and SCH 28080 were added after a 30-min preincubation period in the absence (○) or presence (●) of 3 mM imidazole. Values are mean ± SE of three to seven preparations. (Data from Wallmark et al.[10] and Fryklund.[44])

action, since the rate of acid-catalyzed conversion of omeprazole falls drastically[23] (Fig. 6).

Pepsinogen Release

Since the rabbit gastric gland contains a large proportion of chief cells, methodology has also been developed to measure pepsinogen release in the preparation.[24,25]

Human Gastric Glands

H^+,K^+-ATPase inhibitors such as timoprazole[16] and omeprazole inhibit AP uptake concentration dependently, with IC_{50} values of about 50 nM, irrespective of the agonist used.[26] At high concentrations, i.e., 10^{-5} M, omeprazole also reduces the basal accumulation. Cimetidine, on the other hand, does not suppress the basal uptake, nor does it counteract AP uptake stimulated with other agonists than histamine. The IC_{50} value for cimetidine is about 45 μM. The pA_2 value has been estimated to be 5.7–6.1.[27]

Isolated Parietal Cells

Studies of intracellular second messenger levels or of receptor binding to a specific cell require enrichment of the cell in question. Dispersion of gastric mucosal cells and the subsequent enrichment of parietal cells has been described for the dog,[28] rabbit,[29,30] rat,[31] mouse,[32] guinea pig,[33] pig,[34] and frog.[35] The techniques involve blunt separation of the gastric mucosa

---

[24] H. R. Koelz, S. J. Hersey, G. Sachs, and C. S. Chew, *Am. J. Physiol.* **243**, G218 (1982).

[25] E. Skoubo-Kristensen and J. Fryklund, *Naunyn-Schmiedebergs Arch. Pharmakol.* **330**, 37 (1985).

[26] B. Elander, E. Fellenius, R. Leth, L. Olbe, and B. Wallmark, *Scand. J. Gastroenterol.* **21**, 268 (1986).

[27] R. Leth, B. Elander, U. Haglund, L. Olbe, and E. Fellenius, *Am. J. Physiol.* **253**, G497 (1987).

[28] A. H. Soll, *J. Clin. Invest.* **61**, 381 (1978).

[29] J. Fryklund, B. Wallmark, H. Larsson, and H. F. Helander, *Biochem. Pharmacol.* **33**, 273 (1984).

[30] C. S. Chew, *Am. J. Physiol.* **245**, G221 (1983).

[31] W. J. Thompson, L. K. Chang, and G. S. Rosenfeld, *Am. J. Physiol.* **240**, G76 (1981).

[32] L. J. Romrell, M. R. Coppe, D. R. Munroe, and S. Ito, *J. Cell Biol.* **65**, 428 (1975).

[33] K. F. Sewing, P. Harms, G. Schulz, and H. Hanneman, *Gut* **24**, 557 (1983).

[34] S. Mårdh, Y. Song, C. Carlsson, and T. Björkman, *Acta Physiol. Scand.* **131**, 589 (1987).

[35] F. Michelangeli, *in* "Gastric Hydrogen Ion Secretion" (D. K. Kasbekar, W. S. Rehm, and G. Sachs, eds.), p. 212. Dekker, New York, 1976.

from underlying tissue, enzyme treatment, mechanical disruption, and cell separation. Cell separation is performed either by using velocity sedimentation in an elutriator rotor, or separation by density in a gradient medium or a combination of both. The parietal cells are enriched to 50-90%, depending on the species and the technique used.

Isolated Gastric Membranes Containing H^+,K^+-ATPase

Microsomal membrane fractions containing H^+,K^+-ATPase have been purified both from mammals and lower species, and gastric ouabain-insensitive K^+-stimulated ATPase activity has been found in humans, dogs, pigs, rabbits, and rats.[36-40] The most utilized species has been the pig. The preparation of microsomal membrane fractions from the pig gastric mucosa has been developed on a large scale and the enzyme has been purified to near homogeneity as judged by SDS-PAGE.

Effect of H^+,K^+-ATPase Blockers on Proton Transport and on ATPase Activity

Acid-Activated Sulfhydryl Group-Directed Inhibitors of H^+,K^+-ATPase

Investigations of this class of drugs at the level of H^+,K^+-ATPase are best carried out in isolated sided vesicle preparations in order to catalyze conversion of the compounds selectively by the acid in the lumenal space of the vesicle.

Figure 7 illustrates the effect of omeprazole on K^+-stimulated ATPase activity in a sided ion-tight vesicle preparation containing H^+,K^+-ATPase. In this experiment, ATPase was activated and a lumenal oriented proton gradient was generated by the addition of 2 mM Na$_2$ATP to 5.7 μg of vesicle protein, 2 mM HEPES/NaOH buffer, pH 7.4, containing 2 mM MgCl$_2$, 175 mM KCl, and 1 μg valinomycin/μg protein suspended in a total volume of 1 ml. Omeprazole was dissolved in methanol and added in increasing concentrations to the assay medium 5 min after addition of ATP. The final methanol concentration was 0.5%, which, per se, had no

---

[36] G. Saccomani, H. H. Chang, A. A. Mihas, S. Crago, and G. Sachs, *J. Clin. Invest.* **64**, 627 (1979).
[37] J. Nandi, Z. Meng-Ai, and T. K. Ray, *Biochemistry* **26**, 4264 (1987).
[38] G. Saccomani, H. B. Stewart, D. Shaw, M. Lewin, and G. Sachs, *Biochim. Biophys. Acta* **465**, 311 (1977).
[39] J. M. Wolosin and J. G. Forte, *FEBS Lett.* **125**, 208 (1981).
[40] W. B. Im and D. P. Blakeman, *Biochem. Biophys. Res. Commun.* **180**, 635 (1982).

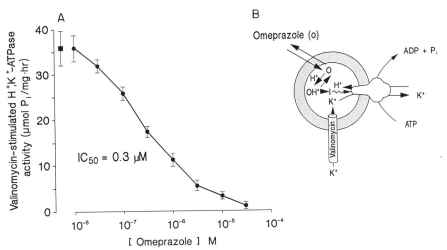

FIG. 7. (A) Effect of omeprazole on H+,K+-ATPase activity in isolated gastric vesicles. (B) The assay was performed in the presence of valinomycin in order to allow entry of K+ into the vesicle lumen and thereby activate the H+,K+-ATPase. OH+, The protonated form of omeprazole; I, the inhibitor generated from omeprazole in acid, the sulfenamide.

effect on the ATPase activity. The assay was performed for 30 min at 37° and the ATPase activity was determined by the release of P_i from ATP, as described elsewhere.[41] It was shown that omeprazole concentration-dependently inhibited ATPase activity, the IC_{50} value being about 0.3 μM.

The effect of omeprazole on H+,K+-ATPase activity was investigated under acidic and nonacidic conditions (Fig. 8), and inhibition of the enzyme activity was shown to require proton transport. The design of these experiments is based on the fact that H+,K+-ATPase is activated by and transports NH_4^+, in addition to K+. Furthermore, NH_4^+ acts as a permeable buffer, due to the rapid permeation of NH_3 into the vesicle lumen. NH_3 is subsequently protonated and thereby effectively neutralizes the vesicle lumen (Fig. 8B). The NH_4^+ generated in the lumen is transported back into the cytosolic medium by H+,K+-ATPase.

In the experiment shown in Fig. 8, 5 μg of vesicle protein is incubated in a medium containing 2 mM Na_2ATP, 2 mM HEPES/NaOH buffer, pH 7.4, 2 mM $MgCl_2$, and 200 mM KCl or 200 mM NH_4Cl. Omeprazole is added in methanol (final concentrations: omeprazole, 10 μM; methanol, 0.5%). The assay is performed for 30 min at 37°. ATPase activity is determined by the release of P_i from ATP. P_i is analyzed as described elsewhere.[41] The experiment demonstrates that NH_4^+ is able to stimulate

---

[41] H. Yoda and L. E. Hokin, *Biochem. Biophys. Res. Commun.* **40**, 880 (1976).

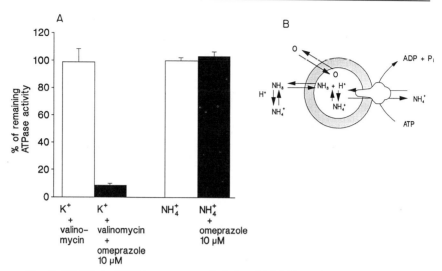

FIG. 8. (A) Effect of NH_4^+ on omeprazole-induced inhibition of gastric H^+,K^+-ATPase. (B) Mechanism for neutralization of the acidic vesicle lumen by NH_4^+. O, Omeprazole.

ATPase to the same maximal level as K^+.[42] In contrast to the inhibition observed with omeprazole in the presence of K^+, no inhibitory effect is found with NH_4^+. These data show that the inhibition of omeprazole is observed only when acidic conditions prevail in the lumen, since under those circumstances no accumulation or acid-catalyzed conversion of omeprazole occur.

The compound converted from omeprazole in acid, the sulfenamide, has been shown to bind to lumenally accessible sulfhydryl groups of the gastric H^+,K^+-ATPase, thereby forming a disulfide bond linkage between the enzyme and the omeprazole-derived inhibitor.[24,42] Binding studies can be performed in isolated gastric vesicles or *in vivo*.[42–44] In the former preparation, a linear relationship has been found between inhibition of both ATPase activity and proton transport and binding of the [^{3}H]omeprazole-derived inhibitor to H^+,K^+-ATPase. At maximal inhibition of ATPase activity, a binding level of about 1.3 nmol/mg of protein is obtained. This corresponds to about two binding sites per phosphorylation site of H^+,K^+-ATPase.

[42] P. Lorentzon, R. Jackson, B. Wallmark, and G. Sachs, *Biochim. Biophys. Acta* **897**, 41 (1987).
[43] D. J. Keeling, C. Fallowfield, and A. H. Underwood, *Biochem. Pharmacol.* **36**, 339 (1987).
[44] J. Fryklund, K. Gedda, and B. Wallmark, *Biochem. Pharmacol.* **37**, 2543 (1988).

Interaction of Reversible K^+-Site Antagonists with H^+,K^+-ATPase

The drugs in this class differ from the omeprazole-type drugs in their mechanism of action inasmuch as they reversibly and competitively interact with the extracellular K^+-activation site of H^+,K^+-ATPase, as described in Fig. 1.

Since these compounds do not require acid-induced conversion for their activity, studies can be performed in ion-permeable gastric vesicles containing H^+,K^+-ATPase. The effect of SCH 28080 on H^+,K^+-ATPase activity is shown in Fig. 9. In this experiment, 10 μg membrane protein/ml is incubated in the presence of 5 mM piperazine-N,N'-bis-2-ethanesulfonic acid (PIPES)/Tris buffer, pH 7.4, 2 mM Na$_2$ATP, 2 mM MgCl$_2$, and KCl concentrations varying between 3 and 0.3 mM. The assays are performed for 10 min at 37°. The ATPase activity is determined as described by Yoda and Hokin.[41] The results are represented as an Eadie–Hofstee plot. SCH 28080 was found to increase the observed K_m value for K^+, whereas no effect was found on the observed maximal velocity, V_{max}. These data are consistent with a competitive interaction between SCH 28080 and the K^+ activation of H^+,K^+-ATPase. The K_i value for SCH 28080 estimated from this type of analysis was found to be 56 nM. Since SCH 28080 is a competitive ligand with respect to K^+, it is important to observe that the

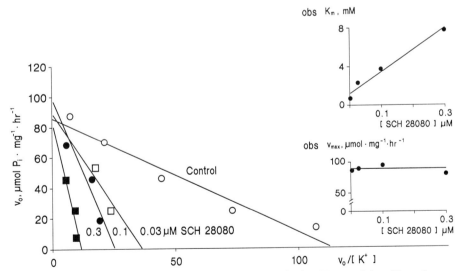

FIG. 9. Effect of SCH 28080 on K^+-stimulated H^+,K^+-ATPase activity. (Data from Wallmark et al.[10])

apparent potency of this type of compound varies with the medium K^+ concentrations.

As in the case of omeprazole, the binding stoichiometry for radiolabeled SCH 28080 has been investigated.[45] At maximal binding levels, which occur in the presence of Mg^{2+} or at low ATP concentrations, about two binding sites exist per phosphorylation site. The reported K_d for binding, 45 nM, correlates well with the above-mentioned K_i for inhibition of K^+-stimulated ATPase activity.

SCH 28080 and related compounds all appear to inhibit H^+,K^+-ATPase in the extracellular (lumenal) domain. This observation is based on a series of experiments involving membrane-impermeable analogs of SCH 28080, which exhibit high-affinity K^+-competitive interaction in ion-permeable preparations, but are essentially without effect in ion-tight preparations.[46] In addition, experiments with radiolabeled SCH 28080 have shown that binding of the drug can be displaced only by lumenal K^+, and not by medium K^+, strongly suggesting that the inhibitory binding site for these drugs is located in the lumenal K^+-activating sector of H^+,K^+-ATPase.

The reversible K^+ site antagonists are thought to bind to an E2 conformation of the H^+,K^+-ATPase, since both the binding affinity and the binding level were favored by ligands that forced the enzyme into an E2 conformation. The binding of SCH 28080 is thought to occur to either the E2 dephosphoenzyme form, as suggested earlier, or to the E2 phosphoenzyme form, as recently suggested based on binding experiments.[10,45]

Relation between Inhibition of Gastric Acid Secretion and Blockade of Gastric H^+,K^+-ATPase

When omeprazole is administered *in vivo* it has been shown that a good correlation exists between inhibition of H^+,K^+-ATPase and gastric acid secretion.[3] For inhibitors of the omeprazole type, which bind covalently to H^+,K^+-ATPase, the enzyme can be isolated from the omeprazole-treated animals without loss of inhibition. However, for inhibitors of the reversible type, this will prove more difficult since these drugs bind reversibly to the H^+,K^+-ATPase and the inhibition of the enzyme will be largely lost during purification.

Acknowledgment

The authors are grateful to Dr. Pia Lorentzon for the use of data shown in Figs. 7 and 8.

[45] D. J. Keeling, A. G. Taylor, and C. Schudt, *J. Biol. Chem.* **264**, 5545 (1988).
[46] C. Briving, B.-M. Andersson, P. Nordberg, and B. Wallmark, *Biochim. Biophys. Acta* **946**, 185 (1988).

[42] Cation Transport Probes: The Amiloride Series

By THOMAS R. KLEYMAN and EDWARD J. CRAGOE, JR.

Introduction

Amiloride and a large number of amiloride analogs were synthesized in the mid-1960s while in search of diuretic agents which possessed both natriuretic as well as antikaliuretic properties.[1,2] Amiloride inhibits the Na^+ channel present in urinary epithelia, and both the natriuresis and antikaliuresis observed clinically are a direct result of inhibition of the Na^+ channel.[3] Amiloride and its analogs have been subsequently shown to inhibit a variety of other ion transporters and enzymes, to interact with specific drug or hormone receptors, and to inhibit DNA, RNA, and protein synthesis[4,5] (Table I).

Amiloride is a pyrazinoylguanidine bearing amino groups in the 3- and 5-positions and a chloro group in the 6-position of the pyrazine ring (Table II). Over 1000 analogs have been synthesized.[1] This chapter reviews the use of amiloride analogs as probes for the characterization of ion transport proteins.

Ion Transport Systems

Amiloride and a large number of amiloride analogs have been used as reversible inhibitors of ion transport.[4,5] Amiloride analogs have a guanidino moiety, and its protonated form is required for inhibition of ion transport.[6-8] The pK_a of amiloride is 8.8 in H_2O (8.7 in 30% ethanol) and decreases with substitution of electron-withdrawing groups on the terminal nitrogen of the guanidino moiety.[4] The concentrations of amiloride required to achieve half-maximal inhibition (IC_{50}) of the epithelial Na^+ channel, Na^+/H^+ exchanger, Na^+/Ca^{2+} exchanger, Na^+,K^+-ATPase, volt-

---

[1] E. J. Cragoe, Jr., *in* "Diuretics" (E. J. Cragoe, Jr., ed.), p. 303. Wiley, New York, 1983.
[2] E. J. Cragoe, Jr., O. W. Woltersdorf, Jr., J. B. Bicking, S. F. Kwong, and J. H. Jones, *J. Med. Chem.* **10**, 66 (1967).
[3] T. R. Kleyman and E. J. Cragoe, Jr., *Semin. Nephrol.* **8**, 242 (1988).
[4] T. R. Kleyman and E. J. Cragoe, Jr., *J. Membr. Biol.* **105**, 1 (1988).
[5] D. J. Benos, *Am. J. Physiol.* **242**, C131 (1982).
[6] D. J. Benos, S. A. Simon, L. J. Mandel, and P. M. Cala, *J. Gen. Physiol.* **68**, 43 (1976).
[7] G. L'Allemain, A. Franchi, E. Cragoe, Jr., and J. Pouyssegur, *J. Biol. Chem.* **259**, 4313 (1984).
[8] G. J. Kaczorowski, F. Barros, J. K. Dethmers, and M. J. Trumble, *Biochemistry* **24**, 1394 (1985).

TABLE I
ION TRANSPORTERS, ENZYMES, AND DRUG AND HORMONE RECEPTORS THAT ARE INHIBITED OR INTERACT WITH AMILORIDE OR AMILORIDE ANALOGS

Ion transporters	Enzymes	Drug and hormone receptors
Epithelial Na^+ channel	Protein kinases	α_1-Adrenergic receptor
Na^+/H^+ exchanger	Type I cAMP-dependent protein kinase	α_2-Adrenergic receptor
Na^+/Ca^{2+} exchanger	Type II cAMP-dependent protein kinase	β-Adrenergic receptor
Na^+, K^+-ATPase	Epidermal growth factor receptor protein kinase	Atrial natriuretic receptor
Na^+-D-glucose cotransporter	Platelet-derived growth factor receptor protein kinase	Guanine nucleotide regulatory proteins
Na^+-L-alanine cotransporter	Insulin receptor protein kinase	Muscarinic acetylcholine receptor
Na^+-PO_4^{3-}-cotransporter	Protein kinase C	Nicotinic acetylcholine receptor
Voltage-gated Na^+ channel	Acetylcholinesterase	
Voltage-gated Ca^{2+} channel	Adenylate cyclase	
Delayed rectifier K^+ channel	DNA topoisomerase II	
Nicotinic acetylcholine receptor	Monoamine oxidase	
	Renal kalekrein	
	Urokinase-type plasminogen activator	

TABLE II
Concentrations of Amiloride Required to Achieve Half-Maximal Inhibition (IC_{50}) of the Epithelial Na^+ Channel, Na^+/H^+ Exchanger, Na^+/Ca^{2+} Exchanger, Na^+, K^+-ATPase, Voltage-Gated Na^+ Channel, and Voltage-Gated Ca^{2+} Channel (L Type)[a]

Amiloride Analog	R¹	R²	R³	Epithelial Na⁺ Channel	Na⁺/H⁺ Exchanger	Na⁺/Ca²⁺ Exchanger	Na⁺/K⁺-ATPase	Voltage-gated Na⁺ Channel	Voltage-gated Ca²⁺ Channel
Amiloride	H—	H—	—H	(0.35) 1	(84) 1	(1100) 1	(4000) 1	(600) 1	(90) 1
Phenamil	H—	H—	—C₆H₅	17	<0.01	5			10
Benzamil	H—	H—	—CH₂C₆H₅	9	<0.08	11	5	16	14
3',4'-Dichlorobenzamil	H—	H—	—CH₂C₆H₃Cl₂	4		37			25
5-(N,N-Dimethyl)amiloride	CH₃—	CH₃—	—H	<0.04	12	2	1.4		5
5-(N-Ethyl-N-isopropyl)amiloride	C₂H₅—	(CH₃)₂CH—	—H	<0.04	220	8	14	100	56
5-(N-Methyl-N-isobutyl)amiloride	CH₃—	(CH₃)₂CHCH₂—	—H		190	8			75

IC_{50} (μM) for Amiloride and Potency Relative to Amiloride

[a] Amiloride analogs that have been commonly used are included in this table along with their potency, relative to amiloride, in inhibiting the ion transporters listed.[4,8,17,19,20,25,29,33] To determine the IC_{50} of a specific amiloride analog for a transport protein listed above, divide the IC_{50} for amiloride by the relative potency of the amiloride analog. The IC_{50} values for the voltage-gated Ca^{2+} channel (L type) were derived from binding experiments.

age-gated Na$^+$ channel, and voltage-gated Ca^{2+} channel (L-type) are listed in Table II. Amiloride analogs that have been commonly used are included in this table, along with their potency, relative to amiloride, in inhibiting the ion transporters listed.

Factors that Influence the IC_{50} of Amiloride and Its Analogs

The IC_{50} values of amiloride and amiloride analogs are not absolute, but vary depending on assay conditions. A decrease in the extracellular (or extravesicular) Na$^+$ concentration results in a decrease in the IC_{50} for the Na$^+$ channel, Na$^+$/H$^+$ exchanger, and Na$^+$/Ca^{2+} exchanger.[4,8–10]

Amiloride and a large number of its analogs (in particular, a number of analogs that have been shown to be potent inhibitors of Na$^+$/H$^+$ exchange) bear hydrophobic substituents and are permeable weak bases. These amiloride analogs may collapse pH gradients across membranes and decrease the driving force for Na$^+$/H$^+$ exchange.[11,12] Amiloride has also been shown to accumulate within cells and may indirectly effect ion transport through inhibition of intracellular enzymes (such as protein kinases and adenylate cyclase) which may be involved in the regulation of ion transport proteins (see Enzymes, Receptors, and Cellular Metabolism, below).

The extent of inhibition of the sodium channel by amiloride has been shown to be dependent on the apical plasma membrane potential. Amiloride interacts with the channel in a manner that is sensitive to the membrane potential.[13,14]

Epithelial Na$^+$ Channel

Sodium ion channels are present on the apical plasma membrane of high-resistance (or "tight") epithelia that transport Na$^+$.[15,16] This channel is the only ion transporter inhibited by amiloride with an IC_{50} of less than 1 μM.[4] In the presence of a physiologic Na$^+$ concentration, the IC_{50} for inhibition of Na$^+$ transport is in the range of 0.1 to 0.5 μM.[4,16] The most specific inhibitors of the epithelial Na$^+$ channel are amiloride analogs bearing hydrophobic substituents on the terminal nitrogen atom of the

---

[9] A. W. Cuthbert and W. K. Shum, *Naunyn-Schmiedebergs Arch. Pharmakol.* **281**, 261 (1974).
[10] J. L. Kinsella and P. S. Aronson, *Am. J. Physiol.* **241**, F374 (1981).
[11] G. Kaczorowski, M. Garcia, T. Kleyman, and E. J. Cragoe, Jr., unpublished observations.
[12] W. P. Dubinsky and R. A. Frizzel, *Am. J. Physiol.* **245**, C157 (1983).
[13] L. G. Palmer, *J. Membr. Biol.* **80**, 153 (1984).
[14] K. L. Hamilton and D. C. Eaton, *Am. J. Physiol.* **245**, C200 (1985).
[15] P. J. Bently, *J. Physiol. (London)* **195**, 317 (1986).
[16] S. Sariban-Sohraby and D. J. Benos, *Am. J. Physiol.* **250**, C175 (1986).

guanidino moiety, such as the benzyl (benzamil) and phenyl (phenamil) analogs. These analogs inhibit Na$^+$ transport with IC$_{50}$ values of approximately 10 nM.[4,17] This is more than three orders of magnitude lower than that reported for inhibition of Na$^+$/H$^+$ and Na$^+$/Ca^{2+} exchangers, Na$^+$/K$^+$-ATPase, and Na$^+$-glucose and Na$^+$-alanine contransporters[4,7,8] (see the next three sections).

Na$^+$/H$^+$ Exchanger

A plasma membrane transporter that exchanges Na$^+$ for H$^+$ in an electroneutral fashion has been described in a variety of cells[18] and is involved in regulation of intracellular pH. Amiloride and its analogs have been used as inhibitors of both Na$^+$/H$^+$ exchange and subsequent events that might be altered through regulation of this transporter.[7,18]

The IC$_{50}$ for inhibition of the exchanger by amiloride is 84 μM in human neutrophils,[19] and as low as 3 to 7 μM in other cells.[4,7,20] This variation is likely to be due, in part, to differences in the concentration of Na$^+$ used to assay Na$^+$/H$^+$ exchange. A recent study also suggests that the Na$^+$/H$^+$ exchanger present on the basolateral plasma membrane of epithelia is more sensitive to 5-(N-ethyl-N-isopropyl)amiloride than is the Na$^+$/H$^+$ exchanger present on the apical plasma membrane.[20a] Amiloride analogs bearing hydrophobic substituents on the 5-amino group located on the pyrazine ring have both the highest activity and specificity for the Na$^+$/H$^+$ exchanger. Several analogs with hydrophilic substituents in this position are potent inhibitors of the exchanger.[4,19] When studied in the presence of a low [Na$^+$], IC$_{50}$ values less than 100 nM have been observed for a number of amiloride analogs.[4,7,20] This is 100-fold less than the IC$_{50}$ observed with these analogs for inhibition of the epithelial Na$^+$ channel,[4,17] Na$^+$/Ca^{2+} exchanger,[8] Na$^+$,K$^+$-ATPase,[20] and Na$^+$-D-glucose, Na$^+$-L-alanine, and Na$^+$-PO$_4^{3-}$ cotransporters.[21]

Amiloride and 5-(N,N-dimethyl)amiloride are 9 and 100 times more potent, respectively, in inhibiting Na$^+$/H$^+$ exchange than in inhibiting protein kinase C.[22] Amiloride inhibits a number of other kinases with IC$_{50}$

---

[17] A. W. Cuthbert and G. M. Fanelli, *Br. J. Pharmacol.* **63**, 139 (1978).
[18] J. L. Seifter and P. S. Aronson, *J. Clin. Invest.* **78**, 859 (1986).
[19] L. Simchowitz and E. J. Cragoe, Jr., *Mol. Pharmacol.* **30**, 112 (1986).
[20] Y. Zhuang, E. J. Cragoe, Jr., T. Shaikewitz, L. Glaser, and D. Cassel, *Biochemistry* **23**, 4481 (1984).
[20a] J. G. Haggerty, N. Argarwal, R. F. Reilly, E. A. Adelberg, and C. W. Slayman, *Proc. Natl. Acad. Sci. U.S.A.* **85**, 6797 (1988).
[21] R. C. Harris, R. A. Lufburrow III, E. J. Cragoe, Jr., and J. L. Seifter, *Kidney Int.* **27**, 310 (abstr.) (1985).
[22] J. M. Besterman, W. S. May, Jr., H. Le Vine III, E. J. Cragoe, Jr., and P. Cuatrecasas, *J. Biol. Chem.* **260**, 1155 (1985).

values similar to that reported for protein kinase C.[23] Amiloride inhibits adenylate cyclase with an IC_{50} similar to the IC_{50} for inhibition of Na^+/H^+ exchange.[24]

Na^+/Ca^{2+} Exchanger

The Na^+/Ca^{2+} exchanger is a reversible electrogenic transporter and is inhibited by amiloride with an IC_{50} of approximately 1 mM.[8] Although proper substituents on either (or both) the 5-amino or the terminal guanidino nitrogen atoms increase the activity of these analogs by up to 100-fold, the IC_{50} is only in the range of 10 μM.[4,8] With the exception of analogs with substituents at both sites, this is a considerably greater concentration than is require for inhibition of the epithelial Na^+ channel and the Na^+/H^+ exchanger. Of the amiloride analogs examined, those active against the Na^+/Ca^{2+} exchanger are equal or more potent inhibitors of the Ca^{2+} channel[25] (see Table II).

Na^+,K^+-ATPase

The inhibition of the Na^+,K^+-ATPase by amiloride has been studied in intact cells, membrane vesicles, and with partially purified Na^+,K^+-ATPase.[20,21,26,27] The IC_{50} for inhibition of ATPase activity by amiloride is greater than 3 mM. Appropriate substitution of the 5-amino moiety of amiloride decreases the IC_{50} to as low as 200 μM.[20] 5-(N,N-Dimethyl)amiloride inhibits the Na^+,K^+-ATPase with an IC_{50} of 3 mM, which is four orders of magnitude greater than the IC_{50} for inhibition of Na^+/H^+ exchanger.[20] The effect of introduction of substituents on the terminal nitrogen atom of the guanidino moiety has not been studied except for benzamil, which has an IC_{50} of less than 1 mM.

Na^+-Coupled Solute Transport

The transport of a number of solutes across the apical plasma membrane of epithelia is tightly coupled to Na^+. The Na^+–D-glucose, Na^+–L-alanine, and Na^+–PO_4^{3-} transporters are inhibited by amiloride at concentrations greater than 1 mM.[21,28] Benzamil and 5-(N,N-dimethyl)

[23] R. J. Davis and M. P. Czech, *J. Biol. Chem.* **260**, 2543 (1985).
[24] Y. Mahe, F. Garcia-Romeu, and R. Montais, *Eur. J. Pharmacol.* **116**, 199 (1985).
[25] G. Kaczorowski, personal communication (1988).
[26] S. P. Soltoff and L. J. Mandell, *Science* **220**, 957 (1983).
[27] E. L. Renner, J. R. Lake, E. J. Cragoe, Jr., and B. F. Scharschmidt, *J. Hepatol.* **6**, 1193 (1986).
[28] J. S. Cook, C. Shaffer, and E. J. Cragoe, Jr., *Am. J. Physiol.* **253**, C199 (1987).

amiloride inhibit the Na$^+$–D-glucose transporter in rabbit kidney brush border membranes with an IC$_{50}$ between 0.5 and 1 mM. The IC$_{50}$ of amiloride is approximately 2 mM.[21] The Na$^+$-coupled glucose transporter in LLC-PK$_1$ cells is inhibited by amiloride analogs bearing hydrophobic substituents on the 5-amino moiety or on terminal nitrogen of the guanidino moiety with IC$_{50}$ values (measured using a low [Na$^+$]) between 0.1 and 0.3 mM.[28]

Na$^+$-dependent L-alanine and PO$_4^{3-}$ transporters are inhibited by amiloride at concentrations greater than 1 mM. Introduction of hydrophobic groups on the 5-amino moiety or the terminal nitrogen atom of the guanidino moiety decreases the IC$_{50}$.[21]

Voltage-Gated Na$^+$/Channel

A voltage-gated Na$^+$-selective ion channel is present in electrically excitable cells. Amiloride inhibits this channel in both synaptosomes and heart membrane vesicles with an IC$_{50}$ of 600 μM.[29] Amiloride analogs bearing hydrophobic substituents on the 5-amino moiety or the terminal nitrogen of the guanidino moiety have lower IC$_{50}$ values [6 μM for 5-(N-ethyl-N-isopropyl)amiloride and 37 μM for benzamil]. Amiloride and analogs with hydrophobic substituents on the 5-amino moiety and terminal nitrogen atom of the guanidino moiety inhibit [^{3}H]batrachotoxin-A 20-α-benzoate and [^{3}H]tetracaine binding to the channel.

Voltage-Gated Ca^{2+} Channels

Amiloride and 3′,4′-dichlorobenzamil inhibit L-type, N-type, and T-type voltage-gated Ca^{2+} channels.[25,30–34] Amiloride binds to the L-type Ca^{2+} channel with an IC$_{50}$ of 90 μM. Amiloride analogs bearing hydrophobic substituents on the 5-amino moiety or the terminal nitrogen of the guanidino moiety have IC$_{50}$ values 5- to 75-fold lower than amiloride[25,33] (Table II). Amiloride analogs bind to a cation-binding site in the pore of the Ca^{2+} channel and allosterically alter binding of dihydropyridines, aralkylamines, and benzothiazepines.[33]

---

[29] J. Velly, M. Grima, N. Decker, E. J. Cragoe, Jr., and J. Schwartz, *Eur. J. Physiol.* **149**, 97 (1988).
[30] C.-M. Tang, F. Presser, and M. Morad, *Science* **240**, 213 (1988).
[31] G. Suarez-Kurtz and G. Kaczorowski, *J. Pharmacol. Exp. Ther.* **247**, 248 (1988).
[32] P. Feigenbaum, M. L. Garcia, and G. Kaczorowski, *Biochem. Biophys. Res. Commun.* **154**, 298 (1988).
[33] M. L. Garcia, V. F. King, J. L. Shevell, R. S. Slaughter, G. Suarez-Kurtz, R. J. Winquist, and G. J. Kaczorowski, *J. Biol. Chem.* **265**, 3763 (1990).
[34] D. R. Bielefeld, R. W. Hadley, P. M. Vassilev, and J. R. Hume, *Circ. Res.* **59**, 381 (1986).

K⁺ Channel

3′,4′-Dichlorobenzamil inhibits the delayed rectifier K⁺ channel in intact frog atrial myocytes with 30 to 40% inhibition of the K⁺ current at a concentration of 5 μM.[34]

Nicotine Acetylcholine Receptor

Amiloride inhibits the nicotinic acetylcholine receptor isolate from *Torpedo*. The IC_{50} is approximately 100 μM.[35]

Enzymes, Receptors, and Cellular Metabolism

Protein Kinases

Amiloride inhibits a number of protein kinases, including type I and type II cAMP-dependent protein kinases,[36] protein kinase C,[22] and protein kinase activity associated with the insulin receptor, epidermal growth factor (EGF) receptor, and the platelet-derived growth factor receptor.[23] The IC_{50} for inhibition of EGF receptor protein kinase is 350 μM. Amiloride inhibits types I and II cAMP-dependent protein kinases with an IC_{50} of approximately 1 mM.[36] Both amiloride and 5-(N,N-dimethyl)amiloride inhibit purified protein kinase C with an IC_{50} of approximately 1 mM.[22] The inhibition of protein kinase activity associated with the EGF receptor is competitive with ATP, suggesting that amiloride binds to an ATP-binding site. Amiloride is a noncompetitive inhibitor of substrate (histone) phosphorylation.[23]

Adenylate Cyclase

The generation of cAMP in fish erythrocytes is inhibited by amiloride in a dose-dependent manner, with an IC_{50} of 6 μM.[24] The effect of amiloride analogs in this system have not been examined.

α- and β-Adrenergic Receptors

Amiloride is a competitive inhibitor of [³H]prazocin binding to α_1 receptors in membrane vesicles from rat renal cortex or bovine carotid artery with an IC_{50} of 24 to 33 μM.[37,38] Amiloride is also a competitive

---

[35] A. Karlin, personal communication (1986).
[36] R. K. Ralph, J. Smart, S. M. Wojcik, and J. McQuillan, *Biochem. Biophys. Res. Commun.* **104**, 1054 (1982).
[37] M. J. Howard, M. D. Mullen, and P. A. Insel, *Am. J. Physiol.* **253**, F21 (1987).
[38] R. C. Bhalla and R. V. Sharma, *J. Cardiovasc. Pharmacol.* **8**, 927 (1986).

inhibitor of [³H]rauwolscine binding to α_2 receptors and [¹²⁵I]iodocyanopindolol binding to β-adrenergic receptors in rat renal cortical membranes with IC_{50} values of 14 and 84 μM, respectively.[37] Both benzamil and 5-(N-ethyl-N-isopropyl)amiloride are between 2- to 25-fold more potent than amiloride in inhibiting specific ligand binding to α_1-, α_2-, and β-adrenergic receptors.

Atrial Natriuretic Factor (ANF) Receptor

ANF is a peptide hormone secreted by atrial myocytes and which binds to specific cell surface receptors. Recent studies have shown that ANF binds to both low- and high-affinity receptors, and that amiloride (100 μM) increases the number of high-affinity sites in membranes obtained from adrenal zona glomerulosa.[39] This change in the number of binding sites is associated with a conformational change in the 150,000-Da receptor.

Guanine Nucleotide Regulatory Proteins

Pertussis toxin catalyzed ADP ribosylation of the guanine nucleotide regulatory proteins G_o and G_i and is inhibited by 10^{-4} M amiloride. Amiloride did not inhibit cholera toxin-catalyzed ADP ribosylation of G_o.[40]

Muscarinic Acetylocholine Receptor

Amiloride inhibits amylase secretion induced by carbachol with an IC_{50} of 40 μM,[45] Ca^{2+} efflux induced by carbachol with an IC_{50} of 80 μM, and is a competitive inhibitor of [³H]quinuclidinyl benzylate binding to the muscarinic acetylcholine receptor.[41]

Inhibition of Cellular Metabolism

Several amiloride analogs with substituents on the 5-amino moiety (at concentrations > 30 μM) deplete intracellular levels of ATP,[20,42] whereas amiloride and 5-(N,N-dimethyl)amiloride at concentrations of 2 mM had no effect.[20] The depletion of intracellular ATP levels may be due in part to inhibition of oxidative phosphorylation. Amiloride analogs bearing hydrophobic substituents on the 5-amino moiety inhibit O_2 consumption in

---

[39] S. Meloche, H. Ong, and A. De Léan, J. Biol. Chem. 262, 10252 (1987).
[40] M. B. Anand-Srivastava, J. Biol. Chem. 264, 9491 (1989).
[41] G. A. J. Kuijpers, J. DePont, I. van Nooy, A. Fleuren-Jakobs, S. Bonting, and J. Rodrigues de Miranda, Biochim. Biophys. Acta 804, 237 (1984).
[42] S. P. Soltoff, E. J. Cragoe, Jr., and L. J. Mandel, Am. J. Physiol. 250, C744 (1986).

carbonyl cyanide *p*-(trifluoro methoxy)-phenylhydrazone (FCCP)-treated renal proximal tubular cells in suspension with IC_{50} values of 250 to 500 μM. The IC_{50} for amiloride is greater than 1 mM.

Others

Amiloride is a noncompetitive inhibitor of acetylcholinesterase, with an IC_{50} in the range of 20 to 60 μM,[43] and is a noncompetitive inhibitor of rat and human renal kallekrein with an IC_{50} in the 85 to 230 μM range.[44] Amiloride is a competitive inhibitor of monoamine oxidase activity measure in rat brain homogenate.[45] Amiloride inhibits urokinase-type plasminogen activator with an IC_{50} 7 μM.[46] Amiloride does not inhibit tissue-type plasminogen activator.

DNA, RNA, and Protein Synthesis

DNA and RNA Synthesis

Amiloride inhibits growth factor-induced DNA, RNA, and protein synthesis.[7,47] The effect of amiloride on DNA and RNA synthesis may be indirect. Lowering the extracellular Na^+ concentration inhibits DNA and RNA synthesis, suggesting that entry of Na^+ into cells may be a requirement for these events. In addition, amiloride analogs inhibit DNA replication with IC_{50} values similar to that observed for inhibition of Na^+/H^+ exchange.[7] Amiloride may also have a direct inhibitory effect on DNA and RNA synthesis. Amiloride and several of its analogs have been shown to intercalate into DNA and to inhibit DNA topoisomerase II.[48]

Protein Synthesis

Amiloride inhibits protein synthesis, as measured by incorporation of radiolabeled amino acids into proteins in both intact cells and cell-free reticulocyte lysate[20,47,49,50] with an IC_{50} between 100 and 400 μM. The IC_{50} varies with the cell type studied. The IC_{50} for inhibition of synthesis of individual proteins may also vary.[49] Amiloride lowers both the initial rate

---

[43] D. Dannenbaum and K. Rosenheck, *Biophys. J.* **49**, 370a (abstr.) (1986).
[44] H. S. Margolius and J. Chao, *J. Clin. Invest.* **65**, 1343 (1980).
[45] V. Palaty, *Can. J. Physiol. Pharmacol.* **63**, 1586 (1985).
[46] J.-D. Vassalli and D. Belin, *FEBS Lett.* **214**, 187 (1987).
[47] K. S. Koch and H. L. Leffert, *Cell (Cambridge, Mass.)* **18**, 153 (1979).
[48] J. M. Besterman, L. P. Elwell, E. J. Cragoe, Jr., C. W. Andrews, and M. Cory, *J. Biol. Chem.* **265**, 2324 (1989).
[49] H. L. Leffert, K. S. Koch, M. Fehlmann, W. Heiser, P. J. Lad, and H. Skelly, *Biochem. Biophys. Res. Commun.* **108**, 738 (1982).
[50] M. Lubin, F. Cahn, and B. A. Countermarsh, *J. Cell. Physiol.* **113**, 247 (1982).

of incorporation of [^{35}S]methionine as well as plateau levels of [^{35}S]methionine-labeled proteins.[49] In cell-free systems, amiloride analogs bearing hydrophobic substituents on the 5-amino moiety have similar IC$_{50}$ values for inhibition of protein synthesis. However, in intact cells the IC$_{50}$ of these analogs varies over 25-fold, suggesting that amiloride analogs may inhibit protein synthesis in intact cells indirectly through inhibition of ion transport.[20] The mechanism by which amiloride directly inhibits protein synthesis in cell-free systems is unclear.

Amiloride and Amiloride Analogs as Probes for Characterizing Transport Proteins

Solubility Characteristics

Amiloride and many of its analogs are soluble in aqueous solutions at concentrations less than 1 to 10 mM. Stock solutions of most amiloride analogs may be made in dimethyl sulfoxide (DMSO) at concentrations of 1 M. We generally store stock solutions at 10^{-2} M protected from light at $-20°$.

Absorption and Fluorescence Characteristics

Three major absorption peaks have been observed for amiloride and several analogs at approximately 360 to 370 nm, 265 to 290 nm, an 215 to 235 nm. Extinction coefficients at these wavelengths are in the range of 10,000 to 25,000 M^{-1} cm^{-1}. The absorption peaks are broad and vary slightly among the different analogs and with the solvent system used. Amiloride analogs are highly fluorescent aromatic compounds. Amiloride has excitation maxima at 286 and 360 nm, and an emission maximum at 410 to 415 nm.[23] The fluorescence and absorption properties of the analogs may interfere with techniques which utilize fluorescent probes to measure intracellular pH and intracellular Ca^{2+}. Amiloride may quench the fluorescence of both acridine orange and carboxyfluorescein, probes which have been used to measure intravesicular and intracellular pH. The absorption peak of amiloride at 360 nm may interfere with the absorption of quin2 (excitation at 342 nm) and Fura-2 (excitation at 340 and 380 nm), probes which have been used to measure intracellular calcium ions.

Intracellular Accumulation

Several studies have shown that amiloride accumulates within cells. Amiloride an its analogs may diffuse or be transported across membranes. Amiloride has been shown to be transported across the plasma membrane of hepatocytes and reach an intracellular concentration 10-fold greater than the extracellular concentration,[49] assuming that cellular amiloride was free in solution and not bound to lipid or compartmentalized. Amiloride accumulates within frog skin epithelial cells[51] and within A431 cells[23] at intracellular concentrations greater than the extracellular concentration. Amiloride diffuses across red blood cell and neutrophil plasma membranes with a permeability coefficient of approximately 10^{-7} cm·sec^{-1}.[52,53] In neutrophils 75% of the intracellular amiloride was considered to be in the lysosomal compartment.[53] In other cell types, the extent to which intracellular amiloride is compartmentalized is uncertain.

Radiolabeled Amiloride Analogs

[^{14}C]Amiloride has been synthesized with a specific activity of 54 mCi/mmol by the reaction of [^{14}C]guanidine with methyl 3,5-diamino-6-chloropyrazinecarboxylate.[2] This procedure has been used to synthesize other [^{14}C]-labeled analogs.

A number of tritium-labeled amiloride analogs have been synthesized, including [phenyl-^{3}H]phenamil, [benzyl-^{3}H]benzamil,[54] and [benzyl-^{3}H]-6-bromobenzamil, with specific activities between 2 and 21 Ci/mmol. These analogs have been synthesized by the reaction of 1-methyl-2-(3,5-diamino-6-chloropyrazinoyl) pseudothiourium iodide or its 6-Br analog with [^{3}H]aniline or [^{3}H]benzylamine.[4,55] 6-[methyl-^{3}H]Bromomethylamiloride has been synthesized by the reaction of bromoamiloride with [^{3}H]CH$_3$I with a specific activity of 1 Ci/mmol.[56] The methyl group is attached to the guanidino nitrogen atom adjacent to the carbonyl moiety. Both 5-([N-ethyl-N-^{3}H]propyl)amiloride and 5-([N-methyl-N-^{3}H]isobutyl)amiloride have been synthesized with specific activities of 2 and 28 Ci/mmol, respectively.[57,58]

6-[^{125}I]Iodoamiloride and a number of 6-[^{125}I]iodoamiloride analogs have been synthesized by reacting 6-H amiloride analogs with ^{125}ICl.[59] A

---

[51] J. V. Briggman, J. S. Graves, S. S. Spicer, and E. J. Cragoe, Jr., *Histochem. J.* **15**, 239 (1983).
[52] D. J. Benos, J. Reyes, and D. G. Shoemaker, *Biochim. Biophys. Acta* **734**, 99 (1983).
[53] L. Simchowitz, O. W. Woltersdorf, Jr., and E. J. Cragoe, Jr., *J. Biol. Chem.* **262**, 15875 (1987).
[54] A. W. Cuthbert and J. M. Edwardson, *J. Pharm. Pharmacol.* **31**, 382 (1979).
[55] E. J. Cragoe, Jr., O. W. Woltersdorf, Jr., and S. J. deSolms, U.S. patent 4,246,406 (1981).
[56] K. Lazorick, C. Miller, S. Sariban-Sohraby, and D. Benos, *J. Membr. Biol.* **86**, 69 (1985).
[57] P. Vigne, C. Frelin, M. Audinot, M. Borsotto, E. J. Cragoe, Jr., and M. Lazdunski, *EMBO J.* **3**, 2647 (1984).
[58] 5-(N-Methyl-N-^{3}H-isobutyl)amiloride was prepared by New England Nuclear by the catalytic tritiation of 5-[N-methyl-N-(2-methylallyl)amiloride.
[59] D. Cassel, M. Rotman, E. J. Cragoe, Jr., and P. Igarashi, *Anal. Biochem.* **170**, 63 (1988).

new amiloride analog bearing a 4'-hydroxyphenethyl moiety may also prove useful in synthesis of radioiodinated amiloride analogs.[4]

Binding Assays Using Radiolabeled Amiloride Analogs

Several recent studies have used radiolabeled amiloride analogs to identify and characterize binding sites of amiloride analogs in plasma membranes or microsomal membranes derived from cells known to have amiloride-sensitive transporters. Analogs bearing substituents on the 5-amino group, including 5-([N-methyl-N-^{3}H]isobutyl)amiloride and 5-([N-ethyl-N-^{3}H]propyl)amiloride have been used to characterize binding to the putative Na$^+$/H$^+$ exchanger.[54,60] [^{3}H]Benzamil, [^{3}H]phenamil, and 6-[^{3}H]bromomethylamiloride have been used to characterize binding to the epithelial Na$^+$ channel and to follow the channel during solubilization and purification.[61-66] Methods for assaying the binding of amiloride analogs to membranes and to detergent-solubilized proteins are outlined below.

Equilibrium Dialysis. Binding of [^{3}H]benzamil to the epithelial Na$^+$ channel in renal cortical microsomal membranes is measure by equilibrium dialysis with M_r 12,000–14,000 M_r cutoff dialysis tubing. Membrane vesicles (approximately 250 μg of protein) in a phosphate buffer and [^{3}H]benzamil are placed in dialysis tubing, which is then placed in a test tube with 7.5 ml of buffer containing the same concentration of [^{3}H]benzamil. The tubes are stirred on a flat-bed rotary mixer for 16 hr at 4, by which time equilibrium is achieved. Aliquots are removed from the dialysis bag and dialysated and counted. Protein determinations on aliquots from the dialysis bag are performed. Nonspecific binding is determined in parallel experiments in which 1 μM unlabeled benzamil is added to the microsomes and dialysate.

Filtration Assay. Measurement of the binding of [^{3}H]benzamil to membrane vesicles may also be performed using filtration to separate bound from free drug. Glass fiber filters have been used to bind the

---

[60] S. J. Dixon, S. Cohen, E. J. Cragoe, Jr., and S. Grinstein, *J. Gen. Physiol.* **88,** 19a (abstr.) (1986).
[61] T. R. Kleyman, T. Yulo, C. Ashbaugh, D. Landry, E. Cragoe, Jr., A. Karlin, and Q. Al-Awqati, *J. Biol. Chem.* **261,** 2839 (1986).
[62] A. W. Cuthbert and J. M. Edwardson, *Biochem. Pharmacol.* **30,** 1175 (1981).
[63] T. Kleyman, T. Yulo, E. J. Cragoe, Jr., and Q. Al-Awqati, *Kidney Int.* **29,** 400 (abstr.) (1986).
[64] P. Barby, O. Chassande, P. Vigne, C. Frelin, C. Ellory, E. J. Cragoe, Jr., and M. Lazdunski, *Proc. Natl. Acad. Sci. U.S.A.* **84,** 4836 (1987).
[65] S. Sariban-Sohraby and D. J. Benos, *Biochemistry* **25,** 4639 (1986).
[66] D. J. Benos, G. Saccomani, B. M. Brenner, and S. Sariban-Sohraby, *Proc. Natl. Acad. Sci. U.S.A.* **83,** 8525 (1986).

vesicles, with minimal nonspecific trapping of the drug on the filter paper. Nitrocellulose and cellulose acetate filter paper should not be used, as there is considerable nonspecific trapping of the radiolabeled amiloride analog on the filter.

Rapid Gel Filtration Assay. The binding of [^{3}H]benzamil to the putative Na$^+$ channel using detergent-solubilized membrane proteins is measured by a rapid gel filtration assay. Sephadex G-25 (fine grade) is swollen with water and washed with buffer used for solubilization (200 mM sucrose, 1 mM EGTA, and 10 mM Tris-Cl, pH 8.0, with the addition of protease inhibitors). A saturating concentration of [^{3}H]benzamil (final concentration = 50 nM) is added to solubilized proteins (total volume = 200 μl) and incubated on ice for > 10 min, by which time equilibrium has been achieved. The mixture is placed over a 1.5-ml column in a 3-ml plastic syringe (using Whatman grade 1 filter paper for support) which is suspended in a plastic test tube, and then spun at 500 g for 30 sec (the speed and duration of the spin must be optimized). The eluate is weighed to determine the total volume and aliquots are removed for counting and for protein determination. In parallel assays, 5 μM benzamil is added to the incubation mixture to determine nonspecific binding (see also Refs. 64 and 65).

Photoaffinity Labels

Two major photoreactive groups, arylhalides and aromatic ethers, have been used in the development of photoactive amiloride analogs. Photolysis of 6-bromo, 6-iodo, or 6-chloro analogs of amiloride can lead to the formation of a free radical, and subsequent covalent incorporation into adjacent proteins. This approach has been used to identify putative subunits of the epithelial Na$^+$ channel, using 6-[^{3}H]bromobenzamil and 6-[^{3}H]bromomethylamiloride as photoreactive amiloride analogs that bind to the Na$^+$ channel with both high affinity and specificity.[61,67] Amiloride and 6-bromobenzamil have a major absorption peak at 360 nm, and we have used this wavelength of light to photoactivate 6-bromobenzamil.[61] 5-(*N*-Ethyl-*N*-isopropyl)-6-[^{14}C]bromoamiloride has been used to label the Na$^+$/H$^+$ exchanger.[68]

Aromatic ethers have been shown to undergo photoactivation and photoincorporation into proteins by the mechanism of aromatic nucleophilic photosubstitution.[69] Photoreactive amiloride analogs have been syn-

[67] D. J. Benos, G. Saccomani, and S. Sariban-Sohraby, *J. Biol. Chem.* **262**, 10613 (1987).
[68] T. Friedrich, J. Sablotni, and G. Burckhardt, *J. Membr. Biol.* **94**, 253 (1986).
[69] J. Cornelisse and E. Havinga, *Chem. Rev.* **75**, 353 (1975).

thesized with a 2'-methoxy-5'-nitrobenzyl moiety located either on the terminal nitrogen atom of the guanidino moiety or on the 5-amino moiety (which also bears an ethyl group). These drugs undergo photoactivation with 313-nm wavelength light. 2'-Methoxy-5'-nitrobenzamil has been used to photolabel and identify the amiloride-binding subunit of the epithelial Na^+ channel.[70] An analog with the photoreactive group on the 5-amino moiety has been used to label the Na^+/H^+ exchanger.[71] Anti-amiloride antibodies were used to detect these photolabels after photoincorporation into the channel or exchanger.[70-72] Radioactive counterparts of these analogs have recently been synthesized.[59,73]

Methods for Photoaffinity Labeling the Binding Sites of Amiloride Analogs. Sodium ion channel-containing microsomes (150 μg/ml) are preequilibrated for 1 hr in a phosphate buffer containing protease inhibitors and 20 nM 6-[^{3}H]bromobenzamil (the concentration of the photolabel will depend on the affinity of the amiloride analog for the particular transporter) and then gassed for 5 min with N_2. The vesicles are photolyzed with a 50-W high-pressure mercury arc lamp with 300–400 band pass and > 345 long pass filters. The solution is constantly stirred, cooled in a water jacket to 4°, and the surface gassed with N_2. The vesicles are initially photolyzed for varying times to determine the optimal duration of photolysis by measuring photoincorporation of tritium into trichloroacetic acid-precipitible protein. To identify labeled proteins following photolysis, the microsomes are collected by centrifugation an then analyzed by SDS-PAGE and autofluorography. Photolabeling with 2'-methoxy-5'-nitrobenzamil is performed under similar conditions using a 313-nm narrow band pass filter.

Affinity Matrices

Amiloride has been coupled to support matrices through either the terminal nitrogen atom of the guanidino moiety or through the 5-amino group of the pyrazine ring. Three separate methods have been used to couple amiloride to a matrix through the guanidino moiety. One method has been used to couple through the 5-amino moiety.

Amiloride has been coupled directly to cyanogen bromide-activated Sepharose.[66] Alternatively, a reactive amiloride precursor, 1-methyl-2-

---

[70] T. R. Kleyman, E. J. Cragoe, Jr., and J. P. Kraehenbuhl, *J. Biol. Chem.* **264,** 11995 (1989).
[71] D. Warnock, T. Kleyman, and E. J. Cragoe, Jr., *FASEB J.* **2,** A753 (abstr.) (1988).
[72] T. R. Kleyman, R. Rajagopalan, E. J. Cragoe, Jr., B. F. Erlanger, and Q. Al-Awqati, Am. J. Physiol. **250,** C165 (1986).
[73] T. R. Kleyman, unpublished observations.

(3,5-diamino-6-chloropyrazinoyl) pseudothiourium iodide,[4,55] was allowed to react with aminohexyl-Sepharose in the presence of a hindered base (triethylamine). The product was amiloride coupled to Sepharose through a six-carbon spacer arm.[63] A mixed anhydride of an amiloride analog bearing a 5'-carboxypentyl group on the terminal guanidino nitrogen atom was synthesized using isobutyl chloroformate, and then coupled to albumin. Coupling to aminohexyl-Sepharose by this method should be straightforward. The amiloride–albumin complex was subsequently coupled to Sepharose, and has been used to affinity purify anti-amiloride antibodies.[72] An amiloride analog bearing a 5-[N-(3-isothiocyanatophenyl)] moiety has been coupled directly to dextran and aminohexyl-Sepharose.[74,75]

Anti-Amiloride Antibodies

An amiloride analog bearing a 5-carboxypentyl group on the terminal nitrogen of the acylguanidino moiety of amiloride was coupled to albumin by generation of a mixed anhydride. Approximately 10 mol of amiloride was bound/mol of albumin. The amiloride–bovine serum albumin was used to raise anti-amiloride antibodies in rabbits, which were subsequently affinity purified with an amiloride–rabbit serum albumin affinity column.[72] Utilizing amiloride coupled to albumin as an immunogen and rabbit anti-amiloride antibodies in the screening assay, monoclonal anti-idiotypic antibodies that recognize the epithelial Na^+ channel have recently been raised.[76] 5-[N-(3-Isothiocyanatophenyl)]amiloride has been coupled to albumin and used to raise anti-amiloride antibodies.[75] Antibodies raised against amiloride coupled to albumin through the acylguanidino group or the 5-amino group {using 5-[N-(3-isothiocyanatophenyl)]amiloride} recognize distinct epitopes on amiloride.[75]

Summary

The use of amiloride and its analogs in the study of ion transport requires a knowledge of the pharmacology of inhibition of transport proteins, and of effects on enzymes, receptors, and other cellular processes, such as DNA, RNA, and protein synthesis, and cellular metabolism. We have reviewed the pharmacology of inhibition of these processes by ami-

---

[74] D. Cassel, E. J. Cragoe, Jr., and M. Rotman, *J. Biol. Chem.* **262,** 4587 (1987).

[75] T. R. Kleyman, J. P. Kraehenbuhl, B. C. Rossier, E. J. Cragoe, Jr., and D. G. Warnock, Am. J. Physiol. 257, C1135 (1989).

[76] T. R. Kleyman, B. Rossier, B. F. Erlanger, and J. P. Kraehenbuhl, *Kidney Int.* **35,** 160 (abstr.) (1989).

loride an its analogs, as well as the use of amiloride analogs as potential probes for the characterization of ion transport systems.

Acknowledgments

The authors are grateful to Drs. A. George an G. Kaczorowski for reviewing the manuscript, and to Drs. G. Fanelli, G. Kaczorowski, A. Karlin, and L. Simchowitz for providing unpublished data on inhibition of Na^+ transport systems by amiloride analogs. This work was supported by the Zyma Foundation and by Grant AM34742 from the United States Public Health Service. T.R.K. is a recipient of Clinician-Scientist Award from the American Heart Association.

[43] Photoaffinity–Labeling Analogs of Phlorizin and Phloretin: Synthesis and Effects on Cell Membranes

By DONALD F. DIEDRICH

Introduction[1]

The inhibitory effects of the glucoside phlorizin (**I**, Fig. 1) and its aglycone, phloretin (**IA**, Fig. 1), on the sugar transport systems in a variety of cell types have been examined by many workers. The Na^+-coupled, D-glucose cotransporter in renal and intestinal brush border epithelial membranes is especially sensitive to phlorizin at low micromolar levels while phloretin is almost inactive. This vulnerability to the glycoside distinguishes the brush border system from the Na^+-independent, equilibrating sugar transporter in the basolateral membrane of these cells; the latter resembles the erythrocyte transporter in being inhibited at low micromolar levels of phloretin (by a strictly competitive mechanism), but not the glucoside, which is at least 100 times less potent. This is somewhat puzzling since one intuitively might think that phlorizin's glucosidic moiety should compete for the sugar-binding site on both transporters. Yet, even though the receptor site of the equilibrating system possesses a high-affinity inhibitory site for phloretin, the β-glucosidic group of phlorizin either prevents the binding or negates the inhibitory effect of the aglycone moiety. Con-

---

[1] Abbreviations include p-AmBPht and p-AzBPht (p-aminobenzyl phloretin and p-azidobenzylphloretin, respectively), p-AmBPhz and p-AzBPhz (p-aminobenzylphlorizin and p-azidobenzylphlorizin, respectively); TLC, thin-layer chromatography; BBMV, brush border membrane vesicles; SDS–PAGE, sodium dodecyl sulfate–polyacrylamide gel electrophoresis.

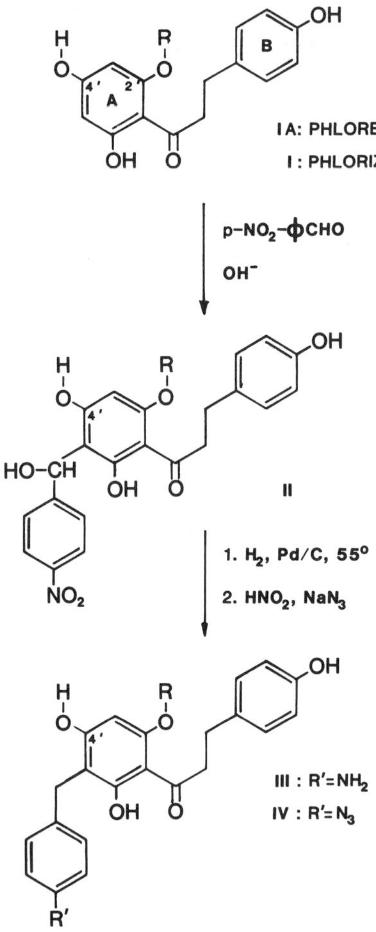

SCHEME A

FIG. 1. Scheme A: Route for the synthesis of *p*-azidobenzylphloretin and *p*-azidobenzylphlorizin. Compounds II, III, and IV are the 2'-β-glucosides (R = glc) and IIA, IIIA, and IVA are the corresponding aglycones (R = H).

versely, occupancy of the phloretin site on the brush border transporter by the aglycone alone causes insufficient perturbation to significantly interfere with glucose binding and transport[2]; the intact phloretinyl-β-glucoside is required for blockade. In ongoing attempts to gain information about the construction and biochemistry of the receptor sites on the respective transporters, some potential photoaffinity-labeling derivatives of these two

[2] D. F. Diedrich, *Am. J. Physiol.* **209,** 621 (1965); H. Vick, D. F. Diedrich, and K. Baumann, *Am. J. Physiol.* **224,** 552 (1973).

inhibitors have been prepared and examined for their membrane effects. The work of my students and contributions of my collaborators are identified in the following sections and are gratefully acknowledged.

Aryl azides are the most widely used photochemical reagents in the strategy of photoaffinity labeling receptor sites on proteins. Their utility and limitations have been reviewed and their disadvantages have been identified.[3] For example, photoactivation yields several reactive intermediates, including singlets (half-life on the order of milliseconds) plus a small fraction of the relatively long-lived triplet nitrenes. Special attention must therefore be given to minimize spurious labeling attendant with reactive species that can migrate from the original binding site. Nitrenes are also thought to react preferentially with nucleophilic groups and rarely are involved in random, near-neighboring hydrogen extractions and insertion into C—H bonds. However, this feature is not necessarily a disadvantage in the employment of the phlorizin or phloretin aryl azide derivatives. For example, the receptor for 4-azidophlorizin (IXA, Fig. 2) probably includes a polar group that serves as the specific interaction site for the $-N_3$ moiety.

Synthesis of the phlorizin/phloretin azides was a natural extension of our early work[4] in which the precursor amine derivatives were coupled to insoluble support media for affinity chromatography of the red cell sugar transporter. P-Aminobenzylphloretin (IIIA, Fig. 1) was the first of several derivatives designed for attachment to agarose without affecting the ligand's high affinity for the transporter. A matrix studded with this aglycone was found to bind intrinsic membrane proteins too tightly; the more polar p-aminobenzylphlorizin derivative (III, Fig. 1) is probably a preferred ligand.[5]

Synthesis of the Analogs: Scheme A (Fig. 1)

3'-(4-Nitrophenylcarbinol)-2',4',6',4-tetrahydroxy-dihydrochalcone (IIA)

Phloretin IA, Fig. 1) is commercially available (Aldrich Chemical Co., Milwaukee, WI) but expensive. It is easily prepared by hydrolyzing 5 g of

---

[3] H. Bayley and J. R. Knowles, this series, Vol. 46, p. 69; H. Bayley and J. V. Staros, in *"Azides and Nitrenes"* (E. F. V. Scriven, ed.), pp. 433–490. Academic Press, New York, 1984.
[4] F. F. Fannin and D. F. Diedrich, *Arch. Biochem. Biophys.* **158**, 919 (1973).
[5] The phloretin derivative actually bound band 4.5 (plus 3 and 7) too avidly [J. Weber, D. A. Warden, G. Semenza, and D. F. Diedrich, *J. Cell. Biochem.* **27**, 83 (1985)]. Only denaturing conditions would partially elute these proteins. The corresponding phlorizin ligand has a lower affinity for the transporter, is much more hydrophilic, and is less apt to be involved in nonspecific binding.

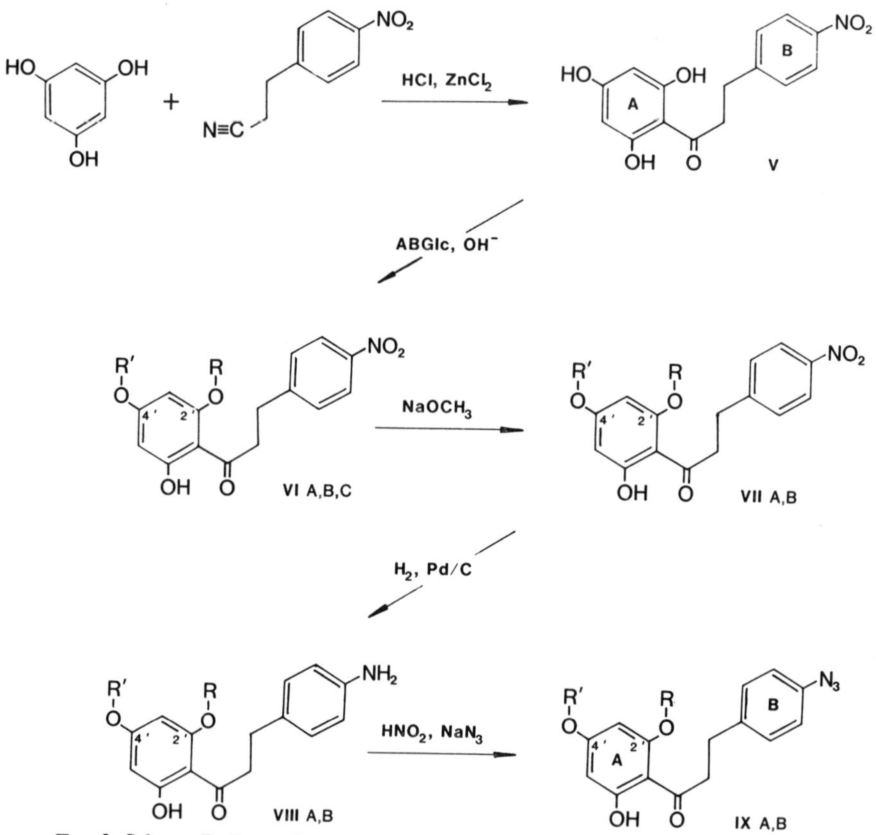

FIG. 2. Scheme B: Route for the synthesis of 4-azidophlorizin. Compounds **VIA – IXA** are 2'-glycosides (R = glc; R' = H); Compounds **VIB–IXB** are 4'-glycosides (R' = glc; R = H); Compound **VIC** is the 2',4'-diglycoside (R' = R = glc).

phlorizin (**I**, Fig. 1) in 25 ml ethanol with 50 ml of 1.25 N HCl at 90° for a few hours under nitrogen. On cooling, 2 g of crude **IA** precipitates. Phloretin can be purified by solubilizing it in hot ethylacetate (avoid direct light), adding 5 vol of benzene and allowing the mixture to cool.

Ten mmol of **IA** and 5 mmol of 4-nitrobenzaldehyde are dissolved in 35 ml methanol in an evacuation flask fitted with a pressure-equalizing, addition funnel containing 40 ml of cold 0.5 N sodium methoxide in methanol. Both solutions are deoxygenated (evacuated and flushed with N_2 twice), and then slowly mixed with cooling and kept for an additional 10–15 min at 0° with constant magnetic stirring. The orange mixture is neutralized under N_2 with cold 1.7 N acetic acid (Congo Red paper must just turn blue to ensure complete neutralization, otherwise oxidation occurs in alkali). The mixture is then poured into 600 ml of ice slurry to precipitate crude, yellow **IIA** and unreacted **IA**. The washed mixture is dissolved in 40% ethanol/water and chromatographed on a 400-ml column of Sephadex G-10 (preequilibrated in 40% ethanol) to which both compounds are avidly bound. After eluting phloretin with the same solvent [monitored by thin-layer chromatography (TLC)], **IIA** can be obtained with a higher (85%) ethanol eluant. Evaporation of the alcohol yields product (in 40–50% yield) which is recrystallized from aqueous methanol (mp >240°, with decomposition). The unstable carbinol cannot be crystallized to absolute purity and it is reduced without delay.

3′-(4-Aminobenzyl)2′,4′,6′,4-tetrahydroxy-dihydrochalcone (p-Aminobenzylphloretin) (p-AmBPht; **IIIA***)*

The semipure **IIA** (2.4 mmol) dissolved in 200 ml methanol is hydrogenated at 30 cm water pressure[6] with 200 mg of 10% palladium/carbon catalyst for 24 hr at 25° with magnetic stirring and then for 24 hr at 55°. The elevated temperature is required to completely reduce the carbinol group; at lower temperatures, an unstable, orange product is obtained. The catalyst is removed (*caution:* see Ref. 7), the solvent is evaporated, and the residue (dissolved in 160 ml 12% methanol in freshly distilled benzene) is chromatographed on a 250-ml silica gel SG-32 column packed in the same solvent. Elution is continued with 18% methanol/benzene to elute the amine within 200–360 ml. The solvent is evaporated and pure product is obtained from 60% ethanol/water at −15° as off-white crystals in about 50% yield (mp 219–220°). Its molar extinction coefficient is $\epsilon_{332} = 21,000$ in 0.025 M Na$_2$B$_4$O$_7$, pH 9.3; the ϵ_{325} of phloretin is 25,500.

3′-(4-Azidobenzyl)-2′,4′,6′,4-tetrahydroxy-dihydrochalcone (p-Azidobenzylphloretin)(p-AzBPht; **IVA***)*

All subsequent steps are conducted in subdued light to avoid photodecomposition of the azide. The amine, **IIIA** (38 mg, 0.1 mmol), is dissolved in 5 ml acetone and 2 ml 0.5 N HCl is added with constant stirring at −2° followed by 0.2 ml of 1.0 M sodium nitrite. After 5–6 min, 0.5 ml of 2.0 M sodium azide is added and the reaction is stopped after 10 min by adding excess urea to decompose the remaining nitrous acid and bring the pH to about 6.5. The acetone is evaporated and the aqueous solution is extracted twice with 10-ml portions of ethylacetate. This extract is evaporated to a red syrup which is applied as a 10% methanol/CHCl$_3$ solution to a silica

---

[6] A. I. Vogel, "A Textbook of Practical Organic Chemistry," 3rd ed., pp. 471–473. Longman, London, 1961.

[7] Spontaneous combustion occurs if the hydrogenated catalyst is allowed to dry on the paper.

gel 60 column (1.5 × 40 cm; in CHCl$_3$). Elution, with stepwise increases of 2–6% methanol in CHCl$_3$, gives pure azide in 290–335 ml eluate volume, preceded by a pink, and followed by a yellow contaminant. It crystallizes as pale yellow needles from aqueous methanol, mp 155–157°. ϵ_{250} = 19,000 and ϵ_{332} = 21,000 cm^{-1} M^{-1}, in 0.025 M borate, pH 9.3. Infrared spectra taken on a Beckman (Fullerton, CA) IR-8 spectrometer (KBr pellets) showed maxima at 2120 and 1290 cm^{-1} which are azide signals. The proton nuclear magnetic resonance (NMR) spectra of **IVA** (and **IIIA**) are comparable to phloretin [100 MHz, on a Varion (Sunnyvale, CA) HA-100 using tetramethyl-Si (δ = 0.000) as internal standard and D$_6$-acetone or CD$_3$OD as solvents]; the A$_2$B$_2$ pattern at 2.82 and 3.26 ppm arising from the —CH$_2$—CH$_2$—CO— moiety in phloretin also appears at 2.84 and 3.32 ppm in both the amine and azide derivatives. The new —CH$_2$— group exhibits a singlet at 3.85 ppm and there is a concomitant decrease in the A-ring singlet intensity from ^{2}H in phloretin to ^{1}H in compounds **IIIA** and **IVA**, whose spectra also show the expected AA'BB' patterns for the new para-substituted aromatic ring.

2'-(β-D-Glucopyranosyl)-4',6',4-trihydroxy-5'-(4-nitrophenylcarbinol)-dihydrochalcone (**II**)

The glucoside is prepared as described above for **IIA** with a few modifications. Phlorizin dihydrate (5 g, 10.6 mmol) and 4-nitrobenzaldehyde (3.2 g, 21 mmol) in 65 ml methanol are condensed anaerobically with 11.6 ml of 1 N KOH for 1.5 hr at 25° and an additional 2 hr at 55°. The reaction mixture is cooled, neutralized, and reduced in volume until slightly turbid. Impurities are then extracted four times with 50 ml of benzene. Crude **II** is extracted from the aqueous layer into ethylacetate (3 × 74 ml) which is washed once with water. The organic solution is then dried with MgSO$_4$; solvent evaporation gives the unstable, nitrocarbinol glucoside ready for reduction.

2'-(β-D-Glucopyranosyl-4',6',4-trihydroxy-5'-(4-aminobenzyl)-dihydrochalcone (p-Aminobenzylphlorizin; p-AmBPhz) (**III**)

The catalytic hydrogenation of **II** is carried out as described above for the aglycone, **IIA**, to yield the amine, **III**. The crude product is dissolved in minimal acetone, mixed with a little silica gel, and applied as a thick slurry to a 3 × 80 cm column of 200÷400-mesh silica gel packed in 5% methanol in CHCl$_3$. The column is washed with 600 ml of 5%, then 1 liter 15% methanol; the eluates are discarded. Elution with 20% methanol/CHCl$_3$ delivers pure **III** in the 550- through 750-ml fraction. The solvent is removed to yield 1.4 g of product (54%) which crystallizes from aqueous methanol as the monohydrate; after drying *in vacuo* at 80°, its melting point was 197–200°. The NMR spectrum of **III** is comparable to that of phlorizin; both compounds possess the A_2B_2 pattern at 2.88 and 3.50 ppm arising from the $-CH_2-CH_2-CO-$ moiety (the triplet at 3.5 ppm is partially obscured by the proton signals from the glucose group). Both B ring AA'BB' systems at 6.72 and 7.10 ppm are identical, but phlorizin's A ring AB system at 5.99 and 6.25 ppm is converted to a ^1H singlet at 7.42 ppm in compound **III**. The AA'BB' system of the new ring gives signals at 6.54 and 6.98 ppm and the $-CH_2-$ group is seen as a ^2H singlet at 3.72 ppm.

P-Azidobenzylphlorizin (p-AzBPhz) (**IV**)

The compound is prepared exactly as described above for the aglycone, **IVA**, but the synthesis can be scaled up 20-fold. After urea addition and acetone evaporation, nearly pure **IV** crystallizes from the cold solution in 90% yield. It is recrystallized from aqueous, 60% methanol, giving a cream-colored product; it does not have a sharp melting point. It softens at about 180° and then decomposes between 195 and 215°. It displays an infrared (IR) spectrum having sharp 2120 and 1280 cm^{-1} azide absorption peaks. Like phlorizin, the compound exhibits two ultraviolet (UV) maxima in water, one at 225 nm and the second at 283 nm, which shifts to 334 nm in borate, pH 9.3 ($\epsilon_{334} = 29,500$ and $\epsilon_{255} = 23,500$ cm$^{-1}M^{-1}$).[8]

Synthesis of the Analogs: Scheme B (Fig. 2)

2',4',6'-trihydroxy-4-nitro-dihydrochalcone (4-Nitrophloretin) (**V**)

The starting material, 3-(4-nitrophenyl)propionitrile,[9] is prepared by adding 52 ml (400 mmol) of hydrocinnamonitrile (Aldrich Chemical Co.) to 220 ml concentrated nitric acid at room temperature with constant stirring. After 24 hr, the mixture is poured onto crushed ice and the product is washed with water; suction filtration helps remove an oily, yellow contaminant. Crystals melting at 78–80° are isolated from methanol in about 40% yield. This nitrated nitrile (26.5 g, 150 mmol) and 19 g (152 mmol) anhydrous phloroglucinol (dried *in vacuo* at 95°) are dissolved in 2 liters anhydrous diethyl ether contained in a two-necked, 4-liter flask fitted with a CaCl$_2$ drying tube and an inverted thistle tube for the intro-

---

[8] A bathochromic shift in absorption maximum is characteristic of analogs having a free phenolic group with a pK_a of 7.2 in the para-position on the A ring [see J. O. Evans and D. F. Diedrich, *Arch. Biochem. Biophys.* **199**, 342 (1980)].

[9] G. Zemplen, Z. Csuros, A. Gerecs, and S. Aczel, *Ber. Dtsch. Chem. Ges.* **61**, 2486 (1928).

duction of dry HC1.[10] After adding 6.0 g of anhydrous $ZnCl_2$, the flask contents are cooled in an ice/salt bath and shaken occasionally while HC1 gas (Matheson Gas Products, Secaucus, NJ) is admitted via Teflon tubing. Gas must be added at a rate sufficient to prevent the reaction mixture from being drawn into the gas tank since the escaping HC1 cools the container. After about an hour, the gas intake tube is sealed off and the flask is allowed to stand in the cold; any HC1 escaping through the drying tube is trapped over water in a vented filter flask. After a day, the reaction mixture, now containing the bulky orange ketimine hydrochloride ($>$C$=$NH·HC1), is again charged with HC1 for an hour. If the reaction has not been kept anhydrous, a deep red–orange oil will have formed, but this eventually solidifies. After cold storage for a week, the HC1/ether is decanted into water in the hood (*not rapidly!* heat generation causes ether to boil) and the solid is washed with 50 ml portions of dry ether. The ether-free product is dissolved in 4 liters of water and the solution is boiled for about an hour in a large beaker. After standing overnight, **V** is harvested as pale yellow crystals (mp 268–272°) in a consistent 35–50% yield, about 18–22 g. The product can be recrystallized from aqueous methanol (no charcoal), mp 274–276°. Compared to phloretin, the proton NMR spectrum of **V** showed that the B ring AA'BB' system was shifted from δ 6.85 ($v_A v_{A'} = 7.01$, $v_B v_{B'} = 6.68$, $J_{AB} = 9$ Hz) to 7.80 ($v_A v_{A'} = 8.12$, and $v_B v_{B'} = 7.48$, $J_{AB} = 9$ Hz). The AB system H—C (3',5') at 5.80 remained unchanged. Acetates and benzoates of *I* have been characterized.[11]

2'-O-(β-D-Glucopyranosyl)-4',6'-dihydroxy-4-aminodihydrochalcone (4-Nitrophlorizin) (VIIA)

Compound **V** (4.5 g, 15 mmol) and 9.3 g (22.6 mmol) acetobromo-D-glucose[12] (Sigma Chemical Co., St. Louis, MO) are dissolved in 120 ml acetone and added to a suction flask fitted with a pressure-equalizing, addition funnel containing 90 ml of cold 0.25 *N* KOH. Both solutions are deoxygenated and then slowly mixed (keep cold). After 24 hr in the dark at 20°, the mixture is added to 2 liters of crushed ice containing 5 ml acetic acid. The sticky product solidifies after a few hours. It is washed with water, air dried, and then extracted twice into 50 ml $CHCl_3$. Most of the insoluble residue (2.5–3.0 g) is unreacted **V** while the extract contains various tri-

---

[10] It is essential that this gas entry tube be wide mouthed and remain barely submerged during and after HC1 addition; otherwise, the gassing port will become clogged as the product crystallizes.

[11] E. M. Gibbs, M. Hosang, B. F. X. Reber, G. Semenza, and D. F. Diedrich, *Biochim. Biophys. Acta* **688**, 547 (1982).

[12] M. E. Krahl and C. F. Cori, *Biochem. Prep.* **1**, 33 (1949).

and tetraacetates of glucose, some **V**, and the tetraacetylglucosides **VIA**, **VIB**, and **VIC**. The derivatives can be separated by conventional chromatography on a silica gel 60 column using increasing levels (up to 4%) of *n*-propanol in chloroform as solvent. However, flash chromatography[13] gives a better fractionation; the concentrated $CHCl_3$ extract (3.1 g solid) is added to a 4.5 × 26 cm silica gel 60 column that is pressure packed with $CHCl_3$, and pressure elution is carried out with 1 liter of 1.5%, 500 ml of 1.8%, and finally 1 liter of 3% *n*-propanol in $CHCl_3$. Eluates (25 ml) are monitored by TLC; volumes between 275 and 350, 375 and 525, and 675 and 1600 ml are pooled, concentrated to dryness, and the residues crystallized from methanol to yield 0.2 g of **VIC** (mp 197–198°), 0.4 g of **VIB** (mp 185–186°), and 0.3 g of **VIA** (mp 88–92°), respectively.

The proton NMR spectrum of **VIA** in $CDCl_3$ reveals that the AB system singlet signal of **V** at δ 5.80 is shifted to δ 6.03 [$v_A = 6.03$, $v_B = 5.98$, $J = 2$ Hz, H—C (3′, 5′)] which indicates substitution at the 2′-position. **VIC** displays signals of the 2′,4′-disubstituted analog with an AB system at δ 6.13 [$v_A = 6.23$, $v_B = 6.03$, H—C (3′, 5′)]. Although the NMR spectrum of **VIB** has not been obtained, the saponified **VIIB** has a signal for a 4′-monoglucoside substitution, i.e., an H—C (3′, 5′) singlet at $\delta = 6.08$.

Saponification of both **VIA** and **VIB** (1 mmol) proceeds smoothly (*in vacuo*) in 2 ml cold 1 *M* methanolic sodium methoxide in subdued light. After 15 min at room temperature, the solution is neutralized under nitrogen with 0.3 *N* HCl to give crystalline deacetylated glucoside in 90% yield. Trihydrates of both **VIIA** and **VIIB** crystallize from 20% methanol; the analytical sample of **VIIB** is obtained from ethylacetate (white plates, mp 200–202°) whereas **VIIA** crystallizes from methanol as hard yellow prisms, mp 225°.

2′-O-(β-D-Glucopyranosyl)-4′,6′-dihydroxy-4-aminodihydrochalcone (4-Aminophlorizin) **(VIIIA)**

The trihydrate of **VIIA** (200 mg, 0.38 mmol) is reduced as described above for the hydrogenation of **IIIA** except that an elevated temperature is not required. Efficiency of magnetic stirring determines the incorporation rate of the 3 Eq of H_2. The catalyst is removed (*Caution:* see Ref. 7) and after evaporating most of the methanol, solid **VIIIA** is obtained by adding water. Recrystallization from dilute methanol yields the tetrahydrate which can be dried at 60° *in vacuo* (mp 139–142°).

The isomeric *4- amino-phlorizin* **(VIIIB)** is prepared by hydrogenation of **VIIB** as described above for **VIIA**; it crystallizes from aqueous methanol as an off-white trihydrate which can be dried at 100° *in vacuo* (mp 159–

---

[13] W. C. Still, M. Kahn, and A. Mitra, *J. Org. Chem.* **43**, 2923 (1978).

162°). $\epsilon_{283} = 19,000$ and $17,000$ cm^{-1} M^{-1} at pH 6.5 and 8.0 (5 mM phosphate, 5% ethanol), respectively. No shift to higher wavelength occurs in alkali.[8]

When mixtures of **VIIIA** and **VIIIB** (plus contaminating diglucoside) are occasionally obtained due to inadequate prepurification of **VIIA, VIIB**, and **VIIC**, the deacetylated glucosides are readily separated by chromatography on a chloroform-packed silica gel 60 column. The mixture is applied as a 35% methanol/CHCl$_3$ solution and elution is performed with stepwise 20–40% methanol mixtures. Alternatively, fractionation with aqueous alcohol can also be obtained on Sephadex G-10 columns to which 4-aminophlorizin is avidly bound. Those derivatives having highest affinity for this matrix (but less for G-25 and nearly none for G-50) possess phenolic and/or amino groups on both the A and B ring systems;[14] 4-amino-*p*-phlorizin (and the diglucoside) are eluted with 20% aqueous methanol whereas 60–70% will desorb the 4-aminophlorizin.

2'-O-(β-D-Glucopyranosyl)-4',6'-dihydroxy-4-azidodihydrochalcone (4-Azidophlorizin) **(IXA)**

The synthetic method resembles that used to prepare **IVA** and **IV**. The 4-aminophlorizin (218 mg) is dissolved in 25 ml acetone to which 10 ml of 0.5 N HCl is added (keep at $-2°$). Cold 0.54 N sodium nitrite (2.5 ml) is added over a 15-sec period and after 6 min of stirring at subzero temperature in subdued light, 2.5 ml of a 2.0 N sodium azide solution is added dropwise followed in 10 min by the addition of excess urea. Acetone is evaporated from the neutral solution and azide is extracted into ethylacetate (2 × 25 ml), which is then dried with anhydrous sodium sulfate. Solvent removal gives crude **IXA** in 70–85% yield. The product is recrystallized from dilute methanol and the monohydrate is dried over CaCl$_2$ at 25°; after softening at 108–110°, it forms a red melt at 150–155°. Anhydrous azide is not obtained by drying at higher temperature without decomposition. **IXA** had an IR spectrum essentially the same as phlorizin except for sharp 2130 and 1280 cm^{-1} azide absorption peaks. The derivative has a UV absorption maximum at 251 nm but otherwise its spectrum is similar to phlorizin; at pH 6.5 or lower, both have absorption maxima at 285 nm ($\epsilon = 15,000$ versus 16,500 for **IXA**) which shift in alkali (0.025 M sodium borate, pH 9.3) to 325 nm (328 nm for **IXA**) with $\epsilon = 25,500$ and $27,000$ cm$^{-1}M^{-1}$ for phlorizin and **IXA**, respectively.

4-Azidophlorizin can also be fractionated on a high-performance liquid chromatography (HPLC) Nucleosid 10-C8 column (250 × 4.6 mm diameter) connected to solvent delivery pumps for the creation of a linear elution

---

[14] J. O Evans and D. F. Diedrich, *Arch. Biochem. Biophys.* **199**, 342 (1980).

gradient. Initial elution conditions are 90% solution A (0.1% trifluoroacetic acid in water) mixed with 10% solution B (85% acetonitrile in water containing 0.08% trifluoroacetic acid). Elution of the azide is detected by its absorbance at 285 nm and occurs when the mixture consists of about 45% solution B.

4'-O-(β-D-Glucopyranosyl)-2',6'-dihydroxy-4-azidodihydrochalcone (4-Azido-p-phlorizin) **(IXB)**

Essentially the same procedure described above for the preparation of **IXA** is followed to convert the isomeric **VIIIB** to **IXB**. The reaction is performed in aqueous 15% methanol with 5 Eq of HCl; timing of the nitrite and azide additions are unchanged. After neutralization and methanol removal, crude **IXB** precipitates. It is either recrystallized as colorless needles from 50% methanol or it can be ultrapurified by silica gel 60 chromatography (CHCl$_3$/methanol). Yield is about 50%. The monohydrate (dried at 25°) melts at 168–170° after discoloring at about 160°.

Thin-Layer Chromatography

Compounds were identified chromatographically on silica gel GF thin-layer plates using phloretin and phlorizin as standards. R_F values of all the compounds in the solvent systems used here are listed in Table I.

Preparation of ^{3}H-Labeled Amines and Azides

The Amersham Corporation (Arlington Heights, IL) employed their TR.7 procedure to catalytically exchange the benzylic hydrogens on two of the amines. P-Aminobenzylphlorizin (0.035 mmol) in 2 ml 4% KOH was exposed to 10 Ci tritium gas with 10% Pd/CaCO$_3$ catalyst for 1 hr with stirring. The mixture was then filtered directly into 1 ml iced glacial acetic acid and labile tritium was removed by repeated rotary evaporation with methanol to dryness. Two separate preparations of **III** had relatively low specific activity of 0.5 and about 2 Ci/mmol by this TR.7 procedure. On the other hand, 10 times greater activity was incorporated into 4-aminophlorizin **(VIIIA)** when this compound was tritiated by this method. The degree of tritium incorporation appears to depend greatly upon the activity of the exchange catalyst employed. Since extremely high specific activity [^{3}H]phlorizin can be prepared by an initial mild bromination followed by catalytic debromination in the presence of tritium gas,[15] the 4-aminophlorizin **(VIIIA)** was subjected to this treatment and the TR.3 procedure by

---

[15] J.-T Lin and K.-D. Hahn, *Anal. Biochem.* **129**, 337 (1983).

TABLE I
THIN-LAYER CHROMATOGRAPHY OF PHLORIZIN DERIVATIVES[a]

Compound	R_f values in solvent systems			
	A	B	C	D
IA	0.45	0.64	0.86	
I	0.17	0.35	0.42	
IIA	0.49	0.70		
II		0.44		
IIIA		0.57		
III		0.26		
IVA		0.70	0.92	
IV		0.46	0.57	
V	0.56			
VIA	0.69			0.14
VIB	0.74			0.29
VIC	0.80			0.40
VIIA	0.25	0.46	0.60	
VIIB	0.21	0.34		
VIIIA	0.12	0.26	0.30	
VIIIB		0.15		
IXA		0.54	0.67	
IXB		0.20	0.50	

[a] All solvents were chloroform plus one of the following: 15% n-propanol (solvent A); 22% methanol (solvent B); 30% methanol (solvent C); and 3% n-propanol (solvent D).

the Amersham Corporation. Upon purification, a product with about 18 Ci/mmol was obtained by this method. The other two amines (**III** and **IIIA**) should also survive this procedure.

Tritiation by either process leads to considerable destruction and the labeled amines must be purified before being converted to azides. Chromatography on silica gel in chloroform/methanol causes an intolerable amount of decomposition when the microamounts of radioactive amines are purified in this manner. Recovery may be improved by using methylene chloride instead of chloroform as eluant solvent and by conducting the chromatography in the absence of oxygen and in subdued light. On the other hand, the high specific activity contaminants can be separated in the aqueous Sephadex G-10 system (a 450 × 8 mm diameter column) with good product recovery if the stepwise increase of methanol in the eluant is gradual enough.

The stability of the labeled derivatives varies. Although azides prepared from moderately tritiated amines are stable for many months when stored in methanol/benzene or aqueous alcohol at $-20°$ or lower, more than 50% deterioration of the high specific activity IXA occurred after about a year at $-20°$ in methanol/chloroform. Cold storage of aqueous ethanol solutions appears to be a better choice of conditions.

Subtle structural changes that occur upon storage are easily overlooked. Rearrangement or decay products having very similar R_f values on TLC are not easily separated by column chromatography and the care required in examining the autoradiograms of the developed TLC monitoring plates cannot be overemphasized. We routinely prespot some nonlabeled derivative at the TLC origin before applying the radiolabeled sample within the boundary of the dried carrier spot. This avoids the significant chemistry (oxidation?) that otherwise occurs on the reactive absorbent surface and minimizes the telltale blackening that appears on the developed radiogram film corresponding to the spot origin. The unlabeled derivative also serves as an internal standard; after developing the plate, the fluorescence-quenching spot of amine or azide is circumscribed with a soft carbon pencil before spraying with Enhance (New England Nuclear, Boston, MA) and preparing the autoradiogram. The pencil mark will appear on the film background as a clear ring. Any radioactive decomposition product with an R_f nearly identical to the standard will be recognizable as a spot off-center to the ring.

The Azides as Membrane Perturbants and Photoaffinity-Labeling Agents

Renal and Intestinal Brush Border Membrane Vesicle (BBMV) Systems

The work presented here is that of E. M. Gibbs and our Swiss collaborators, M. Hosang, B. F. X. Reber, and G. Semenza.

Affinity for the Na^+-Dependent Glucose Transporter. The glucose-binding site of the Na^+/glucose cotransporter complex also serves as the subsite to which the glycosidic group of phlorizin binds.[16] The aglycone portion of the inhibitor presumably also interacts with this subunit of the multimeric transport system,[17] rather than some neighboring protein or lipid component in the membrane. Based on structure–activity studies and molecular model constructs, it seems likely that phlorizin's A ring interacts with a

[16] D. F. Diedrich, *Biochim. Biophys. Acta* **71**, 688 (1963); G. Toggenburger, M. Kessler, A. Rothstein, G. Semenza, and C. Tannenbaum, *J. Membr. Biol.* **40**, 269 (1978); M. Silverman and J. Black, *Biochim. Biophys. Acta* **394**, 10 (1975).
[17] H. Koepsell and A. Madrala, *Top. Mol. Pharmacol.* **4**, 169 (1987).

receptor site that is coplanar to the sugar-binding locus. On the other hand, the phenolic B ring forms a strong hydrogen bond with a membrane component that is out of the plane and about 12Å removed from the glucose fixation site.[18] Those derivatives capable of completing these minimal binding and alignment requirements are the most potent *competitive* inhibitors of glucose transport. Thus, phloretin is a weak inhibitor and acts noncompetitively (allosterically[19]), and phlorizin chalcone (whose A and B rings are kept coplanar by an intervening unsaturated —CO—CH=CH— chain) is inactive. These clues about receptor structure led us to reason that substitution of an electronegative azide group for the —OH on the B ring would not compromise the derivative's high-affinity binding. Evidence for this view is presented in Table II. The ability of phlorizin and its analogs to interact with the transporter in BBMV prepared from rabbit intestine and rat kidney is indicated by three different analyses, all conducted in dim light: (1) capacity to inhibit Na^+-coupled glucose transport (K_i'), (2) specific, Na^+-dependent binding affinity (K_d'), and (3) ability to inhibit the specific binding of tritiated phlorizin (K_i'). All three criteria indicate that 4-azidophlorizin binds to the rat kidney membrane transporter as well as phlorizin.[20] Other results (not tabulated here) indicate that the azide is also almost as active as phlorizin in BBMV from rat intestine (the respective K_i' values are 10 and 4.2 μM). However, the affinities of the two compounds for the transporter in rabbit tissues differ: The K_i' with kidney membranes is 7.5 μM for phlorizin and 140 μM for the 4-azido analog. This same 15-fold difference was found with BBMV from rabbit intestine and the results are consistent with the view that the cotransporter is not identical in the two tissues.[17]

Table II displays other pertinent results. The 4-aminophlorizin is as potent a transport inhibitor as phlorizin while the 4-nitro derivative is 10 times less active. Since the pK_a of aromatic amines is about 5, the —NH_2 group is unionized under our conditions (pH 7.5), which allows it to form the essential hydrogen bond. Bulkiness or electronic differences of the NO_2 group significantly interfer with its interaction. However, since the equally large —N_3 moiety is tightly bound, other factors (perhaps lipid solubility) apparently allow for its size. Conversely, 4-azido-*p*-phlorizin (**IXB**), even at 200 μM, inhibits transport less than 50%. This finding resembles my earlier

---

[18] D. F. Diedrich, *Biochim. Biophys. Acta* **71**, 688 (1963).

[19] F. Alvarado, *Biochim. Biophys. Acta* **135**, 483 (1967); D. F. Diedrich, *Arch. Biochem. Biophys.* **117**, 248 (1966).

[20] The cited values are based on nominal concentrations of the derivatives. The small difference between the azide and phlorizin may be due to the derivative's greater hydrophobicity (seven or eight times larger oil/water partition coefficient) and/or nonspecific binding (M. Hosang and E. M. Gibbs, unpublished observations).

TABLE II
HALF-SATURATION CONSTANTS OF PHLORIZIN AND DERIVATIVES FOR THE GLUCOSE TRANSPORTER IN RAT RENAL AND RABBIT INTESTINAL BRUSH BORDERS[a]

Inhibition of glucose transport by	Rat kidney	Rabbit intestine
	K'_i values (μM)	
Phlorizin	2.1 ± 1.2	8 ± 1.2
4-Azidophlorizin (**IXA**)	5.2 ± 1.9	139 ± 9
4-Aminophlorizin (**VIIIA**)	2.8	44 ± 7
4-Nitrophlorizin (**VIIA**)	17	445 ± 50
4-Azido-p-phlorizin (**IXB**)	> 200	—
p-Azidobenzylphlorizin (**IV**)	> 270; > 300	—

Specific, Na$^+$-dependent binding of	Rat kidney	Rabbit intestine
	K'_d values (μM)	
Phlorizin	1.2 ± 0.3	4.6 ± 0.85
4-Azidophlorizin	3.2 ± 0.7	—

Inhibition of specific, [^{3}H]phlorizin binding by	Rat kidney	Rabbit intestine
	K''_i values (μM)	
Phlorizin	0.9 ± 0.08	12 ± 2
4-Azidophlorizin	3.8 ± 1.7	156 ± 40

[a] Brush border membrane vesicles were incubated in the presence of varying concentrations of phlorizin and its derivatives to measure (1) their inhibitory action on the 2-sec (intestine) and 7-sec (kidney) rates of Na$^+$ gradient-coupled D-glucose uptake yielding K'_i ± SD values from two to four separate Dixon plots; (2) the Na$^+$, ψ-dependent binding of each tritiated ligand (K'_d values) from Scatchard plots at pH 6.5 at the same time periods indicated in (1) above. The K'_d of the 4-azidophlorizin could not be determined with BBMVs from rabbit intestine because of its very high non-specific binding in this relatively transporter-poor tissue (20 pmol/mg BBMV protein compared to 100 pmol for the renal vesicles); (3) their ability to competitively block specific [^{3}H]phlorizin binding at pH 6.5 and 7 sec (kidney) and pH 7.5 and 2 sec (intestine) to obtain an estimate for K''_i. K'_i and K''_i are the respective inhibition constants of (Na$^+$, $\Delta\psi$)-dependent D-glucose uptake and D-glucose-protectable [^{3}H]phlorizin or 4-azido[^{3}H]phlorizin binding. K'_d is the apparent dissociation constant at the indicated pH. All refer to the total (ionized + unionized) ligand concentrations. (See Gibbs et al.[11] for other details.)

observation that transposition of the glucosyl moiety to the para-position in phlorizin leads to a loss of virtually all affinity for the cotransporter.[2] Finally, placement of a benzylazide group on phlorizin's A ring to form *p*-AzBPhz (**IV**) leads to a profound loss in apparent affinity for renal BBMV transporters ($n = 2$). B. Addison and I have recently found that the *p*-aminobenzyl derivative (**III**) at a nominal concentration of $100\mu M$ also fails to block glucose uptake from the rat intestine *in situ* (unpublished observations). However, high nonspecific binding (which was not measured) may have reduced the effective free concentration of both analogs in these experiments.

Photolabeling Studies. Since the calculated K'_d for 4-azidophlorizin was relatively small, we reasoned that the offrate of its photogenerated nitrene from the receptor would be slow enough to allow its insertion into the site with minimal spurious labeling. Photolysis studies were therefore performed with our Zürich collaborators using transporter-enriched membrane vesicles from rabbit intestine and tritiated 4-azidophlorizin (405 mCi/mmol) at its K'_i (100 μM) in the presence of a NaSCN gradient, pH 7.0, and with *p*-aminobenzoate added to scavenge unbound photogenerated intermediates. Irradiation was for 1 min at 20° with 315-nm and longer light generated by a 450-W mercury lamp.[21] Although much nonspecific labeling occurred under these conditions, one of the tagged proteins, with an apparent M_r of 72,000, was identified as the cotransporter (or a part thereof) in SDS-PAGE patterns based on the following criteria: (1) its labeling was reduced when phlorizin (but not the low-affinity *p*-phlorizin) or choline (instead of Na$^+$) was present; (2) the extent of labeling was proportional to membrane purification and specific activity of the transporter in the vesicle preparation; and (3) it was recognized by a monoclonal antibody which specifically inhibited Na$^+$/glucose cotransport and Na$^+$-dependent phlorizin binding in these membranes.[22]

In some preliminary experiments with BBMV from rat kidney, Gibbs and Reber (unpublished) found that a limited number of polypeptides, having M_r values of 130,000, 65,000–72,000 48,000–50,000, 37,000, and 27,000, were labeled when 4-[^{3}H]azidophlorizin was photolyzed under a variety of conditions. However, no Na$^+$ dependency or protection by phlorizin could be shown. Use of the high specific activity azide under optimal and more discriminating conditions should enhance specific labeling (see later) and help solve the problem of how the cotransporter is functionally inserted into the membrane.

[21] M. Hosang, E. M. Gibbs, D. F. Diedrich, and G. Semenza, *FEBS Lett.* **130**, 244 (1981).
[22] U. M. Schmidt, B. Eddy, C. M. Frazer, J. C. Venter, and G. Semenza, *FEBS Lett.* **161**, 279 (1983).

The Human Erythrocyte and K-562 Cell Transporter
  Apparent Affinity of p-AzBPht and p-AzBPhz for the Sugar Transporter. The work presented here is that of F. F. Fannin, J. O. Evans, J. Dozier, and C. Maynard.

  Equilibrium exchange of 3-methoxyglucose in the red cell: Freshly drawn erythrocytes are washed four times in ice-cold pH 7.4 buffer (27 mM glycylglycine in 150 mM NaCl) or pH 6.2 buffer (150 mM NaCl, 20 mM sodium phosphate) to remove white cells and deplete intracellular glucose. The red cells are equilibrated for 1 hr (20°) in either buffer containing 3.5, 7, and 21 mM 3-[^{3}H]methoxyglucose.[23] The cells are then centrifuged for 5 min at 3000 g; the supernatant as well as the residual fluid obtained after a second spin are removed to obtain RBCs packed to about 85% hematocrit. Flux studies are performed using a filter technique[25] in which a 20-μl aliquot of loaded cells is injected into 10 ml of a vigorously stirred wash buffer (10°) containing the same concentration of unlabeled sugar originally loaded in the cells plus inhibitor (added as a dilute ethanol solution; final alcohol less than 0.5%. See Ref. 26). At least four samples of the cell suspension are withdrawn at 7-to10-sec intervals into 10-ml plastic syringes through Swinnex filter assemblies fitted with 3-μm pore SS filters and AP prefilters (Millipore Corporation, Bedford, MA). The time each filtrate enters the syringe ($\pm$0.5 sec) is registered orally on a cassette recorder while the ticks of an adjacent metronome (set to indicate the expired time

[23] Thin-layer chromatography and autoradiography indicated that significant chemical decomposition of especially this sugar can occur when it possesses high specific activity (e.g., 80 Ci/mmol). Over the course of our experiments, the compound was purified either by preparative TLC (methylene chloride:methanol:water, 65:25:4) or by removing anionic oxidation products on minicolumns of Dowex-1 exchange resin.[24]

[24] D. F. Diedrich, J. C. Dozier, and S. Turco, *J. Cell. Biochem., Suppl.* **6**, 152 (1982); J. C. Dozier, D. F. Diedrich, and S. Turco, *J. Cell. Physiol.* **108**, 77 (1981).

[25] R. B. Gunn and O. Fröhlich, *J. Gen. Physiol.* **74**, 351 (1979).

[26] These azides are not stable for more than 1 h or so in buffers, especially if pH is much above 6.5. They are therefore stored as 30–50% ethanol stock solutions and are added to the medium just before the procedure. Furthermore, *p*-azidobenzylphlorizin and especially *p*-azidobenzylphloretin have high affinity for plastic unless they are dissolved in relatively high levels of alcohol and incubations must be conducted in glass vessels. Delivery of an aqueous solution having a low concentration of *p*-AzBPht with an automatic pipet can result in as much as 40–50% binding to the plastic pipet tip. The more polar *p*-AzBPhz is minimally but also significantly bound under these conditions. Quantitative transfer can be performed if sufficient alcohol (>35%) is present in the inhibitor stock solutions. For this reason, the apparent association and inhibition constants cited here reflect the minimal affinities of the agents since they are usually uncorrected for this random binding. Furthermore, the aglycone undergoes unspecific nonsaturable binding to cells which can amount to 70–85% at the hematocrits examined in these studies, whereas *p*-AzBPhz binding to cells is saturable (about 10–15%).

in seconds) are simultaneously recorded. An "infinite time" sample is taken to establish the amount of tritium originally delivered with the packed cell suspension to the efflux medium. The mean value of the efflux rate constant (variance about 10–15%; $n = 3$ to 7 experiments) is estimated from the first order time course of tritium appearance in each cell-free filtrate. The K_t and V_{max} values for the transport process determined from double-reciprocal plots were 10 mM and 22.2 nmol/min/μl cells. Dixon plots ($1/V_o$ versus nominal [I] plus secondary plots of 1/[3-methoxyglucose] versus slopes) were constructed to determine the K_i' and type of inhibition of each derivative at minimally two concentrations.

Sugar transport in the K-562 cell: An original K-562 cell culture was obtained by Dr. S. Turco as passage No. 211 from Prof. C. B. Lozzio and cells were used before passage 240. Logarithmically growing cells were washed twice in Krebs–Ringer phosphate (KRP) buffer, pH 7.4, and harvested by centrifugation at 2000 g for 2–5 min at ambient temperature.

For the zero trans influx experiments, 20 μl of a 5% cell suspension is rapidly mixed at zero time with an equal volume of KRP buffer containing 2 to 40 mM 3-[^{3}H]methoxyglucose and tracer amounts of L-[^{14}C]glucose as extracellular marker plus three concentrations of inhibitor. At a specified time (usually 15 sec), 4 ml of ice-cold stop solution is added and the mixture is rapidly filtered through 5-μm cellulose nitrate filters using not more than 10 cm water vacuum. Leakage of tritiated sugar from the cells in the cold diluent is not detectable for at least 4 min. The filter disk holding the collected cells is promptly removed from the suction (air flow will cause cell lysis) and its underside is rapidly blotted with adsorbent paper to minimize fluid entrapment (amounting to 1.35% of the ^{14}C present). Other details of the procedure and results have been published.[24] The apparent half-saturation constant for zero trans influx, K_t, was 2.5 mM and V_{max} was 2.9 nmol/min/μl cells; incorrect values had been reported in our earlier work.

For the zero trans efflux experiments, aliquots of a washed 50% cell suspension in KRP buffer are transferred to 1.5-ml plastic tubes which are then centrifuged at less than 2000 g (high speed in a microfuge will cause cell damage). After removing all supernatant, the cells are loaded with tritiated 3-methoxyglucose at the desired concentration for 45 min at ambient temperature. After a low-speed spin, the packed cells are stored on ice. More than 96% of the tritium in the cells remains unmetabolized sugar[24] and from 70 to 90% of the cells remain viable (impermeable to Trypan Blue) for up to 2 hr after these manipulations. Sugar efflux is determined as described above for 3-methoxyglucose equilibrium exchange in erythrocytes except that the 10 ml of efflux medium contained no sugar. Since the loaded K-562 cells after packing are too sticky to be rapidly and completely transferred with a plastic tip pipet, a 20-μl droplet of cells is collected on a tared, glass inoculation loop, quickly weighed, and then plunged into the rapidly stirred medium. First order efflux rate constants are determined (20% variance, linear regression) for each experimental condition and double-reciprocal and Dixon plots are constructed to estimate kinetic parameters. K_t and V_{max} for uninhibited efflux are 11.7 mM and 23 nmol/min/μl cell volume, respectively. Comparison of the values determined for the equivalent influx reaction (see above) indicates that the characteristic transporter asymmetry in the mature RBC also exists in the K-562 precursor cell. Its exterior conformation has about five times greater affinity for the sugar than the cytoplasmic assembly, but the maximal entry mechanism operates only about one-eighth as fast as the exit process.

Inhibition of sugar transport: In earlier erythrocyte transport inhibition studies,[27] initial zero trans influx or efflux rates were measured to determine the membrane side at which phloretin and p-AzBPht exerted their effects. When either inhibitor was initially present only in the medium, both appeared to act exclusively on outwardly oriented transporter units; inhibition was competitive under sugar influx conditions and noncompetitive when efflux was tested. However, if the cells were preincubated with either of these membrane-permeable agents, they attained significant free intracellular concentrations and both exhibited mixed inhibition kinetics, suggesting that they could also block from the cytoplasmic side. Unambiguous equilibrium exchange experiments demonstrated that phloretin, phlorizin, and their benzylazides act strictly competitively. The results summarized in Table III illustrate that at pH 7.4 (when each compound is 50% protonated), p-AzBPht inhibits at least 10 times more effectively than phloretin. The most surprising finding was that when the benzylazide group was attached to the phlorizin molecule, inhibitory potency was increased more than 100-fold over the parent glucoside. Furthermore, p-AzBPhz was even more potent when the pH of the medium was adjusted to 6.2, consistent with the idea that the unionized form of these agents preferentially binds to the exteriorized transporter site. The effective concentration of the protonated p-AzBPhz increases as pH is lowered because it fails to penetrate the cell[28]; this is not so for p-AzBPht because the

---

[27] F. F. Fannin, J. O Evans, E. M. Gibbs, and D. F. Diedrich, *Biochim. Biophys. Acta* **649**, 189 (1981).

[28] Other evidence that the glucoside does not penetrate the membrane includes the following: (1) binding of p-[³H]AzBPhz to erythrocytes at the hematocrit used to study transport (usually 0.2%) is saturable and easily reversed simply by washing with fresh buffer, unlike that seen for phloretin and p-AzBPht; and (2) in preliminary photolabeling studies, when up to 5% of the radiolabeled glucoside became covalently bound to red blood cells, no hemoglobin labeling occurred, again unlike the results of parallel experiments with tritiated p-AzBPht.

TABLE III
INHIBITION OF 3-METHOXYGLUCOSE TRANSPORT IN ERYTHROCYTES AND K-562 CELLS BY PHLORETIN, PHLORIZIN, AND THEIR AZIDOBENZYL DERIVATIVES[a]

Type of transport system	Inhibition type and K_i' values (μM) (uncorrected for nonspecific binding)			
	Phloretin	p-Azidobenzylphloretin	Phlorizin	p-Azidobenzylphlorizin
Erythrocytes				
Equilibrium exchange	Competitive	Competitive	Competitive	Competitive
pH 7.4	1.7	0.4	180	1.7
pH 6.2	—	0.30, 0.53	—	0.35, 0.59
K-562 cells				
Zero trans influx	Competitive 4.1	—	Competitive 225	—
Zero trans efflux	—	Noncompetitive 0.37	—	Noncompetitive 2.2

[a] Details of methods used to determine inhibition type and potency are described in the text.

unionized aglycone is highly membrane permeable, enters the red cell, and becomes preferentially bound to hemoglobin. Free intracellular concentrations sufficient to interact with the cytoplasmic face of the transporter are apparently not attained since the aglycone still behaves as a competitive inhibitor.

The results with K-562 cells shown in Table III also fit this model. Zero trans influx of 3-methoxyglucose is inhibited by phloretin and phlorizin in a strictly competitive manner (no preincubation) and their relative effectiveness resembles that found in the red cell studies. When the nonpenetrating p-AzBPhz was examined under zero trans efflux conditions, it produced a noncompetitive blockade (Fig. 3), as would be expected if it bound only to the exterior face of the transporter. However, the inhibitory effects of p-AzBPht are more complex under these conditions. A hyperbolic curve was obtained in the Hunter–Downs plot which identifies a special form of mixed inhibition, coupling, or uncompetitive inhibition. The aglycone appears to be most effective when the transporter is loaded, suggesting that a substrate$_{in}$–transporter–inhibitor$_{out}$ complex is preferentially formed. This is not the mechanism by which p-AzBPht acts on the red cell membrane, where it is strictly a noncompetitive inhibitor of zero trans galactose efflux.[27] A coupling mechanism as described above would

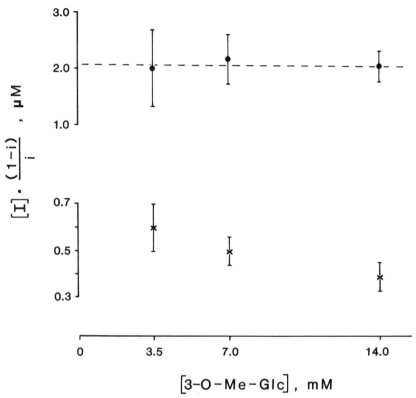

FIG. 3. Hunter–Downs plot of the 3-methoxyglucose zero trans efflux inhibition data. Values represent means ± SD of eight observations from two cell preparations of K-562 cells, preloaded with tritiated sugar at concentrations of 3.5, 7.0, and 14 mM, which were injected into sugar-free efflux media containing either (×) p-azidobenzylphloretin (at 0, 0.4, or 0.8 μM) or (●) p-azidobenzylphlorizin (at 0, 1.5, or 4.0 μM). The appearance of labeled 3-methoxyglucose in at least three consecutive cell-free filtrates obtained in 7- to 10-sec intervals was measured. [I], Inhibitor concentration (not corrected for nonspecific binding); i, the fractional inhibition of flux, equals $1 - (v_i/v)$. A straight line with no slope indicates noncompetitive inhibition, with the ordinate intercept being the respective K_i' value; a hyperbolic curve (as obtained for the p-AzBPht inhibition data) in this plotting method for enzyme kinetics suggests that an EIS complex is formed (noncompetitive inhibition). In this case, the inhibition constant is equal to the intercept on the ordinate of an imaginary line parallel to the x axis and intersecting the data point at the highest substrate concentration.

also be inconsistent with current thinking (but see Ref. 29) that the internal and external sugar-binding sites of the red cell transporter are not simultaneously exposed. The different inhibitory effects of p-AzBPht in these two cells may be caused by its ability to reach intracellular inhibitory concentrations in the hemoglobin-free K-562 cell but not the erythrocyte.

[29] A. L. Helgerson and A. Carruthers, *J. Biol. Chem.* **262**, 5464 (1987).

Interaction of p-AzBPht and p-AzBPhz with the Anion Transporter. The work presented here is that of C. Maynard in collaboration with Dr. Otto Fröhlich, Emory University, Atlanta, Georgia.

Although phloretin has been long recognized as the premier inhibitor of sugar transport in the red cell, Fröhlich and Gunn[30] have shown phloretin to be a mixed inhibitor of chloride exchange having a K_i of 1-2 μM. They propose that it binds to the outward facing transporter in both the unloaded and loaded state, although not necessarily with the same affinity. On the other hand, at the low concentrations which block exchange, phloretin has no effect on *net* chloride efflux.[31] That p-AzBPht and p-AzBPhz also have high affinity for this phloretin receptor on the anion transporter was discovered in some early unpublished experiments performed with Robert Gunn and colleagues. Both of these derivatives blocked chloride exchange; inhibition was either mixed or noncompetitive and their respective K_i values, compared to phloretin, were 0.8-1.5, 0.75-1.4, and 2.1-2.5 μM (corrected for nonspecific binding). We have recently repeated these experiments using the Swinnex filter/syringe technique as originally described by Gunn and Fröhlich[25] and have confirmed that p-azidobenzylphlorizin exhibits a half-maximal inhibitory effect on tracer-chloride exchange at a nominal concentration of 1.5 μM (final hematocrit of 1-2%, 0°, pH = 7.0). The membrane-impermeable p-AzBPhz may therefore represent a useful probe to examine the outer face of the RBC anion transporter. It is interesting that the derivative inhibits the transport mechanisms for glucose (competitively) and anion (noncompetitively) at about the same low micromolar concentrations.

Effect of p-AzBPhz on Salt and Water Flux.. While studying the inhibitory effect of p-AzBPhz on sugar transport (in dim light), we unexpectedly found that it also caused red cell swelling *in isosmotic medium.* The effect was drug dose dependent and high levels of the azide (greater than 12 μM, nominal) could cause lysis. Cell vulnerability depended on cold storage time and medium pH; no lysis occurred when freshly isolated cells were exposed to a nominal concentration of 7.5 μM. We conducted the following early experiments: Blood collected by venipuncture in heparinized vacutainers was stored at 2-4° for up to 10 days. At selected times, cells were isolated and washed three times in cold PBS (isotonic saline buffered with 5 mM potassium phosphate, pH 6.2) and made to a 0.2% hematocrit

---

[30] O. Fröhlich and R. Gunn, *Am. J. Physiol.* **252**, 153 (1987).

[31] The interpretation given by O. Fröhlich [*J. Gen. Physiol.* **84**, 877 (1984)] for this paradoxical observation is that the band 3 protein mediates anion flux by serving as a channel without undergoing a conformational change, a process he calls "tunneling." Other authors (e.g., Kaplan and Passow, Knauf and co-workers) have also studied and described this process.

in a spectrophotometer cuvette. Time of cell exposure in this medium was also varied before addition of p-AzBPhz. Cell turbidity was continuously monitored at 610 nm at the desired temperature for 700 sec; under these conditions, no detectable change in absorbance occurred unless drug was added. Erythrocyte suspensions that had been cold stored for 110 hr and preincubated for 75 min in cold, drug-free PBS showed 30% lysis within 700 sec when exposed to 12 μM of the agent. Results of kinetic analyses using a modified Coulter counter-sizing technique as well as direct phase and electron microscopic examination led us to conclude that the agent promoted salt and water entry into the cell.[32]

That high micromolar levels of p-AzBPhz increases both anion and cation permeability was directly measured in experiments conducted with Dr. Fröhlich. Drug-induced loss of K^+ from intact erythrocytes was determined under two experimental conditions; with cells that had been pretreated with gramicidin or with untreated cells. Since gramicidin introduces a large cation permeability in the red cell membrane, any increase in the K^+ efflux rate caused by p-AzBPhz would reflect an azide-induced increase in anion permeability (which is the rate-limiting flux in the presence of gramicidin). Conversely, K^+ loss from untreated cells is limited by the normally very low cation permeability and any increase in K^+ efflux would suggest that the azide had a direct effect on this process. Potassium ion efflux rate was determined by rapidly injecting normal and gramicidin-treated cells (30 μg gramicidin/ml of packed cells) into efflux media (1–2% Hct) containing 4 mM K^+ (to attain about -90 mV membrane potential) and 150 mM Tris-Cl.[33] At varying times, aliquots of the cell suspension were diluted into ice-cold, isotonic Tris-Cl/100 μM 4,4'-dinitrostilbene-2,2'-disulfonate (DNDS) stopping solution contained in a centrifuge tube having a bottom layer of n-butyl phthalate. Packed cells, having passed through this oil layer and thus separated from the efflux medium, were lysed in distilled water to measure K^+ and hemoglobin. Our results indicated that when p-AzBPhz in the efflux medium was 50 μM (nominal), the rate of K^+ (and Cl^-) loss from the gramicidin-treated cells was increased 10-fold over a control rate of 5 mmol/kg hemoglobin/min. Thus, at a concentration nearly two log units more than that needed to *inhibit* exchange, the azide increases chloride permeability across the gramicidin-modified cell membrane. Furthermore, cation flux was also enhanced at this higher concentration; the azide increased K^+ efflux from gramicidin-untreated cells to more than 60 mmol/kg hemoglobin/min (control rate was zero in normal cells under these conditions).

[32] C. L. Maynard and D. F. Diedrich, *Prog. Clin. Biol. Res.* **258**, 235 (1988).
[33] O. Fröhlich, C. Leibson, and R. Gunn, *J. Gen. Physiol.* **81**, 127 (1983).

How can we account for these results? At low nominal concentrations, when only about 1–2 million molecules of *p*-AzBPhz are bound to each red cell, the azide selectively binds to and interferes with both the sugar and anion transporters; results from an electron spin resonance (ESR) investigation with two lipid-specific spin labels support the idea that the first azide molecules react with the transport proteins and not the bulk lipid phase.[34] However, at much higher doses, it alters membrane integrity to allow salt and water entry, causing the cell to swell and eventually burst. It seems that the amphiphilic agent then acts nonspecifically to form a pore or channel in the membrane. If passages for cations are formed, they must be very small since slightly larger molecules (such as *N*-methylglucamine) still do not penetrate and glucose efflux is *inhibited*. Furthermore, drug-induced lysis that otherwise occurs in 700 sec can be completely inhibited by sucrose when it is present at a level which just offsets the colloid-osmotic pressure of hemoglobin, viz. 25 mM.[32]

It is also important to record that lytic action within 10–15 min is exhibited only by *p*-AzBPhz; phlorizin, phloretin, *p*-AzBPht, and 4-azidophlorizin are all inactive at 12 μM and under conditions when *p*-AzBPhz caused 54% lysis. These limited results suggest that the most active molecule has the lipophilic aromatic azide group precisely oriented in relation to the hydrophilic glucosidic moiety. The azido group might anchor the molecule by inserting into a hydrophobic pocket, causing the glucosidic moiety to either enlarge an aqueous channel already formed by an integral protein or create a new seam between phase boundries. Creation of pores that span the bilayer within the bulk lipid seems unlikely since the maximal distance separating the azide group from the hydrophilic glucosyl moiety is only 10–12 Å.

Photolabeling of Red Cells with p-[³H]Azidobenzylphloretin. Several years ago, F. Fannin found that the reversible sugar transport inhibition caused by *p*-AzBPht in the dark was converted after photolysis to an irreversible blockade. The effect was dependent on inhibitor dose and irradiation time.[35] Photolysis conditions were unsophisticated; cells in temperature-controlled media (0.2% hct) were mixed with the azide (10 μM, nominal) in shallow glass dishes so that a large surface area was exposed to a Sylvania 250-W sun lamp from a distance of 15 cm. After photolysis, noncovalently bound inhibitor was adsorbed onto Sephadex G-10 beads and the recovered red cells were reexamined for their transport capacity. Transporter activity was more than 50% irreversibly lost after a single

---

[34] J. W. Wyse, M. E. Blank, C. L. Maynard, D. F. Diedrich, and D. A. Butterfield, *Biochim. Biophys. Acta* **979**, 127 (1989).

[35] F. F. Fannin, J. O Evans, and D. F. Diedrich, *Biochim. Biophys. Acta* **684**, 228 (1982).

4-min irradiation period. When cells were irradiated without inhibitor or when the cells were exposed to the azide at the same concentration but not irradiated, transport activity remained unaltered (after free azide removal). Under these photolysis conditions, about 3% of the tritiated azide was covalently bound to membrane components, including massive amounts of lipid as well as some hemoglobin. Figure 4 shows the results of an

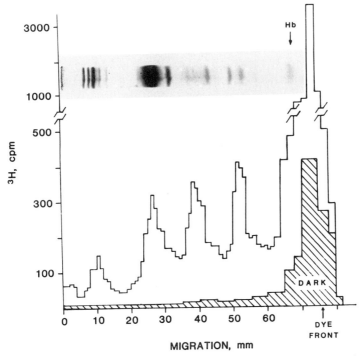

FIG. 4. Sodium dodecyl sulfate polyacrylamide gel electrophoresis analysis of the p-[^{3}H]AzBPht photolabeled erythrocyte membrane components. Intact red cells were labeled as described in the text and then passed through a Sephadex G-10 column which removed most of the unbound ligand and photolytic products. The ghosts prepared from the recovered cells were dissolved in SDS/dithiothreitol (DTT) and electrophoresed on 100-mm polyacrylamide gel rods (8.5%) until the tracking dye migrated 75 mm. One of the rods was stained with Coomassie Blue G and photographed (inset; gel origin is at the left and Hb represents the migrated hemoglobin band). The experimental second rod (unstained) was sliced into 1-mm disks which were then analyzed for tritium. Three major radioactivity peaks were found corresponding to band 3 (24–31 mm), band 4.5 (36–41 mm), and 7 or 8 (50–53 mm). The massive radioactive peak near the dye front (65–80 mm) represents photolytic by-products not adsorbed by Sephadex G-10, labeled membrane lipids, and perhaps labeled small peptide fragments. The hatched bar histogram represents the tritium pattern of electrophoresed membranes obtained from cells incubated with the same azide concentration but which were not irradiated.

SDS-polyacrylamide gel rod analysis of the radiolabeled membrane components from this experiment. Although the resolution is poor, tritium peaks occurred at positions occupied by bands 3, 4.5, and 7 (or hemoglobin dimer).

Since performing this experiment several years ago, we have come to realize that the experimental conditions used were actually the least likely to yield specific labeling. Future photolabeling work will take the following into consideration:

1. Nonspecific binding and labeling can be minimized if photolysis is conducted with azide concentrations below the apparent K_d. The ligand's specific activity must therefore be greater than 10 Ci/mmol. Very short irradiation times (5 sec or less) with UV limited to the azide's absorption maximum (the 254-nm Hg line) may reduce the photoactivation of irrelevant proteins that then interact with unbound ligand and cause spurious labeling.

2. Because of its undetectable membrane penetration, p-azidobenzylphlorizin rather than p-AzBPht is the better photoreagent at low levels as long as irradiation time is kept short and fresh cells are used (to avoid lysis). Addition of sucrose to the medium could block any secondary swelling phenomenon.

3. Photolysis should be executed at pH 6.2 or lower since the protonated form of the azide binds to the transporter. Furthermore, the light used to activate the azides must not be above 320 nm (as was the case in most of our early work) since the ionized, unbound form of the ligand (with a maximal A at 325 nm) is likely to be preferentially activated.

Acknowledgements

Portions of this work were accomplished during a fruitful sabbatical year at ETH-Zürich and association with G. Semenza and colleagues. It is a pleasure to dedicate this paper to Prof. Semenza on his sixtieth birthday. Most of these studies were supported by Grant AM 06878 from NIAMDD and in part with funds from the National Science Foundation (Grant RII-8610671) and the Commonwealth of Kentucky through the Kentucky EPSCOR program. More recent information about the synthesis and purification of the ultrahigh-specific activity 4-azidophlorizin was obtained while I was enjoying a sabbatical leave with R. Kinne, J. Deutscher, and their colleagues at the Max Planck Institüt-Systemphysiologie in Dortmund; supported by the Alexander von Humboldt Stiftung.

[44] Diuretic Compounds Structurally Related to Furosemide

By Scott M. O'Grady, Mark W. Musch, and Michael Field

Diuretic-type drugs can be divided into 3 major categories; (1) carbonic anhydrase inhibitors, (2) thiazide-type diuretics, and (3) loop or high ceiling diuretics.[1] This chapter will focus on a group of loop diuretics that are chemically similar to furosemide. More specifically, issues regarding the chemistry, structure–activity relationships and methods for evaluating the activity of loop diuretics are discussed. In addition, we also address the mechanism of action of these compounds in blocking Cl⁻ transport and their use in the molecular characterization of the $Na^+/K^+/2Cl^-$ cotransport system which is specifically inhibited by this group of diuretics.

Chemistry

The loop diuretics are a group of substituted sulfamyl benzoic acid derivatives which include furosemide, bumetanide, and related compounds.[2] These drugs are potent inhibitors of Cl⁻ transport in both epithelial and nonepithelial tissues. In contrast to carbonic anhydrase inhibitors and thiazide compounds, which block electrolyte reabsorption in proximal and distal segments of the nephron, furosemide, bumetanide and related cogeners inhibit NaCl transport in the thick ascending limb of Henle's loop (TALH). From a structural point of view, diuretics from each category have one chemical group in common, the sulfamyl group (SO_2NH_2). Chemically, carbonic anhydrase inhibitors, thiazides, and loop diuretics are related to sulfanilamide, which, from earlier observations, was shown to possess weak diuretic activity and inhibited carbonic anhydrase (Fig. 1). Attempts to improve the diuretic activity of sulfanilamide resulted in the synthesis of chlorothiazide, the parent compound of the thiazide diuretics.[3] Some of the thiazide diuretics are particularly interesting since they share common structural features with furosemide, including (1) an unsubstituted sulfamyl group, (2) a chlorine atom, and (3) an electronegative CO-containing group. Furosemide also bears some similarity to *p*-sulfamylbenzoic acid (namely the carboxyl and sulfamyl groups), which inhibits

[1] J. B. Puschett, *J. Clin. Pharmacol.* **21,** 564 (1981).
[2] M. Cohen, *J. Clin. Pharmacol.* **21,** 537 (1981).
[3] P. W. Feit, *J. Clin. Pharmacol.* **21,** 531 (1981).

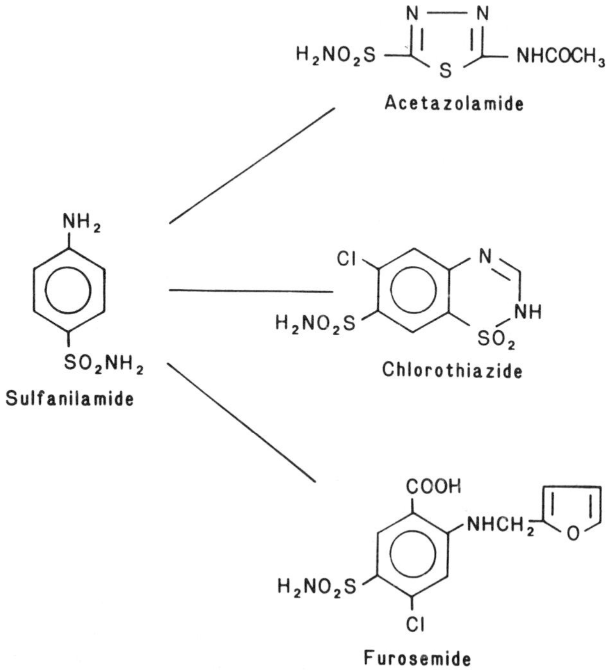

FIG. 1. Structural comparison between representative drugs from the three major categories of diuretics and sulfanilamide, which possesses weak diuretic activity.

carbonic anhydrase activity.[3] In spite of these similarities furosemide appears to represent an intermediate structure between the thiazide-type diuretics and a more specific structure which possesses unique loop diuretic activity. In their search for such a structure, Feit and co-workers synthesized a number of benzoic acid derivatives which resulted in the discovery of a new series of high ceiling diuretics with greater diuretic activity then furosemide.[4,5] The most potent compounds of this series have the general structure shown in Fig. 2. Bumetanide is probably the best known compound of this series. It differs from sulfanilamide, thiazides, and furosemide in that a phenoxy group is located at position 4 and a butylamino group forms the substituent at position 3. The synthesis of bumetanide and related compounds is described in detail by Feit.[4] The general scheme begins with 4-chloro-3-nitro-5-sulfamylbenzoic acid as the starting compound. It is then converted to 4-R_1-3-NH-R_2-5-sulfamylbenzoic acid by an initial alkylation step which substitutes the choro group

[4] P. W. Feit, J. Med. Chem. **14**, 432 (1971).
[5] P. W. Feit, O.B.T. Nielsen, and H. Bruun, J. Med. Chem. **15**, 437 (1972).

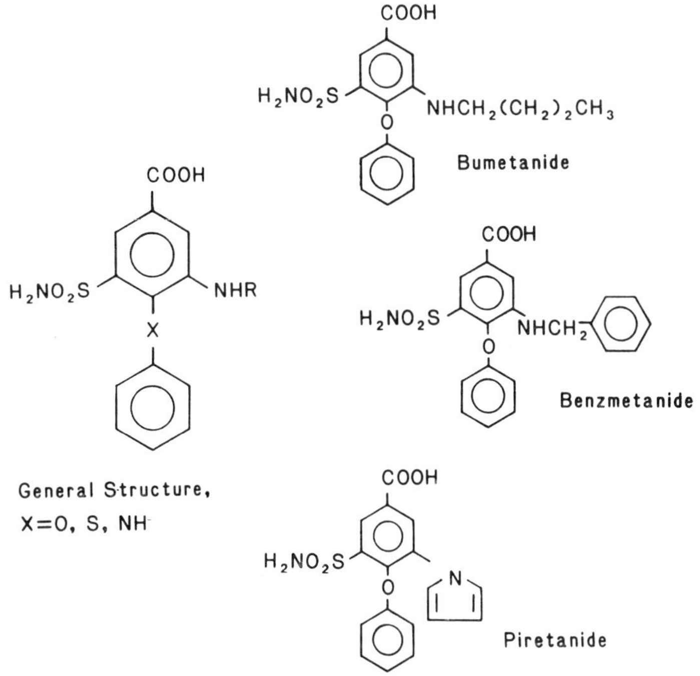

FIG. 2. General structure of 3-aminosulfamylbenzoic acid derivatives, with some specific examples shown to the right.

with a phenoxy group (or various other compounds including thiophenols, amines, or alcohols) followed by reduction of the NO_2 group to the corresponding amino acid (NH_2) at position 3. Additions of constituents such as butyl or benzyl groups at R_2 (position 3) were achieved by reductive alkylation with the appropriate aldehyde. The diuretic activity of bumetanide and related compounds is very dependent upon the nature of the groups at positions 1, 3, and 4.

Structure–Activity Relationships

Structure–activity studies using 3-aminobenzoic acid derivatives have focused on substitutions at three positions on the molecule. It appears that a carboxyl group, or similar group possessing a negative charge at physiological pH (ie., SO_3^-) is important for diuretic activity. It has been suggested that this anionic site may bind to a region of the $Na^+/K^+/2Cl^-$ cotransporter that normally binds Cl^-.[6] Aromatic substitutions at position 4 result in compounds with greater diuretic activity than those with halide

[6] H. C. Palfrey, P. W. Feit, and P. Greengard, *Am. J. Physiol.* **238**, C139 (1980).

substitutions. Compounds with phenylthio substitutions at position 4 were found to have greater efficacy than those with phenoxy or anilino substitutions.[4] Variations of the substituent at position 3 also markedly altered the potency of these derivatives. The presence of a primary amine at this position was totally ineffective but substitutions of either a butyl or benzyl group result in compounds with high diuretic activity.[4,6,7]

Schlatter et al.[8] have tested a number of furosemide-like derivatives in isolated perfused rabbit TALH. The compounds fell into three groups with respect to their effects on short-circuit current: furosemide-like, piretanide-like, and bumetanide-like. The furosemide-like compounds, just as with the 3-aminobenzoic acid derivatives, have increased inhibitory potency when the halide group at position 4 is substituted with an aromatic group. The furfuryl group at position 2 is not essential for activity and can be replaced by benzylamine. However, branching or lengthening the furfuryl group causes a decrease in diuretic activity. Studies with piretanide-like compounds showed that minor modifications of the aromatic ring at position 4 had little effect on activity. Only when bulky substituents were added to the ring did diuretic potency decrease. Substitution of pyrrole for pyrrolidine at position 3 was also found to block the inhibitory activity of these compounds. One of the important conclusions from the study by Schlatter et al.[8] was that the effects of single group substitutions are dependent on other groups on the molecule. For example, the diuretic activity of an anionic substituent at position 1 will vary with substitutions at positions 2 and 3. Hence, it becomes difficult to quantitatively model requirements for inhibitory activity.

Mechanism of Action

Originally, it was proposed that furosemide and related compounds inhibited either the Na^+,K^+-ATPase or some type of Cl^- pump present in the TALH.[9] Later studies in avian[10] and human[11,12] erythrocytes and in Ehrlich cells,[13] however, showed that the loop diuretics inhibited an $Na^+/K^+/2Cl^-$ cotransport process which has been subsequently identified in a wide variety of cell types. Experiments with isolated kidney tubules extended the observations made in red cells to the TALH and showed that

[7] H. C. Palfrey, P. Silva, and F. H. Epstein, *Am. J. Physiol.* **246,** C242 (1984).
[8] E. Schlatter, R. Greger, and C. Weidtke, *Pfluegers Arch.* **396,** 210 (1983).
[9] M. B. Burg, L. Stoner, J. Cardinal, and N. Green, *Am. J. Physiol.* **225,** 119 (1973).
[10] H. C. Palfrey and P. Greengard, *Ann. N.Y. Acad. Sci.* **372,** 291 (1981).
[11] P. Brazy and R. B. Gunn, *J. Gen. Physiol.* **68,** 583 (1976).
[12] J. C. Ellory and G. W. Stewart, *Br. J. Pharmacol.* **75,** 183 (1982).
[13] P. Geck, C. Pietvgyk, B. C. BurckHardt, B. Pfeiffer, and E. Heinz, *Biochim. Biophys. Acta* **600,** 432 (1980).

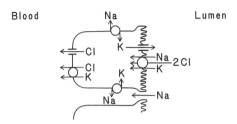

FIG. 3. Model of Na^+ and Cl^- absorption in the thick ascending limb of Henle's loop (TALH) of the kidney.

loop diuretics block secondary active Cl^- transport by inhibiting an $Na^+/K^+/2Cl^-$ cotransport system which is located in the apical membrane.[14]

Effects on Na^+, K^+, and Cl^- Transport in the Kidney

A model that describes the mechanism of Na^+ and Cl^- reabsorption in the TALH of mammalian kidney is presented in Fig. 3. Transport of Na^+, K^+, and Cl^- across the apical membrane is mediated by an electroneutral $Na^+/K^+/2Cl^-$ cotransport pathway. The K^+ that enters the cell with Na^+ and Cl^- recycles back to the lumen through K^+ channels in the apical membrane. Sodium is pumped out of the cell across the basolateral membrane by the Na^+,K^+-ATPase. Chloride, which is present in the cell at a concentration above electrochemical equilibrium, exits the cell by an electroneutral KCl cotransport pathway and to a greater extent through Cl^- channels located in the basolateral membrane. The net transport of Cl^- by this mechanism results in a serosa-negative transepithelial potential that drives paracellular Na^+ reabsorption. Addition of diuretic compounds like furosemide or bumetanide to the lumenal surface of the cell inhibits the uptake of Na^+,K^+, and Cl^- across the apical membrane, resulting in complete inhibition of Na^+ and Cl^- transport across the epithelium. Inhibition of the $Na^+/K^+/2Cl^-$ cotransport system by bumetanide causes hyperpolarization of both apical and basolateral membranes, a decrease in steady state intracellular $[Cl^-]$ to a value near equilibrium, an increase in basolateral membrane resistance due to reduced Cl^- conductance, and a decrease in Na^+,K^+-ATPase activity.[15] The overall result of inhibiting Na^+ and Cl^- reabsorption in this portion of the nephron is to increase urinary flow rate and Na^+ and Cl^- excretion.[16]

Diuretics like furosemide and bumetanide also produce kaliuresis,

---

[14] R. Greger and E. Schlatter, *Klin. Wochenschr.* **61**, 1019 (1983).
[15] R. Greger and P. Wangemann, *Renal Physiol.* **10**, 174 (1987).
[16] B. M. Hendry and J. C. Ellory, *Trends Pharmacol. Sci.* **9**, 416 (1988).

which can lead to a significant loss of K^+ and consequently hypokalemia. This effect can occur by two mechanisms. First, a decrease in Na^+ and Cl^- reabsorption in the TALH causes an increase in the amount of NaCl delivered to the distal tubule and collecting ducts. Sodium ion reabsorption and associated K^+ secretion increases in response to the increase in Na^+ load, resulting in a significant loss of K^+. Second, K^+ loss may occur as a consequence of reduced K^+ reabsorption in the distal portion of the TALH. As the tubular fluid becomes dilute with respect to the peritubular fluid, a dilution potential is generated (20 mV, lumen positive) which decreases the electrogenic efflux of K^+ across the apical membrane and Cl^- across the basolateral membrane of these cells. The result is net K^+ reabsorption in the distal portion of the loop as compared to proximal regions, where peritubular and lumenal fluid compositions are approximately symmetrical.[17] Treatment with furosemide or related compounds abolishes K^+ reabsorption in this portion of the TALH which inhances overall K^+ excretion by the kidney.

Identification and Purification of the $Na^+/K^+/2Cl^-$ Cotransport System Using [³H]Bumetanide

The efficacy and order of potency of loop diuretic compounds serve as important criteria in distinguishing the $Na^+/K^+/2Cl^-$ cotransport system from other coexisting transport mechanisms present in cell membranes (Fig. 4).[18] Comparison of dose–response results with several loop diuretics indicate that the IC_{50} and order of potency of these compounds are remarkably similar from one cell type to another. Binding experiments with [³H]bumetanide to brush border membrane vesicles from flounder intestine showed that saturable binding occurs from 0.025 to 1 μM with a $K_d = 1.2 \times 10^{-7}$ M and $B_{max} = 7.3$ pmol/mg protein (Fig. 5).[19] The K_d from the binding experiments is in close agreement with the IC_{50} for bumetanide shown in Fig. 4. Similar findings have been reported for Ehrlich ascites tumor cells,[20] duck red cells,[21] HT-29 tumor cells,[22] vascular smooth muscle cells,[23] and Madin–Darby canine kidney (MDCK) cells.[24] Simultaneous measurements of cotransport activity and bumetanide binding in Ehrlich ascites cells[20] have provided an estimate of the number of cotran-

[17] R. Greger and H. Velazquez, *Kidney Int.* **31**, 590 (1987).
[18] S. M. O'Grady, H. C. Palfrey, and M. Field, *Am. J. Physiol.* **253**, C177 (1987).
[19] S. M. O'Grady, H. C. Palfrey, and M. Field, *J. Membr. Biol.* **96**, 1 (1987).
[20] E. K. Hoffmann, M. Schiodt, and P. Dunham, *Am. J. Physiol.* **250**, C688 (1986).
[21] M. Haas and B. Forbush, *J. Biol. Chem.* **261**, 8434 (1986).
[22] C. C. Franklin, J. T. Turner, and H. D. Kim, *J. Biol. Chem.* **264**, 6667 (1989).
[23] M. E. O'Donnell and N. E. Owen, *Am. J. Physiol.* **255**, C169 (1988).
[24] E. M. Giesen-Crouse and J. A. McRoberts, *J. Biol. Chem.* **262**, 17393 (1987).

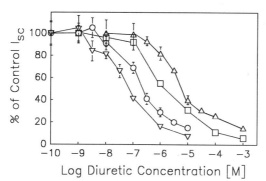

FIG. 4. Inhibition of Cl⁻ transport (as reflected by changes in I_{sc}) across the isolated intestinal epithelium of the winter flounder after mucosal addition of benzmetanide (▽), bumetanide (○), piretanide (□), and furosemide (∇). IC$_{50}$ values are 5×10^{-8}, 3×10^{-7}, 3×10^{-6}, and 7×10^{-6} M, respectively.

sport sites/cell (2×10^6) and a turnover number of 50 Cl⁻/sec. Rugg et al.[25] has reported a range of turnover numbers (120–850 K⁺/sec) for the Na⁺/K⁺/2Cl⁻ cotransporter in MDCK cells while O'Donnell and Owen[23] report 70 K⁺/sec for the Na⁺/K⁺/2Cl⁻ cotransport system in rat vascular smooth muscle cells. In duck red cells stimulated with norepinephrine, approximately 1000 bumetanide binding sites/cell have been estimated with a turnover number of 4000 Na⁺/sec.[21] The lowest estimate of turnover number has been reported in rabbit distal colon epithelial cells (6 K⁺/sec, with 21,000 binding sites/cell).[26] Recently, [³H]bumetanide binding studies in HT-29 cells[22] have shown that activation of protein kinase C with phorbol-12-myristate-13-acetate causes a decrease in the number of bumetanide-binding sites/cell without altering the kinetic properties of the cotransport mechanism. Thus differences in the number of cotransport sites/cell may reflect differences in regulation of the cotransporter between cell types.

The results presented in Fig. 6 show the effects of Cl⁻ concentration on bumetanide binding to brush border membrane vesicles from winter flounder intestine.[19] Maximal binding was observed at a concentration of 5 mM. At concentrations above or below 5 mM, submaximal binding was observed. This result was originally reported in microsomal membranes from canine kidney outer medulla by Forbush and Palfrey[27] and appears to be a consistant finding in other cells types where the Na⁺/K⁺/2Cl⁻ cotrans-

[25] E. L. Rugg, N. L. Simmons, and D. R. Tivey, Q. J. Exp. Physiol. **71**, 165 (1986).
[26] H. Wiener and C. H. van Os, J. Membr. Biol. **110**, 163 (1989).
[27] B. Forbush and H. C. Palfrey, J. Biol. Chem. **258**, 11787 (1983).

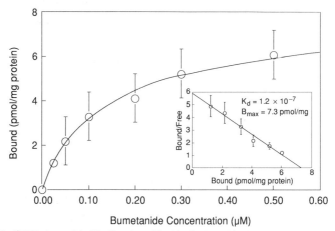

FIG. 5. [^{3}H]Bumetanide binding (specific binding) to brush border membrane vesicles from the winter flounder intestine. A Scatchard plot of the data is shown in the inset graph. The $K_d = 1.2 \times 10^{-7}$ and the $B_{max} = 7.3$ pmol/mg protein.

port system is present (duck red blood cells,[21] bovine kidney microsomal membranes,[19] and HT-29 cells[22]). It has been proposed that two binding sites exist for Cl$^-$ on the cotransporter (a high- and a low-affinity site) which exhibit cooperativity with respect to Cl$^-$ binding.[27] Bumetanide is thought to compete with Cl$^-$ for the low-affinity binding site, which could explain why a decrease in specific binding is observed at high Cl$^-$ concentrations. The submaximal binding observed at concentrations below 5 mM suggests that Cl$^-$ binding to the high-affinity site may cause a conformation change that permits binding of bumetanide to the low-affinity site. Similar effects of Na$^+$ and K$^+$ at low concentrations on bumetanide binding have also been reported, suggesting that conformational changes induced by binding of Na$^+$ and K$^+$ to sites on the cotransporter are required for optimal bumetanide binding.[27]

Bumetanide and furosemide are presently being used to isolate and purify the Na$^+$/K$^+$/2Cl$^-$ cotransporter. Photoaffinity-labeling studies with [^{3}H]4-benzoyl-5-sulfamyl-3-(3-thenyloxy)benzoic acid ([^{3}H]BSTBA) have identified a 150-kDa protein in microsomal membranes from dog kidney that has binding properties consistant with Na$^+$/K$^+$/2Cl$^-$ contransport.[28] A furosemide affinity gel has been used by Zeuthen et al.[29] to identify a 280-kDa protein in bovine kidney microsomal membranes that binds

[28] M. Haas and B. Forbush, Am. J. Physiol. 253, C243 (1987).
[29] T. Zeuthen, P. M. Andersen, K. E. Eskesen, and B. D. Cherksey, Comp. Biochem. Physiol. A 90A, 687 (1988).

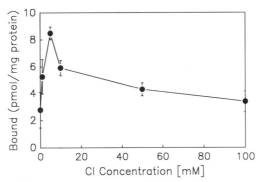

FIG. 6. Effects of Cl⁻ concentration on [³H]bumetanide binding to brush border membrane visicles from winter flounder.

bumetanide at two sites ($K_d = 0.014$ and $0.19 \mu M$). Reconstitution of this protein into a planar lipid bilayer revealed a Cl⁻ channel (12 pS) and a K⁺ channel of about 130 pS. Whether these channel activities represent copurification of channel proteins or actual components of the Na⁺/K⁺/2Cl⁻ cotransporter is unclear at the present time. Recently, a bumetanide affinity column was used to isolate binding proteins from Ehrlich ascites cells.[30] The column was constructed with bumetanide as the ligand by covalent binding of photoactive analog (4-azidobumetanide) to Sepharose. Two major proteins (76 and 36 kDa) were eluted from the column. Future reconstitution into liposomes or lipid bilayers should verify whether the transport properties of these proteins mimick the activity of the cotransporter.

Metabolism

Loop diuretics such as bumetanide have important clinical uses in the regulation of fluid and electrolyte balance. The effectiveness of bumetanide as a diuretic appears to be species dependent when administered orally or intravenously. The oral potency of bumetanide in laboratory mammals was found to decrease in the following order: dog, rabbit, mouse, and rat.[31] The diuretic activity of bumetanide in the dog was directly associated with a high level of unchanged drug excreted in the urine whereas the relatively weak diuretic activity observed in the rat paralleled the excretion of low

[30] P. W. Feit, E. K. Hoffmann, M. Schiodt, P. Kristensen, F. Jessen, and P. D. Dunham, *J. Membr. Biol.* **103**, 135 (1988).
[31] M. P. Magnussen and E. Eilertsen, *Naunyn-Schmiedebergs Arch. Pharmakol.* **282**, R61 (1974).

amounts of bumetanide and large amounts of its metabolites. Studies by Kolis et al.[32] and by Halladay et al.[33] showed that the liver microsomal mixed function oxygenase system was involved in the oxidative metabolism of bumetanide. Inhibitors of this system such as piperonyl butoxide or 2,4-dichloro-6-phenyl phenoxyethyl diethylamine cause an increase in the diuretic response to bumetanide in the rat. The metabolites of bumetanide result from hydroxylation of either the β, γ, or δ carbon atoms [or combination as in the case of 3-(γ, δ-dihydroxybutylamino)-4 phenoxy-5-sulfamylbenzoic acid) of the butylamino group at position 3. The fifth major metabolite of bumetanide is N-desbutyl bumetanide (chemically, 3-amino-4-phenoxy-5-sulfamylbenzoic acid), which results from oxidation of the C_α of the butylamino group. All five metabolites have been shown to be inactive in inhibiting $Na^+/K^+/2Cl^-$ cotransport activity.[34] In dogs a unique metabolite is produced which is excreted in the bile. This metabolite was found to be considerably more polar than bumetanide. It could be converted to bumetanide by alkali treatment, suggesting that an alkali-labile glucuronide of bumetanide is produced by dogs and excreted into the bile. In humans the major metabolites of bumetanide are β- and γ-hydroxylated compounds. The glucuronide metabolite is not a major constituent in either rats or humans.[34]

Methods

The methods outlined and referenced below describe various means of evaluating the activity of sulfamylbenzoic acid derivatives *in vitro*.

Isotopic Influx Measurements

Measurements of initial rates of $^{22}Na^+$, $^{86}Rb^+$, and $^{36}Cl^-$ influx into isolated cells or across the apical border of epithelial sheets provide a direct means of measuring the effectiveness of loop diuretics on $Na^+/K^+/Cl^-$ cotransport activity. Loop diuretics have been shown to be potent cotransport inhibitors in a number of isolated cell systems, including duck,[35] turkey,[6] human,[11] and ferret[36] erythrocytes, human fibroblasts,[37] tracheal epithelial cells,[38] LLC-PK$_1$ cells,[39] MDCK cells,[40] cultured endothelial

---

[32] S. J. Kolis, T. H. Williams, and M. A. Schwartz, *Drug Metab. Dispos.* **4**, 169 (1976).
[33] S. C. Halladay, D. E. Carter, and I. G. Sipes, *Drug Metab. Dispos.* **6**, 45 (1978).
[34] M. A. Schwartz, *J. Clin. Pharmacol.* **21**, 555 (1981).
[35] M. Hass and T. J. McManus, *J. Gen. Physiol.* **85**, 649 (1985).
[36] A. C. Hall and J. C. Ellory, *J. Membr. Biol.* **85**, 205 (1985).
[37] N. E. Owen and M. L. Prastein, *J. Biol. Chem.* **260**, 1445 (1985).
[38] J. H. Widdicombe, I. T. Nathanson, and E. Highland, *Am. J. Physiol.* **245**, C388 (1983).
[39] C. D. Brown and H. Murer, *J. Membr. Biol.* **87**, 131 (1985).

cells,[41] vascular smooth muscle,[42] and squid giant axon.[43] The references given above will allow the reader to find the specific method for tracer uptake measurement in the cells of interest.

Short-Circuit Current (I_{sc}) across Isolated Epithelial Sheets

In epithelial tissues such as flounder intestine,[44] canine tracheal epithelium,[38] and the teleost operculum[45] the I_{sc} provides a convenient measure of net Cl⁻ transport across the tissue. In flounder intestine, for example, mucosa stripped of muscle and mounted in Ussing-type chambers produce a current of approximately 60–100 $\mu A/cm^2$ which was inhibited over 90% by loop diuretics.[44] In a typical experiment, the intestine is removed from the animal and flushed with ice-cold Ringer's solution (150 mM NaCl, 5 mM KCl, 1 mM MgSO$_4$, 1 mM CaCl$_2$, 3 mM Na$_2$HPO$_4$, 5 mM EPPS, 20 mM glucose, pH8.0). The muscle layers are then removed and the mucosa mounted in the chamber with identical Ringer solutions on both sides of the tissue. After a 45-min stabilization period the tissue resistance is measured by passing a 100-μA current pulse across the tissue and recording the change in transepithelial potential difference (PD). The I_{sc} can then be determined and is equal to the amount of current required to nullify the spontaneous PD. Once the initial I_{sc} is recorded, the diuretic compound is then added to the mucosal solution in increasing concentration at 15- to 20-min intervals and the I_{sc} recorded before the next addition.[44] Since the effects of loop diuretics are reversible, the initial I_{sc} can be recovered at the end of the experiment by washing the chamber several times with Ringer's solution.

Perfused Tubule Experiments

Perfusion of the TALH segment of the rabbit nephron has been used to evaluate the effects of furosemide and related compounds on Cl⁻ transport.[9,46] Essentially the method originally outlined by Burg, *et al.*[9] involves the dissection of TALH in saline buffer solution 115mM NaCl, 5mM KCl, 25mM NaHCO$_3$, 10mM sodium acetate, 1mM CaCl$_2$, 1.2mM MgSO$_2$, 1.2mM NAH$_2$PO$_4$, 5mM glucose, 5% rabbit serum) at room temperature

---

[40] J. A. McRoberts, S. Erlinger, M. J. Rindler, and M. H. Saier, Jr., *J. Biol. Chem.* **257**, 2260 (1982).
[41] T. A. Brock, C. Brugnara, M. Canessa, and M. A. Gimbrone, Jr., *Am. J. Physiol.* **250**, C888 (1986).
[42] N. E. Owen, *Biochem. Biophys. Res. Commun.* **125**, 500 (1984).
[43] J. M. Russell, *J. Gen. Physiol.* **81**, 909 (1983).
[44] S. M. O'Grady, M. W. Musch, and M. Field, *J. Membr. Biol.* **91**, 33 (1986).
[45] O. Eriksson and P. J. Wistrand, *Acta Physiol. Scand.* **126**, 93 (1986).
[46] H. O. Berleithner, G. Giebisch, F. Lang, and W. Wang, *Klin. Wochenschr.* **60**, 1173 (1982).

and bubbled with 95% O_2/5% CO_2. The perfusate contained 150 mM NaCl, 2.5 mM K_2HPO_4, 1 mM $CaCl_2$, 1.2 mM $MgSO_4$, pH 7.4 with HCl. The experimental set-up consists of (1) concentric glass pipets which allow for changing the perfusion fluid with the tubule in place, (2) a Sylgard liquid dielectric interface on the outside of the tubule at both ends, and (3) a volumetric constriction pipet for collecting tubule fluid. The transepithelial potential difference (μ, 5mV) was measured between calomel electrodes connected to the fluids in the bath and perfusion pipet with agar bridges (0.16 M NaCl-agar). Transepithelial electrical resistance is calculated by cable analysis using the voltage change at both ends of the tubule that results from passing 0.05 μA of current from the perfusion pipet to a ground connected to the bath. Addition of loop diuretics inhibits the spontaneous transepithelial PD in a dose-dependent manner and causes an increase (about 15%) in the transepithelial resistance.

[^{3}H]Bumetanide-Binding Studies

[^{3}H]Bumetanide-binding experiments have been carried out in intact cells (Ehrlich cells,[20] duck erythrocytes,[21] HT-29 cells,[22] smooth muscle cells,[23] and MDCK cells[24]) and isolated membrane vesicles.[19,26-29] Vesicle-binding studies with microsomal membranes from kidney outer medulla began with homogenzation of the tissue in ice-cold buffer [250 mM sucrose, 5 mM Tris, 1 mM EGTA, 1 mM dithiothreitol (DTT), 1 mM $KHCO_3$, pH 7.4] in a Teflon-glass homogenizer. The homogenate was centrifuged at 5000 g for 15 min at 4° to pellet mitochondria and cell debris. The supernatant was then centrifuged at 48,000 g (4°) for 1 hr. The final pellet was suspended in 100 mM mannitol, 2 mM Tris-HEPES, 0.1 mM PMSF, pH 7.4, and stored at −70° until use. For binding experiments, 50–100 μg of vesicle protein was suspended in binding buffer (20 mM K_2HPO_4, 10 mM NaCl, 5 mM NaH_2PO_4, 2 mM Tris base, pH 7.4) as a 1:1 dilution. For determination of total binding, membranes were incubated with [^{3}H]bumetanide (10 Ci/mmol) over a concentration range of 10^{-8} to $10^{-6} M$. Nonspecific binding was determined with the same concentration of [^{3}H]bumetanide with 10 μM unlabeled bumetanide. Total volume for both total and nonspecific conditions was 100 μl. Binding reaches equilibrium in this preparation after 60 min at 4°. Separation of bound bumetanide from free material was accomplished by filtration using HAWP filters (0.45) presoaked with wash buffer (100 mM Na_2HPO_4, pH 7.4) containing 0.1% bovine serum albumin (BSA). The binding reaction was terminated by addition of 2.0 ml ice-cold wash buffer (without BSA) followed by a second 2.0-ml wash of the filter. Filters were dried and counted by liquid scintillation counting. Specific binding was determined by subtracting nonspecific binding from total binding.[19]

[45] Chloride Channel Blockers

By RAINER GREGER

After a brief introduction into the chloride-transporting membrane proteins, this chapter will focus on a new class of chloride channel blockers. It will be shown that these substances belong to a larger family of agents, of which so-called loop diuretics are just one member.

Introduction

This contribution restricts itself to the blocking of chloride channels and not to chloride permeation in general. The ability to distinguish between permeability in general and permeation through a channel depends on the methods used. If radiolabeled chloride is the measured signal, all permeation pathways for chloride existing in the membrane under study will contribute to the signal. If the chloride flux is measured electrically, only systems transporting chloride with its negative charge will be monitored. It should be clear that this latter chloride flux should only be called "electrogenic," and it cannot, *a priori,* be equated with the flux of chloride through ionic channels. Fortunately, this distinction appears puristic from a practical point of view since only very few rheogenic chloride-transporting carrier systems have been described thus far, whereas chloride channels are present in many different cell types. In the last 10 years or so an identification of chloride channels has become possible by the patch-clamp technique.[1] Still, however, most of our knowledge about blockers of chloride current comes from electrophysiological studies where macroscopic currents or voltages have been measured. In this chapter only chloride transport systems and, more specifically, chloride channels in a few well-defined preparations, mostly of vertebrate origin, are considered.

Chloride channel blockers with very little molecular modification, or generally at higher concentrations, also inhibit other chloride-transporting proteins. This finding has two ramifications on which I will expand. (1) Blockers are usually far less specific than one wishes them to be, and much caution is needed to deduce the existence of a transport system solely on the basis of inhibitor interaction. This consideration, as trivial as it is, is still ignored frequently, and the simple rule that a pharmacological effect must always be looked at on the basis of dose–response curves is neglected in many reports. (2) The fact that a blocker interferes with different affini-

---

[1] B. Sakmann and E. Neher, "Single-Channel Recording," Plenum, New York, 1983.

ties with several chloride-transporting epithelia may also indicate that the transport proteins, even though they may appear functionally different, may share in common some part(s) of the molecule. Therefore, I will briefly describe various chloride transport systems.

The bicarbonate/chloride exchange system is present in the red cell membrane in the form of the band 3 protein (for a review, see Passow[2]). This system or modifications of this system are found in many cells, and may serve pH regulation,[3] bicarbonate transport (for a review, see Novak and Greger[4]), or chloride transport.[5] Another, recently detected[6] chloride transport system is the $Na^+/2Cl^-/K^+$ carrier. This system is present in many polar and apolar cells (for a review see Ellory et al.[7]). It is responsible for the cellular uptake of chloride. If present in epithelia,[8] it serves the transepithelial transport of NaCl, and is usually found together with chloride channels on the opposite cell membranes.[9] Historically, this $Na^+/2Cl^-/K^+$ carrier system is remarkable since blockers for this system, the loop diuretics of furosemide type, had been known for more than 20 years before the system itself was recognized. In fact, the loop diuretic-sensitive transport system was long believed to be an energy-dependent chloride pump. Whereas it is clear now that what had been believed to be a primary pump sensitive to furosemide-like substances turned out to be a secondarily active chloride carrier, chloride pumps are still postulated for insect epithelia.[10] Another chloride-dependent carrier system has been recently detected, namely a chloride/formate exchange.[11]

General Properties of Chloride Channels

The introduction of the patch-clamp technique has now made it possible to study individual chloride channels, to measure their conductances, permselectivities, opening kinetics, activation mechanisms, and the mechanism of action of inhibitors (for a review, see Gögelein[12]). Models for the

[2] H. Passow, *Rev. Physiol. Biochem. Pharmacol.* **103**, 61 (1986).
[3] R. C. Thomas, *Curr. Top. Membr. Transp.* **13**, (1980).
[4] I. Novak and R. Greger, *Pfluegers Arch.* **411**, 546 (1988).
[5] C. M. Liedtke and U. Hopfer, *Am. J. Physiol.* **242**, G272 (1982).
[6] P. Geck, C. Pietrzyk, B.-C. Burckhardt, B. Pfeiffer, and E. Heinz, *Biochim. Biophys. Acta* **600**, 432 (1980).
[7] J. C. Ellory, P. B. Dunham, P. J. Logue, and G. W. Stewart, *Philos. Trans. R. Soc. London, B* **299**, 483 (1982).
[8] R. Greger, *Physiol. Rev.* **65**, 760 (1985).
[9] R. Greger, E. Schlatter, and H. Gögelein, *News in Physiol. Sci.* **1**, 134 (1986).
[10] G. A. Gerencser, *in* "Chloride Transport Coupling in Biological Membranes and Epithelia" (G. A. Gerencser, ed.), p. 183. Elsevier, Amsterdam, 1984.
[11] L. Schild, G. Giebisch, L. Karniski, and P. S. Aronson, *Pfluegers Arch.* **407**, 156 (1986).
[12] H. Gögelein, *Biochim. Biophys. Acta* **947**, 521 (1988).

channels have been derived from such data. At the same time, ligands are used to purify certain chloride channel proteins and to sequence their amino acids. This area of research has exploded during the last few years. I will restrict myself to summarizing the published work on single channels merely in table form. Arbitrarily, at least at this point, the chloride channels may be divided into those of neuronal origin, those present in smooth and striated muscle, and those found in the various epithelia.

As is apparent from Table I,[13-42] it is difficult to find common denominators for the various channels. For example, the single-channel slope conductance varies between less than 1 and 500 pS. Even if one considers that the composition of the bathing solutions may have been different, this range appears surprising. One possible clue to the origin of such a wide range may be the fact that chloride channels appear in clusters within one patch. Such clusters can be taken as subunits of one single channel, which then would have a rather large unitary conductance. Alternatively, clusters

[13] D. G. Owen, M. Segal, and J. L. Barker, *Nature (London)* **311**, 567 (1984).
[14] O. P. Hamill, J. Bormann, and B. Sakmann, *Nature (London)* **305**, 805 (1983).
[15] R. N. McBarney, S. M. Smith, and R. Zorec, *J. Physiol. (London)* **367**, 68P (1985).
[16] A. J. Martin and M. R. Gold, *Biophys. J.* **41**, 62 (1983).
[17] V. I. Geletyuk and V. N. Kazachenko, *J. Membr. Biol.* **86**, 9 (1985).
[18] D. Chesnoy-Marchais and M. G. Evans, *J. Physiol. (London)* **357**, 64P (1984).
[19] M. M. White and C. Miller, *J. Biol. Chem.* **254**, 10161 (1979).
[20] P. T. A. Gray, S. Bevan, and J. M. Ritchie, *Proc. R. Soc. London, B* **221**, 395 (1984).
[21] P. T. A. Gray and J. M. Ritchie, *Proc. R. Soc. London, B* **228**, 267 (1986).
[22] S. S. Kolesnikov, A. L. Lyubarsky, and E. E. Fesenko, *Vision Res.* **24**, 1295 (1984).
[23] A. L. Blatz and K. L. Magleby, *Biophys. J.* **43**, 237 (1983).
[24] A. L. Blatz and K. L. Magleby, *Biophys. J.* **47**, 119 (1985).
[25] W. Schwarze and H.-A. Kolb, *Pfluegers Arch.* **402**, 281 (1984).
[26] A. Coulombe and H. Duclohier, *J. Physiol. (London)* **350**, 52P (1984).
[27] R. Shoemaker, J. Naftel, and J. Farley, *Biophys. J.* **47**, 465a (1985).
[28] G. P. H. Young, J. D.-E. Young, A. K. Deshpande, M. Goldstein, S. S. Koide, and Z. A. Cohn, *Proc. Natl. Acad. Sci. U.S.A.* **81**, 5155 (1984).
[29] D. J. Nelson, J. M. Tang, and L. G. Palmer, *J. Membr. Biol.* **80**, 81 (1984).
[30] H. A. Kolb, C. D. A. Brown, and H. Murer, *Pfluegers Arch.* **403**, 262 (1985).
[31] R. Greger, *Physiol. Aktuel.* **2**, 47 (1986).
[32] R. Greger, M. Bleich, and E. Schlatter, *Renal Physiol. Biochem.* **13**, 37-50 (1990).
[33] H. Gögelein and R. Greger, *Pfluegers Arch.* **407** (Suppl. 2), S142 (1986).
[34] W. Hanrahan, W. P. Alles, and S. A. Lewis, *J. Gen. Physiol.* **84**, 30a (1984).
[35] A. Marty, Y. P. Tan, and A. Trautmann, *J. Physiol. (London)* **357**, 293 (1984).
[36] I. Findlay and O. H. Petersen, *Pfluegers Arch.* **403**, 328 (1985).
[37] R. Greger, E. Schlatter, and H. Gögelein, *Pfluegers Arch.* **409**, 114 (1987).
[38] H. Gögelein, E. Schlatter, and R. Greger, *Pfluegers Arch.* **409**, 122 (1987).
[39] M. J. Welsh, *Pfluegers Arch.* **407** (Suppl. 2), S116 (1986).
[40] J. P. Hayslett, H. Gögelein, K. Kunzelmann, and R. Greger, *Pfluegers Arch.* **410**, 487 (1987).
[41] R. Greger and K. Kunzelmann, *in* "Epithelial Secretion of Water and Electrolytes" (J. A. Young and P. D. Wong, eds.). Springer-Verlag, New York, 1989.
[42] M. E. Krouse, G. T. Schneider, and P. W. Gage, *Nature (London)* **319**, 58 (1986).

TABLE I
Chloride Channels[a]

Organ	Agonist	Antagonist	Voltage dependence	Ca^{2+} dependence	Conductance (pS)	Inhibition	Permselectivity	Reference
Mouse spinal neuron	—	—	Depolarized gates	+	?	—	—	Owen et al.[13]
Mouse spinal neuron	GABA, glycine, diazepam	—	—	—	20–45, multiple states	—	Br > I > Cl > F > SO$_4$ = acetate = O	Hamill et al.[14]
Rat spinal neuron	Cs$^+$ activated	—	—	—	—	—	—	McBarney et al.[15]
Lamprey brain skin	Glycine	—	—	—	10–75	K$^+$	—	Martin and Gold[16]
Molluscan neurons	K$^+$ (cytosolic)	—	Depolarized gates	+	12, 5n, n = 16	—	—	Geletyuk and Kazachenko[17]
Aplysia neurons	—	—	Hyperpolarized gates	—	55 [NaCl, 480 mM]	—	Cl = Ise = NO$_3$ > Cs	Chesnoy-Marchais and Evans[18]
Torpedo electric organ	pH	—	Hyperpolarized gates	—	14 (100 mM)	SITS, DIDS[b]	—	White and Miller[19]
Rat Schwann cell	—	—	—	0	440 (NaCl, 180 mM)	—	I > Br > Cl > CH$_3$SO$_4$ > SO$_4$ > acetate = Ise Cl/Na = 5	Gray et al.[20]
Rat astrocytes	—	—	Depolarized gates	0	—	SITS, DIDS	Br > Cl > CH$_3$SO$_4$ > SO$_4$ > Ise = acetate	Gray and Ritchie[21]
Frog rod	—	—	—	—	200 (NaCl, 100 mM)	—	Cl > F > NO$_3$; propionate	Kolesnikov et al.[22]
Rat myotube	—	—	—	—	45, 61, 430	pH	Cl/K ~ 5; SO$_4$ = CH$_3$SO$_4$ = 0	Blatz and Magleby[23,24]

Tissue	Activation		Gating		Conductance (pS)	Blocker	Selectivity	Reference
Chick and mouse myotube	—	—	Bell shaped	0	210, 340	—	Cl/Na ~ 5	Schwarze and Kolb[25]
Rat heart muscle	—	—	—	—	400–450	—	—	Coulombe and Duclohier[26]
Rat artery	—	—	Depolarized gates	—	12, 150	—	—	Shoemaker et al.[27]
Xenopus oocyte	—	—	Bell shaped	+ (10^{-6} M)	380 (NaCl, 1 M)	—	I, Br > Cl	Young et al.[28]
Frog kidney cells (A6)	—	—	Bell shaped	—	360 (NaCl, 100 mM)	SITS	Cl/Na ~ 10	Nelson et al.[29]
Dog kidney cells (MDCK)	—	—	Spontaneously little action	—	480 (attached)	—	—	Kolb et al.[30]
Rabbit thick ascending limb	—	—	Depolarized gates	—	30–40	DPC, NPPB	—	Greger[31]; Greger et al.[32]
Rabbit proximal tubule	—	—	Depolarized gates	+	30–40	DPC	Cl ~ Na	Gögelein and Greger[33]
Rabbit urinary bladder	—	—	—	0	63 (KCl, 150 mM)	—	Cl = J = Br = NO_3 > gluconate > acetate; Cl/K = 5	Hanrahan et al.[34]
Rat lacrimal gland	Nicotinic	—	—	+	1–2 (NaCl, 150 mM)	—	—	Marty et al.[35]
Mouse lacrimal gland	Nicotinic	—	—	+ (10^{-6} M)	—	—	—	Findlay and Petersen[36]
Shark rectal gland	cAMP	—	Depolarized gates	0	10, 35	DPC	Cl > Br > J > gluconate	Greger et al.[37]
Trachea	cAMP	—	Depolarized gates	0	29	A9C, DPC,	—	Gögelein et al.[38] Welsh[39]
Colon carcinoma cell (HT29)	cAMP	—	Depolarized gates	0	10, 35	DPC, NPPB	Cl = NO_3 > gluconate	Hayslett et al.[40] Greger and Kunzelmann[41]
Colon carcinoma cell (T84)	cAMP	—	Depolarized gates	0	10, 35	DPC	Cl > gluconate	Greger and Kunzelmann[41]
Mouse alveolar type II	—	—	—	—	(60–70)n	—	—	Krouse et al.[42]

[a] —, No information available.
[b] SITS, 4-Acetamido-4′-isothiocyanostilbene-2,2′-disulfonate; DIDS, 4,4′-diisothiocyanostilbene-2,2′-disulfonate; DPC, diphenylamine-2-carboxylic acid; NPPB, 5-nitro-2-(3-phenylpropylamino)benzoic acid; Ise, equivalent short circuit current; A9C, anthracene-9-carboxylate.

are interpreted as individual channels, which then leads to rather small values for the unitary conductance. According to accepted theory, a distinction between both alternative explanations can be made on the basis of whether the single conductance events (or sublevels) are dependent on or independent of each other as tested by statistical methods. In case of independence of events, the term subunit would be misleading if it was not qualified as *independent* subunit. If the conductance events are dependent on each other, independent single channels are excluded and the term subunit is appropriate. The situation may be even more complex if one looks closely at original recordings. Frequently, substates of differing current amplitude are observed, and even one individual chloride channel may vary its slope conductance to some extent as the experiment proceeds. To make it even worse, in one patch several different chloride channels can be found. This has been recognized in myotubes,[23,24] and in epithelia.[12,35,37,38] Usually, large and small channels are distinguished, but if there were no additional criteria for distinction, the slope conductance alone would clearly not be sufficient. Finally, it should not be ignored that the patch-clamp technique for the recording of single ionic channels has its restrictions. It will "see" a channel only if it is open for some limited time and at some limited current level. Our routine conditions (signal-to-noise ratio for a filter frequency of 0.5–1 kHz) would set the limit such that a channel will be seen only if it is open at least some 0.2–0.5 msec and if it has a conductance above 5 pS. Thus, in single-channel recordings, we will ignore any very small chloride channel whether it was physiologically relevant or not. It should be clear, however, that even very small channels can be found in whole cell recordings. To summarize this issue, it seems very dangerous to classify chloride channels according to their conductance. Inherent pitfalls include the substate and subunit dilemma, and, probably even more serious, the variability of the conductance for otherwise identical channels.

A similar dilemma in classification of chloride channels is apparent if one considers the heterogeneity of the voltage dependence. Three patterns are found with almost equal frequency: (1) increase in open probability by depolarization, (2) increase in open probability by hyperpolarization, and (3) bell-shaped curves with the maximum around 0 mV clamp potential. These different patterns cannot be ascribed to the three subclasses of channels chosen for this table. In fact, for neuronal chloride channels all three patterns have been reported. For the epithelial chloride channels it seems to be the "rule" that depolarization increases the open probability and, in many instances, also the slope conductance.

Where determined, the listed channels are clearly anion selective, and large anions are usually excluded. At least in some types of channels I^- and to a smaller extent Br^- are excluded. Chloride channels may be regulated by Ca^{2+}, but the calcium dependence is also heterogeneous and ranges from Ca^{2+} dependent to Ca^{2+} inhibited. However, as with selectivity, too few data have been reported to arrive at any firm conclusion. To summarize Table I, I should like to conclude that chloride channels in the various cells show a large variability. In this respect, other ionic channels, e.g., the acetylcholine receptor, the Na^+ channels, and the K^+ channels, seem to be more homogeneous. As devastating as the data in Table I may be, I am optimistic that subclasses of channels are emerging with far less heterogeneity, as, for example, the various chloride channels in secretory epithelia. Most research groups would agree that the cAMP-dependent chloride channels present in these epithelia have a narrow conductance range (around 40–50 pS), a similar selectivity, and a comparable voltage dependence.

Chloride Channel Blockers

According to their targets, the various agents may be grouped into substances interfering with the γ-aminobutyric acid (GABA) and glycine receptor chloride channels, and agents which interfere with chloride channels in muscle membrane, apolar cells such as the red blood cell, and in epithelial cells. This type of classification is not at all satisfactory since, on the one hand, some agents appear to act in several of the above-described channel types. On the other hand, the representatives for one group of blockers act quite differently. Some may interfere with the activation of the channel, e.g., GABA antagonists, others may act at the chloride-binding site, and yet others are not at all defined in their mode of action. Therefore, this chapter will list the different blockers, provide reference to pertinent recent work, but will not go into details of the mechanism of blocking. For one newly detected group of agents, namely the arylaminobenzoates, sufficient data are available to discuss some hypothetical views of the mechanism of interaction.

Agents Acting at the Neuronal Chloride Channels

Well-studied examples are the GABA receptor and the glycine receptor chloride channels. GABA binding gates the former channel and that of glycine the latter.[14] The GABA interaction is modified by diazepam and related agonists (central type of receptor) such that the GABA effect is enhanced. Diazepam antagonists have the opposite effect and, therefore, reduce the GABA effect on the chloride current. These substances then act as convulsants. It is interesting to note that shortly after their detection in the brain, peripheral diazepam receptors were also found. The function of these receptors, e.g., in the nephron[43] is still obscure. This appears puzzling as the benzodiazepam binding coincides with the site of active NaCl reabsorption in the nephron. This has prompted us to examine several of these agonists and antagonists (kindly provided by Dr. P. Skolnick) in isolated perfused thick ascending limb segments of rabbit and mouse kidney. We are unable to demonstrate an effect on active NaCl reabsorption (unpublished observations).

Other, and chemically not at all related, substances have been reported as inhibitors of the GABA-activated chloride channel: pentylenetetrazole,[44] picrotoxin and bicuculline,[45] and bicyclic phosphates and several insecticides.[46]

The glycine-operated chloride channel is inhibited by strychnine and related compounds. In fact, these antagonists have been used to purify this channel protein.[47]

Agents Acting at the Muscular Chloride Channel

No experiments are available at the level of single chloride channels (patch-clamp analysis). Therefore, the data on putative blockers of chloride fluxes may reflect the properties of systems other than chloride channels. In the striated barnacle muscle cAMP-stimulated chloride fluxes were inhibited by rather low doses of sulfonic acid stilbene derivatives[48]: 4-acetamido-4'-isothiocyanostilbene-2,2'-disulfonate (SITS) and 4,4'-diisothiocyanostilbene-2,2'-disulfonate (DIDS). This result documents the similarity of this chloride transport system in muscle cell membrane to that in the red blood cell (vide infra). Also, furosemide in a rather high dose (0.6 mmol/liter) was effective as a reversible inhibitor in this muscle preparation. As will be pointed out later, it is not pertinent, and may even be incorrect, to conclude from this finding that chloride transport occurred via the $Na^+/2Cl^-/K^+$ cotransport system.

A recent review[49] summarizes the work with chloride channel blockers in striated muscle and subdivides the agents into four groups: (1) polycyclics, with anthracene-9-carboxylate as one example, (2) benzoates, with furosemide as an example, (3) phenoxyacetates, with ethacrynic acid as

---

[43] D. Butlen, *FEBS Lett.* **169**, 138 (1984).
[44] B. A. Weissmann, J. Cott, D. Hommer, R. Quirion, S. Paul, and P. Skolnick, in "Benzodiazepine Recognition Site Ligands: Biochemistry and Pharmacology" (G. Biggio and E. Costa, eds.), p. 139. Raven, New York, 1983.
[45] K. G. Thampy and E. M. Barnes, Jr., *J. Biol. Chem.* **259**, 1753 (1984).
[46] J. R. Bloomquist and D. M. Soderlund, *Biochem. Biophys. Res. Commun.* **133**, 37 (1985).
[47] D. Graham, F. Pfeiffer, and H. Betz, *Eur. J. Biochem.* **131**, 519 (1983).
[48] J. M. Russel and M. S. Brodwick, *J. Gen. Physiol.* **78**, 499 (1981).
[49] A. H. Bretag, *Physiol. Rev.* **67**, 618 (1987).

one example, and (4) sulfonates and sulfamides, with disulfonic stilbenes as an example. It is difficult to accept the first three groups as different entities, since it will be shown below that these agents have far more in common than may be apparent at first glance. The methodology, as it has been applied to date, does not permit any conclusion as to how these substances reduce chloride fluxes or chloride conductance. However, recent studies with these substances in epithelial chloride channels[50] (cf. next section) may also be used to extrapolate these data to the muscle chloride channel.

In the smooth muscle cell most evidence favors the view that cytosolic chloride activity is above Nernstian equilibrium.[51] Many data are available on anion replacements on the extracellular compartment and their effect on membrane voltage, as are many data on disulfonate stilbene as well as on furosemide effects. Little, however, is known on the properties of the respective chloride channels.[27] It has been argued that SITS or DIDS, in analogy to their effects in *Torpedo* electric organ chloride channel,[52] may also interfere with chloride channels in smooth muscle cells. Also, the effects of organic "nitro" compounds, such as nitroglycerin, have been ascribed to the interference of these compounds with chloride channels.[51]

Agents Acting in Apolar Cells

One specific, but obviously very complex, chloride transport system is that of the red blood cell. There, the vast majority of the chloride flux occurs via the band 3 protein.[2] However, chloride channels are also present in these cells. A recent survey has compared several classes of agents in their ability to block the anion shift via the band 3 protein and the chloride movement through single chloride channels.[53] It was concluded that both routes of chloride movement take the same pathway, namely, via the band 3 protein. In other words, this protein may perform electroneutral anion exchange (quantitatively most important), and may also permit the channeling of chloride ions. From this point of view, older data using furosemide[2] and flufenamic acid[54] may be looked at from a new perspective. Flufenamic acid and related substances belong to the group of chloride channel blockers, cf below. The band 3 protein may thus bind blockers of three different classes: the sulfonic acid stilbene derivatives, compounds related to furosemide, and compounds related of the arylaminobenzoate

---

[50] P. Wangemann, M. Wittner, A. Di Stefano, H. C. Englert, H. J. Lang, E. Schlatter, and R. Greger, *Pfluegers Arch.* **407** (Suppl. 2), S128 (1986).
[51] V. A. W. Kreye and F. W. Ziegler, *Adv. Microcirc.* **11**, 114 (1982).
[52] C. Miller and M. M. White, *Proc. Natl. Acad. Sci. U.S.A.* **81**, 2772 (1984).
[53] W. Schwarz and H. Passow, *Proc. Int. Union Physiol. Sci. 30th Congr.* **16**, 544 (1986).
[54] J.-L. Cousin and R. Motais, *Biochim. Biophys. Acta* **687**, 156 (1982).

class. It is important to realize that the affinities to these three classes of substances are quite different from those needed in other preparations. Micromolar concentrations are needed to block the band 3 protein by sulfonic acid stilbenes, yet much higher concentrations are usually required in other systems. Conversely, the concentrations needed of furosemide and even more so of bumetanide are much higher for the band 3 protein when compared to that needed to interfere with the $Na^+/2Cl^-/K^+$ carrier system.[8] Similarly, rather high concentrations of arylaminobenzoates are necessary to inhibit the band 3 protein.

A great deal of effort has been directed toward the pharmacology of glial cell volume regulation, and it was recognized some 4 years ago that several transport systems in which chloride participates are probably responsible for the glial volume increase, which in turn may be the pathophysiological basis of traumatic brain edema. In a recent study, several classes of substances have been tested in brain slices and it was found that the most effective inhibitors were derived from the phenoxyacetic acid group. Specifically (2,3,9,9α-tetrahydro-3-oxo-9α-substituted-1H-fluoren-7-yl)oxyalkanoic acids were effective in doses as low as 10^{-12} mol/liter. However, the authors[55] concluded only that this class of inhibitors blocks Cl^- *uptake* and have no evidence that this uptake is via chloride channels.

Agents Acting in Epithelia

Soon after the recognition of the existence of chloride channels in epithelia such as the thick ascending limb, the colon, and the trachea, a search for agents blocking these channels was initiated in our laboratory. The focus was on anthracene-9-carboxylate and diphenylamine-2-carboxylate since these two substances, previously used in the striated muscle chloride channel[56] and in the red blood cell,[54] showed some inhibitory activity on the chloride conductive pathways in the thick ascending limb[57] and in the amphibian diluting segment.[58] Our survey covered a broad spectrum of substances and used a more systematic approach for the substances showing strong activity. The bioassay used in this survey, namely, the *in vitro* perfused thick ascending limb of the loop of Henle, has the great advantage that chloride conductance is present only at the blood side of the cell, whereas the $Na^+/2Cl^-/K^+$ carrier is present only in the

[55] E. J. Cragoe, Jr., O. W. Woltersdorf, Jr., N. P. Gould, A. M. Pietruszkiewicz, C. Ziegler, Y. Sakurai, G. E. Stokker, P. S. Anderson, R. S. Bourke, H. K. Kimelberg, L. R. Nelson, K. D. Barron, J. R. Rose, D. Szarowski, A. J. Popp, and J. B. Waldman, *J. Med. Chem.* **29,** 825 (1986).
[56] P. T. Palade and R. L. Barchi, *J. Gen. Physiol.* **69,** 879 (1977).
[57] A. Di Stefano, M. Wittner, E. Schlatter, H. J. Lang, H. Englert, and R. Greger, *Pfluegers Arch.* **405** (Suppl. 1), S95 (1985).
[58] H. Oberleithner, M. Ritter, F. Lang, and W. Guggino, *Pfluegers Arch.* **398,** 172 (1983).

lumenal membrane.[8] Loop diuretics of the furosemide type interfere with this carrier system but have little effect on chloride channels. A simple experiment leads to this distinction: whereas loop diuretics inhibit the active transport of chloride and sodium instantaneously when added to the lumenal perfusate, they are devoid of effect when added to the basolateral perfusate. Conversely, blockers of the arylaminobenzoate type, as they are listed in Table II, block the active transport of chloride at low concentrations when added to the basolateral perfusate but not when added at the same concentration to the lumenal perfusate. Nevertheless, if the furosemide concentration on the basolateral cell pole is increased to very high concentrations of 10^{-3} mol/liter or more, some inhibition will also be observed. Similarly, if the chloride channel blocker is added to the lumenal perfusate at very high concentration, it may also exert an effect from this side. These effects may be due to a diffusion of furosemide or the chloride channel blocker to the other cell membrane or to effects other than that on the $Na^+/2Cl^-/K^+$ carrier or on the chloride channel. The first explanation becomes more likely the higher the lipid solubility of the tested compound is. In case of furosemide and related compounds, the lipid solubility at physiologic pH is rather low, but for the chloride channel blockers the opposite is true. Some 30–80% distribute into the lipid phase (e.g., CH_2Cl_2) at pH 7.4. It is not surprising then that these substances also act usually at 10–100 times higher concentrations from the lumenal cell side.[50] The second explanation given above comprises a heterogeneous group of possible effects. Of course, it must be considered that one individual substance may not only inhibit chloride channels, but may also interact with the $Na^+/2Cl^-/K^+$ carrier. We have several examples available.[59,60] It should be clear that preparations other than the intact epithelium, e.g., the isolated membrane patch, must be used to prove this potential effect of an $Na^+/2Cl^-/K^+$ carrier blocker on the chloride channel. Another potential complication comes from the fact that the tested compound may interfere with other systems. For high concentrations of furosemide and related compounds, an interference with carbonic anhydrase has been shown.[61] It should also be noted that the loop diuretics of furosemide type have originated from studies with carbonic anhydrase inhibitors. Furthermore, lipophilic anions will also find access to the mitochondria and will short circuit pH gradients across thinner mitochondrial membrane, thus inter-

[59] R. Greger, P. Wangemann, M. Wittner, A. Di Stefano, H. J. Lang, and H. C. Englert, in "International Conference on Diureties, Sorento, 1986" (T. Andreucci, ed.), p. 33. Nijhoff, Boston, Massachusetts, 1987.
[60] R. Greger, H. J. Lang, H. C. Englert, and P. Wangemann, in "Diuretics" (J. Puschett, ed.), p. 131. Elsevier, New York, 1987.
[61] T. H. Maren, "Renal Physiology: Men and Ideas." Am. Physiol. Soc., Bethesda, Maryland, 1987.

TABLE II[a]

	Blocker of the $Na^+/2Cl^-/K^+$ carrier		Blocker of chloride channel	
	Furosemide	Lumen: $3 \cdot 10^{-6}$M Bath: $> 10^{-4}$M	2-(2-Furfurylmethylamino)-5-nitrobenzoic acid	Lumen: $> 10^{-4}$M Bath: $3 \cdot 10^{-5}$M
	Piretanide	Lumen: $1 \cdot 10^{-6}$M Bath: $> 10^{-4}$M	4-Chloro-3-N-pyrrolidino-benzoic acid	Lumen: $> 10^{-4}$M Bath: $5 \cdot 10^{-5}$M
	Triflocine	Lumen: $3 \cdot 10^{-5}$M Bath: $> 10^{-4}$M	Diphenylamine-2-carboxylic acid	Lumen: $> 10^{-4}$M Bath: $2.6 \cdot 10^{-5}$M
	Torasemide	Lumen: $3 \cdot 10^{-7}$M Bath: $3.4 \cdot 10^{-5}$M	3-((2-Anilino-5-nitrophenyl)sulfonyl)-1-isopropyl urea	Lumen: $\gg 10^{-4}$M Bath: $3 \cdot 10^{-4}$M

[a] Comparison of the blockers of the $Na^+/2Cl^-/K^+$ carrier and of chloride channels in the thick ascending limb of the loop of Henle. The $Na^+/2Cl^-/K^+$ carrier is present only in the lumenal membrane, and Cl^- channels only in the basolateral membrane. The data are taken from Wangemann et al.[50] and from E. Shlatter, R. Greger, and C. Weidthe, Pfluegers Arch. **396**, 210 (1983). The values corespond to IC_{50} values as defined in the footnote to Table III.

fering with mitochondrial function. This putative mechanism may be responsible for the poorly reversible inhibitory effect seen with arylaminobenzoates at high concentrations. This effect is usually followed by a cell depolarization.[50] The chloride channel blockers of the arylaminobenzoate type are also known to block cyclooxygenase,[62] and this may generate a dilemma in tissues in which the chloride conductance is increased by prostaglandins. Hence, the reduction in chloride conductance observed in airway epithelial cells after the addition of very high concentrations of diphenylamine-2-carboxylate[39] may, in part, be caused by cyclooxygenase inhibition.[63]

From our studies we have determined that furosemide is not a chloride channel blocker,[64] and this contrasts to other studies on the cornea[65] and on the tear gland.[66] It should be noted, however, that in the intact preparation, furosemide will lead to a reduction of the chloride conductance simply because it reduces cytosolic chloride activity.[8] In isolated chloride channels, we have never observed any effect of furosemide.[37,64]

Table III summarizes several potent chloride channel blockers as documented by their inhibitory effect on the chloride conductance in isolated perfused thick ascending limbs. Some of these compounds have also been tested in isolated chloride channels of the rectal gland,[37] in airway epithelial cells,[64] in the thick ascending limb of the loop of Henle,[32] and in colonic carcinoma cells.[40] It is apparent that all listed compounds possess several common features. All are anionic at physiological pH. The anionic group is a carboxylate. All compounds possess an amino group, and all compounds have an apolar residue. Within one subclass, e.g., the derivatives of 5-nitro-2-(3-phenylpropylamino)benzoic acid (NPPB), it has been shown that the distances between the carboxylate group and the amino group, between the nitro group and the carboxylate group, and between the phenyl ring and the amino group are very critical. For example, the optimal spacer between the phenyl ring and the amino group was a propyl group. Changing this spacer to butyl or ethyl led to dramatic reductions in the potency. Also the sidedness of substituents on one of the chiral centers of this spacer led to discrepant results. Whereas a D form of one compound was active, the L form was entirely ineffective.[50] Data of this kind have led us to suggest that chloride channel blockers interact with several sites. In the case of NPPB (Fig. 1), these sites may be summarized as follows. (1) An

---

[62] S. H. Ferreira and J. R. Vane, *Annu. Rev. Pharmacol.* **14,** 57 (1974).
[63] M. J. Stutts, D. C. Henke, and R. C. Boucher, *Pfluegers Arch.* **415,** 611 (1990).
[64] K. Kunzelmann, H. Pavenstädt, and R. Greger, *Pfluegers Arch.* **415,** 172 (1989).
[65] R. Patarca, O. A. Candia, and P. S. Reinach, *Am. J. Physiol.* **245,** F660 (1983).
[66] M. G. Evans, A. Marty, Y. P. Tan, and A. Trautmann, *Pfluegers Arch.* **406,** 65 (1986).

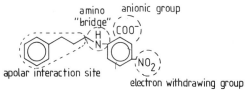

FIG. 1. The chloride channel blocker NPPB.

anionic site is mandatory; this site is ionized at physiological pH. In all the potent compounds it is a carboxylate group. Other anionic groups such as sulfonyl urea[67] are less effective or have very little effect, as in the case of sulfonates (unpublished). This is remarkable since in the case of loop diuretics of the furosemide type, the comparable site can be any of these anionic groups; (2) the amino "bridge" is necessary; it cannot be replaced by oxygen, phosphorus, or carbon[50]; (3) an m-nitro substitution or a p-chloro substitution on the benzoate moiety increases potency; (4) finally, an apolar site of interaction is required. Further details on charge effects and on optimal distances have been summarized recently.[50]

Table II summarizes several chloride channel blockers and compares them to closely related blockers of the $Na^+/2Cl^-/K^+$ carrier.[3] It is apparent from this figure that minute changes in the molecule shift the affinity of one given structure from the chloride channel to the $Na^+/2Cl^-/K^+$ carrier. A general principle seems to be that the blockers of the $Na^+/2Cl^-/K^+$ carrier contain a sulfonamide group or a pyridine nitrogen where the blockers of the chloride channel contain a nitro substitution. I have already stated above that the anionic group is common to both classes of substances. Also the amino nitrogen is common to both classes. Finally, the apolar residue also appears essential in both kinds of compounds. However, the special restrictions on this group are more marked for the chloride channel blockers.

This similarity of blockers for two very distinct chloride permeation pathways stresses that the respective membrane proteins may have molecular similarities as expressed by their "receptor" sites for these blockers. It would not be surprising at this stage if certain amino acid sequences in both membrane proteins would show a high degree of homology. Nevertheless, I wish to emphasize that such an assumption is very indirect and speculative.

[67] M. Wittner, A. Di Stefano, P. Wangemann, J. Delarge, J. F. Liegeois, and R. Greger, *Pfluegers Arch.* **408**, 54 (1987).

TABLE III[a]

Structure	Systematic name	IC$_{50}$ Bath (Cl-channel)	IC$_{50}$ Lumen (Na2ClK-carrier)	Effect on isolate chloride channel	References
	4-Chloro-3-N-Pyrrolidino benzoic acid	$2.5 \cdot 10^{-5}$	$> 10^{-4}$		Wangemann et al.
	4-(4-Benzoylanilino) butyric acid	$3.2 \cdot 10^{-6}$	$5.6 \cdot 10^{-5}$		Wangemann et al.
	5-Chlorodiphenylamine-2-carboxylic acid	$9 \cdot 10^{-6}$	$2 \cdot 10^{-5}$	HT29	Wangemann et al. unpublished
	5-Nitro-2-(3-phenylpropyl-amino)-benzoic acid	$8 \cdot 10^{-8}$	$3 \cdot 10^{-5}$	T84, TAL RGT, HT29	Wangemann et al., Greger et al., Hayslett et al.
	4-Methyl-3-phenylamino-thiophene-2-carboxylic acid	$5.6 \cdot 10^{-6}$	$5 \cdot 10^{-5}$		Wangemann et al.

[a] The IC$_{50}$ values correspond to the concentration needed to block active chloride transport in isolated in vitro perfused cortical thick ascending limbs of the loop of Henle from the peritubular side (IC$_{50}$ bath) or from the lumenal side (IC$_{50}$ lumen). The effect on isolated chloride channels was obtained in excised membrane patches in colonic carcinoma cells (HT29, T84) or in rectal gland tubule (RGT) apical cell membranes or in isolated perfused rat thick ascending limbs (TAL).[37]

Several of the above blockers have meanwhile been tested in other epithelia. Most of this work is still preliminary. However, there is evidence that these compounds act in a variety of chloride channels in, e.g., rat thick ascending limb,[32] frog skin (W. Nagel, personal communication), pancreatic duct,[4] sweat gland duct,[68] rectal gland of the dogfish (*Squalus acanthias*),[69] collecting duct,[70] colonic carcinoma cells,[40] rabbit colon (unpublished from our laboratory), and airway epithelial cells.[64] When compared to data obtained from the intact rabbit thick ascending limb,[50] it is important to note that the sensitivity to a given blocker may be quite different. For example, in sweat duct or in frog skin the compound 5-chlorodiphenylamine-2-carboxylic acid was far more potent than NPPB, the opposite was true for the thick ascending limb of the loop of Henle. This heterogeneity in structure–activity relations for the chloride channel contrasts to the monotonous sequences for the blockers of the $Na^+/2Cl^-/K^+$ carrier, where generally the sequence bumetanide > piretanide > furosemide holds true for any tissue possessing this carrier system.[8,71] This may indicate that there is nothing like "the epithelial chloride channel" but that chloride channels show differences in the various epithelia, and these differences may also express themselves in the sensitivity to blockers.

The mechanism of chloride channel blocking by these compounds is now being studied in the rectal gland,[37] in colonic carcinoma cells,[12,40,41] and in airway epithelial cells.[64] The present data indicate that the blockers act from ouside the channel but not from the cytosolic side. The evidence for this is threefold: (1) In inside-out patches the added blocker acts only with some delay, and also the recovery after removal of the blocker is delayed. We have interpreted these findings in the following way.[37] The blocker must permeate the membrane to reach the channel from the inside of the patch pipet. Also, after removing the blocker from the bulk solution, the blocker must diffuse out of the pipet before the blocking effect is reversed; (2) more direct proof comes from the finding that the effect of a low dose of NPPB, unlike in inside-out patches, was instantaneous in outside-out membrane patches of colonic carcinoma cells[72]; (3) recently we examined derivatives of NPPB coupled to a macromolecular (5 kDa)

---

[68] J. Bijman, H. C. Englert, H. J. Lang, R. Greger, and E. Frömter, *Pfluegers Arch.* **408**, 511 (1987).

[69] R. Greger and E. Schlatter, *Pfluegers Arch.* **402**, 364 (1984).

[70] K. Tago, D. H. Warden, V. L. Schuster, and J. B. Stokes, *Am. J. Physiol.* **251**, F1109 (1986).

[71] R. Greger and E. Schlatter, *Klin. Wochenschr.* **61**, 1019 (1983).

[72] R. Greger, L. Gerlach, and K. Kunzelmann, *in* "SKF Symposium: Ion Transport" (T. Rink, ed.). Academic Press, San Diego, California, in press, 1989.

residue and found that these compounds block only in outside-out patches (unpublished observations).

We have also examined the kinetics of the chloride channel during the blocking process, and we find that low doses of a blocker ($\leq 10^{-6}$ mol/liter) lead to a reduction in the open probability, mainly by a reduction in the number of long lasting (time constant 5–20 msec) open states. Thus, the channel transits through a rapidly flickering state before it appears completely closed at higher blocker concentrations. Given the complexity of the kinetics of chloride channels it appears premature to deduce a certain model of action for the blockers from these data.

In a recent study in vesicles prepared from renal cortex a variety of putative blockers of chloride conductance was tested,[73] and it was found that NPPB was effective in the micromolar concentration range, but that an entirely unrelated compound derived from phenoxyacetic acid[55] was at least equally effective. This latter compound was used as an affinity probe to isolate the chloride channel protein. Similarly, the ability of stilbenes to block the chloride channel of airway cells[74] was used to isolate a SITS-binding protein which is supposed to correspond to the chloride channel.[75]

Conclusion

Chloride channel blockers comprise very heterogeneous compounds. The antagonists of the neuronal chloride channels have nothing in common with, e.g., the blockers for the epithelial chloride channel. On the other hand, fairly simple structures such as the diphenylamine-2-carboxylates and the related arylaminobenzoates block chloride channels in various tissues. These compounds are closely related to the known blockers of other chloride-transporting pathways such as the band 3 protein (e.g., in red blood cells) and the $Na^+/2Cl^-/K^+$ carrier system. The purification and isolation of the membrane proteins transporting chloride is a very active field of current research. It is expected that a few years from now the amino acid sequences of most of these transport proteins will be available, and we will understand more about the mechanisms of blocker interaction. The blockers available now will probably be powerful tools in the purification process.

[73] D. W. Landry, M. Reitman, E. J. Cragoe, and Q. Al-Awqati, *J. Gen. Physiol.* **90**, 779 (1987).
[74] R. J. Bridges, R. T. Worrell, R. A. Frizzell, and D. J. Benos, *Am. J. Physiol.* **256**, C902 (1989).
[75] D. J. Benos, *Proc. Forefront Symp. Nephrol., 3rd* p. 62 (1989).

Acknowledgments

Work from the author's laboratory cited in this chapter has been supported by Deutsche Forschungsgemeinschaft (Gr 480/8) and by the "Kommission der Europäischen Gemeinschaften" [ST2J-0095-2-D (CD)]. The secretarial assistance by Mrs. E. Viereck is gratefully acknowledged.

Section IV

Targeting and Intracellular Trafficking in Epithelial Cells

[46] *In Vitro* Recovery of Exocytic Transport Vesicles from Polarized MDCK Cells

By MARK K. BENNETT, ANGELA WANDINGER-NESS, ANDRÉ W. BRÄNDLI, and KAI SIMONS

Membrane transport through the secretory pathway from the endoplasmic reticulum to the plasma membrane is mediated by vesicular carriers. These carriers shuttle cargo proteins and lipids from one membrane compartment in the pathway to the next. In addition to their cargo, these transport vesicles must contain proteins which are involved in vesicle-specific functions such as budding, targeting, and fusion. By studying the composition of transport vesicles derived from different compartments along the secretory pathway one can hope to define at a molecular level the common as well as the unique components required for each vesicular transport step. Recently transport vesicles have been isolated from animal cells for both the intra-Golgi transport steps[1] and the Golgi to plasma membrane transport step.[2] These studies have resulted in the identification of vesicle-specific proteins, the functions of which can now be investigated.

This chapter will describe a method for the isolation and characterization of transport vesicles derived from the trans-Golgi network (TGN) of polarized epithelial cells [using the Madin–Darby canine kidney (MDCK) cell line]. These vesicles are destined for two different plasma membrane domains, apical and basolateral, having distinct protein and lipid compositions. In MDCK cells the sorting of plasma membrane components occurs intracellularly in the TGN prior to the formation of the transport vesicles.[3,4] This suggests that transport vesicles derived from the TGN would contain the receptor(s) involved in the sorting of plasma membrane components and proteins required for specific targeting, in addition to the vesicle machinery required for the functions common to all vesicular transport steps along the secretory pathway. Therefore, this represents a promising system for the identification of common and unique components involved in vesicular transport.

Cell perforation with a nitrocellulose filter is used to recover transport

---

[1] V. Malhotra, T. Serafini, L. Orci, J. C. Shepherd, and J. E. Rothman, *Cell (Cambridge, Mass.)* **58**, 329 (1989).
[2] I. de Curtis and K. Simons, *Cell (Cambridge, Mass.)* **58**, 719 (1989).
[3] A. Wandinger-Ness and K. Simons, *in* "Intracellular Trafficking of Proteins" (J. Hanover and C. Steer, eds.), Cambridge Univ. Press, London, in press, 1990.
[4] E. Rodriguez-Boulan and W. J. Nelson, *Science* **245**, 718 (1989).

vesicles from MDCK cells.[5] This method is one of many that have been introduced in recent years with the aim of gaining access to the cytoplasmic compartment while leaving intracellular organization intact.[6] These methods have allowed the *in vitro* reconstitution and molecular analysis of a variety of membrane transport steps. The isolation of transport vesicles thus represents a specialized application of a more general technique.

Transport Markers

We have used three different markers to monitor the formation and release of transport vesicles from MDCK cells. Two of these markers are viral glycoproteins that have been extensively used to study polarized traffic in MDCK cells. The spike glycoprotein G of vesicular stomatitis virus (VSV) is delivered preferentially to the basolateral domain whereas the hemagglutinin (HA) of influenza virus is delivered to the apical domain.[3,4] The use of viral proteins provides the advantage that viral infection inhibits host protein synthesis so that the viral glycoprotein is the major protein in transit through the secretory pathway. Since sorting is known to continue following viral infection, other proteins which are identified in the transport vesicle fraction are likely to be proteins that recycle and, therefore, comprise an important part of the transport vesicle machinery. The third marker we have used is the fluorescent lipid analog N-6(7-nitro-2,1,3-benzoxadiazol-4-yl) (C6-NBD) ceramide. This is converted to C6-NBD-sphingomyelin and C6-NBD-glucosylceramide in the Golgi complex.[7] These metabolites are then delivered to both plasma membrane domains by vesicular traffic. The NBD metabolites can be monitored by fluorescence measurements, providing a simple assay for the behavior of vesicles during the different purification steps.

All three markers can be accumulated in the TGN by incubating at 20°, causing a block in membrane traffic. The strategy for the recovery of transport vesicles is to perforate filter-grown MDCK cells in which the transport markers have been accumulated in the TGN at 20°.[5] The cells are then incubated under conditions in which transport vesicles form from the TGN but fail to fuse with the plasma membrane. Instead the vesicles are released into the incubation buffer from which they can be isolated. The emphasis of this chapter will be on the methods used to isolate and biochemically characterize these transport vesicles with the hope that similar methods can be applied to other vesicular carriers.

[5] M. K. Bennett, A. Wandinger-Ness, and K. Simons, *EMBO J.* **7**, 4075 (1988).
[6] A. M. Tartakoff (ed.), *Methods Cell Biol.* **31** (1989).
[7] G. van Meer, E. H. K. Stelzer, R. W. Wijnaendts-van-Resandt, and K. Simons, *J. Cell Biol.* **105**, 1623 (1987).

Cell Methods

Cell Culture

Media and reagents for cell culture were purchased from Gibco Biocult (Eggenheim, FRG), and Seromed (Berlin, FRG). Growth medium consisted of Eagle's minimal essential medium with Earl's salts (E-MEM) supplemented with 10 mM HEPES, pH 7.3, 10% (v/v) fetal calf serum, 100 U/ml penicillin, and 100 µg/ml streptomycin. Infection medium was composed of E-MEM supplemented with 10 mM HEPES, pH 7.3, 0.2% (w/v) bovine serum albumin (BSA), 100 U/ml penicillin, and 100 µg/ml streptomycin. Metabolic labeling medium consisted of E-MEM containing 10 mM HEPES, pH 7.3, 2.5% (v/v) fetal calf serum, and 1.5 mg/liter methionine (one-tenth the normal concentration). Water bath medium was E-MEM supplemented with 10 mM HEPES, pH 7.3, and 0.35 g/liter sodium bicarbonate (instead of the usual 2.2 g/liter).

MDCK strain II cells were grown and passaged as previously described.[8] The cells from one confluent 75-cm^2 flask were resuspended in 20 ml growth medium and seeded on a 10-cm-diameter, 0.4-µm pore size, premounted polycarbonate filter (Transwell 3419; a kind gift from Costar, Cambridge, MA). The filter cultures were maintained at 37° and 5% CO_2 for 2–3 days in a 14-cm Petri dish containing 140 ml of growth medium.

Metabolic Labeling and Viral Infection

Isolates of influenza WSNts61 were obtained from Peter Palese, New York, NY.[9] A stock of phenotypically mixed VSV (Indiana strain) was prepared in Chinese hamster ovary C15.CF1 cells which express HA on their plasma membrane and titered on MDCK cells as previously described.[5]

Metabolic labeling of MDCK filter cultures with [^{35}S]methionine was initiated after 2 days in culture. The cultures were rinsed twice with phosphate-buffered saline (PBS) containing 0.5 mM Mg^{2+} and 0.9 mM Ca^{2+} (PBS+) and placed in a 10-cm culture dish. Ten milliliters of metabolic labeling medium was added to the apical side of the filter and 15 ml metabolic labeling medium supplemented with 1 mCi [^{35}S]methionine (800 Ci/mmol; Amersham, Arlington Heights, IL) was added to the basal side and the culture incubated at 37° for 12 hr. To initiate viral infection the apical medium was aspirated and the basal medium was collected and

---

[8] K. S. Matlin, H. Reggio, A. Helenius, and K. Simons, *J. Cell Biol.* **91,** 601 (1981).
[9] S. Nakajima, D. J. Brown, M. Ueda, K. Nakajima, A. Sugiura, A. K. Pattnaik, and D. P. Nayak, *Virology* **154,** 279 (1986).

saved. The apical surface was rinsed twice with infection medium and an aliquot of virus (5–10 pfu/cell for WSNts61 or 50 pfu/cell of phenotypically mixed VSV) in 1 ml infection medium was added. The cultures were incubated for 1 hr with rocking at 31° for WSNts61 or 37° for VSV. The virus was removed and 10 ml metabolic labeling medium was added to the apical side and the basal medium collected above was supplemented with an additional 250 µCi [^{35}S]methionine and added to the basal side. The infection was continued for an additional 3.5 hr at 39° for WSNts61 or 2.5 hr at 37° for VSV.

C6-NBD-Ceramide Labeling

Phosphatidylcholine liposomes containing C6-NBD-ceramide (Molecular Probes, Eugene, OR) were prepared by octylglucoside dialysis as previously described.[5] Following viral infection the filter cultures were rinsed twice with PBS+ and placed on a 2-ml drop of water bath medium on a piece of Parafilm in a humid chamber. Three milliliters of water bath medium containing C6-NBD-ceramide liposomes (26 nmol C6-NBD ceramide/ml) was added to the apical surface and the culture was incubated in a 20° water bath for 1 hr. The liposomes were removed and the culture placed in a 10-cm culture dish. Ten and 15 ml of water bath medium containing 0.2% (w/v) BSA were added to the apical and basal sides of the filter culture, respectively, and the 20° incubation continued for an additional 1 hr.

Vesicle Formation

A detailed discussion of the cell perforation procedure has been published[10] and is, therefore, only briefly described here. MDCK cells were cultured on polycarbonate filters for 2.5 days and infected with either VSV or WSNts61 as described above. The cultures were then incubated at 20° for 2 hr to accumulate the markers in the TGN. During the 20° incubation the cells were also labeled with C6-NBD-ceramide. The cells were then perforated with a nitrocellulose filter using the following procedure (all steps at 4°): The filter culture was cut from the holder with a scalpel, rinsed briefly in KOAc buffer (25 mM HEPES–KOH, pH 7.4, 115 mM potassium acetate, 2.5 mM MgCl$_2$), and placed (cell side up) in a dry Petri dish in a 20° water bath. A 10-cm nitrocellulose filter (HATF, 0.45-µm pore size; Millipore) previously soaked in KOAc buffer was blotted dry between two pieces of Whatman filter paper for 1 min and placed on top of the filter

[10] M. K. Bennett, A. Wandinger-Ness, I. de Curtis, C. Antony, K. Simons, and J. Kartenbeck, *Methods Cell Biol.* **31**, 103 (1989).

culture. Excess moisture was removed by placing a Whatman filter on top of the nitrocellulose and smoothing with a bent Pasteur pipet. Binding was allowed to proceed for 90 sec and was terminated by the addition of 2 ml KOAc buffer. Excess buffer was aspirated and the two filters separated. The perforated filter culture was incubated in GGA transport buffer [25 mM HEPES–KOH, pH 7.4, 38 mM potassium gluconate, 38 mM potassium glutamate, 38 mM potassium aspartate, 2.5 mM MgCl$_2$, 1 mM dithiothreitol (DTT), 2 mM EGTA, 1 mM ATP, 8 mM creatine phosphate, and 50 μg/ml creatine kinase; 10 ml/filter] at 37° for 1 hr. Following the incubation the transport buffer was collected and used as the starting material for the vesicle isolation. Under these conditions 25–40% of the transport markers were released from the cells while markers for other cellular organelles [Golgi, endoplasmic reticulum (ER), endosomes] and lysosomes remained associated with the cells (<10% released). The release of transport vesicles required the presence of ATP and EGTA.

Vesicle Isolation

The isolation of transport vesicles released from perforated cells required a three-step procedure. First, sedimentation of the vesicles through a sucrose cushion was used to remove soluble cytosolic protein. The second step was an equilibrium flotation gradient to resolve the transport vesicles from other membranes and protein aggregates. The final step of the vesicle purification was immunoisolation using antibodies generated against the cytoplasmic domain of either VSV G or WSN HA.

Differential Centrifugation

The medium fraction from perforated MDCK cells was collected and subjected to sedimentation at 500 g for 10 min to remove any intact cells and large cellular debris. The resulting supernatant ($\approx$9 ml) was loaded onto a 2-ml cushion of 0.25 M sucrose (all sucrose solutions were made in 10 mM HEPES–KOH, pH 7.4, 1 mM EGTA, 1 mM DTT) and subjected to centrifugation at 100,000 g for 3 hr in an SW 40 rotor. The supernatant was carefully aspirated and the pellet resuspended in 1.5 M sucrose.

The resuspended membrane pellet was placed in the bottom of an SW 60 centrifuge tube and overlayed with either a continuous linear gradient (0.3–1.5 M sucrose) or a discontinuous step gradient (0.8 and 1.2 M sucrose) and subjected to centrifugation at 100,000 g for 13 hr. The use of a continuous gradient allowed for the resolution of HA and G into two distinct peaks with HA peaking at a density of 1.099 g/ml and G at 1.113 g/ml. The discontinuous gradient was used to obtain a more concentrated

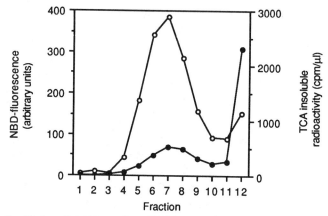

FIG. 1. Equilibrium flotation gradient. Vesicles were isolated from [^{35}S]methionine- and C6-NBD-ceramide-labeled cells as described in the text and resolved on a continuous linear sucrose gradient (0.3–1.5 M sucrose). The gradient was fractionated and aliquots analyzed for NBD fluorescence (○) and acid-insoluble radioactivity (●). TCA, Trichloroacetic acid.

vesicle fraction for use in immunoisolation experiments. Figure 1 shows an example of a continuous equilibrium flotation gradient used to isolate vesicles obtained from perforated cells labeled with both [^{35}S]methionine and C6-NBD-ceramide. The vesicles (NBD fluorescence) float to their equilibrium density of ≈1.1 g/ml while most of the protein (measured as acid-insoluble radioactivity) remains at the bottom of the gradient. This step results in a 10-fold enrichment of the transport vesicles.

Immunoisolation

Immunoisolation was used as the final vesicle purification step, allowing for the separation of transport vesicles destined for either the apical or the basolateral plasma membrane domains. We have used antibodies directed against the cytoplasmic domains of either VSV G or influenza HA for these studies, but antibodies directed against other markers with cytoplasmically exposed epitopes (e.g., mannose 6-phosphate receptor, clathrin, synaptophysin) could be used to isolate different types of vesicles. The following immunoisolation methods represent a modification of previously published procedures. The vesicle pool from the flotation gradient was divided into three equal aliquots (usually 300 μl each). The first aliquot was diluted fourfold with 10 mM HEPES–KOH, pH 7.4, 1 mM EGTA (to reduce the sucrose concentration to ≈0.25 M) and subjected to centrifugation at 55,000 rpm (200,000 g) in a TLS55 rotor in a Beckman TL-100 centrifuge. The resulting pellet served as a control for the total particulate material in the starting vesicle fraction. The remaining two aliquots were used for immunoisolation using either specific or control monoclonal antibodies. Antibody P5D4 directed against the cytoplasmic domain of VSV G[11] served as the specific antibody for vesicles obtained from VSV-infected cells and as the control antibody for vesicles obtained from WSN-infected cells. Antibody 2D1 directed against the cytoplasmic domain of WSN HA[12] served as the specific antibody for vesicles obtained from WSN-infected cells and as the control antibody for vesicles obtained from VSV-infected cells. One-half volume (~150 μl) of PBS containing 0.1% (w/v) gelatin and a protease inhibitor cocktail (10 μg/ml each of chymostatin, leupeptin, antipain, and pepstatin) was added to each aliquot. The monoclonal antibody (20 μl of a 10-fold concentrated hybridoma culture supernatant) was then added and the sample incubated at 4° with rotation for either 3 hr for VSV samples or overnight for WSN samples. Cellulose fibers (prepared as previously described[13]) to which an affinity-purified sheep antibody generated against the Fc domain of mouse IgG was covalently coupled were used as the solid support to recover the antibody–vesicle complexes.[14] Following the antibody incubation 1 mg of the cellulose fibers [500 μl of 2 mg/ml fibers in PBS containing 0.1% (w/v) gelatin] was added to each sample and the incubation continued for 1.5 hr at 4° with rotation. The fibers were pelleted in the microfuge at 1500 rpm for 10 min and washed three times with PBS containing 0.1% (w/v) gelatin and once with PBS. Under these conditions greater than 50% of the viral glycoproteins were recovered on the specific fibers while less than 10% were recovered on the control fibers.

The immunoisolation conditions used were optimized to obtain maximal specific recovery of the desired marker with minimal nonspecific binding. One factor important for maximal recovery was to carry out antibody binding in solution (i.e., in the absence of the solid support) to increase the possibility of antibody–antigen interaction. The HA-containing vesicles proved to be more difficult to recover quantitatively than the VSV G-containing vesicles, perhaps because of the short cytoplasmic domain of WSN HA[9] (only 11 amino acids) or because of differences in antibody affinity for the respective antigens. Two modifications of the VSV G protocol were used which increased the recovery of HA-containing vesicles. The first was to increase the antibody binding time from 3 hr to overnight. The second was the use of a temperature-sensitive mutant of WSN (ts61). The mutant HA fails to be transported out of the ER at the

[11] T. E. Kreis, *EMBO J.* **5**, 931 (1986).
[12] A. Wandinger-Ness and K. Simons, unpublished observations.
[13] J. P. Luzio, *Methodol. Surv. Biochem.* **6**, 131 (1977).
[14] I. de Curtis, K. Howell, and K. Simons, *Exp. Cell Res.* **175**, 248 (1988).

nonpermissive temperature and therefore accumulates there when the infection is carried out at 39°. When the cells are subsequently shifted to 20° the HA transported out moves out of the ER in a synchronous wave and accumulates in the TGN at higher concentrations than would be obtained with wild-type virus. This presumably results in a higher concentration of the antigen in the transport vesicles and therefore a more efficient recovery by immunoisolation.

Nonspecific binding can be a serious problem when dealing with small amounts of material and with the mild washing conditions required to maintain membrane integrity. The minimization of nonspecific binding required screening a number of different solid supports (cellulose fibers, magnetic beads, fixed *Staphylococcus aureus* cells) and the use of gelatin as a blocking reagent. In our experience the cellulose fibers gave the lowest nonspecific binding.

Vesicle Analysis

Total Protein

The total protein composition of the immunoisolated transport vesicles was analyzed following metabolic labeling with [^{35}S]methionine by two-dimensional gel electrophoresis.[15] The filter cultures were metabolically labeled with [^{35}S]methionine for 12 hr prior to viral infection to label all cellular proteins. The cells were then infected with either VSV or WSNts61 and the metabolic labeling continued during the infection. The cells were then perforated and the vesicles isolated as described above. These labeling conditions ensured that both endogenous cellular proteins and the viral glycoproteins would be detected in the vesicle fraction.

Following immunoisolation, each sample (total vesicles, specific fibers, and control fibers) was solubilized in two-dimensional (2D) gel lysis buffer [4% (w/v) NP-40, 2% (v/v) pH 7–9 ampholines (LKB, Freiburg, FRG), 9.8 M urea, 100 mM DTT], resolved by 2D gel electrophoresis, and the labeled proteins detected by fluorography. The use of a highly reproducible 2D gel system was of critical importance in this analysis to allow the comparison of sets of gels run at different times. The resolution and reproducibility of the system described by Bravo[15] were ideal. The first dimension entailed isoelectric focusing with a mixture of ampholines producing very good resolution of proteins with pI values between 4.5 and 8.0.

[15] R. Bravo, *Proc. Natl. Acad. Sci. U.S.A.* **81**, 4848 (1984).

The second dimension comprised SDS/PAGE using 15% (w/v) acrylamide and 0.075% (w/v) bisacrylamide. The quantitative analysis of the 2D gel patterns was carried out by densitometric scanning. The recovery of the viral proteins on the specific and control fibers was determined relative to the total fraction. Each protein on the gel considered to be a potential vesicle-specific protein was then analyzed to determine if it was recovered to the same extent as the viral proteins. A protein was considered specifically associated with the transport vesicles if it was recovered to the same extent as the viral glycoprotein on the specific fibers and was not recovered on the control fibers. This analysis has resulted in the identification of one class of proteins associated specifically with apical vesicles, a second class of proteins associated with basolateral vesicles, and of common proteins which appear to be associated with both classes of vesicles.[16]

We have used several techniques in order to further characterize the putative vesicle-specific proteins. We have determined whether they are integral or peripheral membrane proteins by Triton X-114 (TX-114) phase partitioning.[17] In order to evaluate whether the proteins have domains exposed on the cytoplasmic or lumenal domain we have used selective biotinylation in experiments designed to label either the vesicle surface or the two plasma membrane domains.

Triton X-114 Phase Partitioning

Partitioning of proteins between the detergent and aqueous phases has been used to define membrane–protein interactions. In general, integral membrane proteins are recovered in the detergent phase and peripheral membrane proteins are recovered in the aqueous phase.[17] For this analysis, a vesicle fraction was prepared from metabolically labeled cells by equilibrium flotation and concentrated by sedimentation. The vesicle pellet was dissolved in 50 μl cold PBS containing 1% Triton X-114 and left on ice for 15 min. The sample was then incubated at 37° for 5 min and subjected to centrifugation in a microfuge for 2 min to separate the phases. The aqueous (supernatant) and detergent (pellet) phases were carefully separated. The detergent phase was dissolved directly in 2D gel sample buffer. The aqueous phase was first concentrated by precipitation with 10 vol. −20° acetone, dried, and then dissolved in 2D gel sample buffer. The 2D gel patterns were then analyzed to determine the detergent solubility of the candidate vesicle-specific proteins previously identified by immunoisolation.

---

[16] A. Wandinger-Ness, M. K. Bennett, and K. Simons, unpublished observations.
[17] C. Bordier, *J. Biol. Chem.* **256,** 1604 (1981).

Surface Biotinylations

The surface biotinylation experiments were designed to determine whether a particular protein has a domain exposed on the cytoplasmic surface of the vesicle or on either plasma membrane surface (equivalent to the vesicle lumen). Sulfosuccinimidyl-2-(biotinamido) ethyl-1 3'-dithiopropionate (BSSNHS; Pierce, Rockford, IL) which reacts with free amino groups was used to derivatize proteins with biotin. BSSNHS contains a disulphide bond, allowing recovery of biotinylated proteins after isolation on a streptavidin solid support by cleavage of the disulphide bond with DTT (present in the 2D gel sample buffer).[18,19] Due to the sulfo group, BSSNHS is water soluble and does not penetrate cellular membranes.[20]

Vesicles. A vesicle fraction was prepared from metabolically labeled cells by equilibrium flotation. An aliquot (210 μl) was mixed with 25 μl BSSNHS (2 mg/ml) in PBS and incubated on ice for 30 min. Subsequently, 25 μl of 0.5 M glycine in 10 mM HEPES, pH 7.4, was added to quench the remaining BSSNHS and the sample was diluted with 3 vol of 10 mM HEPES, pH 7.4, 1 mM EGTA. The labeled vesicles were then recovered by centrifugation at 200,000 g for 3 hr in a TLS55 rotor and the pellet dissolved in 550 μl solubilization buffer [PBS containing 2% (w/v) NP-40, 0.2% (w/v) SDS, 35 μg/ml phenylmethylsulfonyl fluoride (PMSF), and a protease inhibitor cocktail consisting of 10 μg/ml each of chymostatin, leupeptin, antipain, and pepstatin]. Streptavidin–agarose (Sigma, St. Louis, MO) was washed three times with wash buffer A (PBS with 1% NP-40) and 30 μl of a 50% suspension of streptavidin–agarose was added to the lysate. The mixture was incubated with rotation for 2 hr at 4° and the streptavidin–agarose pelleted by centrifugation in a microfuge. The streptavidin–agarose was washed three times with wash buffer A, twice with wash buffer B (PBS with 0.1% NP-40 and NaCl to 0.5 M), and once with 50 mM Tris-HCl, pH 7.5. Adsorbed proteins were solubilized in 2D gel sample buffer.

Plasma Membrane Domains. MDCK cells were grown on 24-mm polycarbonate filters for 2 days and metabolically labeled with [^{35}S]methionine for 12 hr. The filter cultures were then washed three times with PBS and incubated with 250 μl of 1.5 mg/ml BSSNHS in PBS for 30 min on ice. For apical biotinylation the BSSNHS was added directly to the apical chamber of the filter. For basolateral biotinylation the filter was placed on top of a drop of BSSNHS on a piece of Parafilm. The filters were washed

---

[18] D. R. Gretch, M. Suter, and M. F. Stinski, *Anal. Biochem.* **163**, 270 (1987).
[19] A. Brändli and K. Simons, unpublished observations.
[20] M. Sargiacomo, M. Lisanti, L. Graeve, A. Le Bivic, and E. Rodriguez-Boulan, *J. Membr. Biol.* **107**, 277 (1989).

four times with PBS and the cells scraped from the filter with a rubber policeman. The cells were transferred to an Eppendorf tube, pelleted by centrifugation, and solubilized in 500 μl solubilization buffer. The cell lysate was cleared by centrifugation at 13,000 rpm for 10 min. Streptavidin-agarose was added to the cleared lysates and the samples processed as described above. Again the resultant 2D gels were compared to the protein patterns obtained following the immunoisolation. Together, the specific biotinylation of the cytoplasmic vesicle surface and of the exocytoplasmic plasma membrane surface permits one to assign a membrane topology to the candidate vesicle proteins. The localization of proteins detected on the cell surface by biotinylation was consistent with the recovery of these proteins in apical, basolateral, or both vesicle fractions.

Discussion

This chapter describes methods for the isolation of transport vesicles derived from the TGN of polarized MDCK cells. Similar methods could possibly be applied to the isolation of other vesicles not only along the secretory pathway, but along the endocytic or transcytotic pathway as well. Two important factors have permitted the isolation of transport vesicles in a form pure enough for biochemical characterization. The first factor was the establishment of a perforated cell system. Such perforated cells maintain their intracellular organization and most organelles remain intact. This is a tremendous advantage over preparing a cellular homogenate that causes many organelles to vesiculate, thus making the recovery of authentic transport vesicles extremely difficult. The conditions required for the formation of transport vesicles in perforated cells were established using a biochemical assay. The HA of fowl plague virus undergoes a proteolytic processing step late in the biosynthetic pathway, probably between the TGN and the plasma membrane.[5] This processing step was monitored to optimize the conditions for transport vesicle formation. A similar proteolytic processing step of Semliki forest virus glycoprotein was used to establish conditions for the formation of transport vesicles from the TGN of Baby Hamster Kidney (BHK) cells both in a perforated cell system[21] and in a postnuclear supernatant.[2] The failure of transport vesicles to undergo fusion with the plasma membrane, resulting in their recovery in the medium, may be caused by the rapid dilution of the cytosol in the perforated MDCK cells upon transfer to the GGA transport buffer. The putative transport vesicles derived from other stages of the secretory pathway seem to remain largely within the cell, perhaps tightly associated with a donor or

[21] I. de Curtis and K. Simons, *Proc. Natl. Acad. Sci. U.S.A.* **85**, 8052 (1988).

acceptor membrane compartment. The end result is that the membranous material released from perforated cells under the conditions described is highly enriched for the TGN-derived transport vesicles.

The second factor which made it possible to purify the vesicles was the availability of specific antibodies allowing the application of immunoisolation techniques. As a result, we were able to separate two classes of transport vesicles derived from the same membrane compartment and having very similar physical properties. The immunoisolation procedure described here could also be used to recover other vesicular compartments. The two main requirements are an antibody that recognizes a cytoplasmically exposed epitope, and a restricted localization of the antigen. It should also be possible to use antibodies against endogenous proteins such as clathrin, mannose 6-phosphate receptor (both types), polymeric IgA receptor, or synaptophysin to characterize other vesicular transport pathways in MDCK or other cell types.

The biochemical characterization of the TGN-derived transport vesicles made possible by the methods described here should lead to a better understanding of the TGN-to-plasma membrane transport step. The identification of vesicle-specific proteins will now allow analysis of protein function. Of primary interest to the area of cell polarity is the problem of sorting of membrane components prior to transport vesicle formation and the targeting of transport vesicles to the appropriate plasma membrane domain. The process of sorting is likely to involve receptors that divert proteins into one (or both) of the exocytic pathways in MDCK cells. The prediction is that the sorting receptor would be included in the transport vesicle, delivered to the plasma membrane, and then recycled back to the TGN. We have been able to identify several apical proteins which recycle between the plasma membrane and the TGN using the ricin-resistant MDCK cell line.[22] We are currently attempting to determine whether these correspond to any of the proteins identified in the apical vesicle fraction isolated from wild-type MDCK cells. Vesicle delivery may in part be mediated by microtubules, at least to the apical plasma membrane.[23] Vesicular proteins (as well as cytosolic ones) are likely to be required for such microtubule interactions. *In vitro* reconstitution assays have been used to demonstrate the specific binding of TGN-derived vesicles to microtubules[24] and can be used to identify components which mediate the interaction. Finally, recent evidence has implicated GTP-binding proteins in vesicular transport and it has been proposed that these proteins may

[22] A. W. Brändli and K. Simons, *EMBO J.* **8**, 3207 (1989).
[23] U. Eilers, J. Klumperman, and H.-P. Hauri, *J. Cell Biol.* **108**, 13 (1989).
[24] P. van der Sluijs, M. K. Bennett, K. Simons, and T. Kreis, unpublished observations.

provide information for the targeting of transport vesicles to the appropriate acceptor membranes.[25] The TGN-derived transport vesicles can also be used to study the role of such GTP-binding proteins. In conclusion, the further characterization of the putative vesicle-specific proteins we have now identified should provide insight into the molecular machinery required both for protein sorting and vesicle formation, as well as for vesicle targeting and fusion.

[25] H. R. Bourne, *Cell (Cambridge, Mass.)* **53**, 669 (1988).

[47] Biogenesis of the Rat Hepatocyte Plasma Membrane

By JAMES R. BARTLES and ANN L. HUBBARD

Model systems which exploit the behavior of exogenous (usually viral) plasma membrane glycoproteins in cultured epithelial cell lines, such as the Madin–Darby canine kidney (MDCK) cell line, have been of great value in studying protein sorting.[1-5] However, these types of studies must be extended to encompass the sorting of endogenous plasma membrane proteins in epithelial cells organized into tissues. Recent developments in methodology have permitted a thorough analysis of the pathways of plasma membrane protein sorting in rat hepatocytes *in situ*.[6-8] Interestingly, these studies have revealed that the sorting pathway may be different for plasma membrane proteins in MDCK cells and hepatocytes. In MDCK cells integral membrane proteins destined for apical and basolateral domains appear to be sorted before ever reaching the plasma membrane.[1-5] In contrast, in hepatocytes, the plasma membrane proteins—apical and basolateral alike—are first shipped to the basolateral plasma membrane. Then the apical proteins are sorted from the basolateral ones and delivered to the apical plasma membrane.[6-8]

The approach utilized to elucidate the pathway of plasma membrane

[1] E. Rodriguez-Boulan, *Mod. Cell Biol.* **1**, 119 (1983).
[2] K. S. Matlin, *J. Cell Biol.* **103**, 2565 (1986).
[3] M. Caplan and K. S. Matlin, *Mod. Cell Biol.* **8**, 71 (1989).
[4] E. Rodriguez-Boulan and W. J. Nelson, *Science* **245**, 718 (1989).
[5] K. Simmons, this volume [24].
[6] J. R. Bartles, H. M. Feracci, B. Stieger, and A. L. Hubbard, *J. Cell Biol.* **105**, 1241 (1987).
[7] J. R. Bartles and A. L. Hubbard, *Trends Biochem. Sci.* **13**, 181 (1988).
[8] A. L. Hubbard, B. Stieger, and J. R. Bartles, *Annu. Rev. Physiol.* **51**, 755 (1989).

protein sorting in rat hepatocytes employed a combination of pulse–chase radiolabeling and subcellular fractionation techniques.[6] The success of the approach relied heavily upon the development of procedures to purify intact hepatocyte plasma membrane[9,10] (referred to as plasma membrane sheets) and to separate vesicles derived from these plasma membrane sheets into apical and basolateral vesicle subpopulations.[11] These procedures are described in detail below in conjunction with a number of accessory techniques which may prove useful in studies of plasma membrane protein sorting in rat hepatocytes *in situ*.

Hepatocyte Plasma Membrane Sheets

Isolation[9]

This isolation procedure is an extensively modified version of a method published by Toda *et al.*[12] It offers several advantages: (1) it is fast, taking a total of only 4–5 hr; (2) it is simple, requiring only four low-speed centrifugations and a single flotation through a one-step sucrose gradient; (3) it is effective, with routine yields of 10–20% and relative enrichments of plasma membrane marker enzymes in the range of 30- to 60-fold; and (4) it yields a preparation of intact hepatocyte plasma membrane which contains apical, lateral, and basal domains in surface area ratios approximating those observed *in situ* (see the section, Characterization, below).

All sucrose solutions should be prepared 24–48 hr before use and their densities determined at room temperature with an Abbe refractometer. The solutions are Millipore (Bedford, MA) filtered (0.22 μm for 0.25 M and 1.2 μm for 1.42 and 2.0 M) and stored at 4°. Male Sprague–Dawley rats (125–150 g; Charles River Breeding Laboratories, Wilmington, MA) are fasted for 18–24 hr and killed by decapitation while under ether anesthesia. Livers are excised and perfused through the portal vein with ice-cold 0.15 M NaCl until the lobes are blanched. All subsequent procedures are carried out at 4°. Individual livers are weighed and added to 4.5 vol of 0.25 M STM (0.25 M sucrose, 5 mM Tris-HCl, 1 mM MgCl$_2$, pH 7.3–7.4) containing protease inhibitors (1 mM phenylmethylsulfonyl fluoride, added fresh from a 0.2 M stock in 100% ethanol; 1 μg/ml each of antipain and leupeptin, added fresh from a 1 mg/ml stock in 10% dimethyl sulfoxide; and 100 KIU/ml of Trasylol or aprotinin, added from a 10,000 KIU/ml aqueous stock solution). The livers are minced into small (30- to

[9] A. L. Hubbard, D. A. Wall, and A. Ma, *J. Cell Biol.* **96**, 217 (1983).
[10] A. L. Hubbard and A. Ma, *J. Cell Biol.* **96**, 230 (1983).
[11] J. R. Bartles, L. T. Braiterman, and A. L. Hubbard, *J. Cell Biol.* **100**, 1126 (1985).
[12] G. Toda, H. Oka, T. Oda, and Y. Ikeda, *Biochim. Biophys. Acta* **413**, 52 (1975).

120-mm³) pieces with scissors and homogenized in 10 gentle up-and-down strokes in a 40-ml Dounce-type glass homogenizer (Wheaton, Millville, NJ) with the loose-fitting pestle (B; nominal clearance 0.0025 – 0.0035 in.). The pestle should be wiped off after the second or third stroke to remove any adherent connective tissue. The homogenate is adjusted to 20% (liver wet wt to total volume) and filtered through four layers of moistened gauze (three to four livers only per gauze pad to prevent clogging). The filtrate is centrifuged (25 – 30 ml/50-ml plastic conical tube) at 280 g for 5 min [1100 rpm in a Beckman (Fullerton, CA) TJ-6R refrigerated table-top centrifuge]. The supernatant is carefully decanted off and saved. The pellet is resuspended in one-half the volume of the homogenate by three strokes of the loose Dounce and centrifuged again as above. The first and second supernatants are combined and centrifuged (25 – 30 ml/50-ml plastic conical tube) at 1500 g for 10 min (2800 rpm in Beckman TJ-6R). The supernatant is removed by aspiration and the resulting pellets are pooled and resuspended with three strokes of the loose-fitting Dounce and one stroke of the tight-fitting Dounce (A; nominal clearance 0.001 – 0.003 in.) in 1 – 2 ml of 0.25 M STM/g of liver (initial wet wt). 2.0 M STM is added to obtain a density of 1.18 g/cm³ (1.42 M sucrose, $n = 1.4016$), and sufficient 1.42 M STM is added to bring the volume to approximately twice that of the original homogenate (i.e., 10%). Aliquots (35 ml) of the sample are added to 25 × 89 mm Ultraclear tubes (Beckman), overlaid with 2 ml of 0.25 M sucrose (containing neither Tris-HCl nor $MgCl_2$), and centrifuged at 113,000 g_{max} for 60 min in a swinging bucket rotor with no brake (25,000 rpm in a Beckman SW-28 rotor and L5-65 ultracentrifuge). The pellicle at the interface is collected with a blunt-tipped plastic transfer pipet and resuspended in a loose Dounce homogenizer in sufficient 0.25 M sucrose to obtain a density of ≤ 1.05 g/cm³. The suspension is centrifuged at 1500 g for 10 min as above. The final pellet is resuspended to a protein concentration of 1 – 2 mg/ml (i.e., in about 0.5 – 1 ml/g of initial liver) in 0.25 M sucrose containing the protease inhibitors by three strokes of the loose-fitting Dounce homogenizer. The plasma membrane sheets can be frozen in aliquots at $-70°$. One can anticipate a recovery of about 1 mg of plasma membrane protein/g of starting liver (wet wt).

Several features of the isolation procedures are essential to the successful preparation of highly purified plasma membrane sheets. (1) To reduce aggregation of subcellular organelles by glycogen, the rats should be fasted for at least 18 hr; (2) livers from relatively small young adult rats (125 – 150 g) should be used to avoid the increased amounts of connective tissue and plasma membrane-associated filaments found in the preparations from the livers of older rats; (3) to minimize autolysis, the homogenization of the perfused livers should be performed within 5 min of excision; (4) to reduce vesiculation and/or production of small membrane fragments, the homogenization should be gentle, with minimal generation of bubbles or vacuum during both the up and down strokes; and (5) to reduce contamination with membranes derived from the endoplasmic reticulum, it is helpful to resuspend the first 1500 g, 10-min pellet in a volume approximately twice that of the initial homogenate in preparation for the flotation step.

Characterization[9-11]

Hepatocyte plasma membrane sheets prepared in this way exhibit a number of desirable properties. First, they are highly purified. Relative to the gauze-filtered liver homogenate, the recovery of typical plasma membrane marker enzymes is usually in the range of 10-20% with 30- to 60-fold enrichments (Table I). In addition, a variety of immunoblotting, radioligand-binding, and marker enzyme assays have indicated substan-

TABLE I
RECOVERIES AND ENRICHMENTS OF VARIOUS ORGANELLAR MARKERS DURING THE ISOLATION OF HEPATOCYTE PLASMA MEMBRANE SHEETS[a]

Marker	Recovery[b] (%)	Enrichment[b] (-fold)
Protein	0.4 ± 0.13	—
5'-Nucleotidase (PM)[c]	17.4 ± 6	39 ± 10
Alkaline phosphodiesterase I (PM)	17.0 ± 5.6	40 ± 9
Asialoglycoprotein binding activity (PM)	16.3 ± 6	47 ± 18
NADH-cytochrome c reductase (ER)[d]	0.3 ± 0.2	1 ± 0.5
Glucose-6-phosphatase (ER)	0.5	0.8
β-N-Acetylglucosaminidase (lysosomes)	0.22 ± 0.13	0.66 ± 0.3
Cytochrome oxidase (mitochondria)	0.12 ± 0.07	0.22 ± 0.1
DNA (nuclei)	0.26	0.3
Galactosyltransferase (Golgi)	1.0	1.9
Sialyltransferase (Golgi)	0.7	1.3
[^{125}I]Asialoorosomucoid-containing endosomes	0.5	0.9
[^{125}I]Mannosyl-BSA receptor (nonhepatocyte PM)[e]	3.1 ± 0.3[e]	5.8 ± 0.5[e]

[a] Compiled from Bartles et al.,[6] Hubbard et al.,[9] and Bartles et al.[11] by permission of the Rockefeller University Press.
[b] Reported as mean ± SD ($n \geq 3$), when available.
[c] PM, Plasma membrane.
[d] ER, Endoplasmic reticulum.
[e] Data obtained from livers which had been labeled with 500 μg of wheat germ agglutinin by isolated perfusion at 4°. These numbers may represent overestimates because, under these conditions, the recovery and enrichment of a typical hepatocyte plasma membrane marker such as alkaline phosphodiesterase I are nearly doubled (35 ± 3% and 68 ± 7-fold, respectively).[11]

tially lower recoveries of membranes derived from other organelles, including the Golgi complex, endoplasmic reticulum, mitochondria, endosomes, lysosomes, and nuclei, and even of plasma membrane derived from the nonhepatocyte cell types of the liver (the Kupffer and endothelial cells) (Table I). This high degree of purity helps to ensure that any newly synthesized plasma membrane protein is truly at the plasma membrane instead of in a contaminating membrane vesicle derived from an intracellular organelle (e.g., the Golgi). Perhaps the simplest way to routinely check recovery and purity of the plasma membrane sheet fraction is to employ the relatively simple colorimetric assays for the plasma membrane marker enzyme alkaline phsophodiesterase I[13] and protein.[14] The freshly isolated plasma membrane sheets should also be examined by phase-contrast microscopy at a magnification of ×25. One should observe relatively few small vesicles and membrane fragments and an abundance of relatively large sheetlike structures characterized by a phase-dense line (lateral surface) with appended ruffles of membrane (predominantly basal surface).

Upon electron microscopic examination it is apparent that the plasma membrane sheets are actually formed by the joining together of large pieces of plasma membrane derived from two or sometimes three adjacent hepatocytes (Fig. 1, bottom left). The pieces of membrane are joined along the lateral surfaces which they shared in the tissue (Fig. 1, top). The plasma membrane sheets retain all three domains—basal [or sinusoidal (sf)], lateral (ls), and apical [or bile canalicular (bc)]—in near-normal surface area ratios, a full complement of intercellular junctions, and an appended subplasmalemmal cytoskeletal network. And, what is most important for subsequent resolution into apical and basolateral subfractions (see below), representative domain-specific integral proteins of the rat hepatocyte plasma membrane have been found to retain their proper localizations on the isolated plasma membrane sheets by immunogold electron microscopy.[11,15]

Apical and Basolateral Vesicles

Vesiculation[6,11]

Isolated hepatocyte plasma membrane sheets are thawed and diluted to a protein concentration of 1.0 mg/ml in 0.25 M sucrose containing pro-

---

[13] O. Touster, N. N. Aronson, Jr., J. T. Dulaney, and H. Hendrickson, *J. Cell Biol.* **47**, 604 (1970).
[14] M. A. K. Markwell, S. M. Haas, L. L. Bieber, and N. E. Tolbert, *Anal. Biochem.* **87**, 206 (1978).
[15] L. M. Roman and A. L. Hubbard, *J. Cell Biol.* **98**, 1488 (1984).

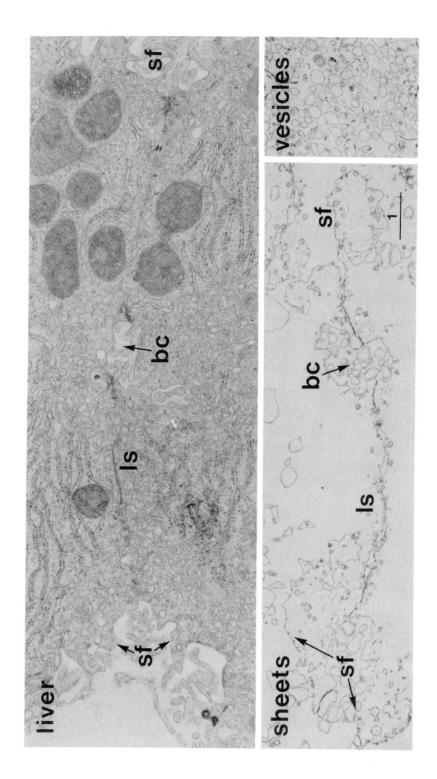

tease inhibitors. Aliquots (2 ml) of membrane suspension are placed in capped 14-ml polystyrene conical tubes and vesiculated at 0° in a bath-type sonicator capable of delivering an output of 80 W at 80 kilocycles (Laboratory Supplies Co., Hicksville, NY). Ten-second bursts are used, interrupted by 1-min periods with no sonication. It is best to monitor the vesiculation process by light microscopy. After each burst or two, a 25-μl drop of the suspension should be placed on a microscope slide, covered with a coverslip, and examined immediately using a ×25 phase-contrast objective. Ideally, one should sonicate for just enough bursts (usually four to six) to convert a majority of the sheets into vesicles which are no longer visible at this magnification (see Fig. 1, bottom right, for electron micrograph of the vesicles). If a bath-type sonicator is not available, then satisfactory vesiculation may also be achieved using a Polytron homogenizer (Brinkman, Westbury, NY) fitted with a 12-mm tip at a setting of 8.

Sucrose Density Gradient Centrifugation[6,11]

Vesiculated plasma membrane suspension (3.9 ml) is loaded onto 32.1-ml linear sucrose gradients (1.06–1.17 g/cm^3) buffered with 10 mM Tris-HCl, pH 7.5. After centrifugation at 72,000 g_{max} for 16–20 hr with no brake (20,000 rpm, Beckman SW 28 rotor and L5-65 centrifuge), 3-ml fractions are collected from the top using an automatic gradient filler/drainer (Auto Densi-Flow IIC, Haake Buchler Instruments, Saddlebrook, NJ). Fraction densities are determined with an Abbe refractometer, and the fractions are stored at −70°.

With this procedure, the vesicles derived from the apical and basolateral domains are partially resolved on the basis of differences in equilibrium density (e.g., see Fig. 2C). The density profile for the apical vesicles is characterized by a single peak centered at a density of 1.10 g/cm^3; the distribution of the basolateral vesicles is bimodal, with a peak in the body of the gradient centered at a density of 1.14 g/cm^3 and smaller (variable) amount found near the bottom of the gradient (density ≥ 1.17 g/cm^3). Poor resolution of the apical and basolateral vesicles is most often the result of either over- or undersonication. We routinely establish the distributions of the apical and basolateral vesicles by immunoblotting for representative apical- and basolateral-specific hepatocyte plasma membrane

---

FIG. 1. Electron microscopic analysis of an isolated plasma membrane sheet (bottom left), the same regions of the hepatocyte plasma membrane revealed in a section of liver tissue (top), and the vesicles generated from the plasma membrane sheets by sonication (bottom right) (Bar, 1 μm.) Shown are the basal domain (sf, sinusoidal front), lateral domain (ls, lateral surface), and apical domain (bc, bile canaliculus).

proteins,[16] such as HA 4 and CE 9 (Fig. 2C). For this purpose, aliquots of the gradient fractions are resolved in 7.5% polyacrylamide-SDS gels, and the proteins are electrophoretically transferred to nitrocellulose (4 hr at constant 400 mA in 20% methanol, 0.1% SDS, 0.19 M glycine, and 0.1 M Tris). After quenching with 2% gelatin, the blots are labeled with either ^{125}I-labeled antibodies or antibodies plus [^{125}I]Protein A, all in 0.2% Triton X-100, 0.15 M NaCl, 10 mM Tris-HCl, 0.05% NaN$_3$, pH 7.4 and subjected to autoradiography for 1–48 hr using Kodak XAR-5 X-ray film and a Du Pont (Wilmington, DE) Cronex Lightning Plus intensifying screen at $-70°$. Marker distributions are then established by laser densitometric scanning (Biomed Instruments, Fullerton, CA, or Ultroscan XL, LKB-Pharmacia, Piscataway, NJ) of autoradiograms. However, there are a number of alternatives available for establishing these distributions. One of the simplest ways is to use marker enzyme assays for representative domain-specific plasma membrane enzymes, such as K$^+$-stimulated p-nitrophenylphosphatase[17] (a partial reaction of the Na$^+$,K$^+$-ATPase) for the basolateral vesicles and 5′-nucleotidase[18] for the apical vesicles. In addition, it is possible to selectively label the basolateral vesicles by prior isolated liver perfusion with [^{125}I]wheat germ agglutinin at 4°.[11]

Pulse–Chase Metabolic Radiolabeling

Injection of Isotope[6]

To tag the subset of newly synthesized plasma membrane proteins and to monitor their intracellular transit as a cohort, we recommend pulse–chase metabolic radiolabeling with either L-[^{35}S]methionine or with Tran^{35}S-label (mixture of L-[^{35}S]methionine and -cysteine, ICN Radiochemicals, Irvine, CA). The radioisotope (preferably 5–10 mCi of ^{35}S in < 0.5 ml) is injected via a 26-gauge × ⅜-in. needle into the saphenous vein of a rat maintained under mild ether anesthesia. To gain access to the vessel requires the prior surgical removal of a 1-cm^2 flap of skin and the overlying fascia. Ten minutes after the initial injection, the rat is administered a chase consisting of 2.5 ml of 20 mg/ml of unlabeled L-methionine (and 7 mg/ml of unlabeled L-cysteine if Tran^{35}S-label is used) in sterile-filtered phosphate-buffered saline, pH 7.4, by intraperitoneal injection. Additional chases, 0.5 ml instead of 2.5 ml, are then administered at 0.5-hr intervals as needed. After various periods of chase, the rats are killed by

[16] A. L. Hubbard, J. R. Bartles, and L. T. Braiterman, *J. Cell Biol.* **100**, 1115 (1985).
[17] B. Stieger, A. Marxer, and H.-P. Hauri, *J. Membr. Biol.* **91**, 19 (1986).
[18] C. C. Widnell and J. C. Unkeless, *Proc. Natl. Acad. Sci. U.S.A.* **61**, 1050 (1968).

decapitation under ether anesthesia, and their livers are excised and perfused with ice-cold saline in preparation for subcellular fractionation (see above).

Immunoprecipitation[6]

To determine recoveries in the plasma membrane sheet fraction and to establish their domain localizations, the ^{35}S-labeled plasma membrane proteins are immunoprecipitated from liver homogenates, hepatocyte plasma membrane sheets, and gradient fractions under conditions known to give near-quantitative recoveries by immunoblotting. Liver homogenates and hepatocyte plasma membrane sheets (at protein concentrations of 25–50 and 0.75–1.5 mg/ml, respectively) are extracted for 45 min at 4° with 20 mM octyl-β-D-glucopyranoside, 0.5% (w/v), Triton X-100, 0.3 M NaCl, 0.025 M NaP$_i$, 3 mM NaN$_3$, pH 7.4. [An exception is made for the apical protein dipeptidylpeptidase IV (DPP IV), which is efficiently solubilized from the liver homogenates only when extracted at lower protein concentrations in the range of 7.5–15 mg/ml.] The mixtures are centrifuged at 150,000 g_{max} for 60 min at 4° (38,000 rpm, Beckman 50 Ti rotor and L5-65 ultracentrifuge). Aliquots of the supernatants (0.3 and 1 ml, respectively) are incubated at 4° with 70 μl of the designated monoclonal or polyclonal antibody–Sepharose 4B slurry (either 1 mg of mouse monoclonal antibody or 2–3 mg of rabbit polyclonal IgG fraction/ml of slurry, conjugated by the cyanogen bromide technique[19]) in a final volume of 1 ml of the same buffer containing 5 mg/ml of bovine serum albumin. After 5–15 hr (depending upon the antibody) of incubation on a platform shaker or rotating wheel, the beads are sedimented (13,000 g, 0.5 min; top speed Brinkman microfuge 5415) and washed six times: three times in the same buffer plus 5 mg/ml of albumin, twice in the same buffer without albumin, and once with 0.15 M NaCl, 20 mM Tris-HCl, 0.02% NaN$_3$, pH 7.4.

Since the resolution of apical and basolateral vesicles is not complete, we prefer to immunoprecipitate from all of the fractions and then make domain assignments on the basis of distributions within the gradient. To immunoprecipitate the proteins from the density gradient fractions, it is necessary to first concentrate the fractions by centrifugation at 150,000 g_{max} for 60 min after a sevenfold dilution in ice-cold water containing the protease inhibitors (see above). The equivalent of one-half of the detergent extract produced from each fraction is used in the immunoprecipitation procedure described above. For economic reasons, we often perform the immunoprecipitations in sequential fashion, with the supernatant from the

[19] S. C. March, I. Parikh, and P. Cuatrecasas, *Anal. Biochem.* **60**, 149 (1974).

first immunoprecipitation being challenged by a second antibody–Sepharose conjugate, etc.

The immunoprecipitates are released from the washed beads by heating for 3 min at 100° in the presence of 75 μl of gel sample buffer containing 55 mM Tris-HCl, 6 mM EDTA, and 4.5% SDS, pH 8.9. The beads are sedimented by centrifugation at 13,000 g for 0.5 min. The supernatants are collected and reduced by the addition of 2.5 μl of 130 mg/ml dithiothreitol, heated for an additional 2 min, and then alkylated by the addition of 20 μl of 36 mg/ml iodoacetamide in 2 M sucrose. The samples are subjected to SDS-polyacrylamide gel electrophoresis in 7.5% gels, electrophoretically transferred to nitrocellulose (constant 400 mA for 4 hr in 20% methanol, 0.1% SDS, 0.19 M glycine, and 0.1 M Tris). The blots are exposed to Kodak XAR-5 X-ray film for 2–6 weeks at −70°. This direct blot autoradiography consistently gives improved band resolution and higher signal-to-noise ratios when compared to fluorography of dried gels impregnated with fluorophores. The amount of radioactivity in the relevant protein bands can then be quantified by laser densitometric scanning of the autoradiograms, and the resulting gradient distributions of the ^{35}S-labeled proteins compared directly to those of the selected domain markers (see above). For each of the plasma membrane proteins which we routinely examine, it is possible to resolve the mature molecular mass (terminally glycosylated) forms of the proteins from their corresponding immature (high-mannose) precursors. The latter are readily detected when the proteins are immunoprecipitated from liver homogenates immediately after pulse labeling or after short times of chase.

Using this type of approach, we found that the three apical proteins [HA 4, dipeptidylpeptidase IV (DPP IV), and aminopeptidase N (AP)] and two basolateral proteins [CE 9 and the asialoglycoprotein receptor (ASGP-R)] examined were delivered to the hepatocyte plasma membrane with similar kinetics: 45 min of chase was required for the terminally glycosylated forms of these five pulse-labeled proteins to reach maximum specific radioactivity in the isolated hepatocyte plasma membrane sheet fraction. A startling finding resulted from our first attempts to assign domain locations for these newly delivered plasma membrane proteins by immunoprecipitation from the fractions of the sucrose density gradient: At 45 min of chase the terminally glycosylated forms of all five of the newly synthesized proteins—apical and basolateral alike—exhibited density profiles coincident with that of the basolateral marker when immunoprecipitated from the sucrose gradient fractions (Fig. 2A and B). It thus appeared that even the apical plasma membrane proteins were delivered first to the basolateral domain of the hepatocyte plasma membrane. When analyzed in a similar fashion at later times of chase, a progressively larger percentage of the

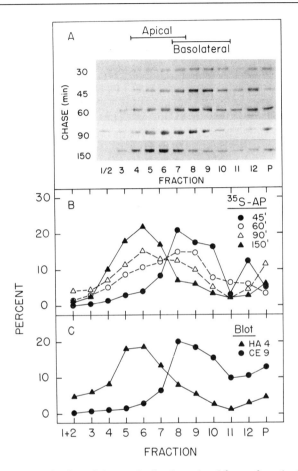

FIG. 2. Domain localization of the terminally glycosylated form of a pulse-labeled apical plasma membrane protein, aminopeptidase N (AP), as a function of time of chase. Hepatocyte plasma membrane sheets were isolated from rats after a 10-min pulse labeling with L-[^{35}S]methionine and the designated period of chase. They were vesiculated by sonication and resolved into apical and basolateral subpopulations in sucrose density gradients. The apical protein AP was immunoprecipitated from the gradient fractions, and the distribution of the terminally glycosylated form of radiolabeled AP was determined by blot autoradiography. (A) Autoradiogram of the blots of the AP immunoprecipitates. Little, if any, of the lower molecular mass high-mannose precursor of AP can be seen. The brackets defined the positions of the apical (HA 4) and basolateral (CE 9) domain markers [see (C)]. (B) Graph of the data obtained from the densitometric analysis of the autoradiograms shown in (A), with the distribution of the terminally glycosylated radiolabeled AP plotted as a percentage of that recovered in the gradient for each time of chase. (C) Distributions of the apical protein HA 4 and the basolateral protein CE 9 in these gradients were determined by immunoblotting and are plotted as a percentage of the recovered antigen. (From Bartles et al.[6] by permission of the Rockefeller University Press.)

newly synthesized apical proteins was found to distribute within the apical region of the gradient (Fig. 2A and B). After 150 min of chase, the pulse-labeled AP exhibited a distribution almost identical to that of the steady state apical marker. Interestingly, the three apical proteins appeared to move into the apical region of the gradient at different rates, with AP the fastest and HA 4 the slowest; after 150 min of chase, only 15–20% of pulse-labeled HA 4 had become apical.[6] These observations form the basis of our current model of plasma membrane sorting in the rat hepatocyte *in vivo*.[6-8]

Accessory Techniques

Both free-flow electrophoresis and vesicle immunoadsorption were used to substantiate the finding that the vesicles containing the newly synthesized apical proteins at 45 min of chase were basolateral plasma membrane.[6] The basics of these techniques are outlined below in the hopes that these methods will prove useful in the examination of other protein-sorting phenomena in this and in other systems.

Free-Flow Electrophoresis[6]

Free-flow electrophoresis can be used as an independent method to resolve the apical and basolateral vesicles derived from the hepatocyte plasma membrane sheets. Though the method offers no real improvement in resolution of the apical and basolateral plasma membrane vesicles, the separation relies on a different physical property, i.e., on net charge density instead of on equilibrium density in sucrose gradients. In addition, free-flow electrophoresis does offer a significant improvement in the resolution of basolateral plasma membrane vesicles from the residual trans-Golgi-derived contaminants (see below).

In preparation for free-flow electrophoresis, hepatocyte plasma membrane sheets are prepared in the presence of protease inhibitors as described above except that the final 1500 g, 10-min wash step is performed using chamber buffer (0.1 M mannitol, 0.2 mM Na$_4$EDTA, 10 mM acetic acid, 10 mM triethanolamine, adjusted to pH 7.4 with NaOH). Vesicles prepared from the plasma membrane sheets by sonication tend to undergo aggregation under the buffer conditions employed. Therefore, better results are obtained when the membrane pellet from one 5- to 7-g rat liver is resuspended in 4 ml of chamber buffer and vesiculated for 45 sec at setting 8 using a Polytron homogenizer (Brinkman) fitted with a 12-mm tip. The suspension is centrifuged for 10 min at 1500 g, and the supernatant is introduced into a free-flow apparatus (Elphor VAP 22; Bender-Hobein, Munich, FRG) using a chamber flow rate of 198 ml/hr (setting 100). The chamber should be precoated overnight with 3% (w/v) bovine serum albumin in water and rinsed thoroughly with water and chamber buffer immediately before the run. The electrode buffer composition is as follows: 2 mM Na$_4$EDTA, 0.1 M acetic acid, and 0.1 M triethanolamine adjusted to pH 7.4 with NaOH. The temperature (4°) and voltage (195 V/cm) are kept constant throughout the separation. Ninety 3.5-ml fractions are collected. Pairs of adjacent fractions are usually pooled and renumbered 1, 2, etc.) starting from the anode.

The distributions of the apical and basolateral plasma membrane can be determined either by immunoblotting or by using enzyme assays (see above). The apically derived plasma membrane vesicles distribute nearer to the anode than do those derived from the basolateral domain (Fig. 3B). To immunoprecipitate the radiolabeled plasma membrane proteins from the free-flow electrophoresis fractions, it is necessary to first concentrate the fractions by centrifugation at 150,000 g_{max} for 60 min. The detergent extract produced from each 7-ml fraction can be used in the immunoprecipitation procedure described above. In agreement with the results from the sucrose gradient separation, at 45 min of chase the terminally glycosylated form of a pulse–labeled apical plasma membrane protein such as DPP IV is found to codistribute with the basolateral marker (Fig. 3A). The remaining trans-Golgi contaminants—marked by the enzyme sialyltransferase[20]–are significantly resolved from the basolateral plasma membrane vesicles and are observed to distribute as a sharp peak near the cathode (Fig. 3B).

Vesicle Immunoadsorption[6,21]

Vesicle immunoadsorption can be used to demonstrate that a particular component of interest is present in the same membrane vesicle as an integral membrane protein against which an antibody exists. In our case, we used vesicle immunoadsorption to help confirm that the terminally glycosylated forms of the radiolabeled apical hepatocyte plasma membrane proteins at 45 min of chase are in ectoplasmic-side-out plasma membrane (PM) vesicles which contain a bonafide basolateral PM protein, the ASGP-R. For these experiments, we had to use a different starting PM preparation—one derived from a rat liver microsomal density gradient—because those vesicles derived from the hepatocyte PM sheets by sonication consistently gave unacceptably high levels of nonspecific immunoadsorption.

---

[20] J. Roth, D. J. Taatjes, J. M. Lucocq, J. Weinstein, and J. C. Paulson, *Cell (Cambridge, Mass.)* **43**, 287 (1985).
[21] S. C. Mueller and A. L. Hubbard, *J. Cell Biol.* **102**, 932 (1986).

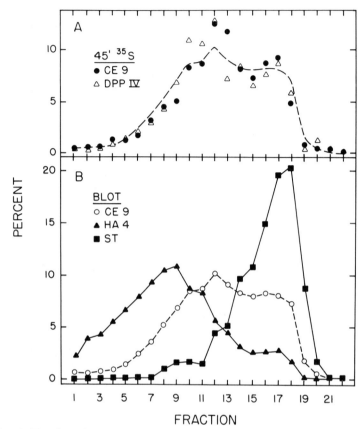

FIG. 3. Free-flow electrophoretic analysis of the hepatocyte plasma membrane sheet fraction after 45 min of chase. Hepatocyte plasma membrane sheets were isolated from pulse-labeled rats at 45 min of chase, vesiculated by Polytron homogenization, and resolved by free-flow electrophoresis. (Fraction 1 was nearest the anode.) The apical protein dipeptidylpeptidase (DPP IV) and the basolateral protein CE 9 were immunoprecipitated from the fractions. (A) The distributions of terminally glycosylated radiolabeled DPP IV and CE 9 were determined by autoradiography and plotted as a percentage of that recovered in the experiment. The broken line shows the steady state distribution of CE 9 as determined by immunoblotting [replotted from (B)]. (B) The distributions of the apical protein HA 4, the basolateral protein CE 9, and the trans-Golgi marker sialyltransferase (ST) were determined by immunoblotting and are plotted as a percentage of the recovered antigen. (From Bartles et al.[6] by permission of the Rockefeller University Press.)

Microsomal Gradients[22]

Unless indicated otherwise, all procedures should be performed at 4°, and all sucrose solutions should be Millipore filtered prior to use. Livers are perfused with ice-cold 0.15 M NaCl, weighed, and homogenized (Potter–Elvehjem, serrated pestle, six or seven strokes) in 5 vol of 0.25 M sucrose containing 3 mM imidazole-HCl buffer, pH 7.4, and the protease inhibitors (see above). The homogenate is then filtered through four layers of moistened gauze. The filtered homogenate is centrifuged at 12,000 g_{max} for 10 min (11,000 rpm, Beckman 60 Ti rotor and L5-65 ultracentrifuge). The supernatant is collected and recentrifuged at 150,000 g_{max} for 90 min (38,000 rpm, Beckman 60 Ti rotor and L5-65 centrifuge). The supernatant is discarded and the pellets resuspended by homogenization (Potter–Elvehjem, smooth pestle, three strokes) in 10–15 ml of 0.25 M sucrose, 3 mM imidazole-HCl, pH 7.4, for every 7–8 g wet wt of starting liver. Aliquots (3–5 ml) of the resuspended microsomal pellet are applied to the tops of 32-ml linear sucrose gradients (density 1.10–1.25 g/ml) buffered with 3 mM imidazole, pH 7.4, and centrifuged for 12–15 hr at 83,000 g in a swinging bucket rotor with no brake (25,000 rpm, Beckman SW 28 rotor and L5-65 ultracentrifuge). Thirty 1.2-ml fractions are collected from the top using an automatic gradient former/drainer (Haake Buchler Auto Densi-Flow IIC). The position of the plasma membrane peak can be deduced by immunoblotting or by colorimetric assay for a plasma membrane marker enzyme such as alkaline phosphodiesterase I[13] or 5′-nucleotidase.[18] Typically, a pool of fractions 13–16 (top = fraction 1) can serve as the source of plasma membrane vesicles for the immunoadsorptions. The lower numbered fractions (6–10) are enriched in Golgi- and endosome-derived vesicles, whereas vesicles derived from the endoplasmic reticulum and lysosomes are found in the higher numbered fractions (≥ 20).[6,22,23]

Immunoadsorption Using Staphylococcus Aureus Cells[6,21]

Vesicle immunoadsorptions are best performed with fixed *Staphylococcus aureus* cells because of their high capacity for binding rabbit polyclonal antibody.[24] Aliquots (50 µl) of the pooled microsomal gradient fractions 13–16 prepared from metabolically radiolabeled rats (see above) are mixed with 3 µg of the designated rabbit preimmune IgG or affinity-purified

---

[22] W. A. Dunn and A. L. Hubbard, *J. Cell Biol.* **98,** 2148 (1984).
[23] D. A. Wall and A. L. Hubbard, *J. Cell Biol.* **101,** 2104 (1985).
[24] A. L. Hubbard, W. A. Dunn, S. C. Mueller, and J. R. Bartles, in "Cell-Free Analysis of Membrane Traffic" (J. Morre, ed.), p. 115. Liss, New York, 1988.

anti-plasma membrane protein antibody (e.g., against the ASGP-R, see below) in 1 ml of PBS (0.15 M NaCl, 20 mM NaP$_i$, 0.02% NaN$_3$, pH 7.4) containing 1% (w/v) bovine serum albumin for 1 hr at 4° on a gyratory shaker. One hundred and twenty-five microliters of a 10% (w/v) suspension of heat-killed, fixed,[25] and PBS/albumin prewashed *S. aureus* cells is added and the incubation continued for an additional 1 hr. The incubation mixture is then diluted to 10 ml with 0.25 M sucrose, 3 mM imidazole-HCl, pH 7.4, and centrifuged at 1600 g for 17 min at 4° (3000 rpm, Beckman TJ6-R). The supernatants are aspirated off and the pellets extracted for analysis of immunoadsorption efficiency and specificity using a combination of immunoblotting, enzyme assays, or even trapped ^{125}I-labeled ligand (for endosomes).[21]

To obtain a significant ^{35}S-autoradiographic exposure to examine the behavior of the ^{35}S-labeled plasma membrane proteins, it is necessary to perform 15–20 replicate immunoadsorptions. The combined *S. aureus* pellets are then detergent extracted in a volume of 3 ml and analyzed by immunoprecipitation as described above. For our studies we used affinity-purified rabbit polyclonal antibodies directed against the intact 42-kDa subunit of the rat liver ASGP-R[21] (rat hepatic lectin 1[26]) and against an extracellular fragment of this polypeptide prepared by prior papain digestion of an endosome-enriched population of membrane vesicles.[6] The advantage of this latter antibody was that it allowed specific adsorption of only those vesicles with an ectoplasmic-side-out orientation, i.e., presumably those derived from the plasma membrane. By immunoblotting, both antibodies gave basolateral-specific immunoadsorption, with about 65–76% immunoadsorption of basolateral proteins (ASGP-R and CE 9) and only 9–13% immunoadsorption of the apical proteins (AP, DPP, IV, and HA 4). In contrast to these immunoblotting results, we found anywhere from 35 to 65% of the terminally glycosylated ^{35}S-labeled apical plasma membrane proteins after 45 min of chase to be specifically immunoadsorbed.[6] Thus, at 45 min of chase, the pulse-labeled apical plasma membrane proteins were enriched to varying degrees in ectoplasmic-side-out vesicles which contained a bona fide basolateral plasma membrane protein, the ASGP-R.

Concluding Remarks

The rationale for these studies of hepatocyte plasma membrane biogenesis stemmed from our desire to examine the sorting of endogenous

---

[25] S. W. Kessler, *J. Immunol.* **115**, 1617 (1975).
[26] J. O. Leung, E. C. Holland, and K. Drickamer, *J. Biol. Chem.* **260**, 12523 (1985).

domain-specific integral plasma membrane proteins in epithelial cells *in vivo*. In the case of the hepatocyte, the alternatives remain limited by the inability to observe significant proliferation and polarization of hepatocytes in tissue culture. *In vivo* studies of plasma membrane protein sorting, whether in liver or in other organs and tissues, will undoubtedly continue to utilize combinations of metabolic radiolabeling (pulse-chase) and subcellular fractionation.[8] However, because of the limitations inherent in subcellular fractionation, it is important to remain alert to the ways in which likely contaminants can influence the interpretation of experimental results. Furthermore, since the yields of the organelles in the subcellular fractions that are finally analyzed are always less than 100% and often less than 10%, extrapolation to the intact cell will continue to be problematic. These limitations notwithstanding, we believe that we have learned a great deal about the pathways of plasma membrane protein sorting in rat hepatocytes *in vivo* using the methods described above. We anticipate that most of these methods will prove adaptable in future studies of hepatocyte plasma membrane biogenesis both *in vivo* in pathological situations and in the isolated perfused rat liver. This latter system represents a good compromise between the *in vivo* situation and tissue culture; it offers the advantage of allowing the controlled manipulation of environmental conditions and introduction of reagents without the attendant disruption of cell polarity and tissue organization.[27]

[27] W. A. Dunn, D. A. Wall, and A. L. Hubbard, this series, Vol. 98, p. 225.

[48] Oligomerization and Protein Transport

By ROBERT W. DOMS

Membrane and secretory proteins are inserted into the endoplasmic reticulum (ER) in a largely unfolded state. Folding begins cotranslationally and is aided by a number of enzymes in the ER, including protein disulfide isomerase and proline isomerase.[1] In addition to attaining their secondary and tertiary structures, many membrane and secretory proteins assemble into oligomeric complexes. Proteins which fail to fold or assemble correctly are usually retained in the ER and eventually degraded.[2] The means by

[1] R. B. Freedman, *Nature (London)* **329**, 196 (1987).
[2] J. Lippincott-Schwartz, J. S. Bonifacino, L. C. Yuan, and R. D. Klausner, *Cell (Cambridge, Mass.)* **54**, 209 (1988).

which misfolded or unassembled proteins are recognized, retained, and degraded is an active area of research and has recently been reviewed.[3,4] Characterizing protein folding and assembly is central to understanding the quality control mechanisms which govern transport from the ER. In addition, three-dimensional "signal patches" may well play an important role in the sorting of proteins in polarized epithelial cells, so that the effects of tertiary and quaternary structure on transport need not be limited to the ER.[5]

It can be exceedingly difficult to ascertain the oligomeric structure of a protein complex consisting of noncovalently associated subunits. It is not uncommon for oligomers to dissociate following detergent solubilization, making analysis difficult at best. In this chapter I will briefly discuss several of the most commonly used methods to determine a protein's quaternary structure and the application of these techniques to correlate folding and assembly with transport. While studies on membrane proteins will be emphasized, the techniques are of course applicable to secretory proteins as well.

Velocity Gradient Sedimentation

Velocity gradient centrifugation is perhaps one of the most useful techniques for isolating and analyzing oligomeric complexes. For membrane proteins, detergent solubilization is a necessary first step. The choice of a detergent is not trivial since oligomers may dissociate or even aggregate in the presence of certain detergents.[6] There being no way of telling in advance which detergent will best preserve native structure, several must be tried under a variety of conditions. Obviously, the choices of detergents and buffers are almost unlimited, so a somewhat systematic sampling method should be used. The nonionic and zwitterionic detergents are those most commonly used for protein solubilization and reconstitution. Representatives from these classes include the nonionic polyoxyethylene detergents Triton X-100 (TX100) and NP-40, the nonionic detergent octylglucoside (which because of its exceptionally high critical micellar concentration is easily removed by dialysis), and the zwitterionic detergent 3-[(3-cholamidopropyl)dimethylammonio]-1-propane sulfonate (CHAPS). These widely used and readily available detergents have been employed to solubilize numerous oligomeric proteins without disrupting subunit–subunit interactions or biological activity. However, not all multisubunit

[3] J. K. Rose and R. W. Doms, *Annu. Rev. Cell Biol.* **4**, 257 (1988).
[4] S. M. Hurtley and A. Helenius, *Annu. Rev. Cell Biol.* **5**, 277 (1989).
[5] S. R. Pfeffer and J. E. Rothman, *Annu. Rev. Biochem.* **56**, 829 (1987).
[6] K. Simons, A. Helenius, and H. Garoff, *J. Mol. Biol.* **80**, 119 (1973).

proteins will be stable to these detergents. In some cases, it may be possible to stabilize weakly associated complexes by using a mixture of detergents or a mixture of detergent and phospholipid.[7] Wagner and Helenius[8] have found that a combination of 0.5% TX100 and 1 mM phosphatidylcholine stabilizes certain types of influenza hemagglutinin (HA) trimers which dissociate in the presence of TX100 alone. The polyoxyethylene detergent $C_{12}E_8$ has also been found to stabilize several proteins which likewise dissociate in TX100.[8] Additional nonionic and zwitterionic detergents may sometimes have to be used, in which case several comprehensive reviews on detergents may be consulted.[9-11]

Additional factors which may affect the stability of proteins in detergent solution include pH, the presence of cations, salt concentration, and temperature. As an example, Ca^{2+} is known to stabilize influenza C virus HA trimers.[12] In the absence of Ca^{2+} the trimers dissociate to monomers. Thus, MNT buffer [20 mM 2-(N-morpholino) ethanesulfonic acid (MES), 30 mM Tris, 100 mM NaCl] with 1 mM Ca^{2+} at slightly elevated pH (7.6–7.8, to help prevent protein aggregation) is a good choice when beginning studies on a protein's quaternary structure. Since some oligomers are more stable at 4° than at elevated temperatures, solubilizations should at least initially be performed on ice. For effective solubilization, an excess of detergent over lipid on the order of 10:1 is required.[11] The detergent concentration must be maintained above the critical micellar concentration, with 0.1% TX100, 40 mM BOG, and 25 mM CHAPS being suitable concentrations to include in sucrose gradients. Higher concentrations may be required when solubilizing large amounts of lipids.

A protein's conformation may also affect its stability to detergent solubilization and centrifugation.[13] An example of this is provided by the trimeric vesicular stomatitis virus (VSV) G protein. The VSV G protein exists in two distinct, pH-dependent conformations. The protein exists in a neutral pH conformation on the surface of the virus. Once the virus is internalized by the host cell and delivered to endosomes, the mildly acid pH in this compartment triggers a conformational change in G protein which leads to fusion between the viral and the endosomal membranes. In this regard G protein is like the acid-activated fusion factors of other enveloped viruses, like influenza HA and the Semliki Forest (SFV) enve-

[7] N. Epstein and E. Racker, *J. Biol. Chem.* **253**, 6660 (1978).
[8] K. Wagner and A. Helenius, personal communication.
[9] A. Helenius and K. Simons, *Biochim. Biophys. Acta* **415**, 29 (1975).
[10] C. Tanford and J. A. Reynolds, *Biochim. Biophys. Acta* **457**, 133 (1976).
[11] A. Helenius, D. R. McCaslin, E. Fries, and C. Tanford, this series, Vol. 56, p. 734.
[12] F. Formanowski and H. Meier-Ewert, *Virus Res.* **10**, 177 (1988).
[13] R. W. Doms, D. S. Keller, A. Helenius, and W. E. Balch, *J. Cell Biol.* **105**, 1957 (1987).

lope proteins.[14] As it turns out, the acid pH conformation of G protein is far more stable than its neutral pH counterpart. This fortuitous result made it possible to quantitatively recover G protein in trimeric form and characterize its assembly pathway. A similar approach may be tried for other proteins which exist in two or more defined conformations, such as studying the oligomeric structure of receptors in the presence or absence of ligand.

To separate newly synthesized monomers from oligomers, relatively shallow, low-density sucrose gradients are employed, classically 5 to 20% (w/w). Within this concentration range differences in solvent viscosity are negligible. As a result, proteins sediment at nearly a constant rate from the top of the gradient to the bottom. Since the sedimentation rate is proportional to molecular weight for molecules with a similar shape, proteins can be conveniently sized on these gradients under the appropriate run conditions. Specifically, centrifugation times should be adjusted so that the largest protein of interest sediments from one-half to two-thirds the distance of the tube, thus providing maximum separation as well as the greatest accuracy in determining sedimentation coefficients and molecular weight. The time required to sediment a protein half-way through a gradient at 5° can be estimated from the equation $T = k'/2S$, where T is the time in hours at the rotor's maximum speed, k' is a constant for any given rotor,[15] and S is the sedimentation coefficient of the protein. Thus, it takes approximately 16 hr to sediment human immunodeficiency virus (HIV)-1 env protein dimers ($S_{20,w} = 10.8$) half-way through a 5–20% sucrose gradient at 5° in the SW40 rotor.[16]

There are several ways to estimate a protein's sedimentation coefficient and molecular weight following velocity gradient centrifugation. The sedimentation coefficient of a protein can be estimated by the method of McEwen,[17] which requires knowledge of the sucrose concentration at the top and bottom of the gradient as well as the peak fraction. Alternatively, proteins with known $S_{20,w}$ values may be employed as standards either in the same or parallel gradients. A list of proteins which can be used for sedimentation standards is shown in Table I. Since migration through a 5–20% (w/w) sucrose gradient is relatively constant, it is possible to plot distance travelled from the miniscus versus either $S_{20,w}$ value or molecular weight to provide a crude estimate for any unknown protein. In addition,

---

[14] T. Stegmann, R. W. Doms, and A. Helenius, *Annu. Rev. Biophys. Biophys. Chem.* **18**, 187 (1989).
[15] Values of k' when sedimenting proteins in swinging bucket rotors: SW50.1 = 162, SW41 = 335, SW40 = 368.
[16] P. L. Earl, R. W. Doms, and B. Moss, *Proc. Natl. Acad. Sci. U.S.A.* **87**, 648, (1990).
[17] C. R. McEwen, *Anal. Biochem,* **20**, 114 (1967).

TABLE I[30]
PROTEINS WHICH CAN BE USED FOR SEDIMENTATION STANDARDS[a]

Protein	Stoke's radius	$S_{20,w}$	Molecular weight
Lysozyme		1.9	14,100
Bovine serum albumin	35	4.6	66,500
Human transferrin		4.9	76,000
Human IgG		7.1	154,000
Aldolase	48	7.3	161,000
Catalase	52	11.3	247,000
Horse spleen ferritin	61	16.5	460,000
Jackbean urease		18.6	483,000
Thyroglobulin (dimer)	85	19.3	660,000
Influenza HA trimers[a,b]		8.9	240,000
VSV G trimers[13,b]		8.0	200,000

[a] References for the values shown in this table can be found in G. D. Fasman (ed.), "Handbook of Biochemistry and Molecular Biology," 3rd ed. CRC Press, Cleveland, Ohio, 1976.

[b] HA, Hemagglutinin; VSV, vesicular stomatitis virus. Influenza HA and VSV G are both membrane proteins which bind an unknown amount of detergent.

the molecular weight of a protein can be estimated from its $S_{20,w}$ value by the relationship[18] $S_1/S_2 = (M_{r1}/M_{r2})^{2/3}$. Ideally, several standards should be used and the results averaged.

Several factors conspire to make sedimentation values and, in particular, molecular weight as determined by velocity gradient sedimentation somewhat uncertain, most notably the effects of detergent binding on membrane proteins and particle shape. The partial specific volume of a protein–detergent complex can be considerably higher than that of the protein alone. Therefore, using soluble proteins as sedimentation standards will likely underestimate the size of a detergent-binding membrane protein.[9] While using other membrane proteins as sedimentation standards can help compensate for this, anomalous detergent binding can nevertheless greatly affect a protein's sedimentation value. If necessary, sedimentation equilibrium centrifugation can be employed to determine the molecular weight of a membrane protein without knowing how much detergent it binds.[19]

The shape of a molecule also affects its sedimentation rate. The standards shown in Table I are essentially globular molecules. As can be seen, there is reasonably good correlation between molecular weight and sedi-

[18] R. G. Martin and B. N. Ames, *J. Biol. Chem.* **236**, 1372 (1961).
[19] M. A. Bothwell, G. J. Howlett, and H. K. Schachman, *J. Biol. Chem.* **253**, 2073 (1978).

mentation coefficient. However, proteins with more elongated shapes sediment much more slowly than would be expected judging by their molecular weight alone. For example, red blood cell spectrin dimers (M_r 485,000) sediment at only 8.4S, much more slowly than catalase (11.3S), which is only 247 kDa. Since membrane proteins are often elongated, and can be made even more assymetrical by detergent binding, their $S_{20,w}$ values may be considerably less than expected. As a result, molecular weight estimated from $S_{20,w}$ values or by cosedimentation with globular standards may be grossly underestimated. Thus, while sedimentation provides a reliable way to separate monomers from oligomers, it is somewhat less useful for estimating the absolute size of a complex without knowledge of the molecule's Stoke's radius and the amount of detergent it binds.

If conditions are found in which newly synthesized protein monomers can be separated from the mature, oligomeric species, then velocity gradient sedimentation may be used to monitor assembly in conjunction with pulse–chase experiments. An example of this approach is shown in Fig. 1. The VSV G protein is initially synthesized as a 4S monomer which assembles posttranslationally into trimers.[13] Cells expressing VSV G protein were pulse labeled for 3 min, and chased for the indicated periods of time. The cells were lysed and the lysates centrifuged on gradients under conditions designed to separate the 4S monomers from 8S trimers. As seen in Fig. 1, most of the G protein sedimented as 4S monomers immediately after the pulse label. With time, the protein began to accumulate in the 8S region as trimers were formed. The experiment showed that the $t_\frac{1}{2}$ for trimer formation was approximately 5 min, and digesting monomers and trimers with endoglycosidases revealed that trimers were formed before arrival at the cis Golgi. Furthermore, the carbohydrates of monomers were never processed by Golgi enzymes. This observation, coupled with other experiments, showed that trimer formation occurred in the ER and was a prerequisite for transport to the Golgi.[13,20] This example demonstrates some of the information which can be obtained from an assay which reproducibly and quantitatively separates monomers from oligomers. An important point to keep in mind, however, is that the assay monitors assembly of only those oligomers which are stable to centrifugation. There are examples in the literature of proteins which assemble in the ER but fail to acquire stability to ultracentrifugation until after subsequent posttranslational modifications.[8,21] Thus, it is important to correlate the kinetics of assembly obtained by this technique with others whenever possible. This is particularly im-

[20] R. W. Doms, A. Ruusala, C. Machamer, J. Helenius, A. Helenius, and J. K. Rose, *J. Cell Biol.* **107,** 89 (1988).
[21] C. S. Copeland, R. W. Doms, E. M. Bolzau, R. G. Webster, and A. Helenius, *J. Cell Biol.* **103,** 1179 (1986).

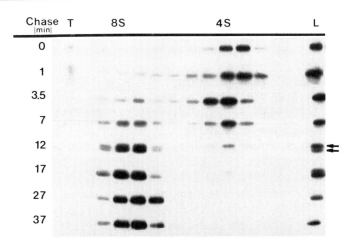

FIG. 1. Kinetics of VSV G trimerization. Chinese hamster ovary cells expressing VSV G protein were labeled with [^{35}S]methionine for 3 min, after which they were chased in the presence of excess cold methionine for the indicated period of time. The cells were lysed in MNT buffer containing 1.0% TX100 at pH 5.5 and transferred to ice. The lysates were centrifuged on 5–20% (w/w) sucrose gradients in order to separate 4S monomers from 8S trimers. Twenty percent of the material loaded onto the gradient was run in the lane marked L in order to monitor recovery. Arrowheads at the 12-min chase point show the distinction between G trimers which have been trimmed by enzymes in the cis Golgi and those which have not yet reached that compartment.

portant when the kinetics of assembly derived from a gradient sedimentation assay suggest that assembly occurs in a post-ER compartment, since the preponderance of data shows that assembly is clearly an ER event.[3,4]

Chemical Cross-Linking

Chemical cross-linkers are powerful reagents for characterizing the oligomeric structure of proteins, particularly those not stable to detergent solubilization. Cross-linkers consist of two reactive groups separated by a spacer arm of variable length. The reactive groups present in the most commonly used cross-linkers are either imidoesters or N-hydroxysuccinimide esters. Cross-linkers may be further subdivided into cleavable and noncleavable reagents. Cleavable reagents contain a disulfide in the spacer arm, and so can be broken upon reduction. Cross-linking reagents also differ in their solubility properties, with some being membrane permeable. Since the imidoesters and N-hydroxysuccinimide esters react only with primary amino groups, the success of cross-linking is contingent upon there being accessible amino groups in adjoining protein subunits which are also the correct distance apart. Thus, it is likely that only a subset of reagents will successfully cross-link any given oligomeric protein. Since there is no way to know *a priori* which cross-linker under which conditions will provide the best results, it is important to screen several reagents with different properties. In a recent study of the oligomeric structure of the HIV-1 env protein, Earl *et al.*[16] screened several types of cleavable and noncleavable reagents, including DMS, DSP, DTBP, DTSSP, and EGS. Each cross-linker was initially tested at concentrations of 0.1, 1.0, and 10.0 mM at room temperature for 15 min at pH 8.0. Generally, a concentration of around 1.0 mM was found to provide the best cross-linking.

Since cross-linkers react with primary amines, it is necessary to use a slightly alkaline pH in order to keep the primary amines in an unprotonated state. Also, the reaction buffer should be free of extraneous amines, such as glycine or Tris, and the cross-linker made fresh before use since hydrolysis is significant competing reaction. Generally, cross-linking reactions proceed more quickly the higher the pH and the higher the temperature. However, efficient cross-linking can be achieved at slightly basic pH (7.4–8.0) and at 4°. The reaction is terminated by addition of a substantial molar excess of primary amines in the form of glycine or other simple compound containing primary amino groups.

The interpretation of chemical cross-linking results is fraught with difficulties.[22] As alluded to above, the inability to cross-link a protein with any single agent is by no means proof that it exists as a monomer. Only if several reagents fail to cross-link a given protein may any degree of confidence be imparted to such a conclusion. If cross-linking is observed, it may be specific and reflect a relevant structure, or it may be nonspecific. It is perhaps easiest to initially cross-link under conditions which have the greatest chance of giving positive results—even at the risk of obtaining nonspecific association—and then cross-link under progressively more rigorous conditions to ascertain whether the result is biologically relevant. Given the instability of many oligomers to solubilization, this means that cross-linking might best be performed on intact cells with a relatively high concentration of reagent, keeping in mind whether a membrane-permeable or -impermeable cross-linker is required. While cross-linking on the intact cell increases the probability of detecting otherwise unstable oligomers, it also increases the probability of nonspecifically cross-linking proteins together. The concentration of proteins in membranes is considerable, thus increasing the frequency of random collisions which may give rise to nonspecific cross-linking.[22] If cross-linking is observed, the assay

---

[22] J. B. C. Findlay, *in* "Biological Membranes: A Practical Approach" (J. B. C. Findlay and W. H. Evans, eds.), p. 179. IRL Press, Oxford, England, 1987.

may then be repeated with reduced concentrations of reagent or for shorter periods of time and/or at reduced temperature to detect those proteins which are preferentially cross-linked. If an identical result is obtained over a broad concentration range of cross-linker or with several reagents, it is much more likely to be specific. Cleavable reagents are particularly useful here, since the monomeric species may be easily reconstituted and identified when electrophoresed under reducing conditions. One way to perform such an analysis is to identify the cross-linked complex by autoradiography following SDS-PAGE. The region of the dried gel which contains the band is cut out and placed in the well of a second polyacrylamide gel. The dried gel slice is then allowed to hydrate by addition of 60 mM Tris, pH 6.8, with reducing agent to the well. After 1 hr running buffer is added and current applied to the gel as usual.[16]

There are other ways to prepare samples for cross-linking which should be considered, particularly when dealing with relatively stable complexes. Cross-linking may be performed after solubilization, which greatly reduces the frequency of random collisions which can result in nonspecific coupling. In addition, the sample can be serially diluted, which should have little effect on the efficiency with which authentic oligomers are cross-linked. Cross-linking can also be performed on proteins which have been partially purified. The risk of these approaches are, of course, that the oligomer may dissociate during purification and thus prior to addition of cross-linker. In addition, the concentration of cross-linker may have to be varied as samples are diluted.

A problem that may be encountered when cross-linking large complexes is that their combined molecular weight may be beyond the range of commonly used molecular weight markers. We encountered this problem when cross-linking the HIV-1 env protein, which consists of 160-kDa subunits. The protein forms stable 320-kDa dimers which in turn assemble into tetramers, probably a dimer of dimers with a final molecular weight of approximately 640,000. To analyze a protein of this size, we used 4% SDS gels which provided a sufficient degree of resolution. For molecular weight markers, we cross-linked β-galactosidase, a 135-kDa protein which exists in tetrameric form. β-Galactosidase (Calbiochem, Los Angeles, CA) at a concentration of 10 mg/ml was cross-linked with 5 mM EGS (a noncleavable reagent) for 15 min at room temperature at pH 8.0. Due to the high concentration of β-galactosidase, forms larger than tetramers were obtained. On application to 4% SDS gels, monomeric through hexameric bands were observed. The banding pattern was far superior to commercially available cross-linked standards and its preparation has proved to be highly reproducible. A point to remember when cross-linking proteins is that heavily cross-linked species may migrate more quickly in SDS-PAGE than expected. This is presumably due to the fact that intrachain cross-links prevent the protein from being completely denatured. As a result, it can maintain a more globular conformation in SDS and migrate more quickly as a result. This situation is analogous to the observation that nonreduced proteins often migrate more quickly in SDS-PAGE than their fully reduced counterparts.

Chemical cross-linking is more useful for determining the subunit structure of an oligomer than it is in correlating assembly with transport. However, cross-linking can be performed following a pulse–chase experiment in several ways. At the end of the pulse, the cells can be shifted to ice and a membrane-permeable cross-linker such as DTBP added. Such an approach was used to monitor the assembly of VSV G protein trimers, though the cross-linking was with low efficiency and the results somewhat ambiguous.[23] A more successful application of cross-linking to monitor assembly was performed by Yewdell et al.,[24] who lysed cells at the end of different chase times and then cross-linked the resulting lysates. Since their cross-linking was complete at later times of chase, they were able to quantitatively follow the assembly of influenza HA trimers.

SDS Resistance

Some membrane proteins display partial resistance to SDS-induced dissociation, even at 95° with reducing agents. Thus, several strains of influenza HA give a "ladder" pattern in gels consisting of monomer, dimer, and trimer bands.[25] Other proteins which display partial resistance to SDS include VSV G,[18] the env proteins of HIV-1,[16] HIV-2,[26] and simian immunodeficiency virus (SIV),[26] and glycophorin.[27] The degree of resistance is often greater when samples are boiled for short periods of time or incubated at reduced temperatures. The concentration of reducing agents can also affect stability. For example, the resistance of the HIV-1 env protein dimers to SDS-induced dissociation is greater when 0.1% 2-mercaptoethanol is used rather than 1.0%.[28] Thus, it is often worthwhile to examine a protein of interest under a variety of sample preparation conditions for SDS-PAGE, since resistance to dissociation provides an easy and rapid way to monitor assembly. Such an approach could conceivably be

---

[23] T. E. Kreis and H. F. Lodish, *Cell (Cambridge, Mass.)* **46**, 929 (1986).
[24] J. W. Yewdell, A. Yellen, and T. Bächi, *Cell (Cambridge, Mass.)* **52**, 843 (1988).
[25] R. W. Doms and A. Helenius, *J. Virol.* **60**, 833 (1986).
[26] M.-A. Rey, B. Krust, A. G. Laurent, L. Montagnier, and A. G. Hovanessian, *J. Virol.* **63**, 647 (1989).
[27] M. Silverberg and V. T. Marchesi, *J. Biol. Chem.* **253**, 95 (1978).
[28] P. L. Earl, personal communication.

taken with any membrane protein which displays a relative degree of SDS resistance. The disadvantages of this technique include the fact that most oligomers are SDS sensitive, while others display only partial resistance, making quantitation difficult. The advantages include the simplicity of the approach and, if resistance is observed, there is a high probability that the interaction represents a relevant structure.

Antibodies

Monoclonal and polyclonal antibodies can provide a convenient means to monitor folding and assembly at both the biochemical and morphological levels. In the case of influenza hemagglutinin, some antibodies are specific for the trimeric form of the protein — they do not immunoprecipitate monomeric HA.[21,24,29] Before attempting to generate oligomer-specific antibodies, any preexisting panels of monoclonals should be screened against monomeric and oligomeric forms of the protein in the hopes of finding some which recognize folding or assembly intermediates. Given the kinetics with which proteins fold and assemble, it is necessary to label cells for short periods of time in order to obtain incompletely folded and assembled molecules. In screening antibodies to the VSV G protein, Machamer et al.[30] performed immunoprecipitations against lysates of virus infected cells which had been labeled for 5 min or labeled for 5 min and chased for 60 min. Under these conditions, it was found that some monoclonals failed to recognize early folding intermediates, specifically G proteins which had not yet acquired their full complement of disulfide bonds.[30] Inclusion of an alkylating agent in the cell lysate was important to prevent folding in vitro.[30] In addition, lysing cells on ice under conditions in which ATP is depleted was found to be important in stabilizing interactions between incompletely folded G protein monomers and GRP78 (BiP), a protein which transiently associates with the folding intermediates of a large number of proteins.[30] Antibodies may also be screened against monomers and oligomers separated on sucrose gradients, or by immunofluoresence. If an antibody reacts only with protein in the Golgi or on the cell surface, it may be oligomer specific. Likewise, an antibody that gives only an ER pattern may react only with monomers or folding intermediates.[21,24,29] When coupled with pulse–chase experiments, oligomer specific antibodies represent a simple, quantitative, and specific way to correlate folding and assembly with transport at both the biochemical and ultrastructural levels.

[29] M.-J. Gething, K. McCammon, and J. Sambrook, *Cell (Cambridge, Mass.)* **46,** 939 (1986).
[30] C. Machamer, R. W. Doms, D. G. Bole, A. Helenius, and J. K. Rose, *J. Biol. Chem.* **265,** 6879 (1990).

Protease Resistance

Membrane proteins often display considerable resistance to proteases. However, newly synthesized proteins, being in a largely unfolded state, are exquisitely sensitive. The acquisition of protease resistance may provide a relatively simple assay to follow folding and assembly. Basically, a protease must be found to which the mature protein is resistant in detergent solution. It is then possible to pulse–label cells for a short period of time followed by detergent solubilization and addition of protease to determine if folding and assembly intermediates are protease sensitive. If these conditions are met, then it is straightforward to determine the kinetics with which a protein becomes protease resistant and to correlate this with folding and assembly. It should be noted, however, that protease resistance is a rather stringent assay for correct assembly and final folding. Molecules may remain protease sensitive for some period of time even after assembly. Likewise, some mutations may alter folding enough to maintain the protein in a protease-sensitive state even though assembly and transport are unaffected.[31]

Summary

As noted above, there are a number of methods available for ascertaining a protein's quaternary structure and correlating assembly with transport, though each has its own set of artifacts. Almost any technique can give false negative results due to oligomer instability, while others can give false positive results, with nonspecific cross-linking or detergent-induced aggregation being obvious examples. Thus, it is extremely important to corroborate results obtained by one technique with those from another. Chemical cross-linking and velocity gradient sedimentation are two techniques which complement each other nicely. A protein cross-linked on intact cells can be applied to sucrose gradients, after which it is recovered from fractions and analyzed by SDS-PAGE, thereby making it possible to correlate molecular weight with $S_{20,w}$ value. Comparing the distribution of cross-linked and non-cross-linked material across a gradient will reveal if instability to solubilization is a significant problem. Conversely, cross-linking may be performed after gradient centrifugation. This also allows correlation of molecular weight with $S_{20,w}$ value, and greatly reduces the risk of nonspecifically cross-linking the protein of interest to other molecules.

An example in which velocity gradient centrifugation was correlated with chemical cross-linking is shown in Fig. 2. Cells expressing the HIV-1

---

[31] M. G. Roth, C. Doyle, J. Sambrook, and M.-J. Gething, *J. Cell Biol.* **102**, 1271 (1986).

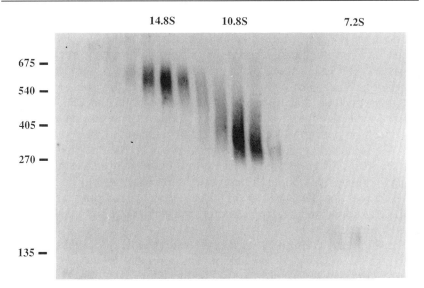

FIG. 2. Sedimentation and cross-linking of the HIV-1 env protein. Cells expressing the HIV-1 env protein were harvested and lysed as described[16] and loaded onto a 5–20% (w/w) sucrose gradient and centrifuged for 20 hr at 4° at 40,000 rpm in the SW40 rotor. Fractions were collected from the gradient and cross-linked with 5 mM EGS (Pierce) for 15 min at room temperature. After acid precipitation, the proteins were analyzed by SDS-PAGE and Western blotting using an antibody to gp120. *Escherichia coli* β-galactosidase (Calbiochem), cross-linked with 1 mM EGS as described in the text, was used as a molecular weight marker.

env protein were lysed and the lysates subjected to centrifugation.[16] Each fraction recovered from the gradient was then incubated with 5 mM EGS for 15 min at room temperature. The proteins were recovered by acid precipitation and analyzed by SDS-PAGE and Western blotting. As can be seen, a ladder pattern was obtained.[16] At the top of the gradient, monomeric gp160 (with an S value of 7.2) was recovered and could not be cross-linked to higher order forms, confirming that the 7.2S material was indeed monomeric. The 10.8S gp160 was quantitatively cross-linked into dimeric form with a molecular weight of approximately 320,000, while 14S–16S material was cross-linked into a yet higher order structure, probably a tetramer.[16] Similar results were obtained when cross-linking was performed before centrifugation, indicating that cross-linking in the lysate was specific, that non-cross-linked molecules were relatively stable to lysis and centrifugation, and confirming the identity of the 7.2S, 10.8S, and 14S–16S material.[16,28]

If the kinetics of assembly obtained by two techniques are in good agreement, it is then possible to ascertain the cellular location of oligomerization. For most proteins, this occurs in the ER.[3,4] Identification of the site of assembly is easiest when oligomer-specific antibodies are available, in which case immunogold labeling or immunofluorescence may be employed.[21,24,29,32] In the absence of a morphological assay, transport from the ER may be blocked by a variety of methods. If assembly still occurs, then one can conclude that oligomerization can occur in the ER. However, most of the techniques available to block transport may also have deleterious effects on assembly—incubation at low temperature or depleting cells of ATP, for example. Recently, however, it has become possible to block exit from the ER in a rapid, efficient, and entirely reversible manner by using the fungal metabolite Brefeldin A.[33,34] Brefeldin A may be applied at 37° and does not affect cellular ATP levels, thereby providing more appropriate conditions under which assembly may take place. Its effects are also rapidly reversible.

In summary, commonly used biochemical techniques are more than adequate for studying protein folding and assembly *in vivo*, provided an appropriate degree of caution is exercised in interpreting the results. Studies directed along these lines will undoubtedly help elucidate the means by which proteins fold and assemble as well as the mechanisms which exist to govern transport of molecules from the ER.

Acknowledgments

I would like to thank Dr. Kevin Gardner for his copious supply of information on sedimentation values and Stoke's radii of proteins, Dr. Pat Earl for providing unpublished data and very helpful conversations, and Drs. Krystn Wagner and Ari Helenius for providing unpublished results on stabilizing recalcitrant proteins with detergent:phospholipid combinations.

---

[32] C. S. Copeland, K. P. Zimmer, K. R. Wagner, G. A. Healy, I. Mellman, and A. Helenius, *Cell (Cambridge, Mass.)* **53**, 197 (1988).
[33] J. Lippincott-Schwartz, J. S. Bonifacino, L. C. Yuan, and R. D. Klausner, *Cell (Cambridge, Mass.)* **56**, 209 (1989).
[34] R. W. Doms, G. Russ, and J. W. Yewdell, *J. Cell Biol.* **109**, 61 (1989).

Author Index

Numbers in parentheses are footnote numbers and indicate that an author's work is referred to although the name is not cited in the text.

A

Abboud, H. E., 135
Abboud, H., 135
Abdel-Latif, A. A., 677
Aboolian, A., 411, 421(10)
Abramow, M., 210, 227, 250, 289, 344
Aczel, S., 761
Adachi, H., 622
Adam, W. R., 524, 525(37)
Adams, D. J., 695
Adams, G., 339
Adelberg, E. A., 405, 406(74), 500, 743
Adelstein, R. S., 664
Adler, E. M., 427
Adrian, T. E., 654
Agarwal, N., 500
Agus, Z. S., 65
Ahearn, G. A., 488
Ahnen, D. J., 460
Ait-Mohamed, A., 335, 338
Akabas, M. H., 582
Akaiwa, Y., 622
Akiba, I., 637, 638(117)
Al-Awqati, Q., 255, 256(1), 257, 258, 259(3), 260(3), 262, 263(3), 265(13), 271, 414, 420(44), 425(44), 506 751, 752(61), 753, 754(63,72), 809
Al-Zahid, G., 250
Alavi, N., 410, 411, 418(4,9), 422(4)
Alderton, J., 693
Alexander, E. A., 81, 94(22), 415, 420(64)
Algren, A., 412, 416(28), 417(28), 424(28)
Alhenc-Gelas, F., 335, 339
Allegrini, J., 339
Allen, D. P., 272, 396, 400(42)
Allen, K. P., 266
Allen, M. L., 328, 412, 414, 416(39), 418(39), 422(39)
Alles, W. P., 795, 797(34)
Alleyne, G. A. O., 113
Alminger, T., 721
Alorn, D., 31
Alpert, B. E., 134
Altman, L. G., 278
Alvarado, F., 768
Alvey, T. R., 335
Ames, B. N., 845, 850(18)
Amgel, L. M., 198
Amiel, C., 91, 93(48), 94, 295, 496
Amin, D., 678
Amiranoff, B., 612
Amirian, D. A., 621, 627
Amsler, K., 404, 405, 406(58,76)
Amsterdam, A., 686
Amzel, L. M., 376
Anagnostopoulos, T., 8, 9
Anan, M., 585
Anand-Srivastava, M. B., 747
Andersen, P. M., 788, 792(29)
Anderson, B.-M., 738
Anderson, H. C., 500
Anderson, P. S., 802, 809(55)
Anderson, R. J., 412, 414, 416, 417(25), 418(37,25,71), 422(37,71)
Anderson, T., 683
Andersson, G. L., 304
Andreoli, T. E., 87, 248, 249, 250, 251(10), 252, 289, 356, 369
Andreucci, V. E., 73, 112, 121(11), 122(11)
Andrews, C. W., 748
Andrews, P. M., 152
Andrews, P., 68
Angel, L. M., 622
Änggard, E. E., 664, 707, 715(5)
Angielski, S., 495, 684
Antoniotti, H., 612, 620(36)
Antony, C., 816

Aoshima, H., 585
Apell, H.-J., 479
Aperia, A., 107, 121, 122
Apfeldorf, W., 671
Appel, R. G., 415, 423(54)
Applebaum, S. W., 174
Arais, J. M., 335
Arataki, M., 108
Ardaillou, R., 135, 304
Ardito, T., 266, 272(21)
Arend, L. J., 328, 414, 416(39,42), 418(39,42), 422(39,42), 422(45)
Arendhorst, W. J., , 112
Argarwal, N., 743
Armanini, D., 524, 525(37)
Armstrong, F. B., 167
Arnaout, M. A., 561, 562
Arnaud, C. D., 303
Aronson, N. N. Jr., 829
Aronson, P. S., 454, 491, 742, 743, 794
Arriza, J. L., 521
Aruffo, A., 639
Asano, Y., 306, 307(48), 323(48)
Ashbaugh, C., 751, 752(61)
Ashkenazi, A., 637, 638(118,121)
Ashwell, G., 633
Asselin, J., 620, 621(54)
Atkins, R. C., 142, 143(4)
Atlas, D., 622
Auchus, R. J., 415, 422(50)
Audinot, M., 750
Auhus, R. J., 305
Aurbach, G. D., 622
Aurbach, J. D., 303
Ausiello, D. A., 147, 415, 425(67), 468, 501, 504(32), 556, 557(25), 559, 561, 563(25), 564(25), 567, 568, 569(59,61), 570(59), 571(59)
Avison, M. J., 405
Aviv, H., 441
Avron, M., 474, 475(12)
Axelrod, J., 682
Ayalon, A., 621, 627(60)
Azzone, G. F., 474

B

Baack, R., 31
Baba, S., 612, 620(37,38)
Babinowitz, L., 182
Bach, G. G., 46, 71(25)
Bächi, T., 850
Bachmann, S., 266, 271(18), 272(18), 286(18)
Bacskai, 385, 386(21)
Badie-Dezfooly, B., 411, 418(5,7), 420(7), 421(8)
Badr, K. F., 112
Baer, P. G., 76, 77(13), 78(13), 79(13)
Bagnasco, S., 331
Bahr, D. P., 478
Bailey, J. R., 223
Bailie, M. D., 159
Bailly, C., 94, 295, 324
Baines, A. D., 81, 93(17)
Baines, C. J., 81, 93(17)
Bainton, D. F., 431
Bakalyar, H. A., 569
Baker, C. J., 715
Baker, R., 677
Balaban, R. S., 10, 399, 401
Balaban, R., 331
Balch, W. E., 843, 846
Balla, T., 715
Ballard, P. L., 524, 525(32)
Ballermann, B. J., 135, 148, 149(19)
Bankir, L., 265, 273(12), 286(13), 287, 410
Bannister, W. H., 726
Banschbach, M. W., 676, 685
Baraban, J. M., 701
Barac-Nieto, M., 305
Barajas, L., 154
Barasch, J., 257, 258(3), 259(3), 260(3), 263(3), 271
Barbry, P., 325
Barby, P., 751
Barcenas-Ruiz, L., 700
Barchi, R. L., 802
Barfuss, D. W., 203, 209(66), 219(66), 220(66), 236, 249, 355, 366
Barker, G., 429
Barker, J. L., 795, 796(13)
Barlet, C., 324, 335, 338
Barlett, C., 549
Barnard, E. A., 639
Barnes, A. D., 342
Barnes, E. M., Jr., 800
Barnett, R., 135, 137(35)
Barratt, T. M., 109, 110
Barrett, L., 399

Barrett, P. Q., 664, 665, 666(18), 673
Barritt, G. J., 684
Barron, K. D., 802, 809(55)
Barros, F., 739, 741(8), 742(8), 743(8), 744(8)
Barsuhn, C., 622
Bartles, J. R., 825, 826, 828(11), 831(6, 11), 832, 833(6), 835(6)
Bartles, J. T., 836(6, 7, 8), 838(6), 839, 840(6), 841(8)
Bartlett, G. R., 688
Bartter, F. C., 522, 523(20), 524(20)
Bassingthwaighte, J. B., 36, 45(9), 46(9), 67, 68(9), 71(9)
Bataille, D., 612, 634(28)
Batty, I. R., 707, 708(4)
Bauduin, H., 685
Baukal, A. J., 715
Baum, B. J., 483
Bauman, K., 342
Baumann, K., 87, 100, 102, 343
Baverel, G., 326
Bawab, W., 612
Baxter, J. D., 522, 523(20), 524(20)
Bayerdörffer, E., 401
Bayley, H., 757
Baylis, C., 113, 115
Baylor, S. M., 701
Bear, J., 45
Bearer, E. L., 556
Beason-Griffin, C., 128, 129(54)
Beaven, M. A., 627
Beck, F. X., 327
Beck, F., 96
Becker, J. E., 432
Becker, P. L., 700
Beckmann, C., 343
Beeuwkes, R., III, 39, 71(17)
Beibisch, G.,290
Beierwaltes, W., 135
Beil, W., 722
Beinfeld, M. C., 627
Belcher, M., 305, 306, 307(47), 323(47)
Belin, D., 748
Bell, J. R., 638
Bell, P. D., 236
Bello Reuss, E., 10
Bello-Reuss, E., 411, 417(18), 422(18), 424(18)
Ben Abdelkhalek, M., 335, 338

Benavides, J., 319, 324(80)
Bendayan, M., 281
Bender, P., 683
Bengele, H. H., 81, 94(22)
Bennett, C. D., 637
Bennett, M. K., 814, 815(5), 816, 821, 823(5), 824
Bennett, V., 272, 396, 400(42)
Benos, D. J., 10, 739, 742, 750, 751, 752, 753(63), 809
Benos, D., 750
Benovic, J. L., 637
Bens, M., 135
Bentley, K., 207
Bentley, M. D., 70
Bentley, S. K., 201, 202, 203(57,60), 204, 205(57), 206(57), 207(57,60), 215, 216(81,83)
Bently, P. J., 742
Benveniste, J., 680
Berek, J. S., 536
Bereziat, G., 681
Berg, M. B., 209
Berg, M. M., 175, 180, 181
Berglindh, T., 470, 474, 475(14), 640, 641, 647(5,6), 648(4), 649(6), 660(6), 724, 726, 727(17), 728(17)
Bergmann, J. E., 255, 261(2), 264(2)
Bergstrand, A., 508
Bergstrom, S., 676
Berkowitz, J. M., 625
Berl, T., 415, 416(62), 423(62)
Berleithner, H. O., 791
Berliner, R. W., 81, 89(16), 92(16), 93(16), 90
Bernard, C., 167
Bernard, L. M.,231
Berner, W., 508
Bernotat-Danielowski, S., 585, 598(29), 600(29), 601(29), 602(29), 603(29), 604(29)
Berridge, M. J., 664, 691, 705, 707, 708(1, 2), 709(2), 711(1, 2)
Berry, C. A., 250, 565
Bersch, B., 479
Bershengorn, M. C., 684
Berthold, R., 303
Berthon, G., 612, 620(36)
Besterman, J. M., 743, 746(22), 748
Betz, H., 800

Bevan, C., 387, 403(31), 407(31)
Bevan, S., 795, 796(20)
Beyenbach, K. W., 173, 174, 175, 176, 178, 179, 180, 181, 183, 184, 185, 186, 188, 189, 190, 191, 192(30,31,32,33,34), 193(32), 194(33,34,40), 202, 220, 221(65,92), 223, 240, 289, 299, 301(44)
Bhalla, R. C., 746
Biagi, B., 295, 299
Bianchi, C., 303, 304(8)
Biava, C. G., 155
Biber, J., 451, 459(10), 495, 496, 498, 501, 503(35), 504(35)
Bicking, J. B., 739
Bidet, M., 412, 416(22), 417(22), 416(26), 417(26), 422(26), 424(26), 425
Bieber, L. L., 829
Bielefeld, D. R., 745, 746(34)
Biemesderfer, D., 147, 266, 268, 270(27), 271(26,27), 272(27)
Bijman, J., 808
Billah, M. M., 678, 679, 687
Binns, G., 638
Birabean, M. A., 632
Birardet, M., 540
Birnir, B., 584
Bissonnette, B. M., 622
Bitar, K. N., 612, 615(24)
Bjornsson, O. G., 699
Black, J., 590
Blackburn, J. G., 417, 424(73)
Blakman, D. P., 734
Blanchard, R. F., 65
Blanchardie, P., 529
Blank, M. E., 778
Blantz, R. C., 113
Blas, S. D., 65
Blatz, A. L., 795, 796(23,24), 798(23,24)
Bleich, M., 293, 299, 300, 301(19,49), 302(19), 795, 797(32), 808(32)
Bligh, E. G., 686
Blissard, D., 475
Bloomfield, D. A., 36
Bloomquist, J. R., 800
Blouch, K., 171, 173(5), 241
Bluck, S., 258
Blumenthal, R., 450, 633
Blumenthal, S. S., 411, 415, 420(55)
Boardman, L., 428
Boatman, D. E., 688

Bockaert, J., 303
Bockhorn, H., 335
Bockus, B. J., 259
Bode, F., 342
Bodmer, M., 501, 503(35), 504(35)
Boeynaems, J. M., 685
Bohman, S.-O., 277
Bolanowski, M. A., 637
Bole, D. G., 851
Bolinder, R. P., 282, 284(97)
Bollack, C., 130, 134(8)
Bolzau, E. M., 846, 851(21)
Bonifacino, J. S., 854
Bonifacion, J. S., 841
Bonjour, J. P., 306
Bonjour, J.-P., 495, 496
Bonnard, M., 326
Bonner, T. I., 637, 638(119,120)
Bonsib, S. M., 396
Bonting, S. L., 209
Bonting, S., 747
Bonvalet, J. P., 306, 520, 522, 523(21)
Bonvalet, J.-P., 330
Bonventre, J. V., 171, 173(5), 241
Booth, A. G., 450, 586
Booth, I. R., 490
Booth, J., 66
Bordier, C., 821
Bormann, J., 795, 796(14), 799(14)
Boron, W. F., 197, 404, 406(58)
Boron, W., 290
Borsotto, M., 750
Borst, J. G. G., 526
Bothwel, M. A., 845
Bott, P. A., 73, 75(9), 92(9), 94(9), 102
Bouby, N., 265, 286(13)
Boucher, R. C., 15, 805
Boulanger, M., 405
Boulanger, Y., 405
Boulpaep, E. L., 4, 8, 197, 265, 266, 270(11), 271(11), 293, 294, 296(26)
Boulpaep, E., 299, 301(45)
Bourdeau, J. E., 343
Bourguet, J., 553
Bourke, R. S., 802, 809(55)
Bourne, H. R., 825
Bourne, J. A., 281
Bowen-Pope, D. F., 148
Bowman, R. L., 95, 96, 357
Boylan, J. G., 335

Boylan, J. W., 184
Bradford, 508
Bradley, T. J.,182
Brädström, A., 721, 724(4), 729(4), 730(4), 731(4)
Braell, W., 441
Braiterman, L. T., 826, 828(11), 831(11), 832
Brand, P. H., 217
Brandli, A. W., 824
Brändli, A., 822
Brann, M. R., 637, 638(119,120)
Branton, D., 553
Braun, E. J., 223
Braun, G., 84
Bravo, R., 655, 820
Brazy, P., 784, 790(11)
Breckon, R., 412, 414, 416, 417(25), 418(37,71), 422(25,37,71)
Brehe, J. E., 335
Brendel, K., 131
Brennan, J. C., 184
Brenner, B. M., 51, 52, 75, 76, 77(10,11), 90, 108, 113, 114, 115, 123(7,21,31), 125, 128, 130, 135, 137(1), 387, 390(25), 391(25), 392(25), 394(25), 401, 407(25), 408(25,28), 751
Bresler, E. H., 52
Bretag, A. H., 800
Breton, M., 681
Brett, L., 523
Breuer, W. V., 404
Brezis, M., 31, 385, 387(13,14)
Bridges, C. R., 52, 128
Bridges, K., 633
Bridges, R. J., 809
Briggman, J. V., 750
Briggs, A. P., 115
Briggs, J. P., 120, 124
Briggs, J., 119
Brink, H. M., 32
Brion, L. P., 415, 420(65)
Briving, C., 722, 729(10), 730(4, 10), 731(10), 732(10), 737(10), 738
Broberger, O., 122
Brock, T. A., 791
Brodwick, M. S., 800
Brokl, O. H., 201, 202, 204, 208(58), 209(58), 215, 216(60,64,82,84), 217(84), 218(58), 219(58)

Bron, C., 540
Brooker, G., 311
Brooks, B., 158, 162, 163, 167
Brooks, C. A., 335
Brooks, W. L., 153
Brosius, W. L., 162, 163
Bross, T. E., 335
Brown, C. D. A., 435, 496, 498, 501, 503(35), 504(35), 795, 797(30)
Brown, C. D., 790
Brown, D. B., 438
Brown, D. J., 815, 819(9)
Brown, D. M., 717, 718(18)
Brown, D., 258, 271, 279, 396, 400(38), 468, 551, 552(6), 553, 555, 556, 557, 558(36), 559, 561, 563(25), 564(25)
Brown, G. B., 162
Brown, M. R., 641, 642(10), 654, 659(10), 660(10)
Brown, M. S., 638
Brown, P., 158, 167
Brown, R. H. J., 95, 357
Brown, R. S., 31
Brunette, M. G., 95, 305, 307(38)
Brunner, F. P., 88
Bruun, H., 782
Brzeski, H., 408
Buckley, N. J., 637, 638(119,120)
Buday, M. A., 411
Bulger, R. E., 268
Bullivant, M., 32
Bullock, G. R., 281
Burch, H. B., 325, 331, 332, 335, 336(49)
Burch, R. M., 682
BurckHardt, B. C., 784
Burckhardt, B.-C., 104
Burckhardt, G., 453, 506, 508(4,5), 511(4), 512(4), 513(5), 514(5), 518, 583, 752
Burg, M. B., 61, 192, 207, 209, 210, 219, 227, 242, 270, 291, 326, 330(1a), 331, 335, 344, 349, 363, 367, 415, 416(68), 420(68), 520, 784, 791(9)
Burg, M., 289, 331, 412, 416(27,28), 417(27,28), 424(27,
Burgnara, C., 791
Burke, C., 414, 418(37), 422(37)
Burke, G., 678
Burleson, G. R., 405
Burnatowska-Hledin, M. A., 305, 385, 391(19)

Burnatowska-Hledin, M., 412, 414, 416(42), 418(42), 422(42)
Burnham, C., 463
Burrowes, M., 401
Burt, D. R., 639
Burt, D., 523, 524, 525(36)
Bush, I. E., 524, 525(33)
Butcher, R. W., 662, 663
Butlen, D., 306, 307, 312(72), 313(72), 314(72), 315(72), 316, 317(54), 319(78), 320(52), 321(52), 322(54), 322(56), 323(50,52,56), 324(52,78), 800
Butler, T. C., 470
Butterfield, D. A., 778
Büttner, D., 98
Button, D. C., 700
Byers, L. W., 167

C

Cahn, F., 748
Cala, P. M., 739
Calamia, J., 405, 406(72)
Calayag, M. C., 639
Calcagno, P. L., 306, 307(47), 323(47)
Camargo, M. J. F., 340, 341(2), 342, 343(2)
Cambon, N., 335
Campen, T. J., 538
Candia, O. A., 805
Canessa, M., 791
Cannon, C., 258, 414, 420(44), 425(44)
Cantiello, H., 405, 406(75)
Cantin, M., 303, 304(8), 305
Cantley, L. C., 691, 701(2)
Cantley, L., 437, 585, 604(27)
Caplan, M. J., 500
Caplan, M., 268
Capon, D. J., 637, 638(121)
Capon, D., 637, 638(118)
Capparelli, A., 412, 417(23), 421(23), 422(23)
Capponi, A. M., 632
Card, W. I., 526
Cardinal, J., 300, 327, 784, 791(9)
Carlemalm, E., 278, 280
Carlson, E. C., 131
Carlsson, C., 733
Carlsson, E., 723, 725(12)
Carmines, P. K., 126

Caron, M. G., 622, 637
Carone, F. A., 343
Carretero, O. A., 414, 416(41), 418(41), 422(41), 423(41)
Carriere, B., 327
Carriere, S., 95, 327
Carroll, M. A., 412, 416(31), 424(31)
Carruthers, A., 775
Carter, D. E., 790
Carter, N. W., 86, 92(32), 97, 521
Caruso, D., 685
Casellas, D., 126
Cassel, D., 743, 744(20), 747(20), 748(20), 750, 754
Cassola, A. C., 295
Castellot, J. J., 148, 149(18)
Catt, K. J., 663, 715
Caverzasio, J., 495, 496
Celis, J. E., 655
Cereijido, M., 434
Cerelli, G., 521
Chabardès, D., 130, 227, 305, 306(35), 307, 308, 309(37), 310(37), 312(66,72), 313(66,72), 314(66,72,74), 315(72), 324, 330, 331(28), 335, 338(28), 380, 400(2)
Chaffee, R. R., 478
Chaillet, J. R., 404, 406(58)
Chamberlin, M. E., 327, 391, 394(32), 407(32), 461
Chamness, G. C., 529
Chan, A. W. K., 335
Chan, C. C., 683
Chan, I. S., 67
Chan, K. M., 684
Chance, B., 469
Chang, H. H., 721, 734
Chang, H., 474, 475(15), 477(15)
Chang, K.-J., 529, 530(48)
Chang, L. K., 733
Chang, R. L. S., 51, 52(26,27)
Chang, R. S. L., 621, 633(58)
Chansel, D., 135
Chao, J., 748
Charbonnel, B., 136
Charlton, C. G., 622
Chassande, O., 751
Chastre, E., 612
Chatelet, F., 328, 385, 388(15)
Chaudhari, A., 305, 411, 421(8)

Cheah, A. M., 684
Chekal, M. A., 305, 307(45), 323(45), 523, 525(24)
Chen, PJ. Y., 564, 565(46)
Chen, E. R. Y., 343
Chen, E. Y., 638
Chen, L. G., 259
Chen, T. B., 621, 633(58)
Chen, T. R., 430
Chen, W. S., 638
Cherksey, B. D., 788, 792(29)
Cherner, J. A., 611
Cherry, W. R., 426, 495
Chesnoy-Marchais, D., 795, 796(18)
Chevalier, J., 553
Chew, C. S., 611, 640, 641, 642(10), 644(7), 647(6,7), 649(6), 651, 653(12), 654, 659(10), 660(6,10), 733, 736(24)
Chi, M. M-Y., 335
Chi-Rosso, G., 683
Chiang, H.-C., 618, 619(52), 630(52), 631(52)
Chiba, T., 612
Chignard, M., 680
Chinard, F. P., 36, 37, 40, 43, 58(19)
Chirgwin, J. M., 441
Chirito, E., 76, 77(13), 78(13), 79(13)
Choi, S., 331, 335
Chonko, A. M., 61, 232, 234(1)
Chorneyko, K. A., 278
Chowdhury, J. R., 625
Christensen, E. I., 304, 343
Christophe, J. P., 628
Christophe, J., 620, 635
Chu, C., 331, 335
Chu, L. L. H., 530
Chui, F., 121
Chuman, L. M., 438
Chuman, L., 427
Chung, D. S., 410, 418(4), 422(4)
Chung, S. D., 496
Churg, J., 260
Cieplak, W., 426
Cirksena, W. J., 81, 89(16), 92(16), 93(16)
Civan, M. M., 13
Civitelli, R., 305
Claire, M., 306, 522, 529, 530(13)
Clapham, D. E., 683
Clapp, W. L., 265
Clapp, W. L., III, 343

Clark, A. J. L., 638
Clarke, N., 717
Clarkson, T. W., 4
Claude, P., 268
Clausen, C., 14, 18
Clerc, P., 612, 620(39)
Cliff, W. H., 186, 188, 189, 190, 191, 192(31,32,33), 193(32), 194(33,40)
Clique, A., 305, 307(38,41), 308, 314(74), 324, 330, 331(28), 335, 338(28)
Cluzeaud, F., 328, 385, 388(15)
Coady, M. J., 584, 603(22), 604(22)
Cochran, W. G., 287
Cogan, M. C., 117, 123(31)
Cohen, A. J., 31
Cohen, D., 385, 386(22)
Cohen, M., 781
Cohen, P., 663
Cohen, R. L., 259
Cohen, S., 751
Cohn, Z. A., 795, 797(28)
Coifier, E., 680
Cojocel, C., 343
Colard, O., 681
Colb, C. E., 663
Cole, B. R., 335, 416
Cole, J. A., 496
Coleman, R. A., 555
Coleman, R., 677
Coll, K. E., 699
Collen, M. J., 615, 616(45), 622
Colonna, R., 474
Conlon, T. P., 628
Conti-Tronconi, B. M., 634, 638(102)
Cook, D. I., 9, 296
Cook, J. S., 744, 745(28)
Cooke, A. M., 705
Coon, H., 415, 416(68), 420(68)
Copeland, C. S., 846, 851(21), 854
Coppe, M. R., 733
Corbin, J. D., 651, 653, 663
Corda, D., 682
Cori, C. F., 762
Cornejo, M., 412, 416(35), 417(35)
Cornelisse, J., 752
Cornu, G., 529
Corradi, A., , 112, 121(11), 122(11)
Corrie, J. E. T., 523
Cortney, M. A., 343
Corvol, P., 522, 524, 525(35), 530(13)

Cory, M., 748
Coseki, C., 335
Costanzo, L. S., 94
Cotelli, F., 12
Cotran, R. S., 80, 147, 148
Cott, J., 800
Coulombe, A., 795, 797(26)
Countermarsh, B. A., 748
Cousin, J.-L., 801, 802(54)
Coussen, F., 569
Coussens, L., 638
Coverstone, M., 622
Coy, D. H., 618
Crago, S., 734
Cragoe, E. J., 405, 406(74), 809
Cragoe, E. J., Jr., 325, 384, 388(7), 403(7), 406(7), 739, 741(4), 742, 743, 744, 745, 746(22), 747, 748, 750, 751, 753, 754, 802, 809(55)
Cragoe, E., 475
Cragoe, E., Jr., 751, 752(61), 754(63)
Craig, J. M., 107
Crane, R. K., 583, 587(4,5)
Cremaschi, D., 12
Creutz, C. A., 144, 146(6)
Crews, F. T., 305
Croghan, P. C., 95, 357
Crone, C., 56
Crooke, S. T., 683
Csuros, Z., 761
Cuatreasas, P., 833
Cuatrecasas, P., 529, 530(48), 678, 679, 743, 746(22)
Currie, M. G., 416
Curthoys, N. P., 495
Cuthbert, A. W., 742, 743, 750, 751
Czech, M. P., 744, 746(23), 749(23), 750(23)

D

d'Armagnac de Castanet, E., 326
Da Cruz, M. E. M., 583, 584(6), 587(6), 596(6), 598(6)
Dacus, S. C., 426
Dahl, R., 460
Dal Canton, A., , 112, 121(11), 122(11)
Daniels, E. G., 153, 163
Danielsson, M., 676

Dannenbaum, D., 748
Dantzler, W. H., 200, 201, 202, 203, 204, 205(57,63), 206(57), 207, 208(55,58), 209(58,66), 210(55), 211(55,61), 214, 215, 216(55,60,61,63,64,80,81,82,83,84), 217, 218(58,59), 219(58,59,66), 220, 221(65,92), 223
Darbon, J.-M., 635
Darby, N. B., Jr., 438
Darlson, M. G., 639
Darnell, J. E., Jr., 438
Daugharty, T. M., 75, 76, 77(10,11)
Davidson, R. S.,280
Davis, A., 633
Davis, J. M., 120
Davis, P. J., 65
Davis, R. J., 744, 746(23), 749(23), 750(23)
Davis, R. P., 536
Dawson, R. M. C., 683, 707, 708(2), 709(2), 711(2), 717
Day, R. P., 156
Dayer, J. H., 501, 504(32)
de Curtis, I., 813, 816, 819, 823
De Kruijff, B., 556
De Léan, A., 305, 747
De Mello, G., 32
De Mey, J., 144, 146(6)
De Pierre, D., 157
De Rouffignac, C., 76, 77(12), 78(12), 81, 82, 93(17), 94, 95(19), 96, 108, 123, 266, 273(22), 295, 307
De Vries, L. A., 526
Dean, J., 345
Dean, N. M., 707
Dechatelet, L. R., 685
Dechenne, C. A., 144, 146(6)
Decker, N., 745
Decodt, P., 129
Deen, W. M., 51, 52, 76, 77(11), 108, 115, 123(7), 125, 128
Deetjen, P., 83, 89(26), 100, 583
DeFronzo, R. A., 65
DeFronzo, R., 266, 270(28), 272(28), 273(28)
Dehaye, J.-P., 620
DeKloet, E. R., 523
Delaney, R., 197
Delarge, J., 806
Dell'Antone, P., 474

DeLuca, M. A., 338
Demel, R. A., 556
Demharter, R., 682
DeMoura, J. L. C., 18
Demura, H., 304
Dence, C. N., 335
DePont, J., 747
DeSchryver-Kecskemeti, K., 416
Deshpande, A. K., 795, 797(28)
Desnuelle, P., 454
deSolms, S. J., 750, 754(55)
DeSousa, R. C., 553, 555
Dethmers, J. K., 739, 741(8), 742(8), 743(8), 744(8)
Detriasac, C. J., 417, 424(73)
Dev, B., 120
Deveney, C. W., 625
deVerteuil, F., 592
Dew, M. E., 384, 385(8,9), 386(8,9), 388(8,9), 396(8,9), 399(9)
Di Bella, F. P., 303
Di Stefano, A., 293, 295, 801, 802, 803, 805(50), 806, 808(50)
Diamond, J. M., 8, 10(10), 11(10), 14(10), 18
DiBona, D. R., 268, 275
Dibona, D. R., 474, 475(14)
Dickman, K. G., 411, 421(19)
Didet, M., 412, 425(22)
Diedrich, D. F., 756, 757, 762, 764, 767, 768, 769(11), 770, 771, 772(24), 773, 774(27), 777, 778
Diezi, J., 538, 539(65)
Dillingham, M. A., 412, 416(25), 417(25), 422(25)
Ding, G., 272
Dirks, J. H., 73, 76, 77(13), 78(13), 79(13), 81, 84(6), 89(16), 91, 92(16), 93(16), 94
DiScala, V. A., 553, 555
Dix, J. A., 562
Dixon, B. S., 414, 416, 418(37,71), 422(37,71)
Dixon, J. F., 677, 678, 685(9), 686(9), 687(9), 688(9), 689(9), 691(9)
Dixon, R. A. F., 637
Dixon, S. J., 751
Djabali, K., 496
Dodge, W., 683
Dohlman, H. G., 637
Dolan, W. J., 434

Doms, R. W., 842, 843, 844, 846, 847(3), 848(16), 849(16), 850, 851, 853(16), 854
Donin, C. L. L., 12
Doninska, B., 436
Dörge, A., 14, 96
Dorup, J., 266, 271(15), 282(15)
Doucet, A., 231, 270, 303, 305, 307(44), 316, 318(44), 319(79), 320(79), 323(44), 324, 330, 331, 335, 338, 396, 400(37), 497, 522, 525(15), 530(15), 549
Douglas, J. G., 135, 137(33)
Dousa, T. P., 130, 134(10), 135, 303, 305, 307, 330, 335, 338
Downes, C. P., 664, 707, 708(1, 2), 709(2), 711(1, 2), 715
Downward, J., 638
Dozier, J., 771, 772(24)
Dragstein, P. R., 450
Dray, F. B., 136
Draznin, B., 684
Drenckhahn, D., 272, 396, 400(42)
Drescher, C., 120
Drickamer, K., 638, 840
Droz, B., 303, 304(7)
Drugge, E. D., 412, 416(31), 424(31)
Drungis, A. M., 625
Du Bois, R., 129
Dubach, U. C., 270, 328, 335, 336(26), 337, 340, 380, 396, 400(1), 464
Dubinsky, W. P., 548, 742
Dublineau, I., 307
DuBois, R., 250
Dubroeucq, M. C., 319, 324(80)
Ducis, I., 584, 587(8), 589(8), 590, 591(8,39), 592, 596(8), 596(8,43), 598(8)
Duclohier, H., 795, 797(26)
Duddy, S., 703, 704(25)
Dufau, M. L., 663
Duhamel, L., 683
Dulaney, J. T., 829
Dull, T. J., 638
Dulley, J. R., 499
Dumont, Y., 620, 621(54)
Duncan, R. L., 528
Dunham, P. B., 794, 789
Dunham, P., 786, 792(20)
Dunn, M. J., 135, 137(33,34), 163, 415, 416(51), 420(61), 423(51,54)
Dunn, W. A., 839, 841

Duong, L. T., 636
Durr, J., 306, 307(55), 323(55)
Dürrenberger, W., 280
Dussault, J. H., 304
Duval, D., 163
Dworkin, L. D., 130, 137(1)
Dworzack, D. L., 387, 392(30), 394(30)
Dzelzkalns, M., 625
Dzhandzhugazyan, K. N., 464, 465(59)

E

Earl, P. L., 844, 848(16), 849(16), 850, 853(16,18)
Earle, A. M., 71
Eaton, D. C., 8, 10, 11(10), 14(10,20,21), 742
Eber, S. L., 496
Eddy, A. A., 481
Eddy, B., 770
Edelman, A., 8
Edelman, I. S., 327, 522, 523, 524, 525(22,29), 527, 530, 532, 533(56), 534(56), 536, 537
Edelmann, C. M., Jr., 118, 121(32)
Edwards, C. R. W., 523, 524, 525(36)
Edwards, H. E., 405
Edwards, J. D., 414, 420(44), 425(44)
Edwards, R. M., 130, 134(10), 307, 330, 335, 338
Edwardson, J. M., 750, 751
Effros, R. M., 68
Effros, R., 36
Eiksen-Olsen, M. J., 70
Eilers, U., 824
Eilertsen, E., 789
Eisenbach, G. M., 121
Eisenberg, E., 664
Eisner, G. M., 305
Eknoyan, G., 325
El mernissi, G., 316, 319(79), 320(79), 324
El-Refai, M. F., 632
Elalouf, J. M., 94, 266, 273(22), 307
Elander, B., 724, 734(16)
Elashoff, J. D., 627
Elger, B., 453, 454(29)
Elger, M., 182, 184(20)
Elias, P. M., 556
Eliat, D., 345
Ellis, E. N., 68

Ellison, D. H., 520
Ellory, C., 751
Ellory, J. C., 784, 785, 790, 794
Elwell, L. P., 748
Emanuel, J. R., 437, 585, 604(27)
Emmanouel, D. S., 306, 307(49), 323(49)
Emmanouel, S., 65, 66(45)
Emslie, K. R., 31
Endo, M., 495
Endou, H., 306, 307(51), 323(51), 330, 331, 335, 386(18)
Engel, S. L., 625
Englert, H. C., 293, 801, 803, 805(50), 806(50), 808
Englert, H., 293, 802
Enns, T., 40, 58(19)
Epstein, F. H., 31, 385, 387(13,14), 784
Epstein, N., 843
Eriksson, O., 791
Erlanger, B. F., 753, 754
Erlichman, J., 652
Erlij, D., 268
Erlinger, S., 791
Ernst, S. A., 268, 335, 340
Ernster, L., 508
Esbrit, P., 305
Escoubet, B., 496
Eskensen, K. E., 788, 792(29)
Essig, A., 8, 10(15)
Esteve, J. P., 612, 620(36)
Esteve, J.-P., 612, 620(39)
Evan, A. P., 245, 269
Evans, J. O., 764, 771, 773, 774(27), 778
Evans, M. G., 795, 796(18), 805
Evans, R. M., 521, 638
Eveloff, J. L., 405, 406(72)
Eveloff, J., 327, 387, 390(24), 391(24), 392(24), 394(24), 396(24), 397(24), 399(24), 401, 451, 453, 461
Evenou, J. P., 303
Evers, C., 451, 460(16)

F

Fabritius, J., 413
Fagioli, S., 331, 335
Fallowfield, C., 736
Fambrough, D. M., 553
Fanelli, G. M., 743
Fanestil, D. D., 324, 536, 538

Fannin, F. F., 757, 771, 773, 774(27), 778
Fara, J. W., 625
Farin, F. M., 151
Farley, J., 795, 797(27)
Farman, N., 306, 330, 520, 522, 523(21)
Farquhar, M. G., 280, 450
Farraggiana, T., 260
Fasold, H., 84, 92(30), 105, 106(18)
Faure, R., 304
Faust, R. G., 583, 584, 587(7,9), 596(7,9), 598(7,9)
Faustil, D. D., 505
Fawcett, D. W., 154
Fay, F. S., 612, 700, 701(18)
Fedak, S. A., 622
Fehlmann, M., 748, 749(49), 750(49)
Feig, P. U., 14
Feigenbaum, P., 745
Feinstein, M. B., 671, 682
Feinstein, P. O., 569
Feit, P. W., 781, 782, 783, 784(4), 789, 790(6)
Fejes-Tóth, A., 414, 416(40,41), 418(40,41), 420(40), 421(40), 422(41), 423(41), 425(43)
Fejes-Toth, G., 385, 386(20)
Felder, R. A., 305, 306, 307(47), 323(47)
Feldman, D., 522, 523, 524, 525(22,29), 532, 533(56), 534(56), 537
Fellenius, E., 721, 724, 729(4), 730(4), 731(4), 734(16)
Feltkamp, C. A., 556
Feracci, H. M., 825, 826(6), 831(6), 833(6), 835(6) 836(6), 838(6), 839(6), 840(6)
Ferguson, D., 584, 587(14), 589(14), 592(14), 596(14), 596(14), 598(14), 600(14)
Fernandez, J. M., 18
Ferriera, S. H., 805
Fesenko, E. E., 795, 796(22)
Field, M. J., 272
Field, M., 451, 786, 788(19), 791, 792(19)
Figueiredo, J. F., 349
Filburn, C., 496, 699
Fimognari, G. M., 536
Findlay, I., 795, 797(36)
Findlay, J. B., 848
Fine, L. G., 335, 384, 388(7), 403(7), 406(7), 411, 418(5,7), 420(7), 421(8), 496
Fink, E., 335, 339
Finkelstein, A., 582

Finn, A. L., 7, 9(9), 21(9)
Fischer, C., 265, 273(12)
Fischer, S., 265, 273(12)
Fisher, E. R., 154
Fisher, R. S., 12, 16(32)
Flamier, A., 319, 324(80)
Fleischer, N., 652
Fleischer, S., 508, 567
Fleuren-Jakobs, A., 747
Flockhart, D. A., 651
Fluharty, S. J., 704
Fluk, L. J., 427
Fluss, S. R., 472, 473(9)
Foellmer, H. G., 147
Fogo, A., 120, 121
Foidart, J. B., 144, 146(6)
Folkers, K., 622, 629
Folkert, V. W., 135, 136(18)
Fondacaro, J. D., 673
Fong, P., 404
Forbe, U., 290
Forbush, B., 268, 786, 787, 788, 792(21,27,28)
Forbush, B., III, 461
Ford, T. C., 391
Ford, T., 391
Ford-Hutchinson, A., 683
Forgo, J., 496
Formanowski, F., 843
Forrest, J. N., Jr., 183, 289, 299
Forssmann, W. G., 268
Forster, R. P., 188
Forte, J. G., 654, 721, 734
Forte, L. R., 496
Fortune, J., 416, 418(71), 422(71)
Foulkes, E. C., 583
Fourmy, D., 635
Fowell, D. M., 115
Franchi, A., 739, 743(7), 748(7)
Frandeleur, P., 325
Franki, N., 272
Franklin, C. C., 786, 788(22), 792(22)
Franz, H. E., 385, 386(22), 403
Franzen-Sieveking, M., 343
Frazer, M. C., 770
Frazier, H. S., 14
Fredrich, T., 752
Freedman, R. B., 841
Frelin, C., 325, 750, 751
Freshney, R. I., 141
Freund-Mercier, M. J., 303, 304(10)

Frey-Wyssling, A. J., 553
Freychet, P., 305
Fried, T. A., 130
Friedlander, G., 496
Friedman, 385, 386(21)
Friedman, P. A., 349
Frielle, T., 637
Friend, D. S., 556
Fries, E., 843
Fries, J. W. U., 68
Friese, W., 585, 603(31), 604(31)
Friesen, H. G., 633
Frindt, G., 293, 299(21), 301(21)
Fritsch, E. F., 443
Fritzsch, G., 105, 106(18), 598, 604(49)
Frizzel, R. A., 742
Frizzell, R. A., 9, 12, 809
Fröhlich, O., 771, 776, 777
Frommer, J. P., 325
Frömter, E., 4, 5, 7(4), 9, 10, 12(16), 18, 19(48,49), 88, 100, 104, 176, 184, 185, 290, 296, 299, 356, 414, 424(47), 448, 485, 808
Frucht, H., 618
Fryklund, J., 722, 724, 726, 727(18), 728(18), 729(10,18), 730, 731, 732, 734, 736, 737(10)
Fuchs, R., 505
Führ, J., 117
Fujita, N., 639
Fujita, T., 496
Fujiwara, T., 639
Fukami, K., 706
Fukase, M., 496
Fukuda, K., 637, 638(116)
Fukuhara, Y., 483, 485, 486(6), 491
Fuller, P., 522, 523(17)
Fuller, S. D., 450
Funder, J. W., 522, 523, 524, 525(22,27,29,30), 528(30), 532, 533(56), 534(56) 537
Funk, B., 328, 336(26)
Furnelle, J., 635
Furusawa, M., 385, 386(18)
Furuya, M., 303, 304(9), 305(9)

G

Gachot, B., 412, 416(22), 417(22), 416(26), 417(26), 422(26), 424(26), 425(22)
Gage, P. W., 795, 797(42)
Gaginella, T. S., 611, 618(9,10), 629(9)
Gagnan-Brunette, M., 308
Galand, N., 685
Galla, H.-J., 592
Gallagher, D. V., 696, 706(9)
Galli, G., 685
Ganote, C. E., 270, 343
Ganser, A. L., 721
Gapstur, S. M., 305, 307(42)
Garancis, J. C., 415, 420(55)
Garavito, R. M., 278
Garcia Diaz, J. F., 8, 10(15)
Garcia, M. L., 745
Garcia, M., 742
Garcia, R., 303, 304(8)
Garcia-Perez, A., 385, 388(16), 411, 414(14), 416(14), 422(14), 424(14)
Garcia-Romeu, F., 744
Gardner, J. D., 609, 610(1), 611, 612, 613(19,20), 614(1,19,20), 615, 616, 617(48), 618, 619, 620(1,19,30,42,53), 621, 622, 623(25), 625 627(19,20), 628, 629, 630(52), 631(52), 634(30)
Garg, L. C., 227, 305, 520
Garg, L. E., 335, 338
Garoff, H., 842
Garrels, J. I., 655
Gärtner, K., 98
Gartner, R., 682
Garvin, J. L., 242
Gascard, P., 707
Gassée, J. P., 129
Gassner, B., 438, 439(7)
Gastineau, M., 412, 425(22)
Gattone, V. H., 269
Gausch, C. R., 426
Gaush, C. R., 495
Gavin, J. R., III, 66
Gebler, B., 9, 10(16), 12(16), 100
Geck, P., 471, 784, 794
Gedda, K., 726, 727(18), 728(18), 729(18), 731, 734(23)
Geering, K., 540
Geibel, J., 242, 290
Geison, R. L., 676
Geletyuk, V. I., 795, 796(17)
Gellai, M., 307
Genest, J., 303, 304(8)
Geniteau, M., 385, 386(17), 396(17)
Genty, N., 431

George, J. N., 483
Gerecs, A., 761
Gerencser, G. A., 794
Gerger, R., 290, 293, 295(25)
Gerlach, L., 808
Germain, G. S., 162, 167
Germain, G., 153, 163
Gerozissis, K., 328, 414, 416(39), 418(39), 422(39)
Gerritsen, W. J., 556
Gerten, J., 157
Gertz, K. H., 84, 102
Gesbach, C., 612
Gesek, F. A., 304
Gespach, C., 612, 634(28)
Gething, M.-J., 851, 852
Ghouse, A. M., 184
Gibbs, E. M., 762, 769(11), 770, 773, 774(27)
Giebisch, G. H., 272
Giebisch, G., 8, 73, 88, 100, 102, 266, 270(27,28), 271(27,29), 272(28), 273(28), 294, 295, 296(26), 299, 301, 357, 376, 405, 791, 794
Giebisch, J., 198
Giesen-Crouse, E. M., 786
Giesen-Crouse, E., 325
Gilford, L. A., 683
Gill, G. N., 638
Gill, J. E., 259
Gilman, A. G., 566, 569, 662
Gilmore, J. P., 71
Gimbrone, M. A., 147
Gimbrone, M. A., Jr., 791
Ginkinger, K., 456
Ginsberg, B. H., 632
Giocondi, M.-C., 327
Giraud, F., 707
Gjöry, A. Z., 84, 86(28)
Glaser, L., 743, 744(20), 747(20), 748(20)
Glaser, T. M., 521
Glasgow, E. F., 113, 142, 143(4)
Gleich, R., 414, 416(38), 418(38), 419, 422(38), 424(38)
Gleischer, B., 508
Glencorse, T. A., 639
Glick, A. D., 120, 121
Glickman, J., 572
Glomset, J. A., 682
Glossmann, H. P., 303
Gluck, S., 258, 271, 506

Gmaj, P., 451, 459(10), 469(12), 495, 675, 684
Gnionsahe, A., 306
Gocke, D. J., 157
Goerig, M., 682
Goerke, J., 556
Goetzl, E. J., 683
Gögelein, H., 293, 296(18), 299, 300(46), 301, 794, 795, 797(33,37,38,40), 798(12), 798(37,38), 805(37), 806(37), 808(37,40)
Gold, M. R., 795, 796(16)
Goldberg, A. M., 65
Goldenring, J. R., 654
Goldfarb, D., 384, 388(7), 403(7), 406(7)
Goldfine, I. D., 625, 627, 633, 635, 636
Goldman, D. W., 683
Goldrich, M., 625
Goldshleger, E., 478
Goldstein, E. J., 335
Goldstein, J. L., 638
Goldstein, L., 453, 454(29)
Goldstein, M., 795, 797(28)
Goldstein, R. S., 31
Goldwasser, P., 135, 137(35)
Goldyne, M., 670
Goligorsky, M., 411, 416(16), 417(16), 418(17), 421(16,17), 422(17), 424(16)
Goltzman, D., 303, 304(13)
Gomez-Sanchez, C. E., 529
Gomez-Sanchez, E. P., 529
Gonick, H. C., 108
Gontcharevskaia, O., 308
Good, D. W., 331, 335, 357
Good, D., 331
Goodenough, D. A., 267, 268
Goodfriend, T. L., 158
Goodman, D., 527
Goodman, R. R., 622
Goodpasture, C., 495
Goodwin, D., 280
Goodwin, F. I. C., 627
Gordon, L. G. M., 10
Goresky, C. A., 36, 40, 45(9), 46, 67(9,10), 68(9), 71(9), 71(25)
Gottschalk, C. W., 73, 81, 86, 92(18), 94(23)
Gougoux, A., 327, 386, 411
Gould, N. P., 802, 809(55)
Gould, R. J., 632
Gown, A. M., 148
Goyal, R. K., 611, 625(18), 627(18)

Graeve, L., 822
Gragoe, E. J., Jr., 415, 420(55)
Graham, B., 522, 523(20), 524(20)
Graham, D., 800
Graham, J., 391
Graham, R. C. J., 557
Graham, R. M., 303
Granström, E., 670
Grantham, F. F., 270
Grantham, J. J., 61, 189, 248, 387, 392(30), 394(30)
Grantham, J., 210, 227, 289, 291, 344
Grassi, J., 136
Grassl, S. M., 453
Graves, J. S., 750
Gray, A., 638
Gray, P. T. A., 795, 796(20,21)
Green, N., 301, 412, 415, 416(27,28,68), 417(27,28), 420(68), 424(27,28), 425(33,34), 784, 791(9)
Green, R., 100
Greengard, P., 651, 652(13), 783, 784, 790(6)
Greensporn, S. A., 131, 142
Greger, R., 10, 73, 82, 86, 87(24), 89(25), 98, 183, 289, 291, 293, 294(14, 16), 295, 296(16,18), 297, 298, 299, 300, 301, 302(19), 357 461, 572, 784, 785, 786, 794, 795, 797(31,32,33,37,38,40,41), 798(37,38), 801, 802, 803, 805, 806, 808
Grenier, F. C., 391, 392(35), 414, 418(48), 423(48), 425(48), 425(48)
Gretch, D. R., 822
Grick-Ghannam, C., 307
Grider, J. R., 611, 612, 615(24), 625(16)
Griel, W., 682
Grieve, P. A., 499
Griewel, B., 405, 406(76)
Griffiths, N. M., 305, 307(43)
Grigas, E. V., 625
Grima, M., 745
Grinstein, S., 751
Grogan, W. M., 528
Gronwald, R., 682
Groome, L. S., 52
Gross, J. B., 520
Gross, P., 414, 424(47)
Grosso, A., 553, 555
Groth, D. F., 451, 455(13)
Gruder, W. G., 497

Grulich, J., 682
Grundke, E. W., 592
Grunewald, R. W., 387, 403(31), 407(31)
Grunfeld, C., 638
Grupp, C., 385, 386(22), 387, 391(29), 392(29), 394(29), 400(29), 403, 407(55), 408(55), 465
Grybowski, D. M., 611
Grynkiewics, G., 696
Grynkiewicz, G., 405, 661
Gstraunthaler, G. J. A., 380
Gstraunthaler, G., 326, 438, 439(7)
Guder, W. G., 305, 306, 326, 327, 328, 330, 331, 332, 335, 336(26), 338, 339
Gueremy, C., 319, 324(80)
Guerrant, R. L., 673
Guggino, S. E., 301, 412, 416(35), 417(35), 425(33,34)
Guggino, W. B., 8, 198, 301, 371, 374, 376, 412, 416(35), 417(35), 425(33,34)
Guggino, W., 802
Guilllemette, G., 715
Gullans, S. R., 405
Gumbiner, D., 435
Gunn, R. B., 771, 776, 777, 784, 790(11)
Gupta, R. K., 405
Gurdon, J. B., 437
Gurich, R. W., 506, 519(3), 520(3)
Gusovsky, F., 414, 422(45)
Gutkowska, J., 303, 304(8)
Gyory, A. Z., 102

H

Haas, J. A., 91
Haas, M., 786, 787(21), 788, 792(21,28)
Haase, W., 105, 106(18), 327, 387, 390(24), 391(24), 392(24), 394(24), 397(24), 399(24), 407(24), 451, 455, 460(16), 461, 495, 584, 585, 587(14), 589(14), 590, 592(14), 594(39), 596(14), 598(14), 600(14), 601(30), 603(31), 604(30,31)
Habenicht, A. J. R., 682
Haber, E., 156
Hackabath, I., 722
Hackbarth, H., 98
Hackenthal, E.,156
Hackenthal, R., 156
Hadar, E. M., 625, 636

AUTHOR INDEX

Haden, R. L., 537
Hadley, R. W., 745, 746(34)
Haeggstrom, J., 684
Hafezi, A., 384, 388(7), 403(7), 406(7)
Haga, K., 637, 638(116,117)
Haga, T., 637, 638(116,117)
Hagedorn, H. H., 170, 173, 179, 180, 181
Hagenbuch, B., 459, 460(43)
Haggerly, J. G., 405, 406(74), 500, 743
Hagiwara, H., 304
Hagiwara, Y., 304
Haglund, U., 724, 733, 734(16)
Hajo, K., 335
Halenda, S. P., 682
Hall, A. C., 790
Hallac, R., 31
Halladay, S. C., 790
Hallbach, J., 335, 339
Hallden, G., 636
Hallem, T., 702
Hamada, H., 260
Hamburg, M., 670
Hamill, O. P., 445, 795, 796(14), 799(14)
Hamilton, K. L., 742
Hamilton, W. A., 490
Hamilton, W. F., 115
Hammarström, S., 670
Hammer, R., 611, 618(7), 629(7)
Hammerman, M. R., 66, 495
Hampel, W., 95, 100
Hamzeh, A., 411, 418(7), 420(7)
Hanano, M., 67
Handa, B. K., 622
Handelin, B. L., 521
Handler, J. S., 428, 435, 450, 496, 498, 503, 551, 555, 559
Handler, J., 412, 416(27), 417(27), 422(45), 424(27)
Handler, S. A., 272
Handwerger, S., 683
Hanh, K.-D., 765
Hanley, M. J., 127
Hanley, M. R., 707, 708(1), 711(1)
Hanneman, H., 733
Hannig, K., 456, 458(40), 459(40)
Hanrahan, W., 795, 797(34)
Hansen, C. A., 699
Hansen, G. P., 266
Hansson, A., 684
Hanzel, D. K., 654

Harada, Y., 639
Haramati, A., 91
Hard, W. C., 495
Hard, W. L., 426
Harmanci, M. C., 553, 555
Harms, P., 733
Harootunian, A., 405
Harper, J. F., 311
Harper, P. A., 151
Harris, C., 76, 77(13), 78(13), 79(13)
Harris, H. W., 555, 559(24)
Harris, H. W., Jr., 272, 559
Harris, R. C., 405, 415, 416, 418(70), 420(70), 421(70), 423(63), 743, 745(21)
Harris, S. I., 399
Harrison, J. F., 342
Harrison, L. C., 636
Harrison, M., 163
Harrisson, H. E., 538
Hartroft, P. M., 155
Hartwig, J., 271
Harty, R. F., 612
Hasamura, S., 335
Haskell, J. F., 305
Hass, M., 790
Hass, S. M., 829
Hasse, W., 506, 508(4), 511(4), 512(4)
Hassid, A., 135
Hauri, H.-P., 824, 832
Hauser, D., 622
Havinga, E., 752
Hawk, C. T., 202, 218(59), 219(59)
Hawkins, P. T., 715
Hayashi, K., 663
Hayashi, Y., 303, 304(9), 305(9)
Hayat, M. A., 281
Hayflick, J. S., 638
Hayhurst, R. A., 396
Hayhurst, R. W., 270
Haynes, R. C., 701
Hays, A. E., 335
Hays, R. M., 272, 551, 555
Hays, S. E., 627
Hayslett, J. P., 266, 271(26), 272(21), 795, 797(40), 805(40), 808(40)
Hazen-Martin, D. J., 417, 424(73)
Heald, J. I., 135
Healy, D. P., 303, 304(12)
Healy, G. A., 854
Hediger, M. A., 584, 603, 604(22,51), 605

Hedin, L., 652
Heidrich, H. G., 456, 458(40), 459(40)
Heidrich, H.-G., 327, 330, 335, 338(35), 384, 385(8,9), 386(8,9), 388(8,9), 396(8,9), 399(9)
Heinle, H., 335
Heinz, E., 453, 471, 784, 794
Heinz-Erian, P., 611
Heiser, W., 748, 749(49), 750(49)
Heitmann, K., 585, 603(31), 604(31)
Helander, H. F., 724, 726, 727(17), 728(17), 733
Helander, H., 641, 647(5)
Helenius, A., 505, 815, 842, 843, 844, 845(9), 846, 847(4), 850, 851, 854
Helgerson, A. L., 775
Heller, B. I., 115
Hellfritzsch, M., 304
Helman, S. I., 6, 10, 12, 16(32), 177, 220, 221(92), 299, 301(44)
Helwig, J. J., 130, 134(8)
Henderson, L. S., 673
Hendrickson, H., 829
Hendry, B. M., 785
Henin, S., 12
Henke, D. C., 805
Hensen, C. A., 705
Hentschel, H., 182, 184(20)
Herawi, M. G., 611
Herin, P., 121, 122
Herrera, R., 638
Hersey, S. J., 611, 641, 644(7), 647(6,7), 649(6), 660(6), 731, 733, 736(24)
Herzlinger, D., 414, 420(44), 425(44)
Hesketh, T. R., 698
Heslop, J. P., 707, 708(2), 709(2), 711(2)
Hevert, F., 289
Hicks, K. O., 32
Hierholzer, K., 94, 335, 338, 524, 528(40)
Higashihara, E., 521
Highland, E., 537, 790, 791(38)
Hilberman, M., 121
Hildman, B., 455, 456(38), 460(38)
Hilgenfeld, V., 156
Hill, J. J., 245
Hiller, J. M., 622
Hiller, S., 410, 418(4), 422(4)
Hiller-Chung, S. D., 496
Hiller-Grohol, S., 496
Hinckle, P. C., 579

Hinman, J. W., 153, 163
Hirano, T., 639
Hirata, Y., 639
Hirose, S., 304
Hirose, T., 637, 638(116,117)
Hirsch, D., 266
Hirsch, S., 258
Hirschmann, W., 15
Hirsh, D. J., 266, 272(21)
Hjelle, J., 342
Hjelmeland, L. M., 632
Hoffman, E. A., 70
Hoffman, E. K., 786, 792(20)
Hoffman, K., 683
Hofman, J. F., 478
Hofmann, E. K., 789
Höhmann, B., 335, 337
Hoilien, C. A., 460
Hokin, L. E., 676, 677, 678, 682(2), 684(2), 685(9), 686(9), 687(9), 688, 689(9), 691(9), 735, 737(41)
Hokin-Neaverson, M., 676, 685
Holborow, D. J., 280
Holdsworth, S. R., 142, 143(4)
Holland, E. C., 840
Holliday, M. A., 109, 110
Hollingworth, S., 701
Hollis, T. M., 135
Holmberg, S., 416
Holmgren, J., 612, 634(31)
Holter, H., 328
Holthöfer, H., 396
Holub, B. J., 676, 679
Homma, T., 121
Hommer, D., 800
Honeyman, T., 612
Hook, J. B., 31
Hooper, J., Jr., 154
Hoover, R. L., 135, 148, 149(18), 151
Hopfer, U., 482, 794
Horacek, M. J., 71
Horan, P. K., 259
Hori, R., 455, 457(39)
Horisberger, J.-D., 538, 539(65)
Horn, K., 682
Horster, M. F., 413, 414, 416(36,38), 418(38), 419, 422(38), 424(38)
Horster, M., 107, 118(3), 120, 335, 411, 412(21), 416(21), 418(21)
Hosang, M., 762, 769(11), 770

Hoshi, T., 8, 584, 587(11), 596(11), 598(11)
Hostetter, T. H., 113, 115
Housman, D. E., 521
Housman, D., 437, 585, 604(27)
Hovanessian, A. G., 850
Howard, M. J., 746, 747(37)
Howell, K., 819
Howlett, G. J., 845
Hoyer, J., 412, 416(28), 417(28), 424(28)
Hruska, K. A., 305, 411, 415, 416(16), 417(16), 418(17), 421(16,17), 422(17,49), 424(16), 495
Huang, S. C., 618, 619(52), 630(52), 631(52)
Huang, Y.-H., 455, 456(38), 460(38)
Hubbard, A. L., 825, 826, 828(9, 10, 11), 829, 831(6, 11), 832, 833(6), 835(6), 836(6), 837, 838(6), 839, 840(6, 20), 841
Huett, M., 428
Huf, E. G., 435
Hughes, A. R., 703
Hughes, M. L., 123, 125
Hugon, J. S., 553
Hui, Y. S. F., 135
Hull, R. N., 426, 495
Humbert, F., 268, 555
Humble, L., 721, 738(3)
Hume, J. R., 745, 746(34)
Hunkapiller, M. W., 634, 638(102)
Hunter, M., 299, 301
Hunter, S. A., 524, 525(33)
Hunter, T., 637
Hurley, S., 154
Hurtley, S. M., 842, 847(4)
Hus-Citharel, A., 324, 331
Huskey, M., 305
Husted, R. F., 430
Hwang, T.-C., 572
Hyatt, P. J., 669

Im, W. B., 583, 584(7), 587(7), 596(7), 598(7), 734
Imai, M., 192, 195(43), 223, 225(97), 301, 303, 304(9), 305(9), 306, 307(48), 323(48), 369, 520
Imbault, F., 319, 324(80)
Imbert, M., 130, 227, 305, 307(37), 309(37), 310(37), 330, 331(28), 335, 338(28), 380, 400(2)
Imbert-Teboul, M., 305, 306(35), 307, 308, 312(66), 313(66), 314(66,74), 324, 335, 339
Imbs, J. L., 325
Inagami, T., 154, 158, 159(19), 160(19), 161(19), 162(19), 163(19), 164(19), 165(19), 304
Ingelman-Sundberg, M., 684
Innis, R. B., 625, 627(79)
Insel, P. A., 303, 304(12), 427, 746, 747(37)
Inui, K., 455, 457(39)
Irish, J. M., III, 232, 234(1)
Irvine, R. F., 664, 677, 683, 691, 696, 706(9), 707, 708, 709(2, 12), 711(2), 712, 714(3), 715(5)
Irwin, R. C.,189
Irwin, R. L., 248
Isaacson, L. C., 14
Isaacson, L., 291
Isac, T., 592
Ishic, S., 638
Ishii, H., 585
Ishikawa, S., 415, 416(52), 422(59), 423(52,53,56,59)
Ishitoya, J., 683
Itin, A., 636
Ito, S., 474, 475(14), 733
Itoh, N., 67
Iyengar, R., 467

I

Ichikawa, I., 112, 113, 114, 115, 116, 120, 121, 123, 124, 125, 130, 137(1)
Ichiyama, A., 637, 638(116,117)
Iga, T., 67
Igarashi, P., 750
Ikeda, T. S., 584, 603(22), 604(22)
Ikeda, Y., 826, 829(12), 832(12)
Illsley, N. P., 404

J

Jackson, B. A., 307, 330, 335, 338
Jackson, R., 722, 729(10), 730(10), 731(10), 732(10), 736, 737(10)
Jacob, R., 702
Jacobs, S., 529, 530(48)
Jacobsen, C., 453, 461(23,25,26)
Jacobson, H., 520
Jacobson, K., 592

Jacobson, W. E., 115
Jahnsen, T., 652
Jakoby, W. B., 141
Jakubowski, Z., 591, 596(42)
Jamieson, J. D., 454, 500, 635, 686
Jamison, R. L., 90, 93, 520
Jang, H. J., 293
Jans, A. W. H., 405, 406(71,76,77)
Janzen, A., 266, 270(28), 272(28), 273(28)
Jarausch, K. H., 81, 95(19)
Jarck, D., 98
Jard, S., 303
Jaresten, B.-M., 721, 724(4), 729(4), 730(4), 731(4)
Jehart, J., 326
Jen, P., 65
Jenkins, J. S., 524
Jennings, M. L., 396
Jennische, E., 304
Jensen, R. T., 609, 610(1), 611, 612, 613(19,20), 614(1,19,20), 615, 616, 617(48), 618, 619, 620(1,19,53), 621, 622, 625, 627, 628(1,53,57), 629, 630(52), 631(52)
Jessen, F., 789
Jiang, N.-Y., 618
Jim, K., 135
Jo, O. D., 412, 417(23), 421(23), 422(23)
Joacobs, W. R., 404
Jobin, M., 524, 525(31)
Jockim, H., 197
Joelsson, I., 122
Joha, H., 438, 439(7)
Johanson, R. A., 699
Johnson, A., 148
Johnson, G. D., 280
Johnson, J. P., 435, 498
Johnson, L. V., 259
Johnson, V., 31, 342
Jones, F., 162
Jones, J. H., 739
Jones, S. W., 612, 620(42), 622, 629
Jørgensen, K. E., 452, 461(20,21,22)
Jørgensen, P. L., 463, 464, 465(59)
Jose, P. A., 305, 306, 307(47), 323(47)
Joseph, S., 682
Journet, E., 707
Jukkala, K., 265, 273(12)
Jung, C. Y., 411, 418(9), 567, 568, 586, 598(32,33), 603(32)
Jung, K. Y., 306, 307(51), 323(51)

K

Kaback, H. R., 469, 470, 476
Kachadorian, W. A., 553, 555
Kachar, B., 267
Kaczmarczyk, J., 117
Kaczorowski, G. J., 739, 741(8), 742(8), 743(8), 744(8)
Kaczorowski, G., 742, 745
Kahn, M., 763
Kaiser, N., 536
Kaissling, B., 260, 265, 266, 270, 271(18), 272(18, 20), 273(22), 275, 286(18), 396, 461, 530
Kakizuka, A., 639
Kakkar, V. V., 682
Kakuno, K., 335
Kampmann, L., 18
Kan, F. W. K., 281
Kaneko, K., 622
Kangawa, K., 415, 637, 638(116, 117)
Kano-Kameyama, A., 584, 587(11), 596(11), 598(11)
Kapitza, H. G., 592
Karbowiak, I., 524, 525(37)
Karlin, A., 746, 751, 752(61)
Karlish, S. J. D., 463, 478
Karlson, K. H., 414, 415, 425(43, 66)
Karniski, L. P., 454
Karniski, L., 794
Karnovsky, M. J., 135, 144, 147, 148, 149(18), 151, 557
Karnovsky, M. M., 205
Kartenbeck, J., 816
Kasahara, M., 579
Kasbekar, D. K., 611, 615, 616(46), 618(4)
Kashgarian, M. C., 147
Kashgarian, M., 266, 268, 271(26), 272(21)
Kass, G., 703, 704(25)
Katsuta, H., 495
Katz, A. I., 65, 66(45), 231, 270, 305, 306, 307(44, 45, 49, 53), 318(44), 323(44, 45, 49, 53), 324(44), 330, 335, 338, 396, 400(37), 522, 523, 525(15, 24), 530(15)
Kau, S. T., 31
Kauffman, S., 306, 307(53), 323(53)
Kaufman, D. L., 639
Kawaguchi, Y., 412, 416(24), 417(23, 24), 421(23, 24), 422(23, 24)
Kawahara, K., 8, 301
Kawai, S., 706

Kawanishi, Y., 639
Kawashima, H., 335
Kazachenko, V. N., 795, 796(17)
Kazorowski, G., 744
Keeling, D. J., 722, 736, 738
Keenen, R. W., 676
Keiser, H. R., 163
Kellenbach, E. R., 405, 406(71)
Kellenberger, E., 280
Keller, D. S., 843, 846
Keller, T. M., 415, 425(60)
Kelvie, S. L., 305
Kempner, E. S., 567, 571, 586, 598(34)
Kempson, S. A., 495
Kennedy, B. G., 503
Kennedy, M. B., 665
Kenny, A. J., 450, 586
Kent, R. B., 437, 585, 604(27)
Kepner, G. R., 567
Kerjaschki, D., 280
Kersting, U., 438, 439(7), 448
Kessler, R. J., 505
Kessler, S. W., 840
Keutmann, H. T., 303
Khadouri, C., 335, 338
Khalil, T., 627
Khatra, B. S., 663
Kiely, J. M., 148
Kikeri, D., 401
Killen, P., 151
Kim, D., 683
Kim, H. D., 786, 788(22), 792(22)
Kim, J. K., 306, 307(55), 323(55)
Kim, J., 460
Kim, S. H., 618
Kimelberg, H. K., 802, 809(55)
King, V. F., 745
Kinne, R. K. H., 387, 391(29), 392(29), 394(29), 400(29), 401, 403, 405, 406(71, 76, 77), 407(24, 55), 408(55), 465
Kinne, R., 60, 327, 337, 342, 387, 390(24), 391(24), 392(24), 394(24), 397(24), 399, 401, 408, 451, 452, 453, 454, 455, 456, 458(17, 40), 459(40), 460(16), 461, 462, 463(52, 53, 54), 466, 468, 469, 508, 583, 584(6), 586, 587(6), 596(6), 598(6, 32), 603(32), 672
Kinne-Saffran, E., 60, 451, 456, 458(40), 459(40), 462, 463(52, 54), 466, 468, 469
Kinne-Safran, E., 672
Kinoshita, Y., 496
Kinsella, J. L., 491, 742
Kinskey, S. C., 556
Kinter, W. B., 188
Kirk, K. L., 236, 275, 306
Kirschenbaum, M. A., 305
Kirsten, R., 539
Kishimoto, T., 639
Kitazawa, S., 455, 457(39)
Klaerke, D. A., 463
Klahr, S., 113
Klausner, R. D., 633, 841, 854
Klee, W. A., 632
Kleeman, C. R., 108
Klein, K. L., 135, 307
Klein-Robbenhaar, G., 266, 270(28), 272(28), 273(28)
Kleinman, J. G., 411, 415, 420(55)
Kleinzeller, A., 188, 427
Kletzein, R. F., 432
Kleyman, T. R., 739, 741(4), 742(4), 743(4), 744(4), 751, 753, 754
Kleyman, T., 742, 750(4), 751, 752(61), 753, 754(63)
Kline, R., 495
Klinman, B., 156
Klose, R. M., 102, 357
Klöss, S., 84, 86, 92(30), 103, 104
Klueppelberg, U. G., 636
Klumperman, J., 824
Klyce, S. D., 14
Knepper, M. A., 227, 231, 242, 307, 357, 520
Knoth, J., 490
Knowles, J. R., 757
Knox, F. G., 70, 73, 80, 82, 86, 89(25), 91, 357
Kobilka, B. K., 637
Koch, K. S., 748, 749(49), 750(49)
Koechlin, N., 266, 273(22), 306, 412, 416(26), 417(26), 422(26), 424(26)
Koelz, H. R., 611, 731, 733, 736(24)
Koeppen, B. M., 220, 221(92), 260, 265, 299, 301(44), 396
Koepsell, H., 584, 585, 587(8, 14), 589(8, 14), 590, 591, 592, 596(8, 14, 18, 42, 43), 598, 599, 600(14, 28, 29), 601(29, 30), 602(29), 603(28, 29, 31, 50), 604(29, 31, 49), 605, 767, 768(17)
Koerner, D. R., 524, 525(34)
Koerner, T., 156
Koeschke, K., 102

Kohn, L. D., 682
Koide, S. S., 795, 797(28)
Kojima, I., 665, 666(18), 669, 671
Kojima, K., 665, 666(18), 669
Kojima, M., 637, 638(116)
Kokko, J. P., 206, 207, 218(73), 265, 369, 520, 521, 540
Kokko, J., 520
Kolb, H. A., 795, 797(30)
Kolb, H.-A., 795, 797(25)
Kolesnick, R. N., 684
Kolesnikov, S. S., 795, 796(22)
Kolis, S. J., 790
Kon, V., 115, 116, 123
Koniatsu, H., 622
Konieczkowski, M., 135, 415, 420(61)
König, B., 462, 463(53)
Konoblauch, C., 404
Köpfer-Hobelsberger, B., 327
Korenstein, R., 479
Korn, K., 584, 585, 587(14), 589(14), 590, 592(14), 596(14), 596(14), 598, 600(14, 29), 601(29), 602(29), 603(29), 604(29, 49)
Koseki, C., 303, 304(9), 305(9), 331, 335, 385, 386(18)
Koski, G., 632
Kostelek, M., 678
Kotanko, P., 326
Kottra, G., 10, 18, 19(48, 49)
Koushanpour, E., 35, 39(1), 49(1)
Kouznetzova, B., 135
Koyama, H., 495
Kraehenbuhl, J. P., 540, 753, 754
Kragh-Hansen, U., 453, 461(23)
Krahl, M. E., 762
Krakower, C. A., 131, 142
Kramer, L. B., 528
Krause, K. H., 683
Kraut, J. A., 415, 420(57)
Krebs, E. G., 663
Krebs, H. A., 32
Kreis, T. E., 819, 850
Kreis, T., 824
Kreisberg, J. I., 143, 144, 147, 151, 501
Kretchevsky, D., 687
Kreutter, D. K., 674
Kreye, V. A. W., 801
Krijnen, E. S., 405, 406(77)
Krishnamurthi, S., 682

Kristensen, P., 789
Kriz, W., 31, 35, 39(1), 49(1), 68, 265, 266, 270, 271(18), 272(18), 273(12), 286(13, 18), 287, 385, 387(12), 396, 410, 414, 418(46), 424(47), 461, 530
Krouse, M. E., 795, 797(42)
Krozowski, Z. S., 524, 525(30), 528(30)
Krozowski, Z., 524, 525(37)
Kruiger, W., 638
Krupinski, J., 569
Krust, B., 850
Krüttengen, C. D., 117
Kubo, T., 637, 638(116, 117)
Kubota, T., 295
Kuff, E. L., 639
Kuhn, K., 268
Kuhn, L. C., 638
Kuijpers, G. A. J., 747
Kulpa, C. F., 405
Kumar, A. M., 405
Kung, H., 706
Kunkel, K. A., 621
Kuno, M., 639
Kuntziger, H. E., 91, 93(48)
Kunz, U., 585, 590, 598(28), 600(28), 603(28)
Kunzelmann, K., 795, 797(40, 41), 805, 808
Kurihara, S., 412, 416(24), 417(24), 421(24), 422(24)
Kurokawa, K., 65, 135, 307, 335
Kurtz, I., 411, 421(8)
Kurzyca, J., 436
Kusano, E., 306, 307(48), 323(48)
Kuwahara, M., 565, 566
Kuylenstierna, L., 508
Kuzuya, T., 415, 416(52), 422(59), 423(52, 53, 56, 59)
Kwan, C., 702, 703(23)
Kwong, S. F., 739

L

L'Allemain, G., 739, 743(7), 748(7)
La Maout, S., 412, 416(22), 417(22), 416(26), 417(26), 422(26), 424(26), 425
Laburthe, M., 612
Lachman, H. M., 415, 420(65)
Lad, P. J., 748, 749(49), 750(49)
Laemmli, U. K., 652, 655(14)

Lahiti, R. A., 622
Laing, S. M., 722
Lake, J. R., 744
Lakskmi, V., 524
Lamb, J. F., 428
Lambert, M., 635
Lambert, P. P., 129
Lambrecht, G., 611
Lan, M. C., 522, 523(20), 524(20)
Landry, D. W., 573, 574(4), 576, 578(5), 809
Landry, D., 751, 752(61)
Lane, A. L., 622
Lane, C. D., 437, 639
Lang, F., 73, 82, 86, 89(25), 242, 275, 357, 791, 802
Lang, H. J., 293, 801, 802, 803, 805(50), 806(50), 808
Langhans, S., 385, 386(22)
Lapetina, E. G., 678, 679, 683, 687
Laragh, J. H., 157
Laris, P. C., 478
Larose, L., 620, 621(54)
Larsson, H., 721, 723, 724(4), 725(12), 733, 738(3)
Larsson, L., 107, 343
Lassiter, W. E., 73
Lassiter, W., 86
Latif, S., 527
Latta, H., 154
Laub, J. F., 501
Laurent, A. G., 850
Laychock, S. G., 682, 683(27)
Lazdunski, M., 325, 750, 751
Lazorick, K., 750
Le Bivic, A., 822
Le Couedic, J. P., 680
Le Fur, G., 319, 324(80)
Le Grimellec, C., 76, 77(12), 78(12), 90, 91
Le Hir, M., 260, 265, 266, 270, 328, 335, 337, 340, 396
Le Maout, S., 412, 425(22)
Le Vine, H., III, 743, 746(22)
Leach, B. E., 153, 162, 163, 167
Leaf, A., 539
Lear, S., 387, 390(25), 391(25), 392(25), 394(25), 407(25), 408(25)
LeBouffant, F., 331
Lechene, C., 73, 86, 171, 173(5), 241, 242, 357, 405, 416, 418(70), 420(70), 421(70)
Leder, P., 441

Lee, H.-S., 584
Lee, J. C., 154
Lee, J., 335, 338, 638
Lee, S. M. K., 305, 307(45), 323(45), 523, 525(24)
Lee, T., 684
Leffert, H. L., 748, 749(49), 750(49)
Lefkowith, J. B., 133
Lefkowitz, R. J., 622, 637
LeFurgey, A., 274, 327, 391, 394(32), 407(32), 461
LeGrimellec, H., 327
Leibson, C., 777
Leitner, J. W., 684
Lemieux, G., 327, 386, 411
Lemp, G. F., 621, 625
Lencer, W. I., 468, 556, 557(25), 559, 561, 563(25), 564
Lerner, R. L., 65, 135
Letcher, A. J., 664, 683
Leth, R., 733
Lether, A. J., 707, 715(5)
Leung, J. O., 638, 840
Levasseur, S., 678
Levenson, J. A., 112
Levenson, R., 437, 585, 604(27)
Lever, J. E., 451, 453, 503, 584, 596(16), 598(16)
Levine, D. Z., 120
Levine, S. D., 555
Levinson, G., 182
Levitzki, A., 622
Lew, D. P., 683
Lewand, D. L., 411, 415, 420(55)
Lewin, H., 721
Lewin, M., 734
Lewis, R. A., 396, 400(41)
Lewis, S. A., 8, 10, 11(10), 14(10, 20, 21), 18, 272, 551, 559(3), 675, 795, 797(34)
Lianos, E., 135, 137(34)
Liao, Y.-C., 638
Libermann, T. A., 638
Liedtke, C. M., 794
Liegeois, J. F., 806
Light, D. B., 414, 415, 425(43, 60, 66, 67)
Lin, C. R., 638
Lin, J. T., 583, 587(6), 596(6), 598(6)
Lin, J.-T., 453, 584, 586, 587(12), 596(12), 598(12, 32), 603(32), 765
Lin, P.-F., 438

Lindberg, P., 721
Lindemann, B., 15, 18(40)
Linderstrom-Lang, K., 328
Lindstrom, J. M., 634, 638(102)
Ling, K. Y., 583, 584(7), 587(7), 596(7), 598(7)
Ling, K.-Y., 584, 587(9), 596(9), 598(9)
Lingappa, V. R., 639
Linke, R., 335, 339
Lippincott-Schwartz, J. S., 841, 854
Lisanti, M., 822
Liu, A. Y. C., 651, 652(13)
Livingstone, D., 410, 418(4), 422(4)
Lodish, H. F., 441, 850
Loewenstein, W. R., 7
Logsdon, C. D., 639
Logsdon, C., 612, 620(39)
Logue, P. J., 794
Lohmann, S. M., 652
Lombes, M., 529
London, G. M., 112
Lonnroth, I., 612, 634(31)
Lookman, T., 592
Lopes, A. G., 374, 376
Lopes, A., 299, 301(45)
Lord, J. A. H., 622
Loreau, N, 135
Lorentzon, P., 736
Lorenz, P. E., 345
Loth, U., 682
Lotti, V. J., 621, 633(58)
Louie, D. S., 612
Louvard, D., 431, 454, 501
Lovelace, C., 453
Lovett, D. H., 147
Low, P. S., 266
Lowe, A. G., 411, 418(5, 7), 420(7)
Lowry, O. H., 227, 328, 331, 332, 335, 542
Lu, Y.-A., 622, 629
Lubin, M., 748
Luchtel, D., 148
Lucocq, J. M., 837
Lucoq, J. M., 281
Ludt, J., 404
Luer, K., 88
Lufburrow, R. A., III, 743, 745(21)
Luig, J., 405, 406(77)
Luini, A., 682
Lukeman, D. S., 324

Lumsden, C. J., 36, 37, 45(16), 46(16), 48(16), 49(12, 16), 51(16), 52, 53(12), 54(12), 55(12), 56(31), 57, 63, 65, 66, 67(49), 71(16)
Luo, M., 304
Luskey, K. L., 638
Lustenberger, P., 529
Luzio, J. P., 819
Lyubarsky, A. L., 795, 796(22)

M

Ma, A., 826, 828(9, 10)
Ma, D. P., 638
Maach, T., 343, 345(18), 346(18), 349(18), 351(18), 352(18), 353(18)
Maack, T., 32, 340, 341(1, 2, 3), 342, 349
McArdle, S., 305
McAvoy, E. M., 188
McBurney, R. N., 795, 796(15)
McCammon, K., 851
McCann, F. V., 414, 415, 425(43, 60)
McCarthy, W. Z., 335
McCarthy, W., 331, 335
McCaslin, D. R., 843
McChesney, D. J., 639
McClelland, A., 638
McCoy, R. N., 130
McCreary, M. D., 335
McDaniel, M. L., 684
McDermott, M., 527
MacDonald, C., 408
MacDonald, R. J., 441
McDonogh, P. M., 700
McEwens, C. R., 844
Macey, R. L., 567
McGiff, J. C., 412, 416(31), 424(31)
McGuigan, J. E., 612
McGuire, W. L., 529
Machamer, C., 846
Machen, T. E., 405, 472
McHugh, P. P., 625
McIntyre, M. A., 523
Mack, E., 404
McKnight, A. D. C., 501, 504(32)
Maclouf, J., 136
McManus, T. J., 790
McMurchie, E. J., 592

McNamara, E. R., 81, 94(22)
McNamara, P. D., 456
McNamee, K. C., 280
McQuillan, J., 746
Macrae, W. A., 157
McRoberts, J. A., 429, 786, 791
Maddox, D. A., 114, 117, 123(21, 31)
Maddrell, S. H., 178
Madin, S. H., 438
Madison, L. D., 635
Madrala, A., 590, 591, 596(42), 598, 599, 603(50), 604(49), 767, 768(17)
Madsen, K. M., 265, 266, 269, 271(7, 24), 275(47), 343
Maeda, A., 637, 638(116, 117)
Magleby, K. L., 795, 796(23, 24), 798(23, 24)
Magnussen, M. P., 789
Mahadevappa, V. G., 676, 679
Mahamer, C., 851
Mahe, Y., 744
Mahieu, P., 144, 146(6)
Mailman, D. S., 467
Mainwaring, W. I. P., 535
Mais, D. E., 682
Majerus, P. W., 677, 679
Makhail, N., 411, 418(5)
Makhlouf, G. M., 611, 612, 615(24), 625(16)
Malathi, P., 583, 584, 586, 587(4, 5, 10), 596(10), 598(10, 33)
Malchiodi, F., 260
Malco, D. G., 612
Maleque, A., 331, 335
Malhotra, S., 687
Malhotra, V., 813
Mall, A., 335
Malmsten, C., 670
Malmstrom, K., 495, 498, 499(25), 501(25), 504(25)
Malmstrom, L., 496
Malnic, G., 102, 357
Malone, B., 684
Mamelok, R. D., 451, 455(13)
Mamon, J. F., 638
Mandel, G. S., 411, 415, 420(55)
Mandel, L. J., 10, 327, 384, 391, 394(32), 399, 401, 404, 407(32), 411, 421(19), 461, 739, 747
Mandel, N. S., 411, 415, 420(55)

Mandel, P., 130, 134(8)
Mandell, L. J., 744
Manganiello, V. C., 307
Mangos, J. A., 84
Maniatis, T., 443
Manillier, C., 306
Manlunowska, D. H., 731
Mantey, S. A., 611, 618
Mantey, S., 618
Manyard, C., 771, 777, 778
Marbaix, G., 437
March, S. C., 833
Marchand, G. R., 80
Marchand, G., 82, 89(25)
Marchesi, V. T., 850
Marchetti, J., 335, 339
Marcus, D. C., 18
Márdh, S., 733
Maren, T. H., 803
Margaroli, A., 701
Margolius, H. S., 748
Marie, J.-C., 612
Markakis, D., 376
Markwell, M. A. K., 829
Maroux, S., 454
Marshall, P. J., 688
Marshall, W. S., 14
Marsy, S., 335, 338
Martin, A. J., 795, 796(16)
Martin, R. G., 845, 850(18)
Martinez, J., 619, 620(53), 628(53)
Martinez-Maldonado, M., 113
Martinez-Palomo, A., 268
Marty, A., 445, 795, 797(35), 798(35), 805
Marver, D., 521, 522, 523, 527, 530, 532, 533(56), 534(56), 539, 540, 541(11), 542, 547(69), 548(69), 549(69), 550(69)
Marx, S. J., 303
Marxer, A., 832
Mason, A., 638
Mason, J., 96
Massry, S. G., 411, 418(6), 421(6)
Masu, Y., 639
Masur, S. K., 562
Matheravidather, 731
Matlin, K. S., 815, 825
Matlin, K., 431
Maton, P. N., 611, 613(20), 614(20), 618, 619(52), 627(20), 630(52), 631(52)

Matozaki, T., 612, 620(37, 38)
Matsuo, H., 415, 637, 638(116, 117)
Matsuoko, K., 706
Maude, D. L., 88
Mauer, S. M., 68
Maunbach, A. B., 342, 343(5)
Maxfield, F. R., 404, 472, 473(9)
Maxwell, M. H., 108, 280
May, W. S., Jr., 743, 746(22)
Mayes, E. L. V., 638
Mayr, G. W., 708
Mayron, B. R., 209
Mays, J. M., 366
Mead, J. A. R., 337
Medow, M. S., 456
Meek, J. L., 708
Meezan, E., 131, 305
Meier, K. E., 427
Meier-Ewert, H., 843
Meigs, R. A., 524, 525(33)
Meister, R., 687
Melcion, C., 385, 386(17), 396(17)
Meldolesi, J., 454, 701
Mellman, I., 505, 854
Meloche, S., 747
Menard, J., 339, 524, 525(35)
Mendlein, J., 722, 729(10), 730(10), 731(10), 732(10), 737(10)
Mendrick, D. L., 411, 418(12), 420(12)
Meng-Ai, Z., 734
Menozzi, D., 611, 613(20), 614(20), 627(20)
Menton, D. N., 411, 416(16), 417(16), 421(16), 424(16)
Menuhr, H., 584, 587(8, 14), 589(8, 14), 590, 591(8, 39), 592(8, 14), 596(8, 14), 598(8, 14), 600(14)
Merlet-Benichou, C., 108
Merlino, G. T., 638
Merot, J., 412, 416(22, 26), 417(22, 26), 422(22, 26), 424(22, 26), 425
Merriam, P., 612
Merritt, J., 702
Merz, W. A., 283
Messow, C., 98
Metcalfe, J. C., 698
Mettrick, D. F., 687, 688(66)
Metz, S. A., 684
Meyer, G., 12
Meyer, M. M., 562, 563(44), 564(44)
Meyer, P., 303, 304(7)

Meyer, T. W., 68
Michaels, A. S., 121
Michaud, A., 522, 529, 530(13)
Michelangeli, F., 733
Micheletti, R., 611
Migone, L., 112, 121(11), 122(11)
Mihas, A. A., 734
Mikami, A., 637, 638(116, 117)
Mikhail, N., 384, 388(7), 403(7), 406(7)
Milavec-Dritzman, M., 303
Milholland, R. J., 536
Miliani, D., 701
Miller, C., 750, 795, 796(19), 801
Miller, J. H., 571
Miller, J. S., 639
Miller, J., 683
Miller, L. J., 625, 635, 636
Miller, M. M., 495
Miller, S. G., 665
Miller, S. S., 303
Mills, G. R., 268
Mills, J. W., 501, 504(32)
Mills, R. J., 113
Milton, A., 304
Mimoune, D., 135, 304
Mims, S., 415, 416(68), 420(68)
Minta, A., 405
Minuth, W. W., 265, 286(13), 414, 418(46), 424(47)
Misfeldt, D. S., 485
Mishina, M., 637, 638(116)
Mistaoui, M., 306, 307, 312(72), 313(72), 314(72), 315(72), 316(54), 317(54), 318(54), 322(54), 323(54)
Mitchell, P., 469, 470(2)
Mitchell, R. H., 677
Mitchell, W. J., 490
Mitchell, W., 526
Mitra, A., 763
Miura, K., 31
Miwa, T., 223, 225(97), 226(98)
Miyahara, T., 412, 416(24), 417(23, 24), 421(23, 24), 422(23, 24)
Miyanoshita, A., 306, 307(51), 323(51)
Miyauchi, A., 496
Mizgala, C. L., 495
Modlin, I. M., 654
Moewes, B., 518
Mogan, R. O., 715
Mohandas, T. K., 605

AUTHOR INDEX

Molhuysen, J. A., 526
Molitoris, B. A., 460
Monck, J. R., 691, 693(3), 695, 699, 704
Monder, C., 523, 524
Mong, S., 683
Monnens, L., 123
Monod, A., 683
Montagnier, L., 850
Montais, R., 744
Montégut, M., 305, 307, 308, 312(66, 72), 313(66, 72), 314(66, 72, 74), 315(72), 324, 330, 331(28), 335, 338(28)
Montesano, R., 555, 556
Montrose, M. H., 404
Montrose-Razifadeh, C., 371
Moody, F. G., 726
Moody, T., 622
Moons, W. M., 32
Moor, H., 553
Moor, R. M., 707, 714(3)
Moore, G., 703, 704(25)
Moore, J. P., 698
Mooseker, M. S., 267
Morad, M., 745
Moradeshagi, P., 384, 388(7), 403(7), 406(7)
Moran, A., 61, 452, 459(19), 461(18, 19), 482, 483(3), 491, 492(19), 503
Moran, T., 611, 625
Morean, J.-P., 618
Morel, F., 81, 82, 84, 87(24), 91, 94(23, 24), 95(19), 130, 227, 231, 303, 304(7), 305, 306, 307, 308, 309(37), 310(37), 312(66, 72), 313(66, 72), 314(66, 72, 74), 315(72), 316(54), 317(54), 318(52, 54), 320(52), 321(52), 322(54), 323(52), 324, 330, 331, 335, 338, 339, 380, 396, 400(2, 37), 497
Morgan, B. A., 622
Morgan, J. P., 664
Morgan, K. G., 664
Morin, J.-P., 342
Morisset, J., 620, 621(54)
Moroder, L., 612, 634(28)
Morris, A. P., 696, 706(9)
Morris, D. J., 527, 536
Morris, J. F., 555
Morrison, A. R., 305, 415, 422(49, 50)
Morrison, T. K., 126
Morton, J. J., 157
Moscarello, M. A., 553

Moser, U., 611
Moses, H. L., 270
Moskowitz, D., 305, 411, 418(17), 421(17), 422(17)
Moss, B., 844, 848(16), 849(16), 850(16), 853(16)
Motaba, T., 154, 158(2)
Motais, R., 801, 802(54)
Moyer, J. D., 707
Mroz, E. A., 357
Muallem, S., 415, 420(57), 475
Mueller, S. C., 837, 839, 840(20)
Muhlethaler, K., 553
Muirhead, E. E., 153, 158, 159(19), 160(19), 161(19), 162, 163, 164(19), 165(19), 167
Muirhed, K. A., 259
Mujais, S. K., 270, 306, 307(53), 323(53)
Mukherjee, C., 622
Mukherjee, S. K., 217
Mullen, M. D., 746, 747(37)
Müller, J., 555
Muller-Esteri, W., 335, 339
Mullin, J. M., 427
Mumford, R. A., 637
Munroe, D. R., 733
Munson, K., 722, 729(10), 730(10), 731(10), 732(10), 737(10)
Munson, P. J., 529, 625, 628(76), 631(76)
Münzel, P. A., 303, 304(12)
Murakami, K., 154, 158(2)
Murayama, N., 305, 307(42)
Murayama, Y., 84
Murer, H., 60, 404, 451, 455, 456(38), 459, 460(38, 16, 43), 469(12), 488, 495, 496, 498, 499(25), 501, 502(24), 503(35), 504(25, 35), 583, 675, 790, 795, 797(30)
Murphy, P., 438
Musacchio, I., 684
Musch, M. W., 791
Mustschler, E., 611
Myer, B. D., 115
Myers, B. D., 115, 128, 121
Mylle, M., 81, 86, 92(18), 94(23)

N

Naftel, J., 795, 797(27)
Nagai, Y., 683
Nagao, M., 612, 620(37, 38)

Nagata, N., 326
Nagel, W., 8, 10(15), 13, 15
Nagle, R. B., 204
Nahorski, S. R., 707, 708(4)
Naik, D., 551
Nakada, J., 330
Nakada, M., 496
Nakajima, K., 815, 819(9)
Nakajima, S., 815, 819(9)
Nakamura, R., 306, 307(48, 49), 323(48, 49)
Nakanishi, O., 706
Nakanishi, S., 639
Nakao, A., 412
Nakayama, K., 639
Nakumura, M., 622
Nanci, A., 281
Nandi, J., 734
Narang, N., 335, 338
Náray-Fejes-Toth, A., 385, 386(20), 414, 416(40, 41), 418(40, 41), 420(40), 421(40), 422(41), 423(41), 425(43)
Narins, R. G., 331, 335
Naruse, K., 154, 304
Naruse, M., 304
Nathanson, I. T., 790, 791(38)
Navar, L. G., 93, 126
Nayak, D. P., 815, 819(9)
Neeb, M., 585, 590, 598(28), 599, 600(28), 603(28, 50)
Needleman, P., 335, 416
Negulescu, P. A., 405
Neher, E., 299, 445, 793
Neilsen, O. B. T., 782
Nellans, H. N., 12
Nelson, D. J., 795, 797(29)
Nelson, L. R., 802, 809(55)
Nelson, N. F., 426
Nelson, W. J., 813, 814(4), 825
Neufeld, E. J., 677, 679
Newsholme, E. A., 338
Newton, J. P., 428
Nir, S., 592
Nishiitsutsuji-Uwo, J. M., 32
Nishimura, H., 192, 195(43), 223, 225(97), 226(98)
Nishizuka, Y., 664, 684
Nitschke, R., 290, 295
Nitta, K., 304
Nivez, M. P., 135
Njus, D., 490

Noguchi, M., 611, 612, 613(19), 614(19), 620(19, 42), 627(19), 628(19)
Noinel, N., 96
Nolan, M. F., 40, 58(19)
Nonoguchi, H., 307, 330, 331, 335
Nord, E. P., 384, 388(7), 403(7), 406(7), 411, 421(8, 10)
Nordberg, P., 721, 738
Norgaard, J. O. R., 144
Norman, J., 411, 421(8)
Noronha-Blob, L., 496
Northrup, T. E., 130, 134(10)
Northrupt, T. E., 135
Novak, I., 289, 794, 808(4)
Noweicka, C., 684
Nozawa, Y., 678
Numa, S., 637, 638(116, 117)

O

O'Donnell, M. E., 786
O'Donohue, T. L., 622
O'Farrell, P. H., 655
O'Grady, S. M., 786, 788(19), 791, 792(19)
O'Neal, S., 517
O'Neil, R. G., 10, 265, 270, 271(11), 293, 396, 548
Oberleithner, H., 198, 295, 376, 438, 439(7), 448, 802
Öbrink, K. J., 640, 641, 647(5), 648(4), 724, 726, 727(17), 728(17)
Ochs, T. J., 198
Oda, M., 622
Oda, T., 826, 829(12), 832(12)
Oddsdottir, M., 654
Oesterhelt, D., 580
Ogata, E., 496
Ogawa, M., 192, 195(43), 223
Ogden, P., 501
Ogino, T., 405
Ogura, Y., 412, 417(23), 421(23), 422(23)
Ohkubo, H., 639
Ohkuma, S., 261
Ohlson, A. S., 328
Ohnuma, N., 303, 304(9), 305(9)
Oka, H., 826, 829(12), 832(12)
Okada, K., 415
Okada, T., 612
Okamura, T., 158

AUTHOR INDEX

Okano, T., 455, 457(39)
Okano, Y., 678
Oken, D. E., 87
Okumura, H., 495
Olbe, L., 723, 724, 734(16)
Olefsky, J., 65
Oliver, J., 73
Ollig, D., 584, 585, 587(14), 589(14), 592(14), 596(14), 598(14, 29), 600(14, 29), 601(29), 602(29), 603(29), 604(29)
Olsnes, S., 404
Olson, D. M., 683
Olson, P., 135
Ong, H., 747
Oparil, S., 159, 343
Ophir, I. J., 605
Oppenhoff, I., 157
Orci, L., 268, 279, 396, 400(38), 427, 555, 556, 557, 558(36), 559(36), 813
Ordway, R. W., 683
Orloff, J., 207, 210, 227, 270, 289, 291, 326, 330(1a), 344, 349
Orrenius, S., 703, 704(25)
Orth, K., 569
Osborne, M. J., 303, 304(7)
Osgood, R. W., 127, 130
Othmani, L., 412, 416(22), 417(22), 422(22), 424(22), 425(22)
Oudinet, T. P., 135
Owen, D. G., 795, 796(13)
Owen, N. E., 786, 790, 791
Owyang, C., 612

P

Padan, E., 605
Page, L., 156
Pagel, H. D., 84
Pagenelli, C. V., 294
Paing, M., 163
Pajor, A. M., 605
Palacios, J. M., 303, 304(10)
Palade, G. E., 454, 500
Palade, P. T., 802
Palaty, V., 748
Palfrey, H. C., 461, 783, 784, 786, 787, 788(19, 27), 790(6), 792(19, 27)
Palmer, J. S., 622

Palmer, L. G., 293, 299(21), 301(21), 742, 795, 797(29)
Pan, G. Z., 615, 616(45)
Paoletti, R., 685
Papahadjopoulos, D., 592
Papahadjopoulous, D., 553
Papermeister, D. S., 278
Pappenheimer, J. R., 18, 52
Paradiso, A. M., 472
Parent, L., 300
Parikh, I., 833
Paris, K., 328, 336(26)
Pariza, M. W., 432
Park, C. H., 340, 341(2), 343, 345(18), 346(18), 349(18), 351(18), 352(18), 353(18)
Park, J., 611, 612, 627
Park, S., 664
Parry, G., 335
Parsa, B., 184
Pasino, D. A., 31
Passonneau, J. V., 227, 332, 335, 542
Passow, H., 437, 585, 794, 801
Pastan, I., 141, 638
Patarca, R., 805
Patel, P. Y., 147
Paterson, B. M., 639
Patlak, C. S., 61, 209, 210(77), 248, 251(10), 252(10), 356
Pattnaik, A. K., 815, 819(9)
Patton, M., 127
Paul, S. M., 627
Paul, S., 800
Paule, J., 108
Paulson, J. C., 837
Pavenstädt, H., 805
Pavenstädt-Grupp, I., 387, 403, 407(55), 408(55)
Pazanelli, C. V., 197
Pearce, D., 564, 565
Pearce, P. T., 523, 525(27)
Pearson, J. D., 431
Pearson, R. K., 636
Peerce, B. E., 584, 603(19, 20, 21), 604(10, 21)
Pegram, S. E., 118
Pellet, M., 326
Peralta, E. G., 637, 638(118, 121)
Perfetto, M., 147
Perl, W., 36, 37

Perrelet, A., 268, 556
Perrier, M. L., 319, 324(80)
Perrin, F., 524, 525(31)
Pert, C., 622
Pesonen, M., 431
Peter, K., 328
Peterkin, B. M., 535
Petersen, O. H., 293, 795, 797(36)
Peterson, D. R., 343
Peterson, G. L., 637, 638(118)
Peterson, L. H., 36
Peterson, O. H., 696, 706(9)
Petrusz, P., 281
Petruzzelli, L. M., 638
Pettinger, W. A., 303, 307
Petty, K. J., 540
Petzel, D. H., 175, 179, 180, 181, 188, 190, 192(33), 194(33)
Pfaller, W., 71, 270, 282(60), 284(60), 326, 438, 439(7)
Pfeffer, J. M., 115
Pfeffer, M. A., 115
Pfeffer, S. R., 842
Pfeifer, B., 794
Pfeiffer, A., 611
Pfeiffer, B., 784
Pfeiffer, F., 800
Phillips, J. E., 172, 289
Picking, R. A., 701
Pietruszkiewicz, A. M., 802, 809(55)
Pietrzyk, C., 794
Pietvgyk, C., 784
Pillion, D. J., 305
Pindo Da Silva, P., 267
Pines, M., 174
Pink, D. A., 592
Pisano, J. J., 305, 307(46), 323(46)
Pitcock, J. A., 153, 158, 159(19), 160(19), 161(19), 162, 163, 164(19), 165(19), 167
Pizzonia, 385, 386(21)
Poelling, R. E., 496
Poenie, M., 405, 661, 693, 696
Poggioli, J., 707
Poiree, J. C., 584, 587(13, 15), 596(13, 15), 598(13, 15)
Poirer, G. G., 620, 621(54)
Pollack, J. M., 281
Pollard, J. W., 655
Pollock, W. K., 707, 708, 714(3)
Poloczek, P., 620
Polson, J., 428

Pool, C. W., 304
Poole, B., 261
Popowicz, J., 436
Popowicz, P., 436
Popp, A. J., 802, 809(55)
Portoukalian, J., 687
Pothmann, M., 98
Potter, B. V., 705
Potter, E. L., 107
Potter, V. R., 432
Poujeol, P., 76, 77(12), 78(12), 327, 328, 384, 385, 386(17), 388(6, 15), 396(17), 403(6), 406(6), 412, 416(22), 417(22), 416(26), 417(26), 422(26), 424(26), 425
Pouyssegur, J., 739, 743(7), 748(7)
Powers, S. P., 636
Pozzan, T., 698, 701
Pradayrol, L., 635
Pradelles, P., 136, 306, 307
Prado, A., 260
Prager, M. D., 426
Praissman, M., 625
Prastein, M. L., 790
Preiser, H., 583, 584, 586, 587(4, 5, 10), 596(10), 598(10, 33)
Presser, F., 745
Preston, A. S., 428, 435, 498
Pretlow, T. G., 143
Pretlow, T. P., 143
Pricam, C., 268
Price, R. G., 130
Printz, R., 663
Prusiner, S. B., 451, 455(13)
Pryor, J. L., 265, 270(11), 271(11)
Przbyla, A. E., 441
Pucacco, L., 521
Pumplin, D. W., 553
Purkerson, M. L., 113
Purnode, A., 156
Pürschel, S., 326, 327, 335, 338, 339(60, 63)
Puschett, J. B., 781
Putney, J. W., 702, 703
Putney, J. W., Jr., 682, 683(27)

Q

Qualizza, P. B., 189, 248
Quamme, G. A., 73, 84(6), 91, 94, 495
Quinton, P. M., 173, 289
Quirion, R., 800

R

Rabinovitch, H., 135
Rabito, C. A., 498, 501, 502(34)
Rabito, C., 405, 406(75)
Rabon, E., 474, 475(15), 477(15), 721, 722, 729(10), 730(104), 731(10), 732(10), 737(10)
Racker, E., 517, 843
Radnik, R. A., 147
Radtke, H. W., 103
Rae-Venter, B., 627
Rafaeli, A., 174
Rafestin-Oblin, M. E., 529
Rafestin-Oblin, M.-E., 522, 530(13)
Raftery, M. A., 634, 638(102)
Raison, J. K., 592
Rajagopalan, R., 753, 754(72)
Rajerison, R., 303
Ralph, R. K., 746
Ramachandran, J., 639
Ramchandran, J., 637, 638
Ramos, S., 470, 476
Ramsay, J. A., 95, 170, 357
Rana, R. S., 676, 677(2), 682(2), 684(2)
Rance, M. J., 622
Randle, H. W., 202, 205(63)
Rands, E., 637
Rasmussen, H., 326, 664, 665, 666, 669, 671, 673, 674
Rastegar, A., 266, 271(26)
Raszeja-Specht, A., 585, 598(29), 600(29), 601(29), 602(29), 603(29), 604(29)
Ratan, R. R., 404
Rater, M., 414, 420(44), 425(44)
Rattan, S., 611, 625(18), 627(18)
Rau, N. V., 57
Raufman, J.-P., 615, 616(46)
Ravid, R., 304
Ray, T. K., 734
Razgaitis, K. A., 683
Reale, E., 268
Reale, V., 639
Reaven, G., 65
Reber, B. F. X., 762, 769(11)
Rector, F. C., 265
Rector, F. C., Jr., 86, 88, 92(32), 113
Reed, R. R., 569
Reese, T. S., 267
Reetz, K. L., 415, 420(55)
Reggio, H., 815

Rehm, A. G., 682
Reid, I. A., 156
Reif, M. C., 250
Reilly, R. F., 500, 743
Reimann, E. M., 653
Reinach, P. S., 18, 805
Reineck, H. J., 127
Reisine, T., 625
Reitman, M., 809
Remington, J. W., 115
Renault, C., 319, 324(80)
Renda, D., 707
Renda, F., 707
Renfro, L. J., 190
Renner, E. L., 744
Rennke, H. G., 68, 80, 129, 411, 418(12), 420(12)
Reshkin, S., 496, 498, 502(24)
Reuss, L., 7, 9(9), 21(9)
Rey, M.-A., 850
Reyes, J., 750
Reynolds, E. E., 695
Reynolds, J. A., 843
Reynolds, R. A., 456
Rhee, L. C., 639
Rhee, S., 706
Rhim, J. S., 414, 422(45)
Rhoads, D. B., 517
Ribadeneira, M., 236
Ribet, A., 612, 620(36, 39), 635
Ricapito, S., 462, 463(53)
Ricciardi, R. T., 639
Richards, A. N., 204
Richards, J. S., 652
Richardson, J. C. W., 434
Richardson, W., 128, 129(54)
Richet, G., 91, 93(48)
Rick, R., 14, 96
Rickwood, D., 391
Riedel, S., 583, 584(6), 587(6), 596(6), 598(6)
Riehl, R. E., 637
Rigaud, M., 135
Riggs, A. D., 495
Rightsel, W. A., 158, 159(19), 160(19), 161(19), 162(19), 163(19), 164(19), 165(19)
Rindler, M. J., 432, 438, 791
Rink, T. J., 698
Rithie, J. M., 795, 796(20, 21)
Ritman, E. I., 70

Rittenshouse, S. E., 677
Ritter, M., 802
Rivier, J. E., 622
Rizzoli, R., 496
Robberecht, P., 628
Robbins, E. S., 434
Robenek, H., 556
Roberts, B. E., 639
Robertson, C. R., 51, 52(26, 27), 76, 77(11), 108, 123(7), 125(7)
Robinson, D., 130
Robinson, G. A., 662
Robinson, J. M., 151
Robinson, J., 338
Robinson, P., 625
Robinson, R. R., 266
Roczniak, S., 135, 136(18)
Rodbard, D., 529, 625, 628(76), 631(76)
Rodbell, M., 662
Rodman, J. S., 450
Rodrigues de Miranda, J., 747
Rodriguez, H., 639
Rodriguez-Boulan, E., 813, 814(4), 822, 825
Roe, B. A., 638
Roghstein, A., 404
Rogulja, I., 704
Røigaard-Petersen, H., 453, 461(23, 24, 25, 26)
Roinel, N., 91, 94, 173, 266, 273(22), 295
Rollins, T. E., 414, 418(48), 423(48), 425(48)
Roman, I., 684
Roman, L. M., 829
Roman, R. J., 357
Romero, C., 70
Romero, T., 135
Romrell, L. J., 733
Ronco, P., 328, 385, 386(17), 388(15), 396(17)
Rose, C. P., 36, 67(10)
Rose, J. K., 842, 846, 847(3), 851
Rose, J. R., 802, 809(55)
Roseau, S., 306, 307(50, 56), 322(56), 323(50, 56), 335, 339
Rosen, F., 536
Rosen, O. M., 638
Rosen, S., 31, 385, 387(13, 14)
Rosenfeld, G. S., 733
Rosenfeld, M. G., 638
Rosenheck, K., 748
Rosenthal, H. E., 529

Rosenweig, S. A., 635
Rosenzweig, S. A., 635
Ross, B. D., 32, 306
Ross, E. M., 662
Ross, R. D., 31
Ross, R., 148
Rosselin, G., 612, 634(28)
Rossier, B. C., 540, 754
Rossier, B., 754
Roth, J., 279, 281, 396, 400(38), 837
Roth, K. S., 456
Roth, M. G., 851
Roth, M., 157
Rothman, J. E., 559, 813, 842
Rotman, M., 750, 754
Rottenberg, H., 474, 475(12)
Rottman, A. J., 612, 620(30), 634(30)
Rotunno, C. A., 434
Rouleau, M. F., 303, 304(13)
Roy, C., 147, 303
Rubin, B., 625
Rubin, C. S., 652
Ruddle, F. H., 438, 638
Rugg, E. L., 429, 435, 787
Ruggles, B. T., 305, 307(42)
Rumrich, G., 84, 86, 87, 92(30), 103, 104, 105, 106(18), 356
Rüppel, D. A., 592
Rupprecht, A., 331
Russ, G., 854
Russel, G., 280
Russel, J. M., 800
Russell, D. W., 638
Russell, J. M., 791
Russo-Marie, F., 135, 163
Rutter, W. J., 441
Ruusala, A., 846
Ryan, G. B., 31
Ryberg, B., 721, 723, 724(4), 729(4), 730(4), 731(4)
Ryu, S., 706

S

Sabatini, D. D., 434
Sablotni, J., 752
Sabolić, I., 453, 506, 508(4, 5), 511(4), 512(4), 513(5), 514(5), 518

Saccomani, G., 721, 734, 751, 752
Sachs, G., 60, 454, 467, 474, 475, 477(15), 611, 640, 647(6), 649(6), 660(6), 721, 722, 726, 727(18), 728(18), 729(10, 18), 730(10), 731, 732(10), 733, 734, 736, 737(10)
Sackin, H., 8
Sackmann, E., 592
Sacktor, B., 301, 412, 425(33, 34), 496
Safar, M. E., 112
Sagalowsky, A., 303
Sage, S. O., 695
Saheki, T., 304
Saier, M. H., 427, 429, 432
Saier, M. H., Jr., 438, 791
Saito, A., 627
Saito, T., 415, 416(52), 422(59), 423(52, 53, 56, 59)
Sakai, F., 331, 335
Sakai, T., 68
Sakamoto, C., 612, 620(37, 38), 635
Sakauye, C., 537
Sakhrani, L. M., 411, 418(5, 6), 421(6), 496
Sakmann, B., 299, 445, 793, 795, 796(14), 799(14)
Sakurai, Y., 802, 809(55)
Sala, G. B., 663
Salehmoghaddam, S., 411, 418(7), 420(7)
Salomonsson, M., 297
Sambroock, J., 443
Sambrook, J., 851, 852
Samuel, D., 479
Samuelson, B., 670
Samuelsson, B., 676, 684
Sanaka, T., 304
Sanders, M. J., 485
Sando, J. J., 673
Sands, J. M., 231
Sandstrom, D. J., 68
Sandvig, K., 404
Sangren, W. C., 71
Sankaran, H., 625, 627
Sargiacomo, M., 822
Sariban-Sohraby, S., 742, 750, 751, 752, 753(63)
Sarkar, D., 652
Sasai, Y., 639
Sasaki, S., 299
Satlin, L. M., 255, 259, 261(2), 263, 264(2), 273, 279(70)

Sato, G., 149, 427
Sato, K., 103, 104, 289, 495
Sato, M., 415, 416(51), 423(51)
Satriano, J. A., 135, 136(18)
Sautebin, L., 685
Sauve, R., 300
Savada, H., 280
Savin, V. J., 128, 129(54), 132
Sawada, Y., 67
Sawin, L. L., 343
Sawyer, D. B., 178, 184, 186, 188(32), 192(32), 193(32)
Scalera, V., 455, 456(38), 460(38), 495
Scanlon, M., 700, 701(18)
Scatchard, G., 529
Scemama, J.-L., 635
Schachmann, H. K., 845
Schachmann, R., 721
Schaefer, A., 455, 495
Schaeffer, W. I., 410, 425(3)
Schafer, J. A., 87, 90, 248, 249, 250, 251(10), 252, 275, 289, 355, 356, 366, 369
Scharschmidt, B. F., 744
Scharschmidt, L. A., 135, 137(33, 34, 35)
Schatzmann, H. J., 87
Schechter, A. N., 345
Schenk, E. A., 396, 400(41)
Schettino, T., 12
Schettler, G., 682
Schild, H. O., 616, 617(49)
Schild, L., 794
Schiller, A., 268
Schimerlik, M., 637, 638(118)
Schiodt, M., 786, 789, 792(20)
Schlatter, E., 10, 183, 293, 295, 297, 298, 299, 300, 301, 302(19), 572, 784, 785, 794, 795, 797(32, 37, 38), 798(37, 38), 801, 802, 803(50), 805(37, 50), 806(37, 50), 808
Schlegel, W., 567, 683
Schlessinger, J., 638
Schlondorff, D., 116, 130, 134, 135, 136(18), 137(2), 304, 305(20)
Schlüter, K., 272, 396, 400(42)
Schmer, G., 148
Schmid, H., 335, 340
Schmidt, H., 335
Schmidt, M., 325
Schmidt, S. L., 559
Schmidt, U. M., 770

Schmidt, U., 328, 335, 336(26), 380, 400(1), 464
Schmidt-Nielsen, B., 182
Schmitz, J. M., 303
Schmolke, M., 413, 414, 416(38), 418(38), 419, 422(38), 424(38)
Schneider, B. S., 278
Schneider, G. T., 795, 797(42)
Schneider, W. J., 638
Schnermann, J., 93, 120, 542
Schofield, P. R., 639
Scholer, D. W., 327
Schölermann, B., 462, 463(52, 54)
Scholz, M., 335
Schoumacher, R. A., 572
Schreiner, G. F., 147, 148
Schreiner, G., 133
Schreyver, 135
Schrier, R. W., 306, 307(55), 323(55)
Schudt, C., 738
Schuldiner, S., 474, 475(12), 605
Schultz, S. G., 9, 12, 368, 370(20)
Schulz, G., 733
Schulz, J., 437, 585, 604(27)
Schulz, W., 338
Schurek, H. J., 31
Schurek, H.-J., 385, 387(12)
Schuster, V. L., 396, 808
Schütz, H., 462, 463(52, 54)
Schwab, A. J., 67
Schwarc, K., 584, 587(12), 596(12), 598(12)
Schwartz, G. J., 255, 256(1), 257, 258(3), 259, 260(3), 261(2), 263, 264(2), 271, 415, 420(65), 506, 520
Schwartz, G. S., 273, 279(70)
Schwartz, I. L., 466, 468, 672
Schwartz, J. H., 415, 420(64, 65)
Schwartz, J., 325, 745
Schwartz, M. A., 790
Schwartz, M. J., 539, 547(69), 548(69), 549(69), 550(69)
Schwartz, R. H., 396, 400(41)
Schwarz, W., 437, 585, 801
Schwarze, W., 795, 797(25)
Schwiebert, E. M., 414, 415, 425(43, 66)
Scoble, J., 411, 418(17), 421(17), 422(17)
Scott, C. K., 722
Scott, D. M., 380, 408, 412, 413(30), 416(30), 417(30), 422(30)
Scott, D., 726, 727(18), 728(18), 729(18)
Scratchard, G., 624
Sealey, J. E., 158
Sedor, T. R., 135
Seeburg, P. H., 638, 639
Seed, B., 639
Segal, M., 795, 796(13)
Segal, S., 456
Segel, I. H., 486
Seidman, L., 450
Seifter, J. L., 387, 390(25, 26), 391(25, 26), 392(25, 26), 394(25, 26), 405, 407(25, 26), 408(25, 27, 28), 411, 416, 418(12, 70), 420(12, 70), 421(70), 562, 743, 745(21)
Sekar, M. C., 678
Seldin, D. W., 86, 88, 92(32), 113
Seldin, D., 520
Semenza, G., 762, 769(11), 770
Semmekrot, B., 306, 307(56), 322(56), 323(56)
Semple, P. F., 157
Senn-Bilfinger, J., 722
Sens, D. A., 417, 424(73)
Sepulveda, F. V., 431
Serafini, T., 813
Serhan, C. N., 684
Sesselmann, E., 14
Seto, J., 496
Severi, C., 611, 612
Severs, N. J., 556
Sewing, K. F., 733
Sewing, K.-F., 722
Shackleton, C. H. L., 523, 524, 525(36)
Shaffer, C., 744, 745(28)
Shaffer, L., 438
Shah, S. V., 130, 134(10), 135
Shahak, Y., 478
Shaikewitz, T., 743, 744(20), 745(21), 747(20), 748(20)
Shaltz, L. J., 672
Shannon, J. A., 183
Shanser, A., 479
Shapiro, J. P., 170
Sharma, R. V., 746
Sharp, C. M., 619, 620(53), 628(53)
Sharp, G. W., 539
Shaver, J. L. F., 464
Shaw, D., 734
Shaw, M. J., 625
Shayman, J. A., 305, 415, 422(49, 50)

Shculz, W., 335
Sheikh, M. I., 452, 453, 461(20, 21, 22, 23, 24, 25, 26)
Shelanski, M. L., 404
Shepherd, J. C., 813
Sheppard, C. W., 71
Sheridan, E., 103
Sherman, W. R., 684
Sherwood, I. M., 157
Shi, L. B., 564
Shi, L.-B., 556
Shiegemoto, R., 639
Shields, G. I., 555
Shigemoto, R., 639
Shiigai, T., 299, 331, 335
Shimomura, A., 61
Shimonaka, M., 304
Shiraga, H., 120, 121(38)
Shiu, R. P., 633
Shizume, K., 304
Shlatz, L. J., 466
Shlosser, J., 411, 421(8)
Shoemaker, D. G., 750
Shoemaker, R., 795, 797(27)
Shuldiner, S., 476
Shulman, R. G., 405
Shum, W. K., 742
Siaume-Perez, S., 305, 307
Siebe, H., 335, 338, 524, 528(40)
Siebens, A. W., 415, 417(69), 420(69)
Siegel, F. L., 683
Siemems, I. R., 705
Siess, W., 683
Sigal, I. S., 637
Sigworth, F. J., 445
Silbernagl, S., 448, 583
Siliiano, J. D., 267
Silva, P., 31, 385, 387, 390(25, 26), 391(25, 26), 392(25, 26), 394(25, 26), 401, 407(25, 26), 408(25, 27, 28), 784
Silver, P. J., 664
Silverberg, M., 850
Silverblatt, F. J., 135
Silverman, M., 35, 36, 37, 41(15), 42, 43(22), 44(21), 45(16), 46(16), 48(16), 49(16, 21), 51(16, 21), 52(11, 21), 53(21), 56, 57(32), 58(14), 59(14), 61, 63, 65, 67(14, 15, 49), 71(16), 584, 590, 591(17), 596(17), 598(17)
Simchowitz, L., 743, 750

Simmonds, W. F., 632
Simmons, N. I., 305, 307(43), 429, 434, 435, 501, 787
Simon, A. C., 112
Simon, E. J., 622
Simon, F. R., 416, 418(71), 422(71)
Simon, G. T., 278
Simon, M., 100
Simon, S. A., 739
Simone, J. N., 205
Simons, K., 431, 435, 450, 460(4, 5), 813, 814, 815, 816, 819, 821, 822, 823, 824, 842, 843, 845(9)
Sims, P. J., 478
Singer, J. J., 683
Singh, P., 627
Sipes, I. G., 790
Sistare, F. D., 701
Sjöstrand, S.-E., 723
Skelly, H., 748, 749(49), 750(49)
Skolnick, P, 800
Skorecki, K. L., 567, 568, 569(59, 61), 570(59), 571(59)
Skottner, A., 304
Skoubo-Kristensen, 733
Slater, E. E., 637
Slaughter, O., 569
Slayman, C. W., 405, 406(74), 500, 743
Slegers, J. F., 32
Slotki, I. N., 415, 420(64)
Smaill, B. H., 32
Smart, J., 746
Smith, A. I., 523, 525(27)
Smith, C. F. C., 622
Smith, D. H., 637, 638(118, 121)
Smith, G. A., 698
Smith, J. C., 435
Smith, J. N., 337
Smith, M. R., 706
Smith, P. M., 707, 714(3), 716
Smith, R., 523, 525(27)
Smith, S. M., 795, 796(15)
Smith, T. F., 426, 495
Smith, W. L., 328, 385, 388(16), 391, 392(35), 411, 412, 414, 416(14, 39), 418(39, 48), 422(14, 39), 423(48), 424(14), 425(48)
Smyth, D. D., 307
Snedecor, G. W., 287
Snyder, F., 684

Snyder, S. H., 622, 625, 627(79), 701
Soderland, C., 148
Soderlund, D. M., 800
Sohraby, S., 412, 416(27), 417(27), 424(27)
Sohtell, M., 295, 299
Soll, A. H., 609, 610(2), 611, 615, 620(2), 621, 627, 629(2), 640, 641, 647(6), 733
Solomon, A. K., 87
Solomon, T. E., 686
Soltoff, S. P., 290, 384, 401, 691, 701(2), 744, 747
Sommons, K., 825
Song, Y., 733
Sonnenberg, H., 83, 89(26), 95, 100
Sonnenberg, W. K., 328, 412, 414, 416(39), 418(39), 422(39)
Sono, H., 304
Soreg, H., 437, 585, 604(26)
Sottocasa, G. L., 508
Souquet, J.-C., 612, 615(24)
Soyle, C., 851
Spangler, R. A., 411, 418(9)
Specht, H.-C., 265, 273(12)
Spector, A. A., 632
Speight, P., 584, 591(17), 596(17), 598(17)
Speilman, W. S., 385, 391(19)
Spicer, S. S., 750
Spielman, W. S., 305, 328, 412, 414, 416(39, 42), 418(39, 42), 422(39, 42, 45)
Spitzer, A., 118, 121(32), 405
Spokes, K., 31, 385, 387(13)
Spring, K. R., 197, 198, 223(54), 275, 294, 415, 417(69), 420(69)
Sraer, J. D., 135, 304
Sraer, J., 135, 304
Stadtman, E. R., 469
Staehelin, L. A., 267
Staley, R. S., 163
Stambo, G. W., 305
Stanbury, R. S., 217
Stanton, B. A., 265, 272, 414, 415, 416(13), 425(43, 60, 66, 67)
Stanton, B., 266, 270(27, 32), 271(27)
Stanton, R. C., 411, 418(12), 420(12)
Stanziale, R., 112, 121(11), 122(11)
Starita-Geribaldi, M., 584, 587(13, 15), 596(13, 15), 598(13, 15)
Stark, H. A., 619, 620(53), 628(53)
Steele, R. E., 428, 435, 498

Steele, R., 412, 416(27), 417(27), 424(27)
Steer, M. L., 622
Steffes, M. W., 68
Steffgen, J., 385, 386(22)
Stegmann, T., 844
Steigerwalt, R. W., 625
Steigner, W., 438, 439(7), 448
Stein, J. H., 127, 130
Steiner, D. F., 66
Steiner, L., 731
Steinhardt, R., 693
Steinhart, H., 63
Steinhausen, M., 113, 121
Stelzer, E. H. K., 814
Stendahl, O., 683
Stephens, L. R., 715
Stephenson, F. A., 639
Stephenson, J. L., 265
Stephenson, P., 153, 162, 163
Stern, P. S., 174
Stern, P., 553, 555
Sterzel, R. B., 147
Stetson, D. L., 266, 271(29), 272, 551, 559(3), 675
Stevenson, B. R., 267
Stewart, G. W., 784, 794
Stewart, H. B., 734
Stewart, J. C., 717, 718(18)
Stewart, J., 532, 533(56), 534(56)
Stewart, P. M., 523, 524, 525(36)
Stieger, B., 451, 459(10), 825, 826(6), 831(6), 832, 833(6), 835(6), 836(6, 8), 838(6), 839(6), 840(6), 841(8)
Still, W. C., 763
Stinebaugh, B. J., 325
Stinski, M. F., 822
Stirling, C., 464
Stirman, J. A., 162
Stoeckel, M. E., 303, 304(10)
Stoeckenius, W., 580
Stoff, J. S., 31
Stokes, J. B., 387, 391(29), 392(29), 394(29), 400(29), 430, 465, 808
Stokes, J. B., III, 387, 403(31), 407(31)
Stokker, G. E., 802, 809(55)
Stokols, M. F., 412, 416(29), 417(29), 424(29)
Stolarsky, L. S., 638
Stoll, A. P., 303

Stone, D., 520
Stoner, L. C., 197
Stoner, L., 784, 791(9)
Storey, J. M., 401
Strader, C., 637
Strader, D. J., 637
Strandhoy, J. W., 304
Strange, G., 495
Strange, K., 198, 223(54), 275, 289, 559
Strasheim, A., 415, 416(62), 423(62)
Strassberg, J., 108
Strauss, W., 557
Streaty, R. A., 632
Striker, G. E., 130, 144, 146(5), 148, 151
Striker, L. J., 130, 144, 146(5), 148
Stroh, A., 584, 587(12), 596(12), 598(12)
Strong, J. A., 526
Strum, J. M., 523
Stukowski, B., 326
Stull, J. T., 664
Stuttis, M. J., 805, 808(64)
Suarez-Kurtz, G., 745
Sudaka, P., 584, 587(13, 15), 596(13, 15), 598(13, 15)
Suenaga, K., 8
Sugimoto, K., 637, 638(117)
Sugino, N., 304
Sugiura, A., 815, 819(9)
Sugiyama, Y., 67
Suh, P., 706
Suki, W. N., 325
Sullivan, M., 638
Sulpice, J.-C., 707
Summer, S. N., 306, 307(55), 323(55)
Summers, R. J., 303, 304(11)
Sumpio, B. E., 342
Sun, F., 135
Sundaresan, P. R., 305
Sundell, E., 722
Sundell, G., 723, 725(12)
Suppattapnoe, S., 701
Suresh, K. R., 411, 418(20), 421(20), 422(20)
Susini, C., 612, 620(36, 39)
Sussman, K. E., 684
Sutanto, W. S., 523
Suter, M., 822
Sutherland, C. A., 678
Sutherland, D. E. R., 68

Sutherland, E. W., 662
Sutherland, E., 416, 418(71), 422(71)
Sutliff, V. E., 611, 615, 616(45), 619, 620(53), 628(53)
Suzuki, K., 18
Suzuki, M., 412, 417(23), 416(24), 421(23, 24), 422(23, 24)
Svennerholm, L., 612, 634(31)
Svoboda, M., 612, 620(39), 635
Swaab, D. F., 304
Swaneck, G. E., 530
Swartz, R. D., 31
Syer, W. J., 686
Szarowski, D., 802, 809(55)
Szecowka, J., 633, 636
Szwarc, K., 586, 598(32), 603(32)

T

Taatjes, D. J., 279, 837
Tack, F. F., 345
Taga, T., 639
Taggart, B., 523
Tago, K., 808
Tahiri-Jouti, N., 612, 620(39)
Tait, J. F., 669
Tait, S. A. S., 669
Takahashi, H., 637, 638(117)
Takahashi, M., 586, 598(33)
Takaichi, K., 307
Takakeshi, H., 637, 638(116)
Takano, M., 455, 457(39)
Takaoka, T., 495
Takehara, Y., 330
Takemura, H., 703
Takenaka, M., 496
Takenawa, T., 683, 706
Takeuchi, J., 299
Takii, Y., 154, 158
Takuwa, Y., 496
Talor, Z., 65, 66(45)
Tam, A. W., 638
Tamaki, H., 639
Tan, Y. P., 795, 797(35), 798(35), 805
Tanaka, K., 639
Tanford, C., 843
Tang, C.-M., 745
Tang, J. M., 795, 797(29)

Tang, M. J., 411, 418(20), 421(20), 422(20)
Tang, W. J., 569
Tange, J. D., 31
Taniguchi, J., 301
Taniguchi, T., 639
Tannen, R. L., 411, 418(20), 421(20), 422(20)
Tanner, G. A., 113
Tartakoff, A. M., 814
Taub, M., 149, 410, 411, 418(4, 5), 422(4), 427, 432, 496
Tauc, M., 328, 385, 388(15), 412, 416(22), 417(22), 416(26), 417(26), 422(26), 424(26), 425
Taugner, R., 268
Taylor, A. G., 738
Taylor, C. W., 705
Taylor, D. G., 130
Taylor, J. E., 618
Taylor, K., 338
Taylor, M. V., 698
Taylor, N. R. W., 526
Taylor, W. R., 40, 58(19)
Tchao, R., 501, 502(34)
Tein, X.-Y., 611, 618(9), 629(9)
Teitelbaum, I., 415, 416(62), 423(62)
Temmar, M. A., 112
Ten Holt, S. P., 526
Teplitz, R. L., 495
Terada, Y., 231
Terreros, D. A., 132
Terris, S., 66
Tessitore, N., 411, 418(6), 421(6)
Teulon, J., 8
Thampy, K. G., 800
Thastrup, O., 703
Thaw, C., 684
Thees, M., 60
Thibault, G., 303, 304(8)
Thomas, A. P., 695
Thomas, J. A., 278, 405
Thomas, J. K., 405
Thomas, L. P., 611, 627
Thomas, L., 621, 627(60)
Thomas, M. J., 685
Thomas, M., 683
Thomas, R. C., 299, 794, 806(3)
Thompson, I., 405, 406(75)
Thompson, J. C., 627
Thompson, S. M., 6
Thompson, T. E., 580
Thompson, W. J., 733
Thompson, W., 677
Thomson, N. H., 142, 143(4)
Thorne, P. K., 496
Thurau, K., 93, 96
Tillack, T. W., 556
Tisher, C. C., 207, 265, 266, 269, 271(7, 24), 274, 275(47)
Tisher, C. G., 343
Tivey, D. R., 787
Tobin, A. J., 639
Toda, G., 826, 829(12), 832(12)
Tokuyasu, K. T., 279
Tolbert, N. E., 829
Tomita, K., 305, 307(46), 323(46)
Tompsett, S. L., 526
Tong, Y. A., 304
Tonnessen, T. I., 404
Torikai, S., 303, 304(9), 305(9), 307, 335
Torres, V. E., 130, 134(10), 135
Touster, O., 829
Townsend, C. M., 627
Toyoshima, Y., 580
Trainor, C., 36, 37, 52(11), 58(14), 59(14), 67(14)
Trautmann, A., 795, 797(35), 798(35), 805
Triche, T., 412, 415, 416(28, 68), 417(28), 420(68), 424(28)
Trinh-Trang-Tan, M. M., 265, 286(13)
Trizna, W., 411, 418(5)
Trong-Trang, C., 429
Trouet, A., 342
Troutman, S. L., 248, 250, 251(10), 252
Troxler, T., 622
Troy, J. L., 51, 52(27), 75, 76, 77(10, 11), 108, 113, 123(7), 125
Trumble, M. J., 739, 741(8), 742(8), 743(8), 744(8)
Tsiakiras, D., 524, 528(40)
Tsien, R. Y., 405, 470, 472, 473(4), 692, 693, 696, 698
Tsien, R., 661
Tsubokawa, M., 638
Tsuchida, K., 639
Tsukada, N., 622
Tsunoo, H., 633
Tsutsumi, M., 411, 418(17), 421(17), 422(17)
Tucker, D. R., 189

Tune, B. M., 209, 210(77), 219
Tune, M., 61
Turco, S., 771, 772(24)
Turk, E., 603, 604(51), 605
Turk, J., 684
Turnell, R. W., 536
Turner, J. T., 786, 788(22), 792(22)
Turner, R. J., 35, 61, 452, 459(19), 461(18, 19), 479, 480(1), 482, 483, 485, 486, 487(1), 491, 492(19), 493(1), 494(8), 503, 586, 598(34), 590,

U

Uchida, S., 331, 335, 415, 416(68), 420(68)
Ueda, M., 815, 819(9)
Ueki, I. F., 51, 52(27)
Ullrich, A., 638
Ullrich, K. J., 81, 84, 86, 87, 92(30), 95, 98, 100, 103, 104, 105, 106(18), 291, 293(15), 583
Ulmann, A., 524, 525(35)
Ulreich, J., 131
Ulrich, K., 356
Umeda, Y., 612
Umemura, S., 307
Unanue, E. R., 147, 148
Underwood, A. H., 736
Unkeless, J. C., 832, 839(18)
Uno, I., 651, 652(13)
Urushidani, T., 654
Ussing, H. H., 4, 175
Ussing, H.-H., 295
Uzan, A., 319, 324(80)

V

Vadrot, S., 306, 307(50), 323(50)
Vail, W. J., 553
Valentich, J. D., 412, 416(29), 417(29), 424(29)
Valentino, R., 524, 525(36)
Vallotton, M. B., 632
Valtin, H., 107, 118(3), 551, 553, 555
van Adelsberg, J., 262, 265(13), 414, 420(44), 425(44)
van Breeman, C., 695
Van Deenen, L. L. M., 556
Van Den Bosch, H., 679
van der Sluijs, P., 824
Van der Waarden, A. W. M., 556
Van Driessche, W., 15
van Meer, G., 450, 460(4), 814
van Noorden, S., 281
van Nooy, I., 747
van Os, C. H., 787, 792(26)
Vanderhoek, J. Y., 682
Vandewalle, A., 306, 327, 328, 330, 332, 335, 338, 384, 385, 386(17), 388(6, 15), 396(17), 403(6), 406(6), 520, 522, 523(21)
Vane, J. R., 805, 808(64)
Vanhoutte, P. M., 68
Vannier, C., 454
Vasan, R., 503
Vassalli, D. B., 427
Vassalli, J.-D., 748
Vassalli, P., 556
Vasseliv, P. M., 745, 746(34)
Vassent, G., 306, 307(56), 322(56), 323(56)
Vaucher, N., 319, 324(80)
Vauclin, N., 612, 634(28)
Vaughn, D. A., 505, 538
Vaysse, N., 612, 620(36, 39)
Vaystub, S., 384, 388(7), 403(7), 406(7)
Velazquez, H., 520, 786
Velly, J., 745
Venkatachalam, M. A., 80, 129, 147
Venkatachalam, M., 127
Venter, J. C., 770
Verhoeven, A., 699
Verkleij, A. J., 556
Verkman, A. S., 404, 468, 473, 551, 556, 557(25), 559, 561, 562, 563(25, 44), 564, 565, 566, 568, 569(59, 61), 570(59), 571(59)
Verlander, J. W., 265, 266
Verma, I. M., 638
Vernier, R. L., 109, 110
Verniory, A., 129, 250
Verroust, P., 328, 385, 386(17), 388(15), 396(17)
Verry, F., 540
Ververgaert, P. H. J., 556
Viets, J. W., 125
Vigne, P., 325, 750, 751
Vigurie, N., 612, 620(39)

Villanueva, M. L., 611, 612, 613(19), 614(19), 620(19, 42), 627(19), 628(19)
Villiger, E., 280
Villiger, W., 278
Vinay, P., 305, 327, 386, 405, 411
Viogneault, N., 95
Vlasuk, G. P., 636
Vogel, A. I., 759
Vogel, U., 448
Völkl, H., 242, 275
von Schulthess, G. K., 70
Von Voightlander, P. F., 622
Von-Schrenck, T., 611
Vurek, G. G., 95, 96, 118, 357

W

Wade, J. B., 260, 265, 266, 270(11, 27), 271(11, 27, 29), 272, 396, 551, 553, 555, 559(3, 24)
Wade, J., 266, 270(28), 272(28), 273(28), 675
Wagar, M. A., 496
Wagemann, P., 803
Waggoner, A. S., 259, 478
Waggoner, A., 476, 478
Wagner, H., 303
Wagner, K. R., 854
Wagner, K., 843, 846
Wahawisan, C., 611, 618(10)
Wahawisan, R., 611, 618(9), 629(9)
Waite, B. M., 685
Waite, M., 679
Wald, H., 456
Waldman, J. B., 802, 809(55)
Waldner, H., 553
Walker, A. M., 73, 75(9), 92(9), 94(9)
Wall, D. A., 340, 341(3), 826, 828(9), 839, 841
Wall, S. M., 415, 420(57)
Wallace, A. M., 524, 525(36)
Wallace, L. J., 611, 618(9, 10), 629(9)
Wallmark, B., 721, 722, 724, 726, 727(18), 728(18), 729(4, 10, 18), 730, 731, 732, 734, 736, 737, 738
Walsh, C. E., 685
Walsh, J. V., Jr., 683
Walsh, M. L., 259

Walter, U., 651, 652
Wandinger-Ness, A., 813, 814, 815(5), 816, 819, 821
Wang, C. H., 67, 478
Wang, F., 183, 289, 299
Wang, L. H., 618
Wang, M. S., 135, 307
Wang, W., 791
Wangemann, P., 293, 295, 785, 801, 805(50), 806, 808(50)
Wank, S. A., 618, 619(52), 621, 628(57), 629(57), 630(52), 631(52)
Warden, D. H., 808
Warncke, J., 15, 18(40)
Warner, R. R., 242, 357
Warnock, D. G., 506, 519(3), 520(3), 522, 530(12), 754
Warnock, D., 753
Warshawsky, H., 303, 304(13)
Waterfield, M. D., 638
Watkins, M. L., 252, 355
Watlington, C. O., 435, 528
Waugh, M. H., 625
Wearn, J. T., 73
Weaver, G. W., 426, 495
Webb, R. A., 687, 688(66)
Weber, M. R., 411, 417(18), 422(18), 424(18)
Weber, W. M., 437
Weber, W., 638
Weber, W.-M., 585
Webster, R. G., 846, 851(21)
Weibel, E. R., 205, 282, 284(96, 97)
Weidtke, C., 295, 784
Weikel, C. S., 673
Weinberger, C., 521
Weinstein, J. N., 633
Weinstein, J., 837
Weiss, D. D., 343
Weissmann, B. A., 800
Weller, P. F., 209, 363, 367
Weller, P. W., 192
Welling, D. J., 198, 204, 205, 232, 234(1), 245, 269
Welling, L. W., 198, 204, 205, 232, 245, 248, 269
Wells, J., 411, 418(7), 420(7)
Welsh, M. J., 430, 795, 797(39), 805(39)
Wendel, A., 335

Werness, J. L., 305, 307(42)
Weskamp, P., 18, 19(50)
West, M., 155
Westbook, S., 305
Westbrook, S., 411, 418(17), 421(17), 422(17)
Wetzel, G. D., 14
Weyer, P., 557, 558(36), 559
Wheeler, N. C., 115
White, M. M., 795, 796(19), 801
White, R. H. R., 113
Whiteside, C. I., 36, 40, 42, 43(22), 44(21), 49(12, 21), 51(21), 52(11, 12, 21), 53(12, 21), 54(12), 55(12), 56, 57(32), 63, 65, 67(49)
Whiting, J. A., 684
Whittembury, G., 87
Whittier, F. C., 189
Whittle, N., 638
Widdicombe, J. H., 404, 790, 791(38)
Widnell, C. C., 832, 839(18)
Wiederholt, M., 94
Wieland, O., 326
Wiener, H., 787, 792(26)
Wier, W. G., 700
Wiesner, W., 326
Wiessner, J. H., 415, 420(55)
Wigglesworth, V. B., 169, 174(2)
Wight, D., 501
Wijnaendts-van-Resandt, R. W., 814
Willem, R., 405, 406(71)
Williams, D. A., 700, 701(18)
Williams, J. A., 625, 627, 633, 635, 636, 639
Williams, J. C., 178
Williams, J. C., Jr., 173, 174, 175, 248, 249
Williams, R. T., 337
Williams, T. H., 790
Williamson, J. R., 691, 693(3), 695, 699, 704, 705
Williamson, M. T., 699
Williamson, R. E., 704
Willingham, M. C., 559
Willis, W. T., 113
Wills, N. K., 10, 14, 18(38)
Willumsen, N. J., 15
Wilson, C. B., 113
Wilson, D. B., 677
Wilson, D. R., 113
Wilson, J. M. G., 526

Wilson, P. D., 412, 414, 416(25, 36), 417(25), 422(25), 460
Wilson, R. K., 638
Wilson, V. S., 701
Wilton, P., 107
Winand, J., 620
Windhager, E. E., 4, 8, 72, 84(1), 87, 88, 94, 100, 175, 349, 453
Wingo, C. S., 270
Winslow, J. A., 115
Winslow, J. W., 637, 638(118, 121)
Wirthensohn, G., 305, 326, 327, 330, 331, 332, 335, 338, 339
Wirthenson, G., 497
Wissmüller, T. F., 584, 587(8), 589(8), 590, 591(8, 39), 592(8), 596(8), 596(8), 598(8)
Wittner, M., 293, 295, 801, 802, 803, 805(50), 806, 808(50)
Witz, H., 81
Wlstrand, P. J., 791
Wohlwend, A., 427
Wojcik, S. M., 746
Wolf, B. A., 684
Wolff, D. W., 304
Wolff, N. A., 451, 454(29)
Wolosin, J. M., 734
Woltersdorf, O. W., Jr., 739, 750, 754(55), 802, 809(55)
Wong, K. Y., 635
Wong, K., 564
Wong, K.-Y., 625
Wong, P. Y., 335
Woodland, H. R., 437
Woodward, C. H., 303
Woodward, C. J., 622
Work, J., 90, 520
Worley, P. F., 701
Worrell, R. T., 809
Wreggett, K. A., 707, 708, 709(12), 712, 714(3)
Wright, E. M., 584, 603, 604(10, 21, 22, 51), 605
Wright, F. S., 94, 95(56), 120, 124, 357, 520
Wright, T. C., 151
Wu, J.-S. R., 584, 596(16), 598(16)
Wunderlich, P., 120
Wünsch, E., 612, 634(28)
Wunsch, E., 612, 620(36)

Wurtz, M., 280
Wyse, J. W., 778

X

Xu, J.-C., 622, 629
Xu, Y.-L., 638
Xu, Z.-J., 453

Y

Yamada, H., 330
Yamada, K., 678
Yamada, T., 611, 612, 627
Yamagata, R., 337
Yamaguchi, Y., 385, 386(18)
Yamaizumi, M., 438
Yamakawa, A., 706
Yamamoto, T., 638
Yamasaki, K., 639
Yamashiro, P. D., 472, 473(9)
Yanagawa, N., 412, 417(23), 421(23), 422(23)
Yano, K., 678
Yarden, Y., 638
Yared, A., 115
Yawata, H., 639
Ye, R., 564
Yellen, A., 850
Yeo, P., 134
Yewdell, J. W., 850, 854
Yiacas, E., 625
Yoda, H., 735, 737(41)
Yokota, Y., 639
Yokoyama, R., 8
Yonath, J., 13
Yoshida, Y., 120, 121
Yoshitomi, K., 299, 301
Young, A. L., 637, 638(119, 120)
Young, G. P. H., 795, 797(28)
Young, J. D.-E., 795, 797(28)
Yu, D.-H., 611, 613(19), 614(19), 616, 617(48), 620(19), 627(19), 628(19)
Yu, J., 553
Yuan, L. C., 841, 854
Yulo, T., 751, 752(61), 754(63)
Yunis, M., 135

Z

Zahidi, A., 635
Zallakian, M., 490
Zalups, R. K., 266
Zavilowitz, B. J., 415, 420(65)
Zavoico, G. B., 682
Zawalich, K., 665, 666(18), 673
Zdon, M. J., 654
Zeidel, M. L., 387, 390(25, 26), 391(25, 26), 392(25, 26), 394(25, 26), 401, 407(25, 26), 408(25, 27, 28)
Zeiske, W., 15
Zemplen, G., 761
Zerhan, I., 295
Zeuthen, T., 788, 792(29)
Zhang, L., 618, 619(52), 630(52), 631(52)
Zhou, X.-M., 437, 585, 604(27)
Zhou, Z.-C., 611, 612, 613(19), 614(19), 620(19, 42), 627(19), 628(19)
Zhuang, Y., 743, 744(20), 747(20), 748(20)
Ziegler, C., 802, 809(55)
Ziegler, F. W., 801
Ziegler, K., 605
Ziegler, W. H., 46, 71(25)
Zierler, K. L., 36
Zietler, P., 683
Zilberstein, D., 605
Zimmer, K. P., 854
Zorec, R., 795, 796(15)
Zoubek, E., 590
Zubay, G., 636
Zucker, K. A., 654
Zuckermann, M. J., 592
Zusman, R. M., 163
Zwiebel, R., 337
Zwingelstein, G., 687

Subject Index

A

4-Acetamido-4'-isothiocyanostilbene-2,2'-disulfonate, as chloride channel blocker in muscle, 800–801
Acetylcholine receptor, 799
Acetylcholinesterase, inhibition by amiloride and/or its analogs, 748
N-Acetyl-β-D-glucosaminidase, measured in single isolated nephron segments, 334
Acid–base balance, normal, maintenance of, 92
Acid–base transporters, localization, 263
Acid phosphatase, measured in single isolated nephron segments, 334
Acridine orange
  cell studies, to determine presence and location of acid spaces, 475
  in measurement of proton pump activity in endocytotic vesicles, 508–509, 512–514
  in optical measurements of pH gradients in vesicles, 474–475
Actinomycin D, effects on binding of [^{3}H]aldosterone-labeled receptor complexes to chromatin, 532
Active transport, 1–2, 268–269, 290
  in glomerular filtration, 35
  primary, 2
  rheogenic, 20
  secondary, 2
Active transport potential, 290
Active transport voltage, 290, 293
Adenine nucleotides, measured in single isolated nephron segments, 333
Adenosine triphosphate, measured in isolated nephron segments, 338
  with enzymatic cycling, 336
Adenylate cyclase
  activation of, 615
    sustained phase of, 663
  activity
    dual control of, 662
    in gastric parietal cells, measurement of, 650
  inactivation, in lyophilized or freeze-dried tissue, 331
  inhibition by amiloride and/or its analogs, 746
  measured in single isolated nephron segments, 334
    radiochemical methods, 338
  microassay of, in permeabilized tubules, 307
  vasopressin-sensitive, target analysis studies of, 566–571
Adenylate kinase, measured in single isolated nephron segments, 333
α-Adrenergic receptor, inhibition by amiloride and/or its analogs, 746–747
β-Adrenergic receptor
  identification of
    radiolabeled ligands used in binding studies for, 621
    tissues tested, 610
  inhibition by amiloride and/or its analogs, 746–747
  mammalian, amino acid sequence of, 637
Aedes aegypti
  diuretic hormone. *See also* Mosquito natriuretic factor
    bioassay for, 178–181
    sources, 170
Affinity cross-linking, 634
Affinity labeling, 4
Aggrephores, 272
Agonist
  pathways mediating action of, 613–615
  potentiation of response to second agonist, 615–616
  relative potencies of, in receptor identification, 613
  response to, 611–613
Agonist–receptor complex, 611
Alanine aminotransferase, measured in single isolated nephron segments, 333

Albumin. *See also* Bovine serum albumin
  endocytosis
    by proximal convoluted tubules of rabbit, uptake curve of, 350, 352
    studies of, inulin leak in, 349
  fate of, in isolated perfused proximal convoluted tubules, 347–349
  filtration of, 340–341
  iodinated, as volume marker, 242
  radioiodination, 345–346
  reductive methylation procedure for, 345–346
  as tracer, in renal multiple indicator dilution, 40
  triatiation, 345–346
  tubular absorption and metabolism of, 340–341
    in isolated perfused proximal convoluted tubules of rabbit, 350–351
Aldehydes, in fixative solution, 276
Aldosterone
  clearance studies, 536–539
  in cytoplasmic mineralocorticoid receptor assay, 529–531
  metabolism, in kidney, 526–528
  renal target sites, 520
  secretion from adrenal glomerulosa cell, Ca^{2+}-cAMP interactions in, 668–669
  tritiated, tubular binding
    conditions adopted in studies using, 317–319
    microassay, 324
    sites, 324
Alkaline phosphatase
  in apical membranes isolated from renal cells, 502
  cellular localization, in renal cells, 501
  measured in single isolated nephron segments, 334
Alkylacyl phospholipids, as source of arachidonate, 680–681
Amiloride, 539–541
  absorption and fluorescence characteristics, 749
  affinity matrices using, 753–754
  antibiotics, 754
  antikaliuretic properties, 739
  concentrations required to achieve half-maximal inhibition of epithelial ion transporters, 739–742
  coupled to cyanogen bromide-activated Sepharose, 753–754
  coupled to support matrices, 753–754
  effect on cell membrane resistance, 10, 12–15
  inhibition of sodium channel, in urinary epithelia, 739
  intracellular accumulation, 749–750
  natriuretic properties, 739
  as probe for characterizing transport proteins, 749–755
  radiolabeled, 750
  solubility characteristics, 749
  structure, 739, 741
  transport across plasma membrane, 749–750
Amiloride–albumin complex, coupled to Sepharose, to affinity purify anti-amiloride antibodies, 754
Amiloride analogs
  absorption and fluorescence characteristics, 749
  binding sites of, photoaffinity labeling, methods for, 753
  IC_{50} values of, 741–742
    factors that influence, 742
  photoactive, 752–753
  photoaffinity labels, 752–753
  potency relative to amiloride, in inhibiting ion transporters, 739–742
  as probes for characterization of ion transport proteins, 739, 749–755
  radiolabeled, 750–751
    binding assays using, 751–752
  solubility characteristics, 749
  stock solutions of, 749
D-Amino-acid oxidase, measured in single isolated nephron segments, 333
Amino acids, cotransport, with sodium transport by proximal tubule epithelium, 92
9-Aminoacridine
  emission wavelength, 475
  excitation wavelength, 475
  in intact cells, 475
  for optical measurements of pH gradients in vesicles, 474–475
3-Aminobenzoic acid derivatives, structure–activity studies using, 783–784
p-Aminobenzylphloretin, 757
  synthesis of, 756, 759

p-Aminobenzylphlorizin, 757
  synthesis of, 756, 760–761
3′-(4-Aminobenzyl)-2′,4′,6′,4-tetrahydroxy-dihydrochalcone. *See p*-Aminobenzylphloretin
γ-Aminobutyric acid antagonists, 799
γ-Aminobutyric acid receptor, amino acid sequencing of, 639
Amino[^{14}C]pyrine accumulation, measurement of, as measure of parietal acid secretory activity, 647–649
p-Aminohippuric acid
  in multiple indicator dilution experiments to study organic anion transport in renal proximal tubule, 61–65
  secretion, from isolated, perfused snake tubules, 214–216
4-Aminophlorizin, synthesis of, 758, 763–764
3-Aminosulfamylbenzoic acid derivatives, 782–783
Ammoniagenesis, measured in single isolated nephron segments, 333
Amphibian renal tubules
  in vitro microperfusion of, 197–198
  isolation, 197
  unique tubular functions, 196
Amphibians
  sources, 196–197
  vasopressin-sensitive epithelia, studies of, 551–571
Amphiuma
  diluting segment, complex double-cylinder model of, 376–378
  source, 196–197
Amphotericin B, effect on cell membrane resistance, 10, 15
Amplification. *See also* Cell membrane amplification
Amylase, secretion
  cholecystokinin-8-stimulated, dose–response curve for, 612–614
  vasoactive intestinal polypeptide-stimulated, dose–response curve for, 612–613
ANF. *See* Atrial natriuretic factor
Angiotensin II, effect on single-nephron glomerular filtration rate, 116
Angiotensin II receptor, solubilization, 632
3-((2-Anilino-5-nitrophenyl)sulfonyl)-1-isopropyl urea, as blocker of chloride channel, 804
Anion ATPase, measured in single isolated nephron segments, 334
Anionic dyes, in monitoring of intravesicular potential, 478–479
Anthracene-9-carboxylate, 802
  as chloride channel blocker in muscle, 800–801
Antibodies, oligomer-specific, in monitoring of protein folding and assembly, 851
Antidiuretic hormone
  cellular action of, 551–571
  transcellular water transport, 272
Apical membrane vesicles, isolation of, 501–503
Arachidonic acid
  activation of Ca^{2+} channels by, 683
  and Ca^{2+} mobilization, 684
  metabolism, 670–671
  phosphatidylinositol-derived, 676–691
  possible compartmentalization of, 685
  possible involvement in phospholipase C action, 683
  release, 677
    mechanisms of, 677–682
    protein kinase C in, 682
  released
    role of, 682–685
    second messenger functions of, 677
  source of, for eicosanoid synthesis, 670–671
Arachidonic acid metabolites, as mitogens, 682–683
Arginine vasopressin. *See also* Vasopressin
  action on tubular H$_2$O reabsorption, 675–676
  effect of
    on immunodissected CCD cells in culture, 422–423
    on second messenger production, 423
Arylaminobenzoates, 799
  as chloride channel blockers, 803–805
Aryl azides. *See also specific compound*
  photoactivation, 757
Ascending thin limbs, 289
Asialoglycoprotein receptor, amino acid sequencing of, 638
Aspartate, measured in single isolated nephron segments, 333

Aspartate aminotransferase, measured in single isolated nephron segments, 333
ATPase, 2. *See also* Proton ATPase
anion, measured in single isolated nephron segments, 334
Ca^{2+}-, 692
measured in single isolated nephron segments, 334
H^+,K^+-
activity, effect of SCH 28080 on, 737–738
as gastric proton pump, 721
inhibition of, and gastric acid secretion, 738
inhibitors, 721–722
classes of, 721–722
effect of
on human gastric glands, 733
on proton transport, 734–736
isolated gastric membranes containing, 734
K^+-site antagonists, 721–722
reversible, interaction with extracellular K^+-activation site of H^+,K^+-ATPase, 737–738
Mg^{2+}-, measured in single isolated nephron segments, 333
Na^+,K^+-, 399
activity
and aldosterone, 540–541
in basolateral plasma membranes of thick ascending limb cells, 464
and surface area of basolateral cell membranes, 270–271
and transport rate for sodium, 270–271
in apical membranes isolated from renal cells, 502
cellular localization, in renal cells, 501
in cultured renal cells, 422
hormone effects on, in cultured CCD cells, 422
inhibition by amiloride and/or its analogs, 739–742, 744
measured in single isolated nephron segments, 333–334, 340
microassay, 547–548
molecular activity, along nephron, 324
in renal $Na^+/K^+/2Cl^-$ cotransport, 785
in renal tubule cells, 268
ultramicroassay, 541–547

Atrial natriuretic factor, effect on second messenger production, 423
Atrial natriuretic factor receptor, inhibition by amiloride and/or its analogs, 747
Atrial natriuretic peptide 1–28, ^{125}I-labeled α-rat, binding conditions adopted in studies using, 317–319
Atrial natriuretic peptide 1–28, ^{125}I-labeled α-rat, binding to glomeruli, dose-dependent inhibition by unlabeled atrial natriuretic peptide analogs, 322
Atrial natriuretic peptide 1–28, ^{125}I-labeled α-rat, binding to glomeruli, effect of temperature on kinetics of, 317
Atriopeptin II, cyclic GMP production induced in glomeruli by, 313–315
dose-dependency of, 314–315
Autoradiographic studies, of renal hormone receptors, 303–304
AVP. *See* Arginine vasopressin
p-AzBPht. *See p*-Azidobenzylphloretin
p-AzBPhz. *See p*-Azidobenzylphlorizin
p-Azidobenzylphloretin
apparent affinity for sugar transporter, 771–775
interaction of, with anion transporter, 776–778
synthesis of, 756, 759–760
p-Azidobenzylphlorizin
apparent affinity for sugar transporter, 771–775
interaction of, with anion transporter, 776–778
synthesis of, 756, 761
3′-(4-Azidobenzyl)-2′,4′,6′,4-tetrahydroxy-dihydrochalcone. *See p*-Azidobenzylphloretin
4-Azidophlorizin, 757
synthesis of, 758, 764–765
4-Azido-*p*-phlorizin, synthesis of, 758, 765

B

Band 3 protein, 396, 794, 801–802
Basolateral plasma membranes
isolated by Percoll density gradient centrifugation, enrichment of lumenal membranes in, 460
phosphate transport in, sodium dependence of, 460
purification of, from rat kidney cortex

using free-flow electrophoresis, 457–458
using Percoll density gradient centrifugation, 456–457
of thick ascending limb cells, Na^+,K^+-ATPase activity, 464
Basolateral plasma membrane vesicles
inside-out oriented, purification of, from outer medulla, using Hypaque gradient centrifugation, 464–465
isolation of, from rat kidney cortex, 455–459
BBMV. See Brush border membrane vesicles
BCECF, 472
as indicator of intracellular pH, 472
staining of cortical collecting tubule with, 262
double-fluorescence labeling with rhodamine-peanut lectin, 263–264
Benzamil, 743–744, 747
potency relative to amiloride, in inhibiting ion transporters, 739–742
[benzyl-^{3}H]Benzamil, 750
Benzmetanide, structure, 783
2,5-di-(*tert*-butyl)-1,4-Benzohydroquinone, intracellular Ca^{2+} mobilization with, 703–704
4-(4-Benzoylanilino)butyric acid, as chloride channel blockers, 807–809
Bessel function, 7
Betaine formation, measured in single isolated nephron segments, 334
Bicarbonate, secretion of, by cultured intercalated cells, 420
Bicarbonate/chloride and solute concentrations, in control of proximal fluid absorption, 88–89
Bicarbonate/chloride exchange, 794. See also Chloride/bicarbonate exchangers
Bicuculline, inhibition of GABA-activated chloride channel, 800
Bicyclic phosphates, inhibition of GABA-activated chloride channel, 800
Biionic potential differences, in epithelial layers, 435–436
Bioluminescence techniques, 338
Bird renal tubules
bird sources, 224
chloride transport, 226
isolation, 224–225
perfusion, 224–225

reptilian-type nephrons, 223
isolation of, 225
perfusion of, 225
thick ascending limbs, 223
isolation of, 224–225
transepithelial potential, 225–226
unique tubular functions, 223
water transport, 226
Biscarboxyethyl fluorescein. See BCECF
Black box approach, 1–2, 37–38
BMcycline, 430
Bombesin receptor, 617
radiolabeled ligands used in binding studied to identify, 621
tissues tested to identify, 610
Bony fish. See also Teleost renal tubules
Boundary length *B*
definition of, 282
measurement, 283–284
Bovine serum albumin, with fluorescein or rhodamine isothiocyanate, as marker for endocytosis, 255
Bowman's capsule, micropuncture of, 76–78
Bowman's space
hydraulic pressure, changes in, 112
oncotic pressure within, 108–109
Bradykinin, radioimmunoassay, 339
Branched chain amino acid aminotransferase, measured in single isolated nephron segments, 333
Brefeldin A, 854
10-*N*-(Bromoacetyl)amino-1-decyl-β-D-glucopyranoside, 585
[benzyl-^{3}H]-6-Bromobenzamil, 750
6-[methyl-^{3}H]Bromomethylamiloride, 750
Brush border, 269
Brush border membranes
aggregated mixed micelles
procedure to remove extraneous proteins from, 601–602
protein patterns of, 600
deoxycholate-solubilized, aggregated mixed micelles from, 599
enrichment of lysosomes and peroxisomes in, by differential precipitation, 460
intact, protein patterns of, 600
isolated, in study of stimulus-response coupling, 675–676
isolation of, from S_3 segments (or pars recta) of proximal tubules, 461

phosphorylation events in, and regulation of transcellular transport, 675
from proximal tubule, enzyme composition of, 673
purification of
from hog kidney cortex, 454–455
on preparative scale, by differential precipitation and alkaline treatment, 454–455
from rat kidney cortex, 451–452
addition of divalent cations, 453–454
final steps of suspension, 454
preparation of crude fraction, 454
tissue selection and pretreatment, 452–453
solubilization of proteins in, 587–590
Brush border membrane transport proteins, renal, 583–605
Brush border membrane vesicles
intestinal, binding of phlorizin azide derivative to, 767–770
renal
binding of phlorizin azide derivative to, 767–770
isolation of, 586–587
renal and intestinal, photolabeling studies, with phlorizin azide derivatives, 770
from winter flounder intestine, [^{3}H]bumetanide binding to, 786–789
Bumetanide, 781–782
inhibition of $Na^+/K^+/2Cl^-$ cotransport system by, 785
kaliuretic effect of, 785–786
metabolism, 789–790
metabolites, 790
structure, 783
synthesis of, 782–783
tritiated, binding sites with, 792

C

Calcitonin, effect on intracellular calcium, in cultured renal cells, 421–422
Calcitonin-gene-related peptide receptor, identification of
radiolabeled ligands used in binding studies for, 621
tissues tested for, 610
Calcium. *See also* Cytosolic free calcium concentration

intracellular, in cultured inner medullary collecting duct, 422
intracellular free, measurement of, 691–706
intracellular stores of
depletion of, 703
Ins(1,4,5)P$_3$-sensitive, 703–704
as messenger in cells capable of transcellular Ca^{2+} transport, 674
in PCT cultures, 421
proximal tubule fluid concentration, 90
transcellular transport, 674
Calcium-binding protein, in calcium messenger system, 701
Calcium channels, voltage-gated, inhibition by amiloride and/or its analogs, 739–742, 745
Calcium indicators. *See* Fluorescent calcium indicators
Calcium messenger system, 661–662, 664–666, 691–692
and cAMP system
integration of, in cellular response, 667–669
interrelationships between, 666
study of
cations that substitute for Ca^{2+} in, 702–703
in Fura-2-loaded cells, 701–706
selective inactivation of Ca^{2+} entry and Ins(1,4,5)P$_3$-sensitive Ca^{2+} stores, 703–704
two-branch model of, 664
Calcium signaling. *See* Calcium messenger system
Calmodulin, 664
cAMP
accumulation, in gastric parietal cells, measurement, 649–650
cellular, agonists that increase, 613–615
effect on killifish proximal tubule secretion, *in vitro* electrophysiological studies of perfused systems, 194–195
intracellular, radioimmunoassays, 307, 310–314
in Malpighian tubule secretory function, 173–175, 177–178
measured in single isolated nephron segments, 334

microassay
    reliability of, 313–314
    in single pieces of living tubules incubated *in vitro*, 308
    regulation of, in cultured CCD cells, 422
cAMP-dependent protein kinase, 663, 666
    activation, 672
    assay of, in gastric parietal cells, 650–654
    hormonal activation of, detection of, 652–654
    inhibition by amiloride and/or its analogs, 746
    isozymes, identification of, 651–652
cAMP messenger system, 661–664
    and calcium system, interrelationships between, 666–667
    in renal epithelia, 423
Carbenoxolone, affinity for mineralocorticoid receptor, 526–527
Carbon dioxide formation
    measured in single isolated nephron segments, 334
        radiochemical methods, 338
    as parameter to determine sodium-dependent energy expenditure, 401
Carbonic anhydrase
    cytoplasmic, 262
    inhibitors, 781
6-Carboxyfluorescein acetate, as indicator of ion flux in cells, 404
Carriers, 35
Cationic dyes, in monitoring of potentials in cells and vesicles, 476–477
Cation transport probes, amiloride series, 739–755
CCD. *See* Cortical collecting duct
CCK-JMV-180, 618–619
Cell culture
    media. *See* Media
    primary, of defined tubule cells, 409–426
    substratum, 417–418
    support, 417–418
    techniques, 410–418
Cell membrane amplification, forms of, in renal tubular cells, 269–270
Cell membrane resistances, 5
    direct current (DC), and DC transepithelial resistance, quantitative comparison of, 7–8
    effect of amiloride on, 10, 12–15

effect of amphotericin B on, 10, 15
effect of gramicidin D on, 10, 15
effect of novobiocin on, 15
effect of nystatin on, 15
effect of silver ions on, 15
effect of vasopressin on, 15
Cell monolayers
    permeability, tests for, 500
    phosphate transport studies in, 497–501
Cells
    measurement of potential in, 476–479
    pH gradient measurements in, principles of, 470
Cell-to-cell fusion, 439
Cellular impalement, 296–297
    successful, criteria for, 297–298
Cellular metabolism, inhibition by amiloride and/or its analogs, 747–748
Cellular response. *See also* Stimulus–response coupling
    autocrine and paracrine inputs in, 670–671
    sustained phase of
        Ca^{2+} action in, 665, 669
        cAMP metabolism during, 663
        synarchic regulation of, 669
Cell volume
    determination of, in monolayer cultures, 432
    measurement
        double-cylinder model for complex, 375–379
        simple, 374–375
    in isolated and perfused renal tubules, 371–379
    overall strategy for, 373–374
    and perturbation in extracellular osmolality, 371
    regulatory mechanisms, 371
cGMP
    intracellular, radioimmunoassays, 307, 310–314
    in single pieces of living tubules incubated *in vitro*, micromethods for measuring, 308
cGMP phosphodiesterase, measured in single isolated nephron segments, 333
Chemical cross-linking
    interpretation of, 848–849
    of protein

correlated with velocity gradient sedimentation, 852–853
for structural characterization, 847–850
Chemiluminescence, 338
Chinchilla, tubule fluid-to-ultrafilterable potassium (TF/UF$_K^+$), 90
Chloride
 in mosquito hemolymph and fluid secreted by isolated Malpighian tubules, 173–174
 secretion
  by fish proximal tubules, 190–192
  in shark proximal tubules, electrophysiological evidence for, 184–186
 transport
  in bird renal tubules, 226
  in kidney, 785
 transport system, in apolar cells, agents acting on, 801–802
Chloride/bicarbonate exchangers
 apical, 263–264
 in isolated renal cells, 406–407
 localization, to apical or basolateral membrane, 263
 polar distribution of, 263–264
Chloride channel blockers, 293, 793–809
 acting at muscular chloride channel, 801–802
 acting at neuronal chloride channels, 798–799
 acting in apolar cells, 801–802
 acting in epithelia, and related Na$^+$/2Cl$^-$/K$^+$ carrier inhibitors, 802–809
 classification, 799
 specificity, 793
Chloride channels
 abnormalities of, diseases associated with, 572
 affinity purification of, IAA ligands for, 576–577
 agonists, 796–797
 antagonists, 796–797
 calcium dependence, 796–797, 799
 cAMP-dependent, 799
 classification of, 798
 conductance, 796–797
 coreconstitution of, with bacteriorhodopsin, 580–581
 electrophysiological behaviors, 572–573
 in epithelia, agents acting in, 802–809

function of, 572
general properties of, 794–799
inhibition, 796–797
muscular, agents acting at, 800–801
neuronal, agents acting at, 799–800
permselectivity, 796–797
purification of, 573–574
reconstitution of, 579–582
 into planar bilayers, 581–582
regulation of, 572
solubilization of, 575–576
voltage dependence, 796–797
Chloride exchange, in red blood cells, inhibition by phloretin, phlorizin, and their azido derivatives, 776–778
Chloride flux
 electrogenic, 793
 through ionic channels, 793
Chloride/formate exchange, 794
5-Chlorodiphenylamine-2-carboxylic acid, as chloride channel blocker, 807–809
4-Chloro-3-N-pyrrolidino-benzoic acid, as chloride channel blocker, 804, 807–809
1-(2-Chlorophenyl)-N-methyl-N-(1-methylpropyl)-3-isoquinoline carboxamide, tritiated. See PK 11195, tritiated
Chlorothiazide, 781
Cholecystokinin
 radioiodinated, cross-linking of, to binding proteins, 635
 short probes of, cross-linking studies using, 635
Cholecystokinin receptors, 618–619. See also Gastrin receptors
 central cortical, 627
 identification of
  radiolabeled ligands used in binding studies for, 621
  tissues tested for, 610
 interaction with, relative potencies of CCK-related peptides in, 625–627
 pancreatic, 625–627
 on pancreatic acini, 613–614
 solubilized
  affinity for antagonists, 633
  affinity for CCK receptor agonists, 633
Cholecystokinin-related peptides, inhibition of binding to [^{125}I]CCK-8, in pancreatic acini, dose–response curves for, 628–629

Cholera, 572
Cholera toxin receptor, identification of
  radiolabeled ligands used in binding
    studies for, 621
  tissues tested for, 610
Choline kinase, measured in single isolated
  nephron segments, 334
Cimetidine
  effect of
    on gastric glands, 729–730
    on isolated gastric mucosa, 724
  inhibitors, effect on human gastric glands,
    733
Citrate synthase
  activity, aldosterone-dependent elevations
    in, 539–541
  measured in single isolated nephron
    segment, 333
  enzymatic cycling used for, 332–336
  ultramicroassay, 541–542, 547–548
C kinase. See Protein kinase C
Clathrin-coated pits, on surface of collecting
  duct principal cells, IMP aggregates in,
  556–557
Cl^-/HCO_3^- exchangers. See Chloride/bicarbonate exchangers
Collagenase
  digestion of gastric mucosa, 640–643
  disintegration of renal connective tissue
    by, 330
  and enzyme activities, 332
  source, 643
  and tubular membrane permeabilities, 332
Collagenase perfusion, for fresh rat kidney,
  227–228
Collecting duct, 288. See also Cortical
  collecting duct; Inner medullary collecting duct; Medullary ray collecting
  ducts; Outer medullary collecting duct;
  Papillary collecting duct
  acidic cytoplasmic vesicles in, 258
  isolated perfused, fluorescent techniques
    to measure water permeability of, 552
Collecting duct cells, 287–288
  isolation of, 387
  lumenal membrane recycling, 272
  lumenal microfold formation, 271
  markers for, 400
  modulation of lumenal membrane area
    by membrane recycling, 271

time course of structural changes, 272–273
Collecting tubule. See also Cortical
  collecting tubule; Inner medullary
  collecting tubules; Medullary collecting
  tubule; Outer medullary collecting
  tubules
  function of, 254
  intercalated cells. See Intercalated cells
  principal cells. See Principal cells
Collecting tubule cells, isolation, for primary
  culture, 413–415
Colloidal gold, labeling, of ultrathin frozen
  sections, 281
Colloid osmotic pressure, of plasma entering
  and leaving glomerular capillaries, 125
Colon, shunt analysis in, 14
Congestive heart failure, glomerular
  filtration rate in, 115
Connecting segment, aldosterone sensitivity,
  520
Contralumenal plasma membranes. See
  also Basolateral plasma membranes
  from canine inner medulla, isolation of,
    467–468
  preparation, purity of, 460–461
Cornea, apical cell membrane conductance
  in, 10
Cortical collecting duct, 288
  histotopographical criteria for, 285
  microdissection of, 229–230
  pH of, 257
Cortical collecting duct cells
  culture
    growth media and supplements, 416
    in vectorial transport studies, 418–419
  cultured
    intracellular signal systems, 422
    transepithelial voltage in, 424–425
    vectorial ion transport in, 420
  immunodissection of, 414
  isolation, for primary culture, 413–415
  transfected with adenovirus 12–SV40
    hybrid, 414
Cortical collecting tubule, 326
  aldosterone sensitivity, 520
  endocytosis in, 256–257
  segments, in rabbit, 310
  stained with acridine orange, 258–259
  staining of

with BCECF, 262
and rhodamine-peanut lectin, double-fluorescence labeling with, 263–264
with 5,6-dicarboxyfluorescein, 262
with 3,3′-dipentyloxacarbocyanine, 259–260
with rhodamine–peanut lectin, 260–261
type III binding site in, 523–524
Cortical collecting tubule cells
culture, support, 417–419
immunodissection of, 411
Cortical connecting tubule
histotopographical criteria for, 287
microdissection of, 229–230
Cortical distal cells, sodium transport-related O_2 consumption, 401–402
Cortical thick ascending limb of loop of Henle, 326
histotopographical criteria for, 287
isolation of cells from, 412
microdissection of, 229–230, 309
Cortical tubule fragments, purification of, by gradient centrifugation, 411
Cortical tubules, suspensions, 326–327
Corticosteroid binding globulin, renal, 522–524
Corticosteroid binding sites, renal
type I, 521–523
steroid specificity of, 528
type II, 521–523
type III, 522–524
Corticosteroid dehydrogenase, measured in single isolated nephron segments, 334
Corticosteroid receptors
assay, 529
bound to agonists and antagonists, glycerol density gradients, 535–536
Cortisol-binding globulin, renal, 528
Cotransporter, 3. *See also specific cotransporter*
Cotransport systems, in membrane vesicle preparations, stoichiometries of, 479–494
Coturnix coturnix. See Japanese quail
Counter-current theory, 92
Countertransport systems, in membrane vesicle preparations, stoichiometries of, 479–494

Coupled transport
A:S coupling stoichiometry, 480
flux measurements, by rapid filtration technique, 481–483
multiple activators, and charge stoichiometry, 481
stoichiometry of
measurement, 483–494
activation method, 483, 485–488
direct method, 483–485
static head method, 491–494
steady state method, 488–490
in vesicles, 479–494
thermodynamics of, 480–481, 485
Covalent cross-linking, in identifying cell surface receptors, 634–636
Creatinine, as tracer, in renal multiple indicator dilution, 40
Cross-linking
affinity, 634
chemical
interpretation of, 848–849
of protein, for structural characterization, 847–850
covalent, in identifying cell surface receptors, 634
of human immunodeficiency virus-type 1 env proteins, 849, 852–853
Cross-linking agents, 847–848
in receptor identification, 634
CTAL. *See* Cortical thick ascending limb of loop of Henle
Cyanocobalamine, as volume marker, 242
Cyclic adenosine monophosphate. *See* cAMP
Cyclic guanosine monophosphate. *See* cGMP
Cyclooxygenase pathway, 670–671, 685, 805
Cystic fibrosis, 572
Cytochrome *P*-450, measured in single isolated nephron segments, 334
Cytosolic free calcium concentration
calculation of, from Fura-2 fluorescence signals, potential problems, 700–701
change in, as intracellular signaling mechanism, 691–692
measurement of, with fluorescent indicators, 692–701
calibration for, 696–699

cell preparations, 694–696
  in cell suspension and monolayers, 694–695
  instrumentation required for, 693–694
  in single cells, 695–696

D

DCCD. See N,N'-Dicyclohexylcarbodiimide
DCT. See Distal convoluted tubule
Decane, in fluid column of constant-bore pipet, to prevent evaporative loss, 235, 238
Descending limbs of short loops, 288
Descending thin limbs, 288. See also Thin descending limb of Henle's loop
Descending thin limbs of long loops, 288–289
Desmosomes, 267
Detergents, for protein solubilization and reconstitution, 842–843
Dextran, fluid-phase endocytosis of, 341
Diabetes insipidus
  hypothalamic, 551
  nephrogenic, 551
Diacylglycerol
  in arachidonate release, 677–681
  in calcium messenger system, 701
  in cellular response, 665, 673–674
  via Ca^{2+} messenger system, 670
Diacylglycerol lipase, 677–678
Diacylphospholipids, as source of arachidonate, 678
Diarrhea, secretory, 572
Diazepam, antagonists, 799
Diazepam receptors, peripheral, 800
5,6-Dicarboxyfluorescein, staining of cortical collecting tubule with, 262
3',4'-Dichlorobenzamil, 745
  in intact frog atrial myocytes, inhibition by amiloride and/or its analogs, 746
  potency relative to amiloride, in inhibiting ion transporters, 739–742
N,N'-Dicyclohexylcarbodiimide, inhibition of ATP-driven proton pumps, 505, 517–518, 520
DIDS. See 4,4'-Diisothiocyanostilbene-2,2'-disulfonate
Diethyl oxocarbodicyanine, in monitoring of potential in cells and vesicles, 477

Diethylstilbestrol, inhibition of ATP-driven proton pumps, 517–518, 520
Differential interference optics, 372–373
Differential solute clearances, 128
4,4'-Diisothiocyanostilbene-2,2'-disulfonate, as chloride channel blocker in muscle, 800–801
Diluting segment, Amphiuma, complex double-cylinder model of, 376–378
Dilution potential difference, in epithelial layer, 436
5-(N,N-Dimethyl)amiloride, 743–747
  inhibition of Na^+,K^+-ATPase by, 744
  potency relative to amiloride, in inhibiting ion transporters, 739–742
Di-O-C_5(3). See 3,3'-Dipentyloxacarbocyanine
3,3'-Dipentyloxacarbocyanine
  intercalated cells functionally identified using, 258–260
  staining of cortical collecting tubule with, 259–260
Diphenylamine-2-carboxylic acid, 802
  as blocker of chloride channel, 804
3,3'-Dipropylthiadicarbocyanine iodide, in monitoring of intracellular and intraesicular potentials, 477–478
DiS-C_3(5). See 3,3'-Dipropylthiadicarbocyanine iodide
Discontinuous density gradient centrifugation, purification of cortical epithelial fragments by, 410–411
Distal convoluted tubule, 326
  aldosterone sensitivity, 520
  function of, 94
  histotopographical criteria for, 287
  microdissection of, 229–230, 309
  portions of, in rabbit, 309–310
Distal tubule, cortical segments of, isolation of cells from, 381–386
Distal tubule cells
  compounds used in identification of, 396
  markers for, 400
  time course of structural changes, 272–273
Disuccinimidyl suberate, 634–635
Disulfonic stilbenes, as chloride channel blocker in muscle, 801
Diuretics
  high ceiling, 782–783

related to furosemide, 781–792
structure–activity relationships, 783–784
types of, 781
DNA, synthesis, inhibition by amiloride and/or its analogs, 748
DNA topoisomerase II, inhibition by amiloride and/or its analogs, 748
DOCC. *See* Diethyl oxocarbodicyanine
Dog
  tubule fluid-to-ultrafilterable (TF/UF) magnesium ratio, 91
  tubule fluid-to-ultrafilterable potassium (TF/UF$_K^+$), 90
Dogfish shark. *See also* Shark
  rectal gland tubules
    isolation, 183
    perfusion, 183
Double-cylinder model
  complex, of tubule, 375–379
  simple, of tubule, 374–375
Double perfusion, in intact mammalian kidney, 98
(D-Pro^4,D-Tryp^{7,9,10})substance P-411, action of, on stimulation of pancreatic enzyme secretion caused by substance P and bombesin, 618
Drug receptors, that are inhibited by or interact with amiloride or amiloride analogs, 740
Dual-channel perfusion pipet, 292, 294

E

Ehrlich ascites tumor cells
  cotransport activity and bumetanide binding in, 786–788
  [^{3}H]bumetanide-binding experiments in, 792
Eicosanoids, 676
  functions of, 682–685
  relationship to Ca^{2+} and cAMP messenger systems, 670–671
  synthesis, 670
Elasmobranchs. *See also* Shark
  sources of, 183
  tubular epithelia of, 182–186
    unique tubular functions, 182–183
Electrolyte transport
  in renal epithelia, correlation of certain structures with, 266

transepithelial
  regulation of, 265
  structural basis for, 266–273
Electron microprobe analyses
  of solid tissue samples, 96
  in study of ion fluxes in epithelial cells, 405–408
Electron microprobe spectrophotometry, 96
Electrophysiological study
  of chloride transport systems, 793
  of defined tubule cells, in primary culture, 424
  of isolated perfused tubules, 289–302
  measurements, 289
  of mRNA-injected MDCK cells, 448–449
  to quantify epithelial shunt conductances, 5
  techniques, 289–290
Endocytosis
  in cortical collecting tubule, 256–257
  fluorescent probes of, 559–561
  in isolated nephron segments, studies using fluorescent dyes, 255–257
  in kidney, tracer studies using horseradish peroxidase and fluorescent probes, 557–561
  markers for, 255
  of proteins, in proximal tubules, 340–354
  in proximal straight tubule, 255–257
Endocytotic vesicles
  ATP-driven proton pump, effect of inhibitors on, 515–517
  enzymatic characteristics of, 515–516
  isolation of
    buffers and solutions for, 507
    procedure, 509–512
  from kidney, as model to study characteristics of vacuolar proton pump, 506
  preparation of
    animals for, 506
    equipment for, 506
    from rat renal cortex, 506
  recovery of protein and ATP-driven proton pump activity during, 514–515
  proton ATPase activity, 517–518
    coupled optical assay, 517–518
    measurement of, 509
      buffers and solutions for, 507

proton pump activity, measurement of, 508, 512–514
  buffers and solutions for, 507
  renal, characterization of, 512–517
Endoplasmic reticulum, protein transport from, 841–842
Endothelial cells. *See also* Glomerular endothelial cells
  isotopic influx measurements in, effect of loop diuretics on, 791
Enzymatic cycling analyses, 332–336
Enzyme inhibition, in isolated single nephron segments, 340
Enzymes, that are inhibited by or interact with amiloride or amiloride analogs, 740
Epidermal growth factor receptor
  amino acid sequencing of, 638
  identification of
    radiolabeled ligands used in binding studies for, 621
    tissues tested for, 610
  in inner medullary collecting duct, 423
  primary structure of, 637
  structural/functional domains, 637
Epidermal growth factor receptor protein kinase, inhibition by amiloride and/or its analogs, 746
Epinephrine, effect on single-nephron glomerular filtration rate, 116
Epithelia
  definition of, 1
  distributed equivalent circuit model of, 20
  equivalent circuit representations of, 5–6
  ionic pathways of, 291
  leaky, 1, 268, 293
    effects of amiloride on, 10
    impedance analysis of, 19
    vs. tight, 1
  paracellular shunt conductance in, 4–27
    distributed model, 6
    lumped model with capacitance C_a and C_{bl} for apical and basal cell membrane, 6
    lumped model with electron motive forces representing 0-current potentials E_i for each membrane element, 6
    simple lumped model, 5–6
      determination methods based on, 7–11

    permeability coefficients, calculation of, 368–370
    shunt pathways, 4–5
  tight, 1, 268, 293
    effects of amiloride on, 10
    shunt resistance of, 12–15, 20
    tubular, shunt analysis in, 5, 8–9
  with two distinct apical permeation pathways, model of, 25
Epithelial cell layers, 2
  growth of
    on collagen-coated nylon mesh, 434
    in Marbrook-type chambers, 434–435
    on Millipore substrates, assessment of confluency in, 433
  Millipore supports, 432–433
  on permanent supports
    demonstration of epithelial polarity of transport or bindings in, 432
    preparation of, 432–435
    transepithelial electrical properties, measurement of, 435–436
Epithelial cells. *See also* Glomerular epithelial cells
  cell-to-cell communication among, 264
  functional activity in, and cAMP metabolism, 640
  patch-clamp study in, 301–302
  pH in, measurement of, 263
  polarity, 450
  stimulus–response coupling, special problems in, 671–676
Epithelial transport, 1–4
  physiological function of, 2
  research, black box approach, 1–2
Epithelial volume per given tubular length (V_L (epi/tub))
  calculation of, 283–286
  definition of, 283
Equivalent short-circuit current, 293–295
Erythrocytes
  anion transporter, effect of *p*-AzBPht and *p*-AzBPhz on, 776–778
  duck
    cotransport activity and bumetanide binding in, 786–788
    [³H]bumetanide-binding experiments in, 792
    equilibrium exchange of 3-methoxyglucose in, 771–772

inhibition of 3-methoxyglucose transport in, by phloretin, phlorizin, and their azidobenzyl derivatives, 773–775
isotopic influx measurements in, effect of loop diuretics on, 790
photolabeling of, with p-[^{3}H]azidobenzylphloretin, 778–780
salt and water flux in, effect of p-AzBPhz on, 776–778
Ethacrynic acid, as chloride channel blocker in muscle, 801
Ether phospholipids, as source of arachidonate, 680–681
Ethidium bromide, effects on binding of [^{3}H]aldosterone-labeled receptor complexes to chromatin, 532
5-(N-Ethyl-N-isopropyl)amiloride, 743, 747
potency relative to amiloride, in inhibiting ion transporters, 739–742
N'-Ethylmaleimide, inhibition of ATP-driven proton pumps, 505, 517–518, 520
Exchangers. *See also specific exchanger*
localization, to apical or basolateral membrane, 263
Excretory ducts, epithelia, 1
Exocytosis, in isolated nephron segments, studies using fluorescent dyes, 257
Expression systems
for expression of receptor protein, 639
giant MDCK cells as, 437–449
Xenopus oocytes, 437
Expression vectors, 639
Extracellular fluid expansion, and salt and water reabsorption in proximal tubule, 89

F

Ferrocyanide, radiolabeled with ^{14}C, used to quantitate single-nephron glomerular filtration rate, 123
Ferrocyanide injection technique, measurement of single-nephron glomerular filtration rate by, 122–123
Fibroblasts, isotopic influx measurements in, effect of loop diuretics on, 790
Filament borosilicate glass, 296
Filipin, as morphological probe, 556
Filipin–sterol complexes, 554–557

Filtration pressure disequilibrium, glomerular, 114
Filtration pressure equilibrium, glomerular, 113
Fish. *See also* Flounder; Teleost renal tubules
euryhaline, renal tubular function in, 186–187
sources, 187
FITC. *See* Fluorescein isothiocyanate
Fixation. *See* Renal tissue, fixation of; Renal tissue, fixed
Fixative, 276–277
osmolality, 277
retrograde vascular perfusion of, 275–276
vehicle, 277
washout of, 276
Flounder, proximal tubules
detection of low-capacity solute transport system in, 193–194
isolated, fluid secretion in, 189–191
measurement of transepithelial volume flow in, 192
Fluid, reabsorption
in isolated perfused proximal convoluted tubules of rabbit, 350–351
in tubule perfusions, measurement and calculation of, 350–352
Fluid droplet technique, for investigating water and solute flux within nephron, 84
Fluid-phase endocytosis, determination of, in isolated perfused nephron segments, 352–353
Fluo-3, 692
Fluorescein
excitation spectrum of, pH sensitivity, 261
as indicator of intracellular pH, 472
permeant derivatives of, cell pH measured by excitation ratio fluorometry using, 261–263
Fluorescein dextran, used to measure pH$_i$ in endocytosed material, 473
Fluorescein isothiocyanate
fading inhibition, 280
as marker for endocytosis, 255
Fluorescein isothiocyanate–dextran, as probe for morphological and functional study of endosomes in kidney, 559–561

Fluorescence
  dual-wavelength, used to measure intracellular pH, 473
  single-wavelength, used to measure intracellular pH, 472–473
Fluorescence microscopy, 254–255
Fluorescent calcium indicators, 692–701.
  See also specific indicator
  loading cells with, 693
Fluorescent dyes
  advantages of, 253
  concentration needed, 253
  derivatization of, 253
  identification and study of specific cell types in isolated nephron segments using, 253–265
  as label on semithin sections, 280
Fluorescent indicator techniques, 4
  to measure water permeability of isolated perfused collecting ducts, 552
Fluorometry, 95–96, 333–334, 337
Flux coefficient, 368
Freeze–fracture
  after filipin treatment of tissue, 552–557
  methods, 274
  of renal tissue, 552–556
Frog skin
  apical cell membrane conductance in, 10
  shunt analysis in, 5, 8, 14–15
Fructokinase, measured in single isolated nephron segments, 333
Fructose-1,6-bisphosphatase, measured in single isolated nephron segments, 333
Fructose-1,6-bisphosphate aldolase, measured in single isolated nephron segments, 333
Fructose 1-phosphate, measured in single isolated nephron segments, 333
Fructose-1-phosphate aldolase, measured in single isolated nephron segments, 333
Fumarase, measured in single isolated nephron segments, 333
Fundulus heteroclitus. See Killifish
Fura-2
  absorption of, amiloride interference with, 749
  Ca^{2+} bound, 700
  Ca^{2+} free, 700
  Ca^{2+} insensitive, 700–701
  as calcium indicator, 692–693, 695–701

  calibration procedures with, 696–699
  loading cells with, 693
  microinjection of, directly into cell, 693
  photobleaching of, 700
Fura-2/AM. *See* Fura-2 tetraacetoxymethyl ester
Fura-2 loaded cells
  Ca^{2+} signaling in, investigation of biochemical mechanisms regulating, 701–706
  microinjection of $Ins(1,4,5)P_3$ into, and monitoring cytosolic Ca^{2+} concentration in, 704–706
Fura-2 tetraacetoxymethyl ester, loading cells with, 693
2-(2-Furfurylmethylamino) 5-nitrobenzoic acid, as blocker of chloride channel, 804
Furosemide, 803–809
  as blocker of $Na^+/2Cl^-/K^+$ carrier, 804
  as chloride channel blocker in muscle, 800–801
  diuretic compounds related to, 781–792
  effects on Na^+, K^+, and Cl^- transport in kidney, 785–786
  mechanism of action, 784–789
  structure–activity relationships, 783–784
  kaliuretic effect of, 785–786
Fusion factors, of enveloped viruses, 843–844

G

GABA. *See* γ-Aminobutyric acid
β-Galactosidase, measured in single isolated nephron segments, 334
Gall bladder
  apical cell membrane conductance in, 10
  Necturus, shunt analysis in, 19
Gap junctions, 267
Garter snake
  distal tubules
    as diluting segment, 221–223
    isolation of, 202–203
    sodium reabsorption, amiloride-sensitive, 220–221
    sodium transport
      active reabsorptive, electrophysiological evidence for, 220–221
      self-inhibition of, 221

transepithelial water permeabilities, 221–223
water permeability of basolateral membrane, 221–223
proximal tubules
glucose transport, 214, 219–220
in vitro microperfusion of, 203–220
epithelial cell water content determination, 209–210
fluid absorption determination, 203–204
fluid absorption mechanism, 201, 204
hydraulic conductivity calculation, 206–207
intracellular organic solute concentration determination, 208–209
membrane permeability estimates, 210–212
morphometric analysis with, 204–205
reflection coefficients for experimental solutes, 205–207
transepithelial organic solute fluxes, 207–208
transepithelial organic solute transport, 207–214
transepithelial voltage evaluation, 207
volume markers used in, 203
isolation of, 199–200
lactate transport, 214, 217–218
lumenal membrane
apparent permeability of, 210–211
efflux coefficients, 213
net reabsorption of N-methylnicotinamide, 214, 218–219
net secretion of PAH, 214–216
net secretion of tetraethylammonium, 214, 218
net secretion of urate, 214, 216–217
organic cation transport, mechanism of, 214, 218–219
organic ion transport, cellular mechanisms of, 214–218
perfusion of, 199–200
Ringer solution for, 200–202
peritubular membrane
apparent permeability of, 211–212
efflux coefficients, 212–213
permeability from bath to cell, determination, 213–214

permeability from lumen to cell, determination, 213–214
segments of, isolation and perfusion of, 200
transepithelial transport of neutral organic solute, mechanism of, 214, 219–220
renal tubules
perfusion, Ringer solution for, 200–202
unique tubular functions, 198–199
source of, 199
Gastric acid, formation, in gastric glands, assessment of, 726–728
Gastric acid secretion, pharmacological agents of, 721–738
Gastric glands
acid formation in, determination of, 726–728
by measurement of [^{14}C]aminopyrine accumulation, 727–728
by measurement of glucose oxidation, 727
by measurement of oxygen consumption, 726–727
aminopyrine accumulation, 727–728
glucose oxidation, 727–729
human, studies of acid secretion and inhibition in, 733
isolation of, 724–726
oxygen consumption, 726–729
pepsinogen release, measurement of, 733
permeable
effect of different inhibitors, 731
preparation, 731
preparation
responsiveness to different inhibitors, 729–730
responsiveness to different stimuli in, 728–729
respiration in, 726–729
stimulation and inhibition of, 728–729
Gastric mucosa
isolated
effect of cimetidine and omeprazole on, 724–725
preparation, 722–723
responsiveness to different secretagogues, 723–724
perfusion of, 644–645

Gastric parietal cells
  adenylate cyclase activity in, 650
  amino[^{14}C]pyrine accumulation, 647–649
  cAMP-dependent protein kinase in, 650–654
  cAMP-dependent protein phosphorylation in, 654–659
  cAMP in, determination of, 649–650
  cAMP metabolism in, 654–659
    correlation with function, 659–661
  cyclic nucleotide-related measurements in, 649–654
  enrichment technique, 641–642, 647
  function, correlation of, with cAMP metabolism, 659–661
  functional measurements in, 647–649
  hydrochloric acid secretion by, 647–649
  isolation
    procedure, 643–647
    techniques, 640–643
  purification, 641, 646–647
  respiration, measurement of, 647
Gastric parietal glands, isolation of, 645–646
Gastrin receptors, 627–628
  on gastric smooth muscle cells, 613–614
  identification of
    radiolabeled ligands used in binding studies for, 621
    tissues tested for, 610
GFR. *See* Glomerular filtration rate
Glial cell volume regulation, 802
Glomerular capillary
  hydraulic pressures, 108–109, 123–124
    measurement of, 76
    transcapillary, 75–76
  oncotic pressure within, 108–109
  ultrafiltration pressure, 108
Glomerular capillary network, 108
Glomerular capillary tuft, 39–40
Glomerular capillary ultrafiltration coefficient (K_f), 115–116, 125
Glomerular capillary wall
  charge-selective properties, 52–55
  forces operating across, 75–78
  hydraulic conductivity, morphological measurements, 129
  properties, 108
  surface area, morphological measurements, 129

  transcapillary hydraulic pressure, 75–76
  transcapillary oncotic pressure, 75
  ultrafiltration at
    determinants, 49
    and solute size, 51–52
Glomerular cells. *See also* Mesangial cells
  cloning, 143–144, 146
  culture, 141
    techniques, 144
  homogeneous, methods to obtain, 144
  isolation, 141–152
Glomerular endothelial cells
  bovine, isolation, 148–149
  cloning, 149–150
  culture, 148–150
  fluorescence-activated cell sorting, 150
  human, isolation, 148–149
  isolation, 148–150
  rat, isolation, 148–149
Glomerular epithelial cells, 144
  cloning, 150–151
  culture, 144–146, 150–151
    assessment of purity, 152
  cytotoxic response to puromycin, 151
  identification, 151
  isolation, 150–152
  shape, 151
Glomerular filtrate, composition of, 76–78
Glomerular filtration
  with increased systemic oncotic pressure, 110–111
  prevention, methods, 31–32
  with reduced glomerular plasma flow rate, 110–111
  with reduced mean transcapillary hydraulic pressure difference, 110–111
  with reduced ultrafiltration coefficient, 110–111
  schematic portrayal of process of, 109
Glomerular filtration barrier, sieving characteristics of, 52
Glomerular filtration rate, 107
  assessment of, 116–123
  autoregulatory maintenance of, in face of reduction in renal perfusion pressure, 112
  mathematical model of, 108–116
  single-nephron, 107
    assessment of, 117–123

with congestive heart failure, 115
  determinants of, 108–110
    assessment of, 123–129
    selective effect of, 110–116
  equations for, 76, 110
  measurement of, in free-flow micropuncture, 79
  whole kidney, 116–117
Glomerular function, direct evaluation of, 75
Glomerular hemodynamics, 123–129
  derivation from macromolecular sieving data, 128–129
  heterogeneity of, 123
Glomerular membrane, ultrafiltration of potassium across, 90
Glomerular mesangium, bone marrow-derived cells, 147–148
Glomerular permselectivity, investigation of, using multiple indicator dilution, 49–55
Glomerular plasma flow rate, 125
  and single-nephron glomerular filtration rate, 113–114
Glomerular polyanions, 52
Glomerular reflection coefficient σ, 52–53
Glomerular–tubular balance, 86–87
Glomerular ultrafiltrate, 34–35, 39
Glomerular ultrafiltration, 75
Glomerular vasculature, hydraulic pressure profile along, 126
Glomerulus, 39
  centrifugation, 133
  function, control of, 130
  incubation of, for determination of prostaglandin synthesis, 135–136
  isolated, 130–140
    continuous superfusion, determination of planar surface area during, 139–140
    evaluation of purity, 133
    incubation of
      K_f determination by, 128
      used to estimate single-nephron glomerular filtration rate, 128
    in vitro biochemical and morphological experiments using, 130
    in vitro contractility of, determined by planar surface area, 138–140
    microperfusion, 126–128
    planar surface area of, determination of, 138–140
    superfusion system of, 136–138
  isolation of
    alternative methods of, 134–140
    from other than adult rats, 134
    from rat kidney, 131–134
    by sequential sieving, 131–134, 142–144
    techniques for, 130–131, 141–146
  microdissection, 130, 229–230
  micropuncture of, 75–78
  transcapillary fluid transport in, 107–129
  washing, 133
Gluaminase, measured in single isolated nephron segments, 334
Glucagon, radioiodinated
  binding, to medullary thick ascending limbs
    dose dependency of, 321–322
    dose-dependent inhibition of by unlabeled glucagon, 321–322
    effect of tubule length on, 320
    binding conditions adopted in studies using, 317–319
  bound to rat medullary thick ascending limbs, separation of, from free labeled hormone, 315–316
Glucagon-binding sites, radioiodinated, and glucagon-sensitive adenylate cyclase, distribution of, along rat nephron, 323–324
Glucocorticoid receptors, renal, 521–528
  equilibrium dissociation constants for [^{3}H]aldosterone binding to, 522
  equilibrium dissociation constants for [^{3}H]dexamethasone binding to, 522–523
Gluconeogenesis, measured in single isolated nephron segments, 333
2'-O-(β-D-Glucopyranosyl)-4',6'-dihydroxy-4-azidodihydrochalcone. See 4-Azidophlorizin
4'-O-(β-D-Glucopyranosyl)-2',6'-dihydroxy-4-azidodihydrochalcone. See 4-Azido-p-phlorizin
2'-(β-D-Glucopyranosyl)-4',6',4-trihydroxy-5'-(4-aminobenzyl)-dihydrochalcone. See p-Aminobenzylphlorizin
2'-(β-D-Glucopyranosyl)-4',6',4-trihydroxy-

5′-(4-nitrophenylcarbinol)-dihydrochalcone, synthesis of, 756, 760
Glucose
  cotransport, with sodium transport by proximal tubule epithelium, 92
  measured in single isolated nephron segments, 333
  transport, in isolated, perfused snake tubules, 214, 219–220
D-Glucose
  radioactive, fluxes, interpreting, 355
  sodium-dependent transport in renal outer medullary brush border membrane vesicles, 491–493
  transepithelial transport of, 2
L-Glucose, as tracer, in renal multiple indicator dilution, 40
Glucose-6-phosphatase, measured in single isolated nephron segments, 333
Glucose 6-phosphate, measured in single isolated nephron segments, 333
Glucose-6-phosphate dehydrogenase, measured in single isolated nephron segments, 333–334
Glucose transporter, sodium-dependent, in renal and intestinal brush border, binding of phlorizin azide derivative to, 767–770
Glutamate, measured in single isolated nephron segments, 333
Glutamate dehydrogenase, measured in single isolated nephron segments, 333–334
Glutamine, measured in single isolated nephron segments, 333
Glutamine synthetase, measured in single isolated nephron segments, 333
α-Glutamylcysteine synthetase, measured in single isolated nephron segments, 334
γ-Glutamyl transferase, in apical membranes isolated from renal cells, 502
γ-Glutamyl transpeptidase, measured in single isolated nephron segments, 334
Glutaraldehyde
  in fixative solution, 276
  stock solution, 276
Glutathione, measured in single isolated nephron segments, 333
Glutathione-S-transferase, measured in single isolated nephron segments, 334

Glycerol density gradients, of ^{3}H-labeled steroid–receptor complexes, 535–536
Glycerol kinase, measured in single isolated nephron segments, 334
  radiochemical technique, 337–338
Glycerol phosphate, measured in single isolated nephron segments, 333
Glycerol-3-phosphate dehydrogenase, measured in single isolated nephron segments, 333
Glycerophosphorylinositol, 678
Glycerophosphorylinositol mono- or bisphosphate, 678
Glycine receptor chloride channels, 799–800
Glycogen, measured in single isolated nephron segments, 333
Glycophorin, SDS resistance, 850
Glycyrrhetinic acid, affinity for mineralocorticoid receptor, 526–527
Goldman–Hodgkin–Katz equation, 370
G proteins, 662, 664
  in calcium messenger system, 701
  inhibition by amiloride and/or its analogs, 747
Gramicidin D, effect on cell membrane resistance, 10, 15
Guanine nucleotide regulatory proteins, pertussis toxin catalyzed ADP ribosylation of, inhibition by amiloride and/or its analogs, 747
Guanine nucleotides, measured in single isolated nephron segments, 333
Guanylate kinase, measured in single isolated nephron segments, 333

H

Hamster, tubule fluid-to-ultrafilterable (TF/UF) magnesium ratio, 91
H^+-ATPase. *See* Proton ATPase
HCO_3^-. *See* Bicarbonate
H^+/dopamine countertransporter, in chromaffin granule ghosts, coupling and charge stoichiometries of, 489–490
Helium glow photometry, 95–96
Hemagglutinin
  of fowl plague virus, as marker of transport vesicle formation, 823

of influenza virus
  oligomer-specific antibodies, 851
  SDS resistance, 850
  as transport marker, in MDCK cells, 814
Hepatocyte plasma membrane
  biogenesis, in rat, 825–841
  microsomal gradients, 839
  vesiculated suspension, sucrose density gradient centrifugation, 831–832
Hepatocyte plasma membrane proteins
  domain localizations, determination of, 833–836
  immunoprecipitation, 833–836
  pulse–chase metabolic radiolabeling, 832–836
  sorting, pathways, 825
Hepatocyte plasma membrane sheets
  apical and basolateral vesicles, 829–832
  free-flow electrophoresis of, 836–838
  vesicle immunoadsorption, 837–840
  characterization, 828–829
  electron microscope analysis of, 829–831
  isolation, 826–828
    recoveries and enrichments of various organellar markers during, 828
  marker enzymes in, 828
  vesiculation, 829–831
HETEs. See Hydroxyeicosatetraenoic acids
Hexokinase, measured in single isolated nephron segments, 334
Hexose transport, sodium ion-dependent, in cultured PCT cells, 418
H^+/HCO_3^- pumps. See Proton/bicarbonate pumps
Histamine receptor, tissues tested to identify, 610
Histochemical staining, in isolated single nephron segments, 340
Histotopographical criteria, for recognition of tubular segments, 287–289
Histotopographical relationships, as landmarks, 285
H^+,K^+-ATPase. See ATPase, H^+,K^+-
Hormonal supplements, for primary cultures of renal cells, 416
Hormone, effects on Na^+,K^+-ATPase, in cultured CCD cells, 422
Hormone receptors
  antibodies against, 304
  identification, prerequisites for, 322
  number of, 306–307
  renal, 303–325
  that are inhibited by or interact with amiloride or amiloride analogs, 740
Horseradish peroxidase
  as marker for fluid-phase endocytosis, 341
  as tracer for endocytosis, 557–559
HT-29 tumor cells
  cotransport activity and bumetanide binding in, 786–788
  [^{3}H]bumetanide-binding experiments in, 792
Human immunodeficiency virus-type 1 env proteins
  cross-linking, 849
  SDS resistance, 850
  sedimentation and cross-linking of, 852–853
Human immunodeficiency virus-type 2 env proteins, SDS resistance, 850
Hybridization arrest of translation system, 639
Hydraulic pressure, and net salt and water absorption in proximal tubule, 90
Hydrochloric acid secretion, by gastric parietal cells, 647–649
Hydrolases, determination of, by fluorimetric measurement of liberated methylumbelliferon, 337
3-Hydroxyacyl-CoA dehydrogenase, measured in single isolated nephron segments, 333–334
3-Hydroxybutyrate dehydrogenase, measured in single isolated nephron segments, 334
5-Hydroxyeicosatetraenoic acid, as mediator of cellular response, 670–671
Hydroxyeicosatetraenoic acids, as mitogens, 683
11α-Hydroxysteroid dehydrogenase
  11-oxidase activity, 524
  11-reductase activity, 524
  renal, 523–528
Hypertension
  induced by transplant of cultured juxtaglomerular cells, 158–162
  juxtaglomerular cell-induced, vascular disease in, 161, 165

I

IAA-94. *See* Indanyloxyacetic acid (IAA)-94
IMCD. *See* Inner medullary collecting duct
Imidazole, neutralization of gastric gland, 731–733
Immunocytochemical studies, 274
  of hormonal receptors, 304
  procedures, 279
Immunostaining
  postembedding techniques, 280
  preembedding techniques, 279–280
IMP clusters, 554–557
  on apical plasma membrane, and transepithelial water flow, 553–555
Impedance measurements, 6–7
Indanyloxyacetic acid (IAA)-94
  chloride transport inhibition, 573
  tritiated, specific activity, 574
  tritiation of, 574–575
Indanyloxyacetic acids, chloride transport inhibition, 573
Indanyloxyacetic affinity column, synthesis of, 576–578
Indo-1
  as calcium indicator, 692–693
  loading cells with, 693
  ratiometric estimation of Ca^{2+} with, 699
Initial collecting ducts, 288
Inner medulla (renal papilla)
  histotopographical criteria for, 288–289
  isolation of plasma membrane vesicles from, 465–469
  microdissection of, 229, 231
Inner medullary collecting duct, 288–289
Inner medullary collecting duct cells
  culture
    growth media and supplements, 416–417
    support, 418
    for vectorial ion transport, 418
  cultured
    hydrogen ion secretion by, 420
    intracellular messenger systems in, 423
    ion channels in, 425
    pH measurements, 420
    uptake of $^{22}Na^+$ into, 420
  intracellular calcium messenger system, 422
  isolated, morphological appearance of, 398–399
  isolation of, 387
    by hypotonic treatment, 414–415
    for primary culture, 413–415
  organic solute membrane transport in, 421
Inner medullary collecting tubules, 310
Inner medullary tubules, suspensions, 327
Inner stripe, 288
Inorganic phosphate, as indicator of intracellular pH, 404–405
Inositol phosphate
  formation, in receptor-stimulated tissued prelabeled with [^{3}H]inositol, studies of, 707
  metabolism, complexity of, 707
Inositol phosphate isomers
  high-performance liquid chromatography, on Partisil SAX or WAX columns, 715
  isocratic ion-exchange high-pressure liquid chromatography of, 711–714
  radiolabeled, two-stage analysis of, 707–718
    advantages of, 708, 714–715
    first stage of, 708–712
    second stage of, 711–715
  separation of
    internal standards for, 715–718
    on ion-exchange resins, 708–712
D-*myo*-Inositol 1,4,5-trisphosphate
  in calcium messenger system, 701
  generation of, receptor-mediated Ca^{2+} signaling initiated by, 691
  microinjection of, into Fura-2-loaded hepatocytes, calcium ion transients induced by, 704–705
1,4,5-Inositol trisphosphate, messenger function, 665
Insect
  Malpighian tubules, 168–182
  renal excretion in, 169
  sources, 170
  urine formation in, 169
Insect gut segments, isolated tubules, *in vitro* perfusion of, 289
Insecticides, inhibition of GABA-activated chloride channel, 800
Ins(1,4,5)P$_3$. *See* D-*myo*-Inositol 1,4,5-trisphosphate
Insulin
  antinatriuretic and antiphosphaturic effect, 63

interaction with antilumenal renal tubular cell membrane, 65–66
ligand–receptor kinetics in renal proximal tubule, 65–66
Insulin receptor
amino acid sequencing of, 638
solubilization, 632
Insulin receptor protein kinases, inhibition by amiloride and/or its analogs, 746
Integuments, epithelia, 1
Intercalated cells, 254, 287–288
accumulation of acridine orange in, 258
band 3 protein for HCO_3^-/Cl^- exchange, 272
collecting tubule, 254
compounds used in identification of, 396
cultured, transepithelial HCO_3^- transport, 420
electrophysiology, 425
endocytosis in, 255
isolation, 414
lumenal microfold formation, 271
markers for, 400
microfolds, 270
mitochondrial strain, 258–260
modulation of lumenal membrane area by membrane recycling, 271
monolayers of, transepithelial electrical resistance across, 424–425
peanut lectin binding to, 260–261
proton ATPase of, associated with studs on cytoplasmic membrane face, 271
proton secretion in, 258
recycling of membrane, 271
rod-shaped intramembrane particles, 271
types of, 263–264
Intermediate junction, 267
Intestinal mucosa cells, polar, activation of C kinase in, 673
Intestine, epithelia, 1
Intracellular ionic activities, 298–299
Intracellular voltage measurements, 295–298
in isolated perfused tubules, 295
purposes, 296
in small cells, 295
Inulin
[$carboxyl$-^{14}C]-, as volume marker, 203
^{14}C-labeled, as volume marker, 242
in extracellular space, for measurement of ion fluxes, 403–404

fluid-phase endocytosis of, 341
in isolated perfused nephron segments, 352–353
as marker of leaky tubules in protein endocytosis studies, 349
[$methoxy$-^3H]-
dialysis, 242–243
as volume marker, 203, 242
6-[^{125}I]Iodoamiloride, 750–751
Ion fluxes, in isolated renal cells
indicators for, 404–405
measurements of, 403–404
Ionic channels, 97
Ion permeability coefficients, of epithelium, 369–370
Ion-selective electrodes, 96–97, 366
manufacture of, 299
measurements with, 298–299
Ion-selective resins, 97
Ion transporters, that are inhibited by or interact with amiloride or amiloride analogs, 740
Ion transport systems
inhibition, by amiloride and amiloride analogs, 739–742
reversible inhibitors of, 739
Isocitrate dehydrogenase, measured in single isolated nephron segments, 333
Isoelectric focusing, study of cAMP-dependent protein phosphorylation, in gastric parietal cells, 655–659
5-[N-(3-Isothiocyanatophenyl)]amiloride, coupled to albumin, used to raise anti-amiloride antibodies, 754

J

Japanese quail, renal tubules, *in vitro* studies, 223–226
JGCs. *See* Juxtaglomerular cells
Junctional complexes, 267
Juxtaglomerular cells
biologic attributes, 152
cloning, by limiting dilution method, 153
culture, 152–153
cultured
appearance, 153–154
electron microscopy, 154–155
fluorescent microscopy, 153–154
light microscopy, 154–155

functions, 156–162
hypertension induced by, 158–162
isolation, 152–153
renin–angiotensin system in, 156–158
Juxtamedullary glomeruli, microdissection of, 309
Juxtamedullary nephrons, *in situ* microperfusion study, 126

K

Kallikrein, inhibition by amiloride and/or its analogs, 748
Kallikrein (kininogenin), measured in single isolated nephron segments, 334
K-562 cells
  inhibition of 3-methoxyglucose transport in, by phloretin, phlorizin, and their azidobenzyl derivatives, 773–775
  sugar transport in, 772–773
KCl cotransport, 785
Kidney. *See also entries under* Renal; Glomerulus
  afferent arteriole blood, colloid osmotic pressures, 124–125
  aglomerular, 186
  arteriolar resistances, 125–126
    resistance per single afferent arteriole, 125
  blood flow rate per single afferent arteriole or glomerulus, 125
  bovine, membranes, preparation of, 574
  concentration–dilution potential of, 93
  corticosterone-metabolizing enzymes in, 524
  efferent arteriolar blood
    collection, measurement of single-nephron glomerular filtration rate by, 121–122
    colloid osmotic pressures, 124–125
    flow rate, 125
  efferent arterioles, hydraulic pressure, 124
  human, microdissection of glomeruli and tubules from, 308
  in vivo transport kinetics, quantitative analysis of, 36
  isolated perfused, 31–34
    advantages of, 31
    apparatus for, 33–34
    arterial cannulas for, 34
    glomerular capillary pressure in, 32
    oncotic pressure of perfusate, 33
    perfusion medium, 33
    proximal tubular pressure in, 32
    pumps for, 34
  mammalian, regional architecture of, 228
  medulla, 39
  membrane transport in, specificity and kinetics of, 35
  microcirculation, 39–40
  micropuncture techniques for, 72–97
  molecular recognition, transport, and mobility within, mechanisms of, 35
  mouse, microdissection of glomeruli and tubules from, 308
  multiple indicator dilution and, 36–72
  Necturus, stop-flow microperfusion of proximal tubules of, 87
  net ultrafiltration pressure, 75–76
  nonfiltering isolated, 31–34
    use of, 31
  perfusion pressure, reduction, to prevent glomerular filtration, 32
  rabbit
    cell isolation from, 381
    microdissection of glomeruli and tubules from, 308
    microdissection of renal tubule segments, 226
    neonatal, explants of S-shaped bodies and of collecting duct anlagen from nephrogenic zone of, 414, 418
    tubule segments isolated from, 254
  rat
    anatomical regions, 39
    blood-free perfusion of, 131
    capillary beds, 39–40
    cortical dissection, 132
    double-microperfusion technique, 98–107
    glomerular isolation from, 131–134, 141–142
      without prior perfusion, 131–134
    microdissection of glomeruli and tubules from, 226–231, 308
    microdissection of tubule segments from, 226–231
    perfusion of
      in nephrectomy, 141–142
      preparation, 32–33
    structural organization of, 326

solute and water fluxes in, 35
structural and functional units, 325
transport kinetics in, 35–36
transport reactions in, 35–36
zones, volume of, determination of, 286
Killifish
  fluid secretion in proximal tubules isolated from, 189–191
  proximal tubules, *in vitro* electrophysiological studies of perfused systems, 190–191, 194–195
Kininases
  measured in single isolated nephron segments, 334
  radioimmunoassay, 339
Kininogen, radioimmunoassay, 339
Kininogenin (kallikrein), radioimmunoassay, 339
Krebs–Henseleit solution, 33

L

L-364, 718
  affinities of, for two subclasses of CCK receptors, 630–631
  in identification of classes of receptors, 618–619
  interaction with CCK receptors, 629–630
  interaction with gastric receptors, 630
L-365,260
  affinities of, for two subclasses of CCK receptors, 630–631
  in identification of classes of receptors, 618–619
Labyrinth, 287
Lactate
  formation
    as indicator of metabolic activity, 401–403
    measured in single isolated nephron segments, 333
  transport, in isolated, perfused snake tubules, 214, 217–218
Lactate dehydrogenase, measured in single isolated nephron segments, 333
Landis–Pappenheimer equation, 33
Lectin binding, 396
Lectin histochemistry, 279
Leucine aminopeptidase
  in apical membranes isolated from renal cells, 502

cellular localization, in renal cells, 501
  measured in single isolated nephron segments, 334
Leukocyte common antigen, 146–148
Leukotriene B_4, and activation of phospholipase, 683
Leukotrienes, 676
  as mediator of cellular response, 670–671
LIGAND (computer program), 625, 628
Light microscopy, for morphological investigation of renal tissue, 273–274
Liposomes, reconstitution of functional transport into, 4
Lipoxin A, 684
Lipoxygenase pathway, 670–671
Lissamine Green, renal transit time, 74
LLC-PK$_1$ cells, 426–427
  adenylate cyclase in
    assay, 568
    unstimulated and vasopressin-stimulated states of, 568–571
  apical membranes isolated from, 501–503
  apical membrane vesicles
    isolation of, 503
    time course of phosphate transport into, 504
  enzyme activities in, cellular localization, 501–502
  isotopic influx measurements in, effect of loop diuretics on, 790
  Na^+-coupled glucose transporter in, inhibited by amiloride analogs, 745
  Na/P_i cotransport in, 496
  radiation inactivation experiments with, 568
  transport systems in, 406–407
Loop diuretics, 793–794
  chemistry of, 781–783
  as cotransport inhibitors, and isotopic influx measurements, 790–791
  mechanism of action, 784–789
  metabolism, 789–790
  related to furosemide, 781–792
Loop of Henle, reabsorption of salt and water, 92
Low-density lipoprotein receptor, amino acid sequencing of, 638
Low-molecular-weight proteins, metabolic clearance, 340

Luciferase
 in measurement of picomolar quantitites of ATP and NAD(P)H, 338
 from *Photobacterium fisherii*, used to measure NAD(P)⁺-dependent enzymes in isolated nephron structures, 339
Lumenal and contralumenal membranes, simultaneous purification of
 from bovine papilla, by free-flow electrophoresis, 466–467
 by free-flow electrophoresis, 459
Lumenal plasma membranes
 from canine inner medulla, isolation of, 467–468
 isolated, purity of, 460–461
 from proximal tubules, isolation of, 451–455
 from red outer medulla, preparation of, using metrizamide density gradient centrifugation, 463–464
 from TALH cells, enrichment of, by differential centrifugation, 461–463
Luminometric methods, 334, 338–339
Lysophosphatidylinositol, 678

M

Macula densa, 309
Madin–Darby canine kidney cells, 426–427, 495, 825
 clonal cell lines of, phenotypic properties, 428–429
 cloning, 428–429
 C6-NBD-ceramide labeling, 816
 cotransport activity and bumetanide binding in, 786–788
 culture, 815
 defined serum-free medium for, 427
 enzyme activities in, cellular localization, 501–502
 epithelial layers of
  electrical properties, measurement of, 436
  preparation, 432–435
 fusion, 439
 giant, 437–449
  attachment to coverslips, 441–442
  enrichment, 439–441
  as expression system, 437–449
  advantages of, 449
  disadvantages of, 449
  microinjection, 443–448
  pressure-induced, 447–448
 mRNA-injected
  electrical properties of, 448–449
  experimental yield with, 449
  survival, 449
 preparation of
  for microinjection, 443–445
  for microinjections, 439–441
 size, 439
 growth medium, 427
 [³H]bumetanide-binding experiments in, 792
 influx of Na⁺ in, determinations of, 431–432
 isotopic influx measurements in, effect of loop diuretics on, 790
 K⁺ transport, measurement, 431
 metabolic labeling, 815–816
 Na/P$_i$ cotransport in, 496
 perforation procedure, 816–817
 plasma membrane domains, biotinylation experiments, 822–823
 plasma membrane proteins, sorting, pathways, 825
 polarized, *in vitro* recovery of exocytic transport vesicles from, 813–825
 single-cell suspension for serial passage of stock Roux bottles, formation of, 428
 stock cultures, 427–428
  freezing of, 429–430
 strains, phenotypic characteristics, variation in, 429
 trans-Golgi network of, transport vesicles derived from, 813
 transport vesicles, 813–825
  analysis, 820–823
  differential centrifugation, 817
  formation, 816–817
  formation and release of, markers, 814
  immunoisolation, 817–820, 824
  isolation, 817–818, 823–824
  recovery of, 813–814, 823
  surface biotinylation experiments, 822
  TGN-derived, vesicle-specific proteins, functional characterization of, 824–825
  total protein composition of, 820–821

Triton X-114 phase partitioning, 821
vesicle-specific proteins, identification of, 821–823
unidirectional transport studies in, 431–432
viral infection, 815–816
Magnesium
proximal tubule fluid concentration, 90–91
reabsorption, in proximal tubule, 91
secretion, by fish proximal tubules, 190–191
transport
in fish renal tubules, 193–194
in teleost renal tubules, 186
Malate dehydrogenase, measured in single isolated nephron segments, 333
m-Maleimidobenzoic acid N-hydroxysuccinimide ester, 634
Malic enzyme, measured in single isolated nephron segments, 333
Malignant hypertension, morphologic expressions of, 161, 165
Malpighian tubules, 168–182
elastase treatment, and formation of blebs in cell membranes, 182
in vitro studies of, 170–182
electron probe analysis of secreted fluid, 173–175
electrophysiological methods for, 175–176
investigation of sodium secretion, 176–178
microperfusion methods for, 174–176
osmotic pressure measurements, 172–173
patch-clamp studies of ion channels, 181–182
Ramsay method, 170–172
storage of small fluid volumes for, 171
voltage response, as bioassay for diuretic hormone, 178–181
isolated, *in vitro* perfusion of, 289
isolation of, 170
organic solute secretion, analysis of, 174
secreted fluids, elemental analysis of, 173–175
secretory function, and cAMP, 173–175, 177–178
Marbrook chambers, culture of epithelial layers in, 434–435

Marine elasmobranchs. *See* Elasmobranchs
MDCK. *See* Madin–Darby canine kidney cells
Mean glomerular transcapillary hydraulic pressure difference, in alteration in glomerular filtration rate, 110–113
Media
basal nutrient, 415–416
supplements, 415–416
Medical imaging, multiple indicator dilution-based, 67–71
Medullary cells
isolated, structural markers for identification of, 396–397
isolation of
dissociation methods, 391–393
perfusion solutions and conditions, 387, 390
separation methods, 391
Medullary collecting duct cells, transport systems in, 407
Medullary collecting tubule, 326. *See also* Inner medullary collecting tubules; Outer medullary collecting tubules
pH of, 257
Medullary ray collecting ducts, 288
Medullary rays, 287
Medullary thick ascending limb of loop of Henle, 326
aldosterone sensitivity, 520
isolation of plasma membrane vesicles from, 461–465
microdissection, 309
Medullary thick ascending limb of loop of Henle cells
centrifugal elutriation, 412
cultured, electrophysiological studies of, 424–425
immunodissection of, 412
isolated, morphological appearance of, 398–399
isolation of, 387
for primary culture, 412
linear density gradient centrifugation, 412
microdissection of, 412
transfected with early region SV40 genes, differential transport functions in, 422
transfected with plasmid pSV2-neo DNA plus DNA from early region of Simian virus 40, 413

Medullary tubules, suspensions, 327
Membrane amplification
　assessment of, 282
　definition of, 282
Membrane resistance, measurement of, dye probes for, 479
Membranes, binding of amiloride analogs to, 751–752
Membrane vesicles, 2
Membrane voltage, stability, and impalement, 297–298
Membrane water permeability, biophysical parameters of, 561–562
Mesangial cells
　contractile, 146–147
　culture of, 145–148
　hillocks, 147
　in homogeneous culture, 144
　isolation, 146–148
　types, 146
Messenger RNA
　microinjection of, into intact mammalian cells, 437–438
　from rat intestine, preparation of, 441–443
5-(N-Methyl-N-isobutyl)amiloride
　potency relative to amiloride, in inhibiting ion transporters, 739–742
　tritiated, 750
N-Methylnicotinamide, net reabsorption of, from isolated, perfused snake tubules, 214, 218–219
4-Methyl-3-phenylaminothiophene-2-carboxylic acid, as chloride channel blockers, 807–809
Mg^{2+}-ATPase, measured in single isolated nephron segments, 333
Microassays, applied to renal tissue, 332–340
Microcatheterization, renal, 95
Microdissection, 305, 328–330, 380
　from fresh tissue, 330
　of glomerulus, 130
　of individual cortical nephron segments, for primary culture, 411–413
　from lyophilized tissue sections, 328–330
　of nephron segments, 308–310
　pros and cons of, 331–332
　of renal tubule segments, 226–231
Microelectrodes, 295–296. *See also* Ion-selective electrodes

cellular impalement with, 297
double-barreled, 299
resistance, during impalement, 297
tip, sealing of, 297
tip potential, 297
Microfold formation, 270
Microinfusion, renal, 81–82
Microinjection
　renal, 81–82
　　of distal convoluted tubule, 94–95
　techniques, to introduce nonpermeable compounds into cells, 704–706
Microperfusion, 325
　continuous, in renal tubule, 102–103
　　with simultaneous capillary perfusion, 104
　in vitro
　　of amphibian renal tubules, 197–198
　　of isolated glomeruli, 130
　　for studies of Malpighian tubules, 174–176
　of juxtamedullary glomeruli, 126
　of peritubular capillaries in renal cortex, 105–106
　of proximal tubules, 343
　renal, 82–84
　　double-perfused tubule *in situ*, 98–107
　　animals for, 98
　　capillary microperfusion only, 105–106
　　composition of perfusates for, 107
　　experimental set-up for, 98–101
　　lumenal and simultaneous capillary perfusion, 104–105
　　lumenal stationary and continuous microperfusion, 102–103
　stationary, 87
　renal, 84
　stop-flow, 87
　of teleost renal tubules, 192–194
　　detection of low-capacity solute transport systems, 193–194
　　detection of small transepithelial volume flow, 192
　　perfusion rates, 192
Micropipets
　binding of protein to, 344–345
　constant-bore, 235–238, 240–241
　filling, with mRNA, 446–447
　fixed volume, 235–236, 238–241

advantages and disadvantages of, 240–241
calibration, 238–240
nomogram for approximating volume of, 238–240
glass, 99–100
device to sharpen, 100–101
double-barrelled, 100
for holding collecting end of isolated perfused tubule, 233–236
preparation of, 445–446
protein coating of, 344–345
for sampling collectate from isolated perfused tubule, 233–235
siliconization of, 344–345
volumetric, 235–236
Micropuncture, 325
free-flow, 78–81, 86
oil block for, 78–79
recollection technique, 80
sites, 78–79
glomerular, 75–78
allowing measurement of single-nephron glomerular filtration rate and its determinants, 123–126
in vivo, 130
in vitro, 73
in vivo, 73
recollection, 89
renal, 123–126
contributions of, 85–86
distal tubule fluid collection, 119–120
early proximal tubule fluid collection, 120–121
future of, 96–97
proximal tubule fluid collection, 117–119
in renal research, 70, 72–97
technical problems in, 85
technique, 73
tubular, 78–97
Microscope
inverted, differential interference optics for, 372–373
level of focus of, to assess cell volume, 373
Microvilli
brush border, 269
stubby, 269–270
MID. *See* Multiple indicator dilution

Millipore filters
cell layers grown on, assessment of confluency in, 433
collagen-coated, as support for epithelial layers, 433
uncoated, as support for epithelial layers, 433
Mineralocorticoid receptors
assay, 529
kinetics, 521–522
renal, 521–528
equilibrium dissociation constants for [^{3}H]aldosterone binding to, 522
equilibrium dissociation constants for [^{3}H]dexamethasone binding to, 522–523
steroid specificity of, 521
Mitochondrial transmembrane potential, in intercalated cells, 259
Molecular weight, determined by velocity gradient sedimentation, 845
Monkeys, tubule fluid-to-ultrafilterable (TF/UF) magnesium ratio, 91
Monoacylglycerol lipase, in arachidonate release, 677–681
Morphological studies, 265–289
microscopic approaches, 273–274
of nephron segments, 253–254
Mosquito. *See also* Malpighian tubules
sources, 170
Mosquito natriuretic factor, 173–174
bioassay for, 178–181
effects on Malpighian tubule secretory function, 179–181
Motlin receptor, tissues tested to identify, 610
MTAL. *See* Medullary thick ascending limb of loop of Henle
Multiple indicator dilution
applications, 67
to transient and time-dependent phenomena *in vivo*, 67–68
definition of, 36
dynamics of, 37
hypothesis underlying, 37–38
as instrument for *in vivo* kinetics, 66–67
limitations of, 67–72
mathematical models of, 68
methodological restriction to (impulse) input–output measurements, 68–70

modeling, progress in, 71–72
principle of, 36
regional studies in, augmenting input–output measurements at whole-organ level, 68–71
renal, 36–72
  dual output design of, 40
  experimental design, 40–42
  investigation of glomerular permselectivity, 49–55
  investigation of peritubular capillary permselectivity, 55–57
  investigation of transport and receptor kinetics in proximal tubule, 57–66
  mathematical approaches to, 37, 43–49
  method, 39–42
  modeling approach to analysis of local tissue physiology and dynamic kidney imaging, 68–71
  multiple simultaneously injected indicators used in, 40
  protocol for, 40
  separation of glomerular and postglomerular events, 42–43
  studies of renal pathophysiology, 72
  stepwise comparison of test substance ⇌ extracellular reference ⇌ vascular reference, 67–71
Muscarinic acetylcholine receptor
  amino acid sequencing of, 638
  inhibition by amiloride and/or its analogs, 747
Muscarinic cholinergic receptor
  radiolabeled ligands used in binding studied to identify, 621
  tissues tested to identify, 610
Mycoplasma, contamination, of established cell lines, detection of, 430
Myosin light chain kinase, 664
Myotonia, 572

N

Na^+/Ca^+ exchanger. See Sodium/calcium exchanger
$Na^+/2Cl^-/K^+$ carrier. See Sodium/potassium/chloride cotransport
NADH, measured in isolated nephron segments, 336

NAD(P)H, measured in isolated nephron segments, 338
Na^+/H^+, localization, to apical or basolateral membrane, 263
Na^+/H^+ antiporter, in rabbit and rat PCT cultures, 418–420
Na^+/H^+ exchanger
  characterization, with radiolabeled amiloride analogs, 751
  inhibition by amiloride and/or its analogs, 739–744
  in isolated renal cells, 406–407
Na^+,K^+-ATPase. See ATPase, Na^+,K^+-
$Na^+/K^+/Cl^-$ cotransport. See Sodium/potassium/chloride cotransport
C6-NBD-ceramide
  labeling of MDCK cells, 816
  as transport marker, in MDCK cells, 814
C6-NBD-glucosylceramide, as transport marker, in MDCK cells, 814
C6-NBD-sphingomyelin, as transport marker, in MDCK cells, 814
Necturus ambyostoma, source, 196–197
NEM. See N'-Ethylmaleimide
Nephron
  acid intracellular compartments in density, 258
  distribution, 258
  of avian kidney, 223
  concentration profile along, 80, 86
  rabbit, isolated structures of, 329
Nephron cells
  cultured
    electrophysiology, 424–425
    requirements for growth and differentiation, 415–418
  immunodissection of, 411
  monolayer cultures, substrata and support of, 417–418
  segmental epithelial transport functions, in culture, 418–426
  solute transport and metabolism, in culture, 418–421
  types of
    correlation of structure with transport, 270–273
    surface area of basolateral celll membranes, 270
Nephronogenesis, 107–108
Nephron segments, 325–326

axial heterogeneity, 284
cell types in, 253
collected volume of, 283
defined
  isolation of homogeneous cell populations from, 380–381
  isolation procedures for, 326–332
  primary cultures of cells derived from, 409–426
  primary homogeneous cultures of, 409–426
  transport in isolated cells from, 380–409
differences in surface area of basolateral membranes among, 270–271
fractional volume of, 283
  determination of, 286
identification of different cells in, using response of pH_i to various maneuvers, 263–264
isolated, cell types in, 253–265
isolation, 254
measurement of cell pH in, 261–263
microdissected
  hormone and drug-binding microassays, 315–325
  primary culture of cells derived from, 411–413
  receptor studies using, general aspects, 306–308
morphological studies, 253–254
perfusion, 254
Netropsin, effects on binding of [^{3}H]aldosterone-labeled receptor complexes to chromatin, 532
Neutral lipids
  chromatography, 687–688
  extraction, from pancreatic preparations, 686–687
Nicotinic acetylcholine receptor
  amino acid sequencing of, 638
  from *Torpedo*, inhibition by amiloride and/or its analogs, 746
N-6(7-Nitro-2,1,3-benzoxadiazol-4-yl). *See* C6-NBD-ceramide
Nitroglycerin, as chloride channel blocker in muscle, 801
3′-(4-Nitrophenylcarbinol)-2′,4′,6′,4-tetrahydroxy-dihydrochalcone, synthesis of, 756–759

5-Nitro-2-(3-phenylpropylamino)benzoic acid, as chloride channel blockers, 807–809
4-Nitrophloretin, synthesis of, 758, 761–762
4-Nitrophlorizin, synthesis of, 758, 762–763
Nonelectrolyte permeabilities, of epithelium, 368–369
Novobiocin, effect on cell membrane resistance, 15
Nuclear binding assays, of ^{3}H-labeled steroid–receptor complexes, 532–535
5-Nucleotidase, measured in single isolated nephron segments, 333
Nystatin, effect on cell membrane resistance, 15

O

22α-OHSD. *See* 11α-Hydroxysteroid dehydrogenase
OK cells, 495
  apical membranes
    isolated, 501–503
    isolation of, 503–504
  apical membrane vesicles prepared from, time course of phosphate transport into, 504
  Na/P_i cotransport in, 496
Oligomerization
  in endoplasmic reticulum, 854
  and protein transport, 841–842
Omeprazole, 721–722
  effect of
    on gastric glands, 729–731
    on isolated gastric mucosa, 724
  inhibition of H^+,K^+-ATPase, and gastric acid secretion, 738
  inhibitors, effect on human gastric glands, 733
Onion skin lesion, 161, 165
Opiate receptor
  radiolabeled ligands used in binding studied to identify, 621
  solubilization, 632
  tissues tested to identify, 610
Optical methods, for transport studies, 471–475
Optical probes, weak base, for measurements of changes of intravesicular pH, 473–475

SUBJECT INDEX

Organic solutes, transport, in teleost renal tubules, 186
Osmotic pressure measurements, in *in vitro* studies of Malpighian tubules, 172–173
Ouabain, tritiated
  binding studies in rabbit nephron segments, 549
  binding to isolated tubules, standard assay conditions for measurements of, 319–320
Outer medulla
  histotopographical criteria for, 288
  isolation of cells from, 381
  microdissection of, 229–230
  plasma membrane vesicles from, isolation of, 461–463–465
Outer medullary collecting duct, 288
Outer medullary collecting duct cells, sodium transport-related O_2 consumption, 401–402
Outer medullary collecting tubules, 310
  aldosterone sensitivity, 520
  arginine vasopressin-induced cyclic AMP accumulation in, inhibition by norepinephrine, 314
Outer medullary tubules, suspensions, 327
Outer stripe, 288
3-Oxoacid transferase, measured in single isolated nephron segments, 333–334
2-Oxoglutarate dehydrogenase, measured in single isolated nephron segments, 334
Oxonol VI
  difference spectrum of, 478
  in monitoring of intravesicular potential, 478–479
Oxygen consumption
  in cultured PCT cells, 421
  as indicator of sodium movement through cell, 399–403

P

^{32}P. *See* Phosphorus-32
Pancreas, isolated tubules, *in vitro* perfusion of, 289
Papillary collecting duct, aldosterone sensitivity, 521
Papillary collecting duct cells. *See* Inner medullary collecting duct cells

compounds used in identification of, 396
sodium transport-related O_2 consumption, 401–402
transport systems in, 407
Paracellular pathway, 267, 291–293, 436
  permeability of, for backflux of ion, 370
Paracellular resistance, 1. *See also* Paracellular shunt conductance; Transepithelial resistances
Paracellular shunt conductance, in epithelia, 4–27. *See also* Transepithelial resistances; Voltage divider ratio
  calculation of
    errors made when either R_{sh} or R_{bl} do not remain constant, 21–23
    when $I-V$ relations of all cell membranes are linear and 0-current potentials constant, 23–27
  determination of R_{sh} from break point in transepithelial current voltage relation, 16
  distributed model, 19
  impedance techniques, 16–20
  macroscopic methods, 11–16, 20
  measurement of epithelial and cell membrane potentials in response to alteration of one membrane resistance, 11–12
  measurement of R_T, VDR, and relationship between cell membrane potential and I_{SC}, 12–13, 23–27
  measurement of transepithelial impedance and voltage divider ratio
    in frequency domain, 18–20
    in time domain, 17–18
  methods based on lumped model circuits, 20
  plot of G_T vs. voltage divider ratio in response to alteration of one cell membrane resistance, 11
  plots of epithelial conductance vs. I_{SC} in response to alteration of one cell membrane resistance, 13–15
  steady state measurements of intracellular activity of ion species for which one cell membrane is permselective, 15–16
  steady state measurements of V_T, 15–16
Paracellular shunt pathways, 2
  in proximal tubules, 87

Paraformaldehyde
  in fixative solution, 276
  solution, 277
Parathyroid hormone
  effect of
    on cAMP, in PCT cells, 422
    on intracellular calcium, in cultured renal cells, 421–422
    on renal tubular phosphate reabsorption, 675–676
  regulation of Na/P_i-cotransport system by, 495
Parietal cells
  enrichment of, 733–734
  isolated, 733–734
Pars recta, 326
  cortical portion of, 309
  terminal portion of, 309
Passive transport, 1
  in proximal tubular reabsorption of sodium, 88
Patch-clamp technique, 4, 97, 425
  application of, to isolated tubule, 299–302
  identification of chloride channels by, 793
  for Malpighian tubule ion channels, 181–182
  for recording of single ionic channels, 798
  with *in vitro* perfused tubule preparation, advantages of, 301–302
PCT. *See* Proximal convoluted tubule
Peanut lectin. *See also* Rhodamine-peanut lectin
  binding to intercalated cells, 260–261
  coupled to fluorescein or rhodamine, 260
Pentylenetetrazole, inhibition of GABA-activated chloride channel, 800
Perognathus penicillatus
  tubule fluid-to-ultrafilterable (TF/UF) magnesium ratio, 91
  tubule fluid-to-ultrafilterable potassium (TF/UF$_K^+$), 90
Peroxidase techniques, 280–281
pH
  of endocytic vesicle, 255
  intracellular, indicators, 404–405
  measurements
    in cells, 470–471
    in vesicles, 470–471
  probes, 472
Phenamil, 743
  potency relative to amiloride, in inhibiting ion transporters, 739–742
[phenyl-^{3}H]Phenamil, 750
pH gradient, measurement, in cells and vesicles, 470–471
Phloretin
  aryl azide derivatives, 757
  inhibition of sugar transport systems, 755–756
  photoaffinity-labeling analogs
    azides, as membrane perturbants and photoaffinity-labeling agents, 767–780
    synthesis, 756–765
    thin-layer chromatography, 765–766
    tritiated, preparation of, 765–767
  preparation, 757–758
  source, 757
Phlorizin, 584
  aryl azide derivatives, 757
  inhibition of sugar transport systems, 755–756
  photoaffinity-labeling analogs
    azides, as membrane perturbants and photoaffinity-labeling agents, 767–780
    synthesis, 756–765
    thin-layer chromatography, 765–766
    tritiated, preparation of, 765–767
Phlorizin binding
  in brush border, 586
  renaturation of, after solubilization of Na$^+$-D-glucose cotransporter, 590–591
Phosphate, reabsorption, in proximal tubule, 91
Phosphate-dependent glutaminase, measured in single isolated nephron segments, 333
Phosphate transport
  in cells grown in petri dishes, 498–499
  in cells grown on permeant filter supports, 499
  in established renal epithelial cell lines, 494–505
  studies with intact cell monolayers, 497–501
Phosphatidylcholine, arachidonate in, 677
Phosphatidylethanolamine, arachidonate in, 677

Phosphatidylinositol, arachidonate in, 676–677
Phosphatidylinositol breakdown
　balance sheet analysis of, 685–691
　　prelabeling of whole pancreata, 685
　　preparation and incubation of minilobules for, 686
　　results, 679–681, 688–691
　in [^{14}C]arachidonate- or [^{14}C]stearate-prelabeled pancreata
　　increases in labeling of arachidonate on stimulation of, 688–689
　　increases in labeling of DAG on stimulation of, 688–689
　　increases in labeling of PA on stimulation of, 688–689
　　increases in labeling of stearate on stimulation of, 688–689
　　increases in labeling of TAG on stimulation of, 688–689
　[^{3}H]glycerol-prelabeled, increases in labeling in PA, DAG, TAG, and free glycerol on stimulation of, 680, 689
　in minilobules prelabeled with [^{3}H]glycerol, radioactivity in lysophosphatidylinositol on stimulation of, 690–691
1-Phosphatidylinositol-4-phosphate kinase, preparation of, 715–716
Phosphatidylinositol phosphorus, determination of, 688
Phosphodiesterases, 662–663
　activation of, 663
　measured in single isolated nephron segments, 334
Phosphoenolpyruvate carboxykinase, measured in single isolated nephron segments, 333–334
Phosphofructokinase, measured in single isolated nephron segments, 333
6-Phosphogluconate, measured in single isolated nephron segments, 333
Phosphoinositidase C, activation of, study of, 707
Phosphoinositide breakdown
　and liberation of arachidonate and/or eicosanoids, 677
　stimulation of, with caerulein, 678–681
Phospholipase A$_2$, 670
　activation of, relationship to phosphoinositide cascade, 682

Ca^{2+} dependency of, 682
　pathway, in arachidonate release, 678
　in release of arachidonate from phosphatidic acid, 678
Phospholipase C, activation of, 613, 615, 664, 691
　possible role of arachidonate and/or Arachidonic acid metabolites in, 683
Phospholipase C–DAG–lipase pathway, of arachidonate release, 677–681
Phospholipids
　chromatography, 687–688
　extraction, from pancreatic preparations, 686–687
Phospholipid turnover, measured in single isolated nephron segments, 334
Phosphoprotein phosphatase, 662–663
　inhibition of, intracellular cAMP in, 663–664
　regulation of, 663–664, 666
Phosphorus-32
　nonspecific absorption to filters, blank value which corrects for, 504
　stock solution, contaminated by pyrophosphate, 504–505
Phosphorylase b kinase, 666
Photoaffinity labeling, 4, 634
Photometry, 334, 337
Physiology
　scalar, 71
　vector, 71
Picrotoxin, inhibition of GABA-activated chloride channel, 800
Pipets. See Micropipets
Piretanide
　as blocker of $Na^+/2Cl^-/K^+$ carrier, 804
　structure, 783
PK 11195, tritiated
　binding to isolated tubules, standard assay conditions for measurements of, 319–320
　mapping of peripheral benzodiazepine-binding sites along rate nephron using, 324
Planar area, definition of, 282
Planimetric measurements
　techniques of, 283–284
　test grids, 283–284
Plasma colloid osmotic pressure, systemic,

and changes in single-nephron glomerular filtration rate, 113
Plasma membrane vesicles, renal
   from inner medulla, isolation of, 465–469
   lumenal and contralumenal, isolation of, 450–469
   from outer medulla, isolation of, 461–465
   from thick ascending limb of Henle's loop, lumenal, 461
Platelet-derived growth factor receptor protein kinase, inhibition by amiloride and/or its analogs, 746
Platelet release reaction, 670–671
Platelets, arachidonate release in, 678
Polyethylene glycol
   in extracellular space, for measurement of ion fluxes, 403–404
   as volume marker, 203, 242
Polypeptide hormones, metabolic clearance, 340
Polypeptides, endocytosis, by proximal tubular cells, 340–354
Polyphosphoinositides, arachidonate in, 676–677
Positive hybridization selection system, 639
Positron emission tomography, 70
Postglomerular (peritubular) capillary, permselectivity, 55
Potassium, in mosquito hemolymph and fluid secreted by isolated Malpighian tubules, 173–174
Potassium channel, 799
   in apical cell membrane, 15
   Ba^+-sensitive, 422
   blockers, 293
   in intact frog atrial myocytes, inhibition by amiloride and/or its analogs, 746
   revealed by patch-clamp analysis, 301–302
Potassium ferrocyanide, as volume marker, 203
Potassium handling, in proximal tubule, 90
Principal cells, 254
   aldosterone-mediated Na^+ reabsorption in, 538–540
   clathrin-coated pits on surface of, IMP aggregates in, 556–557
   water channel recycling in, proposed pathway of, 551–552
   water permeability of apical plasma membrane of, 551
Proflavine sulfate, effects on binding of [^{3}H]aldosterone-labeled receptor complexes to chromatin, 532
Proglumide analog 10 (CR 1409), effect of, on CCK-8-stimulated amylase secretion from dispersed pancreatic acini, 616–617
Proline isomerase, 841
Pronase
   digestion of gastric mucosa, 641–643
   source, 643
5-([N-ethyl-N-^{3}H]Propyl)amiloride, 750
Prostacyclin, 671
   inhibition of phosphoinositide hydrolysis, 683
   as mediator of cellular response, 670–671
Prostaglandin E_2, 671, 685
   inhibition of phosphoinositide hydrolysis, 683
Prostaglandin $E_{2\alpha}$, 685
Prostaglandins, 676
   as mediator of cellular response, 670–671
   synthesis, study of isolated glomeruli, 135–136
Protease resistance, acquisition of, by newly synthesized proteins, 852
Protein, 751
   absorption, by fluid-phase endocytosis, 341
   assembly, 842
      monitoring, with velocity gradient sedimentation in conjunction with pulse–chase experiments, 846–847
   chemical cross-linking
      correlated with velocity gradient sedimentation, 852–853
      for structural characterization, 847–850
   endocytosis, by proximal tubular cells, 340–354
   labeling of
      with ^{125}I, 345–346
      with tritium, 345–346
   lysosomal hydrolysis, 342
   in proximal tubule, 340
   oligomeric structure of, chemical cross-linkers for characterizing, 847–850
   protease resistance, as assay of protein folding and assembly, 852

reabsorption, in tubule perfusions, measurement and calculation of, 350–352
synthesis, inhibition by amiloride and/or its analogs, 748–749
tertiary and quarternary structure, and transport from ER, 841–842
tubular absorption and metabolism of, 341–342
velocity gradient sedimentation, 842–847
correlated with chemical cross-linking, 852–853
Protein disulfide isomerase, 841
Protein folding, 841–842
Protein kinase A, activation of, 613
Protein kinase C, 746
activation of, 613, 673, 704
and arachidonate metabolism, 684
in arachidonate release, 682
in calcium messenger system, 701
Ca^{2+}-sensitive, 665–666
inhibition by amiloride, 743–744
Protein kinase (cAMP), measured in single isolated nephron segments, 334
Protein kinases
calcium/calmodulin-dependent, 666
calmodulin-dependent, 665
free catalytic units, 663
inhibition by amiloride and/or its analogs, 746
regulatory units, 663
Protein sorting
plasma membrane, in rat hepatocytes *in situ*, 825
studies, 825
Proton ATPase (H^+-ATPase), 258, 572
activity, in endocytotic vesicles, 517–518
coupled optical assay, 517–518
measurement of, 509
buffers and solutions for, 507
of intercalated cells, associated with studs on cytoplasmic membrane face, 271
in isolated renal cells, 407
measured in single isolated nephron segments, 334
vacuolar, 505
Proton/bicarbonate pumps, localization, to apical or basolateral membrane, 263
Proton pump
ATP-driven, 505

in endocytotic vesicles, 505–520
of stomach, 721
Proton transport, transepithelial, 255–258
Proximal convoluted tubule, 326
isolated perfused, endocytic uptake and lysosomal hydrolysis of proteins, 344–354
microdissection of, 229–230, 309
rabbit, procedure for perfusion of, 344
Proximal convoluted tubule cells
culture
media for, 415–416
support, 417
cultured
electrophysiological studies, 424
intracellular calcium concentrations, 421
ion channel characteristics of apical cytoplasmic membrane of, 425
solute transport and metabolism in, 418–420
isolation of, for culture, 410–412
Proximal straight tubule, 326
endocytosis in, 255–257
pH of, 257
Proximal tubule, 40
apical cell membrane conductance in, 10
convoluted segment (pars convoluta), 85–86
endocytosis in, 340–354
determination of intact protein and metabolites, 353–354
experiments, criteria for adequacy of, 349–350
methodological approaches to, 342–343
study of
clearance techniques for, 342–343
differential centrifugation techniques for, 342–343
general protocol for, 347–349
microperfusion techniques for, 342–343
morphological techniques for, 342–343
epithelium, reabsorptive net transport of sodium across, 86–87
histotopographical criteria for, 287
hydraulic pressures, 123–124
isolated perfused, analysis of endocytosis and metabolism of proteins in, 343–354

ligand–receptor kinetics, investigation using renal multiple indicator dilution, 65–66
lumenal membranes from, purification of, 451–455
pars recta, 86
passive sodium chloride flux, 88
salt and water transport within, controls of, 89
segments, structural differences of, related to specific functional variations, 86
simple double-cylinder model of, 374–375
S_1 segments, 287
S_2 segments, 287
S_3 segments, 287
superficial, concentration profile of electrolytes along, 85
transport kinetics, quantification of, using renal multiple indicator dilution, 61
transport pathways, enumeration of, using renal multiple indicator dilution, 58–60
Proximal tubule cells
endocytic uptake of filtered proteins, 341
immunodissection of, 411
isolated, morphological appearance of, 398–399
isolation of, 381–386
perfusion conditions for, 381–384
perfusion solutions for, 381–384
separation methods, 386, 388–389
lumenal surface of, 269
markers for, 400
sodium transport-related O_2 consumption, 401–402
transporters in, 3–4
transport systems in, 406
Psammomys obesus
tubule fluid-to-ultrafilterable (TF/UF) magnesium ratio, 91
tubule fluid-to-ultrafilterable potassium (TF/UF$_K^+$), 90
Pseudopleuonectes americanus. See Flounder
Pyranosides, interaction of, with opposing surfaces of tubular epithelium, competitive inhibition experiments *in vivo*, 60

Pyruvate carboxylase, inactivation, in lyophilized or freeze-dried tissue, 331
Pyruvate kinase, measured in single isolated nephron segments, 333

Q

Quartz fiber balance
sensitivity of, 542–543
to weigh pieces of freeze-dried tubules, 542–543
Quin2, 692
absorption of, amiloride interference with, 749

R

Rabbit. *See also* Kidney, rabbit
tubule fluid-to-ultrafilterable (TF/UF) magnesium ratio, 91
tubule fluid-to-ultrafilterable potassium (TF/UF$_K^+$), 90
Radiation inactivation
of enzyme systems, 566–567
experiments, with Na$^+$-D-glucose cotransporter, 585–586
Radiochemical procedures, 334, 338
Radioimmunoassay, 334, 339
of bradykinin, 339
cyclic nucleotide
experimental procedures, 311–312
preliminary control experiments, 312–313
sensitivity of, 313
of intracellular cAMP, 307, 310–314
of intracellular cGMP, 307, 310–314
of kininases, 339
of kininogen, 339
of kininogenin (kallikrein), 339
Radiolabeled ligand, binding of, as function of concentration of radiolabeled ligand, 622–623
Raffinose, as volume marker, 241–242
Ramsay method, for *in vitro* studies of Malpighian tubules, 170–172
Rapid filtration technique, to measure flux of radiolabeled ligands, 481–483
Rat. *See also* Kidney, rat

SUBJECT INDEX

tubule fluid-to-ultrafilterable (TF/UF) magnesium ratio, 91
tubule fluid-to-ultrafilterable potassium (TF/UF$_K^+$), 90
Receptor antagonists, used to identify and characterize various classes of receptors, 616–619, 628–630
Receptor identification, 609–639
  by binding of radiolabeled ligands, 619–631
    materials, 619–620
    methods, 620–630
  by biological activity, 609–619
  by covalent cross-linking, 634–636
  protein and amino acid sequencing in, 636–637
Receptors
  binding of ligand to, 623–624
  effector unit or transducer, 609
  gene sequencing of, 637–639
  purification of, 636
  regulatory unit or binding site, 609
  sequencing, using recombinant DNA technology, 637
  solubilized
    characterization of, 632
    determining saturable binding to, 632
    in receptor reconstitution studies, 633
Receptor solubilization, 631–634
Rectal gland tubules, shark. *See* Shark, rectal gland tubules
Red blood cell, chloride flux in, 801
Renal blood flow, autoregulatory maintenance of, in face of reduction in renal perfusion pressure, 112
Renal cell lines. *See also* LLC-PK$_1$ cells; Madin–Darby canine kidney cells; OK cells
  Caki-1, 426
  Cur, 426
  GRB-MAL1, 417
  GRB-MAL2, 417
  GRB-PAP1, 417
  GRB-PAP-HT25, 420
  JTC, 426
  JTC-12, 495
    Na/P$_i$ cotransport in, 496
  LLC-PK, 495
  MAL-1, 412
  MAL1 clone A3, 412, 417

MTAL-1C, 417
MTAL-1P, 417
mycoplasma in, elimination of, 430
PAP-HT25, 417
phenotypic heterogeneity, 426–427
phosphate transport in, 494–505
PK$_1$, 426
RCCT-28A, 414
tissue culture of, 426–436
Renal cells
  cultured, transcellular signal systems, 421–423
  isolated
    cellular components and cell surface markers, 396, 400
    characterization of, 391–399
    of defined segmental origin, primary culture of, 409–426
    morphology, 391–399
    oxygen consumption studies on, 401–403
    transport in, 380–409
    transport-related parameters in, 399–409
    transport systems in, 406–408
  isolation of, 381
    for culture, 410–415
  swelling, nonuniform shape changes during, 375–377
  volume of. *See also* Cell volume
    determining, 371
Renal clearance, concept of, 116–117
Renal cortex, 39. *See also* entries under Cortical
  histotopographical criteria for, 287
  isolation of cells from, 381
  microdissection of, 229–230
Renal cortical cells, isolation of, 381–386
  dissociation methods, 385–386
  perfusion solutions and perfusion conditions for, 381–384
  separation methods, 386, 388–389
Renal cortical tissue, enzymatic disaggregation of, 410–411
Renal epithelia
  investigation of, *in situ*, strategies for, 273–281
  quantitative evaluation of, 281–287
  structural modulation of, 265
  study of, *in situ*, 266

Renal physiology, history of, 167–168
Renal tissue
  embedding
    into conventional epoxy resin, 278
    at low temperatures, 278–279
  fixation of, 274–275. *See also* Fixative
    procedure, 275–276
  fixed
    freezing, 279
    hardening of, 277–278
    localization of antigen in, 280–281
    processing of, for sectioning, 277–278
  fractionation, 304–305
  freeze-dried sections, microdissection of,
    328
  freeze-fracture, 552–556
  frozen, semithin and ultrathin sections of,
    279–280
  functional morphology, 265–289
    sample size for statistical analysis,
      286–287
  immunostaining. *See* Immunostaining
  processing of
    for microscopic investigation, 274–281
    shrinkage or swelling in, 284
  sampling procedures, 284–287
    and cellular parameters, 285
    definition of functionally corresponding
      locations of cells within a segment,
      284–285
    and fractional volumes of cells within
      segments, 285
    population of cells to be analyzed, 284
    randomness of sampling, 285
    recognition of cell type for, 284
    and tubular parameters, 285
  structural preservation of, 275
  ultrathin sections, cellular parameters
    measured on, 285–286
Renal transport studies, by optical methods,
  469–479
Renal tubule cells
  isolated, suspensions, 327–328
  mitochondrial density and transport rate,
    272
Renal tubule fluid, ultramicroanalytical
  approaches to chemically define, 95
Renal tubules, 39
  with cellular heterogeneity, proportion of
    each cell type calculated as fractional
    volume, 283
  double-perfused, 98–107
  extracellular compartments in, 266–267
  functional morphology of, *in situ*,
    265–289
  intact, water permeability, measurement,
    565–566
  interstitial (serosal) compartment, 267
  isolated
    effect of loop diuretics on, 791–792
    electrophysiological study of, 289–302
    hormonal receptors in, 303–325
    patch-clamp analysis in, 299–302
    perfusion, 372
    studies of volume, shape changes in
      level of focus for, 373
      optical and video apparatus for,
        372–373
      tubule preparation for, 372
    in vitro perfusion, advantages of, 290
    volume and shape changes in,
      measurements of, 371–379
  isolated perfused
    advantages of, 302
    flux measurements in, 354–370. *See
      also* Solute flux
  lumenal compartment, 267
  nonmammalian, 168–226. *See also*
    Amphibian renal tubules; Bird renal
    tubules; Elasmobranchs, tubular
    epithelia of; Malpighian tubules;
    Reptilian renal tubules; Teleost renal
    tubules
  perfusate bathing solutions, for protein
    endocytosis studies, 346–347
  preparation, 372
  structural data, correlation of, with
    transport data, 282–283
  structural study of, 265
  surface area of basolateral cell membranes
    of, and Na^+,K^+-ATPase activity and
    sodium transport rate, 270–271
  suspension, 326–327
    advantages and limitations of, 330–331
Renal tubule segments
  distribution of, in regional sections of
    kidney, 229
  enzyme assays, permeabilization step,
    231
  from fresh kidney, viability, and
    incubation time, 228–229
  isolated, metabolism in, 325–340

isolated, perfused
  collection rate, 232–233
    measurement of, 233–241
    normalization of water flows, 244–245
  osmotic water permeability
    measurement, 249–252
    and perfusion rate, 250–252
  perfusion rate, 232–233
    effect of leakage of perfusate, 244
    measurement of, 241–244
  surface area estimates for, 244–245
  volume absorption measurements
    errors in, 245–248
    and rate of perfusion, 246–248
  volume flow measurements in, 232–252
    crimped end method, 248–249
    by quantitative collection of absorbate, 248–249
  volume flow rate in, 232–233
microdissection of, 226–231
  apparatus for, 227
  collagenase perfusion for, 227–228
  from cortex, 229–230
  from inner medulla, 229, 231
  from outer medulla, 229–230
transfer, 231
Renin–angiotensin system, within cultured juxtaglomerular cells, 156–158
Renomedullary interstitial cells
  antihypertensive action, 161, 163–164
  appearance, 166
  biologic attributes, 152
  culture, 162–166
  function, 167
  isolation, 162–166
Reptiles. *See also* Garter snake
  kidneys, sexual dimorphism, 199
  source of, 199
Reptilian renal tubules. *See also* Garter snake, distal tubules; Garter snake, proximal tubules
  distal tubules, *in vitro* microperfusion of, 220–223
  isolation of, 199–203
  proximal tubules, *in vitro* microperfusion of, 203–220
  unique tubular functions, 198–199
Resistance, of epithelial layers, measurement of, 435
RHC 80267, inhibition of conversion of DAG to arachidonate, stearate, and glycerol by, 678–681, 690
Rhodamine–peanut lectin
  simultaneous staining of cortical collecting tubule with BCECF and, 263–264
  staining of cortical collecting tubule with, 260–261
Rhodopsin receptors, 637–638
RICs. *See* Renomedullary interstitial cells
Ringer solution, for perfusion of reptilian renal tubules, composition of, 200–202
RNA, synthesis, inhibition by amiloride and/or its analogs, 748

S

Salamander, source, 196–197
Salmon gairdneri. See Trout
Salt
  body content, 1
  proximal tubular reabsorption, 86
Saralasin, effect of, on hypertension induced by juxtaglomerular cells, 160–161
Scanning electron microscopy, 274
SCH 28080, 721–722
  effect of
    on gastric glands, 729–731
    on H^+,K^+-ATPase activity, 737–738
SDS resistance, 850–851
Secretin preferring receptor, identification of
  radiolabeled ligands used in binding studies for, 621
  tissues tested for, 610
Sedimentation rate, and shape of molecule, 845–846
Sedimentation standards, proteins which can be used for, 844–845
Sedimentation value, of protein, factors affecting, 844–846
Semliki forest virus glycoprotein, as marker of transport vesicle formation, 823
Shark
  proximal tubules
    chloride secretion in, electrophysiological evidence for, 184–186
    isolated, fluid secretion in, 189
  rectal gland tubules
    isolated, *in vitro* perfusion of, 289
    isolation, 183
    perfusion, 183

renal tubules
　isolation, 184
　perfusion, 184
　source of, 183
　tubular functions in, 182–183
Short-circuit current
　across isolated epithelial sheets, effects of loop diuretics on, 791
　in epithelial layer, 436
Shrinking droplet method, for determination of isotonic fluid reabsorption in mammalian kidney, 84
Silver ions, effect on cell membrane resistance, 15
Simian immunodeficiency virus, env proteins of, SDS resistance, 850
SITS. *See* 4-Acetamido-4′-isothiocyanostilbene-2,2′-disulfonate
Small intestine, apical cell membrane conductance in, 10
Smooth muscle cells
　cytosolic chloride activity in, 801
　[^{3}H]bumetanide-binding experiments in, 792
SNGFR. *See* Glomerular filtration rate, single-nephron
Sodium
　in mosquito hemolymph and fluid secreted by isolated Malpighian tubules, 173–175
　proximal tubule fluid concentration, 90
　proximal tubule transport, role in other tubular transport processes, 91–92
　reabsorption of, along distal convoluted tubule, 94
　receptor, 799
　secretion
　　by fish proximal tubules, 190–192
　　in *in vitro* studies of Malpighian tubules, electrophysiological methods for, 176–178
　transport
　　active reabsorptive, in snake distal tubules, electrophysiological evidence for, 220–221
　　self-inhibition of, in snake distal tubules, electrophysiological evidence for, 221
　transport rate, and membrane surface area, 270–271

uptake systems, identification of, by determination of metabolic activity, 399–403
Sodium/calcium exchanger, inhibition by amiloride and/or its analogs, 739–742, 744
Sodium channel, 799
　amiloride-sensitive, in isolated renal cells, 407
　blockers, 293
　epithelial
　　binding of [^{3}H]benzamil to, 751–752
　　inhibition by amiloride and/or its analogs, 739–742
　　inhibition by amiloride or amiloride analogs, 742–743
　　subunits, identification using photoreactive amiloride analogs, 852
　voltage-gated
　　inhibition by amiloride and/or its analogs, 745
　　inhibition of, by amiloride or amiloride analogs, 739–742
Sodium chloride
　renal transport of, 93–94
　secretion
　　by fish proximal tubules, 190–192
　　from killifish proximal tubules, *in vitro* electrophysiological studies of perfused systems, 194–195
　transport, in teleost renal tubules, 186
Sodium cotransporters, 3–4
　cross-reactivity of antibodies with, 604
　functional coupling of, 604
　polypeptide components of, possible homologies, 603–605
　precipitation of, 587–590
　reconstitution of
　　after solubilization of transport proteins, 591–597
　　into proteoliposomes, 591–597
　solubilization of, 587–590
Sodium-coupled solute transport, inhibition by amiloride and/or its analogs, 744–745
Sodium-coupled transporters
　activities of, in intact membrane vesicles, 587
　in brush border membranes, 583
Sodium ferrocyanide, as volume marker, 203

Sodium-D-glucose cotransporter, 583–605
  components of, 603
  freeze–thaw reconstitution, 592–597
  functional molecular weight of, 585–586, 603
  inhibition by phlorizin, 755
  inhibitor, 584
  intestinal, 584–585
    component of, homology with polypeptide of renal Na^+-D-glucose cotransporter, 603
  in isolated renal cells, 406–407
  partially purified, protein patterns of, 600
  partial purification of, 596–603
  polypeptides, cloned, 584, 603–604
  renal
    components of, 585
    partial purification of polypeptide components of, by ion-exchange chromatography, 603
  structure of, 604
Sodium/phosphate cotransport
  across microvillar brush border membrane, 494–495
  established renal epithelial cell lines, basic characteristics of, 496
  in isolated renal cells, 406–407
  in LLC-PK_1 cells, as function of days in culture, 497
  in OK cells, as functin of days in culture, 498
  regulation of
    studied in established renal epithelial cells, 495–497
    studied in isolated renal cortex brush border membrane vesicles, 495
    studied in primary cultures of proximal tubular epithelial cells, 495
Sodium/potassium antiporter, differential hormonal dependence of, 422
Sodium/potassium/chloride cotransport, 422, 794, 800, 802–803
  blockers, 803–804
  effect of loop diuretics on, 790
  identification and purification of, using [^{3}H]bumetanide, 786–789
  inhibition, by loop diuretics, 784–785
  in isolated renal cells, 407
Solute
  intracellular concentration of, measurement of, 366–368
  quantification of, in small sample volumes, 357
Solute flux
  across membrane, 51
  bath-to-lumen (secretory), 354
    measurement of, 363–366
    perfusion chamber used for, 364
  bidirectional transepithelial, measurement of, in tissue culture of renal cell lines, 436
  lumen-to-bath (absorptive), 354
    measurement of, 359–363
      by measurement of lumenal disappearance of tracer, 360
      by measurement of tracer appearance in bathing solution, 361–363
  measurement
    in isolated perfused tubule segments, 354–370
    and transepithelial volume flow, 355–356
    using radioactive labels, 355–356
  net, 354–355
    methods for measuring, 356–359
    normalized in terms of apparent or morphometrically measured lumenal membrane surface area, 356
    and rate of perfusion, 357–359
    relationship to its electrochemical potential driving force, 368–370
  unidirectional, 354–355
    measurement of, 359–366
    and volume flow measurements, 232
Solute transport
  pathways, 267–270
  transmembrane, measurements of, in cell cultures, 430–432
Somatostatin receptor
  radiolabeled ligands used in binding studied to identify, 621
  tissues tested to identify, 610
Sorbitol, uptake, in cultured nephron segment cells, 420–421
Spectrophotometry, 95–96
Spirolactone SC 26304, dose-dependent antagonist effect of
  on aldosterone-mediated increase in urinary K^+/Na^+ ratio, 532–533
  on binding of [^{3}H]aldosterone to renal receptors, 532–534

Squalus acanthias. *See* Dogfish shark
Squid giant axon, isotopic influx measurements in, effect of loop diuretics on, 791
Standing droplet technique, and simultaneous perfusion of peritubule capillaries, to determine disappearance rates of solutes across epithelial membranes, 84
Staphylococcus aureus cells, vesicle immunoadsorption using, 839–840
Stereologic measurements
  techniques of, 283–284
  test grids, 283–284
Stereology, basic parameters in, 281–283
Stimulus–response coupling. *See also* Cellular response
  in epithelial cells, 671–676
  general models of, 661–671
  in nonpolar cells, 661–671
  in polar cells, 661, 671–672
  and spatial separation of messenger generation and action, 672–674
  study of, isolated brush border membrane vesicle in, 675–676
  synarchic regulation of, 666–667
Substance K receptor, amino acid sequencing of, 639
Substance P receptor
  amino acid sequencing of, 639
  identification of
    radiolabeled ligands used in binding studies for, 621
    tissues tested for, 610
Succinate, sodium-dependent uptake into renal outer cortical brush border membrane vesicles, 484–486
Succinate dehydrogenase
  measured in isolated nephron segments, 340
  measured in single isolated nephron segments, 334
Sucrose, in extracellular space, for measurement of ion fluxes, 403–404
p-Sulfamylbenzoic acid, 781–782
Sulfamylbenzoic acid derivatives, activity of, *in vitro*, evaluation of, 790–792
Sulfanilamide, 781–782
  diuretic activity of, 781–782
Sulfate, sodium-dependent uptake into renal brush border membrane vesicles, 487

Sulfosuccinimidyl-2-(biotinamido)ethyl-13′-dithioproprionate, in localization of proteins by cell surface biotinylization, 822–823
Sulfur, secretion, by fish proximal tubules, 190–191
Superficial glomeruli, microdissection of, 309
Surface density, definition of, 282
Sweat glands, isolated tubules, *in vitro* perfusion of, 289

T

TALH. *See* Thick ascending limb of loop of Henle
Tamm–Horsfall glycoprotein, 396, 412
Target theory, 567–568
Teleost renal tubules, 186–196
  in vitro studies of, 188–196
    electrophysiological studies of perfused systems, 194–195
    microperfusion, 192–194
    nonperfused single tubule method, 189–192
    teased tubules method, 188–189
    uptake of radiolabeled solutes in, 188–189
  isolation of, 187–188
  proximal NaCl and fluid secretion in, 195–196
  unique tubular functions, 186–197
Terminal bars, 1–2
Tetraethylammonium, net secretion of, from isolated, perfused snake tubules, 214, 218
Thamnophis. *See* Garter snake
Thapsigargin, intracellular Ca^{2+} mobilization with, 703–704
Thiazide diuretics, 781
  common structural features with furosemide, 781
Thick ascending limb cells, transport systems in, 407
Thick ascending limb of loop of Henle, 288. *See also* Cortical thick ascending limb of loop of Henle; Medullary thick ascending limb of loop of Henle
  Na^+ and Cl^- absorption in, 785
  NaCl transport in, inhibitors, 781
  perfused, chloride transport assay in, 802

Thick ascending limb of loop of Henle cells
  compounds used in identification of, 396
  culture
    growth media and supplements, 416
    support, 417
  cultured, electrophysiological studies, 424
  markers for, 400
  sodium transport-related O_2 consumption, 401–402
Thin ascending limb, microdissection, 309
Thin descending limb, microdissection, 309
Thin descending limb of Henle's loop, 326
  complex double-cylinder model of, 377–378
Thin-layer chromatography, for quantifying labeled metabolites of solutes, in flux measurements, 355
ThiolP$_3$, microinjection of, into Fura-2-loaded hepatocytes, calcium ion transients induced by, 705–706
Thromboxane A$_2$, 671
Thromboxanes, 676
  as mediator of cellular response, 670–671
Tight junction, 267
  length of, 268
  selective permeability qualities of, 267
  structure, 267–268
  transjunctional resistance, 268
Timoprazole, inhibitors, effect on human gastric glands, 733
Tissue culture, of renal cell lines, 426–436
  and measurement of transmembrane solute transport, 430–432
Torasemide, as blocker of Na$^+$/2Cl$^-$/K$^+$ carrier, 804
Tracer, 4
Tracheal epithelial cells, isotopic influx measurements in, effect of loop diuretics on, 790
Transcellular transport, 450
  study of, isolated brush border membrane vesicle in, 675–676
Transcellular transport pathway, 2, 267–269, 291–293
  permeability of, 5
Transepithelial ionic transport, 290–291
Transepithelial measurements, 290–292
Transepithelial organic solute transport, in reptilian proximal tubules, *in vitro* microperfusion studies, 207–214

Transepithelial resistances, 1, 5
  direct current, and direct current cell membrane resistances, quantitative comparison of, 7–8
  of isolated perfused tubule segment, 291–294
  measurement of, in response to alteration of one membrane resistance, 11–12
  steady state measurement of, 15–16
  and voltage divider ratio, paired measurement of, in response to alteration of one membrane resistance, 9–11, 21–23
Transepithelial transport. *See* Epithelial transport
Transepithelial voltage
  in cultured cells, 424
  measurements, in isolated tubules, 291–292
  in reptilian proximal tubules, *in vitro* microperfusion studies, 207
Transepithelial volume flow, detection, 189–190
  in teleost renal tubules, 192
Transferrin receptor, amino acid sequencing of, 638
Transmission electron microscopy, in structural investigation of kidneys, 273–274
Transporters, 35
Transport proteins
  identification, 4
  purification, 4
Transport studies, by optical methods, 469–479
Transport systems, in isolated renal cells, 408
Triacylglycerol synthesis, measured in single isolated nephron segments, 334
  radiochemical methods, 338
Triflocine, as blocker of Na$^+$/2Cl$^-$/K$^+$ carrier, 804
2′,4′,6′-Trihydroxy-4-nitro-dihydrochalcone. *See* 4-Nitrophloretin
Trout, distal tubules, *in vitro* electrophysiological studies of perfused systems, 195
Tubuloglomerular feedback, 93, 119–120
Tunneling, 776
Two-dimensional cable analysis, 7

U

Ultramicro-enzyme assay, 541–542
Urate, secretion, from isolated, perfused snake tubules, 214, 216–217
Urea, passive reabsorptive movement, from tubule lumen into blood, 92
Ureteral ligation, 31–32
Urinary bladder
  apical cell membrane conductance in, 10
  Necturus, shunt analysis in, 12
  shunt analysis in, 14
Urinary osmolality, 92
Urinary tract, epithelia, 1
Urokinase-type plasminogen activator, inhibition by amiloride and/or its analogs, 748

V

Vacuolar system, 505
Valinomycin, effect on mitochondrial retention of Di-O-C_5(3) in cortical collecting tubule, 259–260
Vascular smooth muscle cells
  cotransport activity and bumetanide binding in, 786–788
  isotopic influx measurements in, effect of loop diuretics on, 791
Vasoactive intestinal peptide receptors
  on pancreatic acini, subclasses of, 628
  radiolabeled ligands used in binding studied to identify, 621
  tissues tested to identify, 610
Vasopressin. *See also* Antidiuretic hormone
  cellular action of, 551–571
    morphological endpoint of, 557
    morphological studies, 552–561
  effect of
    on cell membrane resistance, 15
    on single-nephron glomerular filtration rate, 116
Velocity gradient centrifugation
  for isolating and analyzing oligomeric complexes, 842–847
  of protein, correlated with chemical cross-linking, 852–853
Vesicle immunoadsorption, 837–840
Vesicles
  coupled transport in, stoichiometry of, 479–494
  from kidney cortex, ATP-driven proton transport in, 505–520
  measurement of potential in, 476–479
  pH gradient measurements in, principles of, 470
  transport studies with, 70
Vesicle technology, 469
Vesicular stomatitis virus
  glycoprotein G of, as transport marker, in MDCK cells, 814
  G protein
    assembly, 846–847
    conformation, and stability to detergent solubilization and centrifugation, 843–844
  SDS resistance, 850
Vibrating probe techniques, 2
Video position analyzer, 373
Vitamin D hydroxylation, measured in single isolated nephron segments, 334
Voltage divider ratio (α), 292, 295–296
  direct current (DC), 8
  measurement of, in response to alteration of one membrane resistance, 11–12
  steady state measurement of, 15–16
Voltage scanning, 2
Volume density
  definition of, 282
  measurement, 283
Volume flow measurements, in isolated, perfused renal tubule segments, 232–252
Volume markers
  concentration in collectate, errors in determination of, 245–246
  leakage, in isolated tubule segments, 243–244
  lumen-to-bath flux of, measurement, 243–244
  in measurement of perfusion rate of isolated tubule, 241–244
  in proximal tubule perfusate, in studies of reptiles, 203

W

Water
  body content, 1
  proximal tubular reabsorption, 86
  transepithelial flow, in snake distal tubules, 221–223

transmural flow, in isolated, perfused renal tubule segments, 232–252
transport, in bird renal tubules, 226
Water channels
  endocytosis of, during vasopressin-induced membrane recycling in kidney collecting ducts and toad urinary bladders, 555–556
  in membranes, identification of, 561–562
  recycling, in collecting duct principal cells, proposed pathway of, 551–552
Water permeability
  of apical plasma membrane of kidney collecting duct principal cells, 551
  measurement of
    across isolated membrane vesicles and intact epithelia, 561
    in intact kidney tubules, 565–566
    in isolated vesicles, 561–565
    optical, in vesicles and tubules, 561–566
Whole-organ physiology, black box approach to, 37–38
Winter flounder, intestine, [^{3}H]bumetanide binding to brush border membrane vesicles from, 786–789

X

Xenopus laevis, oocytes
  endogenous transport systems in, 437
  expression system, 437

250732